KIRK-OTHMER ENCYCLOPEDIA OF

CHEMICAL TECHNOLOGY

Fifth Edition

VOLUME 5

KIRK-OTHMER ENCYCLOPEDIA OF CHEMICAL TECHNOLOGY, FIFTH EDITION
EDITORIAL STAFF

KIRK-OTHMER ENCYCLOPEDIA OF

CHEMICAL TECHNOLOGY

Fifth Edition

VOLUME 5

Kirk-Othmer Encyclopedia of Chemical Technology
is available Online in full color and with additional content at
http://www3.interscience.wiley.com/cgi-bin/mrwhome/104554789/HOME.

A John Wiley & Sons, Inc., Publication

Copyright © 2004 by John Wiley & Sons, Inc. All rights reserved.

Published by John Wiley & Sons, Inc., Hoboken, New Jersey.
Published simultaneously in Canada.

For general information on our other products and services please contact our Customer Care
Department within the U.S. at 877-762-2974, outside the U.S. at 317-572-3993 or fax 317-572-4002.

Wiley also publishes its books in a variety of electronic formats. Some content that appears in print,
however, may not be available in electronic format.

Library of Congress Cataloging-in-Publication Data:

Kirk-Othmer encyclopedia of chemical technology. – 5th ed.
        p. cm.
Editor-in-chief, Arza Seidel.
"A Wiley-Interscience publication."
Includes index.
    ISBN 0-471-48494-6 (set) – ISBN 0-471-48518-7 (v. 5)
    1. Chemistry, Technical–Encyclopedias. I. Title: Encyclopedia of
chemical technology. II. Kroschwitz, Jacqueline I.
    TP9.K54   2004
    660'.03–dc22   2003021960

Printed in the United States of America

10 9 8 7 6 5 4 3 2 1

CONTENTS

CONTRIBUTORS

Marvin O. Bagby, *United States Department of Agriculture, Peoria, IL,* Carboxylic Acids

Calvin H. Bartholomew, *Brigham Young University, Provo, UT,* Catalyst Deactivation and Regeneration

Noelie Bertoniere, *Southern Regional Research Center, New Orleans, LA,* Cellulose

R. Malcolm Brown, *The University of Texas at Austin, Austin, TX,* Cellulose

Relva C. Buchanan, *University of Cincinnati, Cincinnati, OH,* Ceramics as Electrical Materials

James C. Burnett, *Virginia Commonwealth University, Richmond, VA,* Cardiovascular Agents

E. P. Butler, *National Institute of Standards and Technology, Gaithersburg, MD,* Ceramic-Matrix Composites

David S. Butts, *DSB International Inc., St. George, UT,* Chemicals from Brine

Henri Chanzy, *CNRS-CERMAV, Grenoble Cedex, France,* Cellulose

Hsin-Tien Chiu, *National Chiao Tung University, Hsinchu, Taiwan,* Chemical Vapor Deposition

James H. Clark, *University of York, York, United Kingdom,* Catalysts, Supported

Robert R. Contrell, *Union Camp Corporation, Wayne, NJ,* Carboxylic Acids

Nick Corner-Walker, *Alan Letki, Alfa Laval Inc., Warminster, PA,* Centrifugal Separation

R. W. Daniels, *Union Camp Corporation, Wayne, NJ,* Carboxylic Acids

Tom J. Evans, *Cubic Defense Applications Group, White City, OR,* Chemical Warfare

Kevin G. Ewsuk, *Sandia National Laboratories, Albuquerque, NM,* Ceramics, Processing

Richard Fengi, *Eastman Chemical Company, Kingsport, TN,* Cellulose Esters, Organic Esters

William Ferguson, *Tanco, Lac du Bonnet, Canada,* Cesium and Cesium Compounds

Alfred D. French, *Southern Regional Research Center, New Orleans, LA,* Cellulose

E. R. Fuller, Jr., *National Institute of Standards and Technology, Gaithersburg, MD,* Ceramic-Matrix Composites

Bruce C. Gates, *University of California – Davis, CA,* Catalysis

Steven Gedon, *Eastman Chemical Company, Kingsport, TN,* Cellulose Esters, Organic Esters

Christine George, *Air Liquide America Corporation, Houston, TX,* Carbon Monoxide

S. Jill Glass, *Sandia National Laboratories, Albuquerque, NM,* Ceramics, Mechanical Properties

Wolfgang Glasser, *Virginia Polytechnic Institute and State University, Blacksburg, VA,* Cellulose

Dena Gorrie, *Tanco, Lac du Bonnet, Canada,* Cesium and Cesium Compounds

Derek Gray, *McGill University, Montreal, Canada,* Cellulose

Kazuyuki Hattori, *Kitami Institute of Technology, Kitami, Japan,* Cellulose

William Howard, *Consultant,* Chelating Agents

R. W. Johnson, Jr., *Union Camp Corporation, Wayne, NJ,* Carboxylic Acids

Gajanan S. Joshi, *ISBDD / Virginia Biotechnology Research Park, Richmond, VA,* Cardiovascular Agents

Subhash Karkare, *Scientia Consulting, Thousand Oaks, CA,* Cell Culture Technology

M. J. Keenan, *Exxon Chemical Company, Baton Rouge, LA,* Carboxylic Acids

Barry T. Kilbourn, *Molycorp Inc., Brea, CA,* Cerium and Cerium Compounds

Steven Kosmatka, *Portland Cement Association, Skokie, IL,* Cement

M. A. Krevalis, *Exxon Chemical Company, Baton Rouge, LA,* Carboxylic Acids

Alan Letki, *Alfa Laval Inc., Warminster, PA,* Centrifugal Separation

Duncan Macquarrie, *University of York, York, United Kingdom,* Catalysts, Supported

Thomas G. Majewicz, *Aqualon Company, Palatine, IL,* Cellulose Ethers

Geoff Moggridge, *University of Cambridge, Cambridge, United Kingdom,* Chemical Product Design

Kenichiro Nakashima, *Nagasaki University, Nagasaki, Japan,* Chemiluminescence, Analytical Applications

Thomas J. Podlas, *Aqualon Company, Palatine, IL,* Cellulose Ethers

Rodney D. Roseman, *University of Cincinnati, Cincinnati, OH,* Ceramics as Electrical Materials

E. T. Sauer, *The Procter & Gamble Company, Cincinnati, OH,* Carboxylic Acids

Giuseppe Spoto, *Universita di Catania, Catania, Italy,* Chemical Methods in Archaeology

Michael C. Shelton, *Eastman Chemical Company, Kingsport, TN,* Cellulose Esters, Inorganic Esters

Rajan Tandon, *Sandia National Laboratories, Albuquerque, NM,* Ceramics, Mechanical Properties

David Wilson, *The Dow Chemical Company, Freeport, TX,* Chelating Agents

CONVERSION FACTORS, ABBREVIATIONS, AND UNIT SYMBOLS

SI Units (Adopted 1960)

The International System of Units (abbreviated SI), is implemented throughout the world. This measurement system is a modernized version of the MKSA (meter, kilogram, second, ampere) system, and its details are published and controlled by an international treaty organization (The International Bureau of Weights and Measures) (1).

SI units are divided into three classes:

BASE UNITS

length	meter[†] (m)
mass	kilogram (kg)
time	second (s)
electric current	ampere (A)
thermodynamic temperature[‡]	kelvin (K)
amount of substance	mole (mol)
luminous intensity	candela (cd)

SUPPLEMENTARY UNITS

plane angle	radian (rad)
solid angle	steradian (sr)

DERIVED UNITS AND OTHER ACCEPTABLE UNITS

These units are formed by combining base units, suplementary units, and other derived units (2–4). Those derived units having special names and symbols are marked with an asterisk in the list below.

[†]The spellings "metre" and "litre" are preferred by ASTM; however, "-er" is used in the *Encyclopedia*.

[‡]Wide use is made of Celsius temperature (*t*) defined by

$$t = T - T_0$$

where T is the thermodynamic temperature, expressed in kelvin, and $T_0 = 273.15$ K by definition. A temperature interval may be expressed in degrees Celsius as well as in kelvin.

Quantity	Unit	Symbol	Acceptable equivalent
*absorbed dose	gray	Gy	J/Kg
acceleration	meter per second squared	m/s^2	
*activity (of a radionuclide)	becquerel	Bq	1/s
area	square kilometer	km^2	
	square hectometer	hm^2	ha (hectare)
	square meter	m^2	
concentration (of amount of substance)	mole per cubic meter	mol/m^3	
current density	ampere per square meter	A/m^2	
density, mass density	kilogram per cubic meter	kg/m^3	g/L; mg/cm^3
dipole moment (quantity)	coulomb meter	$C \cdot m$	
*dose equivalent	sievert	Sv	J/kg
*electric capacitance	farad	F	C/V
*electric charge, quantity of electricity	coulomb	C	$A \cdot s$
electric charge density	coulomb per cubic meter	C/m^3	
*electric conductance	siemens	S	A/V
electric field strength	volt per meter	V/m	
electric flux density	coulomb per square meter	C/m^2	
*electric potential, potential difference, electromotive force	volt	V	W/A
*electric resistance	ohm	Ω	V/A
*energy, work, quantity of heat	megajoule	MJ	
	kilojoule	kJ	
	joule	J	$N \cdot m$
	electronvolt[†]	$eV^†$	
	kilowatt-hour[†]	$kW \cdot h^†$	
energy density	joule per cubic meter	J/m^3	
*force	kilonewton	kN	
	newton	N	$kg \cdot m/s^2$

[†]This non-SI unit is recognized by the CIPM as having to be retained because of practical importance or use in specialized fields (1).

Quantity	Unit	Symbol	Acceptable equivalent
*frequency	megahertz	MHz	
	hertz	Hz	1/s
heat capacity, entropy	joule per kelvin	J/K	
heat capacity (specific), specific entropy	joule per kilogram kelvin	$J/(kg \cdot K)$	
heat-transfer coefficient	watt per square meter kelvin	$W/(m^2 \cdot K)$	
*illuminance	lux	lx	lm/m^2
*inductance	henry	H	Wb/A
linear density	kilogram per meter	kg/m	
luminance	candela per square meter	cd/m^2	
*luminous flux	lumen	lm	$cd \cdot sr$
magnetic field strength	ampere per meter	A/m	
*magnetic flux	weber	Wb	$V \cdot s$
*magnetic flux density	tesla	T	Wb/m^2
molar energy	joule per mole	J/mol	
molar entropy, molar heat capacity	joule per mole kelvin	$J/(mol \cdot K)$	
moment of force, torque	newton meter	$N \cdot m$	
momentum	kilogram meter per second	$kg \cdot m/s$	
permeability	henry per meter	H/m	
permittivity	farad per meter	F/m	
*power, heat flow rate, radiant flux	kilowatt	kW	
	watt	W	J/s
power density, heat flux density, irradiance	watt per square meter	W/m^2	
*pressure, stress	megapascal	MPa	
	kilopascal	kPa	
	pascal	Pa	N/m^2
sound level	decibel	dB	
specific energy	joule per kilogram	J/kg	
specific volume	cubic meter per kilogram	m^3/kg	
surface tension	newton per meter	N/m	
thermal conductivity	watt per meter kelvin	$W/(m \cdot K)$	
velocity	meter per second	m/s	
	kilometer per hour	km/h	
viscosity, dynamic	pascal second	$Pa \cdot s$	
	millipascal second	$mPa \cdot s$	
viscosity, kinematic	square meter per second	m^2/s	
	square millimeter per second	mm^2/s	

Quantity	Unit	Symbol	Acceptable equivalent
volume	cubic meter	m^3	
	cubic diameter	dm^3	L (liter) (5)
	cubic centimeter	cm^3	mL
wave number	1 per meter	m^{-1}	
	1 per centimeter	cm^{-1}	

In addition, there are 16 prefixes used to indicate order of magnitude, as follows

Multiplication factor	Prefix	symbol	Note
10^{18}	exa	E	
10^{15}	peta	P	
10^{12}	tera	T	
10^9	giga	G	
10^6	mega	M	
10^3	kilo	k	
10^2	hecto	h^a	[a]Although hecto, deka, deci, and
10	deka	da^a	centi are SI prefixes, their use
10^{-1}	deci	d^a	should be avoided except for SI
10^{-2}	centi	c^a	unit-multiples for area and
10^{-3}	milli	m	volume and nontechnical use of
10^{-6}	micro	μ	centimeter, as for body and
10^{-9}	nano	n	clothing measurement.
10^{-12}	pico	p	
10^{-15}	femto	f	
10^{-18}	atto	a	

For a complete description of SI and its use the reader is referred to ASTM E380 (4) and the article UNITS AND CONVERSION FACTORS which appears in Vol. 24.

A representative list of conversion factors from non-SI to SI units is presented herewith. Factors are given to four significant figures. Exact relationships are followed by a dagger. A more complete list is given in the latest editions of ASTM E380 (4) and ANSI Z210.1 (6).

Conversion Factors to SI Units

To convert from	To	Multiply by
acre	square meter (m^2)	4.047×10^3
angstrom	meter (m)	1.0×10^{-10}[†]
are	square meter (m^2)	1.0×10^{2}[†]
astronomical unit	meter (m)	1.496×10^{11}

[†]Exact.

To convert from	To	Multiply by
atmosphere, standard	pascal (Pa)	1.013×10^5
bar	pascal (Pa)	$1.0 \times 10^{5\dagger}$
barn	square meter (m^2)	$1.0 \times 10^{-28\dagger}$
barrel (42 U.S. liquid gallons)	cubic meter (m^3)	0.1590
Bohr magneton (μ_B)	J/T	9.274×10^{-24}
Btu (International Table)	joule (J)	1.055×10^3
Btu (mean)	joule (J)	1.056×10^3
Btu (thermochemical)	joule (J)	1.054×10^3
bushel	cubic meter(m^3)	3.524×10^{-2}
calorie (International Table)	joule (J)	4.187
calorie (mean)	joule (J)	4.190
calorie (thermochemical)	joule (J)	$4.184^\dagger$
centipoise	pascal second (Pa·s)	$1.0 \times 10^{-3\dagger}$
centistokes	square millimeter per second (mm^2/s)	$1.0^\dagger$
cfm (cubic foot per minute)	cubic meter per second (m^{3}s)	4.72×10^{-4}
cubic inch	cubic meter (m^3)	1.639×10^{-5}
cubic foot	cubic meter (m^3)	2.832×10^{-2}
cubic yard	cubic meter (m^3)	0.7646
curie	becquerel (Bq)	$3.70 \times 10^{10\dagger}$
debye	coulomb meter (C·m)	3.336×10^{-30}
degree (angle)	radian (rad)	1.745×10^{-2}
denier (international)	kilogram per meter (kg/m)	1.111×10^{-7}
	tex‡	0.1111
dram (apothecaries')	kilogram (kg)	3.888×10^{-3}
dram (avoirdupois)	kilogram (kg)	1.772×10^{-3}
dram (U.S. fluid)	cubic meter (m^3)	3.697×10^{-6}
dyne	newton (N)	$1.0 \times 10^{-5\dagger}$
dyne/cm	newton per meter (N/m)	$1.0 \times 10^{-3\dagger}$
electronvolt	joule (J)	1.602×10^{-19}
erg	joule (J)	$1.0 \times 10^{-7\dagger}$
fathom	meter (m)	1.829
fluid ounce (U.S.)	cubic meter (m^3)	2.957×10^{-5}
foot	meter (m)	$0.3048^\dagger$
footcandle	lux (lx)	10.76
furlong	meter (m)	2.012×10^{-2}
gal	meter per second squared (m/s^2)	$1.0 \times 10^{-2\dagger}$
gallon (U.S. dry)	cubic meter (m^3)	4.405×10^{-3}
gallon (U.S. liquid)	cubic meter (m^3)	3.785×10^{-3}
gallon per minute (gpm)	cubic meter per second (m^3/s)	6.309×10^{-5}
	cubic meter per hour (m^3/h)	0.2271

†Exact.
‡See footnote on p. ix.

To convert from	To	Multiply by
gauss	tesla (T)	1.0×10^{-4}
gilbert	ampere (A)	0.7958
gill (U.S.)	cubic meter (m^3)	1.183×10^{-4}
grade	radian	1.571×10^{-2}
grain	kilogram (kg)	6.480×10^{-5}
gram force per denier	newton per tex (N/tex)	8.826×10^{-2}
hectare	square meter (m^2)	$1.0 \times 10^{4\dagger}$
horsepower (550 ft·lbf/s)	watt (W)	7.457×10^2
horsepower (boiler)	watt (W)	9.810×10^3
horsepower (electric)	watt (W)	$7.46 \times 10^{2\dagger}$
hundredweight (long)	kilogram (kg)	50.80
hundredweight (short)	kilogram (kg)	45.36
inch	meter (m)	$2.54 \times 10^{-2\dagger}$
inch of mercury (32°F)	pascal (Pa)	3.386×10^3
inch of water (39.2°F)	pascal (Pa)	2.491×10^2
kilogram-force	newton (N)	9.807
kilowatt hour	megajoule (MJ)	$3.6^\dagger$
kip	newton (N)	4.448×10^3
knot (international)	meter per second (m/S)	0.5144
lambert	candela per square meter (cd/m^3)	3.183×10^3
league (British nautical)	meter (m)	5.559×10^3
league (statute)	meter (m)	4.828×10^3
light year	meter (m)	9.461×10^{15}
liter (for fluids only)	cubic meter (m^3)	$1.0 \times 10^{-3\dagger}$
maxwell	weber (Wb)	$1.0 \times 10^{-8\dagger}$
micron	meter (m)	$1.0 \times 10^{-6\dagger}$
mil	meter (m)	$2.54 \times 10^{-5\dagger}$
mile (statue)	meter (m)	1.609×10^3
mile (U.S. nautical)	meter (m)	$1.852 \times 10^{3\dagger}$
mile per hour	meter per second (m/s)	0.4470
millibar	pascal (Pa)	1.0×10^2
millimeter of mercury (0°C)	pascal (Pa)	$1.333 \times 10^{2\dagger}$
minute (angular)	radian	2.909×10^{-4}
myriagram	kilogram (Kg)	10
myriameter	kilometer (Km)	10
oersted	ampere per meter (A/m)	79.58
ounce (avoirdupois)	kilogram (kg)	2.835×10^{-2}
ounce (troy)	kilogram (kg)	3.110×10^{-2}
ounce (U.S. fluid)	cubic meter (m^3)	2.957×10^{-5}
ounce-force	newton (N)	0.2780
peck (U.S.)	cubic meter (m^3)	8.810×10^{-3}
pennyweight	kilogram (kg)	1.555×10^{-3}
pint (U.S. dry)	cubic meter (m^3)	5.506×10^{-4}

†Exact.

To convert from	To	Multiply by
pint (U.S. liquid)	cubic meter (m^3)	4.732×10^{-4}
poise (absolute viscosity)	pascal second (Pa·s)	0.10[†]
pound (avoirdupois)	kilogram (kg)	0.4536
pound (troy)	kilogram (kg)	0.3732
poundal	newton (N)	0.1383
pound-force	newton (N)	4.448
pound force per square inch (psi)	pascal (Pa)	6.895×10^3
quart (U.S. dry)	cubic meter (m^3)	1.101×10^{-3}
quart (U.S. liquid)	cubic meter (m^3)	9.464×10^{-4}
quintal	kilogram (kg)	1.0×10^{-2}[†]
rad	gray (Gy)	1.0×10^{-2}[†]
rod	meter (m)	5.029
roentgen	coulomb per kilogram (C/kg)	2.58×10^{-4}
second (angle)	radian (rad)	4.848×10^{-6}[†]
section	square meter (m^2)	2.590×10^6
slug	kilogram (kg)	14.59
spherical candle power	lumen (lm)	12.57
square inch	square meter (m^2)	6.452×10^{-4}
square foot	square meter (m^2)	9.290×10^{-2}
square mile	square meter (m^2)	2.590×10^6
square yard	square meter (m^2)	0.8361
stere	cubic meter (m^3)	1.0[†]
stokes (kinematic viscosity)	square meter per second (m^2/s)	1.0×10^{-4}[†]
tex	kilogram per meter (kg/m)	1.0×10^{-6}[†]
ton (long, 2240 pounds)	kilogram (kg)	1.016×10^3
ton (metric) (tonne)	kilogram (kg)	1.0×10^3[†]
ton (short, 2000 pounds)	kilogram (kg)	9.072×10^2
torr	pascal (Pa)	1.333×10^2
unit pole	weber (Wb)	1.257×10^{-7}
yard	meter (m)	0.9144[†]

[†]Exact.

Abbreviations and Unit Symbols

Following is a list of common abbreviations and unit symbnols used in the Encyclopedia. In general they agree with those listed in *American National Standard Abbreviations for Use on Drawings and in Text (ANSI Y1.1)* (6) and *American National Standard Letter Symbols for Units in Science and Technology (ANSI Y10)* (6). Also included is a list of acronyms for a number of private and government organizations as well as common industrial solvents, polymers, and other chemicals.

Rules for Writing Unit Symbols (4):

1. Unit symbols are printed in upright letters (roman) regardless of the type style used in the surrounding text.
2. Unit symbols are unaltered in the plural.
3. Unit symbols are not followed by a period except when used at the end of a sentence.
4. Letter unit symbols are generally printed lower-case (for example, cd for candela) unless the unit name has been derived from a proper name, in which case the first letter of the symbol is capitalized (W, Pa). Prefixes and unit symbols retain their prescribed form regardless of the surrounding typography.
5. In the complete expression for a quantity, a space should be left between the numerical value and the unit symbol. For example, write 2.37 lm, *not* 2.37 lm, and 35 mm, *not* 35 mm. When the quantity is used in an adjectival sense, a hyphen is often used, for example, 35-mm film. *Exception:* No space is left between the numerical value and the symbols of degree, minute, and second of plane angle, degree Celsius, and the percent sign.
6. No space is used between the prefix and unit symbol (for example, kg).
7. Symbols, not abbreviations, should be used for units. For example, use "A," not "amp," for ampere.
8. When multiplying unit symbols, use a raised dot:

$$N \cdot m \text{ for newton meter}$$

In the case of W·h, the dot may be omitted, thus:

$$Wh$$

An exception to this practice is made for computer printouts, automatic typewriter work, etc, where the raised dot is not possible, and a dot on the line may be used.
9. When dividing unit symbols, use one of the following forms:

$$m/s \quad or \quad m \cdot s^{-1} \quad or \quad \frac{m}{s}$$

In no case should more than one slash be used in the same expression unless parentheses are inserted to avoid ambiguity. For example, write:

$$J/(mol \cdot K) \quad or \quad J \cdot mol^{-1} \cdot K^{-1} \quad or \quad (J/mol)/K$$

but *not*

$$J/mol/K$$

10. Do not mix symbols and unit names in the same expression. Write:

$$\text{joules per kilogram} \quad or \quad \text{J/kg} \quad or \quad \text{J} \cdot \text{kg}^{-1}$$

but *not*

$$\text{joules/kilogram} \quad nor \quad \text{Joules/kg} \quad nor \quad \text{Joules} \cdot \text{kg}^{-1}$$

ABBREVIATIONS AND UNITS

A	ampere		AOAC	Association of Official Analytical Chemists
A	anion (eg, HA)			
A	mass number		AOCS	American Oil Chemists' Society
a	atto (prefix for 10^{-18})			
AATCC	American Association of Textile Chemists and Colorists		APHA	American Public Health Association
			API	American Petroleum Institute
ABS	acrylonitrile–butadiene–styrene		aq	aqueous
abs	absolute		Ar	aryl
ac	alternating current, *n*.		*ar*-	aromatic
a-c	alternating current, *adj*.		*as*-	Asymmetric(al)
ac-	alicyclic		ASHRAE	American Society of Heating, Refrigerating, and Air Conditioning Engineers
acac	acetylacetonate			
ACGIH	American Conference of Governmental Industrial Hygienists			
			ASM	American Society for Metals
ACS	American Chemical Society		ASME	American Society of Mechanical Engineers
AGA	American Gas Association			
Ah	ampere hour		ASTM	American Society for Testing and Materials
AIChE	American Institute of Chemical Engineers		at no.	atomic number
AIME	American Institute of Mining, metallurgical, and Petroleum Engineers		at wt	atomic weight
			av(g)	average
			AWS	American Welding Society
			b	bonding orbital
AIP	American Institute of Physics		bbl	barrel
			bcc	body-centered cubic
AISI	American Iron and Steel Institute		BCT	body-centered tetragonal
			Bé	Baumé
alc	alcohol(ic)		BET	Brunauer-Emmett-Teller (adsorption equation)
Alk	alkyl			
alk	alkaline (not alkali)		bid	twice daily
amt	amount		Boc	*t*-butyloxycarbonyl
amu	atomic mass unit		BOD	biochemical (biological) oxygen demand
ANSI	American National Standards Institute			
			bp	boiling point
AO	atomic orbital		Bq	becquerel

C	coulomb	dil	dilute
°C	degree Celsius	DIN	Deutsche Industrie
C-	denoting attachment to		Normen
	carbon	*dl*-; DL-	racemic
c	centi (prefix for 10^{-2})	DMA	dimethylacetamide
c	critical	DMF	dimethylformamide
ca	circa (Approximately)	DMG	dimethyl glyoxime
cd	candela; current density;	DMSO	dimethyl sulfoxide
	circular dichroism	DOD	Department of Defense
CFR	Code of Federal	DOE	Department of Energy
	Regulations	DOT	Department of
cgs	centimeter-gram-second		Transportation
CI	Color Index	DP	degree of polymerization
cis-	isomer in which	dp	dew point
	substituted groups are	DPH	diamond pyramid
	on some side of double		hardness
	bond between C atoms	dstl(d)	distill(ed)
cl	carload	dta	differential thermal
cm	centimeter		analysis
cmil	circular mil	(*E*)-	entgegen; opposed
cmpd	compound	ϵ	dielectric constant
CNS	central nervous system		(unitless number)
CoA	coenzyme A	*e*	electron
COD	chemical oxygen demand	ECU	electrochemical unit
coml	commerical(ly)	ed.	edited, edition, editor
cp	chemically pure	ED	effective dose
cph	close-packed hexagonal	EDTA	ethylenediaminetetra-
CPSC	Consumer Product Safety		acetic acid
	Commission	emf	electromotive force
cryst	crystalline	emu	electromagnetic unit
cub	cubic	en	ethylene diamine
D	debye	eng	engineering
D-	denoting configurational	EPA	Environmental Protection
	relationship		Agency
d	differential operator	epr	electron paramagnetic
d	day; deci (prefix for 10^{-1})		resonance
d	density	eq.	equation
d-	*dextro*-, dextrorotatory	esca	electron spectroscopy for
da	deka (prefix for 10^{-1})		chemical analysis
dB	decibel	esp	especially
dc	direct current, *n.*	esr	electron-spin resonance
d-c	direct current, *adj.*	est(d)	estimate(d)
dec	decompose	estn	estimation
detd	determined	esu	electrostatic unit
detn	determination	exp	experiment, experimental
Di	didymium, a mixture of all	ext(d)	extract(ed)
	lanthanons	F	farad (capacitance)
dia	diameter	*F*	fraday (96,487 C)

f	femto (prefix for 10^{-15})	hyd	hydrated, hydrous
FAO	Food and Agriculture Organization (United Nations)	hyg	hygroscopic
		Hz	hertz
		^{i}(eg, Pr^i)	iso (eg, isopropyl)
fcc	face-centered cubic	i-	inactive (eg, i-methionine)
FDA	Food and Drug Administration	IACS	international Annealed Copper Standard
FEA	Federal Energy Administration	ibp	initial boiling point
		IC	integrated circuit
FHSA	Federal Hazardous Substances Act	ICC	Interstate Commerce Commission
fob	free on board	ICT	International Critical Table
fp	freezing point		
FPC	Federal Power Commission	ID	inside diameter; infective dose
FRB	Federal Reserve Board		
frz	freezing	ip	intraperitoneal
G	giga (prefix for 10^9)	IPS	iron pipe size
G	gravitational constant $= 6.67 \times 10^{11} \text{N} \cdot \text{m}^2/\text{kg}^2$	ir	infrared
		IRLG	Interagency Regulatory Liaison Group
g	gram		
(g)	gas, only as in H_2O(g)	ISO	International Organization Standardization
g	gravitatonal acceleration		
gc	gas chromatography		
gem-	geminal	ITS-90	International Temperature Scale (NIST)
glc	gas–liquid chromatography		
g-mol wt; gmw	gram-molecular weight	IU	International Unit
		IUPAC	International Union of Pure and Applied Chemistry
GNP	gross national product		
gpc	gel-permeation chromatography		
		IV	iodine value
GRAS	Generally Recognized as Safe	iv	intravenous
		J	joule
grd	ground	K	kelvin
Gy	gray	k	kilo (prefix for 10^3)
H	henry	kg	kilogram
h	hour; hecto (prefix for 10^2)	L	denoting configurational relationship
ha	hectare		
HB	Brinell hardness number	L	liter (for fluids only) (5)
Hb	hemoglobin	l-	levo-, levorotatory
hcp	hexagonal close-packed	(l)	liquid, only as in NH_3(l)
hex	hexagonal	LC$_{50}$	conc lethal to 50% of the animals tested
HK	Knoop hardness number		
hplc	high performance liquid chromatography	LCAO	linear combnination of atomic orbitals
		lc	liquid chromatography
HRC	Rockwell hardness (C scale)	LCD	liquid crystal display
HV	Vickers hardness number	lcl	less than carload lots

LD_{50}	dose lethal to 50% of the animals tested	N	newton (force)
LED	light-emitting diode	N	normal (concentration); neutron number
liq	liquid	N-	denoting attachment to nitrogen
lm	lumen		
ln	logarithm (natural)	n (as n_D^{20})	index of refraction (for 20°C and sodium light)
LNG	liquefied natural gas		
log	logarithm (common)		
LOI	limiting oxygen index	n (as Bu^n),	normal (straight-chain structure)
LPG	liquefied petroleum gas	n-	
ltl	less than truckload lots	n	neutron
lx	lux	n	nano (prefix for 10^9)
M	mega (prefix for 10^6); metal (as in MA)	na	not available
		NAS	National Academy of Sciences
M	molar; actual mass		
$\overline{M}_w$	weight-average mol wt	NASA	National Aeronautics and Space Administration
$\overline{M}_n$	number-average mol wt		
m	meter; milli (prefix for 10^{-3})	nat	natural
		ndt	nondestructive testing
m	molal	neg	negative
m-	meta	NF	*National Formulary*
max	maximum	NIH	National Institutes of Health
MCA	Chemical Manufacturers' Association (was Manufacturing Chemists Association)	NIOSH	National Institute of Occupational Safety and Health
MEK	methyl ethyl ketone	NIST	National Institute of Standards and Technology (formerly National Bureau of Standards)
meq	milliequivalent		
mfd	manufactured		
mfg	manufacturing		
mfr	manufacturer		
MIBC	methyl isobutyl carbinol	nmr	nuclear magnetic resonance
MIBK	methyl isobutyl ketone		
MIC	minimum inhibiting concentration	NND	New and Nonofficial Drugs (AMA)
min	minute; minimum	no.	number
mL	milliliter	NOI-(BN)	not otherwise indexed (by name)
MLD	minimum lethal dose		
MO	molecular orbital	NOS	not otherwise specified
mo	month	nqr	nuclear quadruple resonance
mol	mole		
mol wt	molecular weight	NRC	Nuclear Regulatory Commission; National Research Council
mp	melting point		
MR	molar refraction		
ms	mass spectrometry	NRI	New Ring Index
MSDS	material safety data sheet	NSF	National Science Foundation
mxt	mixture		
μ	micro (prefix for 10^{-6})	NTA	nitrilotriacetic acid

NTP	normal temperature and pressure (25°C and 101.3 kPa or 1 atm)	pwd	powder
		py	pyridine
		qv	quod vide (which see)
NTSB	National Transportation Safety Board	R	univalent hydrocarbon radical
O-	denoting attachment to oxygen	(*R*)-	rectus (clockwise configuration)
o-	ortho	*r*	precision of data
OD	outside diameter	rad	radian; radius
OPEC	Organization of Petroleum Exporting Countries	RCRA	Resource Conservation and Recovery Act
o-phen	*o*-phenanthridine	rds	rate-determining step
OSHA	Occupational Safety and Health Administration	ref.	reference
		rf	radio frequency, *n.*
owf	on weight of fiber	r-f	radio frequency, *adj.*
Ω	ohm	rh	relative humidity
P	peta (prefix for 10^{15})	RI	Ring Index
p	pico (prefix for 10^{-12}	rms	root-mean square
p-	para	rpm	rotations per minute
p	proton	rps	revolutions per second
p.	page	RT	room temperature
Pa	Pascal (pressure)	RTECS	Registry of Toxic Effects of Chemical Substances
PEL	personal exposure limit based on an 8-h exposure	*s*(eg, Bu*s*); *sec*-	secondary (eg, secondary butyl)
pd	potential difference	S	siemens
pH	negative logarithm of the effective hydrogen ion concentration	(*S*)-	sinister (counterclockwise configuration)
		S-	denoting attachment to sulfur
phr	parts per hundred of resin (rubber)	*s*-	symmetric(al)
p-i-n	positive-intrinsic-negative	S	second
pmr	proton magnetic resonance	(s)	solid, only as in $H_2O(s)$
p-n	positive-negative	SAE	Society of Automotive Engineers
po	per os (oral)		
POP	polyoxypropylene	SAN	styrene-acrylonitrile
pos	positive	sat(d)	saturate(d)
pp.	pages	satn	saturation
ppb	parts per billion (10^9)	SBS	styrene–butadiene–styrene
ppm	parts per milion (10^6)	sc	subcutaneous
ppmv	parts per million by volume	SCF	self-consistent field; standard cubic feet
ppmwt	parts per million by weight		
PPO	poly(phenyl oxide)	Sch	Schultz number
ppt(d)	precipitate(d)	sem	scanning electron microscope(y)
pptn	precipitation		
Pr (no.)	foreign prototype (number)	SFs	Saybolt Furol seconds
pt	point; part	sl sol	slightly soluble
PVC	poly(vinyl chloride)	sol	soluble

soln	solution	*trans-*	isomer in which
soly	solubility		substituted groups are
sp	specific; species		on opposite sides of
sp gr	specific gravity		double bond between
sr	steradian		C atoms
std	standard	TSCA	Toxic Substances Control
STP	standard temperature and		Act
	pressure (0°C and	TWA	time-weighted average
	101.3 kPa)	Twad	Twaddell
sub	sublime(s)	UL	Underwriters' Laboratory
SUs	Saybolt Universal seconds	USDA	United States Department
syn	synthetic		of Agriculture
^{t} (eg, Bu^t),	tertiary (eg, tertiary	USP	*United States*
t-, tert-	butyl)		*Pharmacopeia*
T	tera (prefix for 10^{12}); tesla	uv	ultraviolet
	(magnetic flux density)	V	volt (emf)
t	metric to (tonne)	var	variable
t	temperature	*vic-*	vicinal
TAPPI	Technical Association of	vol	volume (not volatile)
	the Pulp and Paper	vs	versus
	Industry	v sol	very soluble
TCC	Tagliabue closed cup	W	watt
tex	tex (linear density)	Wb	weber
T_g	glass-transition	Wh	watt hour
	temperature	WHO	World Health Organization
tga	thermogravimetric		(United Nations)
	analysis	wk	week
THF	tetrahydrofuran	yr	year
tlc	thin layer chromatography	(Z)-	zusammen; together;
TLV	threshold limit value		atomic number

Non-SI (Unacceptable and Obsolete) Units		Use
Å	angstrom	nm
at	atmosphere, technical	Pa
atm	atmosphere, standard	Pa
b	barn	cm^2
bar†	bar	Pa
bbl	barrel	m^3
bhp	brake horsepower	W
Btu	British thermal unit	J
bu	bushel	m^3; L
cal	calorie	J
cfm	cubic foot per minute	m^3/s
Ci	curie	Bq
cSt	centistokes	mm^2/s
c/s	cycle per second	Hz
cu	cubic	exponential form

†Do not use bar (10^5 Pa) or millibar (10^2 Pa) because they are not SI units, and are accepted internationally only in special fields because of existing usage.

Non-SI (Unacceptable and Obsolete) Units		Use
D	debye	$C \cdot m$
den	denier	tex
dr	dram	kg
dyn	dyne	N
dyn/cm	dyne per centimeter	mN/m
erg	erg	J
eu	entropy unit	J/K
°F	degree Fahrenheit	°C; K
fc	footcandle	lx
fl	footlambert	lx
fl oz	fluid ounce	m^3; L
ft	foot	m
ft·lbf	foot pound-force	J
gf den	gram-force per denier	N/tex
G	gauss	T
Gal	gal	m/s^2
gal	gallon	m^3; L
Gb	gilbert	A
gpm	gallon per minute	(m^3/s); (m^3/h)
gr	grain	kg
hp	horsepower	W
ihp	indicated horsepower	W
in.	inch	m
in. Hg	inch of mercury	Pa
in. H_2O	inch of water	Pa
in.-lbf	inch pound-force	J
kcal	kilo-calorie	J
kgf	kilogram-force	N
kilo	for kilogram	kg
L	lambert	lx
lb	pound	kg
lbf	pound-force	N
mho	mho	S
mi	mile	m
MM	million	M
mm Hg	millimeter of mercury	Pa
$m\mu$	millimicron	nm
mph	miles per hour	km/h
μ	micron	μm
Oe	oersted	A/m
oz	ounce	kg
ozf	ounce-force	N
η	poise	$Pa \cdot s$
P	poise	$Pa \cdot s$
ph	phot	lx
psi	pounds-force per square inch	Pa
psia	pounds-force per square inch absolute	Pa
psig	pounds-force per square inch gage	Pa
qt	quart	m^3; L
°R	degree Rankine	K
rd	rad	Gy
sb	stilb	lx
SCF	standard cubic foot	m^3
sq	square	exponential form
thm	therm	J
yd	yard	m

BIBLIOGRAPHY

1. The International Bureau of Weights and Measures, BIPM (Parc Saint-Cloud, France) is described in Ref. 4. This bureau operates under the exclusive supervision of the International Committee for Weights and Measures (CIPM).
2. *Metric Editorial Guide (ANMC-78-1)*, latest ed., American National Metric Council, 900 Mix Avenue, Suite 1 Hamden CT 06514-5106, 1981.
3. *SI Units and Recommendations for the Use of Their Multiples and of Certain Other Units (ISO 1000-1992)*, American National Standards Institute, 25 W 43rd St., New York, 10036, 1992.
4. Based on IEEE/ASTM-SI-10 *Standard for use of the International System of Units (SI): The Modern Metric System* (Replaces ASTM380 and ANSI/IEEE Std 268-1992), ASTM International, West Conshohocken, PA., 2002. See also www.astm.org
5. *Fed. Reg.*, Dec. 10, 1976 (41 FR 36414).
6. For ANSI address, see Ref. 3. See also www.ansi.org

C

Continued

CARBON MONOXIDE

1. Introduction

Carbon monoxide [630-08-0], CO, gaseous in normal atmospheric conditions (15°C and 101.3 kPa), is a colorless, odorless, and highly toxic gas. Lassonne discovered carbon monoxide in 1776 by heating a mixture of charcoal and zinc oxide. It was a source of heat for industrial and domestic purposes, and was used as a primary raw material in the manufacturing of German synthetic fuel during World War II.

Carbon monoxide is produced by the incomplete combustion of carbon in solid, liquid and gaseous fuels. Industrially produced carbon monoxide is used in the chemical and metallurgical industries, for the synthesis of several compounds (eg, acetic acid, polycarbonates, polyketones, etc.), and the creation of reducing atmospheres, respectively. The demand for carbon monoxide as a raw material for chemical synthesis is expected to continue a rapid growth over the next ten years.

Carbon monoxide is also a by-product of highway vehicle exhaust, which contributes about 60% of all CO emissions in the U.S. In cities, automobile exhaust can cause as much as 95% of all CO emissions. These emissions can result in high concentrations of CO, particularly in local areas with heavy traffic congestion. Other sources of CO emissions include industrial processes and fuel combustion in boilers and incinerators. Despite an overall downward trend in concentrations and emissions of CO, most U.S. metropolitan areas still experience high levels of CO. Household appliances fueled with gas, oil, kerosene, or wood may produce CO emissions that if not monitored, can accumulate to dangerous levels. In urban areas, carbon monoxide emissions are strictly controlled

1

and monitored for CO-producing industries, commercial vehicles and private cars. The Environmental Protection Agency (EPA) is the U.S. governmental body responsible for establishing the emission standards.

2. Physical Properties

Gaseous carbon monoxide is colorless, odorless, tasteless, flammable and highly toxic (1,2). It becomes a liquid at 81.62 K. Carbon monoxide is flammable in air over a wide range of concentration: lower limit of 12.5%, and upper limit of 74% at 20°C and 101.3 kPa. Carbon monoxide is moderately soluble in water at low temperatures, and virtually insoluble above 70°C.

Selected physical properties are listed in Table 1. Solubility data are listed in Table 2.

Table 1. **Physical Properties of Carbon Monoxide**

Property	Value
mol wt	28.011
melting point, K	68.09
boiling point, K	81.65
ΔH, fusion at 68 K, kJ/mol[a]	0.837
ΔH, vaporization at 81 K, kJ/mol[a]	6.042
density at 273 K, 101.3 kPa[b], g/L	1.2501
sp gr[c], liquid, 79 K	0.814
sp gr[d], gas, 298 K	0.968
critical temperature, K	132.9
critical pressure, MPa[b]	3.496
critical density, g/cm^3	0.3010
triple point	
temperature, K	68.1
pressure, kPa[e]	15.39
$\Delta G°$ formation at 298 K, kJ/mol[a]	−137.16
$\Delta H°$ formation at 298 K, kJ/mol[a]	−110.53
$S°$ formation at 298 K, kJ/(mol·K)[a]	0.1975
$C°_p$ at 298 K, J/(mol·K)[a]	29.1
$C°_v$ at 298 K, J/(mol·K)[a]	20.8
autoignition temperature, K	882
bond length, nm	0.11282
bond energy, kJ/mol[a]	1070
force constant, mN/m = (dyn/cm)	1,902,000
dipole moment, C·m[f]	0.374×10^{-30}
ionization potential, eV	14.01
flammability limits in air[g]	
upper limit, %	74.2
lower limit, %	12.5

[a]To convert J to cal, divide by 4.184.
[b]101.3 kPa = 1 atm; to convert MPa to atm, multiply by 9.87.
[c]With respect to water at 277 K.
[d]With respect to air at 298 K.
[e]To convert kPa to torr, multiply by 7.5.
[f]To convert C·m to debye, multiply by 2.99×10^{29}.
[g]Saturated with water vapor at 290 K.

Table 2. **Aqueous Solubility of CO at STP, L/L**

Temperature, °C	Bensen coefficient
0	0.03516
5	0.03122
10	0.02782
15	0.02501
20	0.02266
25	0.02076
30	0.01915
40	0.01647
50	0.01420
60	0.01197
70	0.00998
80	0.00762
90	0.00438

3. Chemical Properties

Chemically, carbon monoxide is stable with respect to decomposition (1). The bond energy of 1070 kJ/mol is illustrative of the triple bond configuration described molecular by orbital theory and is the highest observed bond energy for any diatomic molecule. At temperatures of 310–500°C, CO reduces many metal oxides to lower metal oxides, or metals, or produces metal carbides. Carbon monoxide absorbs at 2143 cm^{-1} in the infrared spectrum. The bonding between CO and transition metal atoms weakens the C–O bond, thus allowing the metal bonded CO to react more readily. Carbon monoxide is a reducing agent that reacts with oxidizers and salts such as iodic anhydride, palladium salts, and red mercuric oxide. Catalytic reduction of carbon monoxide produces methane. Catalytic oxidation of carbon monoxide leads to carbon dioxide. Carbon monoxide reacts violently with oxygen difluoride, chlorine produced by phosgene decomposition, and barium peroxide. Hydrogenation of carbon monoxide yields products that vary with catalysts and conditions: methane, benzene, olefins, paraffin waxes, hydrocarbon high polymers, methanol, higher alcohols, ethylene glycol, glycerol, have all been produced.

3.1. Metal Compatibility. Carbon monoxide is compatible with all commonly used metals at pressures below 3.5 MPa. However, at higher pressures, carbon monoxide reacts with nickel, iron, cobalt, manganese, chromium and gold, to form small quantities of metal carbonyls, which are unstable and highly toxic. The following metals and alloys may be employed with carbon monoxide under pressure: carbon steels up to 10 MPa, aluminum alloys, copper and copper alloys, low carbon stainless steel, and nickel-base alloys. The presence of moisture and sulfur-containing impurities in carbon monoxide increases its corrosive action on steel at any pressure (2). High-pressure plant equipment is often lined with copper for increased resistance to carbon monoxide attack.

3.2. Refractory Material Compatibility. Table 3 gives the behavior of refractory materials likely to be employed in high temperature furnaces.

Table 3. **Compatibility of Refractory Materials with Carbon Monoxide**

Material	Compatibility[a]
magnesia	A
zircon	A
bonded alumina	A
fused cast alumina	A
bubble alumina	A
stabilized zirconia	silicon carbide A: $t < 810°C$; B: $t > 1150°C$

silicon nitrite +
   silicon carbide
	A: $t < 810°C$;
	   B: $t > 1150°C$

magnesite	B–C
fosterite	A
synthetic mullite	A
converted mullite	B–C
silica	A
superduty fireclay	A

[a]A = no reaction, material satisfactory; B = slight reaction, material generally satisfactory; and C = reaction, material satisfactory in certain cases.

Table 4. **Compatibility of Elastomers and Plastics with Carbon Monoxide**

Material	Compatibility[a]
Teflon	A
natural rubber	C
neoprene	C
Hypalon	B
butyl rubber	C
Kel-F	A
Buna N	C

[a]A = good resistance; B = fair resistance; and C = poor resistance.

3.3. Compatibility with Elastomers and Plastics. The chemical resistance of certain elastomers and plastics to carbon monoxide is given in Table 4.

4. Reactions

4.1. Industrially Significant Reactions of Carbon Monoxide. *Reppe Chemistry.* Chemicals including acetic acid, acetic anhydride, formic acid, propionic acid, dimethyl carbonate, and methyl methacrylate are examples of final products derived from Reppe chemistry (3–6).

Acetic Acid. Acetic acid is synthesized by carbonylation of methanol by the following reaction:

$$CH_3OH + CO \xrightarrow{\text{catalyst}} CH_3COOH \tag{1}$$

The catalyst can be cobalt iodide, rhodium iodide, or iridium iodide. Other syntheses can be used to manufacture acetic acid, such as oxidation of n-butane or naphtha, oxidation of acetaldehyde, and terephthalic acid coproduct. However, methanol carbonylation has been the preferred process for new capacity over the last ten years because of its favorable raw material (methanol) and energy costs. The CO purity required for methanol carbonylation is in the range of 98 to 99% pure, with low concentrations of methane, hydrogen, nitrogen and argon. The feed pressure of CO in the reactor is approximately 35 MPa The primary use for acetic acid is feedstock for the production of vinyl acetate monomer (VAM). It is also a solvent for the air-based oxidation of p-xylene to terephthalic acid (see also ACETIC ACID AND DERIVATIVES).

Acetic Anhydride. Acetic anhydride is the largest commercially produced carboxylic acid anhydride. Its main industrial application is for acetylation reactions. Over 85% of acetic anhydride production goes into cellulose acetate flake, which is in turn used to make filament yarn, cigarette paper and cellulose ester plastics. Eastman Chemical also uses it as an intermediate to make photographic film base, Tenite cellulose plastics, textile chemicals, and coating chemicals. Acetic anhydride (qv) can be made by carbonylation of methyl acetate, by methanol carbonylation, as follows:

$$CH_3COOH + CH_3OH \longrightarrow CH_3COOCH_3 + H_2O \qquad (2)$$

$$CO + CH_3COOCH_3 \longrightarrow (CH_3CO)_2O \qquad (3)$$

The catalyst system is rhodium and iodide complexes and chromium metal powder on an alumina support.

The CO feed for equation 3 must be anhydrous, of high purity, and pressured at 15–18 MPa.

Vinyl Acetate Monomer (VAM). VAM is currently the most important vinyl ester. It is used mainly for the production of polymers and copolymers for paints, adhesives, textiles, and for the production of poly(vinyl alcohol), and poly(vinylbutyral). VAM can be produced by reacting methyl acetate with CO and hydrogen:

$$CH_3COOH + CH_3OH \longrightarrow CH_3COOCH_3 + H_2O \qquad (2)$$

$$2\,CH_3COOCH_3 + 2\,CO + H_2 \longrightarrow CH_3CH(OOC{-\!}CH_3)_2 + CH_3COOH \qquad (4)$$

$$CH_3CH(OOC{-\!}CH_3)_2 \longrightarrow CH_3COOCH{=\!=}CH_2 + CH_3COOH \qquad (5)$$

Although viable, this synthetic method has not yet been used in industrial applications. It could become more attractive depending on future raw material prices.

Formic Acid. Formic acid like acetic acid, is produced by methanol carbonylation followed by methyl formate hydrolysis:

$$CO + CH_3OH \longrightarrow HCOOCH_3 \qquad (6)$$

$$HCOOCH_3 + H_2O \longrightarrow HCOOCH + CH_3OH \qquad (7)$$

Carbon monoxide specifications for this reaction is typically 98% pure, with low levels of methane, hydrogen, chlorine, nitrogen, and a very low sulfur content of less than 1 ppm. The CO feed pressure in the reactor is around 1.5 MPa. Formic acid is a medium volume commodity chemical that has a variety of applications, including processing of natural rubber, textile finishing, production of dyes, flavors and fragrances, and as a chemical intermediate (see FORMIC ACID AND DERIVATIVES).

Propionic Acid.   Propionic acid is used in the production of cellulose esters, plastic dispersions, herbicides, and to a limited extent in pharmaceuticals, and in flavors and fragrances. It is gaining importance for the preservation of forage cereals, and animal feeds because many putrefying and mold-forming micro-organisms cannot survive in its presence. Propionic acid can be commercially produced by carbonylation of ethylene, a one-step Reppe process catalyzed with nickel propionate ($Ni(CO)_4$), at 300°C and 22 MPa.

$$CO + H_2C\!\!=\!\!CH_2 + H_2O \longrightarrow CH_3CH_2COOH \tag{8}$$

Dimethyl Carbonate (DMC).   DMC is an extremely versatile chemical. It is used as an organic solvent, additive for fuels, reagent, as a substitute for phosgene, and in the synthesis of other alkyl or aryl carbonates used as synthetic lubricants, solvents, and in methylation and carbonylation reactions for the preparation of isocyanates, urethanes, and polycarbonates. DMC can be commercially produced by oxycarbonylation of methanol catalyzed by copper salts at 150 °C and 1–5 MPa.

$$CO + CH_3OH + \frac{1}{2}O_2 \longrightarrow (CH_3O)_2CO + H_2O \tag{9}$$

Methyl Methacrylate (MMA).   MMA polymerizes to form a clear plastic that has excellent transparency, strength, and outdoor durability. The automotive and construction markets create the largest demand for acrylic sheet. It is also used in the manufacturing of acrylic paints, including latex paints, and lacquers. MMA can be commercially produced from acetone, methanol and high purity CO (99.8%) at approximately 4 MPa.

Koch Carbonylation.   The Koch carbonylation is of an olefin in a two-stage reaction. Two main categories of compounds fall under the Koch carbonylation: trialkylacetic acids (monoacids from olefins), and adipic acid (two acidic functions from carbonylation of α-diene) (7).

Trialkylacetic Acids.   The lowest member of the series R=R′=R″=CH_3 is the C5 acid, trimethylacetic acid, also called neopentanoic acid or pivalic acid. The principal commercial products are the C5 acid and the C10 acid (also known as Exxon's neodecanoic acid, or Shell's Versatic 10). The trialkylacetic acids have a number of uses in areas such as polymers, pharmaceuticals, agricultural chemicals, cosmetics, and metal-working fluids. Commercially important derivatives of these acids include acid chlorides, peroxyesters, metal salts, vinyl esters and glycidyl esters. Pivalic acid (C5), for example, is prepared via Koch's reaction:

$$(CH_3)_2\!-\!C\!\!=\!\!CH_2 \longrightarrow (CH_3)_3\!-\!C^+ \tag{10}$$

where the strong acid catalyst is either a Bronsted acid (H_2SO_4, H_3PO_4, HF), or a Lewis acid such as BF_3.

$$(CH_3)_3\text{—}C^+ + CO \longrightarrow (CH_3)_3\text{—}C\text{—}CO^+ \qquad (11)$$

$$(CH_3)_3\text{—}C\text{—}CO^+ + H_2O \longrightarrow (CH_3)_3\text{—}C\text{—}COOH + H^+ \qquad (12)$$

The C10 tryalkylacetic acid is manufactured using the same process and catalysts. For the C10 acids, a branched C9 olefin is typically used. The resulting C10 acid is typically a mixture of isomers due to chemical rearrangement, olefin dimerization, and oligomerization (see also CARBOXYLIC ACIDS, TRIALKYLACETIC ACIDS).

Adipic Acid. Adipic acid, also known as hexanedioic acid, is the most significant commercially of all the aliphatic dicarboxylic acids. Appearing in nature in only minor amounts, it is synthesized on a very large scale worldwide. The principal use of adipic acid is to produce nylon 6/6, a linear polyamide made by condensing adipic acid with hexamethylene diamine, HMDA. The market for nylon 6/6 is predominantly in fibers. The other uses of adipic acid are in plasticizers, unsaturated polyesters, and polyester polyols (for polyurethane resins). Adipic acid can be synthesized by carbonylation of 1,3-butadiene. This process is attractive from a raw material cost, but requires high operating CO pressure (see also ADIPIC ACID).

Phosgenation. *Phosgene.* Phosgene is an inorganic, intermediate produced by the catalytic reaction of chlorine and carbon monoxide (8). It is a gaseous product, that cannot be stored or conveniently shipped owing to its extreme toxicity. As a result, it is usually produced on demand for intermediate use. Phosgene is an important starting compound in the production of intermediates and end-product in many branches of large-scale industrial chemistry. Most phosgene (80%) is used for the production of diisocyanates. The next largest phosgene application is the production of polycarbonates. The commercial production of phosgene is by the following reaction:

$$CO + Cl_2 \longrightarrow COCl_2 \qquad (13)$$

an activated carbon catalyst is used. This reaction is strongly exothermic. Because of toxicity and corrosiveness of phosgene product, strict and extensive safety procedures are incorporated in plant and operation design. CO purity requirements for this reaction vary considerably depending on the end-products. However, a low content in methane and hydrogen is always required for safety reasons to prevent spontaneous exothermic HCl formation when mixing CO and chlorine. Typically, low sulfur impurities (COS, $CSCl_2$) are also required, as those compounds affect the quality of the end-products.

Diisocyanates. Diisocyanates, including toluene diisocyanate (TDI) and 4,4′-methylene diphenyl diisocyanate (MDI), have become large-volume raw materials for addition polymers, such as polyurethanes, polyureas, and polyisocyanurates. By varying the reactants (isocyanates, polyols, polyamines, and others) for polymer formation, a myriad of products have been developed ranging from flexible and rigid insulation foams to the high modulus automotive exterior parts to high quality coatings and abrasion-resistant elastomers unmatched by any other polymeric material. The most common method of preparing isocyanates on a commercial scale is the reaction of phosgene and aromatic or aliphatic amine precursors. The overall reaction is shown below:

$$R\text{—}NH_2 + COCl_2 \longrightarrow R\text{—}NHCOCl + HCl \tag{14}$$

$$R\text{—}NHCOCl \longrightarrow R\text{—}N{=}C{=}O + HCl \tag{15}$$

Nonphosgene routes to isocyanate production have been developed, but none has been commercialized. The term nonphosgene route is primarily used in conjunction with the conversion of amines to isocyanates via the use of carboxylation agents. These approaches are becoming more attractive to the chemical industry as environmental or toxicological restrictions involving chlorine or phosgene are increasingly enforced.

Polycarbonates. Polycarbonates are an unusual and extremely useful class of polymers. The vast majority of polycarbonates (qv) are based on bisphenol A (BPA). The economically most important polycarbonate is the Bisphenol A

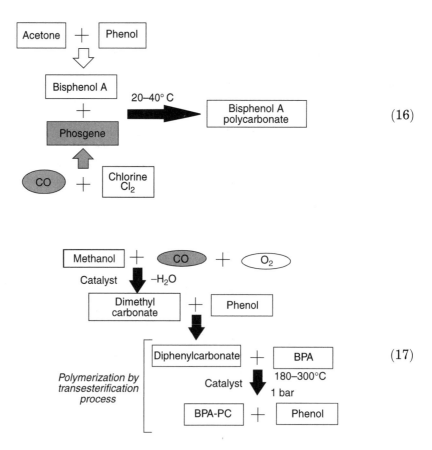

Fig. 1. Polycarbonate production by the phosgene route (eq. 16) and by the base-catalyzed transesterification of a bisphenol (eq. 17).

polycarbonate (BPA–PC). Its great commercial success is owing to its unique combination of properties: extreme toughness, outstanding transparency, excellent compatibility with several polymers, and high heat distortion resistance. Some of its main uses include coatings, films, fibers, resin for shatterproof windows, computer housings, tape reel housings, gas meter covers, lenses for automobiles and appliances, and pipe. Polycarbonates can be commercially produced by the Schotten-Bauman reaction (interfacial polycondensation): the phosgene route (eq. 16) (Fig. 1) or by base-catalyzed transesterification of a bisphenol with a monomeric carbonate (eq. 17) (Fig. 1).

Ethylene–CO Copolymers or Polyketones. These compounds are defined as polymers with 1:1 molar ratio of carbon monoxide to ethylene (9). Carilon was the first polyketone commercially produced by Shell Chemicals in 1995: it is a new family of polymers called aliphatic polyketone, and results from the polymerization of perfectly alternating structures of CO and olefins (such as ethylene). The product's great strength lies in its broad range of high performance characteristics, allied with its ease of processing. It has a number of properties which make it ideally suited for use in the automotive sector, including: superior energy absorption capacity, dimensional stability at elevated temperature, good impact resistance over a broad temperature range, insensitivity to moisture, and excellent resistance to automotive fuels and their vapors. The polymerization is catalyzed by a single-site late transition-metal (ie, palladium) as follows:

$$CH_2{=}CH_2 + CO \longrightarrow -(-CH_2CH_2C-)_x-\ -(-CH_2CH_2-)_y- \qquad (18)$$

4.2. General Reactions of Carbon Monoxide. *With Hydrogen.* In addition to the reactions already discussed, other products may be obtained from synthesis gas depending on the catalyst used. In a liquid-phase high pressure reaction (60 MPa or 600 atm), a rhodium cluster complex catalyzes the direct formation of ethylene glycol, propylene glycol (see Glycols), and glycerol (qv) from synthesis gas (eq. 19) (10). Mixtures of methanol, ethanol (11), acetaldehyde, and acetic acid (12) are formed by using supported rhodium catalysts at 598 K and 17 MPa (168 atm). Rates of reaction for this latter route appear to be too slow for commercial application.

$$2\,CO + 3\,H_2 \longrightarrow HOCH_2CH_2OH \qquad (19)$$

With Alcohols, Ethers, and Esters. Carbon monoxide reacts with alcohols, ethers, and esters to give carboxylic acids. The reaction yielding carboxylic acids is general for alkyl (13) and aryl alcohols (14). It is catalyzed by rhodium or cobalt in the presence of iodide and provides the basis for a commercial process to acetic acid.

Strong base catalyzes the formation of derivatives of formic acid in the reaction between alcohols and carbon monoxide (15). Methyl formate is made at 443–463 K and 1–2 MPa (10–20 atm) (eq. 20).

$$CH_3OH + CO \xrightarrow{\ NaOH\ } HCOOCH_3 \qquad (20)$$

Methanol reacts with carbon monoxide and hydrogen to form ethanol in the homologation reaction. Cobalt carbonyl catalyzes the transformation at 473 K and 30 MPa (300 atm) pressure, and gives yields of less than 75% ethanol. The greatest activity in the homologation reaction is observed for methyl and benzyl alcohols (eq. 21) (16). Reaction between methyl acetate, carbon monoxide, and hydrogen at 408–433 K and up to 10 MPa (100 atm) pressure using a palladium or rhodium iodide catalyst leads to the production of ethylidene diacetate (17) (eq. 22). Ethylidene diacetate can be pyrolyzed to vinyl acetate (eq. 23).

$$CH_3OH + CO + 2\,H_2 \longrightarrow C_2H_5OH + H_2O \tag{21}$$

$$2\,CH_3COOCH_3 + 2\,CO + H_2 \longrightarrow CH_3CH(OOCCH_3)_2 + CH_3COOH \tag{22}$$

$$CH_3CH(OOCCH_3)_2 \longrightarrow CH_2 = CHOOCCH_3 + CH_3COOH \tag{23}$$

With Formaldehyde.   The sulfuric acid catalyzed reaction of formaldehyde with carbon monoxide and water to glycolic acid at 473 K and 70 MPa (700 atm) pressure was the first step in an early process to manufacture ethylene glycol. A patent (18) has described the use of liquid hydrogen fluoride as catalyst, enabling the reaction to be carried out at 298 K and 7 MPa (70 atm) (eq. 24).

$$HCHO + CO + H_2O \longrightarrow HOCH_2COOH \tag{24}$$

With Unsaturated Compounds.   The reaction of unsaturated organic compounds with carbon monoxide and molecules containing an active hydrogen atom leads to a variety of interesting organic products. The hydroformylation reaction is the most important member of this class of reactions. When the hydroformylation reaction of ethylene takes place in an aqueous medium, diethyl ketone is obtained as the principal product instead of propionaldehyde (19). Ethylene, carbon monoxide, and water also yield propionic acid under mild conditions (448–468 K and 3–7 MPa or 30–70 atm) using cobalt or rhodium catalysts containing bromide or iodide (20,21).

Carbon monoxide also reacts with olefins such as ethylene to produce high molecular weight polymers. The reaction of CO with ethylene can be initiated by an x-ray irradiator (22) or transition-metal catalyzed reactions (23). The copolymerization of ethylene with carbon monoxide is catalyzed by cationic Pd (II) complexes such as $\{Pd[P(C_6H_5)_3]_n(CH_3CN)_{4-n}\}(BF_4)_2$ where $n = 1$–3. With this catalyst, copolymerization can be carried out at 25°C and pressures as low as 2.1 MPa.

Oxidative Carbonylation.   Carbon monoxide is rapidly oxidized to carbon dioxide; however, under proper conditions, carbon monoxide and oxygen react with organic molecules to form carboxylic acids or esters. With olefins, unsaturated carboxylic acids are produced, whereas alcohols yield esters of carbonic or oxalic acid. The formation of acrylic and methacrylic acid is carried out in the liquid phase at 10 MPa (100 atm) and 110°C using palladium chloride or rhenium chloride catalysts (eq. 25) (24,25).

$$CH_2{=}CH_2 + CO + \frac{1}{2}\,O_2 \longrightarrow CH_2 = CHCOOH \tag{25}$$

Dimethyl carbonate and dimethyl oxalate are both obtained from carbon monoxide, oxygen, and methanol at 363 K and 10 MPa (100 atm) or less. The choice of catalyst is critical; cuprous chloride (26) gives the carbonate (eq. 26); a palladium chloride–copper chloride mixture (27,28) gives the oxalate, (eq. 27). Anhydrous conditions should be maintained by removing product water to minimize the formation of by-product carbon dioxide.

$$2\,CH_3OH + CO + \frac{1}{2}\,O_2 \longrightarrow (CH_3O)_2CO + H_2O \tag{26}$$

$$2\,CH_3OH + 2\,CO + \frac{1}{2}\,O_2 \longrightarrow CH_3O_2C\!-\!CO_2CH_3 + H_2O \tag{27}$$

Isocyanate Synthesis. In the presence of a catalyst, nitroaromatic compounds can be converted into isocyanates, using carbon monoxide as a reducing agent. Conversion of dinitrotoluene into toluenediisocyanate (TDI) with carbon monoxide (eq. 28), could offer significant commercial advantages over the current process using phosgene. The reaction is carried out at 473–523 K and 27–41 MPa (270–400 atm) with a catalyst consisting of either palladium chloride or rhodium chloride complexed with pyridine, isoquinoline, or quinoline and yields are in excess of 80% TDI (29,30).

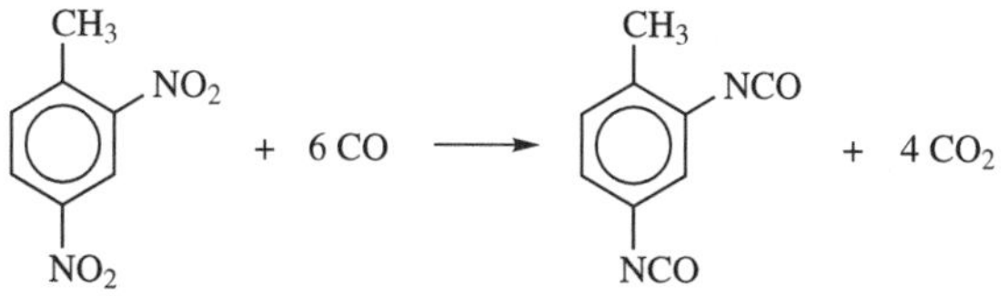

$$\text{(structure: 2,4-dinitrotoluene)} + 6\,CO \longrightarrow \text{(structure: toluene diisocyanate)} + 4\,CO_2 \tag{28}$$

Dimethylformamide. The industrial solvent dimethylformamide is manufactured by the reaction between carbon monoxide and dimethylamine.

$$(CH_3)_2NH + CO \longrightarrow (CH_3)_2NCHO \tag{29}$$

The reaction is carried out in the liquid phase using a sodium methoxide catalyst at 60–130°C and 0.5–0.9 MPa (5–9 atm) (31).

Aromatic Aldehydes. Carbon monoxide reacts with aromatic hydrocarbons or aryl halides to yield aromatic aldehydes (see ALDEHYDES). The reaction of equation 24 proceeds with yields of 89% when carried out at 273 K and 0.4 MPa (4 atm) using a boron trifluoride–hydrogen fluoride catalyst (32), whereas conversion of aryl halides to aldehydes in 84% yield by reaction with CO + H$_2$ requires conditions of 423 K and 7 MPa (70 atm) with a homogeneous palladium catalyst (33) and also produces HCl.

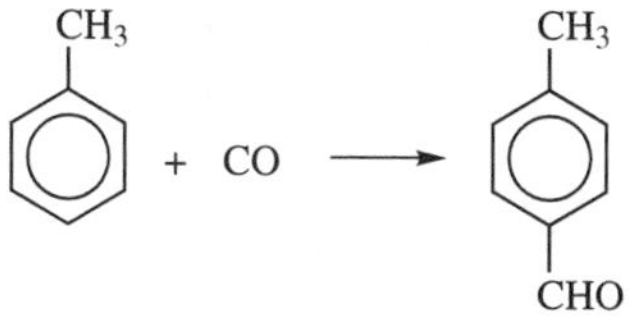

$$\text{(structure: toluene)} + CO \longrightarrow \text{(structure: p-tolualdehyde)} \tag{30}$$

Metal Carbonyls.   Carbon monoxide forms metal carbonyls or metal carbonyl derivatives with most transition metals (34) (see COORDINATION COMPOUNDS).
Metal carbonyls are used in a variety of industrial applications in addition to
their use as catalysts. Methylcyclopentadienylmanganesetricarbonyl (MMT),
$[CH_3C_5H_4Mn(CO)_3]$ was sold as an antiknock additive, but its use in unleaded
gasoline was banned by the Environmental Protection Agency (EPA) in 1978;
tungsten and molybdenum hexacarbonyls are thermally decomposed to obtain
very pure metal films; and numerous carbonyls are used as reagents in organic
synthesis (see CARBONYLS).

5. Manufacture

Commercial carbon monoxide is a co-product, along with hydrogen, of synthetic
gas (syngas) production. Several technologies, based on steam reforming or partial oxidation processes, are used to produce syngas, with a hydrogen-to-carbon
monoxide mole ratio varying from 2 to 0. A ratio of 2 indicates that two parts
hydrogen are produced for one part of carbon mononoxide; a ratio of zero indicates pure carbon monoxide without hydrogen. The principal components of
the resulting syngas, hydrogen and carbon monoxide are then separated and
purified by pressure swing adsorption and/or cryogenic distillation. The purity
of the final carbon monoxide product typically ranges from 97% to 99.9% (35).
The nature and level of the impurities remaining in the final carbon monoxide
product are usually more critical than the total purity for chemical synthesis
applications.

5.1. Syngas Technologies.   The principal technologies employed today
to produce syngasare (*1*) steam methane reforming, (*2*) naphtha reforming,
(*3*) autothermal reforming, (*4*) oxygen secondary reforming, (*5*) partial oxidation
of hydrocarbons, petroleum coke and coal, and (*6*) reverse shift of hydrogen and
carbon dioxide (36–39).

Steam Methane Reforming.   This process involves the catalyzed reaction
of steam and methane to yield a mixture of carbon monoxide, hydrogen, and carbon dioxide. The chemical reactions are as follows:

$$CH_4 + H_2O \iff CO + 3\,H_2 \tag{31}$$

$$CO + H_2O \iff CO_2 + H_2 \tag{32}$$

Equation 31 is the reforming reaction at $900°C$, 3 MPa using a nickel-based catalyst. Equation 32 is the water-shift reaction. The reforming reaction is
endothermic and takes place in a primary reforming furnace. The water-shift
reaction takes place in a separate reformer, and can be eliminated if a CO-rich
mixture is required. The typical hydrogen to carbon monoxide ratio for this technology varies from 3 to 4.9, thus producing a hydrogen-rich stream. Table 5 illustrates typical reformer furnace ($890°C$, 2.4 MPa) outlet syngas composition.

Naphtha Reforming.   This process substitutes naphtha for methane in the
reformer. This process presents the advantage of higher hydrogen to carbon
monoxide feed content; thus resulting in lower hydrogen to CO rates.

Table 5. **Typical Reformer Furnace Outlet Syngas Composition**

Component	Concentration, vol %
hydrogen	51.0
carbon monoxide	10.4
carbon dioxide	5.0
methane	2.0
water vapor	31.6

Autothermal Reforming. This process is a combination of partial oxidation and steam reforming in one reactor. The exothermic heat of reaction of the partial oxidation of the hydrocarbon feedstock provides the energy required for the (endothermic) steam methane reforming reaction to take place. The chemical reactions taking place in an autothermal reactor are as follows:

In the catalytic steam-reforming zone: reforming (eq. 31) and water-shift (eq. 32) reactions using a nickel catalyst.

In the combustion zone (ca 1,200 °C):

$$C_n H_m + n/2\, O_2 \;\Rightarrow\; n\, CO + m/2\, H_2 \tag{33}$$

$$CH_4 + \frac{1}{2}\, O_2 \;\Rightarrow\; CO + 2\, H_2 \tag{34}$$

$$H_2 + \frac{1}{2}\, O_2 \;\Rightarrow\; H_2O \tag{35}$$

$$CO + \frac{1}{2}\, O_2 \;\Rightarrow\; CO_2 \tag{36}$$

In the steam-reforming zone:

$$CH_4 + H_2O \;\Longleftrightarrow\; CO + 3\, H_2 \tag{37}$$

$$CO + H_2O \;\Longleftrightarrow\; CO_2 + H_2 \tag{38}$$

Some of the benefits of autothermal reforming are that no external feed is required, there is flexibility in feedstock selection (from methane-rich natural gas to naphtha), and the ability to produce syngas with low hydrogen to CO ratios. Typically, capital investment for autothermal reforming is lower than for steam reforming. However, operating costs are similar to slightly higher due to the added cost of pure oxygen.

Oxygen Secondary Reforming. This process is a conventional steam methane reformer with a secondary reformer reactor and direct-contact water quench downstream of the primary reactor. Pure oxygen is introduced in the secondary reactor to produce CO-rich syngas. The heavier the hydrocarbon feedstock, the lower the hydrogen to CO ratio.

Partial Oxidation. This process is the nonscatalytic reaction of hydrocarbons, petroleum, coke or coal with steam and oxygen at a high temperature and pressure to produce syngas. The hydrogen to CO ratio depends on the carbon to hydrogen ratio of feedstock. Reactions in partial oxidation are extremely complex. Simplistically, thermal cracking at high temperatures in the reactor produce lowmolecular-weight hydrocarbon fragments. The fragments then react with pure oxygen follows:

$$CH + 1/2\,O_2 \;\Rightarrow\; CO + 1/2\,H_2 \tag{39}$$

Additionally, many of the hydrocarbon fragments are completely oxidized to carbon dioxide and water as follows:

$$CH + 5/4\,O_2 \;\Rightarrow\; CO_2 + 1/2\,H_2O \tag{40}$$

Concurrently, the reversible water-gas shift reaction (eq. 32) takes place, but is incomplete as oxygen is less than stoichiometric, leading to a maximum carbon dioxide content of 2 vol% in the reactor effluent. Reactions 32 and 40 are highly exothermic.

The typical reactor temperature is 1250–1500°C and the pressure is between 2.5 and 8 MPa. For heavy hydrocarbon feedstock, a temperature moderator such as carbon dioxide or steam must be used to control the reactor temperature and adjust the hydrogen to CO ratio. Reactor effluent contains carbon monoxide, hydrogen, carbon dioxide (from the water-gas shift reaction), steam, and trace amounts of argon and nitrogen, which enter the system with the oxygen feedstock. If the hydrocarbon feed contains sulfur, hydrogen sulfide and carbonyl sulfide appears in the raw syngas. The high temperature and highly reducing atmosphere in the reactor prevent the formation of NOx and SOx.

The partial oxidation process presents great flexibility with respect to the feedstock (from natural gas to petroleum residue and petroleum coke), as well as the advantage of producing minimal gaseous emissions of NOx and SOx. However, both capital investment and operating costs are high for this process, which makes it prohibitively expensive for light feedstock, from natural gas to naphtha.

Reverse Shift of Hydrogen and Carbon Dioxide.   This process uses the reverse water-gas shift reaction:

$$CO_2 + H_2 \;\Longleftrightarrow\; H_2O + CO \tag{41}$$

This reaction can be carried out in: (*1*) a conventional primary reformer with a nickel based reforming catalyst at a furnace outlet temperature of 950°C and 1.4 MPa pressure; (*2*) a simple catalyst-filled reactor that takes advantage of the exothermic methanation reaction to provide the heat needed to drive the simultaneous reverse shift reaction. The down sides of each type of reactor are respectively: (*1*) a fuel stream is required to supply heat to the reaction, and produces export steam; (*2*) methane is a by-product in the syngas effluent. In either case, the production of CO by reverse water-shift reaction requires a large, low-cost source of hydrogen and carbon dioxide feedstock. High purity carbon dioxide is available from ethylene oxide and vinyl acetate monomer units. However, the carbon dioxide produced by these two processes is only available at low pressure and must be compressed for use as feedstock in the reverse-shift process. Typically, the reverse-shift process is not a primary source of carbon monoxide, but rather a supplemental syngas production process that allows petrochemical complexes to adjust the hydrogen to CO ratio to a required level.

5.2. Purification Technologies. All sources of CO are essentially gas mixtures with two primary components, hydrocarbon and CO, as well as other gases including nitrogen, carbon dioxide, methane, and moisture. Purification of carbon monoxide from the syngas effluent is therefore required prior to being used in applications such as organic synthesis. The cost of separation of CO or H_2 contributes significantly to the total production cost of high purity CO. Purification techniques fall into four main categories: cryogenic processes, adsorption process, membrane, and liquid absorption processes (40–44). The choice of the most attractive process is based on the feed conditions, and the final product specifications.

Cryogenic Processes. These processes essentially consist of liquefaction of part of the fuel stream, followed by a phase separation and distillation of the remaining liquid components. In general cryogenic processes are suitable for large capacity high purity CO plants where the nitrogen content of the purified CO stream is tolerated by the downstream application. This method is the oldest, and two principal methods can be used in large-scale processes depending on the required purity: partial condensation cycle; methane wash cycle.

Partial Condensation. Partial condensation allows the liquefaction of CO and methane in several cooling steps, leaving a residual gas stream containing approximately 98% hydrogen and 1 to 2% CO. The liquid phase CO, containing methane and hydrogen impurities is then flashed and distilled to produce a high purity CO stream along with a CO–methane fuel stream. Nitrogen present in the feed remains in the purified hydrogen and CO streams, with a majority staying in the CO product. Figure 2 illustrates a typical process flow schematic.

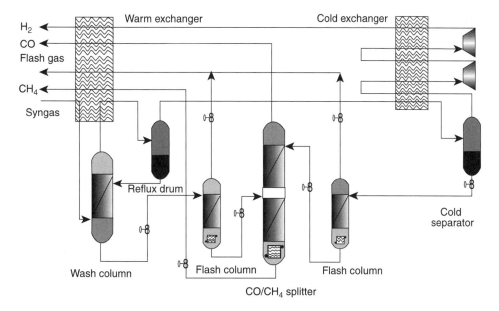

Fig. 2. Process flow schematic for CO/H_2 separation by partial condensation.

The feed gas is compressed and dried to remove residual water and carbon dioxide (that would otherwise freeze in the cold box). The feed gas is cooled against the product stream in the warm exchanger and used to provide heat to the re-boiler of the CO–methane splitter column. Liquid CO and methane are removed in the wash column. Vapor from the wash column is further cooled in the cold exchange. Most of the remaining CO is condensed in the cold exchanger and separated in the cold separator. The final product purity is critically dependent upon the temperature reached at this step. The liquid from the cold separator is a high-purity CO stream used to reflux the CO–methane splitter column. Liquid from the wash column is flashed in the flash column to remove dissolved hydrogen. The vapor from the wash column is rewarmed, compressed and recycled to the feed to recover the contained CO. The liquid from the flash column is directed to the CO–methane column. The CO overhead is rewarmed and recovered as product. The CO–methane liquid is rewarmed and is then available as reformer fuel. Hydrogen from the cold separator is warmed in the cold exchanger and expanded to provide refrigeration for the cycle. Rewarmed in the cold and warm exchangers, hydrogen exits the process at 97–98% purity. The partial condensation is process of choice with high-pressure feed, high H_2/CO ratio, and low hydrogen product purity and pressure requirements.

Methane Wash Cycle. This cycle uses liquid methane to absorb CO from the feed, resulting in a hydrogen stream containing parts per million of CO but 2 to 3% methane. A process flow schematic is presented in Figure 3.

The feed gas is purified by adsorption to remove residual carbon dioxide and water, cooled in the main exchanger and fed to the bottom of the wash column. The column is refluxed with liquid methane to produce a hydrogen stream containing parts per million of CO but saturated with methane (2–3%). The liquid stream from the wash column is preheated and flashed to the flash column where

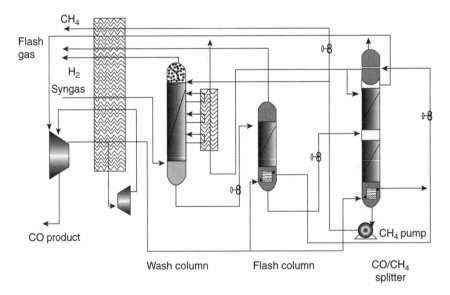

Fig. 3. Process flow schematic CO/H_2 separation by methane wash.

hydrogen dissolved in methane is rejected for use as fuel. To minimize CO losses, the flash column is also refluxed with liquid methane. The hydrogen-free liquid from the flash column is then heated and flashed to the CO−methane splitter column. The CO overhead from the column is rewarmed and compressed. Part of this gaseous stream of CO leaves the process as product, the remainder is cooled and used to reboil the splitter column and preheat the column feeds before being flashed for refrigeration with the liquid used as reflux to the sputter column. The liquid methane from the sputter column is pumped to the wash and flash columns, the excess being vaporized for fuel. The process refrigeration is provided by liquid nitrogen evaporation or by a nitrogen or CO turboexpander recycle system. Methane wash is the process of choice with low feed pressure, low H_2/CO ratio, and a high hydrogen purity with respect to CO.

Removal of nitrogen from the CO stream requires an additional column and cryogenic distillation step. Separation of nitrogen and CO by cryogenic process is difficult as the difference in their boiling point is only 6°C.

Adsorption Process.   Also known as vacuum swing adsorption (VSA), the adsorption process is suited to the production of high purity product with a high yield of CO. Adsorption is the binding of molecules from a gaseous phase to the surface of a solid (the adsorbent). Physical forces that are dictated by the type of molecules and the characteristics of the solid cause this binding. Generally, the adsorbent load capacity increases with the partial pressure of the gas component and with a decreasing temperature. The process consists of passing the syngas mixture (H_2/CO) sequentially through first and second adsorptive beds, each of which adsorbs H_2 more readily than CO. Typically, a minimum of three adsorptive beds are used, as the process involves three cyclical steps: production (selective adsorption), regeneration of the adsorber (evacuation of the adsorbed gas), and repressurization of the bed with a portion of the purified stream. The number of beds is increased for higher flow rates or higher CO recovery rates. In general, CO-selective absorbents are available to only a very limited extent. Sodium-type mordenite active-carbon-supported carbon and activated carbon are the most common absorbents. Active carbon copper is a chemical absorbent based on the selective binding capability of Cu^+ to CO. Several adsorbents consisting of porous carriers, such as activated carbon or zeolite, and supported Cu^+, have been developed for CO adsorption. These adsorbents exhibit selectivity for CO; however, their adsorption capacities and selectivity depends significantly on the properties of the carrier. Sample performance data of a commercial VSA plant for recovery of CO is pesented in Table 6. The VSA cycle yield decreases as the desired CO purity is increased.

Table 6. **Example of Operating Performance of a CO VSA Plant**

Feed gas components	Concentration, (%)	Purified CO stream
H_2	2	flow rate: 150 Nm^3/h
CO	68	purity: 99–99.9%
CO_2	16	yield: 80–90%
N_2	13	
O_2+ Ar	1	

Membrane Processes.   Membrane gas separation was first introduced in the mid-1970s. The first application of membrane processes in syngas separation was to adjust the H_2/CO ratio of natural gas-reformed syngas using the PRISM polysulfone separator (45). A membrane system selectively removes hydrogen and moisture from a compressed syngas feed, leaving a carbon monoxide rich stream as the primary product. This separation is accomplished utilizing hollow fiber polymeric membranes housed in modules with no moving parts. An individual fiber is about the size of a human hair (approximately 0.1 mm in diameter) and each module houses thousands of these fibers. Syngas molecules fed under pressure to the center of the hollow fibers partially dissolve in the polymeric membrane material, diffuse to the low pressure side outside of the fibers, and desorb at the lower pressure. Each component in the syngas stream dissolves in the polymer to a different extent and permeates at a different rate. "Fast" components with a high permeation rate, such as hydrogen and moisture, diffuse through the membrane, flow out through the hollow fiber interior and are channeled into the residue stream. "Slow" molecules, such as carbon monoxide, methane, carbon dioxide, and nitrogen, are held in the high-pressure stream, flow around the walls of the fibers and are removed from the membrane as the dry product gas. Feed gas is sent to a coalescing filter to remove liquids, and is preheated before entering the permeator. Membrane systems offer several advantages over competing separation technologies, such as eliminating: (*1*) the need for a drier and extended equipment; (*2*) cooling time required in cryogenic processes. Also, variations in flow, pressure and composition associated with the VSA technology are nonexistent with membrane processes. One key element in the membrane process is the selection of a selective material. For both syngas H_2:CO ratio adjustment and pure CO production, cellulose acetate membranes were found to provide higher selectivity and permeability compared to polysulfone membranes under similar operating conditions. In membrane processes, impurities such as methane, nitrogen and carbon dioxide, remain with the treated CO, while hydrogen and water are permeated. Subsequent purification of CO to remove the methane and carbon dioxide impurities can be made by VSA processes. New developments in membrane technology focus on improving H_2:CO selectivity. Recent studies (46,47) show that polyimide membranes exhibit competitive H_2:CO selectivity as high as 350. A new strategy to increase selectivity is to cross-link a transition metal complex (ruthenium and niobium-based) to the cellulose acetate membrane to decrease CO permeability.

Liquid Absorption Processes.   These processes are based on the selective and reversible complexation of CO with metal-based complexing compounds in a liquid solution. Liquid absorption processes include: ammoniacal copper liquor process:

$$CO + Cu(NH_3)_2+ \iff Cu(NH_3)_2(CO)^+ \tag{42}$$

COSORB process:

$$CO + ArX \iff COX + Ar \tag{43}$$

Where ArX = copper(I)tetrachloroaluminate(III) aromatic solution (benzene or toluene).

Other gases, such as hydrogen, carbon dioxide, methane, and nitrogen are chemically inert to this solvent, but are slightly soluble in benzene or toluene. Water, ammonia, hydrogen sulfide, sulfur dioxides, and oxides of nitrogen are poisons to the complex, and therefore must be removed in a pretreatment step (ie, molecular sieve adsorption system). The COSORB process works well to produce high-purity CO from a variety of gas mixtures, including a nitrogen-rich feed. Separation of nitrogen from CO by the COSORB process is easier and more economical that by cryogenic distillation.

In a typical process, CO produced by COSORB technology is 99% pure with a 99% yield. COSORB is also less capital intensive than the ammoniacal copper liquor process. However, due to degradation of the absorbent in the COSORB process, this process is seldom used.

New CO-complexing compounds have been studied for liquid absorption applications, including iron complexes, palladium complexes, selenium and secondary amine systems (48–50).

6. Production

Carbon monoxide and hydrogen are generally produced simultaneously by syngas plants. Carbon monoxide production for relatively large users typically falls into one of three cases. They are (1) a plant on the user's property owned and operated by the user; (2) a plant that is on or adjacent to the user's property, owned and operated by an industrial gas company per a long-term contract between the industrial gas supplier and the CO user; and (3) a plant owned and operated by an industrial gas company that supplies carbon monoxide to several users. In the first two categories, carbon monoxide is generally supplied to the user's site via a pipe. In the last case, carbon monoxide is distributed via a pipeline with branches to several users' sites. In most cases, pure hydrogen produced from a syngas plant is distributed similarly to CO. For most applications, an average capacity of a CO plant is approximately 4000 to 8000 Nm^3/h. Some of the largest plants built can produce up to around 25,000 Nm^3/h.

7. Shipment

Gas by pipeline is a cost-effective way to manufacture and supply CO to the user. Losses and distribution costs are minimized. Carbon monoxide pipeline networks are found in heavily industrialized areas such as the Gulf Coast in the United States, and in the Rotterdam area in Europe.

When the syngas plant belongs to an industrial gas company, then excess carbon monoxide and hydrogen roan be stocked in liquid and/or gaseous phase and supplied to smaller customers by cryogenic trucks (liquid phase), or high pressure tube trailers (gas phase). Cryogenic liquid tanks of appropriate size are permanently installed at the customer's site. Tank and piping isolation and design are critical to minimize the inevitable boil-off of the liquid CO to reduce losses. Extreme precautions must be taken when handling liquid CO. Refer to the *Handbook of Compressed Gases* by the Compressed Gas Association

(2) for more detail on liquid CO handling, and to the ASME Boiler and Pressure Vessel Code and ANSI/ASME B31.3 for proper selection and maintenance of cryogenic CO vessels. The Type 300 series stainless steels, 9% nickel steel, and aluminum alloys are suitable for inner vessel material construction.

For small-volume customers involved with applications such as CO-lasers, high-pressure cylinders are the supply mode of choice. The maximum pressure authorized for carbon monoxide cylinders is 6.9 MPa at 21°C if the gas is dry and sulfur-free, the cylinders can be charged to 5/6 the service pressure but never more than 13.7 MPa at 21°C. A high pressure carbon-steel cylinder holds approximately 175 scf of pure CO. Under present regulations, the cylinders authorized for carbon monoxide service, per TC/DOT specifications, must be requalified by hydrostatic test every five years.

8. Economic Aspects

Carbon monoxide is manufactured as a syngas mixture or as purified gas by a number of chemical and industrial gas plants. A large majority of the carbon monoxide produced is used immediately downstream and at the plant site for chemical synthesis, or steel manufacturing. Consequently, published production data are not readily available. Based on the development and growth of some of the applications in chemical synthesis (see Uses), carbon monoxide production has grown over the last few years and is expected to continue to grow over the next ten years.

Carbon monoxide pricing is dependent upon several factors: the price of by-product hydrogen, feedstock price, purity requirement (which determines the manufacturing and purification processes, thus determining capital and operational costs), location (which impacts the distribution cost), mode of supply, and volume. On-site carbon monoxide is the most economical supply method.

Carbon monoxide can also be delivered via high pressure tube trailers, typically containing 50,000 to 100,000 scf each. Liquid carbon monoxide is available only from a very small number of suppliers due to the safety and health risks associated with handling and stocking the product under liquid form.

Prices for bulk CO in the Gulf Cost area would typically range from $0.35/m³ ($1.00/100 scf) for "over the fence" supply of approximatly 25,000 m³/day to $0.85/m³ ($2.40/100 scf) for high pressure tube trailers containing.

For smaller volume requirements, carbon monoxide can be supplied in high pressure steel cylinders with top pressures of (11–13.7 Mpa) and at purities ranging from 99.0% to 99.995%. The price variance between the lowest and highest grades can be in the 1:20 ratio, as extra purification steps are required.

9. Specifications

Typically the purity of the carbon monoxide stream from a commercial production and purification unit is in the range of 97 to 99.9%. Higher purity, of up to 99.995% can be achieved for electronic grades and applications. For most applications involving carbon monoxide, the nature and concentration of the

CO stream impurities are more critical than the total purity value. The typical impurities are hydrogen, methane, carbon dioxide, moisture, nitrogen, oxygen, and argon. The level of these impurities is dependent upon the manufacturing process, the feedstock and the purification process. The chemical reaction using the CO stream drives the purity specifications: ie, in addition to the impurities mentioned above, the commercial process to synthesize formic acid demands less than one part per million mole of sulfur contamination. Consequently, the manufacturing and purification process design must be geared towards the particular application(s) the carbon monoxide is to be used for.

For electronic applications, minimum purity of 99.98–99.995% is required. At 99.995% purity, the typical levels of contaminants are as follows: N_2 <10 ppm, O_2 <3 pm, Ar <10 ppm, CO_2 <1 ppm, H_2 <1 ppm, H_2O <3 ppm, CH_4 <2 ppm (51).

As gas purity and grade names are not standardized across the industrial gas industry, it is important for the end-user to consider the supplier's published guaranteed specifications when performing purity versus price value comparisons.

10. Analytical Methods

Carbon monoxide can be analyzed by a number of procedures based on the reducing properties of CO. Qualitative detection of CO is made by passing the gas through palladium chloride ($PdCl_2$): black metallic palladium appears in the presence of CO. This technique is able to detect levels down to 100–1000 ppm. However, hydrogen, hydrogen sulfide, ethylene, and acetylene also reduce palladium chloride, thus interfering with CO detection. Detection and quantification of CO is also possible via infrared spectrophotometry. The CO infrared stretching frequency is at 2143 cm^{-1}. Electrochemical oxidation of CO to CO_2 is another viable method. Gas chromatography is a method of choice to analyze CO, as it allows both separation of the CO product from its impurities, and quantification. Thermal conduction detectors (TCD) are appropriate for high level concentrations (< 500 ppm). Flame ionization detectors (FID) or discharge ionization detectors (DID) are necessary for low part per million concentration levels. Analysis by a FID necessitates the reduction of CO to CH_4 on a palladium-based catalyst in the presence of high purity hydrogen. Atmospheric CO emissions are measured by continuous emission monitoring systems (CEMS), typically based on infrared, or gas chromatography methods.

Impurities in carbon monoxide are typically analyzed using the following gas chromatography techniques: nitrogen, argon, carbon dioxide, and hydrogen by TCD, methane by FID, moisture via electrical hygrometers based on the direct amperometric method on the piezoelectric sorption detector.

11. Health and Safety Factors

Hazard associated with the use of carbon monoxide derive primarily from: its toxicity; and its flammability.

11.1. Toxicity. Carbon monoxide is a chemical asphyxiant and acts toxically by combining with the hemoglobin of the red blood cells to form a stable compound called carbon monoxide–hemoglobin. This stable compound prevents the hemoglobin from taking up oxygen, thus depriving the body of the oxygen needed for metabolic respiration. The affinity of carbon monoxide for hemoglobin is approximately 300 times than affinity of oxygen for hemoglobin. The inhalation of concentration as low as 0.04% causes headaches and discomfort within 2 to 3 hours. Inhalation of a 0.4% concentration in air is fatal in less than 1 hour. Carbon monoxide is odorless and colorless, which gives no warning of its presence, and inhalation of heavy concentration can cause sudden, unexpected collapse. The current eight-hour time-weighted average threshold limit value (TLV) adopted by the U.S. Occupational Safety and Health Administration is 35 ppm (or 40 mg/m^3) for exposure to carbon monoxide, and a ceiling limit of 200 ppm (229 mg/m^3) (2,52).

According to the *Journal of the American Medical Association*, carbon monoxide is the leading cause of poisoning death in the United States. In concentrations of 12,800 parts per million (ppm) or 1.28 vol% unconsciousness is immediate with the danger of death in 1 to 3 minutes if not rescued. Domestic sources of CO are typically associated with home gas appliances (ovens, water heaters, clothes dryers), generators, furnace, fireplaces, charcoal grills, automobile exhaust fumes, power tools, etc. Only carbon monoxide detectors can detect lethal levels of CO in households. Industrial environments where carbon monoxide is used or stored should also be monitored for CO concentrations with CO detectors and alarms.

11.2. Flammability. Carbon monoxide is flammable in air over a wide range of concentration: lower limit of 12.5%, and upper limit of 74% at 20°C and 101.3 kPa. In an industrial environment, special care should be taken to avoid storing carbon monoxide cylinders with cylinders containing oxygen or other highly oxidizing or flammable materials. It is recommended that carbon monoxide cylinders in use be grounded. Additionally, areas in which cylinders are in use must be free of all ignition sources and hot surfaces.

12. Environmental Concerns

Carbon monoxide is highly toxic and with the single exception of carbon dioxide, its total yearly emissions of CO exceed all other atmospheric pollutants combined. Some of the potential sources of CO emission and exposure are foundries, petroleum refineries, kraft pulp mills, carbon black manufacturers, steel mills, formaldehyde manufacturers, coal combustion facilities, fuel oil combustion operations (ie, power plants, industrial, commercial and domestic uses, charcoal manufacturer, sugarcane processing operations, motor vehicles). In the U.S., two-thirds of the carbon monoxide emissions come from transportation sources, with the largest contribution coming from highway motor vehicles. In urban areas, the motor vehicle contribution to carbon monoxide pollution exceeds 90%. In 1992, carbon monoxide levels exceeded the Federal air quality standard in 20 U.S. cities, home to more than 14 million people.

The Clean Air Act of 1990 gives state and local government primary responsibility for regulating pollution from power plants, factories, and other "stationary sources". The U.S. Environmental Protection Agency (EPA) has primary responsibility for "mobile sources" pollution control (53).

The EPA motor vehicle program has achieved considerable success in reducing carbon emissions from the 1970s until 1990. EPA standards in the early 1970s prompted automakers to improve basic engine design. By 1975 catalytic converters, designed, to convert CO to CO_2, appeared and reduced CO emissions upwards of 80%. In the early 1980s, automakers introduced more sophisticated converters, plus on-board computers and oxygen sensors to help optimize the efficiency of the catalytic converter. Today's passenger cars are capable of emitting 90% less carbon monoxide over their lifetimes than their uncontrolled counterparts in the 1960s. As a result, ambient carbon monoxide levels have dropped, despite large increases in the number of vehicles on the road and the number of miles they travel. However, with continued increases in vehicle travel projected, the increasing number of more pollutant vehicles (utility vehicles, pickup trucks), it is expected that CO levels will climb again. This increase has already started in the most populated urban areas. CO emissions from automobiles increase dramatically in cold weather because cars need more fuel to start at cold temperatures. The addition of oxygen-containing compound to gasoline in cold temperature improves the air-to-fuel ratio, thereby promoting complete fuel combustion, and reduced CO emission. The 1990 Clean Air Act requires oxygenated fuels in designated CO non-attainment areas where mobile sources are a significant source of CO emissions. In other urban areas of Europe and industrialized Asia, automotive CO emissions create similar issues, but local environmental protection agencies have not been as active over the past decade. However, the historically high fuel gas prices in these geographic areas lead to the manufacturing of cars and trucks that are more fuel efficient than in the U.S., which in turn contributed to less CO emissions on a per vehicle basis.

CO emitted from stationary sources, such as refineries, is under the EPA Clean Air Act 1990 regulation but enforced by state and local environmental agencies. For example, refineries are required to monitor CO stack emissions on a periodic basis, ie, daily. Typically two certified calibration standards (one of higher concentration that the expected value, and one of lower value than the expected level), are required to validate the concentration of emitted CO. Additionally, on a quarterly basis, a calibration gas audit with EPA Protocol gases is required. EPA Protocol gases are gas standards manufactured by industrial gas suppliers in accordance with an established U.S. EPA Standards that dictate how to manufacture and certify the gaseous $CO-N_2$ mixture. Plants that emit more than permitted are penalized by fines or have the option of buying emission credits from the plants that have emitted less CO than allowed. Authorized CO emission levels are lower in nonattainment areas.

13. Uses

13.1. Chemical Synthesis. Pure carbon monoxide is used in a number of chemical syntheses: eg, acetic acid, acetic anhydride, polycarbonate and diisocyanates (via phosgene), formic acid, propionic acid, methyl methacrylate, polyketones (see Fig. 4). Commercial petrochemical processes using pure CO are based on four principal classes of chemical reaction: Reppe chemistry, Koch carbonylation, phosgenation, and ethylene–CO copolymers (see section Industrially

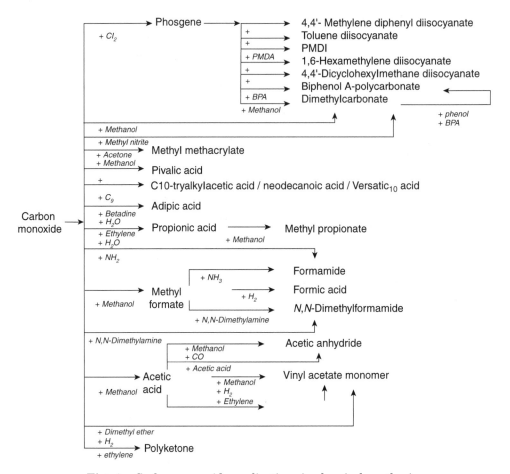

Fig. 4. Carbon monoxide applications in chemical synthesis.

Significant Reactions of Carbon Monoxide, for specific information on these processes and uses for the chemicals produced). Figure 5 provides a summary of the end-use of the various chemicals produced by synthesis involving carbon monoxide.

13.2. Other Application.   With applications of carbon monoxide include the following:

Fuel gas alone or in mixes as waste gas or producer gas.

Metallurgy as a reagent for manufacturing special steels, as a reagent for reducing refractory oxides.

As a reagent to make high grade zinc white pigment for paints and varnishes.

Electronics in dielectric etch recipes: a new chemistry, $CF_4–CO–CHF_3–N_2$, was developed for enhanced selectivity to underlayer and/or overlayer film (54).

Infrared gas lasers (CO laser wavelength range is 5.2–6.0 μm) used in solid state and molecular spectroscopy, nonlinear optics, laser studies.

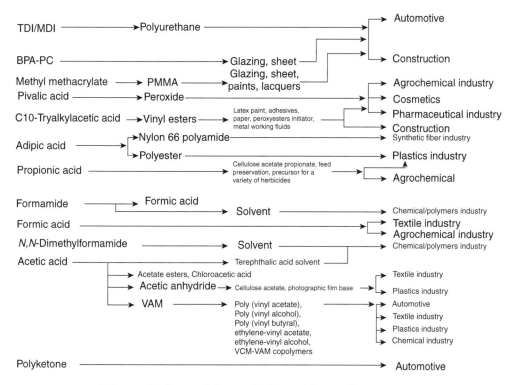

Fig. 5. End-use of chemicals derived from CO synthesis.

BIBLIOGRAPHY

"Carbon Monoxide" in *ECT* 1st ed., Vol. 3, pp. 179–191, by D. D. Lee, E. I. du Pont de Nemours & Co., Inc.; in *ECT* 2nd ed., Vol. 4, pp. 424–445, by R. V. Green, E. I. du Pont de Nemours & Co., Inc.; in *ECT* 3rd ed., Vol. 4, pp. 772–793, by C. M. Bartish and G. M. Drissel, Air Products and Chemicals, Inc.; "Carbon Monoxide" in *ECT* 4th ed., Vol. 5, pp. 97–122, by Ronald Pierantozzi, Air Products and Chemicals, Inc; "Carbon Monoxide" in *ECT* (online), posting date: December 4, 2000, by Ronald Pierantozzi, Air Products and Chemicals, Inc.

CITED PUBLICATIONS

1. *Gas Encyclopedia*, L'Air Liquide Division Scientifique, Elsevier, New York, 1976.
2. *Handbook of Compressed Gases*, 3rd ed., Compressed Gas Association, Inc., Van Nostrand Reinhold, New York, 1989.
3. Eur. Pat., EP 0 643 034 A1 (March 15, 1995), C. Garland, M. Giles, and J. Sunley (to BP Chemicals).
4. Eur. Pat., EP 0 752 406 A1 (Aug. 1, 1997), M. Giles, C. Garland, and M. Muskett (to BP Chemicals).
5. CHEM SYSTEMS reports: ≪ *Acetic acid / anhydride* ≫, report 91-1, (July 4, 1996).

6. U.S. Pat., 5,523,452 (April 4, 1996), Z. Kricsfalussy and co-workers, (to Bayer Lever-kusen).
7. CHEM SYSTEMS reports: ≪ *Adipic acid* ≫, report 92-4, March 1994.
8. Eur. Pat., EP 0 796 819 A1 (Sept. 24, 1997), N. Kunisi, N. Murai, and H. Kusama (to Idemitsu Petrochemical Co, Ltd.).
9. V. Macho, M. Kralik, and L. Komora; *Pet. Coal* **39**(1), 6–12 (1997).
10. U.S. Pats, 3,833,634 (Sept. 3, 1974); 3,957,857 (May 18, 1976), R. L. Pruette and W. E. Walker (to Union Carbide Corp.).
11. Ger. Offen. 2,503,204 (July 31, 1975), M. M. Bhasin (to Union Carbide Corp.).
12. Ger. Offen. 2,503,233 (July 31, 1975), M. M. Bhasin and G. L. O'Connor (to Union Carbide Corp.).
13. U.S. Pat. 3,769,329 (Oct. 31, 1973), F. E. Paulik, A. Hershman, J. F. Roth, and W. R. Knox (to Monsanto Co.).
14. U.S. Pat. 3,769,324 (Oct. 31, 1973), F. E. Paulik, A. Hershman, J. F. Roth, and W. R. Knox (to Monsanto Co.).
15. U.S. Pat. 3,928,435 (Dec. 23, 1975), Y. Awane, S. Otsuka, M. Nagata, and F. Tanaka (to Mitsubishi).
16. M. M. T. Khan and A. E. Martell, *Homogeneous Catalysis by Metal Complexes*, Vol. 1, Academic Press, Inc., New York, 1974.
17. Ger. Offen. 2,610,035 (Sept. 23, 1976), N. Rizkalla and C. N. Winnick (to Halcon).
18. U.S. Pat. 3,911,003 (Oct. 7, 1975), S. Suzuki (to Chevron).
19. U.S. Pat. 3,923,904 (Dec. 2, 1975), H. Hara (to Nippon Oil Co.).
20. U.S. Pat. 3,852,346 (Dec. 3, 1974), D. Forster, A. Hershman, and F. E. Paulik (to Monsanto Co.).
21. U.S. Pats. 3,989,747 and 3,989,748 (Nov. 2, 1976), F. E. Paulik, A. Hershman, J. F. Roth, and J. H. Craddock (to Monsanto Co.).
22. A. Sen, *Adv. Polym. Sci.* **73/74**, 125 (1986).
23. A. Sen, *Chemtech*, 48 (Jan. 1986) and references therein.
24. U.S. Pats. 3,346,625 (Oct. 10, 1967); 3,349,119 (Oct. 24, 1967), D. M. Fenton and K. L. Olivier (to Union Oil Co.).
25. U.S. Pat. 3,907,882 (Sept. 23, 1976), W. Ganzler, K. Kabs, and G. Schroder (to Rohm GmbH).
26. U.S. Pats. 3,846,468 (Nov. 5, 1974); 3,980,690 (Sept. 14, 1976), E. Perrotti and G. Cipriani (to Snam Progetti SPA).
27. Ger. Offen. 2,213,435 (Oct. 11, 1973), W. Ganzler, K. Kabs, and G. Schroder (to Rohm GmbH).
28. U.S. Pats. 3,992,436 (Nov. 16, 1976); 4,005,128-131 (Jan. 25, 1977), L. R. Zehner (to Atlantic Richfield).
29. U.S. Pat. 3,576,835 (Apr. 27, 1971), E. Smith and W. Schnabel (to Olin).
30. U.S. Pat. 3,832,372 (Aug. 27, 1974), R. D. Hammond, W. M. Clarke, and W. I. Denton (to Olin).
31. U.S. Pat. 2,866,822 (Dec. 30, 1958), H. T. Siefen and W. R. Trutna (to E. I. du Pont de Nemours & Co., Inc.).
32. U.S. Pat. 3,948,998 (Apr. 6, 1976), S. Fujiyama, T. Takahashi, S. Kozao, and T. Kasahara (to Mitsubishi).
33. U.S. Pat. 3,960,932 (June 1, 1976), R. F. Heck (to University of Delaware).
34. E. W. Abel and F. G. A. Stone, *Quart. Rev.* **24**, 498 (1970).
35. W. Forg, *Linde Rep. Sci. Technol.*, (15), 20–21 (1970).
36. H. H. Gunardson and J. M. Abrardo; *Hydrocarbon Processing* (Intern. Ed.) **78**(4) 87–90, April 1999.
37. U.S. Pat. 4,564,513; (Jan. 14, 1986), D. Becher and co-workers (to Bayer).
38. U.S. Pat. 5,538,706, (July 13, 1996), A. Kapoor, R. Krishnamurthy and D. L. MacLean (to The BOC Group Inc.).

39. U.S. Pat. 5,496,530 (March 5, 1996), R. Vannby, and C. S. Nielsen (to Haldor Topsoe A/S).
40. N. N. Dutta and G. S. Patil, *GAS separation Purification* **9**(4), p. 277 (Dec 1995).
41. U.S. Pat. 5,632,162 (May 27, 1997), Billy; (to L'Air Liquide).
42. U.S. Pat. 5,096,470, (March 17, 1992), R. Krishnamurty; (to The BOC Group, Inc.)
43. U.S. Pat. 5,073,356 (Dec. 17, 1991), O. Guro and co-workers (to Air Products and Chemicals, Inc.).
44. U.S. Pat. 5,167,125 (Dec. 12, 1992), R. Agrawal; (to Air Products and Chemicals, Inc.).
45. R. W. Spillman, *Chem. Eng. Progr.* **85**, 41–62 (1989).
46. K. Tanaka and co-workers, *Polym. J.* **22**, 381–385 (1990).
47. K. Tanaka, H. Kita, K. I. Okamoto, *Kobunshi Ronbunshu* **47**(12), 945–951 (1990).
48. C. S. Sarma and N. N. Dutta, *Chem. Soc. Chem. Commun.* (1996).
49. S. S. Lyke and co-workers, *Ind. Eng. Chem. Prod. Res. Dev.* **25**, 517–521 (1986).
50. N. Sonodo and co-workers, *Chem. Lett.* 1873–1876 (1990).
51. Air Liquide High Purity Specialty Gases and Equipment Catalog.
52. OSHA Method 210; CO Analysis.
53. EPA 400-F-92-005; Automobiles and Carbon Monoxide, Jan. 1993.
54. R. Lindley and co-workers, *Solid State Technology*, 93–99 (Aug. 1997).

CHRISTINE GEORGE
Air Liquide America Corporation

CARBOXYLIC ACIDS

1. Introduction

Carboxylic acids from the smallest, formic, to the 22-carbon fatty acids, eg, erucic, are economically important; several million metric tons are produced annually. The shorter-chain aliphatic acids are colorless liquids. Each has a characteristic odor ranging from sharp and penetrating (formic and acetic acids), or vinegary (dilute acetic), to the odors of rancid butter (butyric acid), and goat fat (the 6–10-carbon acids). At room temperature, the cis-unsaturated acids through C_{18} are liquids, and the saturated unbranched aliphatic acids from decanoic through the higher acids and trans-unsaturated acids are solids. The latter are higher melting because of their higher degree of linearity and greater degree of crystallinity eg, elaidic acid.

Although vegetable oils and animals fats were commonly used in ancient times, most higher acids were not known until the beginning of the nineteenth century. Then the nature of the naturally occurring 18-carbon fatty acids was established, and hundreds of long-chain fatty acids have been isolated from natural sources and characterized.

Both odd and even numbered alkanoic acids of molecular formula $C_nH_{2n}O_2$ occur naturally. Until chromatographic techniques provided means to identify minor components in natural mixtures, it was believed that only the even numbered higher acids, most often the C_{18} acids, occurred naturally. Formic acid (qv), acetic acid (qv), propionic, and butyric acids are manufactured in large quantities from petrochemical feedstocks. The higher fatty acids are derived from animal fats, vegetable oils, or fish oils. Some higher saturated fatty acids with significant industrial applications are caprylic, pelargonic, capric, lauric, myristic, palmitic, stearic, and behenic acids (Table 1).

In the alkenoic series of molecular formula $C_nH_{2n-2}O_2$, acrylic, methacrylic, undecylenic, oleic, and erucic acids have important applications (Table 2). Acrylic and methacrylic acids have a petrochemical origin, and undecylenic, oleic, and erucic acids have natural origins (see ACRYLIC ACID AND DERIVATIVES; METHACRYLIC ACID AND DERIVATIVES).

The polyunsaturated aliphatic monocarboxylic acids having industrial significance include sorbic, linoleic, linolenic, eleostearic, and various polyunsaturated fish acids (Table 3). Of these, only sorbic acid (qv) is made synthetically. The other acids, except those from tall oil, occur naturally as glycerides and are used mostly in this form.

The shorter-chain alkynoic or acetylenic acids are common in laboratory organic syntheses, and several long-chain acids occur naturally (Table 4).

Many substituted, ie, branched, fatty acids, particularly methacrylic, 2-ethylhexanoic, and ricinoleic acids, are commercially significant. Several substituted fatty acids exist naturally (Table 5). Fatty acids with a methyl group in the penultimate position are called iso acids, and those with a methyl group in the antepenultimate position are called anteiso acids (1). However, the term iso is often used in a broader sense to mean branched or mixtures of branched-chain industrial acids.

Some naturally occurring fatty acids have alicyclic substituents such as the cyclopentenyl-containing chaulmoogra acids (1), notable for their use in treating leprosy (see ANTIPARASITIC AGENTS, ANTIMYCOTICS), and the cyclopropenyl (2) or sterculic acids (Table 6).

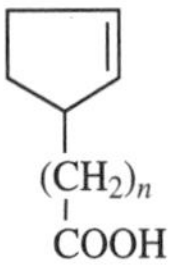

(1) 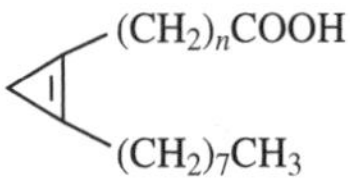   (2) 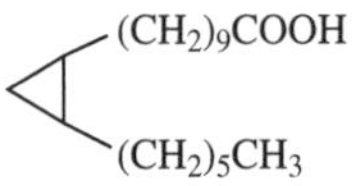   (3)

The prostaglandins (qv) constitute another class of fatty acids with alicyclic structures. These are of great biological importance and are formed by *in vivo* oxidation of 20-carbon polyunsaturated fatty acids, particularly arachidonic acid [27400-91-5]. Several prostaglandins, eg, PGE_1 [745-65-3], have different degrees of unsaturation and oxidation when compared to the parent compound, prostanoic acid [25151-18-9].

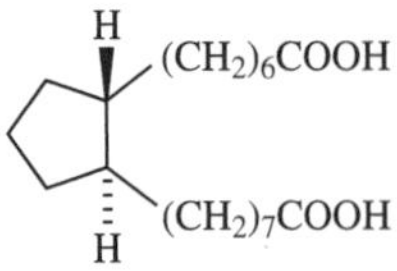

prostanoic acid 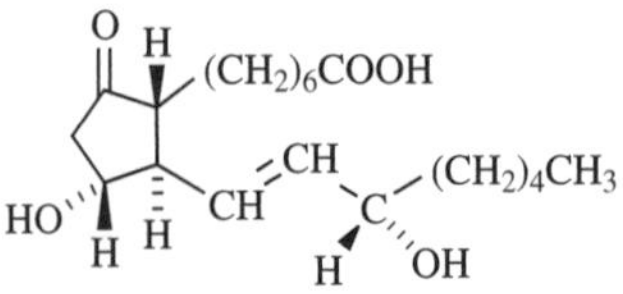 PGE_1

Table 1. Physical Properties of the Straight-Chain Alkanoic Acids, $C_nH_{2n}O_2$

n	Systematic name (common name)[a]	CAS Registry Number	Mol wt	Mp, °C	Bp[b], °C	Density,[c] d_4^{20}	Refractive index, n_D[d]	ΔH_f° at 25°C, kJ/mol[e]	Specific heat, J/g[e]	Heat of fusion, kJ/mol[e]	Surface tension[f] mN/m (=dyn/cm)	Viscosity,[d] mPa·s (=cP)	ΔH_R^{25} kJ/mol[e]
1	methanoic (formic)	[64-18-6]	46.03	8.4	100.5	1.220	1.3714						
2	ethanoic (acetic)	[64-19-7]	60.05	16.6	118.1	1.049	1.3718						
3	propanoic (propionic)	[79-09-04]	74.08	−22	141.1	0.992	1.3874	−511.2 (l)	2.34 (l)		20.7	1.099	−1,536
4	butanoic (butyric)	[107-92-6]	88.11	−7.9	163.5	0.959	1.39906	−534.1 (l)	2.16 (l)		21.8	1.538	−2,194
5	pentanoic (valeric)	[109-52-4]	102.13	−34.5	187.0	0.942	1.4086	−559.2 (l)				2.30	−2,837.8
6	hexanoic ([caproic])	[142-62-1]	116.16	−3.4	205.8	0.929	1.4170	−584.0 (l)	2.23 (l)	15.1	23.4	3.23	−3,492.4
7	heptanoic ([enanthic])	[111-14-8]	130.19	−10.5	223.0	0.922	1.4230	−609.1 (l)		15.0		4.33	−4,146.9
8	octanoic ([caprylic])	[124-07-2]	144.21	16.7	239.7	0.910	1.4280	−635.2 (l)	2.62 (s)	21.4	23.7	5.74	−4,799.9
9	nonanoic (pelargonic)	[112-05-0]	158.24	12.5	255.6	0.907	1.4322	−658.5 (l)	2.91 (s)	20.3		8.08	−5,456.1
10	decanoic ([capric])	[334-48-5]	172.27	31.6	270.0	0.895^{30}	1.4169^f	−685.2 (l)		28.0	25.0	4.30	−6,108.7
11	undecanoic ([undecyclic])	[112-37-8]	186.30	29.3	284.0	0.9905^{25}	1.4202^f	−736.7 (s)		25.1		7.30	−6,762.3
12	dodecanoic (lauric)	[143-07-7]	200.32	44.2	298.9	0.883	1.4230^f	−775.6 (s)	1.80 (s)	36.6	26.6	7.30	−7,413.7
13	tridecanoic ([tridecylic])	[638-53-9]	214.35	41.5	312.4	0.8458^{80}	1.4252^f						
14	tetradecanoic (myristic)	[544-63-8]	228.38	53.9	326.2	0.858^{60}	1.4273^f	−835.9 (s)	1.60 (s)	44.8	27.4	5.83^f	−8,721.4
15	pentadecanoic ([pentadecyclic])	[1002-84-2]	242.40	52.3	339.1	0.8423^{80}	1.4292^f						
16	hexadecanoic (palmitic)	[57-10-3]	256.43	63.1	351.5	0.8534^{60}	1.4309^f	−892.9 (s)	1.80 (s)	54.4	28.2	7.80^f	−10,030.6
17	heptadecanoic (margaric)	[506-12-7]	270.46	61.3	363.8	0.853^{60}	1.4324^f						
18	octadecanoic (stearic)	[57-11-4]	284.48	69.6	376.1	0.847^f	1.4337^f	−949.4 (s)	1.67 (s)	63.2	28.9	9.87^f	−11,342.4
19	nonadecanoic ([nonadecyclic])	[646-30-0]	298.51	68.6	299_{100}	0.8771^{25}	1.4512^{25}	−1013.3 (s)		71.0			−12,646.2
20	eicosanoic (arachidic)	[506-30-9]	312.54	75.3	203_1	0.8240^{100}	1.4250^{100}	−1063.7 (s)		78.7			−13,976

Table 1 (*Continued*)

n	Systematic name (common name)[a]	CAS Registry Number	Mol wt	Mp, °C	Bp[b], °C	Density,[c] d_4^{20}	Refractive index, n_D^d	ΔH°_f at 25°C, kJ/mol[e]	Specific heat, J/g[g]	Heat of fusion, kJ/mol[e]	Surface tension[f] mN/m (=dyn/cm)	Viscosity,[d] mPa·s (=cP)	ΔH_R^{25} kJ/mol[e]
22	docosanoic (behenic)	[112-85-6]	340.59	79.9	306₂₀	0.8221^{100}	1.4270^{100}						
24	tetracosanoic (lignoceric)	[557-59-5]	368.65	84.2		0.8207^{100}	1.4287^{100}						
26	hexacosanoic (cerotic)	[504-46-7]	396.70	87.7		0.8198^{100}	1.4301^{100}						
28	octacosanoic (montanic)	[506-48-9]	424.75										
30	triacontanoic (melissic)	[506-50-3]	452.81										
33	tritriacontanoic (psyllic)	[38232-03-0]	494.89										
35	pentatriacontanoic (ceroplastic)	[38232-05-2]	522.94										

[a]Brackets signify a trivial name no longer in use.
[b]At 101.3 kPa = 1 atm unless otherwise noted in kPa as a subscript.
[c]At 20°C unless otherwise noted by a superscript number (°C).
[d]At 20°C unless otherwise noted.
[e]To convert J to cal, divide by 4.184.
[f]At 70°C.
[g]Heat of combustion (liquid).
[h]At 50°C.
[i]To convert kPa to mm Hg, multiply by 7.5.

30

Table 2. **Physical Properties of the Straight-Chain Alkenoic Acids, $C_n H_{(2n-2)}O_2$**

n	Systematic name (common name)	CAS Registry Number	Mol wt	Mp, °C	Bp, °C[a]	Density,[b] d_4^{20}	Refractive index,[b] n_D^{20}
3	propenoic (acrylic)	[79-10-7]	72.06	12.3	141.9	1.0621[16]	1.4224
4	trans-2-butenoic (crotonic)	[107-93-7]	86.09	72	189	1.018	1.4228[79.7]
4	cis-2-butenoic (isocrotonic)	[503-64-0]	86.09	14	171.9	1.0312[15]	1.4457
4	3-butenoic (vinylacetic)	[625-38-7]	86.09	-39	163	1.013[15]	1.4257[15]
5	2-pentenoic (β-ethylacrylic)	[626-98-2]	100.12				
5	3-pentenoic (β-pentenoic)	[5204-64-8]	100.12				
5	4-pentenoic (allylacetic)	[591-80-0]	100.12	-22.5	188-189	0.9809	1.4281
6	2-hexenoic (isohydroascorbic)	[1191-04-4]	114.14				
6	3-hexenoic (hydrosorbic)	[4219-24-3]	114.14	12	208	0.9640[23]	1.4935
7	trans-2-heptenoic	[10352-88-2]	128.17				
8	2-octenoic	[1470-50-4]	142.20				
9	2-nonenoic	[3760-11-0]	156.23				
10	trans-4-decenoic[c],	[57602-94-5]	170.25				
	cis-4-decenoic[c]	[505-90-8]					
10	9-decenoic (caproleic)	[14436-32-9]	170.25	24.5		0.9075[25]	1.4464
11	10-undecenoic (undecylenic)	[112-38-9]	184.28		275		
12	trans-3-dodecenoic (linderic)	[4998-71-4]	198.31				
13	tridecenoic	[28555-21-7]	212.33				
14	cis-9-tetradecenoic (myristoleic)	[544-64-9]	226.36				
15	pentadecenoic	[26444-04-2]	240.39				
16	cis-9-hexadecenoic (cis-9-palmitoleic)	[373-49-9]	254.41				
16	trans-9-hexadecenoic (trans-9-palmitoleic)	[10030-73-6]	254.41				
17	9-heptadecenoic	[10136-52-4]	268.44				
18	cis-6-octadecenoic (petroselinic)	[593-39-5]	282.47	30		0.8681[40]	1.4533[40]
18	trans-6-octadecenoic (petroselaidic)	[593-40-8]	282.47	54			
18	cis-9-octadecenoic (oleic)	[112-80-1]	282.47	13.6	234[d]	0.8905	1.4582[3]
18	trans-9-octadecenoic (elaidic)	[112-79-8]	282.47	43.7	234[d]	0.8568[70]	1.4405[70]
18	cis-11-octadecenoic	[506-17-2]	282.47	14.5			
18	trans-11-octadecenoic (vaccenic)	[693-72-1]	282.47	44		0.8563[70]	1.4406[70]
20	cis-5-eicosenoic	[7050-07-9]	310.52				
20	cis-9-eicosenoic (godoleic)	[506-31-0]	310.52				
22	cis-11-docosenoic (cetoleic)	[506-36-5]	338.58				

Table 2 (*Continued*)

n	Systematic name (common name)	CAS Registry Number	Mol wt	Mp, °C	Bp, °C[a]	Density,[b] d_4^{20}	Refractive index,[b] n_D^{20}
22	*cis*-13-docosenoic (erucic)	[112-86-7]	338.58	34.7	281[e]	0.85321[70]	1.44438[70]
22	*trans*-13-docosenoic (brassidic)	[506-33-2]	338.58	61.9	265[d]	0.85002[70]	1.44349[70]
24	*cis*-15-tetracosenoic (selacholeic)	[506-37-6]	366.63				
26	*cis*-17-hexacosenoic (ximenic)	[544-84-3]	394.68				
30	*cis*-21-triacontenoic (lumequeic)	[67329-09-3]	450.79				

[a] At 101.3 kPa = 1 atm unless otherwise noted.
[b] Superscript numbers indicate measurement at a temperature other than 20°C.
[c] The common name for both *cis*- and *trans*-4-decenoic is obtusilic.
[d] At 15 kPa = 113 mm Hg.
[e] At 30 kPa = 225 mm Hg.

Table 3. **Some Polyunsaturated Fatty Acids**

Total number of carbon atoms	Systematic name (common name)	CAS Registry Number	Mol wt	Mp, °C	Bp at kPa[a]	Refractive index, n_D^{20}
	Dienoic acids, $C_nH_{2n-4}O_2$					
5	2,4-pentadienoic (β-vinylacrylic)	[626-99-3]	98.10			
6	2,4-hexadienoic[b] (sorbic)	[22500-92-1]	112.13	134.5		
10	trans-2,4-decadienoic	[3036-33-2]	168.24			
12	trans-2,4-dodecadienoic	[24738-48-5]	196.29			
18	cis-9,cis-12-octadecadienoic (linoleic)	[60-33-3]	280.45	–5	202 at 1.4	1.4699
18	trans-9,trans-12-octadecadi-enoic (linolelaidic)	[506-21-8]	280.45	28–29		
18	5,6-octadecadienoic (laballenic)	[5204-84-2]	280.45			
22	5,13-docosadienoic	[26764-24-9]	336.56			
	Trienoic acids, $C_nH_{2n-6}O_2$					
16	6,10,14-hexadecatrienoic (hiragonic)	[4444-12-6]	250.38			
18	cis-9,cis-12,cis-15-octadeca-trienoic (linolenic)	[463-40-1]	278.44	–10–11.3	157 at 0.001	1.4800
18	cis-9,trans-11,trans-13-octa-decatrienoic (α-eleostearic)	[13296-76-9]	278.44	48–49	235 at 12	1.5112
18	trans-9,trans-11,trans-13-octa-decatrienoic (β-eleostearic)	[544-73-0]	278.44	71.5		1.5002
18	cis-9,cis-11,trans-13-octade-catrienoic (punicic)	[544-72-9]	278.44			
18	trans-9,trans-12,trans-15-octadecatrienoic (linolenelaidic)	[28290-79-1]	278.44			
	Tetraenoic acids, $C_nH_{2n-8}O_2$					
18	4,8,12,15 octadecatetraenoic (moroctic)	[67329-10-6]	276.42			
18	cis-9,trans-11,trans-13,cis-15-octadecatetraenoic (α-parinaric)	[593-38-4]	276.42			
18	trans-9,trans-11,trans-13,trans-15-octadecate-traenoic (β-parinaric)	[18841-21-9]	276.42			
20	5,8,11,14-eicosatetraenoic (arachidonic)	[27400-91-5]	304.47			
	Pentaenoic acids, $C_nH_{2n-10}O_2$					
22	4,8,12,15,19-docosapentae-noic (clupanodonic)	[2548-85-8]	330.51			

[a]To convert kPa to mm Hg, multiply by 7.5.
[b]$\Delta H_{298} = -393.5$ kJ/mol ($\frac{94.05}{4.184}$ kcal/mol); flash point (OC) = 127°C.

Table 4. **The Acetylenic Fatty Acids**

Total number of carbon atoms	Systematic name (common name)	CAS Registry Number	Mol wt
3	propynoic (propiolic, propargylic)	[471-25-0]	70.05
4	2-butynoic (tetrolic)	[590-93-2]	84.07
5	4-pentynoic	[6089-09-4]	98.10
6	5-hexynoic	[53293-00-8]	112.13
7	6-heptynoic	[30964-00-2]	126.16
8	7-octynoic	[10297-09-3]	140.18
9	8-nonynoic	[30964-01-3]	154.21
10	9-decynoic	[1642-49-5]	168.24
11	10-undecynoic (dehydro-10-undecylenic)	[2777-65-3]	182.26
18	6-octadecynoic (tariric)	[544-74-1]	280.45
18	9-octadecynoic (stearolic)	[506-24-1]	280.45
18	17-octadecene-9, 11-diynoic (isanic, erythrogenic)	[506-25-2]	274.40
18	*trans*-11-octadecene-9-ynoic (ximenynic)	[557-58-4]	278.44
22	13-docosynoic (behenolic)	[506-35-4]	336.56

Aromatic carboxylic acids are produced annually in amounts of several million metric tons. Several aromatic acids occur naturally, eg, benzoic acid (qv), salicylic acid (qv), cinnamic acid (qv), and gallic acids, but those used in commerce are produced synthetically. These acids are generally crystalline solids with relatively high melting points, attributable to the rigid, planar, aromatic nucleus (see PHTHALIC ACIDS).

2. Nomenclature

Acyclic monocarboxylic acids are named substitutively by dropping the -e from the parent hydrocarbon name and adding -oic acid. Cyclic carboxylic acids are named substitutively by adding the suffix carboxylic acid to the name of the cyclic hydrocarbon. Positions of substituents, eg, amino-, chloro-, hydroxy-, methyl-, oxo-, in acyclic alkanoic acids are indicated by number locants, with the numbering starting from the carbon of the carboxyl group; in common names, Greek letters are used beginning at C-2 = the α-position. It is preferable to name substituted derivatives systematically and not on the basis of the common name for the parent compound, eg, 12-hydroxyoctadecanoic rather than 12-hydroxystearic acid. Acid halides of acyclic acids are named by dropping the ending -ic acid and adding the suffix -oyl or -yl to either the hydrocarbon name or the acid common name, eg, acetyl, hexanoyl, and stearoyl chlorides from the respective names acetic acid, hexanoic acid, and stearic acid.

Esters are named by replacing the ending -ic acid with the suffix -ate. The alcohol portion of the ester is named by replacing the -ane ending of the parent hydrocarbon name with the suffix -yl. The alkyl radical name of an ester is separated from the carboxylate name, eg, methyl formate for $HCOOCH_3$. Amides are named by changing the ending -oic acid to -amide for either systematic or common names, eg, hexanamide and acetamide. Mono-, di-, and triunsaturated fatty

Table 5. **Some Substituted Acids**

Total number of carbon atoms	Systematic name (common name)	CAS Registry Number	Mol wt	Mp, °C	Bp, °C[a]	Density[b] d_4^{20}	Refractive index[b] n_D^{20}
4	2-methylpropenoic (methacrylic)	[79-41-4]	86.09	16	163	1.0153	1.4314
4	2-methylpropanoic (isobutyric)	[79-31-2]	88.10	-47	154.4	0.9504	1.3930
5	2-methyl-cis-2-butenoic (angelic)	[565-63-9]	100.12	45	185	0.9539[76]	1.4434[47]
5	2-methyl-trans-2-butenoic (tiglic)	[80-59-1]	100.12	65.5	198.5	0.9641[76]	1.43297[76]
5	3-methyl-2-butenoic (β,β-dimethyl acrylic)	[541-47-9]	100.12				
5	2-methylbutanoic	[116-53-0]	102.13				
5	3-methylbutanoic (isovaleric)	[503-74-2]	102.13	-37.6	176	0.93319[17.6]	1.40178[22.4]
5	2,2-dimethylpropanoic (pivalic)	[75-98-9]	102.13	35.5	163	0.905[50]	
8	2-ethylhexanoic	[149-57-5]	144.21		220	0.9031[25]	1.4255[28]
14	3,11-dihydroxytetradecanoic (ipurolic)	[36138-54-2]	260.37				
16	2,15,16-trihydroxyhexadecanoic (ustilic)	[557-44-8]	304.43				
16	9,10,16-trihydroxyhexadecanoic (aleuritic)	[533-87-9]	304.43				
16	16-hydroxy-7-hexadecenoic (ambrettolic)	[506-14-9]	270.41				
18	12-hydroxy-cis-9-octadecenoic (ricinoleic)	[141-22-0]	298.47	5.0,7.7,16.0	226[c]	0.9496[15]	1.4145[15]
18	12-hydroxy-trans-9-octadecenoic (ricinelaidic)	[540-12-5]	298.47	52–53	240[c]		
18	4-oxo-9,11,13-octadecatrienoic (licanic)	[17699-20-6]	292.42				
18	9,10-dihydroxyoctadecanoic	[120-87-6]	316.48	90			
18	12-hydroxyoctadecanoic	[106-14-9]	300.48	79			
18	12-oxooctadecanoic	[925-44-0]	298.47	81.5			
18	18-hydroxy-9,11,13-octadecatrienoic (kamlolenic)	[4444-93-3]	294.43	77–78			
18	12,13-epoxy-cis-9-octadecenoic (vernolic)	[31263-20-4]	296.45				
18	8-hydroxy-trans-11-octadecene-9-ynoic (ximenynolic)	[2724-58-5]	294.43				
18	8-hydroxy-17-octadecene-9,11-diynoic (isanolic)	[64144-78-1]	290.40				
20	14-hydroxy-cis-11-eicosenoic (lesquerolic)	[4103-20-2]	326.52				
	mixed isomers (isononanoic) acid	[26896-18-4]	158	ca 70	232–246		
	2,2-dimethyloctanoic (neodecanoic) acid	[129662-90-6]	172	<40	147–150[d]		
	mixed isomers (isostearic) acid	[2724-58-5]	284	ca 7	192–204[e]		

[a]At 101.3 kPa = 1 atm unless otherwise noted.
[b]Superscript numbers indicate measurement at a temperature other than 20°C.
[c]At 1.3 kPa = 9.75 mm Hg.
[d]At 20 kPa.
[e]At 5 kPa.

Table 6. **Some Fatty Acids with Alicyclic Substituents**

Total number of carbon atoms	Common name	CAS Registry Number	n in (1) or (2)	Mol wt
		Cyclopentenyl compounds[a]		
6	aleprolic	[2348-89-2]	0	112.13
10	aleprestic	[2348-90-5]	4	168.24
12	aleprylic	[24874-21-3]	6	196.29
14	alepric	[2519-24-6]	8	224.34
16	hydnocarpic	[459-67-6]	10	252.40
18	chaulmoogric	[502-30-7]	12	280.45
		Cyclopropenyl compounds[b]		
18	malvalic (halphenic)	[503-05-9]	6	280.45
19	sterculic	[738-87-4]	7	294.48
19	lactobacillic[c]	[503-06-0]		296.49

[a]See structure (1).
[b]See structure (2).
[c]Saturated ring; see structure (3).

acids are named with Arabic numeral locants for the unsaturated positions and with the suffix -enoic, -adienoic, and -atrienoic acid in place of the -ane ending of the saturated hydrocarbon name, eg, 10,12,14-octadecatrienoic acid.

Shorthand notations have been developed to avoid repetitive systematic names of unsaturated fatty acids. For example, linolenic or *cis*-9,*cis*-12-,*cis*-15-octadecatrienoic acid can be represented by 18:3(9c,12c,15c). The Greek letter Δ has been used to indicate presence and position of double bonds, eg a $\Delta^{9,12,15}$ fatty acid, but it should never be used in a systematic name. An equally inappropriate but popular designation is derived by counting from the methyl terminus to the nearest double bond. For example, 9,12,15-octadecatrienoic, 9,12-octadecadienoic, and 9-octadecenoic acids are referred to as ω-3, ω-6, and ω-9, respectively.

3. Physical Properties

Melting points, boiling points, densities, and refractive indexes for carboxylic acids vary widely depending on molecular weight, structure, and the presence of unsaturation or other functional groups (Tables 1, 2, 3, and 5). In addition, some useful constants for alkanoic acids are listed in Table 1. Some constants for selected unsaturated and substituted acids are given in Table 7.

Equations for the heat capacities in J/g for solid and liquid states of palmitic acid are, respectively (4)

$$C_p = 1.604 + 0.00544t \quad (-73 \text{ to } 40°C)$$

$$C_p = 1.936 + 0.00734t \quad (63 \text{ to } 92°C)$$

Table 7. **Some Constants for Selected Unsaturated and Substituted Acids**

Acid	Heat of formation,[a] kJ/mol[b]	Flash point, °C	Heat of combustion,[c] kJ/mol[b]
	Unsaturated acids		
acrylic	−384.3 (l)	49[d]	
crotonic	−431.6 (s)	88[d]	
2-pentenoic	−447.5 (s)		
3-pentenoic	−435.3 (s)		
4-pentenoic	−431.6 (l)		
undecenoic	−572.2	146[d]	−6614
elaidic	−786.1 (s)		−11,154
oleic	−802.9 (l)	189	−11,228
brassidic	−938.9		−23,775
erucic	−854.9 (s)		−13,797
	Substituted acids		
methacrylic	−435.0 (l)		
isobutyric	−534.1 (l)		−2166
pivalic	−565.1 (s)		−2834

[a]ΔH°_f at 298 K.
[b]To convert kJ to kcal, divide by 4.184.
[c]ΔH^{25}_R (liquid).
[d]OC = open cup.

For stearic acid, the equations are (5)

$$C_p = 1.7886 + 0.00754t \quad (-120 \text{ to } 65°C)$$

$$C_p = 1.7861 + 0.00754t \quad (70 \text{ to } 78°C)$$

Tall oil fatty acids (TOFA) consist primarily of oleic and linoleic acids and are obtained by the distillation of crude tall oil. Crude tall oil, a by-product of the kraft pulping process, is a mixture of fatty acids, rosin acids, and unsaponifiables (2). These components are separated from one another by a series of distillations (3). Several grades of TOFA are available depending on rosin, unsaponifiable content, color, and color stability. Typical compositions of tall oil fatty acid products are shown in Table 8 (see TALL OIL).

The viscosity of a mixture of fatty acids depends on the average chain length n and can be calculated from the equations (6):

$$\log \eta = -0.602802 + 0.134844n - 0.00259(n)^2 \quad (70°C)$$

$$\log \eta = -0.510490 + 0.101571n - 0.001628(n)^2 \quad (90°C)$$

where η = viscosity in mPa·s(=P).

Heats of combustion for liquid alkanoic acids at 25°C are given by the equation (7):

$$-\Delta H^{25} = 654.4n - 430.4 \text{J/mol} \quad (n > 5)$$

Table 8. Typical Fatty Acid Composition of Tall Oil Products

	CAS Registry Number	Crude tall oil, %	Crude fatty acid, %	<2% Rosin in fatty acid, %	Distilled tall oil, %
		Fatty acids normalized to 100%			
$C_{16}H_{32}O_2$	[57-10-3]	6.3	1.6	0.4	
$C_{17}H_{34}O_2$	[506-12-7]	1.5	0.7	0.7	
$C_{18}H_{36}O_2$	[57-11-4]	1.5	2.2	2.3	1.4
$C_{18}H_{34}O_2$	[112-80-1]	39.8	42.3	46.4	22.9
$C_{18}H_{32}O_2$	[60-33-3]	34.0	34.8	36.3	22.0
$C_{18}H_{32}O_2$ (isomers)	[26764-25-0]	10.2	12.7	10.3	24.6
$C_{19}H_{38}O_2$	[646-30-0]	1.2	1.1	1.1	1.1
$C_{20}H_{36}O_2$	[25448-01-5]	4.6	4.7	2.4	28.0
		Rosin acids			
		40	7	1	30
		Unsaponifiables			
		8	2.5	1.5	2

Crystallographic properties of solid alkanoic acids significantly affect many of their other properties (8–10). For example, heat of crystallization, melting point, and solubility depend on whether the acid is even or odd numbered and vary alternately in a homologous series. Other physical properties such as boiling point, density (liquid), and refractive index depend on molecular rather than crystal structure. Thus these colligative properties change in a regular manner according to molecular weight.

The long-chain alkanoic acids and their derivatives are polymorphic with the unit cell containing dimers formed by hydrogen bonding between carboxyl groups.

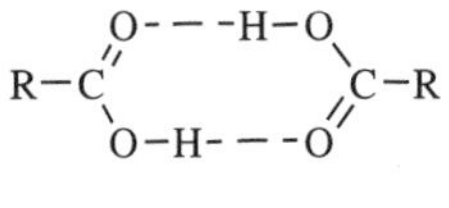

$$R-C\begin{matrix} O\text{-}\text{-}\text{-}\text{-}H-O \\ \diagup \qquad\qquad \diagdown \\ \diagdown \qquad\qquad \diagup \\ O-H\text{-}\text{-}\text{-}\text{-}O \end{matrix}C-R$$

In crystallizing fatty acids, solvent polarity does not influence crystal form as much as temperature and concentration (11). Infrared (11,12) and wide-line nmr spectra (13) as well as x-ray methods (14,15) can be used to detect the various crystalline forms.

Alkenoic acids also have polymorphic crystalline forms. For example, both oleic and elaidic acids are dimorphic with melting points of 13.6 and 16.3°C for oleic, and 43.7 and 44.8°C for elaidic acid (16).

The higher fatty acids undergo decarboxylation and other undesirable reactions when heated at their boiling points at atmospheric pressure. Hence they are distilled at reduced pressure (17,18). Methyl esters boil at lower temperatures than acids at the same pressure as the result of the absence of hydrogen bonding (19). A procedure for calculation of the vapor pressures of fatty acids at various temperatures has been described (20).

Table 9. **Solubilities of Alkanoic Acids in Water and Organic Solvents**

Number of carbon atoms in RCOOH	Solubility at 20°C, g/100 g solvent				
	Water	Acetone	Benzene	Cyclohexane	n-Hexane
4	∞				
5	3.7				
6	0.968				
7	0.244				
8	0.068				
9	0.026	∞	∞	∞	∞
10	0.015	407	398	342	290
11	0.0093	706	663	525	
12	0.0055	60.5	93.6	68	47.7
13	0.0033	78.6	117	100	
14	0.0020	15.9	29.2	21.5	11.9
16	0.00072	5.38	7.3	6.5	3.1
18	0.00029	1.54	2.46	2.4	0.5

Formic, acetic, propionic, and butyric acids are miscible with water at room temperature. Solubility in water decreases rapidly for the higher alkanoic acids as the chain length increases (Table 9) (21). The solubility in water at pH 2–3 for unionized acids is given by the following relationship:

$$\log S = -0.6n + 2.32$$

where S = solubility in mol/L and n = number of carbon atoms (22).

The hydrophilic nature of the carboxyl group balanced against the hydrophobic nature of the hydrocarbon chain allows long-chain fatty acids to form monomolecular films at aqueous liquid–gas, liquid–liquid, or liquid–solid interfaces (20).

The solubility of water in fatty acids, 0.92% for stearic acid at 68.7°C, is greater than the solubility of the acid in water, 0.0003% for stearic acid at 20°C, and this solubility tends to increase with increasing temperature (23). Solubilities of aliphatic acids in organic solvents demonstrate another example of the alternating effect of odd vs even numbered acids (Table 9).

An important chemical characteristic of unsaturated acids is the iodine value (IV), which indicates the average degree of unsaturation. It is equal to the number of grams of iodine absorbed under standard conditions by 100 g of the unsaturated acid.

Unsaturation in a fatty acid increases its solubility in organic solvents, and the differences in solubilities between saturated and unsaturated acids can be used to separate these acids (Table 10).

Formic acid is the most acidic straight-chain alkanoic acid. Solubility in water of alkanoic acids containing more than nine carbon atoms is too low to permit accurate measurement of dissociation (Table 11). The acidity of 2-chloroalkanoic acids is much greater than that of formic acid, and trichloroacetic acid is comparable to the mineral acids in acid strength.

Table 10. **Solubilities of Fatty Acids in Organic Solvents at Various Temperatures**

Fatty acid	Temperature, °C	Solubility, g/100 g solvent		
		Acetone	Toluene	n-Heptane
16:0	10	1.60	1.41	0.30
	0	0.66	0.36	0.08
	−10	0.27	0.086	0.02
	−20	0.10	0.018	0.005
18:0	10	0.54	0.390	0.080
	0	0.11	0.080	0.018
	−10	0.023	0.015	0.004
	−20	0.005	0.003	
18:1 (9c)	−20	5.2		2.25
	−30	1.68	3.12	0.66
	−40	0.53	0.96	0.19
	−50	0.17	0.28	0.05
18:1 (9t)	−10		0.86	0.19
	−20	0.26	0.20	0.06
	−30	0.092	0.056	0.019
18:2 (9c,12c)	−50	4.10		0.98
	−60	1.20		0.20
	−70	0.35		0.042

Table 11. **Dissociation Constant K_a for Straight-Chain and Chlorinated Alkanoic Acids at 25°C**

Acid	Unsubstituted	Cl at C-2
formic	2.1×10^{-4}	
acetic	1.81×10^{-5}	1.4×10^{-3}
propionic	1.32×10^{-5}	1.6×10^{-3}
butyric	1.50×10^{-5}	1.4×10^{-3}
pentanoic	1.56×10^{-5}	
hexanoic	1.40×10^{-5}	

4. Chemical Properties

The alkanoic acids, with the exception of formic acid, undergo typical reactions of the carboxyl group. Formic acid has reducing properties and does not form an acid chloride or an anhydride. The hydrocarbon chain of alkanoic acids undergoes the usual reactions of hydrocarbons except that the carboxyl group exerts considerable influence on the site and ease of reaction. The alkenoic acids in which the double bond is not conjugated with the carboxyl group show typical reactions of internal olefins. All three types of reactions are industrially important.

Reactions of the carboxyl group include salt and acid chloride formation, esterification, pyrolysis, reduction, and amide, nitrile, and amine formation.

Salt formation occurs when the carboxylic acid reacts with an alkaline substance (24):

$$RCOOH + MOH \longrightarrow RCOO^- + M^+ + H_2O$$

where $M = Li$, Na, K, NH_4, R_3HN etc. The alkaline substance can also be an oxide, hydroxide, or carbonate of a metal of higher valence such as Ca, Mg, Zn, or Al. The saponification of fats and oils with caustic soda or potash gives water-soluble soaps. Water-insoluble, metallic soaps are prepared by fusion, precipitation, or direct solution of metal (see DRIERS AND METALLIC SOAPS). Fusion gives fine, dense, but slightly off-color metallic soaps useful as driers. The precipitation method forms fluffy, finely divided soaps of excellent color. Lithium 12-hydroxyoctadecanoate [2918-92-5] is an important constituent of many greases.

Acid chlorides are prepared with reagents such as PCl_3, $SOCl_2$, $(COCl)_2$, and $COCl_2$ (25); preparation with thionyl chloride follows the reaction:

$$RCOOH + SOCl_2 \longrightarrow RCOCl + SO_2 + HCl$$

Fatty acid chlorides are very reactive and can be used instead of conventional methods to facilitate production of amides and esters. Imidazoles are effective recyclable catalysts for the reaction with phosgene (qv) (26).

Esterification is one of the most important reactions of fatty acids (27). Several types of esters are produced including those resulting from reaction with monohydric alcohols, polyhydric alcohols, ethylene or propylene oxide, and acetylene or vinyl acetate. The principal monohydric alcohols used are methyl, ethyl, propyl, isopropyl, butyl, and isobutyl alcohols (28) (see ESTERIFICATION; ESTERS, ORGANIC).

Sulfuric acid or hydrogen chloride may be used to catalyze esterification, and weight ratios of alcohol to fatty acid as high as 2 to 4:1, corresponding to molar ratios of 10 to 20:1, may be used to drive the equilibrium reaction to completion. Stoichiometric quantities of acid and alcohol can be used with hexyl and higher alcohols when the water of reaction is removed by azeotropic distillation with toluene or xylene. With long-chain fatty alcohols, water may be removed by azeotropic distillation, sparging with an inert gas, or subjecting the reaction to reduced pressure. Esters are also prepared by alcoholyses of animal fats or vegetable oils in the presence of alkaline or acidic catalysts (29,30). Esters of monohydric alcohols are used for plasticizers, as lubricants, and in cosmetics.

Esterification with polyhydric alcohols such as ethylene-, propylene-, diethylene-, and polyethylene glycols, glycerol, pentaerythritol, and certain carbohydrates is a more complex reaction because of immiscibility problems (see ALCOHOLS, POLYHYDRIC). A good reaction requires temperatures of 230–235°C and vigorous agitation in contrast to the milder conditions used for the simple alcohols. Temperatures higher than 235°C cause polyols to condense to ethers and to decompose. Nearly stoichiometric quantities of glycol and fatty acid are needed to make either monoesters or diesters. Product water is removed by reduced pressure, azeotropic distillation, or sparging. Monoesters are usually formed by reaction of ethylene or propylene oxide with the fatty acid:

$$RCOOH \ + \ n\,CH_2\!-\!CH_2 \ \longrightarrow \ RCOO\!-\!(CH_2CH_2O\,)_{\overline{n}}\!-\!H$$
$$\underset{O}{\diagdown\diagup}$$

The product is a mixture of various polyoxyethylene chain lengths (31–33). Glycol diesters are used as vinyl plasticizers; the monoesters as surface-active agents and viscosity modifiers for alkyd resins (qv).

Glycerol esterifications are still more complex (34–37). Even with excess glycerol (qv), a mixture of mono-, di-, and triglycerides is formed because of the limited solubility of glycerol in the reaction product. Nearly complete reaction of glycerol can occur at reduced pressure or in inert atmosphere and with 5–20% excess acid. Compositions of the reaction mixture can be calculated on a statistical basis assuming equivalence of the three hydroxyl groups and no isomer formation (35). Glycerides are important as surface-active agents; triolein [122-32-7] is used to some extent as a plasticizer (see SURFACTANTS; PLASTICIZERS).

Pentaerythritol with its four primary hydroxyl groups is used for the preparation of tetraesters and presents little difficulty except for its high melting point of 263°C, when pure. Pentaerythritol tetraesters are used in aircraft lubes, synthetic drying oils, and alkyds. Esters derived from trimethylolalkanes and dipentaerythritol are also used in alkyd resins (qv). Esterification may take place *in situ* during preparation of the alkyd.

Sorbitol is the most important higher polyol used in direct esterification of fatty acids. Esters of sorbitans and sorbitans modified with ethylene oxide are extensively used as surface-active agents. Interesterification of fatty acid methyl esters with sucrose yields biodegradable detergents, and with starch yields thermoplastic polymers (38).

Vinyl esters are prepared by the reaction of a fatty acid with either acetylene in direct condensation or vinyl acetate by acidolysis.

Reduction of fatty acids to alcohols is done by catalytic hydrogenation over a copper chromite catalyst at high temperatures of 325°C and pressures of 24 MPa (3500 psig) (39):

$$RCOOH + 4\,H_2 \longrightarrow RCH_2OH + H_2O$$

The yield of fatty alcohol is ca 90%. Fatty alcohols may also be prepared by high pressure catalytic hydrogenolysis of either a glyceride or a methyl ester. A copper chromite catalyst is used at 270–300°C and 34.6 MPa (5000 psig) of hydrogen pressure. If a glyceride is used, the yield of glycerol is relatively low because of hydrogenolysis to propylene glycol and isopropyl alcohol. The saturated fatty alcohols thus produced are used primarily in the production of detergents.

Reduction of glycerides or other esters with sodium and a secondary alcohol such as cyclohexanol or 4-methyl-2-pentanol was at one time a second commercial method for producing fatty alcohols. Though more costly than hydrogenolysis, this method gives high yields of glycerol and unsaturated fatty alcohols if the original fatty ester is unsaturated.

Selective hydrogenation of the carboxyl or ester group in preference to the olefinic unsaturation also produces unsaturated alcohols. Copper–cadmium and zinc–chromium oxides seem to provide most selectivity (40–42). Copper chromite catalysts are not selective. Reduction of red oil-grade oleic acid has been accomplished in 60–70% yield and with high selectivity with Cr–Zn–Cd, Cr–Zn–Cd–Al, or Zn–Cd–Al oxides (43). The reduction may be a homogeneously catalyzed reaction as the result of the formation of copper or cadmium soaps (44).

Pyrolysis of either saturated or unsaturated fatty acids leads to mixtures of hydrocarbons, olefins, and cyclic compounds (45). Pyrolysis of fatty acid salts or vegetable oils has been used to make hydrocarbon fractions suitable as petroleum substitutes. Various products are obtained by pyrolysis of fatty acid salts depending on the metal salt or catalyst used. Calcium salts generally form symmetrical ketones, aldehydes are formed in the presence of calcium formate, and hydrocarbons are formed in the presence of excess calcium hydroxide. Magnesium and lead salts produce ketones in improved yield over the calcium salts. Zinc oxide promotes hydrocarbon rather than ketone formation. Vapor-phase pyrolysis of fatty acids or esters over thorium or cesium oxides produces ketones in high yields. Homogeneous decarbonylation of fatty acid chlorides with $PdCl_2$ catalyst occurs readily at 185–200°C to make olefins in high yield (46).

Pyrolysis is used to produce undecenoic acid from ricinoleic acid:

$$CH_3(CH_2)_5CHOHCH_2CH{=}CH(CH_2)_7COONa \longrightarrow CH_3(CH_2)_5CHO$$
$$+ CH_2{=}CH(CH_2)_8COONa$$

Undecenoic acid is the starting point for making 11-aminoundecanoic acid [2432-99-7] and nylon-11 (see CASTOR OIL; POLYAMIDES).

Reactions of ammonia and amines with carboxylic acids result in the formation of a variety of products (47,48). Ammonium salts, $RCOONH_4$, are prepared with or without solvent, by reaction with anhydrous ammonia. Ammonium salts readily decompose to acid ammonium salts, $RCOONH_4 \cdot RCOOH$, particularly at temperatures higher than 50°C. Amides are formed at 150–200°C by reaction of the acid with ammonia at reduced pressures:

$$RCOOH \ + \ NH_3 \ \longrightarrow \ \overset{\overset{\text{O}}{\|}}{RC}{-}NH_2 \ + \ H_2O \uparrow$$

Amides are also formed by the reaction of an acid chloride with ammonia or an amine:

$$\overset{\overset{\text{O}}{\|}}{RC}{-}Cl \ + \ R'NH_2 \ \longrightarrow \ \overset{\overset{\text{O}}{\|}}{RC}{-}NHR' \ + \ HCl$$

Ammonolysis or aminolysis of an ester can be used to make the respective amide or *N*-substituted amide.

Ammonium acetate and sodium methoxide are effective catalysts for the ammonolysis of soybean oil (49). Polyfunctional amines and amino alcohols such as ethylenediamine, ethanolamine, and diethanolamine react to give useful intermediates. Ethylenediamine can form either a monoamide or a diamide depending on the mole ratio of reactants. With an equimolar ratio of reactants and a temperature of >250°C, a cyclization reaction occurs to give imidazolines with ethylenediamine (48):

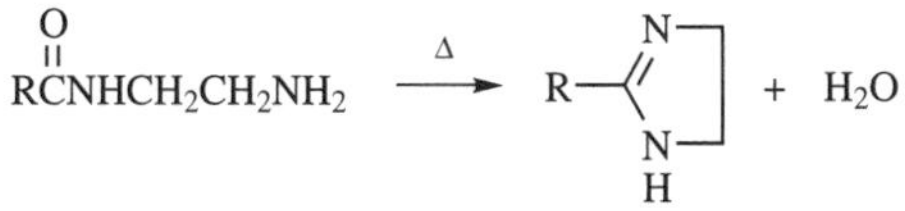

$$\underset{RCNHCH_2CH_2NH_2}{\overset{\overset{\text{O}}{\|}}{}} \ \xrightarrow{\Delta} \ R{-}\left[\begin{array}{c} N{-} \\ {\Big\langle} \\ N{-} \\ H \end{array}\right] \ + \ H_2O$$

Ethanolamine produces oxazolines.

Fatty amines are made by dehydration of amides to nitriles at 280–330°C, followed by hydrogenation of the nitrile over nickel or cobalt catalysts:

$$\underset{\text{RCNH}_2}{\overset{\overset{\text{O}}{\underset{||}{}}}{}} \xrightarrow{\text{–H}_2\text{O}} \text{RCN} \xrightarrow[\text{NH}_3]{\text{Ni,H}_2} \text{RCH}_2\text{NH}_2$$

The presence of ammonia during hydrogenation suppresses formation of secondary amines and inhibits hydrogenation of double bonds in unsaturated nitriles. Fatty amines are used as corrosion inhibitors, flotation agents, quaternary salts for sanitizing agents and textile fabric softeners, and surface-active agents.

Acyl aminimides have proved useful as surface-active and antimicrobial agents and as an intermediate for isocyanate preparation (50).

Reactions of the hydrocarbon chain in alkanoic acids include α-sulfonation and halogenation (51–54). The α-sulfonated fatty ester salts have excellent lime-dispersing properties and are valuable surface-active agents.

Reactions of the double bonds include isomerization and conjugation, cyclization, various addition reactions including hydrogenation, pyrolytic and oxidative cleavage, metathesis, and various polymerization reactions (51,55).

Geometrical isomerization of cis- to trans-alkenoic acids occurs by photosensitization (56) or with heat treatment in the presence of catalysts such as SO_2, I_2, or HNO_2.

Positional isomerization occurs most often during partial hydrogenation of unsaturated fatty acids; it also occurs in strongly basic or acidic solution and by catalysis with metal hydrides or organometallic carbonyl complexes. Concentrated sulfuric or 70% perchloric acid treatment of oleic acid at 85°C produces γ-stearolactone from a series of double-bond isomerizations, hydration, and dehydration steps (57).

Conjugation as well as geometric and positional isomerization occur when an alkadienoic acid such as linoleic acid is treated with a strong base at an elevated temperature. Cyclic fatty acids result from isomerization of linolenic acid in strong base at about 250°C (58). Conjugated fatty acids undergo the Diels-Alder reaction with many dienophiles including ethylene, propylene, acrylic acid, and maleic anhydride.

Addition to the double bond occurs readily with hydrogen halides, hypohalous, sulfuric, or formic acids (53):

$$-\text{CH}{=}\text{CH}- + \text{HX} \longrightarrow -\text{CH}_2\text{CHX}-$$

where X = F, Cl, Br, I, OCl, OBr, OSO_3H, HCOO, etc.

Addition of a weaker acid such as acetic acid takes place if the reaction is catalyzed by a sulfonic acid ion-exchange resin (59). Addition of halogens, mixed halogens, eg, the Wijs reagent ICl, or halogenlike compounds, eg, NOCl, and thiocyanogen, SCN_2, occurs easily and is the basis of several analytical methods for determining total unsaturation (60,61). Addition of hydrogen occurs only in the presence of an active catalyst such as nickel at moderately elevated temperatures and pressures (62). Other reagents that add to the double bond include carbon monoxide and hydrogen, eg, the oxo reaction (63,64); carbon monoxide and water, eg, hydrocarboxylation (64); various carbon free-radical compounds (65); dialkyl-phosphonates (66); formaldehyde (67); alcohols (68); rhodium-catalyzed olefin addition (69); and mercuric acetate (70). Dihydroxylation of a double bond may be brought about by peroxy acids or treatment with alkaline permanganate (71). Addition of oxygen is carried out with peroxy acids in the presence of strong acid catalysts to make epoxidized compounds (72).

Cleavage of an alkenoic acid can be carried out with permanganate, a permanganate–periodate mixture, periodate or with nitric acid, dichromate, ozone, or, if the unsaturation is first converted to a dihydroxy compound, lead tetraacetate (71,73). Oxidative ozonolysis is a process for the manufacture of azelaic acid [123-99-9] and pelargonic acid (74).

$$CH_3(CH_2)_7CH{=}CH(CH_2)_7COOH \xrightarrow{\ O_3\ } CH_3(CH_2)_7COOH + HOOC(CH_2)_7COOH$$

          oleic acid   pelargonic acid   azelaic acid

Alkali fusion of oleic acid at about 350°C in the Varrentrapp reaction causes double-bond isomerization to a conjugated system with the carboxylate group followed by oxidative cleavage to form palmitic acid (75). In contrast, alkali fusion of ricinoleic acid is the commercial route to sebacic acid[111-20-6]:

$$CH_3(CH_2)_5CHOHCH_2CH{=}CH(CH_2)_7COOH \xrightarrow{\ NaOH\ } CH_3(CH_2)_5CHOHCH_3$$

                sebacic acid $+HOOC(CH_2)_8COOH$

Metathesis of oleic acid to produce a C_{18} straight-chain dibasic acid can be carried out at 70°C with a $WCl_6 \cdot Sn(CH_3)_4$ catalyst or with rhenium heptoxide promoted by $Sn(CH_3)_4$ (55,77).

Polymerization takes a variety of forms including dimerization or trimerization to polybasic acids. The internal double bond in oleic acid and other unsaturated fatty acids can participate, but only to a limited extent, in copolymerization with ethylene (78). Conjugated linoleate esters form co-oligomers with styrene by cationic catalysis (79). The resulting dibasic acid or ester can be condensed with ethyleneamine oligomers to make reactive polyamides useful for producing frothed epoxy compositions and as a catalyst for cross-linking rigid urethane foams (80). Conjugated linoleate esters are readily copolymerized with styrene and acrylonitrile by free-radical catalysis (81). Methyl eleostearate, however, inhibits copolymerization with either styrene or acrylonitrile. Polymerization of unsaturated fatty acids in drying oils through autoxidative mechanisms is the basis for a significant use of these materials in the paint industry (82) (see DRYING OILS).

5. Manufacture

Carboxylic acids having 6–24 carbon atoms are commonly known as fatty acids. Shorter-chain acids, such as formic, acetic, and propionic acid, are not classified as fatty acids and are produced synthetically from petroleum sources (see ACETIC ACID; FORMIC ACID AND DERIVATIVES; OXO PROCESS). Fatty acids are produced primarily from natural fats and oils through a series of unit operations. Clay bleaching and acid washing are sometimes also included with the above operations in the manufacture of fatty acids for the removal of impurities prior to subsequent processing.

The composition of common fats and oils are found in Table 12. The most predominant feedstocks for the manufacture of fatty acids are tallow and grease, coconut oil, palm oil, palm kernel oil, soybean oil, rapeseed oil, and cottonseed oil. Another large source of fatty acids comes from the distillation of crude tall oil obtained as a by-product from the Kraft pulping process.

5.1. Fatty Acids from Natural Fats and Oils.

There are essentially four steps or unit operations in the manufacture of fatty acids from natural fats and oils: (1) batch alkaline hydrolysis or continuous high pressure hydrolysis; (2) separation of the fatty acids usually by a continuous solvent crystallization process or by the hydrophilization process; (3) hydrogenation, which converts unsaturated fatty acids to saturated fatty acids; and (4) distillation, which separates components by their boiling points or vapor pressures. A good review of the production of fatty acids has been given (83).

Hydrolysis. Saponification or hydrolysis involves converting the fat or oil (a triglyceride) to a fatty acid and glycerol. This can be done in a number of ways including Twitchell splitting, autoclave batch splitting, continuous high pressure splitting, and enzymatic splitting. The feed should be of quality commensurate with the quality of fatty acid desired. Also, the feedstock generally is given a pretreatment such as clay bleaching or acid washing prior to splitting that allows hydrolysis to take place more efficiently and to give a higher quality product in some cases (84).

Twitchell splitting is an acid catalyzed hydrolysis that uses sulfuric acid as the catalyst along with surface-active agents such as petroleum sulfonates or sulfonated oleic and naphthenic acids (Twitchell reagent). The splitting is carried out in acid-resistant vessels using about 60% fat or oil, 40% water, 0.5% sulfuric acid, and 1% petroleum sulfonate. The advantages of Twitchell splitting are its simple process and equipment required compared with the other methods. The disadvantages are darker colored products that usually contain traces of sulfur-containing materials, disposal of the aqueous acid layers, inefficient use of heat, and long cycle times. Twitchell splitting is used on a relatively small scale (85,86).

Autoclave batch splitting is generally carried out at 1.03–3.45 MPa (150–500 psig). By using certain metal oxides, such as zinc oxide, as catalysts, lower pressures can be used. In a typical batch the amount of water is about 30–60% of the fat weight. Headspace air is removed with steam to minimize oxidation and the autoclave is heated to the desired temperature and pressure. After 6–10 h a split of about 92% is obtained at the lower pressure whereas at higher pressures splits as high as 95% have been realized. After the desired degree of split is obtained the lower aqueous glycerol layer is separated from the upper fatty acid layer. If metal catalysts have been used the fatty acids are treated with sulfuric acid to decompose the soaps, and finally, any residual mineral acid is removed from the fatty acids with a hot water wash (87).

Table 12. Fatty Acid Composition (%) and Significant Properties of Important Fats and Oils

Name	Chain length	Double bonds	Cotton oil	Coconut oil	Palm kernel oil	Corn oil	Palm oil	Castor oil	Rapeseed oil (low erucic)	Rapeseed oil (high erucic)	Soybean oil	Sunflower oil	Herring oil	Sardine oil	Tallow	Tall oil
caproic acid	C_6	0		0–1	a											
caprylic acid	C_8	0		5–10	3–6											
capric acid	C_{10}	0		5–10	3–5											
lauric acid	C_{12}	0		43–53	40–52								a			
myristic acid	C_{14}	0	0–2	15–21	14–18	0–1	0–2					a	5–10	4–6	1–6	
palmitic acid	C_{16}	0	17–29	7–11	6–10	8–19	30–48	2–3	3–6	0–5	7–12	3–10	11–16	9–11	20–37	1–3
stearic acid	C_{18}	0	1–4	2–4	1–4	0–4	3–6	2–3	0–3	0–3	2–6	1–10	0–3	1–3	6–40	0–1
arachidic acid	C_{20}	0	0–1				0–1		0–2	0–2	0–3	0–1			a	0–1
behenic acid	C_{22}	0	a						a	0–2	a	0–1				
palmitoleic acid	C_{16}	0–2							a	a	a	0–1	5–12	10–15	1–9	
oleic acid	C_{18}	1	13–44	6–8	9–16	19–50	38–44	4–9	50–66	9–25	20–30	20–40	8–15	15–25	20–50	9–16
gadoleic acid	C_{20}	1	a						0–5	5–15	0–1	a				
erucic acid	C_{22}	1							0–5	30–60		a				
ricinoleic acid	C_{18}	1						80–87								
linoleic acid	C_{18}	2	33–58	1–3	1–3	34–62	9–12	2–7	18–30	11–25	48–58	50–70	2–4	3–8	0–5	20–32
linolenic acid	C_{18}	3				0–2			6–14	5–12	4–10	0–1	0–2	1–3	0–3	
unsaturated fatty acids	C_{20}	2–6							a	a			20–30	15–30	a	a
unsaturated fatty acids	C_{22}	2–6								0–2			10–28	15–20		
rosin acids																23–37
Properties																
iodine value, g I$_2$/100 g			96–112	8–12	14–23	103–128	44–54	81–91	105–120	91–108	120–140	120–140	120–145	170–193	35–55	
saponification value, mg KOH/g			190–198	250–264	245–255	188–193	194–206	174–186	185–198	170–185	190–195	186–194	178–194	189–193	190–200	130–170

a Trace.

Continuous high pressure splitting was developed by Colgate-Emery and by Procter & Gamble (88,89). Temperatures of 240–270°C are preferred, giving pressures of 4.8–5.2 MPa (700–750 psig). The splitting is carried out in a cylindrical-shaped tower, 18.3–24.4 m high and 0.51–1.22 m in diameter of 316 L stainless steel or 316 L cladding on carbon steel. Splitting coconut oil requires better corrosion resistance because of the shorter-chain fatty acids present. Corrosion-resistant linings such as Carpenter 20 Cb or Incoloy 825 can be utilized. The tower is operated with countercurrent flow, with water being introduced into the top part of the tower and fat at the bottom of the tower. The tower contains disengaging zones where fatty acids are collected at the top of the tower and aqueous glycerol (sweet water) at the bottom (Fig. 1). Heat is conserved by the use of heat exchangers that cool the existing fatty acid while heating the incoming water (steam); the exiting sweet water is cooled by the entering fat, which in turn is heated in the exchanger. Make-up heat is applied to the center

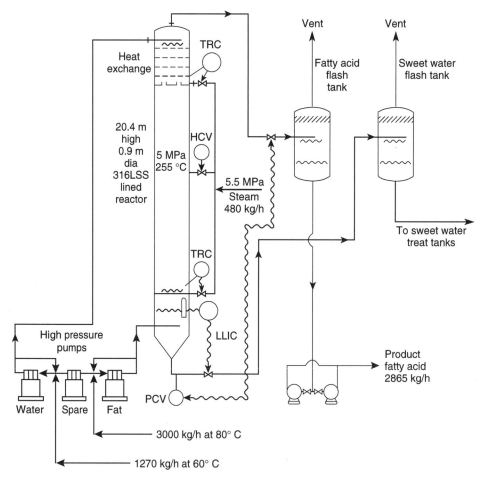

Fig. 1.   Fat splitter. TRC, temperature recorder controller; LLIC, liquid level indicator controller; PCV, pressure control valve; and HCV, heat control valve. To convert MPa to psi, multiply by 145.

of the tower (the hydrolysis section containing the continuous fat phase) using internal steam coils, electric heating, or direct superheated steam. The pressure in the tower is controlled by a backpressure valve in the fatty acid discharge line, whereas the fat/sweet water interface is controlled by the rate of sweet water discharged. About 98–98.5% split is usually obtained. The sweet water contains 10–15% glycerol and is purified in a series of steps involving removal of any dissolved salts, fat, and oil impurities, and then concentration by evaporation of water and/or distillation (90). In some continuous high pressure splitting units, approximately 0.05% of ZnO catalyst is used to speed the reaction rates and raise conversion to >99.0%.

Enzymatic fat splitting was developed as a means to minimize energy costs and where a high order of specificity of splitting is desired. Further details on enzymatic fat splitting are available (91).

Separation Techniques.   Current methods for separating fatty acids are by solvent crystallization or by the hydrophilization process. Other methods that have been used in the past, or perhaps could be used in the future, are panning and pressing, solvent extraction, supercritical fluid extraction, the use of metal salts in assisting in separation, separations using urea complexes, and adsorption/desorption.

Panning and pressing is no longer used because of high labor costs, but the terms single-pressed, double-pressed, and triple-pressed stearic acid come from these processes. These terms are widely used to denote the quality of stearic acid. These commercial "stearic" acids are actually mixtures of palmitic acid and stearic acid.

There are two commercial solvent crystallization processes. The Emersol Process, patented in 1942 by Emery Industries, uses methanol as solvent; and the Armour-Texaco Process, patented in 1948, uses acetone as solvent. The fatty acids to be separated are dissolved in the solvent and cooled, usually in a double-pipe chiller. Internal scrapers rotating at low rpm remove the crystals from the chilled surface. The slurry is then separated by means of a rotary vacuum filter. The filter cake is sprayed with cold solvent to remove free liquid acids, and the solvents are removed by flash evaporation and steam stripping and recovered for reuse (92).

When tallow fatty acids are the feed, stearic acid (actually 60/40 C16/C18) and oleic acids are the products. Solvent separation is also used to separate stearic acid from isostearic acid when hydrogenated monomer is the feed, and oleic acid from linoleic acid when using tall oil fatty acids as feed.

Several processes have been evaluated in attempts to lower the cost of the low temperature refrigeration required in the above solvent separation systems. The use of adiabatic cooling under vacuum to a temperature below the equilibrium temperature for crystallization has been described in which the resulting supercooled solution was transferred to a crystallization vessel where nucleation started and crystallization was reported to be complete within a few minutes. Spheroid-shaped crystals were obtained, which were more easily separated and washed than the usual crystals produced from other solvent processes (93). Two processes have been described in which crystallization occurs at a higher temperature. One uses methyl formate containing 5–10% water as solvent, where crystallization occurs about 15°C higher than when using methanol or acetone as solvent (94). The other process uses 2-nitropropane as solvent and gives a good separation of oleic acid from linoleic acid at $-15°C$ instead of $-30°C$ when using methanol or acetone as solvents (95).

To avoid the hazards and costs of using solvents, the Hydrophilization Process was developed by Henkel in Germany. The steps in the process include (1) cooling to obtain a slurry; (2) addition of aqueous solution of a wetting agent; (3) high speed agitation to form a dispersion of the solid fatty acid in the liquid; (4) addition of an electrolyte to stabilize the dispersion; and (5) separation of the solid fatty acid from the liquid fatty acid by means of a centrifuge. The quality of the products are generally not as good as when solvent separation is used (96–99).

Liquid–liquid extraction can be used to obtain high purity linoleic acid from safflower fatty acids or linoleic acid from linseed fatty acids using furfural and hexane as solvents (100). High purity linoleic acid has been obtained from sunflower fatty acids using a dimethylformamide and hexane solvent system (101).

Supercritical fluid extraction (SFE) has been investigated on oleochemical separations. This method uses a gas in the supercritical state, which generally means working at pressures of 8.3–48.5 MPa (1200–7000 psi), and at relatively low temperatures. Initial costs for commercial equipment can be quite high and will probably limit the use of SFE to higher priced specialty products (102). Methyl esters of Menhaden oil have been separated using supercritical CO_2 into docosa-hexaenoic acid [6217-54-5] (DHA) and eicosapentaenoic acid [10417-94-4] (EPA). These omega-3 fatty acids have been found to be important dietary factors, beneficial in reducing the development of atherosclerotic lesions. Accordingly, there is great interest in obtaining the polyunsaturated fatty acids present in fish oils in a more concentrated form (103).

Adsorption processes have recently been described to separate fatty acids into high purity products. Lauric acid was separated from myristic acid using crystalline silica as the adsorbent and was desorbed using a ketone such as acetone or methyl ethyl ketone (104). Another system, using cross-linked polystyrene as the adsorbent, separated a mixture of palmitic and stearic acids (105). Separation of saturated fatty acid, such as palmitic/stearic acid from an unsaturated fatty oleic acid, using a molecular sieve plus crystalline clay as the adsorbent, has been described (106). The desorbent was acetone. The separation of oleic acid from linoleic acid has been described using as an adsorbent either crosslinked polystyrene, or a molecular sieve-silicate as the adsorbent with a variety of desorbents listed (107). Separation of fatty acids from rosin acids can be accomplished using molecular sieves that have been modified with silicalite and a phosphorus-modified alumina. The preferred desorbents were methyl ethyl ketone–acetic acid or short-chain acids or esters with less than six carbon atoms (108). The separation of fatty acids from unsaponifiables has been carried out using a molecular sieve, comprising a crystalline silica with a silica to alumina ratio of at least 12, as the adsorbent. The desorbent was acetone (109).

Hydrogenation.   In the manufacture of fatty acids hydrogenation is used to saturate the ethylenic linkage to produce a more saturated acid. Generally, the feedstocks for hydrogenation are (1) the solid fractions from crystallizations where the iodine value (IV) is 4–15 and the desired IV after hydrogenation may be as low as 0.5; (2) tallow fatty acids to give a rubber-grade stearic acid having an IV of about 5–12; (3) coconut, palm, and palm kernel fatty acids to give hydrogenated products, which are then fractionated to give saturated fatty acids of six- to sixteen-carbon atoms; (4) fish fatty acid, such as menhaden, and high erucic rapeseed fatty acids to give arachidic (C-20) and behenic (C-22) acids; and (5) castor oil fatty acids (ricinoleic) to give 12-hydroxystearic acid. These and other feedstocks can be partially hydrogenated to give products having iodine values of about 20–80 that have found use in specific areas of application. Monomer acids, by-products from the dimerization of unsaturated fatty acids (primarily tall oil fatty acids), are hydrogenated to give a mixture of primarily stearic and isostearic acid, which are then separated from solvent by crystallization (110).

Hydrogenations can be carried out in batch reactors, in continuous slurry reactors, or in fixed-bed reactors. The material of construction is usually 316 L stainless steel because of its better corrosion resistance to fatty acids. The hydrogenation reaction is exothermic and provisions must be made for the effective removal or control of the heat; a reduction of one IV per g of C_{18} fatty acid releases 7.1 J (1.7 cal), which raises the temperature 1.58°C. This heat of hydrogenation is used to raise the temperature of the fatty acid to the desired reaction temperature and is maintained with cooling water to control the reaction.

The size of a typical batch reactor is 18–23 t. Some reactors have a hydrogen recycle system, whereas in others the hydrogen is internally recycled by the use of a hollow agitator shaft that allows hydrogen to be drawn from the head space to below the impeller blades where it mixes with the liquid fatty acids. This recycle of the hydrogen gives better contact with the liquid and effective mixing does not have to depend solely on agitation. It is also an advantage to vent some of the hydrogen periodically, which allows removal of inert material and possibly catalyst poisons (111).

Continuous slurry reactors are generally either of one of two designs. One type uses a reactor loop, generally known as a Buss loop design; the other is a co-current hydrogen/fatty acid/catalyst system mainly marketed by Lurgi. Continuous slurry reactors are more popular in Europe, Asia, and South America than in the United States.

Fixed-bed reactors have been described in detail and their advantages and disadvantages listed (111). It is reported that only one manufacturer uses fixed-bed hydrogenation for fatty acids (111).

Dry reduced nickel catalyst protected by fat is the most common catalyst for the hydrogenation of fatty acids. The composition of this type of catalyst is about 25% nickel, 25% inert carrier, and 50% solid fat. Manufacturers of this catalyst include Calsicat (Mallinckrodt), Harshaw (Engelhard), United Catalysts (Süd Chemie), and Unichema. Other catalysts that still have some place in fatty acid hydrogenation are so-called wet reduced nickel catalysts (formate catalysts), Raney nickel catalysts, and precious metal catalysts, primarily palladium on carbon. The spent nickel catalysts are usually sent to a broker who sells them for recovery of nickel value. Spent palladium catalysts are usually returned to the catalyst supplier for credit of palladium value.

The most important reaction variables in the hydrogenation of fatty acids are temperature, pressure, agitation, catalyst loading, and catalyst addition. Temperature is normally in the range of 150 to 210°C. Below 150°C the nickel catalyst is not activated sufficiently; however, above 210°C some degradation may occur, which could inactivate the catalyst thus slowing the reaction rate. Pressures of industrial hydrogenations are about 2.08–3.47 MPa (300–500 psig). Pressures below 2.08 MPa require longer times to reach the same IV if the hydrogenation is carried out at 3.47 MPa. Pressure higher than 3.47 MPa has little or no effect on reaction rate. Increasing the agitation in the reactor increases the reaction rate. Determination of the optimum level of agitation is dependent on the configuration inside each specific reactor. Because the reaction is mass and diffusion limited, increasing the speed of the agitator increases the dispersion of catalyst in the fatty acid until it reaches a maximum that corresponds to a maximum reaction rate, assuming everything else is constant. Proper agitation also helps to keep the solid catalyst in suspension, to help maintain temperature control, and to bring the hydrogen in contact with the catalyst/fatty acid. Increasing the catalyst loading increases the reaction rate. However, when extremely large amounts of catalysts are used it may cause a rapid decrease in hydrogen concentration in the fatty acid resulting in a dehydrogenation reaction. The optimum time of catalyst addition is near the reaction temperature and immediately thereafter hydrogen is introduced. Prolonged exposure of a nickel catalyst to hot fatty acids leads to catalyst deactivation (111).

The quality of the feedstock is important since it affects not only the product quality but the rate of hydrogenation. Some of the impurities that affect the rate are sulfur, phosphorus, halides, polyethylene, and moisture. Impurities are usually removed by clay treatment or by distillation (112).

Distillation.   Most fatty acids are distilled to produce high quality products having excellent color and a low level of impurities. Distillation removes odor bodies and low boiling unsaponifiable material in a light ends or heads fraction, and higher boiling material such as polymerized material, triglycerides, color bodies, and heavy decomposition products are removed as a bottoms or pitch fraction. The middle fractions sometimes can be used as is, or they can be fractionated (separated) into relatively pure materials such as lauric, myristic, palmitic, and stearic acids.

Because fatty acids, and especially unsaturated fatty acids, have limited stability when subjected to high temperatures, most distillations are carried out in continuous distillation columns as opposed to batch-type distillations. Almost all distillations are carried out under vacuum and sometimes with the injection of steam to further reduce the temperature at which the fatty acid will distill. Boiling points of various fatty acids at different pressures are shown in Table 13.

The crude fatty acid feed is usually preheated and degassed to remove air and water. If a vapor feed is desired it is then flashed in a vaporizer and fed to the distillation column as a vapor or vapor–liquid. A continuous distillation using one column will generally produce a satisfactory product if a suitable feedstock is used. If the feedstock contains large amounts of low boilers, color, and odor bodies, it is then preferable to use a two-column system. The low boilers and odor are removed as a heads fraction in the first column; the remaining material is removed at the bottom and then fed to a second column where a residue cut is taken off the bottom and the desired fatty acid fraction is removed as a side stream. It is possible to use a one-column system if the side stream is taken in

Table 13. **Boiling Points of Fatty Acids,**[a] °C

Pressure, kPa[b]	Caproic acid	Caprylic acid	Capric acid	Lauric acid	Myristic acid	Palmitic acid	Stearic acid	Oleic acid	Linoleic acid
0.133 kPa	61.7	87.5	110.3	130.2	149.2	167.4	183.6	177.6	178.5
0.267 kPa	71.9	97.9	121.1	141.8	161.1	179.0	195.9	189.5	190.1
0.533 kPa	82.8	109.1	132.7	154.1	173.9	192.2	209.2	202.6	202.8
1.07 kPa	94.6	121.3	145.5	167.4	187.6	206.1	224.1	217.0	216.9
2.13 kPa	107.3	134.6	159.4	181.8	202.4	221.5	240.0	232.9	232.6
4.27 kPa	120.8	149.2	174.6	197.4	218.3	238.4	257.1	250.6	250.0
8.53 kPa	136.0	165.3	191.3	214.6	236.3	257.1	276.8	270.3	269.7
17.1 kPa	152.5	183.3	209.8	234.3	257.3	278.7	299.7	292.5	291.9
34.1 kPa	171.5	203.0	230.6	256.6	281.5	303.6	324.8	317.7	317.2
68.3 kPa	192.5	225.6	254.9	282.5	309.0	332.6	355.2	346.5	346.5
101.3 kPa	205.8	239.7	270.0	298.9	326.2	351.5	376.1	364.9	365.2

[a]Ref. 113.
[b]To convert kPa to mm Hg, multiply by 7.5.

the stripping section below the feed and taken as a vapor and condensed. Doing this will eliminate any high boilers in the side steam.

Most distillation columns in the past used sieve trays or bubble cap trays where the pressure drop could be substantial (1–2 mm/tray) or about 30–50 mm in a column of 20–30 trays. Because of this pressure drop the distillation temperature must be higher and this higher temperature causes some decomposition of the fatty acids. A packing called structured packing has low pressure drop per theoretical plate thus making it ideal for use in the distillation of heat-sensitive fatty acids. It is available throughout the world as either Flexipac or Mellapac (113). Wiped-film evaporators can be used in the distillation of heat-sensitive material because the contact time with a hot surface is extremely short. They have also been used in depitching crude tall oil (113).

Fatty acids are corrosive at high temperatures and selection of materials of construction for distillation systems is critical. Stainless steels with various contents of molybdenum have proved satisfactory. For example, 316 L has 2% Mo and is satisfactory for service up to 260°C; 317 L has 3% Mo and can be used satisfactorily up to 285°C, whereas 904 L can be used up to 310°C (113).

5.2. Synthetic Routes to Fatty Acids from Petroleum. These synthetic processes have been reviewed in detail (114).

Catalytic Oxidation for Straight-Chain Paraffinic Hydrocarbons. Synthetic fatty acids (SFA) are produced by Eastern European countries, Russia, and China using a manganese-catalyzed oxidation of selected paraffinic streams. The technology is based on German developments that were in use during World War II. The oxidation is highly exothermic and is carried out at about 105–125°C, mostly in continuous equipment.

Oxidation of Straight-Chain 1-Olefins. Oxidation of α-olefins has been thoroughly studied using ozone, peracids, nitric acid, chromic acid, and others.

Carboxylation/Oxidation of Straight-Chain 1-Olefins. Selective carboxylation of α-olefins to predominately straight-chain aldehydes is realized through specific catalyst systems and by careful control of reaction conditions. The aldehyde produced is then air-oxidized to the acid using a Mn catalyst. Heptanoic acid [111-14-8] and pelargonic acid [112-05-0] are produced commercially in this manner.

Carboxylation of Straight-Chain 1-Olefins.   Carboxylation is the selective addition of CO and water to an olefin to give either a straight-chain or branched-chain acid. The use of specific catalysts and reaction conditions has given a straight-chain/branched-chain isomer ratio as high as 98:2. In spite of a one-step method to predominately straight-chain isomers of fatty acids, no commercialization using this route has yet occurred in the United States.

Oxidation of Straight-Chain Alcohols.   Two methods have been developed. One uses an air oxidation catalyzed by a metal, eg, copper, platinum, etc, whereas the other is a caustic oxidation. Generally, however, fatty alcohols are priced higher on the world market than their corresponding fatty acids and, consequently, these conversions are uneconomical.

Branched-Chain Carboxylic Acids.   Branched-chain acids such as 2-methylbutyric, 3-methylbutyric, isooctanoic, and isononanoic acids are produced by the oxo reaction, giving first the corresponding aldehyde, which is then oxidized to the acid. 2-Ethylhexanoic acid is produced by the aldol route from butyaldehyde in three steps: aldol condensation; hydrogenation of the carbon–carbon double bond; and oxidation of the branched-chain saturated aldehyde to 2-ethylhexanoic acid.

Manufacturing procedures for most branched-chain acids are well known. For example, oxo process acids are manufactured from branched-chain olefins using hydroformylation followed by oxidation (115) (see OXO PROCESS).

$$CH_3CHCH=CH_2 \ + \ CO \ \xrightarrow{H_2} \ CH_3CHCH_2CH_2CHO \ \xrightarrow{[O]} \ CH_3CHCH_2CH_2COOH$$
$$\quad\ | \qquad\qquad\qquad\qquad\qquad\qquad\ | \qquad\qquad\qquad\qquad\qquad\ |$$
$$\quad CH_3 \qquad\qquad\qquad\qquad\qquad CH_3 \qquad\qquad\qquad\qquad CH_3$$

(primary product)

Isostearic acid is produced from dimerization and reduction of monomeric acids (116).

6. Economic Aspects and Applications

Aliphatic carboxylic acids produced on a reasonably significant commercial scale range from acetic acid (two carbons or C2) through stearic acid (C18). Lesser amounts of commercially available shorter chain-length acids, such as formic (C1), and longer chain-length acids, such as erucic (unsaturated C22) and behenic (saturated C22), are also produced. As a general rule, all of the even chain-length, nonisomeric acids from C6 to C22 are produced from naturally occurring fats and oils. A significant proportion of the lower chain-length (C1–C6) and longer isomeric chain-length (C7–C10) acids are made synthetically.

6.1. Nonoleo-Based Carboxylic Acids.   Some of the more prominent carboxylic acids that are not fat- or oil-based include acetic, acrylic, and olefin-based propionic, butyric/isobutyric, 2-ethylhexanoic, heptanoic, pelargonic, neopentanoic, and neodecanoic. Table 14 summarizes the production, pricing, primary producers and applications of these acids.

Table 14. **U.S. Statistics on Non-Fat and Oil-Based Acids**[a,b]

Acid	Production, 10^3 t	Price, $/kg	Producers	Applications
acetic (C2)	3354	0.19–0.27	Millennium, Eastman, Hoechst Celanese, Sterling, Air Products, Du Pont Primester	vinyl acetate; acetic anhydride; acetate esters; TPA/DMT plastics
acrylic (C3)	1313	0.39	BASF, Hoechst Celanese, Rohm & Haas, Dow	acrylate esters (latex coatings, adhesives, polishes, etc)
propionic (C3)	200	0.24	Eastman, Hoechst Celanese, Dow	grain and feed preservative; propionate salts; pesticides; cellulose acetate
butyric/isobutyric (C4)			Eastman, Hoechst Celanese	cellulose acetate butyrate (plastic); pesticides (isobutyric)
2-ethylhexanoic (C8)			Eastman, Union Carbide	paint driers; heat stabilizers for PVC
benzoic (C7)	129	0.28	Kalama, Pfizer, Velsicol	benzoate plasticizers; benzoate salts (preservative), alkyd resins
heptanoic/pelargonic (C7/C9)			Hoechst Celanese	polyol esters for synthetic lubricants; tetraethylene glycol diheptanoate plasticizer
neoacids (C5–C10)			Exxon	peroxy esters (polymerization initiators); pivaloyl chloride (polymerization, pharmaceuticals)

[a]Refs. 117–119.
[b]Production and prices given when available.

55

6.2. Oleo-Based Carboxylic Acids. Typically, fatty acids make up between 87 and 90% of the fat or oil from which they are made; the remaining 10–13% is glycerol. The most often used raw materials are coconut or palm kernel oil (lauric oils) for C8, C10, C12, and C14 acids; tallow, lard, and palm stearine for C16 and C18 acids; and soybean, sunflower, canola, and tall oil for whole cut unsaturated (lower melting point) acids. Fully hardened soya, canola, or edible tallow are usually used when high C18 food-grade stearic acid is needed, whereas edible tallow and tall oil are the primary raw materials for food-grade oleic. C22 acid is derived from rapeseed and/or marine oil (menhaden). Tall oil, a by-product of the kraft pulp and paper industry, is not a triglyceride and therefore does not contain glycerol. Castor oil (qv) is the primary source for ricinoleic acid or 12-hydroxystearic acid when hardened. The compositions of some of these fats and oils are outlined in Table 15.

In 2001, production of 8-22 carbon acids in the United States, Western Europe, and Japan was 2.5×10^6 t. North American production fell in 2001. New fatty acid plants were built in Southeast Asia. Major sources of raw materials, eg, coconut, palm, are found in this area of the world. Capacity of these plants was 1.5×10^6 t in 2001. There was a ca 9% growth in production and consumption in North America, but declined to 4% in 2001. A 1.6%/yr growth rate is expected throuth 2006.

Tall oil fatty acids are in decline because of reduced refinery output. Growth in Western Europe is expected at the rate of 1.2% throuth 2006. Japan

Table 15. **Typical Composition of Specific Fats and Oils,** %

Acid (chain length)	Lauric oils		Tallow[a]	Palm stearine	Soybean	Tall Oil[b]	High erucic rapeseed[c]
	Coconut	Palm kernel					
Saturated acids							
caproic (C6)							
caprylic (C8)	7	3					
capric (C10)	6	3					
lauric (C12)	50	50			0.5		
myristic (C14)	18	18	3		0.5		
palmitic (C16)	8.5	8	24	51	12		3
stearic (C18)	3	2	20	5	4	2	1
Unsaturated acids[d]							
palmitoleic (C16:1)			2.5				
oleic (C18:1)	6	14	43	33	25	59	18
linoleic (C18:2)	1	2	4	9	52	37	14
linolenic (C18:3)	0.5		0.5		6		7
arachidonic (C20:1)						1	10
erucic (C22:1)							46

[a]Also contains ∼1% of myristoleic acid (C14:1)d.
[b]Also contains ∼1% of arachidic acid (C20).
[c]Also contains ∼0.5% each of arachidic and behenic acid (C22).
[d]The number following the colon indicates the number of sites of unsaturation.

consumption is expected to decline at the rate of 1%/yr because of economic conditions (121).

Table 16 provides application information on various fatty acids as well as listing U.S. producers.

Because they are made from renewable natural raw materials, oleo-based fatty acids are completely biodegradable and find widespread usage in a variety of applications and industries.

7. Health and Safety Factors

Carboxylic acid dust and vapors are generally described as being destructive to tissues of the mucous membrane, eyes, and skin. The small molecules such as formic, acetic, propionic, butyric, and acrylic acids tend to be the most aggressive (Table 17) (123). Formic, acetic, propionic, acrylic, and methacrylic acids have time weighted-average exposure limits of 20 ppm or lower. Acrylic acid showed an LD_{50} of 33.5 mg/kg from oral administration to rats.

The hazards of handling branched-chain acids are similar to those encountered with other aliphatic acids of the same molecular weight. Eye and skin contact as well as inhalation of vapors of the shorter-chain acids should be avoided.

8. Environmental Aspects

Environmental regulation in the oleochemical industry addresses pollution of air, surface, and groundwater, along with land pollution and solid waste disposal. This is administered by the Environmental Protection Agency (EPA) on the national level, an equivalent agency on the state level, and sometimes local agencies also deal with various aspects of pollution abatement.

In-plant controls are perhaps the best approach to eliminate waste generation and pollution problems, and many times good payback exists on recovery of products lost because of poor process controls. If the production department is responsible for the generation and in-plant control of wastes, this will help ensure that initial standards for water use and process loss are reasonable and that they are maintained (124).

9. Trialkylacetic Acids

Trialkylacetic acids are characterized by the following structure:

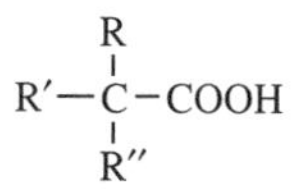

$$R' - \underset{\underset{R''}{|}}{\overset{\overset{R}{|}}{C}} - COOH$$

in which R, R', and R'', are $C_xH_{2x}+1$ with $x \geq 1$. The lowest member of the series (R = R' = R'' CH_3) is the C_5 acid, trimethylacetic acid or 2,2-dimethylpropanoic

Table 16. **Oleo-Based Carboxylic Acids**

Acid	Price[a], $/kg	U.S. producers	Applications
		Oil specific acids	
canola	0.22–0.48	Henkel/Emery, Procter & Gamble, Sherex Caschem, Union Camp	surfactants
castor oil acids (ricinoleic, 12-hydroxystearic)			lubricating greases
coconut oil acids	0.11–0.13	Dial, Henkel/Emery, Procter & Gamble, Witco, Karlshamns	surfactants, soap
hydrogenated and/or separated tallow-based acids		Akzo, Henkel/Emery, Lonza, Procter & Gamble, Sherex, Synpro, Unichema, Witco	metallic stearates (plastic lubricants), tires, candles, crayons, cosmetics alkyd resins (paint)
soybean oil acids		Henkel/Emery, Procter & Gamble, Witco	
tall oil acids			
2% or more rosin	0.17	Arizona, Georgia-Pacific, Hercules, Union Camp, Westvaco	alkyd resins, ore flotation, chemical intermediates, soaps
less than 2%	0.27		
tallow fatty acids		Akzo, Dial, Henkel/Emery, Lonza, Procter & Gamble, Sherex, Synpro. Unichema, Witco	soap, lubricants, fabric softeners, asphalt emulsifiers, synthetic rubber, plastics
		Chain-length specific acids	
capric		Akzo, Henkel/Emery, Procter & Gamble, Witco	synthetic lubricants, medium-chain triglycerides
caprylic		Akzo, Henkel/Emery, Procter & Gamble, Witco	synthetic lubricants, medium-chain triglycerides
caprylic-capric blend		Akzo, Dial, Henkel/Emery, Procter & Gamble, Witco	synthetic lubricants, medium-chain triglycerides
lauric, 95% (dodecanoic)	0.31–0.33	Akzo, Henkel/Emery, Procter & Gamble, Witco	surfactants, soap
myristic, 95% (tetradecanoic)	0.31–0.32	Henkel/Emery, Procter & Gamble, Witco	esters for cosmetics, lotions
oleic	0.22–23	Henkel/Emery, Hercules, Unichema, Witco	surfactants, lubricants, plasticizers
palmitic, 90%		Henkel/Emery, Procter & Gamble, Sherex, Witco	esters for personal care products
pelargonic (nonanoic), 90%		Henkel/Emery	synthetic lubricants, plasticizers
stearic, 90%	0.15–0.18	Akzo, Henkel/Emery, Lonza, Procter & Gamble, Sherex, Synpro, Unichema, Witco	alkyd resins, ore flotation

[a]Ref. 122.

Table 17. **Safety and Toxicity**

Total number of carbon atoms	Systematic name (common name)[a]	Flash point, °C	RTECS number	ACGIH[b] TWA, ppm	Rat[c] LD$_{50}$, mg/kg	DOT/IMO[d] label
1	methanoic (formic)		LQ4900000	5	1,100	corrosive
2	ethanoic (acetic)		AF1225000	10	3,530	corrosive flammable
3	propanoic (propionic)	54	UE5950000	10	2,600	corrosive
4	butanoic (butyric)	66	ES5425000		2,000	corrosive
5	pentanoic (valeric)	96[e]	YV6100000		600[f]	corrosive
6	hexanoic ([caproic])	102[e]	MO5250000		3,000	corrosive
7	heptanoic ([enanthic])		MJ1575000		7,000	corrosive
8	octanoic ([caprylic])	132[e]	RH0175000		10,080	
9	nonanoic (pelargonic)		RA6650000		3,200	
10	decanoic ([capric])		HD9100000			
11	undecanoic ([undecyclic])		YQ2275000		140[f]	
12	dodecanoic (lauric)		OE9800000		12,000	
13	tridecanoic ([tridecyclic])		YD3850000			
14	tetradecanoic (myristic)		QH4375000			
15	pentadecanoic ([pentadecylic])		RZ1925000			
16	hexadecanoic (palmitic)		RT4550000			
17	heptadecanoic (margaric)		MI3850000			
18	octadecanoic (stearic)	196	WI2800000		4,640	
3	propenoic (acrylic)		AS4375000	2	33.5	corrosive
5	4-pentenoic (allylacetic)		SB2800000		470	
4	2-methylpropenoic (methyacrylic)		OZ2975000	20	1,600[f]	corrosive
4	2-methylpropanoic (isobutyric)		NQ4375000	280		corrosive, flammable
5	3-methyl-2-butenoic (β, β-dimethyl acrylic)		GQ5425000		3,560	
5	3-methylbutanoic (isovaleric)		NY1400000		2,000	
5	2,2-dimethyl-propanoic (pivalic)		TO7700000		900	
8	2-ethylhexanoate		MO7700000		3,000	
11	10-undecenoic (undecylenic)		YQ2975000		2,500	
18	cis-9-octadecenoic (oleic)		RG2275000		74,000	
3	propynoic (propiolic)		UD9300000		100	

[a]Bracket signify a trivial name no longer in use.
[b]ACGIU, American Conference of Governmental Industrial Hygienists; TWA, time-weighted average (8-h day or 40-h week).
[c]LD$_{50}$, lethal dose 50% kill (oral).
[d]DOT, Department of Transportation; IMO, International Maritime Organization.
[e]Open cup.
[f]Mouse.

acid (also, neopentanoic acid, pivalic acid). For higher members in the series, the products are typically mixtures of isomers, resulting from the use of mixed isomer feedstocks and the chemical rearrangements that occur in the manufacturing process.

Trialkylacetic acids have been produced commercially since the early 1960s, in the United States by Exxon and in Europe by Shell, and have been marketed as neo acids (Exxon) or as Versatic Acids (Shell). The principal commercial products are the C_5 acid and the C_{10} acid (neodecanoic acid, or Versatic 10), although smaller quantities of other carbon numbers, such as C_6, C_7, and C_9, are also produced.

The trialkylacetic acids have a number of uses in areas such as polymers, pharmaceuticals, agricultural chemicals, cosmetics, and metal-working fluids. Commercially important derivatives of these acids include acid chlorides, peroxyesters, metal salts, vinyl esters, and glycidyl esters.

9.1. Trimethylacetic Acid.   *Physical Properties.*   2,2-Dimethylpropionic acid [75-98-9], $(CH_3)_3CCOOH$, also referred to as neopentanoic acid or pivalic acid, is a solid at room temperature with a pungent odor typical of many lower molecular weight carboxylic acids. It is commercially available at a purity greater than 99.5%. Neopentanoic acid is a single isomer with a high degree of symmetry and, thus, has a relatively high melting point ($+34°C$, compared to $-34.5°C$ for *n*-pentanoic acid). Physical properties of a typical commercial sample are given in Table 18.

Chemical Properties.   Neopentanoic acid [75-98-9] undergoes reactions typical of carboxylic acids. Reactions often proceed less readily than with straight-chain acids because of the steric hindrance around the carbonyl group. However, this steric hindrance at the α-carbon results in derivatives that are typically more resistant to hydrolysis and oxidation.

Acid Chloride Formation.   Neopentanoic acid can be converted to neopentanoyl chloride [3282-30-2] by reaction with thionyl chloride (126), phosgene (127), phosphorus pentachloride, phosphorus trichloride, or by the reaction with

Table 18. **Physical Properties of Commercially Available Neopentanoic Acid**[a]

Property	Value
mp, °C	34.4
bp, °C	163–165
acid value, mg KOH/g	550
color, Pt/Co (Hazen) of molten material	50
specific gravity at 38/38°C	0.913
viscosity at 60°C, mm^2/s (=cSt)	1.7
flash point, °C (Tag closed cup)	63
water, wt %	0.05
vapor pressure, kPa^b at 60°C	1.33
solubility in water, g 100 mL H_2O at 25°C	2.1
heat of vaporization, $kJ/kg,^c$ at the boiling point and 101.3 kPa^b	423
ionization constant, $K_a \times 10^{-6}$ at 25°C	9.3

[a]Ref. 125.
[b]To convert kPa to mm Hg, multiply by 7.5.
[c]To convert kJ to kcal, divide by 4.184.

benzotrichloride in the presence of Friedel-Crafts catalysts (128). A laboratory procedure using tetramethyl-α-halogenoenamines at room temperature has also been reported (129).

Commercially, neopentanoyl chloride is often the preferred starting material for the synthesis of peroxyesters, agricultural chemicals, pharmaceuticals, esters, and other fine chemicals because the reactivity of the acid halide is generally greater than that of the acid.

Esterification. Esters of neopentanoic acid can be prepared either from the chloride or directly from the acid and alcohol. An example of the former reaction is that between neopentanoyl chloride and *tert*-butyl hydroperoxide to give *tert*-butyl peroxyneopentanoate [927-07-1], which is used as a free-radical initiator in polymerizations. For direct esterification, acid catalysts, such as sulfuric acid or toluene sulfonic acid, are used, although higher catalyst concentrations are generally required because of the lower reactivity of the acid. Methyl neopentanoate [598-98-1] has been prepared using an acid catalyst (130) or sulfonic acid cation-exchange resin (131); aromatic neopentanoate esters can be made by standard esterification procedures (132). Vinyl neopentanoate [3377-92-2] is prepared from neopentanoic acid and acetylene using zinc neopentanoate [15827-10-8] as catalyst and zinc chloride as cocatalyst (133). It can also be prepared from the acid and vinyl acetate [108-05-4] using ruthenium carbonyl as the catalyst (134). The glycol monoester can be prepared in high yields, essentially free of the diester, by reaction of the acid with ethylene oxide using an hydroxy-alkylamine as the catalyst (135).

The neo acids are generally less reactive than their straight-chain counterparts. For example, neopentanoic acid reacts approximately 15 times slower than *n*-pentanoic acid (136). Greater steric hindrance, as brought about by ethyl groups, for example, results in even slower esterification rates. Once formed, however, the esters of neopentanoic acid are more resistant to hydrolysis than the corresponding linear acid esters. For example, under basic conditions, hexyl neopentanoate [5434-57-1] is hydrolyzed approximately 20 times more slowly than hexyl valerate [1117-59-5]. Under acid conditions, the difference in hydrolysis rate is a factor of 160 (137).

Reduction. 2,2-Dimethylpropanal [630-19-3] can be prepared by the reduction of neopentanoic acid using various catalysts, such as iron (138), tin or zirconium oxides (139,140), iron–chromium (141), and other reagents (142,143). The reduction of neopentanoic acid to 2,2-dimethylpropanol [75-84-3] (neopentyl alcohol) has been accomplished using supported osmium and rhenium (144), copper–zinc or nickel (145), metal oxides (146), or sodium borohydride (147). Reduction to the alkane, 2,2-dimethylpropane [463-82-1] (neopentane), has been claimed using a copper oxide–zinc oxide catalyst (148).

Other Reactions. The anhydride of neopentanoic acid, neopentanoyl anhydride [1538-75-6], can be made by the reaction of neopentanoic acid with acetic anhydride (149). The reaction of neopentanoic acid with acetone using various catalysts, such as titanium dioxide (150) or zirconium oxide (151), gives 3,3-dimethyl-2-butanone [75-97-8], commonly referred to as pinacolone. Other routes to pinacolone include the reaction of pivaloyl chloride [3282-30-2] with Grignard reagents (152) and the condensation of neopentanoic acid with acetic acid using a rare-earth oxide catalyst (153). Amides of neopentanoic acid can be prepared directly from the acid, from the acid chloride, or from esters, using primary or secondary amines.

Manufacture. Trialkylacetic acids are prepared using variants of the Koch reaction (154), a two-stage reaction for the preparation of carboxylic acids. In the first stage, olefin, carbon monoxide, and a strong acid catalyst react to give what is commonly referred to as the complex. In the second stage, the complex is hydrolyzed to give the carboxylic acid and to regenerate the catalyst. A number of Brønsted acid catalysts have been used, including H_2SO_4, H_3PO_4, HF, and Lewis acids such as BF_3.

Temperatures used depend on the choice of catalyst, and can range from -20 to $+80°C$, pressures used can be as high as 10 MPa (100 atm). Koch reactions have been reviewed (155). The mechanism of reaction is believed to proceed by the formation of a carbenium ion from the olefin, followed by addition of carbon monoxide to give an acylium cation. The acylium cation then reacts with water to give the carboxylic acid. This is illustrated using isobutylene [115-11-7] as the olefin, which gives neopentanoic acid as the product.

$$
\begin{array}{c}
CH_3 \\
\overset{|}{\underset{|}{C}}=CH_2 \\
CH_3
\end{array}
\xrightarrow{H^+}
\begin{array}{c}
CH_3 \\
CH_3-\overset{|}{\underset{|}{C}}{}^+ \\
CH_3
\end{array}
\xrightarrow{CO}
\begin{array}{c}
CH_3\ O \\
CH_3-\overset{|}{\underset{|}{C}}-\overset{||}{C}{}^+ \\
CH_3
\end{array}
\xrightarrow{H_2O}
\begin{array}{c}
CH_3\ O \\
CH_3-\overset{|}{\underset{|}{C}}-\overset{||}{C}-OH \\
CH_3
\end{array}
+\ H^+
$$

Neopentanoic acid has also been produced commercially from diisobutylene [18923-87-0], in which the first step in the reaction sequence is a cracking or depolymerization of the olefin to give isobutylene.

Commercial production of these acids essentially follows the mechanistic steps given. This is most clearly seen in the Exxon process of Figure 2 (156). In the reactor, catalyst, olefin, and CO react to give the complex. After degassing, hydrolysis of this complex takes place. The acid and catalyst are then separated, and the trialkylacetic acid is purified in the distillation section. The process postulated to be used by Shell (Fig. 3) is similar, with additional steps prior to dis-

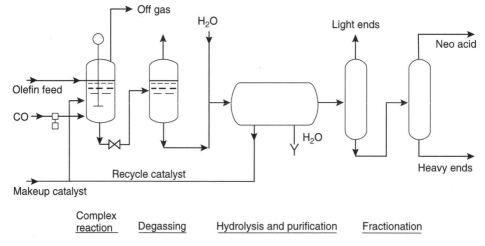

Fig. 2. Neo acid production schematic (156). Courtesy of *Hydrocarbon Processing*.

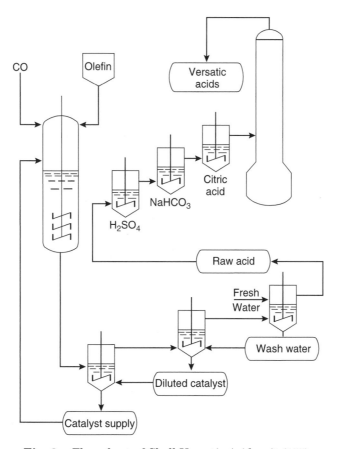

Fig. 3. Flow sheet of Shell Versatic Acid unit (157).

tillation being used. In 1980, the conditions used were described as ca 40–70°C and 7–10 MPa (70–100 bar) carbon monoxide pressure with H_3PO_4–BF_3–H_2O in the ratio 1:1:1 (Shell) or with $BF_3 \cdot 2H_2O$ (Enjay) as catalyst (157).

Work on the process for the production of these acids has continued in recent years. One patent discloses the use of zeolite catalysts (158) for the synthesis of neopentanoic acid from isobutylene. The use of a copper catalyst in a strong acid, such as sulfuric acid, operating at lower pressures, has also been claimed (159).

Shipment. Neopentanoic acid is shipped in heated tank cars, heated tank trucks, and drums.

Health and Safety Factors. Neopentanoic acid possesses low toxicity, either by ingestion (oral LD_{50} in rats is 2.0 g/kg) or by skin absorption (dermal LD_{50} in rabbits is 3.16 g/kg). The principal hazards associated with neopentanoic acid at ambient temperatures are from eye and skin irritation. At elevated temperatures, where concentrations of the vapor are significant, irritation of the respiratory tract can also occur. Contact with the material should be avoided.

Eye contact should be followed by flushing the eyes with large amounts of water. If irritation persists, medical attention should be obtained. Skin contact should be followed by flushing with water, using soap if available. Neopentanoic acid is combustible and will burn. Fire should be extinguished with foam, dry chemical, or water spray.

Uses. Polymers and Resins. *tert*-Butyl peroxyneopentanoate and other peroxyesters of neopentanoic acid can be used as free-radical initiators for the polymerization of vinyl chloride [75-01-4] (160) or of ethylene [74-85-1]. These peresters have also been used in the preparation of ethylene–vinyl acetate copolymers [24937-78-8] (161), modified polyester granules (162), graft polymers of aminoalkyl acrylates with vinyl chloride resins (163), and copolymers of *N*-vinyl-pyrrolidinone [88-12-0] and vinyl acetate [108-05-4] (164). They can also be used as curing agents for unsaturated polyesters (165).

Vinyl neopentanoate is used in the preparation of adhesives and binders (166–168), optical materials for plastic lenses (169), gas permeable membranes for oxygen enrichment (170), and in coating applications (171,172).

Pharmaceuticals. Neopentanoic acid derivatives are widely used in the preparation of pharmaceuticals, eg, as a means of introducing the *tert*-butyl group into a molecule. More frequently, however, derivatives have been prepared that exploit the enhanced hydrolytic stability of the neopentanoate group. For example, when salmon calcitonin is treated with *N*-hydroxysuccinimide pivalate [42014-50-6], the resulting derivative retains the biological activity of the precursor, but gives an extended duration of activity (173).

Chloromethyl 2,2-dimethylpropionate [18997-19-8] has also been used to prepare a number of ester derivatives with improved properties compared to the nonderivatized compound. The pivaloyloxymethyl ester derivative of piperacillin [61477-96-1] is an orally administrable antibiotic agent (174). The pivaloyloxymethyl ester [77372-61-3] of 2-propylpentanoic acid [99-66-1] shows comparable anti-epileptic and anticonvulsant activity to the acid, but is more rapidly and uniformly absorbed in the intestinal tract than the acid itself (175). The pivaloyloxymethyl ester of 2-anilinonicotinic acid retains the analgesic and anti-inflammatory activity without the ulcerogenic activity of the free acid (176). Similarly, pivaloyloxymethyl salicylate [66195-29-7], prepared from sodium salicylate and chloromethyl pivalate, has all the favorable properties of aspirin without undesirable gastric irritation (177).

Neopentanoyl chloride has been used in the preparation of AZT (178), which is used in the treatment of acquired immune deficiency syndrome (AIDS) (see ANTIVIRAL AGENTS).

The effect of acid structure on skin penetration and skin irritation has been studied (179).

Agricultural Applications. One of the largest uses for neopentanoic acid is in the preparation of agricultural chemicals. Neopentanoic acid or its derivatives are used in the preparation of a number of commercial herbicides, such as tebuthiuron [39014-18-1], metamitron [41394-05-2], metribuzin [21087-64-9], clomazone [81777-89-1], and oxadiazon [19666-30-9], and others (180–185). As with pharmaceuticals, the pivalate ester of a herbicide has been prepared to increase the resistance to hydrolysis, allowing the preparation of aqueous dispersions (186). A combination of a pivalic acid amide and metribuzin results in a synergism that broadens the range of applicability of the herbicide (187). The amides themselves also show nematocidal and fungicidal activity. The preparation of organotin compounds for use as miticides also involves the use of neopentanoic acid (188).

Cosmetics. Esters of neopentanoic acid are used as perfumes or perfume precursors (132,189), as liquid binders (190), and in emollient and moisturizing compositions (191).

Fuels, Lubricants, and Transmission Fluids. Polyol esters of neopentanoic acid have been used as high vacuum pumping liquids that are stable in chemically aggressive environments (192). Esters such as 6-(*p*-anilinophenoxy)-hexyl pivalate are used as antioxidants for synthetic ester lubricants (193). Pivalic anhydride [1538-75-6] has been claimed as an antiknock additive for gasoline (194).

Miscellaneous Applications. The fruity odor of a series of esters, including a number of esters of neopentanoic acid, has been related to the structure of the ester, with an emphasis on pearlike odor (195). Methyl pivaloylacetate [55107-14-7], $(CH_3)_3CCOOCH_2COOCH_3$, prepared by condensation of methyl pivalate and methyl acetate, is used as an intermediate in the preparation of photographic chemicals (196). Degradable organic material such as aviation turbo-kerosene can be preserved by using basic zinc carboxylates, such as zinc pivalate [15827-10-8], for their biocidal activity (197) (see AVIATION AND OTHER GAS TURBINE FUELS).

9.2. C₁₀ Trialkylacetic Acids. *Physical Properties.* The C_{10} trialkylacetic acids, referred to as neodecanoic acid [26896-20-8] or as Versatic 10 [52627-73-3], are liquids at room temperature. Typical physical properties for commercially available material are given in Table 19. These materials are typically mixtures of isomers, hence no structures are given throughout this section.

Chemical Properties. Like neopentanoic acid, neodecanoic acid, $C_{10}H_{20}O_2$, undergoes reactions typical of carboxylic acids. For example,

Table 19. **Physical Properties of Commercially Available Neodecanoic Acid**[a]

Property	Value
mp, °C	<−40
bp, °C	250−257
acid value, mg KOH/g	325
color	100 (Pt/Co)
specific gravity at 20/20°C	0.915
viscosity, mm²/s (=cSt)	
at 20°C	35.7
at 60°C	7
flash point, °C (Tag closed cup)	105
water, wt%	0.05
vapor pressure, kPa[b] at 60°C	0.012
solubility in water, g/100 mL H_2O at 25°C	0.017
heat of vaporization, kJ/kg[c], at the boiling point and 101.3 kPa[b]	249.5
ionization constant, $K_a \times 10^{-6}$ at 25°C	4.2

[a] Ref. 125.
[b] To convert kPa to mm Hg, multiply by 7.5.
[c] To convert kJ to kcal, divide by 4.184.

neodecanoic acid is used to prepare acid chlorides, amides (198), and esters (131,135,199,200), and, like neopentanoic acid, is reduced to give alcohols and alkanes (145,148). One area of reaction chemistry that is different from the C_5 acids is the preparation of metal salts. Both neopentanoic acid and neodecanoic acid, like all carboxylic acids, can form metal salts. However, in commercial applications, metal salt formation is much more important for neodecanoic acid than it is for neopentanoic acid.

Metal Salt Formation.   At least three methods are commonly used to prepare metals salts. The first of these is known as the double decomposition method.

$$CoCl_2 \text{ (aq)} + 2Na(neodec) \text{ (org)} \longrightarrow Co(neodec)_2 \text{ (org)} + 2\,NaCl \text{ (aq)}$$
$$\text{where neodec} = \text{the anion of neodecanoic acid}$$

Typically, a slight excess of acid is used, resulting in salts that are neutral or slightly acidic. This method has been applied to the preparation of zirconium salts (201).

A second method for preparing acid salts is termed fusion, represented by

$$Co(OH)_2 + 2 \text{ neodecanoic acid} \longrightarrow Co(neodec)_2 + 2\,H_2O$$

In this method, a metal oxide or hydroxide is slurried in an organic solvent, neodecanoic acid is slowly added, and the mixture is refluxed to remove the water. Salts that are basic can be prepared by using less than stoichiometric amounts of acid. This method has been used in the preparation of metal salts of silver (202) and vanadium (203). The third method of preparation is similar to the fusion process, the difference is the use of finely divided metal as the starting material instead of the metal oxide or hydroxide. This method has been applied to the preparation of cobalt neodecanoate (204). Salts of tin (205) and antimony (206) have been prepared by the fusion method, starting with lower carboxylic acids, then replacing these acids with neodecanoic acid.

Manufacture.   The C_{10} trialkylacetic acids are prepared using the same process and catalysts as are used for the preparation of neopentanoic acid. For the C_{10} acids, a branched C_9 olefin stream is typically used. Because the reaction proceeds by means of a carbenium ion mechanism, rearrangement of the olefin occurs, resulting in a C_{10} acid composed of a large number of isomers. In addition to the carbonylation reaction, olefin dimerization and oligomerization, along with olefin disproportionation also occur, resulting in trialkylacetic acids with carbon numbers less than and greater than 10.

Shipment.   The C_{10} acids are shipped in bulk sea vessels, tank cars, tank trucks, and drums.

Health and Safety Factors.   The C_{10} trialkylacetic acids have toxicities similar to those for other neo acids: oral LD_{50} in rats is 2.0 g/kg, and dermal LD_{50} in rabbits is 3.16 g/kg.

The primary hazard associated with C_{10} trialkylacetic acids is eye irritation. In contact with the eyes, the material is irritating and may injure eye tissue if not removed promptly. Any contact with the eyes should be immediately flushed with large amounts of water. Medical attention should also be obtained.

For skin contact, flush with large quantities of water, using soap if available. To extinguish fires, use foam, dry chemical, or water spray.

Uses. *Polymers, Resins, and Coatings.* Peroxyesters of neodecanoic acid, such as *tert*-butyl peroxyneodecanoate [26748-41-4] and α-cumyl peroxyneodecanoate [26748-47-0], constitute one of the most important uses for neodecanoic acid. These materials are used as free-radical initiators in the polymerization of vinyl chloride (207), acrylates (208), ethylene (209), styrene [100-42-5] (209), and in the copolymerization of vinyl chloride with other monomers, such as propylene [115-07-1] (210), or acrylates (211). The peroxyesters are also used as curing agents for resins (212).

Metal salts of neodecanoic acid have also been used as catalysts in the preparation of polymers. For example, bismuth, calcium, barium, and zirconium neodecanoates have been used as catalysts in the formation of polyurethane elastomers (213,214). Magnesium neodecanoate [57453-97-1] is one component of a catalyst system for the preparation of polyolefins (215); vanadium, cobalt, copper, or iron neodecanoates have been used as curing catalysts for conjugated-diene butyl elastomers (216).

The metal salts of neodecanoic acid have found wide usage as driers for paints and inks (217,218). Metal neodecanoates that are used include silver (202), cobalt (204), and zirconium (201), along with lead, copper, manganese, and zinc (see DRIERS AND METALLIC SOAPS).

Neodecanoic acid is also used as the carrier for metals in poly(vinyl chloride) heat stabilizers (qv). Metals used in this application include barium, cadmium, and zinc. Tin as the neodecanoate salt has also been claimed as a heat stabilizer for maleic anhydride (219).

Adhesion Promoters. One of the growing uses for neodecanoic acid has been in the preparation of adhesion promoters for radial tires. In this application, cobalt or nickel neodecanoate, along with other components, is used during tire manufacture to promote the adhesion or bonding of the rubber to the steel cord. The result is high adhesive strength, good thermal aging resistance and improved resistance to moisture aging (220–222).

Metal-Working and Hydraulic Fluids. In the preparation of fluids for metal-working and hydraulics, the trend has been to replace organic-based materials with aqueous-based materials. Neodecanoic acid has found application in these newer fluids as a corrosion inhibitor and a viscosity improver. For example, neodecanoic acid is used in an aqueous hydraulic fluid concentrate for corrosion inhibition and improved antiwear properties (223), in the preparation of a thickened aqueous hydraulic fluid to reduce viscosity loss (224), and in a water-soluble metal working oil to reduce corrosion (225). In a similar vein, neodecanoic acid has been used in antifreeze concentrates for corrosion inhibition (226,227).

Metal Extraction. As with other carboxylic acids, neodecanoic acid can be used in the solvent extraction of metal ions from aqueous solutions. Recent applications include the extraction of zinc from river water for determination by atomic absorption spectrophotometry (228), the coextraction of metals such as nickel, cobalt, and copper with iron (229), and the recovery of copper from ammoniacal leaching solutions (230).

Fuels and Lubricants. Rare-earth neodecanoates have been claimed as additives for diesel fuels that reduce the precipitation of particles and gum (231).

Neodecanoic acid has also been used in the preparation of ashless detergent additives for fuels and lubricants that reduce engine deposits in internal combustion engines (232).

Electrical and Electronic Applications.   Silver neodecanoate [62804-19-7] has been used in the preparation of a capacitor-end termination composition (233), lead and stannous neodecanoate have been used in circuit-board fabrication (234), and stannous neodecanoate has been used to form patterned semiconductive tin oxide films (235). The silver salt has also been used in the preparation of ceramic superconductors (236). Neodecanoate salts of barium, copper, yttrium, and europium have been used to prepare superconducting films and patterned thin-film superconductors. To prepare these materials, the metal salts are deposited on a substrate, then decomposed by heat to give the thin film (237–239) or by a focused beam (electron, ion, or laser) to give the patterned thin film (240,241). The resulting films exhibit superconductivity above liquid nitrogen temperatures.

Miscellaneous Applications.   Polyamides, prepared from polyamines and neodecanoic acid, are used as wash-cycle antistatic agents (qv) (198). Salts of neodecanoic acid have been used in the preparation of supported catalysts, such as silver neodecanoate for the preparation of ethylene oxide catalysts (242), and the nickel soap in the preparation of a hydrogenation catalyst (243). Metal neodecanoates, such as magnesium, lead, calcium, and zinc, are used to improve the adherence of plasticized poly(vinyl butyral) sheet to safety glass in car windshields (244). Platinum complexes using neodecanoic acid have been studied for antitumor activity (245). Neodecanoic acid and its esters are used in cosmetics as emollients, emulsifiers, and solubilizers (199,246,247). Zinc or copper salts of neoacids are used as preservatives for wood (248).

Table 20. Physical Properties of Commercially Available Glycidyl Neodecanoates

Property	Cardura E10[a]	GLYDEXX N-10[b]
color (Pt/Co scale)	<60	45
density at 20°C, g/mL	0.958–0.968	
water content	<0.1% mass/mass	0.05 wt %
epoxy equivalent weight,[c] g	244–256	250
flash point, °C	126[d]	128[e]
residual epichlorohydrin	<10 mg/kg	<5 ppm
viscosity, mPa·s(=cP)		
25°C	7.13	
100°C	1.31	
150°C	0.72	
vapor pressure at 37.8°C, kPa[f]	0.9	
boiling range at 101.3 kPa,[f] °C	251–278	
freezing point, °C	<−60	
solubility in water at 20°C	0.01 % mass/mass	

[a]Ref. 249.
[b]Ref. 250.
[c]Grams of resin containing 1 g-equivalent of epoxide.
[d]PMCC = Pensky–Martin closed cup.
[e]Tagliabue closed cup.
[f]To convert kPa to mm Hg, multiply by 7.5.

Table 21. **Physical Properties of VeoVa 10**[a]

Property	Value
color (Pt/Co)	<15
density at 20°C, g/mL	0.875–0.885
water content, % mass/mass	<0.1
acid value, mg KOH/g	<1.0
vinyl unsaturation, mol/kg	4.85–5.10
kinematic viscosity at 20°C, mm^{2}s(=cSt)	2.2
vapor pressure, kPa[b]	
at 30°C	<0.1
at 110°C	4.3
at 210°C	101
boiling range, °C at 13.3 kPa[b]	133–136
flash point,[c] °C	75
freezing point, °C	<−20
solubility in water at 20°C, % mass/mass	<0.1

[a]Ref. 251.

[b]To convert kPa to mm Hg, multiply by 7.5.

[c]PMCC = Pensky–Martin closed cup closed cup.

Glycidyl and Vinyl Esters. Glycidyl neodecanoate [26761-45-5], sold commercially as GLYDEXX N-10 (Exxon) or as Cardura E10 (Shell), is prepared by the reaction of neodecanoic acid and epichlorohydrin under alkaline conditions, followed by purification. Physical properties of the commercially available material are given in Table 20. The material is a mobile liquid monomer with a mild odor and is used primarily in coatings. For example, it is used as an intermediate for the production of a range of alkyd resins (qv) and acrylics, and as a reactive diluent for epoxy resins (qv).

Glycidyl neodecanoate is shipped in bulk or in drums and must be protected from contact with atmospheric water during storage.

Vinyl neodecanoate [26544-09-2] is prepared by the reaction of neodecanoic acid and acetylene in the presence of a catalyst such as zinc neodecanoate. Physical properties of the commercially available material, VeoVa 10 from Shell, are given in Table 21. The material is a mobile liquid with a typical mild ester odor used in a number of areas, primarily in coatings, but also in construction, adhesives, cosmetics, and a number of miscellaneous areas. Copolymerization of vinyl neodecanoate with vinyl acetate gives coating materials with excellent performance on alkaline substrates and in exterior weathering conditions.

Vinyl neodecanoate is shipped in bulk or in lined drums, stabilized with 5 ppm of the monomethyl ether of hydroquinone [95-71-6] MEHQ.

BIBLIOGRAPHY

"Acids, Carboxylic" in *ECT* 1st ed., Vol. 1, pp. 139–151, by E. F. Landau, Celanese Corporation of America; in *ECT* 2nd ed., Vol. 1, pp. 224–239; "Carboxylic Acids (Survey)" in *ECT* 3rd ed., Vol. 4, pp. 814–834, by E. H. Pryde, United States Department of

Agriculture; in *ECT* 4th ed., Vol. 5, pp. 147–168, by M. O. Bagby, United States Department of Agriculture; "Fatty Acids, Survey (Manufacture from Fats)" in *ECT* 1st ed., Vol. 6, pp. 231–236, by H. J. Harwood and E. F. Binkerd, Armour and Co.; "Fatty Acids (Manufacture)" in *ECT* 2nd ed., Vol. 8, pp. 825–830, by W. C. Ault, U.S. Department of Agriculture; "Carboxylic Acids (Manufacture)" in *ECT* 3rd ed., Vol. 4, pp. 835–845, by R. H. Potts, Armak Co.; in *ECT* 4th ed., Vol. 5, pp. 168–178, by R. W. Johnson and R. W. Daniels, Union Camp Corporation; "Fatty Acids (Economic Aspects)" in *ECT* 2nd ed., Vol. 8, pp. 839–845, by E. H. Pryde, U.S. Department of Agriculture; "Carboxylic Acids (Economic Aspects)" in *ECT* 3rd ed., Vol. 4, pp. 853–859, by E. H. Pryde, U.S. Department of Agriculture; in *ECT* 4th ed., Vol. 5, pp. 179–187, by E. T. Sauer, The Procter & Gamble Company; "Fatty Acids from Tall Oil" under "Fatty Acids" in *ECT* 2nd ed., Vol. 8, pp. 845–847, by W. C. Ault, U.S. Department of Agriculture; "Fatty Acids from Tall Oil" under "Carboxylic Acids" in *ECT* 3rd ed., Vol. 4, pp. 859–861, by R. W. Johnson, Jr., Union Camp Corp.; in *ECT* 4th ed., Vol. 5, pp. 187–189, by Robert W. Johnson, Jr., and Robert R. Cantrell, Union Camp Corporation; "Branched-Chain Saturated Acids" under "Fatty Acids (Branched-Chain)" in *ECT* 1st ed., Vol. 6, pp. 259–262, by M. D. Reiner (in part), and J. A. Field (in part), Union Carbide and Carbon Corp.; "Branched-Chain Acids" under "Fatty Acids" in *ECT* 2nd ed., Vol. 8, pp. 849–850, by W. C. Ault, U. S. Department of Agriculture; "Branched-Chain Acids" under "Carboxylic Acids" in *ECT* 3rd ed., Vol. 4, pp. 861–863, by R. W. Johnson, Jr., Union Camp Corp.; in *ECT* 4th ed., Vol. 5, pp. 189–192, by Robert W. Johnson, Jr., and Robert R. Cantrell, Union Camp Corporation; "Trialkylacetic" under "Fatty Acids" in *ECT* 2nd ed., Vol. 8, pp. 851–856, by E. J. Wickson, Enjay Laboratories; "Trialkylacetic Acids" under "Carboxylic Acids" in *ECT* 3rd ed., Vol. 4, pp. 863–871, by J. W. Parker, M. P. Ingham, R. Turner, and J. H. Woode, Shell International Chemical Co., Ltd. in *ECT* 4th ed., Vol. 5, pp. 192–206 by M. J. Keenan and M. A. Krevalis, Exxon Chemical Co.; "Carboxylic Acids, Survey" in *ECT* (online), posting date: December 4, 2000, by M. O. Bagby, United States Department of Agriculture; "Carboxylic Acids, Trialkylacetic Acids" in *ECT* (online), posting date: December 4, 2002, by M. J. Keenan, M. A. Krevalis, Exxon Chemical Co.; "Carboxylic Acids, Fatty Acids from Tall Oil" in *ECT* (online), posting date: December 4, 2000, by Robert W. Johnson, Jr., Robert R. Cantrell, Union Camp Corporation; "Carboxylic Acids, Economic Aspects" in *ECT* (online), posting date: December 4, 2000, by E. T. Sauer, The Procter & Gamble Company; "Carboxylic Acids, Manufacture" in *ECT* (online), posting date: December 4, 2000, by R. W. Johnson, R. W. Daniels, Union Camp Corporation; "Carboxylic Acids, Branched-Chain Acids" in *ECT* (online), posting date: December 4, 2000, by Robert W. Johnson, Jr., Robert R. Cantrell, Union Camp Corporation.

CITED PUBLICATION

1. S. Abrahamsson, S. Stallberg-Stenhagen, and E. Stenhagen, in R. T. Holman, ed., *Progress in the Chemistry of Fats and Other Lipids*, Vol. 7, Pt. 1, Pergamon Press, Oxford, UK, 1964, pp. 1–164.
2. E. E. McSweeny, "Sulfate Naval Stores," in D. F. Zinkel and J. Russell, eds., *Naval Stores*, Pulp Chemical Assoc., New York, 1989, 160–186.
3. D. T. A. Huibers and E. Fritz, "Distillation of Fatty Acids" in R. W. Johnson and E. Fritz, eds., *Fatty Acids in Industry*, Marcel Dekker, Inc., New York, 1989, 85–105.
4. T. L. Ward and W. S. Singleton, *J. Phys. Chem.* **56**, 696 (1952).
5. W. S. Singleton, T. L. Ward, and F. G. Dollear, *J. Am. Oil Chem. Soc.* **27**, 143 (1950).
6. F. Fernendez-Martin and F. Montes, *J. Am. Oil Chem. Soc.* **53**, 130 (1976).
7. N. Adriaanse, H. Dekker, and J. Coops, *Rec. Trav. Chim. Pays-Bas* **84**, 393 (1965).

8. R. T. O'Connor, in K. S. Markley, ed., *Fatty Acids*, 2nd ed., Pt. 1, Interscience Publishers, Inc., New York, 1960, 285–378.

9. E. S. Lutton, in Ref. 6, Pt. 4, pp. 2583–2641.

10. K. Larsson, *J. Am. Oil Chem. Soc.* **43**, 559 (1966).

11. A. V. Bailey and co-workers, *J. Am. Oil Chem. Soc.* **49**, 419 (1972).

12. D. Mitcham, A. V. Bailey, and V. W. Tripp, *J. Am. Oil Chem. Soc.* **50**, 446 (1973).

13. A. V. Bailey and R. A. Pittman, *J. Am. Oil Chem. Soc.* **48**, 775 (1971).

14. A. V. Bailey and co-workers, *J. Am. Oil Chem. Soc.* **52**, 196 (1975).

15. D. Chapman, *The Structure of Lipids*, John Wiley & Sons, Inc., New York, 1965, 221–315.

16. J. A. Harris, *J. Am. Oil Chem. Soc.* **44**, 737 (1967).

17. W. S. Singleton, in Ref. 6, pp. 499–607.

18. E. Jantzen and W. Erdmann, *Fette Seifen* **54**, 197 (1952).

19. H. Stage, *Fette Seifen* **55**, 217 (1953).

20. E. L. Lederer, *Seifensieder-Ztg.* **57**, 67 (1930).

21. W. S. Singleton, in Ref. 6, pp. 609–678.

22. G. H. Bell, *Chem. Phys. Lipids* **10**, 1 (1973).

23. C. W. Hoerr, W. O. Pool, and A. W. Ralston, *Oil Soap* **19**, 126 (1942).

24. J. Levy, in E. S. Pattison, ed., *Fatty Acids and Their Industrial Applications*, Marcel Dekker, Inc., New York, 1968, 209–220.

25. N. O. V. Sonntag, in Ref. 6, Pt. 2, pp. 1127–1151.

26. C. F. Hauser and L. F. Theiling, *J. Org. Chem.* **39**, 1134 (1974).

27. K. S. Markley, in Ref. 6, Pt. 2, pp. 757–984.

28. V. Sreeramulu and P. B. Rao, *Ind. Eng. Chem., Prod. Res. Dev.* **12**, 483 (1973).

29. B. Freedman, R. O. Butterfield, and E. H. Pryde, *J. Am. Oil Chem. Soc.* **63**, 1375 (1986).

30. M. Mittelbach and B. Trathnigg, *Fette Wissenschaft Techn.* **92**, 145 (1990).

31. H. Grossmann, *Tenside Deterg.* **12**, 16 (1975).

32. M. Bares and co-workers, *Tenside Deterg.* **12**, 155 (1975).

33. *Ibid.*, p. 162.

34. R. O. Feuge, *J. Am. Oil Chem. Soc.* **39**, 521 (1962).

35. R. O. Feuge and A. E. Bailey, *Oil Soap* **23**, 259 (1946).

36. R. O. Feuge, E. A. Kraemer, and A. E. Bailey, *Oil Soap* **22**, 202 (1945).

37. A. T. Gros and R. O. Feuge, *J. Am. Oil Chem. Soc.* **41**, 727 (1964).

38. M. L. Rooney, *Polymer* **17**, 555 (1976).

39. K. S. Markley, in Ref. 6, Pt. 2, pp. 1187–1305.

40. H. Bertsch, H. Reinheckel, and E. Konig, *Fette, Seifen, Anstrichm.* **69**, 387,731 (1967).

41. H. Bertsch, H. Reinheckel, and K. Haage, *Fette, Seifen, Anstrichm.* **71**, 357,785 (1969).

42. *Ibid.*, p. 851.

43. R. S. Klonowski and co-workers, *J. Am. Oil Chem. Soc.* **47**, 326 (1970).

44. B. Stouthamer and J. C. Vlugter, *J. Am. Oil Chem. Soc.* **42**, 646 (1965).

45. N. O. V. Sonntag, in Ref. 6, Pt. 2, pp. 985–1072.

46. T. A. Foglia, I. Schmeltz, and P. A. Barr, *Tetrahedron* **30**, 11 (1974).

47. S. H. Shapiro, in Ref. 22, pp. 77–154.

48. N. O. V. Sonntag, in Ref. 6, Pt. 3, pp. 1551–1715.

49. W. L. Kohlhase, E. H. Pryde, and J. C. Cowan, *J. Am. Oil Chem. Soc.* **48**, 265 (1971).

50. W. J. McKillip, ed., *Advances in Urethane Science and Technology*, Vol. 3, Technomic Publishing Co., Inc., Westport, Conn., 1974, pp. 81–107.

51. H. J. Harwood, *Chem. Rev.* **62**, 99 (1962).

52. W. Stein and H. Baumann, *J. Am. Oil Chem. Soc.* **52**, 323 (1975).
53. N. O. V. Sonntag, in Ref. 6, Pt. 2, pp. 1073–1185.
54. M. Chals and R. Perron, *J. Am. Oil Chem. Soc.* **48**, 595 (1971).
55. C. Boelhouwer and J. C. Mol, *Prog. Lipid Res.* **24**, 243 (1985).
56. K. S. Markley, in Ref. 6, pp. 251–283.
57. J. S. Showell, D. Swern, and W. R. Noble, *J. Org. Chem.* **33**, 2697 (1968).
58. J. P. Friedrich and R. E. Beal, *J. Am. Oil Chem. Soc.* **39**, 528 (1962).
59. L. T. Black and R. E. Beal, *J. Am. Oil Chem. Soc.* **44**, 310 (1967).
60. H. A. Boekenoogen, *Analysis and Characterization of Oils, Fats, and Fat Products*, Interscience Publishers, Inc., New York, 1964, 30–46.
61. W. E. Link, ed., *Official and Tentative Methods of the American Oil Chemists' Society*, American Oil Chemists' Society, Champaign, Ill., 1976, Cd 1–25 and Cd 2–38.
62. E. N. Frankel and H. J. Dutton, in F. D. Gunstone, ed., *Topics in Lipid Chemistry*, Vol. 1, Logos Press, London, 1970, pp. 161–276.
63. E. H. Pryde, E. N. Frankel, and J. C. Cowan, *J. Am. Oil Chem. Soc.* **49**, 451 (1972).
64. E. N. Frankel and E. H. Pryde, *J. Am. Oil Chem. Soc.* **54**, 873A (1977).
65. E. Roe, D. A. Konen, and D. Swern, *J. Am. Oil Chem. Soc.* **42**, 457 (1965).
66. R. Sasin and co-workers, *J. Am. Chem. Soc.* **81**, 6275 (1959).
67. E. P. DiBella, *J. Am. Oil Chem. Soc.* **42**, 199 (1965).
68. R. V. Madrigal, M. O. Bagby, and E. H. Pryde, *J. Am. Oil Chem. Soc.* **65**, 1508 (1988).
69. A. Behr and A. Laufenberg, *Fette Wissenschaft Techn.* **93**, 20 (1991).
70. M. Naudet and E. Ucciani, in Ref. 62, Vol. 2, 1971, pp. 99–158.
71. D. Swern, in Ref. 6, Pt. 2, pp. 1307–1385.
72. D. Swern, in D. Swern, ed., *Organic Peroxides*, Vol. 2, Wiley-Interscience, New York, 1971, 355–533.
73. L. A. Goldblatt, in K. S. Markley, ed., *Fatty Acids*, Pt. 5, Interscience Publishers, New York, 1968, 3657–3684.
74. E. H. Pryde and J. C. Cowan, in Ref. 70, pp. 1–98.
75. M. F. Ansell, A. N. Radziwill, and B. C. L. Weedon, *J. Chem. Soc. C*, 1851 (1971).
76. E. H. Pryde and J. C. Cowan, in J. K. Stille and T. W. Campbell, eds., *Condensation Monomers*, Wiley-Interscience, New York, 1972, pp. 74–89.
77. E. Verkuijlen, *J. Chem. Soc. Chem. Commun.*, 198 (1977).
78. C. J. Vetter Jr., *Aust. Chem. Eng.*, 3 (1968).
79. J. Baltes, *Fette, Seifen, Anstrichm.* **66**, 942 (1964).
80. W. E. Richardson and C. H. Smith, *J. Am. Oil Chem. Soc.* **51**, 499 (1974).
81. F. R. Mayo and C. W. Gould, *J. Am. Oil Chem. Soc.* **41**, 25 (1964).
82. H. Wexler, *Chem. Rev.* **64**, 591 (1964).
83. R. H. Potts and V. J. Muckerhoide, in E. S. Pattison, ed., *Fatty Acids and Their Industrial Applications*, Marcel Dekker, Inc., New York, 1968, 21–46.
84. L. H. Wiedermann, *J. Am. Oil Chem. Soc.* **58**(3), 159 (1981).
85. S. D. Vaidya and J. G. Kane, *J. Oil Technol. Assoc. (India)* **18**, 200 (1963).
86. A. J. Stirton, E. M. Hammaker, S. F. Herb, and E. T. Roe, *Oil Soap* **21**, 148 (1944).
87. N. O. V. Sonntag, in R. W. Johnson and E. Fritz, eds., *Fatty Acids in Industry*, Marcel Dekker, Inc., New York, 1989, 40–42.
88. U.S. Pat. 2,139,589, (Dec. 6, 1938), M. H. Ittner (to Colgate-Palmolive-Peet Co.); revised 22,006 (1942).
89. U.S. Pat. 2,156,863 (May 2, 1939), V. Mills (to Procter & Gamble Co.).
90. N. O. V. Sonntag, in Ref. 87, pp. 43–47.
91. *Ibid.*, 51–59.

92. E. Fritz, in Ref. 87, pp. 76–77.

93. F. B. White, *Rau Crystallization Process*, American Oil Chemists' Society Meeting, Chicago, Ill., 1983.

94. U.S. Pat. 3,755,389 (Aug. 28, 1973), T. L. Blaney (to Procter & Gamble Co.).

95. U.S. Pat. 3,345,389 (Oct. 3, 1967), K. T. Zilch (to Emery Industries).

96. W. Stein, *J. Am. Oil Chem. Soc.* **45**, 471–474 (1968).

97. U.S. Pat. 3,541,122 (1970), G. R. Payne, W. B. Campbell, and N. S. Yanick (to Kraftco Corp.).

98. U.S. Pat. 3,870,735 (Mar. 11, 1975), W. Stein and H. Hartmann (to Henkel and Cie, GmbH).

99. U.S. Pat. 3,549,676 (Dec. 22, 1970), H. Hartmann and W. Stein (to Henkel and Cie, GmbH).

100. U.S. Pat. 3,052,699 (Sept. 4, 1962), R. E. Beal (to U.S. Department of Agriculture).

101. U.S. Pat. 3,892,789 (July 1, 1975), A. M. Parsons (to Lever Brothers Co.).

102. M. McHugh and V. Krukonis, *Supercritical Fluid Extraction Principles and Practice*, Butterworths, Boston, Mass., 1986.

103. U.S. Pat. 4,675,132 (June 23, 1987), V. F. Stout and J. Spinelli (to U.S. Department of Commerce).

104. U.S. Pat. 4,578,223 (Mar. 25, 1986), M. T. Cleary (to UOP, Inc.).

105. U.S. Pat. 4,353,839 (Oct. 12, 1982), M. T. Cleary and S. Kulprathipanja (to UOP, Inc.).

106. U.S. Pat. 4,524,029 (June 18, 1985), M. T. Cleary, S. Kulprathipanja, and R. W. Neuzil (to UOP, Inc.).

107. U.S. Pat. 4,353,838 (Oct. 12, 1982) M. T. Cleary and S. Kulprathipanja (to UOP, Inc.); U.S. Pat. 4,210,594 (July 1, 1980), T. J. Logan and D. C. Underwood (to Procter & Gamble Co.).

108. U.S. Pat. 4,495,106 (Jan. 22, 1985), M. T. Cleary and W. C. Laughlin (to UOP, Inc.).

109. U.S. Pat. 4,519,952 (May 28, 1985), M. T. Cleary, S. Kulprathipanja, and R. W. Neuzil (to UOP, Inc.).

110. U.S. Pat. 2,812,342 (Nov. 5, 1957), R. M. Peters (to Emery Industries).

111. G. Buehler, in Ref. 87, pp. 126–127.

112. W. Zschau, *J. Am. Oil Chem. Soc.* **61**, 214 (1984).

113. D. T. A. Huibers and E. Fritz, in Ref. 87, pp. 94–96.

114. N. O. V. Sonntag, in Ref. 87, pp. 277–326.

115. *Oxo Synthesis Products*, Farbwerke Hoechst AG, Frankfurt, Germany, 1971.

116. U.S. Pat. 2,812,342 (Nov. 5, 1957), R. M. Peters (to Emery Industries).

117. "Acetic Acid," *Chemical Profile*, Chemical Market Reporter, Feb. 19, 2001.

118. "Acrylic Acid," Chemical Profile, *Chemical Market Reporter*, April 1, 2002.

119. "Propionic Acid," Chemical Profile, *Chemical Market Reporter*, March 17, 2003.

120. "Benzoic Acid," Chemical Profile, *Chemical Market Reporter*, March 4, 2002.

121. *Chemical Economics Handbook*, SRI International, Menlo Park, Calif., 2002.

122. *Chem. Mark. Rep.* (June 2, 2003).

123. *Chemical Data*, Canadian Centre of Occupational Health and Safety, Hamilton, Ontario, Canada, June 30, 1991.

124. M. J. Boyer, in Ref. 87, pp. 587–591.

125. *Neo Acids: Properties, Chemistry, and Applications*, SC 82-134, Exxon Chemical Company.

126. J. C. Castellvi and A. P. Coll, *Afinidad* **44**, 333–336 (1987).

127. Ger. Offen. 2,657,734 (July 7, 1977), J. C. M. Deweerdt and B. P. J. M. Colomby (to Société Nationale des Poudres et Explosifs).

128. Jpn. Kokai Tokkyo Koho 82 165,341 (Oct. 12, 1982) (to Mitsubishi Gas Chemical Co., Inc.).

129. A. Devos and co-workers, *J. Chem. Soc. Chem. Commun.*, 1180–1181 (1979).

130. Eur. Pat. Appl. 286,981 (Oct. 19, 1988), J. Weber, P. Lappe, and H. Springer (to Hoechst A. G.).

131. U.S. Pat. 4,332,738 (June 1, 1982), F. M. Benitez and M. F. English (to Exxon Research and Engineering Co.).

132. Eur. Pat. Appl. 56,500 (July 28, 1982), D. Kastner, E. Sundt, and R. Choppaz (to Firmenich S.A.).

133. U.S. Pat. 3,455,998 (July 15, 1969), H. J. Arpe (to Shell Oil Co.).

134. Eur. Pat. Appl. 351,603 (Jan. 24, 1990), R. E. Murray (to Union Carbide Corp.).

135. Eur. Pat. Appl. 178,913 (April 23, 1986), A. D. Godwin (to Exxon Research and Engineering Co.).

136. K. L. Loening, A. B. Garrett, and M. S. Newman, *J. Am. Chem. Soc.* **74**, 3929 (1952).

137. M. Fefer, *J. Am. Oil Chem. Soc.* **55**, 342A–345A (1978).

138. Eur. Pat. Appl. 304,853 (March 1, 1989), L. Wambach, M. Irgang, and M. Fischer (to BASF A.G.).

139. Jpn. Kokai Tokkyo Koho 88 214,353 (Sept. 7, 1988), M. Shibagaki, K. Takahashi, and H. Matsushita (to Japan Tobacco, Inc.).

140. Jpn. Kokai Tokkyo Koho 87 108,832 (May 20, 1987), T. Maki, M. Nakajima, T. Yokoyama, and T. Setoyama (to Mitsubishi Chemical Industries Co., Ltd.).

141. Eur. Pat. Appl. 178,718 (April 23, 1986), C. S. John (to Shell Internationale Research Maatschappij B.V.).

142. J. S. Cha, J. E. Kim, M. S. Yoon, and Y. S. Kim, *Tetrahedron Lett.* **28**, 6231–6234 (1987).

143. H. C. Brown, J. S. Cha, N. M. Yoon, and B. Nazer, *J. Org. Chem.* **52**, 5400–5406 (1987).

144. Jpn. Kokai Tokkyo Koho 87 210,056 (Sept. 16, 1987), Y. Kajiwara, Y. Inamoto, and J. Tsuji (to Kao Corp.).

145. Eur. Pat. Appl. 180,210 (May 7, 1986), S. A. Butter, and I. Stoll (to Air Products and Chemicals, Inc.).

146. Eur. Pat. Appl. 285,786 (Oct. 12, 1988), H. Matsushita, M. Shibagaki, and K. Takahashi (to Japan Tobacco, Inc.).

147. B. T. Cho and N. M. Yoon, *Synth. Commun.* **15**, 917–924 (1985).

148. Eur. Pat. Appl. 180,240 (May 7, 1986), S. A. Butter and I. Stoll (to Air Products and Chemicals, Inc.).

149. Ger. Offen. 3,037,301 (June 3, 1982), M. Lenthe, K. Fendeisen, G. Dankert, and G. Grah (to Bayer A.G.).

150. Ger. Offen. 3,825,873 (Feb. 1, 1990), C. Schommer and co-workers (to BASF A.G.).

151. Jpn. Kokai Tokkyo Koho 83 13,537 (Jan. 26, 1983) (to Daicel Chemical Industries, Ltd.).

152. Ger. Offen. 3,744,619 (July 13, 1989), L. Sproesser, K. Sperling, W. Trautmann, and H. Smuda (to BASF A.G.).

153. Eur. Pat. Appl. 85,996 (Aug. 17, 1983), F. Wattimena (to Shell Internationale Research Maatschappij B.V.).

154. H. Koch, *Brenn. Chem.* **36**, 321 (1955).

155. H. Bahrmann in J. Falbe, ed., *New Syntheses with Carbon Monoxide*, Springer-Verlag, Berlin, Heidelberg, New York, 1980.

156. W. J. Ellis and C. Roming, Jr., *Hydrocarbon Process.* **44**(6), 139–141 (1965).

157. Ref. 155, pp. 406–407.

158. Ger. Offen. 3,620,581 (Dec. 23, 1987), W. Hoelderich, J. G. Reuvers, R. Kummer, and L. Hupfer (to BASF A.G.).

159. Jpn. Kokai Tokkyo Koho 86 76,434 (April 18, 1986), A. Yamada, H. Miwa, and H. Kawasaki (to Idemitsu Petrochemical Co., Ltd.).

160. Jpn. Kokai Tokkyo Koho 86 141,703 (June 28, 1986), G. Nobuki and M. Nakahara, (to Shin-Etsu Chemical Industry Co., Ltd.).

161. Eur. Pat. Appl. 307,755 (March 22, 1989), K. P. Meurer and co-workers (to Bayer A.G.).

162. Jpn. Kokai Tokkyo Koho 84 215,307 (Dec. 5, 1984) (to Mitsubishi Petrochemical Co., Ltd.).

163. Jpn. Kokai Tokkyo Koho 87 39,609 (Feb. 20, 1987), R. Tsuruta and co-workers (to Mitsui Toatsu Chemicals, Inc.).

164. Eur. Pat. Appl. 104,042 (March 28, 1984), E. S. Barabas and J. R. Cho (to GAF Corp.).

165. Eur. Pat. Appl. 219,900 (April 29, 1987), J. Meijer, H. Westmijze, and E. Boelema (to AKZO).

166. Eur. Pat. Appl. 301,420 (Feb. 1, 1989), W. Zeibig and H. Hueskes (to Deutsche Texaco A.G.).

167. Eur. Pat. Appl. 282,710 (Sept. 21, 1988), P. R. Mudge (to National Starch and Chemical Corp.).

168. Ger. Offen. 3,538,983 (May 14, 1987), E. Penzel and co-workers (to BASF A.G.).

169. Jpn. Kokai Tokkyo Koho 84 219,314 (Dec. 10, 1984) (to Asahi Glass Co., Ltd.).

170. Jpn. Kokai Tokkyo Koho 89 30,620 (Feb. 1, 1989), H. Nakamura, H. Yamamoto, and K. Takahashi (to Asahi Glass Co., Ltd.).

171. Jpn. Kokai Tokkyo Koho 86 293,273 (Dec. 24, 1986), K. Mori and K. Soga (to Kansai Paint Co. Ltd.).

172. Jpn. Kokai Tokkyo Koho 86 261,367 (Nov. 19, 1986), T. Mitani, T. Shimomura, and I. Mihata (to Dainippon Ink and Chemicals, Inc.).

173. U.S. Pat. 4,804,742 (Feb. 14, 1989), E. S. Neiss and co-workers (to Rorer Pharmaceutical Corp.).

174. Jpn. Kokai Tokkyo Koho 88 297,388 (Dec. 5, 1988) (to Daiichi Seito K.K.).

175. Neth. Appl. 8,003,292 (Jan. 4, 1982), J. E. Maltz (to Lagap S.A. Pharmaceuticals).

176. Ger. Offen. 2,928,546 (Jan. 29, 1981), M. A. Las (to Laboratorios Bago S.A.).

177. Ger. Offen. 2,732,144 (Feb. 2, 1978), C. Swithenbank and J. N. Moss (to Rohm and Haas Co.).

178. Eur. Pat. Appl. 295,090 (Dec. 14, 1988), J. D. Wilson, M. R. Almond, and J. L. Rideout (to Wellcome Foundation Ltd.).

179. B. J. Aungst, *Pharm. Res.* **6**, 224–227 (1989).

180. Eur. Pat. Appl. 339,964 (Nov. 2, 1989), R. K. Achgill and L. W. Call (to Eli Lilly and Co.).

181. U.S. Pat. 4,283,543 (Aug. 11, 1981), D. L. Booth and R. M. Rodebaugh (to Morton-Norwich Products, Inc.).

182. Ger. Offen. 3,226,425 (May 19, 1983), K. Fendeisen, and E. Kranz (to Bayer A.G.).

183. Ger. Offen. 3,209,472 (Sept. 29, 1983), K. H. Drauz, A. Kleemann, and E. Wolf-Heuss (to Degussa A.G.).

184. Eur. Pat. Appl. 57,550 (Aug. 11, 1982), E. Vipingtao (to Eli Lilly and Company).

185. U.S. Pat. 4,405,357 (Sept. 20, 1983), J. H. Chang (to FMC Corp.).

186. Fr. Demande 2,618,980 (Feb. 10, 1989), J. Schapira and co-workers (to Compagnie Française de Produits Industriels).

187. Ger. Offen. 3,540,360 (May 21, 1987), M. Schwamborn and co-workers (to Bayer A.G.).

188. Jpn. Kokai Tokkyo Koho 87 39,590 (Feb. 20, 1987), K. Matsumoto and co-workers (to Sankyo Co., Ltd.).

189. Eur. Pat. Appl. 292,889 (Nov. 30, 1988), K. H. Grossebrin, A. Paul, and F. Lanzendoer (to BASF AG).

190. Patent Cooperation Treaty (PCT) Int. Appl. 88 00,039 (Jan. 14, 1988), C. Mercado and D. Verdon (to Charles of the Ritz Group Ltd.).

191. U.S. Pat. 4,005,210 (Jan. 25, 1977), J. Gubernick (to Estée Lauder, Inc.).

192. Eur. Pat. Appl. 342,827 (Nov. 23, 1989), F. B. Waddington and A. Kallinicos (to Micanite and Insulators Co., Ltd.).

193. U.S. Pat. 4,141,848 (Feb. 27, 1979), M. Braid (to Mobil Oil Corporation).

194. U.S. Pat. 4,647,292 (March 3, 1987), P. J. Jessup, S. G. Brass and M. C. Croudace (to Union Oil Co. of Calif.).

195. S. C. Sell, *Dev. Food Sci.* (Flavors, Fragrances) **18**, 777–795 (1988).

196. Jpn. Kokai Tokkyo Koho 87 293,243 (Dec. 12, 1987), T. Nakamura, M. Sakagami, and Y. Ichijima (to Fuji Photo Film Co., Ltd.).

197. Brit. Pat. Appl. 2,104,781 (March 16, 1983), H. B. Silver and D. Berry (to British Petroleum Co. PLC).

198. U.S. Pat. 4,715,970 (Dec. 29, 1987), R. J. Steltenkamp and M. A. Camara (to Colgate-Palmolive Co.).

199. U.S. Pat. 4,125,549 (Nov. 14, 1978), M. Coopersmith and L. Z. Jasion (to Exxon Research and Engineering Co.).

200. Ger. Offen. 3,004,330 (Aug. 7, 1980), R. Venkatram, C. J. Smith, and V. R. Kamath (to Pennwalt Corp.).

201. Eur. Pat. Appl. 234,649 (Sept. 2, 1987), L. A. Filachek and S. E. Whitehead (to AKZO N.V.).

202. U.S. Pat. 4,723,024 (Feb. 2, 1988), M. F. DePompei (to Mooney Chemicals, Inc.).

203. U.S. Pat. 4,374,777 (Feb. 22, 1983), R. A. Henry and A. Adicoff (to U.S. Department of the Navy).

204. U.S. Pat. 3,723,152 (March 27, 1973), A. Alkaitis and G. A. Thomas (to Mooney Chemicals, Inc.).

205. U.S. Pat. 4,495,105 (Jan. 22, 1985), R. F. Miller (to Atlantic Richfield Co.).

206. U.S. Pat. 4,488,997 (Dec. 18, 1984), R. F. Miller and J. Link (to Atlantic Richfield Co.).

207. Jpn. Kokai Tokkyo Koho 87 86,005 (April 20, 1987), H. Ishimi, Y. Hirai, and S. Toyonishi (to Kayaku Noury Corp., Unitika Chemical Co.).

208. Jpn. Kokai Tokkyo Koho 88 72,706 (April 2, 1988), T. Takahashi, K. Shimoda, and T. Shimomura (to Terumo Corporation).

209. U.S. Pat. 4,057,567 (Nov. 8, 1977) (to Argus Chemical Corp.).

210. Eur. Pat. Appl. 96,365 (Dec. 21, 1983), P. A. Mango (to Air Products and Chemicals, Inc.).

211. Jpn. Kokai Tokkyo Koho 87 73,417 (April 4, 1987) (to Sekisui Chemical Industries K.K.).

212. Jpn. Kokai Tokkyo Koho 87 179,518 (Aug. 6, 1987), S. Yanagi, E. Kuronuma, and Y. Hirai (to Kayaku Noury Corp.).

213. U.S. Pat. 4,742,090 (May 3, 1988), D. L. Hunter and D. E. Schiff (to The Dow Chemical Co.).

214. U.S. Pat. 4,584,362 (April 22, 1986), A. R. Leckart and H. V. Hansen, (to Cosan Chemical Corp.).

215. Eur. Pat. Appl. 143,978 (June 12, 1985), J. H. Johnson (to Hercules, Inc.).

216. U.S. Pat. 4,038,472 (July 26, 1977) (to Exxon Research and Engineering Co.).

217. M. Fefer and A. J. Lauer, *J. Am. Oil Chem. Soc.* **45**, 479–480 (1968).

218. U.S. Pat. 4,199,492 (April 22, 1980), H. T. Roth (to Inmont Corp.).

219. U.S. Pat. 3,985,776 (Oct. 12, 1976), C. Samans and M. R. Spatz (to Standard Oil Co. (Indiana)).

220. Jpn. Kokai Tokkyo Koho 89 85,228 (March 30, 1989), Y. Ogino, K. Yamaguchi, and T. Usui (to Nippon Mining Co., Ltd.).

221. Jpn. Kokai Tokkyo Koho 89 139,970 (June 11, 1988), T. Umeda and H. Kondo (to Bridgestone Corp.).

222. Eur. Pat. Appl. 148,782 (July 17, 1985), P. E. R. Tate (to Manchem Ltd.).

223. U.S. Pat. 4,493,780 (Jan. 15, 1985), E. S. Schwartz and C. A. Tincher (to BASF Wyandotte Corporation).

224. Eur. Pat. Appl. 62,891 (Oct. 20, 1982), E. S. Schwartz, J. F. Maxwell, and P. Davis (to BASF Wyandotte Corp.).

225. Jpn. Kokai Tokkyo Koho 90 16,192 (Jan. 19, 1990) (to New Japan Chemical Co., Ltd.).

226. U.S. Pat. 4,657,689 (April 14, 1987) J. W. Darden (to Texaco Inc.).

227. U.S. Pat. 5,741,436 (April 21, 1998) A. Gershun, W. Mercer, and P. H. Woyciesjes (to Prestone Producers Inc.).

228. S. S. Bhattacharyya and A. K. Das, *J. Indian Chem. Soc.* **65**, 303–306 (1988).

229. F. M. Doyle, D. Pouillon, and E. A. Villegas, *Hydrometallurgy* **19**, 289–308 (1988).

230. Can. Pat. 1,223,242 (June 23, 1987), G. M. Ritcy and E. W. Wong (to Canada, Minister of Energy, Mines and Resources).

231. Eur. Pat. Appl. 261,002 (March 23, 1988), P. S. Gradeff, J. F. Davison, and N. A. Sullo (to Rhone-Poulenc Chimie).

232. U.S. Pat. 4,125,383 (Nov. 14, 1978), H. D. Holtz (to Phillips Petroleum Co.).

233. Eur. Pat. Appl. 289,239 (Nov. 2, 1988), J. P. Maher and P. H. Nguyen (to Engelhard Corp.).

234. Ger. Offen. 3,716,640 (Nov. 26, 1987), M. Kono and co-workers (to Harima Chemicals, Inc.).

235. U.S. Pat. 4,752,501 (June 21, 1988), D. B. Hicks, A. L. Micheli, and S. C. Chang (to General Motors Corp.).

236. Jpn. Kokai Tokkyo Koho 89 48,326 (Jan. 18, 1989), J. T. Anderson, V. K. Nagesh, and R. C. Rugby (to Yokogawa-Hewlett Packard, Ltd.).

237. Eur. Pat. Appl. 310,247 (April 5, 1989), J. V. Mantese, A. H. Hamdi, and A. L. Micheli (to General Motors Corp.).

238. K. F. Teng and P. Wu, *IEEE Trans. Compon. Hybrids Manuf. Technol.* **12**(1), 96–98 (1989).

239. R. W. Vest and co-workers, *J. Solid State Chem.* **73**(1), 283–285 (1988).

240. J. V. Mantese and co-workers, *Appl. Phys. Lett.* **53**(14), 1335–1337 (1988).

241. Eur. Pat. Appl. 320,148 (April 5, 1989), J. V. Mantese, A. L. Micheli, A. H. Hamdi, and A. B. Catalan (to General Motors Corp.).

242. U.S. Pat. 4,663,303 (May 5, 1987), M. Becker and K. H. Liu (to Halcon SD Group, Inc.).

243. U.S. Pat. 4,317,748 (March 2, 1982), J. Torok, E. F. McCaffrey, R. H. Riem, and W. S. Cheung (to Emery Industries, Inc.).

244. Ger. Offen. 2,904,043 (Aug. 9, 1979), H. K. Inskip (to E. I. du Pont de Nemours & Co., Inc.).

245. A. R. Khokhar, S. Al-Baker, I. H. Krakoff, and R. Perez-Soler, *Cancer Chemother. Pharmacol.* **23**(4), 219–224 (1989).

246. Ger. Offen. 3,313,002 (Oct. 20, 1983), M. Ochiai and H. Sagitani (to Pola Chemical Industries, Inc.).

247. Jpn. Kokai Tokkyo Koho 80 45,609 (March 31, 1980) (to Nisshin Oil Mills K.K.).

248. Eur. Pat. Appl. 5361 (Nov. 14, 1979), E. A. Hilditch, R. E. Hambling, C. R. Sparks, and D. A. Walker (to Cuprinol Ltd.).

249. *Cardura*^TM *Technical Manual*, CA 1.1, 3rd ed., Data Sheet, Shell Chemical Company, Reprinted June, 1989.

250. *Data Sheet NA 0014, 2M/9-89, GLYDEXX*^TM *N-10: Reactive Diluent for Epoxy Formulators*. Exxon Chemical Company.

251. *VeoVa*™ *Technical Manual*, VV 1.1, 5th ed., Technical Bulletin, Shell Chemical Company.

M. O. Bagby
United States Department of Agriculture

R. W. Johnson, Jr.
R. W. Daniels
Robert R. Contrell
Union Camp Corporation

E. T. Sauer
The Procter & Gamble Company

M. J. Keenan
M. A. Krevalis
Exxon Chemical Co.

CARDIOVASCULAR AGENTS

1. Introduction

Over the last several decades, tremendous advances in basic and clinical research on cardiovascular disease have greatly improved the prevention and treatment of the nation's number one killer of men and women of all races (it is estimated that ~40% of Americans (~60 million between the ages of 40–70, suffer from some degree of this disease) (1–3). During the second half of the twentieth century, the problem of treating heart disease was at the forefront of the international medical communities' agenda. This was reflected in the World Health Organizations (WHO) 1967 classification of cardiovascular disease as the world's most serious epidemic. The problem of cardiovascular disease continues to be the leading cause of death in the United States and other industrialized countries as we progress into the new millennium. According to the American Heart Association, heart disease and stroke cost an estimated $329.2 billion in medical expenses and lost productivity in the United States in 2002– more than double the economic cost of cancer.

Cardiovascular disease encompasses a wide range of disorders, and the methods of its treatment are both vast and diverse. As such, this article covers a range of key therapeutic agents including antiarrhythmic agents, antianginal agents, antilipemic agents, thrombolytic agents, agents used in the treatment of congestive heart failure, antiatherosclerotic agents, and antihypertensive agents. The cardiac physiology, pathophysiology, and the causes of common cardiac diseases are reviewed before considering the drugs used in their treatment.

2. The Circulatory System and Cardiac Physiology

The cardiovascular system consists of the heart and a complex network of blood vessels—arteries, veins, and capillaries. The blood vessels provide nutrients to the body's cells and remove waste products. The heart functions as the pump that maintains the constant movement of blood through the blood vessels, ensuring that nutrient access and waste removal are constantly in homeostasis. During cardiac disease, this homeostasis is compromised by a malfunction in any one or several of the components of the cardiovascular system. Such disorders or diseases can lead to irreversible damage, and if untreated, death. Over the past several decades, great scientific strides have been made to provide therapeutic agents that restore normal function to malfunctioning components of the cardiovascular system.

The human heart and physiological processes that are altered during cardiovascular diseases are reviewed as background to the mechanisms of action of therapeutic agents. However, for in-depth details about heart anatomy, physiology, and electrophysiology, the reader is referred to textbooks and reviews (4).

2.1. Heart Anatomy. The human heart consists of four chambers: the right and left atria, and the right and left ventricles. Blood returning from the body collects in the right atrium, passes into the right ventricle, and is pumped to the lungs. Blood returning from the lungs enters the left atrium, passes into the left ventricle, and is pumped into the aorta. Valves in the heart prevent the backflow of blood from the aorta to the ventricle, the atrium, and the veins.

Heart muscle (the myocardium) is composed of three types of fibers or cells. The first type of muscle cell, found in the sinus and atrioventricular node, is weakly contractile, autorhythmic, and exhibits slow intercellular conduction. The second type of cell, located in the ventricles, is specialized for fast impulse conduction. The third type of myocardial cell is highly contractile, and composes the bulk of the heart.

Muscle cells in the heart abut tightly from end to end and form fused junctions known as intercalated disks. These tight junctions serve two functions. First, when one muscle cell contracts, it pulls on cells attached to its ends. Second, the close contacts facilitate excitation wave propagation. In addition, large channels, referred to as gap-junctions, pass through intercalated disks, and connect adjacent cells. These connections play an important role in transmitting and excitation wave from one cell to another.

Myocardial cells receive nutrients from coronary arteries that branch from the base of the aorta. Blockage of sections of these coronary arteries occurs during coronary artery disease (CAD). This leads to myocardial ischemia, which is the cause of myocardial infarction (heart attack) and angina pectoris.

2.2. Electrophysiology. With the exception of differences in calcium ion uptake and release, the mechanisms of contraction of human skeletal and cardiac muscle are comparable. However, unlike skeletal muscle, which requires neuronal stimulation, heart muscle contracts automatically. A heartbeat is composed of a contraction and relaxation of the heart muscle mass, and is associated with an action potential in each cell. The constant pumping action of the heart depends on the precise integration of electrical impulse generation, transmission, and myocardial tissue response.

A single heartbeat involves three principal electrical events: (*1*) an electrical signal to contract is initiated; (*2*) the impulse signal propagates from its point of origin over the rest of the heart; and (*3*) the signal abates or fades away. Cardiac arrhythmias develop when any of these three events are disrupted or impaired.

Figure 1 shows the principle components of the heart involved in cardiac impulse generation and conduction. In a healthy heart, the electrical impulse signal to contract is initiated in the sinoatrial (SA) node, which is located at the top of the right atrium (Fig. 1). Following depolarization of the SA node, the impulse spreads from cell to cell into the atria membrane. The atria contract first. Following, the impulse is focused through specialized automatic fibers in the atria known as the atrioventricular (AV) node (Fig. 1). At this node, the impulse is slowed so that the atria finish contracting before the impulse is propagated to myocardial tissue of the ventricles, which allows for the rhythmic pumping action that allows blood to pass from the atria to the ventricles.

After the electrical impulse emerges from the AV node, it is propagated by tissue known as the bundle of His, which passes the signal onto fast-conducting myocytes known as Purkinje fibers. These fibers conduct the impulse to surrounding, myocardial cells. The transmission of the impulse results in a characteristic electrocardiographic pattern.

Following contraction, heart muscle fibers enter a refractory period during which they will not contract or accept a signal to contract. Without this resting period, the initial contraction impulse originating in the SA node would not fade away, but would continue to propagate over the heart, leading to disorganized contraction (known as fibrillation).

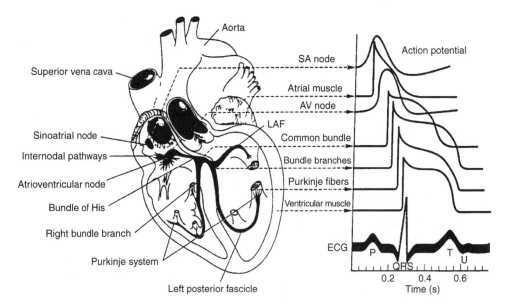

Fig. 1. Action potentials and the conducting system of the heart. Shown are typical transmembrane action potentials of the SA and AV nodes, specialized conducting myocardial cells, and nonspecialized myocardial cells. Also shown is the ECG plotted on the same time scale. Courtesy of Appleton & Lange.

2.3. Excitation and Contraction Coupling. Myocardial pacemaker cells, usually in the SA node, initiate an action potential that travels from cell to cell through the intercalated disks. This opens calcium channels and leads to a small influx of extracellular calcium ions, which triggers events leading to muscle contraction.

The biophysical property that connects excitation impulse and muscle contraction is based on the electrical potential differences that exist across cell membranes. These potentials arise due to several factors: (*1*) intracellular fluid is rich in potassium (K^+) and low in sodium (Na^+) (the reverse is true of extracellular fluid); (*2*) the cell membrane is more permeable to K^+ than it is to Na^+; (*3*) anions in the intracellular fluid are mostly organic and fixed, and do not diffuse out through the membrane; and (*4*) cells use active transport to maintain gradients of Na^+ and K^+.

Stimulation, either electrical or chemical, can depolarize the cell membranes by causing conformational changes that open membrane ion channels. This allows Na^+ to flow into the cell, which produces an action potential that is transmitted in an all-or-none fashion along the cellular membrane. As the action potential travels along the cell membrane, it induces a rise in the levels of free, or activator, calcium (Ca^{2+}) within the cell. This, in turn, initiates the interaction between actin and myosin, which leads to muscle contraction.

The action potential of a nonautomatic ventricular myocyte is shown in Figure 2. It is divided into phases 0–4. Phase 0, rapid membrane depolarization, results from the opening of fast sodium channels, and is augmented by Ca^{2+} entering via calcium channels. Phase 1 follows depolarization where there is a brief initial repolarization due to the closing of the sodium channels, and a brief outward movement of K^+ ions (ie, decrease in potassium conductance by inward rectifying current). Phase 2, is a plateau period, during which the slow influx of Ca^{2+} via an opening of the L-type calcium channel occurs (Fig. 2). During this phase, a prolonged refractory period exists during which the muscle cannot be reexcited. Phase 3, the repolarization period, is due primarily to the opening of and outward-rectifying K^+ channel and the closure of the calcium channels. The repolarization that occurs during this phase involves the interplay of several different types of potassium channels. Following phase 3, the transmembrane potential is restored to its resting value (phase 4, Fig. 2).

Cells of the nodal tissue and specialized conducting myocytes, such as Purkinje fibers, can spontaneously depolarize and generate action potentials that propagate over myocardial tissue. This is referred to as automaticity, and hence, all of these cells have pacemaker potential. In automatic cells, the outward leak of K^+ slows after repolarization; however, Na^+ continues to leach into the cell, which results in a steady increase in intracellular cations, and leads to depolarization. The action potential phase 4 of such cells is not flat, as observed in Figure 2, but becomes less negative until it reaches a threshold that triggers the opening of an L-type calcium channel in nodal tissue, or the sodium channel in conducting tissue. Thus, phase 0 in nodal tissue is due to the influx of Ca^{2+} and not Na^+. Figure 1 displays the action potentials for several cardiac cells having spontaneous and nonspontaneous depolarizability. The electrical activity of myocardial cells produces a current that can be recorded as an electrocardiogram (ECG).

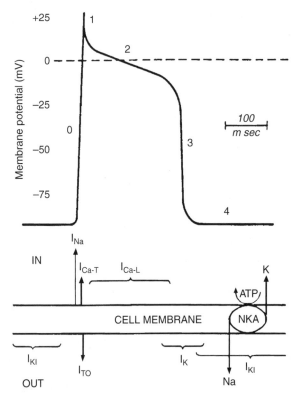

Fig. 2.   Diagrammatic representation of an action potential of a nonautomatic ventricular cell, showing the principal ion fluxes involved in membrane depolarization and repolarization. The membrane potential in millivolts is given on the vertical axis. This denotes the electrical potential of the inner-face of the membrane relative to the outer-face. Phases of the potential are numbered 0,1,2,3,4, and are described in detail in the text.

Under normal physiological conditions, the SA node is the pacemaker for the rest of the heart. However, if the impulse from the SA node is slowed or blocked, or if the process of depolarization is accelerated in other automatic cells, non-SA node cells may initiate a wave of depolarization that either replaces the SA node impulse, or interferes with it. Heartbeats that originate from non-SA pacemaker activity are referred to as ectopic beats. However, not all ectopic beats result from altered pacemaker activity. Reentrant rhythms also are a major cause of ectopic beats. It is also possible to alter pacemaker activity without inducing an ectopic beat (ie, accelerated pacemaker in the atria).

The spontaneous impulse rate of automatic cells depends on the slope of action potential phase 4, the magnitude of the maximum diastolic potential, and the threshold potential. Changes in any of these values can occur during disease states, or from the effects of small "drug" molecules. B_1-Adrenergic receptor agonists increase heart rate by increasing phase 4 of the pacemaker cell action potential. Cholinergic drugs, which are agonists of muscarinic receptors, not only slow the heart by decreasing the phase 4 slope, but also hyperpolarize the cells.

Thus, compounds that block muscarinic receptors (atropine-like) increase heart rate, while compounds that block β-receptors slow the heart.

2.4. Ion Channels. The basic ion channel transmembrane protein consists of subunits designated as α, β, γ, and δ. The α subunit is the major component of Na^+, K^+, and Ca^{2+} channels and is tetrameric in nature. Each unit of this tetramer is designated as a domain, and each domain is made up of six segments designated: S1, S2, S3, S4, S5, and S6. The S5 and S6 segments are linked to each other in a specific arrangement so as to form the lining of the ion channels. The S4 segment of each domain contains many lysine and arginine residues that act in response to changes in the membrane potential, and are thus involved in the opening (voltage gating) of the channel. It is believed that the S4 segment constitutes the "*m*" gate (5–11). While a polypeptide chain that links the S6 segment of domain III to the S1 segment of domain IV constitutes the "*h*" gate (12,13).

Many of the drugs used to treat heart disease exert their therapeutic effects by blocking Na^+, K^+, and Ca^{2+} ion channels. In the case of sodium channel blockers, this results in a decrease in the slope of phase 0 of the action potential, and thereby decreases the V_{max}, or rate of conduction of the impulse. Sodium channel blockade can also prolong the refractory period by increasing the time that the channel is in the inactivated state, before returning to the resting state. Potassium channel blockers increase the duration of the action potential, as potassium currents are responsible for repolarizing the membrane during the action potential. Calcium channel blockers slow impulse conduction through the SA and AV nodes.

2.5. Channel Gates. The term gating refers to the process during which external stimuli cause conformational changes in membrane proteins, leading to the opening and closing of ion channels. It has been theorized that ion channels have at least two gates, referred to as *m* and *h*, and that both gates must be open for ions to pass through the channel (14). According to this model, channel gates cycle through three states: (*1*) closed resting (R), (*2*) open active (A), and (*3*) closed inactive (I).

The gating model shown in Figure 3, is a basic outline of the channel gating mechanism, and is useful for describing drug action (15,16). In the closed resting state, the *h* gate is open and the *m* gate is shut. During depolarization the *m* gate switches to the open position and the channel is activated, allowing the fast passage of ions through the channel. Depolarization also initiates the channel inactivation so that the channel begins to move from the open to the closed inactive state. In the closed inactive state both the *m* and *h* gates are shut, and the channel does not respond to further repolarization until it has moved back to the closed resting state, during which the *h* gate is again open and the *m* gate is shut.

2.6. Sodium Channels. There is strong evidence indicating that the amino acid sequence of sodium channels has been conserved over a long period. As indicated earlier, the inward voltage dependent Na^+ channel consists of four protein subunits designated as α, β, γ, and δ. The α subunit, the major component of Na^+ channel made up of 200 amino acids, is subdivided into four covalently bound domains and contains binding sites for a number of antiarrhythmic compounds and other drugs. The Na^+ channels are found in neurons, vertebrate

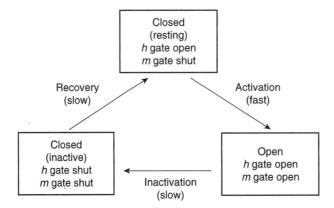

Fig. 3. Simplified representation of the gating mechanism in voltage-activated sodium, potassium, and calcium channels. The model hypothesizes three states: closed resting, open active, and closed inactive; and two gates h and m. The figure depicts what are generally considered to be the essential features of gating, which include a closed "resting" state that is capable of rapidly opening in response to changes in membrane potential followed by a refractory period in which the channel slowly returns to the resting state.

skeletal muscle, and cardiac muscle. Electrophysiological studies indicating that Na^+ channels favor the passage of Na^+ over K^+, point to the fact that Na^+ channels must be narrow and ion conductance depends on the ions size (ionic radius of Na^+ is 0.95 in. as compared to that of K^+ being 1.33 in.) and possibly steric factors (7). There is also some evidence indicating that voltage dependent cardiac Na^+ channels exist in two isoforms: fast and slow (17). Activation of the fast Na^+ channels in cardiac cells produces the rapid influx of Na^+ and depolarizes the membrane in all cardiac myocytes, except nodal tissue, where Na^+ channels are either absent or relatively few in number (18). The Na^+ channels are almost all voltage-gated, and gates open in response to changes in membrane potential.

2.7. Potassium Channels. Potassium channels are outward voltage-dependent channels. Similar to Na^+ channels, the major K^+ channel subunit consists of four domains, but unlike the α subunit of Na^+ channels, the domains of the K^+ channels are not covalently linked. At least 10 genes code for the K^+ channel domains, which means that hundreds of combinations of four-domain channels can be constructed. Recently, it has been reported that some of the K^+ channels contain only two transmembrane segments (9,10). Many K^+ channels are classified as rectifying, which means that they are unidirectional (or transport ions in one direction only), and that their ability to pass current varies with membrane potential.

The inward-rectifying K^+ current (usually designated I_K) allows K^+ to move out of the cell during phase 4 of the action potential, but is closed by depolarization; the outward-rectifying K^+ current (usually designated I_K) is opened by depolarization. Hence, the I_{KI} makes a substantial contribution to the value of the resting (phase 4) membrane potential; the I_K is the major outward current contributing to repolarization (19). Adenosine Triphosphate (ATP) sensitive K^+ channels are activated if the ATP level in the heart decreases as observed in myocardial ischemia (20). This leads to the inward flow of Ca^{2+} ions, thereby reducing myocardial contractility and conserving energy for basic cell survival processes.

The K^+ channels are highly selective and are 100-fold more permeable to K^+ than to Na^+ (21). Certain types of molecules bind extracellularly and block the voltage gated K^+ channels (16). These include several peptide toxins, as well as small charged organic molecules such as tetraethylammonium, 4-aminopyridine, and quinine. Other molecules have been found to be K^+ channel openers. These compounds act on the ATP sensitive K^+ current, and provide cardioprotection during ischemia. The K^+ channel opener molecules also relax smooth muscle cells, and may increase the coronary blood flow during angina (22–28). The activation of ATP-sensitive K^+ channels does result in an increase inward flow of Ca^{2+} and outward flow of K^+ resulting in reduction of action potential duration and net Ca^{2+} entry.

2.8. Calcium Channels. Calcium ions are essential for the chain of events that lead to myocardial contraction. The role of calcium in the cardiac cycle has been studied extensively for years. Four types of voltage-dependent calcium channels with specific function and location have been identified. These include (1) the L type (found mainly in skeletal, cardiac and smooth muscle cells); (2) the T type (located in pacemaker cells); (3) the N type (found in neuronal cells), and (4) the P type (located at neuromuscular junctions) (29–31). L-type Ca^{2+} channels are formed by a complex arrangement of five protein subunits designated as the $\alpha 1$, $\alpha 2$, β, γ, and δ subunits, which are comprised of polypeptide chains of different lengths. The arrangement of these subunits is shown in Figure 4. The tetrameric $\alpha 1$ subunit is the most important functional component forming the Ca^{2+} channel, and is responsible for producing the pharmacological effects of calcium channel blockers. The Ca^{2+} channels play an important role in cellular excitability by allowing the rapid influx of Ca^{2+}, which depolarizes the cell. The resulting increase in intracellular Ca^{2+} is essential for the regulation

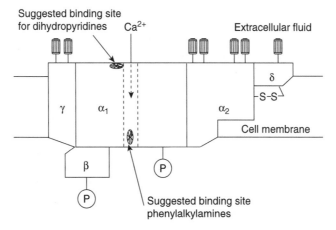

Fig. 4. Suggested structure of an L-type calcium channel from skeletal muscle, showing the five protein subunits that comprise the channel. Phosphorylation sites are indicated by P. Binding sites for phenylalkylamine and dihydropyridine calcium channel blockers. Courtesy of *Trend. Pharmacol. Sci.*

of Ca^{2+} dependent processes including excitation–contraction coupling, excitation–secretion coupling, and gene regulation. It has been suggested that dihydropyridine calcium channel blockers exert their effects by binding at the top of the channel near the cytoplasmic entrance, while arylalkylamines bind at the bottom of the channel between domains III and IV of the $\alpha 1$ subunit. The other hydrophobic $\alpha 2$, β, γ, and δ subunits may play a role in positioning the $\alpha 1$ subunit in the membrane. L-type channels are activated slowly by partial depolarization of the cell membrane, and inactivated by full depolarization and by increasing Ca^{2+} concentration. In nodal tissues, where fast sodium channels are absent or sparse, L- and T-type Ca^{2+} channels are responsible for depolarization.

3. Antiarrhythmic Agents

3.1. Mechanisms of Cardiac Arrhythmias.

The pumping action of the heart involves three principle electrical events: the generation, the conduction or propagation, and the fading away of the signal. When one or more of these events is disrupted, cardiac arrhythmias may arise.

In a healthy heart, cells located in the right atrium, referred to as the SA node or pacemaker cells, initiate a cardiac impulse. The spontaneous electrical depolarization of the SA pacemaker cells is independent of the nervous system; however, these cells are innervated by both sympathetic and parasympathetic fibers, which can cause increases or decreases in heart rate as a result of nervous system stimulation. Other special cells in the heart also possess the ability to generate an impulse, and may influence cardiac rhythm, but are normally surpassed by the dominant signal generation of SA pacemaker cells. When normal pacemaker function is suppressed due to pathological changes occurring from infarction, digitalis toxicity, or excessive vagal tone or when excessive release of catecholamines from sympathomimetic nerve fibers occurs, these other automatic cells (including special atrial cells, certain AV node cells, the bundle of His, and Purkinje fibers) have the potential to become ectopic pacemakers, which can dominate cardiac rhythm, and consequently lead to arrhythmias.

Disorders in the transmission of the electrical impulse can lead to conduction block and reentry phenomenon. Conduction block may be complete (no impulses pass through the block), partial (some impulses pass through the block), and bidirectional or unidirectional. During bidirectional block, an impulse is blocked regardless of the direction of entry; a unidirectional block occurs when an impulse from one direction is completely blocked, while impulses from the opposite direction are propagated (although usually at a slower than normal rate).

During another condition known as heart block, the impulse signal from the SA node is not transmitted through either the AV node or lower electrical pathways properly. Heart block is classified by degree of severity: (1) first degree heart block: all impulses moving through the AV node are conducted, but at a slower than normal rate; (2) second degree heart block: some impulses fully transit the AV node, whereas others are blocked (as a result, the ventricles fail to beat at the proper moment); (3) third degree heart block: no impulses reach the ventricles (automatic cells in the ventricles initiate impulses, but at a slower rate, and as a result the atria and ventricles beat at somewhat independent rates).

The most serious cause of life-threatening cardiac arrhythmias results from a condition known as reentry, which occurs when an impulse wave circles back through the heart, reenters previously excited tissue, and reactivates the cells. Under normal conditions, reentry does not occur, as cells become unable to accept an excitation impulse for a period of time that is sufficient for the original signal to abate. Hence, the cells will not contract again until a new impulse emerges from the SA node. However, there are certain conditions during which this does not happen, and the impulse continues to circulate. The essential condition for reentry to occur involves the development of a cellular refractory period that is shorter than the conduction velocity. Consequently, any circumstance that either shortens the refractory period, or lengthens the conduction time, can lead to reentry. Various types of alteration in automacity (enhanced and triggered automacity) can trigger cardiac arrhythmias.

Nearly all tachycardias, including fibrillation, are due to reentry. The length of the refractory period depends mainly on the rate of activation of the potassium current; the rate of conduction depends on the rate of activation of the calcium current in nodal tissue, and the sodium current in other myocytes. The channels controlling these currents are the targets for suppressing reentry.

While many conditions can lead to reentry, the most common is shown in Figure 5. The conditions needed for this type of reentry are as follows: first, the existence of an obstacle, around which the impulse wave front can propagate,

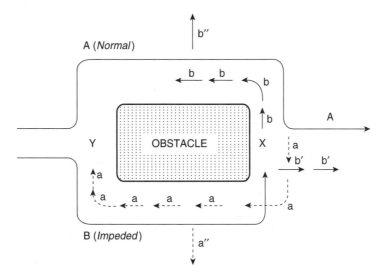

Fig. 5. Model for reentrant activity. A depolarization impulse approaches an obstacle (nonconducting region of the myocardium) and splits into two pathways (A and B) to circumvent the obstacle. If pathway B has impeded ability to conduct the action potential the following may occur. (*1*) If impulse B is slowed and arrives at cross junction X after the absolute refractory period of cells depolarized by A, the impulse may continue around the obstacle as shown by path b and/or follow A along path b′. In both cases, the impulse is said to be reflected. (*2*) If pathway B shows unidirectional block, impulse A may continue around the obstacle as shown by path a. If the obstacle is large enough, so that cells in cross region Y are repolarized before the return of a or b, then a circus movement may be established. Both (*1*) and (*2*) may propagate daughter impulses (a″ and b″) to other parts of the myocardium. These effects can give rise to coupled beats and fibrillation.

is needed. The obstacle may be infarcted or scarred tissue that cannot conduct the impulse. The second condition needed is the existence of a pathway that allows conduction at the normal rate around one side of the obstacle, while the other side of the pathway is impaired. The impairment may be such that it allows conduction in only one direction (unidirectional block) or it may allow conduction to proceed at a greatly reduced rate, such that when the impulse emerges from the impaired tissue, the normal tissue is no longer refractory. These pathways are usually localized, eg, within the AV node or the end branches of part of the Purkinje system. Alternatively, they can be more extensive and may give rise to daughter impulses capable of spreading to the rest of the myocardium.

Unidirectional block occurs in tissue that has been impaired, such that its ability to conduct an impulse is completely blocked in one direction but only slowed in the other. As a result of unidirectional block, the impulse cannot proceed forward along path B (Fig. 5), and the cells on this path remain in a polarized state. However, when the impulse traveling along path A reaches a suitable cross-junction (point X in Fig. 5), the impulse proceeds back along path a (Fig. 5), although at a slower rate than normal (indicated by the dotted path line, Fig. 5). When this impulse reaches another cross-junction (Y), it is picked up by path A and conducted around the circle. If the obstacle is large enough, the cells in path A will have repolarized and the path will again be followed, giving rise to a continuous circular movement. When reentry occurs randomly in the myocardium, it results in random impulses that lead to cardiac fibrillation. Thus, the reentry is critically dependent on two factors: unidirectional conduction block and decremental conduction.

3.2. Types of Cardiac Arrhythmias.

Arrhythmias can be divided into two categories: ventricular and supraventricular arrhythmias. Within these two categories, arrhythmias are further defined by the pace of the heartbeats. Bradycardia indicates a very slow heart rate of <60 beats/min; tachycardia refers to a very fast heart rate of >100 beats/min. Fibrillation refers to fast, uncoordinated heartbeats.

Listed below are common forms of arrhythmias grouped according to their origin in the heart. Supraventricular arrhythmias include (1) sinus arrhythmia (cyclic changes in heart rate during breathing); (2) sinus tachycardia (the SA node emits impulses faster than normal); (3) sick sinus syndrome (the SA node fires improperly, resulting in either slowed or increased heart rate); (4) premature supraventricular contractions (a premature impulse initiation in the atria causes the heart to beat prior to the time of the next normal heartbeat); (5) supraventricular tachycardia (early impulse generation in the atria speed up the heart rate); (6) atrial flutter (rapid firing of signals in the atria cause atrial myocardial cells to contract quickly, leading to a fast and steady heartbeat); (7) atrial fibrillation (electrical impulses in the atria are fired in a fast and uncontrolled manner, and arrive in the ventricles in an irregular fashion); and (8) Wolff-Parkinson-White syndrome (abnormal conduction paths between the atria and ventricles cause electrical signals to arrive in the ventricles too early, and subsequently reenter the atria).

Arrhythmias originating in the ventricles include (1) premature ventricular complexes (electrical signals from the ventricles cause an early heartbeat, after which the heart seems to pause before the next normal contraction of the ventricles occurs); (*2*) ventricular tachycardia (increased heart rate due to ectopic signals from the ventricles); and (*3*) ventricular fibrillation (electrical impulses in the ventricles are fired in a fast and uncontrolled manner, causing the heart to quiver).

3.3. Classification of Antiarrhythmic Drugs. The classification of antiarrhythmic agents is important for clinical application; however, there is no single classification system that has gained universal endorsement. At this time, the method proposed by Singh and Vaughan Williams (32) continues to be the most widespread classification scheme. Since its initial conception, this classification method has undergone several modifications—calcium channel blockers have been added as a fourth class of compounds (33), and class I agents have been subdivided into three groups to account for their sodium channel blocking kinetics (34). Table 1 lists examples of drugs in each of these classes and provides other pertinent information such as CAS registry number, molecular formula, and various trade names. Note that miscellaneous drugs (35) have been added to Table 1 to account for compounds with mechanisms of action that do not fit within the four standard classes (I–IV).

3.4. Perspective: Treatment of Arrhythmias. In recent years, there have been many changes in the way the arrhythmia is treated. New technologies, including radiofrequency ablation and implantable devices for atrial and ventricular arrhythmias, have proven to be remarkably successful mechanical treatments. In addition, cardiac suppression trials (CAST) and numerous other studies have provided evidence indicating drugs that act mainly by blocking sodium ion channels—Class I agents under the Singh and Vaughn Williams system of classification—may have the potential to increase mortality in patients with structural heart disease (36,37). Since the CAST results were released, the use of Class I drugs has decreased, and attention has shifted to developing new Class III agents, which prolong the action potential and refractoriness by acting on potassium channels. Of the Class III antiarrhythmic agents, amiodarone has been studied extensively, and has proven to be a highly effective drug for treating life-threatening arrhythmias.

In addition, new studies have indicated that combination therapies, eg, administration of amiodarone and Class II β-blockers, or concomitant treatment with implantable mechanical devices and drug therapies are effective avenues for treating arrhythmias.

3.5. Class I Agents. Antiarrhythmic agents in this class bind to sodium channels and inhibit or block sodium conductance. This inhibition interferes with charge transfer across the cell membrane. Investigations into the effects of Class I antiarrhythmics on sodium channel activity have resulted in the division of this class into three separate subgroups—referred to as IA, IB, and IC (38).

The basis for dividing the Class I drugs into subclasses resulted from measured differences in the quantitative rates of drug binding to, and dissociation from, sodium ion channels (38). Class IB drugs, which include lidocaine, tocainide, and mexiletine, rapidly dissociate from sodium channels, and consequently have the lowest potencies of the Class I drugs—these molecules produce little to no change in the action potential duration, and shorten repolarization. Class IC drugs, which include encainide and lorcainide, are the most potent of the Class I antiarrhythmics; drugs in this class display a characteristically slow dissociation

Table 1. Antiarrhythmic Agents

Chemical/ generic name	CAS Registry Number	Molecular formula	Trade name	Structure
			Class I Antiarrhythmic Agents	
			Class IA	
quinidine	[56-54-2]	$C_{20}H_{24}N_2O_2$	Cardioquin, Galactoquin, Duraquin, Quinaglute, Quinora, Quincardine, Kinidin, Quiniduran	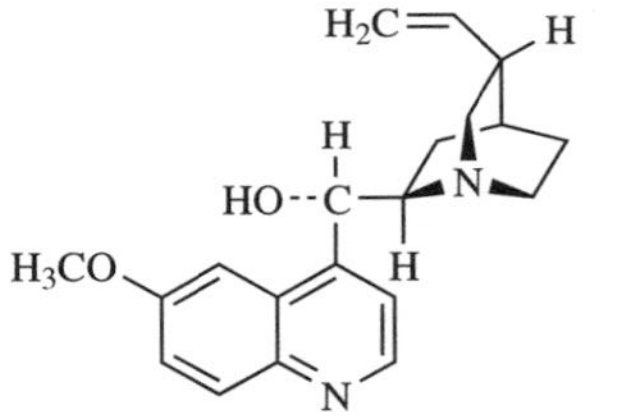
procainamide	[614-39-1]	$C_{13}H_{22}ClN_3O_3$	Amisalin, Novocamid, Procamide Procanbid, Procapan, Pronestyl	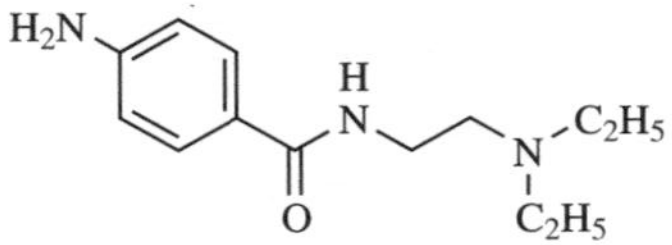
disopyramide	[3737-09-5]	$C_{21}H_{29}N_3O$	Dicorantil, Isorythm, lispine, Rythmodan, Ritmodan	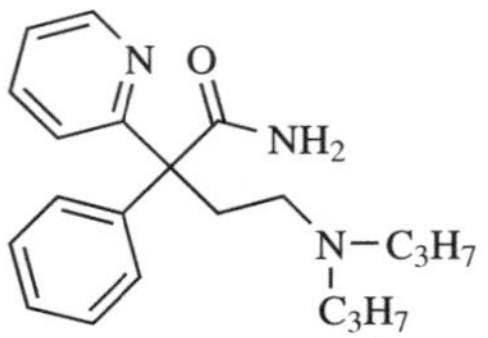

90

lidocaine	[6108-05-0]	$C_{14}H_{23}ClN_2O$	Lidesthesin, Odontalag, Sedagul, Xylocard, Xyloneural 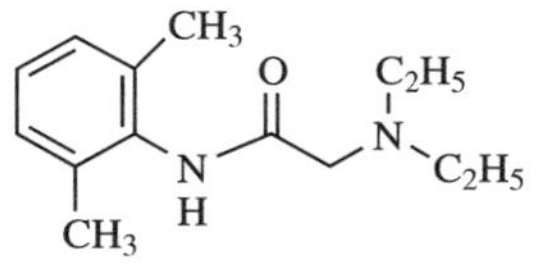
tocainide	[71395-14-7]	$C_{11}H_{17}ClN_2O$	Tonocard, Xylotocan, Taquidil 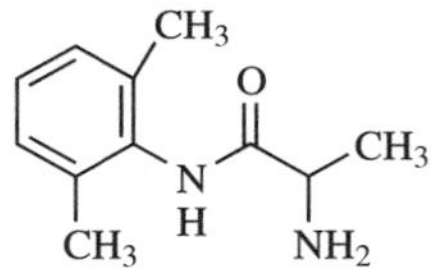
mexiletine	[5370-01-4]	$C_{11}H_{18}ClNO$	Mexitil, Ritalmex 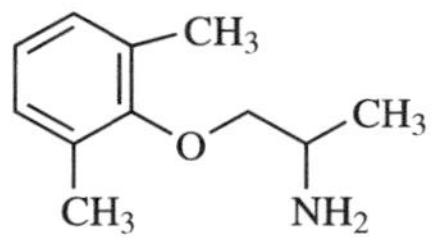
phenytoin	[630-93-3]	$C_{15}N_{11}N_2NaO_2$	Aurantin, Epanutin, Phenhydan, Tacosal, Pyoredol 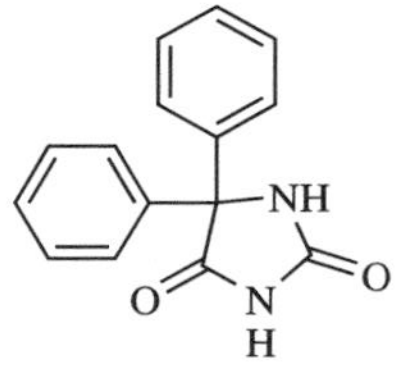

Table 1 (*Continued*)

Chemical/ generic name	CAS Registry Number	Molecular formula	Trade name	Structure
encainide	[66794-74-9]	$C_{22}H_{28}N_2O_2$	*Class IC* Enkaid	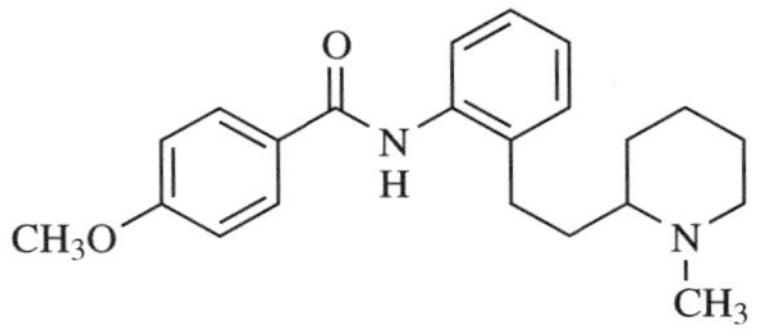
flecainide	[54143-56-5]	$C_{19}H_{24}F_6N_2O_5$	Apocard, Flecaine, Tambocor	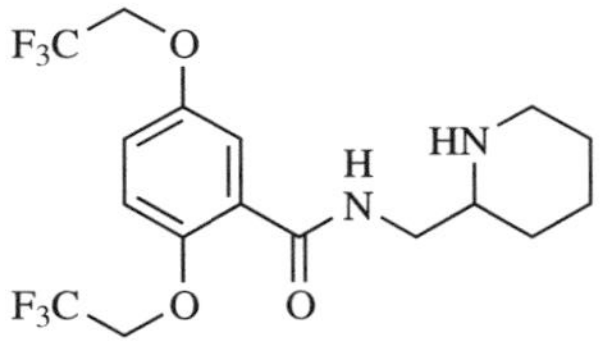
loracainide	[58934-46-6]	$C_{22}H_{28}Cl_2N_2O$	Lopantrol, Lorivox, Remivox	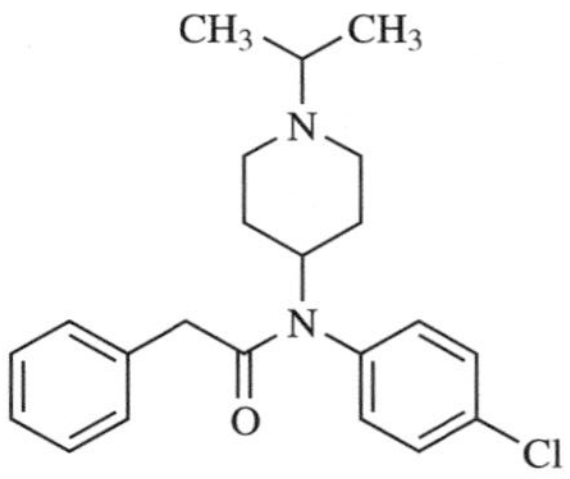

propafenone [34183-22-7] $C_{21}H_{28}ClNO_3$ Rythmol, Pronon, Rythmonorm

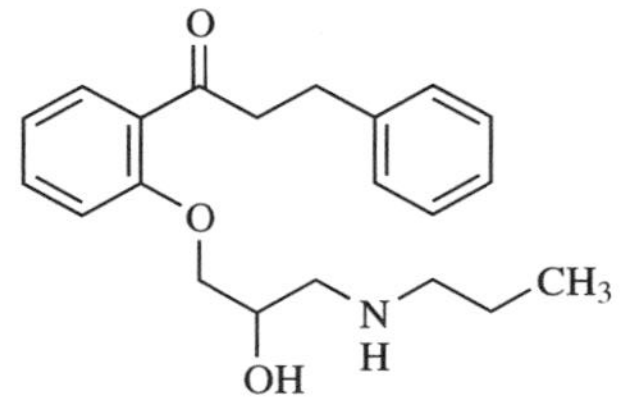

moricizine [29560-58-5] $C_{22}H_{26}ClN_3O_4S$ Ethmozine

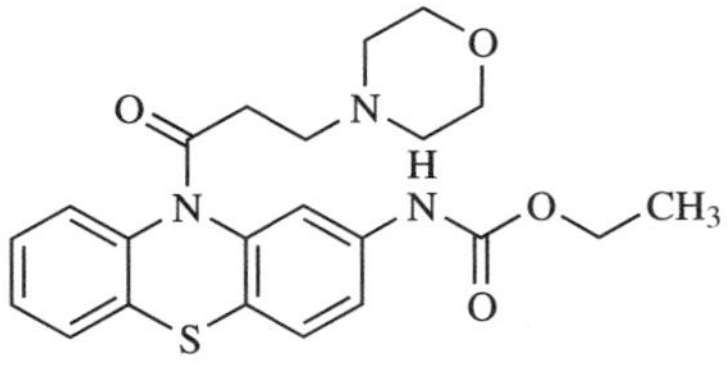

Class II Antiarrhythmic Agents

propranolol HCl [318-98-9] $C_{16}H_{22}ClNO_2$ Inderal, Inderide, Angilol, Bedranol, Duranol, Elbrol, Oposim, Tensol

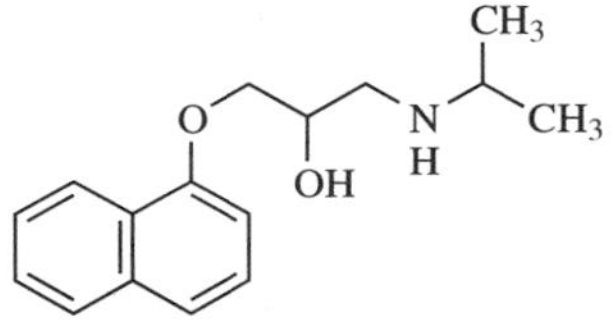

Nadolol [42200-33-9] $C_{17}H_{27}NO_4$ Corgard, Solgol, Corzide, Nadolol

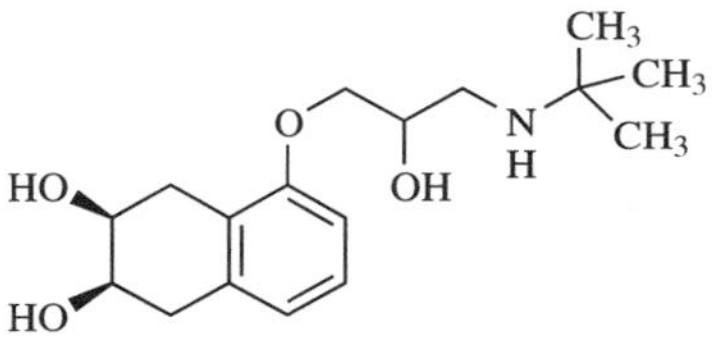

Table 1 (Continued)

Chemical/ generic name	CAS Registry Number	Molecular formula	Trade name	Structure
l-sotalol	[959-24-0]	$C_{12}H_{21}ClN_2O_3S$	Beta-cardone, Betapace, Darob, Sotacor, Sotalex	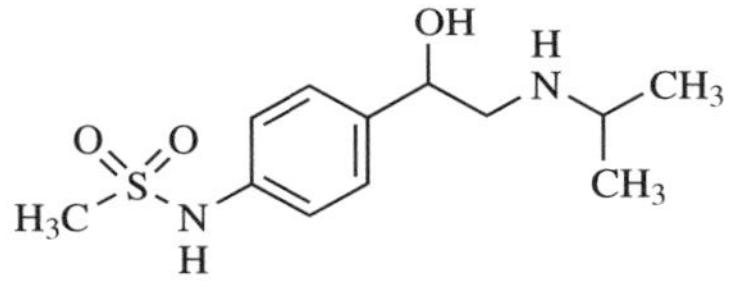
atenolol	[29122-68-7]	$C_{14}H_{22}N_2O_3$	Tenormin, Tenoretic, Atenol, Tenoblock, Basan	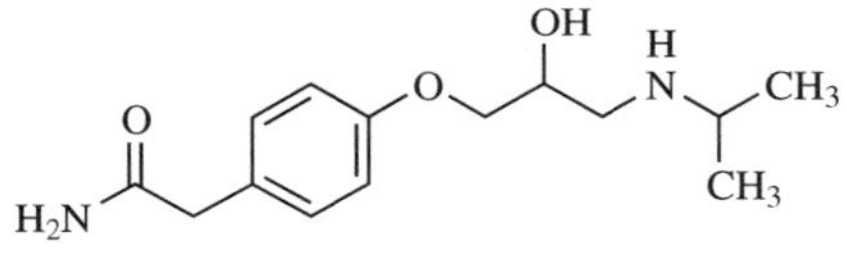
acebutolol	[34381-68-5]	$C_{18}H_{29}ClN_2O_4$	Acecor, Acetanol, Neptal, Prent, Sectral	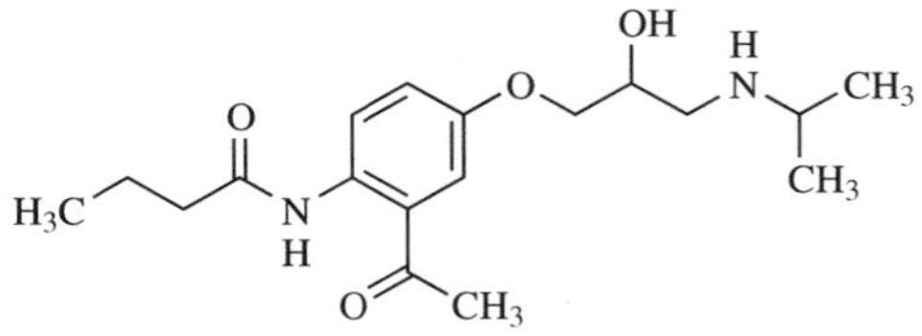
esmolol	[81161-17-3]	$C_{16}H_{26}ClNO_4$	Brevibloc	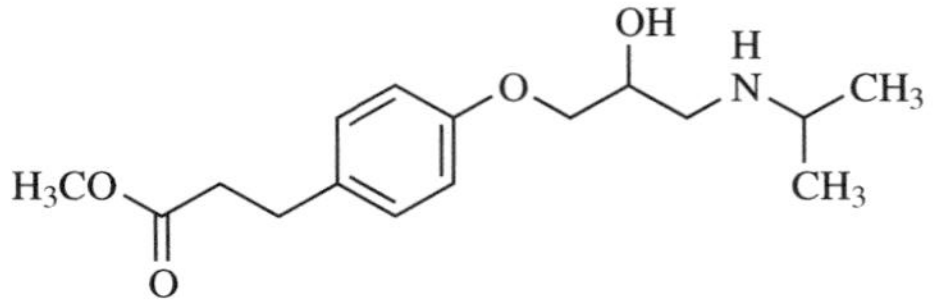

94

Name	CAS	Formula	Trade names
metoprolol succinate	[98418-47-4]	$(C_{15}H_{25}NO_3)_2 \cdot C_4H_6O_4$	Toprol-XL 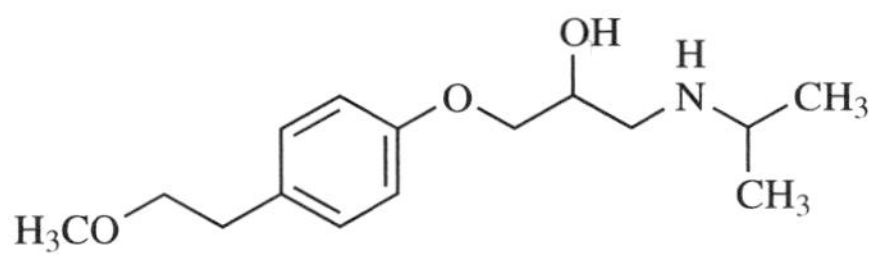
metoprolol tartarate	[56392-17-7]	$(C_{15}H_{25}NO_3)_2 \cdot C_4H_6O_6$	Beloc, Betaloc, Lopressor 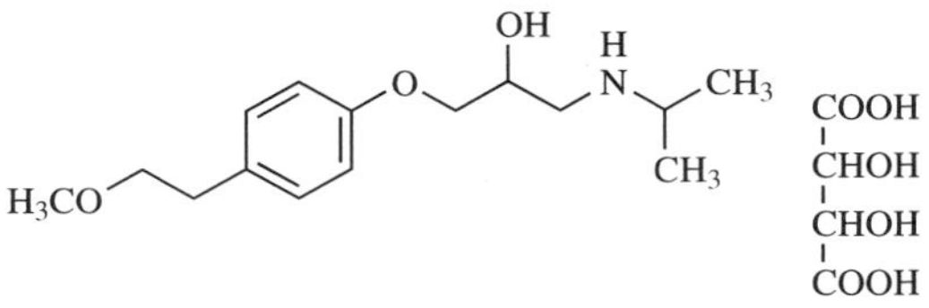

Class III Antiarrhythmic Agents

Name	CAS	Formula	Trade names
sotalol (d,l)	[3930-20-9]	$C_{12}H_{20}N_2O_3S$	Sotalex 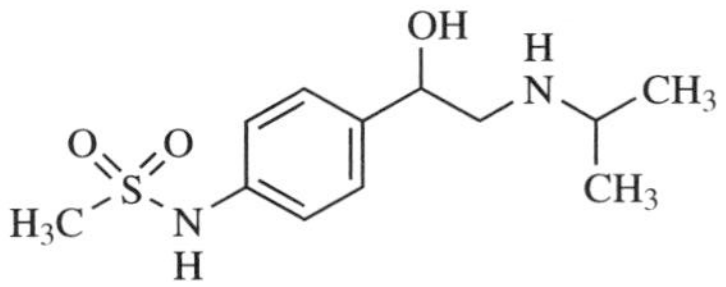
amiodarone	[19774-82-4]	$C_{25}H_{30}ClI_2NO_3$	Amiodar, Anacorn, Cordarex, Cordarone, Ortacrone, Pacerone, Tachydaron, Trangorex 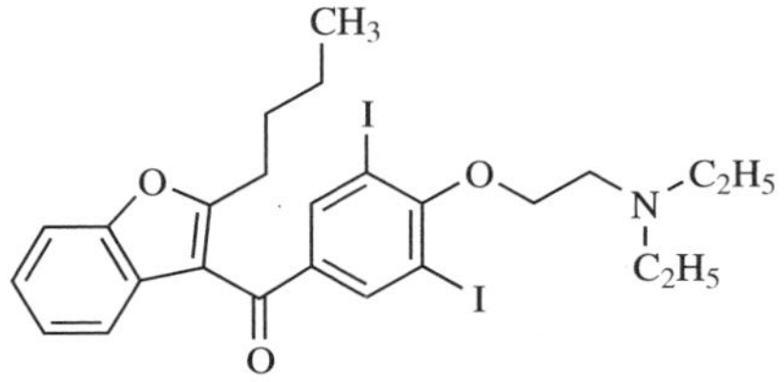

Table 1 (*Continued*)

Chemical/ generic name	CAS Registry Number	Molecular formula	Trade name	Structure
bretylium	[61-75-6]	$C_{18}H_{24}BrNO_3S$	Bretylan, Bretylate, Bretylol, Darenthin, Ornid	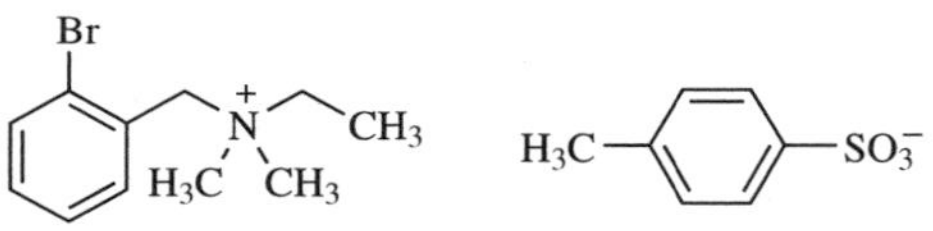
ibutilide	[122647-32-9]	$(C_{20}H_{36}N_2O_3S)_2 \cdot C_4H_4O_4$	Corvert	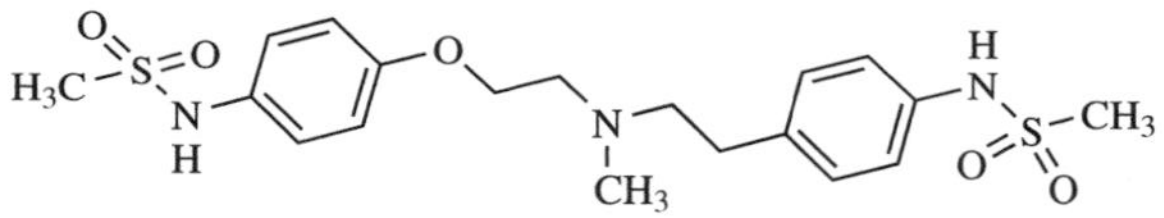
dofetilide	[115256-11-6]	$C_{19}H_{27}N_3O_5S_2$	Tikosyn	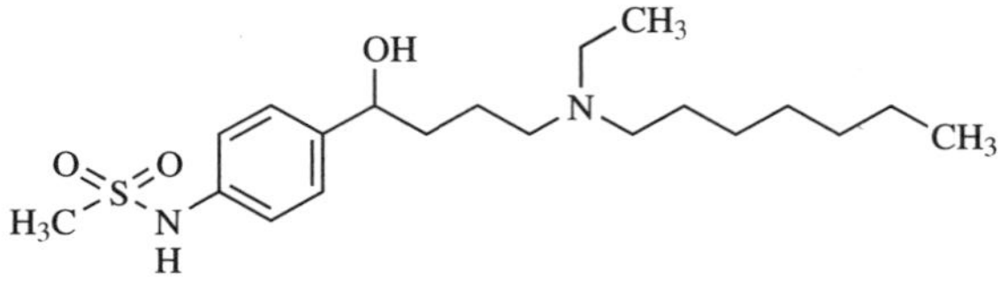
azimilide	[149888-94-8]	$C_{23}H_{30}Cl_3N_5O_3$	Stedicor	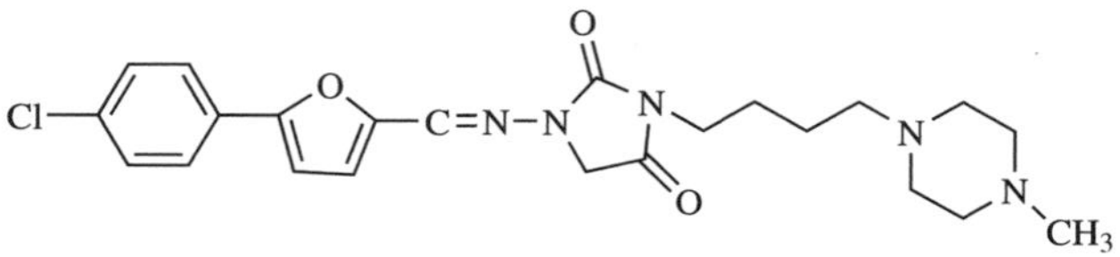

96

Class IV Antiarrhythmic Agents

bepridil	[74764-40-2]	$C_{24}H_{35}ClN_2O$	Angopril, Bepadin, Cordium, Vascor	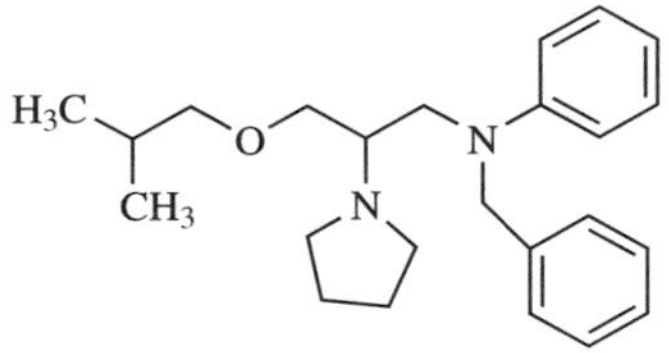
diltiazem	[33286-22-5]	$C_{22}H_{27}ClN_2O_4S$	Adizem, anginyl, Angizem, Bruzem, Calcicard, Cardizem, Citizem, Deltazen, Diladel, Dilzem, Masdil, Tildiem	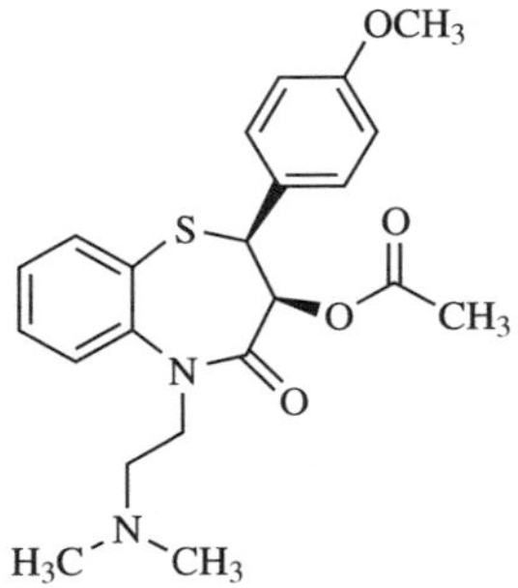
verapamil	[152-11-4]	$C_{27}H_{38}N_2O_4$	Apramyl, Calan, Cordilox, Dignover, Isoptin, Securon, Univer, Vasolan, Veraptin, Verelan, Verexamil, Veramix	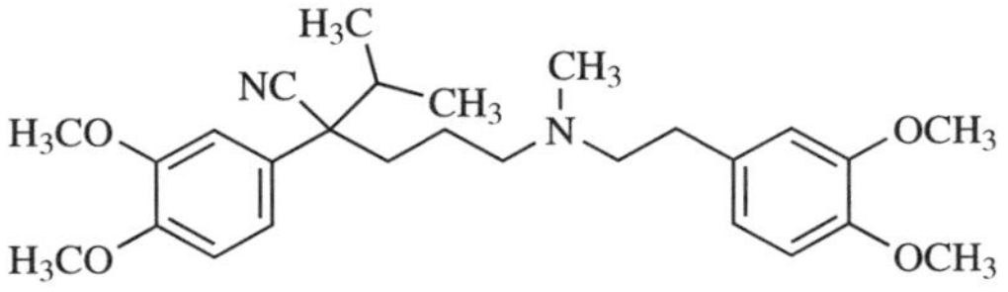

Table 1 (*Continued*)

Chemical/generic name	CAS Registry Number	Molecular formula	Trade name	Structure
			Miscellaneous Agents	
adenosine	[58-61-7]	$C_{10}H_{13}N_5O_4$	Adenocard, Adenocor, Adenoscan	
digoxin	[20830-75-5]	$C_{41}H_{64}O_{14}$	Digacin, Dilanacin, Eudigox, Lanacordin, Lanicor, Lanoxicaps, Rougoxin	

rate from sodium ion channels, causing a reduction in impulse conduction time. Agents in this class have been observed to have modest effects on repolarization. Drugs in Class IA—quinidine, procainamide, and disopyramide—have sodium ion channel dissociation rates that are intermediate between Class IB and IC compounds.

The affinities of the Class I antiarrhythmic agents for sodium channels vary with the state of the channel or with the membrane potential (39). As indicated earlier, sodium channels exist in at least three states: R = closed resting, or closed near the resting potential, but able to be opened by stimulation and depolarization; A = open activated, allowing Na^+ ions to pass selectively through the membrane; and I = closed inactivated, and unable to be opened (38). Under normal resting conditions, the sodium channels are predominantly in the resting or R state. When the membrane is depolarized, the sodium channels are active and conduct sodium ions. Next, the inward sodium current rapidly decays as the channels move to the inactivated (I) state. The return of the I state to the R state is referred to as channel reactivation, and is voltage and time dependent. Class I antiarrhythmic drugs have a low affinity for R channels, and a relatively high affinity for both the A or I channels (40).

3.6. Class IA Agents. *Quinidine.* Quinidine is obtained from species of the genus *Cinchona*, and is the *d*-isomer of quinine. This molecule contains two basic nitrogens: one in the quinoline ring and one in the quinuclidine moiety. The nitrogen in the quinuclidine moiety is more basic. Three salt formulations are available: quinidine gluconate, quinidine polygalacturonate (41), and quinidine sulfate (42). Of the three, the gluconate formulation is the most soluble in water.

Quinidine binds to open sodium ion channels, decreasing the entry of sodium into myocardial cells. This depresses phase 4 diastolic depolarization (shifting the intracellular threshold potential toward zero), and decreases transmembrane permeability to the passive influx of sodium (slowing the process of phase 0 depolarization, which decreases impulse velocity), and increases action potential duration (43). Physiologically, this results in a reduction in SA node impulse initiation, and depression of the automaticity of ectopic cells. Quinidine is also thought to act, at least in part, by binding to potassium channels (Table 1), and is used to treat supraventricular and ventricular arrhythmias including atrial flutter and fibrillation, and atrial and ventricular premature beats and tachycardias. It is primarily metabolized to a hydroxylated metabolite—2-hydroxyquinidine—that is equal in potency to the parent compound (44).

Procainamide. Procainamide is an amide derivative of procaine. Replacement of the ether oxygen in procaine, with an amide nitrogen (in procainamide), decreases central nervous system (CNS) side effects, rapid hydrolysis, and instability in aqueous solution that results from the ester moiety in procaine. Procainamide is formulated as a hydrochloride salt of its tertiary amine. The metabtabolite of procainamide is *N*-acetylprocainamide, which possesses 25% of the parent drugs activity (45,46). Mechanistically, procainamide has the same cardiac electrophysiological effects as quinidine. It decreases automaticity and impulse conduction velocity, and increases the duration of the action potential (47). This compound may be used to treat all of the arrhythmias indicated for treatment with quinidine, including atrial flutter and fibrillation, and atrial and ventricular premature beats and tachycardias.

Disopyramide. The electrophysiological effects of this drug are similar to those of quinidine and procainamide—decreased phase 4 depolarization and decreased conduction velocity (48). This molecule contains the ionizable tertiary amine that is characteristic of compounds in this class, is formulated as a phosphate salt, and is administered both orally and intravenously (49). Due to its structural similarity to anticholinergic drugs, disopyramide produces side effects that are similar to those types of therapeutics, including dry mouth, urinary hesitancy, and constipation. Clinically it is used to treat life-threatening ventricular tachyarrhythmias.

3.7. Class IB Agents. *Lidocaine.* Lidocaine is formulated as a hydrochloride salt that is soluble in both water and alcohol. It binds to inactive sodium ion channels, decreasing diastolic depolarization, and prolonging the resting period (50,51). Lidocaine is administered intravenously for suppression of ventricular cardiac arrhythmias. The first pass metabolite of this compound—monoethylglycinexylidide—is generated from deethylation of the tertiary amine, and is equipotent to its parent compound (52,53).

Tocainide. Tocainide is an analogue of lidocaine, but differs structurally in that it possesses as primary, versus a tertiary, terminal side-chain amine. In addition, a methyl substituent on the side chain partially protects the amide moiety against hydrolysis. Tocainide has a mechanism of action that is similar to lidocaine (54,55). It is orally active, and the presence of a primary amine allows for formulation as a hydrochloride salt. Therapeutically it is used to prevent or treat ventricular tachycardias.

Mexiletine. Structurally, mexiletine resembles lidocaine and tocainide in that it contains a xylyl moiety. However, it differs in that it possesses an ether moiety (versus an amide moiety (as found in lidocaine and tocainide) in its side chain. As a result, mexiletine is not vulnerable to hydrolysis, and has a longer half-life than lidocaine (56). Mexiletine possesses a primary amine, and is formulated as a hydrochloride salt that is orally active. Its effects on cardiac electrophysiology are similar to that of lidocaine (57). It is used in the treatment of ventricular arrhythmias, however, due to proarrhythmic side effects; it is generally not used with lesser arrhythmias (58).

Phenytoin. Phenytoin is a hydantoin derivative of the anticonvulsant therapeutics that does not possess sedative properties. It is structurally dissimilar to all other class I antiarrhythmic compounds, and is the only member of this family of compounds that does not contain an ionizable amine. However, its effects on cardiac cells are similar to those of lidocaine. Mechanistically, it decreases the force of contraction, depresses pacemaker action, and improves atrioventricular conduction, especially when administered in conjunction with digitalis (59).

3.8. Class IC Agents. *Encainide.* Encainide is a benzanilide derivative containing a piperidine ring, and like other Class I compounds, blocks sodium channels (60). Encainide contains a terminal tertiary amine, and is formulated as a chloride salt. It is used orally to suppress and prevent recurrence of documented life-threatening ventricular arrhythmias. Use of encainide for less severe arrhythmias is no longer recommended. A metabolite of this compound, ODE, which results from demethylation of the methoxy moiety, is more potent than encainide (61). It has been effective for the management of various supraventricular tachyarrhythmias, Wolf-Parkinson white syndrome, and AV nodal reentrant tachycardia.

Flecainide. Flecainide is a benzamide–piperidine derivative. However, it is structurally dissimilar from encainide in that it contains one less benzyl group, possesses two lipophilic trifluoroethoxy substituents at the 1 and 4 positions on the benzamide ring (versus a single methoxy substituent at the 4 position of the benzamide in encainide), and lacks a methyl substituent on the piperidine nitrogen. It is formulated as an acetate salt, and like encainide, its metabolites are active. It possesses cardiac physiological effects that are similar to those of encainide, and it is used orally to suppress and prevent the recurrent of documented life-threatening ventricular arrhythmias and supraventricular tachyarrhythmias (62,63). Limited data suggest that flecainide may be effective for conversion of atrial fibrillation to normal sinus rhythm and for the treatment of ectopic or multifocal atrial tachycardias.

Lorcainide. Lorcainide is a benzamide/piperidine derivative with a mechanism of action that is similar to encainide (64,65). Lorcainide is formulated as a hydrochloride salt and is orally active. Metabolism produces an N-dealkylated derivative—norlorcainide (66). This metabolite is as potent as the parent compound, but possesses a half-life that is approximately three times longer. Lorcainide is used to treat ventricular arrhythmia, ventricular tachycardia, and Wolff-Parkinson-White Syndrome.

Propafenone. Propafenone is structurally unlike other compounds in this subclass—ie, encainide, flecainide, and lorcainide. Instead, it is an ortho substituted aryloxy propanolamine that is similar in structure to β-blockers. The racemic mixture possesses Na^+ channel blocking activity, while the (S) (−) isomer is a potent β-blocker. Mechanistically, propafenone has a stabilizing effect on myocardial membranes, which manifests in a reduction in the upstroke velocity (Phase 0) of the action potential (67). In Purkinje fibers, and to a lesser extent myocardial fibers, propafenone decreases the fast inward current carried by sodium ions, prolongs the refractory period, reduces spontaneous automaticity, and depresses triggered activity (68,69). Propafenone is indicated in the treatment of paroxysmal atrial fibrillation–flutter and paroxysmal supraventricular tachycardia. It is also used to treat ventricular arrhythmias, such as sustained ventricular tachycardias.

Moricizine. Moricizine is a phenothiazine derivative, and is a structurally unique member of the Class IC antiarrhythmic agents. Like other agents in this subclass, it decreases the speed of cardiac conduction by lengthening the refractory period and shortening the length of the action period of cardiac tissue (70). Moricizine is formulated as a hydrochloride salt, and is used to treat ventricular arrhythmias.

3.9. Class II Agents. The inhibitors in this class are all β-adrenergic antagonists that have been found to produce membrane-stabilizing or depressant effects on myocardial tissue. It has been hypothesized that the antiarrhythmic properties of these agents are mainly due to their inhibition of adrenergic stimulation of the heart (71,72) by the endogenous catecholamines, epinephrine and norepinephrine. The principal electrophysiological effects of the β-blocking agents manifest as a reduction in the phase 4 slope potential of sinus or pacemaker cells, which decreases heart rate and slows tachycardias.

With the exception of sotalol (73), the compounds in Class II are all structurally similar to aryloxypropanolamines (Table 1). This name originates from the presence of an $-OCH_2-$ group located between a substituted benzene ring, on one side, and an ethylamino side-chain, on the other side. The aromatic ring and its substituents are the primary determinants of β-antagonist selectivity. Substitution of the para position of the benzene ring, in tandem with the absence of meta-position substitution, appears to confer selectivity for β1 cardiac receptors. Sotalol differs from other members of this class in that it lacks the $-OCH_2-$ group. This results in a shortening of the characteristic ethylamino side chain (Table 1).

Propranolol is the prototype agent for this class of compounds. Due to the substitution pattern on its aromatic ring, it is not a selective β-adrenergic blocking agent. During propranolol-mediated β-receptor block, the chronotropic, ionotropic, and vasodilator responses to β-adrenergic stimulation are decreased. Propranolol exerts its antiarrhythmic effects in concentrations associated with β-adrenergic blockade (74). It has also been shown to possess membrane-stabilizing activity that is similar to quinidine. However, the significance of this membrane action in the treatment of arrhythmias is uncertain, as the concentrations required to produce this effect are greater than required for the observance of its β-blocking effects. Effects of β-adrenergic receptor antagonists to reduce cardiac mortality after MI are well established. The β-adrenergic agents, including amiodarone are the only drugs proven to reduce death in these patients.

Nadolol (75) and *l*-Sotalol (76) are both nonspecific β-blockers (Table 1), while para substitutions on the aromatic rings of atenolol (77), acetobutolol (78), esmolol (79), and metoprolol (80) all confer β1 antagonist selectivity (Table 1). Each of these agents exerts electrophysiological effects that result in slowed heart rate, decreased AV nodal conduction, and increased AV nodal refractoriness.

3.10. Class III Agents. The drugs in this class—amiodarone, bretylium, dofetilide, ibutilide, and (*d,l*) or racemic sotalol—all generate electrophysical changes in myocardial tissue by blocking ion channels, however, some are selective, while others are multi-channel blockers (this is not surprising, as there is a high degree of sequence homology between the different ion channels). Importantly, all Class III drugs have one common effect—that of prolonging the action potential, which increases the effective refractory period without altering the depolarization or the resting membrane potential (81).

Racemic sotalol, dofetilide, and ibutilide are potassium channel blockers. Sotalol also possesses β-adrenergic blocking properties (as indicated above), while ibutilide is also a sodium channel blocker. The mechanisms of action of amiodarone and bretylium, which also prolong the action potential, remain unclear. However, both have sodium channel-blocking properties.

Of the compounds listed in this class, sotalol, dofetilide, and ibutilide are structurally similar (Table 1). All three drugs contain a central aromatic ring with a sulfonamide moiety, and a para-substituted alkylamine side chain. Dofetilide, unlike sotalol and ibutilide, is nearly symmetrical, with two methanesulfonamides at either end of the molecule.

Amiodarone. Amiodarone is structurally unique in this class, and has received much attention over the past several years for its ability to treat arrhythmias. Amiodarone is currently the most utilized drug in patients with life-threatening arrhythmias—approximately one-half of the patients currently receiving antiarrhythmic drug therapy are treated with amiodarone (82).

Amiodarone is a benzofuranyl derivative with a central diiodobenzoyl substituent and an alkyl amine side chain. Mechanistically, this agent prolongs the duration of the action potential and effective refractory period, with minimal effect on resting membrane potential (83–85). Amiodarone exhibits mechanisms of activity from each of the four Singh and Vaughan Williams classes. In addition, it also displays noncompetitive α- and β-adrenergic inhibitory properties. It is effective in the treatment of life-threatening recurrent ventricular arrhythmias and atrial fibrillation (86), and is orally available as a chloride salt. Amiodarone contains two iodine substituents, and consequently affects thyroid hormones (87). However, its most serious side effects involve both the exacerbation of arrhythmias and pulmonary toxicity. An experimental noniodinated benzofuranyl derivative of amiodarone, *dronedarone*, has emerged as a potential new member of the Class III antiarrhythmics. It has been found to have similar electrophysiological effects as amiodarone, but with fewer side effects (88).

(d,l) Sotalol.   Sotalol is classified as both a Class II and a Class III antiarrhythmic agent. The *l*-isomer is classified as a β-blocker, and is 50 times more active than the *d*-isomer in this capacity; the racemic mixture of this drug is considered to be a Class III agent, as it inhibits the component of the potassium channel involved in the rectifier potassium current. Sotalol is used to treat and prevent life-threatening ventricular arrhythmias (89). Additionally, because of its Class II and III activity, it is also effective against supraventricular arrhythmias (90). The mode of action, pharmacokinetics, and therapeutic uses of sotalol have been reviewed extensively (91).

Sotalol is formulated as a hydrochloride salt and is orally available. In terms of efficacy, clinical trials have indicated that this agent is at least as effective or more effective in the management of life-threatening ventricular arrhythmia than other available drugs (92).

Bretylium Tosylate.   Bretylium tosylate is a bromobenzyl quaternary ammonium salt. It is formulated as a tosylate salt and is soluble in water and alcohol. It is administered by intravenous or intramuscular injection, and is used to treat ventricular fibrillation and ventricular arrhythmias that are resistant to other therapy. The mechanism of antiarrhythmic action of this drug has not been determined (93). Research has shown that this agent selectively accumulates in neurons and inhibits norepinephrine release, and it has been suggested that its adrenergic neuronal-blocking properties are responsible for its antiarrhythmic activity (94).

Ibutilide.   Ibutilide is formulated as a fumarate salt, and is administered by intravenous injection (95). This agent prolongs repolarization of cardiac tissue by increasing the duration of the action potential and the effective refractory period in cardiac cells. It blocks both sodium and potassium channels (96–98), but unlike sotalol, does not possess β-adrenergic blocking activity. Ibutilide is used in the treatment of supraventricular tachyarrhythmias, such as atrial flutter and atrial fibrillation (99,100). *Trecetilide* is a congener of ibutilide that is currently under investigation for intravenous and oral treatment of atrial flutter and atrial fibrillation.

Dofetilide. Dofetilide is a recent addition to the Class III antiarrhythmic agents. It prolongs repolarization and refractoriness without affecting cardiac conduction velocity, and is a selective blocking agent of the delayed rectifier potassium current (101,102). Unlike ibutilide and sotalol, this agent does not inhibit sodium channels or β-adrenergic receptors. Dofetilide is formulated as a hydrochloride salt, is administered orally, and is used to treat supraventricular tachyarrhythmias, and to restore normal sinus rhythm during atrial fibrillation and atrial flutter (103,104).

Azimilide. Azimilide is a novel Class III antiarrhythmic agent that has been shown to block both the slow activating and rapidly activating components of the delayed rectifier potassium current (105). Structurally, it is unlike other molecules in this class, containing both imidazolidione and piperazine moieties, and is being evaluated in the treatment of atrial flutter, atrial fibrillation, and paroxysmal supraventricular tachycardia (106–108).

3.11. Class IV Agents. All of the calcium channel blockers in this class of agents—verapamil, diltiazem, and bepridil—also possess antianginal activity. With respect to cardiac arrhythmias, these agents affect calcium ion flux, which is required for the propagation of an electrical impulse through the AV node (109). By decreasing this influx, the calcium channel blockers slow conduction. This, in turn, slows the ventricular rate.

Verapamil. Verapamil blocks the influx of calcium ions across cell membranes (110). Structurally, it is not related to other antiarrhythmic drugs. It is formulated as a hydrochloride salt and is readily soluble in water. Verapamil is used to treat supraventricular arrhythmias, including atrial tachycardias and fibrillations (111), and is administered both orally and via intravenous injection. Verapamil is rapidly metabolized to at least 12 dealkylated metabolites. Norverapamil, a major and active metabolite, has 20% of the cardiovascular activity of verapamil, and reaches plasma concentrations that are almost equal to those of verapamil within 4–6 hours after administration.

Diltiazem. Diltiazem is a benzothiazepine derivative, and is formulated as a hydrochloride salt. It may be administered orally or via injection. Similar to verapamil, diltiazem inhibits the influx of Ca^{2+} during the depolarization of cardiac smooth muscle. Therapeutically, this drug reduces the heart rate during tachycardias. Diltiazem has also been shown to decrease the ventricular rate during atrial fibrillation or atrial flutter (112,113).

Bepridil. Bepridil inhibits the transmembrane influx of Ca^{2+} into cardiac and vascular smooth muscle. Like diltiazem it slows the heart rate by prolonging both the effective refractory periods of the atria and ventricles (114). This agent is formulated as a hydrochloride salt, and is orally available. It is used in treating tachyarrhythmias and in the management of high ventricular rates that are secondary to atrial flutter or fibrillation (115,116).

3.12. Miscellaneous Antiarrhythmic Agents. Two antiarrhythmic agents that do not fall within the Singh and Vaughan Williams classification are adenosine and digoxin.

Adenosine. Adenosine is chemically unrelated to other antiarrhythmic drugs. It is soluble in water, but practically insoluble in alcohol. For the treatment of arrhythmias it is administered via intravenous injection.

Adenosine reduces SA node automaticity, slows conduction time through the AV node, and can interrupt reentry pathways. It is used to restore normal sinus rhythm in patients with paroxysmal supraventricular tachycardia, including Wolff-Parkinson-White syndrome (117–120).

Digoxin and Digitoxin.   Digoxin and digitoxin belong to the family of compounds known as the cardiac glycosides. The natural glycosides are isolated from various plant species: *digitalis purpurea Linne, digitalis lanata Ehrhart, strophanthus gratu,* or *acokanthea schimperi.*

Digoxin and digitoxin inhibit sodium–potassium ATPase, which is responsible for regulating the quantity of sodium and potassium inside cells. Inhibition of this enzyme results in an increase in the intracellular concentration of sodium and calcium. However, reduction in the heart rate associated with digitalis results from at least two different mechanisms. Digitalis increases cardiac parasympathetic activity via CNS actions. In addition, the activation of the reverse mode of the sodium–calcium exchanger leads to a net loss of positive charge (3 sodium out to every calcium in) and results in hyperpolarize of the pacemaker cells and decreased heart rate. Digoxin and digitoxin are available both orally and through intravenous injection, and are used to treat and prevent sinus and supraventricular fibrillation, flutter, and tachycardia (121).

3.13. Current and Future Trends in the Treatment of Arrhythmia. Following the discovery that lidocaine was useful for treating cardiac arrhythmias, early drug discovery and development of antiarrhythmic agents focused on compounds that were structurally similar to lidocaine and possessed similar mechanisms of action—that of blocking sodium channels. This led to the initial identification of lidocaine congeners such as tocainide and mexiletine, and later to encainide and flecainide. The long standing hypotheses for treating arrhythmias with sodium ion channel blockers was based on the belief that these molecules could effectively prevent or suppress the onset of arrhythmias and/or terminate this condition when it became persistent (122). However, CAST (123), which evaluated the effects of well-established sodium channel blockers on mortality in postmyocardial patients (with frequent premature ventricular arrhythmias), dispelled this hypothesis. In fact, these studies found that both encainide and flecainide increased mortality. Since the CAST studies, other trials with mexiletine, propafenone, and moricizine (124) (CAST II) (125) have also shown similar results, and a correlation between increased mortality and the use of sodium ion channel blocking agents in post-myocardial infarction patients has been established.

Based on the findings of CAST and related studies, the treatment of antiarrhythmias has shifted away from Class I sodium channel blockers, and now focuses on Class III drugs (126), which act by prolonging the action potential duration and the refractory period. Class III agents lack many of the negative side effects observed in other classes of antiarrhythmics, affect both atrial and ventricular tissue, and can be administered orally or intravenously. Members of this class, such as amiodarone (which has proven to be a clinically efficient therapeutic for the treatment of a wide variety of arrhythmias) and racemic sotalol, have been the center of much attention in recent years, and have led to the search for new Class III drugs with improved safety profiles (127). New and investigational Class III agents that are more selective for potassium channel subtypes include azimilide (128), dofetilide, dronedarone, ersentilide, ibutilide, tedisamil, and trecetilide (129).

There have also been numerous reports on the synthesis and evaluation of new antiarrhythmic compounds; several of these are briefly described: Matyus and co-workers (130) reported the synthesis and biological evaluation of novel phenoxyalkyl amines that exhibit both Class IB and Class III type electrophysiological properties; Tripathi and co-workers (131) performed synthesis and SAR studies on 1-substituted-n-(4-alkoxycarbonylpiperidin-1-yl) alkanes that showed potent antiarrhythmic activity comparable to quinidine; Bodor and co-workers (132) reported a novel tryptamine analogue that was found to selectively bind to the heart (and within the heart to have tissue specificity), and possessed effects on vital signs of the cardiovascular system that indicated antiarrhythmic activity; Morey and co-workers (133) designed a series of amiodarone homologues that resulted in an SAR that will have implications for the future development of amiodarone-like antiarrhythmic agents; Himmel and co-workers (134) synthesized and evaluated the activities of thiadiazinone derivatives that are potent and selective for potassium ion channels, and show Class III antiarrhythmic activity; Thomas and co-workers (135) developed a novel antiarrhythmic agent—BRL-32872—that inhibits both potassium and calcium channels; and Levy and co-workers (136) described novel dibenzoazepine and 11-oxo-dibenzodiazepine derivatives that are effective ventricular defibrillating drug candidates.

Along with advances in the understanding and development of new therapeutic agents, the development of technological devices to treat arrhythmias has also evolved. One of the most important achievements has been the implantable cardioverter defibrillator (ICD) (82). This device has been an option for treating arrhythmias since the early 1980s, and in the treatment of ventricular tachycardia and fibrillation, no other therapy has been as effective in prolonging patient survival (122). However, an important point regarding ICD treatment is that it is often used in combination with antiarrhythmic drug therapy (82). For frequent symptomatic episodes of ventricular tachycardia, administration of an adjuvant drug therapy is often required to provide maximum prevention and treatment of life-threatening arrhythmias. In particular, combination therapy with ICD and both β-blockers and amiodarone have received the most attention (82).

Finally, new evidence suggests that combinations of therapeutics may be more effective at treating and controlling arrhythmias than using any single agent alone (127). In particular, clinical sources have indicated that the pharmacological properties of amiodarone and β-blockers may be additive or even synergistic for treating arrhythmias (127). Details of the analysis of amiodarone interaction with α-blockers in the European myocardial infarct amiodarone trial (EMIAT) and in the Canadian amiodarone myocardial infarction trial (CAMIAT) have recently been reported (137). Data from randomized patients in these trials were analyzed by multivariate proportional hazard models, and indicated that combination therapy consisting of amiodarone and β-blockers led to a significantly better survival rate. Hence, the possibility of administering combination therapies will be an important aspect in the future development of therapeutic techniques for treating arrhythmias.

4. Antianginal Agents

Angina pectoris is the principal symptom of ischemic heart disease, and is caused by an imbalance between myocardial oxygen demand and oxygen supply by coronary vessels. Such an imbalance could be the result of either increased myocardial oxygen demand due to exercise or decreased myocardial oxygen delivery or both. Angina pectoris is always associated with sudden, severe chest pain and discomfort. The location and character of the pain may vary but often radiates from the sternum to the left shoulder and over the flexor surface of the left arm to the tips of the medial fingers. However, some individuals do not experience pain with ischemia. Angina pectoris can be induced by exercise, anxiety, overeating, or stress and is often relieved quickly by rest. Other factors, such as decreased oxygen carrying capacity of the blood, or reduced aortic pressure may be involved. The attack may be transient and damage to the ischemic myocardium may be minimal or it may result in an acute myocardial infarction (MI) and/or death. It is usually accompanied by ST segment changes in the ECG depending on the condition. Angina occurs because the blood supply to the myocardium via coronary vessels is insufficient to meet the metabolic needs of the heart muscle for oxygen (138) either by a decrease in blood supply or an exceedingly large increase in oxygen requirements of the myocardium or both. For a drug to be efficacious in angina it should improve myocardial oxygen supply (increase blood flow) or reduce myocardial oxygen consumption or have both actions.

Several factors affect mycocardial oxygen supply such as (*1*) *Blood oxygenation and oxygen extraction involving tissue ischemia*: A normal and healthy heart extracts about 75% of blood oxygen at rest, however, increased coronary blood flow and extraction results in an increase in oxygen supply. But, ischemic heart disease develops when there is a deficiency in the supply of blood and oxygen to the heart, typically caused by narrowing of the coronary arteries, a condition known as coronary artery disease (CAD) or coronary heart disease (CHD). CAD is a consequence of a complicated pathological process involving the development of atherosclerotic lesions in the coronary arteries, in which cholesterol, triglycerides, and other substances in the blood deposit in the walls of arteries, narrowing them. The narrowing limits the extraction and flow of oxygen rich blood to the heart. (*2*) *Pulmonary conditions*: Sometimes acute and chronic bronchopulmonary disorders such as pneumonia, bronchitis, emphysema, tracheobronchitis, chronic asthmatic bronchitis, tuberculosis, and primary amyloidosis of the lung affect the oxygen extraction and its supply to the heart, causing severe ischemia. Also, if the heart does not work efficiently as it should, it reduces the cardiac output causing congestion of fluid in the tissues leading to swelling (edema). Occasionally, the fluid collects in the lungs and interferes with breathing, causing shortness of breath at rest or during exertion. Edema is also exacerbated by reduced ability of the kidneys to dispose sodium and water. The retained water further increases the edema (swelling). (*3*) *Coronary vascular conditions*: The various conditions such as coronary collateral blood flow, coronary arterial resistance affected by nervous system, accumulation of local metabolites, tissue death, endothelial function, diastolic blood pressure, and endocardial–epicardial blood flow contribute significantly to the pathogenesis of angina.

The factors that govern myocardial oxygen demand include (*4*) *Heart rate*: A significant change in the regular beat (fast or slow) or rhythm of the heart (arrhythmias) may affect the myocardial oxygen demand. Excessive slowing of heartbeat is called bradycardia and is sometimes associated with fatigue, dizziness, and lightheadedness or fainting. The various types of bradycardia have been categorized as sinus bradycardia, and junctional rhythm and heart block. These symptoms can easily be corrected with an electrical pacemaker that is implanted under the skin and takes over the functioning of the natural pacemaker. Conversely, a rapid heart beat is referred to as tachycardia. Tachycardias are classified into two types: supraventricular and ventricular. Different types of abnormal rapid heart beats have been categorized as sinus tachycardia (normal response to exercise), atrial tachycardia, atrial fibrillation, atrial flutter, AV nodal reentry, AV reciprocating tachycardia, premature atrial contractions, ventricular tachycardia, and premature ventricular contractions. Electrocardiographic monitoring is needed for the correct diagnosis of arrhythmias. Since exercise stimulates the heart to beat faster and more forcefully, more blood, and hence oxygen, is needed by the myocardium to meet this increased workload. Normally, this is accomplished by dilation of coronary blood vessels; however, sometimes atherosclerosis may inhibit the flow of oxygen rich blood, causing ischemia. (*5*) *Cardiac contractility (inotropic state)*: Reduction in the cardiac output causes a reflex activation of the sympathetic nervous system to stimulate heart rate and contractility further leading to greater oxygen demand. If the coronary arteries are occluded and incapable of delivering the needed oxygen an ischemia will occur. (*6*) *Preload-venous pressure and its impact on diastolic ventricular wall tension and ventricular volumes*: It has been suggested that an important strategy in the treatment of cardiac function is reduction of the work load of the heart, by reducing the number of heart beats per minute and the work required per heart beat defined by preload and afterload. Preload is defined as the volume of blood that fills the heart before contraction. Contraction of the great veins increases preload, while dilation of veins reduces preload. (*7*) *Afterload—systolic pressure required to pump blood out*: Afterload is defined as the force that the heart must generate to eject blood from ventricles. It largely depends on the resistance of arterial vessels. Contraction of these vessels increases afterload, while dilation reduces afterload.

There are different types of angina (*1*) *Stable Angina*. Stable angina is also called chronic angina, exertional angina, typical or classic angina, angina of effort or atherosclerotic angina. The main underlying pathophysiology of this most common type of angina is usually atherosclerosis, ie, plaques that occlude the vessels or coronary thrombi that block the arteries. This type of angina usually develops by "exertion", exercise, emotional stress, discomfort, or cold exposure and can be diagnosed using EKG. Therapeutic approaches to treat this type of angina include increasing the myocardial blood flow and decreasing the cardiac preload and afterload. (*2*) *Vasopastic Angina*. It is also called variant angina or Prinzmetal's angina. It is usually caused by a transient vasospasm of coronary blood vessels or atheromas, at the site of plaque. This can easily be seen by EKG changes in ST elevation that tend to occur at rest. Sometimes chest pain develops even at rest. A therapeutic approach to treat this type of angina is to decrease vasospasm of coronary arteries normally provoked by α-adrenergic activation in coronary vasculature. However, α-adrenergic activation is not the only cause of vasospasm. *(3) Unstable Angina*. It is also called preinfarction angina, crescendo angina, or angina at rest. It is usually characterized by recurrent episodes of prolonged attacks at rest resulting from the small platelet clots (platelet aggregation) at the atherosclerotic plaque site that may also induce local vasospasm. This type of angina requires immediate medical intervention such as cardiac bypass surgery or angioplasty since it could ultimately lead to myocardial infarction (MI). Treatment regiments include inhibition of platelet aggregation and thrombus formation, vasodilation of coronary arteries (angioplasty) or decrease in cardiac load.

4.1. Etiology and Causes of Angina and Coronary Heart Disease. The risk factors for the development of CHD and angina pectoris are genetic predisposition, age, male sex, and a series of reversible risk factors. The most important factors include high fat and cholesterol rich diet (138–140), lack of exercise and inability to retain normal cardiac function under increased exercise tolerance (141,142), tobacco and smoking (since nicotine is a vasoconstrictor) (143), excessive alcohol drinking, carbohydrate and fat metabolic disorders, diabetes, hypertension (144,145), obesity (146,147), and use of drugs that produce vasoconstriction or enhanced oxygen demand.

4.2. Treatment. In general, the action of various therapeutic drugs is either by (a) alteration of myocardial contractility or heart rate; (b) modification of conduction of the cardiac action potential; or (c) vasodilatation of coronary and peripheral vessels. Therefore, this article primarily focuses on therapeutics that apply to the treatment of angina, arrhythmia, cardiac heart diasease, thrombolysis, and other cadiac complications.

4.3. Treatment of Angina. The various treatment modalities of different kinds of angina include *(1)* prevention of precipitating factors; *(2)* use of nitrates as vasodilators to treat acute symptoms; *(3)* utilization of prophylactic treatment using a choice of drugs among antianginal agents, calcium channel blockers, and β-blockers; *(4)* surgeries such as angioplasty, coronary stenting, and coronary artery bypass surgery; *(5)* anticoagulants and use of antithromobolytic agents.

4.4. Nitrates as Vasodilators. Some of the simple organic nitrates and nitrites find applications for both short- and long-term prophylactic treatment of angina pectoris, myocardial infarction, and hypertension. Most of these nitrates and nitrites are formulated by mixing inert suitable excipients such as lactose, dextrose, mannitol, alcohol, propylene glycol for safe handling, since some of these compounds are heat sensitive, very flammable and powerful explosives, if used alone. The onset, duration of action and potency of organic nitrates could be attributed to structural differences. However, there is no relationship between the number of nitrate groups and activity (Table 2).

4.5. Mechanism of Action. The nitrates and nitrites are simple organic compounds that metabolize to a free radical nitric oxide (NO) at or near the plasma membrane of vascular smooth muscle cells. The basic pharmacological action of nitrates is a relaxation of most vascular smooth muscle cell. It is a direct effect that is not mediated by adrenergic receptors or endothelium. In 1980, Furchgott and Zawadski first discovered that NO is the most potent endogenous vasodilator (148). Nitric oxide is a highly reactive species with a very short

Table 2. Antianginal Agents: Nitrates as Vasodilators

Chemical/ generic name	CAS Registry Number	Molecular formula	Trade name	Uses	Side effects	Structure
amyl nitrate	[628-05-7]	$C_5H_{11}NO_2$	Inhalant	angina pectoris	tachycardia, CNS	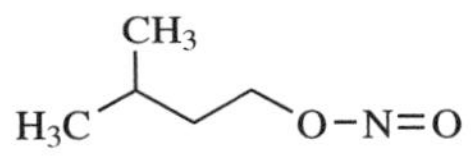
glyceryl trinitrate or nitroglycerin	[55-63-0]	$C_3H_5N_3O_9$	Nitrogard, Nitrolyn, Nitostat, Nitrol	angina pectoris, hypertension, acute MI	CNS	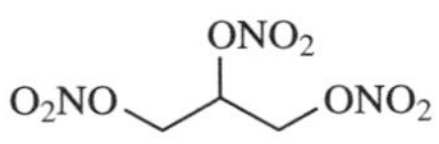
pentaerythritol tetranitrate	[78-11-5]	$C_5H_8N_4O_{12}$		prophylactic anginal attacks	CNS	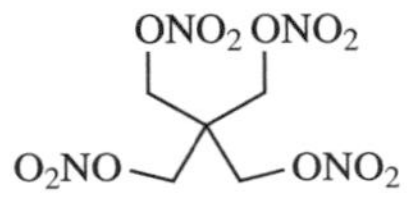
isosorbide dinitrate	[87-33-2]	$C_6H_8N_2O_8$	Sorbitrate, Isordil, Isordil Titradose	angina pectoris, congestive heart failure, dysphasia	reflex tachycardia, CNS	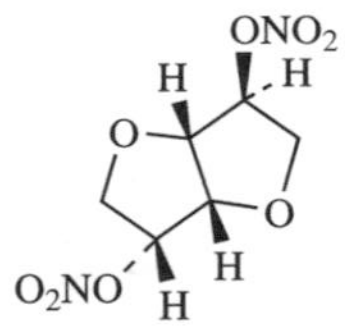
isosorbide mononitrate	[16051-77-7]	$C_6H_9NO_6$	Monoket, Ismo, Imdur, Isotrate ER	angina pectoris, congestive heart failure	CNS, GI intolerance	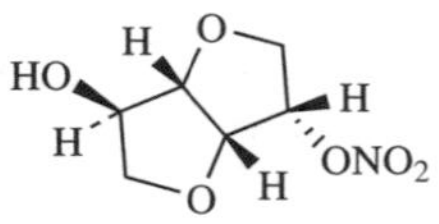

110

Name	CAS	Formula	Trade names	Uses
isoxsuprine	[579-56-6]	$C_{18}H_{23}NO_3$	Dilavase, Duvadilan, Isolait, Navilox, Suprilent, Vadosilan, Vasodilan, Vasoplex, Vasotran	peripheral vascular diseases, Burger's and Raynaud's diseases, arteriosclerosis obliterams / tachycardia, CNS 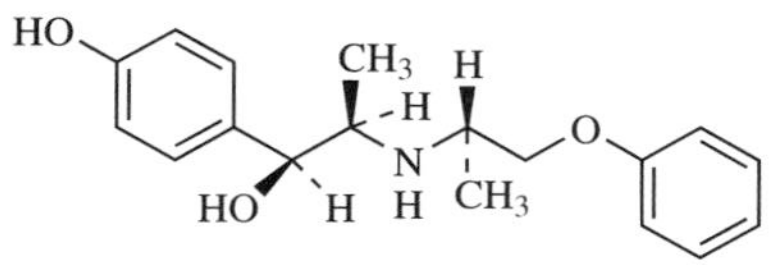
nicorandil	[65141-46-01]	$C_8H_9N_3O_4$		antianginal, hypotension 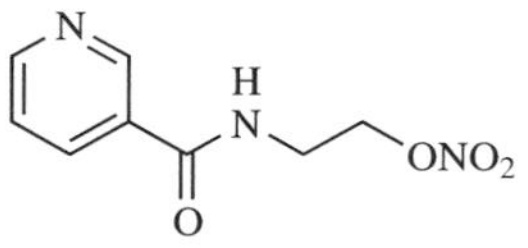
erythrityl tetranitrate	[7297-25-8]	$C_4H_6N_4O_{12}$	Tetranitrol, Tetranitrin, Cardilate, Cardiloid	coronary vasodilator 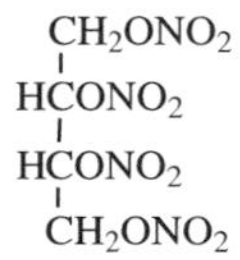

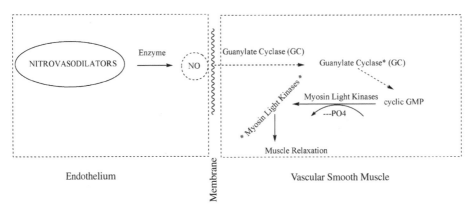

Fig. 6. Suggested mechanism of action of nitrate and nitrites used as vasodilators to generate NO, the most potent (endogenous) vasodilator that induces a cascade of reactions resulting in smooth muscle relaxation and vasodilation.

half-life of a few seconds. It is an endothelium derived relaxing factor (EDRF) that influences vascular tone. Nitric oxide induces vasodilation by stimulating soluble guanylate cyclase to produce cyclic GMP (cGMP) as shown in Figure 6. The latter eventually leads to dephosphorylation of the light chains of myosin (149). The resultant hemodynamic effect is the basis of dilation of epicardial coronary arteries, systemic resistance vessels and veins, always associated with a reduction in the coronary vascular resistance that contributes to the efficacy (150–154). Thus, their main action is peripheral vasodilation, either venous (low doses), or both venous and arterial (higher doses). As a result, pooling of blood in the veins reduces venous return and also the ventricular volume (preload). This reduction in enlargement of the heart wall decreases oxygen demand and the pain of angina is relieved quickly. However, it has been found that nitrates act only on the large coronary vessels to exert their effect. The lack of ability of minor vessels to convert nitrate to NO is the main reason for nitrates being ineffective in minor vessels (Fig. 6).

The use of nitrates leads to reflex activation of the sympathetic nervous system that increases the heart rate spontaneously and the myocardial contractility. Reduction in the ventricular wall tension decreases the myocardial oxygen consumption. At the same time, nitrates improve myocardial oxygen supply by increasing the coronary blood flow to the endocardium. Thus nitrates alter the imbalance of myocardial oxygen consumption and supply, which is the basis of angina pectoris. The main pharmacological effect of organic nitrates is relaxation of vascular smooth muscle that results into vasodilation. Organic nitrate provides an exogenous source of nitric oxide that augments the actions of EDRF, normally impaired with coronary artery diseases. It has been suggested that nitrates may be useful as antiplatelet and antithrombic agents in the management of intracoronary thrombi. Although the exact mechanism of action of nitrates on antiplatelet aggregation is unknown, it is postulated that activation of cGMP inhibits the calcium influx resulting in fibrinogen binding to glycoprotein IIb/IIIa receptors. Nitrates may also release prostacyclin [35121-78-9], $C_{20}H_{32}O_5$, from the endothelial lining of the vasculature and relax vascular smooth muscle. Prostacyclin is a potent endogenous vasodilator and inhibitor of platelet aggregation (155,156).

4.6. Vasodilating Agents. All vasodilators can be divided into three types depending on their pharmacological site of action. These include (1) cerebral; (2) coronary, and (3) peripheral vasodilators. However, in this article, we will focus our attention on various compounds currently used as coronary and peripheral vasodilators represented in Table 2. Some of their properties such as bioavailability, half-life, and some possible side effects are also discussed in the text.

Amyl Nitrate. It is a light sensitive aliphatic compound with an unpleasant odor, volatile and inflammable liquid, immiscible in water. A stabilizer such as diphenylamine or epoxolol is added to the commercial product. Amyl nitrate can be administered to patients with coronary artery disease by nasal inhalation for acute relief of angina pectoris. It has also been used to treat heart murmurs resulting from stenosis and aortic or mitral valve irregularities. Amyl nitrate acts within 30 s after administration and sustantial hemodynamic effects such as increased heart rate and decreased diastolic pressure occurs within 30 s and duration of action persists ~3–5 min. However, this drug has a number of adverse side effects such as tachycardia and headache.

Glyceryl Trinitrate (GTN). Also called nitoglycerin, it was introduced as a drug for the treatment of angina pectoris in 1879. It is a powerful explosive and the undiluted drug occurs as a volatile, white-to-pale yellow, thick flammable liquid with a sweet burning taste. It is slightly soluble in water and soluble in alcohol. Nitroglycerin is diluted with lactose, dextrose, alcohol, propylene glycol, or another inert excipient to permit safe handling. It is administered lingually, sublingually, intrabuccally, orally, topically, by IV infusion, or a transdermal route. However, ~40–80% of the dose is normally lost during IV administration due to the absorption by plastic material used to administer the dose. The transdermal nitroglycerin adhesive or band aid patches overcome problems of handling controlled dosing, ie, eliminate the peaks and valleys of plasma levels associated with conventional therapy, to achieve a prolonged duration of action. Glyceryl trinitrate is a short-acting trinitrate ester of glycerol, with duration of action of 30 min or so. The GTN is easily absorbed through skin and has a strong vasodilating effect. The plasma half-life of nitroglycerin is ~1–4 min and it is rapidly metabolized in the liver and other organs by glutathione organic nitrate reductase and in the blood of some species to pharmacologically inactive 1,3-glyceryl dinitrate, 1,2-glyceryl dinitrate, and glyceryl mononitrate. It is highly bound to plasma protein and has a large volume of distribution suggesting it is widely distributed in the body.

Glyceryl trinitrate is the only vasodilator drug known to stimulate the enhancement of coronary collateral circulation and capable of preventing myocardial infarction induced by coronary occlusion, and therefore is more widely used in preventing attacks of angina than in stopping them once they have instigated. Thus, nitroglycerin remains the drug of choice for treatment of angina pectoris. It has also been found useful for the treatment of congestive heart failure, myocardial infarction, peripheral vascular disease, such as Raynaud's disease, and mitral insufficiency, although the benefits of nitroglycerin in mitral insufficiency have been questioned. The principal side effects of nitroglycerin are headache, dizziness, nausea, vomiting, diarrhea, flushing, weakness, syncope, and tachycardia can result.

Pentaerythritol Tetanitrate (PETN).   PETN is a nitric acid ester of the tetrahydric alcohol namely pentaerythritol. Since PETN is a powerful explosive, it is normally mixed and diluted with other inert materials for safe handling purposes and to prevent accidental explosions. PETN is mainly used in the prophylactic management of angina to reduce the severity and frequency of attacks, since it has a similar mechanism of action, pharmacokinetic and toxic effects on vascular smooth muscle cells to induce vasodilation as other nitrates (156). It is available in regular formulation, however, PETNs duration of action can be prolonged by using a sustained release formulation.

Erythrityl Tetranitrate.   It is a racemic mixture of 1, 2, 3, 4-butanetetrol tetranitrate. It is a less potent antinaginal agent than nitroglycerin but has a more porlonged duration of action. It can be administerared orally or sublingually. The onset time is 5–10 min following sublingual administration and 20–30 min if given orally. The duration of action for the above two routes lasts ~3 and 6 h, respectively. Erythrityl is readily absorbed from the gastrointestinal (GI) tract and undergoes an extensive first-pass metabolism in the liver by glutathione organic nitrate reductase. Adverse effects are similar to those described for nitroglycerin (156).

Isosorbide Dinitrate.   It is a white crystalline rosettes soluble in water. Isosorbide dinitrate can be administrated by oral, sublingual, or intrabuccal routes and the approximate onset and duration of action depends on the administration route and various dosage forms. The approximate onset and duration of action of different dosage forms of isosorbide dinitrate varies, such as oral (1 and 5–6 h), extended release (30 min and 6–8 h), chewable (3 min to 0.5–2 h), and sublingual (within 3 min to 2 h).

Isosorbide dinitrate is metabolized to the corresponding mononitrates (2 and 5 mononitrate) within several minutes to hours depending on the route of administration. Isosorbide dinitrate is routinely used for the treatment and relief of acute angina pectoris as well as in the short and long-term prophylactic management of angina. It can also be used in combination with cardiac glycosides or diuretics for the possible treatment of congestive heart failure (157–159).

Isosorbide Mononitrate.   It is the major active metabolite of isosorbide dinitrate and occurs as a white, crystalline, odorless powder. Similar to dinitrate, mononitrate is freely soluble in water and alcohol. Mononitrate is available commercially as conventional tablets or extended release formulation such as capsules or controlled release coated pellets containing a suitable matrix. Some of the extended release formulations and tablets always should be stored in tight, light resistance containers at room temperature. Isosorbide mononitrate is readily absorbed from the GI tract and is principally metabolized in the liver. But unlike isosorbide dinitrate, it does not undergo first pass hepatic metabolism and therefore the bioavailability of isosorbide mononitrate in conventional or extended release tablets is very high (100–80%, respectively). About 50% dose of isosorbide mononitrate undergoes denitration to form isosorbide, followed by partial dehydration to form sorbitol. Mononitrate also undergoes glucuronidation to form 5-mononitrate glucuronide. None of these metabolites have any apparent pharmacological activity.

Similar to isosorbide dinitrate, mononitrate is used for the acute relief of angina pectoris, for prophylactic management in situations likely to provoke angina attacks, and also for long term management of angina pectoris (160,161).

Isoxsuprine Hydrochloride. This vasodilator is structurally related to nylidrin and occurs as a white crystalline powder, sparingly soluble in water (162). It causes vasodilation by direct relaxation of vascular smooth muscle cells. It acts by decreasing the peripheral resistance, and at high doses is even known to reduce the blood pressure. It also stimulates β-adrenergic receptors. It is used as an adjunct therapy in the management of peripheral vascular diseases such as Burger's disease, Raynaud's disease, arteriosclerosis obliterans, and for the relief of cerebrovascular insufficiency (163–166).

Nicorandil. It is a nicotinamide analogue possessing a nitrate moiety. It is known to exhibit a dual mechanism of action as both nitrovasodilator and potassium channel activator (167,168). It is a balanced arterial and venous dilator and also offers cardioprotection. Nicorandil is used as an antianginal agent, known to improve the myocardial blood flow resulting in decreased systemic vascular resistance and blood pressure, pulmonary capillary wedge and left ventricular end-diastolic pressures (169,170). It is relatively well tolerated when used orally or intravenously in patients with stable angina. However, the use of nicorandil in patients undergoing cardiopulmonary bypass surgery needs further evaluation since severe vasodilation and hypotension requiring significant vasoconstrictor support was observed (171).

4.7. Pharmacokinetics and Tolerance of Organic Nitrates. All organic nitrates exhibit similar pharmacological effects. The foremost factor contributing to the pharmacokinetics of glycerol trinitrates (GTN) and other longer acting organic nitrates is the existence of high capacity hepatic nitrate reductase in the liver that eliminates the nitrate groups in a stepwise process. But in serum, nitrates are metabolized independent of glutathione (151,172,173). In general, the organic nitrates are well absorbed from the oral mucosa following administration lingually, sublingually, intrabuccally, or as chewable tablets. The organic nitrates are also well absorbed from the GI tract and then undergo the first-pass metabolism in the liver. Nitroglycerin is well absorbed through the skin if applied topically as an ointment or transdermal system. Orally administered nitrates and topical nitroglycerin are relatively long acting. However, rapid development of tolerance to the hemodynamic and antianginal effects of these dosage forms is known to occur with continuous therapy. Therefore, an approximately 8 h/day nitrate-free period is needed to prevent tolerance. Slow release transdermal patches of GTN are the most favored form of achieving prolonged nitrate levels. Highly lipophilic nitrates, following intravenous infusion, are widely distributed in to vascular and peripheral tissues, while less lipophilic nitrates are not as widely distributed. At plasma concentrations of 50–500 ng/mL, nitrates are ~30–60% bound to plasma proteins.

4.8. Side Effects. The principal side effects of nitrates include dilation of cranial vessels causing headaches, which can limit the dose used. More serious side effects are tachycardia and hypotension resulting in corresponding increase in myocardial oxygen demand and decreased coronary perfusion, both of which have an adverse effect on myocardial oxygen balance. Another well-documented problem is the development of tolerance to nitrates. Blood vessels become hypo- or nonreactive to the drugs, particularly if large doses, frequent dosing regimes and long-acting formulations are used. To avoid this, nitrates are best used intermittently, allowing a few hours without treatment in each 24-h period.

4.9. Calcium Channel Blockers. Verapamil was the first calcium channel blocker (CCB). It has been used since 1962 in Europe, and then in Japan, for its antiarrhythmic and coronary vasodilatory effects. The CCBs have become prominent cardiovascular drugs during the last 40 years and are widely used in the treatment of various types of angina, hypertension, certain arrhythmias, heart failure, acute myocardial infarction, cardioprotection, cerebral vasospasm, and cardiomyopathy as well. Many experimental and clinical studies have defined their mechanism of action, the effects of new drugs in this therapeutic class, and their indications and interactions with other drugs.

Calcium plays a significant role in the excitation–contraction coupling processes of the heart and vascular smooth muscle cells as well as the conduction cells of the heart and failure to maintain intracellular calcium homeostatis results in cell death. The membranes of conduction cells contain a network of numerous inward channels that are selective for calcium and activation of these channels leads to the plateau phase of the action potential of cardiac muscle cells. Please refer to the earlier section for a detailed discussion on calcium channels, mechanism of action, and their role in cardiovascular diseases.

4.10. Applications. Calcium channel blocking agents are the first drugs of choice for the management of Prinzmetal angina. It has been suggested that extended release or intermediate-long acting calcium channel blocking agents may be useful in the management of hypertension in patients with diabetes mellitus due to their fewer adverse side effects on glucose homeostasis, lipid, and renal function. However, the data from limited clinical studies indicates that patients with impaired glucose metabolism receiving calcium channel blockers are at higher risks of nonfatal MI and other adverse cardiovascular events than those receiving ACE inhibitor or β-adrenergic agents (174).

A number of recent reviews are also available that describe the utility of Ca^{2+} channel blockers in the treatment of hypertension (175). Some of the new drugs have greater selectivity since they can be used to treat hypertension in the presence of concomitant diseases, such as angina pectoris, hyperlipidemia, diabetes mellitus, or congestive heart failure. Sometimes reflex tachycardia and vasodilator headache are the major side effects that limit the use of these agents as antihypertensives (175).

The calcium channel blockers can be divided into four different classes of compounds based upon their pharmacophore and chemical structure. These include (1) arylalkylamines; (2) benzothiazepines; (3) 1–4 dihydropyridines; and (4) Mibefradil, which has been assigned its own class. These drugs have wide applications in cardiovascular therapy due to their effects such as (a) arterial vasodilation resulting in reduced afterload; (b) slowing of impulse generation and conductance in nodal tissue; (c) reduction in cardiac work and sometimes myocardial contractility, ie, negative inotropic effect so as to improve myocardial oxygen balance. We will discuss in detail each of the above class of compounds and their pharmacological action.

Arylalkylamines and Benzothiazepines. These drugs vary in their relative cardiovascular effects and clinical doses, but they have the most pronounced direct cardiac effects (eg, verapamil) (Table 3). Various drugs that are currently in use are discussed next.

Bepridil Hydrochloride.   It is a nondihydropyridine calcium channel blocking agent with antianginal and antiarrhythmic properties. It inhibits calcium ion influx across L-type (slow, low voltage) calcium channels (176). However, unlike other agents, it also inhibits calcium ion influx across receptor operated channels, and inhibits intracellular calmodulin-dependent processes by hindering the release of calcium and sodium influx across fast sodium channels. Thus, it exhibits both calcium and sodium channel blocking activity, and also possesses some electrophysiological properties of class I antiarrhythmic agents by prolonging QT and QTc intervals (177). Although the precise mechanism of action remains to be fully determined, it reduces in a dose-dependent manner heart rate and arterial pressure at rest by dilating peripheral arterioles and reducing total peripheral resistance. This leads to a modest decrease in (<5 mmHg) systolic and diastolic blood pressure and larger decreases in hypertensive patients. If given intravenously, it also reduces left ventricular contractility and increases filling pressure.

Although, bepridil hydrochloride is usually administered orally for the treatment of chronic stable angina, it is not the first choice of drugs due to its arrhythmogenic potential and associated agranulocytosis and is given only if a patient fails to respond to other antianginal agents (178,179). When used alone or in combination with other antianginal agents, it is as effective as β-adrenergic blocking drugs or other dihydropyridine calcium channel blockers. However, bepridil can aggravate existing arrhythmias or induce new arrhythmias to the extent of potentially severe and fatal ventricular tachyarrhythmias, related to increase in QT and QTc interval (180). Bepridil is rapidly and completely absorbed after oral administration and 99% bound to plasma proteins. It has a half-life of 26–64 h. Bepridil is almost completely metabolized in the liver, resulting in 17 various metabolites, including pharmacologically most effective 4-hydroxy-N-phenylbepridil.

Diltiazem Hydrochloride.   Like bepridil, it is also a non-dihydropyridine calcium channel blocker but belongs to a benzothiazepine family of compounds (181,182). The (+)-cis form of the compound is the pharmacologically active isomer. It is a light sensitive crystalline powder, soluble in water and formulated as either a hydrochloride or malate salt. Diltiazem shows pharmacologic actions similar to other calcium channel blockers, ie, by inhibiting the transmembrane influx of extracellular calcium ions across the myocardial cell membrane and vascular smooth muscle cells (183,184). However, unlike dihydropyridine calcium channel blockers, diltiazem exhibits inhibitory effects on the cardiac conduction system mainly at the AV node and minor effects on sinus node. The frequency dependent effect of diltiazem on AV nodal conduction selectively decreases the heart ventricular rate during tachyarrhythmias involving AV node. However, in patients with SA node dysfunction, it decreases the heart rate and prolongs sinus cycle length resulting in a sinus arrest. Diltiazem shows no or little effect on QT interval.

Diltiazem is administered as the hydrochloride salt orally as tablets or extended release capsules for the treatment of Printzmetal angina and chronic stable angina and hypertension. The intravenous infusion is the most preferred

Table 3. Antianginal Agents: Arylalkylamines and Benzothiazepins (Calcium Channel Blockers)

Chemical/ generic name	CAS Registry Number	Molecular formula	Trade name	Uses	Structure
bepridil hydrochloride	[74764-40-2]	$C_{24}H_{35}ClN_2O$	Angopril, Bepadin, Cordium, Vascor	antianginal	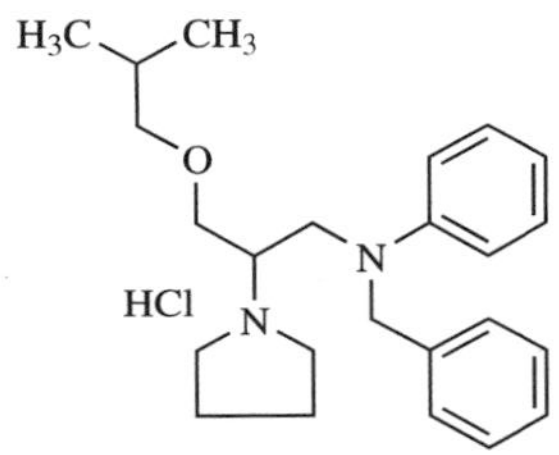
diltiazem hydrochloride	[33286-22-5]	$C_{22}H_{27}ClN_2O_4S$	Adizem, Anginyl, Angizem, Bruzem, Calcicard, Cardizem, Citizem, Deltazen, Diladel, Dilzem, Masdil, Tildiem	antianginal, antihypertension, antiarrhythmic	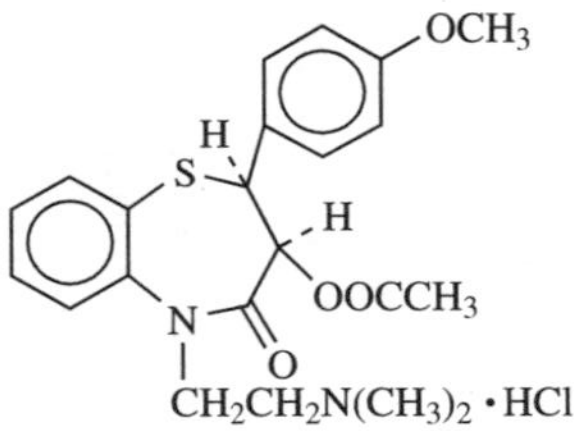
clentiazem	[96125-53-0]	$C_{22}H_{25}ClN_2O_4S$	Logna	antihypertension	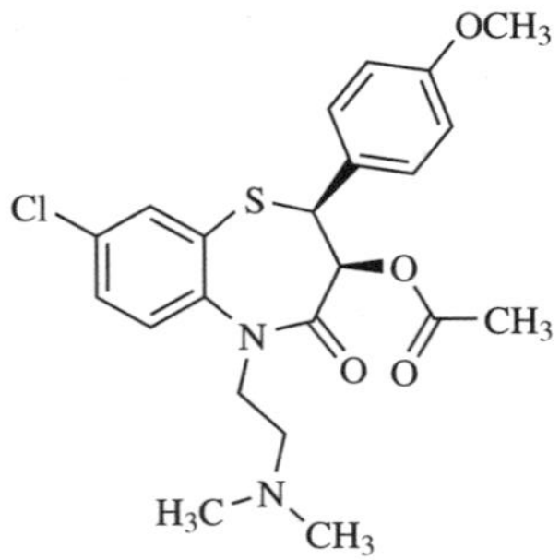

verapamil hydrochloride	[152-11-4]	$C_{27}H_{38}N_2O_4$	Apramyl, Calan, Cordilox, Dignover, Isoptin, Securon, Univer, Vasolan, Veraptin, Verelan, Verexamil, Veramix	antihypertension, antianginal, antiarrhythmic	
gallopamil	[16662-46-7]	$C_{28}H_{40}N_2O_5$	Algocor, Procorum	antianginal	
mibefradil	[116666-63-8]	$C_{29}H_{38}FN_3O_3$	Posicor	antihypertension	
fendiline	[13636-18-5]	$C_{23}H_{25}N \cdot HCl$	Cordan, Fendilar, Sensit	coronary vasodilator	

Table 3 (*Continued*)

Chemical/ generic name	CAS Registry Number	Molecular formula	Trade name	Uses	Structure
prenylamine	[390-64-7]	$C_{24}H_{27}N$	Angormin, Crepasin, Reocorin, Roinin, Sedolatan, Synadrin	coronary vasodilator	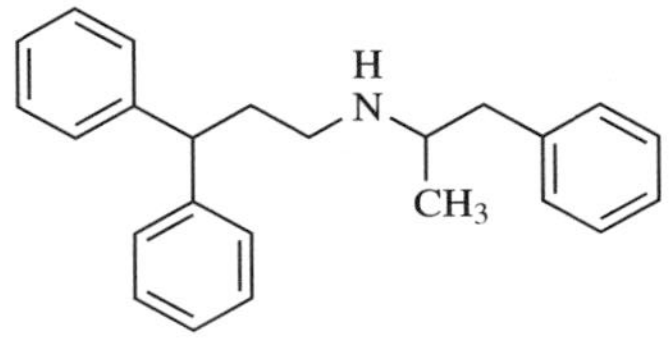
terodiline	[7082-21-5]	$C_{20}H_{27}N$	Bicor, Mictrol, Micturin, micturol	antianginal	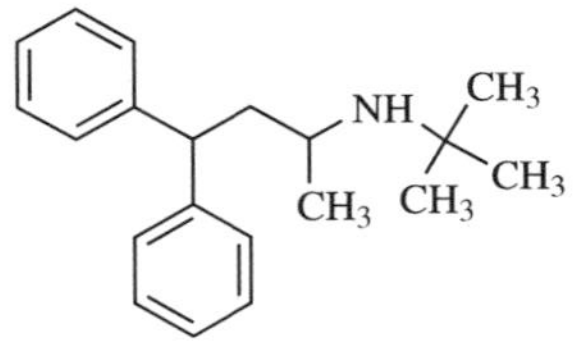

formulation for the treatment of supraventricular tachyarrhythmias. A controlled study also indicates that simultaneous use of diltiazem and a β-adrenergic blocking agent in patients with chronic stable angina reduces the frequency of attacks and increases an exercise tolerance (185). The absorption of diltiazem from the GI tract after po dosing is nearly complete. It undergoes extensive first-pass metabolism in the liver and only 40% of the po dose is bioavailable. About 70–80% of the drug is bound to plasma protein. The onset of action is ~15 min and the peak effect occurs at 30 min. Diltiazem is metabolized by deacetylation, and *N*- and *O*-demethylation. The principal metabolite, desacetyldiltiazem has 25–50% of the pharmacological activity and 10–20% of the plasma levels of the parent compound. About 30–35% of the drug is excreted by the kidneys as metabolites and 2–4% as unchanged drug. Almost 60% is excreted in the feces as metabolites. The elimination half-life of diltiazem is 3–4.5 h. Elimination half-life and plasma levels are increased in patients with liver disease (154,155,186).

The side effects of diltiazem therapy are less than those of verapamil or nifedipine therapy and occur in ~4% of the patients (154,155,186).

Clentiazem. It is a chlorinated derivative of diltiazem and is currently undergoing clinical evaluation for the treatment of angina pectoris and hypertension. The primary mechanism of clentiazem responsible for the antihypertensive effects seems to be reduction in the peripheral arterial resistance due to calcium channel blockade (187,188).

Verapamil Hydrochloride. Like diltiazem, verapamil is also a nondihydropyridine calcium channel blocker. It is available as a racemic mixture and occurs as a crystalline powder, soluble in water. The L-isomer of verapamil, which is 2–3 times more active than the corresponding D-isomer for its pharmacodynamic response on A-V conduction, has been shown to inhibit the ATP-dependent calcium transport mechanism of the sarcolemma (189). It acts by a similar pharmacological mechanism of action to other calcium channel blocking agents by reducing afterload and myocardial contractility. However, verapamil also exerts negative dromotropic effects on the AV nodal conduction and is classified as a class IV antiarrhythmic agent (190). These effects of verapamil on nodal impulse generation and conduction are useful in treating certain types of arrhythmias and its effects on myocardial contractility can be a problem in patients with heart failure. Therefore, verapamil is used in the treatment and prevention of supraventricular tachyaarhythmias and in hypentensive patients, not affected by cardiodepressent effects (191).

It is also used orally in the treatment of Prinzmetal angina or chronic stable angina and is as effective as any other β-adrenergic blocking agent or calcium channel blocker. Intravenous verapamil is the first choice of drugs used in the management of supraventricular tachyarrhythmia, including rapid conversion to sinus rhythm of PSVT (those associated with Wolf-Parkinson-White or Lown-Ganong-Levine syndrome) and temporary relief of atrial fibrillation. It is also used as a monotherapy or in combination with other antihypertensive agents for the treatment of hypertension.

Gallopamil. Gallopamil is a more potent methoxy analogue of verapamil and the drug has demonstrated efficacy in effort and rest angina, hypertension, and supraventricular tachycardia (192–195). However, intracoronary administration of gallopamil may be useful in treating myocardial ischemia during percutaneous transluminal coronary angioplasty (196).

Intrarenal gallopamil has shortened the course of acute renal failure. It has been suggested that the role of inhaled gallopamil in asthma remains to be defined, and well-controlled potential comparisons with verapamil are needed to define the place in therapy of gallopamil for all indications.

Mibefradil. This compound has been assigned its own class and it is a T- and L-type CCB, primarily approved in 1997 by U.S. Food and Drug adminstration (FDA) for management of hypertension and chronic stable angina (197–200). However, postmarketing surveillance discovered potential severe life-threatening drug–drug interactions between mibefradil and β-blockers, digoxin, verapamil, and diltiazem, especially in elderly patients, resulting in one death and three cases of cardiogenic shock with intensive support of heart rate and blood pressure. Therefore, manufacturer voluntarily withdrew mibefradil from the U.S. market in 1998 (201).

Fendiline. Fendiline is used in the long-term treatment of coronary heart disease. Fendiline is a coronary vasodilator and clinical studies have established at least equal therapeutic efficacy in the treatment of angina pectoris as isosorbide dinitrate or diltiazem (202–206). Recently, the action of fendiline on cardiac electrical activity has also been investigated in guinea pig papillary muscle, suggesting that a frequency and concentration dependent block of Na^+ and L-type Ca^{2+} channels occurs in presence of fendiline, leading to inhibition of fast and slow conduction and inactivation of Ca^{2+} channels (207).

Further studies show that fendiline also induces an increase in Ca^{2+} concentration in Chang liver cells by releasing stored Ca^{2+} in an inositol 1,4,5-triphosphate independent manner and by causing extracellular Ca^{2+} influx (208).

Prenylamine. Prenylamine is a homologue of fendiline and normally is used in the treatment of chronic coronary insufficiency and prophylaxis of anginal paroxysms. The latter is recognized by a disturbance in brain blood circulation and sometimes hypertension, but prenylamine is not sufficiently effective in very acute anginal paroxysms (209). Since it is a coronary vasodilator, it acts as a calcium antagonist, but without any substantial effect on the contractility of the myocardium. However, it improves the vascular blood circulation due to vasodilation and thereby oxygen supply of the myocardium. It also decreases the amount of norepinephrine and serotonin in the myocardium and brain and therefore possesses a slight blocking effect on β-adrenergic receptors. Since it enhances the antihypertensive effect of beta blockers, the dose needs to be controlled. But if given in high doses and in cases of tachycardia, it could lead to the deceleration of cardiac related physical activity (210).

Terodiline. It is an alkyl analogue of fendiline and is used as a calcium channel antagonist. However, it also possesses anticholinergic and vasodilator activity (211). When administered twice daily, terodiline has been demonstrated effective in the treatment of urinary urge incontinence (211). Comparative studies with other agents used in urge incontinence are required to determine if the dual mechanism of action and superior absorption of terodiline offer clinical advantages. Safety concerns about ventricular arrhythmias have suspended general clinical investigations.

Dihydropyridine Derivatives. This is an important class of compounds that are widely used as vasodilators. In general, 1,4-Dihydropyridines demonstrate slight selectivity toward vascular versus myocardial cells, and therefore have greater vasodilatory effect than other calcium channel blockers. 1,4-Dyhydropyridines are also known to possess insignificant electrophysiological and negative inotropic effects compared to verapamil or diltiazem. The dihydropyridines have no significant direct effects on the heart, although they may cause reflex tachycardia. Structures and properties of some of the first and second generation dihydropyridines are given in Table 4. Most of the newer drugs have longer elimination half-lives but also higher rates of hepatic clearance and hence low bioavailability. The only exception is amlodidpine, which has a much higher bioavailability (60%) and a long elimination half-life. Several metabolic pathways of DHP-type calcium channel blockers have been identified in humans. However, the most important metabolic pathway appears to be the oxidation of 1,4-Dihydropyridine ring into pyridine, catalyzed by the cytochrome P450 (CYP) 3A4 isoform and the oxidative cleavage of carboxylic acid (212). Calcium antagonists are known to block calcium influx through the voltage-operated calcium channels into smooth muscle cells. Some of the compounds of 1,4-DHP category such as Nifedipine, Nisoldipine, or Isradipine have been demonstrated to be useful in the management of coronary artery diseases. Nevertheless these already available calcium antagonists have some major disadvantages: they are photosensitive and decompose rapidly, they are not soluble in water, and because of their depressive effects on myocardium they have negative inotropic effects. The CCBs account for almost $4 billion in sales and dihydropyridines like lercanidipine are the fastest growing class of CCB. There are 13 derivatives of DHP calcium channel blockers currently licensed for the treatment of hypertension and widely used. Some of the most prescribed drugs include amlodipine, felodipine, isradipine, lacidipine, lercanidipine, nicardipine, nifedipine, and nisoldipine. Currently, thiazide diuretics or β-blockers are recommended as first line therapy for hypertension. Calcium channel blockers, ACE inhibitors or α-adrenergic blockers may be considered when the first line therapy is not tolerated, contraindicated or ineffective.

Amlodipine Besylate. This compound belongs to the 1,4-dihydropyridine family of compounds possessing structural resemblance to nifedipine, felodipine, nimodipine, and others. It is a light sensitive, selective, and potent calcium channel blocking agent with a long duration of action. Amlodipine has a long half-life of >33 h after intravenous dosing and its bioavailability with po doses is 52–88% (213). It is mainly used orally, either alone or in combination with other antihypertensive agents to treat hypertension and Prinzmetal and chronic stable angina, along with other antianginal agents (214,215).

Aranidipine. It is a 1,4-dihydropyridine calcium channel blocker with vasodilating and antihypertensive actions, and therefore used for the treatment of hypertension (216–218). It is used either alone or in combination with a diuretic or β-blocker, for the once-daily treatment of mild-to-moderate essential hypertension. The drug is under investigation for the treatment of angina pectoris, but available data are limited to preclinical animal studies. It decreases T- and L-type calcium currents in a concentration-dependent manner. The duration of aranidipine's antihypertensive effect is longer than that of nifedipine and

Table 4. Antianginal Agents: 1,4-Dihydropyridine Calcium Channel Blockers

Chemical/ generic name	CAS Registry Number	Molecular formula	Trade name[a]	Structure
amlodipine besylate	[111470-99-6]	$C_{26}H_{31}ClN_2O_8$	Antacal, Istin, Monopina, Norvasc	
amlodipine maleate	[88150-47-4]	$C_{24}H_{29}ClN_2O_9$	Lotrel	
aranidipine	[86780-90-7]	$C_{19}H_{20}N_2O_7$	NA	
barnidipine	[104713-75-9]	$C_{27}H_{29}N_3O_6$	NA	

benidipine [105979-17-7] $C_{28}H_{31}N_3O_6$ Coniel

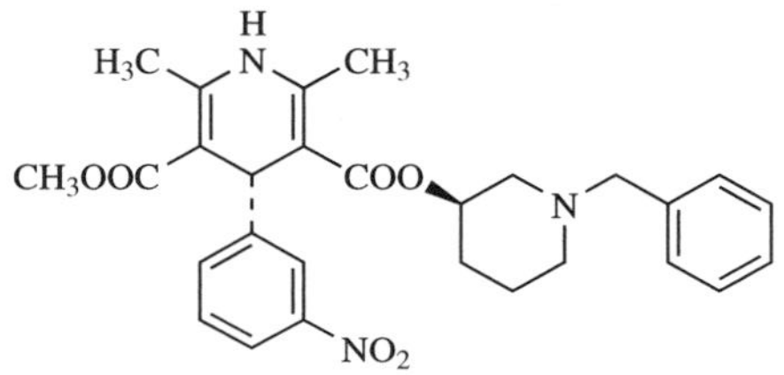

cilnidipine [132203-70-4] $C_{27}H_{28}N_2O_7$ NA

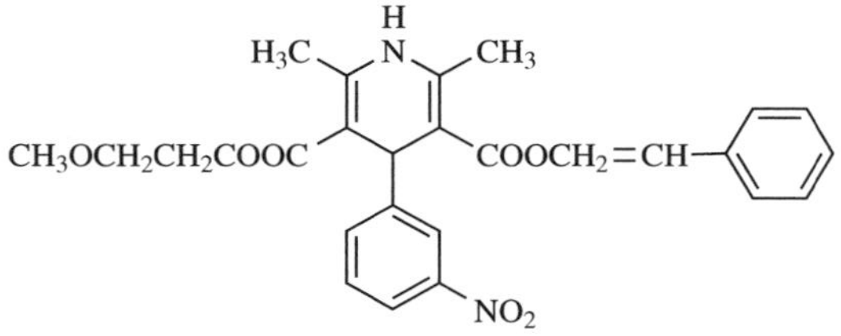

efonidipine [111011-63-3] $C_{34}H_{38}N_3O_7P$ NA

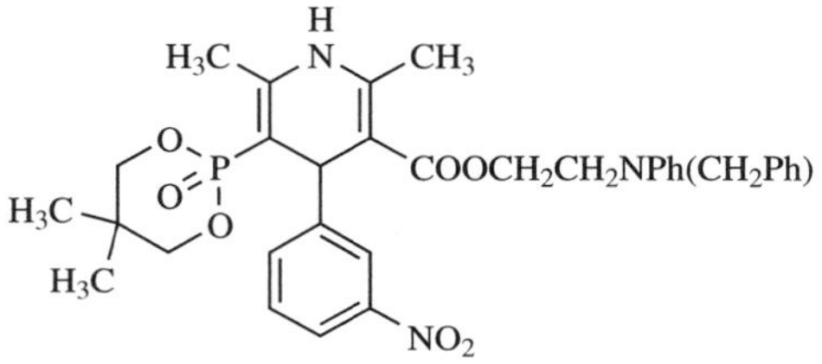

Table 4 (*Continued*)

Chemical/ generic name	CAS Registry Number	Molecular formula	Trade name[a]	Structure
elgonidipine	[119413-55-7]	$C_{29}H_{33}FN_2O_6$	NA	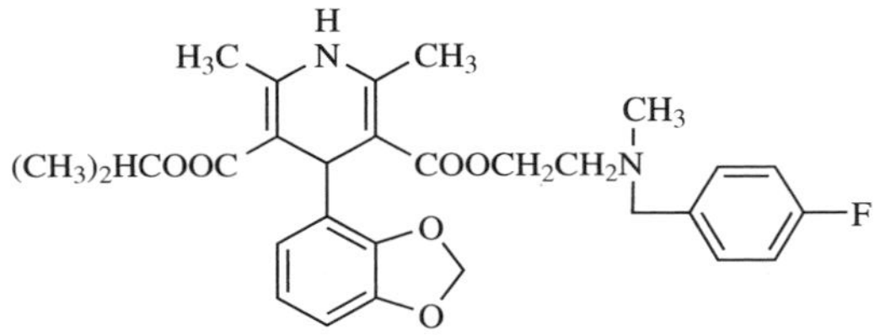
felodipine	[72509-76-3]	$C_{18}H_{19}Cl_2NO_4$	Agon, Hydac, Plendil, Lexxel	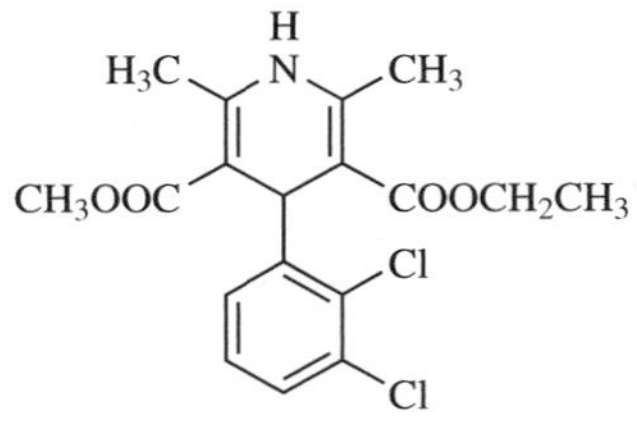
isradipine	[75695-93-1]	$C_{19}H_{21}N_3O_5$	Esradin, Lomir, Prescal	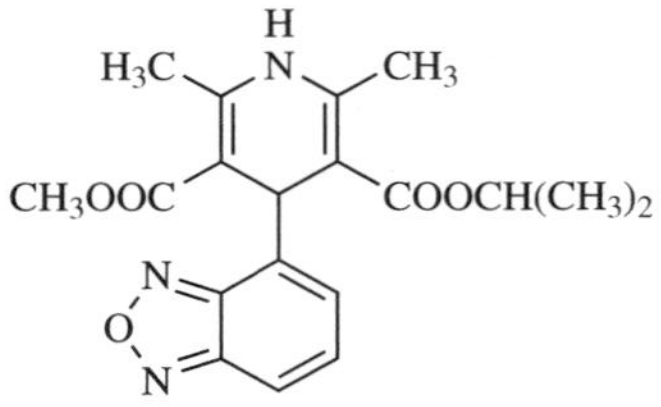

lacidipine

[103890-78-4]

$C_{26}H_{33}NO_6$

Caldine, Motens,
Lacipil, Lacirex

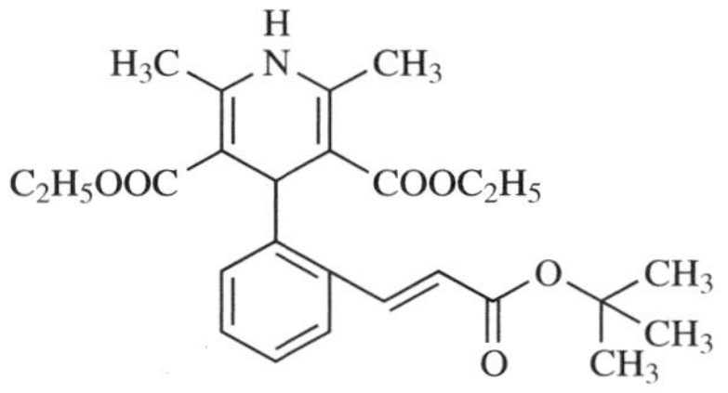

lercanidipine

[100427-26-7]

$C_{36}H_{41}N_3O_6$

Lerdip, Zanidip

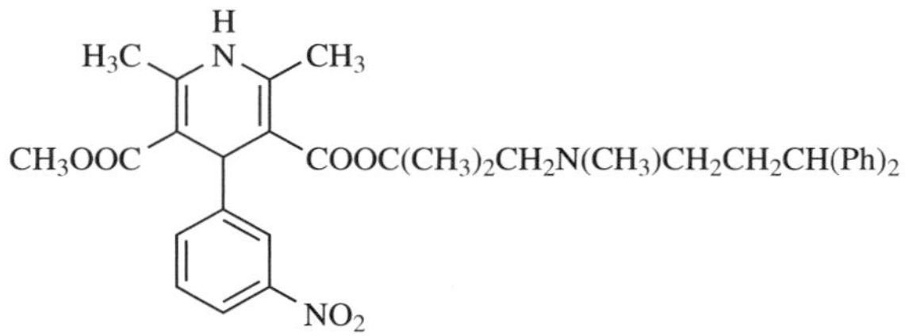

manidipine

[89226-50-6]

$C_{35}H_{38}N_4O_6$

Calslot

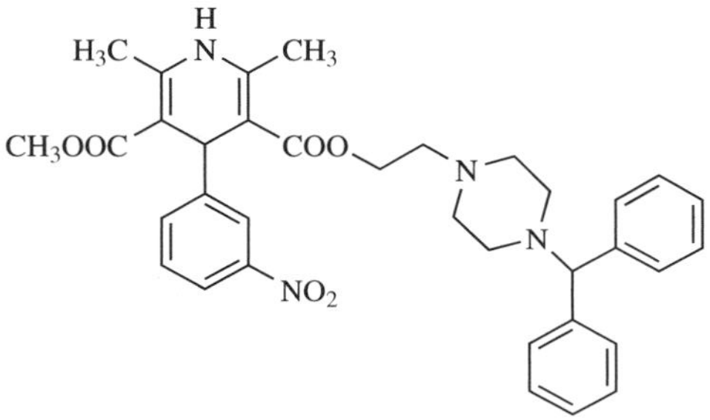

Table 4 (*Continued*)

Chemical/ generic name	CAS Registry Number	Molecular formula	Trade name[a]	Structure
nicardipine	[54527-84-3]	$C_{26}H_{29}N_3O_6$	Barizin, Cardene, Lecibral, Loxen, Nicant, Nicodel, Nimicor, Ranvil, Rydene	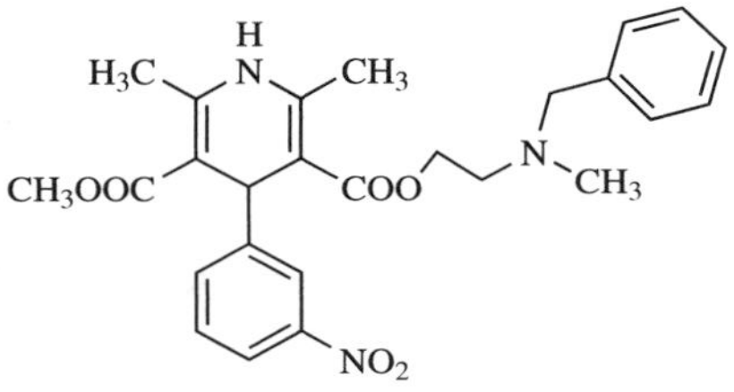
nifedipine	[21829-25-4]	$C_{17}H_{18}N_2O_6$	Adapress, Alfadat, Aprical, Citilat, Introcor, Nifedicor, Tibricol	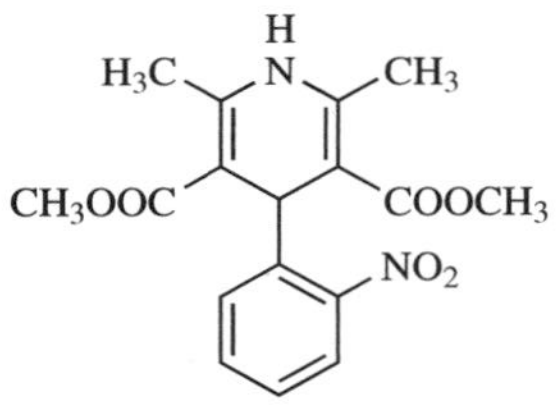
nilvadipine	[75530-68-6]	$C_{19}H_{19}N_3O_6$	Escor, Nivadil	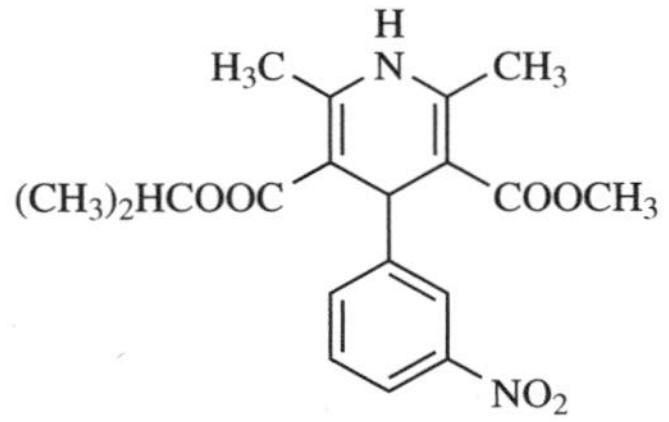

nimodipine	[66085-59-4]	$C_{21}H_{26}N_2O_7$	Admon, Nimotop, Periplum	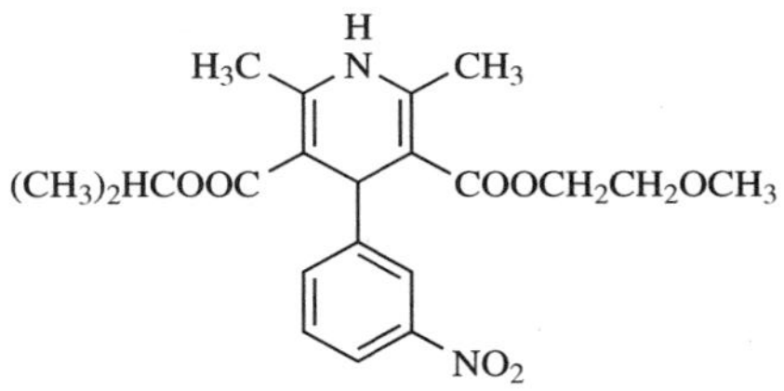
nisoldipine	[63675-72-9]	$C_{20}H_{24}N_2O_6$	Norvasc, Syscor, Zadipina	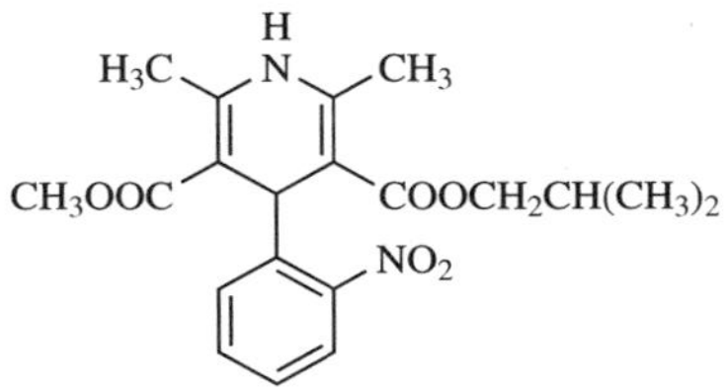
nitrendipine	[39562-70-4]	$C_{18}H_{20}N_2O_6$	Baypress, Bylotensin, Nidrel	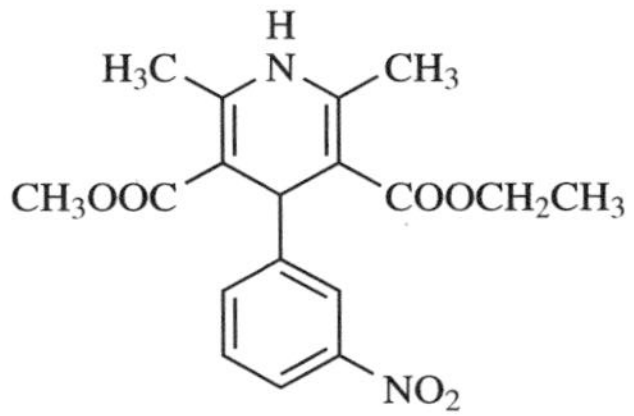

H
H3C
N
CH3
(CH3)2HCOOC
COOCH2CH2OCH3
NO2
H
H3C
N
CH3
CH3OOC
COOCH2CH(CH3)2
NO2
H
H3C
N
CH3
CH3OOC
COOCH2CH3
NO2

[a]NA = not applicable.

nicardipine. It does not significantly affect heart rate, cardiac output, or stroke volume index at rest or after exercising in patients with mild-to-moderate hypertension. However, it significantly increases left ventricular fractional shortening (FS) and left ventricular ejection fraction (EF) at rest. It does not adversely affect hemodynamics or lipoprotein or carbohydrate metabolism and the pharmacokinetics of aranidipine are not altered in the elderly or in patients with renal failure (216–218).

Barnidipine. The long-acting calcium antagonist barnidipine was launched in Japan in September 1992 under the brand name Hypoca(R), and sales of this drug are increasing steadily. Barnidipine is used in Europe under the brand name Vasexten(R). A long-acting calcium antagonist requires administration only once a day for the treatment of angina and hypertension (219–221). It is available as a modified release formulation with a gradual and long duration of action. It is a selective calcium channel antagonist leading to the reduction of peripheral vascular resistance, secondary to its vasodilatory action (222). Recently, it was suggested that Barnidipine administration for a week decreased the blood pressure and made the sodium balance negative by increasing the urinary sodium excretion in patients with essential hypertension. The natriuretic effect of this drug could contribute at least in part to its antihypertensive effect (223). Also, the possible use of barnidipine for protective effects on cerebrovascular lesions in salt-loaded stroke-prone spontaneously hypertensive rats was evaluated by magnetic resonance imaging (mri) (224).

Benidipine. This is a new, potent dihydropyridine and long-lasting calcium antagonist (225). The administration of benidipine once daily effectively decreases blood pressure and attenuates blood pressure response to mental stress. Reflex tachycardia, deterioration of diurnal blood pressure change, and excessive lowering of nighttime blood pressure were not observed after benidipine administration. Therefore, it has been suggested that it may be useful for the treatment of elderly hypertensive patients with cardiovascular disease. It has also been suggested that it also might be useful as an antianginal medication, however, no clinical data is yet available (226).

Cilnidipine. It is a unique calcium antagonist that has both L- and N-type voltage-dependent calcium channel blocking action (227–231). Cilnidipine is under investigation for the treatment of hypertension in Europe.

Recently, cilnidipine, its analogues and other dihydropyridine derivatives were evaluated for their state-dependent inhibition of L-type of Ca^{2+} channels, demonstrating that structurally related DHPs act in distinct ways to inhibit the L-type channel in the resting, open, and inactivated states. Cilnidipine and some related DHPs probably exert their blocking action on the open channel by binding to a receptor distinct from the known DHP-binding site (232). Further effects of cilnidipine on left ventricular (LV) diastolic function in hypertensive patients, as assessed by pulsed doppler echocardiography and pulsed tissue Doppler imaging, has been examined, and suggests that changes in LV diastolic performance in patients with essential hypertension following cilnidipine treatment were biphasic with an initial increase in early diastolic transmitral flow velocity and a later increase in early diastolic LV wall motion velocity. The initial and later changes can be related to an acute change in afterload and a later improvement in LV relaxation (233).

Efonidipine. This is a new long acting dihydropyridine calcium channel blocker derivative used in the treatment of hypertension (234–237). When the use of efonidipine was studied in open-chest anesthetized dogs on Endothelin-1 (ET-1), it was concluded that efonidipine attenuates the ET-1-induced coronary vasoconstriction, and therefore the drug would be useful for some patients with variant angina, in which ET-1 is involved in the genesis of coronary vasoconstriction (238).

Recently, to gain insight into the renoprotective mechanism of efonidipine hydrochloride, the acute effects of efonidipine on proteinuria, glomerular haemodynamics and the tubuloglomerular feedback (TGF) mechanism in anaesthetized 24–25-week-old spontaneously hypertensive rats (SHR) with glomerular injury were evaluated. The results indicate that efonidipine attenuates the TGF response in SHR by dilating the afferent arteriole, thus maintaining the level of renal plasma flow (RPF) and glomerular filtration rate (GFR) despite reduced renal perfusion pressure (239).

Elgodipine. It is a novel type of DHP, very selective, and potent coronary vasodilator calcium channel blocker (240). Elgodipine is very stable to light (2% degradation after 1 year of exposure to room light and temperature) in comparison with other currently available compounds that decompose in 24 h and is water soluble. It is very selective against vascular smooth muscle, in particular coronary vessels. Since elgodipine is >100-folds selective versus vessels than cardiac fibers, it has few negative inotropic effects. Elgodipine seems to be potentially useful as coronary vasodilator during PTCA (percutaneous transluminal coronary angioplasty) because its stability and solubility allows intracoronary administration, in patients with stable angina and because the lack of negative inotropic effects, in patients with moderate heart failure (241–243).

Some of the preliminary electrophysiological data in volunteers have shown that elgodipine differs from other calcium channel blockers in its effects on atria-ventricular conduction.

On the other hand, chemical stability of elgodipine allows its incorporation into suitable polymeric matrices for transdermal administration. Preliminary *in vitro* and *in vivo* data in volunteers have shown that elgodipine penetrates into the skin. Studies are in progress to determine the daily effective dose and therefore, the feasibility of transdermal patches (244).

Felodipine. It is a member of DHP calcium channel blocker family. It is insoluble in water and is freely soluble in dichloromethane and ethanol. It exists as a racemic mixture. Felodipine is used to treat high blood pressure, Raynaud's syndrome, and congestive heart failure (245–246). It reversibly competes with nitrendipine and/or other CCBs for dihydropyridine binding sites, and blocks voltage-dependent Ca^{2+} currents in vascular smooth muscle more than in cardiac muscle.

Following oral administration, felodipine is almost completely absorbed and undergoes extensive first-pass metabolism. However, following intravenous administration, the plasma concentration of felodipine declines triexponentially with mean disposition half-lives of 4.8 min, 1.5, and 9.1 h. Following oral administration of the immediate-release formulation, the plasma level of felodipine also declines polyexponentially with a mean terminal half-life of 11–16 h.

The bioavailability of felodipine is influenced by the presence of high fat or carbohydrate food and increases approximately twofold when taken with grapefruit juice. A similar finding has been seen with other dihydropyridine calcium antagonists, but to a lesser extent than that seen with felodipine (247–249).

Felodipine produces dose-related decreases in systolic and diastolic blood pressure that correlates with the plasma concentration of felodipine. Felodipine can lead to increased excretion of potassium, magnesium, and calcium (250). It has been recommended that in order to prevent side effects of the drug, individuals who are taking felodipine should avoid grapefruit and its juice (251). This is because grapefruit (juice) is an inhibitor of Cytochrome P450 isoforms 3A4 and 1A2, which are needed for the normal metabolism of felodipine.

Isradipine. It is a calcium antagonist available for oral administration. It binds to calcium channels with high affinity and specificity and inhibits calcium flux into cardiac and smooth muscle. It is used in the management of hypertension, either alone or concurrently with thiazide-type diuretics (252–255). In patients with normal ventricular function, isradipine's afterload reducing properties lead to some increase in cardiac output. Effects in patients with impaired ventricular function have not been fully studied.

In humans, peripheral vasodilation produced by isradipine is reflected by decreased systemic vascular resistance and increased cardiac output. In general, no detrimental effects on the cardiac conduction system were seen with the use of isradipine.

Isradipine is 90–95% absorbed and is subject to extensive first-pass metabolism, resulting in a bioavailability of ~15–24%. Isradipine is completely metabolized prior to excretion and no unchanged drug is detected in the urine. Six metabolites have been characterized in blood and urine, with the mono acids of the pyridine derivative and a cyclic lactone product accounting for >75% of the material identified. The reaction mechanism ultimately leading to metabolite transformation to the cyclic lactone is complex.

Lacidipine. It also belongs to a DHP class of calcium channel blockers (256). Recently, it was shown that lacidipine can slow the progression of atherosclerosis more effectively than atenolol, according to the results of the European lacidipine study on atherosclerosis (257). The improvement of focal cerebral ischemia by lacidipine may be partly due to long-lasting improvement of collateral blood supply to the ischemic area (258). When comparative effects of both lacidipine and nifedipine were measured, both drugs reduced blood pressure significantly during the 24-h period with one dosage daily; only lacidipine reduced left ventricular mass significantly after 12 weeks of treatment (259).

Lercanidipine. It is a member of the dihydropyridine calcium channel blocker class of drugs. Recently, a New Drug Application with the FDA to market lercanidipine for the treatment of hypertension has been submitted, although it has been available in European countries for >4 years with an established record of antihypertensive effect and safety in millions of patients (260,261). In fact, it has grown to be the third most prescribed CCB in Italy due to its favorable side effects. Lercanidipine prevents calcium from entering the muscle cells of the heart and blood vessels, which enable the blood vessels to relax, thereby lowering blood pressure. It has a short plasma half-life but its high lipophilicity allows accumulation in cell membranes resulting in a long duration of action. It has been suggested that lercanidipine causes fewer vasodilatory adverse side effects than other CCBs, and therefore is being promoted for the treatment of isolated systolic hypertension (ISH) in elderly patients (262).

Manidipine. It is effective in the treatment of essential hypertension (263,264). When the effect of manidipine hydrochloride on isoproterenol-induced left ventricular hypertrophy and the expression of the atrial natriuretic peptide (ANP) transforming growth factor was evaluated, it was found that manidipine hydrochloride prevented cardiac hypertrophy and changes in the expression of genes for ANP and interstitial components of extracellular matrix induced by isoproterenol (265).

Nicardipine. It belongs to the 1,4-dihydropyridine calcium channel blocking family of compounds. Nicardipine is tissue selective and produces relaxation of coronary vasculature at dose levels that produce little or no inotropic effects. It has been shown to increase exercise tolerance and reduce nitroglycerin consumption and frequency of anginal attacks. It is usually administered orally or by slow continuous intravenous infusion when oral administration is not viable, for the treatment of chronic stable angina and short-term management of hypertension. It is used as a monotherapy or in combination with other antianginal or antihypertensive drugs.

Nicardipine is almost completely absorbed after po administration. Administration of food decreases absorption. It undergoes extensive first-pass metabolism in the liver. Systemic availability is dose-dependent because of saturation of hepatic metabolic pathways. A 30-mg dose is ~35% bioavailable. Nicardipine is highly protein bound (>95%). Peak plasma concentrations are achieved in 0.5–2.0 h. The principal path of elimination is via hepatic metabolism by hydrolysis and oxidation. The metabolites are relatively inactive and exert no pharmacological activity. The elimination half-life is 8.6 h. About 60% of the dose is excreted in the urine as metabolites (<1% as intact drug) and 35% as metabolites in the feces (154,155,186,266).

Nifedipine. The principle physiological action of nifedipine is similar to other 1,4-dihydropyridine derivatives and functions by inhibiting the transmembrane influx of extracellular calcium ions across the myocardial membrane cells and vascular smooth muscle cells, without affecting plasma calcium concentrations. Although the exact mechanism of action of nifedipine is unknown, it is believed to deform the slow channel and hinder the ion control gating mechanism of the calcium channel by interfering with the release of calcium ions from the sarcoplasmic reticulum. The inhibition of calcium influx dilates the main coronary and systemic arteries due to the impediment of the contractile actions of cardiac and smooth muscle. This reduced myocardial contractility, results in increased myocardial oxygen delivery while decreasing the total peripheral resistance associated by a modest lowering of systemic blood pressure, small increase in heart rate and reduction in the afterload ultimately leading to reduced myocardial oxygen consumption.

Unlike verapamil and diltiazem, nifedipine does not exert any effect on SA or AV nodal conduction at therapeutic dosage levels. Nifedipine is administered orally and extended release tablets in various dosage forms. It is rapidly absorbed (90%) from the GI tract following oral administration and its plasma protein binding is concentration dependent. It has a plasma half-life of 2–5 h and is rapidly and completely metabolized in the liver to inactive metabolites. It is mainly used in the treatment of Prinzmetal angina and chronic stable angina. In the latter case, it is as effective as β-adrenergic agents or oral nitrates, but used only when the patient has low tolerance for adequate doses of these drugs. Nifedipine is also used in the management of hypertension, Raynaud's condition and as a second line tocolytic agent. Common adverse side effects include nausea, heartburn, muscle cramps, nervousness and mood changes, and nasal congestion. GI complaints occur occasionally.

Nilvadipine.   It is marketed as a racemic mixture for the treatment of hypertension and angina (267–269). Nilvadipine also provides protective effects against cerebral ischemia in rats having chronic hypertension, and the effects are dependent on the duration of treatment (270). Results of one of the clinical studies in the United States, in which the combination of imidapril with a diuretic, β-adrenoceptor antagonist or calcium channel blocker, such as nilvadipine, indicate a reasonable and safe, treatment option when striving for additive pharmacodynamic effects not accompanied by relevant pharmacokinetic interactions (271).

Nimodipine.   It is a structural analogue of nifedipine and the (S) (−) enantiomer is primarily responsible for the calcium channel blocking activity. The position of the nitro substituent on the aryl ring and planarity of the 1,4-dihydropyridine moiety contribute greatly to the pharmacological effects of nimodipine. Nimodipine is a light-sensitive yellowish crystalline powder. The mechanism of action of nimodipine is similar to other calcium channel blockers; however, the preferential binding affinity of nimodipine toward cerebral tissue is yet to be fully understood. Nimodipine functions by binding to the stereoselective high affinity receptor sites on the cell membrane in or near the calcium channel and inhibiting the influx of calcium ions. The vasodilatory effect of nimodipine also seems to arise partly from the inhibition of the activities of sodium–potassium activated ATPase, an enzyme required for the active transport of sodium across the myocardial cell membranes.

Nisoldipine.   It is a lipophilic, light sensitive, vascular selective dihydropyridine analogue, similar to nifedipine but 5–10 times more potent as a vasodilator, and has a little effect on myocardial contractility. Nisoldipine is available as a long-acting extended release preparation and appears effective in treating mild to moderate hypertension and angina, once-a-day by oral administration (272–274). Nisoldipine selectively relaxes the muscles of small arteries causing them to dilate, but has little or no effect on muscles or the veins of the heart. This indicates that it may not be useful in congestive heart failure.

In vitro studies show that the effects of nisoldipine on contractile processes are selective, with greater potency on vascular smooth muscle than on cardiac muscle. The effect of nisoldipine on blood pressure is principally a consequence of a dose-related decrease of peripheral vascular resistance. While nisoldipine, like other dihydropyridines, exhibits a mild diuretic effect, most of the antihypertensive activity is attributed to its effect on peripheral vascular resistance.

Nisoldipine is highly metabolized into five major urinary metabolites. The major biotransformation pathway appears to be the hydroxylation of the isobutyl ester. A hydroxylated derivative of the side chain, present in plasma at concentrations approximately equal to the parent compound, appears to be the only active metabolite and has ~10% of the activity of the parent compound. Cytochrome P450 enzymes play a key role in the metabolism of nisoldipine. The particular isoenzyme system responsible for its metabolism has not been identified, but other dihydropyridines are metabolized by cytochrome P450 3A4 isozyme. Nisoldipine should *not* be administered with grapefruit juice, as it interferes with nisoldipine metabolism. Since very little information is available about its use in patients with severe congestive heart failure, this calcium channel blocker should be used with caution in any patient with heart failure.

Recently, antianginal and antiischemic effects of nisoldipine and ramipril in patients with Syndrome X (typical angina pectoris, positive treadmill exercise test but negative intravenous ergonovine test and angiographically normal coronary arteries) were evaluated, indicating that they have similar anti-ischemic and antianginal effects in patients with syndrome X (275).

Nitrendipine. It is used to treat mild to moderate hypertension (276,277).

In summary, calcium antagonists inhibit the influx of extracellular calcium ions into the cells, and prevent intracellular calcium from reaching the critical concentration necessary to initiate contraction, resulting in decreased vascular smooth muscle tone and vasodilation and, leading to a reduction in blood pressure. The 1,4-dihydropyridine derivatives (aranidipine, cilnidipine, amlodipine, nisoldipine, nifedipine, felodipine, nitrendipine, nimodipine) differ from the benzothiazepine (eg, diltiazem) and phenylalkylamine (eg, verapamil) classes of calcium antagonists with regard to potency, tissue selectivity, and antiarrhythmic effects. In general, dihydropyridine agents are the most potent arteriolar vasodilators, producing the least negative inotropic and electrophysiological effects; in contrast, verapamil and diltiazem slow AV conduction and exhibit negative inotropic activity while also maintaining some degree of arteriolar vasodilatation.

The calcium antagonists produce beneficial effects in angina by reducing coronary artery spasm, slowing heart rate or decreasing contractile force. Calcium channel blockers are commonly used to treat high blood pressure, angina, and even some forms of arrhythmia. In the treatment of hypertension and chronic heart failure, a combination therapy enhances therapeutic efficacy. Pharmacodynamically, combinations of ACE inhibitor plus a diuretic, β-adrenoreceptor antagonist, or calcium channel blocker are the most promising.

The most common side effects of calcium antagonists are hypotension, facial flushing, dizziness, headache, weakness, sedation, skin rash, edema, and abdominal discomfort such as nausea, vomiting, constipation and, epigastric pressure.

5. Antilipemic Agents

Increased cholesterol levels, due to the consumption of a diet rich in saturated fat, stimulates the liver to produce cholesterol [57-88-5], $C_{27}H_{46}O$, a lipid needed by all cells for the synthesis of cell membranes and in some cells for the synthesis of other steroids. Cholesterol is the principal reversible determinant of risk of heart disease. Low density lipoproteins (LDLs, also referred as "bad" cholesterol) transport cholesterol from the liver to other tissues, whereas high density lipoproteins (HDLs, also referred as "good" cholesterol) transport cholesterol from tissues back to the liver to be metabolized. Triglycerides are transported from the liver to the tissues mainly as very low density lipoproteins (VLDLs). The VLDLs are the precursors of the LDLs. The LDLs are characterized by high levels of cholesterol, mainly in the form of highly insoluble cholesteryl esters. However, there exists a strong relationship between high LDL levels and coronary heart disease, and a negative correlation between HDL and heart disease. Total blood cholesterol is the most common measurement of blood cholesterol and various total blood cholesterol levels and risk factors accepted by most physicians and the American Heart Association are discussed next. In general, for people who have total cholesterol >200 mg/dL, heart attack risk is relatively low, unless a person has other risk factors. If the total cholesterol level is 240 mg/dL, the person has twice the risk of heart attack as people who have a cholesterol level of 200 mg/dL. If total cholesterol level is 240 or more, it is definitely high and the risk of heart attack and, indirectly, of stroke is greater. About 20% of the U.S. population has high blood cholesterol levels. The LDL cholesterol level also greatly affects risk of heart attack and, indirectly, of stroke. Some times the ratio of total cholesterol to HDL cholesterol is used as another measure and the goal is to keep the ratio below 5:1; the optimum ratio is 3.5:1. It is assumed that people with high triglycerides (>200 mg/dL) have underlying diseases or genetic disorders. In such cases, the main treatment is to change the lifestyle by controlling weight and limiting the carbohydrate intake, since they raise triglycerides and lower HDL cholesterol levels.

During the last few years, there has been firm evidence that coronary artery disease (CAD) is a complex genetic disease involving a number of genes associated with lipoprotein abnormalities and genes influencing hypertension, diabetes, obesity, immune, and clotting systems play an important role in atherosclerotic cardiac disorders. Researchers have identified genes regulating LDL cholesterol, HDL cholesterol, and triglyceride levels based upon common apo E genetic variation (278–282). Many genes linked to CAD are involved in how the body removes LDL cholesterol from the bloodstream. If LDL is not properly removed, it accumulates in the arteries and can lead to CAD. The protein that removes LDL from the bloodstream is called the LDL receptor (LDLR). In 1985, Michael Brown and Joseph Goldstein were awarded a Nobel Prize for determining that a mutation in this gene was responsible for familial hypercholesterolemia, or FH. People with FH have abnormally high blood levels of LDL (283).

As with LDLR, mutations in the *apo E* gene affect blood levels of LDL. Although, <30 mutant forms of *apo E* have been identified, people carrying the E4 version of the gene tend to have higher cholesterol levels than the general population, but levels in people with the E2 version are significantly lower. The *apo E* gene has also been implicated in Alzheimer's disease (284). Even though cardiovascular disease due to atherosclerosis remains the leading cause of death in the United States, most of the risk reduction strategies have traditionally focused on detection and treatment of the disease. However, some of the risk factors of cardiac diseases are reversible and changes in the life style could significantly contribute toward decreasing the mortality risk of CHD. One can reduce the risk of hypercholesterolemia by reducing the total amount of fat in diet, being physically active since exercise can help to increase HDL, avoiding cigarette smoking and exposure to second-hand smoke, and also by reducing sodium intake (285). In people whose cholesterol level does not respond to dietary intervention, and for those having genetic predisposition to high cholesterol levels, drug therapy may be necessary.

There are now several very effective medications available for treating elevated cholesterol levels and preventing heart attacks and death. These include hydroxymethylglutartaryl-coenzyme A (HMG-CoA) reductase inhibitors such as statins, namely, atorvastatin, cerevastatin, fluvastatin, lovastatin, pravastatin, simvastatin, and most recently approved rosuvastatin (lowers LDL cholesterol by 30–50% and increases HDL) and Cholestyramine resin (Table 5). Fibrates such as clofibrate, bezafibrate, micronized fenofibrate, and gemfibrozil (Table 5) also lower elevated levels of blood triglycerides and increase HDL. These agents are represented in the Table 2 and some of them are discussed in detail next.

5.1. HMG-CoA Reductase Inhibitors. In humans, biosynthesis of cholesterol from Acetyl CoA in the liver accounts for 60–70% of the total cholesterol pool. The pathway to cholesterol synthesis is shown in Figure 7. Statins are antilipemic agents that are structurally similar to HMG-CoA reductase, the enzyme that catalyzes the conversion of HMG-CoA to mevalonic acid, an early precursor of cholesterol. Statins produce selective and reversible competitive inhibition of HMG-CoA reductase by binding to two separate sites on the enzyme. Statins are either fungus derived (fermentation product of *Aspergillus terreus*) or are synthetically produced by chemical modification of lovastatin (286). Fully synthetic statins exist either as racemic mixtures or as pure enantiomers. All commercially available statins contain a nucleus that interacts with the coenzyme A recognition site of HMG-CoA reductase and a β,δ-dihydroxy acid side chain that competes with HMG-CoA for interaction with the enzyme (287). The nucleus consists of either a hexahydronapthalene moiety, or an indole–pyrrole–pyridine moiety, and any modification alters the lipophilicity of statins. The β,δ-dihydroxy acid chain is essential for catalytic activity, either as an active dihydroxy acid salt or inactive lactone. Compounds possessing the dihydroxy acid salt are orally active, while those with lactone moiety are prodrugs and have little antilipemic activity until hydrolysed *in vivo* to the corresponding dihydroxy form. Statins introduced in the late 1980s, are fast becoming the most widely prescribed drugs to lower cholesterol.

The exact mechanism by which statins reduce serum concentrations of LDL–cholesterol, VLDL–cholesterol, apoB, and triglycerides is complex, but well understood. Normally, the cell synthesizes cholesterol *de novo* for use in cell membrane and steroid—hormone synthesis, or acquires it from circulating LDLs via receptor mediated endocytosis. Statins are rapidly absorbed to different extents following oral administartion and undergo extensive first pass metabolsim in the liver. The absolute bioavailabilities of atorvastatin, cerivastatin, fluvastatin, lovastatin, pravastatin, and simvastatin, are 14, 60, 24, 5, 17, and 5% respectively. Mean peak plasma concentrations of active inhibitors occur at 0.6–4 h following oral administration and appear to be slightly higher in women than in men. All statins are 95–99% plasma protein bound except pravastatin, which is just 50% bound. Atorvastatin, lovastatin, and simvastatin are metabolized by cytochrome P450 (CYP), mainly by the isoenzyme 3A4 (CYP3A4).

Table 5. Antilipemic Agents

Chemical/generic name	CAS Registry Number	Molecular formula	Trade name	Structure
		HMG-CoA Reductase Inhibitors (Statins)		
atrovastatin calcium	[134523-03-8]	$C_{66}H_{68}CaF_2N_4O_{10}$	Lipitor, Sortis, Xarator	
cerivastatin sodium	[143201-11-0]	$C_{26}H_{33}FNNaO_5$	Baycol, Lipobay	

fluvastatin sodium [93957-55-2] $C_{24}H_{25}FNNaO_4$ Lescol, Lescol XL

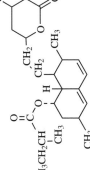

lovastatin [75330-75-5] $C_{24}H_{36}O_5$ Mevacor

pravastatin sodium [81131-70-6] $C_{23}H_{35}NaO_7$ Pravachol

Table 5 (*Continued*)

Chemical/generic name	CAS Registry Number	Molecular formula	Trade name	Structure
simvastatin	[79902-63-9]	$C_{25}H_{38}O_5$	Zocor	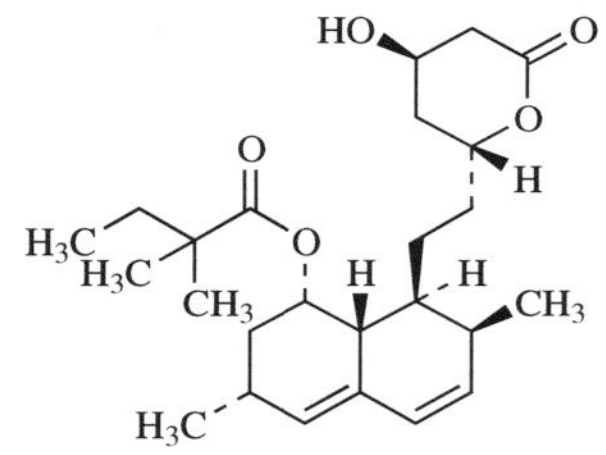
rosuvastatin calcium	[147098-20?]	$2C_{22}H_{27}FN_3O_6S \cdot Ca$	Crestor	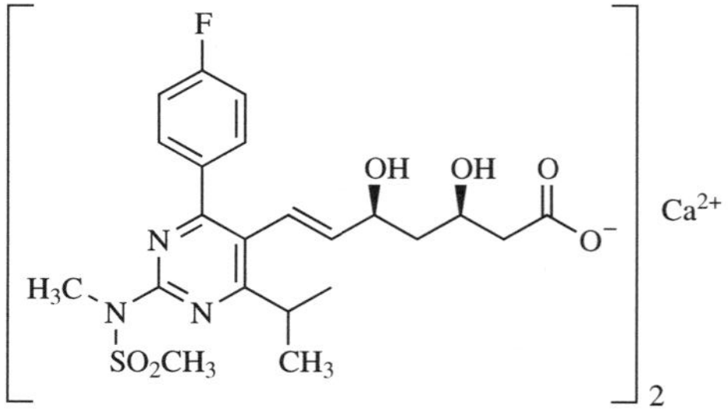
ACAT Inhibitors				
Ezetimibe (zetia)				
SCH 58235	[163222-33-11]	$C_{24}H_{21}F_2NO_3$		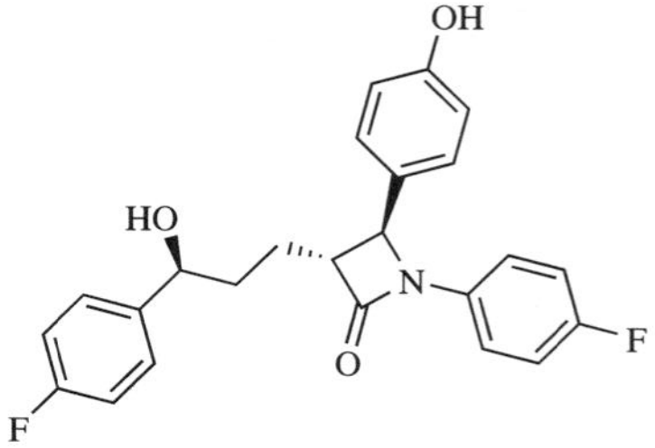

cholestyramine resin	[11041-12-6]	polymer	*Resins* Locholest, Questran, Cholestyramine, Prevalite	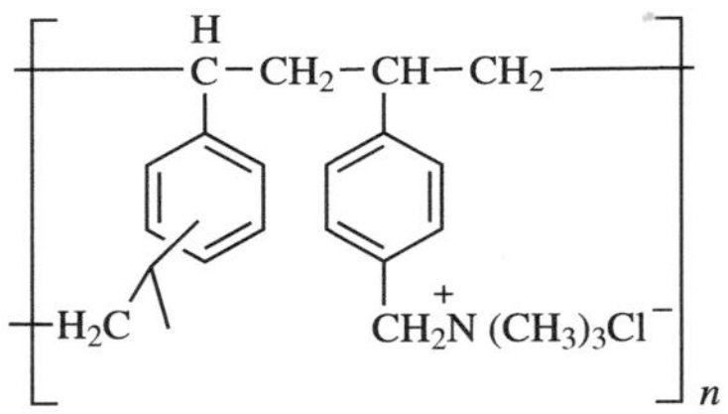 (structure)
colestipol hydrochloride	[37296-80-3]	polymer	Colestid	
clofibrate	[637-07-0]	$C_{12}H_{15}ClO_3$	*Fibric Acid Derivatives* Atromid-S, Clofibrate	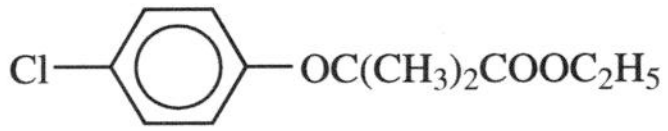 (structure)
bezafibrate	[41859-67-0]	$C_{19}H_{20}ClNO_4$	Bezalip, Bezatol, Cedur,	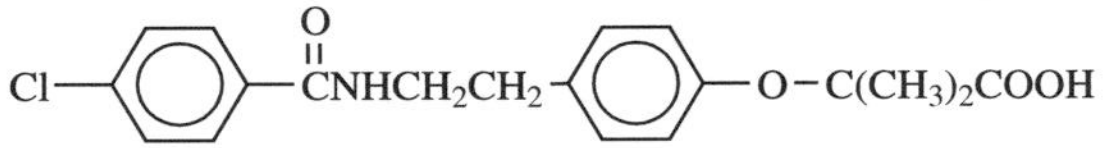 (structure)
fenofibrate	[49562-28-9]	$C_{20}H_{21}ClNO_4$	Fenotard, Lipantil, Lipidil, Tricor	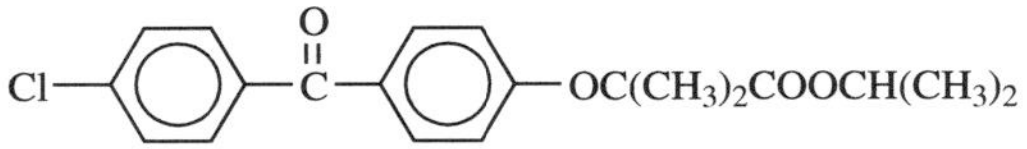 (structure)
gemfibrozil	[25812-30-0]	$C_{15}H_{22}O_3$	Genlip, Lopid, Lipur	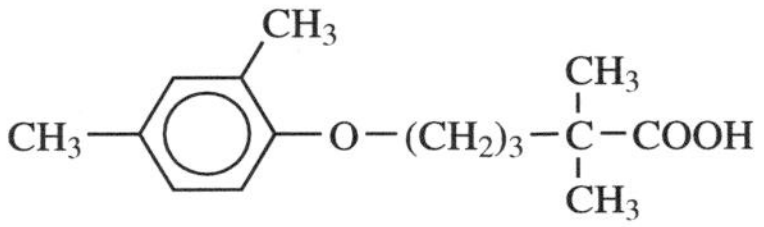 (structure)
niacin			*Other* Niacor, Niaspan	

141

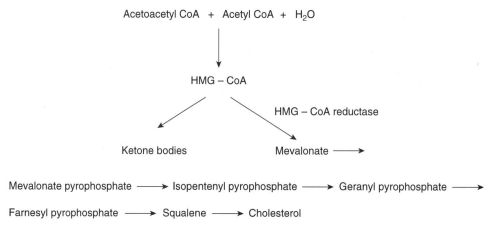

Fig. 7. The pathway to cholesterol synthesis is shown.

Cerevastatin is metabolized by both CYP3A4 and CYP2C8, while fluvastatin is metabolized by isoenzyme 2C9 (CYP2C9). With the exception of atorvastatin (plasma half-life of 14 h) all statins have relatively short half-lives between 0.5 and 4 h.

HMG-CoA reductase inhibitors (Statins) are used as adjuncts to dietary therapy in the management of hypercholestereolemia to reduce the risks of acute coronary events such as CHD, atherosclerosis, MI, or angina. All statins are admistered orally, once a daily, from 10–80 mg/day as per the individual requirements and response. At usual doses, statins are well tolerated and have very few adverse effects. The most common adverse effects are GI disturbances, fatigue, localized pain, and headache. The following five statins are the most prescribed drugs currently on the market.

Atorvastatin Calcium (Lipitor). It is a synthetic antilipemic agent that exists as an active dihydroxy acid. It is the most potent statin for lowering LDL cholesterol (bad cholestrol) and is the most commonly prescribed statin drug. It is used as an adjunct therpy to decrease elevated serum total and LDL, apoB, and triglyceride concentrations and to increase HDL levels. It is also shown to reduces the risk of recurrent ischemic events in patients with CHD having elevated LDL levels. The clinical trials demonstrating the ability of Lipitor to prevent heart attacks and early mortality have not been conducted.

Cerivastatin Sodium (Baycol). It was the most inexpensive statin, however, it was removed from the market in 2001 due to several deaths resulting from severe muscle disease caused by the drug. Other statins are also known to cause similar muscle disease, however, the adverse effects are comparatively mild.

Fluvastatin sodium (Lescol). It is a synthetic and enantiomerically pure antilipemic agent and used as a first line therapy in the treatment of type IIa and IIb hyperlipoproteinemia.

Lovastatin (Mevacor). It is the first statin approved for the use by FDA and the patent expired in 2002 resulting in significantly reduced prices. It is a δ-lactone and produced by fermentation of *Aspergillus terreus* and used similar to other statins to lower bad cholesterol.

Simvastatin (Zocor). It is the second most potent statin drug used to lower LDL cholesterol. Zocor has been found to be the most effective drug as compared to other statins in raising HDL (good) cholesterol levels. It is the first statin used in a study indicating that statins can significantly reduce heart attacks and stroke in high risk patients regardless of cholesterol levels. Similar to Mevacor, zocor is a δ-lactone and produced by the fermentation of *Aspergillus terreus*.

Pravastatin sodium (Pravachol). It is a HMG-CoA inhibitor and differs from other statins where a methyl group on the hexahydronapathalene rings is substituted by a hydroxyl group. The latter change increases the hydrophilicity of Pravastatin making it 100-fold more soluble in water than other statins. A clinical study indicates that Pravachol is also an efficient statin with a capability to prevent heart attacks and early mortality than any other statin. It does not cause drug–drug interactions.

Rosuvastatin (Crestor). Recently a new statin named Rosuvastatin (Crestor), superior to the most widely prescribed statins including atorvastatin, has been approved by the FDA to lower LDL levels. It also increases the HDL significantly more than atorvastatin. Rosuvastatin is an enhanced HMG-CoA reductase inhibitor with a number of advantageous pharmacological properties, relative hydrophilicity, and selective uptake activity by hepatic cells. In one study, crestor (10-mg dose) reduced LDL cholesterol by 49% as compared to 37% with simvastatin (20-mg dose) and 28% with pravastatin (20-mg dose). Similar to other statins, crestor is safe and well tolerated alone or in combination with fenofibrate, extended release niacin and cholstyramine. Crestor undergoes minimal metabolism by P450 (CYP) and is mainly metabolized by 2C9 and 3A4 to a small extent. A new study is underway to examine the efficacy of crestor on atherosclerosis and cardiovascular morbidity and mortality (288).

5.2. New ACAT Inhibitors. Although the statin class of compounds dominate the lucrative market of lipid lowering drugs and are very successful owing to their high efficacy in reducing LDL and total cholesterol, many patients still continue to have higher than normal recommended levels of LDL and total cholesterol. However, in recent years increasing the levels of HDL and reducing levels of triglycerides is becoming the principal focus of antilipemic research and has spurred the development of novel lipid lowering drugs. As discussed earlier, serum cholesterol is regulated by the liver (which produces cholesterol during digestion) and the intestine that absorbs cholesterol from food and bile. Since statins inhibit cholesterol biosynthesis in the liver, new agents that inhibit cholesterol absorption in the intestine would have a synergistic effect if used in combination with statins. Therefore, Acyl coA: cholesterol acyltransferase (ACAT), an enzyme responsible for cholesterol absorption in the intestines, is the target of choice for a new class of compounds that inhibit cholesterol absorption in the intestines.

Ezetimbie is the first member of a new class of cholesterol lowering agents referred to as the cholesterol absorption inhibitors and was recently approved by FDA for the reduction of cholesterol levels in patients with hypercholesterolemia (289). Ezetimibe consists of an azetidione nucleus and an SAR study indicates that essential elements required for inhibition of cholesterol absorption are an N1-aryl group, a (4S)-alkoxyaryl substituent and a C3 arylalkyl sunbstituent

Table 6. **Other Novel Potential Antihyperlipemic Agents**[a]

Class	Compound	Phase status	Company
ACAT inhibitor	avasimibe (CI-1011)	phase III	Pfizer
MTP inhibitor	implitapide	phase III	Bayer
MTP inhibitor	CP-346086	phase II	Pfizer
MTP inhibitor	R-103757	phase I	Janssen
CETP inhibitor	JTT-705	phase II	Tobacco
CETP inhibitor	CP-529414	phase II	Pfizer

[a]Ref. 295.

(290–294). Ezetimibe is administered alone or in combination with statin as an adjunct therapy to diet for lowering elevated levels of LDL, Apo B, and total cholesterol, in patients with primary hypercholesterolemia. Following 10 mg orally/ once daily dose of ezetimibe has been shown to reduce LDL levels, total cholesterol levels apoB, and triglyceride levels accompanied by increases in HDL. Ezetimibe is 18% more effective when used in combination with simvastatin at reducing LDL than simvastatin alone. Table 6 gives other novel antilipemic agents that are under various phases of development and act as ACAT inhibitors, MTP (Microsomal triglyceride transfer protein) inhibitors and CETP (cholesteryl ester transfer protein) inhibitors.

5.3. Bile Acid Sequestrants (BAS). The bile acid binding resins, colestipol [26658-42-4] and cholestyramine, have been in use for ~20 years and are used as an adjunct therapy to decrease elevated serum and LDL–cholesterol levels in the management of type IIa and IIb hyperlipoproteinemia. These drugs are known to reduce the risks of CHD and myocardial infarction (296). Cholestyramine is an anion exchange (stryrene–divinyl benzene) copolymer having Mol. Wt. $>10^6$. It is commercially available as a powder for oral suspension (297).

Colestipol, also an anion-exchange copolymer, is composed of diethylenetriamine [111-40-0], $C_4H_{13}N_3$, and 1-chloro-2, 3-epoxypropane [106-89-8], C_3H_5ClO. Its metabolism and formulation is quite similar to cholestyramine. Following oral administration, these resins lower cholesterol indirectly by binding with bile acids in the intestinal tract. Bile acids are made in the liver from cholesterol and are needed for food digestion. By tying up bile acids, these drugs prompt the liver to make more bile acids. Since the liver uses cholesterol to make the acids, less cholesterol is available to reach the bloodstream. The drug–bile acid complex is eventually excreted along with unchanged resin in the feces, effectively reducing the total cholesterol pool in the body. This results in increased hepatic oxidation and eventual excretion of cholesterol. These agents decrease LDL by up to 20% and do not affect hepatic metabolism; they are, therefore, a good choice in patients with hepatic disease and in young patients. These resins are effective in decreasing both LDL–cholesterol and total serum cholesterol. The drug-induced reduction in cholesterol can be significant; as much as 50% reduction can be observed over long-term therapy (298). Thus control of serum cholesterol can contribute to the antiatherosclerotic attributes of this class of drug.

Because colestipol and cholestyramine are not absorbed, but simply pass through the body by the GI tract, few severe side effects occur. Patients often complain of distaste and constipation, however, more severe side effects such as GI bleeding are relatively uncommon (297). The bile acid sequestrants antilipemic drugs are known to have reduced or no GI absorption and are normally regarded as safe in pregnant patients. One drawback to these agents is a long list of drugs with which they will bind in circulation. It is important to take these agents 1 h prior or at least 4 h after a bile acid dose to reduce the potential for drug interactions.

Colesevelam hydrochloride is a second generation bile acid sequestrant antilipemic agent and is pharmacologically related to other agents in this class such as cholestyramine and colestipol. Colesevelam hydrochloride, alone or in combination with a HMG-CoA reductase inhibitor is used as an adjunct to dietary therapy and exercise to reduce serum LDL–cholesterol in the primary management of hypercholesterolemia. Colesevelam reduces serum total cholesterol, LDL–cholesterol and apolipoprotein B (apo B) and increases high density lipoprotein (HDL) cholesterol concentrations. However, similar to other BAS, colesevelam may cause slight increases in serum triglyceride concentrations and is not absorbed from the GI tract.

It is administered orally, once or twice daily with meals, and at recommended dosages does not appear to interfere with the absorption of fat-soluble vitamins. It is marketed as Welchol from GelTex.

5.4. Fibric Acid Derivatives. Fibrates can be regarded as broadspecturm lipid modulating agents and their main action is to decrease serum triglycerides, LDL–cholesterol, serum VLDL, and to raise HDL–cholesterol. Fibrates are simple small organic molecules. Fibrates reduce the hepatic synthesis and release of VLDL. They also lower triglycerides by up to 50% and raise HDL levels by up to 35%. Fibrates are particularly useful in patients with familial hyperlipidemia and diabetes.

A large number of fibrate analogues have been synthesized and tested for their antihyperlipoproteinemia. The most widely used fibrates include clofibrate, bezafibrate, ciprofibrate, micronized fenofibrate, and gemfibrozil shown in Table 2 (299,300).

All fibrates contain an isobutyric acid side chain, are readily and completely absorbed from the GI tract and 95–98% plasma protein bound. The mechanism of action of fibrates in controlling plasma lipids appears to be primarily through mediation of lipoprotein lipase activity. In addition, gemfibrozil stimulates apolipoprotein AI synthesis (301), presumably leading to increased HDL particles. In fact, the mechanistic effects of the fibrates in lowering serum cholesterol are complex, from indirect regulation of lipoprotein synthesis, to regulation of key enzymes and receptors involved in lipid metabolism. These drugs are well absorbed after oral administration.

The most common adverse side effects of the fibric acid derivatives are similar to those of the bile acid sequestrants. Patients complain of GI effects, including nausea and vomiting.

Clofibrate (Atromid-S). Clofibrate is the ethyl ester of *p*-chlorophenoxy isobutyric acid and is structurally similar to gemfibrozil. Clofibrate is mainly used in the management of severe hypertriglyceridemia (types IV and V). It has been used with good results to prevent or control polydipsia, polyuria, and dehydration in patients with diabetes insipidus. It is rapidly hydrolyzed by serum enzymes *in vivo* to clofibric acid and is subsequently conjugated in the liver with glucuronic acid. Peak plasma clofibric acid concentrations 4–6 h after oral administartion of 500-mg dose average 49–53 µg/mL. Clofirbic acid has an elimination half-life of 12–35 h in normal patients and 29–88 h in patients with renal failure.

   Bezafibrate (Bezalip).   It is used in hyperlipidaemias of types IIa, IIb, IV, and V in patients who have not responded adequately to diet and other appropriate measures.

   Fenofibrate (Tricor).   It is structurally and pharmacologically related to clofibrate and and gemfobrozil. Fenofibrate is a pro-drug and it has no antilipemic activiy until it is hydrolyzed by plasma and serum esterases *in vivo* to fenofibric acid. The antilipemic activity of fenofibrate appears to be related to its effects on the clearnce of triglyceride-rich particles. Data from *in vivo* and *in vitro* studies indicate that fenofibric acid activates lipoprotein lipase and reduces production of apolipoprotein C-III (apo C-III), an inhibitor of lipoprotein lipase activity and increases the clearnce of triglyceride rich particles. This in turn alters the size of LDL–cholesterol from small to dense and larger partciles that can be rapidly catabolized. Fenofibric acid appears to activate peroxisome proliferator activated receptor that induces the synthesis of HDL–cholesterol, apo-A-I and apo-A-II. Further clinical studies with two different formulations of fenofibrate (micronized and nonmicronized) indicate that micornized dosage formulation has gretaer bioavailability and it is the only commercial fenofibrate preparation currently available in the United States.

   Gemfibrozil (Lopid).   Although it is structurally related to clofibric acid, it is a nonhalogenated phenoxypentanoic acid, antilipemic agent. It is mainly used to reduce the risk of developing CHD in patients with type IIb hyperlipoproteinemia and to reduce the risk of recurrent coronary events such as MI and stroke. Gemfibrozil is also used as an adjunct therapy to decrease the elevated serum triglycerides and cholesterol concentrations either alone or in combination with other antilipemic agents, bile acid sequestrants, statins and niacin. Many of the pharmacological effects of gemfibrozil are similar to those of clofibrate. Gemfibrozil also shares the potential adverse side effects of the fibrate class of compounds due to structural similarity. It is rapidly and completely absorbed from the GI tract (97%) and peak plasma concentrations of the drug occur within 1–2 h. The elimination half-life of gemfibrozil is ~1.5 h after a single dose in individuals with normal renal function. Gemfibrozil appears to be metabolized in the liver to four major metabolites produced via three metabolic pathways.

   Niacin (Niacor).   It is a water-soluble, B-complex vitamin that is used as an antilipemic agent. Large doses of niacin are known to lower triglycerides. In addition, niacin can lower LDL–cholesterol and increase HDL–cholesterol, both beneficial effects. Niacin in its extended release form (Niaspan) is better tolerated and also the combination of lovastatin and niacin (Advicor, recently approved by the FDA) is very effective at lowering LDL and TGs while raising HDL. Niacin is rapidly and extensively absorbed following oral administration and peak plasma concentrations of niacin are attained within 30–60 min (Niacor) or 4–5 h (Niaspan).

6. Antihypertensive Agents

Hypertension may result from a single or a combination of several pathogenic states. Physiologically, blood pressure is determined not only by the amount of blood pumped through the heart, but also by the resistance of the blood flow through the arteries, veins, and capillaries, ie, the blood vessels. Abnormalities or malfunctions in the nervous system (the sympathetic-parasympathetic nervous system effects heart rate and blood vessel tone), hormonal system, kidneys, and peripheral vasculature may all contribute to increased blood pressure.

The most common form of high blood pressure is essential hypertension. It is one of the two major factors responsible for cardiovascular diseases, such as CHD, stroke, and CHF. Of the patients diagnosed to have essential hypertension, ~70% have mild hypertension where the diastolic blood pressure is from 90 to 104 mm Hg (12–13.9 Pa), 20% have moderate hypertension, ie, 105–114 mm Hg (14–15.2 Pa), and 10% have severe hypertension, >115 mm Hg (15.3 Pa). According to the National Health and Nutritional Examination Survey III, ~50 million Americans ages 6 and older have high blood pressure and 31.6% do not know that they have it. Of all people with high blood pressure, 27.2% are on adequate medications, 26.2% are on inadequate medications, and 14.8% are not on any therapy. From 1999 to 2000, the death rate in United States due to high blood pressure increased 21.3% , while the actual number of deaths increased by 49.1%. On average, the estimated annual cost of antihypertensive prescriptions in United States is ~\$ 15.5 billion.

The treatment of hypertension is usually tailored to the individual. Initially, nonpharmacological treatments, such as a change in diet and an increase in exercise, may be recommended to lower blood pressure. However, should these measures fail to have an effect, drug therapy is the next option for controlling hypertension. With regard to chemotherapy, the therapeutic regiment is tailored to the individual, and relevant factors affecting each patient are taken into consideration in choosing an antihypertensive drug. The first drug is usually an angiotensin converting enzyme (ACE) inhibitor, a β-adrenoceptor blocker, a calcium channel blocker, or a diuretic. One drug can normalize the blood pressure of ~50–60% of the hypertensive patients, and the combination of two complementary drugs normalizes that of >80% patients. The main goal of hypertension chemotherapy is to give the fewest number of drugs, using the smallest effective amounts having the lowest frequency of dosing and minimum side effects (302,303). An optimal drug regimen should reduce all risk factors of coronary heart diseases, reverse the hemodynamic abnormalities present by preserving cardiac output and tissue perfusion, and lower total peripheral resistance. The physician's challenge is to choose the best antihypertensive drug for concomitant existing diseases while maintaining a good quality of life (304–315).

The drugs used as antihypertensive agents may be classified according to their mechanism of action as follows: (1) agents affecting the renin-angiotensin system (angiotensin converting enzyme inhibitors, angiotensin II receptor blockers, renin inhibitors, aldosterone receptor antagonists, and vasopeptidase inhibitors; (2) agents affecting the adrenergic nervous system (ganglionic blockers, neuronal blockers, neuronal norepinephrine depleting agents, α-adrenoceptor blockers, β-adrenoceptor blockers, and α/β-adrenoceptor blockers); (3) calcium channel blockers; (4) diuretics (low ceiling diuretics, high ceiling diuretics, aldosterone antagonists, and potassium sparing diuretics); (5) centrally acting agents; (6) vasodilators; and potassium channel openers. To view structures of select agents used in the treatment of hypertension, the reader is referred to Table 7.

6.1. Agents Affecting the Renin–Angiotensin System. The renin–angiotensin system (RAS) plays an important role in the regulation of arterial blood pressure in the body (316,317). The enzyme renin, which is synthesized and released from the kidney, catalyzes the formation of angiotensin I, a decapeptide, from the substrate angiotensinogen, a α_2-globulin, synthesized in the liver and released into the blood stream (Fig. 8). ACE, a nonspecific carboxypeptidase, cleaves a dipeptide from the carboxyl terminal of angiotensin I to form angiotensin II, an octapeptide. The blood vessels of the pulmonary circulation are the main site for this conversion. Angiotensin II is further cleaved to the heptapeptide, angiotensin III by the action of tissue aminopeptidase. Angiotensin II is the most potent product of this cascade. Angiotensin I is relatively inactive, whereas angiotensin III is less potent than angiotensin II as a vasoconstrictor, but is as active as angiotensin II in stimulating the release of aldosterone.

The primary effects of angiotensin II are (1) producing vasoconstriction directly or indirectly through activation of the adrenergic nervous system; (2) eliciting cardiac stimulation; (3) stimulating aldosterone and vasopressin secretion; and (4) evoking dipsogenesis. All of these actions, directly or indirectly, elevate blood pressure (316).

6.2. Angiotensin Converting Enzyme Inhibitors. ACE inhibitors are very efficacious in reducing high blood pressure. The antihypertensive activity is better correlated with the vascular tissue ACE inhibition than plasma ACE inhibition. Plasma ACE activity returns to predrug level while significant lowering of the blood pressure is still observed. Teprotide, a nonapeptide ACE inhibitor, was demonstrated to have potent antihypertensive effects after IV administration. The drawback of teprotide is lack of oral activity. Captopril, a sulfhydryl containing small molecule, was the first practical nonpeptide ACE inhibitor in humans (318). Since the advent of captopril, several non-sulfhydryl ACE inhibitors have come on line: benzapril, enalapril, enalaprilat, fosinopril, lisinopril, moexipril, perindopril, quinapril, ramipril, and trandolapril.

ACE inhibitors are used in the first-line therapy of hypertension (319–325). Initially it was thought that ACE inhibitors were efficacious only in patients having high plasma renin activity (PRA) such as in renal or malignant hypertension. Captopril, however, was found to be efficacious in patients having essential hypertension, particularly those with high or moderately high PRA. Only ~20% of patients having essential hypertension have high to moderately high PRA, but ACE inhibitor used as monotherapy is effective in 50–60% of all patients.

ACE inhibitors lower the elevated blood pressure in humans with a concomitant decrease in total peripheral resistance. Cardiac output is increased or unchanged; heart rate is unchanged; urinary sodium excretion is unchanged; and potassium excretion is decreased. ACE inhibitors promote reduction of left ventricular hypertrophy. Other mechanisms of action, such as potentiation of the bradykinin effect, and increased production of prostaglandins have been reported to explain, in part, the antihypertensive effects of ACE inhibitors during chronic therapy (326,327). The inhibition of angiotensin II production is

Table 7. Antihypertensive Agents

Chemical/generic name	CAS Registry Number	Molecular formula	Trade name	Structure
			ACE Inhibitors	
benazepril hydrochloride	[86541-74-4]	$C_{24}H_{29}ClN_2O_5$	Lotensin, Lotrel, Lotensin HCT	
captopril	[110075-07-5]	$C_9H_{15}NO_3S$	Capoten, Capozide	
enalapril maleate	[76095-16-4]	$C_{20}H_{28}N_2O_5 \cdot C_4H_4O_4$	Vasotec, Vaseretic, Lexxel	

149

Table 7 (*Continued*)

Chemical/generic name	CAS Registry Number	Molecular formula	Trade name	Structure
fosinopril sodium	[88889-14-9]	$C_{30}H_{45}N \cdot NaO_7P$	Monopril	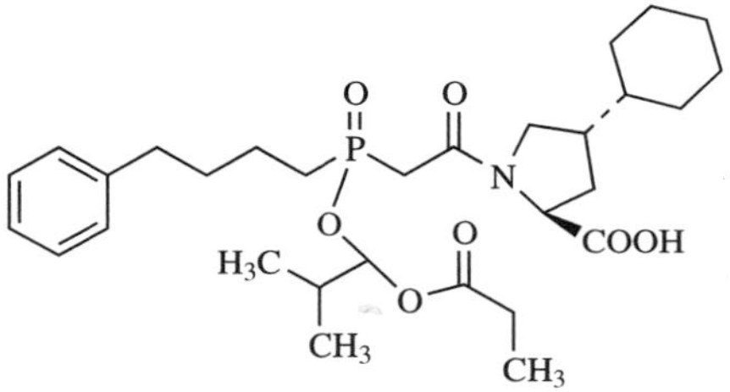
lisinopril	[83915-83-7]	$C_{21}H_{31}N_3O_5 \cdot 2H_2O$	Prinvil, Zestril, Prinzide, Zestoretic	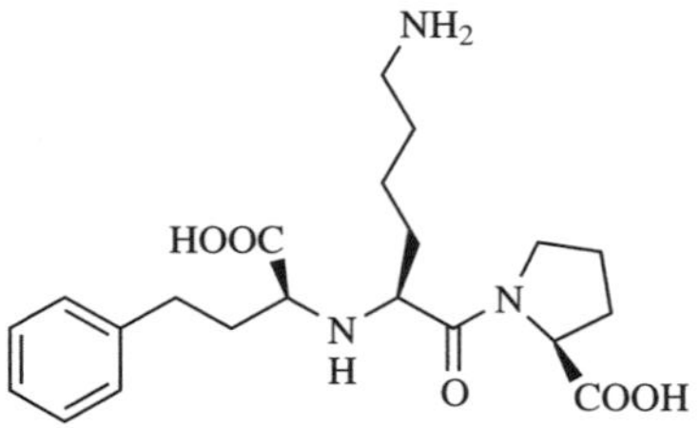
moexipril HCl	[82586-52-5]	$C_{27}H_{34}N_2O_7 \cdot HCl$	Univasc, Uniretic	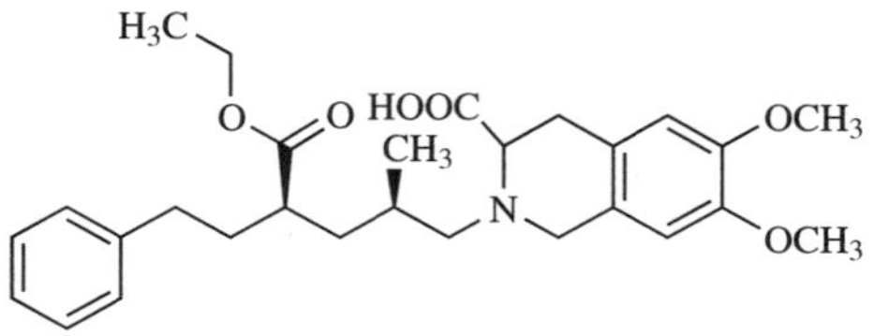

150

perindopril erbumine	[107133-36-8]	$C_{19}H_{32}N_2O_5 \cdot C_4H_{11}N$	Aceon 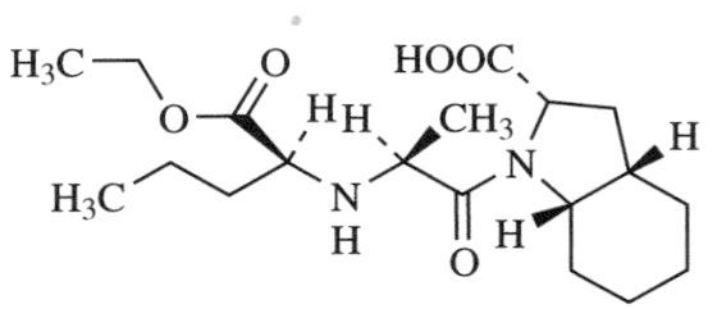
quinapril	[82586-55-8]	$C_{25}H_{30}N_2O_5 \cdot HCl$	Accupril 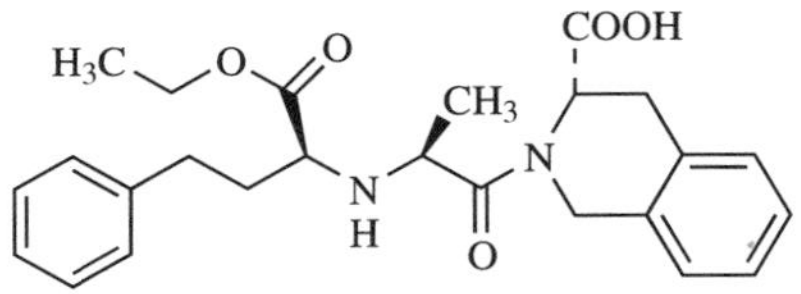
ramipril	[87333-19-5]	$C_{23}H_{32}N_2O_5$	Altace 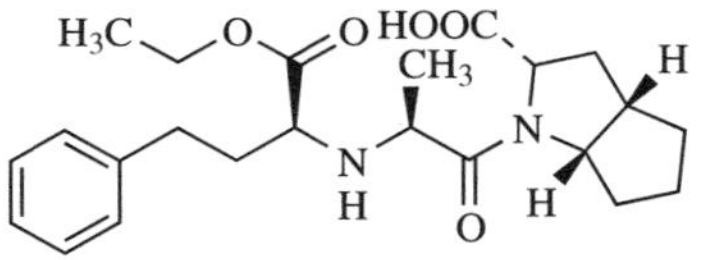
trandolapril	[87679-37-6]	$C_{24}H_{34}N_2O_5$	Mavik, Tarka 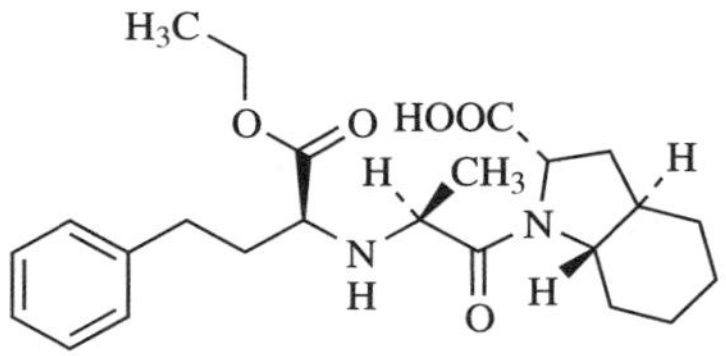

Table 7 (*Continued*)

Chemical/ generic name	CAS Registry Number	Molecular formula	Trade name	Structure
		Angiotensin II Receptor Blockers		
candesartan	[139481-59-7]	$C_{24}H_{20}N_6O_3$	Atacand, Amias, Kenzen	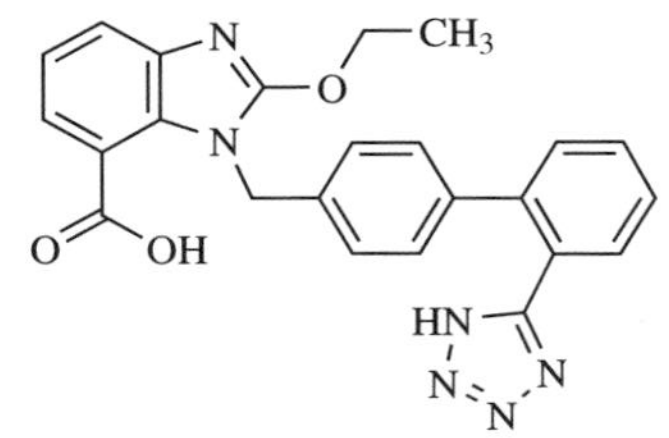
eprosartan	[1333040-01-4]	$C_{23}H_{24}N_2O_4S$	Teveten	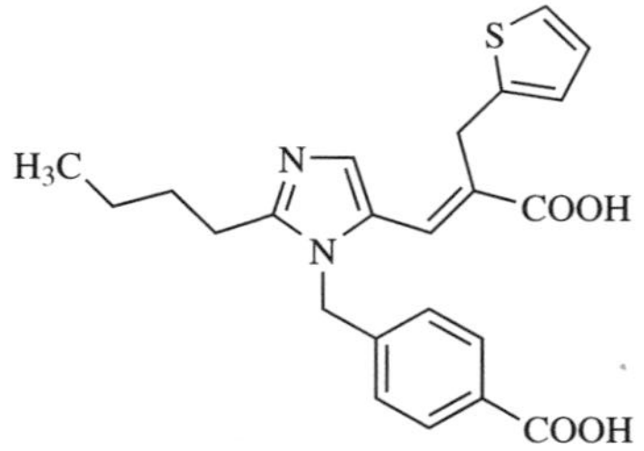
irbesartan	[138402-11-6]	$C_{25}H_{28}N_6O$	Avapro, Avalide	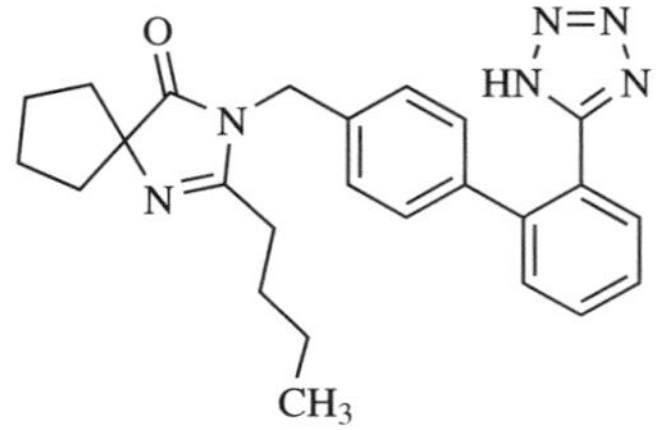

losartan   [124750-99-8] $C_{22}H_{22}ClKN_6O$ Cozaar, Hyzaar

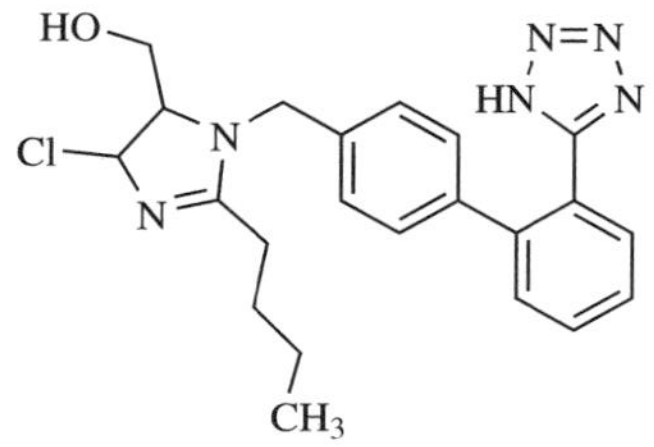

olmesartan   [144689-63-4] $C_{29}H_{30}N_6O_6$ Olmesatran, Medoxomil

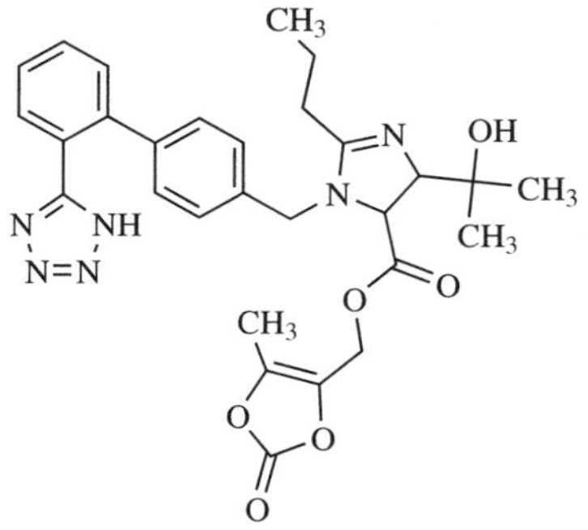

telmisartan   [144701-48-4] $C_{33}H_{30}N_4O_2$ Micardis, Pritor

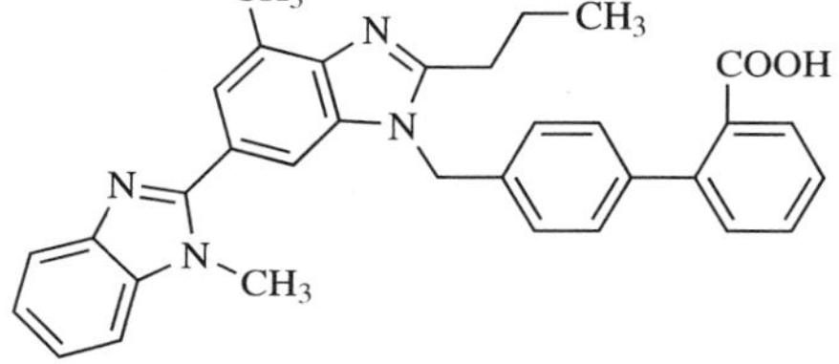

Table 7 (Continued)

Chemical/ generic name	CAS Registry Number	Molecular formula	Trade name	Structure
valsartan	[137862-53-4]	$C_{24}H_{29}N_5O_3$	Diovan, Diovan HCT	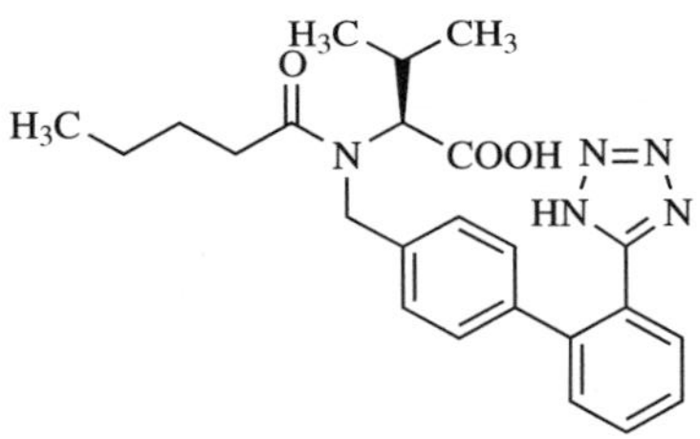
		Aldosterone Receptor Antagonists		
spironolactone	[52-01-7]	$C_{24}H_{32}O_4S$	Verospiron, Xenalon	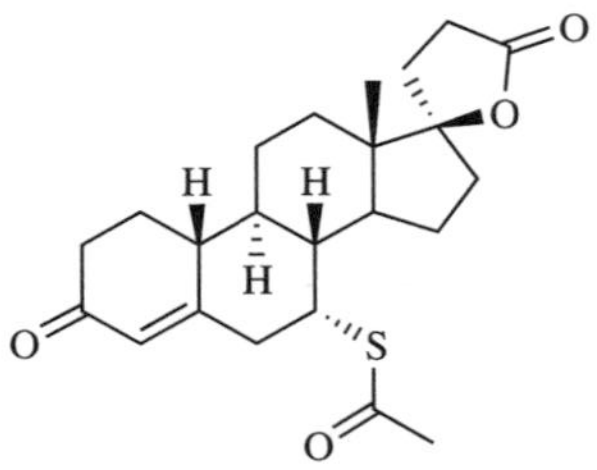
		Vasopeptidase Inhibitors		
omapatrilat	[167305-00-2]	$C_{19}H_{24}N_2O_4S_2$	BMS-186716	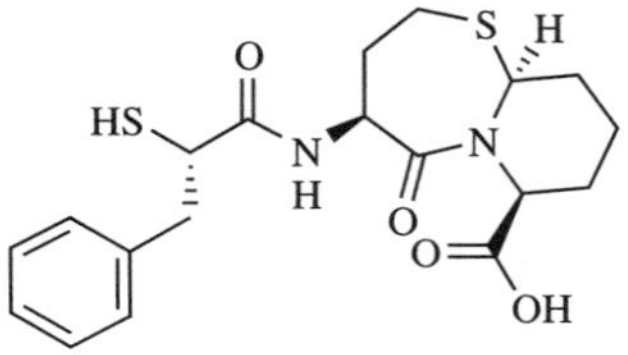

154

Neuronal Norepinephrine Depleting Agents

reserpine [50-55-5] $C_{33}H_{40}N_2O_9$ Rivasin, Sandril, Serfin, Serpasol

Adrenergic Neuronal Blockers

guanethidine monosulfate [60-02-6] $(C_{10}H_{22}N_4)_2 \cdot H_2SO_4$ Ismelin Sulfate

α-Adrenoceptor Blockers

doxazosin mesylate [77883-43-3] $C_{23}H_{25}N_5O_5 \cdot CH_3SO_3H$ Cardura

Table 7 (Continued)

Chemical/generic name	CAS Registry Number	Molecular formula	Trade name	Structure
prazosin HCl	[19237-84-4]	$C_{19}H_{21}N_5O_4 \cdot HCl$	Minipress, Minizide	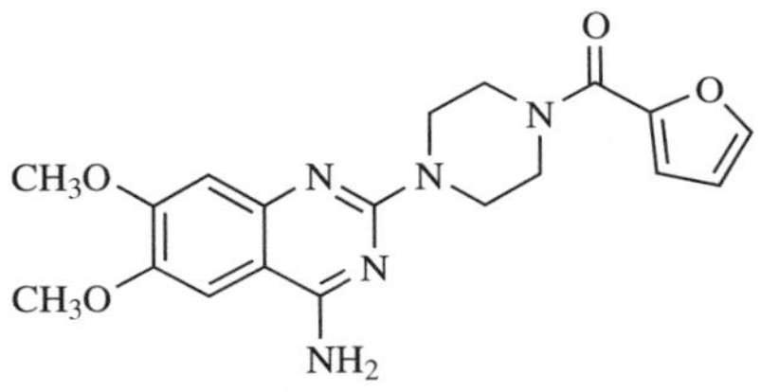
terazosin HCl	[63590-64-7]	$C_{19}H_{25}N_5O_4 \cdot HCl$	Hytrin, Terazosin	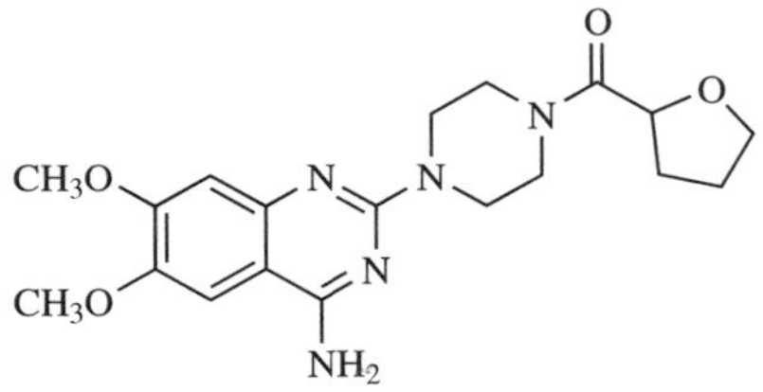
		β-Adrenoceptor Blockers[*]		
bisoprolol fumarate	[104344-23-2]	$C_{18}H_{31}NO_4 \cdot 1/2\ C_4H_4O_4$	Zebeta, Ziac	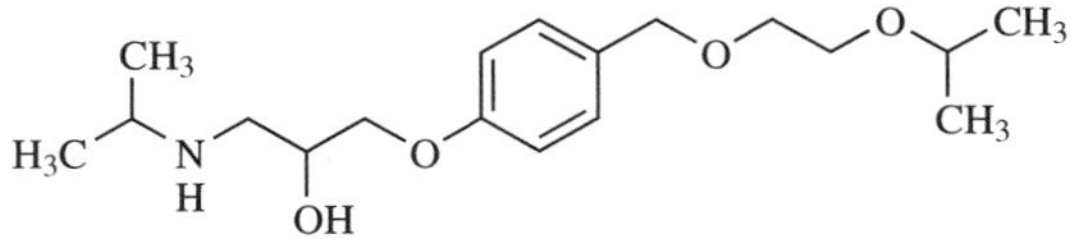
oxprenolol	[6452-71-7]	$C_{15}H_{23}NO_3$	Coretal, Laracor, Paritane, Trasacor	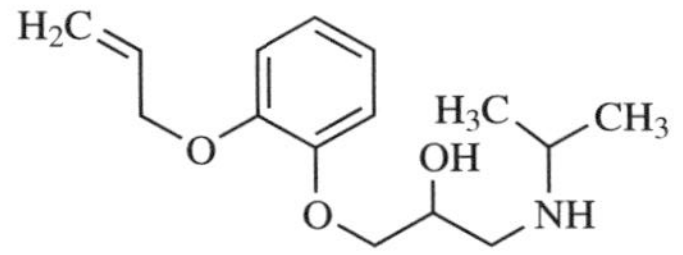

156

pindolol   [13523-86-9] $C_{14}H_{20}N_2O_2$ Visken

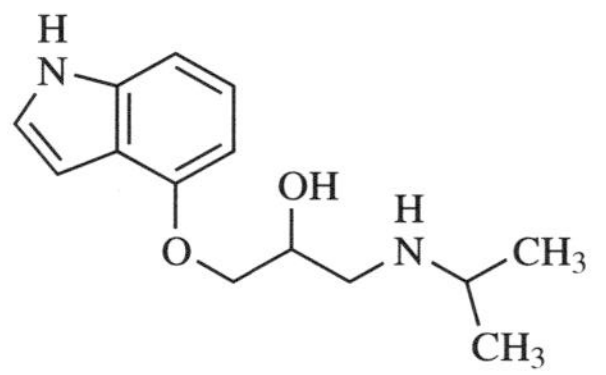

α- and β-Adrenoceptor Blockers

labetalol HCl   [32780-64-6] $C_{19}H_{24}N_2O_3 \cdot HCl$ Normodyne, Trandate

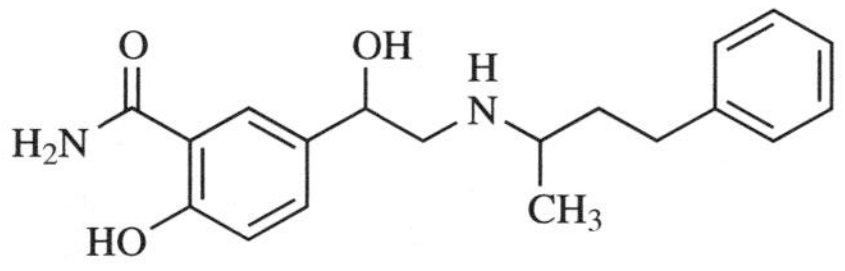

* Other β-blockers that are included as antihypertensive agent in the text are shown in Table 1, as they are also categorized as antiarrhythmic agents.

157

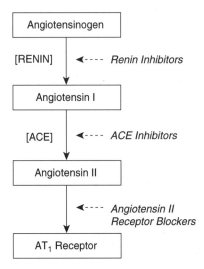

Fig. 8.   The renin–angiotensin system. Renin catalyzes the generation of angiotensin I from angiotensinogen. Angiotensin-converting enzyme, ACE, cleaves angiotensin I to form the active octapeptide–angiotensin II. Angiotensin II activates a G-protein-coupled-receptor—AT_1. AT_1, when activated, produces several effects including vasoconstriction and the stimulation of aldosterone release.

considered the primary, if not the only, mechanism of action of ACE inhibitors, even though the circulating angiotensin II levels may not accurately reflect the situation. Headache, dizziness, fatigue, skin rash, and cough are the most common side effects of ACE inhibitors. Cough in patients treated with an ACE inhibitor is considered to be caused by the build-up of bradykinin and prostaglandins.

6.3. Angiotensin II Receptor Blockers.   Angiotensin II binds to Angiotensin II type 1 (AT_1) receptors to constrict blood vessels. Thus, angiotensin II receptor blockers act not by inhibiting the production of Angiotensin II, but instead, exert their effect by blocking the binding of angiotensin II to AT_1 (Fig. 8), which, in turn, reduces blood pressure (328). However, unlike ACE inhibitors, agents in this class do not inhibit bradykinin metabolism (328), and consequently do not possess the cough inducing side effect that is associated with ACE inhibitors. The FDA has approved losartan, eprosartan, olmesartan, valsartan, irbesartan, candesartan, and telmisartan as therapeutics for hypertension. With the exception of eprosartan, the angiotensin II receptor blockers are structurally similar, containing a characteristic biphenyltetrazole substructure. Eprosartan possesses a carboxylate moiety in place of the tetrazole. Such an interchange of functional groups, with similar capacities to ionize under physiological pH, is a good example of how the principles of isosterism are used during the drug development process. Clinically, the efficacy of these agents has been shown in a number of controlled trials (329–343).

6.4. Renin Inhibitors.   There have been significant efforts to develop compounds that would block the cleavage of angiotensinogen by renin, as such molecules would effectively inhibit the entire RAS (Fig. 8). However, to date only one compound—aliskren—has shown promise in early clinical trials (328). Factors contributing to the lack of success of these compounds include: poor bioavailability, lack of antihypertensive efficacy, and lack of specificity. Furthermore, the success of the ACE inhibitors and angiotensin II blockers has overshadowed the need for renin inhibitors.

6.5. Aldosterone Receptor Antagonists.   Studies have indicated that high levels of aldosterone may contribute to, or cause, hypertension (344,345). Supporting this evidence is the fact that treatment with spironolactone, the only aldosterone receptor antagonist (ARA) that is clinically available, produces decreases in blood pressure in patients with refractory hypertension (346). A drawback of this compound is that it, resembling a steroid, has progestational and antiandrogenic side effects (328). As a result, research to generate ARAs with improved specificity is underway, and has resulted in the generation of eplerenon (347), which has been approved by the FDA.

6.6. Vasopeptidase Inhibitors.   This is a new, still investigational class of agents. The vasopeptidase inhibitors (VPIs) act by inhibiting both ACE and neutral endopeptidase (NEP), which is responsible for the hydrolysis of atrial natriuretic peptide (328). Simultaneous inhibition of ACE and NEP decreases blood pressure with greater efficacy than ACE inhibition alone. Several VPIs are under clinical development: fasidotril, gemopatrilat, omapatrilat, MDL-100240, sampatrilat, and Z-13752A (328). Clinical trials using omapatrilat have indicated that this compound can cause angioedema (328).

6.7. Agents Affecting the Adrenergic Nervous System.   Historically, the vasopressor systems, particularly the adrenergic nervous system, drew the most attention in hypertension research, and many drugs affecting this system have been introduced. Agents antagonize this system either by decreasing the number of nerve impulses traveling down the nerves, by blocking ganglia, by blocking the release of the adrenergic neurotransmitter (norepinephrine), by depleting the norepinephrine stores in the neurons, or by blocking the adrenoceptors in the blood vessels. Vascular tone is maintained mainly by the adrenergic nervous system, and decreasing the tone by any of the mechanisms, listed above, results in a lowering of blood pressure.

6.8. Ganglionic Blockers.   Ganglionic blocking agents such as hexamethonium, trimethaphan, and chlorisondamine, block the autonomic ganglia, and thus decrease sympathetic outflow. Tolerance develops quickly to these agents. Because of such disturbing side effects as severe orthostatic hypotension, sexual dysfunction, dry mouth, and urinary retention, these drugs have become obsolete.

6.9. Adrenergic Neuronal Blockers.   The adrenergic neuronal blocking agents, guanethidin, bretylium, debrisoquin, and guanadrel, produce hypotension by blocking the release of norepinephrine from the nerve terminals of adrenergic neurons. These drugs are taken up by the neurons and decrease sympathetic tone, heart rate, cardiac output, and total peripheral resistance. Some deplete norepinephrine stores, and thus produce an initial sympathomimetic response increasing these various responses. Orthostatic hypotension, severe sexual dysfunction, and impairment, fluid retention, and diarrhea are the primary side effects of this class of agents, which is obsolete and used only when all other therapeutic agents fail to work.

6.10. Neuronal Norepinephrine Depleting Agents.   Reserpine is the most active alkaloid derived from *Rauwolfia serpentina*. The principal antihypertensive mechanism of action primarily results from depletion of nore-pinephrine from peripheral sympathetic nerves and the brain adrenergic neurons. The result is a drastic decrease in the amount of norepinephrine released from these neurons, leading to a decrease in vascular tone and lowering of blood pressure. Reserpine also depletes other neurotransmitters, including epinephrine, serotonin, and dopamine. Reserpine is efficacious in all forms of hypertension, particularly with the concomitant use of a thiazide diuretic. The most serious side effect is mental depression. Other side effects include nasal congestion, sedation, drowsiness, lethargy, decreased libido, impotence, and nightmares.

6.11. α-Adrenoceptor Blockers. Nonselective α-adrenoceptor blockers (Table 8), such as phentolamine, which block both α_1- and α_2-adrenoceptors, produce vasodilation by antagonizing the effects of endogenous norepinephrine. They also produce severe tachycardia and have been replaced by selective α_1-adrenoceptor blockers, such as prazosin, terazosin, and doxazosin, which do not usually cause severe tachycardia.

Prazosin is a selective α_1-adrenoceptor antagonist that exerts its antihypertensive effect by blocking the vasoconstrictor action of adrenergic neurotransmitter, norepinephrine, at α_1-adrenoceptors in the vasculature (348–350). Prazosin lowers blood pressure without producing a marked reflex tachycardia. It causes arteriolar and venular vasodilation, but a significant side effect is fluid retention. Prazosin increases HDL–cholesterol, decreases LDL–cholesterol, and does not cause glucose intolerance.

Doxazosin, a prazosin derivative, is a highly selective α_1-adrenoceptor blocker (351,352). It has no significant antagonistic effects on presynaptic or postsynaptic α_2-adrenoceptors. Blocking the activation of α_1-adrenoceptors prevents the breakdown of phosphatidylinositol, resulting in vasoconstriction. The lowering of blood pressure by doxazosin is not accompanied by an increase in heart rate or cardiac output. Its duration of action is as long as 24 h. Doxazosin in chronic hypertension treatment reduces total cholesterol, LDL–cholesterol, and triglyceride levels, and increases HDL–cholesterol.

Terazosin is a selective α_1-adrenoceptor blocker having hypotensive efficacy equal to that of prazosin. Terazosin has a longer duration of action and better GI absorption profile than prazosin.

6.12. β-Adrenoceptor Blockers. There is no satisfactory mechanism to explain the antihypertensive activity of β-adrenoceptor blockers (Table 8) in humans, particularly after chronic treatment (353–355). Reductions in heart rate correlate well with decreases in blood pressure and this may be an important mechanism. Other proposed mechanisms include reduction in PRA, reduction in cardiac output, and a central action. During long-term treatment, the cardiac output is restored despite the decrease in arterial blood pressure and total peripheral resistance. Atenolol, which does not penetrate into the brain, is an efficacious antihypertensive agent. During short-term treatment, the blood flow to most organs (except the brain) is reduced, and the total peripheral resistance may increase.

Treatment with β-adrenoceptor blockers is effective in patients having high PRA; however, most of these agents are not efficacious in patients having low PRA or in elderly patients. β-Adrenoceptor blockers usually lower arterial

Table 8 Antihypertensive Agents: Diuretics

Chemical/generic name	CAS Registry Number	Molecular formula	Trade name	Uses	Structure
			Thiazide Diuretics		
bendroflumethiazide	[73-48-3]	$C_{15}H_{14}F_3N_3O_4S_2$	Naturetin, Rauzide, Corzide	antihypertensive	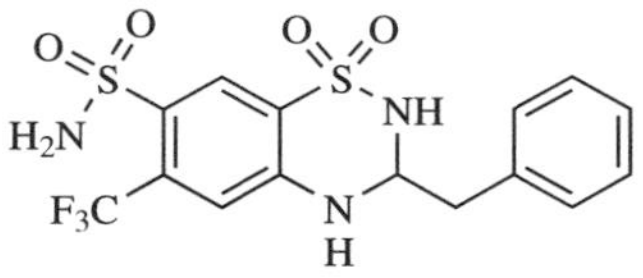
chlorothiazide Na	[7085-44-1]	$C_7H_5ClN_3NaO_4S_2$	Diuril, Aldoclor	antihypertensive	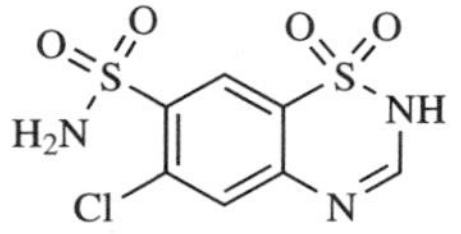
chlorthalidone	[77-36-11]	$C_{14}H_{11}ClN_2O_4S$	Thalitone, Combipress, Tenoretic	antihypertensive	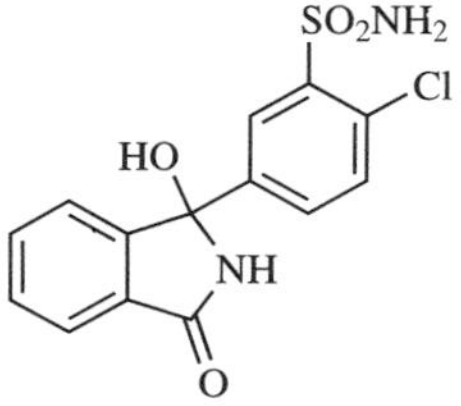
hydrochlorothiazide	[58-93-5]	$C_7H_8ClN_3O_4S_2$	Microzide, Esidrix, HydroDiural, Oretic, Aquazide-H etc.	antihypertensive	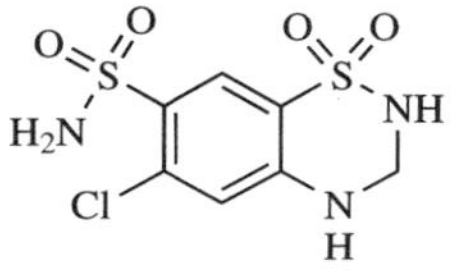

Table 8 (*Continued*)

Chemical/ generic name	CAS Registry Number	Molecular formula	Structure	Uses	Trade name
hydroflumethiazide	[135-09-1]	$C_8H_8F_3N_3O_4S_2$		antihypertensive	Diucardin, Saluron, Salutensin
Indapamide	[26807-65-8]	$C_{16}H_{16}ClN_3O_3S$		antihypertensive	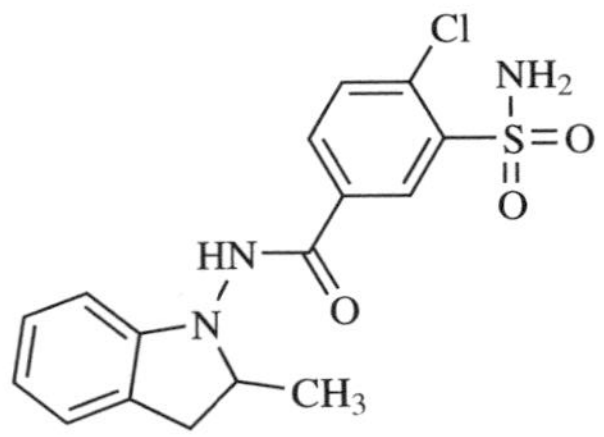 Lozol
methylclothiazide				antihypertensive	Enduron, Aquatensen, Dilutensen
metolazone	[17560-51-9]	$C_{16}H_{16}ClN_3O_3S$		antihypertensive	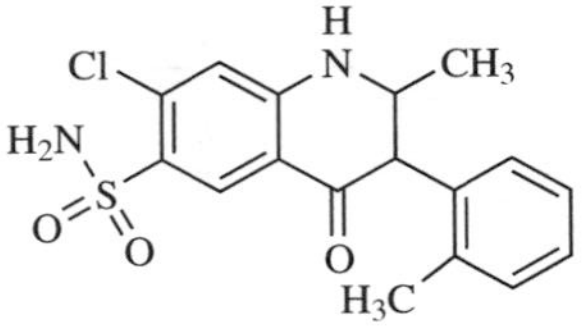 Mykrox, Zaroxolyn

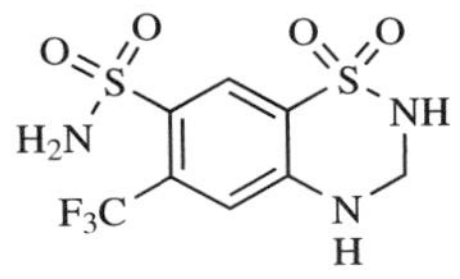

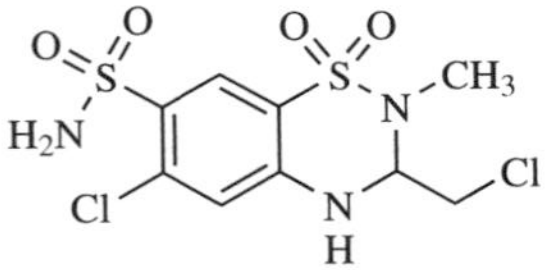

162

polythiazide	[346018-9]	$C_{11}H_{13}ClF_3N_3O_4S_3$	Renese, Minizide		antihypertensive
trichlormethiazide	[133-67-5]	$C_8H_8Cl_3N_3O_4S_2$	Aquazide, Naqua		antihypertensive
bumetanide	[28395-03-1]	$C_{17}H_{20}N_2O_5S$	Bumex		antihypertensive
ethacrynic acid	[6500-81-8]	$C_{13}H_{11}Cl_2NaO_4$	Edecrin		antihypertensive

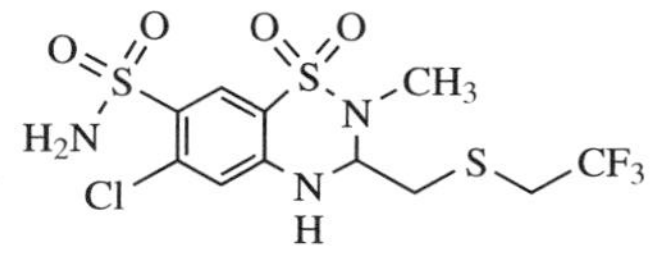

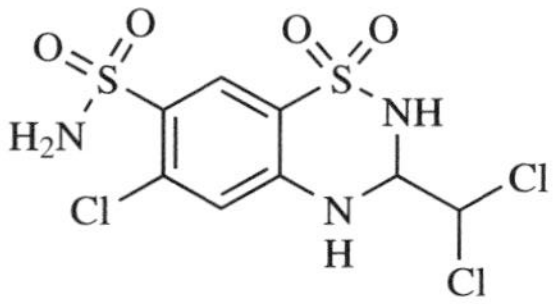

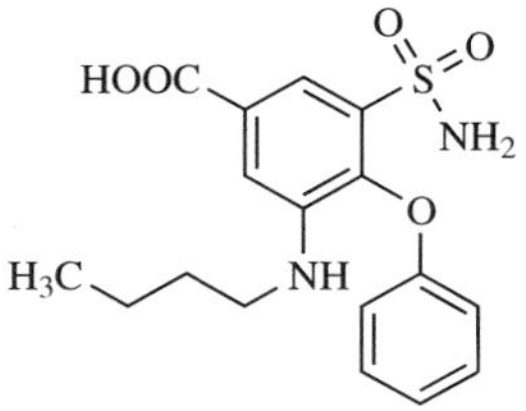

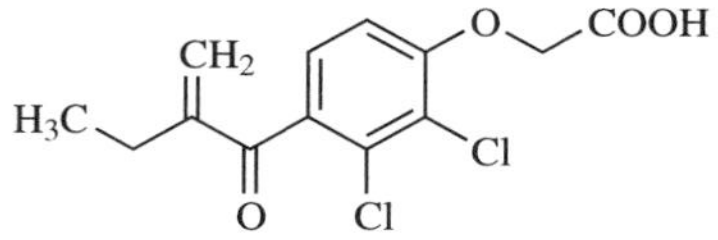

Table 8 (*Continued*)

Chemical/ generic name	CAS Registry Number	Molecular formula	Trade name	Uses	Structure
furosemide	[54-31-9]	$C_{12}H_{11}ClN_2O_5S$	Lasix	antihypertensive	
mannitol	[69-65-8]	$C_6H_{14}O_6$	Osmitrol	antihypertensive	
torsemide	[56211-40-6]	$C_{16}H_{20}N_4O_3S$	Demadex	antihypertensive	
urea	[57-13-6]	CH_4N_2O	Ureaphil	antihypertensive	

Potassium Sparing Diuretics

amiloride HCl [17440-83-4] $C_6H_9Cl_2N_7O \cdot 2H_2O$ Midamor, Moduretic antihypertensive

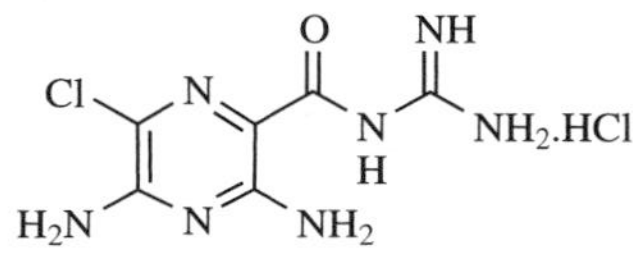

spironolactone [52-01-7] $C_{24}H_{32}O_4S$ Aldactone, Aldactazide antihypertensive

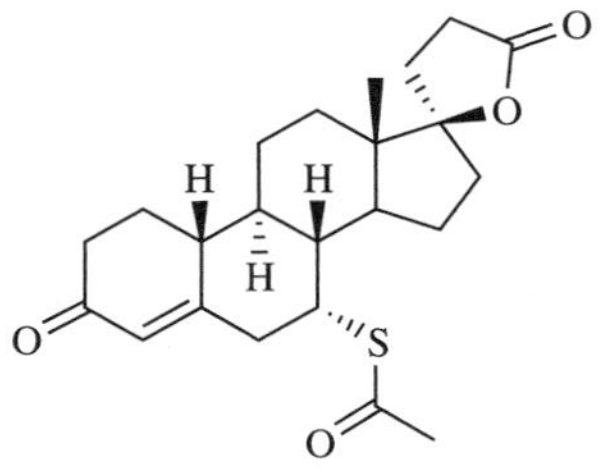

triamterene [396-01-0] $C_{12}H_{11}N_7$ Dyrenium, Dyazide, Maxzide antihypertensive

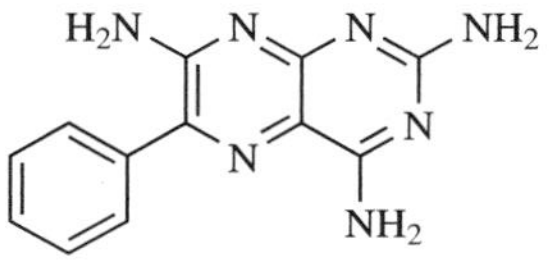

Direct Vasodilators

hydralazine [304-20-1] $C_8H_8N_4 \cdot HCl$ Apresoline, Alphapress

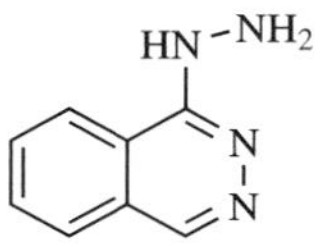

Table 8 (*Continued*)

Chemical/generic name	CAS Registry Number	Molecular formula	Trade name	Uses	Structure
diazoxide	[364-98-7]	$C_8H_7ClN_2O_2S$	Proglycem		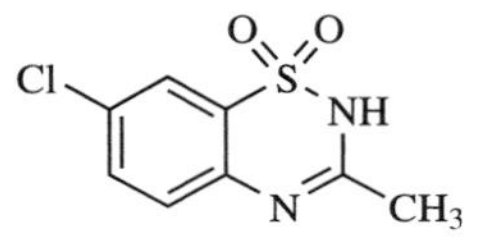
minoxidil	[38304-91-5]	$C_9H_{15}N_5O$	Rogain		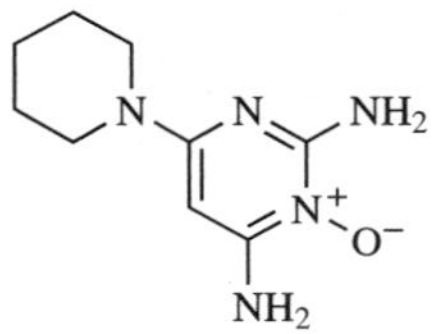
nitroprusside			Nitropress		

166

blood pressure ~10 mm Hg (1.3 kPa). Side effects include lethargy, dyspnea, nausea, dizziness, headache, impotency, cold hands and feet, vivid dreams and nightmares, bronchospasm, bradycardia, and sleep disturbances.

β-Adrenoceptor blockers for the treatment of hypertension include (*1*) the cardioselective β_1-adrenoceptor blockers without intrinsic sympathomimetic activity (ISA), ie, atenolol, bisoprolol, and metoprolol; (*2*) the cardioselective with ISA, ie, acebutolol; (*3*) the noncardioselective without ISA, ie, propranolol and timolol; and (*4*) the noncardioselective with ISA, ie, oxprenolol and pindolol.

6.13. α- and β-Adrenoceptor Blocking Agents. Labetalol possesses both α- and β-adrenoceptor blocking effects, with the (+)-(*SR*) isomer exhibiting more α-blocking activity (356). In particular, this agent possesses β_2-adrenoceptor antagonistic effects, and produces a significant reduction in blood pressure and a slight decrease in heart rate (357). It is efficacious in mild, moderate, and severe hypertension. The cardiac output is maintained due to an increase in stroke volume. Exercise-induced blood pressure increases are blunted and the heart rate is reduced. In chronic treatment, PRA is reduced. Side effects produced by this agent include gastrointestinal discomfort and dizziness.

6.14. Diuretics. Diuretics are drugs that increase the excretion of salts such as sodium chloride (NaCl) since the primary goal and principal mechanism of the hypotensive effects of diuretics is salt and fluid depletion leading to reduction of ECF volume (358,359). Diuretics are someitimes called water pills and are used to treat CHF, high blood pressure or edema (water retention). They are also used for certain types of kidney and or liver diseases. Acute effects of diuretics lead to a decrease in cardiac output and an increase in total peripheral resistance. However, during chronic administration, cardiac output and blood volume return toward normal and total peripheral resistance decreases to below pretreatment values. As a result, the blood pressure falls. The usual reduction in blood volume is ~5%. A certain degree of sustained blood volume contraction has to occur before the blood pressure decreases. The usual decrease in blood pressure achieved using a diuretic is ~20/10 mm Hg (2.7/1.3 kPa) (systolic/diastolic pressures).

Intake of a large amount of sodium chloride negates the antihypertensive effects of diuretics. Other mechanisms, such as direct vasodilating action, decreased responsiveness to vasopressor agents, stimulation of prostacyclin [35121-78-9], $C_{20}H_{32}O_5$, production, and reduction in the intracellular calcium ion concentration are some of the factors that play a trivial role in the overall antihypertensive effects of diuretics.

Diuretics, such as those of the thiazide type, have been the cornerstone of first-line antihypertensive treatments for decades. However, popularity and use have eroded as a result of increases in sudden death in patients on diuretic therapy, and unfavorable effects on blood lipid profiles, ie, increasing cholesterol and triglyceride levels. These effects have been implicated as possible causes for the lack of decrease in the mortality rate resulting from acute MI in patients treated with a diuretic (359,360). However, diuretics do protect against stroke and CHF.

Diuretics are needed to return to normal the expanded extracellular volume that other antihypertensive agents produce, such as fluid retention and blood volume expansion, via compensatory mechanisms of the body. The loss of efficacy of antihypertensive agents can be restored if a diuretic is used concomitantly. In the treatment of hypertension, high ceiling or loop diuretics, such as furosemide, ethacrynic acid, and bumetanide, are no more efficacious than the thiazide-type of diuretics. In fact, these agents cause more side effects, such as dehydration, metabolic alkalosis, etc, and therefore, should not be used except in situations where rapid elimination of fluid volume is clearly indicated.

Diuretics can cause hypokalemia, hyperglycemia, and hyperuricemia. After long-term treatment they may increase serum triglyceride and cholesterol (361). There are three types of diuretic medicines and each type works differently, but they all reduce the amount of salt and water in the body, which helps to lower blood pressure.

Thiazide and Thiazide-like Diuretics.   Thiazide (benzothiadiazine) diuretics are derivatives of 1,2,4-benzothiadiazine-7-sulfonamide-1,1-dioxide (Table 8). Substitution of the thiazide nucleus (R_2 and R_3 positions) increases the potency of the compounds. Most of the thiazids have a saturated 3,4-bond and either Chloro or CF_3 substituent at the 6-position of the thiazide nucleus. Thiazides occur as white crystalline powders and are insoluble in water. Thiazides enhance excretion of sodium, chloride and water by interfering with the transport of sodium ions across the renal tubular epithelium and cortical diluting segment of nephron. It is the only type of diuretic that has a vasodilatory effect. Thiazides also increase the excretion of other electrolytes such as potassium, bicarbonate, and to some extent calcium. These compounds also exercise a weaker diuretic effect as compared to loop diuretics. Most commonly used drugs include chlorothiazide (Diuril), hydrochlorothiazide (Microazide, Esidrix, Oretic), benzthiazide, cyclothiazide, indapamide (Lozol), chlorthalidone (Thalitone), bendroflumethiazide (Naturetin), and metolazone (Mykrox). They are mainly used as a first line therapy for mild to moderate hypertension, edema, congestive heart failure, hypercalciuria and in combination with loop diuretics to treat severe resistant edema.

Thiazides are administered orally (except chlorothiazide which is administered intravenously). They are absorbed from the GI tract in varying degrees. The onset of duration of action following oral administration occurs within 2 h and peak effect occurs 3–6 h after administration. However, chlorothiazide (intravenous route) acts much faster (15 min) and peak effect lasts ~30 min, and therefore it is administered for emergency situations. The duration of diuretic action is determined by individual rates of excretion and varies from 2 to 72 h. Most thiazides are excreted in urine unchanged. One of the most common adverse effects of thiazide therapy is potassium depletion leading to cardiac arrhythmias.

Potassium Sparing Diuretics.   They are weak diuretics that exert their effect mainly on collection tubes and act by blocking the Na^+ channel in the luminal membrane of the *principal cells* of the cortical collecting ducts. This reduces the Na^+ entry through the luminal membrane, and hence the net reabsorption of NaCl. Most commonly used drugs include amiloride (Midamor), triamterene (Dyrenium), spiranolactone (Aldactone), the latter is a competitive aldosterone antagonist at the cystolic receptor level (Fig. 8). Potassium sparing diuretics are used for the prevention of hypokalemia induced by loop or thiazide diuretics, secondary hyperaldosteronism due to hepatic cirrhosis and ascites, and Conn's syndrome.

Loop Diuretics. They are also called "high ceiling" diuretics due to their high diuretic potential since they can eliminate up to 20% of the filtered load of NaCl and H_2O to be excreted in the urine. This class of compounds acts by inhibiting cotransport of Na and K chloride ions in the thick ascending limb of Henle's loop (ALH). These diuretics cause the kidneys to increase flow of urine. They are mainly used for the treatment of hypertension, CHF in the presence of renal insufficiency, ARF, CRF, ascites, and nephritic syndromes. Most of the commonly prescribed loop diuretics (Fig. 8) include: furosemide (Lasix), bumetanide (Bumex), torsemide (Demadex), and ethaycrynic acid (Edecrin).

Carbonic Anhydrase (CA) Inhibitors. This class of compounds acts by inhibiting CA in luminal membrane of proximal tubule, thereby reducing the proximal HCO_3^- reabsorption. They are used to reduce intracellular pressure in glaucoma, to lower mountain sickness and to reduce urine pH in cystinuria. Examples of these drugs include acetazolamide, methazolamide and dichlorphenamide.

Osmotic Diuretics. These are freely filterable nonreabsorbable osmotic agents and primarily act on the proximal tubule to lower the reabsorption of salts and water. They are used to treat or prevent acute renal failure and most commonly used drugs include mannitol (Osmitrol), glycerol and urea (Ureaphil).

6.15. Other Direct Vasodilators. Vasodilators dilate or relax the smooth muscles of the vasculature directly or indirectly by releasing endogenous vasodepressors or by antagonizing the endogenous vasopressors or vasopressor systems (362,363). Vasodilators may interfere with the entry, intracellular release, and utilization of calcium, the activation of the protein kinase C system, cGMP formation, and EDRF turnover.

Hydralazine (Apresoline). Hydralazine reduces perpheral resistance and blood pressure due to its direct vasodilatory effect on vascular smooth muscle cells and the effect is greater on arteries than on veins. It has beeen suggested that cyclic AMP causes relaxation of arterial smooth muscle cells through a calcium binding stimulating mechanism. As a result, diastolic blood pressure decreases more than systolic pressure and it is always accompained by increased heart rate, cardiac output and stroke volume. Hydralazine is readily absorbed from the GI tract and metabolized on first pass through the liver by acetylation. The plasma half-life of hydralazine is ~2–4 h and is sometimes extended up to 8 h in some patients. It is mainly used in the management of moderate to severe hypertension and for the short-term treatment of severe CHF in conjunction with glycosides. Tachycardia, headache, dizziness, water, and sodium retention are principal side effects of hydralazine therapy.

Diazoxide (Proglycem). Diazoxide, chemically related to the thiazide diuretics, is a vasodilator and hyperglycemic agent having no diuretic effects. In fact, diazoxide has marked salt and water retention. It has been suggested that diazoxide has potassium channel opener effects. Because of rapid protein binding, diazoxide has to be given by rapid intravenous injection. Its use is limited to the treatment of hypertensive emergencies or crises. Diazoxide causes marked tachycardia, increases PRA, and causes hyperglycemia.

Minoxidil (Rogain). Minoxidil is a piperidineopyrimidine derivative and is one of the most potent vasodilators available for the treatment of hypertension. It also possesses hair growth stimulant properties. Minoxidil is a potassium channel opener acting directly on the smooth muscle of the arterioles to decrease peripheral resistance and blood pressure. In long-term treatment, minoxidil causes severe fluid retention, tachycardia, and hair growth (hirsutism) in balding areas of androgenetic alopecia. It stimulates the adrenergic nervous system and the renin-angiotensin system. Because of side effects, minoxidil is reserved for use in hypertensive patients having renal failure or in patients having hypertension that is refractory to other antihypertensive drugs.

Nitroprusside (Nitropress).   Sodium nitroprusside is a potent, fast-acting vasodilator that has to be administered intravenously. Nitroprusside exerts direct action on vascular smooth muscle, leading to peripheral vasodilation of arteries and veins. It acts on excitation–contraction coupling of vascular smooth muscle by interfering with both influx and intracellular activation of calcium. However, it has no effect on smooth muscle of the duodenum or uterus and is more active on veins than on arteries. Nitroprusside may also improve CHF by decreasing systemic resistance, preload and afterload reduction, and improved CO. It is used mainly in hypertensive crises and acute management of heart failure and to reduce bleeding during surgeries. Its effects terminate as soon as infusion of the drug is slowed or stopped, and returns to pretreatment levels within 1–10 min. The most common side effect is that it reacts with hemoglobin to produce cyanmethemoglobin and cyanide ion. Therefore, caution must be exercised as nitroprusside injection can result in toxic levels of cyanide. However, when used briefly or at low infusion rates, the cyanide produced reacts with thiosulfate to produce thiocyanate, which is excreted in the urine. It is rapidly metabolized by interaction with sulfhydral groups in the erythrocytes and tissues.

7. Antithrombolytic Agents

It has been well documented that the primary cause of acute myocardial infarction (AMI) is coronary arterial thrombosis (364). Thrombosis formation occurs at sites of ulcerated or fissured atheromatous plaques in the coronary circulation (365,366). Reperfusion of the coronaries to the ischemic–infarcted area within 6 h after the onset of AMI can salvage the myocardium and limit infarct size (367–369). Furthermore, the better the preservation of the ventricular function, the better the survival rate. Progressive irreversible myocardiac damage occurs 30 min after ischemia starts. Therefore, the faster the coronary arterial thrombus is lyzed, the less irreversible damage done on the myocardium. It is imperative that once an AMI is diagnosed, IV thrombolytic agent should be administered as quickly as possible.

The success of thrombus lysis depends mainly on how large the thrombus is and whether any blood flow still remains. The outcome is better if a large surface of the entire thrombus is exposed to the thrombolytic agent. As the clot ages, the polymerization of fibrin cross-linking and other blood materials increases and it becomes more resistant to lysis. Therefore, the earlier the thrombolysis therapy starts, the higher the frequency of clot dissolution. Various available thrombolytic agents used to dissolve blood clots are listed in Table 9 (370–385).

7.1. Glycoprotein IIb/IIIa Receptor Antagonists.   These compounds are also called blood thinners since they thin blood by blocking platelet activity.

Table 9. Antithrmobolytic Agents

Chemical/ generic name	CAS Registry Number	Molecular formula	Trade name	Structure
		Glycoprotein IIb/IIIa Receptor Antagonists		
abciximb	[143653-53-6]		ReoPro	
eptifibatide	[188627-80-7]	$C_{35}H_{49}N_{11}O_9S_2$	Integrillin	
lamifiban	[144412-49-7]	$C_{24}H_{28}N_4O_6$		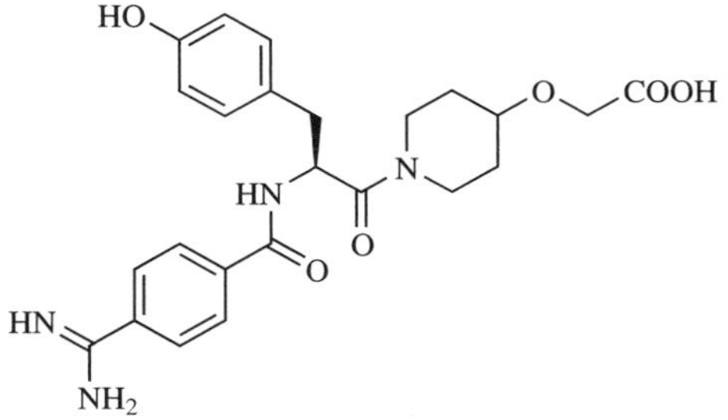
tirofiban	[144494-65-5]	$C_{22}H_{36}N_2O_5S$	Aggrastat	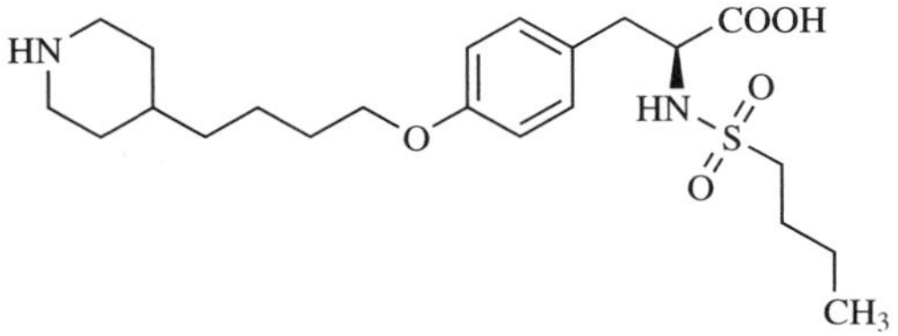

Table 9 (Continued)

Chemical/generic name	CAS Registry Number	Molecular formula	Trade name	Structure
			Anticoagulants	
heparin sodium dalteparin danaparoid enoxaparin tinzaparin	[9041-08-11]		Fragmin Orgaran Lovenox Innohep	
			Thrombolytic Agents	
human t-PA alteplase (TPA) streptokinase	[105857-23-6] [9002-01-1]		Activase Streptase, Kabikinase	
tenecteplase urokinase	[191588-94-0] [9039-53-6]		TNKase Abbokinase, Breokinase, Winkinase	
APSAC-anistreplase prourokinase duteplase	[81669-57-0] [82657-92-9] [120608-46-0]			

Platelet adhesion, activation, and aggregation are key processes leading to the formation of platelet-rich coronary artery thrombi and the development of acute coronary syndromes and ischemic complications during PCI. The platelet glycoprotein IIb/IIIa receptor inhibitor drugs are very beneficial for many patients with angina and do not appear to pose any elevated risk for stroke, including strokes caused by bleeding. Some of the most widely prescribed drugs include Abciximab (ReoPro, Centocor), eptifibatide (Integrelin), lamifiban, and tirofiban (Aggrastat). They are used to reduce the risk for heart attack or death in many patients with unstable angina and non-Q-wave myocardial infarctions when used in combination with heparin or aspirin. Patients with unstable angina showing elevated levels of troponin T factor are good candidates for these drugs. Some of the commonly used drugs are discussed.

Abciximab (ReoPro). It is the Fab fragment of the chimeric human-murine monoclonal immunoglobulin antibody 7E3. It binds selectively to platelet glycoprotein (GP IIb/IIIa) receptors and inhibits platelet aggregation. Abciximab is also known to bind to the vitronectin receptor located on platelets and vascular endothelial and smooth muscle cells. It is used with heparin and aspirin as an adjunct to precutaneous coronary intervention (PCI) for the prevention of acute cardiac ischemic complications in patients undergoing PCI with unstable angina. The GP IIb/IIIa receptor inhibitors are also used for managing unstable angina or non-Q-wave MI. It is administered by direct IV injection or by IV infusion, and the most common adverse effect of abciximab is bleeding.

Eptifibatide (Integrilin). It is a synthetic cyclic heptapeptide and is a selective platelet—aggregation (GP IIb/IIIa receptor) inhibitor. It is structurally similar to barbourin, a peptide constituent of the venom of southeastern pigmy rattlesnake, *Sistrurus m. barbouri*. Eptifibatide is composed of six amino acids and a mercaptopropionyl residue with a disulfide bridge between the cysteine amide and the mercaptopropionyl moiety. The active domain of eptifibatide consists of a modified lysine-glycine-aspartate (KGD) amino acid sequence similar to the physiological arginine-glycine-aspartate (RGD) sequence that binds to GP IIb/IIIa receptor on activated platelet and causes platelet aggregation. The substitution of lysine for arginine in eptifibatide increases the receptor binding selectivity. The pharmacokinetics of eptifibatide is linear, and peak plasma concentrations occur within 5 min after intravenous loading. It has a rapid onset and short duration of action, maximal inhibition of platelet aggregation occurring within 15 min after initiation of therapy. Eptifibatide is ~25% bound to plasma proteins and is eliminated by renal and nonrenal mechanisms. It is metabolized by deamidation to a 41% active metabolite in addition to formation of other polar metabolites. It is mostly used to reduce the risk of acute cardiac ischemic events (MI or death) in patients with unstable angina or non-Q-wave MI syndromes.

Tirofiban (Aggrastat). Tirofiban hydrochloride is a synthetic nonpeptide tyrosine analogue and is a selective, competitive platelet aggregation inhibitor. Similar to abciximab and eptifibatide, tirofiban is a platelet glycoprotein IIb/IIIa receptor inhibitor. It is synthesized by addition of an *n*-butylsulfonyl group to the C-terminus and 4-(piperidin-4-yl) butyloxy group to the N-terminus of tyrosine, resulting in enhancement of the drug's potency. Tirofiban inhibits platelet aggregation by preventing the binding of fibrinogen to activated

GP IIb/IIIa receptors, in a concentration-dependent manner. It also has a rapid onset and short duration of action. However, it is 65% bound to plasma proteins but plasma concentrations decline in biphasic manner, the half-life being 1.2–2 h. Similar to GP IIb/IIIa antagonists, it is used to manage unstable angina or non-Q-wave MI and also during PCI, and coronary artery bypass grafting (CABG).

7.2. Anticoagulants. The most widely accepted indications for anticoagulant therapy include venous thrombosis and pulmonary embolism. Prevention of these conditions in high risk patients or those undergoing certain types of major surgery to prevent postoperative thrombosis is mostly recommended. The indirect acting synthetic anticoagulants include three-substituted derivatives of 4-hydroxy coumarin (dicumarol, warfarin), indan 1, 3-dione derivatives such as antisinidione. These drugs regulate the synthesis of blood coagulation factors II (prothrombin), VII (proconvertin), IX (plasma thromboplastin component) and X (Sturat-Prower-factor) in the liver by interfering with the action of vitamin K. The latter is required for the γ-carboxylation of several glutamic acid residues in the precursor proteins of these coagulation factors. This interference leads to the synthesis of several dysfunctional coagulation factors. Dicumarol, and antisinidione drugs are administered orally, in a single daily dose, while warfarin can be administered both orally as well as by intravenous injections when oral therapy is not feasible. They are mostly used for the prophylaxis and treatment of deep vein thrombosis or any complications associated with atrial fibrillation or coronary occlusion.

Heparin Sodium. It is an anionic, sulfated glycosaminoglycan that occurs in mast cells. Heparin is prepared from either bovine lung tissue or porcine intestinal mucosa and is commercially available as a sodium salt. It is a heterogeneous molecule with an average molecular weight of 12,000 kDa. Heparin acts as a catalyst to enhance the neutralization rate of antithrombin III by thrombin instantaneously and activates coagulation factor X. Heparin is administered by intravenous infusion, injection or deep subcutaneous injection since it is not absorbed by the GI tract. The onset of anticoagulant activity is immediate following direct intravenous injection and is known to bind extensively to LDL, globulins, and fibrinogen. The plasma half-life of heparin is 1–2 h and the metabolic pathway is unclear. The major adverse effect includes hemorrhagic complications. It is used for prophylaxis and treatment of venous thrombosis and as an adjunct therapy in a number of situations.

Dalteparin (Fragmin). It is a depolymerized heparin prepared by nitrous acid degradation of unfractionated heparin of porcine intestinal mucosa. It is commercially available as a sodium salt with a molecular weight of 2000–9000 kDa and therefore also called as low molecular weight heparin. Dalteparin is used for the prevention of postoperative deep vein thrombosis and associated pulmonary embolism in patients undergoing hip replacement or abdominal surgery. It is also used in combination with aspirin to reduce the risk of acute cardiac ischemic events such as unstable angina or non-Q-wave MI.

Danaparoid (Orgaran). It is also isolated from porcine intestinal mucosa and has an average molecular weight of 5500–6000 kDa and structurally it consists of known amounts of three specific glycosaminoglycans (~84% heparin sulfate, 12% dermatan sulfate, and 4% chondroitin sulfate). The mechanism of action is similar to other heparins and is administered as subcutaneous injection. It also finds applications for the prevention of postoperative deep-vein thrombosis.

Enoxaparin (Lovenox). It is a depolymerized heparin produced by alkaline degradation of unfractionated benzylated heparin of porcine intestinal mucosa. The average molecular weight is 4,500 kDa and has a similar mode of action as other heparins. It is mostly used during and following, for the prevention of postoperative thrombosis in patients undergoing hip and knee replacement surgery.

Tinzaprin (Innohep). It is a depolymerized heparin prepared by bacterial enzymatic degradation of unfractionated heparin or porcine intestinal mucosa. The average molecular weight ranges from 5,500–7,500 kDa. The mechanism of action is similar to other heparins and it is generally used for the prevention of thrombosis.

All of these anticlotting agents, either anticoagulants or antiplatelet drugs, are generally used to treat unstable angina, to protect against heart attacks, and prevent blood clots during heart surgery. They can be used alone or in combinations, depending on the severity of the condition. Clopidogrel (Plavix), a platelet inhibitor, has been shown to be more effective than aspirin by 20% for reducing the incidence of a heart attack. Other promising anticlotting drugs comprise argatroban (Novastan) and forms of hirudin (bivalirudin lepidrudin or desirudi), a substance derived from the saliva of leeches. One study suggested that the hirudin agents may be superior to heparin in preventing angina and heart attack, although bleeding is a greater risk with hirudin.

8. Thrombolytic Agents

Streptokinase, prourokinase, and acetylated plasminogen streptokinase activator complex (APSAC) agents are indirect plasminogen activators, t-PA and urokinase are direct plasminogen activators, whereas alteplase and tenecteplase are recombinant human t-PA. All these activators convert the formation of the active proteolytic enzyme plasmin [9001-90-5] from the proenzyme, plasminogen, [9001-91-6] by cleaving the Arg_{560}–Val_{561} bond. After intravenous administration, all five thrombolytic agents achieve approximately the same incidence of reperfusion rate (60–70%) if used optimally. t-PA and prourokinase require simultaneous heparin [9005-49-6] administration when used; the other thrombolytics do not. Reocclusion rate is higher with t-PA if heparin use is not maximized. The half-life is short (5–8 min) using t-PA and prourokinase, medium (16–23 min) using urokinase and streptokinase, and long (105 min) using APSAC. When used at efficacious doses, each agent has similar degrees of bleeding complications. The least expensive thrombolytic agent is streptokinase, which is about one-tenth of the others. No single agent is superior when all factors are taken into consideration. Whereas t-PA is clot selective, it has a very short half-life and simultaneous use of heparin is required. Bleeding is therefore as much of a problem with t-PA as the less clot selective agent, such as streptokinase. Streptokinase causes severe systemic fibrinogen degradation leading to extensive hypofibrinogenemia. Under this condition, reocclusion is less and the dissolution of the thrombus is more complete. This may explain why streptoki-nase has been shown, in most studies, to be as efficacious as t-PA. The most recent comparison study reporting the equal efficacy of t-PA, APSAC, and strep-tokinase tends to confirm this point. Results of the largest clinical trials show that these agents reduce patient mortality from 16 to 30%.

The dosage of each thrombolytic agent should be high enough to produce a lysis state long enough for complete clot dissolution to occur. Aspirin, a platelet aggregation inhibitor, has been demonstrated to decrease mortality of AMI by itself and it is further enhanced by the use of the thrombolytic agents if used con-comitantly (386–388). The use of combinations of aspirin, heparin, vasodilator, and the thrombolytic agent decreases the incidence of rethrombosis.

8.1. Tissue-Type Plasminogen Activator (t-PA). Endogeneous human t-PA is a glycosylated, trypsin-like serine protease secreted mainly by vascular endothelial cells as a single chain polypeptide. It is a physiological activator of plas-minogen in the body. The complete structure of human t-PA consists of a polypeptide of 562 amino acids, that include a leader sequence of 35 amino acids and a mature protein of 527 amino acids having Mol. Wt. ~64,000. It is also made in the liver, spleen, lungs, muscle, and other tissues. The commercial preparation is derived from using DNA recombinant technology and it is a recombinant rt-PA. The t-PA can occur in either a one- or two-chain form. Both forms are active but the two-chain form is longer acting (389–392).

t-PA has a much higher affinity for fibrin and fibrin-bound plasminogen as compared to the circulating plasminogen. For this reason, under normal condi-tions generalized (systemic) fibrinogenolysis in the circulation is thought not to occur. However, in the treatment of AMI, the large amount of t-PA used and infused for a prolonged period causes generalized fibrinogenolysis because of the activation of circulating plasminogen. In the treatment of AMI, a bolus intra-venous injection of 10 mg is followed by an intravenous infusion of 50 mg/h for 1h and 20 mg/h for 2 h. The half-life of t-PA in the circulation is ~4–5 min. It is cleared by the liver. The activation of plasminogen by t-PA is accelerated by heparin. The primary side effect is bleeding or hemorrhage.

Alteplase (Activase). It is a biosynthetic (recombinant DNA) form of the enzyme human t-PA, thrombolytic agent. Alteplase is prepared from the cultures of genetically modified mammalian (chinese hamster ovary, CHO) cells using recombinant DNA technology. Commercially available alteplase is a glycoylated, predominantly (60–80%) one-chain form of t-PA consisting of 527 amino acids and 3 carbohydrate side chains and the molecular weight ranges from 65,000 to 70,000 kDa reflecting the variation in carbohydrate moieties. The amino acid sequence of alteplase is identical to that of human melanoma cell t-PA. Alte-plase is a white lypholyzed powder and practically insoluble in water, but pre-sence of arginine increases the auqeous solubility substantially. Alteplase is administerd by intravenous injection [usual adult dose 100 mg (58 million IU)] and used as a thrombolytic agent in selective cases of MI and coronary artery thrombosis. The most frequent and severe adverse effect of rt-PA is hemorrhage.

Tenecteplase (TNKase). Similar to alteplase, tenecteplase is a rt-PA and is a thrombolytic agent. Although tenecteplase is structurally and pharmacologi-cally related to alteplase, it exhibits higher fibrin selectivity, gerater resistance to PA inhibitors and has a longer plasma half-life than alteplase. It binds to fibrin and converts plasminogen to plasmin. It is administered by intravenous injection over 5 s and should be given as soon as possible or within 30 min after onset of acute MI.

Streptokinase (Streptase). Streptokinase is a nonenzymatic protein produced by a group of C β-hemolytic streptococci. It is a single-chain protein containing 415 amino acids, Mol. Wt. of 45,000–50,000 kDa (393–395). It is available as a lypholized powder and freely soluble in water. It is stable in solution at pH 6–8. In contrast to other anticoagulants, streptokinase promotes thrombolysis. But only after streptokinase combines with plasminogen on a 1:1 basis to form a streptokinase–plasminogen complex, to expose the active site of the plasminogen portion, does it become an active species. This complex then becomes a potent enzyme activator and converts residual plasminogen to plasmin. The complex has a half-life of ~23 min in the circulation. The usual dose used is 1.5 million units to ensure enough streptokinase. It is left after being cleared from the circulation by streptokinase antibodies and inhibitors in the blood stream. Thrombolytic activity lasts for 3–4 h. Streptokinase administration also causes fibrinolysis and fibrinogenolysis, through activation of plasma plasminogen, in the circulation and depletes plasma fibrinogen. Streptokinase is used as a thrombolytic agent in acute MI, coronary artery thrombosis and MI, deep vein thrombosis and pulmonary embolism.

The mortality is usually reduced from 12% in the control group to 9–10% in the streptokinase group. Side effects are bleeding, hemorrhage, fever, and allergy.

Urokinase (Abbokinase). Urokinase is direct plasminogen activator enzyme produced by the kidneys and endothelial cells and excreted in urine. It was first isolated from human urine, but now it is produced by cultured human fetal kidney cells. It is a double chain of 411 amino acids, Mol. Wt. 54,000–55,000 kDa, having a half-life of 16 min. It is metabolized in the liver. Urokinase is given at 4500 units/kg in a loading dose, and then at 4500 units/kg per hour by a continuous intravenous infusion. Urokinase converts plasminogen to plasmin with greater affinity for the fibrin-bound plasminogen. However, during therapy of AMI, the circulating plasminogen is also activated and catalyzes the breakdown of fibrin and also fibrinogen, causing a generalized (systemic) lytic state. Hemorrhage is the most serious side effect. Recently, the U.S. FDA approved the use of urokinase for the tretament of pulmonary embolism.

Acylated (Anisolyated) Plasminogen–Streptokinase Activator Complex (APSAC). After intravenous administration streptokinase becomes active only after forming a complex with plasminogen. This activated form is rapidly inactivated in the plasma and also causes generalized (systemic) fibrinogenolysis. Therefore, most of the active complex cannot reach the arterial thrombus in the coronary circulation. To overcome this, the active site (catalytic center) of the preformed streptokinase–plasminogen complex is protected or blocked by an acyl group and APSAC is formed (396–398). However, the fibrin-binding sites are not changed, therefore APSAC binds rapidly to the fibrin clot. So the local dissolution of the clot is preferentially obtained. Deacylation occurs slowly in plasma.

In the acylated form, the APSAC offers a prolonged half-life (105 min), greater lysis potency, and less systemic lytic state. APSAC is used at 30 U (30 mg) by intravenous injection.

Prourokinase. The single-chain urokinase-type plasminogen activator (scu-PA) urokinase does not have high fibrin specificity and therefore causes systemic fibrinogenolysis. In the early 1980s, a single-chain precursor of urokinase (prourokinase) was isolated and purified from human urine and cultured cell lines. Now, it is produced using recombinant DNA biotechnology. It was found that prourokinase was much more fibrin-specific than urokinase and is not effective in activating plasma plasminogen. Its mechanism of action was distinctly different from that of t-PA.

Prourokinase is a single-chain protein containing 411 amino acids. In clinical use, scu-PA does not bind to fibrin only and its use causes a decreased plasma fibrinogen of 80%. Its half-life in the circulation is 5 min and is cleared by the liver. It is used at 40–70 mg over 1 h and heparin is needed simultaneously. Fibrin specificity and thrombolytic efficacy are similar to that of t-PA.

In general, it has been found that t-PAs reduce 50–90% AMI mortality, 30% of AMI mortality, and 15% AMI mortality if administered in the first hour, within 6 h, and within 6–12 h of symptoms, respectively. The direct benefits of using t-PAs include chest pain relief, diminished ECG changes, lowering of MB creatinine kinase levels and decreased incidences of reinfarction, CHF and shock.

9. New Human t-PA Thrombolytic Agents Under Development

Several new recombinant expressions are currently being explored toward improvement of thrombolytic agents (399). Some of the important methods include construction of mutants of PA, chimeric PA, conjugates of PA with monoclonal antibodies, and PA from either bacterial or animal origin. Some of these thrombolytic agents have shown promise in animal models of venous and arterial thrombosis and are being further investigated in clinical trials. Monteplase is a modified tissue-type plasminogen activator (t-PA) synthesized by substitution of Cys84 by Ser in the epidermal growth factor domain and expressed in baby Syrian hamster kidney cells. This mutation prolongs the half-life by >20 min as compared to 4 min for native t-PA. This mutant has increased thrombolytic activity, slower clearance and enhanced resistance to the inhibitor PAI-1.

Reteplase is a non-glycosylated deletion mutant of wild-type human t-PA containing only kringle 2 and the protease domain, but lacking its kringle 1 and finger and growth factor domains. This structural modification leads to decreased fibrin binding, lower affinity toward endothelial and liver cells resulting into an extended half-life.

Lanoteplase is another deletion mutant of t-PA with >10 times the half-life than alteplase. YM866-is another mutant of human t-PA constructed by deleting 92–173 amino acids of kringle 1and replacing Arg 275 by glutamic acid that confers a longer half-life to the mutant. Also, recombinant glycosylated prourokinase is a rapid acting and safe t-PA, but with greater stability than recombinant unglycosylated pro-urokinase.

Staphylokinase (SAK) produced by *Staphylococcus aureus* induces efficient and rapid recanalization after bolus injection, however it is immunogenic. In patients with acute MI, reteplase, administered as bolus injection. However, the quest continues for plasminogen activators with enhanced potency, specific thrombolytic activity, fibrin selectivity and longer half-life.

The introduction of new clot buster drugs or thrombolytics has revolutionized the treatment of heart attack or AMI, and current research is focused on identifying better thrombolytic and adjunctive agents or mechanical interventions such as the use of angioplasty or stents. A possible future approach that could universally be adopted consists of combination therapy and adjunctive/rescue percutaneous intervention at hospitals.

10. Physiology and Biochemistry of Congestive Heart Failure

The heart, a four-chambered muscular pump, propels blood throughout the cardiovascular system. The left ventricle is the principal pumping chamber and is therefore the largest of the four chambers in terms of muscle mass. The efficiency of the heart as a pump can be assessed by measuring cardiac output, left ventricular pressure, and the amount of work required to accomplish any required amount of pumping.

When the heart begins to fail as a pump, a number of pathophysiologic processes occur. Perhaps the easiest to perceive is simply that some of the blood within the left ventricle is not completely expelled during muscle contraction or systole. This subsequently requires more work during future systoles as additional blood returns from the pulmonary vein. A sort of congestion begins to develop as the heart cannot develop enough pumping ability to provide adequate oxygenated blood to the arterial circulation or removal of metabolic waste through the venous circulation.

There can be a number of underlying causes of CHF. The most prevalent is the lack of oxygenated blood reaching the heart muscle itself because of coronary artery disease with myocardial infarction (401). Hypertension and valvular disease also contribute to CHF as well, but to a lesser extent in terms of principal causes for the disease.

Both high energy containing ATP and calcium are integral components of cardiac muscle contraction. Hydrolysis of ATP is coupled to cross-linking of the actin–myosin contractile proteins, resulting in mechanical shortening of the functional unit of cardiac muscle, the sarcomere. Calcium is involved in binding of the contractile protein troponin, but the quantity of calcium entering the myocardial cell is finely regulated, concomitantly regulating the extent of cardiac muscle contraction. Imbalance in any aspects of the chemical process of contraction can result in reduced pumping ability and contribute to CHF.

10.1. Therapy of Congestive Heart Failure. Many of the drugs used to combat congestive heart failure are inotropic agents, some of which are shown in Table 10. Inotrope is a derivation of the Greek *ino* (fiber) and *tropikos* (changing or turning). A positive inotropic agent is therefore one that increases cardiac muscle contractility associated with CHF.

Table 10. Other Therapeutics

Chemical/ generic name	CAS Registry Number	Molecular formula	Trade name	Uses	Structure
Inotropic Agents: Glycosides					
digitalin	[752-61-4]	$C_{36}H_{56}O_{14}$	Diginorgin	cardiotonic	
digitalis			Digitonin, Digitalin	cardiotonic	
digitogenin	[511-34-2]	$C_{27}H_{44}O_5$		cardiotonic	
digitonin	[11024-24-1]	$C_{56}H_{92}O_{29}$	Digitin	cardiotonic	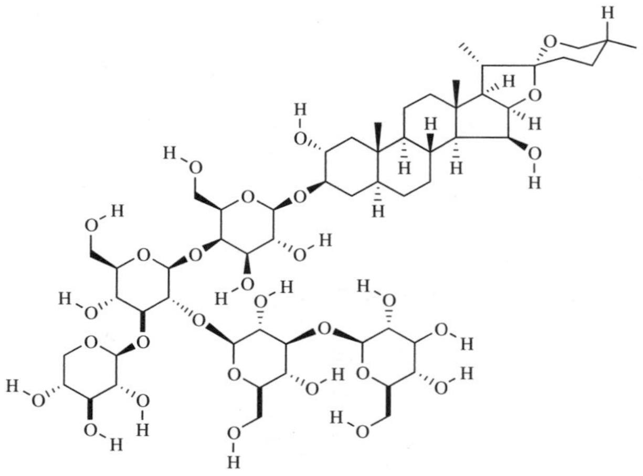
digitoxin	[71-63-6]	$C_{41}H_{64}O_{13}$	Digitalin, Carditoxin	cardiotonic	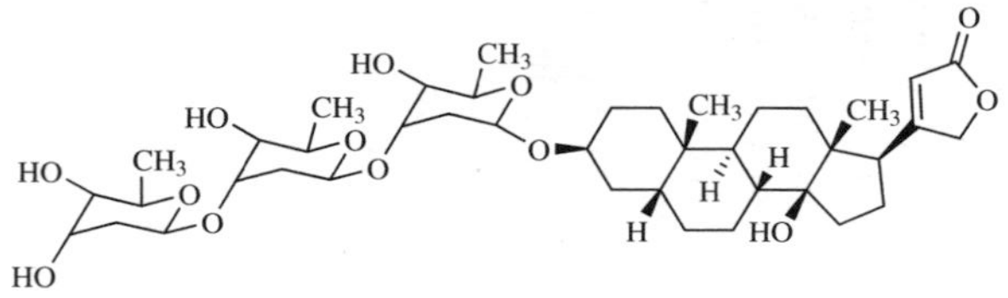

180

Phosphodiesterase Inhibitors

amrinone	[60719-84-8]	$C_{10}H_9N_3O$	Cartonic, Inocor	cardiotonic 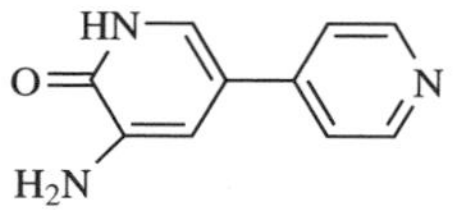
milrinone	[78415-72-2]	$C_{12}H_9N_3O$	Corotrope, Primacor	cardiotonic 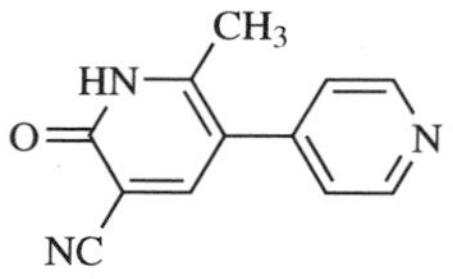
enoximone	[77671-31-9]	$C_{12}H_{12}N_2O_2S$	Perfane, Perfan	cardiotonic 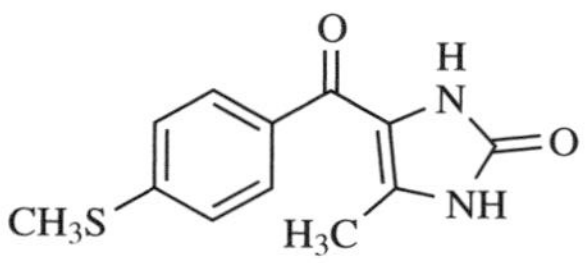
dipyridamole	[58-32-2]	$C_{24}H_{40}N_8O_4$	Cardoxin, Coridil, Prandiol	cardiotonic 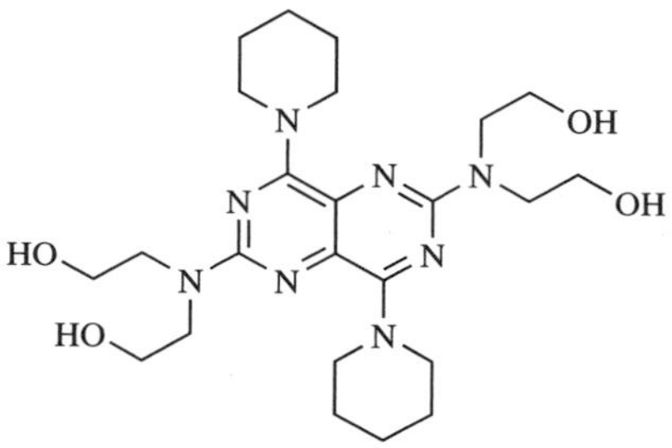

Table 10 (*Continued*)

Chemical/ generic name	CAS Registry Number	Molecular formula	Trade name	Uses	Structure
satigrel	[111753-73-2]	$C_{20}H_{19}NO_4$		antithrombotic	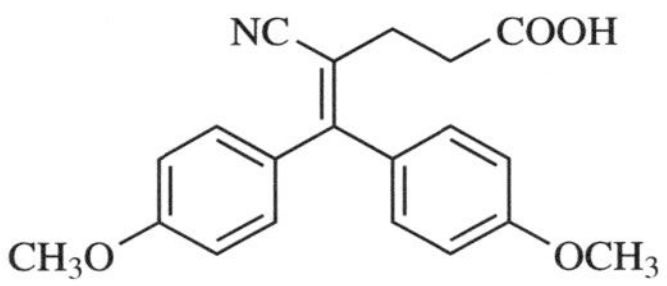
sildenafil	[139755-83-2]	$C_{22}H_{30}N_6O_4S$	Viagra	erectile dysfunction	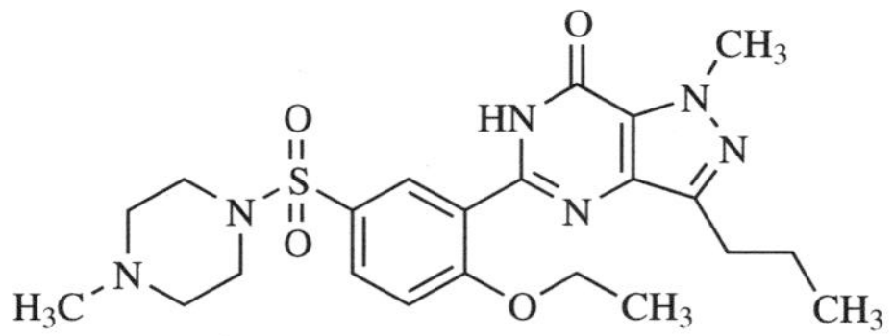
tadalafil	[171596-29-5]	$C_{22}H_{19}N_3O_4$	Cialis	erectile dysfunction	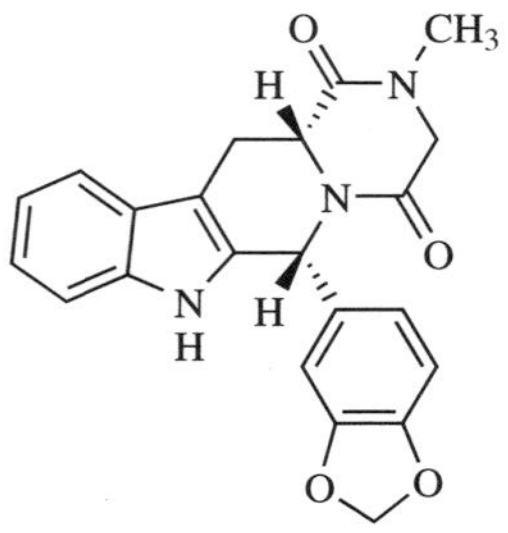

vardenafil

[224785-90-4]

$C_{23}H_{32}N_6O_4S \cdot HC \cdot 3H_2O$

Levitra

erectile
dysfunction

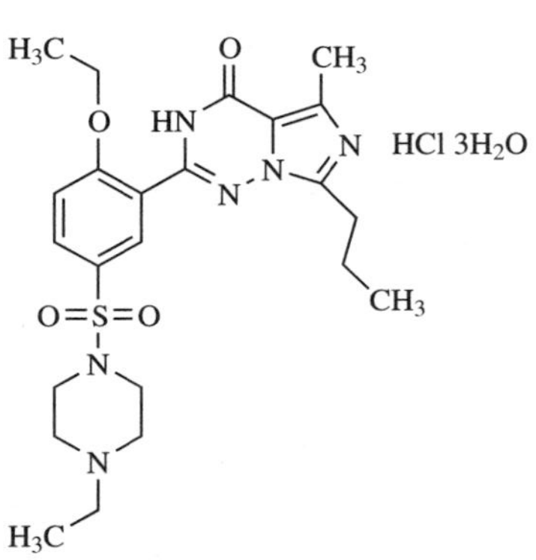

Cardiac Glycosides.   Glycosides are a distinct class of compounds that are either found in Nature or can be synthetically prepared. The natural glycosides are isolated from various plant species, namely, digitalis purpurea Linne, digitalis lanata Ehrhart, strophanthus gratus, or acokanthea schimperi. Therefore, these compounds are also named as digitalis, digoxin, digitoxin. Currently, digoxin is the only cardiac glycoside commercially available in the United States. Glycosides have a characteristic steroid (aglycone) structure complexed with one or more types of sugar moiety at the C3 position of the steroid through the β-hydroxyl group. Increasing, the number of hydroxyl groups on the aglycone or additional sugar moiety increases polarity and decreases lipid solubility. The sugar moiety further affects in part the activity of glycosides by influencing solubility, absorption, distribution, and toxicity. Glycosides are sparingly soluble in water and freely soluble in alcohol. All commercially available glycosides obtained from the *Digitalis* species contain the same basic aglycone but differ in the substitution at the C12 position of the aglycone and in the sugar substituent.

Glycosides are mainly used in the prophylactic management and treatment of congestive heart failure, atrial fibrillation, and recurrent atrial tachycardia. They are also used in conjunction with ACE inhibitors, diuretics, and β-adrenergic blocking agents. They are known to relieve the symptoms of systemic venous congestion (right-sided heart failure or peripheral edema) and pulmonary congestion (left-sided heart failure). However, glycosides also find applications to treat and prevent sinus and supraventricular tachycardia and symptoms of angina pectoris and myocardial infarction, but only in combination with β-adrenergic blocking agents and in patients with congestive heart failure.

The exact mechanism of pharmacological action of glycosides has not been fully elucidated. However, glycosides exhibit a positive inotropic effect accompanied by reduction in peripheral resistance and enhancement of myocardial contractility resulting in increased myocardial oxygen consumption. They also inhibit the activities of sodium–potassium activated ATPase, an enzyme required for the active transport of sodium across the myocardial cell membranes (402).

Glycosides are normally administered via various routes either orally or by intravenous injection and digoxin has a half-life of 36 h while that of digitoxin is to 5–7 days in normal patients. Cardiac glycosides undergo varying degrees of hepatic metabolism, enterohepatic elimination, renal filtration, and reabsorption depending on their polarity and lipid solubility. Cardiac glycosides increase plasma estrogen in women and decrease plasma testosterone in men. Toxic effects of glycosides are mainly GI, CNS, and biochemical in origin.

Nondigitalis Inotropic Agents.   A number of inotropic agents that are not structurally related to the cardiac glycosides have been investigated for treatment of CHF. All are somehow mechanistically related to the complex series of events involved in normal cardiac muscle contraction, and in terms of pumping ability are capable of increasing left ventricular contractility. Many of these drugs do not perform in diseased cardiac tissue in a manner similar to that observed in normal, nondiseased muscle. Many of these agents act via the β-adrenergic receptors of the sympathetic nervous system (SNS).

Catecholamines.   Simplistically, the SNS is composed of nerves capable of synthesizing and releasing various catecholamines that can bind appropriate molecular receptors postsynaptically. Depending on their synthetic structure, they have the ability to bind, and therefore stimulate various postsynaptic events through α-adrenergic, β-adrenergic, and dopaminergic receptors.

Each receptor has a specific role, often vasodilation, which serves to facilitate the ability of a failing heart to pump against reduced systemic pressure. Side effects of catecholamines include arrhythmias and tachyphylaxis. In addition, potent vasodilators such as isoproterenol [7683-59-2], $C_{11}H_{17}NO_3$, have a dual effect, namely increased contractility through direct cardiac muscle stimulation (403). If combined with significant reductions in blood pressure, isoproterenol can produce detrimental effects through a combination of reduced venous return yet increased oxygen requirements of the rapidly contracting cardiac muscle. The orally active catecholamine pirbuterol [38677-81-5], $C_{12}H_{20}N_2O_3$, has many of these classic side effects (403). These drugs have been selectively developed to affect specific receptor subtypes through radioligand binding studies using microsomal membrane preparations from animal tissue. This rapid method of screening for selective agonists and antagonists has produced considerable quantities of chemical structures that have potential as drugs. Each results in positive, yet relatively short-term efficacy in patients having CHF (404).

Dobutamine, a positive inotrope from the sympathomimetic amine family, was introduced in the early 1980s and possesses an asymmetric carbon atom resulting in two enantiomers. The commercial product is a 50:50 racemic mixture of the optical isomers, producing a combined β-adrenergic agonist and α-adrenergic antagonist effect (405) from the (+) enantiomer. The (−) enantiomer is a potent α-adrenoceptor agonist (406). The positive inotropic activity is a result of the two enantiomers, especially as relates to functional, simultaneous effects on cardiac, and vascular smooth muscle. Because of this unique combination of activity, dobutamine is superior to isoproterenol, dopamine, epinephrine, and norepinephrine, primarily because of dobutamine's lesser effect on blood pressure reduction. At doses of 2.5–15 µg/(kg/min) dobutamine increases stroke volume, while reducing peripheral resistance. Even after drug infusion is stopped, the patient can continue to benefit (407). Because it is administered by infusion, the drug is not approved by the FDA for outpatient administration.

Agiotension-Converting Enzyme Inhibitors or α-Blockers. ACE inhibitors prevent the conversion of angiotension I to angiotension II, a potent vasoconstrictor, consequentially reducing plasma concentrations of angiotension II, and hence vasodilation. This results in attenuation of blood pressure. The ACE inhibitors also affect the release of renin from kidneys and increase the plasma renin activity (PRA). It has been suggested that the hypotensive effect of ACE inhibitors may decrease vascular tone due to inhibition of angiotension induced vasoconstriction and increased sympathetic activity. The reduced production of angiotension II lowers the plasma aldosterone concentration (due to less secretion of aldosterone from adrenal cortex) and hence the lower aldosterone excretion. The aldosterone is known to decrease the sodium extraction concentration and water retention resulting in desired hypotensive effect. Controlled clinical studies have shown the ACE inhibitors are at least equal in effectiveness to cardiac glycosides in treatment of CHF (408), and may actually elicit a positive effect on the pathophysiologic mechanism of CHF by inhibiting angiotensin II formation.

ACE inhibitors can be administered with diuretics (qv), cardiac glycosides, β-adrenoceptor blockers, and calcium channel blockers. Clinical trials indicate they are generally free from serious side effects. The effectiveness of enalapril, another ACE inhibitor, in preventing patient mortality in severe (Class IV) heart failure was investigated. In combination with conventional drugs such as vasodilators and diuretics, a 40% reduction in mortality was observed after 6 months of treatment using 2.5–40 mg/day of enalapril (409). However, patients complain of cough, and occasionally rash and taste disturbances can occur. Some of the most commonly prescribed ACE inhibitors were discussed earliet in antihypertensive agents section.

Phosphodiesterase Inhibitors.   cGMP plays a critical role in the modulation of vascular function. Because of the complexity of the biochemical processes involved in cardiac muscle contraction, investigators have looked at manipulation of cGMP pathways for other means of drug intervention for CHF, angina pectoris, MI, and hypertension. Better understanding of regulation and mechanism of cGMP pathways are providing new opportunities through alteration of cGMP levels. Steady-state levels of cGMP are maintained by the rate of formation and degradation of the nucleotide phosphodiesterase (PDE). At least seven classes of phosphodiesterases possessing structural similarities and the amino acid sequences have been identified to date. However, each class is distinguished by its specificty for cAMP (cyclic adenosine phosphate) and cGMP regulation and their response to various agents. One of the areas of investigation involves increased cyclic adenosine monophosphate [60-92-4] (cAMP) through inhibition of phosphodiesterase [9025-82-5] (PDE). This class of compounds includes amrinone, considered beneficial for CHF because of positive inotropic and vasodilator activity. The mechanism of inotropic action involves the inhibition of PDE, which in turn inhibits the intracellular hydrolysis of cAMP (410). In cascade fashion, cAMP-catalyzed phosphorylation of sarcolemmal calcium channels follows, activating the calcium pump (411). A series of synthetic moieties including the bipyridines, amrinone and milrinone, piroximone, and enoximone [77671-31-9], $C_{12}H_{12}N_2O_2S$, all of which have been shown to improve cardiac contractility in short-term studies, were developed (412,413). These drugs initially had a wider therapeutic index than cardiac glycosides, and did not exhibit receptor down-regulation associated with β-adrenoceptor blockers. Questions have since surfaced concerning the detrimental effects of increased contractility upon accelerated deterioration of already diseased cardiac tissue, ie, these drugs can lead to shortened patient survival (414,415). An additional detrimental side effect is arrhythmogenesis (416).

Atrial Natriuretic Peptide.   It is a substance released from the right atrium of the heart and other tissues into the bloodstream that stimulates the kidneys to excrete sodium ("natri-" = sodium) into the urine ("-uretic" = urine). For reasons that are not understood ANP is released in response to subarachnoid hemorrhage. Release of ANP is the cause of the hyponatremia (lower than normal content of sodium in the blood) that frequently complicates subarachnoid hemorrhage. Atrial natriuretic factor (ANP), a polypeptide hormone of 28 amino acids (ANF 28) with potent natriuretic, vasorelaxant, and diuretic properties was originally purified from animal and human atrial extracts (417). The responses evoked from the atrial extracts established the heart as an endocrine organ.

A number of different types of ANP have been described, but the original molecule is a polypeptide of 28 amino acids (qv). The natriuretic peptide family includes A-type (ANP 28), B-type designated as brain natriuretic peptide (BNP 32) and C-type natriuretic peptide (CNP) that are responsible for the body homeostasis and blood pressure control. Both ANP and BNP act on guanylase cyclase-A and CNP on guanylase cyclase-B receptors. The ANPs exert direct and indirect effects on kidneys to alter renal hemodynamics and to enhance salt and fluid excretion and also in part by synthesis and release of renin and aldosterone through renin-angiotensin-aldosterone (RAA) inhibition (418).

It has been shown that ANP s may prevent or attenuate ischemic renal failure (419).Significant benefits were also observed using 4-day infusion of ANP 32 after heart or liver transplantations (420–422). Renal function was improved and all patients developed a strong diuresis and natriuresis with 2–4 h indicating that a continuous low infusion of ANP 32 may present a new concept for treatment of postoperative acute renal failure resistant to conventional therapy (423).

Cardiovascular Disease: Statistics and the Pharmaceutical Industry. Cardiovascular disease (CVD) is a leading cause of death in developed countries, and affects millions of men and women worldwide (328). In the United States alone, 61 million people suffer from some form of CVD, with an estimated 2600 deaths a day attributed to CVD (424). Based on such numbers, it is not surprising that drugs used to treat CVD lead worldwide pharmaceutical sales. In 2003, sales of drugs used to treat CVD are projected to reach $56.6 billion dollars (425). In fact, at a time when new drug approvals are on the decline for most classes of therapeutics, the development times of CVD targeted therapeutics have decreased (ie, new approval times are shortening) (426). Several new classes of therapeutics, including angiotensin II receptor blockers, have contributed significantly to continued growth in this sector of the pharmaceutical industry. In 1997, four new antihypertensive agents were approved by the FDA. These included eprosartan and irbesartan, which are both angiotensin II receptor blockers; mibefradil (a calcium channel blocker), and fenoldopam, which is a selective dopamine1 receptor agonist used intravenously to treat short-term severe hypertension (427). In 2002, the FDA approved two CVD therapeutics used to treat hypertension: olmesartan (an angiotensin II receptor blocker) and eplerenon (a selective aldosterone receptor antagonist). The European AEMP approved telmisartan (an angiotensin II receptor blocker) for the treatment of hypertension, and Bosetan (an endothelin receptor antagonist) for pulmonary arterial hypertension (428). In addition, many new and diverse potential therapeutics are currently in development for the treatment of CVD, and range from phosphodiesterase inhibitors to A1 adenosine receptor agonists (429). The development of diverse new therapeutics to treat new biological pathways associated with CVD, coupled with an ever increasing incidence of CVD worldwide, will ensure that this sector of the pharmaceutical industry will continue to flourish in the foreseeable future.

11. Future Directions

The cardiovascular drug market is one of the larger pharmaceutical markets in the world, with global sales totaling >$50 billion/year and a number of drugs individually exceed $1 billion in annual sales. Even though copious drug classes are used to treat heart failure patients and other cardiovascular diseases, new cases of CHD are growing at >10%/year and the risk of death is also rising incessantly. Therefore, there is an unmet medical need for novel therapeutic agents to treat CHD. Since various kinases are implicated in the catalysis of reversible phosphorylation of direct and indirect signaling pathways in cardiac cells, future discoveries may be made by exploiting such linkages. In particular, protein kinase C (PKC), the mitogen activated protein kinases (MAPKs), and phosphoinositide 3-kinase (PI3K) are associated with various functional responses in cardiac cells such as cell survival, cardiac protection, calcium regulation and contraction, and hypertrophy. Therefore, PKC, MAPK, and PI3K each represent a valid therapeutic target for cardiac function and further in depth elucidation of these kinase pathways is essential (430).

There is increasing evidence of a relationship between apoptosis and pathophysiology of both ischemic and nonischemic cardiomyopathies, and a large number of papers have been published since 1997 suggesting a link between some of the major genetic and biochemical regulators of apoptosis in the heart. There has been a quest for a therapeutical agent that would delay the onset of apoptosis in the ischemic heart. In the future, several therapeutic interventions can be developed to prolong survival of smooth muscle and endothelial cells, and to enhance the vascular contractility, tone, and eventually delay the process of atherosclerosis (431,432).

Elucidation of the phenomenon of myocardial preconditioning may hold the key to the development of a drug for the treatment of ischemic heart disease (433).

Several new therapeutic approaches are under investigation for hypertension. An inhaled prostacyclin analogue, iloprost, may allow ease of administration and minimize some adverse side effects. Oral prostacyclins and inhaled forms of NO therapy are also under study.

NO is a unique moiety implicated in the regulation of various physiological processes including smooth muscle contractility and platelet reactivity. Consequently, it has been suggested that NO may have a significant cardioprotection role in hypercholesterolemia, atherosclerosis, hypertension, and inhibition of platelet aggregation. As a result, the development of selective NOS inhibitors will address the potential beneficial therapeutic outcome of NO modulation to the pathophysiology of these disorders (434,435).

Over the past few years, a number of potent asymmetric aza analogues of the dihydropyrimidine types (DHPM), possessing a similar pharmacological profile to classical dihydropyridine calcium channel blockers, are being studied extensively to evaluate their molecular interactions at the receptor level. Some of the lead compounds (SQ 32926 or SQ 32547) are superior in potency and duration of antihypertensive activity to classical DHP analogues. These compounds compare favorably with second generation drugs such as nicardipine and amlodipine. This class of compounds (DHPM) might be the next generation of CCBs for the treatment of cardiovascular diseases (436).

Clinical administration of drugs with negative inotropic activity is not desirable because of their cardiosuppressive effects, especially in patients with a tendency toward heart failure. Therefore, there has been a search for cardioprotective agents acting through entirely different mechanisms. It has been suggested that reevaluation of dihydropyridine calcium channel blockers might lead to the discovery of therapeutic agents that also have effects on other membrane channels. Efonidipine, possessing inhibitory effects on both L- and T-type Ca^{2+} channels shows potent bradycardic effects through a characteristic prolongation of the phase 4 depolarization, leading to minimum reflex tachycardia or to bradycardia. Both AHC-52 and AHC-93 seem to be interesting prototypes of cardioprotective drugs that act to modify anion homeostatis (437).

Additionally, there are a number of novel potential drug candidates undergoing various clinical studies. One of the most promising candidates is ranolazine. This drug represents the first in a new class of drugs called pFOX inhibitors (partial Fatty Acid Oxidation), which have the potential for treating chronic stable angina. The pFox inhibitors possess a unique method of action, and therefore patients may be able to find relief from the painful attacks of angina without some of the unwanted effects of current antianginal drugs. In particular, energy for cardiac functions such as contraction and relaxation is obtained from ATP breakdown; however, ATP is derived from the oxidation of fatty acids and carbohydrates. In the absence of hypoxia (nonischemic condition), the majority of the energy required for ATP synthesis comes from fatty acids. However, during ischemia fatty acid levels increase further and become a source of cardiac energy other than during nonischemic conditions. Animal studies indicate that pFOX inhibitors directly inhibit fatty acid oxidation of ischemic myocytes.

A new class of hypolipidaemic drugs known as SCAP (escaping high cholesterol) ligands has been proposed (438). Since statins inhibit the rate-limiting enzyme in the cholesterol synthesis pathway, these new compounds act indirectly, by increasing the level of expression of the cell surface LDL receptor (LDLR), which removes cholesterol from circulation. However, LDLR expression is governed by transcriptional activators known as sterol response element-binding proteins (SREBPs), which must travel from the cell's endoplasmic reticulum into the nucleus while undergoing a two-stage cleavage process. Thus, they are chaperoned by another protein called SCAP and new compounds act directly on SCAP to activate it, driving the cleavage and formation of new LDLR and ultimately lowering the LDL levels.

Finally, pharmacogenomics holds the promise that drugs might one day be tailor-made for individual treatment and adapted to each person's own genetic make up.

BIBLIOGRAPHY

"Cardiovascular Agents" in *ECT* 1st ed., Vol. 3, pp. 211–224, by W. Modell, Cornell University Medical College; in *ECT* 2nd ed., Vol. 4, pp. 510–524, by W. Modell, Cornell University Medical College; in *ECT* 3rd ed., Vol. 4, pp. 872–930, by L. Goldman, Lederle Laboratories, American Cyanamid Co.; in *ECT* 3rd ed., Suppl. Vol. pp. 172–190, by P. Cervoni and F. Lai, Lederle Laboratories, American Cyanamid Co.; "Cardiovascular Agents" in *ECT* 4th ed., Vol. 5, pp. 207–300, by Peter Cervoni, David L. Crondall, Peter S. Chan, Lederle Laboratories, American Cyanamid Company; "Cardiovascular Agents" in *ECT* (online), Posting date: December 4, 2000, by Peter Cervoni, David L. Crondall, Peter S. Chan, Lederle Laboratories, American Cyanamid Company.

CITED PUBLICATIONS

1. A. M. Minino and B. L. Smith, *National Vital Statistics Reports* **49**(12), 1 (2001).
2. S. L. Huston, E. J. Lengerich, E. Conlisk, and K. Passaro, *Morbidity Mortility Weekly Review* **47**, 945 (1998).
3. M. A. Marano, *Vital Health Statistics* **10**(199), 1 (1998).
4. R. E. Thomas, in M. E. Wolf, ed., *Burger's Medicinal Chemistry and Drug Discovery*, Vol. 2, 5th ed., 1996, pp. 153–261.
5. A. M. Katz, *N. Engl. J. Med.* **328**, 1244 (1993).
6. W. A. Catterall, *Sci.* **242**, 50 (1988).
7. B. Hille, *J. Gen. Physiol.* **66**, 535 (1975).
8. M. Strong, K. G. Chandy, and G. A. Gutman, *Mol. Bioi. Evol.* **10**, 31 (1993).
9. K. Ho and co-workers, *Nature (London)* **362**, 127 (1993).
10. Y. Kubo, T. J. Baldwin, Y. N. Jan, and L. Y. Jan, *Nature (London)* **362**, 127 (1993).
11. M. Noda, T. Ikeda, T. Kayono, H. Suzuki, H. Takeshima, M. Kurasaki, H. Takahashi, and S. Numa, *Nature (London)* **320**, 188 (1986).
12. W. Sttihmer and co-workers, *Nature (London)* **339**, 597 (1989).
13. R. W. Aldrich, *Nature (London)* **339** 578 (1989).
14. A. L. Hokin and A. F. Huxley, *J. Physiol. (Land.)* **117**, 500 (1952).
15. B. Bean, *Nature (London)* **348**, 192 (1990).
16. O. Pongs, *Trend. Pharmacal. Sci.* **13**, 359 (1992).
17. Task force for the Working Group on Arrhythmias of the European Society of Cardiology, *Eur. Heart J.* **12**, 1112 (1991).
18. H. F. Brown, *Physiol. Rev.* **62**, 505 (1982).
19. W. R. Giles and Y. Imaizumi, *J. Physiol. (Lond.)* **405**, 123 (1988).
20. E. Carmeliet, L. Storms, and J. Vereecke, in D. P. Zipes and H. Jaife, eds., *Cardiac Electrophysiology from Cell to Bedside*, W. B. Saunders, Philadelphia, 1990, pp. 103–108.
21. T. Hoshi, W. N. Zogotta, and R. Aldrich, *Science* **250**, 533 (1990).
22. D. Escande and L. Cavero, *Trend. Pharmacol. Sci.* **13**, 269 (1992).
23. W. C. Cole, C. D. Mcpherson, and D. Sontag, *Circ. Res.* **69**, 571 (1991).
24. G. J. Gross and co-workers, *Am. J. Cardiol.* **63**, IIJ (1989).
25. G. J. Grover and co-workers, *J. Pharmacol. Exp. Ther.* **257**, 156 (1991).
26. J. A. Auchampach, M. Maruyama, I. Cavero, and G. J. Gross, *J. Pharmacol. Exp. Ther.* **259**, 961 (1991).
27. D. Thuringer and D. Escande, *Mol. Pharmacol.* **36**, 897 (1989).
28. V. Mitrovic, E. Oehm, J. Thormann, H. Pitschner, and C. Hamm, *Herz* **25**, 130 (2000).
29. D. Pelzer, S. Pelzer, and T. F. MacDonald, *Rev. Physiol. Biochem. Pharmacal.* **114**, 107 (1990).
30. D. F. Slish, D. Schulz, and A. Schwartz, *Hypertension* **19**, 19 (1992).
31. O. Krizanova, R. Diebold, P. Lory, and A. Schwartz, *Circulation* **87**, VII44 (1993).
32. B. N. Singh and E. M. Vaughan Williams, *Br. J. Pharmacol.* **39**, 675 (1970).
33. B. N. Singh and E. M. Vaughan Williams, *Cardiovasc. Res.* **6**, 109 (1972).
34. B. N. Singh and O. Hauswirth, *Am. Heart J.* **87**, 367 (1974).
35. S. Nattel, *Drugs* **41**, 672 (1991).
36. A. J. Camm and Y. G. Yap, *J. Cardiovasc. Electrophysiol.* **10**, 307 (1999).
37. W. Law, D. Newman, and P. Dorian, *Drugs* **60**, 1315 (2000).
38. E. M. Vaughn Williams, *J. Clin. Pharmacol.* **24**, 129 (1984).
39. T. J. Campbell, *Cardiovasc. Res.* **17**, 251 (1983).
40. T. J. Campbell, *Cardiovasc. Res.* **17**, 344 (1983).

41. A. Sjoerdsma and co-workers, *Circulation* **28**, 492 (1963).
42. M. A. Loutfy and co-workers, *Anal. Profiles Drug Subs.* **12**, 483 (1983).
43. J. W. Mason and L. M. Hondeghem, *Ann. N. Y. Acad. Sci.* **432**, 162 (1984).
44. D. C. Harrison, in J. Morganroth and E. N. Moore, eds., *Cardiac Arrhythmias*, Martinus Nijhoff, Boston, 1985, p. 36.
45. J. Koch-Wester, *Ann. N.Y. Acad. Sci.* **179**, 139 (1971).
46. E. V. Giardinia and co-workers, *Clin. Pharmacol. Ther.* **19**, 339 (1976).
47. R. B. Poet and H. Kadin, *Anal. Profiles Drug Subs.* **4**, 333 (1975).
48. B. Befeler and co-workers, *Am. J. Cardiol.* **35**, 282 (1975).
49. L. A. Vismara and D. T. Mason, *Clin. Pharm. Ther.* **16**, 330 (1974).
50. J. T. Bigger and C. C. Jaffe, *Am. J. Cardiol.* **27**, 82 (1971).
51. M. F. Powell, *Anal. Profiles Drug Subs.* **15**, 761 (1986).
52. G. Hollunger, *Acta Pharmacol. Toxicol.* **17**, 356 (1960).
53. G. Hollunger, *Acta Pharmacol. Toxicol.* **17**, 374 (1960).
54. J. L. Anderson, *Circulation* **57**, 685 (1978).
55. D. M. Roden, R. L. Woolsey, *N. Engl. J. Med.* **315**, 41 (1986).
56. A. H. Beckett and E. C. Chiodomere, *Postgrad. Med. J.* **64** (Suppl. 1), 60 (1977).
57. C. Y. C. Chew and co-workers, *Drugs* **17**, 161 (1979).
58. M. A. Abounassif and co-workers, *Anal. Profiles Drug Subs.* **20**, 433 (1991).
59. R. H. Helfant and co-workers, *Am. Heart J.* **77**, 315 (1969).
60. M. Sami and co-workers, *Am. J. Cardiol.* **44**, 526 (1979).
61. R. E. Kates and co-workers, *Am. J. Cardiol.* **53**, 248 (1983).
62. P. Somani, *Clin. Pharmacol. Ther.* **27**, 464 (1980).
63. J. L. Anderson and co-workers, *N. Engl. J. Med.* **305**, 473 (1981).
64. U. Klotz and co-workers, *Int. J. Clin. Pharmacol. Biopharm.* **17**, 152 (1979).
65. C. E. Erickson and R. N. Brogden, *Drugs* **27**, 279 (1984).
66. R. Woestenborghs, *J. Chromatogr.* **164**, 169 (1979).
67. H. J. Hapke and E. Prigge, *Arzneim. Forsch.* **26**, 1849 (1976).
68. J. Mergenthaler and co-workers, *Naunyn Schmiedebergs Arch Pharmacol.* **363**, 472 (2001).
69. F. Bellandi and co-workers, *Am. J. Cardiol.* **88**, 640 (2001).
70. T. Yamane and co-workers, *Br. J. Pharmacol.* **108**, 812 (1993).
71. F. H. Leenen, *Can. J. Cardiol.* **15** (Suppl A), 2A (1999).
72. K. O. Ogunyankin and B. N. Singh, *Am. J. Cardiol.* **84**, 76R (1999).
73. D. J. MacNeil, *Am. J. Cardiol.* **80** (8A), 90G (1997).
74. J. D. Fitzgerald, in A. Scriabine, ed., *Pharmacology of Antihypertensive Drugs*, Raven Press, New York, 1980, pp. 195–208.
75. L. Slusarek and K. Florey, *Anal. Profiles Drug Subs.* **9**, 455 (1980).
76. B. N. Singh and co-workers, *Drugs* **34**, 311 (1987).
77. M. E. Goldher, ed., *American Pharmacology Association, Washington, D.C.*, 1979, pp. 98–147.
78. B. N. Singh and co-workers, *Drugs* **29**, 531 (1985).
79. R. J. Gorczynski and co-workers, *J. Cardiovasc. Pharmacol.* **5**, 668 (1983).
80. P. Benfield and co-workers, *Drugs* **31**, 376 (1986).
81. B. N. Singh and J. S. Sarma, *Curr. Cardiol. Rep.* **3**, 314 (2001).
82. F. E. Marchlinski and co-workers, *Am. J. Cardiol.* **84**, 69R.
83. T. A. Plomp, *Anal. Profiles Drug Subs.* **20**, 1 (1991).
84. M. Chow, *Ann. Pharmacother.* **30**, 637 (1996).
85. B. N. Singh, *Clin. Cardiol.* **20**, 608 (1997).
86. D. Roy and co-workers, *N. Engl. J. Med.* **342**, 913 (2000).
87. H. C. van Beeren, O. Bakker, and W. M. Wiersinga, *Endocrinology* **137**, 2807 (1996).
88. B. N. Singh and K. Nademanee, *Am. Heart J.* **109**, 421 (1985).

89. J. P. Saul and co-workers, *Clin. Pharmacol. Ther.* **69**, 145 (2001).

90. C. P. Lau and co-workers, *Am. J. Cardiol.* **88**, 371 (2001).

91. A. Fitton and E. M. Sorkin, *Drugs* **46**, 678 (1993).

92. J. Morganroth, *Am. J. Cardiol.* **72**, 3A (1993).

93. R. H. Heissenbuttel and J. T. Bigger, *Ann. Intern. Med.* **91**, 229 (1979).

94. M. R. Rosen and A. L. Wit, *Am. Heart J.* **106**, 829 (1983).

95. K. A. Ellenbogen and co-workers, *J. Am. Coll. Cardiol.* **28**, 130 (1996).

96. J. B. Hester and co-workers, *J. Med. Chem.* **34**, 308 (1991).

97. T. Yang, D. J. Snyders, and D. M. Roden, *Circulation* **91**, 1799 (1995).

98. J. J. Lynch, Jr., *J. Cardiovasc. Pharmacol.* **25**, 336 (1995).

99. L. V. Buchanan and co-workers, *J. Cardiovasc. Pharmacol.* **19**, 256 (1992).

100. G. S. Friedrichs and co-workers, *J. Pharmacol. Exp. Ther.* **266**, 1348 (1993).

101. E. Carmeliet, *Cardiovasc. Drugs Ther.* **7** (Suppl 3), 599, (1993).

102. J. Kiehn, A. E. Lacerda, B. Wible, and A. M. Brown, *Circulation* **94**, 2572 (1996).

103. R. H. Falk, A. Pollak, S. N. Singh, and T. Friedrich, *J. Am. Coll. Cardiol.* **29**, 385 (1997).

104. D. S. Echt and co-workers, *J. Cardiovasc Electrophysiol.* **6**, 687 (1995).

105. R. Karam and co-workers, *Am. J. Cardiol.* **81**, 40D (1998).

106. E. L. C. Pritchett and co-workers, *Eur. Heart J.* **20**, 352 (1999).

107. S. Connelly and co-workers, *Eur. Heart. J.* **20**, 351 (1999).

108. E. L. C. Pritchett and co-workers, *Circulation* **98** (Suppl 1), 633 (1999).

109. K. C. Yedinak, *Am. Pharm.* **33**, 49 (1993).

110. B. N. Singh and co-workers, *Drugs* **25**, 125 (1983).

111. M. J. Niebauer and M. K. Chung, *Cardiol. Rev.* **9**, 253 (2001).

112. A. Gabrielli and co-workers, *Crit. Care Med.* **29**, 1874 (2001).

113. H. F. Tse and co-workers, *Am. J. Cardiol.* **88**, 568 (2001).

114. H. E. Wang and co-workers, *Ann. Emerg. Med.* **37**, 38 (2001).

115. H. Ozaki and co-workers, *J. Cardiovasc. Pharmacol.* **33**, 492 (1999).

116. M. J. Apostolakos and M. E. Varon, *New Horiz.* **4**, 45 (1996).

117. S. Viskin and co-workers, *J. Am. Coll. Cardiol.* **38**, 173 (2001).

118. H. L. Tan and co-workers, *Pacing Clin. Electrophysiol.* **24**, 450 (2001).

119. U. Stark, M. Brodmann, A. Lueger, and G. Stark, *J. Crit. Care* **16**, 8 (2001).

120. S. Luber and co-workers, *Am. J. Emerg. Med.* **19**, 40 (2001).

121. E. O. Robles de Medina and A. Algra, *The Lancet* **354**, 882 (1999).

122. B. N. Singh, *Am. J. Cardiol.* **84**, 3R (1999).

123. The Cardiac Arrhythmia Suppression Tiral (CAST) Investigators, *N. Engl. J. Med.* **321**, 406 (1989).

124. IMPACT Research Group, *J. Am. Coll. Cardiol.* **4**, 1148 (1984).

125. The Cardiac Arrhythmia Suppression Tiral II Investigators, *N. Engl. J. Med.* **327**, 227 (1992).

126. I. Kodama and co-workers, *Am. J. Cardiol.* **84**, 20R (1999).

127. K. O. Ogunyankin and B. N. Singh, *Am. J. Cardiol.* **84**, 76R (1999).

128. S. J. Connolly and co-workers, *Am. J. Cardiol.* **88**, 974 (2001).

129. A. K. Gupta, *Indian Heart J.* **53**, 354 (2001).

130. P. Matyus and co-workers, *Med. Res. Rev.* **20**, 294 (2000).

131. R. C. Tripathi and co-workers, *Bioorg. Med. Chem. Lett.* **9**, 2693 (1999).

132. N. Bodor, H. H. Farag, and P. Polgar, *J. Pharm. Pharmacol.* 889 (2001).

133. T. E. Morey and co-workers, *J. Pharmacol. Exp. Therap.* **297**, 260 (2001).

134. H. M. Himmel and co-workers, *J. Cardio. Pharmacol.* **38**, 438 (2001).

135. D. Thomas and co-workers, *J. Pharmacol. Exp. Therap.* **297**, 753 (2001).

136. O. Lecy, M. Erez, D. Varon, and E. Keinan, *Bioorg. Med. Chem. Lett.* **11**, 2921 (2001).

137. F. Boutitie and co-workers, *Circulation* **99**, 2268 (1999).

138. C. T. Sempos, J. I. Cleeman, M. K. Carroll, and co-workers, *J. Am. Med. Assoc.* **269**, 3009 (1993).
139. R. F. Gillum, *Am. Heart J.* **126**, 1042 (1993).
140. F. M. Sacks, M. A. Pfeffer, L. A. Moye, and co-workers, *New. Eng. J. Med.* **335**, 1001 (1996).
141. A. L. Dunn, B. H. Marcus, J. B. Kampert, M. E. Garcia, H. W. Kohl, and S. N. Blair, *J. Am. Med. Assoc.* **281**, 327 (1999).
142. A. J. Manson, F. Hu, J. W. Rich-Edwards, G. Colditz, M. J. Stampfer, W. H. Willett, F. Speizer, and C. Hennekens, *New. Eng. J. Med.* **341**, 650 (1999).
143. CDC, U.S. 1998, *Morbidiy and Mortality weekly reports* **48** (45) (1999).
144. B. M. Pasty, N. L. Smith, D. S. Siscovick, and co-workers, *J. Am. Med. Assoc.* **277**, 739 (1997).
145. V. L. Burt, J. A. Culter, M. Higgins, and co-workers, *Hypertension* **26**, 60 (1995).
146. D. S. Freedman, W. H. Dietz, S. R. Srinivasan, and G. S. Berenson, *Pediatrics* **103**, 1175 (1999).
147. Clinical guidelines, *J. Obse., Res.* S2, (1998).
148. F. Furchgott and J. V. Zawadzaki, *Nature (London)* **288**, 373 (1980).
149. Moncada, R. M. J. Palmer, and E. A. Higgs, *Pharmacal. Rev.* **43**, 109 (1991).
150. P. Cooke and V. J. Dzau, *Ann. Rev. Med.* **48**, 489 (1997).
151. J. Parker, *Am. J. Cardiol.* **72** (8), 3C (1993).
152. J. Ahlner, R. G. Andersson, K. Torfgard, and K. L. Axelsson, *Pharmacol. Rev.* **43** (3), 351 (1991).
153. H. L. Fung, *Br. J. Clin. Pharmacol.* **34** (1), 5S (1992).
154. J. P. Cooke and V. J. Dzau, *Annu. Rev. Med.* **48**, 489 (1997).
155. J. K. Michelson and co-workers, in Ref. 1, p. 293.
156. F. Murad, in Ref. 2, p. 764.
157. Goldeberg and co-workers, *Acta. Physiol. Scand.* **15**, 173 (1948), *Chem. Abs.* **42**, 5564 (1948).
158. H. Laufen, *Arzneim. Forsch.* **33**, 980 (1983).
159. L. A. Silviveri and N. J. DeAngelis, *Anal. Profiles Drug Subs.* **4**, 225 (1975).
160. *Chem. Abs.* **42**, 5564 (1948).
161. Kochergin and Titkova, *Chem. Abs.* **54**, 8647h (1960).
162. M. V. Dijk, *Rec. Trav. Chim.* **75**, 1215 (1956).
163. U.S. Pat. 3,056,836 (1962), X. Moed (to N. Am Philips).
164. Goldenthal, *Toxicol. App. Pharmacol.* **18**, 185 (1971).
165. H. Wesseling and co-workers, *Eur. J. Clin. Pharmacol.* **20**, 329 (1981).
166. S. H. Skotnicki and co-workers, *Angiology* **35**, 685 (1984).
167. DE Pat. 2,714,713 (1977), U.S. Pat. 4,200,640 (1980) H. Nagano and co-workers.
168. N. Taira and co-workers, *Clin. Exp. Pharmacol. Physiol.* **6**, 301 (1979).
169. F. Yoneyama and co-workers, *Cardiovascu. Drugs Ther.* **4**, 1119 (1990).
170. H. Purcell and K. Fox, *Br. J. Clin. Pract.* **47**, 150 (1993).
171. B. A. Falase, B. S. Bajaj, T. J. Wall, V. Argano, and A. Y. Youhana, *Ann. Thorac. Surg.* **67**, 1158 (1999).
172. H. L. Fung, *Am. J. Cardiol.* **72**, 9C (1993).
173. S. Tsuchida, T. Maki, and T. Sata, *J. Biol. Chem.* **265**, 7150 (1990).
174. V. Burt, P. Whelton, and E. J. Roccella, *Hypertension* **25**, 305 (1995).
175. K. Kato, *Eur. Heart J.* **14** (S B), 40 (1993).
176. U.S. Pat. REISSUED 30577 (1981) N. Busch and co-workers.
177. M. T. Michelin and co-workers, *Therapie* **32**, 485 (1977).
178. S. Vogel and co-workers, *J. Pharmacol. Exp. Ther.* **210**, 378 (1979).
179. C. Labrid and co-workers, *J. Pharmacol. Exp. Ther.* **211**, 546 (1979).
180. M. K. Sharmaa and co-workers, *Am. J. Cardiol.* **611**, 1210 (1988).

181. Ger. Pat. 3,415,035 (1984), U.S. Pat. 4,552,695 (1985) both to Shionogi, K. Igarashi, T. Honma.

182. Sato and co-workers, *Arzneim. Forsch.* **21**, 1338 (1971).

183. T. Nagano and co-workers, *Jpn. J. Pharmacol.* **22**, 467 (1972).

184. M. Chaffmann and R. N. Brogden, *Drugs* **29**, 387 (1985).

185. R. S. Gibson and co-workers, *N. Eng. J. Med.* **315**, 423 (1986).

186. B. R. Lucchesi, in M. Antonacchio, ed., *Cardiovascular Pharmacology*, 3rd ed., Raven Press, New York, 1990, p. 369.

187. S. Kawakita and co-workers, *Clin. Cardiol.* **14**, 53 (1991).

188. H. Narita and co-workers, *Arzneimittel-Forsch* **38**, 515 (1988).

189. L. J. Theodore and W. L. Nelson, *J. Org. Chem.* **52**, 1309 (1987).

190. D. D. Waters and co-workers, *Am. J. Cardiol.* **47**, 179 (1981).

191. D. J. Triggle and V. C. Swamy, *Cir. Res.* **52** (2), 117 (1983).

192. L. J. Theodore and W. L. Nelson, *J. Org. Chem.* **52**, 1309 (1987).

193. H. Haas and E. Busch, *Arzneim. Forsch.* **17**, 257 (1967), *Arzneimittel-Forsch* **18**, 401 (1968).

194. A. Fleckenstein, *Arzneim. Forsch.* **20**, 1317 (1970).

195. N. S. Khurmi and co-workers, *Am. J. Cardio.* **53**, 684 (1984).

196. R. N. Brogden and P. Benfield, *Drugs* **47**, 93 (1994).

197. F. Hefti, *Arzneimittel-Forsch* **40**, 417 (1990).

198. G. Mehrke and co-workers, *J. Pharmacol. Exp. Ther.* **271**, 1483 (1994).

199. M. C. M. Portegies and co-workers, *J. Cardiovasc. Pharmacol.* **18**, 746 (1991).

200. J. P. Clozel and co-workers, *Cardiovasc. Drug Rev.* **9**, 4 (1991).

201. M. E. Mullins, Z. Horowitz, D. H. J Linden, G. W. Smith, R. L. Norton, and J. Stump, *JAMA* **280**, 157 (1998).

202. R. Weyhenmyer and co-workers, *Arzneim. Forsch.* **37**, 58 (1987).

203. W. R. Kukovetz and co-workers, *Arzneim. Forsch.* **26**, 1321 (1976).

204. A. Fleckenstein, *Arzneim. Forsch.* **27**, 562 (1977).

205. Z. Antaloczy and I. Preda, *Ther. Hung.* **27**, 71 (1979).

206. M. Spedding, *Arch. Pharmacol.* **318**, 234 (1982).

207. M. Gautam, A. Tewari, S. Singh, C. Dixit, K. G. Raghu, P. Prakash, and O. Tripathi, *Jpn. J. Pharmacol.* **83**(3), 175 (2000).

208. J. S. Cheng, K. J. Chou, J. L. Wang, K. C. Lee, L. L. Tseng, K. Y. Tang, J. K. Huang, H. T. Chang, W. Su, Y. P. Law, and C. R. Jane, *Clin. Exper. Pharmacol. Physiol.* **28**(9), 729 (2001).

209. DE Pat. 1,100,031 and *Chem. Abs.* **56**, 3413h, (1962), U.S. Pat. 3,152,173 (1964) G. Ehrhart and co-workers.

210. J. E. Murphy, *J. Int. Med. Res.* **1**, 204 (1973).

211. B. Karlen and co-workers, *Eur. J. Clin. Pharmacol.* **23**, 267 (1982).

212. U. Ulmsten and co-workers, *Am. J. Obstet, Gynecol.* **153**, 619 (1985).

213. J. L. Faulkner, *Br. J. Clin. Pharmacol.* **22**, 21 (1986).

214. Fr. Pat. 2,514,761 (1983), U.S. Pat. 4,446,325 (1984) both to Maruko Seiyaku, S. Ohno and co-workers.

215. K. Miyoshi and co-workers, *Eur. J. Pharmacol.* **238**, 139 (1993).

216. A. Kanda and co-workers, *J. Cardiovasc. Pharmacol.* **22**, 167 (1993).

217. S. Suzuki and co-workers, *Arzneim.-Forsch.* **43**, 1152 (1993).

218. S. Nakano and co-workers, *Yakuri Chiryo.* **21**, S931 (1993).

219. DE Pat. 2,904,552 (1979), U.S. Pat. 4,220,649 (1980) T. Kojima, T. Takenaka.

220. K. Tamazawa and co-workers, *J. Med. Chem.* **29**, 2504 (1986).

221. H. Satoh, *Cardiovasc. Drug Rev.* **9**, 340 (1991).

222. H. S. Malhotra and G. L. Plosker, *Drugs* **61**(7), 989 (2001).

223. Y. Ohya, I. Abe, Y. Ohta, U. Onaka, K. Fujii, S. Kagiyama, Y. Fujishima-Nakao, and M. Fujishima, *Int. J. Clin. Pharm Thers.* **38** (6), 304 (2000).

224. H. Ueno, T. Hara, A. Ishi, and K. Shuto, *Jpn. J. Pharmacol.* **84** (1), 56 (2000).

225. S. Muneta1, K. Kohara, and K. Hiwada, *Int. J. Clin. Pharmacol.* **37** (3), 141 (1999).

226. *Arzneim.-Forsch., Drug Res.* **50**, 620 (2000).

227. Eur. Pat. 161,877 (1985), U.S. Pat. 4,672,068 (1987) T. Kutsuma and co-workers.

228. K. Ikeda and co-workers, *Oyo. Yakuri.* **44**, 433 (1992).

229. M. Hosona and co-workers, *J. Pharmacobio.-Dyn.* **15**, 547 (1992).

230. M. Ishi, *Jpn. Pharmacol. Therp.* **21**, 59 (1993).

231. S. Wada, *Chem. Abs.* **118**, 32711 (1992).

232. R. Uchida, J. Yamazaki, S. Ozeki, and K. Kitamura, *Jpn. J. Pharmacol.* **85**(3), 260 (2001).

233. Y. Onose, T. Oki, H. Yamada, K. Manabe, Y. Kageji, M. Matsuoka, T. Yamamoto, T. Tabata, T. Wakatsuki, and S. Ito, *Jpn. Circ. J.* **65**, 305 (2001).

234. WO Pat. 8,704,439 (1987), U.S. Pat. 4,885,284 (1989) K. Seto and co-workers.

235. C. Shudo and co-workers, *Jpn. Pharm. Pharmacol.* **45**, 525 (1993).

236. T. Yamashita and co-workers, *Jpn. J. Pharmacol.* **57**, 331 (1991).

237. T. Saito and co-workers, *Curr. Ther. Res.* **52**, 113 (1992).

238. T. Yokoyama, K. Ichihara, and Y. Abiko, *Jpn. J. Pharmacol.* **72**, 291 (1996).

239. M. Kawabata, T. Ogawa, W. H. Han, and T. Takabatake, *Clin. Exper. Pharmacol. Physio.* **26** (9), 674 (2001).

240. Eur. Pat. 3,02,980 (1989) C. F. Torija, and J. A. G. Ramos, U.S. Pat. 4,952,592 (1990).

241. J. Tamargo and co-workers, *Arzneimittel-Forsch* **41**, 895 (1991).

242. H. Suryapranata and co-workers, *Am. J. Cardiol.* **69**, 1171 (1992).

243. D. U. Acharya and co-workers, *Euro. Heart J.* **15**, 665 (1994).

244. S. Motte, X. Alberich, and F. Harrison, *Int. J. Clin. Pharmacol Therap.* **37**(1), 20 (2001).

245. P. Decoster and co-workers, *Eur. J. Clin. Invest.* **12**, 43 (1982).

246. B. Edgar and co-workers, *Biopharm. Drug Dispos.* **8**, 235 (1987).

247. A. Miniscalco, J. Lundahl, and C. G. Regardh, *J. Pharmacol. Exp. Ther.* **261**, 1195 (1992).

248. D. G. Bailey, J. Malcolm, O. Arnold, and J. D. Spence, *Br. J. Clin. Pharmacol.* **46**, 101 (1998).

249. D. G. Bailey, J. D. Spence, and B. Edgar, *Clin. Invest. Med.* **12**, 357 (1989).

250. U. L. Hulthen and P. L. Katzman, *J. Hypertens.* **6**, 231 (1988).

251. M. R. Werbach, *Foundations of Nutritional Medicine*, Third Line Press, Inc., Tarzana, CA, 1997, p. 208.

252. F. L. S. Tee and J. M. Jaffe, *Eur. J. Clin. Pharmacol.* **32**, 361 (1987).

253. C. E. Handler and E. Sowton, *Eur. J. Clin. Pharmacol.* **27**, 415 (1984).

254. E. B. Nelson and co-workers, *Clin. Pharmacol. Ther.* **40**, 694 (1986).

255. R. P. Hof and co-workers, *J. Cardiovasc. Pharmacol.* **8**, 221 (1986).

256. M. Safar and co-workers, *Clin. Pharmacol. Ther.* **46**, 94 (1989).

257. A. Zanchetti, *Pharm. J.* **266** (7153), 842 (2001).

258. H. Funato, H. Kawano, Y. Akada, Y. Katsuki, M. Sato, and A. Uemura, *Jpn. J. Pharmacol.* **75**(4), 415 (1997).

259. R. H. Hernández, D. M. Castillo, M. J. A. Hernández, M. C. A. Padilla, and J. G. Pajuelo, *Am. J. Hypertens.* **10** (4), 108A, (1997).

260. Eur. Pat. 153,016 (1985) D. Nardi and co-workers, U.S. Pat. 4,705,797 (1987).

261. D. Policicchio, R. Magliocca, and A. Malliani, *J. Cardiovas. Pharmacol.* **29** (S2), S31 (1997).

262. J. A. Staessen and co-workers, *Lancet* **350**, 757 (1997).

263. K. K. Maguro and co-workers, *Chem. Pharm. Bull.* **33**, 3787 (1985).

264. K. Mizuno and co-workers, *Curr. Ther. Res.* **52**, 248 (1992).

265. M. Yoshiyama, K. Takeuchi, S. Kim, A. Hanatani, T. Omura, I. Toda, K. Akioka, M. Teragaki, H. Iwao, and J. Yoshikawa, *Jpn. Cric. J.* **62**, 47 (1998).

266. J. T. Bigger and B. F. Hoffman, in A. G. Gilman and co-eds., *Goodman and Gilman's, The Pharmacologic Basis of Therapeutics*, 8th ed., Pergamon Press, New York, 1990, p. 840.

267. P. A. Molyvdas and N. Sperelakis, *J. Cardiovasc. Pharmacol.* **8**, 449 (1986).

268. G. J. Gross and co-workers, *Gen. Pharmacol.* **14**, 677 (1983).

269. K. Mizuno and co-workers, *Res. Comm.Chem. Pathl. Pharmacol.* **52**, 3 (1986).

270. S. O. Kawamura, Y. Li, M. Shirasawa, N. Yasui, and H. Fukasawa, *Tohoku J. Exp. Med.* **185** (4), (1998).

271. K. B. Grögler, W. Ungethüm, B. M. Witt, and G. G. Belz, *Eur. J. Clin. Pharmacol.* **57** (4), 275 (2001).

272. S. Kazda and co-workers, *Arzneim.-Forsch.* **30**, 2144 (1980).

273. H. A. Friedel and E. M. Sorkin, *Drugs* **36**, 682 (1988).

274. J. Mitchell and co-workers, *J. Clin. Pharmacol.* **33**, 46 (1993).

275. F. Özçelik, A. Altun, and G. Özbay, *Clin. Cardiol.* **22**, 361 (1999).

276. H. Meyer, *Arzneimittel-Forsch* **31**, 407 (1981).

277. U. Brugmann and co-workers, *Hertz* **10**, 53 (1985).

278. J. L. Breslow, *Annu. Rev. Genet.* **34**, 233 (2000).

279. M. A. Austin, *Proc. Nutr. Soc.* **56**, 667 (1997).

280. L. W. Castelini, A. Weinreb, J. Bodnar, A. M. Goto, and M. Doolittle, *Nat. Genet.* **18**, 374 (1998).

281. G. M. Dallinga-Thie, X. D. Bu, M. VLS.Trip, J. I. Rotter, A. J. Lusis, and T. W. de Bruin, *J. Clin. Invest.* **99**, 953 (1997).

282. J. Davingnon, R. E. Gregg, and C. F. Sing, *Arteriosclerosis* **8**, 1 (1988).

283. R. M. Fisher, S. E. Humphries, and P. J. Talmund, *Atherosclerosis* **135**, 145 (1997).

284. H. Knoblauch, B. Muller-Myhsok, A. Busjahn, L. B. Avi, and S. Bahring, *Am. J. Hum. Genet.* **66**, 157 (2000).

285. C. Glaser, *J. Clin. Hypertens.* **2** (3), 204 (2000).

286. A. W. Albertset and co-workers, *Proc. Natl. Acad. Sci.* **77**, 3957 (1980).

287. W. H. Frishman and R. C. Rapier, in Ref. 114, p. 440.

288. A. G. Olsson, F. McTaggart, and A. Raza, *Cardiovas. Rev.* **20** (4), 303 (2002).

289. FDA drug approval list [online] cited Jan 06, 2003 http://www/fda.gov

290. D. A. Burnett, M. A. Caplen, H. R. Davis, R. E. Burrier, and J. W. Clader, *J. Med. Chem.* **39**, 1733 (1994).

291. J. W. Clader and co-workers, *J. Med. Chem.* **39**, 3684 (1996).

292. M. A. Van Heek and co-workers, *J. Pharmacol. Exp. Therapeutic.* **283**, 157 (1997).

293. S. B. Rosenblum and co-workers, *J. Med. Chem.* **41**, 973 (1998).

294. G. Wu, Y. Wong, X. Chen, and Z. A. Ding, *J. Org. Chem.* **64**, 3741 (1999).

295. J. Earl and P. Kirkpatrick, *Nature (London)* **2**, 97 (2003).

296. Lipid Research Clinics Program, *JAMA* **251**, 351 (1984).

297. M. Ast and W. H. Frishman, *J. Clin. Pharmacol.* **30**, 99 (1990).

298. P. T. Kuo and co-workers, *Circulation* **59**, 199 (1979).

299. R. H. Knopp, *Cardiology* **76**(Suppl. 1), 14 (1989).

300. J. P. Monk and P. A. Todd, *Drugs* **33**, 539 (1987).

301. J. M. Cruickshank and J. C. Smith, *Pharmac. Ther.* **42**, 385 (1989).

302. J. H. Laragh, *Hypertensio* **13**(Suppl. 1), 1 (1989).

303. M. C. Houston, *Am. Heart J.* **117**, 911 (1989).

304. K. H. Rahn, *Am. J. Cardiol.* **65**, 82G (1990).

305. M. Burnier and co-workers, *Drugs* **39**(Suppl. 1), 32 (1990).

306. L. Hansson, *J. Hypertens.* (Suppl. 8), S13 (1990).

307. F. H. Messerli, *Postgrad. Med.* **88**, 127 (1990).

308. D. Ballantyne, *J. Hum. Hypertens* **4**(Suppl. 2), 35 (1990).

309. P. Leren, *Clin. Exp. Hypertens. [A]* **12**, 761 (1990).

310. M. H. Weinberger, *Cardiovasc. Drugs Ther.* **4**(Suppl. 2), 379 (1990).

311. R. B. Taylor, *Am. Fam. Physician.* **42**(Suppl. 5), 29S (1990).

312. L. Hansson, *Am. J. Hypertens.* **4**, 84S (1991).

313. A. Breckenridge, *Am. J. Hypertens.* **4**, 79S (1991).

314. R. Gorlin, *Am. Heart J.* **121**, 670 (1991).

315. H. R. Black, *Am. Heart J.* **121**, 2(Pt. 2), 707 (1991).

316. J. J. Wright, in M. J. Antonacchio, ed., *Cardiovascular Pharmacology*, 3rd ed., Raven Press, New York, 1990, p. 201.

317. H. Gavras, *Circulation* **81**, 381 (1990).

318. D. B. Case, E. H. Sonnenblick, and J. H. Laragh, eds., *Captopril and Hypertension*, Plenum Publishing Corp., New York, 1980, pp. 3–230.

319. J. H. Laragh, *Am. J. Hypertens.* **3**, 257S (1990).

320. A. E. Fletcher and C. J. Bulpitt, *J. Hum Hypertens.* **4**(Suppl. 4), 45 (1990).

321. L. E. Rarnsay and W. W. Yeo, *J. Hum. Hypertens.* **4**(Suppl. 4), 1 (1990).

322. J. H. Bauer, *Am. J. Hypertens.* **3**, 331 (1990).

323. H. R. Brunner, *Hosp. Pract.* **25**, 71 (1990).

324. D. McAreavey and J. I. S. Robertson, *Drugs* **40**, 326 (1990).

325. B. Waeber, J. Nussberger, and H. R. Brunner, in J. H. Laragh and B. M. Brenner, eds., *Hypertension: Pathophysiology, Diagnosis and Management*, Raven Press, New York, 1990, p. 2209.

326. P. M. Vanhoutte and co-workers, *Br. J. Clin. Pharmacol.* **28**, 95S (1989).

327. M. R. Ujhelyi, R. K. Ferguson, and P. H. Vlasses, *Pharmacotherapy* **9**, 351 (1989).

328. M. A. Zaman, S. Oparil, and D. A. Calhoun, *Nat. Rev. Drug. Disc.* **1**, 621 (2002).

329. J. Benz, C. Oshrain, and co-workers, *J. Clin. Pharmacol.* **37**, 101 (1997).

330. H. R. Black, A. Graff, and co-workers, *J. Human Hypertension* **11**, 483 (1997).

331. A. D. Bremner, M. Baur, and co-workers, *Clin. Exper. Hypertension* **19**(8), 1263 (1997).

332. P. Chan, B. Tomlinson, and co-workers, *J. Clin. Pharmacol.* **37**, 253 (1997).

333. R. Fogari, A. Zoppi, L. Corradi, and co-workers, *Br. J. Clin. Pharmacol.* **46**, 467 (1998).

334. R. Fogari, A. Zoppi, P. Lazzari, and co-workers, *J. Cardiovasc. Pharmacol.* **32**, 616 (1998).

335. A. H. Gradman, K. E. Arcuri, and co-workers, *Hypertension* **25**, 1345 (1995).

336. N. J. Holwerda and R. Fogari, *J. Hypertension* **14**, 1147 (1996).

337. J. M. Mallion, S. Boutelant and co-workers, *Blood Pressure Monitoring* **2**, 179 (1997).

338. A. Mimran, L. Ruilope, and L. Kerwin, J., and co-workers, *Human Hypertension* **12**, 203 (1998).

339. S. Nielsen, J. Dollerup, B. Nielsen, and co-workers, *Nephrol. Dial. Transplant* **12** Suppl. 2, 19 (1997).

340. A. Pechere-Bertschi, J. Nussberger, and co-workers, *J. Hypertens.* **16**, 385 (1998).

341. N. Perico, D. Spormann, E. Peruzzi, and co-workers, *Clin. Drug Invest.* **14**(4), 252 (1997).

342. D. Ruff, L. P. Gazdick, R. Berman, and co-workers, *J. Hypertens.* **14**, 263 (1996).

343. I. Tikkanen, P. Omvik, and H. A. E. Jensen, *J. Hypertens.* **13**, 1343 (1995).

344. F. Zannad and co-workers, *Circulation* **102**, 2700 (2000).

345. X. Jeunemaitre, and co-workers, *Am. J. Cardiol.* **60**, 820 (1987).

346. D. A. Calhoun, M. A. Zaman, and M. K. Nishizaka, *Curr. Hypertens. Rep.* **4**, 221 (2002).

347. E. Burgess and co-workers, *Am. J. Hypertens.* **15**(Suppl. 1), A57 (2002).

348. P. S. Chan and P. Cervoni, in P. B. Goldberg and J. Roberts, eds., *CRC Handbook on Pharmacology of Aging*, CRC Press, Inc., Boca Raton, Fla., 1983, p. 51.

349. P. A. van Zwieten, *Cardiovasc. Drug Ther.* **3**, 121 (1989).

350. P. A. van Zwieten, in M. J. Antonacchio, ed., *Cardiovascular Pharmacology*, 3rd ed., Raven Press, New York, 1990, p. 37.

351. S. H. Taylor, in F. Messerli, ed., *Cardiovasc. Drug Therapy*, W. B. Saunders, Philadelphia, 1990, p. 714.

352. A. H. Maslowski, *Am. Heart J.* **121**, 1(Pt 2), 323 (1991).

353. L. Hansson, in F. Messerli, ed., *Cardiovascular Drug Therapy*, W. B. Saunders, Philadelphia, 990, p. 408.

354. P. Bolli, P. G. Fernandez, and F. R. Buhler, in J. H. Laragh and B. M. Brenner, eds., *Hypertension: Pathophysiology, Diagnosis and Management*, Raven Press, New York, 1990, p. 2181.

355. J. Conway and A. Bilski, in D. Ganten and P. J. Mulrow, eds., *Pharmacology of Antihypertensive Therapeutics (Handbook Exp. Pharmacology)*, Vol. 93, Springer-Verlag, Berlin, Germany, 1990, p. 65.

356. E. H. Gold and co-workers, *J. Med. Chem.* **25**, 1363 (1982).

357. P. E. Hanna, in J. N. Delgado and W. A. Remers, eds., *Wilson and Gisvold's Textbook of Organic Medicinal Pharmaceutical Chemistry*, 9th ed., J. B. Lippincott, Philadelphia, 1991, p. 441.

358. P. S. Chan and P. Cervoni, in P. B. Goldberg and J. Roberts, eds., *CRC Handbook on Pharmacology of Aging*, CRC Press, Inc., Boca Raton, Fla., 1983.

359. M. C. Houston, *Am. Heart J.* **117**, 911 (1989).

360. W. M. Kirkendall, *Arch. Intern. Med.* **151**, 398 (1991).

361. H. G. Langford and co-workers, *J. Hum. Hyperten.* **4**, 491 (1990).

362. G. B. Weiss, R. J. Winquist, and P. J. Silver, in Ref. 186, p. 75.

363. E. P. MacCarthy, *Cardiovasc. Drug Rev.* **8**, 155 (1990).

364. M. A. DeWood and co-workers, *N. Engl. J. Med.* **303**, 897 (1980).

365. V. Fuster and co-workers, *Circulation* **77**, 1213 (1988).

366. B. Stein and V. Fuster, *Cardiovasc. Drug Ther.* **3**, 797 (1989).

367. R. M. Gunnar and co-workers, *J. Am. Coll. Cardiol.* **16**, 249 (1990).

368. D. C. Collen and H. K. Gold, *Thrombosis Res. Suppl. X*, 105 (1990).

369. C. J. Pepine, *Am. J. Cardiol.* **64**, 2B (1989).

370. S. Sherry, *Cardiovasc. Drug Rev.* **6**, 1 (1988).

371. R. C. Becker and R. Harrington, *Arn. Heart J.* **121**, 1(Pt. 1), 220 (1991), and **121**, 2(Pt. 1), 627 (1991).

372. J. D. Rutherford and E. Braunwald, *Chest.* **97**(Suppl.), 1375 (1990).

373. D. Collen, *J. Cell Biochem.* **33**, 77 (1987).

374. R. W. Holden, *Radiology* **174**, 993 (1990).

375. R. Fears, *Pharmacol. Rev.* **42**, 201 (1990).

376. P. Sleight, *Eur. Heart J.* **11**(Suppl. F), 1 (1990).

377. K. G. Ro and H. V. Anderson, *Postgrad. Med. J.* **88**, 79 (1990).

378. T. Yasuda and H. K. Gold, *Thrombosis Res. Suppl. X* (1990).

379. A. J. Tiefenbrunn and B. E. Sobel, *Fibrinolysis* **3**, 1 (1989).

380. P. Oldershaw, *Postgrad. Med. J.* **64**, 915 (1988).

381. C. L. Grines and A. N. DeMaria, *J. Am. Coll. Cardiol.* **16**, 223 (1990).

382. J. H. Chesebro, L. Badimon, and V. Fuster, *Am. J. Cardiol.* **65**, 12C (1990).

383. J. Loscalzo, *Chest* **97**(Suppl.), 1175 (1990).

384. G. Di Minno and co-workers, *Pharmacol. Res.* **21**, 153 (1989).

385. M. Samama, *Eur. Heart J.* **11**(Suppl. F), 15 (1990).

386. C. H. Hennekens, *Chest* **97**(Suppl.), 151S (1990).

387. C. H. Hennekens and co-workers, *Circulation* **80**, 749 (1989).

388. M. W. Webster, J. H. Chesebro, and V. Fuster, *Hematol. Oncol. Clin. N. Am.* **4**, 265 (1990).

389. G. Agnelli, *Chest* **97**(Suppl.), 161S (1990).

390. S. D. Rogers, L. B. Riemersma, and S. D. Clements, *Pharmacotherapy* **7**, 111 (1987).

391. R. C. Becker and co-workers, *Am. Heart J.* **121**, 220 (1991).

392. M. H. Prins and J. Hirsh, *Am. J. Cardiol.* **67**, 3A (1991).

393. S. Sherry and S. H. Taylor in Ref. 235, p. 1479.

394. S. Sherry and V. J. Marder, *Ann. Intern. Med.* **114**, 417 (1991).

395. K. L. Goa and co-workers, *Drugs* **39**, 693 (1990).

396. R. Fears, *Sem. Throm. Hemostasis* **15**, 129 (1989).

397. S. J. Crabbe, A. M. Grimm, and L. E. Hopkins, *Pharmacotherapy* **10**, 115 (1990).

398. M. A. Mungr and E. A. Forrence, *Clin. Pharmacy* **2**, 530 (1990).

399. M. Verstraete, *Ann. Acad. Med. Singapore* **28**, 424 (1999).

400. R. Gorlin, *Cardiovasc. Rev. Rep.* **4**, 767 (1983).

401. M. R. Sanders, J. B. Kostis, and W. H. Frishman, in W. H. Frishman and S. Charlap, eds., *Medical Clinics of N.A.: Cardiovascular Pharmacotherapy III*, W. B. Saunders, Philadelphia, 1989, p. 293.

402. B. F. Hoffman and J. T. Bigger, in, A. G. Gilman, L. S. Goodman, T. W. Rall, and F. Murad, eds., *The Pharmacologic Basis of Therapeutics*, 7th ed., Macmillan, New York, 1985, pp. 716–747.

403. E. H. Sonnenblick and co-workers, *N. Engl. J. Med.* **300**, 17 (1979).

404. M. R. Bristow and co-workers, *N. Engl. J. Med.* **307**, 205 (1982).

405. T. C. Majerus and co-workers, *Pharmacotherapy* **9**, 245 (1989).

406. R. R. Ruffolo and co-workers, *J. Pharmacol. Exp. Ther.* **224**, 46 (1983).

407. C. V. Leirer and co-workers, *Circulation* **56**, 468 (1972).

408. B. Magnani and co-workers, *Post. Med.* **62**(Suppl. I), 153 (1986).

409. The Consensus Trial Study Group, *N. Engl. J. Med.* **316**, 1429 (1987).

410. P. Honerjager, *Eur. Heart J.* **10** (Suppl. C), 25 (1989).

411. A. A. Alousi and co-workers, *Circulation* **73**, 10 (1982).

412. R. J. Axelrod and co-workers, *J. Am. Coll. Cardiol.* **9**, 1124 (1987).

413. H. C. Herrmann and co-workers, *J. Am. Coll. Cardiol.* **9**, 1117 (1987).

414. C. S. Maskin and co-workers, *Am. J. Med.* **73**, 113 (1982).

415. M. Packer and co-workers, *Circulation* **70**, 1038 (1984).

416. A. M. Katz, *Circulation* **73**, 184 (1986).

417. I. G. Grozier and co-workers, *Clin. Exp. Pharmacol. Physiol.* **15**, 173 (1988).

418. L. S. Lambic, S. Pljesa, V. Stojanov, and D. Avramovic, *Med. Biol.* **5** (1), 6 (1998).

419. P. Widemann, B. Hellmueller, and D. E. Uehlinger, *J. Clin. Endocrinol. Metab.* **62**, 1027 (1986).

420. M. Hummel, M. Kuhn, and A. Bub, *J. Heart Transplant* **12**, 209 (1993).

421. C. Cedidi, E. R. Kuse, and M. Meyer, *Clin. Invest.* **71**, 435 (1993).

422. T. A. Fischer, M. Haass, R. Dietz, R. C. Willenbrock, W. Saggaau, R. E. Lang, and W. Kuber, *Clin. Sci.* **80**, 285 (1991).

423. Cedidi, E. R. Kuse, and M. Meyer, *Eur. J. Clin. Invest.* **4**, 632 (1994).

424. CDC website: http://www.cdc.gov/cvh/aboutcardio.htm

425. Business Communications Company, Inc., 25 Van Zant Street, Norwalk, Conn. 06855. As reported in an online press release (4/21/98): http://buscom.com/editors/RB-124.html.

426. B. M. Bolten and T. DeGergorio, *Nat. Rev., Drug Disc.* **1**, 335 (2002).

427. American Society of Consultant Pharmacists website: http//:www.ascp.com/public/tcp/1998/apr/newdrugreview.shtml, author: J. Wick.
428. S. Frantz and A. Smith, *Nat. Rev., Drug Disc.* **2**, 95 (2003).
429. Cardiology Today website: http://www.cardiologytoday.com/20002/pipelinejump.asap
430. C. J. Vlahos, S. A. McDowell, and A. Clerk, *Nat. Rev. Drug Disc.* **2**, 99 (2003).
431. G. Duque, *Am. J. Ger. Cardiol.* **9** (5), 263 (2000).
432. S. Williams, *N. Eng. J. Med.* **341**, 709 (1999).
433. M. V. Cohen and J. M. Downey, *Annu. Rev. Med.* **47**, 21 (1996).
434. A. J. Hobbs, A. Higgs, and S., Moncada, *Annu. Rev. Pharmacol. Toxicol.* **39**, 191 (1999).
435. J. P. Cooke and V. J. Dzau, *Annu. Rev. Med.* **48**, 489 (1997).
436. C. O. Kappe, *Molecules* **3**, 1 (1998).
437. H. Tanaka and K. Shigenobu, *Cardiovasc. Drug Rev.* **18**, 93 (2000).
438. A. Smith, *Nat. Rev. Drug Disc.* **1**, 8 (2002).

GAJANAN S. JOSHI
Allos Therapeutics, Inc.

JAMES C. BURNETT
Virginia Commonwealth University

CATALYSIS

1. Introduction

Homogeneous catalysis, 324 Heterogeneous catalysis, 340 Catalysis, is the key to efficient chemical processing. Most industrial reactions and almost all biological reactions are catalytic. The value of the products made in the United States in processes that at some stage involve catalysis is approaching several trillion dollars annually, which is more than the gross national products of all but a few nations of the world. Products made with catalysis include food, clothing, drugs, plastics, detergents, and fuels. Catalysis is central to technologies for environmental protection by conversion of emissions.

    This article is an introduction and survey that states the fundamental principles and definitions of catalysis, demonstrates the unity of the subject, and places it in an applied perspective. The selection of industrial catalytic processes discussed has been made for the sake of illustrating principles and representative characteristics of catalysis and catalytic processes. Details of the processes are given in numerous other articles in the *Encyclopedia*.

    A catalyst is a substance that increases the rate of approach to equilibrium of a chemical reaction without being substantially consumed itself. A catalyst changes the rate but not the equilibrium of the reaction. This definition is almost the same as that given by Ostwald in 1895. The term catalysis was coined in ~1835 by Berzelius, who recognized that many seemingly disparate phenomena could be described by a single concept. For example, ferments added in small amounts were known to make possible the conversion of plant materials into alcohol; and there were numerous examples of both decomposition and synthesis reactions that were apparently caused by addition of various liquids or by contact with various solids.

Berzelius attributed catalytic action to ill-defined forces, and the value of Ostwald's more lasting definition is that it identified catalysis as a phenomenon that was consistent with the emerging principles of physical chemistry. Now it is well recognized that catalysts function by forming chemical bonds with one or more reactants, thereby opening up pathways to their conversion into products with regeneration of the catalyst. Catalysis is thus cyclic; reactants bond to one form of the catalyst, products are decoupled from another form, and the initial form is regenerated. The simplest imaginable catalytic cycle is therefore depicted as follows:

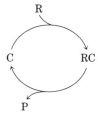

where R is the reactant, P the product, C the catalyst, and RC an intermediate complex. The intermediate complexes in catalysis are often highly reactive and not observable.

Ideally, the catalyst would cycle forever between C and RC without being consumed. But in reality there are competing reactions, and catalysts are converted into species that are no longer catalysts. In practice, catalysts must be regenerated and replaced. Catalyst manufacture is a large industry; catalysts worth some several billion dollars are sold annually in the United States.

Catalysts may be gases, liquids, or solids. Most catalysts used in technology are either liquids or surfaces of solids. Catalysis occurring in a single gas or liquid phase is referred to as homogeneous catalysis (or molecular catalysis) because of the uniformity of the phase in which it occurs. Catalysis occurring in a multiphase mixture such as a gas–solid mixture is referred to as heterogeneous catalysis; usually this is surface catalysis. Biological catalysts are proteins, ie, poly(amino acids), called enzymes, and metalloenzymes, which are proteins incorporating inorganic components, eg, iron sulfide clusters. Some enzymes are present in cytoplasmic solution in cells and others are anchored in cell membranes or on surfaces. Traditionally, homogeneous catalysis, heterogeneous catalysis, and enzymic catalysis have been regarded as almost separate disciplines, and the language and literature have developed without much coherence and overlap.

The performance of a catalyst is measured largely by criteria of chemical kinetics, as a catalyst influences the rate and not the equilibrium of a reaction. The catalytic activity is a property of a catalyst that measures how fast a catalytic reaction takes place and may be defined as the rate of the catalytic reaction, a rate constant, or a conversion (or temperature required for a particular conversion) under specified conditions. The selectivity is a measure of the property of a catalyst to direct a reaction to particular products. There is no single definition of selectivity, but it is sometimes defined as a ratio of activities, such as the ratio of the rate of a desired reaction to the sum of the rates of all the reactions that deplete the reactants. Selectivity may also be represented simply as a product distribution. Because catalysts typically lose activity and/or selectivity during operation, they are also evaluated in terms of stability and lifetime. The stability of a catalyst is a measure of the rate of loss of activity or selectivity. In practical terms the stability might be measured as a rate of deactivation, such as the rate of change of the rate of the desired catalytic reaction or as the rate at which the temperature of the catalyst would have to be raised to compensate for the activity loss. Catalysts that have lost activity are often treated to bring back the activity, ie, reactivated; the regenerability is a measure, often not precisely defined, of how well the activity can be brought back. Technological catalysts are also evaluated in terms of cost; in a typical process, the cost of the catalyst is a small fraction of the total processing cost, often only a few cents or less per kilogram of product.

Liquid-phase catalysts are used in batch and flow reactors. Batch reactors are predominant in small-scale operations and those designed for flexibility such as pharmaceutical and fine chemicals manufacture. Flow reactors predominate in large-scale processes such as commodity chemicals manufacture and petroleum refining. Corrosion, separation, and catalyst recovery and recycle are usually important issues in these processes. Solid catalysts are used in batch and flow reactors; in most large-scale operations, solid particles of catalyst are present in fixed- or fluidized-bed reactors with gaseous or liquid reactants flowing through them (see REACTOR TECHNOLOGY).

The solids used as catalysts are typically robust porous materials with high internal surface areas, typically, hundreds of square meters per gram. Reaction occurs on the internal catalyst surface. The typical solid catalyst used in industry is a composite material with numerous components and a complex structure.

Catalytic processes are classified roughly according to the nature of the product and the industry of application, and to some degree separate literatures have developed. In chemicals manufacture, catalysis is used to make heavy chemicals, eg, ammonia and sulfuric acid; commodity chemicals, eg, acetic acid and n-butyraldehyde; and fine chemicals, eg, fragrances and flavorings. Catalysis is used extensively in the manufacture of pharmaceuticals. In fuels processing, catalysis is used in almost all the processes of petroleum refining and in coal conversion and related synthesis gas (CO and H_2) conversion. Most of the recent large-scale developments in industrial catalysis have been motivated by the need for environmental protection. Many processes for abatement of emissions are catalytic; automobile exhaust conversion catalysts are now the most important of all in terms of sales volume, and catalysts for conversion of nitrogen oxides in stationary power plant effluents are used on a massive scale in Japan and parts of Europe. Recent innovations in pharmaceutical manufacture involve catalysis (usually homogeneous) for asymmetric synthesis. Most of the applications of catalysis in biotechnology (qv) are fermentations, often carried out in stirred reactors with gases, liquids, and solids present; the catalysts are enzymes present in living organisms such as yeasts. There are applications of whole biological cells and of individual enzymes mounted on supports, ie, carriers, and used in fixed-bed reactors (see ENZYMES, INDUSTRIAL APPLICATIONS).

2. History

The science of catalysis is driven by technology, as it has been from the beginning. Some of the earliest known examples of controlled chemical transformations are catalytic. Fermentation (qv) was used in ancient times to make alcoholic beverages, and a number of the earliest examples of chemical technology were also exploitations of catalysis (1). For example, before the sixteenth century ether was made by distilling spirits in the presence of sulfuric acid. In 1746, nitric oxide was used as a catalyst in the lead chamber process for oxidation of sulfur dioxide to give sulfur trioxide in the manufacture of sulfuric acid. In 1781, acids were used to catalyze the conversion of starch into sugar. In 1817, H. Davy discovered that in the presence of platinum, mine gases were oxidized at low temperatures; he designed a safety lamp for miners in which the platinum glowed if the flame was extinguished (2). Many more practical discoveries were made in the 1800s (1,2).

These examples existed prior to Ostwald's 1895 definition (1), ie, before the nature of catalysis was well understood. But in 1850 Wilhelmy (3) had already made the first measurements of kinetics of catalytic reactions in an investigation of sugar inversion catalyzed by mineral acids. In the years following Ostwald's definition, just as the principles of chemical equilibrium and kinetics were becoming known, the field of catalysis became more quantitative and developed rapidly. Kinetics of surface-catalyzed reactions were measured by Bodenstein (4) just after the turn of the century. The defining work that set the stage for modern catalytic technology was the development of the ammonia (qv) synthesis process by Haber, Bosch, Mittasch, and co-workers, beginning ~1908 (1,5).

These pioneers worked out methods of catalyst testing and process development that are essentially the methods of choice today (6,7). They understood the interplay between chemical equilibrium and reaction kinetics; indeed, Haber's research, motivated by the development of a commercial process, helped to spur the development of the principles of physical chemistry that account for the effects of temperature and pressure on chemical equilibrium and kinetics. The ammonia synthesis reaction is strongly equilibrium limited. The equilibrium conversion to ammonia is favored by high pressure and low temperature. Haber therefore recognized that the key to a successful process for making ammonia from hydrogen and nitrogen was a catalyst with a high activity to allow operation at temperatures low enough for a relatively favorable equilibrium.

Bosch and co-workers (6) devised laboratory reactors to operate at high pressure and temperature in a recycle mode. These test reactors had the essential characteristics of potential industrial reactors and were used by Mittasch and co-workers (6) at BASF in Germany to screen some 20,000 samples as candidate catalysts. The results led to the identification of an iron-containing mineral that is similar to catalysts used in industry today. The researchers recognized the need for porous catalytic materials and materials with more than one component, today identified as the support, the catalytically active component, and the promoter. Today's technology for catalyst testing has become more efficient because much of the test equipment is automated and some is miniaturized. The analysis of products and catalysts is also much faster and more accurate today.

3. Homogeneous Catalysis

3.1. Characterization of Solution Processes.
There are many important examples of catalysis in the liquid phase, but catalysis in the gas phase is unusual. From an engineering viewpoint, most of the liquid-phase processes have the following characteristics in common.

Pressure and Temperature. The pressure and temperature are relatively low, typically less than $\sim$2 MPa (20 atm) and 150°C. High pressures are used in some applications, but they require expensive thick-walled reactors. High temperatures are largely avoided because they result in high autogenous pressures and because many of the catalysts are unstable at high temperatures.

Corrosiveness. The catalyst solutions are corrosive, and the reactors, separation devices, etc, that come in contact with them must be made of expensive corrosion-resistant materials.

Separation Processes. Separation of the catalyst from the products is expensive; the process flow diagram and the processing cost are often dominated by the separations. Many soluble catalysts are expensive, eg, rhodium complexes, and must be recovered and recycled with high efficiency. The most common separation devices are distillation columns; extraction is also applied.

Gas Handling. The reactants are often gaseous under ambient conditions. To maximize the rate of the catalytic reaction, it is often necessary to minimize the resistance to gas–liquid mass transfer, and the gases are therefore introduced into the liquid containing the catalyst as swarms of bubbles into a well-stirred mixture or into devices such as packed columns that facilitate gas–liquid mixing and gas absorption.

Exothermicity. The catalytic reactions are often exothermic bond-forming reactions of small molecules that give larger molecules. Consequently, the reactors are designed for efficient heat removal. They may be jacketed or contain coils for heat-transfer media, or the heat of reaction may be used to vaporize the products and aid in the downstream separation by distillation.

There are also a number of generalizations about the chemistry of these processes. Often the reactants are small building blocks, many formed from organic raw materials, namely, petroleum, natural gas, and coal. The reactants include O_2, low molecular weight olefins, and synthesis gas (CO and H_2). In older technology, acetylene was a common building block. Many reactions are catalyzed by acids and bases, usually in aqueous solution. Many reactions are catalyzed by transition-metal complexes, usually in nonaqueous organic solvents, but sometimes in water and in ionic liquids with low melting points such as alkylammonium salts. The transition metal complex catalysts used in technology are often highly selective.

3.2. Influence of Mass Transport on Reaction Rates. When a relatively slow catalytic reaction takes place in a stirred solution, the reactants are supplied to the catalyst from the immediately neighboring solution so readily that virtually no concentration gradients exist. The intrinsic chemical kinetics determines the rate of the reaction. However, when the intrinsic rate of the reaction is relatively high and/or the transport of the reactant is relatively slow, as in a viscous polymer solution, the concentration gradients become significant, and the transport of reactants to the catalyst cannot keep the catalyst supplied sufficiently for the rate of the reaction to be that corresponding to the intrinsic chemical kinetics. Assume that the transport of the reactant in solution is described by Fick's law of diffusion with a diffusion coefficient D, and the intrinsic chemical kinetics is of the following form

$$r = k[C][A] \tag{1}$$

where C is catalyst and A the reactant. Then in the general case in which the diffusion influences the rate (8),

$$\eta = r/r_{\max} = \cfrac{1}{1 + \cfrac{k}{4\pi\left(\frac{R_A + R_C}{2}\right)D}} \tag{2}$$

where η, the ratio of the reaction rate to the maximum value that could occur, ie, the rate in the absence of a transport influence, is called the effectiveness factor, and R_C and R_A are the radii of the catalyst and reactant molecules, respectively. There are two limiting cases; if $k \ll 4\pi(\frac{R_A + R_B}{2})D$, then there is no diffusion influence and the rate is described by equation 1, where the concentrations are the bulk average concentrations in the solution. On the other hand, when the inequality is inverted, the rate becomes

$$r = 4\pi\left(\frac{R_A + R_B}{2}\right)D[C][A] \tag{3}$$

In the former case, the rate is independent of the diffusion coefficient and is determined by the intrinsic chemical kinetics; in the latter case, the rate is independent of the rate constant k and depends on the diffusion coefficient; the reaction is then diffusion controlled. This kind of mass transport influence is different from that characteristic of dissolution of a gas into a liquid phase.

3.3. Examples of Solution Catalysis. *Acid-Base Catalysis.* Inexpensive mineral acids, eg, H_2SO_4, and bases, eg, KOH, in aqueous solution are widely applied as catalysts in industrial organic synthesis. Catalytic reactions include esterifications, hydrations, dehydrations, and condensations. Much of the technology is old and well established, and the chemistry is well understood. Reactions that are catalyzed by acids are also typically catalyzed by bases. In some instances, the kinetics of the reaction has a form such as the following (9):

$$r = (k_0[H_2O] + k_{H^+}[H^+] + k_{HA}[HA] + k_{OH^-}[OH^-] + k_B[B])[R] \tag{4}$$

where r is the reaction rate, [R] the concentration of the reactant, and the terms on the right-hand side represent catalysis by water, hydrogen ions, undissociated acid HA, hydroxide ions, and undissociated base. Often all the terms on the right-hand side but one are negligible.

If the second term on the right-hand side in equation 4 is the only significant one, then catalysis is by hydrogen ions (hydrated protons), and the catalysis is called specific acid catalysis; the rate of the reaction is proportional to the concentration of the hydrogen ions and to the concentration of the reactant. Many examples conform to this pattern. Similarly, if the third term on the right-hand side of equation 4 is the only significant one, then the catalysis is by HA, and general acid catalysis occurs. If the fourth term is the significant one, hydroxide ions are the catalyst and specific base catalysis occurs. If the fifth term is the significant one, general base catalysis occurs.

The chemistry of acid–base catalysis involves the transfer of hydrogen ions, H^+ and sometimes H^-. In general or specific acid catalysis, the undissociated acid molecules or the hydrated protons formed by dissociation of an acid, respectively, donate protons to the reactant. The proton acceptor in an aqueous solution is a good base, often an organic compound including O, N, or S. In catalysis by bases, the reactant donates a proton to the catalyst. The proton donor in general acid catalysis may be a weak or a strong acid, and often the activity of the catalyst increases with its strength as an acid, measured by the acid dissociation constant K_a.

Donation of a proton to the reactant often forms a carbenium ion or an oxonium ion, which then reacts in the catalytic cycle. For example, a catalytic cycle suggested for the conversion of phenol and acetone into bisphenol A, which is an important monomer used to manufacture epoxy resins and polycarbonates, in an aqueous mineral acid solution is shown in Figure 1 (10).

The kinetics of the bisphenol A synthesis has been reported to be of the following form (11)

$$r = k[H^+][A][P] \tag{5}$$

where A is acetone and P is phenol. When the reaction is carried out in the presence of low concentrations of an added compound, thioglycolic acid, $HSCH_2COOH$, the rate is increased, with the kinetics reported to be of the following form (12)

$$r = k'[H^+][A][P](k_1 + k_2[T]) \tag{6}$$

where T is thioglycolic acid, a promoter. A catalyst promoter is a substance that, although it lacks significant catalytic activity itself, increases the activity, or possibly the selectivity, of a catalyst. Promoters are common in catalysis.

In some processes, the reactant bases are too weak to be protonated significantly except in the presence of very strong acids such as fuming sulfuric acid or a mixture of concentrated sulfuric and nitric acids, ie, mixed acid.

The kinetics of reactions catalyzed by very strong acids are often complicated. The exact nature of the proton donor species is often not known, and typically the rate of the catalytic reaction does not have a simple dependence on the

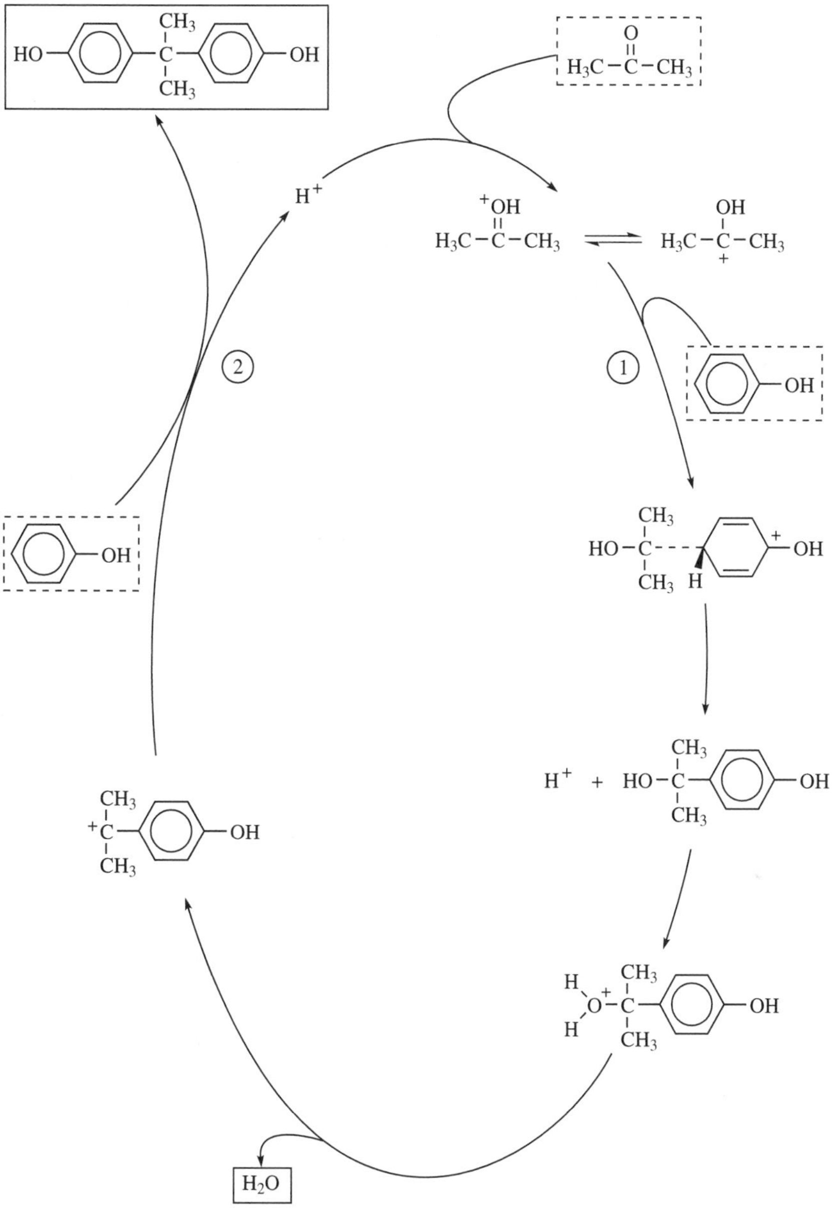

Fig. 1. Catalytic cycle for synthesis of bisphenol A from phenol and acetone in the presence of a dissociated mineral acid (10).

total concentration of the acid. However, sometimes there is a simple dependence of the catalytic reaction rate on some empirical measure of the acid strength of the solution, such as the Hammett acidity function H_0, which is a measure of the tendency of the solution to donate a proton to a neutral base. Sometimes the rate is proportional to h_0 ($H_0 = -\log h_0$). Such a dependence may be expected when the slow step in the catalytic cycle is the donation of a proton by the solution to a neutral reactant, ie, base; but it is not easy to predict when such a dependence may be found.

An important petroleum refining process, alkylation of propylene or butenes with isobutane to give high octane-number branched products, is catalyzed by concentrated sulfuric acid or hydrofluoric acid. The reactions are carried out at subambient temperatures, at which the equilibrium is favorable, in stirred, refrigerated reactors containing two-phase liquid mixtures. Most of the hydrocarbon reactants are in the organic phase, and, when the catalyst is H_2SO_4, most of the catalyst is in the aqueous phase; the reaction takes place near the liquid–liquid interface (see ALKYLATION).

In the alkylation process, the reactants are weak bases. The stronger bases in the mixture, the olefins, are protonated, forming carbenium ions. The carbenium ions are good Lewis acids, and they react with isobutane, a good hydride ion donor, to form paraffins and the relatively stable *tert*-butyl cation, ie, trimethyl carbenium ion. This is a hydride transfer reaction; such reactions are important in many acid-catalyzed hydrocarbon conversions. The *tert*-butyl cation becomes the predominant carbenium ion in the organic solution and reacts with propylene or butenes in Lewis acid–Lewis base reactions to form carbon–carbon bonds, giving C_7 or C_8 products among others. There are many C_7 and C_8 isomers formed because isomerizations are rapid. A partial cycle that illustrates the important characteristics of alkylation is shown in Figure 2 for the ethylene–isobutane alkylation. The cycle is shown for ethylene because it gives a simpler product distribution than propylene or butenes; ethylene is much less reactive than propylene or butenes because on protonation it forms a primary carbenium ion, which is much less stable than the secondary carbenium ion formed by protonation of propylene or straight-chain butenes. Ethylene is not used in the commercial alkylation processes.

Acid-catalyzed reactions of paraffins require much stronger acids than those of olefins or aromatics, because the paraffins are much weaker bases.

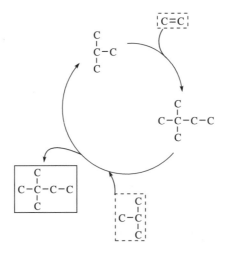

Fig. 2. Catalytic cycle for the ethylene–isobutane alkylation (10).

The only acids that catalyze reactions of paraffins alone at relatively low temperatures are combinations of Brønsted and Lewis acids; examples include $HCl + AlCl_3$, $HF + SbF_5$, and $FSO_3H + SbF_5$ (magic acid). Some of these are called superacids (13), eg, magic acid. Superacids are stronger acids than concentrated sulfuric acid and some have values of H_0 of -20 and less, whereas that of 100% sulfuric acid is only about -12. The Hammett acidity function is logarithmic; thus some superacids are more than eight orders of magnitude stronger proton donors than sulfuric acid. Hence, these superacids can protonate paraffins, even methane. They catalyze isomerization and carbon–carbon bond-forming and bond-breaking reactions at low temperatures but are not found in commercial applications, in part because of their extreme corrosiveness.

Metal Complex Catalysis.   Most of the recent innovations in industrial homogeneous catalysis have resulted from discoveries of transition-metal complex catalysts. Thousands of transition-metal complexes (including organometallics, ie, those with metal–carbon bonds), are known, and the rapid development of organotransition metal chemistry in recent decades has been motivated largely by the successes and opportunities in catalysis. The chemistry of metal complex catalysis is explained by the bonding and reactivity of organic groups (ligands) bonded to the metals. The important reactions in catalytic cycles are those of ligands bonded in the coordination sphere of the same metal atom. Bonding of ligands such as CO or olefin to a transition metal activates them and facilitates the catalysis. The following reactions of transition-metal complexes account for almost all the steps in the catalytic cycles.

Ligand exchange:

$$ML_x + L' \rightleftharpoons ML_{x-1}' + L \qquad (7)$$

where M is metal and L and L' are ligands. This reaction may be thermally or photochemically initiated; it may proceed by dissociation of a ligand to give a coordinatively unsaturated complex, to which another ligand is then bonded, but it may also proceed associatively, without prior dissociation of a ligand. Ligand exchange is one way in which reactants become bonded to catalysts.

Oxidative addition and reductive elimination:

$$ML_{x-2} \ + \ AB \ \underset{\text{reductive elimination}}{\overset{\text{oxidative addition}}{\rightleftharpoons}} \ ML_{x-2}AB \qquad (8)$$

The forward reaction requires an electron-rich metal in a low oxidation state that is coordinatively unsaturated. Oxidative addition provides another way for bonding of reactant ligands to a metal, and reductive elminination provides a way for products to leave a metal.

Migratory Insertion:

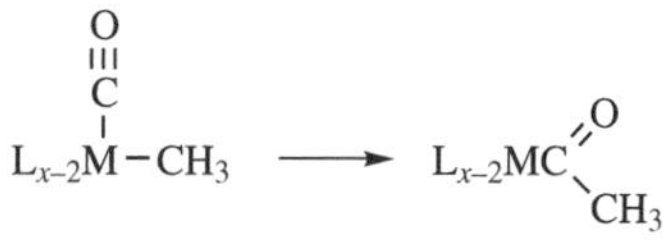

$$L_{x-2}\overset{\displaystyle \overset{O}{\overset{\|||}{\underset{|}{C}}}}{M}-CH_3 \ \longrightarrow \ L_{x-2}MC\overset{\diagup O}{\diagdown CH_3} \qquad (9)$$

In this reaction, one ligand migrates to a neighboring ligand that then becomes formally inserted between the metal and the ligand that has migrated, creating a site of coordinative unsaturation so that another reactant ligand can be associated with the metal. The migratory insertion reaction accounts for chain-growth steps of olefin polymerization reactions.

Wilkinson Hydrogenation.   One of the best understood catalytic cycles is that for olefin hydrogenation in the presence of phosphine complexes of rhodium, the hydrogenation named after the Nobel-laureate Geoffroy Wilkinson (14,15). The reactions of a number of olefins, eg, cyclohexene and styrene, are rapid, taking place even at room temperature and atmospheric pressure; but the reaction of ethylene is extremely slow. Complexes of a number of transition metals in addition to rhodium are active for the reaction.

The Wilkinson hydrogenation cycle shown in Figure 3 (16) was worked out in experiments that included isolation and identification of individual rhodium complexes, measurements of equilibria of individual steps, determination of rates of individual steps under conditions of stoichiometric reaction with certain reactants missing so that the catalytic cycle could not occur, and determination of rates of the overall catalytic reaction. The cycle demonstrates some generally important points about catalysis: the predominant species present in the reacting solution and the only ones that are easily observable by spectroscopic methods, eg, $RhCl[P(C_6H_5)_3]_3$, $RhCl[P(C_6H_5)_3]_2$ (olefin), and $RhCl_2[P(C_6H_5)_3]_4$, are outside the cycle, possibly in virtual equilibrium with species in the cycle, and not involved in the kinetically significant steps of the cycle. Often the only species that can be observed during catalysis are those that are not involved in the cycle. Consequently, elucidation of catalytic cycles is difficult, and inferences about cycles that are based on easily measured compositions of reacting solutions are usually less than well founded. The exceptions to this rule are usually the cycles of least interest, those of slow catalytic reactions.

The Wilkinson hydrogenation illustrates the roles of the oxidative addition, reductive elimination, and insertion reactions, all of which occur in the cycle (Fig. 3). Some of the properties of the rhodium phosphine complexes that explain their success as catalysts include the following: Rhodium exists in two stable oxidation states, Rh(I) and Rh(III), separated by two units. Oxidative addition and reductive elimination reactions formally require changes of two units in metal oxidation state. There are no intermediates that are so stable that they form bottlenecks in the cycle. There is an exquisite dynamic balance, and the cycle turns over rapidly with only low concentrations of the intermediates being present in a steady state. The phosphines are electron-donor ligands that are important in controlling the reactivities of the intermediates and regulating the dynamic balance. Replacement of triphenylphosphine by other phosphines leads to significant changes in the rate of the catalytic hydrogenation. The phosphines can influence the reactivities of the intermediates, and thus the rate of catalysis, by virtue of both electronic and steric effects.

Asymmetric Hydrogenation.   Biological reactions are stereoselective, and numerous drugs must be pure optical isomers. Metal complex catalysts have been found that give very high yields of enantiomerically pure products, and some have industrial application (17,18). The hydrogenation of the methyl ester of acetamidocinnamic acid has been carried out to give a precursor of

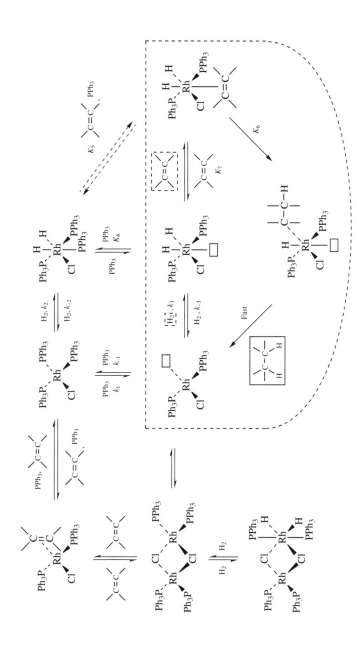

Fig. 3. Catalytic cycle (within dashed lines) for the Wilkinson hydrogenation of olefin (14,16). Ph represents phenyl (C_6H_5). Values of rate constants and equilibrium constants are as follows: $k_1 = 0.68$ s^{-1}; $k_{-1} \geq 7 \times 10^4$ L/(mol·s); $K_1 \leq 10^{-5}$ mol/L; $k_2 = 4.8$ L/(mol·s); $k_{-2} = 2.8 \times 10^{-4}$ s^{-1}; $K_2 = 1.7 \times 10^4$ L/mol; $k_3 \geq 7 \times 10^4$ L/(mol·s); $k_6 = 0.22$ s^{-1}; $k_2 = 3 \times 10^{-4}$; $k_2 = 4.8$ L/(mol·s); $k_{-1}/k_4 \cong 1$; $K_5 = 3 \times 10^{-4}$; $k_2 = 4.8$ L/(mol·s); $K_5 \cong 1$; $K_5 = 3 \times 10^{-4}$.

211

L-dopa, ie, 3,4-dihydroxyphenylalanine, a drug used in the treatment of Parkinson's disease.

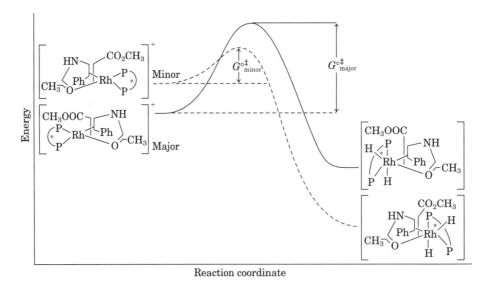

The strategy of the catalyst development was to use a rhodium complex similar to those of the Wilkinson hydrogenation but containing bulky chiral ligands in an attempt to direct the stereochemistry of the catalytic reaction to favor the desired L isomer of the product (17). Active and stereoselective catalysts have been found and used in commercial practice (although there is now a more economical route to L-dopa than through hydrogenation of the prochiral precursor). The 2001 Nobel Prize in Chemistry was awarded to W. S. Knowles and R. Noyori, recognizing their work on these catalysts (and to K. B. Sharpless for work on asymmetric oxidation catalysis).

Two pathways were found for the chiral hydrogenation, and they give products with different stereochemistries (19). One pathway involves the preferred mode of initial binding of the reactant to the catalyst. The other pathway involves an isomer of the reactant–catalyst complex that is formed in only small amounts, but its conversion is energetically favorable and constitutes the kinetically predominant pathway to products (9) (Fig. 4). Thus the chirality of the product is determined not by the preferred mode of the initial binding, but instead by the more favorable energetics of the pathway involving the minor isomer of the reactant–catalyst complex.

Fig. 4. Schematic representation of energy profiles for the pathways for the hydrogenation of a prochiral precursor to make L-dopa (19). The chiral disphosphine ligand is shown schematically as $\overset{*}{P} \frown P$. Ph represents phenyl (C_6H_5).

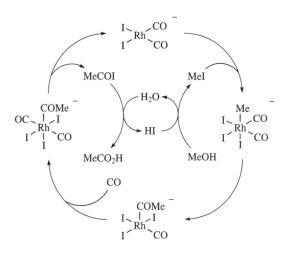

Fig. 5. Catalytic cycle for methanol carbonylation (21).

Methanol Carbonylation. An important industrial process catalyzed by rhodium (and more recently iridium) complexes in solution is methanol carbonylation to give acetic acid.

$$CH_3OH + CO \rightarrow CH_3COOH \tag{10}$$

The catalytic cycle (Fig. 5) (20,21) is well established. The CH_3I works as a cocatalyst or promoter because it undergoes an oxidative addition with $[Rh(CO)_2I_2]^-$, and the resulting product has the CO ligand bonded cis to the CH_3 ligand; these two ligands are then poised for a migratory insertion reaction. When the catalyst is iridium instead of rhodium, a similar cycle pertains, but the reaction of $[H_3CIr(CO)_2I_3]^-$ with CO to give $[H_3CCOIr(CO)_2I_3]^-$ is rate determining (21). The migratory insertion in the iridium complex can be accelerated by the addition of either methanol or a Lewis acid (SnI_2); each appears to facilitate substitution of an iodide ligand by CO, resulting in faster methyl migration. The greater stability of $[H_3CIr(CO)_2I_3]^-$ than of $[H_3CRh(CO)_2I_3]^-$ accounts for the different characters of the reactions catalyzed by the two metals (21).

Methanol carbonylation is one of only a few industrially important catalytic reactions for which the quantitative reaction kinetics is known (22).

$$r = k[CH_3I][\text{Rh complex}] \tag{11}$$

Here, for the rhodium-complex-catalyzed reaction, $k = 3.5 \times 10^6 \exp(-E_{act}/RT)$ and R is the gas constant, T is the absolute temperature, and E_{act} is 61.5 kJ/mol (14.7 kcal/mol). Because the reaction is zero order in the reactants, a stirred reactor has no disadvantage in comparison with a tubular flow reactor with respect to the efficiency of use of the reactor volume. A stirred reactor might therefore be preferred to facilitate rapid heat transfer and good gas–liquid contacting for rapid gas–liquid mass transfer, ie, dissolution of the CO into the reactant solution.

The reaction is carried out in a polar solvent, as the intermediates are anionic. Water is present from the following reaction, which achieves virtual equilibrium:

$$CH_3OH + HI \rightleftharpoons CH_3I + H_2O \tag{12}$$

Water is also formed in the acid-catalyzed dehydration of methanol to give dimethyl ether. The solution is acidic because of the presence of the HI.

The process flow diagram (23) (Fig. 6) is characteristic of those for homogeneous catalytic processes in being dominated by the separations devices. The products are purified by distillation, and several columns are required because the boiling point of the product acetic acid is near the middle of the boiling range of the product mixture; both the light products, including unconverted CO, CH_3I, and dimethyl ether, and the heavy rhodium complex must be recovered and recycled efficiently for the process to be economical. Corrosion is also a concern, and the reactor, separations devices, and lines are made of expensive stainless steel alloys.

For years, researchers have attempted to minimize the problems of separation and corrosion by using catalysts that are solid analogues of the metal complexes used in solution. The early attempts were not successful, in large measure because the rhodium complex was leached from the support (typically a

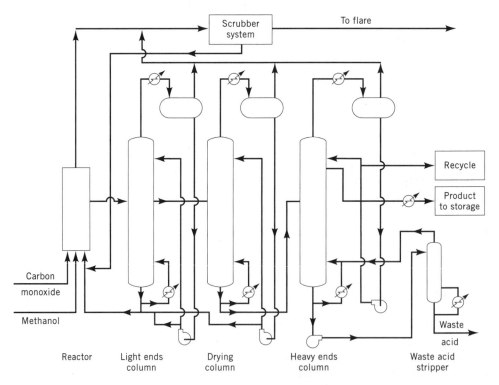

Fig. 6.   Process flow diagram for methanol carbonylation catalyzed by a rhodium complex to make acetic acid (23).

cross-linked polymer with pendant phosphine groups) and the polymer was not sufficiently stable. Recent work with more stable vinyl pyridine resins as supports has led to stable catalysts and apparently to a viable industrial process (24); this is important as perhaps the first successful application of a supported catalyst that is nearly analogous to a molecular metal complex catalyst.

Olefin Hydroformylation (The Oxo Process). One of the most important industrial applications of transition-metal complex catalysis is the hydroformylation of olefins (25), illustrated for propylene:

$$CH_3CH{=}CH_2 \ + \ CO \ + \ H_2 \ \longrightarrow \ CH_3\overset{\overset{\displaystyle CHO}{|}}{C}HCH_3 \ + \ CH_3CH_2CH_2CHO \qquad (13)$$

The linear isomer is more valuable than the branched isomer (see BUTYRALDE-HYDE). The product aldehydes are hydrogenated to give so-called oxo alcohols; long-chain products are converted into sulfonates and used as detergents.

The oxo process was discovered in 1938 by Roelen and co-workers of Ruhrchemie. The first catalysts were cobalt carbonyl complexes formed from [HCo(CO)$_4$]. They are still used on a large scale. The process is carried out at temperatures of ~150°C and pressures of >30MPa (several hundred atm). The high pressures are required to maintain the cobalt in the form of the soluble metal carbonyl complex; at lower pressures, cobalt metal can form and deposit on the reactor walls, possibly in a porous form that adds a significant resistance to heat transfer. Use of cobalt complexes with phosphine ligands led to processes with improved selectivities. The process flow diagram is complicated because it is expensive to separate the catalyst from the products and recycle it in the high pressure process. In the Kuhlmann process (25), the catalyst is extracted into an aqueous stream and then regenerated with acid to give [HCo(CO)$_4$], which is absorbed into the reactant solution.

A significant advance in hydroformylation technology was made with the discovery that phosphine complexes of rhodium, similar to those used in the Wilkinson hydrogenation but incorporating H, CO, and olefin ligands as well as triphenylphosphine, have catalytic activities several orders of magnitude greater than those of cobalt; furthermore, the rhodium complexes are stable enough to be used at low pressures, typically 1.5 MPa (15 atm) at ~150°C, and separated by distillation (25). Rhodium complex catalysts now are used in industrial processes for hydroformylation of propylene. When a large excess of triphenylphosphine is used as the solvent, the selectivity for the desired straight-chain product is high, with >95% of the aldehyde being *n*-butyraldehyde.

The processing costs associated with separation and corrosion are still significant in the low pressure process; for the process to be economical, the efficiency of recovery and recycle of the rhodium must be very high. Consequently, researchers have continued to seek new ways to facilitate the separation and confine the corrosion. Extensive research was done with rhodium phosphine complexes bonded to solid supports, but the resulting catalysts were not sufficiently stable, as rhodium was leached into the product solution. A successful solution to the engineering problem resulted from the application of a two-phase liquid–liquid process (26) . The catalyst is synthesized with polar –SO$_3$Na groups on the phenyl rings of the triphenylphosphine

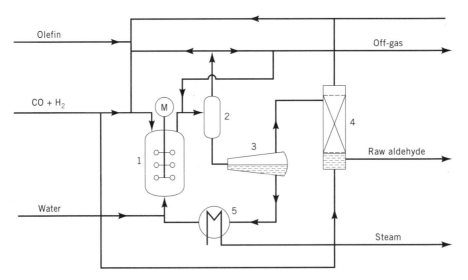

so that it is water soluble. In the reactor, the catalyst is largely confined to the aqueous phase and the reactants are largely confined to the organic phase; reaction likely takes place near the interface. A process flow diagram is shown in Figure 7 (26). The two liquid phases are separated downstream of the stirred reactor and the catalyst is recycled with a high efficiency. The process is applied

Fig. 7. Process flow diagram for the two-phase hydroformylation of propylene where 1 = reactor; 2 = separator; 3 = phase separator; 4 = stripping column; and 5 = heat exchanger (26).

commercially and is competitive with the single-phase process. It is indicative of a trend toward application of more homogeneous catalytic processes with environmentally friendly solvents such as water.

Attempts to anchor transition-metal complex catalysts to supports have been extended to reactions far beyond just methanol carbonylation and hydroformylation, but almost all the resulting catalysts have failed to find practical application. The principal difficulty is that the catalyst at each step of the catalytic cycle must be stably bonded to the support, and this is often too stringent a design criterion. Perhaps a more generally useful way to confine the catalyst in a phase separate from that containing most of the reactants is to use a second liquid phase. Water is one appealing choice, and promising work has also been done with fluorous phases (27,28).

Olefin Metathesis. The olefin metathesis (dismutation) reaction (29,30) converts olefins to lower and higher molecular weight olefins. For example, propylene is converted into ethylene and butene

$$2\,RCH\!=\!CHR' \longrightarrow RCH\!=\!CHR + R'CH\!=\!CHR' \tag{14}$$

This reaction is catalyzed in solution by complexes of tungsten, molybdenum, or rhenium in high oxidation states, eg, Re^{7+}. Examples of catalyst precursors are $[W(CH\text{-}t\text{-}C_4H_9)(OR)_2X_2]$ (31a), where X is halide. A number of catalysts require cocatalysts such as $CH_3Al_2Cl_3$.

The olefins that undergo metathesis include most simple and substituted olefins; cyclic olefins give linear high molecular weight polymers. The mechanism of the reaction is believed to involve formation of carbene complexes that react via cyclic intermediates, ie, metallacycles. An extension of the chemistry has led to efficient ring-opening metathesis polymerization (32).

Industrial olefin metathesis processes are carried out with solid catalysts (29).

The Wacker Oxidation of Ethylene to Acetaldehyde. One of the early industrial examples of organometallic catalysis in solution was described by Smidt and co-workers in 1959 (33). This is the Wacker oxidation of ethylene to give acetaldehyde (qv). It had long been known that Pd(II) complexes in solution would stoichiometrically oxidize ethylene to acetaldehyde. Smidt's insight was to make this reaction part of a catalytic cycle. The researchers created a process by closing the cycle, using an excess of an oxidizing agent, Cu^{2+}, to rapidly reoxidize the palladium that oxidized the ethylene and thereby keep palladium from plating out on the reactor walls as metal. The cycle is completed as O_2 reoxidizes the Cu^+ formed from Cu^{2+}. The sequence of reactions is the following:

$$CH_2\!=\!CH_2 + H_2O + PdCl_2 \rightarrow CH_3CHO + Pd + 2\,HCl \tag{15}$$

$$Pd + 2\,CuCl_2 \rightarrow PdCl_2 + 2\,CuCl \tag{16}$$

$$2\,CuCl + \frac{1}{2}\,O_2 + 2\,HCl \rightarrow 2\,CuCl_2 + H_2O \tag{17}$$

These reactions constitute the cycle; their sum gives the stoichiometry of the Wacker oxidation:

$$CH_2{=}CH_2 + \frac{1}{2}\,O_2 \rightarrow CH_3CHO \tag{18}$$

This process is unusual among those involving organometallic chemistry in that it takes place in aqueous solution. In one process design, there is a single reactor in which all three reactions in the sequence take place, and the feed contains pure O_2. In another process, there are two reactors, with the ethylene oxidation (eq. 15) and the palladium reoxidation (eq. 16) taking place in one reactor and the reoxidation of Cu^+ with air taking place in the other. An advantage of the single-reactor process is the confinement of the corrosive solutions in one vessel, but this is compensated by the need for purified oxygen . The feed gas contains ethylene and oxygen, and the composition must be kept outside the explosive limits. Consequently, the feed gas is not a stoichiometric mixture, and gas recycle is necessary for economic operation. In contrast, the two-reactor process allows the use of air as a feed to the reactor where Cu^+ is reoxidized; the air flows once through the reactor and is almost all converted. Ethylene flows into the other reactor. A disadvantage of the two-reactor process is the need for corrosion-resistant materials in both reactors and the lines between them. Both designs have been successfully applied.

A process similar to the Wacker process has been applied for the oxidation of ethylene with acetic acid to give vinyl acetate, but now the principal applications are with a solid catalyst.

Free-Radical Oxidation of Hydrocarbons. The chemistry of the Wacker reaction is atypical for oxidation. A more common pattern in catalytic oxidation is a chain reaction proceeding through free-radical intermediates; hydroperoxides are common products, and they are often converted themselves into other products (34,35). Free-radical reactions are characterized by complex product distributions because O_2 has a high reactivity with organic reactants, with metal centers, and with many ancillary ligands. Metals typically play a role in an initiation process, helping to start a free-radical chain reaction; free radicals are often generated by metal-catalyzed decomposition of organic hydroperoxides. Some of the important applications of such chemistry are oxidations of petroleum-derived hydrocarbons. Catalysts include iron, manganese, cobalt, and copper in high oxidation states; a typical solvent is acetic acid. The processes include oxidation of *n*-butane to give acetic acid, formation of fatty acids from paraffin wax, and oxidation of cyclohexane to give adipic acid (qv) precursors in nylon manufacture.

A simplified mechanism of the cyclohexane oxidation is shown in Figure 8 (36). A free-radical chain is set up, with R· and ROO· being the chain carriers (R is C_6H_{11}), as shown in the upper right of the figure. One role of Co^{3+} is suggested to be that of an electron-transfer agent undergoing a one-electron change; the role of the metal is that of a redox initiator, as cobalt cycles between the +2 and +3 oxidation states in a process that generates free radicals. As the

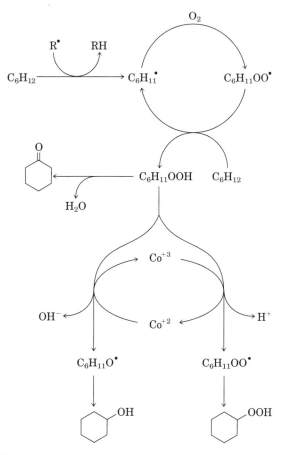

Fig. 8. Simplified mechanism for the free-radical oxidation of cyclohexane (36).

cobalt also reacts in a cycle, its role is catalytic, and only small amounts are needed.

One of the most important oxidation processes is the conversion of p-xylene into terephthalic acid or dimethyl terephthalate. There are numerous commercial processes, generally using air as the oxidant and cobalt and manganese salts as the catalysts (37). In one process, the oxidation takes place at $\sim$225°C and 15 atm, and bromides are involved in the catalysis (37).

CO and H₂ Reactions. Synthesis gas (CO and H_2) can be converted into numerous organic chemicals, including acetic anhydride formed in a commercial process with a rhodium complex catalyst. Soluble rhodium catalysts are also active for synthesis of ethylene glycol along with methanol and other products (38). The process is not applied commercially because extremely high pressures are required. But the process is conceptually important because it is believed that the catalysts may be compounds with metal–metal bonds, called metal clusters. A simple metal cluster is $[Rh_{13}(CO)_{24}H_3]^{2-}$, which has the following structure:

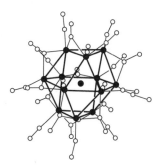

There are many related compounds, including rhodium carbonyl cluster anions that are present in the solutions catalyzing ethylene glycol formation and which may be the catalytically active species or in equilibrium with them (38).

There are only a few well-documented examples of catalysis by metal clusters, and not many are to be expected as most metal clusters are fragile and fragment to give metal complexes or aggregate to give metal under reaction conditions (39). However, the metal carbonyl clusters are conceptually important because they form a bridge between catalysts commonly used in solution, ie, transition-metal complexes with single metal atoms, and catalysts commonly used on surfaces, ie, small metal particles or clusters.

3.4. Phase-Transfer Catalysis.   When two reactants in a catalytic process have such different solubility properties that they can hardly both be present in a single liquid phase, the reaction is confined to a liquid–liquid interface and is usually slow. However, the rate can be increased by orders of magnitude by application of a phase-transfer catalyst (39,40), and these are used on a large scale in industrial processing (see CATALYSTS, PHASE-TRANSFER). Phase-transfer catalysts function by facilitating mass transport of reactants between the liquid phases. Often most of the reaction takes place close to the interface.

Industrial examples of phase-transfer catalysis are numerous and growing rapidly; they include polymerization, substitution, condensation, and oxidation reactions. The processing advantages, besides the acceleration of the reaction, include mild reaction conditions, relatively simple process flow diagrams, and flexibility in the choice of solvents.

4. Heterogeneous Catalysis

4.1. Characterization of Surface Processes.   Most of the largest scale catalytic processes take place with gaseous reactants in the presence of solid catalysts. From an engineering viewpoint, these processes offer the following advantages, in contrast to those involving liquid catalysts: (*1*) wide ranges of temperature and pressure are economically applied. The most direct way to increase the activity of a catalyst is to increase its temperature, and high temperatures can be used economically with solid catalysts as the reactants are gases and there is no liquid phase to require a high pressure. The catalysts must be robust to withstand the high temperatures; solid catalysts are commonly used at temperatures of 500°C, and some endure temperatures several hundred degrees higher; (2) solid catalysts are only rarely corrosive; (3) the separation of gaseous or liquid products from solid catalysts is simple and costs little. It is simplest when the catalyst particles are confined in a fixed-bed reactor; when very small particles are used in a fluidized bed, they are entrained in the product stream, and a gas–solid separator is required; (4) the mixing and mass transport in a fixed- or fluidized-bed reactor are facilitated by the solid catalyst particles through which the reactants and products flow. If no liquid phase is present, a potentially significant mass-transfer resistance is eliminated; mass-transfer resistance in the gas phase is usually small. However, reactions take place on the internal surfaces of porous catalyst particles, and the resistance to transport through the pores is often significant; (5) strongly exothermic and strongly endothermic reactions are routinely carried out with solid catalysts. Fluidized-bed reactors offer the advantage of excellent mixing and high rates of heat transfer. They are commonly used when heat effects are large, eg, for oxidation reactions; alternatively, small tubes are used in bundles to minimize the distance through which the heat is transferred in a fixed bed. The catalyst particles themselves constitute a significant heat-transfer resistance, and the temperature gradients in particles are sometimes large.

4.2. Properties of Solid Catalysts. Most solid catalysts used on a large scale are porous inorganic materials. A number of these and the reactions they catalyze are summarized in Table 1 (10); a list of new processes was compiled by Armor (41).

Catalysis takes place as one or more of the reactants is chemisorbed (chemically adsorbed) on the surface and reacts there. The activity and selectivity of a catalyst depend strongly on the surface composition and structure.

The catalysts with the simplest compositions are pure metals, and the metals that have the simplest and most uniform surface structures are single crystals. Researchers have done many experiments with metal single crystals in ultrahigh vacuum chambers so that unimpeded beams of particles and radiation can be used to probe them. These surface science experiments have led to fundamental understanding of the structures of simple adsorbed species, such as CO, H, and small hydrocarbons, and the mechanisms of their reactions; they indicate that catalytic activity is often sensitive to small changes in surface structure. For example, paraffin hydrogenolysis reactions take place rapidly on steps and kinks of platinum surfaces but only very slowly on flat planes; however, hydrogenation of olefins takes place at approximately the same rate on each kind of surface site.

A few industrial catalysts have simple compositions, but the typical catalyst is a complex composite made up of several components, illustrated schematically in Figure 9 by a catalyst for ethylene oxidation. Often it consists largely of a porous support or carrier, with the catalytically active components dispersed on the support surface. For example, petroleum refining catalysts used for reforming of naphtha have ~1 wt% Pt and Re on the surface of a transition alumina such as γ-Al_2O_3 that has a surface area >100 square meters per gram. The expensive metal is dispersed as minute particles or clusters so that a large fraction of the

Table 1. **Some Large-Scale Industrial Processes Catalyzed by Surfaces of Inorganic Solids**[a]

Catalyst	Reaction
metals (eg, Ni, Pd, Pt, as powders or on supports) or metal oxides (eg, Cr_2O_3)	C=C bond hydrogenation (eg, olefin $+ H_2 \rightarrow$ paraffin)
metals (eg, Cu, Ni, Pt)	C=O bond hydrogenation (eg, acetone $+ H_2 \rightarrow$ 2-propanol)
metal (eg, Pd, Pt) Fe, Ru (supported and promoted with alkali metals)	complete oxidation of hydrocarbons, oxidation of CO_3 $H_2 + N_2 \rightarrow 2NH_3$
Ni	$CO + 3H_2 \rightarrow CH_4 + H_2O$ (methanation) $CH_4 + H_2O \rightarrow 3 H_2 + CO$ (steam reforming)
Fe or Co (supported and promoted with alkali metals)	$CO + H_2 \rightarrow$ paraffins $+$ olefins $+ H_2O + CO_2 +$ oxygen-containing organic compounds) (Fischer-Tropsch reaction)
Cu (supported on ZnO, with other components, eg, Al_2O_3)	$CO + 2H_2 \rightarrow CH_3OH$
Re + Pt (supported on γ-Al_2O_3 and promoted with chloride)	paraffin dehydrogenation, isomerization and dehydrocyclization (eg, n-heptane $\rightarrow$ toluene $+ 4 H_2$) (naphtha reforming)
solid acids (eg, SiO_2-Al_2O_3, zeolites)	paraffin cracking and isomerization; aromatic alkylation; polymerization of olefins
γ-Al_2O_3	alcohol $\rightarrow$ olefin $+ H_2O$
Pd supported on zeolite	paraffin hydrocracking
metal-oxide-supported complexes of Cr, Ti, or Zr	olefin polymerization (eg, ethylene $\rightarrow$ polyethylene)
metal-oxide-supported complexes of W or Re	olefin metathesis (eg, 2-propylene $\rightarrow$ ethylene $+$ butene)
V_2O_5 or Pt	$2SO_2 + O_2 \rightarrow 2SO_3$
Ag (on inert support, promoted by alkali metals)	ethylene $+ \frac{1}{2}O_2 \rightarrow$ ethylene oxide (with $CO_2 + H_2O$)
V_2O_5 (on metal oxide support)	naphthalene $+ \frac{9}{2}O_2 \rightarrow$ phthalic anhydride $+ 2 CO_2 + 2 H_2O$ o-xylene $+ 3 O_2 \rightarrow$ phthalic anhydride $+ 3H_2O$
bismuth molybdate, uranium antimonate, other mixed metal oxides	propylene $+ \frac{1}{2}O_2 \rightarrow$ acrolein propylene $+ \frac{3}{2}O_2 + NH_3 \rightarrow$ acrylonitrile $+ 3 H_2O$
mixed oxides of Fe and Mo	$CH_3OH + O_2 \rightarrow$ formaldehyde (with $CO_2 + H_2O$)
Fe_3O_4 or metal sulfides [Co $-$ Mo$/\gamma-$ Al_2O_3 (sulfided) {Ni $-$ Mo$/\gamma-$ Al_2O_3 (sulfided) Ni $-$ W$/\gamma - Al_2O_3$ (sulfided)]	$H_2O + CO \rightarrow H_2 + CO_2$ (water gas shift reaction) olefin hydrogenation aromatic hydrogenation hydrodesulfurization hydrodenitrogenation

[a] Ref. 10.

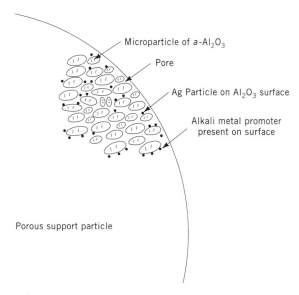

Fig. 9. Schematic representation of a catalyst for ethylene oxide synthesis (not to scale). The porous support particle consists of microparticles held together by a binder.

atoms are exposed at the surface and accessible to reactants (see CATALYSTS, SUPPORTED).

The typical industrial catalyst has both microscopic and macroscopic regions with different compositions and structures; the surfaces of industrial catalysts are much more complex than those of the single crystals of metal investigated in ultrahigh vacuum experiments. Because surfaces of industrial catalysts are very difficult to characterize precisely and catalytic properties are sensitive to small structural details, it is usually not possible to identify the specific combinations of atoms on a surface, called catalytic sites or active sites, that are responsible for catalysis. Experiments with catalyst poisons, substances that bond strongly with catalyst surfaces and deactivate them, have shown that the catalytic sites are usually a small fraction of the catalyst surface. Although most models of catalytic sites rest on rather weak foundations, some are well established. For example, NO decomposition on ruthenium takes place at steps, as shown by scanning tunneling microscopy images of individual reacting atoms on the surface (42).

Important physical properties of catalysts include the particle size and shape, surface area, pore volume, pore size distribution, and strength to resist crushing and abrasion. Measurements of catalyst physical properties (43a) are routine and often automated. Pores with diameters <2.0 nm are called micropores; those with diameters between 2.0 and 5.0 nm are called mesopores; and those with diameters >5.0 nm are called macropores. Pore volumes and pore size distributions are measured by mercury penetration and by N_2 adsorption. Mercury is forced into the pores under pressure; entry into a pore is opposed by surface tension. For example, a pressure of ∼71 MPa (700 atm) is required to fill a pore with a diameter of 10 nm. The amount of uptake as a function of pressure determines the pore size distribution of the larger pores (44). In complementary experiments, the sizes of the smallest pores (those 1–20 nm in diameter) are determined by measurements characterizing desorption of N_2 from the catalyst. The basis for the measurement is the capillary condensation that occurs in small pores at pressures less than the vapor pressure of the adsorbed nitrogen. The smaller the diameter of the pore, the greater the lowering of the vapor pressure of the liquid in it.

Surface areas are determined routinely and exactly from measurements of the amount of physically adsorbed (physisorbed) nitrogen. Physical adsorption is a process akin to condensation; the adsorbed molecules interact weakly with the surface and multilayers form. The standard interpretation of nitrogen adsorption data is based on the BET model (45), which accounts for multilayer adsorption. From a measured adsorption isotherm and the known area of an adsorbed N_2 molecule, taken to be 0.162 nm^2, the surface area of the solid is calculated (see ADSORPTION).

4.3. Influence of Mass Transport on Catalyst Performance.
Reactants must diffuse through the network of pores of a catalyst particle to reach the internal area, and the products must diffuse back. The optimum porosity of a catalyst particle is determined by tradeoffs: making the pores smaller increases the surface area and thereby increases the activity of the catalyst, but this gain is offset by the increased resistance to transport in the smaller pores; increasing the pore volume to create larger pores for faster transport is compensated by a loss of physical strength. A simple quantitative development (46–48) follows for a first-order, isothermal, irreversible catalytic reaction in a spherical, porous catalyst particle.

If there is a significant resistance to transport of the reactant in the pores, a concentration gradient will exist at steady state, whereby the concentration of the reactant is a maximum at the particle periphery and a minimum at the particle center. The product concentration will be higher at the particle center than at the periphery. The concentration gradients provide the driving force for the transport.

As a reactant molecule from the fluid phase surrounding the particle enters the pore structure, it can either react on the surface or continue diffusing toward the center of the particle. A quantitative model of the process is developed by writing a differential equation for the conservation of mass of the reactant diffusing into the particle. At steady state, the rate of diffusion of the reactant into a shell of infinitesimal thickness minus the rate of diffusion out of the shell is equal to the rate of consumption of the reactant in the shell by chemical reaction. Solving the equation leads to a result that shows how the rate of the catalytic reaction is influenced by the interplay of the transport, which is characterized by the effective diffusion coefficient of the reactant in the pores, D_{eff}, and the reaction, which is characterized by the first-order reaction rate constant.

The result is shown in Figure 10, which is a plot of the dimensionless effectiveness factor as a function of the dimensionless Thiele modulus ϕ, which is $R(k/D_{eff})^{1/2}$, where R is the radius of the catalyst particle and k is the reaction rate constant. The effectiveness factor is defined as the ratio of the rate of the reaction divided by the rate that would be observed in the absence of a mass transport influence. The effectiveness factor would be unity if the catalyst were nonporous.

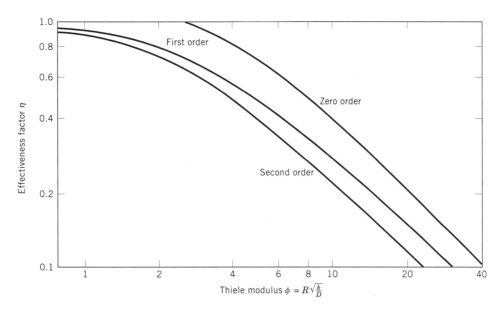

Fig. 10. The Thiele plot accounting for the influence of intraparticle mass transport on rates of catalytic reaction. The dimensionless terms η and ϕ are the effectiveness factor and the Thiele modulus, respectively, and are defined in the text.

Therefore, the reaction rate is

$$r = \eta k [\mathrm{A}]_s \tag{19}$$

where η is the effectiveness factor and $[\mathrm{A}]_s$ the concentration of the reactant A at the peripheral surface of the catalyst particle. The observed reaction rate constant is not the intrinsic reaction rate constant k but the product ηk.

Figure 10 shows that η is a unique function of the Thiele modulus. When the modulus ϕ is small ($\lesssim 1$), the effectiveness factor is unity, which means that there is no effect of mass transport on the rate of the catalytic reaction. When ϕ is greater than ~ 1, the effectiveness factor is less than unity and the reaction rate is influenced by mass transport in the pores. When the modulus is large ($\gtrsim 10$), the effectiveness factor is inversely proportional to the modulus, and the reaction rate (eq. 19) is proportional to k/ϕ, which, from the definition of ϕ, implies that the rate and the observed reaction rate constant are proportional to $(1/R)(D_{\mathrm{eff}}k)^{1/2}[\mathrm{A}]_s$. This result shows that both the rate constant, ie, a measure of the intrinsic activity of the catalyst, and the effective diffusion coefficient, ie, a measure of the resistance to transport of the reactant offered by the pore structure, influence the rate. It is not appropriate to say that the reaction is diffusion controlled; it depends on both the diffusion and the chemical kinetics. In contrast, as shown by equation 3, a reaction in solution can be diffusion controlled, depending on D but not on k.

In the case of a significant intraparticle diffusion influence, the temperature dependence of the observed rate constant, the apparent activation energy, is not that of the intrinsic rate constant, ie, the true activation energy; rather, as the Thiele modulus becomes large (>10), the temperature dependence approaches that of $(D_{eff}k)^{1/2}$. The temperature dependence of a diffusion coefficient is usually small, and the measured activation energy is thus about half the intrinsic activation energy. This is an important and general result: The kinetics is disguised by the transport influence. When the effectiveness factor is unity, there is no disguise, and the true activation energy is measured.

This development has been generalized. Results for zero- and second-order irreversible reactions are shown in Figure 10. Results are given elsewhere (48) for more complex kinetics, nonisothermal reactions, and particle shapes other than spheres. For nonspherical particles, the equivalent spherical radius, three times the particle volume/surface area, can be used for R to a good approximation.

Even when there is a transport disguise, the reaction order remains one for a first-order reaction. But for reactions that are not intrinsically first order, the transport disguise changes the observed reaction order; for an intrinsically zero-order reaction, the observed order becomes 1/2 and for an intrinsically second-order reaction it becomes 3/2 when $\phi \gtrsim 10$. For all reaction orders the apparent activation energy is approximately one-half of the intrinsic value in this limit.

The mass transport influence is easy to diagnose experimentally. One measures the rate at various values of the Thiele modulus; the modulus is easily changed by variation of R, the particle size. Crushing and sieving the particles provide catalyst samples for the experiments. If the rate is independent of the particle size, the effectiveness factor is unity for all of them. If the rate is inversely proportional to particle size, the effectiveness factor is less than unity and $\phi \gtrsim 10$. If the dependence is between these limits, then several experimental points allow triangulation on the curve of Figure 10 and estimation of η ανδ ϕ. It is also possible to estimate the effective diffusion coefficient and thereby to estimate η ανδ ϕ from a single measurement of the rate (48).

If the effectiveness factor is less than unity, the catalyst is not being used efficiently, ie, the central region is starved of reactant. The results of Figure 10 show how to increase the effectiveness factor: decrease the Thiele modulus. This can be done by some combination of the following: (1) decreasing the particle size R. However, if particles are too small, they may cause too great a pressure drop in a flow reactor or be entrained in the product stream; (2) changing particle shape to reduce the transport length. Particles with cross sections resembling clover leaves and wagon wheels have been used, for example; (3) increasing the effective diffusion coefficient. Larger catalyst pores accomplish this, but with a sacrifice in physical strength; and (4) decreasing the activity of the catalyst measured by k. This option is unappealing as researchers strive to make more active catalysts, but it may be economical to reduce the activity by reducing the loading of the catalytically active component on a support. It may also be appropriate to prepare the catalyst with the active component concentrated near the particle periphery and not in the particle interior.

Intraparticle mass transport resistance can lead to disguises in selectivity. If a series reaction A→B→C takes place in a porous catalyst particle with a small effectiveness factor, the observed conversion to the intermediate B is less than what would be observed in the absence of a significant mass transport influence. This happens because as the resistance to transport of B in the pores increases, B is more likely to be converted to C rather than to be transported from the catalyst interior to the external surface. This result has important consequences in processes such as selective oxidations, in which the desired product is an intermediate and not the total oxidation product CO_2.

Rates and selectivities of solid-catalyzed reactions can also be influenced by mass-transport resistance in the external fluid phase. Most reactions are not influenced by external-phase transport, but the rates of some very fast reactions, eg, ammonia oxidation, are determined solely by the resistance to this transport. As the resistance to mass transport within the catalyst pores is larger than that in the external fluid phase, the effectiveness factor of a porous catalyst is expected to be less than unity whenever the external-phase mass transport resistance is significant. A practical catalyst that is used under such circumstances is the ammonia oxidation catalyst. It is a nonporous metal and consists of layers of wire mesh.

4.4. Catalyst Components. Industrial catalysts are typically complex in composition and structure, consisting of catalytically active phases, supports, binders, and promoters.

Catalytically Active Species. The most common catalytically active materials are metals, metal oxides, and metal sulfides. Occasionally, these are used in pure form; examples are Raney nickel, used for fat hydrogenation, and γ-Al_2O_3, used for ethanol dehydration. More often the catalytically active component is highly dispersed on the surface of a support and may constitute no more than ~1% of the total catalyst. The main reason for dispersing the catalytic species is the expense. The expensive material must be accessible to reactants, and this requires that most of the catalytic material be present at a surface. This is possible only if the material is dispersed as minute particles, as small as 1 nm in diameter and even less. It is not practical to use minute particles by themselves, as they would be entrained in products and clog lines and pumps, and their use in a fixed-bed reactor would cause large pressure drops. Dispersion on a support may also help stabilize the catalytically active species.

Supports. The principal component of a typical catalyst is the porous support (49,50) . Most supports are robust solids that can be made with wide ranges of surface areas and pore size distributions. The most widely applied supports are metal oxides; others are carbon, kieselguhr, organic polymers, and zeolites.

The most commonly applied catalyst support is γ-Al_2O_3, one of the transition aluminas (33,51). These are defective metastable solids formed from nonporous $Al(OH)_3$ by heating it to ~500°C or more; continued heating gives the more stable δ-Al_2O_3. As the $Al(OH)_3$ is heated in air, it decomposes into an oxide with a micropore system and a surface area of hundreds of square meters per gram. The solid consists of small, crystalline primary particles; the spaces between these primary particles are micropores and mesopores. As the solid is heated to ~1100°C, there is a series of phase changes and ultimately a collapse of the pore structure and loss of almost all the internal surface area as finally the extremely hard, crystalline α-Al_2O_3 (corundum) is formed. It has a melting point of ~2100°C [see ALUMINUM OXIDE (ALUMINA)].

Transition aluminas are good catalyst supports because they are inexpensive and have good physical properties. They are mechanically stable, stable at relatively high temperatures even under hydrothermal conditions, ie, in the presence of steam, and easily formed in processes such as extrusion into shapes that have good physical strength such as cylinders. Transition aluminas can be prepared with a wide range of surface areas, pore volumes, and pore size distributions.

Macropores can be introduced into γ-Al$_2$O$_3$ by including particles of an organic material such as carbon or sawdust with the Al(OH)$_3$ (51). When the γ-Al$_2$O$_3$ is formed, the organic particles are surrounded by the alumina. The organic material is removed by burning, leaving macropores; the macropore dimensions are determined by the particle size of the organic material.

Most catalyst supports are simply nearly inert platforms that help stabilize the dispersion of the catalytically active phase. Sometimes, however, the supports play a direct catalytic role, as exemplified by the alumina used in supported Pt and RePt catalysts for naphtha reforming.

The surfaces of γ-Al$_2$O$_3$ and most other metal oxides are covered with polar functional groups including OH groups and O^{2-} ions (52). A structural model of γ-Al$_2$O$_3$ is shown in Figure 11 (53). When the solid is heated, it gives off water in a reversible process called dehydroxylation (54). As a result, surface OH groups are lost and Al^{3+} ions are exposed at the surface. The microparticles of the oxide interact strongly with each other through the surface functional groups; eg, hydrogen bonding can occur. As a result, a microporous solid consisting of such particles is a strong material. In contrast, some support materials, exemplified by α-Al$_2$O$_3$, have few surface functional groups, and a solid consisting of microparticles of such a material lacks physical strength.

Binders. To create needed physical strength in catalysts, materials called binders are added (51); they bond the catalyst. A common binder material is a clay mineral such as kaolinite. The clay is added to the mixture of microparticles as they are formed into the desired particle shape, eg, by extrusion. Then the support is heated to remove water and possibly burnout material and then subjected to a high temperature, possibly 1500°C, to cause vitrification of the clay;

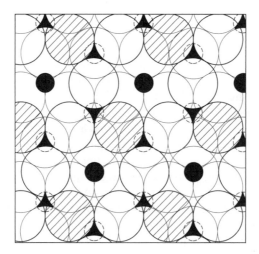

Fig. 11. Structural model of the (111) face of γ-alumina (53). The small solid circles represent Al^{3+}, the large open circles OH groups, and the hatched circles oxygen. The surface is 50% dehydroxylated.

this is a conversion of the clay into a glasslike form that spreads over the microparticles of the support and binds them together.

Promoters. Many industrial catalysts contain promoters, commonly chemical promoters. A chemical promoter is used in a small amount and influences the surface chemistry. Alkali metal ions are often used as chemical promoters, eg, in ammonia synthesis catalysts, ethylene oxide catalysts, and Fischer-Tropsch catalysts (55). They may be used in as little as parts per million quantities. The mechanisms of their action are often not well understood. In contrast, seldom-used textural promoters, also called structural promoters, are used in massive amounts in some catalysts and affect their physical properties. These are used in ammonia synthesis catalysts.

4.5. Catalyst Treatments. Catalysts often require activation or regeneration and their disposal also requires special consideration (56).

Activation. Some catalysts, eg, Ziegler olefin polymerization catalysts (57), are highly reactive in the presence of air, and some, eg, iron catalysts, are even pyrophoric; these must be handled under a blanket of inert gas. The surfaces of most catalysts are reactive and not easily maintained in an active state in the form in which they are conveniently supplied. Thus, many catalysts must be activated prior to use. The activation may be as simple as exposure to reactants under processing conditions, but some catalysts require specialized treatments. For example, catalysts used for hydroprocessing of fossil fuels are usually delivered as supported metal oxides, but in the operating state they are supported metal sulfides. Catalyst suppliers specify detailed procedures for treatment, eg, with a mixture of hydrogen and oil containing organosulfur compounds, to carry out the sulfiding properly. The details are often critical, and the catalyst manufacturer's guarantees may be void if the procedure is not carried out properly.

Deactivation. Catalysts lose activity and selectivity in many ways, and much effort in process development goes into measuring the deactivation and finding means to minimize it. Some catalysts undergo physical changes during normal operation; eg, a catalytically active phase may be transformed into an inactive phase. Catalysts also undergo sintering, which is a coalescence of particles to give larger particles, accompanied by loss of surface area. Iron sinters under conditions of ammonia synthesis, but when a textural promoter such as alumina is present, the sintering is greatly reduced. Catalyst components may be volatile and gradually vaporized during use; examples are silica used as a support and molybdenum oxides present in selective oxidation catalysts.

Catalysts commonly lose activity in operation as a result of accumulation of materials from the reactant stream. Catalyst poisoning is a chemical phenomenon. A catalyst poison is a component such as a feed impurity that as a result of chemisorption, even in small amounts, causes the catalyst to lose a substantial fraction of its activity. For example, sulfur compounds in trace amounts poison metal catalysts. Arsenic and phosphorus compounds are also poisons for a number of catalysts. Sometimes the catalyst surface has such a strong affinity for a poison that it scavenges it with a high efficiency. The poison may then adsorb strongly on the catalyst at the upstream end of a fixed-bed reactor at the beginning of operation, with a wave then moving downstream through the reactor as the upstream surface becomes saturated.

A selective poison is one that binds to the catalyst surface in such a way that it blocks the catalytic sites for one kind of reaction but not those for another. Selective poisons are used to control the selectivity of a catalyst. For example, nickel catalysts supported on alumina are used for selective removal of acetylene impurities in olefin streams (58). The catalyst is treated with a continuous feed stream containing sulfur to poison it to an exactly controlled degree that does not affect the activity for conversion of acetylene to ethylene but does poison the activity for ethylene hydrogenation to ethane. Thus the acetylene is removed and the valuable olefin is barely converted.

Because catalyst surfaces are reactive and often sensitive to their environments, they may be irreversibly changed by exposure to undesired reactants. Upsets in plant operations can lead to catastrophic losses of whole catalyst charges. A large catalyst charge that is ruined can cost hundreds of thousands of dollars as well as the cost of lost operation.

Catalysts are also deactivated or fouled by physical deposition of materials present in or formed from feeds. Sometimes massive deposits form on surfaces and block access of reactants to the catalytic sites. Coke is carbonaceous material of various compositions, often aromatic and with a high molecular weight and a typical composition of approximately CH. Coke forms on every hydrocarbon processing catalyst and on most catalysts used for organic chemical conversions. Inorganic materials are also deposited on catalysts. For example, the organovanadium and organonickel compounds in petroleum residua react to form vanadium and nickel sulfides on the surfaces of hydroprocessing catalysts. The solid deposits reduce activity by covering catalytic sites and by filling pores and restricting the entry of reactants. When the effectiveness factor for the deposition reaction is small, the pore mouths can become blocked and catalysts can suffer near catastrophic failure. Small particles of solid such as dust can also foul catalysts. This may be a problem in processes for cleanup of NO_x emissions from coal-fired power plants, and catalysts are designed in the form of monoliths (honeycombs) to minimize the effect.

Regeneration.   Deactivated catalysts are treated to bring back the catalytic activity in processes called regenerations. Coke deposits are removed by controlled combustion. Often low partial pressures of oxygen are used to keep the rate of combustion and the temperature rise from becoming too large and leading to damage of the catalyst, such as by sintering. Periodic coke burn-off can be carried out many times with little damage to many catalysts. Most catalysts last for months or years between regenerations, but catalysts used for cracking of petroleum are in contact with reactants for only a few seconds; then they are separated and cycled to another reactor where they are regenerated (and then they are cycled back to the cracking reactor) (see CATALYSTS, REGENERATION).

Redispersion.   Expensive catalyst components such as precious metals are used in high dispersions on supports. During operation, the small metal particles tend to sinter, ie, migrate and agglomerate, into larger particles with a loss of metal surface area and thereby a loss of catalytic activity. The metals in such catalysts may be redispersed as part of the catalyst regeneration (59). For example, after the coke is burned off a supported platinum catalyst, the catalyst may be treated with a reactive atmosphere containing chlorine and oxygen to form volatile platinum oxychloride species that are transported through the gas phase and deposited on the pore walls, where they are then treated in H_2 and reduced. The result is an increased dispersion of the platinum and a reactivated catalyst.

Reclamation, Disposal, and Toxicity. Removal of poisons and inorganic deposits from used catalysts is typically difficult and usually uneconomical. Thus, some catalysts are used without regeneration, although they may be processed to reclaim expensive metal components. Used precious metal catalysts, including automobile exhaust conversion catalysts, are treated (often by the suppliers) to extract the metals, and recovery efficiencies are high. Some spent hydroprocessing catalysts may be used as sources of molybdenum and other valuable metals.

Some catalysts are hazardous materials, or they react to form hazardous materials. For example, catalysts used for hydrogenation of carbon monoxide form volatile metal carbonyl compounds such as nickel carbonyl, which are highly toxic. Many catalysts contain heavy metals and other hazardous components, and environmentally safe disposal has become an increasing concern and expense.

4.6. Catalyst Preparation. Catalyst preparation is more an art than a science (60). Many reported catalyst preparations omit important details and are difficult to reproduce exactly, and this has hindered the development of catalysis as a quantitative science. However, the art is developing into a science and there are now many examples of catalysts synthesized in various laboratories that have nearly the same physical and catalytic properties.

Supports are often prepared first and the catalyst and promoter components added later. Metal oxide supports are usually prepared by precipitation from aqueous solutions. Nitrates are commonly used anions; alkalies and ammonium are commonly used cations. Metal oxide supports, eg, silica and alumina, are prepared in the form of hydrogels. Mixed oxides such as silica–alumina are made by cogelation. Careful control of conditions such as pH is important to give uniform products. Sol–gel processes are useful for the preparation of some high-area oxides (61).

Supports are washed at controlled values of pH to remove impurities. Ions and impurities in the preparative solution are easily occluded in the solid and difficult to remove by washing. Therefore, ions that might poison the catalyst, eg, Cl^-, SO_4^{2-}, or alkali metal ions, are avoided. Many of the materials are cation exchangers, and washing does not remove cations from them. Metal ions can be removed by exchange with ammonium ions, which on heating give off ammonia and leave hydrogen ions in surface OH groups. Drying of precipitates and hydrated gels leads to evolution of gases and may generate microporosity, as described above for transition aluminas. Porosity can also be created by reduction of a nonporous oxide; porous iron can be made this way.

Catalyst components are usually added in the form of precursor metal salts in aqueous solutions. In impregnation, the support may be dried, evacuated, and brought in contact with an excess of an impregnating solution containing metal salts. The processes are complex, possibly involving some dissolution of the support and reprecipitation of structures including mixed metal species. The solid is then dried and calcined, ie, brought to a high temperature, usually in air. Alternatively, in the incipient wetness method, just enough of the impregnating solution is used to fill the pores of the support. The chemistry of the interactions of catalyst precursors with metal oxide supports is beginning to be understood. Important parameters that control the adsorption of metal complex precursors from aqueous solution are the isoelectric point of the metal oxide, the pH of the solution, and the nature of the metal complex (62). Depending on the conditions of the contact, cationic or anionic species may be adsorbed. Sometimes these are simple mononuclear (single-metal atom) species, but sometimes they are complicated polynuclear ions. The support may dissolve to some degree in the preparation solution and be redeposited in some form (perhaps with the catalyst precursor) (63). The nature of the initially adsorbed species may significantly affect the structure of the catalytic species in the resultant catalyst. After impregnation, the catalyst may be activated, for example, by drying and calcining. Promoters may be added at various stages, eg, as a final step in the preparation or just prior to operation.

Supported metal catalysts are reduced, eg, by treatment in hydrogen at temperatures in the range of 300–500°C. The reduction temperature may influence the stability of the metal dispersion.

4.7. Examples of Surface Catalysis. *Molecular Catalysis on Supports.* The term molecular catalysis is commonly applied only to reactions in uniform fluid phases, but it applies nearly as well to some reactions taking place on supports. Straightforward examples are reactions catalyzed by polymers functionalized with groups that closely resemble catalytic groups in solution. Industrial examples include reactions catalyzed by ion-exchange resins, usually sulfonated poly(styrene-divinylbenzene) (64). This polymer is an industrial catalyst for synthesis of methyl *tert*-butyl ether (MTBE) from methanol and isobutylene and synthesis of bisphenol A from phenol and acetone, among others. The former application grew rapidly as MTBE became a component of high-octane-number gasoline (see ETHERS).

The polymer has the following structure, which is shown schematically.

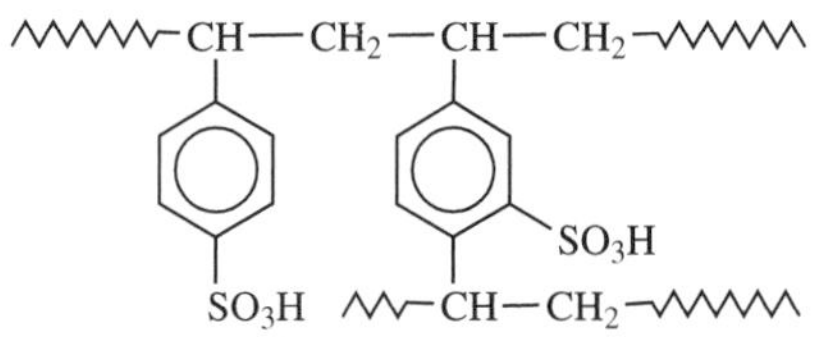

The sulfonated resin is a close analogue of *p*-toluenesulfonic acid in terms of structure and catalyst performance. In the presence of excess water, the SO_3H groups are dissociated, and specific acid catalysis takes place in the swelled resin just as it takes place in an aqueous solution. When the catalyst is used with weakly polar reactants or with concentrations of polar reactants that are too low to cause dissociation of the acid groups, general acid catalysis prevails and water is a strong reaction inhibitor (65).

The polymer-supported catalysts are thus important conceptually in linking catalysis in solutions and catalysis on supports. The acid–base chemistry is fundamentally the same whether the catalytic groups are present in a solution or anchored to the support. The polymer-supported catalysts have replaced acid solutions in numerous processes because they minimize the corrosion, separation, and disposal problems posed by mineral acids.

Polymer-supported methanol carbonylation catalysts incorporating metal complexes, mentioned above, also behave in much the same way as their soluble analogues.

Surfaces of inorganic solids can be functionalized with catalytic groups just as organic polymers can. For example, the hydroxyl groups on the surface of silica can be used for synthesis of the following structure:

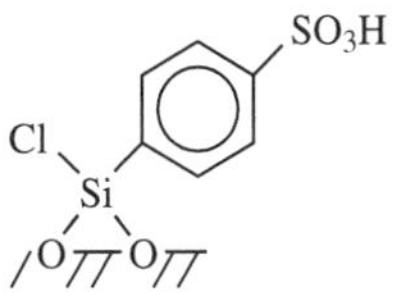

This is an ion-exchanger like the sulfonated polymer. The silica surface can also be functionalized with phosphine complexes; when combined with rhodium, these give anchored complexes that behave like their soluble and polymer-supported analogues as catalysts for olefin hydrogenation and other reactions:

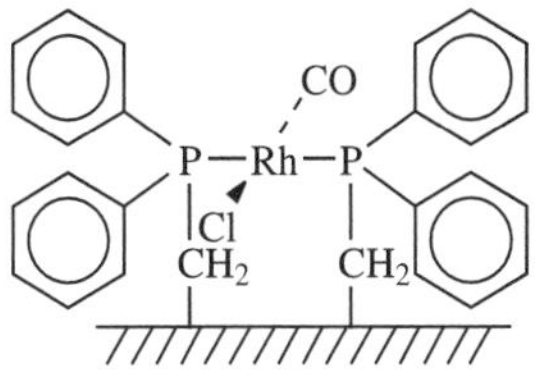

These silica-supported catalysts demonstrate the close connections between catalysis in solutions and catalysis on surfaces, but they are not industrial catalysts. However, silica is used as a support for chromium complexes, formed either from chromocene or chromium salts, that are industrial catalysts for polymerization of α-olefins (66,67). Supported chromium complex catalysts are used on an enormous scale in the manufacture of linear polyethylene in the Unipol and Phillips processes (see OLEFIN POLYMERS). The exact nature of the catalytic sites is still not known, but it is evident that there is a close analogy linking soluble and supported metal complex catalysts for olefin polymerization.

The newest industrial catalysts for olefin polymerization are supported metal complexes, metallocenes (eg, zirconocenes) promoted by aluminum-containing components such as methylalumoxane (68). These are used for polymerization of ethylene and for the stereospecific polymerization of propylene giving high yields of isotactic polypropylene; they are also used for synthesis of copolymers. The discovery and rapid development of the large new class of supported catalysts is one of the major successes of industrial catalysis in recent years.

The conceptual link between catalysis in solution and catalysis on surfaces extends to surfaces that are not obviously similar in structure to molecular species. For example, the early Ziegler catalysts for polymerization of propylene were α-TiCl$_3$. (Today, supported metal complexes are used instead). These catalysts are selective for stereospecific propylene polymerization. The catalytic sites are believed to be located at the edges of $TiCl_3$ crystals. The surface structures have been inferred to incorporate anion vacancies, that is, sites where Cl^- ions are not present and where Ti^{3+} ions are exposed (69). These cations exist in octahedral surroundings. The polymerization has been explained by a mechanism whereby the growing polymer chain and an adsorbed propylene bonded cis to it on the surface undergo a migratory insertion reaction (70). In this respect, there is no essential difference between the explanation of the surface catalyzed polymerization and that catalyzed in solution.

Stereospecific polymerization in solution and on surfaces incorporating analogous metallocenes has been explained in part by the steric restrictions of ligands bonded to the metal center. For example, the following structure, among numerous others, has been postulated as an intermediate in solution catalysis (71):

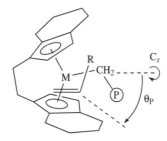

The steric constraints imposed by the bulky ligands cause the propylene to bond almost entirely with a single orientation with respect to the growing polymer chain, $CH_2-\text{\textcircled{P}}$, which leads to the stereoregular product.

The explanation for the stereospecificity of the surface catalysis, which preceded that for the solution catalysis, is based on the structure inferred for the α-$TiCl_3$ crystal edges; the locations of the Cl^- ions at the anion vacancies create an unsymmetrical environment whereby the growing polymer chain and the adsorbed propylene are oriented predominantly in a single, energetically favored way that leads to the stereoregular polymer as a result of a series of insertion reactions. The explanation is simplified, but again there is a strong conceptual link between molecular and surface catalysis. It is difficult to forge many such links because of the lack of detailed understanding of most surface-catalyzed reactions, which is a consequence of the complexity of the surface compositions and structures of solid catalysts.

4.8. Catalysis by Metals.   Metals are among the most important and widely used industrial catalysts (72). They offer activities for a wide variety of reactions (Table 1). Atoms at the surfaces of bulk metals have reactivities and catalytic properties different from those of metals in metal complexes because they have different ligand surroundings. The surrounding bulk stabilizes surface metal atoms in a coordinatively unsaturated state that allows bonding of reactants. Thus metal surfaces offer an advantage over metal complexes, in which there is only restricted stabilization of coordinative unsaturation. Furthermore, metal surfaces provide catalytically active sites that are stable at high temperatures. For example, supported palladium catalysts have replaced soluble palladium for vinyl acetate synthesis; the advantages of the solid include reduced corrosion and reduced formation of by-products.

CO Oxidation Catalyzed by Palladium.   One of the best understood catalytic reactions occurring on a metal surface is the oxidation of carbon monoxide on palladium:

$$2\,CO + O_2 \rightarrow 2\,CO_2 \qquad (20)$$

This reaction takes place similarly in automobile exhaust converters.

Carbon monoxide oxidation catalysis is understood in depth because potential surface contaminants such as carbon or sulfur are burned off under reaction conditions and because the rate of CO oxidation is almost independent of pressure over a wide range. Thus ultrahigh vacuum surface science experiments could be done in conjunction with measurements of reaction kinetics (73). The results show that at very low surface coverages, both reactants are adsorbed randomly on the surface; CO is adsorbed intact and O_2 is dissociated and adsorbed atomically. When the coverage by CO is >1/3 of a monolayer, chemisorption of oxygen is blocked. When CO is adsorbed at somewhat less than a monolayer, oxygen is adsorbed, and the two are present in separate domains. The reaction that forms CO_2 on the surface then takes place at the domain boundaries.

The available results are consistent with the following sequence of steps on the surface (73):

$$O_2 + 2\,S \rightarrow O_{2,ads} \rightarrow\ 2\,O_{ads} \qquad (21)$$

$$CO + S \rightarrow CO_{ads} \qquad (22)$$

$$CO_{ads} + O_{ads} \rightarrow CO_2 + 2\,S \qquad (23)$$

$$CO + O_{ads} \rightarrow CO_2 + S \qquad (24)$$

This depiction is vague because the exact nature of the sites S and their bonding with reactants are not known. The experimental results have led to an approximate potential energy diagram characterizing these elementary steps on the surface (Fig. 12) (73). This shows the role of the surface in providing an efficient pathway for the reaction. Most of the energy is liberated as the reactants are adsorbed; the activation energy for reaction of the adsorbed CO with the adsorbed O is relatively small, and this step is only slightly exothermic.

Ammonia Synthesis.   Another well-understood reaction is the ammonia synthesis:

$$N_2 + 3\,H_2 \rightarrow 2\,NH_3 \qquad (25)$$

This reaction is catalyzed by iron, and extensive research, including surface science experiments, has led to an understanding of many of the details (74,75). The adsorption of H_2 on iron is fast, and the adsorption of N_2 is slow and characterized by a substantial activation energy. N_2 and H_2 are both

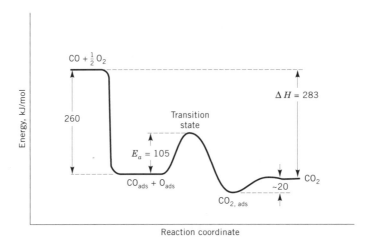

Fig. 12. Schematic potential energy diagram illustrating the changes associated with the individual reaction steps in CO oxidation on Pd (73). $E_{act} = 105$ kJ/mol; $\Delta H = 283$ kJ/mol. To convert kJ to kcal, divide by 4.184.

dissociatively adsorbed. Adsorption of N_2 leads to reconstruction of the iron surface and formation of structures called iron nitrides that have depths of several atomic layers with compositions of approximately Fe_4N. There is a bulk compound Fe_4N, but it is thermodynamically unstable when the surface structure is stable. Adsorbed species such as the intermediates NH and NH_2 have been identified spectroscopically.

The following sequence of steps explains the observations (73):

$$N_2 + 2\,S \rightarrow 2\,N_{ads} \tag{26}$$

$$H_2 + 2\,S \rightarrow 2\,H_{ads} \tag{27}$$

$$N_{ads} + H_{ads} \rightarrow NH_{ads} + S \tag{28}$$

$$NH_{ads} + H_{ads} \rightarrow NH_{2,ads} + S \tag{29}$$

$$NH_{2,ads} + H_{ads} \rightarrow NH_3 + 2\,S \tag{30}$$

where S refers to surface sites, the exact nature of which is unknown. An approximate potential energy diagram for this sequence of steps is shown in Figure 13, which shows how the catalyst facilitates the bond breaking reactions. The energy gain resulting from the formation of the strong metal–nitrogen and metal–hydrogen bonds makes the first steps endothermic. The dissociative adsorption of N_2 is rate determining, not because of a high activation energy barrier but because the frequency factor, (preexponential factor) in the rate

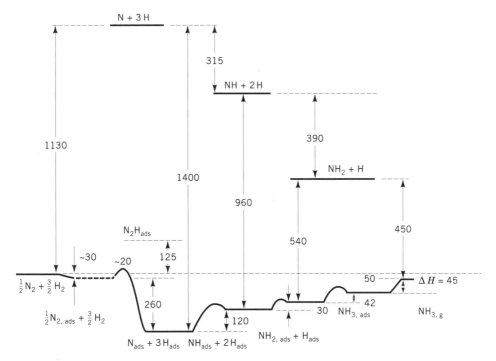

Fig. 13. Schematic potential energy diagram for the catalytic synthesis and decomposition of ammonia on iron. The energies are given in kJ/mol; to convert kJ to kcal, divide by 4.184 (73).

constant is small. The ammonia synthesis mechanism is so well understood that rates of the reaction under practical conditions have been predicted from the rate of adsorption of N_2 measured under low-pressure conditions, far from those of practical catalysis, combined with the equilibria of the other steps (75) . The prediction was within a factor of two of the observed rate of the industrial reaction. This result is a satisfying consolidation of decades of fundamental research motivated by an important catalytic process.

The industrial catalysts for ammonia synthesis consist of far more than the catalytically active iron (76). There are textural promoters, alumina and calcium oxide, that minimize sintering of the iron and a chemical promoter, potassium ($\sim$1 wt% of the catalyst), and possibly present as K_2O; the potassium is believed to be present on the iron surface and to donate electrons to the iron, increasing its activity for the dissociative adsorption of N_2. The primary iron particles are $\sim$30 nm in size, and the surface area is $\sim$15 m^2/g. These catalysts last for years. Some newer catalysts incorporate ruthenium instead of iron.

4.9. Catalysis by Metal Oxides and Zeolites.   Metal oxides are common catalyst supports and catalysts. Some metal oxides alone are industrial catalysts; an example is the γ-Al_2O_3 used for ethanol dehydration to give ethylene. But these simple oxides are the exception; mixed-metal oxides are more common. For example, silica–alumina was used earlier as a catalyst for cracking of petroleum and is still a component of such catalysts, and bismuth molybdates were used for ammoxidation of propylene to give acrylonitrile. Metal oxides supported on metal oxides are also commonly applied. For example, rhenium oxide, Re_2O_7, supported on alumina is used for olefin metathesis, and complicated supported oxides related to bismuth molybdates are used for ammoxidation.

Metal oxide surfaces are more complex in structure and composition than metal surfaces, and they are not so easy to characterize with some ultrahigh vacuum techniques, eg, electron spectroscopies, because they are poor electrical conductors and build up electrical charge when subjected to streams of charged particles. Consequently, understanding of catalysis on metal oxide surfaces is less advanced than understanding of catalysis on metal surfaces, although significant progress has been made (77,78).

Acid–base chemistry of metal oxide surfaces is important in catalysis and is characterized by measurements such as infrared spectroscopy with adsorbed probe molecules, eg, the base pyridine. The surfaces have both basic and acidic character (79). Both OH and O groups have base strengths ranging from weak, eg, in silica gel, to moderate, eg, in γ-Al_2O_3, to strong, eg, in highly dehydroxylated MgO. Surface OH groups are acids with proton donor strengths ranging from weak, eg, in Al_2O_3, to strong, eg, in SiO_2-Al_2O_3. Metal ions exposed at surfaces are Lewis acids. Redox properties are also important in some catalytic applications, as cations in some oxides, eg, V_2O_5, can change oxidation state. Furthermore the principles of organometallic chemistry are useful in describing the interactions of organic ligands with metal ions exposed at surfaces.

Zeolites and Catalytic Cracking.   Some of the best-understood catalysts are zeolites, ie, crystalline aluminosilicates (80–82). They are well understood because they have much more nearly uniform compositions and structures than amorphous metal oxides such as silica and alumina. Here the usage of amorphous refers to results of X-ray diffraction experiments; the crystallites of a metal oxide such as γ-Al_2O_3 that constitute the microparticles are usually so small that sharp X-ray diffraction patterns are not measured; consequently the solids are said to be X-ray amorphous or simply amorphous.

Zeolites contain Si, Al, and O ions and various other cations. The structures are built up of linked SiO_4 and AlO_4 tetrahedra that share O ions. These tetrahedra are arranged in a number of ways to give the different zeolites. The structures are unique in that they incorporate pores as part of the regular crystalline structures. The pores have dimensions of the order of molecular dimensions so that some molecules fit into the pores and some do not. Hence, the zeolites are molecular sieves (qv), and they are applied in industrial separations processes to take advantage of this property. Some zeolites and their pore dimensions are listed in Table 2.

A catalytically important family of zeolites called faujasites (zeolites X and Y) is represented in Figure 14. Here the points of intersection of the lines represent Si or Al ions; oxygen is present at the center of each line. This depiction emphasizes the framework structure of the zeolite and shows the presence of the intracrystalline pore structure in which there are spaces called supercages, each with a diameter of ∼1.2 nm. The pore structure is three dimensional; the supercages are connected by apertures with diameters of ∼0.74 nm. Some rather large molecules can fit through these apertures (Fig. 15) and undergo catalytic reaction in the cages.

Table 2. **Zeolites and Their Pore (Aperture) Dimensions**[a]

Zeolite	CAS Registry Number	Number of oxygens in the ring	$10 \times$ Aperture dimensions, nm
chabazite	[12251-32-0]	8	3.6×3.7
erionite	[12510-42-8]	8	3.6×5.2
zeolite A		8	4.1
ZSM-5 (or silicalite)	[58339-99-4]	10	5.1×5.5; 5.4×5.6
ZSM-11		10	5.1×5.5
heulandite		10	4.4×7.2
ferrierite[b]		10	4.3×5.5
faujasite	[12173-28-2]	12	7.4
zeolite LTL		12	7.1
mordenite	[12173-98-7]	12	6.7×7.0
offretite		12	6.4

[a] The framework oxygen is assumed to have a diameter of 0.275 nm.
[b] There are also apertures with eight-membered oxygen rings in this zeolite.

The zeolite frame is made up of SiO_4 tetrahedra, which are neutral, and AlO_4 tetrahedra, which have a charge of -1. The charge of the AlO_4 tetrahedra is balanced by the charges of additional cations that exist at various crystallographically defined positions in the zeolite, many exposed at the internal surface. Zeolites are thus ion exchangers. The cations may be catalytically active. When the cations are H^+, the zeolites are acidic. Acidic zeolite Y finds enormous industrial application as a component of petroleum cracking catalysts (81). In the following simplified structure, the OH groups located near AlO_4 tetrahedra are moderately strong Brønsted acids and responsible for the catalytic activity for many reactions:

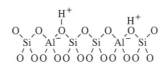

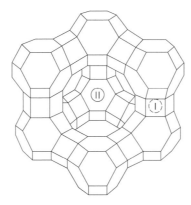

Fig. 14. Schematic representation of the structure of a faujasitic zeolite. I and II indicate cation positions.

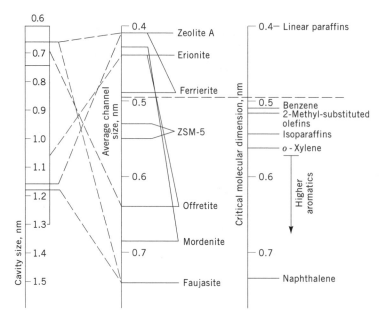

Fig. 15. Pore dimensions of zeolites and critical dimensions of hydrocarbons (82).

Zeolites are named to represent the exchangeable cations in them; eg, zeolite NaY is zeolite Y with sodium ions in the cation exchange positions.

Another catalytically important zeolite is ZSM-5. There is a three-dimensional network of pores in this zeolite, represented in Figure 16. A set of straight parallel pores is intersected by a set of perpendicular zigzag pores. These pores are smaller than those of the faujasites (Fig. 15). ZSM-5 is classified as a medium pore zeolite, the faujasites are large pore zeolites, and zeolite A (Table 2) is a small pore zeolite.

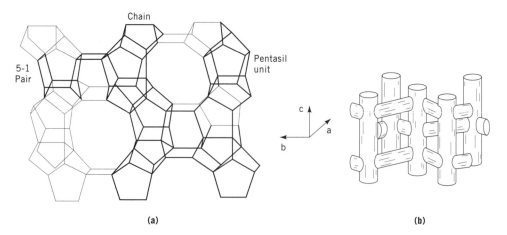

(a)   **(b)**

Fig. 16. Structure of the zeolite ZSM-5 (83): **(a)** framework of the zeolite; **(b)** schematic representation of the pore structure.

Both the faujasites and ZSM-5 in the acidic form catalyze many reactions that are catalyzed by other soluble and solid acids. The zeolites are not very strong acids at low temperatures, but at 500°C they are able to protonate paraffins and initiate their cracking. Almost all the catalytic applications of zeolites take advantage of their acidic properties. Activities of a family of HZSM-5 samples with different Si/Al ratios have been studied (84). When the Al contents are low, the catalytic activity is proportional to the Al content of the zeolite over a wide range of compositions. These results identify the proton-donor sites associated with the Al cations as the catalytic sites for the cracking reaction. At higher concentrations of Al in the zeolite, the dependence is no longer linear.

ZSM-5 is a component of some catalysts for cracking of petroleum (85–87), but the larger pored zeolite Y in an acidic form is the principal catalytic component. Zeolite Y is sometimes used in a form containing hydrogen ions to provide acidity and rare-earth ions such as La^{3+}, which make the structure more stable. The stability is valuable because the zeolites are deactivated in a matter of a few seconds of contact with reactant vapors in the catalytic reactor and are subjected to an atmosphere of oxygen and steam at temperatures as high as ~800°C in the regenerator where coke is burned off.

The feedstocks to crackers are petroleum fractions ranging from gas oil to residuum. They undergo a complicated set of reactions, including cracking, isomerization, disproportionation, and coke formation, that proceed through carbenium ion intermediates and give predominantly lower molecular weight products, including, eg, many in the gasoline boiling range. The patterns in the chemistry of catalytic cracking are consistent with the chemistry of hydrocarbons and carbenium ions in solution at much lower temperatures. For example, cracking of a paraffin takes place predominantly in a cycle that is initiated by formation of a carbenium ion by protonation of an olefin by the catalyst. The chain reaction is the following:

$$R^+ + R'H \rightarrow RH + R'^+ \tag{31}$$

$$R'^+ \rightarrow \text{olefin} + R^+ \tag{32}$$

The second step is a β-scission, the breaking of a carbon–carbon bond β to the charged carbon. The sum of the two reactions is the stoichiometry of the overall cracking reaction $R'H \rightarrow RH + \text{olefin}$. R^+, a relatively stable carbenium ion such as the *tert*-butyl cation, is a chain carrier. The role of the catalyst is to donate the proton to start the chain. This representation is greatly simplified.

Cracking catalysts are complex composites consisting of a support, eg, silica–alumina, and the catalytically active zeolite, present as crystallites roughly 1 μm in size dispersed in a matrix of the amorphous support (85). The catalyst particles are small, roughly 50 μm in diameter on average so that they can be fluidized by the vaporized oil entering the cracking reactor. The reactant vapors carry the catalyst particles with them; such a design is necessary because the catalyst is largely deactivated by coke after only several seconds of operation and must therefore be efficiently transported out of the reactor and into the regenerator where the coke is burned off.

The catalyst may contain ~20 wt% zeolite, and more or less is used, depending on the feedstock and operating goals. The principal component of the catalyst is the matrix, which has a relatively low catalytic activity, but it is active for cracking the molecules that are too large to fit into the zeolite pores. The matrix also plays the role of a heat-transfer medium; the cracking reactions are fast and endothermic, and the temperature of the catalyst and the rates of the reactions fall as the catalyst is carried by the oil vapors from the inlet toward the outlet of the reactor. The thermal mass of the matrix keeps the temperature drop from being too large and causing the cracking rate to fall off too quickly. After leaving the reactor, the catalyst is heated up in the regenerator where the coke is burned off and reenters the reactor at a suitably high temperature.

In addition to the matrix and the catalytically active zeolite, there are small amounts of a supported metal such as platinum on alumina, which catalyzes CO oxidation in the regenerator and minimizes the emissions of CO. There are also metal oxide components that minimize the emission of SO_x formed in the regenerator from combustion of organosulfur compounds from the oil. The metal oxides react with SO_x in the regenerator to make stable metal sulfates. Cycled with the regenerated catalyst to the reducing atmosphere of the cracking reactor, the sulfates are converted into H_2S, which is removed by scrubbing the effluent gas stream.

The discovery of new zeolites continues apace, and the newer ones tend to have larger pores than those discovered earlier (88). Zeolites can be modified by incorporation of cations in the crystalline lattice which are not exchangeable ions, but can play catalytic roles. For example, silicalite, which has the structure of ZSM-5 but without Al, incorporating Ti in the lattice, is a commercial catalyst for oxidation of phenol with H_2O_2 to give diphenols; the catalytic sites are isolated Ti cations (89).

There are numerous structures that are similar to zeolites, such as aluminophosphate molecular sieves, AlPOs, and these are finding increasing catalytic applications. A new process for removal of wax by isomerization to remove straight-chain paraffins reportedly uses a nonzeolitic molecular sieve. Unidimensional pores are selective in producing highly branched products and few cracking products; the products has a low pour point and a high viscosity index (90).

Shape-Selective Catalysis.   The zeolites are unique in their molecular-sieving character, which is a consequence of their narrow, uniform pores (81,91). The transport of molecules in such small pores is different from that in the larger pores of typical catalysts. Figure 17 is a schematic representation of the diffusion coefficients of molecules in pores (92). When the pore diameters are large in comparison with the dimensions of the diffusing molecules, then molecular diffusion occurs, as it does in a fluid phase. When the pores become smaller, the interactions of the molecules with the pore walls become dominant, and Knudsen diffusion occurs. When the pores become so small that the molecules barely fit through them, then configurational diffusion occurs. This diffusion may be characterized by a substantial activation energy; the rate of diffusion is a strong function of the pore and molecule sizes in this regime. In the limit, a pore is too small for a molecule to fit in, and the diffusion coefficient becomes zero.

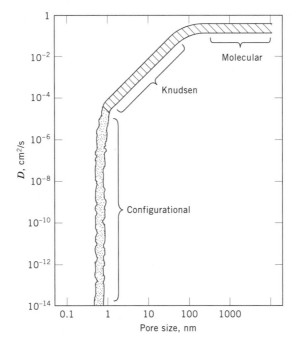

Fig. 17. Schematic representation of the regimes of diffusion in pores (92).

Catalytic processes have been developed to take advantage of the unique transport and molecular sieving properties of zeolites. The zeolite that has found the most applications is the medium-pored HZSM-5 (81,91). The term shape-selective catalysis is applied to describe the unique effects. There are different kinds of shape selectivity. Mass transport shape selectivity is a consequence of transport restrictions whereby some species diffuse more rapidly than others in the zeolite pores. In the simplest kind of shape-selective catalysis, small molecules in a mixture enter the pores and are catalytically converted whereas large molecules pass through the reactor unconverted because they do not fit into the pores where the catalytic sites are located. This statement is slightly oversimplified as there are a few catalytic sites on the outer surfaces of zeolite crystallites. Similarly, product molecules formed inside a zeolite may be so large that their transport out of the zeolite may be very slow, and they may be converted largely into other products that diffuse more rapidly into the product stream. A different kind of shape selectivity is called restricted transition state selectivity (81,91). It is not related to transport restrictions; rather, it is related to the size restriction of the catalyst pore that suppresses the formation of the transition state for a certain reaction, whereas it may not suppress the formation of a smaller transition state for another reaction. A way to diagnose the nature of the shape selectivity is to use zeolite crystallites of various sizes as the catalyst. Mass transport selectivity is influenced by the particle size; restricted transition state selectivity is not.

For example, HZSM-5 is a component of some cracking catalysts. It selectively cracks the straight-chain paraffins rather than the branched paraffins because of restricted transition state selectivity. This is a desired processing goal as the straight-chain isomers have lower octane numbers than the others and are less desirable gasoline components. In this example, the catalyst favors the reaction of the straight-chain compounds, even though they are intrinsically less reactive because they initially form secondary rather than the more stable and easily formed tertiary carbenium ions. Thus the pore size restriction reverses the pattern of selectivity that would be expected on the basis of the intrinsic chemistry alone.

Mass transport shape selectivity is illustrated by a process for disproportionation of toluene catalyzed by HZSM-5 (81,91). The desired product is *p*-xylene; the other isomers are less valuable. The ortho and meta isomers are bulkier than the para isomer and diffuse less readily in the zeolite pores. This transport restriction favors their conversion to the desired product in the catalyst pores; the desired para isomer is formed in excess of the equilibrium concentration. Xylene isomerization is another reaction catalyzed by HZSM-5, and the catalyst is preferred because of restricted transition state shape selectivity (91). An undesired side reaction, the xylene disproportionation to give toluene and trimethylbenzenes, is suppressed because it is bimolecular and the bulky transition state cannot readily form.

Mixed-Metal Oxides and Propylene Ammoxidation. The best catalysts for partial oxidation are metal oxides, usually mixed-metal oxides. For example, phosphorus–vanadium oxides are used commercially for oxidation of *n*-butane to give maleic anhydride, and oxides of bismuth and molybdenum with other components are used commercially for oxidation of propylene to give acrolein or acrylonitrile. A key to the success of a hydrocarbon oxidation catalyst is its ability to convert the hydrocarbon selectively to a partially oxidized product, rather than CO_2; this is difficult because the partially oxidized product is intrinsically more reactive than the hydrocarbon, especially when it is a paraffin.

The surface of a mixed-metal oxide exposes two kinds of metal ions in addition to O^{2-} ions and OH groups. Selective oxidation of hydrocarbons, represented schematically in Figure 18 (10), takes place on surface sites having oxygen atoms of limited reactivity, associated with the metal M_1 in the figure. These react with a hydrocarbon to give water and a partially oxidized organic compound rather than CO_2. The surface sites are reoxidized by other components of the solid catalyst rather than by O_2 directly. A second metal plays the role of an intermediary and oxygen is transported as ions through the bulk of the mixed metal oxide catalyst. A compensating transport of electrons and reaction of O_2 with the surface at sites different from those where the hydrocarbon is adsorbed make the process cyclic. Essentially this same pattern has already been illustrated by the Wacker oxidation, whereby the hydrocarbon reacts with an oxide, H_2O, in a step mediated by palladium, and a second metal, eg, copper, reacts with O_2 and then reoxidizes the palladium.

An important industrial partial oxidation process is the conversion of propylene to acrylonitrile:

$$CH_2\!\!=\!\!CHCH_3 + \frac{3}{2}\,O_2 + NH_3 \rightarrow CH_2\!\!=\!\!CHCN + 3\,H_2O \qquad (33)$$

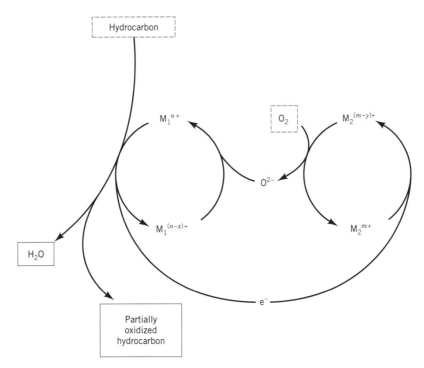

Fig. 18. Schematic representation of the catalytic cycle for ammoxidation of propylene and related reactions. M_1 and M_2 represent the two metals in a mixed-metal oxide catalyst (10).

The first catalysts used commercially to convert the propylene with high selectivity were mixed oxides of bismuth and molybdenum, referred to as bismuth molybdates. Improved catalysts consisting of a number of solid phases have been developed, with each generation becoming more complicated than its predecessor. Among the catalysts cited in a patent is the following: $Co_6^{2+}Ni_2^{2+}Fe_3^{3+}Bi^{3+}(MoO_4)_{12}$ on 50 wt % SiO_2 with some P and K (93). Silica is the support. The other components provide new functions, perhaps making the catalyst more stable. The catalysts are mechanically stable and used in fluidized-bed reactors at low pressures $\sim$100 kPa ($\sim$1 atm), and temperatures of $\sim$400–500°C. Years of development and improvement of catalysts for this process have led to the recognition that lattice oxygen of certain multivalent transition-metal oxides serves as a versatile, selective oxidizing agent, and these sites must be site isolated on the catalyst surface (94).

4.10. Catalysis by Supported Metals. Metals used in industrial catalysis are often expensive, and they are predominantly used in a highly dispersed form. Metal species dispersed on supports may be as small as the mononuclear chromium and zirconium complexes used for olefin polymerization, or they may be clusters containing as few as $\sim$10 metal atoms (95), or they may be larger particles that have three-dimensional structures and resemble small chunks of metal. The surfaces of the latter expose a number of different crystal faces, and if the particles are larger than $\sim$5 nm the distribution of crystal faces may be almost independent of particle size and shape. The smaller clusters, however, are less like bulk metals and are more likely to have unique structures and catalytic properties. The interactions between the metals and the support may be thought of as effects comparable to the ligand effects in molecular catalysis; the catalytic properties are sensitive to the structure and size of the metal cluster (95).

Ethylene Oxidation to Ethylene Oxide. A thoroughly investigated reaction catalyzed by a supported metal is the commercially applied partial oxidation of ethylene to give ethylene oxide (96). The desired reaction is the formation of ethylene oxide, ie, epoxidation; the following reaction scheme is a good approximation:

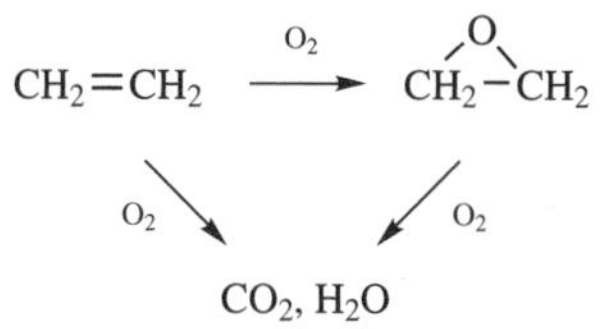

$$CH_2{=}CH_2 \xrightarrow{\ O_2\ } CH_2\underset{O}{\overset{O}{-}}CH_2$$
$$\underset{O_2}{\searrow} \qquad \underset{O_2}{\swarrow}$$
$$CO_2,\ H_2O$$

The selective oxidation is catalyzed by silver, which is the only good catalyst. Other olefins are not converted selectively to the epoxides in the presence of silver . However, propylene epoxidation is applied commercially; the catalysts are either molybdenum complexes in solution or solids containing titanium (see ETHYLENE OXIDE; PROPYLENE OXIDE).

The ethylene epoxidation catalysts (Fig. 9) are multicomponent mixtures consisting of a support (α-Al$_2$O$_3$), the catalytically active component (silver particles), and chemical promoters (alkali metal ions such as Cs$^+$ and perhaps anions such as sulfate), and a binder; the older literature also describes textural promoters. Furthermore, trace amounts of chlorine-containing compounds such as ethylene dichloride are continuously added with the feed; these compounds, like the alkali metal promoter, increase the selectivity of the catalyst for ethylene oxide. The data of Figure 19 illustrate the role of an alkali metal promoter (97). Extremely small amounts of the promoter markedly improve the selectivity of the catalyst; selectivities as high as 80% are reported. Under conditions of the catalytic oxidation, the silver surface is covered with a layer of oxygen and may be more properly described as an oxide than as a metal. There is only an incomplete understanding of the nature of the catalytic sites, the role of promoters, and the bonding of the reactants to the catalyst surface.

The support needs to be inert, which explains the choice of α-Al$_2$O$_3$; most metal oxides, including transition aluminas, catalyze unselective oxidation. The catalyst has a low surface area, $\sim$1 m^2/g, and large pores to minimize the influence of intraparticle diffusion, which would reduce the selectivity.

A new process for epoxidation of butadiene with a supported silver catalyst has recently been announced (98).

Naphtha Reforming and Bifunctional Catalysis. In some supported metal catalysts the support is not just an inert platform but plays an active catalytic role. This point is illustrated by catalysts for reforming of naphtha to make high octane number gasoline, a process that is a classic example of bifunctional

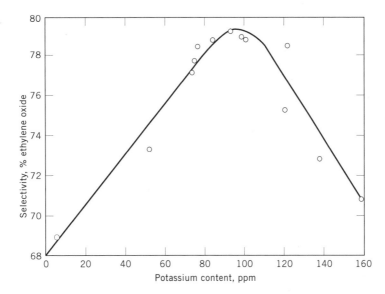

Fig. 19. Promotion of ethylene oxidation by potassium. The selectivity is the percentage of ethylene converted to ethylene oxide (97).

surface catalysis (99). The catalysts consist of metal, originally platinum (100), but now largely rhenium–platinum or tin–platinum, on a transition alumina, γ-Al_2O_3 or $\eta''''\lambda_-O_3$. Platinum is chosen because it is more active than any metal except iridium in a number of reactions that increase the octane number of paraffins without substantially changing their molecular weights. These reactions include dehydrogenation, eg, conversion of methylcyclohexane to toluene, and dehydrocyclization, eg, conversion of n-heptane to toluene. Skeletal isomerization is also desired, but platinum has only a low activity for this reaction. Rather, the reaction is acid catalyzed. Consequently, platinum supported on an amorphous solid acid is a good catalyst.

The support must not be too strongly or too weakly acidic; chlorided Al_2O_3 is optimal. The catalyst works well for reforming although the individual functions, ie, the metal and the acid, alone do not. The metal alone does not catalyze the branching reactions; they require an acidic function to generate carbenium ions, which undergo the desired isomerization. The acidic function alone is not sufficient to generate the carbenium ions; if it were so strongly acidic, the catalyst would be deactivated rapidly by carbonaceous deposits. The metal in the catalyst catalyzes dehydrogenation of paraffins to give olefins, which are easily protonated and thereby converted by carbenium ion routes.

Some reforming processes operate with platinum on alumina catalysts at ~500°C and high pressures, eg, 5 MPa (50 atm), with the reactant stream containing predominantly hydrogen (33,99). It is unusual for a process in which hydrogen is produced to be carried out in the presence of excess hydrogen, but this is done to minimize catalyst deactivation. Coke causes deactivation, but the high partial pressure of hydrogen retards coke formation as aromatic coke precursors and coke are hydrogenated. Catalyst regeneration by controlled coke burnoff may be needed only about twice a year.

However, there is a disadvantage to the high hydrogen partial pressures that became especially apparent with the removal of tetraethyllead from gasoline and an increased need for reforming processes as a source of high octane number hydrocarbons. Aromatic compounds have high octane numbers, and the dehydrogenation of naphthenes, ie, cyclic paraffins, to give aromatics is favored thermodynamically at high temperatures and low hydrogen partial pressures. Therefore, there was an incentive to find a way to operate reforming processes economically at much lower pressures.

Low pressure operation became routine with the application of new catalysts that are resistant to deactivation and withstand the low pressures. The catalysts are bimetallic, eg, incorporating tin or rhenium as well as platinum (101,102). The structures of these catalysts are still not well understood, but they likely involve cationic forms of tin or rhenium, and the latter resists deactivation better than the monometallic catalyst, especially in the presence of sulfur-containing compounds.

There are now a number of applications of supported bimetallic catalysts. In some instances, the role of the added metal is to isolate the atoms of the other metal at the surface, thereby reducing the rates of reactions that require ensembles of the atoms as the catalytic sites. These reactions are called structure-sensitive reactions (103). Some reactions, such as olefin hydrogenation, are structure-insensitive reactions; they seem to be catalyzed by isolated metal centers on surfaces, much as the Wilkinson hydrogenation is catalyzed by mononuclear metal complexes. Iridium–platinum catalysts have been used for naphtha reforming, and one of the roles of the platinum may be to combine with the iridium in alloylike bimetallic clusters to suppress an undesired structure-sensitive reaction, viz, paraffin hydrogenolysis (103). This reaction takes place in reforming and is largely undesired because it reduces the molecular weight range.

4.11. Catalysis by Metal Sulfides. Metal sulfides such as MoS_2, WS_2, and many others catalyze numerous reactions that are catalyzed by metals (104). The metal sulfides are typically several orders of magnitude less active than the metals, but they have the unique advantage of not being poisoned by sulfur compounds. They are thus good catalysts for applications with sulfur-containing feeds, including many fossil fuels.

Metal sulfide catalysts are widely applied in hydroprocessing of petroleum. The reactions include hydrogenation, hydrodesulfurization, and hydrodenitrogenation (105). Hydrodesulfurization is the reaction of organosulfur compounds with hydrogen to give hydrocarbons and H_2S; hydrodenitrogenation gives hydrocarbons and ammonia. Hydrodesulfurization is carried out to remove sulfur from feeds to naphtha reformers because the metals in the reforming catalysts are poisoned by sulfur. Hydrodesulfurization of heavy petroleum fractions is carried out to minimize formation of SO_x resulting from combustion of the fuels. Hydrodenitrogenation is carried out to remove basic nitrogen-containing compounds from fuels that are later hydrocracked, as hydrocracking catalysts have acidic components that are poisoned by the basic nitrogen-containing compounds.

MoS_2 is one of the most active hydroprocessing catalysts, but it is expensive, and the economical way to apply it is as highly dispersed material on a support, γ-Al_2O_3. The activity of the supported catalyst is increased by the presence of promoter ions, Co^{2+} or Ni^{2+}. The structures of the catalysts are fairly well understood; the MoS_2 is present in extremely small layer structures only a few atoms thick on the support surface, and the promoter ions are present at the edges of the MoS_2 layers, where the catalytic sites are located (106,107).

The catalysts are prepared by impregnating the support with aqueous salts of molybdenum and the promoter. In acidic solutions, molybdate ions are present largely in the form of heptamers, $[Mo_7O_{24}]^{6-}$, and the resulting surface species are believed to be present in islands, perhaps containing only seven Mo ions. Before use, the catalyst is treated with H_2 and some sulfur-containing compounds, and the surface oxides are converted into the sulfides that are the catalytically active species.

The applications of supported metal sulfides are unique with respect to catalyst deactivation phenomena. The catalysts used for processing of petroleum residua accumulate massive amounts of deposits consisting of sulfides formed from the organometallic constituents of the oil, principally nickel and vanadium (108). These, with coke, cover the catalyst surface and plug the pores. The catalysts are unusual in that they can function with masses of these deposits that are sometimes even more than the mass of the original fresh catalyst. Mass transport is important, as the deposits are typically formed with effectiveness factors less than unity, and in the extreme case the deposits block the pore mouths. Modeling of the transport–reaction phenomena has guided the preparation of catalysts with tailored pore structures to minimize the detriment of the deposits. These have been some of the most fruitful applications of the principles of chemical engineering in catalyst design and preparation.

5. Catalyst Development, Testing, and Production

Catalysts are discovered to meet processing needs and opportunities, but the discovery of a catalytic application to take advantage of some newly discovered material almost never occurs. Catalyst development is largely a matter of trial and error testing. The methodology was defined by Mittasch in the development of the ammonia synthesis process. Catalyst developers benefit from an extensive and diverse literature and often can formulate good starting points in a search for candidate catalysts by learning what has been used successfully for similar reactions. Deeper insights, such as would arise from understanding of the mechanistic details of a catalytic cycle, are usually not attained; the exceptions to this rule largely pertain to molecular catalysis, usually reactions occurring in solution. Fundamental insights were valuable in guiding the development of the process for chiral hydrogenation and that for methanol carbonylation, among others, but it would be inappropriate to infer that understanding of the fundamental chemistry led to straightforward design of the catalysts. Indeed, the initial working hypothesis about the chiral hydrogenation turned out to be incorrect. The more complicated processes of surface catalysis are for the most part only partially understood even when the processes are established and extensive after-the-fact research has been done. Creative research in catalyst discovery and development is usually the result of intuition and partial understanding combined with efficient testing and serendipity. Researchers who are repeatedly successful in finding new and improved catalysts seem to recognize needs and opportunities and notice significant exceptions to expected patterns and reason inductively by imperfect analogies.

Catalyst testing and evaluation have been revolutionized by computers, automated and miniaturized test reactors, and analytical methods. Combinatorial methods are playing an increasing role in the preparation and testing with modern equipment. Researchers can systematically prepare and screen many catalysts in a short time and efficiently determine, not only the initial catalytic activity and selectivity, but also the stability and the appearance of trace products that may indicate some new catalytic properties worthy of further development.

Catalyst design is in a primitive stage. There are hardly any examples of true design of catalysts. However, development of improved catalysts has been guided successfully in instances when the central issues were the interplay of mass transport and reaction. An example is catalysts used for hydroprocessing of heavy fossil fuels.

Almost all industrial catalysts are developed by researchers who are motivated to improve processes or create new ones. Thus the organization that first uses a new catalyst is usually the one that has discovered it. This organization, however, only rarely becomes the manufacturer of the catalyst used on a large scale. Catalysts are for the most part highly complex specialty chemicals, and catalyst manufacturers tend to be more efficient than others in producing them. Catalyst manufacturing is a competitive industry. Catalyst users often develop close relations with catalyst manufacturers, and the two may work together to develop and improve proprietary catalysts.

6. The Catalysis Literature

Catalysis is a broad, complex subject that is documented in many and varied sources. The field is rich in opportunity, in part because there is so much information that it is possible to find nuggets that competitors miss. Industrial catalysis is a competitive field, and much practical knowledge is proprietary.

The literature consists of patents, books, journals, and trade literature. The examples in patents may be especially valuable. The primary literature provides much catalyst performance data, but there is a lack of quantitative results characterizing the performance of industrial catalysts under industrially realistic conditions. Characterizations of industrial catalysts are often restricted to physical characterizations and perhaps activity measurements with pure component feeds, but it is rare to find data characterizing long-term catalyst performance with impure, multicomponent industrial feedstocks. Catalyst regeneration procedures are scarcely reported. Those who have proprietary technology are normally reluctant to make it known. Readers should be critical in assessing published work that claims a relevance to technology.

Often the catalysts described in the literature are not quite the same as those used in industrial processes, and often the reported performance is for pure single-component feeds. Sometimes the best quantitative approximations that can be made from the available literature are those based on reported kinetics of reactions with pure feeds and catalysts that are similar to but not the same as those used in practice. As a first approximation, one may use the published results and scale the activity on the basis of a few laboratory results obtained with realistic feeds and commercially available catalysts.

Catalyst suppliers are valuable sources of information. They often have extensive experience testing catalysts in long-term operation with real industrial feedstocks and may be willing to share information to improve their chances of selling a catalyst. Also they may work with a potential customer to develop catalysts that they could then supply. There is an extensive literature produced by catalyst manufacturers and organizations that license and market catalytic technology. This trade literature should be read critically as its purpose is to generate sales, but it often contains valuable information. Catalyst manufacturers and those who license and sell technology are motivated to demonstrate their technical knowledge. Their success in marketing depends on their technical reputations and their reliability in supplying catalysts and technology that consistently meet specifications.

BIBLIOGRAPHY

"Catalysis" in *ECT* 1st ed., Vol. 3, pp. 245–272 and Suppl. 1, pp. 144–150, by V. I. Komarewsky, Illinois Institute of Technology, and J. B. Coley, Standard Oil Co. of Indiana; in *ECT* 2nd ed., Vol. 4, pp. 534–586, by G. A. Mills, Houdry Process and Chemical Co.; in *ECT* 3rd ed., Vol. 5, pp. 16–61, by G. A. Mills, U.S. Department of Energy, and J. A. Cusumano, Catalytica Associates, Inc.; in *ECT* 4th ed,. Vol. 5 pp. 320–373, by Bruce C. Gates, University of Delaware; "Catalysis" in *ECT* (online), posting date: December 4, 2000, by Bruce C. Gates, University of Delaware.

CITED PUBLICATIONS

1. A. Mittasch, *Kurze Geschichte der Katalyse in Praxis und Theorie*, J. Springer-Verlag, Berlin, Germany, 1939.
2. A. Mittasch and E. Thies, *Von Davy und Döbereiner bis Deacon, ein Halbes Jahrhundert Grenzflächenkatalyse*, Verlag Chemie, Berlin, Germany, 1932.
3. L. Wilhelmy, *Ann. Physik. Chem. (Poggendorf)* **81**, 413, 419 (1850).
4. M. Bodenstein, *Bericht des V. Internationalen Kongresses für angewandte Chemie zu Berlin 1903*, Sektion X, Band IV, 1904, p. 561.
5. A. Mittasch, *Adv. Catal.* **2**, 81 (1950).
6. S. A. Topham, in J. R. Anderson and M. Boudart, eds., *Catalysis–Science and Technology*, Vol. 5, Springer-Verlag, Berlin, Germany, 1984, p. 119.
7. B. Timm, *Proceeding of the 8th International Congress on Catalysis (Berlin)* **1**, 7 (1984).
8. M. Boudart, *Kinetics of Chemical Processes*, Prentice-Hall, New York, 1968.
9. R. P. Bell, *The Proton in Chemistry*, Cornell University Press, Ithaca, New York, 1973.
10. B. C. Gates, *Catalytic Chemistry*, John Wiley & Sons, Inc., New York, 1992.
11. P. K. Ghosh, T. Guha, and A. N. Saha, *J. Appl. Chem.* **17**, 239 (1967).
12. J. I. de Jong and F. H. D. Dethmers, *Rec. Trav. Chim.* **84**, 460 (1965).

13. G. Olah, G. K. S. Prakash, and J. Sommer, *Superacids*, John Wiley & Sons, Inc., New York, 1985.

14. J. P. Collman, L. S. Hegedus, J. R. Norton, and R. G. Finke, *Principles and Applications of Organotransition Metal Chemistry*, 2nd ed., University Science Books, Mill Valley, Calif., 1987.

15. B. R. James, *Homogeneous Hydrogenation*, John Wiley & Sons, Inc., New York, 1974.

16. J. Halpern, *Trans. Am. Crystallogr. Assoc.* **14**, 59 (1978); J. Halpern, T. Okamoto, and A. Zakhariev, *J. Mol. Catal.* **2**, 65 (1976); J. Halpern and C. S. Wong, *J. Chem. Soc., Chem. Commun.*, 629 (1973).

17. W. S. Knowles, M. J. Sabacky, and B. D. Vineyard, *Adv. Chem. Ser.* **132**, 274 (1974).

18. R. Noyori, *Asymmetric Catalysis in Organic Synthesis*, John Wiley & Sons, Inc., New York, 1994.

19. J. Halpern, *Science* **217**, 401 (1982).

20. D. Forster, *J. Am. Chem. Soc.* **98**, 846 (1976).

21. P. M. Maitlis, A. Haynes, G. J. Sunley, and M. J. Howard, *J. Chem. Soc. Dalton Trans.* **1996**, 2187.

22. J. Hjorkjaer and V. W. Jensen, *Ind. Eng. Chem. Prod. Res. Dev.* **15**, 46 (1976).

23. R. T. Eby and T. C. Singleton, in B. E. Leach, ed., *Applied Industrial Catalysis*, Vol. 1, Academic Press, New York, 1983, p. 275.

24. N. Yoneda, S. Kusano, M. Yasui, P. Pujado, and S. Wilcher, *Appl. Catal. A Gen.* **221**, 253 (2001).

25. H.-W. Bohnen and B. Cornils, *Adv. Catal.*, in press.

26. H. Bach, W. Gick, W. Konkol, and E. Wiebus, *Proceeding of the 9th International Congress on Catalysis (Calgary)* Vol. 1, 1988 p. 254.

27. I. T. Horvath, *Acc. Chem. Res.* **31**, 641 (1998).

28. J. A. Gladaysz, *Pure Appl. Chem.* **73**, 1319 (2001).

29. R. L. Banks, in Ref. 23, Vol. 3, p. 215.

30. H. S. Eleuterio, *J. Mol. Catal.* **65**, 55 (1991).

31. (a) J. Kress, M. Wesolek, and J. A. Osborn, *J. Chem. Soc., Chem. Commun.*, 514 (1982); (b) N. Calderon, E. A. Ofstead, J. P. Ward, W. A. Judy, and K. W. Scott, *J. Am. Chem. Soc.* **90**, 4133 (1968).

32. M. Scholl, S. Ding, C. W. Lee, and R. H. Grubbs, *Org. Lett.* **1**, 953 (2000); C. W. Bielawski and R. H. Grubbs, *Angew. Chem. Int. Ed. Engl.* **39**, 2903 (2000).

33. B. C. Gates, J. R. Katzer, and G. C. A. Schuit, *Chemistry of Catalytic Processes*, McGraw-Hill Book Co., Inc., New York, 1979.

34. J. E. Lyons, in Ref. 23, Vol. 3, p. 131.

35. R. A. Sheldon and J. K. Kochi, *Adv. Catal.* **25**, 272 (1976).

36. G. W. Parshall, *J. Mol. Catal.* **4**, 243, (1978).

37. G. W. Parshall and S. D. Ittel, *Homogeneous Catalysis*, 2nd ed., John Wiley & Sons, Inc., New York, 1992.

38. B. D. Dombek, *Adv. Catal.* **32**, 325 (1983).

39. E. V. Dehmlow and S. S. Dehmlow, *Phase-Transfer Catalysis*, VCH, Weinheim, Germany, 1993.

40. C. M. Starks, C. L. Liotta, and M. Halpern, *Phase-Transfer Catalysis: Fundamentals, Applications, and Industrial Perspectives*, Chapman and Hall, New York, 1994.

41. J. N. Armor, *Appl. Catal. A: General* **222**, 407 (2001).

42. J. Wintterlin, *Adv. Catal.* **45**, 131 (2000).

43. (a) S. J. Gregg and K. S. W. Sing, *Adsorption, Surface Area, and Porosity*, 2nd ed., Academic Press, London, UK, 1967. (b) F. Schüth, K. Sing, and J. Weitkamp, eds., *Handbook of Porous Solids*, Wiley-VCH, Weinheim in press.

44. A. J. Lecloux, in Ref. 6, Vol. 2, p. 171.

45. P. H. Emmett, *Adv. Catal.* **1**, 65 (1948).

46. E. W. Thiele, *Am. Sci.* **55**(2), 176 (1967).

47. P. B. Weisz, *Science* **179**, 433 (1973).

48. C. N. Satterfield, *Mass Transfer in Heterogeneous Catalysis*, MIT Press, Cambridge, Mass., 1970.

49. B. G. Linsen, J. M. H. Fortuin, C. Okkerse, and J. J. Steggerola, eds., *Physical and Chemical Aspects of Adsorbents and Catalysts*, Academic Press, London, 1967.

50. A. B. Stiles, ed., *Catalyst Supports and Supported Catalysts*, Butterworths, Boston, Mass., 1987.

51. R. K. Oberlander, in Ref. 23, Vol. 3, p. 64.

52. H.-P. Boehm and H. Knözinger, in Ref. 6, Vol. 4, p. 39.

53. H. Knözinger and P. Ratnasamy, *Catal. Rev.-Sci. Eng.* **17**, 31 (1978).

54. H. H. Lamb, B. C. Gates, and H. Knözinger, *Angew. Chem., Int. Ed. Engl.* **27**, 1127 (1988).

55. W.-D. Mross, *Catal. Rev. Sci. Eng.* **25**, 591 (1983).

56. J. B. Butt and E. E. Petersen, *Activation, Deactivation, and Poisoning of Catalysts*, Academic Press, San Diego, Calif., 1988.

57. K. B. Tripplett, in Ref. 23, Vol. 1, p. 178.

58. A. B. Stiles, in Ref. 23, Vol. 2, p. 110.

59. E. Ruckenstein, in S. A. Stevenson, J. A. Dumesic, R. T. K. Baker, and E. Ruckenstein, eds., *Metal-Support Interactions in Catalysis, Sintering, and Redispersion*, Van Nostrand Reinhold Co., Inc., New York, 1987, p. 141.

60. G. Ertl, H. Knözinger, and J. Weitkamp, eds., *Preparation of Solid Catalysts*, Wiley-VCH, Weinheim, Germany, 1999.

61. C. G. Brinker and G. W. Scherer, *Sol–Gel Science*, Academic Press, New York, 1989.

62. J. P. Brunelle, *Pure Appl. Chem.* **50**, 1211 (1978).

63. M. Che, *Stud. Surf. Sci. Catal.* **130**, 115 (2000).

64. M. A. Harmer and Q. Sun, *Appl. Catal. A Gen.* **221**, 45 (2001).

65. R. Thornton and B. C. Gates, *J. Catal.* **34**, 275 (1974).

66. M. P. McDaniel, *Adv. Catal.* **33**, 47 (1985).

67. J. P. Hogan, in Ref. 23, Vol. 1, p. 149.

68. G. G. Hlatky, *Chem. Rev.* **100**, 1347 (2000).

69. P. Cossee, *J. Catal.* **3**, 80 (1964).

70. E. J. Arlman and P. Cossee, *J. Catal.* **3**, 99 (1964).

71. R. Waymouth and P. Pino, *J. Am. Chem. Soc.* **112**, 4911 (1990).

72. V. Ponec and G. C. Bond, *Catalysis by Metals and Alloys*, Elsevier, Amsterdam, The Netherlands, 1995.

73. G. Ertl, *Proceedings of the 7th International Congress on Catalysis*, Part A, p. 21, 1981.

74. R. Jennings, ed., *Ammonia Synthesis—Theory and Practice*, Pergamon Press, 1991.

75. P. Stoltze and J. K. Nørskov, *Phys. Rev. Lett.* **55**, 2502 (1985); P. Stoltze, *Phys. Scr.* **36**, 824 (1987); B. Hammer and J. K. Nørskov, *Adv. Catal.* **45**, 71 (2000).

76. J. S. Merriam and K. Atwood, in Ref. 22, Vol. 3, p. 113.

77. V. E. Henrich and P. A. Cox, *The Surface Science of Metal Oxides*, Cambridge University Press, Cambridge, 1994.

78. H. Idriss and M. A. Barteau, *Adv. Catal.* **45**, 261 (2000).

79. B. C. Gates, in *Handbook of Heterogeneous Catalysis*, John Wiley & Sons, Inc., in press.

80. D. W. Breck, *Zeolite Molecular Sieves*, Wiley, New York, 1974.

81. N. Y. Chen, W. E. Garwood, and F. G. Dwyer, *Shape Selective Catalysis in Industrial Applications*, Dekker, New York, 1996.

82. E. G. Derouane, in M. S. Whittington and A. J. Jacobson, eds., *Intercalation Chemistry*, Academic Press, New York, 1982, p. 101.

83. E. M. Flanigen and co-workers, *Nature (London)* **271**, 572 (1978).
84. W. O. Haag, R. M. Lago, and P. B. Weisz, *Nature (London)* **309**, 590 (1984).
85. P. B. Venuto and E. T. Habib, *Catal. Rev. Sci. Eng.* **18**, 1 (1978).
86. B. W. Wojciechowski and A. Corma, *Catalytic Cracking: Catalysts, Chemistry, and Kinetics*, Dekker, New York, 1986.
87. I. E. Maxwell, *CATTECH* **1**, 5 (1997).
88. M. E. Davis and R. F. Lobo, *Chem. Mater.* **4**, 756 (1992).
89. B. Notari, *Adv. Catal.* **41**, 253 (1996).
90. S. J. Miller, *Stud. Surf. Sci. Catal.* **84**, 2319 (1994).
91. W. O. Haag and N. Y. Chen, in L. L. Hegedus, ed., *Catalyst Design, Progress and Perspectives*, John Wiley & Sons, Inc., New York, 1987, p. 163.
92. P. B. Weisz, *CHEMTECH 3.*, 498 (1973).
93. U.S. Pat. 3,414,631 (Dec. 3,1968), R. K. Grasselli, G. Heights, and J. L. Callahan (to Standard Oil Co.); U.S. Pat. 3,642,930 (Feb. 15, 1972), R. K. Grasselli, G. Heights, and H. F. Hardman (to Standard Oil Co.).
94. R. K. Grasselli, *Top. Catal.* **15**, 93 (2001).
95. B. C. Gates, *Adv. Chem. Eng.* **27**, 49 (2001).
96. J. M. Berty, in Ref. 23, Vol. 1, p. 207.
97. U.S. Pat. 4,356,312 (Oct. 26, 1982), R. P. Nielson and J. H. La Rochelle (to Shell Oil Co.).
98. J. R. Monnier, *Appl. Catal. A Gen.* **221**, 73 (2001).
99. G. A. Mills, H. Heinemann, T. H. Milliken, and A. G. Oblad, *Ind. Eng. Chem.* **45**, 134 (1953).
100. M. J. Sterba and V. Haensel, *Ind. Eng. Chem. Prod. Res. Dev.* **15**, 2 (1976).
101. G. Antos, A. M. Aitani, and J. M. Parera, eds., *Catalytic Naphtha Reforming: Science and Technology*, Dekker, New York, 1995.
102. U.S. Pat. 3,415,737 (Dec. 10, 1968), H. E. Klucksdahl (to Chevron Research Co.).
103. J. H. Sinfelt, *Bimetallic Catalysts, Discoveries, Concepts, and Applications*, John Wiley & Sons, Inc., New York, 1983.
104. O. Weisser and S. Landa, *Sulphide Catalysts, Their Properties and Applications*, Pergamon Press, Oxford, UK, 1973.
105. M. J. Girgis and B. C. Gates, *Ind. Eng. Chem. Res.* **30**, 2021 (1991).
106. H. Topsøe, B. S. Clausen, and F. E. Massoth, *Hydrotreating Catalysis, Science and Technology*, Springer, Berlin, 1996.
107. J. V. Lauritsen, S. Helveg, E. Laegsgaard, I. Stensgaard, B. S. Clausen, H. Topsøe, and E. Besenbacher, *J. Catal.* **197**, 1 (2001).
108. R. J. Quann, R. A. Ware, C.-W. Hung, and J. Wei, *Adv. Chem. Eng.* **14**, 95 (1988).

GENERAL REFERENCES

References 10, 14, 18, 23, 33, 37, 39, 40, 48–50, 56, 72, 77, and 80 are general references. Also see B. Cornils, and W. A. Herrmann, eds., *Applied Homogeneous Catalysis with Organometallic Compounds*, Wiley-VCH, Weinheim, Germany, 2002; G. Ertl, H. Knözinger, and J. Weitkamp, eds., *Handbook of Heterogeneous Catalysis*, Wiley-VCH, Weinheim, Germany, 1997. *Advances in Catalysis*, Academic Press, New York, continuing series. Nomenclature used in catalysis is summarized in B. Cornils, W. A. Herrmann, R. Schlögl, and C. H. Wong, eds., *Catalysis from A to Z*, Wiley-VCH, Weinheim, Germany, 2000.

Bruce C. Gates
University of California, Davis

CATALYST DEACTIVATION AND REGENERATION

1. Introduction

Catalyst deactivation, the loss over time of catalytic activity and/or selectivity, is a problem of great and continuing concern in the practice of industrial catalytic processes. Costs to industry for catalyst replacement and process shutdown total billions of dollars per year. Time scales for catalyst deactivation vary considerably; for example, in the case of cracking catalysts, catalyst mortality may be on the order of seconds, while in ammonia synthesis the iron catalyst may last for 5–10 years. It is inevitable, however, that all catalysts will decay.

Typically, the loss of activity in a well-controlled process occurs slowly. However, process upsets or poorly designed hardware can bring about catastrophic failure. For example, in steam reforming of methane or naphtha great care must be taken to avoid reactor operation at excessively high temperatures or at steam-to-hydrocarbon ratios below a critical value. Indeed, these conditions can cause formation of large quantities of carbon filaments that plug catalyst pores and voids, pulverize catalyst pellets, and bring about process shutdown, all within a few hours.

While catalyst deactivation is inevitable for most processes, some of its immediate, drastic consequences may be avoided, postponed, or even reversed. Thus, deactivation issues (ie, extent, rate, and reactivation) greatly impact research, development, design, and operation of commercial processes. Accordingly, there is considerable motivation to understand and treat catalyst decay. Indeed, over the past three decades, the science of catalyst deactivation has been steadily developing, while literature addressing this topic has expanded considerably to include books (1–4); comprehensive reviews (5–8); proceedings of international symposia (9–14); topical journal issues (eg, Ref. 15); and more than 7000 patents for the period of 1976–2001. (In a patent search conducted in April 2001 for the keywords catalyst and deactivation, catalyst and life, and catalyst and regeneration, 1781, 3134, and 5068 patents were found respectively.) This area of research provides a critical understanding that is the foundation for modeling deactivation processes, designing stable catalysts, and optimizing processes to prevent or slow catalyst deactivation.

2. Mechanisms of Deactivation of Heterogeneous Catalysts

There are many paths for heterogeneous catalyst decay. For example, a catalyst solid may be poisoned by any one of a dozen contaminants present in the feed. Its surface, pores, and voids may be fouled by carbon or coke produced by cracking/condensation reactions of hydrocarbon reactants, intermediates, and/or products. In the treatment of a power plant flue gas, the catalyst can be dusted or eroded by and/or plugged with fly ash. Catalytic converters used to reduce emissions from gasoline or diesel engines may be poisoned or fouled by fuel or lubricant additives and/or engine corrosion products. If the catalytic reaction is conducted at high temperatures, thermal degradation may occur in the form of active phase crystallite growth, collapse of the carrier (support) pore structure, and/or solid-state reactions of the active phase with the carrier or promoters. In addition, the presence of oxygen or chlorine in the feed gas can lead to formation of volatile oxides or chlorides of the active phase, followed by gas-phase transport from the reactor. Similarly, changes in the oxidation state of the active catalytic phase can be induced by the presence of reactive gases in the feed.

Thus, the mechanisms of solid catalyst deactivation are many; nevertheless, they can be grouped into six intrinsic mechanisms of catalyst decay: (1) poisoning, (2) fouling, (3) thermal degradation, (4) vapor compound formation and/or leaching accompanied by transport from the catalyst surface or particle, (5) vapor–solid and/or solid–solid reactions, and (6) attrition/crushing. As mechanisms 1, 4, and 5 are chemical in nature while 2 and 5 are mechanical, the causes of deactivation are basically threefold: chemical, mechanical, and thermal. Each of the six basic mechanisms is defined briefly in Table 1. Mechanisms 4 and 5 are treated together, since 4 is a subset of 5.

2.1. Poisoning. *Poisoning* (3,16–22) is the strong chemisorption of reactants, products, or impurities on sites otherwise available for catalysis. Thus, poisoning has operational meaning; that is, whether a species acts as a poison depends upon its adsorption strength relative to the other species competing for catalytic sites. For example, oxygen can be a reactant in partial oxidation of ethylene to ethylene oxide on a silver catalyst and a poison in hydrogenation of ethylene on nickel. In addition to physically blocking adsorption sites, adsorbed poisons may induce changes in the electronic or geometric structure of the surface (17,21).

Mechanisms by which a poison may affect catalytic activity are multifold as illustrated by a conceptual two-dimensional model of sulfur poisoning of ethylene hydrogenation on a metal surface shown in Fig. 1. To begin with, a strongly adsorbed atom of sulfur physically blocks at least one three- or fourfold adsorption/reaction site (projecting into three dimensions) and three or four

Table 1. **Mechanisms of Catalyst Deactivation**

Mechanism	Type	Brief definition/description
poisoning	chemical	strong chemisorption of species on catalytic sites which block sites for catalytic reaction
fouling	mechanical	physical deposition of species from fluid phase onto the catalytic surface and in catalyst pores
thermal degradation	thermal	thermally induced loss of catalytic surface area, support area, and active phase-support reactions
vapor formation	chemical	reaction of gas with catalyst phase to produce volatile compounds
vapor–solid and solid–solid reactions	chemical	reaction of vapor, support, or promoter with catalytic phase to produce inactive phase
attrition/crushing	mechanical	loss of catalytic material due to abrasion loss of internal surface area due to mechanical-induced crushing of the catalyst particle

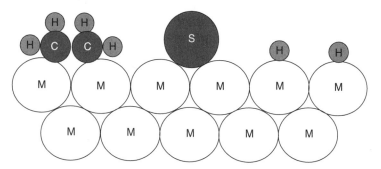

Fig. 1. Conceptual model of poisoning by sulfur atoms of a metal surface during ethylene hydrogenation.

topside sites on the metal surface. Second, by virtue of its strong chemical bond, it electronically modifies its nearest neighbor metal atoms and possibly its next-nearest neighbor atoms, thereby modifying their abilities to adsorb and/or dissociate reactant molecules (in this case H_2 and ethylene molecules), although these effects do not extend beyond about 5 atomic units (21). A third effect may be the restructuring of the surface by the strongly adsorbed poison, possibly causing dramatic changes in catalytic properties, especially for reactions sensitive to surface structure. In addition, the adsorbed poison blocks access of adsorbed reactants to each other (a fourth effect) and finally prevents or slows the surface diffusion of adsorbed reactants (effect number five).

Catalyst poisons can be classified according to their chemical makeup, selectivity for active sites, and the types of reactions poisoned. Table 2 lists four groups of catalyst poisons classified according to chemical origin and their type of interaction with metals. It should be emphasized that interactions of Group VA–VIIIA elements with catalytic metal phases depend on the oxidation state of the former, ie, how many electron pairs are available for bonding and the degree of shielding of the sulfur ion by ligands (16). Thus, the order of decreasing toxicity for poisoning of a given metal by different sulfur species is H_2S, SO_2, SO_4^{2-}, ie, in the order of increased shielding by oxygen. Toxicity also increases with increasing atomic or molecular size and electronegativity, but decreases if the poison can be gasified by O_2, H_2O, or H_2 present in the reactant stream (21); for example, adsorbed carbon can be gasified by O_2 to CO or CO_2 or by H_2 to CH_4.

Table 2. **Common Poisons Classified According to Chemical Structure**

Chemical type	Examples	Type of interaction with metals
Groups VA and VIA	N, P, As, Sb, O, S, Se, Te	through s and p orbitals; shielded structures are less toxic
Group VII A	F, Cl, Br, I	through s and p orbitals; formation of volatile halides
toxic heavy metals and ions	As, Pb, Hg, Bi, Sn, Zn, Cd, Cu, Fe	occupy d orbitals; may form alloys
molecules that adsorb with multiple bonds	CO, NO, HCN, benzene, acetylene, other unsaturated hydrocarbons	chemisorption through multiple bonds and back bonding

Table 3. **Poisons for Selected Catalysts in Important Representative Reactions**

Catalyst	Reaction	Poisons
silica–alumina, zeolites	cracking	organic bases, hydrocarbons heavy metals
nickel, platinum, palladium	hydrogenation/ dehydrogenation	compounds of S, P, As, Zn, Hg, halides, Pb, NH_3, C_2H_2
nickel	steam reforming of methane, naphtha	H_2S, As
iron, ruthenium	ammonia synthesis	O_2, H_2O, CO, S, C_2H_2, H_2O
cobalt, iron	Fischer–Tropsch synthesis	H_2S, COS, As, NH_3, metal carbonyls
noble metals on zeolites	hydrocracking	NH_3, S, Se, Te, P
silver	ethylene oxidation to ethylene oxide	C_2H_2
vanadium oxide	oxidation/selective catalytic reduction	As/Fe, K, Na from fly ash
platinum, palladium	oxidation of CO and hydrocarbons	Pb, P, Zn, SO_2, Fe
cobalt and molybdenum sulfides	hydrotreating of residues	asphaltenes; N, Ni, V compounds

Table 3 lists a number of common poisons for selected catalysts in important representative reactions. It is apparent that organic bases (eg, amines) and ammonia are common poisons for acidic solids such as silica–aluminas and zeolites in cracking and hydrocracking reactions, while sulfur- and arsenic-containing compounds are typical poisons for metals in hydrogenation, dehydrogenation, and steam reforming reactions. Metal compounds (eg, of Ni, Pb, V, and Zn) are poisons in automotive emissions control, catalytic cracking, and hydrotreating. Acetylene is a poison for ethylene oxidation, while asphaltenes are poisons in hydrotreating of petroleum residues.

"Selective" poisoning involves preferential adsorption of the poison on the most active sites at low concentrations. If sites of lesser activity are blocked initially, the poisoning is "antiselective." If the activity loss is proportional to the concentration of adsorbed poison, the poisoning is "nonselective." An example of selective poisoning is the deactivation of platinum by CO for the para-H_2 conversion (23) while Pb poisoning of CO oxidation on platinum is apparently antiselective (24), and arsenic poisoning of cyclopropane hydrogenation on Pt is nonselective (25). For nonselective poisoning the linear decrease in activity with poison concentration or susceptibility (σ) is defined by the slope of the activity versus poison concentration curve. Several other important terms associated with poisoning are defined in Table 4. Poison tolerance, the activity at saturation coverage of the poison, and resistance (the inverse of deactivation rate) are important concepts that are often encountered in discussions of poisoning including those defined in Table 4.

Activity versus poison concentration patterns are based on the assumption of uniform poisoning of the catalyst surface and surface reaction rate controlling, ie, negligible pore-diffusional resistance. These assumptions, however, are rarely

Table 4. **Important Poisoning Parameters**

Parameter	Definition
activity (a)	reaction rate at time t relative to that at $t = 0$
susceptibility (σ)	negative slope of the activity versus poison concentration curve $[\sigma = (a-1)/C(t)]$. Measure of a catalyst's sensitivity to a given poison
toxicity	susceptibility of a given catalyst for a poison relative to that for another poison
resistance	inverse of the deactivation rate, property that determines how rapidly a catalyst deactivates
tolerance ($a(C_{\text{sat}})$)	activity of the catalyst at saturation coverage (some catalysts may have negligible activity at saturation coverage)

met in typical industrial processes because the severe reaction conditions of high temperature and high pressure bring about a high pore-diffusional resistance for either the main or poisoning reaction or both. In physical terms, this means that the reaction may occur preferentially in the outer shell of the catalyst particle, or that poison is preferentially adsorbed in the outer shell of the catalyst particle, or both. The nonuniformly distributed reaction and/or poison leads to nonlinear activity versus poison concentration curves, but do not represent truly selective or antiselective poisoning. For example, if the main reaction is limited to an outer shell in a pellet where poison is concentrated, the drop in activity with concentration will be precipitous.

As sulfur poisoning is a difficult problem in many important catalytic processes (eg, hydrogenation, methanation, Fischer–Tropsch synthesis, steam reforming, and fuel cell power production), it merits separate discussion as an example of catalyst poisoning phenomena. Studies of sulfur poisoning in hydrogenation and CO hydrogenation reactions have been thoroughly reviewed (8,21,26–30). Much of the previous work focused on poisoning of nickel metal catalysts by H_2S, the primary sulfur poison in many important catalytic processes, and thus provides some useful case studies of poisoning.

Previous adsorption studies (27–29) indicate that H_2S adsorbs strongly and dissociatively on nickel metal surfaces. Extrapolation of high temperature data to zero coverage using a Tempkin isotherm (28) yields an enthalpy of adsorption of -250 kJ/mol; in other words, at low sulfur coverages, surface nickel–sulfur bonds are a factor of 3 more stable than bulk nickel–sulfur bonds. The absolute heat of adsorption increases with decreasing coverage and the equilibrium partial pressure of H_2S increases with increasing temperature and increasing coverage. It is expected that H_2S (and other sulfur impurities) will adsorb essentially irreversibly to high coverage in most catalytic processes involving metal catalysts.

Two important keys to reaching a deeper understanding of poisoning phenomena include (1) determining surface structures of poisons adsorbed on metal surfaces and (2) understanding how surface structure and hence adsorption stoichiometry change with increasing coverage of the poison. Studies of structures of adsorbed sulfur on single crystal metals (especially Ni) (3,27,31–34) provide such information. They reveal, for example, that sulfur adsorbs on Ni(100) in an ordered $P(2 \times 2)$ overlayer, bonded to four Ni atoms at $S/Ni_s < 0.25$ and in a

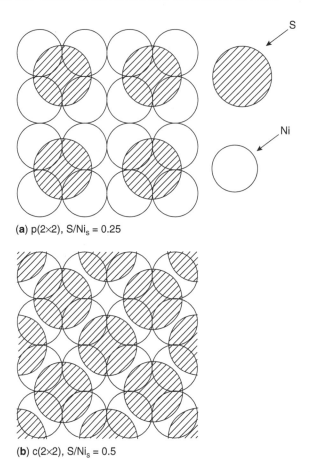

(a) p(2×2), S/Ni$_s$ = 0.25

(b) c(2×2), S/Ni$_s$ = 0.5

Fig. 2. Schematic view of sulfur adsorbed on a Ni(100) surface at a **(a)** S/Ni$_s$ = 0.25 in a p(2 × 2) structure and **(b)** S/Ni$_s$ = 0.50 in a c(2 × 2) structure.

C(2 × 2) overlayer to two Ni atoms for S/Ni$_s$ = 0.25–0.50 (see Fig. 2; Ni$_s$ denotes a surface atom of Ni); saturation coverage of sulfur on Ni(100) occurs at S/Ni$_s$ = 0.5. Adsorption of sulfur on Ni(110), Ni(111), and higher index planes of Ni is more complicated; while the same P(2 × 2) structure is observed at low coverage, complex overlayers appear at higher coverages—for example on Ni(111) in two additional stages (structures) up to saturation at S/Ni$_s$ = 0.5. In more open surface structures such as Ni(110) and Ni(210), saturation coverage occurs at S/Ni$_s$ = 0.74 and 1.09 respectively; indeed, there is a trend of increasing S/Ni$_s$ with decreasing planar density for Ni while the saturation sulfur concentration remains constant at 44 ng/cm^2 Ni (see Table 5).

Reported saturation stoichiometries for sulfur adsorption on polycrystalline and supported Ni catalysts (S/Ni$_s$) vary from 0.25 to 1.3 (27). The values of saturation coverage greater than S/Ni$_s$ = 0.5 may be explained by (1) a higher fractional coverage of sites of lower coordination number, ie, involving more open planes or intersections of planes (Table 5); (2) enhanced adsorption capacity at higher gas phase concentrations of H$_2$S in line with the observed trend of

Table 5. **Sulfur Adsorption Densities on Various Crystal Faces of Nickel**[a]

Crystal face	Sulfur conc. at saturation, ng·S/cm^2	Number of S atoms/cm^2 ($\times 10^{15}$)	Number of Ni atoms/cm^2 ($\times 10^{15}$)	S atoms per surface Ni atoms
(111)	47 ± 1	0.86	1.8	0.48
(100)	43 ± 1	0.80	1.6	0.50
(110)	44.5 ± 1	0.82	1.1	0.74
(210)	42 ± 1	0.78	0.72	1.09
polycrystalline	44.5 ± 1	0.82	—	—

[a]Data from Ref. 31.

increasing saturation coverage with increasing H$_2$S concentration; and/or (3) reconstruction of the surface by adsorbed sulfur at higher adsorption temperatures.

The first effect would be favored, and in fact is observed, for supported catalysts of higher dispersion (27). The second effect may explain the typically lower observed values of S/Ni$_s$ for single crystal Ni, which are measured at extremely low pressures (high vacuum) relative to the higher values of S/Ni$_s$ for polycrystalline and supported Ni, typically measured at orders of magnitude higher pressure; in the case of the single crystal work the surface is not in equilibrium with gas phase H$_2$S/H$_2$. The third effect, reconstruction of nickel surfaces by adsorbed sulfur, has been reported by a number of workers (27); for example, McCarroll and co-workers (33,34) found that sulfur adsorbed at near saturation coverage on a Ni(111) face was initially in a hexagonal pattern but upon heating above 700 K reoriented to a distorted C(2 × 2) structure on a Ni(100) layer. In another study (32), sulfur adsorbed on a Ni(810) caused decomposition to (100) and (410) facets. On the basis of their review of the reconstruction studies, Bartholomew and co-workers (27) concluded that at high temperatures and near saturation coverages, restructuring by sulfur of different facets of Ni to the more stable Ni(100) is probably a general phenomenon. If so, the S/Ni$_s$ ratio at saturation would in principle be 0.5 for the reconstructed surface. In the first example above, restructuring would not affect the S/Ni$_s$ ratio at saturation, since it is 0.5 for both (100) and (111) planes; however, in the second example, the S/Ni$_s$ ratio at saturation would probably decrease, as rough planes transform to smoother ones. Nevertheless, the possibility of increases in the S/Ni$_s$ ratio at saturation due to reconstruction cannot be ruled out.

The nature of reconstruction of a surface by a poison may depend on its pretreatment. For example, in a scanning tunneling microscopy (STM) study of room temperature H$_2$S adsorption on Ni(110), Ruan and co-workers (35) found that the S/Ni structure at saturation varied with the initial state of the surface, ie, whether clean or oxygen covered (see Fig. 3). This study showed that no reconstruction occurs by direct exposure to H$_2$S at room temperature, rather only in the presence of O$_2$ (or air). This emphasizes the complexities inherent in predicting the structure and stability of a given poison adsorbed on a given catalyst during a specified reaction as a function of different pretreatments or process disruptions, eg, exposure to air.

Fig. 3. A series of *in situ* STM images recorded after exposure of Ni(110) to oxygen and then progressively higher exposures of H$_2$S: (**a**) (2×1)O overlayer; (**b**) white islands and black troughs with a C(2×2)S structure after exposure to 3 and 8 L of H$_2$S; (**c**) 25 L, islands transform to low-coordinated rows in the [001] direction; and (**d**) 50 L, stable, well-ordered (4×1)S (35).

It is evident that structure and stoichiometry of sulfur adsorbed on nickel are complex functions of temperature, H$_2$S concentration, sulfur coverage, and pretreatment, phenomena that account at least in part for the complex nature of nickel poisoning by sulfur (27,36). Could one expect similar complexities in the poisoning of other metals? Probably, since poisoning of nickel is prototypical, ie, similar principles operate and similar poisoning behaviors are observed in other poison/metal systems, although none have been studied to the same depth as sulfur/nickel.

Since one of the necessary steps in a catalytic reaction is the adsorption of one or more reactants, investigation of the effects of adsorbed sulfur on the adsorption of other molecules can provide useful insights into the poisoning process (21,27). Previous investigations (27,37–43) indicate that both H$_2$ and CO adsorptions on nickel are poisoned by adsorbed sulfur. Sulfur poisoning can affect reaction selectivity as well as activity (27).

Because sulfur adsorbs so strongly on metals and prevents or modifies the further adsorption of reactant molecules, its presence on a catalyst surface usually effects substantial or complete loss of activity in many important reactions. The steady-state methanation activities of Ni, Co, Fe, and Ru are relative to the fresh, unpoisoned surface activity as a function of gas phase H_2S concentration. Data indicate that Ni, Co, Fe, and Ru all suffer 3–4 orders of magnitude loss in activity at 15–100 ppb of H_2S, ie, their sulfur tolerances are extremely low. Moreover, the sharp drop in activity with increasing H_2S concentration suggests highly selective poisoning. Nevertheless, the rate of sulfur poisoning and hence sulfur resistance varies from catalyst to catalyst and is apparently a function of catalyst composition (27) and reaction conditions (44). Indeed, it is possible to significantly improve sulfur resistance of Ni, Co, and Fe with catalyst additives such as Mo and B that selectively adsorb sulfur. Because the adsorption of sulfur compounds is generally rapid and irreversible, surface sulfur concentrations in catalyst particles and beds are nonuniform, eg, H_2S adsorbs selectively at the entrance to a packed bed and on the outer surface of catalyst particles, making the experimental study and modeling of sulfur poisoning extremely difficult.

There are other complications in the study of sulfur poisoning. For example, the adsorption stoichiometry of sulfur in CO hydrogenation on Ni is apparently a function of the temperature, H_2/CO ratio, and water partial pressure (44). Moreover, at high CO partial pressures sulfur may be removed from the surface as COS, which is not as strongly adsorbed as H_2S. At low temperature conditions, eg, those representative of Fischer–Tropsch synthesis or liquid phase hydrogenations, the gas phase concentration of H_2S in poisoning studies must be kept very low, ie, below 0.1–5 ppm, to avoid formation of bulk metal sulfides — a phenomenon that seriously compromises the validity of the results. Thus, the importance of studying poisoning phenomena *in situ* under realistic reaction conditions, at low process-relevant poison concentrations, and over a process-representative range of temperature and concentration conditions is emphasized.

There are a number of industrial processes in which one intentionally poisons the catalyst in order to improve its selectivity. For example, to minimize unwanted cracking reactions, to improve isomerization selectivity, to minimize coking, etc.

2.2. Fouling, Coking, and Carbon Deposition.

Fouling is the physical (mechanical) deposition of species from the fluid phase onto the catalyst surface, which results in activity loss due to blockage of sites and/or pores. In its advanced stages it may result in disintegration of catalyst particles and plugging of the reactor voids. Important examples include mechanical deposits of carbon and coke in porous catalysts, although carbon- and coke-forming processes also involve chemisorption of different kinds of carbons or condensed hydrocarbons that may act as catalyst poisons. The definitions of carbon and coke are somewhat arbitrary and by convention related to their origin. Carbon is typically a product of CO disproportionation while coke is produced by decomposition or condensation of hydrocarbons on catalyst surfaces and typically consists of polymerized heavy hydrocarbons. Nevertheless, coke forms may vary from high molecular weight hydrocarbons to primarily carbons such as graphite, depending upon the conditions under which the coke was formed and aged. A number of books and reviews treat the formation of carbons and coke on catalysts and the attendant deactivation of the catalysts (1,4,45–50).

The chemical structures of cokes or carbons formed in catalytic processes vary with reaction type, catalyst type, and reaction conditions. Menon (50) suggested that catalytic reactions accompanied by carbon or coke formation can be broadly classified as either coke-sensitive or coke-insensitive, analogous to Boudart's more general classification of structure-sensitive and structure-insensitive catalytic reactions. In coke-sensitive reactions, unreactive coke is deposited on active sites, leading to activity decline, while in coke-insensitive reactions, relatively reactive coke precursors formed on active sites are readily removed by hydrogen (or other gasifying agents). Examples of coke-sensitive reactions include catalytic cracking and hydrogenolysis; on the other hand, Fischer–Tropsch synthesis, catalytic reforming, and methanol synthesis are examples of coke-insensitive reactions. On the basis of this classification Menon (50) reasoned that the structure and location of a coke are more important than its quantity in affecting catalytic activity.

Consistent with Menon's classification, it is also generally observed that not only structure and location of coke vary but also its mechanism of formation varies with catalyst type, eg, whether it is a metal or metal oxide (or sulfide, sulfides being similar to oxides).

Carbon and Coke Formation on Supported Metal Catalysts. Possible effects of fouling by carbon (or coke) on the functioning of a supported metal catalyst are as follows. Carbon may (*1*) chemisorb strongly as a monolayer or physically adsorb in multilayers and in either case block access of reactants to metal surface sites, (*2*) totally encapsulate a metal particle and thereby completely deactivate that particle, and (*3*) plug micro- and mesopores such that access of reactants is denied to many crystallites inside these pores. Finally, in extreme cases, strong carbon filaments may build up in pores to the extent that they stress and fracture the support material, ultimately causing the disintegration of catalyst pellets and plugging of reactor voids.

Mechanisms of carbon deposition and coke formation on metal catalysts from carbon monoxide and hydrocarbons (4,45–49) are illustrated in Figs. 4

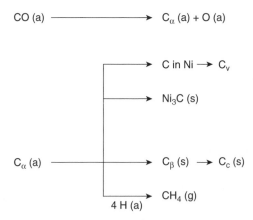

Fig. 4. Formation, transformation, and gasification of carbon on nickel (a, g, s refer to adsorbed, gaseous, and solid states respectively) (48).

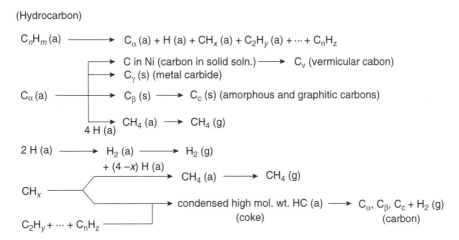

(Hydrocarbon)

Fig. 5. Formation and transformation of coke on metal surfaces (a, g, s refer to adsorbed, gaseous, and solid states respectively); gas phase reactions are not considered (48).

and 5. Different kinds of carbon and coke that vary in morphology and reactivity are formed in these reactions. For example, CO dissociates on metals to form C_a, an adsorbed atomic carbon; C_a can react to C_b, a polymeric carbon film. The more reactive, amorphous forms of carbon formed at low temperatures (eg, C_a and C_b) are converted at high temperatures over a period of time to less reactive, graphitic forms (48).

It should also be emphasized that some forms of carbon result in loss of catalytic activity and some do not. For example, at low temperatures ($<300-375°C$) condensed polymer or β-carbon films and at high temperatures ($>650°C$) graphitic carbon films encapsulate the metal surfaces of methanation and steam reforming catalysts (48). Deactivation of steam reforming catalysts at high reaction temperatures ($500-900°C$) may be caused by precipitation of atomic (carbidic) carbon dissolved in the Ni surface layers to a depth of more than $50-70$ nm (50,51). If it accumulates on the metal surface (at high or low temperatures), adsorbed atomic carbon can deactivate metal sites for adsorption and/or reaction. For example, Durer and co-workers (52) demonstrated that carbon atoms residing in the fourfold hollow sites of Rh(100) block the adsorption of hydrogen (and hence could block sites for hydrogenation). In the intermediate temperature range of $375-650°C$, carbon filaments are formed by precipitation of dissolved carbon at the rear side of metal crystallites, causing the metal particles to grow away from the support (45). Filament growth ceases when sufficient carbon accumulates on the free surface to cause encapsulation by a carbon layer; however, encapsulation of the metal particles does not occur if H_2/CO or H_2O/hydrocarbon ratios are sufficiently high. Thus, carbon filaments sometimes formed in CO hydrogenation or steam reforming of hydrocarbons would not necessarily cause a loss of intrinsic catalyst activity unless they are formed in sufficient quantities to cause plugging of the pores (48) or loss of metal occurs as the carbon fibers are removed during regeneration (53,54). However, in practice, regions of carbon forming potential in steam reforming must be carefully avoided, since once initiated, the rates of filamentous carbon formation are sufficiently high to cause catastrophic pore plugging and catalyst failure within a few hours to days.

The rate at which deactivation occurs for a given catalyst and reaction depends greatly on reaction conditions—especially temperature and reactant composition. A fundamental principle for coke-insensitive reactions on metals (eg, methanation, Fischer–Tropsch synthesis, steam reforming, catalytic reforming, and methanol synthesis) is that deactivation rate depends greatly on the difference in rates of formation and gasification of carbon/coke precursors, ie, $r_d = r_f - r_g$. If the rate of gasification r_g is equal to or greater than that of formation r_f, carbon/coke is not deposited. Rates of carbon/coke precursor formation and gasification both increase exponentially with temperature, although the difference between them varies a great deal with temperature because of differences in pre-exponential factors and activation energies. Thus, carbon/coke formation is avoided in regions of temperature in which precursor gasification rate exceeds deposition rate. A similar principle operates in steam reforming, ie, at a sufficiently low reaction temperature, the rate of hydrocarbon adsorption exceeds the rate of hydrocracking and a deactivating polymer film is formed (55); accordingly, it is necessary to operate above this temperature to avoid deactivation.

In steam reforming filamentous carbon formation rate is a strong function of hydrocarbon structure; for example, it decreases in the order acetylenes, olefins, paraffins, ie, in order of decreasing reactivity, although activation energies for nickel are in the same range (125–139 kJ) independent of hydrocarbon structure and about the same as those observed for formation of filamentous carbon from decomposition of CO (48). This latter observation suggests that the reactions of CO and different hydrocarbons to filamentous carbon proceed by a common mechanism and rate-determining step—probably the diffusion of carbon through the metal crystallites (48).

The rate at which a carbon or coke is accumulated in a given reaction under given conditions can vary significantly with catalyst structure, including metal type, metal crystallite size, promoter, and catalyst support. For example, supported Co, Fe, and Ni are active above 350–400°C for filamentous carbon formation from CO and hydrocarbons; the order of decreasing activity is reportedly Fe, Co, Ni (48). Pt, Ru, and Rh catalysts, on the other hand, while equally or more active than Ni, Co, or Fe in steam reforming produce little or no coke or carbon. This is attributed to reduced mobility and/or solubility of carbon in the noble metals, thus retarding the nucleation process. Thus, it is not surprising that addition of noble metals to base metals retards carbon formation; for example, addition of Pt in Ni lowers carbon deposition rate during methanation, while addition of Cu or Au to Ni substantially lowers carbon formation in steam reforming (48,56). In contrast to the moderating effects of noble metal additives, addition of 0.5% Sn to cobalt substantially increases the rate of carbon filament formation from ethylene (57), an effect desirable in the commercial production of carbon filament fibers.

Since carbon formation and gasification rates are influenced differently by modifications in metal crystallite surface chemistry, which are in turn a function of catalyst structure, oxide additives or oxide supports may be used to moderate the rate of undesirable carbon or coke accumulation. For example, Bartholomew and Strasburg (58) found the specific rate (turnover frequency) of filamentous carbon deposition on nickel during methanation at $350°C$ to decrease in the order Ni/TiO_2, $NiAl_2O_3$, Ni/SiO_2, while Vance and Bartholomew (59) observed C_α hydrogenation rates at $170°C$ to decrease in this same order (the same as for methanation at $225°C$). This behavior was explained in terms of promotional or inhibiting effects due to decoration of metal crystallites by the support, for example silica, inhibiting both CO dissociation and carbon hydrogenation. This hypothesis is consistent with observations (60,61) that silica evaporated on metal surfaces and supported metals inhibits formation of filamentous carbon. Similarly Bitter and co-workers (62) observed rates of carbon formation in CO_2/CH_4 reforming to decrease in the order $Pt/g-Al_2O_3 \rightarrow Pt/TiO_2 > Pt/ZrO_2$; while 90% of the carbon deposited on the support, the authors linked deactivation to carbon accumulated on the metal owing to an imbalance between carbon formed by methane dissociation and oxidation by chemisorbed CO_2. The rate of formation of coke in steam reforming is delayed and occurs at lower rates in nickel catalysts promoted with alkali or supported on basic MgO (63).

Since formation of coke, graphite, or filamentous carbon involves the formation of C−C bonds on multiple atoms sites, one might expect that coke or carbon formation on metals is structure-sensitive, ie, sensitive to surface structure and metal crystallite size. Indeed, Bitter and co-workers (62) found that catalysts containing larger Pt crystallites deactivate more rapidly than those containing small crystallites. Moreover, a crystallite size effect, observed in steam reforming of methane on nickel (48,63), appears to operate in the same direction, ie, formation of filamentous carbon occurs at lower rates in catalysts containing smaller metal crystallites.

In summary, deactivation of supported metals by carbon or coke may occur chemically owing to chemisorption or carbide formation or physically and mechanically owing to blocking of surface sites, metal crystallite encapsulation, plugging of pores, and destruction of catalyst pellets by carbon filaments. Blocking of catalytic sites by chemisorbed hydrocarbons, surface carbides, or relatively reactive films is generally reversible in hydrogen, steam, CO_2, or oxygen. Further details of the thermodynamics, kinetics, and mechanisms of carbon and coke formation in methanation and steam reforming reactions are available in reviews by Bartholomew (48) and Rostrup-Nielsen (55,63).

Coke Formation on Metal Oxide and Sulfide Catalysts. In reactions involving hydrocarbons, coke may be formed in the gas phase and on both non-catalytic and catalytic surfaces. Nevertheless, formation of coke on oxides and sulfides is principally a result of cracking reactions involving coke precursors (typically olefins or aromatics) catalyzed by acid sites (64,65). Dehydrogenation and cyclization reactions of carbocation intermediates formed on acid sites lead to aromatics, which react further to higher molecular weight polynuclear aromatics and condense as coke.

Olefins, benzene and benzene derivatives, and polynuclear aromatics are precursors to coke formation. However, the order of reactivity for coke formation is clearly structure dependent, ie, decreases in the order polynuclear aromatics > aromatics > olefins > branched alkanes > normal alkanes. For example, the weight percent coke formed on silica−alumina at $500°C$ is 0.06, 3.8, 12.5, and 23% for benzene, naphthalene, fluoranthene, and anthracene respectively (66).

Coking reactions in processes involving heavy hydrocarbons are very complex; different kinds of coke may be formed and they may range in composition from CH to C and have a wide range of reactivities with oxygen and hydrogen depending upon the time on stream and temperature to which they are exposed. For example, coke deposits occurring in hydrodesulfurization of residues have been classified into three types (67):

1. Type I deposits are reversibly adsorbed normal aromatics deposited during the first part of the cycle at low temperature.
2. Type II deposits are reversibly adsorbed asphaltenes deposited early in the coking process.
3. Type III deposits result from condensation of aromatic concentrates into clusters and then crystals that constitute a "mesophase." This crystalline phase is formed after long reaction times at high temperature. This hardened coke causes severe deactivation of the catalyst (67).

In addition to hydrocarbon structure and reaction conditions, extent and rate of coke formation are also a function of the acidity and pore structure of the catalyst. Generally, the rate and extent of coke formation increase with increasing acid strength and concentration. Coke yield decreases with decreasing pore size (for a fixed acid strength and concentration); this is especially true in zeolites where shape selectivity plays an important role in coke formation. However, in pores of molecular diameter, a relatively small quantity of coke can cause substantial loss of activity. It should be emphasized that coke yield can vary considerably into the interior pores of a catalyst particle or along a catalyst bed, depending upon the extent to which the main and deactivation reactions are affected by film mass transport and pore diffusional resistance.

The mechanisms by which coke deactivates oxide and sulfide catalysts are, as in the case of supported metals, both chemical and physical. However, some aspects of the chemistry are quite different. The principal chemical loss of activity in oxides and sulfides is due to the strong adsorption of coke molecules on acidic sites. But as discussed earlier, strong acid sites also play an important role in the formation of coke precursors, which subsequently undergo condensation reactions to produce large polynuclear aromatic molecules that physically coat catalytic surfaces. Physical loss of activity also occurs as coke accumulates, ultimately partially or completely blocking catalyst pores as in supported metal catalysts. For example, in isomerization of *cis*-butene on SiO_2/Al_2O_3 (68) catalyst deactivation occurs by rapid, selective poisoning of strong acid sites; coke evolved early in the reaction is soluble in dichloromethane and pyridine and is slightly aromatic. Apparently, the blocking of active sites does not significantly affect porosity or catalyst surface area, as SiO_2/Al_2O_3 contains relatively large mesopores.

In the case of supported bifunctional metal/metal oxide catalysts, different kinds of coke are formed on the metal and the acidic oxide support, eg, soft coke (high H/C ratio) on Pt or Pt–Re metals and hard coke (low H/C ratio) on the alumina support in catalytic reforming (69). In this case coke precursors may be formed on the metal via hydrogenolysis, following which they migrate to the support and undergo polymerization and cyclization reactions, after which the larger molecules are dehydrogenated on the metal and finally accumulate on the support, causing loss of isomerization activity. Mild sulfiding of these catalysts (especially Pt–Re/alumina) substantially reduces the rate of hydrogenolysis and the overall formation of coke on both metal and support; it especially reduces the hard coke, which is mainly responsible for deactivation.

Several studies (65,70–80) have focused on coke formation during hydrocarbon reactions in zeolites including (1) the detailed chemistry of coke precursors and coke molecules formed in zeolite pores and pore intersections (or supercages) and (2) the relative importance of adsorption on acid sites versus pore blockage. The principal conclusions from these studies can be summarized as follows: (1) the formation of coke and the manner in which it deactivates a zeolite catalyst are shape-selective processes, (2) deactivation is mainly due to the formation and retention of heavy aromatic clusters in pores and pore intersections, and (3) while both acid-site poisoning and pore blockage participate in the deactivation, the former dominates at low coking rates, low coke coverages (eg, in Y-zeolite below 2 wt%), and high temperatures, while the latter process dominates at high reaction rates, low temperatures, and high coke coverages. Thus, pore size and pore structure are probably more important than acid strength and density under typical commercial process conditions. Indeed, deactivation is typically more rapid in zeolites having small pores or apertures and/or a monodimensional structure (78). Fig. 6 illustrates four possible modes of deactivation of HZSM-5 by carbonaceous deposits with increasing severity of coking (78).

These conclusions (in the previous paragraph) are borne out, for example, in the study by Cerqueira and co-workers (80) of USHY zeolite deactivation during methylcyclohexane transformation at 450°C, showing the following:

1. Coke is probably mainly formed by rapid transformation of styrenic C_7 carbenium ions with lesser contributions from reactions of cyclopentadiene, C_3–C_6 olefins, and aromatics.

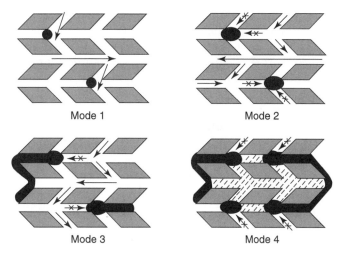

Fig. 6. Schematic of the four possible modes of deactivation by carbonaceous deposits in HZSM-5: (1) reversible adsorption on acid sites, (2) irreversible adsorption on sites with partial blocking of pore intersections, (3) partial steric blocking of pores, and (4) extensive steric blocking of pores by exterior deposits (78).

2. Soluble coke consists of polynuclear aromatic clusters containing three to seven five- and six-membered rings having typical compositions of $C_{30}H_{40}$ to $C_{40}H_{44}$ and having dimensions of 0.9×1.1 nm to 1.1×1.5 nm, ie, sizes that would cause them to be trapped in the supercages of Y-zeolite.

3. At short contact times, coking is relatively slow and deactivation is mainly due to acid-site poisoning, while at long contact times, coking is much faster because of the high concentrations of coke precursors; under these latter conditions coke is preferentially deposited at the outer pore openings of zeolite crystallites and deactivation is dominated by pore-mouth blockage.

That coke formed at large contact times not only blocks pores and/or pore intersections inside the zeolite but also migrates to the outside of zeolite crystallites where it blocks pore entrances has been observed in several studies (74,76,77,80). However, the amount, structure, and location of coke in ZSM-5 depends strongly on the coke precursor, eg, coke formed from mesitylene is deposited on the external zeolite surface whereas coking with isobutene leads to largely paraffinic deposits inside pores; coke from toluene, on the other hand, is polyaromatic and is deposited both on external and internal zeolite surfaces (74).

2.3. Thermal Degradation and Sintering. *Background.* Thermally induced deactivation of catalysts results from (*1*) loss of catalytic surface area due to crystallite growth of the catalytic phase, (*2*) loss of support area due to support collapse and of catalytic surface area due to pore collapse on crystallites of the active phase, and/or (*3*) chemical transformations of catalytic phases to noncatalytic phases. The first two processes are typically referred to as "sintering." Sintering processes generally take place at high reaction temperatures (eg, >500°C) and are generally accelerated by the presence of water vapor.

Most of the previous sintering and redispersion work has focused on supported metals. Experimental and theoretical studies of sintering and redispersion of supported metals published before 1997 have been reviewed fairly extensively (8,81–90). Three principal mechanisms of metal crystallite growth have been advanced: (*1*) crystallite migration, (*2*) atomic migration, and (*3*) (at very high temperatures) vapor transport. Crystallite migration involves the migration of entire crystallites over the support surface, followed by collision and coalescence. Atomic migration involves detachment of metal atoms or molecular metal clusters from crystallites, migration of these atoms over the support surface, and ultimately, capture by larger crystallites. Redispersion, the reverse of crystallite growth in the presence of O_2 and/or Cl_2, may involve (*1*) formation of volatile metal oxide or metal chloride complexes that attach to the support and are subsequently decomposed to small crystallites upon reduction and/or (*2*) formation of oxide particles or films that break into small crystallites during subsequent reduction.

There is controversy in the literature regarding which mechanism of sintering (or redispersion) operates at a given set of conditions. Logically, atomic migration would be favored at lower temperatures than crystallite migration, since the higher diffusivities of atoms or small clusters would facilitate their migration, whereas the thermal energy necessary to induce motion of larger crystallites would only be available at higher temperatures. Moreover, migration of small crystallites might be favorable early in the sintering process but unfavorable as crystallites become larger. However, fixing on only one of the three sintering mechanisms (and two dispersion mechanisms) is a simplification that ignores the possibility that all mechanisms may occur simultaneously and may be coupled with each other through complex physicochemical processes including the following: (1) dissociation and emission of metal atoms or metal-containing molecules from metal crystallites, (2) adsorption and trapping of metal atoms or metal-containing molecules on the support surface, (3) diffusion of metal atoms, metal-containing molecules and/or metal crystallites across support surfaces, (4) metal or metal oxide particle spreading, (5) support surface wetting by metal particles, (6) metal particle nucleation, (7) coalescence of, or bridging between, two metal particles, (8) capture of atoms or molecules by metal particles, (9) liquid formation, (10) metal volatilization through volatile compound formation, (11) splitting of crystallites in O_2 atmosphere owing to formation of oxides of a different specific volume, and (12) metal atom vaporization. Depending upon reaction or redispersion conditions, a few or all of these processes may be important; thus, the complexity of sintering/redispersion processes is emphasized.

In general, sintering processes are kinetically slow (at moderate reaction temperatures) and irreversible or difficult to reverse. Thus, sintering is more easily prevented than cured.

Factors Affecting Metal Particle Growth and Redispersion in Supported Metals. Temperature, atmosphere, metal type, metal dispersion, promoters/impurities and support surface area, texture, and porosity are the principal parameters affecting rates of sintering and redispersion (see Table 6) (8,86–90). Sintering rates increase exponentially with temperature. Metals sinter relatively

Table 6. **Effects of Important Reaction and Catalyst Variables on Sintering Rates of Supported Metals Based on GPLE Data**[a]

Variable	Effect
temperature	sintering rates are exponentially dependent on T; E_{act} varies from 30 to 150 kJ/mol. E_{act} decreases with increasing metal loading; it increases in the following order with atmosphere: NO, O_2, H_2, N_2
atmosphere	sintering rates are much higher for noble metals in O_2 than in H_2 and higher for noble and base metals in H_2 relative to N_2; sintering rate decreases for supported Pt in atmospheres in the following order: NO, O_2, H_2, N_2
metal	observed order of decreasing thermal stability in H_2 is Ru > Ir $\cong$ Rh > Pt; thermal stability in O_2 is a function of (1) volatility of metal oxide and (2) strength of metal oxide–support interaction
support	metal–support interactions are weak (bond strengths of 5–15 kJ/mol); with a few exceptions, thermal stability for a given metal decreases with support in the following order: Al_2O_3 > SiO_2 > carbon
promoters	some additives decrease atom mobility, eg, C, O, CaO, BaO, CeO_2, GeO_2; others increase atom mobility, eg, Pb, Bi, Cl, F, or S; oxides of Ba, Ca, or Sr are "trapping agents" that decrease sintering rate
pore size	sintering rates are lower for porous versus nonporous supports; they decrease as crystallite diameters approach those of the pores

[a]Refs. 8 and 86–90.

rapidly in oxygen and relatively slowly in hydrogen, although depending upon the support, metal redispersion can be facilitated by exposure at high temperature (eg, 500–550°C for Pt/Al_2O_3) to oxygen and chlorine, followed by reduction. Water vapor also increases the sintering rate of supported metals.

Normalized dispersion (percentage of metal exposed at any time divided by the initial percentage exposed) versus time data show that at temperatures of 650°C or higher, rates of metal surface area loss (measured by hydrogen chemisorption) due to sintering of Ni/silica in hydrogen atmosphere are significant, causing 70% loss of the original metal surface area within 50 h at 750°C. In reducing atmosphere, metal crystallite stability generally decreases with decreasing metal melting temperature, ie, in the order Ru > Ir > Rh > Pt > Pd > Ni > Cu > Ag, although this order may be affected by relatively stronger metal–support interactions, eg, the observed order of decreasing stability of supported platinum in vacuum is Pt/Al_2O_3 > Pt/SiO_2 > Pt/C. In oxidizing atmospheres, metal crystallite stability depends on the volatility of metal oxides and the strength of the metal–oxide–support interaction. For noble metals, metal stability in air decreases in the order Rh > Pt > Ir > Ru; formation of volatile RuO_4 accounts for the relative instability of ruthenium (91).

The effect of temperature on sintering of metals and oxides can be understood physically in terms of the driving forces for dissociation and diffusion of surface atoms, which are both proportional to the fractional approach to the absolute melting point temperature (T_{mp}). Thus, as temperature increases, the mean lattice vibration of surface atoms increases; when the Hüttig temperature ($0.3T_{mp}$) is reached less strongly bound surface atoms at defect sites (eg, edges and corner sites) dissociate and diffuse readily over the surface, while at the Tamman temperature ($0.5T_{mp}$) atoms in the bulk become mobile. Accordingly, sintering rates of a metal or metal oxide are significant above the Hüttig temperature and very high near the Tamman temperature; thus, the relative thermal stability of metals or metal oxides can be correlated in terms of the Hüttig or Tamman temperatures (92). For example, sintering of copper catalysts for methanol synthesis is promoted by traces of chlorine in the feed, which react at about 225°C (500 K) with the active metal/metal oxide surface to produce a highly mobile copper chloride phase having a Tamman temperature of only 79–174°C (352–447 K) relative to 405–527°C (678–800 K) for copper metal or metal oxides (93).

Promoters or impurities affect sintering and redispersion by either increasing (eg, chlorine and sulfur) or decreasing (eg, oxygen, calcium, cesium) metal atom mobility on the support; in the latter case this is due to their high resistance to dissociation and migration due to high melting points as well as their hindering dissociation and surface diffusion of other atoms. Similarly, support surface defects or pores impede surface migration of metal particles — especially micropores and mesopores with pore diameters about the same size as the metal crystallites.

Historically, sintering rate data were fitted to a simple power-law expression (SPLE) of the form

$$-d(D/D_0)/dt = k_s(D/D_0)^n \qquad (1)$$

where k_s is the sintering rate constant, D_0 the initial dispersion, and n the sintering order, which for typical catalyst systems may vary from 3 to 15; unfortunately, the SPLE is in general not valid for sintering processes because it assumes that surface area or dispersion ultimately reaches zero, given sufficient time, when in fact, for a given temperature and atmosphere, a nonzero or limiting dispersion is observed after long sintering times. Moreover, the use of the SPLE is further questionable because variations in sintering order are observed as a function of time and temperature for a given catalyst in a fixed atmosphere (88–90); thus, data obtained for different samples and different reaction conditions cannot be quantitatively compared. Nevertheless, it has been shown by Fuentes (94) and Bartholomew (87–89) that the effects of temperature, atmosphere, metal, promoter, and support can be quantitatively determined by fitting sintering kinetic data to the general power-law expression (GPLE)

$$-d(D/D_0)/dt = k_s(D/D_0 - D_{eq}/D_0)^m \tag{2}$$

which adds a term $-D_{eq}/D_0$ to account for the observed asymptotic approach of the typical dispersion versus time curve to a limiting dispersion D_{eq} at infinite time; m, the order of sintering, is found to be either 1 or 2. A recently compiled, comprehensive quantitative treatment of previous sintering rate data based on the GPLE with an order m of 2 (87–89) quantitatively addresses the effects of catalyst properties and reaction conditions on sintering rate (91,95–97).

Sintering studies of supported metals are generally of two types: (1) studies of commercially relevant supported metal catalysts and (2) studies of model metal–support systems. The former type provides useful rate data that can be used to predict sintering rates, while the latter type provides insights into the mechanisms of metal particle migration and sintering, although the results cannot be quantitatively extrapolated to predict behavior of commercial catalysts. There is direct evidence from the previous studies of model-supported catalysts (87,90) for the occurrence of crystallite migration (mainly in well-dispersed systems early in the sintering process), atomic migration (mainly at longer sintering times), and spreading of metal crystallites (mainly in oxygen atmosphere). There is also evidence that under reaction conditions, the surface is dynamic, ie, adsorbates and other adatoms rapidly restructure the surface and slowly bring about faceting; moreover, thermal treatments cause gradual changes in the distribution of coordination sites to minimize surface energy. There is a trend in increasing sophistication of spectroscopic tools used to study sintering and redispersion. Additional insights into atomic and molecular processes during reaction at the atomic scale using STM, analytical HRTEM, and other such powerful surface science tools are expected during the next decade.

Sintering of Catalyst Carriers.   Sintering of carriers has been reviewed by Baker and co-workers (86) and Trimm (98). Single-phase oxide carriers sinter by one or more of the following processes: (1) surface diffusion, (2) solid-state diffusion, (3) evaporation/condensation of volatile atoms or molecules, (4) grain boundary diffusion, and (5) phase transformations. In oxidizing atmospheres, γ-alumina and silica are the most thermally stable carriers; in reducing atmospheres, carbons are the most thermally stable carriers. Additives and impurities affect the thermal properties of carriers by occupying defect sites or forming new phases. Alkali metals, for example, accelerate sintering, while calcium, barium, nickel, and lanthanum oxides form thermally stable spinel phases with alumina. Steam accelerates support sintering by forming mobile surface hydroxyl groups that are subsequently volatilized at higher temperatures. Chlorine also promotes sintering and grain growth in magnesia and titania during high temperature calcination (99). By contrast, sulfuric acid treatment of hydrated alumina (gibbsite) followed by two-step calcination results in a very stable transitional alumina with needle-like particle morphology (98). Dispersed metals in supported metal catalysts can also accelerate support sintering; for example, dispersed nickel accelerates the loss of Al_2O_3 surface area in Ni/Al_2O_3 catalysts.

Effects of Sintering on Catalyst Activity.   Baker and co-workers (86) have reviewed the effects of sintering on catalytic activity. Specific activity (based on catalytic surface area) can either increase or decrease with increasing metal crystallite size during sintering if the reaction is structure-sensitive, or it can be independent of changes in metal crystallite size if the reaction is structure-insensitive. Thus, for a structure-sensitive reaction, the impact of sintering may be either magnified or moderated; while for a structure insensitive-reaction, sintering has in principle no effect on specific activity (per unit surface area). In the latter case, the decrease in mass-based activity is proportional to the decrease in metal surface area. Ethane hydrogenolysis and ethane steam reforming are examples of structure-sensitive reactions, while CO hydrogenation on supported cobalt, nickel, iron, and ruthenium is structure-insensitive.

2.4. Gas/Vapor–Solid and Solid-State Reactions.   In addition to poisoning, there are a number of chemical routes leading to catalyst deactivation: (*1*) reactions of the vapor phase with the catalyst surface to produce (a) inactive bulk and surface phases (rather than strongly adsorbed species) or (b) volatile compounds that exit the catalyst and reactor in the vapor phase; (*2*) catalytic solid-support or catalytic solid-promoter reactions, and (*3*) solid-state transformations of the catalytic phases during reaction.

Gas/Vapor–Solid Reactions.   Reactions of Gas/Vapor with Solid to Produce Inactive Phases.   Dispersed metals, metal oxides, metal sulfides, and metal carbides are typical catalytic phases, the surfaces of which are similar in composition to the bulk phases. For a given reaction, one of these catalyst types is generally substantially more active than the others, eg, only Fe and Ru metals are active for ammonia synthesis, while the oxides, sulfides, and carbides are inactive. If, therefore, one of these metal catalysts is oxidized, sulfided, or carbided, it will lose essentially all of its activity. While these chemical modifications are closely related to poisoning, the distinction here is that rather than losing activity owing to the presence of an adsorbed species, the loss of activity is due to the formation of a new phase altogether.

Examples of vapor-induced chemical transformations of catalysts to inactive phases are listed in Table 7.

Reactions of Gas/Vapor with Solid to Produce Volatile Compounds.   Metal loss through direct vaporization is generally an insignificant route to catalyst deactivation. By contrast, metal loss through formation of volatile compounds, eg, metal carbonyls, oxides, sulfides, and halides in CO, O_2, H_2S, and halogen-containing environments, can be significant over a wide range of

Table 7. **Examples of Reactions of Gases/Vapors with Catalytic Solids to Produce Inactive Phases**

Catalytic process	Gas/vapor composition	Catalytic solid	Deactivating chemical reaction	Ref.
auto emissions control	N_2, O_2, HCs, CO, NO, H_2O, SO_2	Pt–Rh/Al_2O_3	$2\ Rh_2O_3 + \gamma\text{-}Al_2O_3 \rightarrow RhAl_2O_4 + 0.5\ O_2$	100,101
ammonia synthesis and regeneration	H_2, N_2	Fe/K/Al_2O_3	$Fe \rightarrow FeO$ at >50 ppm O_2	8
catalytic cracking	Traces O_2, H_2O HCs, H_2, H_2O	La-Y-zeolite	$Fe \rightarrow FeO$ at >0.16 ppm H_2O/H_2 H_2O induced Al migration from zeolite framework causing zeolite destruction	8
CO oxidation, gas turbine exhaust	N_2, O_2, 400 ppm CO, 100–400 ppm SO_2	Pt/Al_2O_3	$2\ SO_3 + \gamma\text{-}Al_2O_3 \rightarrow Al_2(SO_4)_3$ which blocks catalyst pores	8
diesel HC/soot emissions control	N_2, O_2, HCs (gas and liquid), CO, NO, H_2O, soot, SO_2	Pt/Al_2O_3 and β-zeolite; oxides of CaCuFeVK on TiO_2	formation of $Al_2(SO_4)_3$ or sulfates of Ca, Cu, Fe, or V which block catalysts pores and lower activity for oxidation; Al_2O_3 stabilized by BaO	102–104
Fischer–Tropsch	CO, H_2, H_2O, CO_2, HCs	Fe/K/Cu/SiO_2	$Fe_5C_2 \rightarrow Fe_3O_4$ due to oxidation at high X_{CO} by-product H_2O, CO_2	105
Fischer–Tropsch	CO, H_2, H_2O, HCs	Co/SiO_2	$Co + SiO_2 \rightarrow CoO \cdot SiO_2$ and collapse of SiO_2 by-product H_2O	106
selective catalytic reduction (SCR), stationary	N_2, O_2, NO, PM,[a] H_2O, SO_2	$V_2O_5/WO_3/TiO_2$	formation of $Al_2(SO_4)_3$ if Al_2O_3 is used	107
steam reforming and regeneration in H_2O	CH_4, H_2O, CO, H_2, CO_2	Ni/Al_2O_3	$Ni + Al_2O_3 \rightarrow Ni_2Al_2O_4$	8

[a]Particulate matter.

275

Table 8. **Types and Examples of Volatile Compounds Formed in Catalytic Reactions**

Gaseous environment	Compound type	Example of compound
CO, NO	carbonyls and nitrosyl carbonyls	$Ni(CO)_4$, $Fe(CO)_5$ ($0-300°C$)[a]
O_2	oxides	RuO_3 ($25°C$), PbO ($> 850°C$), PtO_2 ($>700°C$)
H_2S	sulfides	MoS_2 ($>550°C$)
halogens	halides	$PdBr_2$, $PtCl_4$, PtF_6, $CuCl_2$, Cu_2Cl_2

[a]Temperatures of vapor formation are listed in parentheses.

conditions, including relatively mild conditions. Classes and examples of volatile compounds are listed in Table 8.

While the chemical properties of volatile metal carbonyls, oxides, and halides are well known, there is surprisingly little information available on their rates of formation during catalytic reactions. There have been no reviews on this subject and relatively few reported studies to define the effects of metal loss on catalytic activity (27,108–121). Most of the previous work has focused on volatilization of Ru in automotive converters (108–111); nickel carbonyl formation in nickel catalysts during methanation of CO (113,119) or during CO chemisorption at 25°C (27,115); formation of Ru carbonyls during Fischer–Tropsch synthesis (116,117); volatilization of Pt during ammonia oxidation on Pt–Rh gauze catalysts (120,121); and volatilization of Cu from methanol synthesis and diesel soot oxidation catalysts, leading to sintering in the former and better catalyst–soot contact but also metal loss in the latter case (92).

Results of selected studies are summarized in Table 9.

Loss of nickel metal during CO chemisorption on nickel catalysts at temperatures above 0°C is also a serious problem; moreover, this loss is catalyzed by sulfur poisoning (27). In view of the toxicity of nickel tetracarbonyl, the rapid loss of nickel metal, and the ill-defined adsorption stoichiometries, researchers are advised to avoid using CO chemisorption for measuring nickel surface areas; instead, hydrogen chemisorption, an accepted ASTM method with a well-defined adsorption stoichiometry, is recommended (124).

Decomposition of volatile platinum oxide species formed during high temperature reaction may (125–127) lead to formation of large Pt crystallites and/ or substantial restructuring of the metal surface. For example, Wu and Phillips (125–127) observed surface etching, enhanced sintering, and dramatic surface restructuring of Pt thin films to faceted particles during ethylene oxidation over a relatively narrow temperature range (500–700°C). The substantially higher rate of sintering and restructuring in O_2/C_2H_4 relative to that in nonreactive atmospheres was attributed to the interaction of free radicals such as HO_2, formed homogeneously in the gas phase, with the metal surface to form metastable mobile intermediates. Etching of Pt–Rh gauze in a H_2/O_2 mixture under the same conditions as Pt surfaces (600°C, $N_2/O_2/H_2 = 90/7.5/2.5$) has also been reported (123). A significant weight loss was observed in a laminar flow reactor with little change in surface roughness, while in an impinging jet reactor, there was little weight loss, but substantial restructuring of the surface to particle-like structures, 1–10 μm in diameter; these particles were found to have the same

Table 9. Documented Examples of Reactions of Vapor with Solid to Produce Volatile Compounds

Catalytic process	Catalytic solid	Vapor formed	Comments on deactivation process	Ref.
automotive converter	$Pd–Ru/Al_2O_3$	RuO_4	50% loss of Ru during 100-h test in reducing automotive exhaust	111
methanation of CO	Ni/Al_2O_3	$Ni(CO)_4$	$P_{CO} > 20$ kPa and $T < 425°C$ due to $Ni(CO)_4$ formation, diffusion and decomposition on the support as large crystallites	113
CO chemisorption	Ni catalysts	$Ni(CO)_4$	$P_{CO} > 0.4$ kPa and $T > 0°C$ due to $Ni(CO)_4$ formation; catalyzed by sulfur compounds	114
Fischer–Tropsch synthesis (FTS)	Ru/NaY zeolite Ru/Al_2O_3, Ru/TiO_2	$Ru(CO)_5$, $Ru_3(CO)_{12}$	loss of Ru during FTS ($H_2/$ $CO = 1$, 200–250°C, 1 atm) on Ru/NaY zeolite and Ru/ Al_2O_3; up to 40% loss while flowing CO at 175–275°C over Ru/Al_2O_3; for 24 h, rate of Ru loss less on titania-supported Ru and for catalysts containing large metal crystallites (3 nm) relative to small metal crystallites (1.3 nm); surface carbon lowers loss	116,117
ammonia oxidation	Pt–Rh gauze	PtO_2	loss: 0.05–0.3 g Pt/ton HNO_3; recovered with Pd gauze; loss of Pt leads to surface enrichment with inactive Rh	8,120,122
HCN synthesis	Pt–Rh gauze	PtO_2	extensive restructuring and loss of mechanical strength	8,123
methanol synthesis	CuZnO	$CuCl_2$, Cu_2Cl_2	mobile copper chloride phase leads to sintering at reaction temperature (225°C)	92
diesel soot oxidation	oxides of K, Cu, Mo, and trace Cl	$CuCl_2$, Cu_2Cl_2	mobile copper chloride improves catalyst–soot contact; catalyst evaporation observed	92

Pt–Rh composition as the original gauze. The nodular structures of about 10-μm diameter formed in these experiments are strikingly similar to those observed on Pt–Rh gauze after use in production of HCN at 1100°C in 15% NH_3, 13% CH_4, and 72% air. Moreover, because of the high space velocities during HCN production, turbulent rather than laminar flow would be expected as in the impinging jet reactor. While little Pt is volatilized from the Pt–Rh gauze catalyst during HCN synthesis, the extensive restructuring leads to mechanical weakening of the gauze (8).

Other examples of catalyst deactivation due to volatile compound formation include (*1*) loss of the phosphorus promoter from the VPO catalyst used in the fluidized-bed production of maleic anhydride with an attendant loss of catalyst selectivity (8), (*2*) vapor-phase loss of the potassium promoter from steam-reforming catalysts in the high temperature, steam-containing environment (8), and (*3*) loss of Mo from a 12-Mo-V-heteropolyacid due to formation of a volatile Mo species during oxydehydrogenation of isobutyric acid to methacrylic acid (118).

While relatively few definitive studies of deactivation by volatile compound formation have been reported, the previous work does provide the basis for enumerating some general principles. A generalized mechanism of deactivation by formation of volatile metal compounds can be postulated (see Fig. 7). In addition, the roles of kinetics and thermodynamics can be stated in general terms:

1. At low temperatures and partial pressures of the volatilization agent (VA), the overall rate of the process is limited by the rate of volatile compound formation.
2. At intermediate temperatures and partial pressures of the VA, the rate of formation of the volatile compound exceeds the rate of decomposition. Thus, the rate of vaporization is high, the vapor is stable, and metal loss is high.
3. At high temperatures and partial pressures of the VA, the rate of formation equals the rate of decomposition, ie, equilibrium is achieved. However, the volatile compound may be too unstable to form or may decompose before there is an opportunity to be transported from the system. From the previous work, it is also evident that besides temperature and gas phase composition, catalyst properties (crystallite size and support) can play an important role in determining the rate of metal loss.

Solid-State Reactions.   Catalyst deactivation by solid-state diffusion and reaction appears to be an important mechanism for degradation of complex multi-

Generalized Mechanism:

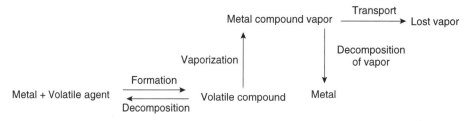

Generalized Kinetics:

(a) rate of volatile compound formation = rate of formation − rate of decomposition

(b) rate of metal loss = rate of vaporizaion − rate of vapor decomposition

Fig. 7.   Generalized mechanisms and kinetics for deactivation by metal loss (8).

component catalysts in dehydrogenation, synthesis, partial oxidation, and total oxidation reactions (8,128–139). However, it is difficult in most of these reactions to know the extent to which the solid-state processes such as diffusion and solid-state reaction are affected by surface reactions. For example, the rate of diffusion of Al_2O_3 to the surface to form an aluminate may be enhanced by the presence of gas-phase oxygen or water or the nucleation of a different phase may be induced by either reducing or oxidizing conditions. Recognizing this inherent limitation, the focus here is nevertheless on processes in which formation of a new bulk phase (and presumably the attendant surface phase) leads to substantially lower activity. There is probably some overlap with some of the examples given under Gas/Vapor–Solid Reactions involving reactions of gas/vapor with solid to produce inactive phases.

Examples from the literature of solid-state transformations leading to catalyst deactivation are summarized in Table 10.

There are basic principles underlying most solid-state reactions in working catalysts that have been enumerated by Delmon (135): (1) the active catalytic phase is generally a high-surface-area defect structure of high surface energy and as such a precursor to more stable, but less active phases and (2) the basic reaction processes may themselves trigger the solid-state conversion of the active

Table 10. Examples of Solid-State Transformations Leading to Catalyst Deactivation

Catalytic process	Catalytic solid	Deactivating chemical reaction	Ref.
ammonia synthesis	$Fe/K/Al_2O_3$	formation of $KAlO_2$ at catalyst surface	138
catalytic combustion	PdO/Al_2O_3, PdO/ZrO_2	$PdO \rightarrow Pd$ at $T > 800°C$	131
catalytic combustion	Co/K on MgO, CeO_2, or La_2O_3	formation of $CoO–MgO$ solid soln., $LaCoO_3$, or K_2O film on CeO_2	139
dehydrogenation of styrene to ethyl benzene	$Fe_2O_3/Cr_2O_3/K_2O$	K migration to center of pellet caused by thermal gradient	8
Fischer–Tropsch	Fe/K, Fe/K/CuO	transformation of active carbides to inactive carbides	136,137
oxidation of SO_2 to SO_3	$V_2O_5/K_2O/Na_2O/$ kieselguhr	formation of inactive V(IV) compounds at $T < 420–430°C$	134
partial oxidation of benzene to maleic anhydride	$V_2O_5–MoO_3$	decreased selectivity due to loss of MoO_3 and formation of inactive vanadium compounds	128
partial oxidation of methanol to formaldehyde	$Fe_2(MoO_4)_3$ plus MoO_3	structural reorganization to β-$FeMoO_4$; reduction of MoO_3	129,135
partial oxidation of propene to acrolein	$Fe_2(MoO_4)_3$	reductive transformation of $Mo_{18}O_{52}$ to Mo_4O_{11}	132,135
partial oxidation of isobutene to methacrolein	$Fe_2(MoO_4)_3$	reduction to $FeMoO_4$ and MoO_{3-x}	130,133

phase to an inactive phase; for example, it may involve a redox process, part of which nucleates the inactive phase.

A well-documented example of these principles occurs in the partial oxidation of propene to acrolein on a $Fe_2(MoO_4)_3$ catalyst (132,135). This oxidation occurs by the *"Mars van Krevelen" mechanism*, ie, a redox mechanism in which lattice oxygen reacts with the adsorbed hydrocarbon to produce the partially oxygenated product; the reduced catalyst is restored to its oxidized state through reaction with gaseous oxygen. In propene oxidation, two atoms of oxygen from the catalyst are used, one for removing two hydrogen atoms from the olefin and the other one in forming the unsaturated aldehyde. The fresh, calcined catalyst MoO_3 consists of corner-sharing MoO_6 octahedra (with Mo at the center and six oxygen atoms at the corners), but upon reduction to MoO_2, octahedra share edges. However, it has been reported (132,135) that only slightly reduced (relative to MoO_3), open structures such as $Mo_{18}O_{52}$ and Mo_8O_{23} are the most active, selective phases; more complete reduction of either of these structures leads to formation of Mo_4O_{11} having substantially lower selectivity. Delmon and co-workers (133,135) have shown that addition of an oxygen donor such as Sb_2O_4 facilitates spillover of oxygen and thereby prevents overreduction and deactivation of the catalyst.

2.5. Mechanical Failure of Catalysts. *Forms and Mechanisms of Failure.* Mechanical failure of catalysts is observed in several different forms, including (*1*) crushing of granular, pellet, or monolithic catalyst forms due to a load; (*2*) attrition, the size reduction, and/or breakup of catalyst granules or pellets to produce fines, especially in fluid or slurry beds; and (*3*) erosion of catalyst particles or monolith coatings at high fluid velocities. Attrition is evident by a reduction in the particle size or a rounding or smoothing of the catalyst particle easily observed under an optical or electron microscope. Washcoat loss is observed by scanning a cross section of the honeycomb channel with either an optical or an electron microscope. Large increases in pressure drop in a catalytic process are often indicative of fouling, masking, or the fracturing and accumulation of attritted catalyst in the reactor bed.

Commercial catalysts are vulnerable to mechanical failure in large part because of the manner in which they are formed; that is, catalyst granules, spheres, extrudates, and pellets ranging in diameter from 50 μm to several millimeters are in general prepared by agglomeration of 0.02–2 μm aggregates of much smaller primary particles having diameters of 10–100 nm by means of precipitation or gel formation followed by spray drying, extrusion, or compaction. These agglomerates have in general considerably lower strengths than the primary particles and aggregates of particles from which they are formed.

Two principal mechanisms are involved in mechanical failure of catalyst agglomerates: (*1*) fracture of agglomerates into smaller agglomerates of approximately $0.2d_0$–$0.8d_0$ and (*2*) erosion (or abrasion) of aggregates of primary particles having diameters ranging from 0.1 to 10 μm from the surface of the agglomerate (140). While erosion is caused by mechanical stresses, fracture may be due to mechanical, thermal, and/or chemical stresses. Mechanical stresses leading to fracture or erosion in fluidized or slurry beds may result from (*1*) collisions of particles with each other or with reactor walls or (*2*) shear forces created by turbulent eddies or collapsing bubbles (cavitation) at high fluid velocities. Thermal stresses occur as catalyst particles are heated and/or cooled rapidly; they are magnified by temperature gradients across particles and by differences in thermal expansion coefficients at the interface of two different materials, eg, catalyst coating/monolith interfaces; in the latter case the heating or cooling process can lead to fracture and separation of the catalyst coating. Chemical stresses occur as phases of different density are formed within a catalyst particle via chemical reaction; for example, carbiding of primary iron oxide particles increases their specific volume and micromorphology leading to stresses that break up these particles (141).

Role of Properties of Ceramic Agglomerates in Determining Strength and Attrition Resistance. Factors Affecting the Magnitude of Stress Required for Agglomerate Breakage and the Mechanisms by Which it Occurs. The extent to which a mechanism, ie, fracture or erosion, participates in agglomerate size reduction depends upon several factors: (*1*) the magnitude of a stress, (*2*) the strength and fracture toughness of the agglomerate, (*3*) agglomerate size and surface area, and (*4*) crack size and radius. Erosion (abrasion) occurs when the stress (eg, force per area due to collision or cavitation pressure) exceeds the agglomerate strength, ie, the strength of bonding between primary particles. Erosion rate is reportedly (140) proportional to the external surface area of the catalyst; thus, erosion rate increases with decreasing agglomerate size.

Most heterogeneous catalysts are complex, multiphase materials that consist in large part of porous ceramic materials, ie, are typically oxides, sulfides, or metals on an oxide carrier or support. When a tensile stress of a magnitude close to the yield point is applied, ceramics almost always undergo brittle fracture before plastic deformation can occur. Brittle fracture occurs through formation and propagation of cracks through the cross section of a material in a direction perpendicular to the applied stress. Agglomerate fracture due to a tensile stress occurs by propagation of internal and surface flaws; these flaws created by external stresses or inherent defects are stress multipliers, ie, the stress is multiplied by $2(a/r)^{0.5}$, where a is the crack length and r is the radius of curvature of the crack tip; since a/r can vary from 2 to 1000, the effective stress at the tip of a crack can be 4–60 times the applied stress. Tensile stress multipliers may be microcracks, internal pores, and grain corners.

The ability of a material to resist fracture is termed fracture toughness. The plain strain fracture toughness K_{Ic} is defined as

$$K_{Ic} = Y\sigma(\pi a)^{0.5} \tag{3}$$

where Y is a dimensionless parameter (often close to 1.0–2.0), the magnitude of which depends upon both specimen and crack geometries, σ is the applied stress, and a is the length of a surface crack or half the length of an internal crack. Crack propagation and fracture are likely if the right hand side of equation 3 exceeds the experimental value of plain strain fracture toughness (left-hand side of eq. 3). Plane strain fracture toughness values for ceramic materials are significantly smaller than for metals and typically below 10 MPa(m)$^{0.5}$; reported values for nonporous, crystalline alumina (99.9%), fused silica, and zirconia (3 mol% Y_2O_3) are 4–6, 0.8, and 7–12 MPa(m)$^{0.5}$ respectively; flexural strengths (analogous to yield strengths for metals) for the same materials are 280–550, 100, and 800–1500 MPa (142). Thus, on the basis of both fracture toughness and flexural strength, nonporous, crystalline zirconia is much stronger toward fracture than alumina, which in turn is much stronger than fused silica.

The introduction of porosity to crystalline or polycrystalline ceramic materials will, on the basis of stress amplification, significantly decrease elastic modulus and flexural strength for materials in tension.

Thus far the discussion has focused mainly on tensile strength, the extent of which is greatly reduced by the presence of cracks or pores. However, for ceramic materials in compression, there is no stress amplification due to flaws or pores; thus ceramic materials (including catalytic materials) in compression are much stronger (approximately a factor of 10) than in tension. In addition, the strength of ceramic materials can be dramatically enhanced by imposing a residual compressive stress at the surface through thermal or chemical tempering. Moreover, introduction of binders such as graphite enables agglomerates of ceramic powders to undergo significant plastic deformation before fracture.

Tensile Strengths and Attrition Resistance of Catalyst Supports and Catalysts.   The strengths cited above for nonporous, annealed crystalline or polycrystalline materials do not necessarily apply to porous catalyst agglomerates even under compression; rather, agglomerate strength is dependent upon the strengths of chemical and physical bonds including the cohesive energy between primary particles. Agglomerate strength would depend greatly on the preparation of the compact. Representative data for catalyst agglomerates (see Table 11) suggest they are generally substantially weaker than polycrystalline ceramic materials prepared by high temperature sintering, such as alumina (140,142,144–148).

From the data in Table 11 it is evident that even subtle differences in preparation and pretreatment also affect agglomerate strength. For example, spheres of γ-Al$_2$O$_3$ prepared by sol–gel granulation are substantially (17 times) stronger than commercial γ-Al$_2$O$_3$ spheres (143). Moreover, 30- and 90-μm diameter particles of TiO$_2$ prepared by thermal hydrolysis or basic precipitation are 30 and 15 times stronger than commercially available 4-mm extrudates (146).

Catalyst attrition is a difficult problem in the operation of moving-bed, slurry-bed, or fluidized-bed reactors. Generally, stronger materials have greater attrition resistance; this conclusion is supported by representative data in Table 11 for γ-Al$_2$O$_3$, showing that the strength of the alumina prepared by sol–gel granulation is 17 times higher, while its attrition rate is 5 times lower.

The mechanism by which attrition occurs (erosion or fracture) can vary with catalyst or support preparation, crush strength, and with reactor environment; it can also vary with the mechanical test method. There is some evidence in the attrition literature supporting the hypothesis that in the presence of a large stress, weaker oxide materials are prone to failure by fracture, while stronger materials tend to erode (149). However, there is also contrary evidence (145), showing that fracture may be the preferred mechanism for strong TiO$_2$ agglomerates, while abrasion is favored for weaker agglomerates. Supporting a third trend, data (140) show that attrition mechanism and rate are independent of agglomerate strength but depend instead on the type of material. 100-μm-diameter agglomerates of precipitated Fe/Cu/K Fischer–Tropsch catalyst

Table 11. **Mechanical Strengths and Attrition Rates of Catalyst Supports Compared to Those of Sintered Ceramic Agglomerates**

Catalyst support or ceramic	Preparation/pretreatment/ properties	Strength, MPa	Attrition index, wt%/h	Ref.
	High surface area catalyst supports			
γ-Al$_2$O$_3$, 1.2– 4.25-mm spheres	sol–gel granulation/dried 10 h at 40°C, calcined 3 h at 450°C/ 389 m^2/g, d_{pore} = 3.5 nm	11.6 ± 1.9	0.033	144
γ-Al$_2$O$_3$, 4.25-mm spheres	Alcoa LD-350	0.7	0.177	144
g-Al$_2$O$_3$, 100 μm	VISTA-B-965-500C	6.2 ± 1.3		140
TiO$_2$ (anatase), 30 μm	thermal hydrolysis/dried 110°C, calcined 2 h at 500°C/92 m^2/g, <10-nm primary crystallites	28[a]		145
TiO$_2$ (anatase), 90 μm	basic precipitation/dried 110°C, calcined 2 h at 500°C/81 m^2/g, 10–14-nm primary crystallites	15[a]		145
TiO$_2$ (75% anatase, 25% rutile)	Degussa P25, fumed/4-mm extrudates/48 m^2/g, V_{pore} = 0.34 cm^3/g, d_{pore} = 21 nm	0.9		146
TiO$_2$ (anatase)	Rhone-Poulenc DT51, ppt./4 mm extrudates/92 m^2/g, V_{pore} = 0.40 cm^3/g, d_{pore} = 8, 65 nm	0.9		146
	Low surface area ceramics			
Al$_2$O$_3$	spray dried with organic binder; plastic deformation observed	2.3		147
Al$_2$O$_3$	heat treated (sintered), 99.9%	282–551		142
TiO$_2$ (Rutile)	partially sintered	194		147
ZrO$_2$ (yttria additive)	commercial samples from three companies, spray-dried	0.035–0.43		148
ZrO$_2$ (3% Y$_2$O$_3$)	heat treated (sintered)	800–1500		142

[a]Rough estimates from break points on relative density versus log[applied pressure] curves; data are consistent with mass distribution versus pressure curves from ultrasonic tests.

[prepared by United Catalyst (UCI)] and having nearly the same strength shown in Table 11 for Vista-B Al$_2$O$_3$ (6.3 vs. 6.2 MPa), were found to undergo substantial fracture to 5–30-μm fragments (an increase from 45 to 85%) as well as substantial erosion to 1 μm or less fragments (increase from 2 to 50%). Under the same treatment conditions, 90-μm-diameter agglomerates of Vista-B Al$_2$O$_3$ underwent by comparison much less attrition, mainly by erosion (20% increase in 0.1–5-μm fragments). The very low attrition resistance of the Fe/ Cu/K UCI catalyst is further emphasized by the unsatisfactory outcome of a test of this catalyst by the U.S. Department of Energy (DOE) in a pilot-scale slurry-phase bubble-column reactor in LaPorte, Tex.; following one day of operation, the filter system was plugged with catalyst fines, preventing catalyst–wax separation and forcing shutdown of the plant (150).

Thus, based on these three representative examples, it follows that which of the two attrition mechanisms predominates depends much more on material composition and type than on agglomerate strength. However, irrespective of mechanism the rate of attrition is usually greater for the weaker material.

The catalyst preparation method can have a large effect on the attrition resistance of an Fe/Cu Fischer–Tropsch catalyst (151). This catalyst, prepared by precipitation, undergoes severe attrition during a 25-min treatment with ultrasonic radiation; indeed the mass fraction finer than 0.1–5 μm increases from 0 to 65%. However, after a spray drying treatment of the same catalyst, less than a 10% increase in the same fractions is evident.

In their review of attrition and attrition test methods, Bemrose and Bridgewater (152) discuss how attrition varies with reactor type, eg, involves mainly particle–wall impacts in moving pellet bed reactors and particle–particle impacts in fluidized-bed reactors of high fluid velocity. In fact, jet attrition of catalyst particles in a gas fluidized bed involving principally abrasion due to collision of high-velocity particles has been modeled in some detail (149,153). Thus, given such important differences in attrition mechanism, realistic attrition test methods should attempt to model reactor operation as closely as possible. In addition, the ideal test would require only a small catalyst sample, a simple, inexpensive apparatus, and a few minutes to complete the test. Relatively quick, inexpensive single-particle crushing tests have been devised (152); however, properties of a single particle are rarely representative of those for the bed; moreover, it is difficult to relate the results of this crushing test to the actual abrasion process. Realistic tests have been devised for two reactor types involving a moving catalyst, ie, an air-jet test for fluidized-bed catalysts (154,155), and a rotating drum apparatus for moving-bed catalysts (156); however, the air-jet test requires a large quantity (eg, 50 g) of catalyst, an expensive apparatus, and about 20 h to run. In the past decade a new jet-cup test has been developed for testing of fluidized-bed catalysts (154,155), which requires only a 5-g sample and about 1 h to complete; comparisons of results for the jet-cup and air-jet tests indicate that the two tests give comparable results (154,155). Nevertheless, the mechanisms for the two tests are different, ie, the air-jet (fluid-bed) test is abrasion- (erosion-) dominant, while the jet-cup test includes both abrasion and fracture mechanisms (155). A 30-min, 10-g ultrasonic attrition test based on cavitation has also been developed in the past decade (145,151,157); while it likewise involves both abrasion and fracture mechanisms, the results appear to correlate with other methods. For example, particle size distributions for the same Co/silica catalyst after ultrasonic, jet-cup, and laboratory-scale, slurry-bed column reactor (SBCR) tests are very similar, indicating that both fracture and abrasion mechanisms operate in the small-scale SBCR. Moreover, the good agreement among the three methods suggests that both the jet-cup and ultrasonic tests may provide data representative of the attrition process in laboratory-scale SBCR reactors. It is evident that these two small-scale methods are especially useful for screening of a series of catalysts to determine relative strength.

Nevertheless, the more realistic large-scale tests are probably needed for accurately determining design attrition rates of a commercial catalyst to be used in a full-scale process. The observation that attrition of a fluid catalytic cracking (FCC) catalyst initially involves fracture of weak agglomerates followed by abrasion of strong agglomerates emphasizes the need to collect and analyze the particle size distribution of attrited fines as a function of time in order to define which mechanism (or mechanisms) operates at startup as well as in the steady-state process. Because the mechanism may be time dependent, rapid, small-scale tests may produce misleading results.

While realistic laboratory-scale tests have been developed for simulating attrition in large moving-bed and fluidized-bed reactors, no such laboratory test has been developed and demonstrated yet for simulation of large-scale SBCR reactors, although recent research has focused on the development of such tests. For example, in laboratory-scale, SBCR tests of supported cobalt catalysts over several days (157), it was observed that the attrition resistance decreases in the order Co/Al_2O_3, Co/SiO_2, Co/TiO_2 (especially the anatase form underwent attrition at a high rate); attrition resistance was observed to increase with increasing cobalt loading from 10 to 40 wt%.

Implications of Mechanistic Knowledge of Attrition for Catalyst Design. The understanding of mechanisms important in attrition of catalyst supports and catalysts, the relationship between strength and attrition rate for a given material, and test data can be used to great advantage in the design of attrition resistant catalysts. Several alternatives follow from the previous discussion for increasing attrition resistance: (*1*) increasing aggregate/agglomerate strength by means of advanced preparation methods, eg, sol–gel granulation, spray drying, and carefully controlled precipitation methods (see Table 11 for examples), (*2*) adding binders to improve strength and toughness, eg, the addition of a polyvinylpyrrolidone binder to agglomerates of quartz sand increases agglomerate strength from 0.1 to 3 MPa (158), (*3*) coating aggregates with a porous but very strong material such as ZrO_2, eg, embedding a fluidized-bed catalyst for partial oxidation of *n*-butane to maleic anhydride in a strong, amorphous matrix of zirconium hydrogen phosphate significantly improves its attrition resistance (159), and (*4*) chemical or thermal tempering of agglomerates to introduce compressive stresses that increase strength and attrition resistance, eg, heating and cooling particles rapidly by passing them through a low-residence-time, high-temperature furnace to harden the agglomerate exterior, while preventing significant sintering of or phase changes in the porous interior. The subject of preventing mechanical degradation and other forms of catalyst deactivation is addressed in greater detail under Prevention of Catalyst Decay.

2.6. Summary of Deactivation Mechanisms for Solid Catalysts. Causes of solid (heterogeneous) catalyst deactivation are basically threefold: (*1*) chemical, (*2*) mechanical, and (*3*) thermal. Mechanisms of heterogeneous catalyst deactivation can be classified into five general areas: (*1*) chemical degradation including volatilization and leaching, (*2*) fouling, (*3*) mechanical degradation, (*4*) poisoning, and (*5*) thermal degradation. Poisoning and thermal degradation are generally slow processes, while fouling and some forms of chemical and mechanical degradation can lead to rapid, catastrophic catalyst failure. Some forms of poisoning and many forms of fouling are reversible; hence, reversibly poisoned or fouled catalysts are relatively easily regenerated. On the other hand, chemical, mechanical, and thermal forms of catalyst degradation are rarely reversible.

It is often easier to prevent rather than cure catalyst deactivation. Many poisons and foulants can be removed from feeds using guard beds, scrubbers, and/or filters. Fouling, thermal degradation, and chemical degradation can be minimized through careful control of process conditions, eg, lowering temperature to lower sintering rate or adding steam, oxygen, or hydrogen to the feed to gasify carbon or coke-forming precursors. Mechanical degradation can be minimized by careful choice of carrier materials, coatings, and/or catalyst particle forming methods.

While treating or preventing catalyst deactivation is facilitated by an understanding of the mechanisms, additional perspectives are provided by examining the route by which each of the mechanisms causes loss of catalytic activity, ie, how it influences reaction rate (92). Thus, catalytic activity can be defined in terms of the observed site-based rate constant k_{obs}, which is equal to the product of the active site density σ (number of sites per area of surface), the site-based intrinsic rate constant k_{intr}, and the effectiveness factor η, ie,

$$k_{obs} = \sigma k_{intr} \eta \qquad (4)$$

Loss of catalytic activity may be due to a decrease in any of the three factors in equation 4, whose product leads to k_{obs}. Thus, catalyst deactivation can be caused by (1) a decrease in the site density σ, (2) a decrease in intrinsic activity (ie, decrease in k_{intr}), and/or (3) lowered access of reactants to active sites (decrease in η). Poisoning, for example, leads to a loss of active sites, ie, $\sigma = \sigma_0(1-\alpha)$, where α is the fraction of sites poisoned; sintering causes loss of active sites through crystallite growth and reduction of active surface area. Fouling can cause both loss of active sites due to blocking of surface sites as well as plugging of pores causing a decrease in the effectiveness η. Moreover, poisoning, as discussed earlier, can also lead to a decrease in intrinsic activity by influencing the electronic

Table 12. How Deactivation Mechanisms Affect the Rate of a Catalyzed Reaction and the Rapidity and Reversibility of Deactivation Process

	Effects on reaction rate		Deactivation process		
Deactivation mechanism	Decrease in number of active sites	Decrease in intrinsic activity (k_{intr})	Decrease in effectiveness factor (η)	Fast or slow[a]	Reversible
chemical degradation	×	×	×[b,c]	varies	no
fouling	×		×	fast	yes
mechanical degradation	×			varies	no
poisoning	×	×		slow	usually
sintering	×	×[b,d]	×[b,e]	slow	sometimes
vaporization/ leaching	×		×[b,f]	fast	sometimes

[a]Generally.

[b]In some cases.

[c]Chemical degradation can cause breakdown of support, pore plugging, and loss of porosity.

[d]If the reaction is structure-sensitive, sintering could either increase or decrease intrinsic activity.

[e]Sintering of the support may cause support collapse and loss of porosity; it may also increase average pore diameter.

[f]Leaching of aluminum or other cations from zeolites can cause buildup of aluminum or other oxides in zeolite pores.

structure of neighboring atoms. Thus, each of the deactivation mechanisms affects one or more of the factors comprising observed activity (see Table 12); all of the mechanisms, however, can effect a decrease in the number of catalytic sites.

3. Homogeneous Catalysts and Enzymes

3.1. Homogeneous Catalysts.
The discussion of the deactivation of homogeneous catalysts has received less attention relative to that of heterogeneous catalysts (160,161). Indeed, the first comprehensive review of homogeneous catalyst deactivation appeared just recently (160). Nevertheless, the vast literature of homogeneous catalysis provides numerous anecdotal accounts of problems with catalyst decomposition and references to homogeneous catalysts having a limited number of turnovers, all testifying to the importance of these phenomena.

Homogeneous catalysts may undergo degradation by routes similar to those of heterogeneous catalysts, eg, by chemical modification, poisoning, and thermal degradation. However, the specific details of these mechanistic routes are generally somewhat different, since the catalyst is a molecule rather than a solid; that is, an organometallic complex is quite different from a metal surface in terms of structure and scale. For example, reaction of impurities with homogeneous catalytic complexes is analogous to poisoning of a heterogeneous catalyst by impurities, although the former is essentially a chemical reaction of two species of similar dimensions while the latter involves adsorption of a molecule on the surface of a crystallite containing hundreds to thousands of atoms.

Homogeneous catalysts are generally metal–ligand complexes. The metal center functions as the active site, while the ligands serve to influence site chemistry through electronic modifications of the metal that influence activity/selectivity and through geometric constraints that enhance selectivity. Hence activity and selectivity properties of homogeneous catalysts can be significantly influenced by processes that change the chemistry either of the metal center or the ligands or both.

Mechanisms (or causes) of homogeneous catalyst degradation can be classified as (1) metal deposition reactions, eg, decarbonylation of carbonyl complexes, loss of protons from cationic species, or reductive elimination of C-, N-, or O-donor fragments; (2) decomposition of ligands attached to a catalytic complex; (3) reactions of metal–carbon and metal–hydride bonds with polar species (eg, water, oxygen, acids, alcohols, olefins, and halides); and (4) poisoning of active sites by impurities, reactants, or products or by dimerization of the catalyst.

Principal features of these mechanisms and examples thereof are summarized in Table 13. It is noteworthy that mechanisms 1 and 2 lead mainly to deactivation by either loss or modification of ligands, while mechanisms 3 and 4 cause deactivation largely by either modifying or poisoning the metal, although ligands are also clearly modified by type 3 mechanisms. Of the four mechanisms, deactivation by metal formation and deposition is the most common, although all are important.

Table 13. **Deactivation Mechanisms for Homogeneous Catalysts**[a]

Deactivation mechanism	Comments	Examples
1. *Metals deposition reactions*	most common decomposition mechanism	
ligand loss	decarbonylation is most common	$2\,HCo(CO)_4 \rightarrow CO_2(CO)_8 \rightarrow CO_4(CO)_{12} \rightarrow Co$ metal
loss of protons from cationic species	reductive elimination as HX; pH dependent; basic media lead to zero-valent metals	in Wacker reaction elimination of HCl from PdHCl leads to Pd(0); in Heck reaction Pd(0) is an intermediate
reductive elimination of C-, N-, O-donor fragments	occurs in cross-coupling reactions to form C-C, C-N, C-O bonds	oxidative addition of aryl halides
2. *Ligand decompositions*	ligands greatly influence activity and selectivity of homogeneous catalysts	
oxidation	phosphorus or sulfur-based ligands are readily oxidized by O_2, H_2O, CO_2, peroxides; nitrogen-based ligands are more stable	$PR_3 + H_2O \rightarrow H_2 + O$ "PR_3; $PR_3 + 1/2\,O_2 \rightarrow O$ "PR_3
oxidative addition	breaking of C-P bond with insertion of a metal	decomposition of Rh and Co hydroformylation catalysts (161)
nucleophilic attack	internal or external attack causing insertion of metal in C-P bond or displacement of metal with Ar	decomposition of triphenyl-phosphines (TPPs) in Pd catalyst by acetate ion
thermal decomposition	depends on temperature and gas composition	decomposition of $RhH(CO)(PPh_3)$ hydrofor-mylation catalysts to stable cluster containing μ_2-PPh_2 fragments in absence of H_2 and CO
reactions with water alcohols; rearran-gements	hydrolysis, alcoholysis, and transesterification of phos-phites, imines, and pyridines	hydrolysis of diphosphites in Rh-catalyzed hydroformyl-ation of alkenes
3. *Reactions of metal–carbon and metal–hydride bonds*		
with water, oxygen, acids, and alcohols	decomposition of reactive metal alkyls with water or oxygen	deactivation of Ziegler catalysts containing alkyl complexes of Ti, Zr, and V
with olefins, halides, and aluminoxanes	formation of metallated transi-tion metal ion complex deacti-vates polymerization catalysts	reaction of propene with zirconium alkyl catalyst forms an alkane and a π-allyl zirconium species
4. *Poisoning of active sites*		
by dienes and alkynes	forms a stable π-allylic complex; these poisons must be removed in polyolefin manufacture	methoxycarbonylation of pro-pyne using Pd-2-pyridyl-DPP is poisoned by buta-diene and 1,2-propadiene

Table 13 (*Continued*)

Deactivation mechanism	Comments	Examples
by polar impurities	such as basic amines	enantioselective isomerization of the allylamine to the asymmetric enamine (in menthol synthesis) is poisoned by a stronger basic amine isomer
by dimer formation	active monomeric catalyst species form dimers	Rh–TPP hydroformylation catalyst dimerizes; Pd(I) dimers in carbonylation catalysts

^{a}Data from 160.

Mechanisms 1 and 4 are reversible to some extent. Mechanisms 2 and 3, involving breaking of active site bonds and formation of stable products, are largely irreversible. Products of ligand oxidation are generally more stable than the complexes from which they were formed.

3.2. Enzymes. *Structural and Catalytic Properties of Enzymes.* Enzymes are globular macromolecular polypeptide proteins (molecular weights of $10^4 - 10^6$) synthesized by living organisms (8,162). Each enzyme has a unique three-dimensional structure with a binding site or pocket that is chemically and geometrically compatible with a single reactant molecule (substrate) or group of chemically related reactants; in other words enzymes have molecular-recognition capability. Enzymes are unique in their ability to catalyze biochemical reactions with high selectivity (essentially 100%) at extraordinarily high rates, ie, 10–10,000 molecules/(enzymes) compared to typical values of 1–10 or less for conventional catalysts. These activities enable enzymes to be effective catalysts at extremely low concentrations, eg, $10^{-5} - 10^{-10}$ mol/L, at substrate (reactant) concentrations of greater than 10^{-6} mol/L.

The high activity of enzymes has been illustrated for urease and catalase (163,164). The stereochemical specificity of enzymes is unmatched and absolute, ie, their sites can distinguish between optical and geometrical isomers, almost always catalyzing only the reaction of one isomer of an enantiomeric pair (162). Nevertheless, some enzymes catalyze reactions of chemically unrelated species; for example, nitrogenase reduces N_2 to NH_3 as well as hydrogenating acetylene to ethylene (165).

In 1976 there were 1800 known enzymes, and new enzymes were being discovered at the rate of about 60 per year (166); accordingly, there were an estimated 3000 known enzymes in 1996. It is estimated that an average cell contains 3000 different enzymes (165), and it is speculated that as many as 25,000 different enzymes exist (167).

While they are synthesized *in vitro* and are only active within a limited range of pH and temperature, enzymes otherwise have properties similar to synthetic homogeneous and polymer-supported catalysts. Moreover, they can be extracted from their biological source, purified, crystallized, and used in laboratory studies or industrial processes. Further, they can be attached to glass or ceramic supports and used as heterogeneous catalysts. And their application in industrial processes is rapidly increasing.

Enzymes are formed in living systems by condensation/dehydration of amino acids to produce peptide (C-N) bonds that constitute the backbone of long protein chains. The active conformation of an enzyme is produced by folding of the protein chain into secondary (helical), tertiary (folded), and quaternary (combined tertiary) structures. The folded layers are held in place by hydrogen bonding and disulfide linkages. There are 20 naturally occurring amino acids, each having the composition H_2N-CHR-COOH, the R group (side chain) having different molecular functions, eg, proton donation, proton removal, and bridge formation; for example, amino acids include glycine, alanine, and serine with side chains of H, CH_3, and CH_2OH respectively. Upon folding, side chains become the functional groups of the active site or ligands for binding of metals ions, which then become functional groups at the active site. Naturally occurring metal ions in enzymes include Mg^{2+}, Zn^{2+}, Ca^{2+}, Ni^{2+}, Fe^{2+}, Fe^{3+}, Co^{3+}, and Mo^{2+}.

Distinctive catalytic characteristics of enzymes (162) include (1) their flexible structure, which facilitates an "induced fit" of the substrate, the making and breaking of bonds, and the departure of products, and (2) their sensitivity to reaction *effectors* (inhibitors or activators), which function similarly to promoters of heterogeneous catalysts. Some enzymes require a *cofactor* that combines with the enzyme to form a catalytic site; metal ions are examples of cofactors. Enzymatic reactions may also require a *coenzyme* that reacts with the reactant to produce an enzyme-compatible substrate. Living organisms control and optimize biological processes using a variety of tools: (1) enzyme effectors, (2) regulation of enzyme growth or activation rates, (3) compartmentalization of enzymes within organs or organelles, and (4) destruction (editing) of undesired intermediates or products (162).

Deactivation of Enzymes. Enzymes generally function only under mild conditions of temperature and pH observed in living organisms. Under typical commercial reaction conditions (40–60°C, 1 atm) enzymes otherwise stable in solution may lose activity rapidly as a result of only slight changes in their environment such as temperature, pressure, pH, and ionic strength that induce small free energy changes from native to denatured states (168); moreover, their separation from the product is generally difficult and may cause further denaturation and loss of catalytic activity. The modest, largely reversible losses of activity resulting from small changes in reaction environment are largely due to modest changes in conformation of the active site. More severe changes in reaction conditions (eg, a 10°C increase in temperature) typically bring about the dissociation and unfolding of the quaternary and tertiary structures, respectively, into primary chains that subsequently order into fibrous protein bundles; in the process active sites are irreversibly destroyed. If further exposed to severe conditions of temperature and pH, the principal chain structure of the protein may undergo loss or modification of functional groups or amino acid residues.

The activity of a typical enzyme increases exponentially with temperature in accordance with the Arrhenius law up to about 50–60°C, passes through a maximum and declines precipitously above about 60–70°C. Thus, catalyst life may be on the order of days to weeks at around 50°C; however, the deactivation rate is extremely high at only slightly higher temperatures, eg, 50% loss of activity in 5 min at 65–70°C is typical. Nevertheless, a few enzymes are active and stable at temperatures exceeding 100°C; for example, α-amylase catalyzes starch liquifaction at 105–115°C. Because their deactivation rates are highly temperature-dependent, enzymes are generally shipped and stored under refrigeration (0–4°C); at these low temperatures they are generally stable for months.

Causes of deactivation can be classified (as in the case of heterogeneous catalysts) as chemical, mechanical, and thermal. However, for enzymes these causes are closely linked, since mechanically and thermally induced routes almost always effect chemical changes. Thermally induced chemical change (at elevated temperature) is the most likely scenario for enzyme deactivation.

Chemical deactivation mechanisms include (1) changes in stereo configuration by protons or hydroxyl ions at or near active sites (169,170), (2) structural modifications in aqueous or nonaqueous solvents (171–173), (3) poisoning of active sites by inhibitors (162,174), including "Trojan-horse inhibitors" that are activated by the target enzyme (162), (4) aggregation (175), (5) unfolding (6) fragmentation due to solvolysis, hydrolysis in water, or self-hydrolysis (autolysis) of proteases, eg, trypsin (176), and (7) oxidation in air (177,178). Mechanisms 1–5 may be reversible, while mechanisms 6 and 7 are generally irreversible. Mechanical deactivation may be caused by hydrodynamic shear forces, eg, by stirring or gas sparging, sometimes leading to fragmentation and/or aggregation (175,178).

Thermal inactivation of enzymes is a well-studied phenomenon (179–185); it may be either reversible or irreversible (180). Potentially reversible changes (due to small, short excursions in temperature near the characteristic unfolding temperature) include light aggregation, conformational changes, folding without further chemical change, disulfide exchange, and/or breaking of hydrogen bonds. Irreversible denaturation (due to prolonged, severe thermal treatment) may be caused by cleavage of disulfide bonds and/or cystinyl cross-links; unfolding followed by chemical change; chemical changes of the primary structure and/or active site, eg, cleavage of the polypeptide chain by hydrolysis or destruction of individual amino acid residues; strong aggregation of inactive unfolded forms; and formation of rubbery, tough fibrous structures due to alignment and bundling of unfolded primary chains (similar to that observed during the boiling of an egg). Chemical bonding of unfolded primary chains to form fibers is thermodynamically favorable because chemical bonding of hydrophobic functions exposed by unfolding lowers the entropy and hence free energy of the system.

Table 14 summarizes representative examples of enzyme deactivation by the various mechanisms.

Methods of enhancing enzyme stability have received considerable attention (168,173,178,180,183,185–187). Strategies to improve both chemical and thermal stability include (1) use of soluble additives, (2) immobilization, (3) protein engineering, and (4) chemical modification. Chemical modification (183, 185–187) and immobilization (164,165,172,188–191) are probably the most successful and widely used methods. As examples of the first kind, modification of protein surfaces by chemical binding with polysaccharides can improve thermostability, while polyol binding increases enzyme solubility in organic solvents with little loss of activity (171,176,183). Enzyme stability can be greatly enhanced and recovery problems obviated by immobilizing (heterogenizing)

Table 14. **Representative Examples of Deactivation Mechanisms for Enzymes**

Deactivation mechanism	Cause(s)/reversibility	Examples	Ref.
1. *Chemical*	generally involve formation or breaking of bonds in enzyme structure		
modest changes in active site configuration	caused by (a) introduction of H^+ or OH^- near active site, (b) small changes in pH or solvent environment/ largely reversible	model of effects of pH on phytases: enzyme is in equilibrium with protonated and hydroxylated forms which are less active or inactive	184
poisoning of active site	adsorption of inhibitor on active site/ sometimes reversible	mechanistic study of the inhibition of crotonase by (methylenecyclopropyl)-formyl-CoA; MCP ring trapping of an active site nucleophile is suggested	174
aggregation	caused by changes in pH or solvent environment with partial unfolding/ sometimes reversible	dimers and trimers of lysozyme are formed and activity is lost in a stirred reactor; mechanism may involve collision-induced conversion of enzyme to inactive state, followed by formation of disulfide bridges	175
unfolding, fragmenta-tion, bundling of primary chains into fibers	cleavage of enzyme bonds due to interaction with solvent, H^+, or OH^- due to medium to large changes in pH/ irreversible	deactivation of peroxidase in organic solvents including DMSO; solvent may strip water from enzyme, leading to reduced conformational mobility and unfolding	168,172,173
2. *Mechanical*	caused by hydrodynamic shear forces, eg, stirring or gas sparging, which can break bonds and cause aggregation of enzymes/usually irreversible	lysozyme is aggregated and irreversibly inactivated in a stirred reactor; the deactivation rate constant is proportional to the impeller power	175
3. *Thermal*			
modest changes in active site configuration and reversible unfolding	caused by small, short excursions in tempera-ture near the transition temperature^a/ reversible	equilibrium measure-ments of the tempera-ture-induced unfolding of bovine ribonuclease; repeated measurements after cooling fall on the same plot of fraction unfolded versus T	179,180

Table 14 (*Continued*)

Deactivation mechanism	Cause(s)/reversibility	Examples	Ref.
irreversible unfolding, fragmentation, bundling of primary chains into fibers	cleavage of enzyme bonds due to interaction with solvent, H^+, or OH^- due to medium to large changes in pH/ irreversible	irreversible thermo-inactivation of hen egg white lysozyme at 100°C and pH 4, 6, 8; inactivation is due to monomolecular changes in coordination, eg, hydrolysis of the Asp-X peptide bonds, deamidation of Asn residues, destruction of cystine residues, and formation of incorrect structures	180

[a]Characteristic temperature for a specific enzyme above which unfolding occurs and below which refolding occurs.

enzymes (164,165,172,188–191) through (1) covalent binding to a support, (2) cross-linking of enzymes using a bifunctional agent, (3) adsorption on a solid surface, (4) entrapment in a gel, or (5) containment in a membrane. Moreover, immobilization enables the catalytic process to be run continuously using a reactor of substantially lower volume, thereby substantially reducing capital and operating costs. These important advantages have stimulated the development of a significant number of commercial immobilized enzyme systems.

4. Prevention of Catalyst Decay

4.1. General Principles of Prevention.

The age-old adage that says "an ounce of prevention is worth a pound of cure" applies well to the deactivation of catalysts in many industrial processes. The catalyst inventory for a large plant may entail a capital investment of tens of millions of dollars. In such large-scale processes, the economic return on this investment may depend on the catalyst remaining effective over a period of up to 3–5 years. This is particularly true of those processes involving irreversible or only partially reversible deactivation (eg, sulfur poisoning or sintering). Some typical industrial catalysts, approximate catalyst lifetimes, and factors that determine their life are listed as examples in Table 15. It is evident that in many processes more than one mechanism limits catalyst life. Moreover, there is a wide variation in catalyst lifetimes among different processes, ie, from 10^{-6} to 15 years. While there is clearly greater interest in extending catalyst lifetimes in processes where life is short, it should be emphasized that great care must be exercised in protecting the catalyst in any process from process upsets (eg, temperature runaway, short-term exposure to impure feeds, or changes in reactant composition) that might reduce typical catalyst life by orders of magnitude, eg, from years to hours.

While complete elimination of catalyst deactivation is not possible, the rate of damage can be minimized in many cases through understanding of the

Table 15. **Typical Lifetimes and Factors Determining the Life of Some Important Industrial Catalysts**[a]

Reaction	Operating conditions	Catalyst	Typical life (years)	Process affecting life of catalyst charge	Catalyst property affected
Ammonia synthesis $N_2 + 3 H_2 \rightarrow 2 NH_3$	450–470°C, 200–300 atm	Fe with promoters (K_2O) and stabilizer (Al_2O_3)	10–15	slow sintering	activity
methanation (ammonia and hydrogen plants) $CO/CO_2 + H_2 \rightarrow CH_4 + H_2O$	250–350°C, 30 atm	supported nickel	5–10	slow poisoning by S, As, K_2CO_3 from plant upsets	activity and pore blockage
acetylene hydrogenation ("front end") $C_2H_2 + H_2 \rightarrow C_2H_4$	30–150°C, 20–30 atm	supported palladium	5–10	slow sintering	activity/selectivity and temperature
sulfuric acid manufacturing $2 SO_2 + O_2 \rightarrow 2 SO_3$	420–600°C, 1 atm	vanadium and potassium sulfates on silica	5–10	inactive compound formation; pellet fracture; plugging by dust	activity, pressure drop, and mass transfer
methanol synthesis $CO + 2 H_2 \rightarrow CH_3OH$	200–300°C, 50–100 atm	copper on zinc and aluminum oxides	2–5	slow sintering; poisoning by S, Cl, and carbonyls	activity
low temperature CO shift $CO + H_2O \rightarrow CO_2 + H_2$	200–250°C, 10–30 atm	copper on zinc and aluminum oxides	2–4	slow poisoning and accelerated sintering by poisons	activity
hydrocarbon hydrode sulfurization $R_2S + 2 H_2 \rightarrow H_2S + R_2$	300–400°C, 30 atm	cobalt and molybdenum sulfides on aluminum oxide	1–10	slow coking, poisoning by metal deposits in residues	activity, mass transfer, and pressure drop
high temperature CO shift $CO + H_2O \rightarrow H_2 + CO_2$	350–500°C, 20–30 atm	Fe_3O_4 on chromia	1–4	slow sintering, pellet breakage due to steam	activity and pressure drop
steam reforming, natural gas $CH_4 + H_2O \rightarrow CO + 3 H_2$	500–850°C, 30 atm	nickel on calcium aluminate or α-alumina	1–3	sintering, sulfur-poisoning, carbon formation, and pellet breakage due to plant upsets	activity and pressure drop
ethylene partial oxidation $2 C_2H_4 + O_2 \rightarrow 2 C_2H_4O$	200–270°C, 10–20 atm	silver on α-alumina with alkali metal promoters	1–3	slow sintering, poisoning by Cl, S	activity and selectivity

294

Reaction	Conditions	Catalyst	Life (yr)	Deactivation	Affected property
butane oxidation to maleic anhydride C_4H_{10} + $3.5\,O_2 \rightarrow C_4H_2O_3 + 4\,H_2O$	400–520°C, 1–3 atm	vanadium phosphorus oxide with transition metal additives	1–2	loss of P; attrition or pellet breakage; S, Cl poisoning	activity and selectivity
reduction of aldehydes to alcohols $RCHO + H_2 \rightarrow RCH_2OH$	220–270°C, 100–300 atm	copper on zinc oxide	0.5–1	slow sintering, pellet breakage (depends on feedstock)	activity or pressure drop
ammonia oxidation $2\,NH_3 + 5/2\,O_2 \rightarrow 2\,NO + 3\,H_2O$	800–900°C, 1–10 atm	Pt–Rh alloy gauze	0.1–0.5	surface roughness, loss of platinum, fouling by Fe	selectivity
oxychlorination of ethylene to ethylene dichloride $2\,C_2H_4 + 4\,HCl + O_2 \rightarrow 2\,C_2H_4Cl_2 + 2\,H_2O$	230–270°C, 1–10 atm	copper chlorides on alumina (fluidized bed)	0.2–0.5	loss by attrition and other causes resulting from plant upsets	fluidized state and activity
catalytic hydrocarbon reforming	460–525°C, 8–50 atm	platinum alloys on treated alumina	0.01–0.5	coking, frequent regeneration	activity and mass transfer
catalytic cracking of oils	500–560°C, 2–3 atm (fluidized bed)	synthetic zeolites	0.000002	very rapid coking (continuous regeneration)	activity and mass transfer

[a] Adapted from Ref. 9.

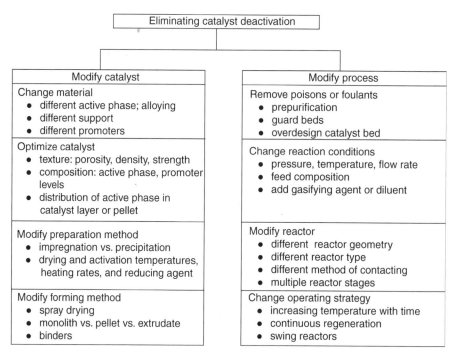

Fig. 8. Approaches to eliminating catalyst deactivation.

mechanisms, thereby enabling control of the deactivation process, ie, prevention is possible through control of catalyst properties, process conditions (ie, temperatures, pressures), feedstock impurities, methods of contacting, and process design. Figure 8 illustrates general approaches to eliminating or moderating deactivation through modifications in catalyst and/or process. Examples of how deactivation can be prevented are discussed below in connection with the most important causes of deactivation: chemical degradation, fouling by coke and carbon, poisoning, sintering, and mechanical degradation. Principles for preventing deactivation by these mechanisms are summarized in Table 16. Representative results from studies focusing on prevention or minimization of catalyst deactivation are found in Refs. 18,48,55,56,192–245.

4.2. Prevention of Chemical Degradation (by Vapor–Solid and Solid–Solid Reactions). The most serious problems-oxidation of metal catalysts, overreduction of oxide catalysts, and reaction of the active catalytic phase with carrier or promoter-can be minimized or prevented by careful catalyst and process design (as enumerated in Table 16). For example, the loss of Rh due to solid-state reaction with alumina in the automotive three-way catalyst can be prevented by supporting Rh on ZrO_2 in a separate layer from Pt and/or Pd on alumina. In Fischer–Tropsch synthesis, the oxidation of the active cobalt phase in supported cobalt catalysts to inactive oxides, aluminates, and silicates can be minimized by employing a two- or three-stage process in which product steam is moderated in the first stage by limiting conversion and in subsequent stages by interstage removal of water. It can also be moderated by addition of noble metal promoters that facilitate and maintain high reducibility of the cobalt

Table 16. **Methods for Preventing Catalyst Decay**

Basic mechanism	Problem	Cause	Methods of minimization
chemical degradation	oxidation of metal catalysts to inactive oxides	oxidation of metal by contaminant O_2 or reactant/product water	(1) purify feed of oxidants; (2) minimize reactant/product water by recycle/separation, staged reactors, and otherwise limiting conversion; (3) incorporate additives that facilitate resistance to oxidation
	transformation of active phase to stable, inactive phase	solid-state reaction of active phase with support or promoters	(1) avoid conditions (eg, oxidizing condition, high steam pressures, and high temperatures) that favor solid-state reactions and (2) select combinations of active phase and promoters/supports that are non-interacting
		overreduction of active oxide phases	(1) stabilize oxidation state using promoters that induce resistance to reduction or that serve as oxygen donors and (2) add steam to the reactants to prevent overreduction
fouling by coke or carbon	loss of catalytic surface sites due to formation of carbon or coke films	free-radical reactions in gas phase	(1) avoid formation of free radicals, lower temp.; (2) minimize free space; (3) free radical traps, diluents; (4) add gasifying agents (eg, H_2, H_2O)
		free-radical reactions at reactor walls	(1) coat reactor with inert material
		formation and growth on metal surfaces	(1) avoid accumulation of coke precursors (eg, atomic carbon, olefins) through careful choice of reactant conditions or membranes; (2) add gasifying agents (eg, H_2, H_2O), diluents; (3) incorporate catalyst additives to increase rate of gasification or to change ensemble size; (4) passivate metal surfaces with sulfur; (5) decrease dispersion; and (6) recycle inerts to flush surface of heavy oligomers and to moderate temperature
		formation and growth on metal oxides, sulfides	(1) utilize measures 1, 2, 3, and 6 for metal surfaces; (2) design catalyst for optimum pore structure and acidity; and (3) use shape-selective, coke-resistant molecular sieves
	loss of catalyst effectiveness; plugging of pores; destruction of catalyst	formation of gas phase coke, vermicular carbons, and liquid or solid cokes in massive quantities	(1) minimize formation of free radicals or coke precursors as above; (2) use gasifying agents; (3) incorporate catalyst additives that lower solubility of carbon in metal or that change ensemble size; (4) use supports with large pores, large pellets

Table 16 (*Continued*)

Basic mechanism	Problem	Cause	Methods of minimization
		hot spots in pellet or bed	(1) use wash coat or small pellets; (2) use slurry- or fluid-bed reactor, gas diluents
mechanical failure	crushing of granules, pellets, or monoliths in a fixed bed	brittle fracture due to a mechanical load	(1) minimize porosity of pellets or monoliths; (2) improve bonding of primary particles in agglomerates that make up pellets or monoliths using advanced forming methods, eg, spray drying and controlled thermal treatments; (3) add binders such as carbon to the support material, which facilitate plastic deformation and thus protect against brittle fracture; and (4) chemically or thermally temper agglomerates
	attrition and/ or erosion in fixed or moving beds	abrasion of catalyst coatings or particles due to mechanical, thermal, or chemical stresses	(1) avoid highly turbulent shear flows and/or cavitation, leading to high erosion rates; (2) avoid thermal stresses in the preparation and use of catalysts that lead to fracture or separation of coatings; and (3) avoid formation of chemical phases of substantially different densities or growth of carbon filaments that cause fracture of primary particles and agglomerates. Choose supports, support additives, and coating materials such as titanates, zirconia, and zirconates, having high fracture toughness
poisoning	loss of catalytic surface sites	blockage of sites by strong adsorption of impurity	(1) purify feed and/or use guard bed to adsorb poison; (2) employ additives that selectively adsorb poison; (3) choose reaction conditions that lower adsorption strength; (4) optimize pore structure and choose mass transfer regimes that minimize adsorption of poison on active sites; and (5) apply coating that serves as diffusion barrier to poison
thermal degradation, sintering	loss of metal area	metal particle or subparticle migration at high temperatures	(1) lower or limit reaction temperature while facilitating heat transfer; (2) add thermal stabilizers to catalyst; and (3) avoid water
	loss of support area	crystallization and/or structural modification or collapse	same as for avoiding loss of metal area

and by coating the alumina or silica support with materials such as ZrO_2 that are less likely to react with cobalt to form inactive phases.

4.3. Prevention of Fouling by Coke and Carbon. Rostrup-Nielsen and Trimm (45), Trimm (47), and Bartholomew (48) have discussed principles and methods for avoiding coke and carbon formation. General methods of preventing coke or carbon formation are summarized in Table 16. Most of these are based on one important fundamental principle, namely that *carbon or coke results from a balance between the reactions that produce atomic carbon or coke precursors and the reactions of these species with H_2, H_2O, or O_2 that remove them from the surface.* If the conditions favor formation over gasification, these species accumulate on the surface and react further to less active forms of carbon or coke, which either coat the surface with an inactive film or plug the pores, causing loss of catalyst effectiveness, pore plugging, or even destruction of the carrier matrix.

Methods to lower rates of formation of carbon or coke precursors relative to their rates of gasification vary with the mechanism of formation (ie, gas, surface, or bulk phase) and the nature of the active catalytic phase (eg, metal or oxide). For example, gas phase formation can be minimized by choosing reaction conditions that minimize the formation of free radicals, by using free-radical traps, by introducing gasifying agents (eg, H_2, H_2O) or gas diluents, and by minimizing the void space available for homogeneous reaction. Similarly, the formation and growth of carbon or coke species on metal surfaces is minimized by choosing reaction conditions that minimize the formation of atomic carbon or coke precursors and by introducing gasifying agents. Selective membranes or supercritical conditions can also be used to lower the gas-phase and surface concentrations of coke precursors. Since carbon or coke formation on metals apparently requires a critical ensemble of surface metal atoms and/or dissolution of carbon into the bulk metal, introduction of modifiers that change ensemble sizes (eg, Cu or S in Ni or Ru) or that lower the solubility of carbon (eg, Pt in Ni) can be effective in minimizing these forms of deactivation.

Coke deposition on oxide or sulfide catalysts occurs mainly on strongly acidic sites; accordingly the rate of coking can be lowered by decreasing the acidity of the support. For example, silanation of HY and HZSM-5 zeolites decreases their activities but improves catalyst life (245). In steam reforming, certain catalyst additives, eg, MgO, K_2O, or U_3O_8, facilitate H_2O or CO_2 adsorption and dissociation to oxygen atoms, which in turn gasify coke precursors (8,48,55).

As in the case of poisoning (see below), there are certain reactor bed or catalyst geometries that minimize the effects of coking on the reaction. For example, specific film-mass transport or pore diffusion regimes favor coke or carbon deposition on either the outside or inside of the catalyst pellet (246,247). Choosing supports with relatively large pores minimizes pore plugging; choice of large-diameter, mechanically-strong pellets avoids or delays reactor plugging. But in view of the rapidity at which coke and carbon can deposit on, plug, and even destroy catalyst particles, the importance of preventing the onset of such formation cannot be overemphasized.

Reforming of naphtha provides an interesting case study of catalyst and process designs to avoid deactivation by coking (8,206,208,248). The classical

Pt/Al$_2$O$_3$ catalyst is bifunctional; that is, the metal catalyzes dehydrogenation while the acid sites of the Al$_2$O$_3$ catalyze isomerization and hydrocracking. Together the two functions catalyze dehydrocylization and aromatization. Addition of Re, Sn, or Ge to Pt and sulfiding of the Pt–Re catalyst substantially reduce coke formation by diluting large Pt ensembles that would otherwise produce large amounts of coke, while addition of Sn and Ir improves selectivity for dehydrogenation relative to hydrogenolysis, the latter of which leads to coke formation. Naphtha reforming processes are designed for (1) high enough H$_2$ pressure to favor gasification of coke precursors while minimizing hydrocracking, (2) maintenance of Cl and S contents throughout the bed to ensure optimum acidity and coke levels, and (3) low enough overall pressure to thermodynamically and kinetically favor dehydrogenation and dehydrocylization. Accordingly, optimal process conditions are a compromise between case 1 and case 3. The above-mentioned improvements in catalyst technologies, especially resistance to coking, have enabled important process improvements such as optimal operation at lower pressure; thus, processes have evolved over the past two to three decades from conventional fixed-bed reactors at high pressure (35 bar) using nonregenerative Pt catalysts to low pressure (3.5 bar), slowly moving-bed, continuously regenerated units with highly selective Pt/Sn catalysts, resulting in substantial economic benefits (248).

4.4. Prevention of Poisoning. Since poisoning is generally due to strong adsorption of feed impurities and since poisoned catalysts are generally difficult or impossible to regenerate, it is best prevented by removal of impurities from the feed to levels that will enable the catalyst to operate at its optimal lifetime. For example, it is necessary to lower the feed concentration of sulfur compounds in conventional methanation and Fischer–Tropsch processes involving base metal catalysts to less than 0.1 ppm in order to ensure a catalyst lifetime of 1–2 years. This is typically accomplished using a guard bed of porous ZnO at about 200°C. In cracking or hydrocracking reactions on oxide catalysts, it is important to remove strongly basic compounds such as ammonia, amines, and pyridines from the feed; ammonia in some feedstocks, for example, can be removed by aqueous scrubbing. The poisoning of catalysts by metal impurities can be moderated by selective poisoning of the unwanted metal. For example, in catalytic cracking of nickel-containing petroleum feedstocks, nickel sites, which would otherwise produce copious amounts of coke, are selectively poisoned by antimony (249). The poisoning of hydrotreating catalysts by nickel and vanadium metals can be minimized by (1) using a guard bed of inexpensive Mo catalyst or a graded catalyst bed with inexpensive, low-activity Mo at the top (bed entrance) and expensive, high-activity catalyst at the bottom (see Fig. 9) and (2) depositing coke prior to the metals since these metal deposits can be physically removed from the catalyst during regeneration (250).

It may be possible to lower the rate of poisoning through careful choice of reaction conditions that lower the strength of poison adsorption (48) or by choosing mass-transfer-limiting regimes that limit deposits to the outer shell of the catalyst pellet, while the main reaction occurs uninterrupted on the interior of the pellet (246). The manner in which the active catalytic material is deposited on a pellet (eg, uniformly or in an eggshell or egg yolk pattern) can significantly influence the life of the catalyst (17,251).

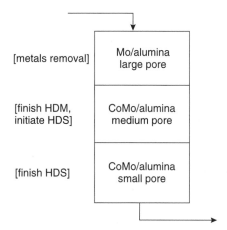

Fig. 9. Staged reactor system with decreasing pore size strategy for HDM/HDS of residues (224).

An example of reducing catalyst poisoning (and oxidation) through process design has been reported in a process patent for staged hydrocarbon synthesis via the Fischer–Tropsch reaction (252). While cobalt catalysts are favored because of their high activities and while it is desirable to achieve high conversions of CO in the process, the one-pass conversion for cobalt is limited by (*1*) its tendency to be oxidized at high partial pressures of product water observed at high CO conversions and (*2*) its tendency to form under these conditions the oxygenated products (eg, alcohols and aldehydes) that poison or suppress its synthesis activity. One alternative is to separate products and recycle the unused CO and H_2, but this requires costly recompression and separation of the oxygenates. Costly separation and/or poisoning can be prevented by operating a first-stage reactor containing a cobalt catalyst to a moderately high conversion followed by reacting the remaining CO and H_2 in a second stage to above 95% conversion on an iron catalyst, which is not sensitive to the oxygenates and which shifts some of the product water to H_2 and CO_2, thus minimizing its hydrothermal degradation.

An example of reducing catalyst poisoning through catalyst design occurs in abatement of emissions for automotive and motorcycle engines (18,222). Application of an alumina or zeolite coating or alternatively preparing the active phase in a sublayer provides a diffusion barrier that prevents or slows the access of poisons from the fuel or oil (eg, phosphorus and/or zinc from lubricating oil or corrosion products) to the catalyst surface. The principle is to optimize the pore size distribution of the diffusion barrier to provide access to the catalytic phase of relatively small hydrocarbon, CO, NO, and O_2 molecules while preventing access of larger molecules such as from lubricating oil and/or particulates.

4.5. Prevention of Sintering. Since most sintering processes are irreversible or are reversed only with great difficulty, it is important to choose reaction conditions and catalyst properties that avoid such problems. Metal growth is a highly activated process; thus by choosing reaction temperatures lower than 0.3–0.5 times the melting point of the metal, rates of metal sintering can be greatly minimized. The same principle holds true in avoiding recrystallization of metal oxides, sulfides, and supports. Of course, one approach to lowering reaction temperature is to maximize activity and surface area of the active catalytic phase.

Although temperature is the most important variable in the sintering process, differences in reaction atmosphere can also influence the rate of sintering. Water vapor in particular accelerates the crystallization and structural modification of oxide supports. Accordingly, it is vital to minimize the concentration of water vapor in high temperature reactions on catalysts containing high surface area supports.

Besides lowering temperature and minimizing water vapor, it is possible to lower sintering rates through addition of thermal stabilizers to the catalyst. For example, the addition of higher melting noble metals (such as rhodium or ruthenium) to a base metal (such as nickel) increases the thermal stability of the base metal (253). Addition of Ba, Zn, La, Si, and Mn oxide promoters improves the thermal stability of alumina (254).

Designing thermally stable catalysts is a particular challenge in high temperature reactions such as automotive emissions control, ammonia oxidation, steam reforming, and catalytic combustion. The development of thermally stable automotive catalysts has received considerable attention, thus providing a wealth of scientific and technological information on catalyst design (eg, Refs. 8 and 225–232). The basic design principles are relatively simple: (1) utilize thermally and hydrothermally stable supports, eg, high-temperature δ- or θ-aluminas or alkaline-earth or rare-earth oxides that form ultrastable spinels with γ-alumina; (2) use PdO rather than Pt or Pt–Rh for high temperature converters, since PdO is considerably more thermally stable in an oxidizing atmosphere because of its strong interaction with oxide supports; and (3) use multilayer strategies and/or diffusion barriers to prevent thermally induced solid-state reactions (eg, formation of Rh aluminate) and to moderate the rate of highly exothermic CO and hydrocarbon oxidations. For example, a typical three-way automotive catalyst may contain alkaline-earth metal oxides (eg, BaO) and rare-earth oxides (eg, La_2O_3 and CeO_2) for stabilizing Pt and/or PdO on alumina and ZrO_2 as a thermal stabilizer for the CeO_2, an oxygen storage material, and as a noninteracting support for Rh in a separate layer or in a separate phase in a composite layer.

4.6. Prevention of Mechanical Degradation. While relatively few studies have focused on this topic, there are nevertheless principles that guide the design of processes and catalysts in preventing or minimizing mechanical degradation. In terms of catalyst design it is important to (1) choose supports, support additives, and coatings that have high fracture toughness, (2) use preparation methods that favor strong bonding of primary particles and agglomerates in pellets and monolith coatings, (3) minimize (or rather optimize) porosity (thus maximizing density), and (4) use binders such as carbon to facilitate plastic deformation and thus protect against brittle fracture. Processes (and to some extent preparation procedures) should be designed to minimize (1) highly turbulent shear flows or cavitation that lead to fracture of particles or separation of coatings, (2) large thermal gradients or thermal cycling leading to thermal stresses, and (3) formation of chemical phases of substantially different densities or formation of carbon filaments leading to fracture of primary particles and agglomerates. Nevertheless, thermal or chemical tempering can be used in a controlled fashion to strengthen catalyst particles or agglomerates.

Examples of catalyst design to minimize attrition can be found in the recent scientific (239,240) and patent (241–244) literature focusing on the Fischer–Tropsch synthesis in slurry reactors. These studies indicate that (1) spray drying of particles improves their density and attrition resistance; (2) addition of silica and/or alumina into titania improves its attrition resistance, while addition of only 2000–3000 ppm of titania to γ-alumina improves alumina's attrition resistance; and (3) preformed alumina spheres promoted with La_2O_3 provide greater attrition resistance relative to silica. Increasing attrition resistance is apparently correlated with increasing density (239,240,244). According to Singleton and co-workers (244), attrition resistance of Co/Al_2O_3 is improved when the γ-alumina support is (1) formed from synthetic boehmite having a crystallite diameter of 4–5 nm and (2) pretreated in acidic solution having a pH of 1–3; moreover, attrition resistance decreases in the order Co/Al_2O_3, Co/SiO_2, Co/TiO_2 and is greater for catalyst prepared by aqueous versus nonaqueous impregnation.

5. Regeneration of Deactivated Catalysts

Despite our best efforts to prevent it, the loss of catalytic activity in most processes is inevitable. When the activity has declined to a critical level, a choice must be made among four alternatives: (1) restore the activity of the catalyst, (2) use it for another application, (3) reclaim and recycle the important and/or expensive catalytic components, or (4) discard the catalyst. The first alternative (regeneration and reuse) is almost always preferred; catalyst disposal is usually the last resort especially in view of environmental considerations.

The ability to reactivate a catalyst depends upon the reversibility of the deactivation process. For example, carbon and coke formation is relatively easily reversed through gasification with hydrogen, water, or oxygen. Sintering on the other hand is generally irreversible, although metal redispersion is possible under certain conditions in selected noble metal systems. Some poisons or foulants can be selectively removed by chemical washing, mechanical treatments, heat treatments, or oxidation (255,256); others cannot be removed without further deactivating or destroying the catalyst.

The decision to regenerate/recycle or discard the entire catalyst depends largely on the rate of deactivation. If deactivation is very rapid, as in the coking of cracking catalysts, repeated or continuous regeneration becomes an economic necessity. Precious metals are almost always reclaimed where regeneration is not possible. Disposal of catalysts containing nonnoble heavy metals (eg, Cr, Pb, or Sn) is environmentally problematic and should be a last resort; if disposal is necessary, it must be done with great care, probably at great cost. Accordingly, a choice to discard depends upon a combination of economic and legal factors (256). Indeed, because of the scarcity of landfill space and an explosion of environmental legislation, both of which combine to make waste-disposal prohibitively expensive, there is a growing trend to regenerate or recycle spent catalysts (257,258). A sizeable catalyst regeneration industry benefits petroleum refiners by helping to control catalyst costs and limiting liabilities (259,260); it provides for *ex situ* regeneration of catalyst and recovery/recycling of metals, eg, of cobalt, molybdenum, nickel, and vanadium from hydroprocessing catalysts (257).

Consistent with its importance the scientific literature treating catalyst regeneration is significant and growing (includes several hundred journal articles since 1990), (eg. refs. 256,261–263).

The patent literature treating catalyst regeneration/reactivation is enormous (more than 2000 patents); the largest fraction of this literature describes processes for regeneration of catalysts in three important petroleum refining processes, FCC, catalytic hydrocarbon reforming, and alkylation. However, a significant number of patents also claim methods for regenerating absorbents and catalysts used in aromatization, oligomerization, catalytic combustion, SCR of NO, hydrocracking, hydrotreating, halogenation, hydrogenation, isomerization, partial oxidation of hydrocarbons, carbonylations, hydroformylation, dehydrogenation, dewaxing, Fisher–Tropsch synthesis, steam reforming, and polymerization.

Conventional methods for regenerating (largely *in situ*) coked, fouled, poisoned, and/or sintered catalysts in some of these processes and representative examples thereof (264–296) are summarized in Table 17, while the basic principles and limitations involved in regeneration of coked, poisoned, and sintered catalysts are briefly treated in the subsections that follow.

5.1. Regeneration of Catalyst Deactivated by Coke or Carbon.

Carbonaceous deposits can be removed by gasification with O_2, H_2O, CO_2, and H_2. The temperature required to gasify these deposits at a reasonable rate varies with the type of gas, the structure and reactivity of the carbon or coke, and the activity of the catalyst. Walker and co-workers (305) reported the following order for rates of uncatalyzed gasification at 10 kN/m^3 and 800°C (relative rates in parenthesis): O_2 (105) > H_2O (3) > CO_2 (1) > H_2 (3×10^{-3}). However, this activity pattern does not apply in general for other conditions and for catalyzed reactions (1). Nevertheless, the order of decreasing reaction rate of $O_2 > H_2O > H_2$ can be generalized.

Rates of gasification of coke or carbon are greatly accelerated by the same metal or metal oxide catalysts upon which carbon or coke deposits.

Because catalyzed removal of carbon with oxygen is generally very rapid at moderate temperatures (eg, 400–600°C), industrial processes typically regenerate catalysts deactivated by carbon or coke in air. Indeed, air regeneration is used to remove coke from catalysts in catalytic cracking (64), hydrotreating processes (261), and catalytic reforming (262).

One of the key problems in air regeneration is avoiding hot spots or over-temperatures which could further deactivate the catalyst. The combustion process is typically controlled by initially feeding low concentrations of air and by increasing oxygen concentration with increasing carbon conversion (261,306); nitrogen gas can be used as a diluent in laboratory-scale tests while steam is used as a diluent in full-scale plant operations (306). For example, in the regeneration of hydrotreating catalysts McCulloch (261) recommends keeping the temperature at less than 450°C to avoid the γ- to α-alumina conversion, MoO_3 sublimation, and cobalt or nickel aluminate formation, which occur at 815, 700, and 500–600°C respectively.

Table 17. **Conventional Methods for and Representative Examples of Catalyst Regeneration from Scientific and Patent Literatures**

Deactivation mechanism/reaction/catalyst	Problem/cause	Method(s) of regeneration/phenomena studied/conclusions	Refs.
Deactivation by coke, carbon			
alkene aromatization oligomerization/zeolites, esp. ZSM-5, -22, -23, beta-zeolite, ferrierite	catalyst fouling by condensation of heavy oligomers to coke	(1) ZSM-5 catalyst for light olefin oligomerization containing 2–3% coke is treated in 8–10% steam/air mixture (1300 kPa, 93°C inlet) in a fluidized bed (2) a coked crystalline alumogallosilicate is contacted with oxygen at a concentration of 0.05–10 vol%, 420–580°C, and 300–4000 h^{-1}	264,265
alkylation of isoparaffins on solid catalysts/sulfated zirconia, USY[a], Nafion, silicalite, ZSM-5	rapid catalyst deactivation due to coke formation; unacceptable product quality, and thermal degradation of catalyst during regeneration	(1) coked zeolite is regenerated in liquid phase ($P > 3500$ kPa) fluid bed with H_2 in two steps: (a) at reaction temperature (20–50°C) and (b) at 25°C above reaction temperature (2) coked Pd- and Pt/Y-zeolite catalysts containing 10–13% coke are regenerated in either air or H_2; H_2 treatment enables removal of most of the coke at low-moderate temperatures; higher temperatures are required for air	266,267
catalytic reforming of naphtha/ Pt/Al$_2$O$_3$ promoted with Re, Sn, Ge, or Ir	poisoning and fouling by coke produced by condensation of aromatics and olefins	(1) coke on Pt bimetallic reforming catalyst is removed off-stream in a moving bed at 300–600°C, followed by oxychlorination (350–550°C) (2) coke on Pt/zeolite is removed in halogen-free oxygen-containing gas at $T < 415°C$ (3) sintering during oxidation of coke on Pt–Ir/Al$_2$O$_3$ catalyst can be minimized at low regeneration temperature (4) study of influence of heating rate, temperature, and time on structural properties of regenerated Pt–Sn/Al$_2$O$_3$ (5) study of effects of Cl, Sn content, and regeneration sequence on dispersion and selectivity of Pt–Sn/Al$_2$O$_3$ (6) regenerated Pt–Re/Al$_2$O$_3$ is more stable than the fresh catalyst in n-heptane conversion and more selective for toluene	268–273
dehydrogenation of propane and butane/Cr$_2$O$_3$/Al$_2$O$_3$, Cr$_2$O$_3$/ZrO$_2$, FeO/K/MgO, Pt/Al$_2$O$_3$, Pt–Sn/Al$_2$O$_3$, Pt–Sn/KL-zeolite	catalyst activity is low owing to equilibrium limitations and build-up of product H_2; rapid loss of activity occurs owing to coke formation	(1) temperatures gradients were measured during burn off of coke formed on a chromia–alumina catalyst during butene dehydrogenation; data were used in developing a mathematical model for predicting temperatures and coke profiles (2) coked supported palladium catalyst used in the dehydrogenation of dimethyltertrahydronaphthalenes to dimethylnaphthalenes is reactivated with an organic polar solvent at a temperature below 200°C	275,275

Table 17 (*Continued*)

Deactivation mechanism/reaction/catalyst	Problem/cause	Method(s) of regeneration/phenomena studied/conclusions	Refs.
Fischer–Tropsch synthesis/ Co/Al_2O_3	loss of activity due to blocking of sites by carbon overlayers and heavy hydrocarbons	(1) carbidic surface carbon deposited on cobalt can be largely removed in hydrogen at 170–200°C and in steam at 300–400°C (2) slurry-phase cobalt catalysts may lose 50% activity during synthesis over a period of a few days; the activity can be rejuvenated *in situ* by injecting H_2 gas into vertical draft tubes inside the reactor	276,277
fluid catalytic cracking (FCC) of heavy hydrocarbons/USY or REO-Y[b] in silica matrix	rapid loss of activity due to poisoning of acid sites and blocking of small zeolite pores by coke	(1) process and apparatus for increasing the coke burning capacity of FCC regenerators; auxiliary regenerator partially burns off the coke at turbulent or fast fluidized-bed conditions (2) multistage fluidized-bed regeneration of spent FCC catalyst in a single vessel by incorporating two relatively dense phase fluidized beds beneath a common dilute phase region	278,279
hydrocracking of heavy naphtha/CoMo, NiW, MoW on Al_2O_3 or SiO_2–Al_2O_3; Pt or Pd on Y-zeolite, mordenite, or ZSM-5	loss of activity due to poisoning of acid sites and blocking of small zeolite pores by coke	(1) regeneration of noble metal/zeolite via progressive partial removal of carbonaceous deposits under controlled oxidizing conditions to maximize sorption of a probe molecule while minimizing metal sintering (2) regeneration of noble metal/zeolite in air at about 600°C, followed by a mild treatment in aqueous ammonia to improve catalytic activity	280,281
hydrotreating of gas oil	loss of activity due to formation of types I, II, and III coke on metal sulfide and alumina surfaces and in pores	(1) TPO studies of oxidative regeneration of CoMo and NiW HDS catalysts; sulfur is removed at 225–325°C, carbon at 375–575°C. Redispersion of NiW was observed by EXAFS (2) physicochemical changes in CoMo and NiCoMo HDS catalysts during oxidative regeneration, including redispersion of Co, Ni, and Mo oxides and surface area loss, were examined (3) changes in NiMo catalyst structure and coke composition during reaction and regeneration were examined and correlated (4) properties of NiMo catalyst deactivated during shale oil hydrogenation and regenerated in O_2 or H_2 were examined. regeneration in 1.6% O_2 was more effective than that in 5% H_2. Ni aluminate spinel was observed after burn off (5) hard and soft cokes formed on CoMo catalysts during HDS of gas oil were characterized. At low coke levels, hard coke was more easily removed in H_2 than in O_2 (6) spent catalysts are washed with solvent and contacted with steam at about 600°C	282,283, 297–300

methanol to olefins or gasoline/ silica–alumina, Y-zeolite, ZSM-5, other zeolites, and aluminophosphate molecular sieves	severe coking and deactivation of silica–alumina and Y-zeolite catalysts observed during high conversions of methanol, also substantial coking of ZSM-5, other zeolites, and aluminophosphate molecular sieves	(1) kinetics of coke burnoff from a SAPO-34 used in converting methanol to olefins were studied; kinetics are strongly dependent on the nature of the coke. Kinetics are slowed by strong binding of coke to acid sites (2) ZSM-34 catalyst used in conversion of methanol to light olefins is effectively regenerated in H_2-containing gas; this approach avoids the formation of catalyst-damaging products such as steam that would be formed during burn off in air	284,285

Poisoning

FCC of residues/USY or REO-Y in silica matrix	(1) poisoning of acid sites by N-containing compounds. (2) deposition of Ni and V metals on acid sites which change selectivity and decrease activity	(1) organometallic solutions of Sb and Bi are added to process steam to passivate Ni by forming inactive Ni–Sb and Ni–Bi species (2) V metal deposits are trapped by reaction with magnesium orthosilicate to form an unreactive magnesium vanadium silicate (3) spent metal-contaminated catalyst is demetallized by chlorinating and washing followed by contacting with NH_4F and one antimony compound (4) metal-contaminated catalyst is contacted with an aqueous solution of a carboxylic acid (eg, formic, acetic, citric, or lactic acid) (5) metal-contaminated catalyst is contacted with HCl, HNO_3, or H_2SO_4 (6) metal contaminated catalyst is contacted with reducing CO gas to form gaseous metal carbonyls that separated from the catalyst	284,285, 301–304
hydrogenation or dechlorination	poisoning of metal sites by arsenic, sulfur, and other poisons	(1) regeneration of Ni/SiO_2 catalyst poisoned by thiophene using a sequence of oxidation–reduction treatments at low PO_2 and 1 atm H_2 respectively (2) regeneration in dilute hypochlorite solution of a Pd/Al_2O_3 catalyst deactivated during the aqueous-phase dechlorination of trichloroethylene in the presence of sulfite or HS^- ions present in ground water	288,289
hydrotreating of residues/ Al_2O_3-supported Mo and CoMo	pore-mouth poisoning and blockage by Ni, V, and Fe sulfides present in feed as organometallics	(1) regeneration of catalysts containing V, Ni, or Fe by contacting with H_2O_2 solution and organic acid (2) following removal of coke by air or solvent wash, catalyst is acid leached to remove undesired metals	290,291

Table 17 (Continued)

Deactivation mechanism/ reaction/catalyst	Problem/cause	Method(s) of regeneration/phenomena studied/conclusions	Refs.
Thermal degradation			
Catalytic reforming of naphtha/Pt/Al$_2$O$_3$ promoted with Re, Sn, Ge, or Ir; Pt/ KL-zeolite	sintering of Pt causing formation of large metal crystallites crystals and loss of active surface area	(1) redispersion of Pt–Ir bimetallic catalysts using a wet HCl/air treatment, since the conventional oxychlorination is not effective (2) redispersion of Pt/KL-zeolite using wet HCl/air treatment followed by brief calcination and reduction (3) redispersion of Pt–Re/ Al$_2$O$_3$ in Cl$_2$ and O$_2$ (4) redispersion of supported Pt, other noble metals, and Ni in Cl$_2$ and O$_2$	270,273, 292,293
hydrocracking of heavy naphtha/CoMo, NiW, MoW on Al$_2$O$_3$ or SiO$_2$–Al$_2$O$_3$; Pt or Pd on Y-zeolite, mordenite, or ZSM-5	sintering of noble metal causing formation of large metal crystallites crystals and loss of active surface area	redispersion of noble metals on molecular sieves including silica-aluminates, ALPOS, SAPOS	294
hydrotreating of gas oil and residues/Al$_2$O$_3$-supported Mo and CoMo	sintering of Mo and Co sulfides causing formation of large sulfide crystals and loss of active surface area	(1) oxidative regeneration of hydroprocessing catalyst at 600°C optimizes surface area and Mo dispersion (2) oxidative regeneration in several steps with a final oxidation at 500–600°C to restore residual catalyst activity	295,296

[a]USY: ultrastable Y-zeolite. [b]REO-Y: rare-earth exchanged Y-zeolite.

308

Because coke burn-off is a rapid, exothermic process, the reaction rate is controlled to a large extent by film heat and mass transfer. Accordingly, burn-off occurs initially at the exterior surface and then progresses inward with the reaction occurring mainly in a shrinking shell consistent with a "shell-progressive" or "shrinking-core" model (307); as part of this same work, Richardson (307) showed how experimental burn-off rate data can be fitted to various coking transport models, eg, parallel or series fouling. Burn-off rates for coke deposited on SiO_2/Al_2O_3 catalysts were reported by Weisz and Goodwin (308); burning rate was found to be independent of initial coke level, coke type, and source of catalyst.

5.2. Regeneration of Poisoned Catalysts. Much of the previous literature has focused on regeneration of sulfur-poisoned catalysts used in hydrogenations and steam reforming. Studies of regeneration of sulfur-poisoned Ni, Cu, Pt, and Mo with oxygen/air, steam, hydrogen, and inorganic oxidizing agents have been reported (27). Rostrup-Nielsen (309) indicates that up to 80% removal of surface sulfur from Mg- and Ca-promoted Ni, steam reforming catalysts occurs at 700°C in steam. The presence of both SO_2 and H_2S in the gaseous effluent suggests that the following reactions occur:

$$Ni - S + H_2O \rightarrow NiO + H_2S \tag{5}$$

$$H_2S + 2\,H_2O \rightarrow SO_2 + 3\,H_2 \tag{6}$$

Although this treatment is partially successful in the case of low-surface-area steam reforming catalysts, the high temperatures required for these reactions would cause sintering of most high-surface-area nickel catalysts.

Regeneration of sulfur-poisoned catalysts, particularly base metal catalysts, in air or oxygen has been largely unsuccessful. For example, the treatment of nickel steam-reforming catalysts in steam and air results in the formation of sulfates, which are subsequently reduced back to nickel sulfide upon contact with hydrogen. Nevertheless, sulfur can be removed as SO_2 at very low oxygen partial pressures, suggesting that regeneration is possible under carefully controlled oxygen or species such as CO_2 or NO that dissociate to oxygen. Apparently, at low oxygen pressures the oxidation of sulfur to SO_2 occurs more rapidly than the formation of nickel oxide while at atmospheric pressure the converse is true, ie, the sulfur or sulfate layer is rapidly buried in a nickel oxide layer. In the latter circumstance, the sulfur atoms diffuse to the nickel surface during reduction, thereby restoring the poisoned surface. Regeneration of sulfur-poisoned noble metals in air is more easily accomplished than with steam, although it is frequently attended by sintering. Regeneration of sulfur-poisoned nickel catalysts using hydrogen is impractical because (1) adsorption of sulfur is reversible only at high temperatures at which sintering rates are also high, and (2) rates of removal of sulfur in H_2 as H_2S are slow even at high temperature.

Inorganic oxidizing agents such as $KMnO_4$ can be used to oxidize liquid phase or adsorbed sulfur to sulfites or sulfates (16). These electronically shielded structures are less toxic than the unshielded sulfides. This approach has somewhat limited application, ie, in partial regeneration of metal catalysts used in low temperature liquid-phase hydrogenation reactions or in liquid-phase destruction of chlorinated organic compounds. For example, Lowrey and

Table 18. **Typical Regeneration Procedure for Reforming Catalysts**[a]

1. *Preliminary operations:*
   cool the catalyst to about 200°C and strip hydrocarbons and H_2 with N_2

2. *Elimination of coke by combustion:*
   inject dilute air (0.5% O_2) at 380°C and gradually increase oxygen content to about 2% by volume while maintaining temperature below 450–500°C to prevent further sintering of the catalyst. To prevent excessive leaching of Cl_2, HCl or CCl_4 may be injected during the combustion step

3. *Restoration of catalyst acidity:*
   Restoration of acidity occurs at 500°C by injection of a chlorinated compound in the presence of 100–200 ppm water in air

4. *Redispersion of the metallic phase:*
   expose the catalyst to a few Torr of HCl or CCl_4 in 2–10% O_2 in N_2 at 510–530°C for a period of about 4 h. After redispersion, O_2 is purged from the unit and the catalyst is reduced in H_2

[a]Ref. 261 and 262.

Reinhard (289) reported successful regeneration in dilute hypochlorite solution of a Pd/Al_2O_3 catalyst deactivated during the aqueous-phase dechlorination of trichloroethylene (TCE) in the presence of sulfite or HS^- ions. These poisons are formed by sulfate-reducing bacteria present in natural groundwater and are apparently adsorbed on the alumina or Pd surfaces more strongly than sulfate ions.

5.3. Redispersion of Sintered Catalysts. During catalytic reforming of hydrocarbons on platinum-containing catalysts, growth of 1-nm platinum metal clusters to 5–20-nm crystallites occurs. An important part of the catalyst regeneration procedure is the redispersion of the platinum phase by a high temperature treatment in oxygen and chlorine, generally referred to as "oxychlorination." A typical oxychlorination treatment involves exposure of the catalyst to HCl or CCl_4 at 450–550°C in 2–10% oxygen for a period of 1–4 h (see details in Table 18). During coke burning some redispersion occurs, eg, D increases from 0.25 to 0.51, while during oxychlorination the dispersion is further increased, eg, from 0.51 to 0.81 (262).

Some guidelines and principles regarding the redispersion process are worth enumerating:

1. In cases involving a high degree of Pt sintering or poisoning, special regeneration procedures may be required. If large crystallites have been formed, several successive oxychlorinations are performed (262).

2. Introducing oxygen into reactors in parallel rather than in series results in a significant decrease in regeneration time (84).

3. Introduction of hydrocarbons present in the reactor recycle after regeneration is said to stabilize the catalyst; solvents such as ammonium acetate, dilute nitric acid containing lead nitrate, EDTA and its diammonium salt are reported to dissolve out metal aggregates without leaching out the dispersed metal (84).

4. The procedures for redispersion of Pt/alumina are not necessarily applicable to Pt on other supports or to other metals. For example, Pt/silica is redispersed at lower temperature and higher Cl_2 concentration (150–200°C

and 25% Cl_2). Pd/alumina can be redispersed in pure O_2 at 500°C. While Pt–Re/alumina is readily redispersed by oxychlorination at 500°C, Pt–Ir/ alumina is not redispersed in the presence of O_2 unless the catalyst is pretreated with HCl (270).

An extensive scientific and patent literature of redisperson describes the use of chlorine, oxygen, nitric oxide, and hydrogen as agents for redispersion of sintered catalysts. Most of the early literature shows positive effects for chlorine compounds in the presence of oxygen in redispersing alumina-supported platinum and other noble metals. Recent literature demonstrates the need for understanding the detailed surface chemistry in order to successfully develop and improve redispersion processes, especially in more complex catalyst systems such as alumina-supported bimetallics. For example, on the basis of a fundamental study of the redispersion surface chemistry, Fung (270) developed a redispersion procedure for Pt–Ir bimetallic catalysts using a wet HCl/ air treatment, since the conventional oxychlorination is not effective for this catalyst.

Redispersion of alumina-supported platinum and iridium crystallites is also possible in a chlorine-free oxygen atmosphere if chlorine is present on the catalyst. The extent of redispersion depends on the properties of the Pt/Al_2O_3 catalyst and temperature. The question whether redispersion of platinum occurs only in oxygen without chlorine present on the catalyst remains controversial.

Two models, "the thermodynamic redispersion model" and "the crystallite splitting model," have been advanced to explain the redispersion in oxygen (84,85,310). The "thermodynamic" redispersion model hypothesizes the formation of metal oxide molecules that detach from the crystallite, migrate to active sites on the support, and form surface complexes with the support. Upon subsequent reduction, the metal oxide complexes form monodisperse metal clusters. In the "crystallite splitting" model, exposure of a platinum crystallite to oxygen at 500°C leads to formation of a platinum oxide scale on the outer surface of the crystallite, which stresses and ultimately leads to splitting of the particle (310). Dadyburjor hypothesizes that the crystallite splitting model is most applicable to the behavior of large crystallites and to all particles at relatively small regeneration times while the thermodynamic migration model is useful for small particles and most particles after longer regeneration times.

6. Summary and Perspective

6.1. Summary

1. The causes of deactivation are basically of three kinds: chemical, mechanical, and thermal. The five intrinsic mechanisms of catalyst decay, (a) poisoning, (b) fouling, (c) thermal degradation, (d) chemical degradation, and (e) mechanical failure, vary in their reversibility and rates of occurrence. Poisoning and thermal degradation are generally slow, irreversible processes while fouling with coke and carbon is generally rapid and reversible by regeneration with O_2 or H_2.

2. Catalyst deactivation is more easily prevented than cured. Poisoning by impurities can be prevented through careful purification of reactants. Carbon deposition and coking can be prevented by minimizing the formation of carbon or coke precursors through gasification, careful design of catalysts and process conditions, and by controlling reaction rate regimes, eg, mass transfer regimes, to minimize effects of carbon and coke formation on activity. Sintering is best avoided by minimizing and controlling the temperature of reaction.

3. Prevention and monitoring are important engineering principles in "standard of care" practice. The prevention of catalyst decay is important in every aspect of a process including design, construction, operation, and regeneration. Careful monitoring of process variables is a necessity in understanding and preventing catalyst decay problems of either a slow or a catastrophic nature.

4. The optimization of a catalytic process considers optimum operation and regeneration policies subject to constraints of catalyst cost, operation cost, regeneration cost, and product value. The optimum operating policy maximizes the rate of formation of product during the operating period.

5. Catalyst deactivation kinetics for reactions involving relatively slow deactivation can be experimentally determined using a laboratory fixed-bed, mixed-fluid (CSTR) reactor. Reactors and processes involving a slowly deactivating catalyst can be designed using relatively simple numerical analysis of the design equations and a pseudo-steady-state approximation for the main reaction.

6. Modeling and experimental assessment of deactivation processes are useful in providing (a) accelerated simulations of industrial processes, (b) predictive insights into effects of changing process variables on activity, selectivity, and life, (c) estimates of kinetic parameters needed for design and modeling, (d) estimates of size and cost for scale-up of a process, and (e) a better understanding of the basic decay mechanisms. It is now possible to develop realistic mathematical models of most catalytic processes, which can be used in conjunction with short-term experimental tests to accurately predict catalyst life in a commercial unit. Proper application of this approach could save companies millions of dollars by alleviating the need for long-term deactivation tests and/or premature shutdown. For details on this aspect of assessment, refer to the expanded version of this article published in the *Encyclopedia of Catalysis* (78,311).

6.2. Perspectives and Trends. Research and development activities in the area of catalyst deactivation have grown steadily in the past three decades. Catalyst deactivation symposia are held annually as part of national meetings of chemical engineering and chemical societies in the United States and Europe. The rising quality of work presented at the international symposium on catalyst deactivation, held every four years, is evident. In view of the importance of deactivation problems in industrial processes, this trend will most probably continue.

Several other trends are evident:

1. The increasing use of more sophisticated analytical tools to investigate the chemistry and mechanisms of deactivation. Surface science tools such as AES, quantitative HRTEM, XPS, and STM are now routinely applied to investigate deactivation mechanisms at very fundamental levels.
2. The increasing development of more sophisticated models of deactivation processes.

These trends are also likely to continue. Moreover, the combination of more sophisticated methods and models will hasten the practical application of models for predicting catalyst/process life. This is already happening in selected companies. For example, for more than a decade now operators at Phillips Petroleum Co. have been using deactivation models (developed at their corporate research) in their refineries to predict when shutdown will be necessary. One of these models enables them to predict accurately the lifetime of hydrotreating catalysts on the basis of catalyst and feedstock properties.

6.3. Future Needs. *Collection of Data.* It is evident from careful examination of the literature that *few deactivation rate data are available* for even the most important large-scale catalytic systems. Accordingly, there is a *critical need for collection of such data at the laboratory, bench, and plant scale.* There is much that could be done with good data. Sophisticated analytical tools and well-designed reactors are available at most companies for collecting and analyzing such data. The field is ripe and ready to harvest. The wise will seize these opportunities.

Data Analysis and Model Development. Much of the previously collected data were analyzed using outdated methods. There is much that could be learned by reanalyzing some of these data using new approaches such as the GPLE and microkinetic modeling. Critical reviews that include collections of carefully selected rate and kinetic data would constitute important contributions to scientific knowledge and technological development. The incorporation of these data into models would enable more sophisticated design of catalysts, reactors, and processes.

BIBLIOGRAPHY

"Catalyst Regeneration, Metal Catalysts" in *ECT* (online), posting date: December 4, 2000, by D. W. Robinson, UOP.

1. J. L. Figuerido, in J. L. Figuerido, ed., *Carbon Formation and Gasification on Nickel*, M. Nijhoff Publishers, Boston, 1982.
2. R. Hughes, *Deactivation of Catalysts*, Academic Press, London, 1984.
3. J. Oudar and H. Wise, eds., *Deactivation and Poisoning of Catalysts*, Marcel Dekker, New York, 1985.
4. J. B. Butt and E. E. Petersen, *Activation, Deactivation, and Poisoning of Catalysts*, Academic Press, San Diego, 1988.

5. P. J. Denny and M. V. Twigg, in Ref. 9, p. 577.

6. C. H. Bartholomew, *Chem. Eng.* **91**, 96 (1984).

7. J. B. Butt, in J. R. Anderson and M. Boudart, eds., *Catalysis—Science and Technology*, Vol. 6, Springer-Verlag, New York, 1984, p. 1.

8. R. J. Farrauto, C. H. Bartholomew, *Fundamentals of Industrial Catalytic Processes*, Kluwer Academic Publishers, London, 1997.

9. B. Delmon and G. F. Froment, eds., *Catalyst Deactivation 1980* (Studies in Surface Science and Catalysis, Vol. 6), Elsevier, Amsterdam, 1980.

10. B. Delmon and G. F. Froment, eds., *Catalyst Deactivation 1987* (Studies in Surface Science and Catalysis, Vol. 34), Elsevier, Amsterdam, 1987.

11. C. H. Bartholomew and J. B. Butt, eds., *Catalyst Deactivation 1991* (Studies in Surface Science and Catalysis, Vol. 68), Elsevier, Amsterdam, 1991.

12. B. Delmon and G. F. Froment, eds., *Catalyst Deactivation 1994* (Studies in Surface Science and Catalysis, Vol. 88), Elsevier, Amsterdam, 1994.

13. C. H. Bartholomew and G. A. Fuentes, eds., *Catalyst Deactivation 1997* (Studies in Surface Science and Catalysis, Vol. 111), Elsevier, Amsterdam, 1997.

14. B. Delmon and G. F. Froment, eds., *Catalyst Deactivation 1999* (Studies in Surface Science and Catalysis, Vol. 126), Elsevier, Amsterdam, 1999.

15. J. A. Moulijn, ed., A series of papers on Catalyst Deactivation, *Appl. Catal., A: Gen.* **212**, 1–255 (2001).

16. E. B. Maxted, *Adv. Catal.* **3**, 129 (1951).

17. L. L. Hegedus and R. W. McCabe, in Ref. 9, p. 47.

18. L. L. Hegedus and R. W. McCabe, *Catalyst Poisoning*, Marcel Dekker, New York, 1984.

19. J. B. Butt, in J. L. Figuerido, ed., *Progress in Catalyst Deactivation* (NATO Advanced Study Institute Series E, No. 54), M. Nijhoff Publishers, Boston, 1982, p. 153.

20. J. Barbier, in Ref. 3, p. 109.

21. C. H. Bartholomew, in Ref. 10, p. 81.

22. J. R. Rostrup-Nielsen, in Ref. 11, p. 85.

23. V. J. Volter and M. Hermann, *Z. Anorg. Allg. Chem.* **405**, 315 (1974).

24. K. Baron, *Thin Solid Films* **55**, 449 (1978).

25. R. D. Clay and E. E. Petersen, *J. Catal.* **16**, 32 (1970).

26. R. J. Madon and H. Shaw, *Catal. Rev. Sci. Eng.* **15**, 69 (1977).

27. C. H. Bartholomew, P. K. Agrawal, and J. R. Katzer, *Adv. Catal.* **31**, 135 (1982).

28. J. R. Rostrup-Nielsen, in J. L. Figuerido, ed., *Progress in Catalyst Deactivation* (NATO Advanced Study Institute Series E, No. 54), M. Nijhoff Publishers, Boston, 1982, p. 209.

29. H. Wise, J. McCarty, and J. Oudar, in Ref. 3, p. 1.

30. J. R. Rostrup-Nielsen and P. E. Nielsen, in Ref. 3, p. 259.

31. M. Perdereau and J. Oudar, *Surf. Sci.* **20**, 80 (1970).

32. J. Oudar, *Catal. Rev. Sci. Eng.* **22**, 171 (1980).

33. J. J. McCarroll, T. Edmonds, and R. C. Pitkethly, *Nature* **223**, 1260 (1969).

34. T. Edmonds, J. J. McCarroll, and R. C. Pitkethly, *J. Cat. Sci. Technol.* **8**, 68 (1971).

35. L. Ruan, F. Besenbacher, I. Stensgaard, and E. Laegsgaard, *Phys. Rev.* **69**, 3523 (1992).

36. J. Hepola, J. McCarty, G. Krishnan, and V. Wong, *Appl. Catal. B* **20**, 191 (1999).

37. W. Erley and H. Wagner, *J. Catal.* **53**, 287 (1978).

38. K. D. Rendulic and A. Winkler, *Surf. Sci.* **74**, 318 (1978).

39. D. W. Goodman and M. Kiskinova, *Surf. Sci.* **105**, L265 (1981).

40. M. Kiskinova and D. W. Goodman, *Surf. Sci.* **108**, 64 (1981).

41. S. Johnson and R. S. Madix, *Surf. Sci.* **108**, 77 (1981).

42. R. J. Madix, M. Thornberg, and S. B. Lee, *Surf. Sci.* **133**, L447 (1983).

43. E. L. Hardegree, P. Ho, and J. M. White, *Surf. Sci.* **165**, 488 (1986).
44. E. J. Erekson and C. H. Bartholomew, *Appl. Catal.* **5**, 323 (1983).
45. J. R. Rostrup-Nielsen and D. L. Trimm, *J. Catal.* **48**, 155 (1977).
46. D. L. Trimm, *Catal. Rev. Sci. Eng.* **16**, 155 (1977).
47. D. L. Trimm, *Appl. Catal.* **5**, 263 (1983).
48. C. H. Bartholomew, *Catal. Rev. Sci. Eng.* **24**, 67 (1982).
49. L. F. Albright and R. T. K. Baker, eds., *Coke Formation on Metal Surfaces* (ACS Symposium Series 202), American Chemical Society, Washington, D.C., 1982.
50. P. G. Menon, *J. Mol. Catal.* **59**, 207 (1990).
51. J. D. Deken, P. G. Menon, G. F. Froment, and G. Haemers, *J. Catal.* **70**, 225 (1981).
52. W. G. Durer, J. H. Craig, Jr., and J. Lozano, *Appl. Surf. Sci.* **45**, 275 (1990).
53. A. D. Moeller and C. H. Bartholomew, *Prepr.—Am. Chem. Soc., Div. Fuel Chem.* **25**, 54 (1980).
54. K. J. Marschall and L. Mleczko, *Ind. Eng. Chem. Res.* **38**, 1813 (1999).
55. J. R. Rostrup-Nielsen, in J. R. Anderson and M. Boudart, eds., *Catalysis—Science and Technology*, Vol. 5, Springer-Verlag, New York, 1984, p. 1.
56. F. Besenbacher, I. Chorkendorff, B. S. Clausen, B. Hammer, A. M. Molenbroek, J. K. Norscov, and I. Stensgaard, *Science* **279**, 1913 (1998).
57. T. Nemes, A. Chambers, and R. T. K. Baker, *J. Phys. Chem.* **102**, 6323 (1998).
58. C. H. Bartholomew, M. V. Strasburg, and H. Hsieh, *Appl. Catal.* **36**, 147 (1988).
59. C. K. Vance and C. H. Bartholomew, *Appl. Catal.* **7**, 169 (1983).
60. R. T. K. Baker and J. J. Chludzinski, *J. Catal.* **64**, 464 (1980).
61. D. E. Brown, J. T. K. Clark, A. I. Foster, J. J. McCarroll, and M. L. Sims, in Ref. 49, p. 23.
62. J. H. Bitter, K. Seshan, and J. A. Lercher, *J. Catal.* **183**, 336 (1999).
63. J. R. Rostrup-Nielsen, *J. Catal.* **33**, 184 (1974).
64. B. C. Gates, J. R. Katzer, and G. C. A. Schuit, *Chemistry of Catalytic Processes*, McGraw-Hill, New York, 1979.
65. C. Naccache, in C. Naccache, ed., *Deactivation of Acid Catalysts*, Marcel Dekker, New York, 1985.
66. W. G. Appleby, J. W. Gibson, and G. M. Good, *Ind. Eng. Chem. Process Des. Dev.* **1**, 102 (1962).
67. H. Beuther, O. H. Larson, and A. J. Perrotta, in Ref. 9, p. 271.
68. A. G. Gayubo, J. M. Arandes, A. T. Aguayo, M. Olazar, and J. Bilbao, *Ind. Eng. Chem. Res.* **32**, 588 (1993).
69. S. M. Augustine, G. N. Alameddin, and W. M. H. Sachtler, *J. Catal.* **155**, 217 (1989).
70. M. Guisnet and P. Magnoux, *Appl. Catal.* **54**, 1 (1989).
71. F. Bauer, V. Karazirev, C. Vlaev, R. Hanisch, and W. Weiss, *Chem. Techn.* **41**, 297 (1989).
72. W. A. Grotten, B. W. Wojciechowski, and B. K. Hunter, *J. Catal.* **138**, 343 (1992).
73. A. Bellare and D. B. Dadyburjor, *J. Catal.* **140**, 510 (1993).
74. M. A. Uguina, D. P. Serrano, R. V. Grieken, and S. Venes, *Appl. Catal.* **99**, 97 (1993).
75. C. Li, Y. Chen, S. Yang, and R. Yen, *Appl. Surf. Sci.* **81**, 465 (1994).
76. J. G. Buglass, K. P. d. Jong, and H. H. Mooiweer, in *Proc. 120th National Meeting of the American Chemical Society*, Aug. 20–24, 1995, p. 631.
77. D. Chen, H. P. Rebo, K. Moljord, and A. Holmen, in *Proc. 14th International Symposium on Chemical Reaction Engineering, Part B*, May 5–9, 1996, p. 2687.
78. M. Gusinet, P. Magnoux, and D. Martin, in Ref. 13, p. 1.
79. T. Masuda, P. Tomita, Y. Fujikata, and K. Hashimoto, in Ref. 14, p. 89.
80. H. S. Cerqueira, P. Magnoux, D. Martin, and M. Gusinet, in Ref. 14, p. 105.
81. S. E. Wanke and P. C. Flynn, *Catal. Rev. Sci. Eng.* **12**, 93 (1975).
82. P. Wynblatt and N. A. Gjostein, *Prog. Solid State Chem.* **9**, 21 (1975).

83. E. Ruckenstein and B. Pulvermacher, *AIChE J.* **19**, 356 (1973).
84. E. Ruckenstein and D. B. Dadyburjor, *Rev. Chem. Eng.* **1**, 251 (1983).
85. S. E. Wanke, in J. L. Figueiredo, ed., *Progress in Catalyst Deactivation* (NATO Advanced Study Institute Series E, No. 54), M. Nijhoff Publishers, Boston, 1982, p. 315.
86. R. T. Baker, C. H. Bartholomew, and D. B. Dadyburjor, *Stability of Supported Catalysts: Sintering and Redispersion, Catalytic Studies Division, 1991.*
87. C. H. Bartholomew, *Catalysis (Spec. Period. Rept.).* **10** (1992).
88. C. H. Bartholomew, *Appl. Catal., A: Gen.* **107**, 1 (1993).
89. C. H. Bartholomew, in Ref. 12, p. 1.
90. C. H. Bartholomew, in Ref. 13, p. 585.
91. C. H. Bartholomew and W. Sorenson, *J. Catal.* **81**, 131 (1983).
92. J. A. Moulijn, A. E. van Diepen, and F. Kapteijn, *Appl. Catal., A: Gen.* **212**, 13–16 (2001).
93. G. W. Bridger, and M. S. Spencer, in M. V. Twigg ed., *Catalyst Handbook*, 2nd ed., Manson Publishing, London, 1996, p. 441.
94. G. A. Fuentes, *Appl. Catal.* **15**, 33 (1985); G. A. Fuentes and F. A. Ruiz-Trevino, in Ref. 11, pp. 637–644.
95. J. P. Bournonville and G. Martino, in Ref. 9, p. 159.
96. G. A. Somorjai, *X-ray and Electron Methods of Analysis*, Plenum Press, New York, 1968.
97. S. R. Seyedmonir, D. E. Strohmayer, G. J. Guskey, G. L. Geoffroy, and M. A. Vannice, *J. Catal.* **93**, 288 (1985).
98. D. L. Trimm, in Ref. 11, p. 29.
99. A. G. Shastri, A. K. Datye, and J. Schwank, *Appl. Catal.* **14**, 119 (1985).
100. L. L. Hegedus and K. Baron, *J. Catal.* **54**, 115 (1978).
101. J. Summers and L. L. Hegedus, *Ind. Eng. Chem. Prod. Res. Dev.* **18**, 318 (1979).
102. U.S. Pat. 6,003,303 (Dec. 21, 1999), J. D. Peter-Hoblyn, J. M. Valentine, B. N. Sprague, and W. R. Epperly (to Clean Diesel Technologies, Inc.).
103. U.S. Pat. 6,013,599 (Jan. 11, 2000), I. Manson (to Redem Corp.).
104. U.S. Pat. 6,093,378 (July 25, 2000), M. Deeba, Y. K. Lui, and J. C. Dettling (to Engelhard Corp.).
105. M. E. Dry, in J. Anderson and M. Boudart, eds., *Catalysis—Science and Technology*, Springer-Verlag, New York, 1981, p. 159.
106. G. W. Huber, C. G. Guymon, B. C. Stephenson, and C. H. Bartholomew, *Catalyst Deactivation 2001* (Studies in Surface Science and Catalysis, Vol. 139), Elsevier, Amsterdam, 2001, p. 423.
107. G. Busca, L. Lietti, G. Ramis, and F. Berti, *Appl. Catal., B: Environ.* **18**, 1–36 (1998).
108. T. P. Kobylinski, B. W. Taylor, and J. E. Yong, in *Proc. SAE*, Detroit, 1974.
109. M. Shelef and H. S. Gandhi, *Platinum Met. Rev.* **18**, 1 (1974).
110. H. S. Gandhi, H. K. Stepien, and M. Shelef, *Mat. Res. Bull.* **10**, 837 (1975).
111. C. H. Bartholomew, *Ind. Eng. Chem. Prod. Res. Dev.* **14**, 29 (1975).
112. R. W. Clark, J. K. Tien, and P. Wynblatt, *J. Catal.* **61**, 15 (1980).
113. W. M. Shen, J. A. Dumesic, and C. G. Hill, *J. Catal.* **68**, 152 (1981).
114. R. B. Pannell, K. S. Chung, and C. H. Bartholomew, *J. Catal.* **46**, 340 (1977).
115. G. Lohrengel and M. Baerns, *Appl. Catal.* **1**, 3 (1981).
116. I. Qamar and J. G. Goodwin, in *Proc. 8th Am. Meeting Catal. Soc.*, Philadelphia, 1983 (Paper C-22).
117. J. G. Goodwin, D. O. Goa, S. Erdal, and F. H. Rogan, *Appl. Catal.* **24**, 199 (1986).
118. O. Watzenberger, T. Haeberle, D. T. Lynch, and G. Emig, in Ref. 11, p. 441.
119. M. Agnelli, M. Kolb, and C. Mirodatos, *J. Catal.* **148**, 9 (1994).
120. H. C. Lee and R. J. Farrauto, *Ind. Eng. Chem. Res.* **18**, 1 (1989).
121. R. J. Farrauto and H. C. Lee, *Ind. Eng. Chem. Res.* **29**, 1125 (1990).

122. F. Sperner and W. Hohmann, *Platinum Met. Rev.* **20**, 12 (1976).

123. J. M. Hess and J. Phillips, *J. Catal.* **136**, 149 (1992).

124. C. H. Bartholomew, *Catalysis Lett.* **7** (1990).

125. N. L. Wu and J. Phillips, *J. Phys. Chem.* **89**, 591 (1985).

126. N. L. Wu and J. Phillips, *Appl. Phys.* **59**, 769 (1986).

127. N. L. Wu and J. Phillips, *J. Catal.* **113**, 129 (1988).

128. A. Bielanski and M. Najbar, in Ref. 9, p. 127.

129. N. Burriesci, F. Garbassi, M. Petrera, and G. Petrini, in Ref. 9, p. 115.

130. Y. L. Xiong, R. Castillo, C. Papadopoulou, L. Dada, J. Ladriere, P. Ruiz, and B. Delmon, in Ref. 11, p. 425.

131. R. J. Farrauto, M. Hobson, T. Kennelly, and E. Waterman, *Appl. Catal.* **81**, 227 (1992).

132. P. L. Gai-Boyes, *Catal. Rev. Sci. Eng.* **34**, 1 (1992).

133. B. Delmon, in Ref. 12, p. 113.

134. K. M. Erickson, D. A. Karydis, S. Boghosian, and R. Fehrmann, *J. Catal.* **155**, 32 (1995).

135. B. Delmon, in Ref. 13, p. 39.

136. N. B. Jackson, A. K. Datye, L. Mansker, R. J. O'Brien, and B. H. Davis, in Ref. 13, p. 501.

137. S. A. Eliason and C. H. Bartholomew, in Ref. 13, p. 517.

138. A. Baranski, R. Dziembaj, A. Kotarba, A. Golebiowski, Z. Janecki, and J. B. C. Pettersson, in Ref. 14, p. 229.

139. C. A. Querini, F. Ravelli, M. Ulla, L. Cornaglia, and E. Miro, in Ref. 14, p. 257.

140. H. N. Pham, J. Reardon, and A. K. Datye, *Powder Technol.* **103**, 95 (1999).

141. D. S. Kalakkad, M. D. Shroff, S. Kohler, N. Jackson, and A. K. Datye, *Appl. Catal.* **133**, 335 (1995).

142. W. D. Callister, *Materials Science and Engineering: An Introduction*, John Wiley & Sons, Inc., New York, 2000.

143. R. L. Coble and W. D. Kingery, *J. Am. Ceram. Soc.* **39**, 381 (1956).

144. S. G. Deng and Y. S. Lin, *AIChE J.* **43**, 505 (1997).

145. S. G. Thoma, M. Ciftcioglu, and D. M. Smith, *Powder Technol.* **68**, 53 (1991).

146. M. Bankmann, R. Brand, B. H. Engler, and J. Ohmer, *Catal. Today* **14**, 225 (1992).

147. V. M. Kenkre and M. R. Endicott, *J. Am. Ceram. Soc.* **79**, 3045 (1996).

148. H. Song and J. R. G. Evans, *J. Am. Ceram. Soc.* **77**, 806 (1994).

149. J. Werther and W. Xi, *Powder Technol.* **76**, 39 (1993).

150. B. L. Bhatt, E. S. Schaub, E. C. Hedorn, D. M. Herron, D. W. Studer, and D. M. Brown, in G. J. Stiegel and R. D. Srivastava, eds. *Proc. of Liquefaction Contractors Review Conference*, U.S. Department of Energy, Pittsburgh, Pa., 1992, p. 403.

151. H. N. Pham and A. K. Datye, *Catal. Today* **58**, 233 (2000).

152. C. R. Bemrose and J. Bridgewater, *Powder Technol.* **49**, 97 (1987).

153. M. Ghadiri, J. A. S. Cleaver, V. G. Tuponogov, and J. Werther, *Powder Technol.* **80**, 175 (1994).

154. S. A. Weeks, P. Dumbill, *Oil Gas J.* **88**, 38 (Apr. 16, 1990).

155. R. Zhao, J. G. Goodwin, K. Jothimurugesan, J. J. Spivey, and S. K. Gangwal, *Ind. Eng. Chem. Res.* **39**, 1155 (2000).

156. P. K. Doolin, D. M. Gainer, and J. F. Hoffman, *J. Testing Evaluation* **21**, 481 (1993).

157. R. Oukaci, A. H. Singleton, D. Wei, and J. G. Goodwin, *Preprints, 217th National Meeting, ACS Division of Petroleum Chemistry*, Anaheim, Calif., 1999, p. 91.

158. M. J. Adams, M. A. Mullier, and J. P. K. Seville, *Powder Technol.* **78**, 5 (1994).

159. G. Emig and F. G. Martin, *Ind. Eng. Chem. Res.* **30**, 1110 (1991).

160. P. W. N. M. van Leeuwen, *Appl. Catal., A: Gen.* **212**, 61 (2001).

161. P. E. Garrou, *Chem. Rev.* **85**, 171 (1985).

162. B. C. Gates, *Catalytic Chemistry*, John Wiley & Sons, Inc., New York, 1992.
163. J. L. Latham and A. E. Burgess, *Elementary Reaction Kinetics*, 3rd ed., Butterworths, London, 1977.
164. W. Hartmeier (translated by J. Wiesner), *Immobilized Biocatalysts*, Springer-Verlag, Berlin, 1988; J. A. Hurlbut and co-workers, *J. Chem. Educ.* **50**, 149 (1973).
165. O. R. Zaborsky, in J. Burton and L. Garten, eds., *Advanced Materials in Catalysis*, Academic press, New York, 1977.
166. S. L. Neidleman, *Catalysis of Organic Reactions*, Marcel Dekker, New York, 1984.
167. S. Kindel, *Technology* **1**, 62 (1981).
168. M. N. Gupta, *Biotechnol. Appl. Biochem.* **14**, 1 (1991).
169. A. W. H. Fersht, *Enzyme Structure and Mechanism*, W. H. Freeman & Co., New York, 1984.
170. J. R. Whitaker, *Principles of Enzymology for the Food Sciences*, 2nd ed., Marcel Dekker, New York, 1994.
171. M. Matsumoto, K. Kida, and K. Kondo, *J. Chem. Technol. Biotechnol.* **70**, 188 (1997).
172. A. M. Azevedo, D. M. F. Prazeres, J. M. S. Cabral, and L. P. Fonseca, *J. Mol. Catal., B: Enzymatic* **15**, 147 (2001).
173. A. M. Klibanov, *Tibtech* **15**, 97 (1997).
174. D. Li, Z. Guo, and H. Liu, *J. Am. Chem. Soc.* **188**, 275 (1996).
175. S. Colombie, A. Gaunand, and B. Lindet, *J. Mol. Catal., B: Enzymatic* **11**, 559 (2001).
176. Z. Zhang, Z. He, and M. He, *J. Mol. Catal., B: Enzymatic* **14**, 85 (2001).
177. G. Toscano, D. Pirozzi, M. Maremonti, and G. Greco, Jr., *Biotechnol. Bioeng.* **44**, 682 (1994).
178. M. Mohanty, R. S. Ghadge, N. S. Patil, S. B. Sawant, J. B. Joshi, and A. V. Deshpande, *Chem. Eng. Sci.* **56**, 3401 (2001).
179. A. Ginsberg and W. R. Carroll, *Biochemistry* **4**, 2159 (1965).
180. T. J. Ahern and A. M. Klibanov, *Meth. Biochem. Anal.* **33**, 91 (1988).
181. K. A. Dill, D. O. V. Alonso, and K. Hutchinson, *Biochemistry* **28**, 5439 (1989).
182. T. J. Hancock and J. T. Hsu, *Biotechnol. Progr.* **12**, 494 (1996).
183. M. Longo and D. Combes, *J. Chem. Technol. Biotechnol.* **74**, 25 (1999).
184. L. M. M. Tijskens, R. Greiner, E. S. A. Biekman, and U. Konietzny, *Biotechnol. Bioeng.* **72**, 323 (2001).
185. V. V. Mozhaev and K. Martinek, *Enzyme Microb. Technol.* **6**, 50 (1984).
186. V. V. Mozhaev, N. S. Melik-Nubarov, V. Siksnis, and K. Martinek, *Biocatalysis* **3**, 189 (1990).
187. M. G. Roig and J. F. Kennedy, *Crit. Rev. Biotechnol.* **12**, 391 (1992).
188. A. M. Klibanov, *Anal. Biochem.* **93**, 1 (1979).
189. P. Monsan and D. Combes, *Methods Enzymol.* **137**, 584 (1988).
190. J. M. Guisan, P. Sabuquillo, R. Fernandez-Lafuent, G. Fernandez-Lorente, C. Mateo, P. J. Halling, D. Kennedy, E. Miyata, and D. Re., *J. Mol. Catal., B: Enzymatic* **11**, 817 (2001).
191. P. Wang, S. Dai, S. D. Waezsada, A. Y. Tsao, and B. H. Davison, *Biotechnol. Bioeng.* **74**, 249 (2001).
192. R. G. Silver, J. C. Summers, and W. B. Williamson, *Catalysis and Automotive Pollution Control II*, Elsevier, Amsterdam, 1991, p. 167.
193. G. B. Fisher, M. G. Zammit, and J. LaBarge, SAE Report 920846, 1992.
194. R. J. Farrauto and R. M. Heck, *Catal. Today* **51**, 351 (1999).
195. G. W. Huber, C. G. Guymon, B. C. Stephenson, and C. H. Bartholomew, *Catalyst Deactivation 2001* (Studies in Surface Science and Catalysis, Vol. 139), Elsevier, Amsterdam, 2001, p. 423.
196. U.S. Pat. 6,169,120 (Jan. 2, 2001), G. L. Beer (to Syntroleum Corp.).
197. C. H. Bartholomew, M. W. Stoker, L. Mansker, and A. Datye, in Ref. 14, p. 265.

198. U.S. Pat. 5,728,894 (Mar. 17, 1998), O. Nagano and T. Watanabe (to Ashahi Kasei Kogyo Kabushiki Kaisha).

199. T. Maillet, J. Barbier, and D. Duprez, *Appl. Catal., B* **9**, 251 (1996).

200. WIPO Pat. 93/16020A3 (Sept. 16, 1993), G. Mathys, L. Martens, M. Baes, J. Verduijn, D. Huybrechts, and C. Renata (to Exxon Chem.).

201. U.S. Pat. 5,672,800 (Sept. 30, 1997), G. Mathys, L. Martens, M. Baes, J. Verduijn, and D. Huybrechts (to Exxon Chem.).

202. U.S. Pat. 6,080,903 (June 26, 2000), L. Stine, B. Muldoon, S. Gimre, and R. Frame (to UOP).

203. B. Subramaniam, V. Arunajatesan, and C. J. Lyon, in Ref. 14, p. 63.

204. WIPO Pat. 99/33769 (July 8, 1999), D. Ginosar, R. Fox, and P. Kong (to Lockheed Martin).

205. F. H. Ribeiro, A. L. Bonivardi, and C. Kim, *J. Catal.* **150**, 186 (1994).

206. D. Ginosar and B. Subramaniam, in Ref. 12, p. 327.

207. E. E. Petersen, in Ref. 13, p. 87.

208. J. W. Gosselink and J. A. R. V. Veen, in Ref. 14, p. 3.

209. L. Lin, T. Zao, J. Zang, and Z. Xu, *Appl. Catal.* **67**, 11 (1990).

210. D. E. Resasco and G. L. Haller, *Catalysis (Spec. Period. Rept.)* **11**, 379 (1994).

211. R. D. Cortright and J. A. Dumesic, *J. Catal.* **148**, 771 (1994).

212. H. Weyten, K. Keizer, A. Kinoo, J. Luyten, and R. Leysen, *AIChE J.* **43**, 1819 (1997).

213. P. Praserthdam, T. Mongkhonsi, S. Kunatippapong, B. Jaikaew, and N. Lim, in Ref. 13, p. 153.

214. WIPO Pat. 00/69993 (May 12, 2000), B. Rose and T. Kiliany (to Mobil Oil Corp.).

215. A. Guerrero-Ruiz, A. Sepulveda-Escribano, and I. Rodriguez-Ramos, *Catal. Today* **21**, 545 (1994).

216. D. Qin and J. Lapszewicz, *Catal. Today* **21**, 551 (1994).

217. S. Stagg and D. Resasco, in Ref. 13, p. 543.

218. K. Fujimoto, K. Tomishige, O. Yamazaki, Y. Chen, and X. Li, *Res. Chem. Intermed.* **24**, 259 (1998).

219. U.S. Pat. 5,191,142 (Mar. 2, 1993), C. Marshall and J. Miller (to Amoco Corp.).

220. A. G. Gayubo, A. T. Aguayo, A. E. S. D. Campo, P. L. Benito, and J. Bilbao, in Ref. 14, p. 129.

221. U.S. Pat. 5,248,647 (Sept. 28, 1993), P. T. Barger (to UOP).

222. WIPO Pat. 99/42202 (Feb. 20, 1998), J. Cox and J. Evans (to Johnson Matthey).

223. WIPO Pat. 98/50487 (May 2, 1997), S. Leviness, C. Mart, W. Behrmann, S. Hsia, and D. Neskora (to Exxon Research and Engineering Co.).

224. C. H. Bartholomew, in M. Oballa and S. Shih, eds., *Catalytic Hydroprocessing of Petroleum and Distillates*, Marcel Dekker, New York, 1993, p. 1.

225. J. Summers and W. B. Williamson, in J. Armor, ed., *Environmental Catalysis 1993*, Vol. 552, American Chemical Society, Washington, D.C., 1993, p. 94.

226. J. Dettling, Z. Hu, Y. K. Lui, R. Smaling, C. Z. Wan, and A. Punke, in *Studies in Surface Science and Catalysis*, Vol. 96, Elsevier, Amsterdam, 1995, p. 461.

227. U.S. Pat. 4,910,180 (Mar. 20, 1990), M. Berndt and D. Ksinsik (to Doduco).

228. U.S. Pat. 4,985,387 (Jan. 15, 1991), M. Prigent, G. Blanchard, and P. Phillippe (to ProCatalyse).

229. U.S. Pat. 5,041,407 (Aug. 20, 1991), W. Williamson, D. Linden, and J. Summers (to Allied-Signal Inc.).

230. U.S. Pat. 5,116,800 (May 26, 1992), W. Williamson, D. Linden, and J. Summers (to Allied-Signal Inc.).

231. U.S. Pat. 5,234,881 (Aug. 10, 1993), C. Narula, W. Watkins, and M. Chattha (to Ford Motor Co.).

232. U.S. Pat. 5,254,519 (Oct. 19, 1993), C. Wan, S. Tauster, and H. Rabinowitz (to Engelhard Corp.).

233. T. Furuya, S. Yamanaka, T. Hayata, J. Koezuka, T. Yoshine, and A. Ohkoshi, in *Proc. Gas Turbine Conference and Exhibition*, Anaheim, Calif., 1987.

234. T. Kawakami, T. Furuya, Y. Sasaki, T. Yoshine, Y. Furuse, and M. Hoshino, in *Proc. Gas Turbine and Aeroengine Congress and Exposition*, Toronto, Ont., June 4–8, 1989.

235. U.S. Pat. 5,250,489 (1993), R. D. Betta, F. Ribeiro, T. Shoji, K. Tsurumi, N. Ezawa, and S. Nickolas (to Catalytica, Inc.).

236. T. Fujii, Y. Ozawa, and S. Kikumoto, *J. Eng. Gas Turbines Power* **120**, 509 (1998).

237. D. O. Borio and N. S. Schbib, *Comput. Chem. Eng.* **19**, S345 (1995).

238. U.S. Pat. 5,028,634 (July 2, 1991), R. Fiato (to Exxon Research and Engineering Co.).

239. R. Zhao and J. G. Goodwin, Jr., K. Jothimurugesan, S. K. Gangwal, and J. J. Spivey, *Ind. Eng. Chem. Res.* **40**, 1065–1075 (2001).

240. R. Zhao, J. G. Goodwin, Jr., K. Jothimurugesan, S. K. Gangwal, and J. J. Spivey, *Ind. Eng. Chem. Res.* **40**, 1076 (2001).

241. U.S. Pat. 5,939,350 (Aug. 17, 1999), A. H. Singleton, R. Oukaci, and J. G. Goodwin (to Energy International Corp.).

242. U.S. Pat. 6,087,405 (July 11, 2000), S. Plecha, C. H. Mauldin, and L. E. Pedrick (to Exxon Research and Engineering Co.).

243. U.S. Pat. 6,124,367 (Sept. 26, 2000), S. Plecha, C. H. Mauldin, and L. E. Pedrick (to Exxon Research and Engineering Co.).

244. W.O. Pat. 00/71253 (Nov. 30, 2000), A. H. Singleton, R. Oukaci, and J. G. Goodwin (to Energy International Corp.).

245. M. Seitz, E. Klemm and G. Emig, in Ref. 14, p. 211.

246. S. Masamune and J. M. Smith, *AIChE J.* **12**, 384 (1966).

247. Y. Murakami, T. Kobayashi, T. Hattori, and M. Masuda, *Ind. Eng. Chem. Fundam.* **7**, 599–605 (1968).

248. W. H. J. Stork, in G. F. Froment, B. Delmon, and P. Grange, eds., *Hydrotreatment and Hydrocracking of Oil Fractions*, (Studies in Surface Science and Catalysis, Vol. 106), Elsevier, New York, 1997, pp. 41–67.

249. G. D. Parks, A. M. Schaffer, M. J. Dreiling, and C. B. Shiblom, *Prepr.—Am. Chem. Soc., Div. Petr. Chem.* **25**, 335 (1980).

250. D. L. Trimm, in J. L. Figueiredo, ed., *Progress in Catalyst Deactivation* (NATO Advanced Study Institute Series E, No. 54), M. Nijhoff Publishers, Boston, 1982, pp. 3–18.

251. E. R. Becker and J. J. Wei, *J. Catal.* **46**, 365–381 (1977).

252. U.S. Pat. 5,498,638 (Mar. 12, 1996), D. C. Long (to Exxon Research and Engineering Co.).

253. C. H. Bartholomew, *Appl. Catal., A: Gen.* **107**, 1–57 (1993).

254. B. R. Powell, *Presented at the Materials Research Society Annual Meeting*, Boston, Nov. 16–21, 1980 (Paper H9).

255. R. Heck and R. Farrauto, *Catalytic Air Pollution Control: Commercial Technology*, Van Nostrand Reinhold, New York, 1995.

256. D. L. Trimm, *Appl. Catal., A: Gen.* **212**, 153 (2001).

257. G. Berrebi, P. Dufresne, and Y. Jacquier, *Environ. Prog.* **12**, 97 (1993).

258. R. L. D'Aquino, *Chem. Eng.* **107**, 32 (2000).

259. T. Chang, *Oil Gas J.* **96**(41), 49 (1998).

260. S. R. Blashka and W. Duhon, *Int. J. Hydrocarbon Eng.* **4**(1), 60 (1998).

261. D. C. McCulloch, in B. E. Leach, ed., *Applied Industrial Catalysis*, Academic Press, New York, 1983, pp. 103–110.

262. J. P. Franck and G. Martino, in J. L. Figueiredo, ed., *Progress in Catalyst Deactivation* (NATO Advanced Study Institute Series E., No. 54), M. Nijhoff Publishers, Boston, 1982, pp. 355–398.

263. J. J. Spivey, G. W. Roberts, and B. H. Davis, eds., *Catalyst Deactivation 2001* (Studies in Surface Science and Catalysis, Vol. 139), Elsevier, Amsterdam, 2001.

264. U.S. Pat. 5,043,517 (Aug. 27, 1991), J. H. Haddad, N. Mohsen, and H. Owen (to Mobil Oil Corp.).

265. U.S. Pat. 5,306,682 (Apr. 26, 1994), M. Ueda, T. Murakami, S. Shibata, K. Hirabayashi, T. Kondoh, K. Adachi, N. Hoshino, and S. Inoue (to Research Association for the Utilization of Light Oil JP).

266. U.S. Pat. 5,675,048 (Oct. 7, 1997), S. Y.-F. Zhang, C. D. Gosling, P. A. Sechrist, and G. A. Funk (to UOP).

267. G. Panattoni and C. A. Querini, in Ref. 263, p. 181.

268. U.S. Pat. 5,854,162 (Dec. 29, 1998), P. Dufresne and N. Brahma (to Eurecat).

269. U.S. Pat. 5,883,031 (Mar. 16, 1999), R. A. Innes, D. L. Holtermann, and B. F. Mulaskey (to Chevron).

270. S. C. Fung, *Chemtech* **24**, 40 (1994).

271. J. C. Alfonso, D. A. G. Aranda, M. Schmal, and R. Frety, *Fuel Proc. Technol.* **50**, 35 (1997).

272. G. J. Arteaga, J. A. Anderson, and C. H. Rochester, *J. Catal.* **187**, 219 (1999).

273. C. L. Pieck, C. R. Vera, and J. M. Parera, in Ref. 263, p. 279.

274. D. R. Acharya, R. Hughes, M. A. Kennard, and Y. P. Liu, *Chem. Eng. Sci.* **47**, 1687 (1992).

275. U.S. Pat. 4,999,326 (Mar. 12, 1991), D. L. Sikkenga, I. C. Zaenger, and G. S. Williams (to Amoco).

276. A. Ekstrom and J. Lapszewicz, *J. Phys. Chem.* **88**, 4577 (1984); *J. Phys. Chem.* **91**, 4514 (1987).

277. U.S. Pat. 5,268,344 (Dec. 7, 1993), L. E. Pedrick, C. H. Mauldin, and W. C. Behrmann (to Exxon Research and Engineering Co.).

278. U.S. Pat. 5,338,439 (Aug. 16, 1994), H. Owen and P. H. Schipper (to Mobil Oil Corp.).

279. U.S. Pat. 5,198,397 (Mar. 30, 1993), M. F. Raterman (to Mobil Oil Corp.).

280. U.S. Pat. 5,393,717 (Feb. 28, 1995), M. R. Apelian, A. S. Fung, G. H. Hatzikos, C. R. Kennedy, C.-H. Lee, T. R. Kiliany, P. K. Ng, and D. A. Pappal (to Mobil Oil Corp.).

281. U.S. Pat. 5,340,957 (Aug. 23, 1994), D. E. Clark (to Union Oil Co.).

282. Y. Yoshimura, T. Sato, H. Shimada, N. Matsubayashi, M. Imamura, A. Nishijima, S. Yoshitomi, T. Kameoka, and H. Yanase, *Energy Fuels* **8**, 435 (1994).

283. E. S. Oh, Y. C. Park, and I. C. Lee, *J. Catal.* **172**, 314 (1997).

284. A. T. Aquavo, A. G. Gayubo, A. Atutxa, M. Olazar, and J. Bilbao, *J. Chem. Tech. Biotech.* **74**, 1082 (1999).

285. U.S. Pat. 4,777,156 (Oct. 11, 1988), N. P. Forbus, M. May-Som Wu (to Mobil Oil Corp.).

286. A. Krishna, C. Hsieh, A. E. English, T. Pecoraro, and C. Kuehler, *Hydrocarbon Process.* 59–66 (Nov. 1991).

287. C. Altomare, G. Koermer, P. Schubert, S. Suib, and W. Willis, *Chem. Mater.* **1**, 459–463 (1989).

288. A. Aguinaga and M. Montes, *Appl. Catal., A: Gen.* **90**, 131 (1992).

289. G. V. Lowry and M. Reinhard, *Environ. Sci. Technol.* **34**, 3217 (2000).

290. U.S. Pat. 4,830,997 (May 16, 1989), D. C. Trinh and A. Desvard (to Institut Français du Petrole).

291. U.S. Pat. 5,230,791 (July 27, 1993), D. E. Sherwood (to Texaco Inc.).

292. S. C. Fung, in Ref. 263, p. 399.

293. U.S. Pat. 5,672,801 (Sept. 30, 1997), B. Didillon (to Institut Français Du Petrole).

294. U.S. Pat. 4,929,576 (May 29k 1990), Y.-Y. P. Tsao and R. von Ballmoos (to Mobil Oil Corp.).

295. P. Dufresne, N. Brahma, and F. Girardier, *Revue de l'Institut Français due Petrole* **50**, 283 (1995).

296. U.S. Pat. 5,275,990 (Jan. 4, 1994), F. T. Clark and A. L. Hensley, Jr. (to Amoco Corp.).

297. A. Brito, R. Arvelo, and A. R. Gonzalez, *Ind. Eng. Chem. Res.* **37**, 374 (1998).

298. V. L. S. Teixeira-da-Silva, F. P. Lima, and L. C. Dieguez, *Ind. Eng. Chem. Res.* **37**, 882 (1998).

299. C. E. Snape, M. C. Diaz, Y. R. Tyagi, S. C. Martin, and R. Hughes, in Ref. 263, p. 359.

300. U.S. Pat. 5,445,728 (Aug. 29, 1995), D. E. Sherwood, Jr., and J. R. Hardee, Jr. (to Texaco Inc.).

301. U.S. Pat. 5,021,377 (June 4, 1991), M. K. Maholland, C.-M. Fu, R. E. Lowery, D. H. Kubicek, and B. J. Bertus (to Phillips Petroleum Co.).

302. U.S. Pat. 5,141,904 (Aug. 25, 1992), D. H. Kubicek, C.-M. Fu, R. E. Lowery, and M. K. Maholland (to Phillips Petroleum Co.).

303. U.S. Pat. 5,151,391 (Sept. 29, 1992), C.-M. Fu, M. Maholland, and R. E. Lowery (to Phillips Petroleum Co.).

304. U.S. Pat. 6,063,721 (May 16, 2000), Y. Hu, B. Luo, K. Sun, Q. Yang, M. Gong, J. Hu, G. Fang, and Y. Li (to China Petro-Chemical Corp.).

305. P. L. Walker, Jr., F. Rusinko, Jr., and L. G. Austin, *Adv. Catal.* **11**, 133–221 (1959).

306. J. W. Fulton, *Chem. Eng.* **96**, 111–114 (1988).

307. J. T. Richardson, *Ind. Eng. Chem. Process Des. Dev.* **11**, 8 (1972).

308. P. B. Weisz and R. B. Goodwin, *J. Catal.* **6**, 227 (1966).

309. J. R. Rostrup-Nielsen, *J. Catal.* **21**, 171–178 (1971).

310. D. B. Dadyburjor, in B. Delmon and G. F. Froment, eds., *Catalyst Deactivation*, Elsevier, Amsterdam, 1980, pp. 341–351.

311. C. Bartholomew in I. I. Harvath, eds., *Encyclopedia of Catalysis*, Vol. 2, John Wiley & Sons, New York, 2003, pp. 182–315.

CALVIN BARTHOLOMEW
Brigham Young University

CATALYSTS, SUPPORTED

1. Introduction

The field of heterogeneous catalysis of organic reactions covers an enormous undertaking, including the conversion of the primary source of almost all the organic compounds known (oil and gas) to a range of functional intermediates that are then used to prepare the vast array of chemical products used in every-day life. The field can be split into two distinct areas—heterogeneous catalysis of gas-phase reactions and of liquid-phase reactions. The former is a well-known and long established field, representing some of the largest chemical processes known—the cracking and other conversions of crude oil and gas which lays the foundation for the remainder of organic chemistry. These processes are typically high temperature, high energy processes, operating on a megaton scale, and typically use zeolites as catalysts. These processes are very well described in a series of books and reviews as well as in other chapters of this work (see *Catalysis, Molecular Sieves, Supported Catalysts*) (1–5).

This contribution, therefore, relates specifically to the use of heterogeneous catalysis to the synthesis of fine chemicals. The vast majority of these systems operate under moderate temperatures and in the liquid phase, although there are a few that take place in the gas phase.

2. Preparation Methods for Inorganic Supported Catalysts

There are several methods for the preparation of supported catalysts such as those described in this article (6). One of the key concepts is the maximization of surface area; this is normally achieved by having a very porous support material, and the methods below aim to either generate materials with high surface areas, or to coat pre-formed high surface area supports with small amounts of catalytic species. In the latter case, it is important to note that the surface area of a typical commercially available support material such as alumina or silica is in the range 100–600 m^2/g^1. On the other hand, a crystalline solid such as potassium fluoride (KF) or zinc chloride will have a surface area of <1 m^2/g^1.

Given that the exposed surface area is where reaction takes place, it is clearly very important not to overload the material to the point where microcrystals are formed on the surface although there may be instances where microcrystals are a necessary feature of the catalyst (eg, in epoxidations using supported silver catalysts) (7). The formation of the high surface area supports (which often may have catalytic activity themselves) is typically carried out by the condensation of inorganic precursors in the presence of a structure directing agent, whose role is to control the formation of a porous or layered structure and thus maximise the surface area.

Mechanical and thermal properties are also very important features of supports (8–10). Mechanical attrition of particles in the aggressive environment of a large chemical reactor must be allowed for and may cause changes in performance on reuse (this is especially important when a solid is used in a high shear stirred tank reactor). Many commonly used support materials have more than adequate thermal stability to cope with the temperatures commonly encountered in liquid phase-organic reactions. Some materials are sensitive, however, notably those that are organic polymer based (eg, acid resins) or have some surface-based organic functionalities (eg, sulfonic acids). The key properties of some of the more widely used support materials are given in Table 1.

2.1. Coating Pre-Formed Supports. The simplest method, and often the most appropriate, is wet impregnation, where a solution of the material to be supported is dissolved in a solvent and mixed with the support. The incipient wetness technique involves filling the pages of the support with a solution of the reagent. Removal of the solvent then leads to the supported catalyst. The degree of dispersion depends on several factors such as the rate of removal of solvent—too

Table 1. **Some Widely Used Support Materials**

Support	Surface area (m^2/g)	Pore size (nm)	Comments
commercial silicas	300–600	2–20	widely available moderate-large pore volumes wide pore-size distribution.
synthetic silicas (micelle-templated)	to >1000	~2 to ~10	can have narrow pore size distribution and large pore volumes
alumina	100–300+	various (broad distributions)	commercial materials are widely available synthetic, narrow poresize distribution materials are as yet, less well established (less stable)
montmorillonites	50–300	<1 to >10	natural clays have small intermamellar-spacings but can swell pillared clays provide large (>1 mm) spacing; acid-treated clays are amorphous and often mesoporous
zeolitic materials (inc. MCMs)	to ~1000	<1 to >1	narrow pore distribution original zeolites are microporous and may be unsuitable for many liquid phase organic reactions larger pore materials including MCMs are more suitable
carbons	to > 1000	various	complex surface chemistry can complicate applications synthetic mesoporous carbons are among recent innovations
organic polymers (Inc. resins)	[a]	various[a]	limited thermal and chemical stability in many cases but often inexpensive newer materials such as Nafion-based have many proven applications

[a]Generally not reported and may be meaningless due to swelling on applications.

rapid can lead to precipitation of clusters of catalyst, leading to reduced accessibility and possibly pore blockage. For many of the smaller pore size catalysts, it may be sensible to allow a period of a few hours contact time to allow ingress of the solution into the pores (incipient wetness), but for many catalysts this is not always necessary. Loading is also a critical factor. Too high a loading can again lead to clustering (microcrystals) and reduced activity, too low can lead to the need for a larger quantity of catalyst than is optimal. In general, 1 mmol catalyst per gram of support is an upper limit, but the situation can be complicated by factors such as a strong interaction with the support (9). In this case, there may be a small number of sites where there is a particularly advantageous interaction between the support and the catalyst, leading to unusually high activity. In these cases, a very low loading may be required, with additional material making very little difference to the overall activity of the supported catalyst.

Choice of solvent is important, and is especially critical in strongly acidic systems such as supported $AlCl_3$, where the extreme sensitivity and reactivity of the catalyst limits the choice of solvent to simple aromatics such as benzene and chlorobenzene. Posttreatment of such catalysts is also vital, partly to remove loosely bound material that can desorb easily during reaction and contaminate the liquid phase, and partly to enhance the activity of the catalyst. Thorough washing with solvents is generally enough, and the problem can be mitigated by appropriate choice of loading. Increases in activity are often brought about by heating of the catalyst (usually to $100-150°C$), which serves to remove the last traces of solvent and water from the catalyst. More extreme heating usually is detrimental, but in the case of clayzic (11) a reaction between the support and the zinc chloride is effected by heating to $280°C$, which leads to a dramatic increase in activity. Similarly, sulfated zirconia is prepared by treating ZrO_2 with sulfuric acid, and then calcining at high temperature.

Organic species can be supported onto oxide surfaces by grafting species such as $(RO)_3SiR^1$, where R^1 is the catalytic site or precursor thereto. The RO groups are typically methoxy or ethoxy, and these form $M-O-Si$ bonds to the surface of the support. It is rare that all three RO groups are replaced with $M-O-Si$ linkages, but singly and doubly bound species are easily formed, with the latter being more stable to thermolysis and solvolysis. Physisorbed material can also be formed, depending on the nature of the R' groups and the surface. This must be removed by appropriate washing before use.

Other techniques have also been used less frequently. These include grinding of two dry powders, or direct addition of both components to the reaction mix, where the supported catalyst forms in situ. Such approaches are occasionally of some value, but generally reproducibility is less good than with a carefully preformed system. Ion exchange can be an important method of preparation but is obviously limited to appropriate materials such as ion-exchange resins and solid acids that may be converted to sources of catalytically useful metal ions.

2.2. Formation of High Surface Area Supports/Catalysts.

Catalysts can also be prepared by (co)-precipitation of oxide precursors, often brought about by a change in pH (eg, in the case of silica, sodium silicate can be partly protonated on decreasing pH from ~12 to ~10; the SiOH units formed readily condense to form SiOSi units and eventually silica) or by hydrolysis of alkoxides $M(OR)_n$. This route is typically carried out in water or aqueous alcohols, and can be used to prepare catalytic systems with one or more M atoms present. When this is the case, it is important to control the relative hydrolysis rates of the two different M alkoxides. Fortunately, the hydrolysis rates of many such species are well understood, and usually a combination of variation of the R group (larger = slower hydrolysis) and/or the use of bidentate systems such as acetylacetone is sufficient to match the hydrolysis rates of the two components so that a uniform material can be prepared.

An extension of this approach allows the incorporation of organic functionality directly into an inorganic framework. Catalysts based on the co-condensation of $Si(OR)_4$ and $R'Si(OR)_3$ can be achieved such that an organic–inorganic hybrid material is formed, where R' represents the active site of the catalyst, or a precursor thereto, or may be a group that modifies the polarity of the surface, improving adsorption–desorption phenomena and improving the activity of the catalyst. In these methods, it is important that the materials formed have a high surface area, in order to allow access to as much surface per unit mass catalyst as possible (in optimum cases surface areas of ~1000 m^2/g can be achieved) and it is also critical that the pores in the material are sufficiently large enough and are open enough to allow free movement of the reactants and products (12). This is generally achieved by the use of templates, molecules that are added to the synthesis mixture and that encourage the formation of the solid around them, and that are subsequently removed. For microporous materials such as zeolites, the templates are normally quaternary ammonium salts, and the pores formed are very small (in general <1 nm). These microporous systems can accommodate small organic molecules readily, but exclude larger organics, limiting their usefulness. Larger pores can be achieved by the use of micellar templates or liquid-crystal phases, using aggregates of molecules such as quaternary ammonium surfactants, long-chain amines, or block polyethers as structure directing agents. In this way, standard, well-understood micelle chemistry can be used to tailor pore sizes of 2–10 nm, allowing access to the majority of organic molecules. For purely inorganic materials, high temperature calcination steps are usually used to remove template, but for organic–inorganic hybrid materials, lower temperature solvent extraction methods are used. These have the advantage of allowing recovery of the template and subsequent reuse, reducing the cost of the material.

While these template assisted formations of high surface area oxides are now well established and lead to a range of materials with various pore systems (such as MCM-41—which has a hexagonally stacked series of relatively straight cylindrical pores, and MCM-48, which has a three-dimensional network of interconnecting pores), nontemplated methods are also known, which lead to structured and regular materials. Hydrotalcites are one such class of compounds, where a layered structure, reminiscent of the clays, is formed—again by alteration of pH.

Many of these methods lead directly to powdered solids that are directly usable in stirred tank reactors and that can be readily separated from reaction mixtures by centrifugation. However, some systems lend themselves to processing as films or monoliths, which may be more appropriate for use in continuous reactors by immobilisation onto, eg, a metal support or even, in some cases, the formation of a membrane.

3. Supported Catalysts as Solid Acids

Solid acids are generally categorised by their Brønsted and/or Lewis acidity, the strength and number of these sites, and the morphology of the support (eg, surface area, pore size). The synthesis of pure Brønsted and pure Lewis solid acids as well as the control of the acid strength are important objectives (13). Pure Lewis acidity is hard to achieve with hydroxylated support materials especially

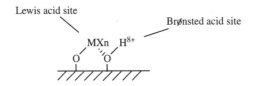

Fig. 1 Brønsted acidity on an hydroxylated support material enhanced by complexation with a neighboring supported Lewis acid.

as a result of polarization of those groups by supported Lewis acid centers such as shown in Figure 1.

Pure Lewis acidity can be important in maximizing selectivity in rearrangement reactions such as those of epoxides whereas Brønsted acidity can be necessary in some Friedel-Crafts reactions and reactions at sterically hindered sites. With acid strength, high acidity is often necessary for alkylations and esterifications while lower strength sites are preferred for acetal formation and hydrolysis reactions. In addition, the steric constraints imposed by the porous structure of many solid acids can influence reaction pathway although actual shape selective catalysis is largely restricted to microporous materials rather than the mesoporous solid acids favored in many liquid-phase organic reactions.

Acids are the most widely used catalysts in organic chemistry with applications covering all major sectors from petrochemicals to pharmaceuticals. The most commonly used include H_2SO_4, HF, $AlCl_3$, and BF_3 and are typically soluble in the organic reaction medium or remain as a separate liquid phase.

At the end of the reaction, such acids are normally destroyed in a water quench stage and require subsequent neutralization thus consuming additional (alkaline) resource and producing salt waste. These increasingly important environmental issues are incentives to utilize solid acids, which stay in a separate and easily recoverable phase from the organic components throughout the reaction. Acidic resins and clays have a proven track record in organic chemicals manufacturing but here we will consider newer solid acids that can substitute environmentally threatening soluble and liquid acids.

3.1. Supported Lewis Acids. The most widely used Lewis acid is aluminium chloride and its immobilization onto typical support materials such as clays, aluminas, and silicas has been widely studied (14–21). This can be achieved from the vapor phase or from a suitable solvent (both aromatic hydrocarbons and chloroaliphatics have been suggested), but in all cases we can expect some combination of physisorbed and chemisorbed, and various levels and types of acidic sites on the surface species (Fig. 2).

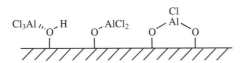

Fig. 2. Possible surface species formed on reaction of $AlCl_3$ with a hydroxylated support material.

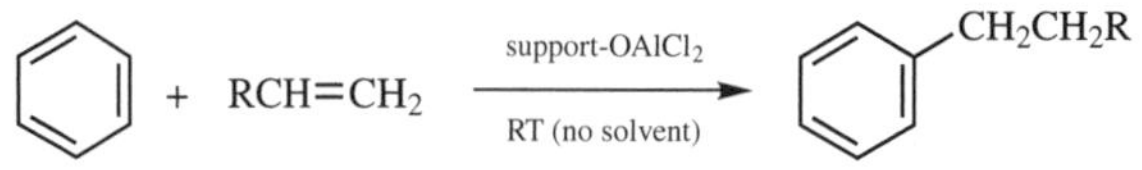

$$\text{C}_6\text{H}_6 + \text{RCH}=\text{CH}_2 \xrightarrow[\text{RT (no solvent)}]{\text{support-OAlCl}_2} \text{C}_6\text{H}_5\text{-CH}_2\text{CH}_2\text{R}$$

Fig. 3. Formation of alkylbenzenes using supported aluminium chloride as catalysts (r.t. = room temperature).

Catalysts of this type have been shown to be active in reactions including *n*-alkanes isomerizations, aromatic alkylations and oligomerizations and polymerization of unsaturated hydrocarbons (21,22). The use of monodisperse mesoporous solids (eg, 24 Å micelle templated silica) has been shown to give enhanced selectivities, eg, in terms of the degree of ring monoalkylation in the reaction of benzene with alkenes (up to 100% with >C14) although it is not clear if this is due to shape selectivity (largely unproven in pores larger than that in microporous zeolite materials) or diffusional factors (Fig. 3) (17). The proportion of the particularly important linear alkylbenzenes compares well with more traditional homogeneous catalytic methods. These are more active catalysts than the solid acids now being introduced into commercial alkyl benzene plants.

Supported Lewis acids, including supported aluminium chloride, has been described as good catalysts for the oligomerization of mixed refinery feedstocks leading to hydrocarbon resin products (Fig. 4) (21,23). This methodology can avoid the need for a water quench step (necessary in the familiar homogeneous Lewis acid continuous stirred tank reaction method) and enable catalyst reuse. These novel heterogeneous systems can also be expected to lead to new products with different and potentially useful properties such as softening point, molecular weights, and a degree of cross-linking.

Supported complexes of boron trifluoride can be used to make aromatic resins in this way (24). These catalysts are actually believed to be largely Brønsted acidic due to the use of alcohols (or other protic molecules) as cocatalysts (Fig. 5). The catalyst has also been reported as being active in other reactions typically catalyzed by homogeneous BF_3 including phenol alkylations and Claisen condensation reactions.

Commercial Friedel-Crafts catalysts based on supported Lewis acids have been developed for use in typical batch reactions. These include acid clay

unsaturated/saturated aliphatic and/or aromatic hydrocarbon $\xrightarrow[\substack{\text{RT or higher} \\ \text{(no solvent)}}]{\text{supported Lewis acid}}$ hydrocarbon resins

Fig. 4. Production of hydrocarbon resins using refinery feedstocks supported Lewis acid catalyst.

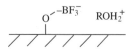

Fig. 5. Active species in supported boron trifluoride complexes.

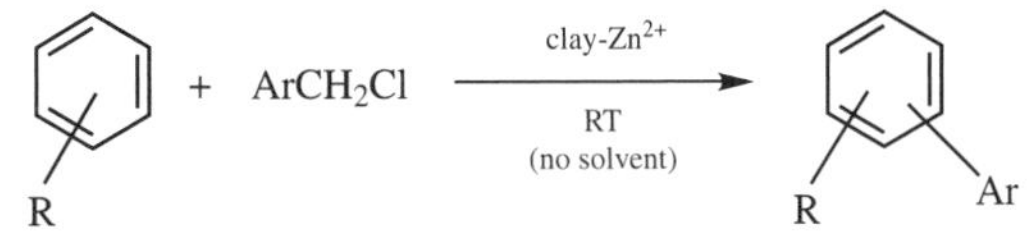

Fig. 6. Benzylations of aromatic substrates catalyzed by commercial acid clay supported zinc(II) catalyst.

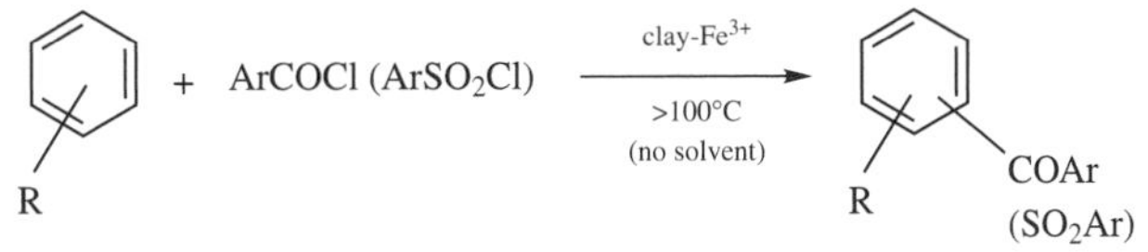

Fig. 7. Benzoylations and sulfonylations of aromatic substrates catalyzed by commercial acid clay supported iron(III) catalyst.

supported zinc(II) salts for use in reactions including benzylations (Fig. 6) and iron salts for the more demanding benzoylations and sulfonylations (Fig. 7) (25).

The reactions are best carried out in the absence of solvent, adding to their environmental credentials. Acid clay supported zinc salts have also been reported as catalysts for aromatic brominations (with some improvements in selectivity compared to normal homogeneous reactions) and the rearrangement of α-pinene oxide to the valuable perfumery intermediate campholenic aldehyde Fig. 8 (26). The last of these is especially demanding on the control of the acidity, since for any Brønsted acidity catalysis, there are several other reactions resulting in a complex product mixture.

3.2. Supported Heteropoly Compounds.
Heteropolyacids supported on silica have been claimed to show commercially viable activity and catalyst lifetime for the manufacture of ethyl acetate from ethylene and acetic acid (27). One of the major current commercial routes to ethyl acetate is based on $PdCl_2/CuCl_2$ catalysis (the Wacker process), but this has a number of drawbacks, including the use and subsequent disposal of high concentrations of HCl. A solid acid catalyst could be a viable option and when comparing a range of materials including clays, acid resins and zeolites to $H_4SiW_{12}O_{40}$-silica, the last of these is clearly the most active (330 $g/L^1/h^1$ space–time yield). Careful choice of the catalyst and associated process conditions has made such a system a genuinely commercially viable option.

The hydrolysis of ethyl acetate can also be achieved using a supported heteropoly compound—$Cs_{2.5}H_{0.5}PW_{12}O_{40}$-silica (28). It is also important to note

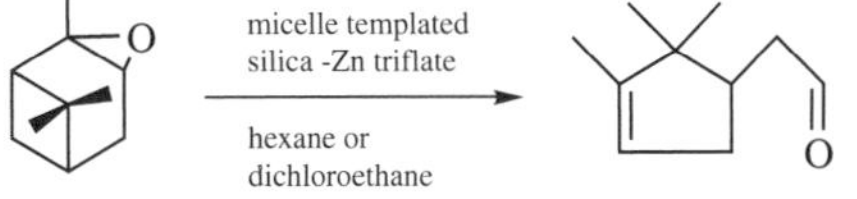

Fig. 8. The rearrangement of α-pinene oxide catalyzed by mesoporous silica supported zinc(III) triflate.

that the catalyst is unchanged at the end of the reaction. Supported heteropoly compounds are also active in the formation of methyl *tert*-butylether (MTBE) (29,30) with a selectivity of 95% particularly using dodecatungtophosphoric acid-K10 clay the dehydration of diols, the oligomerisation of propene, and the etherification of phenethyl alcohol with alkanols (31). Their industrial potential is further illustrated by a substantial recent patent literature that covers reaction applications including the isomerization of nonenes, the formation of *tert*-butylacrylamide (as a replacement for H_2SO_4), the formation of alkyl diphenylamine (as a replacement for $AlCl_3$), and polyisobutylene (as a replacement for BF_3 and $AlCl_3$) (31). Several types of support material have been proposed and while silica (especially mesoporous silica) and acid-treated clays are especially popular, others, including alumina and zirconia have been reported to be effective in some reactions.

3.3. Supported Sulfur(VI) Acids.

The widespread importance of sulfuric acid and other sulfur(VI) acids, notably methanesulfonic and triflic acids, has encouraged a substantial amount of research into heterogeneous analogues and not insignificant commercial application, notably via polymeric acidic resins.

Modern mesoporous silicas and other materials have been successfully functionalized by sulfonic acid and their activity proven in a number of important reactions including esterifications, transeseterification and alkylations (32,33). A good example of this is the use of sulfonic acid functionalized MCM-41 as a catalyst for the liquid-phase condensation of phenol with acetone to form the commercially important bis(phenol)-A with high selectivity (Fig. 9) (34). Reaction occurs at a reasonable rate at temperatures as low as 70°C, which can be contrasted with the same support modified with the heteropolyacid 12-tungstophosphoric acid that is only catalytically active in this reaction at >120°C when reaction selectivity is relatively poor (35). It is also important to note that there is negligible loss of sulfur from the solid sulfonic acid after reaction.

Bisphenol-A is manufactured on a commercial scale using sulfonic acid-modified polystyrenes. Acidic resins have been used for a range of other commercial scale chemical processes including ester hydrolysis, olefin (eg, propylene) hydration, alcohol dehydration and the reaction of olefins with alcohols (notably the reaction of methanol with isobutylene to give MTBE (31). Commercially, MTBE is produced using a macroporous sulfonic acid resin catalyst in the liquid phase (36,37).

More speciality, applications for sulfonic acid resins include the preparation of perfumery-grade methyl anthranilate from the reaction of methanol with anthranilic acid. Numerous other esterifications have also been show to be effectively catalyzed be these solid acids (31).

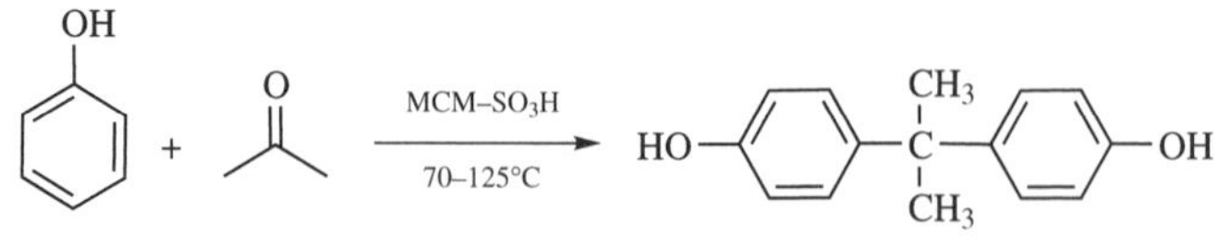

Fig. 9. The preparation of bisphenol-A using a sulfonic acid functionalized mesoporous solid.

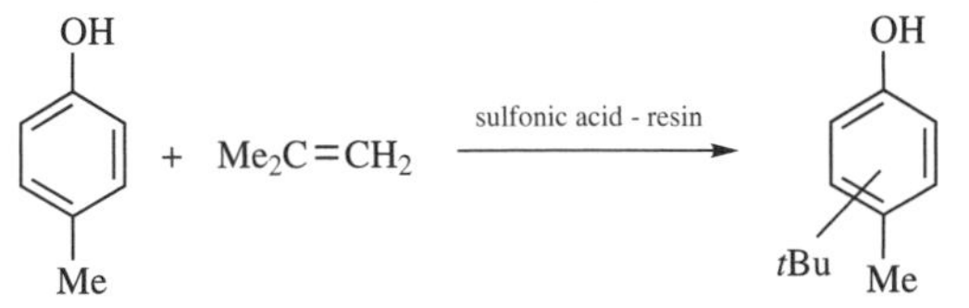

Fig. 10. Alkylation of cresol catalyzed by a sulfonic acid resin.

Polyfluorinated sulfonic acids are considerably stronger Brønsted acids than the conventional styrene-based materials ($\Delta H_0 \sim 10$). They are also more thermally robust. Successful applications of these strong solid acids include the bimolecular conversion of alcohols to ethers. An interesting development in this context is the preparation of composites of silica with fluorinated sulfonic acid resins (eg, Nafion) (38,39). These can be considered as effectively being higher surface area forms of the highly acidic polymer with a commensurate increase in catalytic activity of up to 10^3 and allows catalysis of reactions not normally considered possible for conventional ion exchange resins. Thus a nafion resin/silica catalyst is active for the alkylation of benzene with C_{12} olefins to give the very large scale detergent intermediate dodecylbenzene. Its particularly important to note that very little branched alkylates on other by-products are formed. Selectivity in this application is very good although the catalyst is significantly less active in such alkylations than supported aluminium chloride or sulfated zirconia.

The alkylation of phenols such as the tertiary-butylation of p-cresol is a very important process in the manufacture of numerous chemical products including fragrances, antioxidants, herbicides, and insecticides (40). Sulfonic acid resins have replaced homogeneous and liquid acids in many of these processes especially where the alkylation is not too difficult (Fig. 10).

For more challenging systems, the more acidic fluorinated analogues may be more suitable. Similarly, olefin isomerisations require the more acidic materials to achieve reasonable reaction rates, eg, the Nafion-silica composites 200 times more active than plain Nafion resins and ~ 50 times more active than the zeolitic solid acid H-ZSM-S in the isomerization of 1-dodecene to its linear isomers (31,39).

Rather than fix a sulfur(VI) acid function to a support with a hydrocarbon or perfluorocarbon linker group, attempts have been made to support H_2SO_4 directly onto a support materials. The most successful of these is sulfated zirconia (SZ), which in many practical applications provides a mixture of Brønsted and Lewis acid sites (Fig. 11) (41).

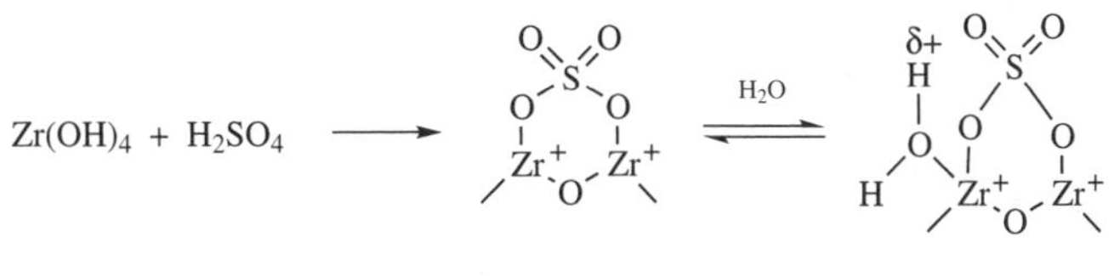

Fig. 11. Formation of sulfonated zirconia and the effect of water.

While many of the supported applications for SZ are in the vapor phase (where Lewis acidity can be expected to dominate) an increasing number of reports show it to be an effective Brønsted solid acid on mixed Brønsted/Lewis solid acid in some liquid phase reactions. The activity of SZ is, however, highly dependant on the temperature of activation. Thus the alkylation of benzene with olefins is readily catalyzed via optimized SZ at room temperature and with excellent selectivity to alkylation (vs digomerization, good selectivity to monoalkylation (>90%), and reasonable selectivity in the formation of the normally preferred 2-isomer (43%). The importance of pore size and pore volume, is very evident. Microporous SZ is blocked by high molecular weight by-products and shows little activity on reuse in a batch reaction. Mesoporous SZ retains its activity and while it will eventually become blocked, it is possible to reactivate the catalyst by a combination of solvent extraction and thermal treatment (Fig. 12) (42).

SZ has also been shown to be active in some Friedel-Crafts acylation reactions (41,43). The SZ will catalyze the benzylation of benzene (and other aromatic substrates) without solvent at moderate temperatures. Unlike $AlCl_3$, the SZ is genuinely catalytic in such reaction although rates or reaction are slow (but SZ does seem to be significantly more active with activated substrates such as anisole). A particularly important advantage is that a water quench is not needed at the end of the reaction (SZ can be removed by filtration, eg).

Loss of activity on moving from a homogeneous acid to a heterogeneous analogues is a frequent cause for concern and has undoubtedly delayed the commercial uptake of many supported catalyst systems. One way around this problem is through the utilization of a moving bed of the liquid catalyst wherebye the catalyst slowly moves over the surface of a support while reactants are fed through the mixed-catalyst bed (Fig. 13) (31).

One example of this reactor technology at work is in the very large scale butane alkylation reaction (n-butene and isobutane to isooctane) using CF_3SO_3H as the (moving) catalyst. Excellent activity can be achieved (100% conversion to the desired alkylates) along with good selectivity (up to 80% C_8). When

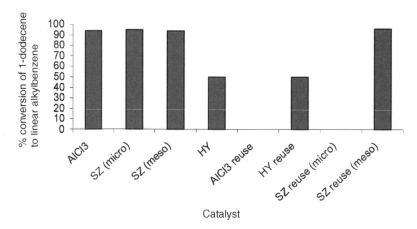

Fig. 12. Comparisons of the effectiveness and reusability of different catalysts for the production of linear dodecyclbenzenes from the reaction of benzene with 1-dodecene at 35°C.

Fig. 13. Moving bed of the liquid catalyst.

the catalyst elutes from the support, it can be recycled. This pseudo-solid-acid technology offers health and safety advantages over the conventional HF and H_2SO_4-based methods.

3.4. Acid Catalysis by Zeolites. The use of zeolites as acid catalysts is one of the most prominent area of gas-phase catalysis, predominantly in the field of catalytic cracking. In recent years, some applications have begun to emerge in the field of fine chemicals synthesis, mostly in the liquid phase, but with one or two important exceptions.

The acid sites of zeolites comes about by the presence of Al atoms in tetrahedral sites in the oxide lattice. This leads to an overall negative charge on the lattice that can be balanced by a range of cations, including the proton. The acid strength of a zeolite is a complex function of the aluminium content, with acid strength ranging from moderate (high number of Al sites giving a large number of moderately acidic sites) to superacidic (low amount of Al leading to a small number of extremely acidic centres). Extension of this concept to the mesoporous "analogues" of zeolites does not appear to be feasible, and mesoporous molecular sieves (eg, MCM-41) seem to have very modest acidity.

The channels of zeolites can accommodate benzenoid systems, given that there are no meta or ortho substituents. These are barely large enough to fit inside (or form inside) the channels of the zeolites. Choosing a zeolite capable of absorbing benzene or a monosubstituted benzene might therefore lead to the possibility of highly selective para substitution under Friedel-Crafts conditions. Such work has been carried out with some success. Nitration using H-ZSM-5, H-ZSM-11, and H-beta is known where benzene, toluene and halobenzenes can be nitrated in excellent yield and para-selectivity (44). Selectivities up to 98% were achieved. Halogenation is also possible in excellent selectivity, with zeolites also being useful in retarding over-halogenation (more of a problem than with nitration) (45). Friedel-Craft alkylations are also documented using zeolites, and here selectivity is also controlled by the pore constraints of the catalysts chosen (46). Friedel-Crafts acylation is possibly the most challenging of these processes, with traditional acylations not being strictly catalytic. Nonetheless, progress has been made, and mildly acidic zeolites have been shown to be sufficiently active to effect the reaction. The acylation of toluene can be carried out using a Ce-exchanged Y-zeolite at 150°C, giving excellent para selectivity and yields that increase from low with short-chain acids (not acid chlorides!) to excellent for longer chain acids (47). The much more activated anisole has been acylated using a range of zeolites, with H-beta and HY being the most active (48). Rhodia curently operate a process that uses this technology to prepare 4-acetylanisole (49). Other activated substrates such as thiophene (50) and 2-methoxynaphthalene (51) have also been extensively studied. 4-Acetyl isobutylbenzene, a key intermediate in the synthesis of ibuprofen, has also been manufactured by the zeolite-catalyzed acylation of isobutylbenzene with H-beta zeolite in excellent yield and selectivity (52).

The Beckmann rearrangement of cyclohexanone oxime has been the subject of a substantial amount of research using zeolitic catalysts (53). Selectivity is excellent and catalyst lifetime can be very long, when a cofeed of carbon dioxide or methanol and water are used.

4. Basic Catalysts

Solid base catalysts are less well developed than solid acids, but nonetheless represent an important class of catalysts. They range from simple solids such as alumina, which has some basic activity, through more complex inorganic species such as zeolites, hydrotalcites, "blue alumina", and to the newer organic–inorganic hybrid materials, based on silica/amine composites. Basicity ranges from very mild, typical of simple amines, through to superbasic ($H^- > 37$).

$$H^- = pK_{BH} + \log_{10}[B^-]/[BH] \tag{1}$$

One generic difficulty inherent in the handling and use of basic catalysts (particularly with the most basic systems) is the presence of acidic impurities in the atmosphere. Both water, and especially CO_2 can poison basic catalysts very rapidly, and possibly irreversibly. In the case of reversible adsorption of CO_2, the temperature-programmed desorption of CO_2 can be used as a measure of base strength. In the case of KF-alumina, the very rapid adsorption of CO_2 from the atmosphere may be the cause of the wildly differing reports of its base strength in the early days of research into its properties. Despite this problem, even the most basic of materials can be and are used on an industrial scale.

4.1. "Blue Alumina".   An interesting and very basic material, which is finding industrial application, is based on the reaction of alumina first with NaOH, and then, in a high temperature step, with sodium metal, to produce the strongly basic "blue alumina".

The basicity of this material is exceptionally high, with H^- >37. Its preparation involves the treatment of gamma alumina with sodium hydroxide, forming $NaAlO_2$. In a second step, this is treated at high temperature with sodium metal, leading to the superbasic solid. The potassium analogues are similarly superbasic. The basicity of these materials are such that they are extremely sensitive to air, water, and CO_2, but nonetheless they are used on a commercial scale for alkene isomerisation and for the side-chain alkylation of alkylaromatics. A range of patents, mostly to Sumitomo, describes such transformations, two of which are discussed below.

Superbases can be used to bring about the thermodynamic isomerization of alkenes, giving the most stable isomer. One example is the isomerization of 5-vinylbicyclo[2.2.1]hept-2-ene to 5-ethylidenebicyclo[2.2.1]hept-2-ene (54) an important component of ethylene–propylene synthetic rubber. The reaction proceeds at $-30°C$ in the presence of catalysts, giving the product cleanly. The endo/exo ratio appears to be related to the temperature of the reaction (lower temperature gives higher selectivity to the endo isomer) but not to the endo/exo ratio of the reactant.

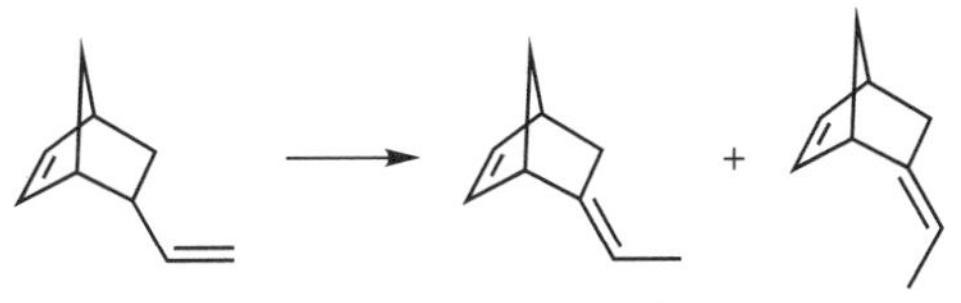

The alkylation of alkyl aromatics at the benzylic position is an important class of reaction, which can often give different isomers when compared to acid catalyzed routes. Such superbase-catalyzed alkylations are used industrially in the production of isobutylbenzene from toluene and propene. The alkylation of cumene with ethylene gives the pentylbenzene shown below in 99% yield under very mild conditions (54), this concept can be extended to the preparation of alkenyl substituted benzenes by using alkenes as alkylating agents (55). Substituted pyridines are also accessible via this route (56).

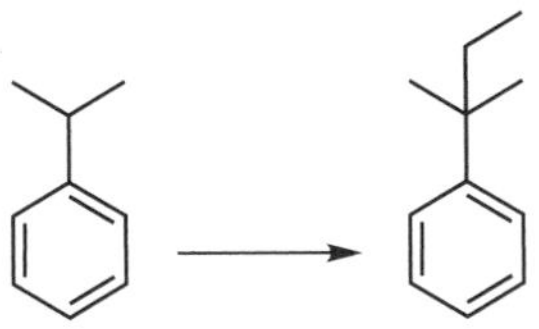

4.2. Zeolites. While the majority of zeolites have pore sizes that are too small to allow all but the smallest molecules to enter, and are consequently of little use in organic synthetic applications, a few examples exist where larger pore zeolites have been used as supports in base catalysis. The limited reaction space of the smaller zeolites becomes even more constrained by the requirement for a (normally) large counterion such as Cs, which is required to increase the basicity of the framework oxyanion sufficiently to allow good activity. Simple ion-exchange allows the formation of ion pairs between the M^+ and the framework O^-. This leads to moderately basic catalysts, with the Cs^+ containing material being most basic (57). More powerfully basic sites are obtained when the zeolites are loaded with salts that, upon thermal decomposition, give small clusters of Group I (I A) or (II A) oxides (58). These much more basic materials must be kept free of water and carbon dioxide, both of which reduce their basicity considerably. The latter appear to be capable of catalyzing reaction types similar to blue alumina, while the former have been studied for their activity in mild-base catalyzed condensation reactions (eg, Knoevenagel).

Reviews on zeolites written with a slant toward their organic chemical applications are available (59). Also see the Bibliography for more background information.

4.3. Hydrotalcites.   Hydrotalcites are another group of catalysts that are proving efficient as medium-to-strong basic catalysts. As catalyst supports, they have been used in the synthesis of the superbases described above (56), but they are also interesting in their own right. Hydrotalcites are lamellar solids consisting of layers of mixed oxides, typically Mg and Al, but others are also known. The structure contains (partly) exchangeable anions such as hydroxide and carbonate that gives basicity to the structure and, combined with their lamellar structure, leads to them being referred to sometimes as anionic clays (60). Surface areas are generally $\sim$100–250 m^2/g^1 depending on thermal pretreatment/rehydration.

A range of C–C bond-forming reactions are known to be catalyzed by hydrotalcites and related materials (61,62). In particular, aldol reactions have been extensively studied (62), and the conversion of n-butyraldehyde to 2-ethylhexenal has been the subject of much work (63). There also exist patents describing the use of hydrotalcites as catalysts for the formation of alkoxylated alcohols and related products, of use as nonionic surfactants, by oligomerization of alkylene oxides (64) and to the preparation of crystalline stereoregular poly(propylene oxide) using a similar route (65).

Ni–Al hydrotalcites have been shown to have excellent properties for the industrially important reduction of nitriles to primary amines. The combination of basic character with reducing centers plays an important role in minimizing the production of secondary amines (66).

4.4. Metal Oxides.   Metal oxides all have the potential for their surface oxygen atoms to act as basic sites. In practice, those with divalent cations are the most basic, and therefore the most studied. Tri- and tetravalent cations can give materials with basic sites, but these are often relatively mildly basic, and acidic sites can also be present. Such materials are very interesting for processes that require the presence of both types of sites, and an example is given below.

Of the divalent oxides, it is found that basicity increases as expected in the order MgO < CaO < SrO < BaO. Some sites with H^- >26 are found on the most strongly basic of these. These catalysts are very good at abstracting protons from allylic positions, making them very effective catalysts for alkene isomerisation, as discussed in the section on "blue alumina".

Zirconium oxide possesses both mildly basic and mildly acidic sites, and this makes it particularly suitable for certain reaction types. Industrial processes exist for the hydrogenation of aromatic carboxylic acids to aldehydes (67), and for the formation of vinylcyclohexene from 1-cyclohexylethanol (68). The formation of diisobutyl ketone from isobutyraldehyde has been practiced industrially for a quarter of a century (69). All of these processes are enhanced by the presence of both acid and basic sites on the surface of the catalyst.

4.5. KF-alumina. KF-alumina is the most studied of a group of MF-support catalysts, most of which are characterized by a strong interaction (or a much more fundamental reaction) between the two components. This leads to novel basic sites; in the case of KF-alumina this can be represented by the equations below:

$$12\,KF + Al_2O_3 + 3\,H_2O \rightarrow 2\,K_3AlF_6 + 6\,KOH$$
$$6\,KF + 2\,Al_2O_3 \rightarrow 3\,K_3AlF_6 + 3\,KAlO_2$$

These catalysts are active in a wide range of organic reactions (70). The basicity is such that many CH acids such as nitroalkanes and methyl ketones can undergo C−C bond-forming reactions such as Knoevenagel, Michael, and aldol reactions.

4.6. Organic−Inorganic Hybrid Catalysts. A range of mildly basic catalysts for fine chemicals synthesis exist, where the basic centre is an amine or diamine group, attached to a silica matrix via a short hydrocarbon spacer. Two routes exist to these compounds, including grafting with $(RO)_3Si(CH_2)_3NH_2$ onto amorphous silica or the newer micelle templated silicas such as MCM-41 (71). The second route involves the direct cocondensation of a silica precursor with the trialkoxyaminopropyl silane under micelle templating conditions (72). This latter route is more efficient with respect to the silane used, and the template can also be recovered and reused, making this method a relatively cheap and simple route to these compounds. The basicity of these materials is relatively mild, and thus these materials are being investigated for the condensation of C-acids such as diketones, dinitriles and ketoesters. Activity in Knoevenagel and Michael reactions, as well as aldol and related condensations is generally good.

4.7. Ion-Exchange Resins. Ion-exchange resins, which are often used in acid catalysis, are much less commonly utilized in base catalyzed processes. This is due to their limited thermal stability, with the quaternary ammonium hydroxide active sites being relatively easily destroyed by Hofmann degradation to the amine, alkene, and water. This finding is reported to occur at temperatures as low as 50°C, although some reports indicate that, in the carbonylation of methanol with CO, little loss of activity is found up to 70°C (73).

5. Supported Metal Complexes

The heterogenization of catalytically useful metal complexes is attractive both economically (by facilitating catalyst recovery and reuse or by enabling use of a fixed-bed catalyst reactor) and in the context of green chemistry (by simplifying reaction work-up and minimizing release of toxic metals into the environment) (74).

In principle, this can be achieved via immobilization typically on to a high surface area support material. However, limitations in the stabilities of supported metal complexes in liquid-phase reactions and uncertainties over possible parallel homogeneous catalysis due to leached metallic species makes this area controversial and still largely unproven (75).

A good example of these "hidden" complexities is the well-researched area of supported chromium complexes and their use in liquid-phase oxidations. While chromium(III) complexes are quite stable toward substitution of the metal, chromium(VI) is very labile and soluble especially in semiaqueous systems. Thus it has been conclusively shown that chromium aluminium phosphate, CRAPO, leaks chromium in reactions using aqueous hydrogen peroxide (76). This can be attributed to the formation of chromium(VI). In determining this a useful test for catalyst stability toward leaching has been developed. The so-called "hot filtration" test involves the filtration of the solid catalyst from the reaction mixture at reaction temperature, and then monitoring the filtrate for any further reaction of the organic substrate. This is an effective check on any homogeneous contribution to the reaction but does not rule out leaking of any inactive metallic species (this can only be determined by analysis of the filtrate but is limited by detectability of the metal). Numerous other attempts have been made to immobilize chromium for oxidation reactions including a commercial catalyst that appears to be based on Cr(VI) chemisorbed on alumina (25) and a chromium Schiffs base complex imprinted on a mesoporous silica surface (77).

While some of the reported activities of these materials in various oxidation reactions (notably the relatively facile ethylbenzene to acetophenone) are quite impressive and are based on good clean synthesis principles (no solvent, air as the only consumable source of oxygen), they are not stable in aqueous peroxide systems, and their stability in nonaqueous reaction mixture is questionable (though probably good enough to give very high turnover systems in moderate temperature alkylbenzene oxidations (75,77). Similar materials based on supported cobalt may have better long term potential (78) and also have value in other reactions such as allylic oxidations (79).

Other noteworthy attempts to develop genuine heterogeneous metallic catalysts for oxidation reactions include metalloporphyrinosilicas (prepared by methods including cocondensation with tetraethoxysilane) (80) and a mononuclear iron carboxylate complex immobilized on modified silica used to mimic the active site of methane monooxygenase. The latter catalyzes the room temperature oxidation of hexane using air although the reaction requires several auxiliaries (a mercaptan, acetic acid, and triphenylphosphine) Several other direct oxidations of alkanes using air, catalyzed by supported metal complexes have been reported including Fe-aluminophosphate (FeAlPU-31) for cyclohexane to adipic acid (81). This provides an interesting contrast with one of the current commercial routes (Table 2). Bifunctional molecular sieve catalysts have been reported as being effective for the one-step solvent-free production of cyclohexaneoxime and caprolactam using a mixture of air and ammonia (Fig. 14) (82).

Epoxidation reactions have been successfully catalyzed by an immobilized version of Jacobsens catalyst using NaOCl as the oxidation (82). This is interesting although long-term catalyst stability is again a likely problem.

Chemically modified (typically porous) solid supports can also make very effective ligands for palladium enabling a range of important reactions including amidations, carbon–carbon bond forming coupling reactions, hydrogenations,

Table 2. **Comparison of Traditional and New Heterogeneous Metal-catalyzed Methods for the Manufacture of Adipic Acid**[a]

	Traditional	New
method	nitric acid oxidation of a mixture of cyclohexanone and cyclohexanol >60°C, ~0.2 MPa; 60% HNO$_3$, V/Cu catalyst	direct conversion of cyclohexane; 100+ °C; 1–5 MPa air; Fe–AlPO catalyst
yield	90%	65%
major products	adipic acid, glutanic acid, succinic acid	adipic acid and cyclohexanone
by-products	NO$_x$.CO$_2$	glutanic and succinic acids
other factors	plant corrosion, health and safety concerns, expensive to recover catalyst health and safety concerns expensive to recover catalyst	slow reaction

[a]See Ref. 81.

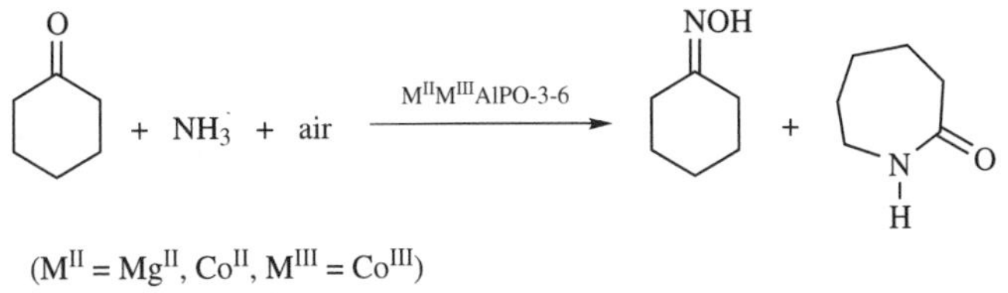

$(M^{II} = Mg^{II}, Co^{II}, M^{III} = Co^{III})$

Fig. 14. The solvent-free production of cyclohexaneoxime and caprolactam using a solid catalyst.

and carbonylations. Apart from organic polymers, silicas and clays have been used as support materials. Silica modified with long chain bis-(diphenylphosphine) ligands can be used as a ligand for Pd(0) in the amidation of aryl halides (Fig. 15) (83).

Silica or clay supported mercaptans can also be used to bind Pd(0) and the resulting materials are catalytically active in the Heck reaction between aryl iodides and activated alkenes (84). Palladium supported on silica modified with pyridyl functions has also been used to catalyse Heck and Suzuki reactions (85,86) (Fig. 16). These catalysts have excellent turnover numbers (several thousand) and pass the "hot filtration test", ie, the hot filtrates show no residual activity indicating that no active soluble palladium species are formed during the reactions.

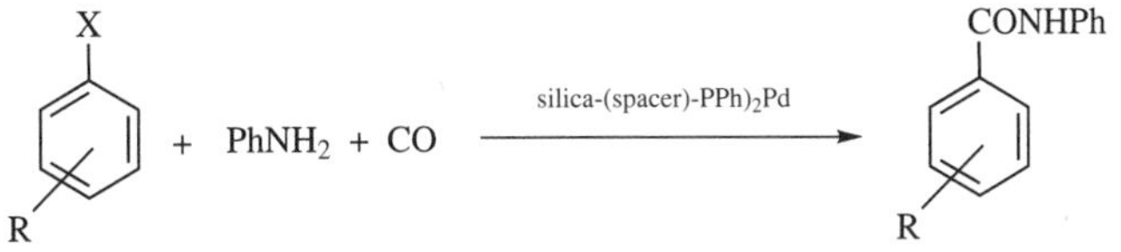

Fig. 15. Amidation of aryl halides catalyzed by supported palladium (0).

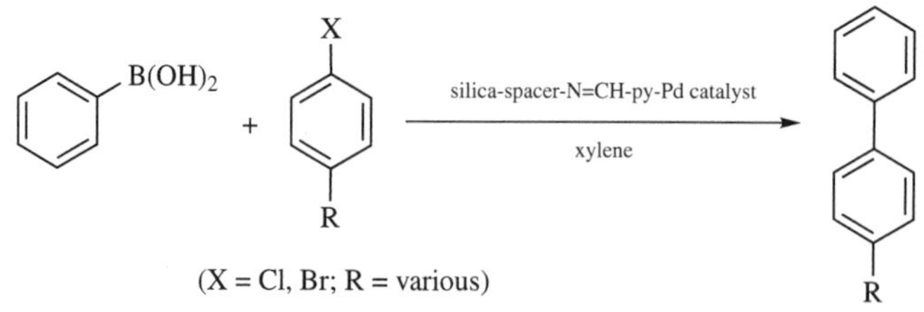

Fig. 16. Suzuki reaction to form biaryls catalyzed by silica supported palladium complexes.

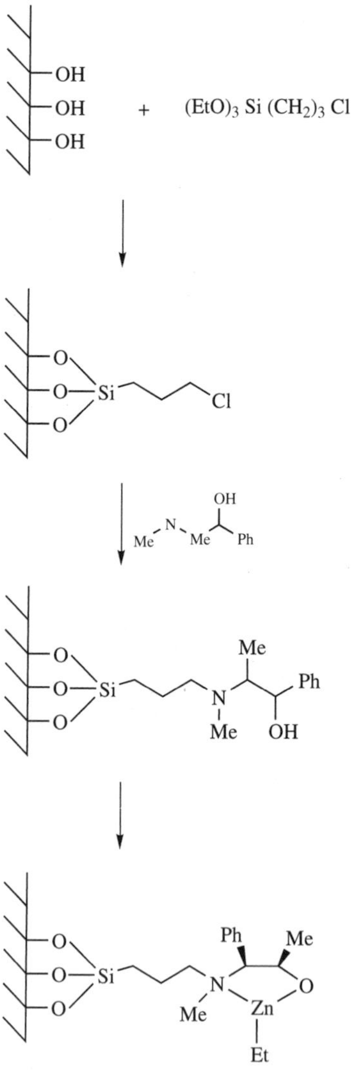

Fig. 17. Preparation of a solid chiral Zn catalyst for enantioselective alkylations.

A chiral palladium catalyst can be prepared using MCM-41 modified with 1,1-bis(diphenyl-phosphino)ferrocene ligand. Some regioselectivity and enantiometric excess is observed when using this catalyst in the allylic amination of cinnamyl acetate; the performance of the catalyst is superior to that of the homogeneous analogue (87). Another solid catalyst for enantioselective reactions is also based on MCM-41, this time modified with (−) or (+) ephedrine (88,89). The solid chiral auxiliary was then used in the enantioselective alkylation of benzaldehyde by diethylzinc to give the corresponding (*R*)- or (*S*)- 1-phenylpropan-1-ol (88). The preparation of the catalyst is shown in Figure 17.

Various other metals can be effectively immobilized on chemically modified supports, silica and clay supported rhodium complexes are good hydrogenation catalysts, eg, (90). Heteropolyacids have been used to enhance the solid-metal binding (91). Silica supported rhodium hydrides are efficient isomerisation catalysts (92). A good example of a supported ruthenium catalyst is that prepared from on sol–gel mixture containing *cis*-Cl(H)Ru(CO) (phosphine)$_3$ with (EtO)$_4$Si and [AllO-i-Pr)$_3$]. The resulting material is quite stable and is active in the hydrogenation of *trans*-crotyl aldehyde with reasonable chemoselectivity to *cis*- and *trans*-crotyl alcohol (93). Osmium has also been effectively immobilised on a silica supported bis(cinchona) alkaloid and the very stable and reusable material that results is catalytically active in the asymmetric aminohydroxylation and dihydroxylation of alkenes (94).

BIBLIOGRAPHY

"Catalysts, Supported" in *ECT* 4th ed., Vol. 5, pp. 383–418, by Robert P. Nielsen, Consulted; "Catalysts, Supported", in *ECT* (online), posting date: December 4, 2000, by Robert P. Nielsen, Consultant.

CITED PUBLICATIONS

1. B. C. Gates, *Catalytic Chemistry*, John Wiley & Sons, Inc., New York, 1991.
2. B. Imelite, C. Naccache, J. Ben Taarit, J. C. Vedrine, G. Coudurier, and H. Praliaud, *Catalysis by Zeolites*, Elsevier, Amsterdam, The Netherlands, 1980.
3. Y. Murakami, A. Iijima, and J. W. Ward, *New Developments in Zeolite Science*, Elsevier, Amsterdam, The Netherlands, 1986.
4. J. B. Nagy, P. Bodari, I. Hannus, and I. Kirischi, *Synthesis, Characterisation and Use of Zeolite*, Microporous Materials, DecaGen, Szeged, 1998.
5. I. E. Maxwell, *Cat. Tech.* **1**, 5 (1997).
6. J. A. Schwarz, C. Contescu, and A. Contescu, *Chem. Rev.* **95**, 477 (1995).
7. R. A. Van Santen, in G. Ertl, H. Knözinger, and J. Weitkamp, eds., *Handbook of Heterogeneous Catalysis*, Wiley–VCH, Vol. 5, 1997, p. 2244.
8. J. M. Thomas and W. J. Thomas, *Principles and Practice of Heterogeneous Catalysis*, VCH, Weinstein, 1997.
9. J. H. Clark, A. P. Kybett, and D. J. Macquarrie, *Supported Reagents Preparation Analysis and Applications*, VCH, New York, 1992.
10. J. H. Clark, *Catalysis of Organic Reactions by Supported Inorganic Reagents*, VCH, New York, 1994.

11. J. H. Clark, S. R. Cullen, S. J. Barlow, and T. W. Bastock, *J. Chem. Soc., Perkin 2*, 1117 (1994).
12. A. Stein, B. J. Melde, and R. C. Schroden, *Adv. Mater.* **12**, 1403 (2000).
13. J. H. Clark, *Pure App. Chem.* **73**, 103 (2001) and K. Wilson and J. H. Clark, *Pure App. Chem.* **72**, 1313 (2000).
14. R. S. Drago, S. C. Petrosius, and C. W. Chronister, *Inorg. Chem.* **33**, 367 (1994).
15. J. H. Clark, K. Martin, A. J. Teasdale, and S. J. Barlow, *Chem. Commun.* 2037 (1995).
16. S. Sato and G. C. Maciel, *J. Mol. Catal. A: Chem.* **101**, 153 (1995).
17. J. H. Clark, P. M. Price, D. J. Macquarrie, K. Martin, and T. W. Bastock, *J. Chem. Res.* 430 (1997).
18. E. E. Getty and R. S. Drago, *Inorg. Chem.* **29**, 1186 (1990).
19. P. M. Price, J. H. Clark, D. J. Macquarrie, K. Martin, and T. W. Bastock, *Org. Proc. Res. Dev.* **2**, 221 (1998).
20. Eur. Pat. Appl.No.98302068.6 (1998) P. M. Price, J. H. Clark, D. J. Macquarrie, and T. W. Bastock.
21. K. Shorrock, J. H. Clark, K. Wilson, and J. Chisem, *Org. Proc. Res. Dev.* **5**, 249 (2001).
22. P. M. Price, J. H. Clark, and D. J. Macquarrie, *J. Chem. Soc. Dalton* 101 (2000).
23. Eur. Pat. Appl., PCT/EP00/07083 (1999), J. H. Clark, J. Chisem, L. Garcia, K. Lewtas, J. K. Shorrock, and K. Wilson.
24. Eur. Pat. Appl., PCT/EP00/07084 (1999), J. H. Clark, L. Garcia, K. Lewtas, and K. Wilson.
25. K. M. Martin, in J. H. Clark and D. J. Macquarrie, eds., *Handbook of Green Chemistry and Technology*, Blackwell Science, Abingdon, U. K., 2002, pp. 321-327.
26. K. Wilson, A. Renson, and J. H. Clark, *Catal. Lett.* **61**, 51 (1999).
27. M. J. Howard, G. J. Sunley, A. D. Poole, R. J. Watt, and B. K. Sharma, *Stud. Surf. Sci. Catal.* **121**, 61 (1999).
28. Y. Izumi, M. Uno, M. Kitagawa, M. Yashida, and K. Urabe, *Microp. Mater.* **5**, 225 (1995).
29. M. Misono, *Korean J. Chem. Eng.* **14**, 427 (1997).
30. N. Mizuno and M. Misono, *Curr. Opin. Solid State. Mater. Sci.* **2**, 84 (1997).
31. M. Harmer, in J. H. Clark, and D. J. Macquarrie eds., *Handbook of Green Chemistry and Technology*, Blackwell Science, Abingdon, U. K., 2002, pp. 86–119.
32. W. M. Van Rhijn, D. E. De Vos, W. D. Bossaert, J. Bullen, B. Wouters, P. Grobet, and P. Jacobs, *Stud. Surf. Sci. Catal.* **117**, 183 (1998).
33. W. M. Van Rhijn, D. E. De Vos, B. F. Sels, W. D. Bossaert, and P. A. Jacobs, *Chem.-Commun.* 317 (1998).
34. D. Das, J. F. Lee, and S. Cheng, *Chem. Commun.* 2178 (2001).
35. K. Nowinska and W. Kaleta, *App. Catal. A* 203 (2001).
36. H. J. Panneman and A. A. C. M. Beenackers, *Ind. Eng. Chem. Res.* **34**, 4318 (2001).
37. D. Parra, and co-workers *Ind. Eng. Chem. Res.* **37**, 3575 (1998).
38. M. A. Harmer, Q. Sun, and W. E. Farneth, *J. Amer. Chem. Soc.* **118**, 7708 (1996).
39. M. A. Harmer, Q. Sun, A. J. Vega, W. E. Farneth, A. Heidehun, and W. F. Hoelderich, *Green Chem.* **6**, 7 (2000).
40. R. H. Rosenwald in M. Grayson, ed., *Kirk-Othmer Encyclopedia of Chemical Technology*, John Wiley & sons, Inc., New York, 1975, p. 50.
41. G. D. Yadav and J. J. Nair, *Microp, Mesop. Mater.* **33**, 1 (1999).
42. J. H. Clark, G. L. Monks, D. J. Nightingale, P. M. Price, and J. F. White, *J. Catal.* **193**, 348 (2000).
43. J. H. Clark, G. L. Monks, D. J. Nightingale, P. M. Price, and J. F. White, in M. Ford, ed., *Catalysis of Organic Reactions*, Marcel Dekker, New York, 2000, pp. 135–143.
44. K. Smith, *Bull. Chem. Soc. Fr.* 272 (1989); K. Smith, A. Musson, and G. A. DeBoos, *I. Org-Chem.* **63**, 8448 (1998).

45. P. Ratnasamy and A. P. Singh, in R. A. Sheldon and H. van Bekkum, eds., *Fine Chemicals through Heterogeneous Catalysis*, 2001, p. 133.
46. S. M. Csicsery, in J. A. Rabo, ed., *Zeolite Chemistry and Catalysis*, ACS Monograph, 161, *Am. Chem. Soc.*, 680 (1976); G. S. Lee, J. J. Maj, C. E. Roche, and J. M. Garces, *Cat. Lett.*, **2**, 234 (1989).
47. B. Chiche, A. Finiels, C. Gauthier, P. Geneste, J. Graille, and D. Poche, *J. Org. Chem.* **51**, 2128 (1986).
48. A. Corma, M. J. Climent, H. Garcia, and J. Primo, *App. Cat.* **49**, 109 (1989); M. Spagnol, L. Gilbert, and D. Alby in *Industrial Chemistry Library*, Vol. 8, Elsevier, Vol. 29, 1996.
49. WO 9E/35655 (1996), M. Spagnol, L. Gilbert, E. Benazzi, and C. Marcilly.
50. A. Finiels, A. Calmettes, P. Geneste, and P. Moreau, *Stud. Surf. Sci. Catal.* **78**, 595 (1993).
51. G. Harvey and G. Mäder, *Coll. Czech. Chem. Comm.* **57**, 862 (1992); E. A. Gunneswegh, S. S. Gopie, and H. van Bekkum, *J. Mol. Catal. A.* **106**, 151 (1996).
52. EP 0701987 (1996), A. Vogt and A. Pfenninger.
53. T. Tatsumi, in R. A. Sheldon and H. van Bekkum, eds., *Fine Chemicals through Heterogeneous Catalysis*, Wiley-VCH, Weinheim, Germany, 2001, p. 185.
54. G. Suzukamo, M. Fukao, T. Hibi, K. Tanaka, and M. Minobe, *Stud. Surf. Sci., Catal.* **108**, 649 (1997).
55. EP 0949230, G. Suzukamo and M. Yamamoto, (to Sumitomo).
56. JP11029554, M. Yamamoto and T. Suzukamo, (to Sumitomo).
57. F. Yogi, H. Tsiyi, and H. Hattori, *Microporous Mater.* **9**, 237 (1997); J. C. Kim, H. X. LI, C. J. Chen, and M. E. Davis, *Microp. Mater.* **2**, 413 (1994).
58. P. A. Anderson, R. J. Singer, and P. P. Edwards, *Chem Commun.* 914 (1991); P. A. Anderson, A. R. Armstrong, and P. P. Edwards, *Angew. Chem. Int. Engl.* **33**, 641 (1994); L. J. Woodall, P. A. Anderson, A. R. Armstrong, and P. P. Edwards, *J. Chem. Soc. Dalton Trans.* 719 (1996).
59. P. B. Venento, *Microp. Mater.* **2**, 297 (1994); P. Espeel, R. Parton, H. Taylor, J. Martens, W. F. Hoelderich, and P. Jacobs, in J. Weitkamp and L. Poppe, eds., *Catalysis and Zeolites*, Springer Verlag, New York, 1999, p. 377.
60. D. D. Tichit, M. H. Lhouty, A. Guida, B. Chiche, F. Figueras, A. Auroux, D. Bartalani, and E. Garrone, *J. Catal.* 151, **50** (1995).
61. B. F. Sels, D. E. De Vos, and P. A. Jacobs, *Catal. Rev.* **43**, 443 (2001).
62. F. Figueras and J. Lopez, in R. A. Sheldon and H. van Bekkum, eds., *Fine Chemicals through Heterogeneous Catalysis*, 2001, p. 327.
63. EP0539002, B. J. Arena and J. S. Holmgren (to UOP); US5258558 and US5254743.
64. EP1156994, A. Behler, A. Folge, and H. D. Schares (to Cognis Deutschland); (WO0061534) GmbH and J. A. Barnhorst, C. W. Blewett, and A. F. Elsasser, Jr. (to Henkel).
65. US4962281, D. E. Laycock (to Dow).
66. D. Tichit, F. Medina, R. Durand, C. Mateo, B. Coq, J. E. Sueiras, and P. Salagre, *Stud. Surf. Sci., Catal.* **108**, 297 (1997).
67. T. Yokoyama, T. Setoyama, N. Fujita, M. Nakajima, T. Maki, and K. Fukii, *Appl. Catal A.* **88**, 149 (1992).
68. T. Yamaguchi, H. Sasaki, and K. Tanabe, *Chem. Lett.* 677 (1976).
69. I. Koga, *Yuki Gouseikagaku* **33**, 702 (1975).
70. H. Hattori, *Chem. Rev.* **95**, 537 (1995).
71. D. Brunel, *Microp, Mesop. Mater.* **27**, 329 (1999).
72. S. L. Burkett, S. D. Sims, and S. Mann, *Chem. Commun.* 1367 (1996); D. J. Macquarrie, *Chem. Commun.* 1961 (1996) and R. Richer and L. Mercier, *Chem. Comun.* 1775 (1998).

73. M. di Girolamo and M. Marchionna, *J. Mol. Cat., A* **177**, 33 (2001).
74. J. H. Clark and C. N. Rhodes, in J. H. Clark, ed., *Clean Synthesis using Porous Inorganic Solid Catalysts and Supported Reagents*, RSC Clean Technology Monograph Series, RSC, Cambridge, 2000.
75. J. S. Rafelt and J. H. Clark, *Catal. Today* **57**, 33 (2000).
76. H. E. B. Lempers and R. A. Sheldon, *J. Catal.* **175**, 62 (1998).
77. I. C. Chisem, J. Rafelt, M. Tantoh Shieh, J. Chisem, J. H. Clark, R. Jachuck, D. J. Macquarrie, C. Ramshaw, and K. Scott, *Chem.Commun.* 1949 (1998).
78. B. K. Das and J. H. Clark, *Chem. Commun.* 605 (2000).
79. J. Salvador and J. H. Clark, *Chem. Commun.* 33 (2001).
80. P. Battioni, E. Cardin, M. Louloudi, B. Schöllhorn, G. A. Spyroulias, D. Mansuy, and T. G. Taylor, *Chem. Commun.* 2037 (1996).
81. J. M. Thomas, R. Raja, G. Sankar, and R. G. Bell, *Acc. Chem. Res.* **34**, 191 (2001).
82. R. Roya, G. Sankar, and J. M. Thomas, *J Am. Chem. Soc.* **123**, 8153 (2001).
83. M. Z. Cai, C. S. Song, and X. Huang, *Syn. Commun.* **27**, 361 (1997).
84. M. Z. Cai, C. S. Song, and X. Huang, *Synthesis* 521 (1997).
85. J. H. Clark, D. J. Macquarrie, and E. B. Mubofu, *Green Chem.* **2**, 53 (2000).
86. E. B. Mubofu, J. H. Clark, and D. J. Macquarrie, *Green Chem.* **3**, 23 (2001).
87. B. F. G. Johnson, S. A. Raynor, D. S. Sheppard, T. Mashmeyer, J. M. Thomas, G. Sanhar, S. Bromley, R. Oldroyd, L. Gladden, and M. D. Mantle, *Chem. Commun.* 1167 (1999).
88. M. Laspéras, N. Bellocq, D. Brunel, and P. Moreau, *Tetrahedron Asymmetry* **9**, 3053 (1998).
89. D. Brunel, N. Bellocq, P. Sutra, A. Cauvel, M. Laspéras, P. Moreau, F. Direnzo, A. Galarneau, and F. Fajula, *Coord. Chem. Rev.* **180**, 1085 (1998).
90. T. J. Pinnavaia, *Science* **220**, 365 (1983).
91. R. Augustine, S. Tanielyan, S. Anderson, and H. Yang, *Chem. Commun.* 1257 (1999).
92. M. D. Ward, T. V. Harris, and J. Schwartz, *Chem. Commun.* 357 (1980).
93. E. Lindner, A. Jafer, M. Kemmler, F. Aver, P. Wegner, H. A. Mayer, and E. Plies, *Inorg. Chem.* **36**, 862 (1997).
94. C. E. Song, C. R. Oh, S. W. Lee, S. G. Lee, L. Canali, and D. C. Sherrington, *Chem. Commun.* 2435 (1998).

GENERAL REFERENCES

R. A. Sheldon and H. van Bekkum, eds., *Fine Chemicals Through Heterogeneous Catalysis*, Wiley-VCH, Weinheim, 2001.
G. Ertl, H. Knözinger, and J. Weitkamp, eds., *Handbook of Heterogeneous Catalysis*, Wiley-VCH, Weinheim, 1997.
H. Hattori, *Chem. Rev.* **95**, 537 (1995).
A. Corma, *Chem. Rev.* **95**, 571 (1995).

JAMES H. CLARK
DUNCAN J. MACQUARRIE
University of York

CELL CULTURE TECHNOLOGY

1. Introduction

Cell culture processes, the *in vitro* growth of animal, insect, or plant cells on a large scale to manufacture biochemicals of commercial importance, have been used for some time for the manufacture of viral vaccines (see VACCINE TECHNOLOGY). Significant growth in this technology, primarily because of the advent of recombinant DNA methods (see GENETIC ENGINEERING) for the production of therapeutic proteins (qv) and hybridoma technology for production of monoclonal antibodies (see IMMUNOASSAY) occurred in the late 1980s and 1990s. The need for cell culture technology stems mainly from the fact that bacteria do not have the capability to perform many of the posttranslational modifications that most large proteins require for *in vivo* biological activity. These modifications include intracellular processing steps such as protein folding, disulfide linkages, glycosylation, and carboxylation.

Historically, large-scale cell culture technology traces its roots to the production of viral vaccines and other therapeutically important secreted products of human and primate cells such as those listed in Table 1. Most of these products were made by cells grown either in suspension or attached to microcarriers. Such cells were cultured in stirred tank reactors adapted from bacterial fermentation (qv) processes. The primary challenges in adapting microbial fermenters to the rigors of cell culture technology were reducing the risk of microbial contamination, modifying the agitation system to provide low shear agitation to the shear-sensitive mammalian cells, and providing adequate oxygenation under the low shear conditions mandated by the cells (see AERATION, BIOTECHNOLOGY).

The major applications for cell culture technology are found in the production of Monoclonal Antibodies (MAbs), recombinant therapeutics and vaccines. Monoclonal antibodies were initially used in small quantities for diagnostic purposes. However, in the past few years several monoclonals have been developed for therapeutic purposes. The first therapeutic Mab (OKT-3) was developed by Johnson & Johnson to prevent rejection of organ transplants. This was produced in mouse ascites as the requirements were relatively small. Several new Mabs have been introduced in the market, that utilize suspension culture technology. For example, Johnson & Johnson has introduced ReoPro for prevention of blood clotting during high risk angioplasty. More recently, IDEC Pharmaceuticals

Table 1. Historical Products of Cell Culture Technology

Product	Cell line	Process[a]	Year introduced	Reference
FMD[b] vaccine	baby hamster kidney	SC	1962	1
rabies vaccine	dog kidney	SM	1978	2
interferon	human namalwa	SC	1979	3
polio vaccine	monkey kidney	SM	1980	4

[a]suspension culture = SC; stirred microcarrier = SM.
[b]Foot-and-mouth disease = FMD.

Table 2. **Examples of Therapeutic Products Manufactured by Cell Culture**

Product	Company	Therapeutic use
OKT-3	Johnson & Johnson	prevention of transplant rejection
Zenapax	Hoffmann	prevention of transplant rejection
ReoPro	Johnson & Johnson	antiplatelet for high risk angioplasty
Rituxan	Idec Pharmaceuticals	non-Hodgkin's lymphoma
Remicade	Johnson & Johnson	rheumatoid arthritis
Herceptin	Genentech	breast cancer
Avakine	Johnson & Johnson	Crohn's disease
Synagis	Medimmune	RSV
Activase	Genentech	heart attacks, strokes
Epogen	Amgen	anemia in kidney dialysis patients
Aranesp	Amgen	anemia in kidney dialysis and cancer chemotherapy patients
Kogenate	Miles/Cutter	hemophelia
Pulmozyme	Genentech	cystic fibrosis
Avonex	Biogen	multiple sclerosis
Enbrel	Amgen	rheumatoid arthritis
Mylotarg	Celltech Chiroscience	acute myeloid leukemia
Zevalin	Idec Pharmaceuticals	non-Hodgkin's lymphoma
Xolair	Genentech	allergic asthma
Humira	Abbott	rheumatoid arthritis

received FDA approval for Rituxan for the treatment of non-Hodgkins lymphoma, Hoffmann La Roche received approval for Zenapax—A humanized monoclonal to prevent transplant rejection, Genentech received approval for Herceptin (for treatment of metastatic breast cancer), and Medimmune received approval for Synagis (a treatment of respiratory synctial virus). Additionally, Johnson & Johnson introduced Avakine for the treatment of Crohn's disease. Some examples of these products are shown in Table 2 (along with the therapeutic Mabs mentioned earlier). Vaccines also continue to be produced by cell culture. A recent example of this is the vaccine against chicken pox (Varivax) introduced by Merck.

Finally, insect cell culture is being utilized increasingly for quickly producing research quantities of new proteins using the baculovirus expression system (5). The strong polyhedrin promoter and the insect's cells ability to perform many posttranslational modifications have made the system useful for the expression of mammalian proteins that cannot be produced in native form in *Escherichia coli*. Technology development for these products has centered around the differences in characteristics of mammalian versus microbial cells, notably, the shear sensitivity and susceptibility to contamination of the mammalian lines.

Although the focus of this article is mainly on mammalian cells, the technologies described herein also apply in principle to insect and plant cells.

2. Characteristics of Mammalian Cells

2.1. Environmental Conditions. Mammalian cells *in vivo* are maintained in a carefully balanced homeostatic environment and thus have evolved

Table 3. **Environmental Parameters for Mammalian Cell Cultivation**

Parameter	Range	Typical value
pH	6.6–7.6	7
temperature, °C	33–39	37
dissolved oxygen, Pa[a]	0.7–40	10
osmolarity, mOsm/kg[b]	280–360	300
dissolved CO_2, Pa[a]	0.9–20	7
tolerable shear rate, s^{-1}	0–3000	1500

[a]To convert Pa to mm Hg, multiply by 7.5.
[b]Milliosmolar or milliosmole = mOsm, where an osmole equals 1 mol of solute divided by the number of ions formed per molecule of the soluble, ie, 1 mol of sodium chloride is equivalent to 2 osmoles of sodium chloride and $1\,M$ NaCl = 2 Osm NaCl.

to require fairly stringent environmental conditions. These cells differ significantly from bacterial cells in that they lack a rigid cell wall, and are hence much more shear sensitive. Many animal cells are also attachment dependent, needing a surface to grow on. Many of the cell culture technologies provide the low shear, high surface area environment needed for the mammalian cells. Another approach, however, is to adapt cells to suspension culture and select cells that are less shear sensitive permitting the use of fermentation technology for the culture of animal cells. The optimum environmental parameters depend on cell type and are specific to cell type. Typical ranges for some of these parameters are listed in Table 3.

2.2. Nutritional Requirements. The nutrient requirements of mammalian cells are many, varied, and complex. In addition to typical metabolic requirements such as sugars, amino acids (qv), vitamins (qv), and minerals, cells also need growth factors and other proteins. Some of the proteins are not consumed, but play a catalytic role in the cell growth process. Historically, fetal calf serum of 1–20 vol% of the medium has been used as a rich source of all these complex protein requirements. However, the composition of serum varies from lot to lot, introducing significant variability in manufacture of products from the mammalian cells.

Serum is expensive, the 2003 price is ∼$470/L, and supply depends on cattle supply. Use of this serum also poses significant difficulties in validating processes for absence of viral contamination. Hence, a goal in cell culture technology is to develop serum-free media for cell culture. Much work has gone into developing serum-free media and a sizable portion of cell culture research is devoted to this project. Several recent publications have reported development of protein-free and animal component free media for cell culture (6–8).

Several generic media formulations have been developed for growth and cultivation of mammalian cells and are commercially available. Each contains amino acids, inorganic salts for providing the right osmolarity, essential minerals, and buffering capacity; vitamins, and energy sources such as glucose. Whereas mixtures of the different formulations are often used to optimize growth and productivity of cell lines, many of the basal media need to be supplemented with serum or other appropriate proteins (or other protein free components such as soy peptone or meat digests) for promoting cell growth. Some of the more commonly used media formulations include: minimum essential medium (MEM) for a broad spectrum of mammalian cells; basal media eagle (BME) for diploid or primary mammalian cells; Dulbecco's modified eagle media (DMEM) for a broad spectrum of cells; CMRL media for Earl's "L" cells and monkey kidney cells; Fischer's media for murine leukemic cells; Iscove's modified Dulbecco's media (IMDM) for rapidly proliferating high density cell cultures; McCoy's media for human lymphocytes; Ham's F10 and F12 for Chinese hamster ovary cells and other mammalian cells; and RPMI 1630/1640 for suspension cells and human leukemic cells. These media are available from Gibco Laboratories (Grand Island, New York), Irvine Scientific (Santa Ana, California), Sigma Chemical Co. (St. Louis, Missouri), Hyclone Laboratories (Logan, Utah), and JRH (Lenexa, Kansas) among others. Most of these companies also offer proprietary serum free and protein free formulations for CHO, hybridoma, and other mammalian cells (eg, Ex-Cell brand from JRH Biosciences).

Another essential nutrient not supplied with the media is oxygen. The oxygen consumption rate of mammalian cell cultures is much lower than that of bacterial ones because cell densities are much lower than those achieved in bacterial cultures. The oxygen consumption rate varies from cell line to cell line, but the range has been reported to be as wide as $0.05-0.5$ mmol/(10^9cells · h) (9). Hence, designing oxygenation systems for mammalian cells is a function of the cells being used. Use of the worst case scenario may lead to costly overdesign. This is especially so if silicone tube oxygenators are being considered. Direct sparging in the reactor to accomplish oxygen transfer often leads to cell damage unless protective agents, such as pluronic polyols, are used (10). However, as long as the pluronic is nontoxic to the cells and is compatible with downstream processing steps, this use is probably the most efficient route to oxygenating cell culture systems. In some cases, oxygenation via sparging can be used without significant damage to cells (and without the use of surfactants) as long as the sparging is resorted to only on demand. Most cell lines also require a small amount of dissolved carbon dioxide for growth, especially at low cell densities. However, at higher cell densities, carbon dioxide may build up in the bioreactors and impact product formation in negative ways. Recently, strategies for carbon dioxide removal from large scale fed batch cultures have been described (11).

2.3. Kinetics of Cell Growth and Product Formation. Mammalian cells grow at a much slower rate than bacterial and yeast cells (see YEASTS). The maximum specific growth rates for mammalian cells range from 0.01 to 0.05 h^{-1}, corresponding to cell doubling times of $14-70$ h depending on the cell line and environmental conditions. Most primary cell lines are anchorage dependent and need a surface to grow on. They are also contact inhibited, ie, they stop growing once the surface is confluent. Alternatively, most of the cell lines used industrially for recombinant products and monoclonal antibodies are not attachment dependent. For example Chinese hamster ovary (CHO) cells are commonly used as host cells for recombinant products. These cells are transformed by tumor viruses and do not require a surface to grow on. They do, however, prefer to grow on surfaces and requirement for serum factors diminishes significantly when they are attached to surfaces. Thus CHO cells can be cultured for several weeks in protein-free media if grown on microcarriers, whereas they require serum proteins such as fetuin or appropriate protein substitutes (such as soy peptone) to grow in suspension. CHO cells can also be adapted to grow in chemically defined serum-free and protein free media which are commercially available from many vendors. Most hybridomas used for making monoclonal antibodies are attachment independent, grow in suspension, and have minimal requirement for serum proteins. It is necessary to adapt these cells to serum-free media for several days before they start growing well in these media.

Cell growth kinetics of mammalian cells can be described by the typical lag, exponential growth, then stationary and death phases. The exponential phase may be described adequately by a Monod type of kinetic model when the growth rate is much larger than the death rate. At low growth rates, it is necessary to include cell death kinetics to account for the lower viability and to predict the cell viability. Toward the end of the exponential culture, cells are also subject to growth inhibition from metabolic by-products such as lactate and ammonia. Hence, for continuous processes, a comprehensive model should contain terms for cell growth, based on the limiting substrate concentration, cell death, and inhibition kinetics.

Product formation kinetics in mammalian cells has been studied extensively for hybridomas. Some studies suggest that monoclonal antibodies are produced at an enhanced rate during the G_0 phase of the cell cycle (12–14). A model for antibody production based on this cell cycle dependence and traditional Monod kinetics for cell growth has been proposed (15). However, it is not clear if this cell cycle dependence carries over to recombinant CHO cells. In fact it has been reported that dihydrofolate reductase, the gene for which is coamplified with the gene for the recombinant protein in CHO cells, synthesis is associated with the S phase of the cell cycle (16). Hence it is possible that the product formation kinetics in recombinant CHO cells is different from that of hybridomas.

3. Cell Culture Processes

A wide variety of mammalian cells are used in industrial practice. The scale of operation and product characteristics also vary considerably. To accommodate this diversity in cell lines, scale and products, several cell culture processes have evolved. Commonly used processes include batch (or fed batch) suspension culture, continuous perfusion culture, and microcarrier systems as well as a few other systems developed to meet specific needs. Figure 1 schematically illustrates the configuration of a few of these culture systems. Table 4 summarizes the pros, cons, and some typical applications of these technologies.

3.1. Batch Suspension Culture. The batch suspension culture is perhaps the simplest technology available. It is adapted from traditional bacterial fermentation (qv) technology by changing the impellers to low shear marine propeller type, thus reducing the shear forces to which cells are subjected. Oxygenation is achieved either by direct sparging, if the cells are not subject to damage by bursting bubbles or are protected by surface-active agents such as pluronic polyols, or by membrane oxygenation where gas-permeable tubing is inserted into the fermentor.

Most hybridomas can be grown in batch suspension culture. Recombinant CHO cells can also be adapted for growth in suspension. However, CHO cells often require serum or expensive serum proteins to grow in this manner. The

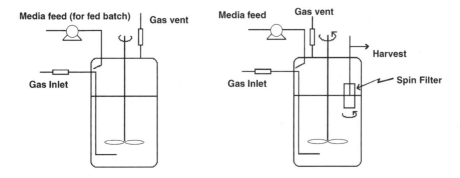

Batch/fed batch suspension culture reactor   Continuous perfusion culture with spin filter

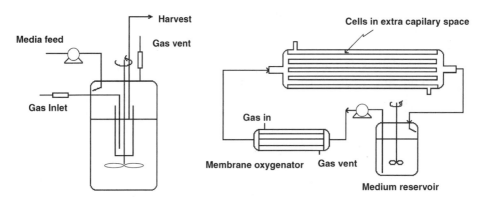

Microcarrier perfusion culture with spin filter   Hollow fiber culture system with membrane oxygenator

Fig. 1.   Commonly used cell culture processes.

applicability of this process hinges on whether an inexpensive serum-free medium can be developed for the cells in question and whether the proteins used for serum-free media development, eg, bovine serum albumin [9048-46-8] can be effectively separated from the product during downstream processing. Viral vaccines are often produced in suspension culture reactors because the product is isolated from the cells and presence of serum is not a hindrance to purification. For example, FMD vaccine is produced in 3000-L batch suspension culture reactors (17). Genentech produces tPA in 12,000-L batch fermenters using recombinant CHO cells. Batch suspension culture is also used for large-scale production of monoclonal antibodies. Celltech has scaled up airlift suspension culture reactors to 2000-L scale for the production of monoclonals (18).

A batch suspension culture reactor consists of a stirred tank typically having a height to diameter ratio of 1:1–3:1, fitted with a low shear impeller. The tank has a hemispherical bottom to avoid stagnant zones, because agitation level is very low, and the agitator shaft is either magnetically coupled to the motor or is coupled with a double mechanical seal to protect the culture from contaminating microorganisms. The vessel is pressure rated for in-place steam sterilization. Medium is filtered into the fermentor through a 0.1-μm absolute filter.

Table 4. **Commonly Used Cell Culture Technologies**

Technology	Pros	Cons	Applications
batch/fed batch suspension culture	simple, scaleable, homogeneous environment, high cell densities and product titers achievable with fed batch	not applicable to attachment dependent cells, *not* suitable for products sensitive to proteases and glycosidases	monoclonal antibodies from hybridomas and transfected myelomas, CHO cell based products, some viral vaccines, insect cells
continuous perfusion culture	useful when waste products are toxic to the cells or when product is sensitive to proteases or glycosidases, smaller reactor size, less expensive media	higher risk of lot failure due to contamination, equipment problems or clogging of filters, *not* always suitable for attachment dependent cells	monoclonal antibodies with hybridomas or transfected myelomas, some recombinant CHO cell products
microcarrier systems	suitable for attachment dependent cells, scaleable, use of traditional stirred tank equipment, low cost media	microcarriers are expensive, shear damage to cells, scaleup train can be cumbersome, clogging of screens can be a problem in continuous systems	large scale viral vaccine production from primary cells, some CHO and BHK cell based products
automated roller bottle systems	suitable for attachment dependent cells, fast implementation for low volume products, high reliability	scale-up is limited, higher cost of production	low volume/high value products including viral vaccines and recombinant CHO cell products
hollow fiber systems	low shear environment, low medium cost, smaller reactor size	scale-up is limited, nonhomogeneous environment, higher risk of lot failure	low volume monoclonal antibodies

Inoculum is added via a steam sterilized connection from the inoculum fermentor. The seed fermentor is typically one fifth to one-tenth the size of the production fermentor. Hence, for large-scale production purposes, a long inoculum train is required. Oxygenation may be effected by either direct sparging into the fementor or by using a coil of gas permeable tubing, either silicone or microporous Teflon tubing. The vessel is fitted with temperature, dissolved oxygen, and pH probes. The pH is controlled by addition of carbon dioxide (qv) to lower the pH and a suitable base, eg, dilute NaOH or $NaHCO_3$, to raise it. Dissolved oxygen is controlled by addition of air and/or pure oxygen. Insulated jackets are used for controlling the temperature. The vessel is manufactured from 316-L stainless steel and polished to a high degree, typically 240 grit followed by electropolishing, for ease of cleaning. Other materials used for seals, etc, are restricted to medical-grade silicone, Teflon, Viton, and borosilicate glass to ensure that toxic materials do not leach into the culture and affect the cells.

The top left schematic in Figure 1 shows a typical cell culture process using batch (or fed batch) suspension culture. The downstream processing of harvest from the fermentor usually consists of a clarification step, either a centrifuge or a microfilter, followed by a concentration/diafiltration step using a tangential flow ultrafiltration membrane of an appropriate molecular weight cutoff. The concentrated protein is further purified by a series of chromatography (qv) steps. The downstream processing steps are usually similar regardless of the cell culture technology being used.

Batch suspension culture has many inherent advantages. It is relatively simple to operate and scale up and less susceptible to microbial contamination. The homogeneous nature of the process makes process control (qv) and optimization easier, and from a regulatory point of view, it is the easiest to define and validate. This is therefore the process of choice if the cells can be grown in suspension in a relatively inexpensive medium. A fed batch suspension culture process is a variation of this wherein concentrated medium is fed to the bioreactor over several days in order to sustain further cell growth and production. This enables the process to reach very high cell dinsities and product titers. A recent paper reviews a generic fed batch process that has been successfully applied to many different antibody producing cell lines (19). A disadvantage of the process is that labile products may degrade during the long batch periods, especially because of the presence of proteases released by lysed cells toward the end of the process. For such products, a continuous process having low reactor residence times may be more suitable.

3.2. Batch Microcarrier Process.

A variation of suspension culture reactors is a system where the cell concentration is increased by perfusing medium through the reactor continuously while retaining the cells in the reactor by means of a spin filter device. This is shown schematically in the top right of Figure 1. Many other types of cell retention devices have been used as well. For example, an acoustic filter was used for perfusion cell culture of CHO cells at 100/L scale (20). A review of commonly used cell retention devices is available (21). A small fraction of cells are either continuously or periodically purged from the reactor to maintain high viability of cells. This type of system provides the benefits of high cell density and maintains a homogeneous environment for the cells. Since the medium is continuously perfused through the reactor, the product does not stay in contact with the cells for a long time. Therefore, this system is more suitable for products that are susceptible to degradation due to proteases and glycosidases. These systems can also be effectively used when there is a need to increase the capacity of existing stirred tank fermentors. Similar to batch suspension cultures, the applicability of the system is limited to attachment independent cells. The spin filter devices are also prone to clogging after several weeks of operation. External cell retention devices make the system more complex to set up and operate.

Applications of perfusion culture include production of factor VIII, and monoclonal antibodies by hybridomas (Bayer and Johnson & Johnson respectively). Production of monoclonal antibodies using a perfusion culture system with real time glucose control has been described (22).

For attachment dependent cells, the batch microcarrier process is the equivalent of the batch suspension process except for the fact that cells are attached to microcarriers. Cells attached on the surface of microcarriers are far more sensitive to shear forces than suspension cells because microcarriers are much larger than suspended cells. The damage to cells is theorized to be caused primarily by turbulent eddies when the eddy size becomes smaller than the particle size of microcarriers. Another mechanism for cell damage, especially at high bead concentration, is the bead to bead collision frequency. These mechanisms of cell damage in microcarrier cultures are discussed in the literature and have been quantified (23,24). Because of this limitation on agitation power input, agitator design is of great importance in microcarrier reactors. High efficiency impellers, which maximize flow and minimize shear, are utilized. The shear effects mentioned here are applicable to solid microcarriers where the cells are attached to the surface. Most recently, some macroporous carriers have been introduced that prevent the shear damage by providing attachment surfaces on the internal pores of the carriers. However, the attachment rate on such microcarriers is slower than that on solid microcarriers. This makes the porous microcarriers more suitable for long-term perfusion cultures.

Many microcarriers are available commercially for mammalian cell culture. The choice of microcarrier depends to some extent on the cell line being used and whether a batch or continuous process is being contemplated. Table 5 lists some of the microcarriers commercially available. The Cytodex family of beads is probably the most widely used. Cytodex 2 is recommended for cells having fibroblast-like morphology; Cytodex 3 is recommended for cells having an epithelial-like morphology. In long-term serum-free cultures, Cytodex 2 tends to retain cells longer than Cytodex 3, whereas the latter is useful when available inoculum density is low. In some cases, productivity is affected by the surface characteristics. For example, some cells have higher productivity on negatively charged polystyrene, eg, Biosilon, than the positively charged dextran. In designing a microcarrier process, it is recommended that a quick screening experiment be conducted to assess the suitability of the microcarriers available. A more extensive review of various types of commercial and noncommercial microcarriers and their applications is available (25).

In addition to attachment dependent cells, a batch microcarrier process may also be used for other cells that can grow in the attached mode because it allows the use of totally protein-free media. Many cells can survive for long periods of time in completely protein-free medium if attached to a surface. However,

Table 5. **Microcarriers for Stirred Tank Reactors**

Trade name	Manufacturer	Diameter, μm	Characteristics
Cytodex 1	Pharmacia, Sweden	147–248	dextran, high positive charge
Cytodex 2	Pharmacia, Sweden	135–200	dextran, positive charge
Cytodex 3	Pharmacia, Sweden	141–211	collagen-coated dextran
Biosilon	Nunc, Denmark	160–300	polystyrene, negative charge
Cultispher	Hyclone, United States	170–270	gelatin, macroporous
Bioglas	Solo Hill, United States	90–150	glass-coated plastic
Bioplas	Solo Hill, United States	90–150	cross-linked polystyrene

microcarrier reactors face several practical difficulties. First, if the cells are attachment dependent, generating sufficient inoculum for large-scale reactors involves trypsinization of microcarriers from a smaller reactor and reattachment in the larger reactor. The exposure to trypsin [9002-07-7] has to be controlled carefully in order to minimize cell damage. Second, the switch to a protein-free medium entails draining the reactor and refilling with fresh medium, requiring settling of the beads for long periods of time, especially in large-scale reactors. During this time, the cells may be deprived of oxygen and other nutrients and may be affected adversely. Hence, batch microcarrier processes are suitable for vaccine manufacture where presence of serum may not be a hindrance. Finally, the low shear requirements make oxygenation much more difficult than in suspension culture. Gas sparging tends to carry all the microcarriers into the foam layer because the beads tend to adhere to the bubbles as they rise through the liquid. One approach to solving this problem is to aerate the liquid in a rotating or vibrating cage, separating the microcarriers from the bubbles. The movement of the cage prevents the cage from getting clogged, especially in a perfusion system, with cells and microcarriers. Alternatively, gas permeable tubing may be used. Many fermentor manufacturers now offer caged aeration and perfusion systems as an option for cell culture fermentors. These include Applikon (The Netherlands), New Brunswick Scientific (United States), Cellex Biosciences (United States), and B. Braun (Germany/United States). B. Braun and Bioengineering (Switzerland) also offer silicone membrane oxygenation as an option. A good overview of microcarrier culture technology and an in-depth discussion of design issues and applications is available (26).

3.3. Microcarrier Perfusion Systems.   Microcarriers may also be used in a continuous perfusion mode. In the perfusion mode, the reactor is constantly fed with fresh, sterile medium and product is harvested from the reactor at an equal rate. Perfusion systems have the advantage that the same cells can be maintained in a productive mode for several days, thus reducing costly downtime and the cost of expensive growth medium. Once the cells attach to the carriers and become confluent, they can be maintained in a productive mode in protein-free medium for several days or, depending on the cell line, for several months. High density perfusion systems are also useful for labile products, because low residence time in the reactor minimizes the exposure of the product to degradative enzymes and conditions. Continuous systems allow steady-state operation, which makes process control and process optimization easier. Continuous processes suffer two principal drawbacks. First, the risk of contamination is higher because of the increased complexity of the process. Additionally, contamination is much more costly for a continuous process than a batch process because the magnitude of product and labor loss is much higher. Second, if the cell line is subject to genetic instability, the reactor may lose productivity over a long period because of slow overgrowth of nonproducing cells. Further, it is harder to define a batch for regulatory purposes when the product is being made by a slowly changing population of cells, making validation of the process that much more difficult.

The bottom left schematic in Figure 1 shows a typical microcarrier perfusion system utilizing a rotating screen to separate the microcarriers from the harvest liquid. Oxygenation is accomplished by sparging gas within this cage.

This eliminates the possibility of damage to the cells from gas bubbles. The medium and harvest tanks are usually sized to hold at least 3 days worth of medium. Peristaltic or steam sterilizable diaphragm metering pumps are commonly used for pumping the medium and harvest. Steam sterilizable connections are used to maintain sterility during medium feed and harvest operations. Although the screen is shown as mounted on the agitation shaft, it is possible to attach the screen to a separate shaft driven by a second motor. The latter system has the advantage that the rotational speed of the screen can be optimized for microcarrier rejection without affecting the mixing and shear characteristics of the agitator. Other methods for retaining microcarriers within the reactor include settling towers, based on gravity settling of microcarriers, and "self-cleaning" static screens (27). In long-term continuous operation, cells tend to clump up and detach from nonporous microcarriers. For this reason, the macroporous carriers may be more suitable for long-term perfusion cultures.

Process design for microcarrier processes involves determination of the surface area of carriers required to accomplish the production in a given time. This can be translated from small-scale T-flask culture experiments. There is a practical limit to the surface area per unit volume that can be accommodated in a reactor. For example, for carriers having an average diameter of 150 µm, 30,000 cm^2/L is a reasonable limit. Beyond this limit the collision severity becomes a factor in cell damage. The reactor volume is determined based on these numbers and provision is made for the necessary oxygen transfer depending on the specific oxygen transfer rate of the cells. The procedures used for designing and scaling up microcarrier reactors have been described (28).

3.4. Fluidized-Bed Systems.   So far we have discussed systems that have been used at large scale in industrial setting. However, there are many other systems that have been used for commercial purposes. For example, erythropoietin is produced commercially with an automated roller bottle system (29). Cells are grown on the surface of roller bottles in a growth medium. Cells are then shifted to a serum-free production medium for harvesting. All operations are carried out aseptically in a clean room environment by an automated machine providing a high degree of reliability and consistency. Although roller bottles are often considered to be obsolete because of the labor and space intensive nature of the process, automation makes them a viable process for products where the volume requirements are not very high. The Technology Partnership (Cambridge, England) offers robotic systems for the production of cell culture-derived products using roller bottles and T-flasks. Automated roller bottle systems can be used when the volume requirements are relatively small. Automation of the various roller bottle handling steps and media and cell manipulation steps provides a high degree of reproducibility and reliability to the process. This technology is also being used commercially for viral vaccine production with primary cells.

Another commonly used technology for small volume production of proteins is the hollow fiber system (shown schematically in the bottom right of Fig. 1). A hollow-fiber device consists of a bundle of hollow fibers, usually made of hollow anisotropic plastic fibers that allow diffusion of molecules smaller than a specified molecular weight cut-off, potted at both ends of a plastic shell (see HOLLOW-FIBER MEMBRANES). The cells are immobilized in the extra capillary space (ecs) of the hollow fibers and the medium recirculates through the lumen of the fibers entering and exiting via headers at either end of the reactor. The cells are held in a static mode and grow to high tissuelike density in the ecs. The medium access to the cells is via diffusion and Starling flow. Thus shear forces on the cells are minimal. This system is especially suitable for cells that are extremely sensitive to shear induced injury. Another advantage is that the fibers can be specified such that the product, often a high molecular weight protein, can be retained within the ecs, while medium components and metabolic by-products can diffuse through. This arrangement allows for *in situ* concentration of the product to very high levels. Similarly, serum usage can be minimized by entrapping the high molecular weight protein components of serum in the ecs and using protein-free medium for perfusion.

Scale-up of this system is limited by the size limitation on hollow-fiber manufacturing. This system is also more suitable for the diagnostic markets where quantities of protein required are relatively low. A disadvantage of the system is that nutrient gradients are set up in the ecs leading to a nonhomogeneous environment making process control difficult. This problem can be solved by pressure induced flow through the ecs (30). However, this increases the complexity of the system. More recently, production of tPA by recombinant BHK cells in a microfiltration hollow fiber bioreactor has been described (31). Hollow-fiber devices are available through Biovest International (Englewood Cliffs, New Jersey). Hollow fiber systems are being utilized by Cytogen and IDEC pharmaceuticals for small volume commercial production of monoclonal antibodies.

In the last 20 years, many different processes were developed that for one reason or another never found significant commercial applications. These include the fluidized-bed system (32) and a ceramic matrix bioreactor developed by Charles River Biotechnical Services (33). Other companies have attempted to scale-up existing T-flask processes linearly by increasing the available surface area in a compact space. An examples of such a system is the cell cube bioreactor (34). A disposable bioreactor using wave induced agitation has also been described (35). These systems are commonly used in the laboratory environment for research purposes—but have not been utilized for commercial production.

4. Economic Aspects

The 2003 market for cell culture-derived products is expected to exceeded $15 billion/year. The market is expected to continue growing substantially throughout the next decade. Cell culture products include erythropoietin and its second generation product ARANESP, 2003 expected sales of ~$7.4 billion, for the treatment of anemia associated with kidney dialysis and chemotherapy, Rituxan, 2003 expected sales of $2.5 billion, for treating non-Hodgkins lymphoma, Remicade, 2003 expected sales of $1.6 billion, for treating rheumatoid arthritis, and Enbrel, 2003 expected sales of $1.3 billion, for treating rheumatoid arthritis.

4.1. Process Economics.   Relative economics of various cell culture processes depend heavily on the performance of the cell line in a system and on the cost of raw materials, particularly the medium. Models are usually developed for the various processes using productivity data obtained from small-scale experiments (see PILOT AND MICROPLANTS). Often, for high value products, the process which ensures the shortest time to market may be the process of choice because of other economic criteria. This is especially true for pharmaceuticals (qv). Reliability concerns also often outweigh economic considerations in choosing a process for a high value product.

Continuous processes have lower labor costs but have higher failure risk. Batch processes can be started back up in a shorter period of time than can a complex continuous process. Batch processes are easier to take through the regulatory process than are continuous processes. Thus batch processes are often chosen for mammalian cell culture systems, even though continuous processes can offer significant cost advantages. Cell culture costs constitute only a small (10–30%) fraction of the overall cost of making a product. A detailed discussion of the economics of biotech products is available (36).

5. Regulations and Standards

Most of the products derived from cell culture technologies are for therapeutic or diagnostic applications and manufacture is regulated by the federal government through the Food and Drug Administration (FDA). The FDA requires that all drugs be manufactured in compliance with current Good Manufacturing Practices (cGMPs). Guidelines for cGMPs are provided through the *Code of Federal Regulations* (CFR) Title 21. Essentially, cGMPs require that all process steps and products be defined in a quantitative manner by the manufacturer, ie, specifications for all important processes must be developed and methods for testing and validating those steps must be identified. The FDA has published a guidance document for submission of relevant chemistry, manufacturing, and controls information for therapeutic recombinant DNA derived products or monoclonal antibody products for *in vivo* use (37). Other relevant guidance documents are also available at the FDA website (*www.fda.gov*) (38,39).

The biotechnology (qv) industry has no formal standards for equipment manufacture and quality control as of this writing. The American Society of Mechanical Engineers (ASME) has an active committee to devleop standards for bioprocess equipment (Bioprocess Equipment Standards Committee—located in www.asme.org).

6. Safety Considerations

The fact that cell culture-derived products are often injected into humans as therapeutic agents makes it imperative that there be no component in the final product that can pose a potential health risk to the patient. Health risks can be introduced into a product from many sources including: the cells themselves; raw materials, such as serum, media components, etc; materials used in purification, eg, antibodies; and external contamination. For a therapeutic product such risk factors are identified at the outset and ways of reducing them to acceptable levels are designed into the process. Before a product is released by the FDA the manufacturer has to demonstrate this risk reduction by rigorous validation of the process.

Some of the cells used in manufacturing are continuous or "immortal." Many of these have been shown to be tumorigenic in immunosuppressed animals. The cells also contain endogenous materials such as retroviruses and nucleic acids (qv), both of which can induce tumorigenesis, and immunogenic foreign proteins. Serum used in media can also introduce adventitious agents such as viruses and mycoplasma into the product. Other process chemicals, including cleaning agents, are low molecular-weight compounds that may be hazardous as well. Purification chemicals, such as monoclonals used for affinity purification, can be immunogenic to humans. Some of the potential health risks in mammalian cell culture processes and the methods used for risk reduction include:

cells	microfiltration
retroviruses	irradiation, sonication, heat, solvents, etc
nucleic acids	chromatography
cellular proteins	chromatography, ultrafiltration
bacterial contamination	microfiltration
process chemicals	diafiltration with appropriate buffers
serum proteins	affinity/ion-exchange chromatography

Most of these methods are commonly employed in the downstream processing of the desired cell culture technology product. Hence, most of the time it is only necessary to demonstrate that the designed process is reducing the putative risk factors to acceptable levels. Validation methods employed for risk reduction are discussed in the literature (40). In recent years, the risk of transmission of BSE via animal derived raw materials has led many companies to source their serum from "BSE-free" countries such as Australia and New Zealand. This risk has also prompted development of animal product free media for cell culture.

BIBLIOGRAPHY

"Cell Culture Technology" in *ECT* 4th ed., Vol. 5, pp. 461–476, by Subhash B. Karkare, Amgen, Inc.; "Cell Culture Technology" in *ECT* (online), posting date: December 4, 2000, by Subhash B. Karkare, Amgen, Inc.

CITED PUBLICATIONS

1. P. B. Capstick, R. C. Telling, W. G. Chapman, and D. L. Stewart, *Nature (London)* **195**, 1163 (1962).
2. A. L. van Wezel and G. Steenis, *Dev. Biol. Standard.* **40**, 69 (1978).
3. M. D. Johnston, G. Christofinis, G. D. Ball, K. H. Fantes, and N. B. Finter, *Dev. Biol. Standard.* **42**, 189 (1979).
4. A. L. van Wezel, C. A. M. van der Velden-de-Groot, and J. A. M. van Herwaarden, *Dev. Biol. Standard.* **46**, 151 (1980).
5. V. A. Luckow, in A. Prokop, R. K. Bajpai, and C. S. Ho, eds., *Recombinant DNA Technology and Applications*, McGraw Hill, New York, 1991, pp. 97–152.

6. R. Heidemann, C. Zhang, H. Qi, J. L. Rule, C. Rozales, S. Park, S. Chuppa, M. Ray, J. Michaels, K. Konstantinov, and D. Naveh, *Cytotechnology* **32**, 157 (2000).
7. K. Kao, J. Ross, A. Albee, B. Fuhr, and M. Caple, in E. Lindner-Olsson, N. Chatzissavidou, and E. Lullau, eds., *Animal Cell Technology: From Target to Market*, Kluwer Academic Press, Dordrecht, The Netherlands, 2001, pp. 189–191.
8. K. Landauer, S. Wiederkum, M. Durrschmid, H. Klug, G. Simic, G. Bluml, and O. Doblhoff-Dier, *Biotechnol. Prog.* **19**, 21 (2003).
9. J. N. Thomas, in A. S. Lubiniecki, ed., *Large Scale Mammalian Cell Culture Technology*, Marcel Dekker, New York, 1990, pp. 93–145.
10. B. Maiorella, D. Inlow, A. Shauger, and D. Harano, *Bio/Technol.* **6**, 1406 (1988).
11. S. S. Mostafa, and X. Gu, *Biotechnol. Prog.* **19**, 45 (2003).
12. R. A. Richieri, L. S. Williams, and P. C. Chau, *Cytotechnology* **5**, 243 (1991).
13. S. J. Kromenaker and F. Srienc, *Biotechnol. Bioeng.* **38**, 665 (1991).
14. N. G. Ray, S. B. Karkare, and P. W. Runstadler, *Biotechnol. Bioeng.* **33**, 724 (1989).
15. E. Suzuki and D. F. Ollis, *Biotechnol. Bioeng.* **34**, 1398 (1989).
16. B. D. Mariani, D. L. Slate, and R. T. Schimke, *Proc. Natl. Acad. Sci.* **78**, 4985 (1981).
17. P. J. Radlett, T. W. F. Pay, and A. H. M. Graland, *Dev. Biol. Standard.* **60**, 163 (1985).
18. M. Rhodes and J. Birch, *Bio/Technol.* **6**, 518 (1988).
19. P. W. Sauer, J. E. Burky, M. C. Wesson, H. D. Sternard, and L. Qu, *Biotechnol. Bioeng.* **67**, 585 (2000).
20. V. M. Gorenflo, L. Smith, B. Dedinsky, B. Persson, and J. M. Piret, *Biotechnol. Bioeng.* **80**, 438 (2002).
21. D. Voisard, F. Meuwly, P. A. Ruffieux, G. Baer, and A. Kadouri, *Biotechnol. Bioeng.* **82**, 751 (2003).
22. S. S. Ozturk, J. C. Thrift, J. D. Blackie, and D. Naveh, *Biotechnol. Bioeng.* **53**, 372 (1997).
23. S. Lakhotia and E. T. Papoutsakis, *Biotechnol. Bioeng.* **39**, 95 (1992).
24. N. Gregoriades, J. Clay, N. Ma, K. Koelling, and J. Chalmers, *Biotechnol. Bioeng.* **69**, 171 (2000).
25. S. Reuveny, in Ref. 9, pp. 271–341.
26. R. Fleischaker, in B. K. Lydersen, ed., *Large Scale Cell Culture Technology*, Hanser, New York, 1987, pp. 59–79.
27. S. B. Karkare, S. T. Cole, R. Sachdev, V. Satyagal, and J. C. Fieschko, *Ann. N. Y. Acad. Sci.* **665**, 371 (1992).
28. K. L. Nelson, *Biopharm.* **34** (Mar. 1988).
29. R. Kunitake, A. Suzuki, H. Ichihashi, S. Matsuda, O. Hirai, and K. Morimoto, *J. Biotechnol.* **52**, 289 (1997).
30. M. D. Hirschel and M. L. Gruenberg, in Ref. 26, pp. 112–144.
31. T. Tanase, Y. Ikeda, K. Iwama, A. Hashimoto, T. Kataoka, and T. Kobayashi, *J. Soc. Ferm. Bioeng.* **74**, 435 (1996).
32. U.S. Pat. 4,978,616 (Dec. 18, 1990), R. C. Dean, Jr. P. V. Grela, S. B. Karkare, and P. W. Runstadler, Jr., (to Verax Corp.).
33. B. K. Lydersen, in Ref. 26, pp. 168–192.
34. J. G. Aunins, B. Bader, A. Caola, J. Griffiths, M. Katz, P. Licari, K. Ram, C. S. Ranucci, and W. Zhou, *Biotechnol. Prog.* **19**, 2 (2003).
35. V. Singh, *Cytotechnol.* **30**, 149 (1999).
36. H. Reisman, in M. C. Flickinger and S. W. Drew, eds., *Encyclopedia of Bioprocess Technology: Fermentatioon, Biocatalysis, and Bioseparation*, John Wiley & Sons Inc., New York, 1999, pp. 863–882.
37. Guidance for Industry: For the submission of Chemistry, Manufacturing, and Controls information for a therapeutic recombinant DNA-derived product or a monoclonal antibody product for *in-vivo* use, August 1996, www.fda.gov.

38. Points to consider—Characterization of Cell Line used to produce Biological Products, 7/12/93, *www.fda.gov*.

39. Guideline for Industry: Quality of Biotechnological Products: Analysis of the expression constructs in cells used for production of r-DNA derived protein products, February 1996, *www.fda.gov*.

40. A. S. Lubiniecki, M. E. Wiebe, and S. E. Builder, in Ref. 9, pp. 515–541.

SUBHASH B. KARKARE
Scientia Consulting

CELLULOSE

1. Introduction

Cellulose [9004-34-6], characterized by Anselme Payen in 1838 (1), is the main molecule in cell walls of higher plants. The name cellulose indicates that it is the sugar (the "ose") from cells, and we now know that cellulose consists of a long chain, or polymer, of glucose units (Fig. 1). Cellulose is also formed by some algae, fungi, bacteria, the ameboid protozoan *Dictyostelium discoideum*, and a group of marine animals, the tunicates. It has even been reported in humans suffering from the rare disease of scleroderma (2). The secondary cell walls of cotton fibers are almost pure (about 94%) cellulose. In other plant sources such as the wood of mature trees, cellulose is enmeshed in as much as 36% lignin, a three-dimensional polymer of several aromatic compounds. About 7.5×10^{10} t of cellulose grow and disappear each year, establishing it as the most abundant regenerated organic material on earth.

Natural cellulosic materials such as grass are eaten by grazing animals, and various species build nests or dens with wood. Cellulose in wood (qv), in animal manure, or in bagasse (the stalks of sugar cane after the juice has been pressed out), serves directly as fuel while scientists strive to develop efficient conversion of cellulose to alcohol and other fuels. After minimal processing of natural cellulosic materials, they are used as lumber, textiles, and cordage. After industrial treatment, with and without chemical derivatization, cellulose is made into diverse products including paper, cellophane films, membranes, explosives, textiles (rayon and cellulose acetate), and dietary fiber (see Cellular Esters; Cellulose Fibers, Regenerated). The U.S. consumption of paper and other products made from wood pulp in 1999 was 340 kg per person (3). Besides its use in relatively simple products, cellulose fiber is being used to reinforce plastics in composites. This is an exciting area that can result in strong, lightweight, economical and biodegradable materials (4).

Cellulose is mostly insoluble in natural environments. Its fibers are relatively strong, with ramie and Fortisan (formerly made by Hoechst-Celanese) both having specific breaking stress values of 0.59 Pa·mm^3/g, compared with

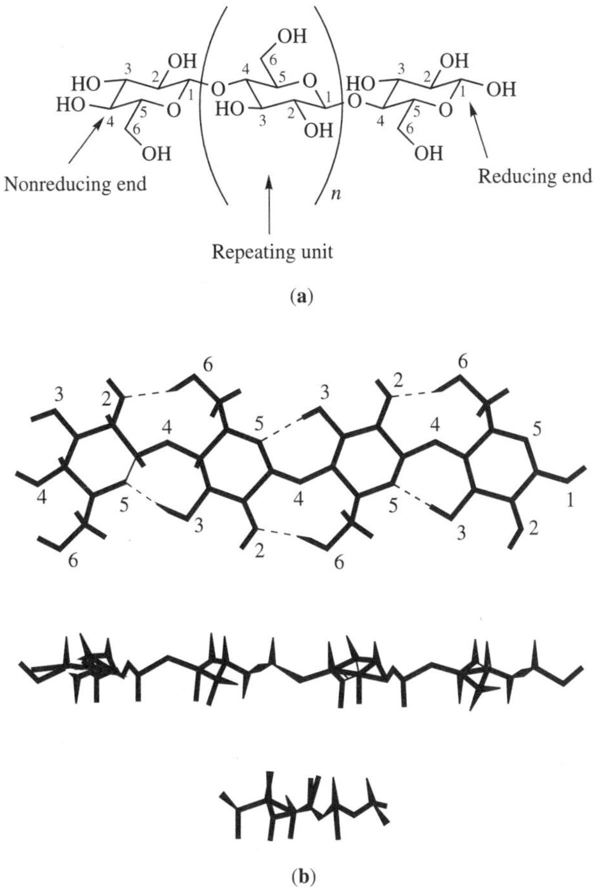

Fig. 1. Drawings of the cellulose molecule. (**a**) The chemist's structural formula for cellulose that shows the β-1,4-linked glucose units, and the numbering of the carbon atoms. (**b**) A segment of a cellulose chain composed of four β-D-glucopyranose residues (cellotetraose) in three different views. The upper image, viewed perpendicular to the flat surface of the molecule, shows the covalent bonds and electron clouds around the atoms. The dotted lines indicate hydrogen bonds between the O6-H and the O2 atoms and between the O3-H and the O5 atoms. Disregarding the ending hydroxyl hydrogen atoms, the molecule has the twofold helical conformation and hydrogen bonding typical of some proposed structures of crystalline cellulose. The middle image shows a view of the long edge of the molecule, and the bottom image shows the end of the molecule.

steel wire, for example, with 0.26 Pa mm^3/g (5). After factoring out their densities, these values can be converted to breaking stress values of 0.9 and 2.0 GPa, respectively. These attributes of strength and insolubility allow cellulose to stabilize the overall structure of plants. The stability of cellulose combined with lignin allows some trees to have very long lives. Some bristlecone pines (*Pinus longaeva (P. aristata)*) in cool, dry mountain regions of Colorado are more than 5000 years old. On the other hand, cellulosic materials in damp, warm conditions are degraded naturally by enzymes collectively known as cellulases that are present in fungi, in bacteria that exist in soil, and in cattle rumen. Cellulases are also found in protozoa in the gut of insects such as termites. Very strong acids can also degrade cellulose. The human digestive system has little effect on cellulose.

One of the reasons for the stability of cellulose is that it is usually in the form of dense crystals that have extensive van der Waals' attractive forces as well as hydrogen bonds. Although its properties, especially its flexibility, are unlike the properties of materials that are recognizable crystals, cellulose molecules have sufficient regularity to meet the criterion for crystallinity. The main reason for the difference in properties between crystals of cellulose and crystals of molecules that do not form fibers is that typical cellulose crystals are very small crosswise while being relatively long. Cellulose is polymorphic, ie, there are a number of different crystalline forms that reflect the history of the molecule. It is almost impossible to describe cellulose chemistry and biochemistry without referring to these different forms. Briefly, cellulose I, with its subclasses Iα and Iβ, is the form that occurs, with limited exception, as the result of biosynthesis. Cellulose II occurs when cellulose is regenerated from solution, such as during the manufacture of rayon, or when cellulose I is treated with strong alkali, washed, and dried. A major part of this chapter on cellulose is devoted to a more detailed explanation of these and other structures.

Plant cell walls are complicated composites, with primary and oftentimes secondary cell walls. The primary walls are next to the outer lipid membrane, and contain substantial amounts of other compounds, including pectin and hemicelluloses. The primary walls of all higher plant cells are thought to be similar in structure. During growth of the cell, expansin proteins loosen the associations of these polymers so that the cell wall can expand (6). The natural cellulosic fibers of commercial interest are typically the walls of elongated cells. For example, the cellulose-rich cotton fiber (see COTTON) is a single cell that develops on the coat of a cottonseed. Other cellulosic fibers are collections of cells called "ultimate fibers" (7).

In industrial terminology, α-cellulose is mostly β-1,4-glucan, although some insoluble hemicelluloses [9034-32-6] may also be present. Hemicelluloses (qv), which occur along with cellulose in plant cell walls, are polysaccharides such as glucomannan and acetylated glucuronoxylan. Other hemicelluloses are natural derivatives of cellulose itself, with side chains of xylose, galactose, or fucose. Holocellulose is an industrial word for delignified cellulose that still contains the hemicellulose. In the older nomenclature, β- and γ-cellulose are fractions of hemicellulose and partially degraded cellulose that are insoluble and soluble, respectively, after their alkaline solution is neutralized. All of this industrial terminology pre-dates explicit knowledge of the chemical and physical structures of these polysaccharides. The word cellulose means β-1,4-D-glucan, regardless of source.

Because of the importance of cellulose and the difficulty in unraveling its secrets regarding structure, biosynthesis, chemistry, and other aspects, several societies are dedicated to cellulose, lignin, and related molecules. These include the Cellulose and Renewable Resources Division of the American Chemical Society, the Cellulose Society of Japan, TAPPI (the Technical Association of the Pulp and Paper Industry), and Cellucon, which has organized numerous international symposia regarding cellulose. Besides the publication of the Cellulose Society of Japan (*Cellulose Communications*), there are two journals dedicated to cellulose (*Cellulose* and *Cellulose Chemistry and Technology*). Research results are published in many other journals as well. The 1638 pages of the proceedings of the Tenth Cellulose Conference that were published in 1989 (8) also indicate the vitality and interest in this subject. Several fairly recent books on cellulose have been published (9–14). The history of the proof of chemical structure of cellulose, one of the milestones in organic chemistry, is reviewed in Reference 15. The Cellulose Society of Japan published an encyclopedia of cellulose science and technology in Japanese (16). Another perspective on the long history of cellulose structure is available (17). An overall chapter is included in an encylcopedia on biopolymers (18), and a 1998 book covers many aspects of importance for those wishing to carry out chemical modifications (19).

2. Sources

Cellulose for industrial conversion comes from wood and scores of minor sources such as kenaf. Paper and rayon are now made mostly from wood pulp. Cotton rags were historically important for paper making, and cotton linters (short fibers not used to spin yarns) are now used in high quality writing and currency papers. The importance of cellulose recycling is increasing, especially for paper products. Some cellulose comes from the hairs (trichomes) on seeds, eg, cotton, kapok, and milkweed. Bast fibers are obtained from the stems of plants such as hemp, kenaf, ramie (a perennial Asian nettle), flax (linen), and jute. Besides the "soft" bast fibers, "hard" cellulose fibers are obtained from the leaves of plants such as agave (especially sisal), banana, and pineapple. In some cases, such as corn stover (the stalks and leaves of maize), the substantial amounts of cellulose present are interesting but not extensively exploited.

Celluloses from algae such as *Valonia ventricosa* are of considerable research interest because they occur in large and well-oriented crystallites. Superior structural data can be obtained by various experimental methods when crystallites are larger. However, ramie is also used for such experiments because it represents textile celluloses better than algal or bacterial cellulose. Ramie fibers contain smaller but highly oriented crystallites. Cellulose from the "tunic" of the tunicates has been processed to yield even larger crystals than those from algal celluloses, and these crystals have allowed the determination of the structure of one important form of native cellulose (see Structure, below). Bacterial cellulose is of research interest because the synthesis of cellulose by an individual bacterium, *Acetobacter xylinium*, can be observed directly with a microscope (20). It forms tangled extracellular masses of cellulose called pellicles. The pellicles resemble a nonwoven fabric and can be grown in shapes as complicated as a glove (21). A single thread of bacterial cellulose can grow to a length of a meter, compared with a few centimeters for cotton fibers.

A commercial bacterial cellulose product (Cellulon) was introduced by Weyerhaeuser (22). For use in foods, the product is called PrimaCel and is available from NutraSweet Kelco. The fiber is produced by an aerobic fermentation of glucose from corn syrup in an agitated fermentor (23,24). Because of its small particle diameter (10 μm), it has a surface area 300 times greater than normal wood cellulose, and gives a smooth mouthfeel to formulations in which it is included. It has an unusual level of water binding and works with other viscosity builders to improve their effectiveness. It is anticipated that it will achieve GRAS (Generally Regarded As Safe) status, and is neutral in sensory quality; microcrystalline cellulose (see below) has similar attributes. Other products made from bacterial cellulose include the oriental dessert, Nata di Coco, high quality loudspeaker cones, and Biofilm, a temporary skin substitute.

Recently, cellulose from sugar beet pulp and from citrus pulp has aroused interest for use as a fat substitute (25). The parenchyma cell walls involved do not contain secondary wall cellulose. The very small microfibrils with some remaining pectin, for example, give this microfibrillated cellulose properties that are similar to Cellulon; but since it is made from low cost by-products, is much less expensive.

3. Biosynthesis

During the past decade or so, we have witnessed major advances in our understanding of the biosynthesis of cellulose. Two areas are of interest, biochemistry and molecular biology.

3.1. Biochemistry.   There has been reasonable progress in the biochemistry of cellulose synthesis, but inherent difficulties lie in the isolation and characterization of the proteins that comprise the membrane-bound terminal complex (TC) first described by Brown and Montezinos (26). This complex is associated with the plasma membrane. The discovery of linear and rosette TCs using the freeze-fracture method opened the pathway for the isolation and characterization of these complexes. The first success, in 1989, was the purification of cellulose synthase from the gram-negative bacterium, *Acetobacter xylinum* (27). Extracts were capable of synthesizing cellulose *in vitro*, but only in the cellulose II crystalline form usually found for regenerated (rayon) or mercerized cellulose, not the normal native cellulose I form (see Structure, below). In 1990, Lin and co-workers (28) identified the UDP–glucose binding subunit of cellulose synthase from *Acetobacter xylinum*, using an azido photoaffinity probe. This allowed for the first identification of an 83-kDa polypeptide as the catalytic subunit of cellulose synthase. The polypeptide was subsequently sequenced for the first isolation of a cellulose synthase gene in 1990 by Saxena and co-workers (29) (see below). A comprehensive series of biochemical investigations with cotton membrane extracts (30) and product analysis led to the in vitro synthesis of cellulose II. Kinetic analyses (31) and identification of UDP–glc binding subunits of β-glucan synthases using photoaffinity labeling (32) were made.

In 1994, the first assembly of cellulose I outside of a living cell was conducted using an artificial system by means of a cellulase-catalyzed polymerization of β-glycosyl fluoride substrate (33). Micelles were postulated to organize the polymerizing sites in such a manner that a laboratory synthesis produced cellulose I microfibrils for the first time. This work (R. M. Brown Jr., unpublished) provided insight into the crystallization mechanisms and suggested that a specific association of the catalytic subunits was necessary to produce synthetic cellulose I. This was achieved by a substantial purification of a minor component of the complete *Trichoderma* cellulase system. Further work (unpublished) showed that a minor 38-kDa endoglucanase is the purified component that synthesizes cellulose I from the artificial cellobiosyl fluoride substrate.

In 1995, the first in vitro synthesis of cellulose I from native plant membrane extracts was achieved (34). The judicious use of MOPs buffer and two independent digitonin solubilization steps led to the synthesis of cellulose I microfibrils in vitro. Cellulose II and callose (β-1,3-glucan) also were assembled in vitro. A comparative study of cellulose I synthesized in vitro from cotton and mung bean revealed apparent differences in crystallinity and resistance to acetic/nitric acid reagent (35). A major breakthrough in the separation of activities leading to cellulose and callose was achieved in 1997 (36). Using native gel electrophoresis under nondenaturing conditions led to the separation of in vitro assembly of cellulose I and callose. Electron micrographs of negatively stained fractions yielded protein complexes of different morphologies involved in the synthesis of cellulose I and of callose, suggesting that the long-standing assumption that the same complex could assemble the two glucans was no longer tenable. This study also initiated a systematic fractionation of the proteins separated on the native gel, and a number of these have been sequenced (R. M. Brown Jr., unpublished). However, other than the cellulose synthases themselves, we do not yet have concrete evidence regarding which polypeptides function or participate in cellulose biosynthesis in vascular plants. Recently, Lai-Kee-Him et al. developed a larger-scale, in vitro synthesis of cellulose microfibrils (37). The cellulose was in the form of cellulose IV (see Structure, below), thought to be a disorganized form of cellulose I. Our understanding of the polymerization events is still fragmentary, as evidenced by a recent paper by Peng and co-workers (38) that discusses sitosterol β-glucoside as a possible primer for cellulose synthesis.

Freeze-fracture labeling studies have given new insight and proven the initial hypothesis that TCs are the sites of cellulose assembly. In 1999, Kimura and co-workers (39) showed the antibody labeling of rosette TCs of cotton fibers. The antibodies were produced against a recombinant polypeptide to the gene sequence of cellulose synthase. Similarly, antibodies to the c-di-GMP-binding protein were produced and tested with freeze-fracture labeling (40), and these localized with the row of the TCs in *Acetobacter* that are responsible for cellulose microfibril assembly. This study provided the first structural evidence for a cellulose-associated polypeptide known to be functional in cellulose biosynthesis in *Acetobacter*. Early work by Mizuta and Brown (41) showed the effects of 2,6-dichlorobenzonitrile and Tinopal LPW on TCs of the alga *Vaucheria*, suggesting that the complex is disrupted by these agents.

Dynamic assembly of cellulose by *Acetobacter* also has provided new understanding of the polymerization and crystallization process. In a series of papers, Cousins and Brown (42–44) showed that cellulose altered during synthesis by the optical brightener, Tinopal, was in the form of monomolecular sheets. Removal of the dye by photoisomerization or acid washing led to the assembly of cellulose I microfibrils. Comparisons of the molecular mechanics energies of different small arrangements of cellulose chain fragments (42) suggested that crystallization involved a two-step association, first to form monomolecular glucan sheets associated with each other by van der Waals' forces, followed by stacking of these sheets by hydrogen bonding to form the three-dimensional microfibril.

3.2. Molecular Biology and Molecular Genetics. Some of the most important advances in cellulose biosynthesis have originated from the combined areas of molecular biology and molecular genetics. Since the first gene for cellulose synthase in *Acetobacter xylinum* was cloned and sequenced by Saxena and co-workers in 1990 (29), many genes for cellulose synthase have been discovered. Selected mutation of *Acetobacter* cellulose synthase components affected crystallization of the product (45). In 1995, Saxena and co-workers (46) subjected the *Acetobacter xylinum* sequence to hydrophobic cluster analysis and revealed a clearly predictable pattern in processive β-glycosylation reactions. These data revealed a conservative DDD QXXRW amino acid motif common to all living organisms, prokaryotic or eukaryotic, plant or animal, that have processive β-glycosyl transferases. Using these data, the structure–function of cellulose synthase has been recently presented in the form of a genetic algorithm (47). In 1996, Pear and co-workers (48), using the data generated from the hydrophobic cluster analysis of Saxena et al., sequenced a cellulose synthase gene from a vascular plant. Independently, Arioli and co-workers (49) found the same gene in *Arabidopsis*. Since then, many other cellulose and cellulose-like synthase genes have been reported (50). Recent advances in the mutation of cellulose synthase and associated genes have provided more interesting data on the complexities of control and regulation in cellulose biosynthesis in growth and development. For example, Taylor and co-workers (51) demonstrated that the irregular xylem3 locus of *Arabidopsis* encodes a cellulose synthase required for secondary wall synthesis. In the same plant, Nicol and co-workers (52) found that a putative *endo*-1,4-β-D glucanase is required for normal cell wall assembly and may be associated with cellulose synthase. Cellulose synthases also have been isolated from forest trees, particularly *Populus* (53). In 2000, Fagard and co-workers demonstrated that PROCUSTE1 encodes for a cellulose synthase required for normal cell elongation (55). Sato and co-workers (56) found KORRIGAN, an endo-1,4-β glucanase in *Arabidopsis* that may be responsible for cell elongation and cellulose synthesis. Along the same lines, Lane and co-workers (57) found temperature-sensitive alleles that link the KORRIGAN glucanase to cellulose synthesis and cytokinesis in *Arabidopsis*. Zuo and co-workers (58) also studied the KORRIGAN endoglucanase and found that it is essential for cytokinesis. Along another line, Gillmor and co-workers (59) found that α-glucosidase I is required for cellulose biosynthesis and morphogenesis in *Arabidopsis*. Recently the question of whether one or more cellulose synthases are required for normal growth and development has been addressed, and it appears that multiple cellulose synthase catalytic subunits are required (60). Recent work from the lab of Zogaj and co-workers (61) has shown that even pathogenic bacteria produce cellulose in the form of biofilms. They conclude that cellulose biosynthesis may have an underlying function in pathogenicity. In addition, the relationship between herbicide resistance and cellulose biosynthesis was recently described by Peng and co-workers (62).

From the genetic evidence has come exciting new insight on the relationship of cellulose synthase between organisms, especially in terms of phylogeny. The recent discovery of cellulose biosynthesis in cyanobacteria, a group of organisms that have been postulated to be one of the most ancient forms of life on earth, has led to some interesting conclusions (63). For example, the cyanobacterial cellulose synthase is more like that of vascular plants than bacteria and other prokaryotes. This supports the endosymbiotic hypothesis for the early transfer of genes for cellulose synthase from cyanobacteria to a primitive eukaryotic cell that eventually led to the evolution of land plants.

3.3. The Future. With rapid advancements in molecular genetics and the ability to create, manipulate, and utilize individual molecules (nanotechnology), many exciting applications for the design and fabrication of synthetic celluloses may be on the horizon. First, however, we must learn from living organisms how a large multi-subunit enzyme complex can simultaneously polymerize up to thousands of glucan chains, then aggregate them into a metastable crystalline microfibril. It is ironic that a biopolymer made only of the sugar, glucose—one which has been around for some billions of years—is so complex in its function and regulation. A few web sites are recommended for further information (64).

4. Preparation

Wood can be used for structures with little preparation. Cotton fiber is also relatively ready for use in textiles. After it is picked, either mechanically or by hand, the fiber is removed from the cottonseed by ginning. With only some additional cleaning, the fiber is ready for textile manufacture. Bast and some leaf fibers are extracted from their surrounding tissues by various processes such as microbiological "retting," the process of rotting the nonfiber parts of the plant. A mechanical process is called decortication. It involves crushing the leaves and scraping the fibers. Other manual or mechanical stripping operations are also used (65).

Most large-scale manufacturers of cellulosic products other than textiles begin with cellulose that is in the form of pulp. Pulping partially separates cellulose from the and hemicelluloses, leaving it in a fibrous form that is more susceptible to chemical treatment than the starting material. After pulping, the pulp is purified and otherwise treated to tailor it to the required specifications. Following drying, the pulp is shipped in large rolls, to serve as feedstock for papermaking, and for the manufacture of rayon fiber, films, and other products.

High purity chemical cellulose, usually in the form of "dissolving pulp," is not only mostly free of lignin and hemicellulose, but the molecular weight of the cellulose, while fairly uniform, is lowered. This increases solubility in alkali and provides desired viscosity levels in solution. These dissolving pulps are used to make derivatives such as sodium cellulose xanthate [9051-13-2], via alkali cellulose [9081-58-7], and various esters and ethers (see Cellulose Esters; CELLULOSE ETHERS). A description of the technical details of bleaching is also available (66). The final use determines the extent of these treatments. Many steps may be used when edible or transparent films are the final product, but most newsprint receives only a single-step reductive brightening that merely whitens the fibers to an acceptable level without causing weight loss or other physical effects.

While not nearly as widely used as pulping, steam explosion (67–76) is another way to break down lignocellulose [11132-73-3] into a fibrous mulch that has substantially increased accessibility to chemical and biological agents.

These render the mulch partly soluble in organic solvents or alkali (to the extent of 40–60%) and degradable by cellulolytic enzymes. Moisture-saturated wood chips, straw, or other materials are subjected to high pressure steam (3.5–4.0 MPa) and temperature (200–250°C), followed by rapid decompression. The temperatures and pressures used depend on both the type of starting material and the intended use of the product, with more severe conditions providing greater separation of the components and greater degradation of the cellulose molecular weight. Several of the highest-energy treatments yield nearly complete separations into cellulose, lignin, and other carbohydrate degradation products. However, this comes at the expense of molecular degradation to low molecular weight sugars, furfural, and phenolics at higher treatment severities. Products such as chemical cellulose, microcrystalline cellulose, cellulose acetate [9004-35-7], sulfur-free (thermoplastic) lignin, vanillin [121-33-5], and xylose [58-86-6] that require high quality starting materials can even be made from waste materials. Milder conditions are used as an alternative to conventional pulping for the manufacture of high yield pulps for the manufacturing of paper and board. Lignocellulose given weaker treatments can be formed into molded building materials. Steam-exploded wood can be fed to cattle because of its enhanced digestibility. It is also a suitable substrate for fermentation into alcohol and other products.

Microfibrillated cellulose has several preparations, either from wood pulp or from sugar beet or citrus pulp. Sugar beet pulp is first extracted with acid or base hydrolysis to extract the pectins and hemicelluloses. After grinding or other shearing operation, a rapid pressure drop of at least 20 MPa is followed by a "rapid decelerating impact" (25).

5. Structure and Its Relation to Chemical and Physical Properties

As noted by Payen (1), both cellulose and starch can both be degraded to glucose. Yet, the properties and suitable uses of these two molecules are very different. For example, we eat starch, and we wear cellulose. If clothes were somehow made from starch without added protection, they would fall apart if they became wet. Dietary cellulose for humans furnishes no energy, although polygastric animals such as cattle do get calories from cellulose. The reason for the differences in the properties of starch and cellulose is largely the difference in the spatial arrangements of their molecules. The importance of structure was recognized at the same time as the discovery of cellulose. Therefore, cellulose structures have been studied intensively. Some basic information from more than 70 years ago has been important in helping to understand many other polymers. Still, only at the time this is being written are the intra- and intermolecular arrangements of some of the important cellulose forms resolved well enough that most scientists can be comfortable with the results. Especially in the case of natural cellulose fibers, there are many different levels of structure. Ultimately, the properties depend on the structures of these levels, and it is important to understand the role of each level.

5.1. Chemical Structure. Figure 1a is a chemist's drawing of the cellulose molecule. On the right is the reducing end, so called because it can reduce

Cu^{2+} ions in Fehlings solution to Cu^+ ions. The nonreducing end is on the left. When the molecular weight is high, the number of reducing groups for the amount of cellulose is relatively small, and the sample may test as nonreducing. Cellulose molecules are unbranched chains of up to 20,000 1,4-linked β-D-glucose [492-61-5] residues (Fig. 1), but shorter chains occur in the primary walls and under other circumstances. This measure of the molecular mass is called the degree of polymerization (DP), a measure more easily interpreted than the molecular weight in daltons. DP may be multiplied by 162 to get the molecular weight in daltons.

Attack of cellulose with strong acid gives hydrocellulose, cellulose with reduced molecular weight. The prefix hydro is used because each cleavage of the cellulose chain is accompanied by addition of the hydrogen and hydroxyl parts of a water molecule to the new fragments. After two hours in boiling 2.5 N HCl, the cellulose reaches the "leveling off degree of polymerization (LODP)". Different sources of cellulose give different values for the LODP, with number-average values of 390 glucose residues for cotton and 200 for Fortisan rayon (77). When cellulose is hydrolyzed with extremely strong acid (eg trifluoroacetic acid) for two hours, molecules with two to seven β-1,4-linked glucose residues (cellobiose to celloheptaose) will remain, along with glucose itself (78). Further attack on these cellodextrins with acid results in just glucose.

The linkage between the glucose rings is not the only location of chemical reactivity. The 2, 3, and 6 hydroxyl groups can be substituted with many different groups, and with widely varying degrees of substitution (see Chemical Reactivity, below). This polyhydroxylic nature of cellulose makes it an attractive polymer for chemical modification. Because cellulose is composed of chiral glucose residues, and because it forms fibers and films, it can be used to separate chiral materials. This is important for various types of organic synthesis, especially for drugs. It can also be used in woven mat form to complex various pollutants.

5.2. Physical Structure. Despite its ultimate degradation to just glucose, one often reads and hears that cellulose is a polymer of cellobiose units. The crystallographic unit cells of numerous forms of cellulose do repeat after two glucose units, but other crystalline cellulose complexes and derivatives have unit cells that repeat after three, four, five (79), or even eight (80), residues. Further, cellulose is not always crystalline. Amorphous and dissolved cellulose do not have geometrically repeated three-dimensional structural units. Therefore, the "polymer of cellobiose units" statement could limit thinking about the range of shapes of this somewhat flexible molecule. Any polymeric molecule that is composed of only β-1,4-linked glucose units is cellulose, regardless of the molecular shape.

Almost all glucose rings in cellulose have chair forms ($^4C_1$) with all of the substituents disposed equatorially. Any of three orientations of the primary hydroxyl oxygen atoms (O6) are possible. When O6 is located as in Figure 1**b**, it is said to be trans (anti) to O5 (the ring oxygen) and gauche (syn) to C4, or tg. When O6 atoms are positioned between O5 and C4, they are in the gauche, gauche (gg) position, and those gauche to O5 and trans to C4 are said to be gt. Other issues include the relative orientations of the adjacent glucose rings in the chain and, if in a crsytal lattice, the screw-axis symmetry.

Cellulose is rarely encountered as isolated individual molecules. Still, this state has been studied because such molecules would be similar to those in the noncrystalline or amorphous state. Also, it should give insight to the intrinsic shape that is preferred by the cellulose molecule. The amorphous form of cellulose is more chemically reactive than others. The range of allowed shapes of the isolated cellulose chain is determined by the fairly rigid glucose rings, which can, to a certain degree, rotate about their interconnecting bonds. An estimate of the flexibility of the linkage can be obtained by studying crystal structures of cellobiose and related molecules. The ranges of rotational orientations about the C1−O1 and O1−C4 bonds for the cellobiose linkages are substantial: 30° for the former and 60° for the latter (81). Compared with other polymers, these fairly large ranges cause relatively little change in the overall shape. There is also one structure with a rotation about the O1−C4 bond nearly 180° from the other conformations that have β-1,4 linkages (82).

Recently, isolated molecules have been imaged with high resolution transmission electron microscopy. Those molecules also have rather extended shapes (83). Computerized molecular modeling studies usually show that the lowest energy for an isolated individual cellulose molecule occurs when it is quite extended, as in Figure 1b (84). When dissolved, cellulose molecules are still fairly extended, but they are described as somewhat flexible random coils with relatively large distances between their ends (Fig. 2) (85). This information was also derived from computerized modeling studies in combination with light-scattering and rheological data from solutions of lightly derivatized cellulose chains. The light substitution is used to keep cellulose in solution for the experiments.

Most cellulose exists in small crystals, or crystallites. In some cases, it is thought that the unit of biosynthesis is a crystallite with only six molecules on each side (this is apparently not the case for cotton). In at least some cases, the length may be equivalent to the molecular weight. Such a crystallite is called an elementary fibril. In a nearly square crystallite with 36 chains, the number of molecules on the surface (twenty) exceeds the number in the interior (sixteen). The crystallites in primary wall material are apparently even smaller. In secondary walls of higher plants such as cotton, a few elementary fibrils are combined to form microfibrils. The details of the separation between the microfibrils and elementary fibrils are not known. Perhaps there is a boundary layer of water in the developing cell. In algal, bacterial, and tunicate cellulose, larger microfibrils are found. In cotton, there are several further levels of organization, with the microfibrils of the secondary cell wall being arranged into complex, reversing helical arrangements. Many properties of cotton are therefore different from those of other, simpler cellulosic substances.

Cellulose can be studied with diffraction crystallography. Many of these studies are reviewed in References 86 and 87. The first objective is to learn the nature of the order of the molecules, if any. If the atoms in the sample do not have a regular repeating pattern, the structure is amorphous and only a diffuse halo will result. Crystalline powders with the particles oriented randomly give a pattern of concentric rings. Powder patterns are distinctive "fingerprints" and effectively distinguish among different crystalline forms. Diffraction patterns of aligned fibers, in which the long crystallites are randomly oriented around

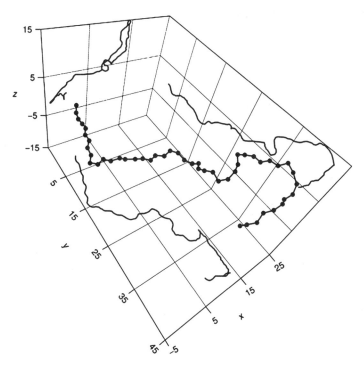

Fig. 2. A representation of the cellulose chain in solution, projected against three two-dimensional surfaces. The circles represent the oxygen atoms that link the individual glucose residues, and the lines take the place of the sugar residues. This result of a modeling study (85) that did not explicitly include solvent molecules indicated that cellulose is somewhat more flexible than found experimentally.

the long axis, give spots on layer lines. The fiber diagrams in Figure 3 show many more spots than those usually obtained from cellulose.

The positions of the diffraction spots depend on the dimensions of the unit cell, the smallest part of the crystal that can reproduce the entire crystal by simple translations of its contents along its edges. (For polymeric samples, the molecules are considered to be infinitely long.) Another objective of the diffraction experiment is to determine the unit cell dimensions. On the basis of the unit cell dimensions, educated guesses can often be made about the arrangement of the molecules. For example, the repeat of 10.38 Å along the meridian is roughly twice the length of a glucose residue. This, along with the missing intensities of the meridional reflections on the odd-numbered layer lines, suggests that the cellulose molecule has twofold screw-axis symmetry and is aligned along the fiber axis.

The final major objective is to determine the positions of the atoms within the unit cell. Until patterns of the quality in Figure 3**a** were obtained, there were always more atomic coordinates to be determined than appearing in diffraction data. Therefore, a completely experimental determination could not be carried out. In such situations, workers rely on computer models to supply information on the positions of the atoms, and calculated intensities from various models are

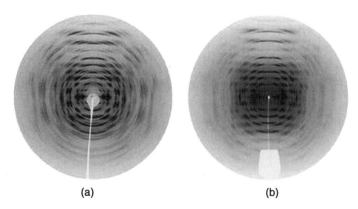

Fig. 3.   X-ray fiber diffraction patterns for (**a**) cellulose Iß and (**b**) II. The fibers and the long molecules are vertical in these patterns. The clear areas in the centers are from the shadow of the "beam stop" that keeps the undiffracted main beam from ruining the detector. A horizontal line through the middle is called the equator and the meridional line, which is vertical, also passes through the center. The patterns both show 10 layer lines. The inverse of the distance between the layer lines gives the fiber repeat distance of 10.38 Å for Iß and 10.36 Å for II. The pattern for Iß is not actually from a fiber but instead from a stack of thin films of tunicate crystallites that were dried from a slurry on the sides of a rotating horizontal vial (88). The pattern for II is for repeatedly mercerized flax. From the archives of Prof. HC working in collaboration with Y. Nishiyama and P. Langan.

compared with the observed data. Often the energies of the trial structures are calculated as well, and selection of the final structure takes into account both the extent of disagreement with the diffraction data and the energetic stability of the model (89). Errors in the measured intensities may be important (90), but a more definitive approach is usually not available.

Amorphous Cellulose.   Some cellulose is amorphous, from either mechanical action (such as ball milling) or chemical treatment (91,92). Chemical and biochemical reactions of less crystalline cellulose are usually more rapid than those of highly crystalline materials. Most samples of cellulose have some amorphous character. Sources of the amorphous halo include differences in the structures of the surface molecules compared with the structures of the interior molecules; external factors such as scattering from the air and moisture; thermal motion of the atoms; Compton scattering; chain ends, bends, or twists of the crystallites; and other departures from perfection in the array of molecules. Another cause of the background is the very small size of the crystallites. This leads to broad diffraction spots and the disappearance of many weaker spots into the background. Most amorphous cellulose probably retains many of the traits of the original crystalline material. Early textbook examples of disordered polymers in "plate of spaghetti" arrangements are quite exaggerated compared with the situation that pertains to most amorphous cellulose.

Different experiments to determine the amount of amorphous material in a given cellulose sample give different answers (93). These discrepancies, amounting to 20% or more, are thought to be due in part to the different roles of molecules on crystallite surfaces. In some methods such as X-ray diffraction, the surface molecules appear to be crystalline, and in other methods, such as

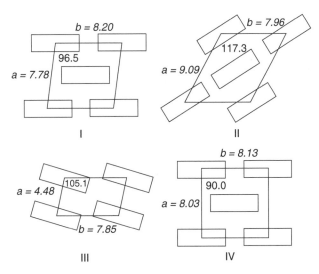

Fig. 4. Comparisons of the unit cells proposed for cellulose I–IV. In all cells, the *c* dimension (perpendicular to the plane of the drawing) is ca 10.31–10.38 Å. The dimensions were taken from References 95–98, respectively.

water sorption, they behave like amorphous material. Even cotton, the most crystalline of the commercial celluloses, has about 20% or more disordered material, including chain ends and crystallite surfaces.

Crystalline Cellulose. There are several different crystalline arrangements of cellulose (Fig. 4), each giving a distinctive diffraction pattern. These polymorphs, or allomorphs, are denoted with Roman numerals I to IV, with some subclasses. Another form is called cellulose x. The particular crystalline form depends on source and treatment. In some cases, more than one form is present in a sample. If so, the fraction of each form can be determined with X-ray diffraction or spectroscopic methods, especially nuclear magnetic resonance (NMR). Again, because they measure different phenomena, different methods will give somewhat different fractions of each form (94). During the allomorphic conversions, many other changes are likely to occur at several levels of structure. Therefore, it is risky to interpret changes in properties solely in terms of crystal form and the underlying changes in chain packing, hydrogen bonding, or molecular conformation. Still, the particular crystal form is an important aspect of solid cellulose.

The microcrystals (microfibrils) are 3–30 nm across (Fig. 5) and perhaps 7 μm in length (99,100). These microcrystals are usually straight, but bent microfibrils have appeared in samples given ultrasonic treatment (101,102). In some cases, such as cotton, the microfibrils organize into macrofibrils 60–300 nm wide. The microfibrils or macrofibrils are then organized into fibers. Diffraction contrast electron microscopy of algal cellulose shows that cross-sections perpendicular to the long axes of microfibrils are nearly square (99,100). This technique produces images that show extended cellulose molecules running parallel to the

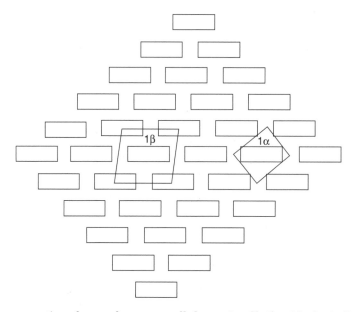

Fig. 5. A cross-section of a nearly square cellulose microfibril, with the individual molecular chains shown as rectangles. Also shown are the one- and two-chain unit cells of Iα and Iβ. This view of the microfibril is parallel to the long axis. The chains are arranged so that the edges of the crystal (microfibril) correspond to diagonals of the two-chain unit cell, or the sides of the one-chain cell (99,100).

long axis of the microfibril (99). Other experiments with atomic force microscopy confirm the conclusions from transmission electron microscopy (103). Cross-sections of other cellulose fibrils have a variety of shapes (104).

Generally speaking, polymer chains can pack efficiently only if their long axes are parallel. The notable differences among cellulose forms relate to the orientations of the rotatable hydroxyl and primary alcohol groups. Those changes permit a variety of hydrogen-bonding and crystal-packing arrangements and result in different availabilities and accessibilities of hydroxyl groups to reagents. There are three ways in which chains could pack in the monoclinic or triclinic crystals found for cellulose (105). If the chains alternate in direction, with half the reducing ends at each end of the microfibril, the packing is "anti-parallel." If all reducing ends are at one of the ends of the microcrystal, the chain packing is "parallel up." If they are all at the other end, the packing is "parallel down." The subtle distinction between the up and down types of parallel packing escaped early workers in the field. Over the years the preferred conventions for describing the unit cell have varied, making the similarity less obvious. For example, the monoclinic angle is currently obtuse, whereas it was formerly acute. Also, the unique monoclinic axis for the older work and small molecule structures is b, whereas it is c for the current fiber structures.

Cellulose I.   Originally, most native cellulose was thought to have the same crystal structure, cellulose I. In the 1950's, it became apparent from infrared spectroscopy (106,107) and electron diffraction (108) that cellulose I from the higher plants was somehow different from that of algal and bacterial cellulose. In

1984, cross polarization magic angle spinning (CPMAS) ^{13}C NMR studies showed that most native celluloses are mixtures of cellulose Iα, prevalent in algae and bacteria, and Iβ, prevalent in higher plants such as flax and cotton (105). Electron diffraction on small areas of large microfibrils from the alga *Microdictyon tenuius* shows that the Iα and Iβ forms both occur in the same microfibril (110). One area of the microfibril gave diffraction spots from a one-chain triclinic unit cell that yields the Iα components of the NMR spectrum. Another area gave the diffraction pattern of the Iβ unit cell, which is essentially the monoclinic, two-chain cell established by Andress (111) and slightly refined by Meyer and Mark (112). At some places along the microfibril, the diffraction pattern is a mixture of the two sets of intensities. In particular, that mixture had led earlier to the proposal of a now-obsolete, eight-chain unit cell for algal and bacterial cellulose I (108).

After many years of controversy, the details of cellulose Iβ (Fig. 6) have been determined to a high level of reliability (113). Extremely crystalline tunicate cellulose was examined by synchrotron X-ray and neutron diffraction analysis. Unlike the 30 or so diffraction spots that can be obtained from samples of ordinary celluloses such as flax, more than 300 spots were recorded (eg Fig. 3a). A complete exchange of the hydroxyl hydrogen atoms with deuterium (114) allowed determination of the hydrogen-bonding system through neutron diffraction.

There are two parallel-up cellulose chains, located at the corners and center, in the monoclinic Iβ unit cell. Chains at the corners are linked by hydrogen bonds into sheets, as are the chains at the centers. There are no hydrogen bonds between the chains at the corners of the unit cell and the chains at the centers.

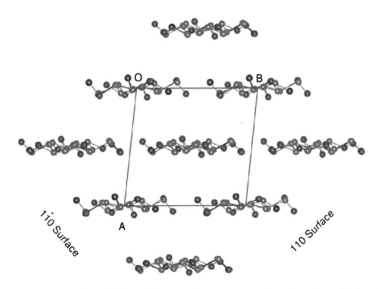

Fig. 6. A segment of the cellulose Iß crystal structure, showing the relationship of the cellulose chains to the unit cell axes a and b (the origin is O) and to the surfaces of the microfibril. The hydrogen atoms are not shown. The interchain hydrogen bonds would be in directions parallel to the b axis. Created from the coordinates in Reference 113.

The chain in the center of the unit cell is shifted along the molecular axis by 0.258 nm relative to the corner chains. Thus, every other sheet of cellulose chains is shifted up from its neighbors. The O6 groups have the unusual tg orientations. Both chains have twofold screw-axis symmetry ($P2_1$ space group) but the structures of the two chains differ. The two unique glucose rings have somewhat different shapes, and the hydrogen bonds for the corner chains and for the center chains are different. The hydrogen atoms on O3 always make hydrogen bonds with O5 on the adjacent glucose ring, but the hydrogen atoms on O2 and O6 are disordered. Those hydroxyls make a variety of different hydrogen bonds. This is thought to allow them to be more reactive than the hydrogen on O3. There is substantial dispersive (van der Waals') attraction between the two different sets of chains, exceeding the strength of the hydrogen-bonding interactions (115).

The chain packing in the Iα crystal structure favored by algal and bacterial cellulose is similar to that in the Iβ crystal. There is only one chain with two glucose residues in the triclinic unit cell, however; so by definition the chains must be packed parallel. Because there is only translational symmetry in the $P1$ space group, the two glucose residues are not identical to each other. There is no symmetry other than translation of the unit cell. The adjacent chains are shifted by approximately the same amount as in Iβ, but the difference is that the shifting is continual, instead of for every other sheet. Although a new crystal structure is being determined (Paul Langan, personal communication), the single chain per unit cell enables a reasonably confident analysis by modeling (116,118). Again the O6 groups have the unusual tg orientation, and disorder of the hydroxyl groups is a possibility. The O6 orientation is not only indicated by modeling, but also found by experimental atomic force microscopy (103).

The Iα crystal structure can be annealed to produce the Iβ structure (119) according to electron diffraction studies, and other studies of the effects of steam on the NMR spectra have been carried out (120). Both cellulose I structures appear in the same microfibrils of most algal and bacterial celluloses (121). The issue of polymorphic composition of native cellulose is further complicated, because sometimes naturally produced cellulose is not in either cellulose I form. Cellulose IV (see below) was found in some primary walls (122), although it can be difficult to distinguish between I and IV when the samples are not very crystalline.

Cellulose II.   Rayon, made from dissolved and regenerated wood and/or cotton, has the cellulose II structure, as does cellulose that has been treated with strong NaOH solutions. The effects of alkali treatments were discovered by John Mercer (123) who found that mercerized cotton had improved luster and ability to take up dyes. It also does not shrink. Treatments with alkali are still used to obtain these benefits, but in the textile industry, more dilute solutions are used at higher temperatures. In the laboratory, concentrated, low temperature solutions are most effective at creating the cellulose II crystal structure. (Cold, highly concentrated NaOH can dissolve many cellulose samples (124).) Other systems, such as 65% nitric acid, also swell cellulose and can convert it to cellulose II (125,126).

A good example of the fundamental importance of the particular crystal form is the difference in rate of digestion by bacteria. Bacteria from cattle rumen rapidly digest cellulose I but degrade cellulose II very slowly (127). Thus, allomorphic form can be an important factor in biochemical reactions of cellulose as well as in some conventional chemical reactions. On the other hand, the improved receptivity of mercerized cotton to dyes may result more from the increased amorphous content, rather than from the packing arrangement of the chains and the hydrogen-bonding system of the new crystal structure.

Besides garments, rayon is used for tire cord and industrial belting. Typical rayon yarns have lower tensile strength than cellulose I yarns, but Fortisan, a heterogeneously saponified cellulose acetate formerly made by Hoechst-Celanese, is a high strength rayon yarn, with its crystallites highly aligned along the fiber axis. The variables in the process of producing rayon fibers allow a large variety of performance characteristics (see Cellulose Fibers, Regenerated).

Some rare bacteria normally create cellulose II instead of I (128,129). Those bacteria apparently have large distances between the synthesizing complexes. This allows the various emerging molecules to fold upon themselves to take advantage of the van der Waals' forces, giving a broad band of cellulose that has its long molecular axes perpendicular to the direction of propagation of the band. Folding often occurs for some synthetic polymers such as polyethylene, but the various proposals of folding for native cellulose I now seem obsolete.

A repeatedly mercerized flax (linen) fiber (which gave an X-ray diffraction pattern similar to Fig. 3b) was studied with neutron diffraction to confirm that, unlike the cellulose I crystal structure, the cellulose II structures are composed of antiparallel chains (130). The O6 groups have the more usual gt disposition, and the interchain hydrogen bonding includes links between the corner and center chains as well as between the corner chains and between the center chains. Cellulose II is the most stable known form (117,131,132). Langan, Nishiyama, and Chanzy reject the previously proposed tg orientations for half of the O6 groups (96,133,134), and show the structure of II to be very similar to the structures of methyl cellotrioside (135), and of cellotetraose (136), both of which give cellulose II type powder diffraction patterns. The crystal densities of cellulose Iβ and II are almost identical.

Transformation from I to II. Knowledge is increasing regarding the transformation from the parallel-chain cellulose I structures to the antiparallel cellulose II. In Mercer's experiments the cellulose I yarns were dipped in alkali, then removed and rinsed. While the yarn remained intact (with some shrinkage), the crystal structure had converted from I to II. In the 1970s, when the combination of parallel I and antiparallel II structures was first widely articulated (96,133), many workers found it counterintuitive that such a fundamental change in the molecular orientation could occur with no substantial revisions in gross structure. Several facts, besides these new, well-determined crystal structures of I and II, support this change in chain packing. The parallel-up nature of Iβ has been confirmed in several different experiments, including an elegant electron microscopy study that lays out the relationship of the crystal structure to the fiber morphology (137). A number of cellulose II structures have chain folding, an antiparallel topology (102,129,138).

In the case of isolated or loosely arranged small microfibrils, such as in primary walls, the fibrous nature of the cellulose is completely lost by treatment in

NaOH, although cellulose II is produced (94). Similarly, while large, isolated, untreated crystalline microfibrils of algal cellulose are resistant to alkali, if treated with acid first and then with alkali, "shish kebab" structures are formed. The "shish" retains the cellulose I structure, and the lamellar "kebab" takes the cellulose II structure (138). In these cases, and in the cases of the bacterial cellulose where normal post-biosynthesis crystallization is thwarted, the antiparallel structure occurs through chain folding. Energy calculations show that such folding can occur with little energy penalty (84).

The current understanding (94) is that the fibrous form is retained during conversion from cellulose I to II only in secondary wall cellulose where microfibrils are adjacent to each other. While the molecules within any given cellulose I microfibril are parallel, the adjacent microfibrils in secondary wall material are themselves antiparallel. Thus a swelling of the crystal lattice, caused by the strong aqueous NaOH, allows the interpenetration, or interdigitation, of chains from neighboring antiparallel microfibrils, resulting in an antiparallel crystal (134,139). Such treatments, even when causing a complete transformation from the cellulose I structure, usually cause an increase in the amorphous cellulose content. Perhaps this is caused by folding of loose chain ends.

Other Polymorphs. There are several other cellulose polymorphs. Cellulose exposed to amines or liquid ammonia forms complexes with the swelling agent. Upon removal of the swelling agent, cellulose III is produced. The actual form of III depends on whether the starting material is cellulose I or II, giving III_I or III_{II}. Their diffraction patterns are similar to each other but the meridional intensities differ. Further, both revert to their parent structures if placed in a heated, high humidity environment. Treatment of the algal cellulose (a mixture of $I\alpha$ and $I\beta$) from *Valonia* in ethylenediamine to give cellulose III_I simultaneously induced subfibrillation in the initial microfibril (140). Thus crystallites 20 nm wide were split into subunits only 3–5 nm wide, even though the length was retained. Conversion of this III_I back to I gave a material with an electron diffraction pattern and NMR spectrum similar to that of cotton cellulose $I\beta$. The unit cell for III_I contains only one chain with two glucose residues (97), rendering obsolete a previously proposed two-chain cell. This cell results in a density of only 1.41 g/cm^3, considerably lower than the values 1.63 and 1.61 for cellulose $I\beta$ and II. As in the case of $I\alpha$ the structure of III_I must be composed of parallel chains. At present, a new crystal structure for III is being determined (Masahisa Wada, personal communication).

Similarly, cellulose IV_I and IV_{II}, formed when cellulose is treated in glycerol at temperatures ca 260°C, can revert to the parent I and II structures. Again, differences between the two diffraction patterns are biggest for the meridional reflections. The proposed structures of IV (98) are not stable in energy calculations (115,132), so details of these structures are not discussed here. Long-chain oligomers, eg, with 20 residues, crystallize from solution at high temperatures (90°C) as the IV_{II} form (141).

Cellulose x results from strong hydrochloric or phosphoric acid treatments (142). The degree of polymerization of their samples was as low as 15 or 20 following the strong acid treatment. No molecular arrangements have been proposed for cellulose x.

Conversions from cellulose I to II are widely considered to be irreversible, although there are some reports of regeneration of cellulose I (143,144). The irreversibility is considered to stem from the greater stability of cellulose II, compared with the cellulose I forms, and from the entropic difficulty in segregating the antiparallel chains of II into groups of parallel chains. Cellulose III and IV are thought to have higher energies than I and II, explaining the reversibility of the I $\rightarrow$ III$_I$ and II $\rightarrow$ III$_{II}$ conversions. The fact of parallel I and antiparallel II also explains the difference between the III$_I$ and III$_{II}$ structures. Cellulose x can also convert to cellulose IV$_{II}$, which can then convert to II.

Cellulose Hydrates. The interaction of cellulose with water is important for many reasons. Initially, cellulose II was incorrectly called cellulose hydrate; cellulose II itself has no crystalline water. Despite the high absorption of moisture, water does not penetrate the crystals in most cases. Instead, it inserts itself into the voids and onto the surfaces of the microfibrils. In some cases, however, there is water inside an otherwise pure cellulose crystal lattice. Cellulose II hydrate is stable at 93% rh (145). It contains four water molecules in a two-chain unit cell. At high hydration, these waters remained after Fortisan was swollen in hydrazine and washed with water. Another cellulose–water complex is called soda cellulose IV. Although soda cellulose I, II, and III (139) all do contain sodium ions, soda cellulose IV is formed as the sodium hydroxide is washed out of the cellulose (146). This structure, with unit cell dimensions different from those of cellulose hydrate, has two water molecules per two-chain unit cell. Two other hydrates were reported earlier (147,148).

General Considerations. All of the above crystal structures repeat in about 10.35 C. This is consistent with two individual glucose residues in the $^4C_1$ conformation. Even if the structures do not possess exact twofold symmetry, such as in Iα, there is only minor distortion from that ideal (except for regions with folds). This corresponds to a ribbon-like shape that is a flattened rectangle in cross-section, as indicated in Figures 1,4,5,6. This shape offers two different types of interaction with neighboring molecules. On the one hand, the short edges of the cross-section present the hydroxyl groups that donate and accept hydrogen bonds with neighboring molecules. The broad edges are the locations of C–H bonds that are perpendicular to the main plane of the glucose rings. These hydrophobic surfaces interact with similar surfaces on other molecules. These properties suit cellulose for its role in nature, with the efficient lateral packing similar to bricks in a wall giving a high density of strong covalent bonds parallel to the fiber axis. Also, solvents must be able to cope with both the network of hydrogen bonds and with the van der Waals' forces. The latter probably exceed those of the hydrogen bonds and contribute substantially to the long-term stability of cellulose.

With so many instances of this twofold ribbon shape, it could be surmised that this is the intrinsically favored shape. Besides the structures already mentioned, there are numerous complexes of cellulose such as those of amines (ethylenediamine, diaminopropane, and hydrazine) (149) in which the cellulose molecule has the same basic shape. The same is true for many chemical derivatives of cellulose, such as the commercially important cellulose triacetate structures (150,151).

However, as mentioned above, other shapes exist. Threefold helical structures, repeating in about 15 C, are the most frequently found alternative. This conformation does not allow an O3–O5′ hydrogen bond. It is frequently found for derivatives such as perethylated cellulose, which because of its ethyl groups is unable to form hydrogen bonds (152). Soda cellulose II is a complex in which the cellulose chain takes a threefold structure (153). Nitrocellulose [9004-70-0] molecules repeat in 25 C, with two helical turns in that distance, each with 2.5 glucose residues (154). When cellulose triacetate complexes with nitromethane, its 40.2 C repeat signifies helices with eight residues in three turns (80). Twofold helices can be considered to be either right- or left-handed, but cellulose structures with more than two residues per turn are apparently left-handed (81,155).

Supramolecular Structure.   Cotton fibers have a complex, reversing, helical arrangement of macrofibrils. The properties of cotton fabrics are therefore distinct from properties of fabrics made of ramie and linen fibers. Those and other bast and leaf fibers have crystallites arranged much more parallel to the fiber axis, but there is also a hierarchy of structures. There is the stem of the plant, bast fiber bundles, technical fiber, and finally the plant cell, which is called the elementary fiber. Within that elementary fiber there are again microfibrils. Among cotton varieties, variations in this architecture are associated with differences in strength and other qualities. In addition to diffraction, NMR, and vibrational spectroscopy, two other categories of methods indicate the extent of crystallinity (156). Those involving chemical reaction include acid hydrolysis, formylation, periodate oxidation, and chemical microstructural analysis (CMA). CMA requires reaction with *N,N*-diethylaminoethyl chloride [100-35-6] and is based on the relative availability of the hydroxyl groups to that of OH-3, which is least accessible to this agent. Other methods involving sorption include deuteration, moisture regain, and iodine sorption. Both chemical reaction and sorption techniques differ from physical measurements because they measure the fraction of the cellulose that is not readily accessible to a specific reagent. The determined fraction of ordered cellulose varies depending on the method.

Pore Structure.   Most cellulosic materials have pores into which reagents must penetrate in order to react. Variations in these pore sizes govern the extent of reaction and quality of finished products (157). In early investigations into pore size distributions, Aggebrandt and Samuelson obtained accessibilities for rayon and cotton fibers with solute exclusion of a series of ethylene glycols (158). Stone and Scallan (159) and Stone and co-workers (160) used individual samples in flasks and conducted the measurements with oligomeric sugars and dextrans of various high molecular weights to measure fiber saturation values, to assess increases in pore sizes with the cell wall during pulping, and to characterize distribution of pore sizes in wood pulps and celluloses. In another paper, Stone and Scallan (161) developed a structural model for the cell wall of water-swollen bleached pine sulfate pulp based on solute exclusion.

Martin and Rowland (162,163) extended the solute exclusion method by developing a chromatographic technique. A series of water-soluble molecules of increasing size are used as molecular probes or "feeler gauges." Sugars of low molecular weight, ethylene glycols, glymes, and dextrans were included because information on their molecular diameters is available. Also, each probe molecule should penetrate the cellulose sample and not be adsorbed on the cellulosic surfaces. Their initial experiments involved decrystallized cotton, and subsequently chopped cottons (which retained most of their crystallinity), whole fibers and fabrics. These studies were extended to include the effects of cotton genotype, fabric pretreatments such as scouring, caustic mercerization, and liquid ammonia treatment. Also studied were the effects of cross-linking with the conventional agent dimethyloldihydroxyethyleneurea (DMDHEU) and the formaldehyde-free reagents 4,5-dihydroxy-1,3-dimethyl-2-imidazolidone (DHDMI) and butanetetracarobxylic acid (BTCA). The work with DHDMI compared residual pore volume with dyeability and the study with BTCA compared the effectiveness of different catalysts. Cotton genotypes differ in their pore size distributions (164). Scouring of fabric increases the accessible internal volume, which is substantially enhanced by caustic mercerization (165). Liquid ammonia treatment also increases the internal volume (not to the extent of caustic mercerization) but the degree is dependent upon the technique used to remove the ammonia (166). Water exchange produced the greatest increase, and removal by dry heat the least. The internal volume is substantially reduced by crosslinking with DMDHEU (167) and, to a lesser degree, with DHDMI (168). Fabrics treated with DHDMI are susceptible to dyeing with small direct dyes. The study of BTCA catalysis showed that the residual internal volume in small pores was inversely related to the resilience level achieved (169). This work was recently reviewed (170).

Similar information is obtained from both the static and chromatographic methods. Results give the internal volume in the water-swollen cellulose that is available as solvent to the probes. After the initial equipment setup, the chromatographic technique is preferred for obtaining large amounts of data.

In an entirely different approach Schurz and coworkers (171) have used small-angle X-ray scattering to study the void system and the inner surfaces of fibers. They propose that the cellulose fiber consists of crystalline portions, "amorphous" regions, and air-filled voids. They obtained the ratio of inner surface to the volume of the void and certain average parameters, which characterize the void size, from the tail-end portion of the scattering curve. In their comparison of the regenerated cellulose fibers modal (a high tenacity viscose rayon) and lyocell (see Cellulose Solvents, below) they conclude that a certain void fraction is required for any fiber. The range lies between 0.0005 and 0.01. This represents the "space reserve," which is indispensable for a good fiber. They conclude that it is a very favorable void system, consisting of elongated and well-oriented voids situated between the compact elementary fibrils, which is responsible for the superior mechanical properties of lyocell fibers. Although this technique has provided information about voids in cellulosic fibers, it does not indicate if these voids are accessible to chemical agents. Fischer et al. found lozenge-shaped voids with small-angle neutron scattering (172) of deuterated ramie cellulose but not Fortisan rayon. They attributed these voids to the packing of kinked microfibrils.

6. Microcrystalline Cellulose

Pulverized forms of woodpulp have been widely used as fillers in some foods and pharmaceuticals. However, their utility is limited because the highly fibrous form results in poor mouthfeel. This problem can be overcome by reducing the woodpulp fibers to colloidal microcrystalline cellulose (173,174). It is made by reducing the particle size and molecular weight by hydrolysis with hydrochloric acid to the point of LODP (see Chemical Structure, above). In aqueous suspensoids, these much finer particles have a smooth texture resembling uncolored butter and exhibit pseudoplastic properties, including stable viscosity, over a wide temperature range. It can therefore be used as a low calorie substitute for fat. Microcrystalline celluloses are important for their heat stability; ability to thicken, with favorable mouthfeel; and flow control. They extend starches, form sugar gels, stabilize foams, and control formation of ice crystals. A few of the foods in which microcrystalline cellulose has been commercially successful are fillings, meringue (cold process), chocolate cake sauce (frozen), cookie fillings, whipped toppings, and imitation ice cream for use as a bakery filling. In the pharmaceutical industry, microcrystalline cellulose is used mostly for tableting. It is used as an excipient to assist in the flow, lubrication, and bonding properties of the ingredients to be tableted, to improve the stability of the drugs in tablet form, and especially to provide for rapid disintegration in the stomach. The determined fraction of ordered cellulose varies depending on the method.

In 2000, 55,000 t of microcrystalline cellulose were sold by the original vendor, FMC, and other companies also sell it. Its utility has led to development of other colloidal polymer microcrystals (see Colloids). For example, polyamides and polyesters from recycled materials can be biodegraded to give microcrystals having a size of 30 nm (175).

7. Chemical Reactivity

Cellulose is chemically like other carbohydrate polymers that consist of pyranose rings bearing hydroxyl groups. These chains of glucose residues include a reducing end unit, a nonreducing end unit, and intermediate units. Most celluloses have a high degree of polymerization; the intermediate glucose residues determine the chemical and physical properties and the end units may be ignored. The glycosidic bonds in cellulose are strong and stable under a variety of reaction conditions. It is a generally insoluble, highly crystalline polymer.

Cellulose can be degraded by acid, or, to a lesser extent, by alkali. The glycosidic bond is susceptible to acid-catalyzed hydrolysis. High yields of glucose can result when hydrolysis proceeds for a long enough time, needing days or weeks. After a few hours under conditions such as room temperature and HCl concentrations around $10 N$ or $0.25 M$ H_2SO_4 at $100°C$, noncrystalline cellulose is lost, leaving a more crystalline material. An LODP of 150 or 180 glucose residues is reached (176), with the length depending on the source of the cellulose. Explanations based on chain folding (177) for the various LODP values from different celluloses are not consistent with evidence from other studies of most cellulose structures. However, except for a proposal of periodic weak zones (178), no explanation has been put forth.

While alkaline degradation (179,180) is usually more subtle, losses during pulping of as much as 25% of alpha cellulose have been reported. The problem is exaggerated when excess oxygen is present. Where oxygen has been excluded, cellulose undergoes endwise degradation, scission of the glycosidic linkages, and, under certain conditions, breakdown to low molecular weight organic acids. Typically, some 50 glucose residues react and are peeled off the end of each cellulose molecule before a stable metasaccharinate (3-deoxy-D-ribohexonate) is formed that blocks further alkaline action.

Oxidation under moderate conditions (181) yields solid products referred to as oxycelluloses. This general term describes various products that must be qualified by indicating the oxidant employed. Among oxidants used are periodate, dinitrogen tetroxide, and sodium hypochlorite. Cellulose is particularly susceptible to oxidation under alkaline conditions.

Industrially important chemical modifications of cellulose generally involve reaction with its 2, 3, and 6 hydroxyl groups. These reactive sites undergo most of the reactions characteristic of alcohols. Etherification and esterification are of particular importance for cellulose (182). Cross-linking of the polymer chains gives durable press properties (183) to cellulosic textiles and dimensional stability to wood products. Reactions with the hydroxyl groups usually take place under heterogeneous conditions because of the insoluble and crystalline nature of cellulose. Under such mild heterogenous conditions, the reactivities of hydroxyl groups may depend on whether they are involved in hydrogen bonds (157). Compared with soluble polysaccharides, therefore, the extents of such reactions are inhibited.

8. Cellulose Solvents

Solvents for cellulose are central to the rayon and cellophane industries as well as being necessary for many analyses. Despite the difficulty of dissolving cellulose in aqueous and organic liquids, several cellulose solvents have been been devised over the last 150 years and reviews have been published (184–187). The solvents fall into several categories; solvents discussed in the following paragraphs do not include processes where cellulose is converted to a derivative that is subsequently dissolved in another medium. For example, cellulose can be gradually dissolved in the mixture of pyridine and acetic anhydride. However, this involves a chemical reaction and the resulting dissolved matter is "cellulose acetate," not "cellulose." The viscose process is the most important industrial method for dissolving cellulose (188). In this process, alkali cellulose [9081-58-7], pulp swollen in NaOH solution, is reacted with CS_2 to give a cellulose xanthate [9032-37-5]. The xanthate is dissolved in aqueous alkali and subsequently spun into the coagulating bath containing sulfuric acid to convert back to cellulose (189). This process, from which cellulose is readily regenerated, is sometimes considered to use a cellulose solution because solvation and derivatization occur simultaneously. Again, the dissolved molecule is a derivative, not pure cellulose. Because of the crystallite size, molecular weight and purity differences in cellulose from various sources, solvents that work well for some celluloses may not work for others. Cellulose subjected to high temperature and pressure during the steam explosion process can be dissolved in strong base (190). It has been exceptionally difficult to find effective solvents that preserve the original molecular weights of high dp samples.

The first solvent systems for cellulose were heavy metal–amine complex solutions. Aqueous solutions of Cu with ammonia (191) or ethylenediamine (192), called "cupra" and "cuen" [111274-71-6], respectively, dissolve cellulose rapidly although cellulose is subjected to progressive, oxidative degradation. The cuproammonium system is used for making high purity dialysis membranes. Instead of Cu, the alternative metals Co, Zn, Ni, and Cd can also be used effectively (193,194). Also discovered early on, some aqueous inorganic salt solutions dissolve cellulose at temperatures above 100°C. However, only a few salts will work, including $ZnCl_2$ [7646-85-7] (195), $Ca(SCN)_2$ [2092-16-2] (196) and NaSCN (197). All of these salt solutions must be highly concentrated to be effective solvents. In addition to the extremely strong, cold NaOH solutions that can dissolve many celluloses; cold, weaker NaOH solutions can also swell cellulose and dissolve 20–30% of the cellulose. Sonication (198) completes the dissolution of these "soda cellulose Q" slurries (199,200).

Reliable nonaqueous salt solutions include thiocyanate/amine and LiCl [7447-41-8]/dimethylacetamide [127-19-5] (DMAc) systems. In the thiocyanate/amine system, ammonia (201), hydrazine (202), and ethylenediamine (203) can be used as the amine. The potent thiocyanates are NH_4SCN in ammonia and NaSCN and KSCN in hydrazine and ethylenediamine. With these thiocyanate/amine solvents, solutions up to about 20% (w/w) of DP210 cellulose can be obtained without heating and/or pretreatment. The LiCl/DMAc system is useful for carrying out chemical modification of cellulose under homogeneous conditions (204) because it has no active functional group to compete with a nucleophilic attack. With DMAc, however, heating to ca 150°C and/or swelling procedures are required for the complete dissolution of cellulose. Both thiocyanate/amine and LiCl/DMAc systems afford liquid crystals of cellulose, suggesting that the chain rigidity of cellulose increases in these solvents. Most other nonaqueous solvent systems, using dipolar aprotic solvents such as dimethyl sulfoxide (203,205) and dimethylformamide (207), cause chemical derivatization even though the derivatives are unstable and readily regenerated to cellulose.

The N-methylmorpholine-N-oxide [7529-22-8] (NMMO)/H_2O system (208) is the only industrialized solvent for the spinning of cellulosic fiber that is used in place of viscose process. The solutions of dissolved cellulose have liquid crystalline properties (209) so that the lyocell (Tencel) fibers spun from this solvent have high tenacity and modulus. A drawback is that the dissolution of cellulose occurs above 130°C, close to the explosive point of NMMO (150°C). Rosenau (210) has reviewed the cellulose/NMMO system and ways to avoid runaway thermal reactions.

Most cellulose solvents are multicomponent; however, N-ethylpyridinium chloride is a single-component system (211). Recently, ionic liquids such as 1-butyl-3-methylimidazolium salts (212) have aroused interest as a single-component solvent system. The primary driving forces behind investigation of new solvents include environmental concerns, the ability to form liquid crystals and single component in the new solvent systems.

9. Liquid Crystals

Many cellulose derivatives form liquid crystalline phases, both in solution (lyotropic mesophases) and in the melt (thermotropic mesophases). The first report (213) showed that aqueous solutions of 30% hydroxypropylcellulose [9004-64-2] (HPC) form lyotropic mesophases that display iridescent colors characteristic of the chiral nematic (cholesteric) state. The field has grown rapidly and has been reviewed from different perspectives (80,214–217). A major reason for the interest in cellulosic liquid crystals is their role in the production of high strength, high modulus fibers. Cellulose fiber spun from an anisotropic phosphoric acid solution had a breaking stress value of 1.7 GPa, about twice that of the highest-strength native fibers (218). Even higher strengths (2.7 GPa) were reported for fibers spun from liquid crystalline cellulose acetate in trifluoroacetic acid, stretched in steam and then saponified (219,220).

The separation of liquid crystals as the concentration of cellulose increases above a critical value (30%) is mostly because of the higher combinatorial entropy of mixing of the conformationally extended cellulosic chains in the ordered phase. The critical concentration depends on solvent and temperature, and has been estimated from the polymer chain conformation using lattice and virial theories of nematic ordering (221–226). The side-chain substituents govern solubility, and if sufficiently bulky and flexible can yield a thermotropic mesophase in an accessible temperature range. Acetoxypropylcellulose [96420-43-8], prepared by acetylating HPC, was the first reported thermotropic cellulosic (227), and numerous other heavily substituted esters and ethers of hydroxyalkyl celluloses also form equilibrium chiral nematic phases, even at ambient temperatures.

Substituted cellulose chains have chiral twists. This leads to chiral nematic liquid crystals, in which the polymer is oriented in macroscopic helicoidal structures. If the pitch of these helicoidal structures is of the same magnitude as the wavelength of visible light, the samples show striking optical properties, in particular the reflection of circularly polarized light with a wavelength related to the pitch. The wavelength of the reflected light depends on factors such as the nature of the side groups, the degree of substitution, the molecular weight of the polymer, temperature, the nature of the solvent, and the polymer concentration. Hydroxypropylcellulose and several of its ether and ester derivatives form right-handed nematic phases (215,227–229). Ethylcellulose [9004-57-3] in glacial acetic acid gives a left-handed nematic phase (230). A change in the nematic chirality may occur with a change in the side-group substituents (231–234) and solvent (235,236). Thermally induced inversions of the twist sense have been reported for oligomers of tri-*O*-2-(2-methoxyethoxy)ethylcellulose [123423-08-5] (TMEC) and tri-*O*-heptylcellulose [100214-73-1] (237).

The helicoidal structure of such liquid crystals can be carried to the solid state by cross-linking (238,239) or by careful evaporation of solvent (240,241). Underivatized cellulose can also form ordered mesophases (209,242), and gel films precipitated from lithium chloride dimethylacetamide retain some mesophase structure (241).

Interest has been growing in liquid crystalline phases where the ordering species is not a molecularly dispersed cellulosic chain, but rather a colloidal particle of cellulose. Surprisingly, colloidal dispersions of cellulose crystallites, produced by careful sulphuric acid hydrolysis of natural cellulose fibres, were found to self-order into a chiral nematic phase above a critical concentration in water (243). The critical concentration depended primarily on the axis ratio of the rodlike cellulose particles, which typically had widths of a few nanometers, and lengths of tens of nanometers (244,245). The phase separation also depended on the ionic strength (246) and on the nature of the counterions (247) of the suspension. The chiral nematic order of the suspensions was maintained in films cast from the suspensions (248). The above suspensions were electrostatically stabilized by the presence of sulfate groups on the surface. Stabilization in water by grafting (249) and in nonpolar solvents (250) has been reported. The surface of the crystallites, termed "cellulose nanocrystals," may be modified chemically (251). Many properties of the suspensions have been examined by small-angle X-ray altering, small-angle neutron scattering (253), and induced CD of suspensions (254) and films (255). Flow properties (256) and the interfacial tension between isotropic and chiral nematic phases (257) have been investigated, and the suspension has been proposed as a medium for nmr dipolar coupling (258).

BIBLIOGRAPHY

"Cellulose" in *ECT* 1st ed., Vol. 3, pp. 342–357, by J. Barsha and P. Van Wyck, Hercules Powder Company; in *ECT* 2nd ed., Vol. 4, pp. 593–616, J. K. Hamilton and R. L. Mitchell, Rayonier, Inc.; in *ECT* 3rd ed., Vol. 5, pp. 70–88, by A. F. Turbak, ITT Rayonier, Inc.; D. F. Durso, Johnson & Johnson; O. A. Battista, Research Services Corp.; H. I. Bolker, Pulp and Paper Research Institute of Caroda; and J. R. Colvin, National Research Council, Canada; in *ECT* 4th ed., Vol. 5, pp. 476–496, by Alfred D. French, Noelie R. Bertoniere, Southern Regional Research Center, O. A. Battista, (Microcrystalline Cellulose) Research Services Corporation, John A. Cuculo, (Cellulose Solvents) North Carolina State University, Derek G. Gray, (Liquid Crystals) Pulp and Paper Research Institute of Canada; "Cellulose" in *ECT* (online), posting date: December 4, 2000, by Alfred D. French, Noelie R. Bertoniere, Southern Regional Research Center, O. A. Battista, (Microcrystalline Cellulose) Research Services Corporation, John A. Cuculo, (Cellulose Solvents) Norther Carolina State University, Derek G. Gray, (Liquid Crystals) Pulp and Paper Research Institute of Canada.

CITED PUBLICATIONS

1. A. Payen, *Compt. Rend.* **7**, 1052 (1838).
2. D. A. Hall, F. Happey, P. J. Lloyd, and H. Saxl, *Proc. Royal Soc. (London)* **151**, 497 (1959).
3. *AF&PA Statistics of Paper, Paperboard and Wood Pulp*, American Forest and Paper Association, Washington, D.C., 2001.
4. A. K. Bledzki and J. Gassen, *Prog. Polym. Sci.* **24**, 221 (1999).
5. W. E. Morton and J. W. S. Hearle, *Physical Properties of Textile Fibres*, 2nd ed., The Textile Institute, Manchester, 1993, p. 282.
6. J. Cosgrove, *Nature* **407**, 321 (2000).
7. R. M. Rowell and H. P. Stout, in M. Lewin and E. M. Pearce, eds., *Handbook of Fiber Chemistry*, 2nd ed., Marcel Dekker, Inc., New York, 1998, p. 465.
8. C. Schuerch, ed., *Cellulose and Wood—Chemistry and Technology*, John Wiley & Sons, Inc., New York, 1989.
9. T. P. Nevell and S. H. Zeronian, eds., *Cellulose Chemistry and Its Applications*, Ellis Horwood, Chichester, U.K., 1985.
10. R. H. Atalla, ed., *The Structures of Cellulose—Characterization of the Solid States* (ACS Symposium Series 340), American Chemical Society, Washington, D.C., 1987.

11. R. A. Young and R. M. Rowell, eds., *Cellulose: Structure, Modification and Hydrolysis*, Wiley-Interscience, New York, 1986.
12. J. F. Kennedy, G. O. Phillips, and P. A. Williams, eds., *Cellulose Structural and Functional Aspects*, Ellis Horwood, Chichester, U.K., 1989.
13. R. M. Brown Jr., ed., *Cellulose and Other Natural Polymer Systems: Biogenesis, Structure, and Degradation*, Plenum Press, New York, 1982.
14. H. A. Krassig, *Cellulos—Structure, Accessibility and Reactivity*, Gordon and Breach Science Publishers, Amsterdam, 1993.
15. R. H. Marchessault and P. R. Sundararajan, in G. O. Aspinall, ed., *The Polysaccharides*, Vol. **2**, Academic Press, New York, 1983, pp. 12–95.
16. Serur-osu Gakkai hensh-u, *Serur-osu no jiten*, Asakura Shoten, Tokyo, 2000.
17. A. D. French, in S.-F. Yang, ed., *Discoveries in Plant Biology III*, World Scientific Publishing Company, Hong Kong, 2000, pp. 163–196.
18. D. Klemm, H.-P. Schmaur, and T. Heinze, *Biopolymers*, Vol., **6**, Wiley-VCH, Weinheim, 2001.
19. T. Heinze and W. Glasser, eds., *Cellulose derivatives: Modification, Characterization and Nanostructures* (ACS Symposium Series 688), American Chemical Society, Washington, D.C., 1998.
20. F. C. Lin, R. M. Brown Jr., J. B. Cooper, and D. P. Delmer, *Science* **230**, 822 (1985).
21. D. G. White and R. M. Brown Jr., in Ref. 8, pp. 573–590.
22. J. Krieger, *Chem. Eng. News* 35–37 (May 21, 1990).
23. U.S. Pat. 4,960,763 (Oct. 2, 1990), R. S. Stephens, J. A. Westland, and A. N. Neogi (to Weyerhaueser Co.).
24. R. A. Kent, R. S. Stephens, and J. A. Westland, *Food Technol.* **45**, 108 (1991).
25. U.S. Pat. 5,964,983 (Oct. 12, 1999), E. Dinand, H. Chanzy, M. R. Vignon, A. Maureaux, and I. Vincent.
26. R. M. Brown Jr., and D. Montezino, *Proc. Natl. Acad. Sci. U.S.A.* **73**, 143 (1976).
27. F. C. Lin and R. M. Brown Jr., in Ref. 8, pp. 473–492.
28. F. C. Lin, R. M. Brown Jr., R. P. Drake Jr., and B. E. Haley, *J. Biol. Chem.* **265**, 4782 (1990).
29. I. M. Saxena, F. C. Lin, and R. M. Brown Jr., *Plant Mol. Biol.* **15**, 673 (1990).
30. K. Okuda, I. Li, K. Kudlicka, S. Kuga, and R. M. Brown Jr., *Plant Physiol.* **101**, 1131 (1993).
31. L. Li and R. M. Brown Jr., *Plant Physiol.* **101**, 1143 (1993).
32. L. Li, R. R. Drake Jr., S. Clement, and R. M. Brown Jr., *Plant Physiol.* **101**, 1149 (1993).
33. J. H. Lee, R. M. Brown Jr., S. Kuga, S. Shoda, and S. Kobayashi, *Proc. Natl. Acad. Sci. U.S.A.* **91**, 7425 (1994).
34. K. Kudlicka, R. M. Brown Jr., L. Li, J. H. Lee, and S. Kuga, *Plant Physiol.* **107**, 111 (1995).
35. K. Kudlicka, J. H. Lee, and R. M. Brown Jr., *Am. J. Bot.* **83**, 274 (1996).
36. K. Kudlicka and R. M. Brown Jr., *Plant Physiol.* **115**, 643 (1997).
37. J. Lai-Kee-Him, H. Chanzy, M. Müller, J.-L. Putaux, T. Imai, and V. Bulone, *J. Biol. Chem.* **277**, 36931 (2002).
38. L. Peng, Y. Kawagoe, P. Hogan, and D. Delmer, *Science* **295**, 147 (2002).
39. S. Kimura, W. Laosinchai, T. Itoh, X. Cui, R. Linder, and R. M. Brown Jr., *Plant Cell* **11**, 2075 (1999).
40. S. Kimura, H. P. Chen, G. Kikuchi, I. M. Saxena, R. M. Brown Jr., and T. Itoh, *J. Bacteriol.* **183**, 5668 (2001).
41. S. Mizuta and R. M. Brown Jr., *Protoplasma* **166**, 200 (1992).
42. S. K. Cousins and R. M. Brown Jr., *Polymer* **36**, 3885 (1995).
43. S. K. Cousins and R. M. Brown Jr., *Polymer* **38**, 897 (1997).

44. S. K. Cousins and R. M. Brown Jr., *Polymer* **38**, 903 (1997).
45. I. M. Saxena, K. Kudlicka, K. Okuda, and R. M. Brown Jr., *J. Bacteriol.* **176**, 5735 (1994).
46. I. M. Saxena, R. M. Brown Jr., M. Fevre, R. Geremia, and B. Henrissat, *J. Bacteriol.* **177**, 1419 (1995).
47. I. M. Saxena, R. M. Brown Jr., and T. Dandekar, *Phytochem.* **57**, 1135 (2001).
48. J. R. Pear, Y. Kawagoe, W. E. Schreckengost, D. P. Delmer, and D. M. Stalker, *Proc. Natl. Acad. Sci. U. S. A.* **93**, 12637 (1996).
49. T. Arioli, L. Peng, A. S. Betzner, J. Burn, W. Wittke, W. Herth, C. Camilleri, H. Höfte, J. Plazinski, R. Birch, A. Cork, J. Glover, J. Redmond, and R. E. Williamson, *Science* **279**, 717 (1998).
50. T. Richmond, *Genome Biol.* **1**, 3001.1 (2000).
51. N. G. Taylor, W. R. Scheible, S. Cutler, C. R. Somerville, and S. R. Turner, *Plant Cell* **11**, 769 (1999).
52. F. Nicol, I. His, A. Jauneau, S. Vernhettes, H. Canut, and H. Höfte, *EMBO J.* **17**, 5563 (1998).
53. F. Sterky, S. Regan, J. Karlsson, M. Hertzberg, A. Rohde, A. Holmberg, B. Amini, R. Bhalerao, M. Larsson, R. Villarroel, M. Van Montagu, G. Sandberg, O. Olsson, T. T. Teeri, W. Boerjan, P. Gustafsson, M. Uhlén, B. Sundberg, and J. Lundeberg, *Proc. Natl. Acad. Sci. U.S.A.* **95**, 13330 (1998).
54. L. Wu, C. P. Joshi, and V. L. Chiang, *Plant J.* **22**, 495 (2000).
55. M. Fagard, T. Desnos, T. Desprez, F. Goubet, G. Refrégier, G. Mouille, M. McCann, C. Rayon, S. Vernhettes, and H. Höfte, *Plant Cell* **12**, 2409 (2000).
56. S. Sato, T. Kato, K. Kakegawa, T. Ishii, Y. G. Liu, T. Awano, K. Takabe, Y. Nishiyama, S. Kuga, Y. Nakamura, S. Tabata, and D. Shibata, *Plant Cell Physiol.* **42**, 251 (2001).
57. D. R. Lane, A. Wiedemeier, L. Peng, H. Höfte, S. Vernhettes, T. Desprez, C. H. Hocart, R. J. Birch, T. I. Baskin, J. E. Burn, T. Arioli, A. S. Betzner, and R. E. Williamson, *Plant Physiol.* **126**, 278 (2001).
58. J. Zuo, Q. W. Niu, N. Nishizawa, Y. Wu, B. Kost, and N. H. Chua, *Plant Cell* **12**, 1137 (2000).
59. C. S. Gillmor, P. Poindexter, J. Lorieau, M. M. Palcic, and C. Somerville, *J. Cell Biol.* **156**, 1003 (2002).
60. N. G. Taylor, S. Laurie, and S. R. Turner, *Plant Cell* **12**, 2529 (2000).
61. X. Zogaj, M. Nimtz, M. Rohde, W. Bokranz, and U. Römling, *Mol. Microbiol.* **39**, 1452 (2001).
62. L. Peng, F. Xiang, E. Roberts, Y. Kawagoe, L. C. Greve, K. Kreuz, and D. P. Delmer, *Plant Physiol.* **126**, 981 (2001).
63. D. Nobles, D. Romanovicz, and R. M. Brown Jr., *Plant Physiol.* **127**, 529 (2001).
64. Websites for additional information on cellulose and cellulose biosynthesis: The Cellulose Electronic Network—A website devoted to all aspects of cellulose: http://128.83.195.51/cen/; The Stanford Cell Wall Page—A website devoted to cellulose synthase genes and other aspects of plant cell walls: http://cellwall.stanford.edu/; The Unrooted N J Tree of Processive Glycosyl Transferases—A website devoted to updated cataloging of processive glycosyl transferases, including cellulose synthases: http://www.botany.utexas.edu/cen/library/tree/main.htm.
65. S. K. Batra, in M. Lewin and E. M. Pearce, eds., *Handbook of Fiber Chemistry*, 2nd ed., Marcel Dekker, Inc., New York, 1998, p. 505.
66. R. P. Singh, ed., *The Bleaching of Pulp*, TAPPI Monograph Series, 3rd ed., Technical Association of the Pulp and Paper Industry, Atlanta, Ga., 1979.
67. R. H. Marchessault and J. St. Pierre, in L. E. St. Pierre and G. R. Brown, eds., *Proceedings of the CHEMRAWN Conference, Toronto, Canada, July 1978*, Pergamon Press, New York, 1980, pp. 613–625.

68. R. H. Marchessault, S. Coulombe, T. Hanai, and H. Morikawa, *Pulp. Pap. Mag. Can. Trans.* **6**, TR52–TR56 (1980).

69. B. K. Avellar and W. G. Glasser, *Biomass and Bioenergy* **14**, 205 (1998).

70. W. G. Glasser and R. S. Wright, *Biomass and Bioenergy* **14**, 219 (1998).

71. S. C. Van Winkle and W. G. Glasser, *J. Pulp Pap. Sci.* **21**, J37 (1995).

72. G. O. Phillips and J. Meadows, in J. F. Kennedy, G. O. Phillips, and P. A. Williams, eds., *Wood Processing and Utilization*, Ellis Horwood, Chichester, U.K., 1989, pp. 3–28; R. P. Overend and E. Chornet, pp. 395–400; T. Yamashiki and co-workers, pp. 401–406; K. Shimizu, K. Sudo, H. Ono, and T. Fujii, pp. 407–411.

73. U.S. Pat. 4,645,541 (Feb. 24, 1987), E. A. DeLong.

74. U.S. Pat. 4,947,743 (Aug. 14, 1990), D. B. Brown and H. Malys (to Stake Technology Ltd.).

75. U.S. Pats. 4,119,025 (Oct. 10, 1978) and 4,186,658 (Feb. 5, 1980), D. B. Brown (to Stake Technology Ltd.).

76. U.S. Pat. 4,136,207 (Jan. 23, 1979), R. Bender (to Stake Technology Ltd.).

77. H. Krassig and W. Kappner, *Makromol. Chem.* **44**/46, 1 (1961).

78. K. B. Hicks, A. T. Hotchkiss Jr., K. Sasaki, P. L. Irwin, L. W. Doner, G. Nagahashi, and R. M. Haines, *Carbohydr. Polym.* **25**, 205 (1994).

79. K. Hess and C. Trogus, *Z. phys. Chem. (B)* **13**, 25 (1931).

80. P. Zugenmaier, in Ref. 11, pp. 221–245.

81. Z. Peralta-Inga, G. P. Johnson, M. K. Dowd, J. Rendleman, E. D. Stevens and A. D. French, *Carbohydr. Res.* **337**, 851 (2002).

82. A. Ernst and A. Vasella, *Helv. Chim. Acta* **79**, 1279 (1996).

83. T. Kondo, E. Togawa, and R. M. Brown Jr., *Biomacromolecules* **2**, 1324 (2001).

84. A. D. French, in Ref. 8, pp. 103–118.

85. D. A. Brant and M. D. Christ, *ACS Symp. Ser.* **430**, 42 (1990).

86. P. Zugenmaier, *Prog. Polym. Sci.* **26**, 131 (2001).

87. A. Isogai, in R. Gilbert, ed., *Cellulosic Polymers*, Hanser Publishers Inc., Cincinnati, Ohio, 1994.

88. Y. Nishiyama, S. Kuga, M. Wada, and T. Okano, *Macromolecules* **30**, 6395 (1997).

89. A. Sarko, C.-H. Chen, B. J. Hardy, and F. Tanaka, *ACS Symp. Ser.* **430**, 345 (1989).

90. A. D. French, W. A. Roughead, and D. P. Miller, *ACS Symp. Ser.* **340**, 15 (1987).

91. L. R. Schroeder, V. M. Gentile, and R. H. Atalla, *J. Wood Chem. Technol.* **6**, 1 (1986).

92. A. Isogai, Y. Akishima, F. Onabe, M. Usuda, and R. H. Atalla, in J. F. Kennedy, G. O. Phillips, and P. A. Williams, eds., *Cellulose*, Horwood, Chichester, U.K., 1990, pp. 105–110.

93. Ref. 14, p. 169.

94. E. Dinand, M. Vignon, H. Chanzy, and L. Heux, *Cellulose* **9**, 7 (2002).

95. C. Woodcock and A. Sarko, *Macromolecules* **13**, 1183 (1980).

96. A. J. Stipanovic and A. Sarko, *Macromolecules* **9**, 851 (1976).

97. M. Wada, L. Heux, A. Isogai, Y. Nishiyama, H. Chanzy, and J. Sugiyama, *Macromolecules* **34**, 1237 (2001).

98. E. S. Gardiner and A. Sarko, *Can. J. Chem.* **63**, 173 (1985).

99. H. Chanzy, in J. F. Kennedy, G. O. Phillips and P. A. Williams, eds., *Cellulose Sources and Exploitation*, Ellis Horwood, New York, 1990, pp. 3–12.

100. A. Frey-Wyssling, *Biochim. Biophys. Acta* **18**, 166 (1955).

101. R. St. John Manley, *J. Polym. Sci. Part A-2* **9**, 1025 (1971).

102. F. J. Kolpak and J. Blackwell, *Textile Res. J.* **45**, 568 (1975).

103. A. A. Baker, W. Helbert, J. Sugiyama, and M. J. Miles, *Biophysical J.* **79**, 1139 (2000).

104. E. A. Bayeer, H. Chanzy, R. Lamed, and Y. Shoham, *Curr. Opin. Struct. Biol.* **8**, 548 (1998).

105. K. H. Gardner and J. Blackwell, *Biopolymers* **13**, 1974 (1975).
106. H. J. Marrinan and J. Mann, *J. Polym. Sci.* **21**, 301 (1956).
107. C. Y. Liang and R. H. Marchessault, *J. Polym. Sci.* **37**, 385 (1959).
108. G. Honjo and M. Watanabe, *Nature* **181**, 326 (1958).
109. R. H. Atalla and D. L. VanderHart, *Science* **223**, 283 (1984).
110. J. Sugiyama, R. Vuong, and H. Chanzy, *Macromolecules* **24**, 4168 (1991).
111. K. R. Andress, *Z. Phys. Chem. A* **136**, 279 (1928).
112. K. H. Meyer and H. F. Mark, *Z. Phys. Chem. B* **2**, 115 (1929).
113. Y. Nishiyama, P. Langan, and H. Chanzy, *J. Am. Chem. Soc.* **124**, 9074 (2002).
114. Y. Nishiyama, A. Isogai, T. Okano, M. Mueller, and H. Chanzy, *Macromolecules* **32**, 2078 (1999).
115. A. D. French, D. P. Miller, and A. Aabloo, *Int. J. Biol. Macromol.* **15**, 30 (1993).
116. A. Aabloo, A. D. French, R.-H. Mikelsaar, and A. J. Pertsin, *Cellulose* **1**, 161 (1994).
117. A. P. Heiner, J. Sugiyama, and O. Telleman, *Carbohydr. Res.* **273**, 207 (1995).
118. R. J. Vietor, K. Mazeau, M. Lakin, and S. Perez, *Biopolymers* **54**, 342 (2000).
119. J. Sugiyama, T. Okano, H. Yamamoto, and F. Horii, *Macromolecules* **23**, 3196 (1990).
120. F. Horii, H. Yamamoto, R. Kitamaru, M. Tanahashi, and T. Higuchi, *Macromolecules* **20**, 2949 (1987).
121. T. Imai and J. Sugiyama, *Macromolecules* **31**, 6275 (1998).
122. H. Chanzy, K. Imada, and R. Vuong, *Protoplasma* **94**, 299 (1978).
123. E. A. Parnell, *The Life and Labours of John Mercer*, Longmans, Green & Co., London, 1886, pp. 175–207, 214–216, 317.
124. A. Isogai and R. H. Atalla, *Cellulose* **5**, 309 (1998).
125. J. R. Katz and K. Hess, *Z. Phys. Chem. (Leipzig)* **122**, 126 (1927).
126. J. Chedin and A. Marsaudon, *Chim. Ind. (Paris)* **71**, 55 (1954).
127. P. J. Weimer, J. M. Lopez-Guisa, and A. D. French, *Appl. Environ. Microbiol.* **56**, 2421 (1990).
128. W. A. Sisson, *Science* **87**, 350 (1938).
129. S. Kuga, S. Takagi, and R. M. Brown Jr., *Polymer* **34**, 3293 (1993).
130. P. Langan, Y. Nishiyama, and H. Chanzy, *J. Am. Chem. Soc.* **121**, 9940 (1999).
131. B. G. Ranby, *Acta Chem. Scand.* **6**, 101 (1952).
132. A. D. French, M. K. Dowd, S. K. Cousins, R. M. Brown Jr., and D. P. Miller, *ACS Symp. Ser.* **618**, 13 (1995).
133. F. J. Kolpak and J. Blackwell, *Macromolecules* **9**, 273 (1976).
134. F. J. Kolpak, M. Weih, and J. Blackwell, *Polymer* **19**, 123 (1978); F. J. Kolpak and J. Blackwell, *Polymer* **19**, 132 (1978).
135. S. Raymond, B. Henrissat, D. T. Qui, A. Kvick, and H. Chanzy, *Carbohydr. Res.* **277**, 209 (1995).
136. K. Gessler, N. Krauss, T. Steiner, C. Betzel, A. Sarko, and W. Saenger, *J. Am. Chem. Soc.* **117**, 11397 (1995).
137. M. Koyama, W. Helbert, T. Imai, J. Sugiyama, and B. Henrissat, *Proc. Natl. Acad. Sci. U.S.A.* **94**, 9091 (1997).
138. H. Chanzy and E. Roche, *Appl. Polym. Symp.* **28**, 701 (1976).
139. A. Sarko, H. Nishimura, and T. Okano, *ACS Symp. Ser.* **340**, 169 (1987).
140. B. Henrissat, M. Vincendon, S. F. Tanner, and P. S. Belton, *Carbohydr. Res.* **160**, 1 (1987).
141. I. Quenin and H. Chanzy, *ACS Symp. Ser.* **340**, 189 (1987).
142. O. Ellefsen and N. Norman, *J. Polym. Sci.* **58**, 769 (1962).
143. R. H. Atalla, R. H., and S. C. Nagel, *Science* **185**, 522 (1974).
144. R. E. Whitmore and R. H. Atalla, *Int. J. Biol. Macromol.* **7**, 182 (1985).
145. D. M. Lee and J. Blackwell, *Biopolymers* **20**, 2165 (1976).
146. H. Nishimura and A. Sarko, *Macromolecules* **24**, 771 (1991).

147. I. Sakurda and K. Hutino, *Kolloid Z.* **77**, 346 (1936).
148. P. H. Hermans and A. Weidinger, *J. Colloid Sci.* **1**, 185 (1946).
149. J. Blackwell, D. Kurz, M.-Y. Su, and D. M. Lee, *ACS Symp. Ser.* **340**, 199 (1987).
150. A. J. Stipanovic and A. Sarko, *Polymer* **19**, 3 (1978).
151. E. Roche, H. Chanzy, M. Boudeulle, R. H. Marchessault, and P. Sundrarajan, *Macromolecules* **11**, 86 (1978).
152. U. Vogt and P. Zugenmaier, *Ber. Bunsenges. Phys. Chem.* **89**, 1217 (1985).
153. P. M. Whitaker, I. A. Nieduszynski, and E. D. T. Atkins, *Polymer* **15**, 125 (1974).
154. D. Meader, E. D. T. Atkins, and F. Happey, *Polymer* **19**, 1371 (1978).
155. P. Zugenmaier, *J. Appl. Polym. Sci., Appl. Polym. Symp.* **37**, 223 (1983).
156. N. R. Bertoniere and S. H. Zeronian, *ACS Symp. Ser.* **340**, 255 (1987).
157. S. P. Rowland and N. R. Bertoniere, in Ref. 9, pp. 112–137.
158. L. Aggebrandt and O. Samuelson, *J. Appl. Polym. Sci.* **8**, 2810 (1964).
159. J. E. Stone and A. M. Scallan, *Pulp. Pap. Mag. Can.* **69**, 69 (1968).
160. J. E. Stone, E. Treiber and B. Abrahamson, *Tappi J.* **52**, 108 (1969).
161. J. E. Stone and A. M. Scallan, *Cell. Chem. Technol.* **3**, 343 (1968).
162. L. F. Martin and S. P. Rowland, *J. Chromatogr* **28**, 139 (1967).
163. L. F. Martin and S. P. Rowland, *J. Polym. Sci., Part A-1* **5**, 2563 (1967).
164. N. R. Bertoniere, W. D. King, and S. E. Hughs, in J. F. Kennedy, G. O. Phillips and P. A. Williams, eds., *Lignocellulosics—Science, Technology, Development and Use*, Ellis Horwood Limited, Chichester, U.K., 1992, p. 457.
165. N. R. Bertoniere and W. D. King, *Textile Res. J.* **59**, 114 (1989).
166. N. R. Bertoniere, W. D. King, and S. P. Rowland, *J. Appl. Polym. Sci.* **31**, 2769 (1986).
167. N. R. Bertoniere and W. D. King, *Textile Res. J.* **60**, 606 (1990).
168. N. R. Bertoniere and W. D. King, *Textile Res. J.* **59**, 608 (1989).
169. N. R. Bertoniere, W. D. King and C. M. Welch, *Textile Res. J.* **64**, 247 (1994).
170. N. R. Bertoniere in M. Raheel, ed., *Modern Textile Characterization Methods*, Marcel Dekker, Inc., New York, 1996, pp. 265–290.
171. J. Schurz, J. Lenz and E. Wrentschur, *Angew. Makromol. Chem.* **229**, 175 (1995).
172. E. W. Fischer, P. Herschenröder, R. St. J. Manley, and M. Stamm, *Macromolecules* **11**, 213 (1978).
173. O. A. Battista, *Microcrystal Polymer Science*, McGraw-Hill, New York, 1975.
174. O. A. Battista, in J. I. Kroschwitz, ed., *Encyclopedia of Polymer Science and Engineering*, Vol. **3**, John Wiley & Sons, Inc., New York, 1985, pp. 86–90.
175. U.S. Pat. 3,536,647 (Oct. 27, 1970), O. A. Battista; U.S. Pat. 3,931,082 (Jan. 6, 1976), M. M. Cruz Jr., N. Z. Erdi, and O. A. Battista.
176. O. A. Battista, *Ind. Eng. Chem.* **42**, 502 (1950).
177. M. Chang, *J. Polym. Sci. C* **36**, 343 (1971).
178. M. Marx-Figini, *J. Appl. Polym. Sci., Appl. Polym. Symp.* **37**, 157 (1982).
179. R. J. Whistler and J. N. BeMiller, *Adv. Carbohydr. Chem.* **13**, 289 (1958).
180. T. P. Nevell, in Ref. 9, pp. 223–242.
181. Ref. 180, pp. 243–265.
182. P. J. Wakelyn, N. R. Bertoniere, A. D. French, S. H. Zeronian, T. P. Nevell, D. P. Thibodeaux, E. J. Blanchard, T. A. Calamari, B. A. Triplett, C. K. Bragg, C. M. Welch, J. D. Timpa, and W. R. Goynes Jr., in M. Lewin and E. Pierce, eds., *Handbood of Fiber Chemistry*, 2nd ed., Marcel Dekker, Inc., New York, 1998, pp. 577–724.
183. S. L. Vail, in Ref. 9, pp. 375–422.
184. S. M. Hudson and J. A. Cuculo, *J. Macromol. Sci., Rev. Macromol. Chem. C* **18**, 1 (1980); A. F. Turbak, R. B. Hammer, R. E. Davis, and H. L. Hergert, *CHEMTECH* **10**, 51 (1980).

185. D. C. Johnson, in Ref. 9, pp. 181–201.

186. B. Philipp, *J. Macromol. Sci., Pure Appl. Chem. A* **30**, 703 (1993).

187. V. V. Myasoedova, in V. V. Myasoedova, ed., *Physical Chemistry of Non-Aqueous Solutions of Cellulose and Its Derivatives*, John Wiley & Sons, Inc., Chichester U.K., 2000.

188. E. E. Treiber, in Ref. 9, pp. 455–479.

189. A. F. Turbak, in J. I. Knoschwitz, ed., *Encyclopedia of Polymer Science and Engineering*, Vol. **14**, John Wiley & Sons, Inc., New York, 1985, pp. 45–72.

190. T. Yamashiki, M. Saitoh, K. Yasuda, K. Okajima, and K. Kamide, *Cell. Chem. Technol.* **24**, 237 (1990).

191. E. Schweizer, *J. Prakt. Chem.* **72**, 109 (1857).

192. W. Traube, *Ber. Dtsch. Chem. Ges.* **44**, 3319 (1911).

193. G. Jayme, in N. M. Bikales and L. Segal, eds., *Cellulose and Cellulose Derivatives*, Vol. **V**, Part IV, John Wiley & Sons, Inc., New York, 1970, pp. 381–411.

194. K. Saalwächter, W. Burchard, P. Klüfers, G. Kettenbach, P. Mayer, D. Klemm, and S. Dugarmaa, *Macromolecules* **33**, 4094 (2000).

195. Br. Pat. 13,296 (1850), J. Mercer.

196. A. Dubose, *Bull. Soc. Ind. Rouen* **33**, 318 (1905).

197. K. Kamide and M. Saito, *Polym. J.* **20**, 447 (1988).

198. U.S. Pat. 5,605,567 (Feb. 25, 1997), E. P. Lancaster (to Weyerhaueser Co.).

199. H. Sobue, H. Kiessig, and K. Hess, *Z. Phys. Chem. (B)* **43**, 309 (1939).

200. A. H. Nissen, G. H. Hungen, and S. S. Sternstein, in N. M. Bikales, ed., *Encyclopedia of Polymer Science and Technology*, Vol. **3**, John Wiley & Sons, Inc., New York, 1965, p. 166.

201. U.S. Pat. 4,367,191 (Jan. 4, 1983), J. A. Cuculo and S. M. Hudson (to Research Corp.).

202. K. Hattori, J. A. Cuculo, and S. M. Hudson, *J. Polym. Sci., Part A: Polym. Chem.* **40**, 601 (2002).

203. K. Hattori, T. Yoshida, and J. A. Cuculo, *224th National Meeting*, American Chemical Society, Boston, 2002. Abstract, BTEC32.

204. T. R. Dawsey and C. L. McCormick, *J. Macromol. Sci., C: Rev. Macromol. Chem. Phys.* **30**, 405 (1990).

205. U.S. Pat. 4,097,666 (June 27, 1978), D. C. Johnson and M. D. Nicholson (to the Institute of Paper Chemistry).

206. B. Philipp, H. Schleicher, and W. Wagenknecht, *CHEMTECH* **7**, 702 (1977).

207. U.S. Pat. 3,669,916 (June 13, 1972), O. Nakao, S. Nakagawa, J. Hirose, S. Yamazaki, T. Amano, T. Nakamura, and H. Yamamoto (to Tomoegawa Paper Manufacturing Co., Ltd.).

208. U.S. Pat. 3,447,939 (June 3, 1969), D. L. Johnson (to Eastman Kodak Co.).

209. H. Chanzy and A. Peguy, *J. Polym Sci., Polym. Chem. Ed.* **18**, 1137 (1980).

210. T. Rosenau, A. Potthast, H. Sixta and P. Kosma, *Prog. Polym. Sci.* **26**, 1763 (2001).

211. U.S. Pat. 1,943,176 (Jan. 9, 1934), C. Graenacher (to Society of Chemical Industry).

212. R. P. Swatloski, S. K. Spear, J. D. Holbrey, and R. D. Rogers, *J. Am. Chem. Soc.* **124**, 4794 (2002).

213. R. S. Werbowyj and D. G. Gray, *Mol. Cryst. Liq. Cryst. (Lett.)* **34**, 97 (1976).

214. D. G. Gray, *J. Appl. Polym. Sci., Appl. Polym. Symp.* **37**, 179 (1983).

215. R. D. Gilbert and P. A. Patton, *Prog. Polym. Sci.* **9**, 115 (1983).

216. D. G. Gray, *Faraday Discuss. Chem. Soc.* **79**, 257 (1985).

217. P. Sixou and A. Ten Bosch, in Ref. 11, pp. 205–219.

218. M. G. Northolt, H. Boerstoel, H. Maatman, R. Huisman, J. Veurink, and H. Elzerman, *Polymer* **42**, 8249 (2001).

219. U.S. Pat. 4464323 (Aug. 7, 1984), J. P. O'Brien (to E. I. Du Pont de Nemours and Company).

220. U.S. Pat. 4501886 (Feb. 26, 1986), J. P. O'Brien (to E. I. Du Pont de Nemours and Company).

221. R. S. Werbowyj and D. G. Gray, *Macromolecules* **13**, 69 (1980).

222. G. Conio, E. Bianchi, A. Ciferri, A. Tealdi, and M. A. Aden, *Macromolecules* **16**, 1264 (1983).

223. S. N. Bhandani, S.-L. Tseng, and D. G. Gray, *Makromol. Chem.* **184**, 1727 (1983).

224. E. Bianchi, A. Ciferri, G. Conio, L. Lanzavecchia, and M. Terbojevich, *Macromolecules* **19**, 630 (1986).

225. G. V. Laivins and D. G. Gray, *Macromolecules* **18**, 1753 (1985).

226. J. M. Mays, *Macromolecules* **21**, 3179 (1988).

227. S.-L. Tseng, A. Valente, and D. G. Gray, *Macromolecules* **14**, 715 (1981).

228. S. Bhandani and D. G. Gray, *Mol. Cryst. Liq. Cryst.* **99**, 29 (1983).

229. A. M. Ritcey and D. G. Gray, *Macromolecules* **21**, 1251 (1988).

230. U. Vogt and P. Zugenmaier, *Ber. Bunsenges. Phys. Chem.* **89**, 1217 (1985).

231. A. M. Ritcey and D. G. Gray, *Macromolecules* **21**, 2914 (1988).

232. J. X. Guo and D. G. Gray, *Macromolecules* **22**, 2086 (1989).

233. W. P. Pawlowski, R. D. Gilbert, R. E. Fornes, and S. T. Purrington, *J. Polym. Sci., Part B: Polym. Phys.* **25**, 2293 (1987).

234. H. Steinmeier and P. Zugenmaier, *Carbohydr. Res.* **173**, 75 (1988).

235. P. Zugenmaier and P. Haurand, *Carbohydr. Res.* **160**, 369 (1987).

236. B. R. Harkness and D. G. Gray, *Can. J. Chem.* **68**, 1135 (1990).

237. T. Yamagishi, T. Fukua, T. Miyamoto, T. Ichizuka, and J. Watanabe, *Liq. Cryst.* **7**, 155 (1990).

238. S. N. Bhandani and D. G. Gray, *Mol. Cryst. Liq. Cryst.* **102**, 255 (1984).

239. S. Suto and H. Tashiro, *Polymer* **30**, 2063 (1989).

240. G. Charlet and D. G. Gray, *Macromolecules* **20**, 33 (1987).

241. J. Giasson, J.-F. Revol, A. M. Ritcey, and D. G. Gray, *Biopolymers* **27**, 1999 (1988).

242. G. Conio, P. Corazza, E. Bianchi, A. Tealdi, and A. Ciferri, *J. Polym. Sci., Polym. Lett. Ed.* **22**, 273 (1984).

243. J.-F. Revol, H. Bradford, J. Giasson, R. H. Marchessault, and D. G. Gray, *Int. J. Biol. Macromolecules* **14**, 170 (1992).

244. J.-F. Revol, L. Godbout, X.-M. Dong, D. G. Gray, H. Chanzy, and G. Maret, *Liq. Crys.* **16**, 127 (1994).

245. X. M. Dong, J.-F. Revol, and D. G. Gray, *Cellulose* **5**, 19 (1998).

246. X. M. Dong, T. Kimura, J.-F. Revol, and D. G. Gray, *Langmuir* **12**, 2076 (1996).

247. X. M. Dong and D. G. Gray, *Langmuir* **13**, 2404 (1997).

248. U.S. Pat. 5,629,055 (May 13, 1997), J.-F. Revol, D. L. Godbout, and D. G. Gray (to Pulp and Paper Research Institute of Canada).

249. J. Araki, M. Wada, and S. Kuga, *Langmuir* **17**, 21 (2001).

250. L. Heux, G. Chauve, and C. Bonini, *Langmuir* **16**, 8210 (2000).

251. W. T. Winter and A. J. Stipanovic, in *221st National Meeting*, Amercian Chemical Society, Orlando, Fla., 2001, Abstract P, CELL-076.

252. T. Furuta, E. Yamahara, T. Konishi, and N. Ise, *Macromolecules* **29**, 8994 (1996).

253. W. J. Orts, L. Godbout, R. H. Marchessault, and J.-F. Revol, *Macromolecules* **31**, 5717 (1998).

254. X.-M. Dong and D. G. Gray, *Langmuir* **13**, 3029 (1997).

255. C. D. Edgar and D. G. Gray, *Cellulose* **23**, 1 (2001).

256. J. Araki, M. Wada, S. Kuga, and T. Okano, *Colloids Surf., A* **142**, 75 (1998).

257. W. Chen and D. G. Gray, *Langmuir* **18**, 633 (2002).
258. K. Fleming, D. Gray, S. Prasannan, and S. Matthews, *J. Am. Chem. Soc.* **122**, 5224 (2000).

ALFRED D. FRENCH
NOELIE R. BERTONIERE
Southern Regional Research Center, USDA
R. MALCOLM BROWN
The University of Texas at Austin
HENRI CHANZY
CNRS-CERMAV
DEREK GRAY
Pulp and Paper Research Centre, McGill University
KAZUYUKI HATTORI
Kitami Institute of Technology
WOLFGANG GLASSER
Virginia Polytechnic Institute and State University

CELLULOSE ESTERS, INORGANIC ESTERS

1. Introduction

Cellulose plays an extremely important structural role in nature. Cellulose is a linear polysaccharide comprised of β-1,4-linked cellobiose repeat units (Fig. 1). The hydroxyl groups present on each anhydroglucose unit provide the biopolymer with the ability to form strong structures through hydrogen bonding. This vast network of hydrogen bonds that link the individual polymer chains into more complex structures greatly reduces the solubility of cellulose in conventional organic and aqueous solvents. Derivatization of cellulose as esters and/or ethers modifies the solubility profile of the cellulosic while maintaining many of the polymeric properties of the molecule. Cellulose esters have found numerous commercial applications, including plastics, coatings, and ion exchange applications. Esters of cellulose can be either inorganic or organic. The focus of this section is on inorganic esters of cellulose. This article covers both the synthesis and use of inorganic esters of cellulose.

The importance of the general structures in Figure 1 and the numbering scheme (the six carbons that make up the anhydroglucose backbone of a cellulosic are represented by a number between C1–C6) described in Figure 2 becomes clear to the reader throughout this article.

Inorganic esters of cellulose have been known for well over 100 years and are still commercially viable products. Cellulose nitrate, commonly called

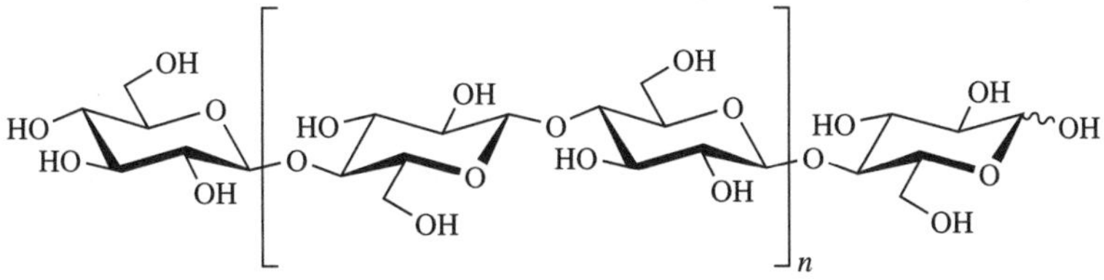

(**1**)

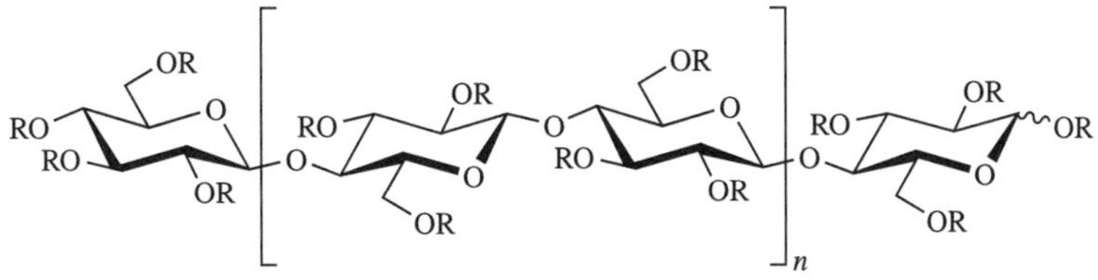

(**2**)

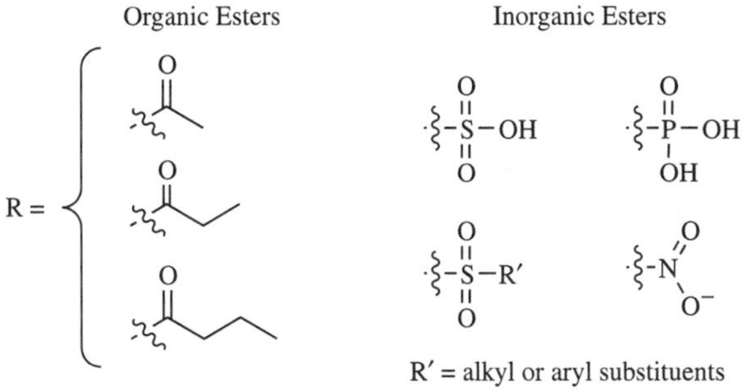

Organic Esters   Inorganic Esters

R' = alkyl or aryl substituents

Fig. 1. General structures of cellulose and cellulose esters.

nitrocellulose, is the ester with the most significant historical importance. Cellulose nitrate led to the birth of the plastics industry, the film and motion picture industry, and major improvements in explosives and coatings applications. Additionally, cellulose nitrate played a prominent role in numerous biochemical discoveries, including elucidation of the genetic code.

Due to its historical importance cellulose, nitrate has traditionally dominated the content of articles devoted to inorganic esters of cellulose. While not understating the importance of cellulose nitrate, this report emphasizes the

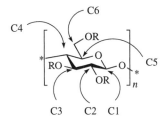

Fig. 2. Description of numbering of anhydroglucose carbons.

value of other inorganic esters of cellulose, including cellulose phosphate, cellulose sulfate, and cellulose sulfonates.

2. Preparation of Inorganic Esters of Cellulose

Inorganic esters of cellulose include all esters where the atoms linked directly to the cellulosic oxygens are non-carbon. The following cellulose derivatives will be discussed in this article, cellulose nitrate, cellulose sulfate, cellulose sulfonate, and cellulose phosphate.

2.1. Nitrogen-Containing Esters of Cellulose. *Cellulose Nitrate.*

Cellulose nitrate, without question the most commercially important inorganic ester of cellulose, was first prepared in 1833 (1). Schonbein prepared the first stable form of cellulose nitrate in 1846 (2). The preparation of cellulose nitrate, most commonly called nitrocellulose, has been reviewed in the literature numerous times (3,4). Several nitration techniques have been reported for the preparation of cellulose nitrate, including $HNO_3/H_2SO_4/H_2O$, HNO_3 (>75%), N_2O_4 (g), HNO_3/H_3PO_4, $HNO_3/H_3PO_4/P_2O_5$, $HNO_3/Mg(NO_3)_2$, HNO_3/CH_2Cl_2, and $HNO_3/HOAc/Ac_2O$ (5–11). Cellulose nitrate is still manufactured using a mixture of nitric acid, sulfuric acid, and water.

A number of versions of cellulose nitrate are commercially available. Variations in nitrogen content and viscosity are the main differences between nitrocellulose product lines. The amount of water present during the preparation of cellulose nitrate directly influences the final nitrogen content of the product (Fig 3). The viscosity of the final product is determined by selecting a starting material with the appropriate degree of polymerization and/or by performing a viscosity reduction step following nitration. Cellulose trinitrate has a degree of substitution of 3 and a nitrogen content of 14.15%. Cellulose dinitrate (DS = 2) and cellulose mononitrate (DS = 1) have nitrogen contents of 11.11 and 6.76%, respectively (12). Degree of substitution is a term commonly used to describe

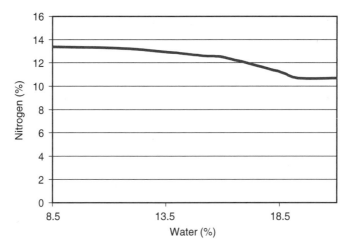

Fig. 3. Influence of water on degree of nitration of cellulose.

the amount of a substituent that is attached to a cellulosic backbone. Each anhydro-glucose unit of a cellulosic backbone contains 3 hydroxyl groups and thus a fully esterified cellulose ester, commonly referred to as a triester, has a degree of substitution of 3 (DS = 3). Degree of substitution is used throughout this report to represent the extent of substitution on the cellulose ester. A subscript is often used to indicate the nature of the substituent, eg, DS_N means the degree of substitution of nitrogen-containing compounds. The terms DS_P and DS_S mean degree of substitution of phosphorus- and sulfur-containing compounds, respectively. The maximum DS_N of commercial nitrocellulose is ∼2.9 (∼13.8% N). The hydroxyl group at C6 is nitrated much faster than those at C2 and C3. Under equilibrium conditions (ie, commercial processes utilizing HNO_3, H_2SO_4, and H_2O), the relative rates of nitration of the hydroxyl groups on cellulose are as follows, C6 = 5.8, C2 = 1.8, C3 = 1.0 (13).

2.2. Sulfur-Containing Esters of Cellulose. Cellulose sulfate and cellulose sulfonates are examples of sulfur-containing esters of cellulose. The sulfur is linked directly to an oxygen atom contained within an anhydroglucose unit of cellulose (see Fig. 1).

Cellulose Sulfate. Sulfation of cellulose has been accomplished via a number of procedures. Three forms of cellulose sulfate have been described in the literature; surface modified cotton fabric or fibers, water-swellable cellulose sulfate, and water-soluble cellulose sulfate. Sulfation of cellulose is commonly accomplished using sulfuric acid, chlorosulfonic acid, or sulfur trioxide/pyridine (eq. 1). Philipp and Wagenknecht (14) compiled a list of 50 sulfation strategies for the preparation of cellulose sulfate (Tables 1–5). The methods described in Tables 1–5 produce cellulose sulfate with various degrees of substitution. A common theme in the synthesis of cellulose esters (both inorganic and organic) is the trade-off between esterification and polymer degradation. Preparation of cellulose sulfate is no exception to this rule. Increasing the temperature during the sulfation reaction produces a more homogeneous mixture and thus more complete sulfation, but the rate of polysaccharide chain degradation is also rapidly

Table 1. **Preparation of Cellulose Sulfate Using H_2SO_4**

Sulfation system				DS
H_2SO_4				1.0–2
H_2SO_4	SO_2			0.9
H_2SO_4	chlorinated hydrocarbon			0.3
H_2SO_4	diethyl ether			0.2–0.4
H_2SO_4	low aliphatic alcohol			0.05–1
H_2SO_4	low aliphatic alcohol	inert solvent		0.35–0.7
H_2SO_4	chlorinated aliphatic alcohol			0.5
H_2SO_4	Ac_2O			0.6–2.8
H_2SO_4	Ac_2O	HOAc	salt	0.1–0.5
H_2SO_4	Ac_2O	HOAc	inert solvent	0.3–1.0
H_2SO_4	Ac_2O	pyridine	dioxane	2
H_2SO_4	aliphatic carboxylic acid C3-C4			NA[a]
H_2SO_4	HOAc	salt		0.3
H_2SO_4	HOAc	benzene		0.6

[a] Not available = NA.

Table 2. **Preparation of Cellulose Sulfate Using ClSO$_3$H**

	Sulfation system		DS
ClSO$_3$H			NA^a
ClSO$_3$H	SO$_2$		1.8
ClSO$_3$H	chlorinated hydrocarbon		NA^a
ClSO$_3$H	pyridine		1.9–2.8
ClSO$_3$H	pyridine propanolamine		1.6
ClSO$_3$H	pyridine	toluene	2.8
ClSO$_3$H	formamide		2

^{a} Not available = NA.

Table 3. **Preparation of Cellulose Sulfate Using SO$_3$**

	Sulfation system			DS
SO$_3$				3
SO$_3$	SO$_2$			2.2
SO$_3$	CS$_2$			3
SO$_3$	chlorinated hydrocarbon			0.5–0.9
SO$_3$	diethyl ether			1.3–2.1
SO$_3$	DMSO^a			1.3–2.0
SO$_3$	DMF^b			1.5–2.6
SO$_3$	pyridine			NA^c
SO$_3$	pyridine	various solvents		0.1–2.2
SO$_3$	aliphatic amine	DMF^b		0.4–2.0
SO$_3$	aliphatic amine	pyridine		1.5
SO$_3$	triethylphosphate			3
SO$_3$	N$_2$O$_4$	DMF^b		0.3–1.1
SO$_3$	N$_2$O$_4$	DMF^b	various solvents	<2.0

^{a} Dimethyl sulfoxide = DMSO.
^{b} Dimethylformamide = DMF.
^{c} Not available = NA.

increased. The rate of sulfation for the following sulfating agents is as follows: $H_2SO_4 < SO_3 < SO_2Cl_2 < HSO_3Cl$, with SO_2Cl_2 generating the most chain degradation during sulfation (14). Chlorosulfonic acid is generally considered the most effective sulfation agent with regard to both sulfation and retention of chain length. Excess sulfuric acid is typically used during sulfation reactions; however,

Table 4. **Preparation of Cellulose Sulfate Using SO$_2$Cl$_2$**

	Sulfation system		DS
SO$_2$Cl$_2$	alkali cellulose		NA^a
SO$_2$Cl$_2$	benzene	alkali cellulose	NA
SO$_2$Cl$_2$	ClCH$_2$COOH	alkali cellulose	NA
SO$_2$Cl$_2$	DMF, formamide		0.2–0.5
SO$_2$Cl$_2$	N$_2$O$_4$	DMF	0.4–0.8

^{a} Not available = NA.

Table 5. Preparation of Cellulose Sulfate Using Miscellaneous Reagents

	Sulfation system			DS
$NaSO_3Cl$	HOAc			NA^a
$NaSO_3Cl$	Ac_2O	HOAc		NA^a
$NaSO_3Cl$	pyridine			NA^a
$C_2H_5SO_3H$	pyridine	nitrobenzene		2.8
NH_2SO_3H	urea			NA^a
FSO_3H	pyridine			NA^a
FSO_3Na	benzene	C_2H_5OH	alkali cellulose	NA^a
FSO_3NH_4	benzene	C_2H_5OH	alkali cellulose	NA^a
Na-acetylsulfate				NA^a
$NOSO_4H$	DMF, DMSO, $DMAc^b$			0.7–1.2
$NOSO_4H$	diethyl ether, various solvents			0.02–0.4

^{a} Not available = NA.
^{b} Dimethylacetamide = DMAC.

stoichiometric amounts of acetylsulfuric acid can be used to prepare cellulose sulfate (15).

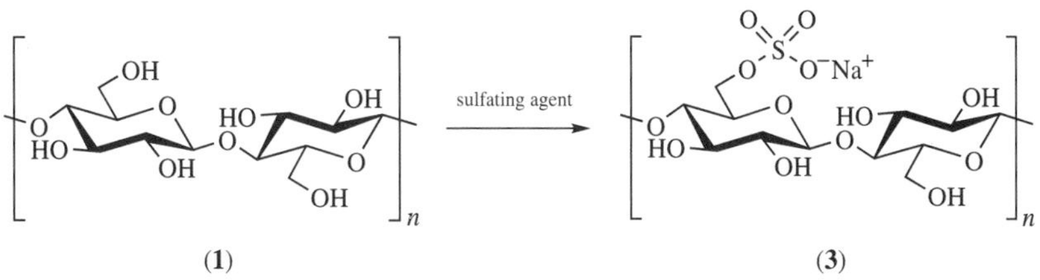

$$(1)$$

Cellulose sulfates with a wide range of degree of substitution can be water soluble, eg, cellulose sulfates with DS_S as low as 0.25 can be water soluble. The solubility of the final product is influenced by the method of preparation. The regioselectivity of sulfation can affect the properties and end uses of cellulose sulfate, eg, the ability of a cellulose sulfate to act as an anticoagulant is directly influenced by the distribution of the sulfates around the cellulosic backbone (16). A heterogeneous process, a homogeneous process, or a protection-deprotection strategy can be used to regioselectively derivatize cellulose (17). Generally, a bulky substituents is required to achieve reasonable regioselectivity in a heterogeneous process, so either a homogeneous process or a protection-deprotection strategy is preferred for the preparation of regioselectively substituted cellulose sulfate. Cellulose sulfates can be prepared under homogeneous conditions by first converting cellulose to the trimethylsilyl (TMS) ethers (18). Water-soluble cellulose sulfates were obtained by applying this strategy. The TMS ethers apparently act as reactive intermediates resulting in SO_3-induced sulfation only at sites previously silylated. The free hydroxyls present in the TMS cellulose (DS_{Si} 1.5–2.5) are unreactive during sulfation. When a similar sulfation strategy is applied to cellulose acetates (DS_{Ac} 1.8–2.5) the free hydroxyls are readily

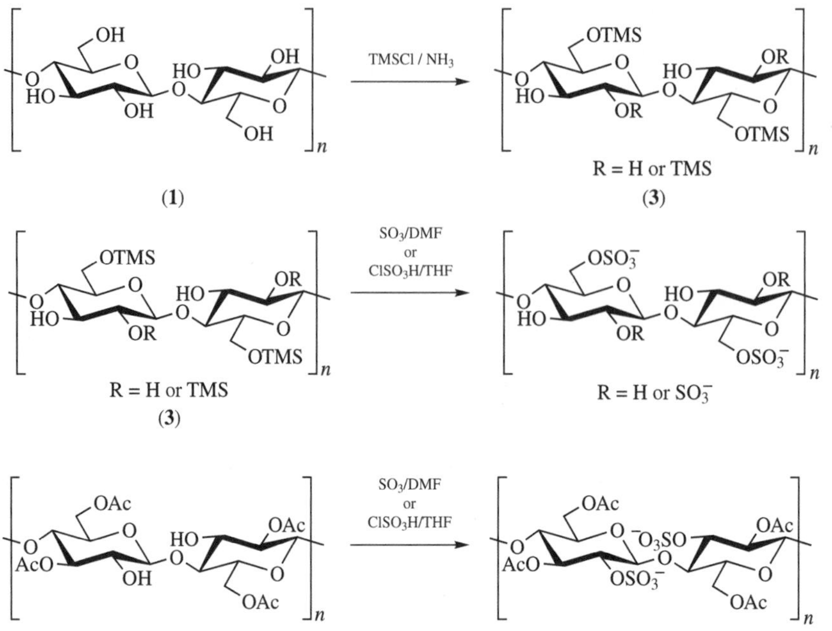

Fig. 4. Synthesis of cellulose sulfate from silylated and acetylated cellulose.

sulfated. It has been suggested that the TMS groups effectively shield the unreacted hydroxyl groups thus preventing sulfation (19). Sulfation of TMS cellulose derivatives with SO_3 or HSO_3Cl in DMF possibly occurs via insertion of SO_3 into the Si–O bond with the sulfate generated by subsequent hydrolysis and neutralization (16,18). Wagenknecht and co-workers (20) and Mischnick and co-workers (21) developed routes to evaluate the substitution pattern of cellulose sulfates.

There are numerous synthetic strategies available for preparing cellulose sulfates with a wide range of properties (see Fig. 4).

Cellulose Sulfonates. Cellulose sulfonates are inorganic esters of cellulose and sulfonic acids in which there is a sulfur–oxygen bond formed to link the cellulose and sulfonic acid moieties. Figure 5 depicts the three most common cellulose sulfonates described in the literature, cellulose toluenesulfonate (tosylate) (**8**), cellulose methanesulfonate (mesylate) (**7**), and cellulose dansylate (**9**) (22–25). Cellulose tosylate is the most commonly described cellulose sulfonate in the literature. Cellulose sulfonates are typically prepared in a homogeneous process by dissolving cellulose in a dimethylacetamide/lithium chloride (DMAc/LiCl) mixture and reacting the cellulose with sulfonylchloride in the presence of triethylamine (eq. 2) (25). Cellulose sulfonates can also be prepared *in situ* for use as reactive intermediates (26,27). The use of cellulose tosylate as a reactive intermediate will be discussed in more detail later in this article.

Cellulose deoxysulfonates have been described in the literature, but will not be discussed further in this report (28).

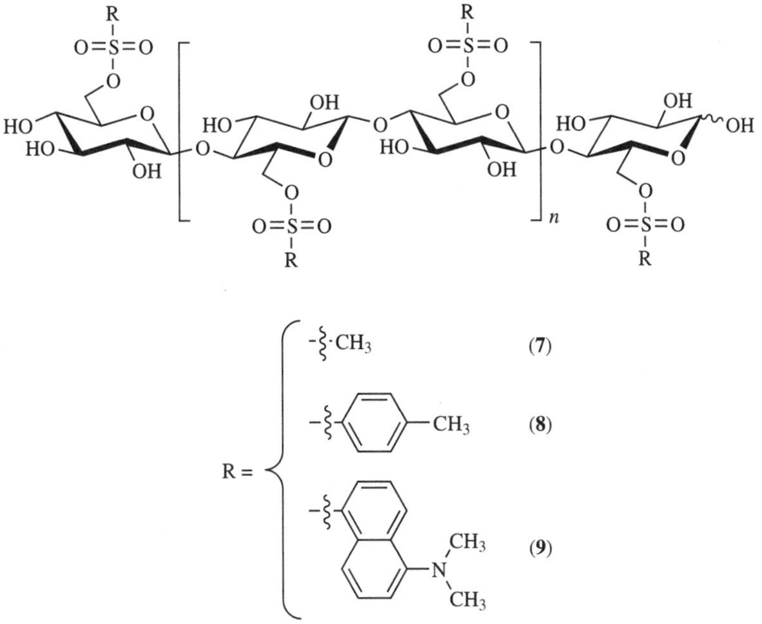

Fig. 5. Structures of cellulose sulfonates.

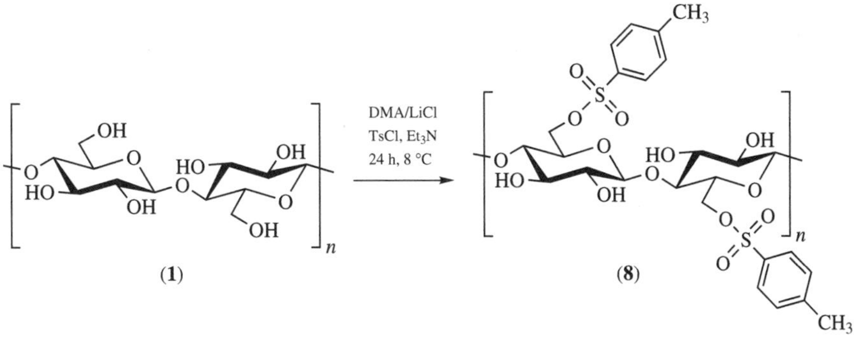

2.3. Phosphorus-Containing Esters of Cellulose. *Cellulose Phosphates.*

There are a number of strategies for preparing cellulose phosphate. Reaction of cellulose with tetrapolyphosphoric acid in the presence of triethylamine produced cellulose phosphate with $DS_P \leq 1.75$ (29). Treatment of cellulose with $POCl_3$ resulted in cellulose phosphate with low DS_P, but also resulted in introduction of chlorine to the cellulosic backbone (Table 6) (30). Phosphorylations of cellulose with *N*-phosphoryl-*N'*-methylimidazole and a mixture of phosphorus pentoxide, triethyl phosphate, and phosphoric acid have been reported (Fig. 6) (31,32).

Table 6. **Phosphorylation and Chlorination of Cellulose with POCl$_3$**

Reaction temp	POCl$_3$ (%)	DS$_P$	DS$_{Cl}$
50	2	0.09	0.02
	4	0.08	0.03
	8	0.03	0.03
75	2	0.19	0.04
	4	0.08	0.08
	8	0.06	0.47
	10	0.08	0.52
	12	0.07	0.55
100	2	0.22	0.04
	4	0.18	0.12
	8	0.20	0.37

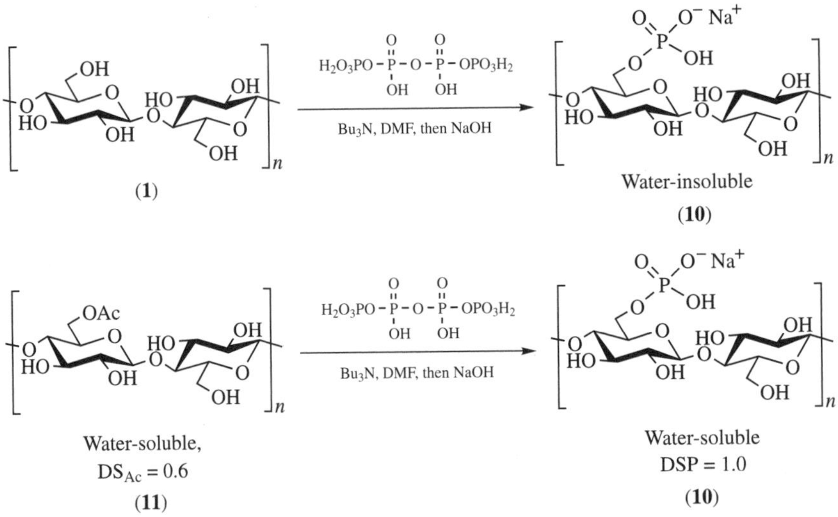

Fig. 6. Synthesis of water-insoluble and water-soluble cellulose phosphate.

3. Properties of Inorganic Esters of Cellulose

In general, inorganic esters of cellulose have significantly different solubility profiles than native cellulose. Cellulose has poor solubility in most organic or aqueous systems, while inorganic esters of cellulose are typically more soluble. Inorganic esters of cellulose that are soluble in organic or aqueous solvents are available. The solubility of inorganic esters of cellulose is dependent on the identity and degree of substitution of the substituents, the number of unreacted free hydroxyls, and the degree of polymerization of the cellulose backbone.

3.1. Solubility. Cellulose sulfate and cellulose phosphate can be either water-soluble or water-swellable depending on the conditions used during preparation (see above). Cellulose nitrate and some cellulose sulfonates are generally more hydrophobic and are typically soluble in organic solvents. In general, the level of esterification of cellulose directly impacts the solubility of the product in various solvents and solvent blends.

3.2. Film Formation and Rheology. The ability to form films is the most important property of inorganic cellulose esters. When cellulose nitrate is dissolved in an organic solvent and then applied to a substrate (eg, wood, metal, glass) a clear film is formed on the substrate once the solvent evaporates. Coatings of this nature are referred to as lacquers. Coatings applications are the largest commercial applications for cellulose nitrate and will be discussed further in the applications section of this article. Clear films of cellulose sulfate (33) and cellulose phosphate (34) have also been reported.

The rheological behavior of cellulose esters, both inorganic and organic, is an important property that contributes to the utility of these cellulosics, particularly in coatings applications. Rheological profiles of cellulose nitrate are well documented (35–38). The semirigid cellulose nitrate displays a flow behavior similar to that of linear flexible polymers (38). Marx-Figini and Gonzalez (35) suggested that the transition from Newtonian to non-Newtonian behavior is likely due to entanglement of molecules above a critical concentration. The higher the degree of polymerization of the cellulose nitrate, the lower the critical concentration at a given shear rate and the lower the critical shear rate at a given concentration. The critical concentrations are much lower for cellulose nitrate than for other synthetic polymers, such as polystyrene or polyacrylamide, with comparable molecular weights (35).

4. Applications of Inorganic Esters of Cellulose

Nitrocellulose is the most commercially important inorganic ester of cellulose. The use of nitrocellulose led to development of the plastics industry, the motion picture industry, and powder-less ammunition. In most cases, nitrocellulose has been replaced over the years by less expensive or less flammable materials. Though considerable market share for nitrocellulose has been lost to alternate products, nitrocellulose remains a commercially viable product and should be for many years.

4.1. Coatings, Adhesives, and Inks. The largest market for cellulose nitrate is in the coatings industry. Relatively clear and workable (ie, sandable or removable) lacquers are generated when cellulose nitrate is used in a coatings application. Cellulose nitrate lacquers dry rapidly, have excellent flow properties and the final coating has excellent appearance, with reasonable strength and durability. Cellulose nitrate became a commercially viable product in the 1920s and rapidly replaced slower drying varnishes that were used in both automotive and wood coatings (39). Cellulose nitrate has played such a significant role in lacquer coatings applications that cellulose nitrate-based lacquers are simply referred to as "lacquers" and it is assumed that a lacquer is cellulose nitrate-based unless stated otherwise. Over time, cellulose nitrate was replaced in automotive topcoats by more durable acrylic lacquers. Cellulose nitrate maintained much of its market share in applications where high temperature curing is not acceptable, for example in wood coatings (39). Cellulose nitrate is still used in some automotive primer refinish applications, but currently its primary use is in wood coatings. Cellulose nitrate provides enhanced wood grain appearance, rapid drying, and easy damage repair to wood coatings. Cellulose nitrate has also been used as an additive to improve the performance, appearance, and drying time of coatings based on other resins, such as acrylics, vinyls, polyamides, epoxies, and polyesters (40). Cellulose nitrate aids in pigment dispersion and acts as a binder following solvent evaporation.

Cellulose nitrate used in coatings applications is commercially available in three general forms based on nitrogen content: SS (10.7–11.3% N), AS (11.3–11.7% N), and RS (11.7–12.2% N) (39). Additionally, each product type is available in low, medium, and high viscosity grades. According to product literature from ICI Nobel Enterprises (Scotland, U.K.) low and medium nitrogen grades are typically used in ink applications. Medium nitrogen content is used for film coatings. High nitrogen content cellulose nitrate is commonly used in wood coatings, general purpose lacquers, and automotive refinish applications. Cellulose nitrate-based lacquers have many advantages over other coatings. They are easily applied, have good adhesion, good solvent release properties and excellent pigment dispersion. The problems with nitrocellulose lacquers include yellowing, flammability, and that they must be sprayed at low solids levels. Nitrocellulose-based lacquers found use in cloth book bindings (40). Nitrocellulose is also used in adhesives and inks (40,41).

4.2. Explosives. Cellulose nitrate with a nitrogen content of 12.3–13.5% N (degree of substitution, $DS_N = 2.25$–2.5) is used in explosive and propellant applications. Cellulose nitrate was initially developed for use in explosives and propellants and this remains one of the largest markets for cellulose nitrate today.

4.3. Plastics. In 1847, it was discovered that cellulose nitrate could be dissolved in a mixture of ether and alcohol to produce a solution called Collodion. Collodion is still in use today in pharmaceutical applications. Combination of nitrocellulose with castor oil and eventually camphor by Parkes (1862) and Hyatt (1870) led to the production of Celluloid and ultimately the birth of the plastics industry.

4.4. Flame Retardants. Flame retardant cotton fibers are prepared by surface modification of cotton cellulose via phosphorylation or phosphorylation–halogenation producing cellulose phosphate and deoxychlorocellulose phosphate, respectively. Surface modification of cotton fibers generates flame retardant fabrics by increasing the char content of the fiber, which in turn produces a lower percentage of flammable volatiles. Since the phosphorylated–halogenated fibers break down to produce acidic products, the overall thermal stability of the fabric is greatly reduced, and the material breaks down at a lower temperature and ultimately reduces the potential of the fabric bursting into flames (42).

4.5. Separations Applications. *Ion Exchange.* Inorganic esters of cellulose have been used in ion exchange applications and as chromatographic adsorbents since the late 1950s (43). Cellulose sulfate and cellulose phosphate are excellent ion exchangers. Exposure of the ester to a salt of a metal ion (eg, NaCl, KCl) or neutralization of the acid with a base (eg, NaOH, KOH) will result in exchange of the metal ion from the salt or base onto the sulfate or phosphate attached to the cellulose backbone (eq. 3).

Cellulose phosphate is the most widely used cellulose inorganic ester for ion exchange. Cellulose phosphate is an effective cation exchanger used in the bioprocessing industry (44). Cellulose phosphate and cellulose polyphosphate effectively remove Al^{3+} contamination from adenosine triphosphate (ATP) (45). A number of products are commercially available, most notably from Whatman, for use as stationary phases in column chromatography and filtration applications. Whatman P1, P11, and P81 are examples of commercially available cellulose phosphate-based ion exchangers.

The ion exchange nature of cellulose sulfate resulted in its use to concentrate viruses (46).

Molecular Weight Determination. Since there are nitration methods available that have minimal impact on the chain length of cellulose, conversion from cellulose to cellulose nitrate is an established method of molecular weight determination of cellulose. This is another situation where the improved solubility of cellulose nitrate is advantageous over the relative insolubility of cellulose. Cellulose nitrate molecular weight can be successfully determined using size-exclusion–gel permeation chromatography due to its improved solubility over cellulose (47–53).

Chiral Chromatography. Cellulose trinitrate-impregnated silica beads was used as a chiral stationary phase for the separation of several racemic aromatic compounds (54).

4.6. Membrane Applications. Cellulose esters are effective in membrane applications. Ultrafiltration membranes based on cellulose nitrate were first described by Collander in 1924 (55). Cellulose nitrate films cast from mixtures of methyl acetate or acetone with glycerine and mixtures of ether and ethanol produce microfiltration and ultrafiltration membranes, respectively (56). Cellulose nitrate microfiltration membranes typically have 0.02–10-μm diameter pores and 4×10^{-3}–15 cm/s·atm permeability. Cellulose nitrate ultrafiltration membranes typically have 0.003–0.03 μm diameter pores and 1–100×10^{-4} cm/s·atm permeability (56). Nitrocellulose membranes are used in numerous biochemical and diagnostic applications and will be discussed later.

Cellulose sulfate, in particular C6 sulfates, can be used for preparing polyanion–polycation complexes for pervaporation membranes (57–59).

4.7. Synthetic Intermediates. Cellulose sulfonates have been evaluated as synthetic intermediates (27,60–63). Cellulose tosylate activates the C6 hydroxyl for substitution with various nucleophiles (eq. 4). Additionally, cellulose sulfonates can also function as a protecting group to prevent reaction of the C6 hydroxyl during esterification reactions (Fig. 7). Cellulose tosylates with degrees of substitution from 0.38 to 2.30 have been reported. Though an effective synthetic tool, the expense of organic solvents utilized and the cost and corrosive

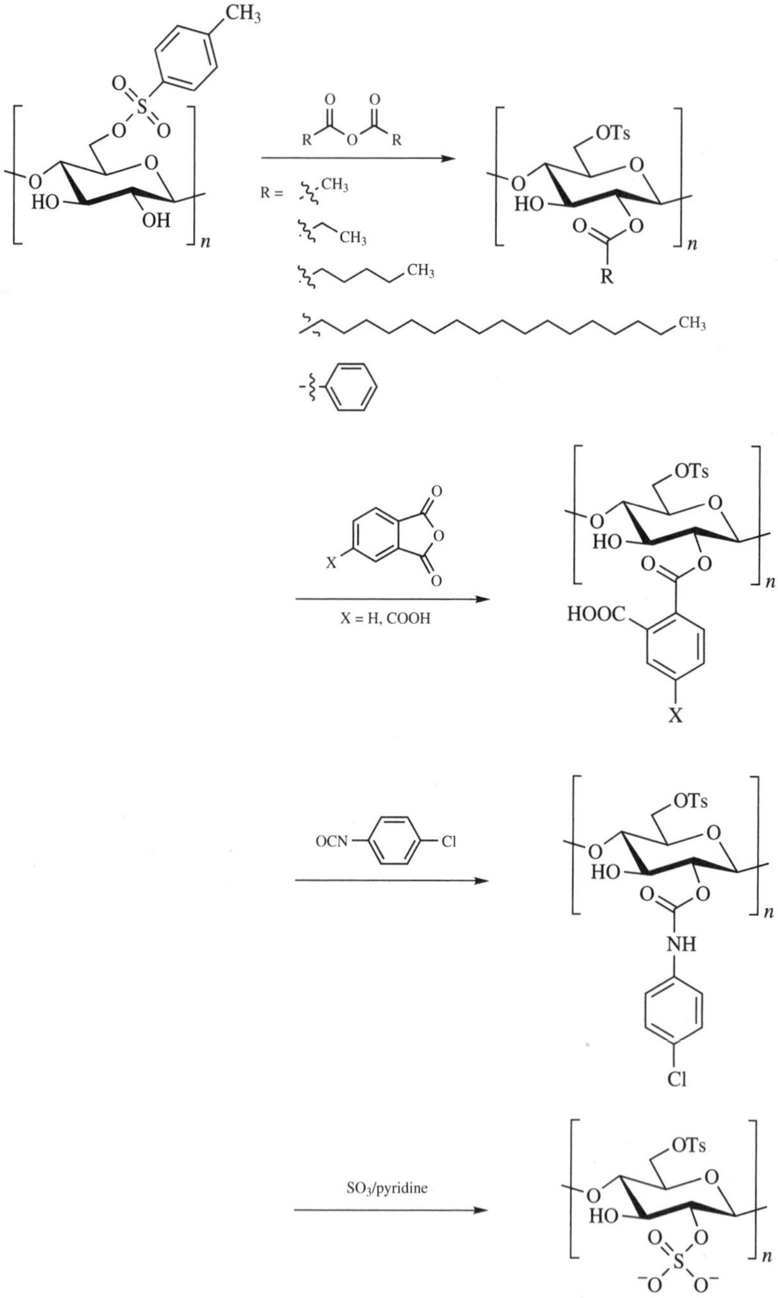

Fig. 7. Subsequent chemical modification of cellulose tosylate.

nature of tosyl chloride have limited the use of this methodology on a commercial scale.

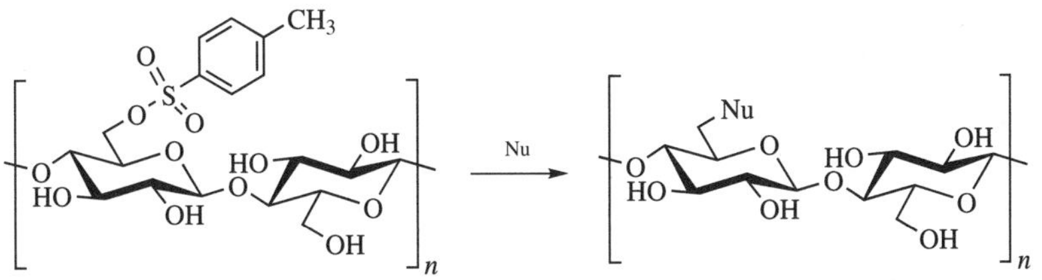

$$Nu- = Cl^-, N_3^-, NH_2-(CH_2)_m-NH_2, \text{ or other nucleophiles}$$

4.8. Medical Applications.   Inorganic esters of cellulose have been evaluated in a number of medical–pharmaceutical applications. A few examples include treatment of hypercalciura, use as antiviral agents, reduction of cholesterol levels, orthopedic applications, and blood stabilization.

Sodium cellulose phosphate is used in the treatment of adsorptive hypercalciura and nephrocalcinosis (both a common cause of kidney stones). The ability of inorganic esters of cellulose to function as ion exchangers, described previously in this report, is the key to the use of sodium cellulose phosphate in the treatment of adsorptive hypercalciuria. Hypercalciuria is an increased level of calcium in a patient's urine. This disorder commonly results in painful kidney stones. Sodium cellulose phosphate has been used to complex the calcium ions and thus remove them from the patient and minimize the risk of stone formation (64–68).

It is well established that sulfated polysaccharides play important biological roles (69–77). Cellulose sulfate displays heparin-like activity and to a lesser extent antiviral activity (78). More specifically, cellulose sulfates with a medium to high degree of sulfation at the C2 and C6 positions of the anhydroglucose units are the most active forms of the ester. High molecular weight cellulose sulfate lowers cholesterol levels in humans by inhibiting pancreatic cholesterol esterase (79,80). Contamination of high molecular weight cellulose sulfate with its low molecular weight counterpart results in a toxic, unusable product. A method to prepare and isolate high molecular weight cellulose sulfate without contamination with low molecular weight cellulose sulfate has been developed (79). Ushercell (a commercial form of sodium cellulose sulfate marketed by Polydex Pharmaceuticals) has been evaluated in the treatment of papilloma virus infection (81). Usher and co-workers (82) also investigated the use of sodium cellulose sulfate as an antimicrobial agent and a contraceptive. Cellulose sulfate with a DS > 1.0 and sodium cellulose sulfate when complexed with (β-hydroxy-γ-trimethylaminopropyl) hydroxyethyl cellulose chloride show excellent anticoagulative property (83,84).

Encapsulation of microorganisms, enzymes, plant cells, and animal cells (including mammalian cells) has been accomplished using a mixture of sodium cellulose sulfate with polydiallyldimethyl ammonium chloride (85,86). Encapsulation of hydridoma cells with cellulose sulfate has provided a means of subcutaneous delivery of monoclonal antibodies. The implant vascularized as early as

15 days after implantation and the "neoorgan" remained an active source of anti-bodies in mice for several months (87).

Cellulose phosphate has been evaluated in femoral implants to improve the mineralization of biomaterials used in orthopedic applications (88). Cellulose phosphate (with or without bound antimicrobial agents) and cellulose phosphate borate effectively stabilize blood for over four years (89).

4.9. Biochemical Applications. Cellulose phosphate paper (P81), also called phosphocellulose paper, has been used for enzyme assay applications (90). As with many other applications of cellulose phosphate, the ability of the material to act as an ion exchanger is the key property that leads to its use in enzyme assays. Cellulose phosphate paper quantitatively binds small peptides that contain at least two basic residues and a free amino terminus, which allows for protein kinase assays to be performed and ^{32}P-labeled cofactors [γ-^{32}P]ATP and non-peptidic by-products [^{32}P]Pi to be washed away (91).

Even though a number of applications have been demonstrated for cellulose phosphate in the biotechnology laboratory, the use of cellulose nitrate dominates the field. Cellulose nitrate filters and membranes have played key roles in numerous biochemistry and molecular biology discoveries and still play important roles in these laboratories today. Protein and glycoprotein immobilization, absorption of single-stranded DNA (Southern blotting), absorption of single-stranded RNA (Northern blotting), binding of transfer RNA linked to ribosomes, tissue printing, and dialysis are common practices accomplished through the use of nitrocellulose membranes or filters (92–97). The binding of ribosomes and tRNAs to nitrocellulose was used to elucidate the genetic code. Cellulose nitrate is an efficient binder of antibodies and can be used as a solid support for immunoaffinity chromatography (98).

4.10. Diagnostic Applications. The binding characteristics of cellulose nitrate membranes have led to their use in numerous biosensor applications, such as in an electrochemical microbial biosensor for EtOH (99), and in glucose oxidase-based glucose sensors (100). Cellulose nitrate continues to play an important role in the development of new biosensors as both a solid support and a method for removal of unbound reagents (101–107).

Additionally, cellulose nitrate has been utilized in nonbiological detectors including an indoor radon gas detector. The cellulose nitrate portion of the film-based detector is the active portion for the detection of α-particles (108).

5. Conclusions

Inorganic esters of cellulose have a valuable place in history, from the birth of the plastics industry to elucidation of the genetic code. It is important not to forget the large number of applications in which these materials have played a role: coatings, explosives, pharmaceuticals, membranes, and synthetic intermediates. Though inorganic esters of cellulose have been around for >100 years, these materials remain important tools for coatings scientists, explosives experts, and biochemists.

BIBLIOGRAPHY

"Cellulose Derivatives, Esters" under "Cellulose Derivatives, Plastics" in *ECT* 1st ed., Vol. 3, pp. 391–411, by W. O. Bracken, Hercules Powder Co.; in *ECT* 2nd ed., Vol. 4, pp. 653–683, by B. P. Rouse, Jr., Tennessee Eastman Co.; "Cellulose Derivatives, Esters" in *ECT* 3rd ed., Vol. 5, pp. 118–143, by R. T. Bogan, C. M. Kuo, and R. J. Brewer, Tennessee Eastman Co.; "Cellulose Esters, Inorganic" in *ECT* 4th ed., Vol. 5, pp. 529–540, by Richard Fengl, Tennessee Eastman; "Cellulose Esters, Inorganic Esters" in *ECT* (online), posting date: December 4, 2000, Richard Fengl, Tennessee Eastman.

CITED PUBLICATIONS

1. H. Braconnot, *Annals* **1**, 242 (1833).
2. C. F. Schönbein, *Philos. Mag.* **31**, 7 (1847).
3. R. T. Bogan, C. M. Kuo, and R. J. Brewer, *Kirk-Othmer Encyclopedia of Chemical Technology*, John Wiley & Sons, Inc., New York, 1979, p. 118.
4. R. J. Brewer and R. T. Bogan, *Encyclopedia of Polymer Science and Engineering*, Vol. **2**, 1998, p. 139.
5. Can. Pat. 536,191 (Jan. 22, 1957), W. C. Ramsey (to Olin Mathieson Chemical Corp.).
6. U.S. Pat. 2,776,965 (Jan. 8, 1957), J. L. Bennett, R. M. Brooks, J. G. McMillan, W. L. Plunkett (to Hercules Powder Co.).
7. U.S. Pat. 3,063,981 (Nov. 13, 1962), J. D. Cochrane III, D. S. Wilt (to Hercules Powder Co.).
8. C. Bennett and T. Timell, *Svensk Paperstidn.* **58**, 281 (1955).
9. D. Goring and T. Timell, *Tappi* **45**, 454 (1962).
10. K. Thinius and W. Tummler, *Makromol. Chem.* **99**, 117 (1966).
11. M. Marx-Figini, *Makromol. Chem.* **50**, 196 (1961).
12. C. W. Saunders and L. T. Taylor, *J. Energ. Mater.* **8**, 3, 149 (1990).
13. T. K. Wu, *Macromolecules* **13**, 74 (1980).
14. B. Philipp and W. Wagenknecht, *Cellul. Chem. Technol.* **17**, 443 (1983).
15. U.S. Pat. 4,480,091 (Oct. 30, 1984) R. J. Brewer (to Eastman Kodak Co.).
16. D. Klemm, T. Heinze, B. Philipp, W. Wagenknecht, *Acta Polymer.* **48**, 277 (1997).
17. B. Philipp, W. Wagenknecht, I. Nehls, D. Klemm, A. Stein, and T. Heinze, *Polymer News* **21**, 155 (1996).
18. W. Wageknecht, I. Nehls, A. Stein, D. Klemm, and B. Philipp, *Acta Polymer.* **43**, 266 (1992).
19. B. Philipp, D. Klemm, and U. Heinze, *Polymer News* **24**, 9, 305 (1999).
20. B. Saake and W. Wagenknecht, *Carbohydr. Polym.* **48**, 7 (2002).
21. M. Gohdes and P. Mischnick *Carbohydr. Res.* **309**, 109 (1998).
22. T. Heinze and K. Rahn, *Macromol. Symp.* **120**, 103 (1997).
23. M. Sakamoto, Y. Yamada, N. Ojima, and H. Tonami, *J. Appl. Polym. Sci.* **166**, 1495 (1972).
24. K. Rahn, M. Diamantoglou, D. Klemm, H. Berghmans, and T. Heinze, *Angew. Makromol. Chem.* **238**, 143 (1996).
25. T. Heinze and K. Rahn, *J. Pulp. Pap. Sci.* **25**, 4, 136 (1999).
26. T. Heinze, K. Rahn, M. Jaspers, and H. Berghmans, *Macromol. Chem. Phys.* **197**, 12, 4207 (1996).
27. A. Koschella and T. Heinze, *Macromol. Biosci.* **1**, 178 (2001).
28. K. I. Furuhata and H. Ikeda, *Reactive Functional Polymers* **42**, 103 (1999).
29. G. A. Towle and R. L. Whistler, *Methods Carbohydr. Chem.* **6**, 408 (1972).

30. T. L. Vigo and C. M. Welch, *Carbohydr. Res.* **32**, 331 (1974).
31. H. Takaku, Y. Shimada, and K. Aoshima, *Chem. Pharm Bull.* **21**, 2068 (1973).
32. P. L. Granja, L. Pouysegu, M. Petraud, B. De Jeso, C. Baquey, and M. A. Barbosa, *J. Appl. Polym. Sci.* **82**, 3341 (2001).
33. U.S. Pat. 2,969,256 (Jan. 24, 1961), G. P. Touey and J. E. Kiefer (to Eastman Kodak Co.).
34. Jpn. Tokkyo Koho JP 42010638 (Sept. 18, 1964), S. Makishima, H. Hirai, and H. Hirokane (to Sumitomo Chemical Co., Ltd.).
35. M. Marx-Figini and R. Gonzalez, *J. Appl. Polym. Sci* **38**, 219 (1989).
36. W. Jian, *International Annual Conference of ICT*, 21st,Vol. **68**, 1990, p. 12.
37. F. S. Baker and G. J. Privett, *Wood Cellul.* 457 (1987).
38. F. R. Gonzalez, M. Marx-Figini, and R. V. Figini, *Makromol. Chem.* **189**, 2409 (1988).
39. E. C. Hamilton, L. W. Early, *Federation Series on Coatings Technology* **Unit 21**, 3 (1973).
40. R. A. Weidener, *Treatise Adhes. Adhes.* **2**, 429 (1969).
41. PCT Int. Appl. WO Pat. 01/90221 (Nov. 29, 2001) G. E. Miller, Jr. (to Arizona Chemical Co.).
42. R. K. Jain, K. Lal, and H. L. Bhatnagar, *J. Appl. Polym. Sci.* **33**, 247 (1987).
43. N. Kabay, M. Demircioglu, S. Yayli, M. Yuksel, M. Saglam, and P. R. Levison, *Sep. Sci. Technol.* **34**, 41 (1999).
44. J. M. Ward, L. J. Wallace, D. Cowan, P. Shadbolt, and P. R. Levison, *Anal. Chim. Acta* **249**, 195 (1991).
45. J. V. Schloss, G. Smith, A. Aulabaugh, and W. W. Cleland, *Anal. Biochem.* **120**, 176 (1982).
46. K. Tamayose, Y. Hirai, and T. Shimada, *Human Gene Therapy* **7**, 507 (1996).
47. B. Saake, R. Patt, J. Puls, K. J. Linow, and B. Philipp, *Makromol. Chem., Macromol. Symp.* **61**, 219 (1992).
48. B. Saake, R. Patt, J. Puls, K. J. Linow, and B. Philipp, *Papier* **45**, 727 (1991).
49. K. Fischer and W. Picker, J. Fritz, *Cellulose Chem. Technol.* **23**, 415 (1989).
50. M. Rinaudo and J. P. Merle *European Polymer J.* **6**, 41 (1970).
51. L. Segal, *J. Polym. Sci, Polym. Lett. Ed.* **4**, 1011 (1966).
52. K. U. Usmanov and T. I. Sushkevich, *J. Polymer Chem.* **58**, 1325 (1962).
53. E. H. Immergut, B. G. Ranby, and H. F. Mark, *Ind. Eng. Chem.* **45**, 2483 (1953).
54. U.S. Pat. 4,714,555 (Dec. 22, 1987) T. Shibata, I. Okamoto (to Daicel Chemical Industries, Ltd.).
55. R. Collander, *Soc. Sci. Fennica, Commentationes Biol.* **2**, 1 (1926).
56. W. Pusch, *Wood Cellul.* 475 (1987).
57. DE Pat. 4229530 (Sept. 08, 1992) M. Aderhold, R. D. Behling, K. V. Peinemann, N. Scharnagel, H. H. Schwarz, D. Paul, K. Richau, R. Apostel, and G. Frigge (to GKSS-Forschungszentrum Geesthacht GmbH, Germany).
58. DE Pat. 4435180 (Sept. 30, 1994) W. Wagenknecht and H. H. Schwarz (to Fraunho fer-Gesellschaftzur Foerderung der Angewenten Forschung e.V., Germany; GKSS-Forschungszentrum Geesthacht GmbH, Germany).
59. DE Pat. 4437869 (May 02, 1995) H. H. Schwarz, K. Richau, R. Apostel, and G. Frigge (to GKSS-Forschungszentrum Geesthacht GmbH, Germany).
60. T. Heinze, *J. Appl. Polym. Sci.* **60**, 1891 (1996).
61. T. Heinze and K. Rahn, *Papier (Darmstadt)* **50**, 721 (1996).
62. G. Siegmund and D. Klemm, *Polymer News* **27**(3), 84 (2002).
63. P. Berlin, D. Klemm, J. Tiller, and R. Rieseler, *Macromol. Chem. Phys.* **201**(15), 2070 (2000).
64. R. Hautmann, R. J. Hering, and W. Lutzeyer, *J. Urol.* **120**, 6, 712 (1978).
65. C. Y. Pak, *J. Clin. Pharm.* **9**, 8–9, 451 (1979).

66. A. Dwarakanathan and W. G. Ryan, *Bone Mineral* **2**(4), 333 (1987).
67. Y. Mizusawa and J. R. Burke, *J. Ped. Child Health* **32**, 350 (1996).
68. U. Backman, B. G. Danielson, G. Johansson, S. Ljunghall, and B. Wilkstrom, *J. Urology*, **123**, 9 (1980).
69. R. Sasisekharan, Z. Shriver, G. Venkataraman, and U. Narayanasami, *Nature Rev. Cancer* **2**, 7, 521 (2002).
70. PCT Int. Appl. WO 02/41901 (May 30, 2002) D. Cullis-Hill (to Arthropharm Pty Ltd, and Cullis-Hill, David).
71. A. Chaidedgumjorn, H. Toyoda, E. R. Woo, K. B. Lee, Y. S. Kim, J. Toida, and T. Imanari, *Carbohydr. Res.* **337**, 10, 925 (2002).
72. C. Fukuda, O. Kollmar, T. Schafer, Y.-H. Tian, and M. K. Schilling, *Transplant Inter.* **15**, 1, 17 (2002).
73. U.S. Pat. 6,372,794 (Apr. 16, 2002) M. E. Nimni.
74. PCT Int. Appl. WO Pat. 02/08295 (Jan. 31, 2002) J. Diaz, C. Pecquet, E. Perrin, and C. Viskov.
75. PCT Int. Appl. WO Pat. 02/02189 (Jan. 10, 2002) T. C. Usher, R. A. Anderson, and L. J. D. Zaneveld (to Polydex Pharmaceuticals Ltd; Rush-Presbyterian-St. Luke's Medical Center).
76. K. Moelling, T. Schulze, and H. Diringer, *J. Virology* **63** 12, 5489 (1989).
77. PCT Int. Appl. WO Pat. 9630027 (Oct. 3, 1996) S. N. Anderson, J. P. Schaller, T. B. Mazer, and S. J. Kirchner (to Abbott Laboratories).
78. I. Yamamoto and co-workers, *Carbohydr. Polym.* **14**, 53 (1991).
79. EP Pat. 0712864 (Oct. 13, 1996) L. G. Lange, III, C. A. Spilburg, and D. T. Rearden (to CV Therapeutics, Inc.).
80. U.S. Pat. 5,378,828 (Sept. 2, 1993) T. C. Usher, N. Patel, and C. G. Tele (to Dextran Products Ltd).
81. WO Pat. 02/02189 (Jan. 10, 2002) T. C. Usher, R. A. Anderson, and L. J. D. Zaneveld (to Polydex Pharmaceuticals Ltd, Bahamas; Rush-Presbyterian-St. Luke's Medical Center).
82. WO Pat. 9712621 (Sept. 27, 1997) R. A. Anderson, L. J. D. Zaneveld, N. Patel, and T. C. Usher (to Novadex Pharmaceuticals, Canada; Rush-Presbyterian-St. Luke's Medical Center).
83. T. Groth and W. Wagneknecht, *Biomaterials* **22**, 2719 (2001).
84. U.S. Pat. 4,708,951 (Nov. 24, 1987) H. Inagaki, T. Miyamoto, H. Ito, and T. Shibata (to Daicel Chemical Industries, Ltd).
85. H. Dautzenberg, U. Schuldt, G. Grasnick, P. Karle, P. Muller, M. Lohr, M. Pelegrin, M. Piechaczyk, K. V. Rombs, W. H. Gunzburg, B. Salmons, and R. M. Saller, *Ann. NY Acad. Sci.* **875**, 46 (1999).
86. S. Yao, *Chem. Eng. J.* **78**, 199 (2000).
87. M. Pelegrin, M. Marin, D. Noel, M. del Rio, R. Saller, J. Stange, S. Mitzner, W. H. Gunzburg, and M. Piechaczyk, *Gene Therapy* **5**, 828 (1998).
88. P. L. Granja, M. A. Barbosa, C. Pouysegu, B. De Jeso, F. Rouais, and C. Bacquey, *J. Mat. Sci.* **36**, 2163 (2001).
89. T. J. M. Sinha and P. Vasudevan, *Biomat., Med. Devices, Artif. Organs* **12**, 273 (1985).
90. M. J. King and R. K. Sharma, *Anal. Biochem.* **199**, 149 (1991).
91. J. E. Casnellie, *Methods Enzymol.* **200**, 115 (1991).
92. M. Schärfke, H. Baumeister, and W. Meyerhof, *Bio Techniques* **30**, 266 (2001).
93. D. J. Thornton, I. Carlstedt, and J. K. Sheehan, *Methods Mol. Biol.* **32**, 119 (1994).
94. A. P. Nygaard and B. D. Hall, *Biochem. Biophys. Res. Commun.* **12**, 98 (1963).
95. B. D. Hall, L. Haarr, and K. Kleppe, *Trends Biochem. Sci.* **28**, 254 (1980).

96. D. Voet and J. G. Voet, *Biochemistry*, 2nd ed., John Wiley & Sons, Inc., New York (1995).

97. R. Taylor, *Methods in Plant Electron Microscopy and Cytochemistry*, 101 (2000).

98. J. Thalhamer, P. Hammerl, and A. Hartl, *Methods Mol. Biol.* **147**, 89 (2000).

99. S. Subrahmanyam, K. Shanmugam, T. V. Subramanian, M. Murugesan, V. M. Madhar, and D. Jeyakumar, *Electoroanalysis* **13**, 11, 944 (2001).

100. G. Cui, J. H. Yoo, J. Yoo, S. W. Lee, H. Nam, and G. S. Cha, *Electroanalysis* **13**, 224 (2001).

101. R. Graf and P. Friedl, *Electrophoresis* **22**, 890 (2001).

102. WO Pat. 01/03721 (Jan. 18, 2001) G. Cullen (to Merlin Biomedical & Pharmaceutical, Ltd.).

103. WO Pat. 01/11363 (Feb. 15, 2001) J. Carlsson and M. Lonnberg (to Pharmacia Diagnostics AB).

104. WO Pat. 01/46464 (June 28, 2001) M. Huber, W. Schmidt, M. Muller, and R. Hiller (to Lion Bioscience AG).

105. WO Pat. 01/44813 (June 21, 2001) R. M. Kaylor, A. B. Choi, M. H. H. Grunze, and C. D. Chidebelu-Eze (to Kimberly-Clark Worldwide, Inc.).

106. WO Pat. 01/81921 (Nov. 1, 2001) R. M. Kaylor, C. D. Chidebelu-Eze, and A. B. Choi, (to Kimberly-Clark Worldwide, Inc.).

107. U.S. Pat. 6,221,678 (4/24/01) H. M. Chandler.

108. G. Jonsson, *Nucl. Tracks Radiat. Meas.* **13**, 1, 85 (1987).

MICHAEL C. SHELTON
Eastman Chemical Company

CELLULOSE ESTERS, ORGANIC ESTERS

1. Introduction

Cellulose (qv) is one of nature's most abundant structural materials, providing the primary framework of most plants. For industrial purposes cellulose is derived from two primary sources, cotton linters and wood pulp. Linters are derived from the machine by the same name used for removing the short fibers adhering to cotton seeds after ginning and consist essentially of pure cellulose (see COTTON). Wood (qv), on the other hand, contains 40–60% cellulose, which must be extracted by the chemical degradation of the wood structure.

The chemical structure of cellulose is relatively simple (Fig. 1). The simplicity lies in the repetitive utilization of the anhydroglucose unit, $C_6H_{10}O_5$, as the building block for chain structure. The term cellulose does not designate a specific chemical or homogeneous substance but serves to characterize the homologous series of compounds having specifically a $(1 \longrightarrow 4)$ β (diequatorial) linkage between each anhydroglucose unit. Many other polyglucoside structural isomers exist (Fig. 2), but few have achieved the widespread commercial applications of cellulose. Thus two samples of cellulose contain the same relative amounts of β-D-Glcp1l 4-β-D-Glcp1l {4-β-D-Glcp1l}$_n$4-β-D-Glcp1l 4-β-D-Glcp1l 4

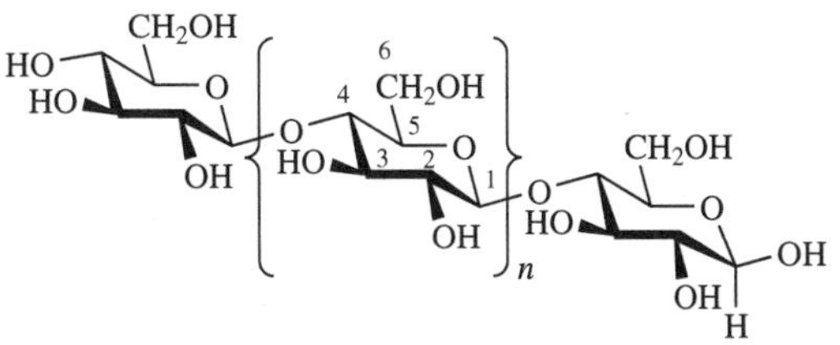

Fig. 1.   Structure of natural cellulose, $C_6H_{12}O_6$.

Starch   4α-D-Glcp 1l {4 α-D-Glcp 1l }$_n$4 α-D-Glcp 1l 4 α-D-Glcp 1l

Pustulan   6 β-D-Glcp 1l {6 β-D-Glcp 1l }$_n$6 β-D-Glcp 1l

Pullulan, m = 1 or 2

$$
\begin{array}{l}
\text{b} \\
6 \\
(\alpha\text{-D-Glcp 1 l}\quad 4\ \alpha\text{-D-Glcp 1l}\quad 4\ \alpha\text{-D-Glcp} \\
\qquad\qquad\qquad\qquad\qquad\qquad\qquad 1 \\
\qquad\qquad\qquad\qquad\qquad\qquad\qquad \text{b} \\
\qquad\qquad\qquad\qquad\qquad\qquad\qquad 6 \\
\qquad\qquad\alpha\text{ - D-Glcp 1l } \{4\ \alpha\text{ - D-Glcp 1}\}_m\text{l}\quad 4\ \alpha\text{ - D- Glcp)}_n \\
\qquad\qquad\qquad\qquad\qquad\qquad\qquad\qquad\qquad 1 \\
\qquad\qquad\qquad\qquad\qquad\qquad\qquad\qquad\qquad \text{b}
\end{array}
$$

Fig. 2.   Structural isomers of cellulose.

carbon, hydrogen, and oxygen but may vary considerably in chemical reactivity and physical properties. Molecular weight and, consequently, the number of anhydroglucose units per molecule or degree of polymerization (DP) vary as a function of the type of cellulose. Molecular weight determinations by the ultra-centrifuge method have assigned a molecular weight average of 570,000 to native cellulose. In the synthesis of cellulose derivatives, however, chain cleavage determines the molecular weight of the product and hence many of the observed physical properties.

Cellulose esters are commonly derived from natural cellulose by reaction with organic acids, anhydrides, or acid chlorides. Cellulose esters of almost any organic acid can be prepared, but because of practical limitations esters of acids containing more than four carbon atoms have not achieved commercial significance.

Cellulose acetate [9004-35-7] is the most important organic ester because of its broad application in fibers and plastics; it is prepared in multi-ton quantities with degrees of substitution (DS) ranging from that of hydrolyzed, water-soluble monoacetates to those of fully substituted triacetate (Table 1). Soluble cellulose acetate was first prepared in 1865 by heating cotton andacetic anhydride at 180°C (1). Usingsulfuric acid as a catalyst permitted preparation at lower temperatures (2), and later, partial hydrolysis of the triacetate gave anacetone-soluble cellulose acetate (3). The solubility of partially hydrolyzed (secondary)

Table 1. **Relationship of Cellulose Acetate DSto Acetyl Content and Combined Acetic Acid**

DS[a]	Acetyl, wt %[b]	Combined acetic acid, wt %[c]
0.5	11.7	16.3
0.75	16.7	23.2
1.0	21.1	29.4
1.5	28.7	40.0
2.0	35.0	48.8
2.5	40.3	56.2
3.0	44.8	62.5

[a] Defined as the average number of acetyl groups in the anhydroglucose unit of cellulose.
[b] Unit molecular weight of acetyl group,CH_3CO, is 43.
[c] Degree of acetylation is often expressed as percent combined acetic acid.

cellulose acetate in less expensive and less toxic solvents such as acetone aided substantially in its subsequent commercial development.

During World War I, cellulose acetate replaced the highly flammablecellulose nitrate coating on airplane wings and the fuselage fabrics. After World War I, it found extensive use in photographic and x-ray films, spun fibers, and molding plastics.

Although cellulose acetate remains the most widely used organic ester of cellulose, its usefulness is restricted by its moisture sensitivity, limited compatibility with other synthetic resins, and relatively high processing temperature. Cellulose esters of higher aliphatic acids, C_3 and C_4, circumvent these shortcomings with varying degrees of success. They can be prepared relatively easily with procedures similar to those used for cellulose acetate. Mixed cellulose esters containing acetate and either the propionate or butyrate moieties are produced commercially in large quantities by Eastman Chemical Co. in the United States (Table 2). Bayer AG discontinued the production of mixed esters at Leverkusen in Germany in mid-1987 citing poor economics as the reason for the closing.

Cellulose esters of aromatic acids, aliphatic acids containing more than four carbon atoms and aliphatic diacids are difficult and expensive to prepare because of the poor reactivity of the corresponding anhydrides with cellulose; little commercial interest has been shown in these esters. Of notable exception, however, is the recent interest in the mixed esters of cellulose succinates, prepared by the sodium acetate catalyzed reaction of cellulose withsuccinic anhydride. The additional expense incurred in manufacturing succinate esters is compensated by the improved film properties observed in waterborne coatings (5).

Mixed cellulose esters containing the dicarboxylate moiety, eg, cellulose acetate phthalate, have commercially useful properties such as alkaline solubility and excellent film-forming characteristics. These esters can be prepared by the reaction of hydrolyzed cellulose acetate with adicarboxylic anhydride in apyridine or, preferably, an acetic acid solvent with sodium acetate catalyst. Cellulose acetate phthalate [9004-38-0] for pharmaceutical and photographic uses is produced commercially via the acetic acid–sodium acetate method.

Table 2. **Domestic and Foreign Producers of Cellulose Acetate**

Producer	Country	Annual capacity, 10^3 t
North America[a]a		
Eastman Kodak Co.[b]b	United States	197
Hoechst Celanese Corp.	United States, Canada, Mexico	241
Western Europe[c]c		
Agfa-Gevaert	Belgium	5
Tubize Plastics[d]d	Belgium	7
Rhône-Poulenc Chemie	France	25
Rhodia AG	Germany	>25
Industrias del Acetato de Celulosa	Spain	2
Courtaulds Fibres Ltd.	United Kingdom	54
Nelsons Acetate Ltd.[e]e	United Kingdom	14
Asia		
Daicel Chemical Industries, Ltd.	Japan	100
Teijin Acetate Ltd.	Japan	11.8

[a]CEH estimates as of Nov. 1, 1988(4).

[b]Also produces cellulose acetate butyrate (CAB) and cellulose acetate propionate (CAP) mixed esters.

[c]CEH estimates as of Jan. 1, 1989(4); Bayer AG (Germany) discontinued CAB/CAP production at Leverkusen in mid-1987; capacity for resin-grade cellulose acetate flake had been reduced from 25,000 t in 1981 to 15,000 t prior to closing.

[d]In 1984, PMC Inc. (United States) purchased Tubize Plastics SA from government owned Fabelta Tubize SA. This facility was sold to Rhône-Poulenc in Oct. 1988.

[e]In 1988, Courtaulds took over Hercules' 50% interest in Nelsons Acetate.

2. Properties

The properties of cellulose esters are affected by the number of acyl groups per anhydroglucose unit, acyl chain length, and the degree of polymerization (DP) (molecular weight). The properties of some typical cellulose triesters are given in Table 3. In this series, with increasing acyl chain length from C_2 to C_6, the melting point, tensile strength, mechanical strength, and density generally decrease, whereas solubilities in nonpolar solvents and resistance to moisture increase. Fewer acyl groups per anhydroglucose unit, ie, increased hydroxyl content, increase the solubility in polar solvents and decrease moisture resistance. The physical and chemical properties of mixed esters vary according to the ratio of the esters used, eg, acetyl to butyl or acetyl to propionyl. General trends of the properties of mixed esters, such as cellulose acetate butyrate [9004-36-8] (CAB), as a function of composition are illustrated in Figure 3, in which increasing butyryl (decreasing acetyl) content increases flexibility, moisture resistance, and nonpolar solubility and decreases melting point and density.

The common commercial products are the primary (triacetate) and the secondary (acetone-soluble, ca 39.5% acetyl, 2.45 DS) acetates; they are odorless, tasteless, and nontoxic. Their properties depend on the combined acetic acid content (acetyl, see Table 1 and Figure 4) and molecular weight. Solubility characteristics of cellulose acetates with various acetyl contents are given in Table 4.

Table 3. **Properties of Cellulose Triesters**[a]

| Cellulose ester | Shrinking point, °C | Mp[b], °C | Water tolerance value | Moisture regain[d], % | | | Density, g/mL | Tensile strength, MPa[e] |
				50% rh	75% rh	95% rh		
cellulose[f]				10.8	15.5	30.5	1.52	
acetate		306	54.4	2.0	3.8	7.8	1.28	71.6
propionate	229	234	26.9	0.5	1.5	2.4	1.23	48.0
butyrate	178	183	16.1	0.2	0.7	1.0	1.17	30.4
valerate	119	112	10.2	0.2	0.3	0.6	1.13	18.6
caproate	84	94	5.88	0.1	0.2	0.4	1.10	13.7
heptylate[g]	82	88	3.39	0.1	0.2	0.4	1.07	10.8
caprate	82	86	1.14	0.1	0.1	0.2	1.05	8.8
caprate[h]	87	88		0.1	0.2	0.5	1.02	6.9
laurate	89	91		0.1	0.1	0.3	1.00	5.9
myristate	87	106		0.1	0.1	0.2	0.99	5.9
palmitate	90	106		0.1	0.1	0.2	0.99	4.9

[a]Ref. 6. Courtesy of the American Chemical Society.
[b]Char point is 315°C or higher unless otherwise noted.
[c]Milliliters of water required to start precipitation of the ester from 125 mL of an acetone solution of 0.1% concentration.
[d]At 25% rh moisture regain for cellulose is 5.4%; for the acetate, 0.6%; for the propionate and butyrate, 0.1%; all others are zero.
[e]To convert MPa to psi, multiply by 145.
[f]Starting cellulose, prepared by deacetylation of commercial, medium viscosity cellulose acetate (40.4% acetyl content).
[g]Char point = 290°C.
[h]Char point = 301°C.

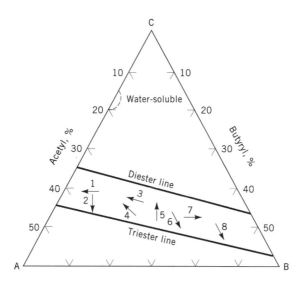

Fig. 3. Effects of composition on physical properties. A, acetyl; B, butyryl; C, cellulose. 1, increased tensile strength, stiffness; 2, decreased moisture sorption; 3, increased melting point; 4, increased plasticizer compatibility; 5, increased solubilities in polar solvents; 6, increased solubilities in nonpolar solvents; 7, increased flexibility; 8, decreased density (7).

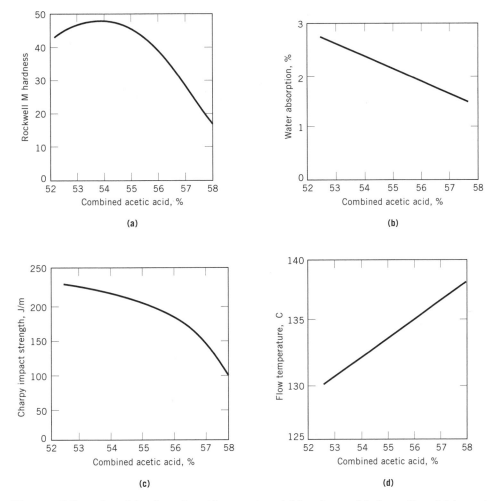

Fig. 4. Effect of combined acetic acid content on (**a**) hardness, (**b**) absorption, (**c**) impact strength, and (**d**) temperature of cellulose acetate (8). To convert J/m to ft·lbs/in., divide by 53.38.

Cellulose triacetate [9012-09-3] has the highest melting point (ca 300°C) of the triesters; melting points generally decrease to a minimum of ca 230°C as the acetyl content decreases to 38–39% (secondary acetate).

Moisture sensitivity and vapor-permeability rate of cellulose acetate increase with decreasing acetyl (increasing hydroxyl) content. Thermoplastic

Table 4. Solubility Characteristics of Cellulose Acetates

Acetyl, %	Soluble in	Insoluble in
43.0–44.8	dichloromethane	acetone
37–42	acetone	dichloromethane
24–32	2-methoxymethanol	acetone
15–20	water	2-methoxymethanol
≤13	none of the above	all of the above

characteristics are greatly improved as the acetyl content is increased from ca 20% (DS(acetyl) = 1) to ca 39% (DS(acetyl) = 2.4) (9).

The bulk density of cellulose acetate varies with physical form from 160 kg/m^3 (10 lb/ft^3) for soft flakes to 481 kg/m^3 (30 lb/ft^3) for hammer-milled powder, whereas the specific gravity (1.29–1.30), refractive index (1.48), and dielectric constant of most commercial cellulose acetates are similar.

In fibers, plastics, and films prepared from cellulose esters, mechanical properties such as tensile strength, impact strength, elongation, and flexural strength are greatly affected by the degree of polymerization and the degree of substitution. Mechanical properties significantly improve as the DP is increased from ca 100 to 250 repeat units.

2.1. Liquid Crystalline Solutions.

Cellulose esters, when dissolved in the appropriate solvents at the proper concentration, show liquid crystalline characteristics similar to those of other rigid chain polymers (10) because of an ordered arrangement of the polymer molecules in solution. Cellulose triacetate dissolved at 30–40 wt% in trifluoroacetic acid, dichloroacetic acid, and mixtures of trifluoroacetic acid and dichloromethane exhibits brilliant iridescence, high optical rotation, and viscosity–temperature profiles characteristic of typical aniostropic phase-containing liquid crystalline solutions (11). Similar observations have been made for cellulose acetate butyrate (12), cellulose diacetate (13), and other cellulose derivatives (14,15). Wet spinning of these liquid crystalline solutions yields fibers with much higher strength properties than fibers normally obtained from cellulose esters (16,17).

3. Manufacture and Processing

Simple triesters such as cellulose formate [9036-95-7] (7), cellulose propionate [9004-48-2] (9,18), and cellulose butyrate [9015-12-7] (19) have been prepared and their properties studied; none of these triesters is produced in large quantities. Cellulose formate esters, prepared by reaction of cellulose with formic acid, are thermally (20) and hydrolytically (7) unstable. Cellulose propionate and cellulose butyrate triesters are synthesized by methods similar to those used in the preparation of cellulose acetate with propionic or butyric anhydride in the presence of an acid catalyst (21). These anhydrides, especially butyric, react more slowly with cellulose than acetic anhydride. Therefore, the cellulose must be activated and the temperature must be controlled to avoid degradation. Esterification rates decrease with increasing acyl chain length, and degradation becomes more severe in the order acetic < propionic < butyric < isobutyric anhydride. Esterification with isobutyric anhydride is normally so slow that highly activated cellulose must be used and thesulfuric acid catalyst must be distributed uniformly. Swelling agents, eg, water, containing dissolved acid catalyst are used to ensure uniform catalyst distribution for the preparation of isobutyrate esters. The swelling agent is removed by solvent exchange, leaving sorbed acid uniformly distributed in the activated cellulose (22).

Cellulose activated with ethylenediamine [107-15-3] is used to prepare high molecular-weight cellulose butyrate (23). Cellulose so activated has a larger measured surface area (120 m^3/g) than cellulose activated with acetic acid (4.8 m^3/g).

The diamine is removed with water, followed by solvent exchange with acetic acid and butyric acid before esterification.

More recently, however, a process for the manufacture of ultrahigh molecular-weight cellulose esters has been developed by reaction of nonactivated, secondary cellulose with trifluoroacetic acid, trifluoroacetic anhydride, and either an organic acid or acid chloride (24). This process is amenable to a larger variety of organic esters not normally available through conventional means. The technique requires less reaction time, less excess solvent, and it is easier to control the extent of the reaction than conventional sulfuric acid activation. Unfortunately, the handling and toxic nature of trifluoroacetic acid and the anhydride currently prevent its use on a large scale.

Cellulose valerates have been synthesized by conventional methods using valeric anhydride and sulfuric acid catalyst (25,26). Alternatively, the cellulose is activated by soaking in water, which is then displaced bymethylene chloride or valeric acid; the temperature is maintained at <38°C to minimize degradation.

Production of cellulose esters from aromatic acids has not been commercialized because of unfavorable economics. These esters are usually prepared from highly reactive regenerated cellulose, and their physical properties do not differ markedly from cellulose esters prepared from the more readily available aliphatic acids. Benzoate esters have been prepared from regenerated cellulose withbenzoyl chloride in pyridine–nitrobenzene (27) or benzene (28). These benzoate esters are soluble in common organic solvents such as acetone or chloroform. Benzoate esters, as well as the nitrochloro-, and methoxy-substituted benzoates, have been prepared from cellulose with the appropriate aromatic acid and chloroacetic anhydride as the impelling agent and magnesium perchlorate as the catalyst (29).

Cellulose chloroacetates (30) and aminoacetates (30,31), acetate sorbates (32), and acetate maleates (33) have been prepared but are not commercially important. These esters are made from hydrolyzed cellulose acetate with the appropriate anhydride or acid chloride in pyridine.

Cellulose esters of unsaturated acids, such as the acetate methacrylate, acetate maleate (34), and propionate crotonate (35), have been prepared. They are made by treating the hydrolyzed acetate or propionate with the corresponding acyl chloride in a pyridine solvent. Cellulose esters of unsaturated acids are cross-linkable by heat or uv light; solvent-resistant films and coatings can be prepared from such esters.

Amine-containing cellulose esters, eg, the acetate N,N-diethylaminoacetate (36) and propionate morpholinobutyrate (35), are of interest because of their unique solubility in dilute acid. Such esters are prepared by the addition of the appropriate amine to the cellulose acrylate crotonate esters or by replacement of the chlorine on cellulose acrylate chloroacetate esters with amines. This type of ester has been suggested for use in controlled release, rumen-protected feed supplements for ruminants (36,37).

Mixed esters, such as cellulose acetate propionate and cellulose acetate butyrate, have desirable properties not exhibited by the acetate or the high acyl triesters. These mixed esters are produced commercially in multiton quantities by methods similar to those for cellulose acetate; they are prepared over a wide range of acyl substitutions and viscosities. The ratio of acetyl to higher acyl

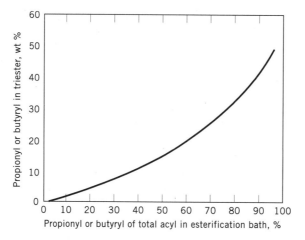

Fig. 5. Composition of cellulose acetate butyrate (propionate) as a function of butyryl (propionyl) content of esterification bath.

in the product is proportional to the concentration of components in the esterification solution (Figs. 5 and 6). Thus it is possible to esterify cellulose with propionic or butyric anhydride in the presence of acetic acid to produce the mixed esters. In a similar manner, acetic anhydride can be used in the esterification with either propionic or butyric acid to produce a cellulose ester containing both acyl moieties. The commercial production of cellulose acetate butyrate

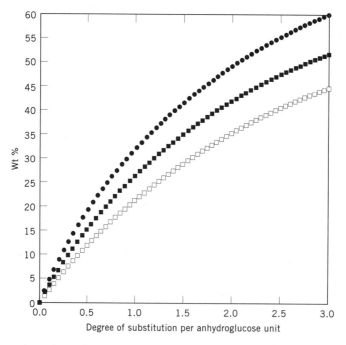

Fig. 6. Conversion of weight percent to degree of substitution per anhydroglucose unit (max = 3.0), where ● is butyryl, ◆ is propionyl, and ◇ is acetyl.

has been described (38), and the different reactivities of lower anhydrides toward cellulose have been investigated in detail. Cellulose butyrate has been prepared in homogenous solution by the reaction of cellulose in dimethyl sulfoxide (DMSO)-paraformaldehyde with butyric anhydride and pyridine catalyst (39). The maximum degree of substitution is ca 1.8, and the products are soluble in common organic solvents. Dichloromethane has been used in the preparation of cellulose acetate butyrate to prevent excessive degradation and provide an ester with higher molecular weight (40).

Mixed esters containing the dicarboxylate moiety, eg, cellulose acetate phthalate, are usually prepared from the partially hydrolyzed lower aliphatic acid ester of cellulose in acetic acid solvent by using the corresponding dicarboxylic acid anhydride and a basic catalyst such as sodium acetate (41,42). Cellulose acetate succinate and cellulose acetate butyrate succinate are manufactured by similar methods as described in reference 43.

Other mixed esters, eg, cellulose acetate valerate [55962-79-3], cellulose propionate valerate [67351-41-1], and cellulose butyrate valerate [53568-56-2], have been prepared by the conventional anhydride sulfuric acid methods (25). Cellulose acetate isobutyrate [67351-38-6] (44) and cellulose propionate isobutyrate [67351-40-0] (45) have been prepared with a zinc chloride catalyst. Large amounts of catalyst and anhydride are required to provide a soluble product, and special methods of delayed anhydride addition are necessary to produce mixed esters containing the acetate moiety. Mixtures of sulfuric acid and perchloric acid are claimed to be effective catalysts for the preparation of cellulose acetate propionate in dichloromethane solution at relatively low temperatures (46); however, such acid mixtures are considered too corrosive for large-scale productions.

Mixed esters are hydrolyzed by methods similar to those used for hydrolyzing cellulose triacetate. The hydrolysis eliminates small amounts of the combined sulfate ester, which, if not removed, affects thermal stability. Sulfuric acid is the preferred catalyst for hydrolysis since it is already present in the esterification mixture. On a large scale, partial neutralization of the catalyst may be necessary before hydrolysis. Increasing the amount of water during hydrolysis reduces the rates of viscosity reduction and acyl hydrolysis of cellulose acetate propionate and acetate butyrate esters (47). Several methods of hydrolyzing cellulose esters of higher aliphatic acids are described (48,49). Acetate phthalate mixed esters preferentially lose acetyl groups when hydrolyzed in aqueous acetic acid media with sulfuric acid catalyst. On the other hand, hydrolysis with a basic catalyst such as sodium or potassium acetate in aqueous acetic acid results in preferential loss of the phthaloyl moiety (50). Ester properties can be modified by changing the DS, ie, removing acyl groups.

3.1. Stabilization. After hydrolysis, precipitation, and thorough washing of the cellulose esters to remove residual acids, the esters must be stabilized against thermal degradation and color development, which may occur during processing, such as extrusion or injection molding. Thermal instability is caused by the presence of oxidizable substances and small amounts of free and combined sulfuric acid (51). The sulfuric acid combines with the cellulose almost quantitatively and most of it is removed during the latter stages of hydrolysis. The remaining sulfuric acid can be neutralized with alkali metal salts, such as sodium, calcium, or magnesium acetate, to improve ester stability. The combined sulfate ester may also be removed by treatment in boiling water or at steam temperatures in an autoclave. Treatment with aqueous potassium or calcium iodide reportedly stabilizes the cellulose acetate against thermal degradation (52).

Dialkyl esters of 3,3'-thiodipropionic acid (53), cyclic phosphonites such as neopentylphenyl phosphite, derivatives of phosphaphenathrene-10-oxide (54), secondary aromatic amines, eg, diphenylamine (55), and epoxidized soybean oils (56) are effective stabilizers for preventing discoloration of cellulose esters during thermal processing.

Exposure to uv radiation may cause chain scission and loss of physical properties in cellulose esters exposed to outdoor environments; esters formulated for such use must be stabilized accordingly. Some resorcinol and benzophenone derivatives, such as resorcinol monobenzoate and 2-hydroxy-4-methoxybenzophenone, are reportedly excellent uv-light stabilizers for cellulose esters (57,58). Other stabilizers include piperidine derivatives (59) and substituted triazole compounds alone (60) and in combination with resorcinol monobenzoate (61).

3.2. Cellulose Acetate.   Almost all cellulose acetate, with the exception of fibrous triacetate, is prepared by a solution process employing sulfuric acid as the catalyst with acetic anhydride in an acetic acid solvent. The acetylation reaction is heterogeneous and topochemical wherein successive layers of the cellulose fibers react and are solubilized in the medium, thus exposing new surfaces for reaction. The reaction course is controlled by the rates of diffusion of the reagents into the cellulose fibers, and therefore the cellulose must be swollen or activated before acetylation to achieve uniform reaction and avoid unreacted fibers in the solution (62).

Cellulose dissolved in suitable solvents, however, can be acetylated in a totally homogeneous manner, and several such methods have been suggested. Treatment in dimethyl sulfoxide (DMSO) with paraformaldehyde gives a soluble methylol derivative that reacts with glacial acetic acid, acetic anhydride, or acetyl chloride to form the acetate (63). The maximum degree of substitution obtained by this method is 2.0; some oxidation also occurs. Similarly, cellulose can be acetylated in solution with dimethylacetamide–paraformaldehyde and dimethylformamide-paraformaldehyde with a potassium acetate catalyst (64) to provide an almost quantitative yield of hydroxymethylcellulose acetate.

Several derivatives of cellulose, including cellulose acetate, can be prepared in solution in dimethylacetamide–lithium chloride (65). Reportedly, this combination does not react with the hydroxy groups, thus leaving them free for esterification or etherification reactions. In another homogeneous-solution method, cellulose is treated with dinitrogen tetroxide in DMF to form the soluble cellulose nitrite ester; this is then ester-interchanged with acetic anhydride (66). With pyridine as the catalyst, this method yields cellulose acetate with DS < 2.0.

In the fibrous acetylation process, part or all of the acetic acid solvent is replaced with an inert dilutent, such astoluene,benzene, orhexane, to maintain the fibrous structure of cellulose throughout the reaction. Perchloric acid is often the catalyst of choice because of its high activity and because it does not react with cellulose to form acid esters. Fibrous acetylation also occurs upon treatment with acetic anhydride vapors after impregnation with a suitable catalyst such as zinc chloride (67).

Table 5. **Typical Specifications for Acetylation-Grade Pulp**[a]

Property	Value
α-cellulose, min %	95.6
moisture, %	5.8[b]
pentosans, max %	2.1
cuprammonium viscosity[c], mPa · s(=cP)	1100–4000
intrinsic viscosity, dL/g	5.5–7.5
ether extractable, max %	0.15
ash, max %	0.08
iron, max ppm	10
trial acetylation	
haze, max ppm	100
color, max ppm	600

[a] Ref. 62.
[b] Off supplier's dryer.
[c] 2.5% solution.

An apparatus for the continuous fibrous acetylation of cellulose in benzene has been described (68, 69). The process involves continuous activation, acetylation, partial saponification of the resulting triacetate, and drying of the product.

Activation of Cellulose. The activation required depends on the source of cellulose (cotton linter or wood pulp), purity, and drying history. Typical specifications for an acetylation-grade cellulose are given in Table 5. Cellulose that has never been dried or has been mildly dried to ca 5% moisture requires little, if any, further activation.

Normally, water or aqueous acetic acid is the activating agent; glacial acetic acid may also be used. Water is more effective because it swells the fibers more than other agents and alters the hydrogen bonding between the polymer chains to provide a greater surface area for reaction. When water or aqueous acids are used, the cellulose must be dehydrated by displacing the water with acetic acid before the start of acetylation. Commercially, it is not unusual to activate cellulose with glacial acetic acid containing a small part of the total required sulfuric acid catalyst; this reduces the molecular weight of the cellulose as needed to obtain a satisfactory product. The efficiency of activation is increased by increased temperature, time, amount of catalyst, and lower acetic acid–cellulose ratio. Several other swelling agents and methods for cellulose activation have been reported but have little commercial value because of cost and performance considerations.

Ethylenediamine (70,71), benzyl alcohol and acetone (72), ethylene glycol (73) and C_2–C_{18} carboxylic acids (74) are claimed to increase the reactivity of cellulose toward acetylation. Sodium hydroxide and liquid ammonia (71) are excellent swelling agents and have been used to activate cellulose before esterification. Ultrasonic treatment of cellulose slurries (75) reportedly swells the fibers and improves reactivity.

In one process to produce highly activated cellulose for acetylation, cellulose is treated with NaOH (mercerization) followed by a hydroxyalkylating agent, eg, ethylene oxide or propylene oxide, to give a cellulose hydroxyalkyl ether with a DS of 0.05–0.3 (76). The resulting water-insoluble material is highly reactive to conventional acetic anhydride–sulfuric acid acetylation.

Catalysts for Acetylation.   Sulfuric acid is the preferred catalyst for esterifying cellulose and is the only known catalyst used commercially for this function. The role of sulfuric acid during acetylation has been discussed (77,78). In the presence of acetic anhydride, sulfuric acid rapidly and almost quantitatively forms the cellulose sulfate acid ester (77). Even in the absence of anhydride, the sulfuric acid is physically or mechanically retained (sorbed) on the cellulose. The degree of absorption is a measure of the reactivity or accessibility of different celluloses.

Sulfuric acid reacts with acetic anhydride to form acetylsulfuric acid (79). This reaction is favored by low temperature and high anhydride concentration. In cellulose acetylation, probably both sulfuric acid and acetylsulfuric acid exist and react with cellulose to form cellulose sulfate acid ester.

Perchloric acid is a well-known acetylation catalyst, especially in the fibrous method of preparing cellulose triacetate. Unlike sulfuric acid, perchloric acid does not combine with cellulose (78), ie, it does not form esters, and therefore virtually complete acetylation (DS 3.0, 44.8% acetyl) occurs. However, the extremely corrosive nature of perchloric acid and explosive nature of its salts have precluded its use industrially as an acetylation catalyst.

Zinc chloride is a Lewis acid catalyst that promotes cellulose esterification. However, because of the large quantities required, this type of catalyst would be uneconomical for commercial use. Other compounds such as titanium alkoxides, eg, tetrabutoxytitanium (80), sulfate salts containing cadmium, aluminum, and ammonium ions (81), sulfamic acid, and ammonium sulfate (82) have been reported as catalysts for cellulose acetate production. In general, they require reaction temperatures above 50°C for complete esterification. Relatively small amounts ($\leq 0.5\%$) of sulfuric acid combined with phosphoric acid (83), sulfonic acids, eg, methanesulfonic, or alkyl phosphites (84) have been reported as good acetylation catalysts, especially at reaction temperatures above 90°C.

Hydrolysis.   The primary functions of hydrolysis are to remove some of the acetyl groups from the cellulose triester and to reduce or remove the combined acid sulfate ester to improve the thermal stability of the acetate.

The acetylation reaction is stopped by the addition of water to destroy the excess anhydride, causing rapid hydrolysis of the combined sulfate acid ester (Fig. 7). This is followed by a much slower rate of hydrolysis of the acetyl ester groups. The rate of hydrolysis is controlled by temperature, catalyst concentration, and, to a lesser extent, by the amount of water. Higher temperatures and catalyst concentrations increase the rate of hydrolysis. Higher water content slightly increases the hydrolysis rate and helps minimize degradation (85). The amount of water also influences the ratio of primary to secondary hydroxy groups in the hydrolyzed cellulose acetate; high water content favors primary hydroxyl formation (86).

In commercial processes, the water content during hydrolysis ranges from 5 to 20 wt % based on total liquids and depends on the temperature and the final product desired. Hydrolysis reactions can be performed at temperatures ranging from ca 38°C to pressurized reactions at 229°C (87). In a continuous process at 129°C, a triacetate solution is passed vertically upward through three consecutive chambers containing rotating disks to maximize plug flow of the solution (88). Kinetics of cellulose triacetate hydrolysis have been investigated (89) and,

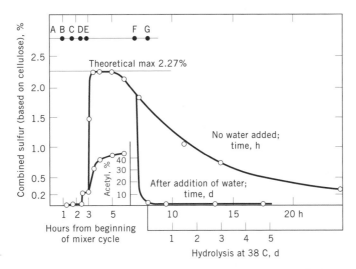

Fig. 7. Combined sulfur during preparation of cellulose acetate; hydrolysis of sulfate and esters (6). Acetylation schedule: A, mixer charged with linters and acetic acid; B, minor portion of catalyst added; C, began cooling to 18°C; D, acetic anhydride added and continued cooling to 16°C; E, significant portion of catalyst added; F–G, water added during 1 h.

with sulfuric acid as the catalyst, the rate constant was found to be linear for both catalyst concentration and water content.

The rate of hydrolysis of cellulose acetate can be monitored by removing samples at intervals during hydrolysis and determining the solubility of the hydrolyzed acetate. When the desired DS is reached, the hydrolysis is stopped by neutralizing the catalyst with magnesium, calcium, or sodium salts dissolved in aqueous acetic acid.

Precipitation and Purification. During the hydrolysis, control tests are made by turbidimetric titration of samples taken intermittently. When the desired degree of hydrolysis is reached, the ester is precipitated from the reaction solution into water. It is important for the precipitate to have the proper texture for subsequent washing to remove acid and salts for thermal stabilization. Before precipitation, the reaction solution is usually diluted with additional aqueous acetic acid to reduce the viscosity. If a flake texture is desired, the solution is poured into a vigorously stirred, 10–15% aqueous acetic acid. To precipitate the acetate in powder form, dilute acetic acid is added to the stirred reaction solution. In both cases, the precipitated ester is suspended in 25–30% aqueous acid solutions and finally washed with deionized water. The dilution, precipitation temperature, agitation, and strength of the acid media must be controlled to ensure uniform texture.

Another method for direct precipitation of cellulose acetate powder suitable for extrusion into plastics is described (90). The reaction solution is precipitated with dilute aqueous acetic acid at 80–85°C in the presence of a coagulant such as isopropyl acetate. The resulting powder particles have a higher bulk density and absorb plasticizers more readily than powders obtained by the usual methods.

Granules can be precipitated and formed by extruding the viscous reaction mixture through a circular die containing several holes over which a knife blade

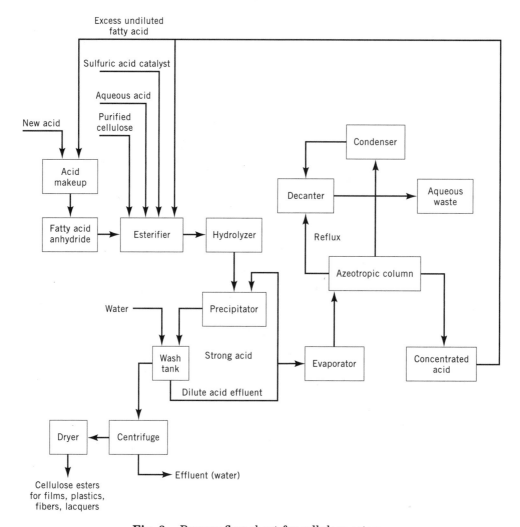

Fig. 8.   Process flow sheet for cellulose esters.

rotates to cut the strands into granules (91). The granules are simultaneously slurried in dilute acetic acid to harden the particles for further washing.

Solution Process.   With the exception of fibrous triacetate, practically all cellulose acetate is manufactured by a solution process using sulfuric acid catalyst with acetic anhydride in an acetic acid solvent. An excellent description of this process is given (85). In the process (Fig. 8), cellulose (ca 400 kg) is treated with ca 1200 kg acetic anhydride in 1600 kg acetic acid solvent and 28–40 kg sulfuric acid (7–10% based on cellulose) as catalyst. During the exothermic reaction, the temperature is controlled at 40–45°C to minimize cellulose degradation. After the reaction solution becomes clear and fiber-free and the desired viscosity has been achieved, sufficient aqueous acetic acid (60–70% acid) is added to destroy the excess anhydride and provide 10–15% free water for hydrolysis. At this point, the sulfuric acid catalyst may be partially neutralized with calcium, magnesium, or sodium salts for better control of product molecular weight.

The cellulose acetate is hydrolyzed in solution at 40–50°C for varying lengths of time (4–20 h) until the desired DS is obtained; at this point, the ester is precipitated with dilute aqueous acetic acid. The precipitate is hardened in 25–30% aqueous acetic acid, which is drained off and recovered.

The ester is washed thoroughly in iron-free water to remove acid and any desirable salts; these wash liquids are sent for acid recovery. The final wash may contain some sodium, calcium, or magnesium ions to stabilize traces of sulfate esters remaining on the cellulose acetate.

Recent Developments. A considerable amount of cellulose acetate is manufactured by the batch process, as described previously. In order to reduce production costs, efforts have been made to develop a continuous process that includes continuous activation, acetylation, hydrolysis, and precipitation. In this process, the reaction mixture, ie, cellulose, anhydride, catalyst, and solvent, pass continuously through a number of successive reaction zones, each of which is agitated (92,93). In a similar process, the reaction mass is passed through tubular zones in which the mixture is forced through screens of successively small openings to homogenize the mixture effectively (94). Other similar methods for continuous acetylation of cellulose have been described (95,96).

Cellulose acetate with improved solubility properties can be prepared from low quality wood pulps by multistage addition of the components (97) or by interrupting the reaction in the early stages, filtering, and continuing the acetylation with fresh reactants (98,99).

In an integrated continuous process, cellulose reacts with acetic anhydride prepared from the carbonylation of methyl acetate withcarbon monoxide. The acetic acid liberated reacts further with methanol to give methyl acetate, which is then carbonylated to give additional acetic anhydride (100,101).

High temperature acetylation of cellulose above 50°C produces cellulose acetate from low purity wood pulp cellulose in shorter reaction times. In a high temperature method recently disclosed (102), cellulose reacts with 200–400% acetic anhydride in the presence of <5% acid catalyst at 68–85°C for 3–20 min. After the acid catalyst is neutralized with magnesium acetate, the cellulose acetate is hydrolyzed at 120°C for two hours (103). Several modified catalyst systems have been developed for acetylation of cellulose above 90°C (89,90).

4. Economic Aspects

From 1980 to 1988, annual cellulose acetate flake production in the United States showed a slight decrease in production from 392,000 t to 323,000 t with an annual decline of −0.4 to −0.1% (Table 6). World demand for cellulose acetate flake has also fallen. A modest recovery has occurred in recent years as a result of the increased demand for cigarette-filter tow; world consumption of cigarette-filter tow has risen about 2.5% per year since 1980 (Tables 7 and 8). In contrast, world demand for textile fibers and cellulose ester plastics decline 4.6% and 4.2% per year, respectively (Fig. 9).

Demand for cellulose acetate flake in the United States is projected to decline slightly from 1988 to 1993. Cigarette-filter tow for export is the only

Table 6. **U.S. Consumption of Cellulose Acetate Flake,**[a] **10^3 t**

Year	Cigarette-filter tow	Textile fibers	Cellulose acetate plastics[b]	Total
1979	164	145	83	392
1980	175	145	48	368
1983	173	105	62	339
1988	205	68	50	323
1993	210–215	59	47	315–321
Average annual growth rate				
1988–1993	0.5–1.0%	−3.0	−1.1%	−0.4 − 0.1%

[a] Data are reported as cellulose acetate flake equivalents; plasticizers not included.
[b] Mixed esters (CAB and CAP) included. Courtesy of CEH Estimates.

Table 7. **World Consumption of Cellulose Acetate Flake,**[a] **10^3 t**

Year	Textile fibers	Cigarette-filter tow	Plastics[b]	Total
1980	335	338	132	805
1985	241	335	172	710
1987	240	401	98	740

[a] Data for textile fibers and cigarette-filter tow include Eastern Europe and the People's Republic of China.
[b] Cellulose acetate ester plastics are produced largely in the United States, Western Europe, and Japan. World consumption is assumed to be approximately equivalent to production of cellulose ester plastics in these three regions. Courtesy of CEH Estimates.

Table 8. **World Capacity for Cellulose Acetate Flake, 1988, 10^3 t**

Region	Capacity
North America	
Canada	30
United States	453
Central America	
Mexico	20
Western Europe	
Belgium	12
France	25
Germany (formerly FRG)	25
Italy	9
Spain	2
United Kingdom	68
Eastern Europe	
Germany (formerly GDR)	6
Russia	70
Asia	
Japan	112
Totals[a]	*832*

[a] In addition to the producers listed above, Victorio Ghisolfiwith state owned Gepi purchased the Taban SpA (a Mon-tedison subsidiary) plant in Pallanza, Italy in 1988 with plans to restart in 1989 (4).

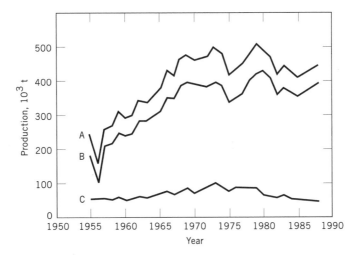

Fig. 9. U.S. production of cellulose esters; A, total cellulose esters production; B, cellulose acetate flake; C, other cellulose esters, ie, cellulose nitrate, cellulose acetate propionate, cellulose acetate butyrate (4).Courtesy of CEH Estimates.

market projected to grow. Cellulose acetate for textile fibers is expected to decline, as will flake demand for plastics, with the growth of photographic films somewhat offsetting declining markets in other plastics end uses.

The price of cellulose ester flake has generally increased with inflation and as of mid-1987 was estimated at ca $3.64–$4.71/kg for cellulose diacetate molding resin and from ca $4.16–$4.71/kg for the mixed esters molding resins depending on purity and the number of propionyl or butyryl esters (4).

From 1946 to mid-1987, Farbenfabriken Bayer AG in Germany was the European producer of cellulose acetate, cellulose acetate butyrate (CAB), and cellulose acetate propionate (CAP) before closing its facilities. Bayer's exit from the cellulose acetate mixed esters business leaves Eastman Chemical Co. in the United States as the sole producer of CAB/CAP resins.

In the United States, plastic and fiber grades of cellulose acetate and mixed cellulose ester flake, ie, acetate butyrate and acetate propionate, are produced by the Eastman Chemical Co.; Hoechst-Celanese manufactures a fiber-grade cellulose acetate flake suitable for plastics applications. Foreign producers of cellulose acetate flake include Gevaert and Fabelta in Belgium; Rhône Poulenc Chemie in France; British Hercules, Courtaulds, and Nelson in the United Kingdom; Mazzucchelli Celluliode SpA and Montedison in Italy; and Daicel and Teijin in Japan.

Cellulose ester flake production for nonfiber applications represents a small part of the total U.S. output. Cigarette-filter tow and cellulose acetate textile fibers consume more than 80% of flake production (104).

With the takeover of the American Celanese Corp. in mid-1986, the new Hoechst Celanese Corp., with Eastman Chemical Co. are the world leaders in cellulose acetate tow and flake production as of this writing. With confidence in both the acetate market and acetic anhydride supply (105), Hoechst Celanese recently announced an expansion that will double its tow and flake capacity from 35,750 metric tons to 60,500 metric tons by 1994 (106). The increased capacity will be achieved through expansions and productivity improvements at the U.S., Belgium, Canada, Mexico, and China operation centers.

In a similar announcement, Eastman Chemical Co. and Rhône-Poulenc S.A. have recently formed a 50/50 joint venture to make 59,090 metric tons/yr of cellulose acetate filter tow and fiber by the fourth quarter of 1993 (107). Eastman Chemical Co. announced a multimillion dollar expansion of its own filter tow capacity by approximately 11,363 metric tons/yr to be completed by mid-1991. The expansion brings the company's worldwide filter tow capacity up to 154,545 metric tons (108). Eastman Chemical Co. announced plans to expand the production of cellulose acetate butryates and cellulose acetate propionates by 20 and 40%, respectively, by mid-1991 (109).

5. Analytical and Test Methods

Standardized test methods for analyzing the chemical composition, viscosity, and physical properties of cellulose esters have been adopted by the ASTM and are described in substantial detail (110).

5.1. Degree of Substitution and DS Distribution.

For cellulose esters, the substitution level is usually expressed in terms of DS; that is, the average number of substituents per anhydroglucose unit (AGU). Cellulose contains three hydroxyl groups in each AGU unit that can be substituted; therefore DS can have a value between zero and three. Because DS is a statistical mean value, a value of 1 does not assure that every AGU has a single substituent. In some cases there can be unsubstituted anhydroglucose units, some with two and some with three substituents, and more often than not the value will be a noninteger. The physical properties commonly associated with commercial cellulose acetates, cellulose acetate butyrates, and cellulose acetate propionates are, in many cases, directly related to the degree of substitution as well as the overall substitution pattern. The degree of substitution or acetyl content of cellulose acetate has traditionally been determined by saponifying a known amount of the ester with an excess of standard sodium hydroxide solution in the presence of swelling agent or solvent. The excess sodium hydroxide is back-titrated with a standard solution of hydrochloric acid to determine the total acetyl content. The relative amounts of acetyl, propionyl, and butyryl in cellulose mixed esters, however, are determined by partition analysis in butyl acetate–water mixtures. The esters are saponified in sodium hydroxide, and phosphoric acid is added to liberate the organic acids from their sodium salts. The acids are partitioned between butyl acetate and water and their mole ratios are determined by comparison to carefully prepared control standards.

The acetyl content of cellulose acetate may be calculated by difference from the hydroxyl content, which is usually determined by carbanilation of the ester hydroxy groups in pyridine solvent withphenyl isocyanate [103-71-9], followed by measurement of uv absorption of the combined carbanilate. Methods for determining cellulose ester hydroxyl content by near-infrared spectroscopy (111) and acid content by nmr spectroscopy (112) and pyrolysis gas chromatography (113) have been reported.

With the advent of high resolution proton and carbon nuclear magnetic resonance spectroscopy, however, determining the DS and the DS distribution of mixed ester systems by aqueous saponification has become obsolete. Numerous studies have shown nmr to be a fundamental tool for probing the microscopic behavior of a wide variety of synthetic polymers and biomacromolecules (114). In the past, however, most spectral assignments required laboriously prepared derivatives containing a trideuterioacetyl or trideuteriomethyl group at a predetermined position or by comparison to spectra of mono- or oligosaccharides (115). These methods proved to be limited in detail because of line broadening and small differences in chemical shifts. Understanding the basic relationships between macroscopic properties and the microstructure of these biopolymer derivatives through nmr is the direction of much cellulose esters research.

Determining the degree of substitution using standard proton nmr relies on the integral ratio between the cellulosic ring protons ($\approx 5.0 - 2.9\delta$) and the ester alkyl protons ($\approx 1.2\delta$ for butyryl and propionyl and $\approx 2.0\delta$ for acetyl methyl groups). This simple procedure is used extensively to determine the extent of esterification and is currently the fastest, easiest way for determining the DS of mixed cellulose esters.

Standard proton nmr techniques provide information on the degree of substitution and ester ratios in mixed ester systems, but it has been the development of two-dimensional techniques that has allowed the greatest insight into the microstructure of the cellulosic polymer (116–119). The combination of ^{1}H and ^{13}C nmr spectroscopy has, for example, been applied to cellulose triacetate for determination of its configuration (120), identification of the chemical shifts of ring protons and carbons (117), and determination of the distribution of acetyl groups over C_2, C_3, and C_6 (118,121). More complex experiments such as insensitive nuclei assigned by polarization transfer (INAPT) and nuclear Overhauser exchange spectroscopy (NOESY) have been successfully applied toward the spectral assignments of most common triesters, the results of which are listed in Table 9. With INAPT, the lack of sensitivity that is commonly associated with two-dimensional nmr techniques is generally avoided. This increased sensitivity circumvents the need for ^{13}C enrichment and decreases the demand on instrument time.

Other two-dimensional techniques, such as COSY (122), DEPT (123), HOHAHA, solid state (124) etc. give varying degrees of success when applied to the structure-property relationship of cellulose triesters. The recent application of ^{1}H–^{13}C multiple-bond correlation (HMBC) spectroscopy for the unambiguous assignment of cellulose mixed esters has successfully demonstrated the utility of nmr for the structure elucidation of complex cellulose esters (125). It is this unique ability to provide detailed information on intermolecular interactions of cellulose esters in coatings (119) or in polymeric blends that continues to put nmr spectroscopy well ahead of other analytical techniques.

5.2. Viscosity. The viscosity of cellulose esters, a measure of the degree of polymerization, is determined by the falling-ball method. The time in Saybolt units (SU) required for an aluminum or stainless-steel ball of specified diameter to go through a specified distance in a solution of the cellulose ester is determined. The choice of solvents used to prepare the solutions for viscosity determination depends on the DS of the cellulose ester; the concentration of ester in the

Table 9. **¹H and ¹³C Nmr Chemical Shifts, ppm, and Coupling Constants, J,[a] for Cellulose Triacetate, Cellulose Tripropionate, and Cellulose Tributyrate**[b]

	CTA DMSO-d₆, 25°C	CTA DMSO-d₆,[c]	CTA CDX₃,[d] d	CTP CDX₃,[d] 25°C	CTB CDX₃,[d] 25°C
			¹H Values		
H-1	4.65 (d, $J_{1,2}$ = 7.9 Hz)	4.65 (d, $J_{1,2}$ = 7.9 Hz)	4.42 (d, $J_{1,2}$ = 7.9 Hz)	4.35 (d, $J_{1,2}$ = 7.9 Hz)	4.65 (d, $J_{1,2}$ = 7.9 Hz)
H-2	4.52 (t, J = 7.3 Hz)	4.55 (t, J = 8.6 Hz)	4.79 (t, J = 8.6 Hz)	4.77 (t, J = 8.6 Hz)	4.76 (t, J = 8.6 Hz)
H-3	5.06 (t, J = 9.2 Hz)	5.04 (t, J = 9.2 Hz)	5.07 (t, J = 9.0 Hz)	5.07 (t, J = 9.1 Hz)	5.06 (t, J = 9.2 Hz)
H-4	3.65 (t, J = 9.2 Hz)	3.68 (t, J = 9.2 Hz)	3.71 (t, J = 9.2 Hz)	3.66 (t, J = 9.1 Hz)	3.61 (t, J = 9.2 Hz)
H-5	3.81 (m)	3.77 (m)	3.53 (m)	3.47 (m)	3.48 (m)
H-6$_s$	4.22 (d, $J_{6s,6r}$ = 10 Hz)	4.26 (d, $J_{6s,6r}$ = 10 Hz)	e	e	e
H-6$_r$	3.98 (m)	4.04 (m)	4.06 (m)	4.03 (m)	4.03 (m)
			¹³C Values		
C-1		99.8 (d, J = 167 Hz)	100.4 (d, J = 165 Hz)	100.3 (d, J = 163 Hz)	100.1 (d, J = 163 Hz)
C-2		72.2 (d, J = 152 Hz)	71.7 (d, J = 153 Hz)	71.7 (d, J = 150 Hz)	71.4 (d, J = 150 Hz)
C-3		72.9 (d, J = 151 Hz)	72.5 (d, J = 148 Hz)	72.2 (d, J = 148 Hz)	71.8 (d, J = 147 Hz)
C-4		76.4 (d, J = 151 Hz)	76.0f	75.8 (d, J = 153 Hz)	75.8f
C-5		72.5 (d, J = 146 Hz)	72.7 (d, J = 139 Hz)	73.0 (d, J = 138 Hz)	73.1 (d, J = 143 Hz)
C-6		62.8 (t, J = 151 Hz)	61.9 (t, J = 151 Hz)	61.9 (t, J = 147 Hz)	61.9 (t, J = 145 Hz)
C-2 acetyl			169.1g		
C-3 acetyl			169.6g		
C-6 acetyl			170.7g		

[a] Digital resolution for ¹H was 0.20–0.26 Hz; For additional information see Ref.116. For¹³C the digital resolution was 0.52 Hz.
[b] All solutions are 30 mg/mL.
[c] At 80°C for ¹H; and 90°C for ¹³C.
[d] X = Cl for ¹H; X = I for ¹³C. 25°C
[e] H-6$_r$ overlaps with H-1.
[f] The coupled resonance overlaps with the solvent peaks. For additional information see Refs. 116 and 117.
[g] Doublet, coupling constant unavailable.

solution is normally 20 wt%. Dilute-solution viscosity (intrinsic viscosity) is determined by using a solution of 0.25-g ester dissolved in 100 mL of solvent. The flow time of the solution through a specially designed capillary viscometer is compared to that of the pure solvent. Methods of calculating molecular weight from intrinsic viscosity of cellulose esters have been reported (126).

5.3. Thermal Properties. The thermal stability of cellulose esters is determined by heating a known amount of ester in a test tube at a specific temperature a specified length of time, after which the sample is dissolved in a given amount of solvent and its intrinsic viscosity and solution color are determined. Solution color is determined spectroscopically and is compared to platinum–cobalt standards. Differential thermal analysis (dta) has also been reported as a method for determining the relative heat stability of cellulose esters (127).

The thermal transitions of a cellulose ester such as the glass-transition temperature and the melting point are usually determined by differential scanning calorimetry (dsc) (Fig. 10), which measures the flow of heat into and out of a sample. Generally a first heating run is necessary in order to erase any previous thermal processing characteristics, ie, regions of crystallinity. A slow cool followed by a reheat above the melt temperature provides information on the glass-transition temperature, crystallization temperature (exothermic), and the melting temperature (endothermic). This information is essential for the development of new thermoplastics.

Similar information can be obtained from analysis by dynamic mechanical thermal analysis (dmta). Dmta measures the deformation of a material in response to vibrational forces. The dynamic modulus, the loss modulus, and a mechanical damping are determined from such measurements. Detailed information on the theory of dmta is given (128).

Determination of the thermal decomposition temperature by thermal gravimetric analysis (tga) defines the upper limits of processing. The tga for cellulose

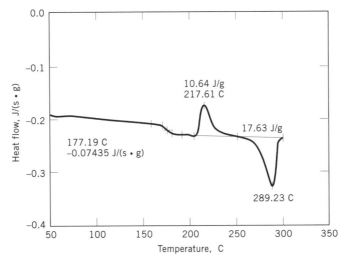

Fig. 10. Differential scanning calorimetry of cellulose triacetate. Second heating at 20°C/min.; glass-transition (T_g) temperature = 177°C; crystallization on heating (T_{ch}) = 217°C; melting temperature (T_m) = 289°C. To convert J to cal, divide by 4.184.

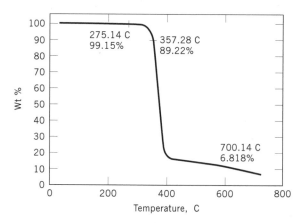

Fig. 11. Thermogravimetric analysis of cellulose triacetate. Method: 20°C/min to 700°C, in (N₂) at 40 mL/min purging rate.

triacetate is shown in Figure 11. Comparing the melt temperature (289°C) from the dsc in Figure 10 to the onset of decomposition in Figure 11 defines the processing temperature window at which the material can successfully be melt extruded or blended.

5.4. Molecular Weight.   The molecular-weight distribution of cellulose esters is normally determined by gel-permeation chromatography (129) in which the ester, dissolved in a suitable solvent, is eluted through a column of porous cross-linked polystyrene. The elution profiles are compared to narrow molecular-weight polystyrene standards to obtain the molecular weight or DP distribution. Other methods, such as fractional precipitation and fractional extraction, although considerably more laborious, may be used to determine DP and DS distribution of cellulose esters (130). The DP of cellulose triacetate has been determined by gpc using a styrene–divinylbenzene copolymer column with an assigned peak molecular weight of approximately 60,000. Other solvents such as THF and NMP have also been shown to solvate secondary cellulose acetates and mixed esters for gpc analysis (131).

6. Health and Safety Factors

The vapors of the organic solvents used in the preparation of cellulose ester solutions represent a potential fire, explosion, or health hazard. Care should be taken to provide adequate ventilation to keep solvent vapor concentrations below the explosive limits. Mixing equipment should be designed to ensure that solvent temperatures do not approach their flash point during the mixing cycle. All equipment must be electrically grounded to prevent static discharge, and appropriate precautions should be followed as recommended by the manufacturer of the solvents.

Mixing cellulose esters in nonpolar hydrocarbons, such as toluene or xylene, may result in static electricity buildup that can cause a flash fire or explosion. When adding cellulose esters to any flammable liquid, an inert gas atmosphere should be maintained within the vessel (132). This risk may be reduced by the use of conductive solvents in combination with the hydrocarbon or by use of an antistatic additive. Protective clothing and devices should be provided.

Cellulose esters, like most dry organic materials in powder form, are capable of creating dust explosions (133). The explosion at Bayer's cellulose acetate plant at Dormagen, Germany in 1976 can attest to the explosive potential of dust. Damage to the plant was estimated at between DM 5–10 million (134).

Cellulose esters are considered nontoxic and may be used in food-contact applications. However, since cellulose esters normally are not used alone, formulators of coatings and films for use in food packaging should ensure that all ingredients in their formulations are cleared by the United States Food and Drug Administration for such use.

7. Uses

The cellulose esters with the largest commercial consumption are cellulose acetate, including cellulose triacetate, cellulose acetate butyrate, and cellulose acetate propionate. Cellulose acetate is used in textile fibers, plastics, film, sheeting, and lacquers. The cellulose acetate used for photographic film base is almost exclusively triacetate; some triacetate is also used for textile fibers because of its crystalline and heat-setting characteristics. The critical properties of cellulose acetate as related to application are given in Table 10.

Large quantities of secondary cellulose acetate are used worldwide in the manufacture of filter material for cigarettes. Because of its excellent clarity and ease of processing, cellulose acetate film is widely used in display packaging and extruded plastic film for decorative signs (see PACKAGING MATERIALS). Injection-molded plastics of cellulose acetate are used in toothbrush handles, computer brushes, and a large variety of other applications (7).

Low viscosity cellulose acetate is used in lacquers and protective coatings for paper, metal, glass, and other substrates and as an adhesive for cellulose photographic film because of its quick bonding rate and excellent bond peel strength (135) (see COATINGS). Heat-sensitive adhesives for textiles have also been prepared from cellulose acetate (136). Extruded cellulose acetate film

Table 10. **Uses and Critical Properties of Cellulose Acetate**[a]

Property	Yarn	Photographic film	Plastics	Lacquers
color, absence of haze	I		C	I
			I	I
false viscosity[c]	I	I		I
filterability	C	I	I	I
adhesion		C		
radioactivecontamination				

[a] Ref. 62.
[b] I = important; C = critical.
[c] Concentrated solution viscosity higher than predicted for the intrinsic viscosity.

makes an excellent base for transparent pressure-sensitive tape (137) (see ADHESIVES).

Cellulose acetate films, specially cast to have a dense surface and a porous substructure, are used in reverse osmosis to purify brackish water (138–141) in hollow fibers for purification of blood (artificial kidney) (142), and for purifying fruit juices (143,144) (see MEMBRANE TECHNOLOGY).

Compaction of cellulose acetate desalination membranes, causing reduction in throughput and performance with time, can be significantly reduced by irrigation grafting of styrene onto the membrane (145).

Eyeglass frames made of cellulose acetate plasticized with diglycerol esters do not exhibit opaqueness at the frame-lens junction with polycarbonate plastic lenses (146,147).

Biodegradable film (148), foam-molding compositions, eg, sponges (149), tobacco substitutes (150), and microencapsulated drug-delivery systems (151) are potentially new and useful applications for cellulose acetate esters.

With the renewed interest in environmentally friendly products, cellulose esters are being re-evaluated as a natural source of biodegradable thermoplastics. Cellulose acetates are potentially biodegradable (152). Films prepared from a cellulose acetate with a DS of 2.5 were shown to require only a 10–12 day incubation period for extensive degradation in an *in vitro* enrichment assay. Similarly, films prepared from a cellulose acetate with a DS of 1.7 saw 70% degradation in 27 days in a waste water treatment facility, whereas films prepared from a cellulose acetate with a DS of 2.5 required approximately 10 weeks for similar degradation to occur. The results of this work demonstrate that cellulose acetate fibers and films are potentially environmentally nonpersistant.

Cellulose acetate propionate and butyrate esters have numerous applications, such as sheeting, molding plastics, film products, lacquer coatings, and melt dip coatings. The properties of propionate and acetate propionate esters ordinarily lie between those of cellulose acetate and acetate butyrate. The acetate propionate mixed esters have traditionally covered a narrow composition range compared to the range of acetate butyrate esters. Table 11 shows uses and a range of commercial compositions of cellulose acetate butyrate esters. Through proper variation of acetyl and butyryl contents, the esters can be adapted to a broad range of applications. Cellulose acetate propionate and acet-

Table 11. **Application and Characteristics of Commercial-Grade Cellulose Acetate Butyrate**[a]

			Degree of esterification[b]			
Acetyl, %	Butyryl, %	Hydroxyl, %	Acetate	Butyrate	Hydroxyl	Application
29.5	17	1	2.1	0.7	0.2	lacquers
20.5	26	2.5	1.4	1.1	0.5	lacquers
13	37	2	0.95	1.65	0.4	plastics, lacquers
6	48	1	0.5	2.3	0.2	melt coatings

[a] Ref. 153.
[b] Ratio of ester groups to glucose residues (see Fig. 6).

ate butyrate are thermoplastic; properly formulated, these esters are processible by methods such as injection molding and extrusion and can be dissolved and cast into films from a variety of solvents (see FILMS). The mixed esters are generally more compatible with various plasticizers and synthetic resins than the acetates, and their films possess excellent clarity and toughness. For example, cellulose acetate butyrate is compatible with polyester, acrylic, vinyl, andalkyd resins (qv), depending on the amount of butyryl substitution and the degree of hydrolysis of the esters.

Cellulose acetate butyrates with high butyryl content and low viscosity are soluble in inexpensive lacquer solvents. They are widely used in lacquers for protective and decorative coatings applied to automobiles and wood furniture.

Higher butyryl esters, formulated with acrylic polymers, provide coatings with excellent weather resistance, good color fastness and dispersibility, and good flow properties (154). Formulations for a typical automotive refinishing lacquer and a wood furniture lacquer are given in Tables 12 and 13, respectively. Low viscosity, high butyryl cellulose esters tolerate substantial amounts of alcohol solvent without appreciable increase in solution viscosity. An alcohol-soluble cellulose acetate butyrate containing ca 50% butyryl and ca 4.5% hydroxy is available commercially.

Low viscosity cellulose propionate butyrate esters containing 3–5% butyryl, 40–50% propionyl, and 2–3% hydroxyl groups have excellent compatibility with oil-modified alkyd resins (qv) and are used in wood furniture coatings (155). Acetate butyrate esters have been used in such varied applications as hot-melt adhesive formulations (156), electrostatically spray-coated powders for fusible, non-cratering coatings on metal surfaces (157–159), contact lenses (qv) with improvedoxygen permeability and excellent wear characteristics (160–162), and as reverse-osmosis membranes for desalination of water (163).

In a relatively new decorative-coating technique called wet-on-wet coatings, cellulose acetate butyrate ester as the pigmented basecoat provides good pigment

Table 12. **CAB–Polyester Automotive Refinishing Lacquer**[a]

Components	Wt %
polyester	11.5
CAB 381-20[b]	11.4
CAB 381-0.5b	5.7
phosphoric acid, 85%	0.3
TiO_2 pigment	4.2
toluene	34.4
xylene	6.1
MIBK	5.0
isopropyl alcohol	9.3
acetone	7.5
n-butyl alcohol	3.2
Ektasolve EB acetate[c]	1.4
Total	*100.0*

[a] Ref. 154.
[b] CAB = cellulose acetate butyrate.
[c] Eastman Kodak Co.

Table 13. **CAB-Based Clear Topcoat Formulation for Wood Furniture**

Components	Wt %
CAB ester[a]	16.1
Unirez 7003 maleic resin[b]	17.4
DOP[c]	4.0
toluene	35.3
isopropyl alcohol	14.0
acetone	4.5
xylene	7.7
SF-69[d]	1.0
Total	*100.0*

[a] CAB = cellulose acetate butyrate.
[b] Union Camp Corp.
[c] DOP = dioctyl phthalate.
[d] Slip aid; 1% xylene (General Electric Co.)

and metal-flake control before applications of the clear topcoat (164,165). Such coatings provide good appearance and excellent resistance to weathering and are expected to find broad use in automotive decorative coatings.

Because of certain properties, such as a high melting point, high tolerance for alcohol solvents, low odor, and excellent surface hardness, cellulose acetate propionates are used in printing inks (flexographic and gravure) (166) (Table 14). Alcohol-soluble cellulose acetate propionate ester tolerates substantial quantities of water in the solvent blend and thus provides an environmentally desirable system for flexographic ink coatings (167).

Acetate propionate esters are nontoxic, exhibit excellent clarity and high tensile strength, and can be formulated into hot-melt dip coatings for food (168). Alternatively, they may be dissolved in volatile solvents and applied to foods in the form of a lacquer coating (169).

Desalination membranes with improved, rigid, and stable surfaces have been prepared from cellulose acetate propionate (170). These films are generally more resistant to hydrolysis than those from cellulose acetate.

Table 14. **Flexographic Ink Formulation Containing Alcohol-Soluble Cellulose Acetate Propionate**

Components	Wt %
cellulose acetate propionate[a]	6.1
sucrose acetate isobutryate (SAIB)	1.5
Kodaflex DOP plasticizer[b]	4.1
Uni-Rez 710 maleic resin[c]	8.2
pigment	5.1
isopropyl alcohol, 99%[a]	56.3
water	18.7
Total	*100.0*

[a] Alcohol-soluble propionate (ASP) CAP 504-0.2.
[b] DOP = dioctyl phthalate (Eastman Kodak Co.)
[c] Union Camp Corp.

Cellulose esters, especially acetate propionate and acetate butyrate mixed esters, have found limited use in a wide variety of specialty applications such as in nonfogging optical sheeting (171), low profile additives to improve the surface characteristics of sheet-molding (SMC) compounds and bulk-molding (BMC) compounds (172,173), and controlled drug release via encapsulation (174).

BIBLIOGRAPHY

"Cellulose Derivatives, Esters" under "Cellulose Derivatives, Plastics" in *ECT* 1st ed., Vol. 3, pp. 391–411, by W. O. Bracken, Hercules Powder Co.; in *ECT* 2nd ed., Vol. 4, pp. 653–683, by B. P. Rousse, Jr., Tennessee Eastman Co.; "Cellulose Derivatives, Esters" in *ECT* 3rd ed., Vol. 5, pp. 118–143, by R. T. Bogan, C. M. Kuo, and R. J. Brewer, Tennessee Eastman Co.

CITED PUBLICATIONS

1. P. Schutzenberger, *Compt. Rend.* **61**, 485 (1865).
2. A. Franchimont, *Compt. Rend.* **89**, 711 (1879).
3. U.S. Pat. 835,350 (Dec. 11, 1906), G. W. Miles.
4. *1982 Directory of Chemical Producers of Western Europe*, Vol. 5, SRI International, Menlo Park, Calif., 1982, p. 1084; *JCW Chemicals Guide 1982/1983*, The Chemical Daily Company, Ltd., Tokyo, Japan, 1982, p. 95; K. Wheeler, W. Cox, and N. Takei, *Cellulose Acetate and Cellulose Ester Plastics*, CEH Marketing Research Report, SRI International, Menlo Park, Calif., Apr. 1989.
5. *Publication No. X-309*, Eastman Chemical Products, Inc., Eastman Kodak Co., Kingsport, Tenn., Aug. 1990.
6. C. J. Malm and co-workers. *Ind. Eng. Chem.* **43**, 688 (1951).
7. C. J. Malm and G. D. Hiatt, in E. Ott and co-eds., *Cellulose and Cellulose Derivatives*, Part II of *High Polymers*, 2nd ed., Vol. 5, John Wiley & Sons, Inc., New York, 1954, p. 766.
8. J. A. Brydson, *Plastic Materials*, D. Van Nostrand Co., Inc., New York, 1966, 369–371.
9. J. J. Creely and co-workers, *J. Appl. Polym. Sci.* **19**, 1533 (1965).
10. D. L. Patel and R. D. Gilbert, *J. Polym. Sci. Polym. Phys. Ed.* **19**, 1449 (1981); B. Yu. Yunusov and co-workers, *Vysokomol. Soedin. Ser. B* **24**, 414 (1982).
11. D. G. Gray and co-workers, *Makromol. Chem.* **184**, 1727 (1983).
12. S. Suto and co-workers, *Rheol. Acta* **21**, 62 (1982).
13. E. D. T. Atkins and co-workers, *TAPPI International Dissolving Pulps Conference, 5th Conference Paper*, Technical Association of the Pulp and Paper Industry, Atlanta, Ga., 1980, 208–213.
14. J. Bheda and co-workers, *Colloid Polym. Sci.* **258**, 1335 (1980).
15. J. Bheda and co-workers, *Technical Paper, Regional Technical Conference of the SPE, May 5–8, 1980*, The Society of Plastics Engineers, Brookfield Center, Conn., 1980, p. 321.
16. Jpn. Kokai Tokkyo Koho JP 82 57729 (Apr. 7, 1982) (to Asahi Chemical Industry Co., Ltd.).
17. J. Bheda and co-workers, *J. Appl. Polym. Sci.* **26**, 3955 (1981).
18. C. J. Malm, *Sven. Kem. Tidskr.* **73**, 523 (1961).

19. C. J. Malm and co-workers, *Ind. Eng. Chem.* **50**, 1961 (1958).
20. G. Tocco, *G. Chim. Ind. Appl.* **13**, 325 (1931).
21. U.S. Pat. 2,208,569 (July 23, 1940), L. W. Blanchard, Jr. (to Eastman Kodak Co.).
22. U.S. Pat. 2,790,794 (Apr. 30, 1957), C. J. Malm and co-workers (to Eastman Kodak Co.).
23. N. I. Klenhova, *Zh. Prikl. Khim. (Leningrad)* **49**, 2701 (1976).
24. U.S. Pat. Appl. No. 495,186 (June 1989), C. M. Buchanan (to Eastman Kodak Co.).
25. J. W. Mench and co-workers, *Ind. Eng. Chem.* **5**, 110 (1966).
26. U.S. Pat. 3,103,506 (Sept. 10, 1963), C. J. Malm and co-workers (to Eastman Kodak Co.).
27. A. Wohl, *Z. Agnew. Chem.* **26**, 437 (1913).
28. K. Atsuji and K. Shimoyama, *Cellul. Ind. Tokyo* **2**, 336 (1926).
29. U.S. Pat. 1,704,283 (Mar. 25, 1929), H. T. Clarke and C. J. Malm (to Eastman Kodak Co.).
30. H. A. Khidoyatov and Z. A. Rogovin, *Polym. Sci. USSR* **11**, 2123 (1969).
31. G. D. Hiatt and co-workers, *Ind. Eng. Chem.* **3**, 295 (1964).
32. E. I. Berenshtein and co-workers, *Strukt. Modif. Khlopk. Tsellyul.* **4**, 91 (1969).
33. Russ. Pat. 659,574 (Apr. 30, 1979), Kh. U. Usmanov and co-workers.
34. R. S. Alimardanov and co-workers, *Azerb. Khim. Zh.* **6**, 105 (1980).
35. J. W. Mench and B. Fulkerson, *Ind. Eng. Chem.* **7**, 2 (1968).
36. S. H. Wu and co-workers, in D. H. Lewis, ed., *Controlled Release Pesticides and Pharmaceuticals, Proceedings of 7th International Symposium (1980)*, Plenum Publishing Corp., New York, 1981, 319–331.
37. Def. Publ. U.S. Pat. Off. T100,404 (Mar. 3, 1981), S. H. Wu (to Eastman Kodak Co.).
38. U.S. Pat. 2,824,098 (Feb. 18, 1958), F. M. Volberg and M. D. Martin (to Eastman Kodak Co.).
39. K. Kenichiro and Y. Ogiwara, *Sen-i Gakkaishi* **36**, 82 (1980).
40. E. P. Grishin and co-workers, *Khim. Tekhnol. Proizvod. Tsellyul.* **98** (1971).
41. U.S. Pat. 2,759,925 (Aug. 21, 1956), G. D. Hiatt and co-workers (to Eastman Kodak Co.).
42. C. J. Malm and co-workers, *Ind. Eng. Chem.* **49**, 84 (1957).
43. V. N. Kryazhev and co-workers, *Plast. Massy* **10**, 12 (1970).
44. U.S. Pat. 2,828,303 (Mar. 25, 1958), C. J. Malm and L. W. Blanchard, Jr. (to Eastman Kodak Co.).
45. U.S. Pat. 2,828,304 (Mar. 25, 1958), C. J. Malm and L. W. Blanchard, Jr. (to Eastman Kodak Co.).
46. L. V. Gurkovskoya and co-workers, *Khim. Tekhnol. Proizvod. Tsellyul.* **93** (1971).
47. C. J. Malm and co-workers, *Ind. Eng. Chem.* **5**, 81 (1966).
48. U.S. Pat. 2,801,240 (July 30, 1957), C. J. Malm and L. J. Tanghe (to Eastman Kodak Co.).
49. U.S. Pat. 2,816,106 (Dec. 10, 1957), C. J. Malm and co-workers (to Eastman Kodak Co.).
50. V. N. Kryazhev and co-workers, *Khim. Tekhnol. Vysokomol. Soedin.* **42** (1969).
51. B. P. Smirnov and co-workers, *Sov. Plast.* **1**, 71 (1972).
52. Russ. Pat. 458,560 (Jan. 30, 1975), A. Maciulis and co-workers.
53. U.S. Pat. 3,723,147 (Mar. 27, 1973), E. L. Wood and R. E. Gibson (to Eastman Kodak Co.).
54. U.S. Pat. 4,137,201 (Jan. 30, 1979), C. M. Kuo and co-workers (to Eastman Kodak Co.).
55. A. Z. Tatarnova and co-workers, *Sov. Plast.* **9**, 11 (1969).
56. U.S. Pat. 4,325,997 (Apr. 20, 1982), R. T. Bogan and R. J. Brewer (to Eastman Kodak Co.).

57. L. M. Malinin and K. F. Yakunina, *Plast. Massy* **47** (1966).
58. V. Shlyapintokh and co-workers, *Kunstst. Fortschrittsberg* **1**, 25 (1976).
59. D. Kho Khalikov and co-workers, *Vysokomol. Soedin.* **19**, 1132 (1977).
60. U.S. Pat. 4,269,629 (May 16, 1981), J. S. Zannucci (to Eastman Kodak Co.).
61. Ger. Offen. 3,007,797 (Sept. 25, 1980), R. J. Brewer and R. T. Bogan (to Eastman Kodak Co.).
62. G. D. Hiatt and W. J. Rebel, in N. M. Bikales and L. Segal, eds., *Cellulose and Cellulose Derivatives*, Part V of *High Polymers*, 2nd ed., Vol. 5, Wiley-Interscience, New York, p. 749.
63. R. B. Seymour and E. L. Johnson, *J. Polym. Sci. Polym. Chem. Ed.* **16**, 1 (1978).
64. R. Leoni and A. Baldini, *Carbohydr. Polym.* **2**(4), 298 (1982).
65. C. L. McCormick and T. S. Chen, in R. Seymour and co-eds., *Macromolecular Solutions: Solvent-Property Related Polymers*, Pergamon Press, Inc., Elmsford, N.Y., 1982, p. 10.
66. P. Mansson and L. Westfelt, *Cellu. Chem. Technol.* **14**(1), 13 (1980).
67. Ref. 59, p. 748.
68. Russ. Pat. 319,227 (Dec. 5, 1975), F. M. Mikhal'skii and co-workers.
69. Ger. Offen. 2,135,735 (Jan. 25, 1973), F. M. Mikhal'skii and co-workers.
70. L. Makova and co-workers, *Zh. Prikl. Khim (Leningrad)* **47**, 610 (1974).
71. A. Koura, G. Faserforsch. *Textiltech.* **29**(6), 414 (1978).
72. Russ. Pat. 479,780 (Aug. 5, 1975), V. J. Sharkov and M. I. Perminova.
73. S. A. Kadyrova and co-workers, *Dokl. Akad. Nauk Uzb. USSR* **26**(10), 29 (1969).
74. U.S. Pat. 4,336,370 (June 22, 1982), V. M. Yasnovaky and D. M. MacDonald (to Inter. Paper Co.).
75. V. V. Safonova and N. I. Klenkova, *Zh. Prikl. Khim. (Leningrad)* **42**, 2636 (1969).
76. K. D. Sears and co-workers, *J. Appl. Polym. Sci.* **27**, 4599 (1982).
77. C. J. Malm and co-workers, *Ind. Eng. Chem.* **38**, 77 (1946).
78. A. J. Rosenthal, *Pure Appl. Chem.* **14**, 535 (1967).
79. L. J. Tanghe and R. J. Brewer, *Anal. Chem.* **40**, 350 (1968).
80. Jpn. Kokai Tokkyo Koho 78 101,083 (Feb. 16, 1978), M. Mishino and co-workers (to Chemical Industry Co., Ltd.).
81. A. Takahashi and S. Takarashi, *Kobunshi Kagaku* **27**, 394 (1970).
82. *Ibid.*, **26**, 485 (1969).
83. U.S. Pat. 4,314,056 (Feb. 2, 1982), R. J. Brewer and B. S. Wininger (to Eastman Kodak Co.).
84. U.S. Pat. 4,329,447 (May 11, 1982), R. J. Brewer and B. S. Wininger (to Eastman Kodak Co.).
85. Ref. 62, p. 756.
86. C. J. Malm and co-workers, *J. Am. Chem. Soc.* **72**, 2674 (1950).
87. U.S. Pat. 2,836,590 (May 27, 1958), H. W. Turner (to Hercules Co.).
88. U.S. Pat. 2,790,796 (Apr. 30, 1957), J. Robin and R. Clevy (to Société Rhodiaceta).
89. V. E. Sabinin and co-workers, *Zh. Prikl. Khim. (Leningrad)* **42**, 1638 (1969).
90. U.S. Pat. 4,228,276 (Oct. 14, 1980), C. M. Kuo and R. T. Bogan (to Eastman Kodak Co.).
91. U.S. Pat. 3,414,640 (Dec. 3, 1968), G. Giuseppe and co-workers (to Rhodiatoce SpA).
92. U.S. Pat. 2,996,485 (Aug. 15, 1961), K. C. Laughlin and co-workers (to Celanese Corp.).
93. Can. Pat. 609,900 (1960), K. C. Laughlin and co-workers (to Celanese Corp.).
94. U.S. Pat. 3,040,027 (Aug. 15, 1961), H. Bates and co-workers (to British Celanese, Ltd.).
95. H. Genevray and J. Robin, *Pure Appl. Chem.* **14**, 489 (1967).

96. U.S. Pat. 3,767,642 (Oct. 23, 1973), K. C. Campbell and co-workers (to Celanese Corp.).

97. Can. Pat. 973,174 (Aug. 19, 1975), K. B. Gibney and co-workers (to Canadian Cellulose Co., Ltd.).

98. Can. Pat. 975,764 (Oct. 7, 1975), K. B. Gibney and co-workers (to Canadian Cellulose Co., Ltd.).

99. U.S. Pat. 3,870,703 (Mar. 11, 1975), K. B. Gibney and co-workers (to Canadian Cellulose Co., Ltd.).

100. U.S. Pat. 4,234,719 (Nov. 18, 1980), C. G. Wan (to Halcon Research and Development Corp.).

101. U.S. Pat. 4,234,718 (Nov. 18, 1980), D. Brown (to Halcon Research and Development Corp.).

102. Jpn. Kokai Tokkyo Koho 81 59,901 (May 23, 1981), I. Yoshiyuki (to Daicel Chemical Industries, Ltd.).

103. U.S. Pat. 4,306,060 (Dec. 15, 1981), I. Yoshiyuki (to Daicel Chemical Industries, Ltd.).

104. K. Wheeler, W. Cox, and N. Takei, *Cellulose Acetate and Cellulose Ester Plastics*, CEH Marketing Research Report, SRI International, Menlo Park, Calif., Apr. 1989.

105. *Chem Mark. Rep.* **241**(12), 42 (Mar. 23, 1992).

106. *Chem Eng. News* **68**(2), 11 (Oct. 15, 1990).

107. *Chem Eng. News* **69**(49), 8 (Dec. 9, 1991); *J. Commer.* 7A (Dec. 6, 1991); *Chem. Mark Rep.* **240**(24), 3, 14 (Dec. 9, 1991); *Eur. Chem.* (31), 9 (July 30, 1990).

108. *Chem Mark. Rep.* **234**(9), 4 (Aug. 29, 1988).

109. *Eur. Past News* **18**(2), 6 (Mar. 1991).

110. ASTM D817-72, *Annual Book of ASTM Standards*, Pt. 06.02, American Society of Testing and Materials, Easton, Md., 1983, 198–217.

111. R. L. Jackson, *Tappi* **51**, 560 (1968).

112. V. W. Goodlett and co-workers, *J. Polym. Sci. Part A-1* **9**, 155 (1971).

113. E. Isobe and T. Nakajima, *Sen'i Gakkaishi* **31**, T-101 (1975).

114. F. A. Bowey, *High Resolution NMR of Macromolecules*, Academic Press, Inc., New York, 1972; K. Wutherich, *NMR in Biological Research: Peptides and Proteins*, American Elsevier Publishing, New York, 1976.

115. D. Horton and J. H. Lauterbach, *Carbohyd. Res.* **43**, 9–33 (1975).

116. C. M. Buchanan, J. A. Hyatt, and D. W. Lowman, *J. Am. Chem. Soc.* **111**, 7312–7319 (1989).

117. C. M. Buchanan, J. A. Hyatt, and D. W. Lowman, *Macromolecules* **20**, 2750–2754 (1987).

118. C. M. Buchanan, J. A. Hyatt, and D. W. Lowman, *Carbohydr. Res.* **177**, 228–234 (1988).

119. K. Kamide, K. Okajima, and M. Saito, *Polym. J.* **13**, 115–125 (1981); K. Kamide, K. Okajima, K. Kowsaka, and M. Saito, *Polym. J.* **19**, 1337–1383 (1987); K. Kamide, K. Okajima, and M. Saito, *Polym. J.* **20**, 1091–1099 (1988).

120. R. U. Lemiux and J. D. Steven, *Can. J. Chem.* **43**, 1059 (1965).

121. D. Gagnaire and M. Vincendon, *Bull. Soc. Chim. Fr.*, 204 (1966); V. W. Goodlett, J. T. Dougherty, and H. W. Patton, *J. Polym. Sci. A-1* **9**, 155 (1971).

122. R. A. Newmark, *Appl. Spectrosc.* **39**, 507–512 (1985).

123. M. Takai, K. Fukuda, and J. Hayashi, *J. Polym. Sco. Part C: Polym. Lett.* **25**, 121–126 (1987).

124. S. Doyle and co-workers, *Polymer* **27**, 19–24 (1986).

125. T. Iwata, J. I. Azuma, K. Okamura, M. Muraoto, and B. Chun, *Carbohydr. Res.* **224**, 277–283 (1992).

126. L. J. Tanghe and co-workers, in R. L. Whistler, ed., *Methods in Carbohydrate Chemistry*, Vol. III, Academic Press, Inc., New York, 1963, p. 210; L. B. Genung, *Anal. Chem.* **36**, 1817 (1964).

127. A. S. Buntyakov and co-workers, *Plast. Massy* **3**, 71 (1969).

128. T. Murayama, *Dynamic Mechanical Analysis of Polymeric Materials*, Elsevier Scientific Publishing Co., New York, 1978.

129. R. J. Brewer and co-workers, *J. Polym. Sci. Part A-1* **6**, 1697 (1968).

130. R. E. Glegg and co-workers, in Ref. 62, Part IV, 491–517.

131. J. F. Kennedy, G. O. Phillips, and P. A. Williams, eds., *Wood Cellulose*, Horwood, Chichester, UK, 203–210 (1987).

132. *Standard on Explosion Prevention Systems, NFPA 69*, and *Static Electricity, NFPA 77*, National Fire Protection Association, Quincy, Me., 1984.

133. *Prevention of Fire and Dust Explosions in the Chemical, Dye, Pharmaceutical, and Plastics Industry, NFPA 654*, National Fire Protection Association, Quincy, Me., 1982.

134. *Chem. Age (London)*, 4 (Oct. 1, 1976); *Eur. Chem. News* **10**, 4 (Sept. 1976).

135. Ger. Offen. 2,104,032 (Aug. 5, 1971), W. Ueno and N. Minagaiwa (to Fuji Photo Film, Ltd.); Brit. Pat. 1,352,605 (May 9, 1974) (to Fuji Photo Film, Ltd.).

136. Brit. Pat. 1,197,570 (July 8, 1970), J. A. Smith and co-workers (to Courtaulds, Ltd.).

137. B. Wright, *Adhes. Age* **14**, 25 (1971).

138. P. Aptel and I. Cabasso, *Desalination* **36**, 25 (1981).

139. R. E. Kesting and co-workers, *Proc. Int. Symp. Fresh Water Sea* **4**, 73 (1976).

140. R. D. Ammons, *Gov. Rep. Announce. U.S.* **79**, 93 (1978).

141. U.S. Pat. 4,145,295 (Mar. 20, 1979), S. Sourirajan and co-workers (to Canadian Patents and Development, Ltd.).

142. Jpn. Kokai Tokkyo Koho JP 82 119,809 (July 26, 1982) (to Daicel Chemical Industries, Ltd.); Ger. Offen. 2,619,250 (Nov. 11, 1976), R. E. Kesting (to Chemical Systems, Inc.).

143. R. L. Merson and co-workers, *Polym. Sci. Technol.* **13**, 405 (1980).

144. A. G. Baxter and co-workers, *Chem. Eng. Commun.* **4**, 471 (1980).

145. H. B. Hopfenberg and co-workers, *Appl. Polym. Symp.* **13**, 139 (1970).

146. Jpn. Kokai 78 58,559 (May 26, 1978), K. Wanatabe and co-workers (to Daicel, Ltd.).

147. Jpn. Kokai Tokkyo Koho JP 79 08,654 (Jan. 23, 1979), M. Hirotake and co-workers (to Daicel Chemical Industries, Ltd.).

148. V. T. Stannett and co-workers, *J. Macromol. Sci. Chem.* **16**, 473 (1981); *J. Polym. Sci. Polym. Lett. Ed.* **11**, 731 (1973).

149. Fr. Demande 2,140,454 (Feb. 23, 1973), W. Fischer (to Bayer AG).

150. Jpn. Kokai 77 79,096 (July 2, 1977), T. Yokota and co-workers (to Daicel Chemical Industries, Ltd.).

151. M. Singh and co-workers, *Makromol. Chem.* **183**, 1897 (1982).

152. C. M. Buchanan and R. M. Gardner, *Cellulose '91*, New Orleans, 1991, p. 228.

153. K. J. Saunders, *Organic Polymer Chemistry*, Chapman and Hall, London, 1973, p. 265.

154. R. L. Smith, *Paint Varn. Prod. Mngr.* **59**, 53 (1969).

155. U.S. Pat. 4,166,809 (Sept. 4, 1979), R. J. Brewer and W. C. Wooten (to Eastman Kodak Co.).

156. Def. Publ. U.S. Pat. Off. T944,005 (Mar. 2, 1976), R. J. Brewer and co-workers (to Eastman Kodak Co.).

157. R. W. Buchanan, *SME Technical Paper FC74-576*, Sept. 17–19, 1974.

158. D. Russell, *Met. Finish.* **74**, 32 (1976).

159. U.S. Pat. 4,133,783 (Jan. 9, 1979), R. J. Brewer and co-workers (to Eastman Kodak Co.).

160. Ger. Offen. 2,856,891 (July 17, 1980), F. Wingler and co-workers (to Farbenfabriken Bayer AG).
161. U.S. Pat. 4,116,549 (Jan. 12, 1976), J. E. Harris and B. D. Parish (to Continuous Curve Contact Lenses, Inc.).
162. Brit. Pat. Appl. GB 1,593,553 (July 15, 1981), S. Loshack and C. M. Shen (to Wesley-J Inc.).
163. H. Ohya and co-workers, *J. Appl. Polym. Sci.* **24**, 663 (1979).
164. K. Walker, *Farbe & Lack* **87**, 198 (1981).
165. K. Walker, *Double Liaison-Chim. Print.* **27**, 258 (1980).
166. C. H. Coney and G. B. Bowen, *Am. Inkmaker* **51** (20), 24 (1973).
167. R. L. Baticle and co-workers, *FATIPEC Fed. Assoc. Techn. Ind. Peint Vernis Emaux Impr. Eur. Cont. Congr.* **12**, 437 (1974).
168. U.S. Pat. 3,313,639 (Apr. 11, 1967), F. M. Ball and J. H. Davis (to Eastman Kodak Co.).
169. Ger. Offen. 2,412,426 (Sept. 18, 1975), M. Stemmler and H. Stemmler (to M&H Stemmler G).
170. S. Sourirajan and co-workers, *Can. J. Chem. Eng.* **54**, 364 (1976).
171. Ger. Offen. 3,148,008 (Apr. 5, 1973), K. Landt and P. Neuber (to Winter-Optik GmbH).
172. W. W. Blount, *Mod. Plast.* **49**, 68 (1972).
173. W. W. Blount and co-workers, *SPI Reinforced Plastics Composites Institute, Proceedings 27th Annual Conference*, Feb. 8–11, Society of the Plastic Industry, New York, 1972, Pt. 12-C.
174. D. L. Gardner and co-workers, *Am. Chem. Soc. Div. Org. Coat. Plast. Chem. Pap.* **36**, 362 (1976).

GENERAL REFERENCES

R. T. Bogan and R. J. Brewer, in J. I. Kroschwitz, ed., *Encyclopedia of Polymer Science and Engineering*, 2nd ed., Vol. 3, John Wiley & Sons, Inc., New York, 1985, 158–181.
For a review of cellulose and wood chemistry see: J. F. Kennedy, G. O. Phillips, and P. A. Williams, eds., *Cellulose: Structure and Functional Aspects*, John Wiley & Sons, Inc., New York, 1989.
J. F. Kennedy, G. O. Phillips, and P. A. Williams, eds., *Wood and Cellulose: Industrial Utilizations, Biotechnology, Structure, and Properties*, Ellis Horwood Limited, Chichester, UK, 1987
R. A. Young and R. M. Rowell, *Cellulose: Structure, Modifications, and Hydrolysis*, Wiley-Interscience, New York, 1986.
M. Yalpani, *Studies in Organic Chemistry 36: Polysaccharides: Synthesis, Modification, and Structure/Property Relations*, Elsevier, New York, 1988.
E. Sjöström, *Wood Chemistry: Fundamentals and Applications*, Academic Press, Inc., New York, 1981.
J. F. Kennedy, *Carbohydrate Chemistry*, Oxford Science Publication, New York, 1988.

STEVEN GEDON
RICHARD FENGI
Eastman Chemical Company

CELLULOSE ETHERS

1. General Considerations

Alkylation of cellulose yields a class of polymers generally termed cellulose ethers. Most of the commercially important ethers are water-soluble and are key adjuvants in many water-based formulations. The most important property these polymers provide to formulations is rheology control, ie, thickening and modulation of flow behavior. Other useful properties include water-binding (absorbency, retention), colloid and suspension stabilization, film formation, lubrication, and gelation. As a result of these properties, cellulose ethers have permeated a broad range of industries including foods, coatings, oil recovery, cosmetics, personal care products, pharmaceuticals, adhesives, printing, ceramics, textiles, building materials, paper, and agriculture.

Estimates from SRI International figures indicate that the world consumption, excluding eastern European countries and the former Soviet Union, of all grades of cellulose ethers in 1987 was about 230,000 t (1). Cellulose ethers represent a mature industry with annual sales of over one billion dollars and annual growth rate averaging 2–3% per year. The U.S. consumption in 1987 was approximately 69,000 t, Western Europe about 116,000 t, and Japan about 22,000 t. Prices for the principal products range from about $1.65/kg for crude grades of sodium carboxymethylcellulose to over $11/kg for purified hydroxypropylcellulose. The highest volume cellulose ethers, the industry workhorses, are sodium carboxymethylcellulose, hydroxyethylcellulose, and hydroxypropylmethylcellulose. Cellulose ethers as a class compete with a host of other materials including natural gums, starches, proteins (qv), synthetic polymers, and even inorganic clays (see CARBOHYDRATES; GUMS). They provide effective performance at reasonable cost and are derived from a renewable, natural resource.

Cellulose ethers are manufactured by reaction of purified cellulose with alkylating reagents under heterogeneous conditions, usually in the presence of a base, typically sodium hydroxide, and an inert diluent. Cottonseed linter fiber and wood fiber are the principal sources for cellulose. Purified cellulose cotton linters, commonly termed chemical cottons, are generally of higher purity and higher maximum molecular weight than purified cellulose from dissolving grades of wood pulp. The base, in combination with water, activates the cellulose matrix by disrupting hydrogen-bonded crystalline domains, thereby increasing accessibility to the alkylating reagent. This activated matrix is commonly termed alkali cellulose (2–6). The base also promotes the etherification reaction. The several purposes of the inert diluent are to suspend/disperse the cellulose, provide heat transfer, moderate reaction kinetics, and facilitate recovery of the product. Crude grades of cellulose ethers, most notably sodium carboxymethylcellulose, may be made in the absence of any diluent. Reactions are typically conducted at elevated temperature, ~ 50 to $140°C$, and under nitrogen to inhibit oxidative molecular weight degradation of the polymer (7,8), if so desired. After reaction, crude grades are simply dried, ground, and packed out; purified grades require removal of byproducts in a separate operation prior to drying. Various additives, such as colloidal silicas, may be added in small amounts to some

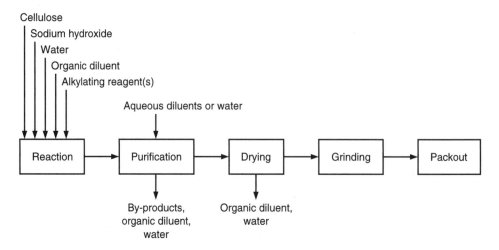

Fig. 1.   Unit operations for the manufacture of purified cellulose ethers.

products prior to drying or before packout to improve dry handling properties. In addition to these unit operations, schematically outlined in Figure 1, a molecular weight reduction operation may be included in the process at any of several points, most notably either in the reaction vessel (before, during, or after the alkylation reaction), after purification before drying, or treatment of dry material (9–11). Hydrogen peroxide is commonly used, though deliberate degradation may also be induced through controlled alkaline-catalyzed autoxidation with oxygen (air) in the reactor. Acids can also be used to cleave the cellulose chain (12,13).

An important characterization parameter for cellulose ethers, in addition to the chemical nature of the substituent, is the extent of substitution. As the Haworth representation of the cellulose polymer shows, it is a linear, unbranched polysaccharide composed of glucopyranose (anhydroglucose) monosaccharide units linked through their 1,4 positions by the β anomeric configuration.

The structurally similar starch amylose polymer is linked through the α anomeric configuration. The three hydroxyl functions per anhydroglucose unit are noteworthy; these hydroxyls are the active sites for ether formation.

The extent of substitution is described as the degree of substitution (DS), defined as the average number of hydroxyl groups substituted per anhydroglucose unit. Excluding the terminal residues, each anhydroglucose moiety has three available hydroxyl groups for a maximum DS value of three. In certain cases the alkylating reagent, such as an alkylene oxide, generates a new hydroxyl group upon reacting which can then further react to give oligomeric chains. The product is then characterized by its molar substitution (MS), the moles of reagent combined per mole of anhydroglucose unit. The ratio of MS to DS is a measure of the average chain length of the oligomeric side chains. Organosoluble ethylcellulose has a DS of 2.3–2.8. Most water-soluble derivatives have DS values of 0.4–2.0. Hydroxyalkyl ethers have MS values typically between 1.5 and 4.0.

MS and DS values are average values, placing no significance on the formal distribution of the substituents within or among polymer chains. Substituent distribution, however, is an important molecular parameter affecting solution rheology and ultimately end use properties. A classic example is offered by sodium carboxymethylcellulose (14). Depending on reaction conditions, sodium carboxymethylcellulose (CMC) of DS ~0.80 can be made to exhibit a varying degree of solution thixotropy, the effect being dependent not so much on degree of substitution but rather on the distribution of substituents.

Thixotropic solutions are characterized by a decrease in viscosity on shearing followed by a time-dependent increase after the shear stress is removed. A plot of shear rate versus shear stress reveals a hysteresis loop. Association among polymer chains via electrostatic, hydrogen-bonding, or hydrophobic effects can lead to thixotropy. Carboxymethyl substituent uniformity along the polymer chain affects thixotropic behavior of CMC because regions of contiguous unsubstituted anhydroglucose units, ie, blocks, among polymer chains tend to associate through hydrogen bonding (15). The extent of blockiness is controlled by reaction conditions. Solution thixotropy is important in applications requiring suspension or stabilization of particulates.

As substituent uniformity is increased, either by choosing appropriate reaction conditions or by reaction to high degrees of substitution, thixotropic behavior decreases. CMCs of DS $\geq$ 1.0 generally exhibit pseudoplastic rather than thixotropic rheology. Pseudoplastic solutions also decrease in viscosity under shear but recover instantaneously after the shear stress is removed. A plot of shear rate versus shear stress does not show a hysteresis loop.

Other examples illustrating the effect of substituent distribution on properties include: (1) enzymatic stability of hydroxyethylcellulose (16,17); (2) salt compatibility of carboxymethylcellulose (18,19); and (3) thermal gelation properties of methylcellulose (20). The enzymatic stability of hydroxyethylcellulose is an example where the actual position of the substituents within the anhydroglucose units is considered important. Increasing substitution at the C2 position promotes better resistance toward enzymatic cleavage of the polymer chain. Positional distribution is also a factor in the other two examples.

[13]C-nmr is the premier method for the compositional and structural characterization of cellulose ethers (21–27). Another analytical method is based on chromatography of hydrolyzates (28). The results show that the reactivity of the three hydroxyl groups can vary significantly depending on the alkylating reagent, the type of reaction, and reaction conditions. For most cellulose ethers, substitution occurs primarily at the C2 and C6 hydroxyl groups (29).

Aside from the chemical nature of the substituent and its DS (MS) and distribution, solution viscosity, ie, polymer molecular weight, is another important characterization parameter. Generally, manufacturers supply cellulose ethers in different viscosity grades, made by choosing a cellulose furnish of appropriate molecular weight or by reducing the molecular weight during processing (9–11). Most manufacturers specify solution viscosity data and not polymer molecular weight in their technical literature; however, molecular weight can be estimated from the classical Mark-Houwink equation:

$$[\eta] = KM^a$$

where $[\eta]$ is intrinsic viscosity, K and a are empirically derived constants, and M is the viscosity-average molecular weight, which approximates the weight-average molecular weight. Values of K and a for various cellulose ethers are available (30–32).

Molecular weight from solution viscosity measurements represents an average value of polymer chain lengths. The distribution of chain lengths making up a polymer fraction is commonly termed molecular weight distribution (MWD), a parameter that can influence performance properties, particularly mechanical properties of films. Size exclusion chromatography (sec), also termed gel-permeation chromatography (gpc), has been the classical method for determining MWD. However, the lack of suitable columns hindered development of aqueous sec methods for water-soluble polymers until the 1980s. Now, because of the development of high performance packings and columns, the MWD of water-soluble cellulose ethers is a measurable parameter (33–36).

Many cellulose ethers contain mixed substituents (cellulose mixed ethers) in order to enhance or modify the properties of the monosubstituted derivative. For example, incorporation of low levels of hydroxypropyl or hydroxyethyl groups into methylcellulose increases its thermal gelation and flocculation temperatures in aqueous media. Carboxymethylation of hydroxyethylcellulose produces a product having excellent tolerance to mono- and divalent metal ions in solution but which readily cross-links with tri- or tetravalent ions to give highly viscoelastic gels. The solubility of ethylcellulose in organic aliphatic solvents is improved by incorporating hydroxyethyl moieties. A fairly new commercial cellulose ether composition is a mixed ether, hydroxyethylcellulose modified with hydrophobic long-chain hydrocarbyl groups (37–39). The hydrophobic groups promote polymer chain association in solution, drastically altering rheology and surface activity properties.

1.1. Health and Safety Factors.   No adverse toxicological or environmental factors are reported for cellulose ethers in general (14,39–50). Some are even approved as direct food additives, including purified carboxymethylcellulose, methylcellulose, hydroxypropylmethylcellulose, and hydroxypropylcellulose.

The only known hazard associated with cellulose ethers is that they may form flammable dusts when finely divided and suspended in air, a hazard associated with most organic substances. An explosion may result if suspended dust contacts an ignition source. Cloud and layer ignition temperatures generally vary between 290 and 410°C (51). Critical air-borne concentrations vary depending on particle size. This hazard can be minimized largely through good housekeeping and proper design and operation of handling equipment.

Another minor hazard is that water-soluble cellulose ether powders form a slippery surface when wet; therefore, spills should be cleaned promptly to avoid slipping accidents.

2. Commercial Cellulose Ethers

2.1. Sodium Carboxymethylcellulose. *Properties.* Sodium carboxymethylcellulose [9004-32-4] (CMC), also known as cellulose gum, is an anionic, water-soluble cellulose ether, available in a wide range of substitution. The most widely used types are in the 0.7 to 1.2 DS range. Water solubility is achieved as the DS approaches 0.6; as the DS increases, solubility increases. The rate at which CMC dissolves depends primarily on its particle size. Finely ground material dissolves faster than coarser grades. The coarse material, however, does not agglomerate as readily when added to water and is therefore easier to disperse. The rate of dissolution also increases with increasing substitution and decreasing molecular weight, ie, viscosity. High molecular-weight grades of CMC have viscosities as high as 12,000 mPa·s(=cP) at 1% solids (as recorded on a Brookfield LVT Viscometer at 30 rpm). Lower molecular-weight CMCs have viscosities in water as low as 50 mPa·s(=cP) at 4% solids.

CMC is soluble in hot and cold water. Solutions may be pseudoplastic or thixotropic depending on molecular weight, DS, and manufacturing process. High molecular-weight, low DS CMCs tend to be more thixotropic. Solutions are viscosity stable at ambient temperature over a wide range of pH. In general, maximum solution viscosity and best stability are obtained at pH 7 to 9. Above pH 10, a slight viscosity decrease is observed. As pH is lowered below 4, viscosity may first increase then decrease as intermolecular associations among free acid groups start affecting solubility. CMC is not soluble in organic solvents, but dissolves in mixtures of water and water-miscible solvents such as ethanol or acetone. Low viscosity CMCs are more tolerant of higher levels of organic solvents.

Monovalent cations are compatible with CMC and have little effect on solution properties when added in moderate amounts. An exception is silver ion, which precipitates CMC. Divalent cations show borderline behavior and trivalent cations form insoluble salts or gels. The effects vary with the specific cation and counterion, pH, DS, and manner in which the CMC and salt are brought into contact. High DS (0.9–1.2) CMCs are more tolerant of monovalent salts than lower DS types, and CMC in solution tolerates higher quantities of added salt than dry CMC added to a brine solution.

CMC is compatible with most water-soluble nonionic gums over a wide range of concentrations. When a solution of CMC is blended with a solution of a nonionic polymer such as hydroxyethylcellulose or hydroxypropylcellulose, a synergistic effect on viscosity is usually observed. Such blends produce solution viscosities considerably higher than would ordinarily be expected. This effect is reduced if other electrolytes are present in the system.

Some typical properties of commercial CMCs are given in Table 1.

Manufacture. Common to all manufacturing processes for CMC is the reaction of sodium chloroacetate [3926-62-3] with alkali cellulose complex

Table 1. **Typical Properties of Purified CMC**^{a}

Property	Value
Powder	
appearance	white to off-white
assay, dry basis, min %	99.5
moisture, max %	8.0
browning temp, °C	227
charring temp, °C	252
bulk density, g/cm^3	0.75
molecular weight, M_w	$9.0 \times 10^4 - 7.0 \times 10^5$
Solution	
viscosity, Brookfield, 30 rpm, mPa·s(=cP)	
at 1% solids (high M_w)	~6000
at 4% solids (low M_w)	~50
sp gr, 2% at 25°C	1.0068
pH, 2%	7.5
surface tension, 1%, mN/m(= dyn/cm)	71
refractive index, 2% at 25°C	1.3355
Film	
refractive index	1.515

^{a}Ref. 14.

represented here as R_{cell} OH:NaOH:

$$R_{cell}OH + NaOH + H_2O \longrightarrow R_{cell}OH : NaOH$$
$$R_{cell}OH : NaOH + ClCH_2COO^-Na^+ \longrightarrow R_{cell}OCH_2COO^-Na^+ + NaCl + H_2O$$
$$\text{sodium carboxymethylcellulose}$$

A by-product is sodium glycolate [2836-32-0] (sodium hydroxyacetate):

$$ClCH_2COO^-Na^+ + NaOH \longrightarrow HOCH_2COO^-Na^+ + NaCl$$

Generally, monochloroacetic acid [79-11-8] (MCA) is added to the reaction slurry containing sufficient excess sodium hydroxide to neutralize the MCA and effect its reaction. The use of esters of MCA has also been reported (52). Common reaction diluents are isopropyl alcohol, *t*-butyl alcohol, or ethyl alcohol (53,54). Dimethoxyethane has also been reported to be effective (55). The product is isolated and washed with aqueous alcohol or acetone to remove by-product salts. Unpurified crude grades are generally prepared in the absence of diluents (56–59).

Economic Aspects.   CMC is the most widely used cellulose ether. Excluding the former Soviet Union and Eastern Bloc countries, from which little data are available, world consumption of crude and purified grades totaled approximately 123,000 metric tons in 1987 (Table 2). Annual growth rate is nominal at 1–2%. The total volume in the United States declined in the 1980s from ~32,000 metric tons in 1981 to ~19,500 in 1987 because of decreased oil well drilling activity, an important outlet.

Table 2. **World Supply and Demand for CMC^a, 10^{3}t**

Region	Capacity^b	Production	Consumption	Imports	Exports
United States	26	19.5	27.5	10.0	4.0
Western Europe	184	99	66	1	33
Japan	44	25	16	<1	9
Canada, Latin America, other			13		
Asia					
Total	*254*	*143.5*	*122.5*		

^{a}In 1987 (1).
^{b}Crude and purified grades, expressed in 100% CMC.

In the United States Aqualon Co., a Hercules Incorporated Company, is the largest producer, followed by Carbose Corp. and MAK Chemical Corp. Western Europe has about 14 companies manufacturing CMC; the five largest are Metsa Serla in Finland, Aqualon France SA in France, Hoechst AG in Germany, Billerud AB in Sweden, and AKZO in The Netherlands and Italy. Among the six producers in Japan, Dai-ichi Kogyo Seiyaku (DKS) and Diacel Chemical Industries are the two largest.

Specifications and Standards; Test Methods. Certain types of purified sodium carboxymethylcellulose meet standards set by the *U.S. Code of Federal Regulations* (CFR) Title 21, Section 182.1745, substances that are generally recognized as safe (GRAS). The FDA defines the direct food additive as the sodium salt of carboxymethylcellulose, not less than 99.5% on a dry weight basis, with a maximum substitution of 0.95 carboxymethyl groups per anhydroglucose unit, and with a minimum viscosity of 25 mPa·s(=cP) in a 2% (by weight) aqueous solution at 25°C.

Cellulose gum is the accepted common name for purified CMC. It may be used in milk products, dressings, jellies, syrups, beverages, and other select products. It is permitted in food contact and packaging applications.

Sodium carboxymethylcellulose is listed in the 1990 *United States Pharmacopeia* under the categories of pharmaceutic aid (suspending agent, tablet binder, viscosity-increasing agent), and cathartic tablets.

Procedures for the analysis of CMC are available in manufacturers' bulletins (14).

Uses. CMC is an extremely versatile polymer, and it has a variety of applications. A sampling of significant applications is given in Table 3. A more extensive listing can be found in reference 14.

2.2. Hydroxyethylcellulose. *Properties.* Hydroxyethylcellulose [9004-62-0] (HEC), is a nonionic polymer. Low hydroxyethyl substitutions (MS = 0.05–0.5) yield products that are soluble only in aqueous alkali. Higher substitutions (MS ≥ 1.5) produce water-soluble HEC. The bulk of commercial HEC falls into the latter category. Water-soluble HEC is widely used because of its broad compatibility with cations and the lack of a solution gel or precipitation point in water up to the boiling point. The MS of commercial HEC varies from about 1.8 to 3.5. The products are soluble in hot and cold water but insoluble in hydrocarbon solvents.

Table 3. **Applications for CMC**[a]

Industry	Application	Function
foods	frozen desserts	inhibit ice crystal growth
	dessert toppings	thickener
	beverages, syrups	thickener, mouthfeel
	baked goods	water-binder, batter viscosifier
	pet food	water-binder, thickener, extrusion aid
pharmaceuticals	tablets	binder, granulation aid
	bulk laxatives	water-binder
	ointments, lotions	stabilizer, thickener, film-former
cosmetics	toothpaste	thickener, suspension aid
	denture adhesives	adhesion promoter
	gelled products	gellant, film-former
paper products	internal additive	binder, improve dry-strength
	coatings, sizes	water-binder, thickener
adhesives	wallpaper paste	adhesion promoter, water-binder
	corrugating	thickener, water-binder, suspension aid
	tobacco	binder, film-former
lithography	fountain, gumming	hydrophilic protective film
ceramics	glazes, slips	binder (promotes green strength)
	welding rods	binder, thickener, lubricant
detergents	laundry	soil antiredeposition aid
textiles	warp sizing	film-former, adhesion promoter
	printing paste, dye	thickener, water-binder

[a]Ref. 14.

HEC swells or becomes partly to mostly soluble in select polar solvents, usually those that are miscible with water.

Commercially, HEC is available in a wide range of viscosity grades, ranging from greater than 500 mPa·s(=cP) at 1% solids to less than 100 mPa·s(=cP) at 5% total solids. Because HEC is nonionic, it can be dissolved in many salt solutions that do not dissolve other water-soluble polymers. It is soluble in most 10% salt solutions and in many 50% (or saturated) salt solutions such as sodium chloride and aluminum nitrate. As a rule, the lower substitution grades are more salt-tolerant.

HEC is soluble in both hot and cold water; however, as with most water-soluble thickeners, the particles have a tendency to agglomerate, or lump, when first wetted with water. This is especially evident when the HEC is added to water with poor agitation. Manufacturers have eliminated the problem of lumping and slow dissolving by surface treating the particles, most commonly with glyoxal [107-22-2] (59–62). When added to water, the particles completely disperse. After an initial induction period, commonly termed the delayed hydration time, the dispersed particles begin to dissolve producing smooth, lump-free solutions. The delayed hydration time can be increased or decreased by lowering or raising, respectively, the pH. Most manufacturers supply dispersible grades.

Solutions of HEC are pseudoplastic. Newtonian rheology is approached by very dilute solutions as well as by lower molecular-weight products. Viscosities change little between pH 2 and 12, but are affected by acid hydrolysis or alkaline oxidation under pH and temperature extremes. Viscosities of HEC solutions change reversibly with temperature, increasing when cooled and decreasing when warmed.

HEC is generally compatible with other cellulosic water-soluble polymers to give clear, homogeneous solutions. When mixed with an anionic polymer such as CMC, however, interactions between the two polymers may result in synergistic behavior, ie, viscosities higher than predicted and calculated. HEC has excellent compatibility with natural gums.

Some typical properties of HEC are given in Table 4.

Manufacture.   Purified hydroxyethylcellulose is manufactured in diluent-mediated processes similar to those used to produce carboxymethylcellulose except ethylene oxide [75-21-8] is used in place of MCA (63,64):

$$R_{cell}OH:NaOH \ + \ x\,CH_2{-}CH_2 \ \longrightarrow \ R_{cell}(OCH_2CH_2)x\,OH \ + \ NaOH$$

hydroxyethylcellulose

A competing reaction that consumes ethylene oxide is hydrolysis to ethylene glycol and oligomeric glycol by-products.

$$NaOH \ + \ x\,CH_2{-}CH_2 \ \longrightarrow \ HO(CH_2CH_2O)x^-Na^+$$

$$HO(CH_2CH_2O)_x^-Na^+ + H_2O \longrightarrow HO(CH_2CH_2O)_xH + NaOH$$

Because of the low boiling point of ethylene oxide, reactions are generally conducted in stirred autoclaves at elevated pressures.

Table 4. **Typical Properties of HEC**[a]

Property	Value
Powder	
appearance	white to light tan
moisture, max %	5.0
ash content (as Na_2SO_4), %	5.5
bulk density, g/cm^3	0.6–0.75
browning temp, °C	205–210
molecular weight, M_w	$9 \times 10^4 - 1.3 \times 10^6$
Solution	
viscosity, Brookfield, 30 rpm, $mPa \cdot s (=cP)$	
at 1% solids (high M_w)	5000
at 5% solids (low M_w)	75
sp gr, 2%, g/cm^3	1.0033
pH	7
surface tension, $mN/m (= dyn/cm)$	
MS 2.5 at 0.1%	66.8
at 0.001%	67.3
refractive index, 2%	1.336
File	
refractive index	1.51
moisture content, %, at 25°C	
50% rh	6
84% rh	29

[a]Ref. 40.

Table 5. **World Supply and Demand For HEC**[a]**, 10³ t**

Region	Capacity	Production	Consumption	Imports	Exports
United States[b]	32	20.5	20.5	1.0	3.5
Western Europe	23	17	13		
Japan	1	1	1		
Canada, Latin America, other Asia			4		
Total	*56*	*38.5*	*38.5*		

[a]In 1987 (1).
[b]Includes carboxymethylhydroxyethylcellulose, which accounts for <5%.

Economic Aspects. A breakdown of salient 1987 world supply and demand figures for HEC is given in Table 5.

Aqualon Co. and Union Carbide Corp. have manufacturing facilities in the United States and Western Europe. Hoechst AG in Europe and Fuji Chemical Co., Ltd. in Japan are the only other procedures of HEC.

Specifications and Standards; Test Methods. Hydroxyethylcellulose is included in the list of materials that are in compliance with requirements of the U.S. FDA for use in adhesives and in resinous and polymeric coatings employed on the food-contact surfaces of metal, paper, or paperboard articles, and other substrates intended for use in food packaging as specified in CFR 21. HEC made dispersible by cross-linking with glyoxal is cleared only as an adhesive and as a component of paper and paperboard in contact with food. It has not been cleared as a direct food additive.

Procedures for determining ash, moisture, solution preparation, and viscosity measurements can be found in manufacturers' product bulletins (40,41) and in ASTM D2364-69 (65).

Uses. HEC is used as a thickener, protective colloid, binder, stabilizer, and suspending agent in a variety of industrial applications. A guide to the principal uses is given in Table 6.

Table 6. **Applications for HEC**[a]

Industry	Application	Function
coatings	latex paints	thickener
	polymer emulsions	protective colloid
construction	cements, mortars	thickener, water-binder, retarder
paper	coatings, sizes	thickener, water-binder
pharmaceuticals	lotions, ointments	thickener, stabilizer, water-binder
cosmetics	toothpastes	thickener
	shampoos	thickener
	creams, lotions	thickener, stabilizer
ceramics	welding rods	water-binder, extrusion aid
	glazes	water-binder (promotes green strength)

[a]Ref. 40.

Mixed Ether Derivatives of HEC. Several chemical modifications of HEC are commercially available. The secondary substituent is generally of low DS (or MS), and its function is to impart a desirable property lacking in HEC.

Carboxymethylhydroxyethylcellulose (CMHEC). This is an anionic modification of HEC manufactured by Aqualon Co. Sodium carboxymethylhydroxyethyl-cellulose [9088-04-4] is manufactured by reaction of alkali cellulose either simultaneously or sequentially with ethylene oxide and sodium chloroacetate. Various grades, with carboxymethyl DS, CM(DS), of 0.3 to 0.5 and hydroxyethyl MS, HE(MS), of 0.7 to 2.0 are available. CMHEC has properties of both HEC and CMC. It is more compatible than CMC with salts because of the presence of nonionic hydroxyethyl groups. In saturated NaCl, only CMC with a CM(DS) ≥ 1.0 is completely soluble. CMHEC, on the other hand, is soluble with a CM(DS) as low as 0.3. CMHEC is also very tolerant of Ca^{2+} and consequently readily dissolves in seawater. Unlike HEC, CMHEC in solution may be cross-linked with trivalent cations such as Fe^{3+} and Al^{3+} to give greatly increased viscosity or three-dimensional viscoelastic gels (66).

CMHEC products are used predominantly in oil recovery applications. The high water binding capability, salt compatibility, and adsorption to clay and mineral surfaces give CMHEC ethers excellent control over high salinity fluids (67,68). Water loss in cement slurries is also reduced (69). CMHEC is also used in hydraulic fracturing fluids. Gels formed by cross-linking with multivalent cations can suspend and transport proppants into a well bore and then fracture (66,70). Some typical properties of CMHEC having a CM(DS) ~0.3 and HE(MS) ~0.7 are listed in Table 7.

Cationic Hydroxyethylcelluloses. These materials are manufactured by Union Carbide Corp. and National Starch and Chemical Corp., marketed under the trade names Polymer JR and Celquat, respectively (47,48). The

Table 7. Typical Properties of Mixed Ether Derivatives of HEC[a]

Property	CMHEC	Cationic HEC	EHEC	HMHEC
		Powder		
appearance	off-white	light yellow	off-white	off-white
bulk density, g/cm^3	0.6	0.48	0.4–0.8	0.55–0.75
ash content (as Na$_2$SO$_4$), %		3	3 (as NaCl)	10 max
volatiles, %	6–8	7	8 max	5 max
		Solution		
pH	6.5–10	7	6–7	6–8.5
flocculation temp in water, °C			~65	
surface tension, mN/m(= dyn/cm)			55	~62
		Film		
tensile strength, MPa^b	69^c	14–22^c	45–55^d	
flexibility^e		60–70^c	25–35^d	
refractive index	1.530		1.49	

[a]Refs. 39, 47, 49, 71.
[b]To convert MPa to psi, multiply by 145.
[c]At 50% rh.
[d]At 65% rh.
[e]MIT double folds.

cationic substituent on Polymer JR is presumably 2-hydroxypropyltrimethylammonium chloride (72). Celquat is presumably the reaction product of HEC with *N,N*-diallyl-*N,N*-dimethylammonium chloride (73). Their primary application is in shampoos and hair conditioners wherein the cationic moiety imparts substantivity to hair. Some typical properties of Celquat resins are given in Table 7.

Hydrophobic Hydroxyethylcelluloses. These materials are produced by stepwise or simultaneous reaction of ethylene oxide and a hydrophobic alkylating reagent. Commercial products include: ethylhydroxyethylcellulose [9004-58-9] (EHEC), manufactured by Berol Kemi AB under the Bermocoll trade name (49), and HEC modified with a long-chain alkyl group, generically termed HMHEC (where HM = Hydrophobically Modified), manufactured by Aqualon Co. and sold under the trade name Natrosol Plus (39). These products are water-soluble. An organo-soluble ethyl modification is also available, which is classified as a derivative of ethylcellulose.

Water-soluble EHEC is a moderate ethyl DS ($\sim$1.0) modification of high hydroxyethyl MS ($\geq$2.0) HEC. Ethyl groups lower the surface and interfacial tensions, thereby increasing surface activity. This group also modifies adsorption properties of the polymer to particulates found in many formulations such as clays, pigments, and latices. Aqueous solutions have pseudoplastic rheology. High viscosity grades are more pseudoplastic than low viscosity materials, which approach Newtonian flow behavior. Viscosities decrease reversibly with increasing temperature. Above 65°C, EHEC precipitates from solution. Salts lower the temperature at which precipitation occurs. Solution viscosities are insensitive to pH between about 3 to 11. Aqueous solutions are miscible with lower alcohols, glycols, and ketones up to equal proportions. Water-soluble EHECs are used to thicken and stabilize a variety of materials, including water-borne paints, plasters, detergents, cosmetics, and pharmaceuticals (49).

HMHEC is a modification with low levels of much longer hydrocarbon chains (hydrophobes) that not only increase surface activity but also impart associative behavior to HEC; this produces dramatic effects on solution viscosity and rheology (37,74). For example, a HEC with a 2% solution viscosity of 10 mPa$\cdot$s(=cP) modified with $\sim$2.5 wt % of a C_{14}-chain hydrocarbyl moiety has a viscosity of 800 mPa$\cdot$s(=cP). The effect has been attributed to micellar aggregation of the hydrophobic groups in solution (37–39). The solubility and rheological properties of HMHEC depend primarily on the molecular weight, the DS and chain length of the hydrophobe, the polymer concentration, and the composition of the aqueous media. It has been found that HMHEC is an efficient rheology control agent in latex paints (39,74). Typical properties of a HMHEC are given in Table 7.

2.3. Methylcellulose and its Mixed Ethers. *Properties.* Methylcellulose [9004-67-5] (MC) and its alkylene oxide derivatives hydroxypropylmethylcellulose [9004-65-3] (HPMC), hydroxyethylmethylcellulose [9032-42-2] (HEMC), and hydroxybutylmethylcellulose [9041-56-9] (HBMC) are nonionic, surface-active, water-soluble polymers. Each type of derivative is available in a range of methyl and hydroxyalkyl substitutions. The extent and uniformity of the methyl substitution and the specific type of hydroxyalkyl substituent affect the solubility, surface activity, thermal gelation, and other properties of the polymers in solution.

These four methylcelluloses are available in a range of substitutions.

	Methyl DS	Hydroxyalkyl MS
methylcellulose	1.4–2.0	
hydroxypropylmethylcellulose	1.1–2.0	0.1–1.0
hydroxyethylmethylcellulose	1.3–2.2	0.06–0.5
hydroxybutylmethylcellulose	≥1.9	≥0.04

Methylcellulose with a methyl DS less than about 0.6 is alkali-soluble. From about 1.6 to 2.4, it is water-soluble (most commercial grades); above 2.4, it is soluble in a wide variety of organic solvents. Methylcellulose solutions in water start to gel at ~55°C, independent of molecular weight. The gelation is a function of the DS, rate of heating, and type and amounts of additives such as salts. As the temperature increases, the viscosity initially decreases (typical behavior). When the gelling temperature is reached, the viscosity sharply rises until the flocculation temperature is reached. Above this temperature, the viscosity collapses. This process is reversible with temperature (75).

The mixed derivatives HEMC, HPMC, and HBMC tend to precipitate rather than gel as the temperature is increased. The higher the hydroxyalkyl substitution, the greater the tendency for precipitation. HEMCs and HPMCs tend to have higher gelation and flocculation temperatures (75). The mixed derivatives are generally more tolerant of added salts than methylcellulose itself. HPMC and HBMC are tolerant of and are soluble in some organic solvents, especially lower alcohols and glycols.

Solutions of methylcelluloses are pseudoplastic below the gel point and approach Newtonian flow behavior at low shear rates. Above the gel point, solutions are very thixotropic because of the formation of three-dimensional gel structure. Solutions are stable between pH 3 and 11; pH extremes will cause irreversible degradation. The high substitution levels of most methylcelluloses result in relatively good resistance to enzymatic degradation (16).

Methylcellulose and its mixed ethers are surface-active cellulose ethers having surface tension values as low as 44 mN/m(= dyn/cm) and interfacial tension values as low as 17 mN/m(= dyn/cm) against paraffin oil.

Typical properties of MC, HPMC, HEMC, and HBMC are given in Table 8.

Manufacture.   Methylcellulose is manufactured by the reaction of alkali cellulose with methyl chloride (76).

$$R_{cell}OH : NaOH + CH_3Cl \longrightarrow R_{cell}OCH_3 + NaCl$$

The reaction is accompanied by side reactions that lead to methanol and dimethyl ether by-products.

$$CH_3Cl + NaOH \longrightarrow CH_3OH + NaCl$$

$$CH_3OH + CH_3Cl + NaOH \longrightarrow CH_3OCH_3 + NaCl + H_2O$$

Table 8. **Typical Properties of Methylcellulose Ethers**[a]

Property	MC	HPMC	HEMC	HBMC
	Powder			
appearance	white	white	white	white
bulk density, g/cm^3		0.25–0.70		
volatiles, %		8 max		
ash content (as Na$_2$SO$_4$), %		2.5 max		
	Solution			
viscosity[b], mPa·s(=cP)	10–15,000	5–70,000	100–70,000	
sp gr, 2% at 20°C		1.0032		
pH, 1%		5.5–9.5		
surface tension, 0.1%, mN/m(= dyn/cm)	47–53	44–56	46–53	49–55
interfacial tension[c], mN/m(= dyn/cm)	19–23	17–30	17–21	20–22
gelation temp, °C	48	54–70		49
flocculation temp, °C	50–75	60–90	60–90	
	Film			
tensile strength, MPa[d], 50% rh	58.6–78.6	58.6–61		
elongation, %, 50% rh	10–15	5–10		
softening point, °C		240		
melting point, °C	290–305	260		
vapor transmission, nmol/(m·s)[e]				
water, 37°C, 90–100% rh		520		
oxygen, 24°C		560		

[a]Refs. 42,43.
[b]2% Solution, Brookfield, 20 rpm.
[c]Against paraffin oil.
[d]To convert MPa to psi, multiply by 145.
[e]To convert nmol/(m·s) to g·mil/100 in.^{2}·d for water, multiply by 3.95; for O$_2$, by 7.02.

Hydroxyalkyl modification is made by simultaneous or staged addition of an alkylene oxide, as exemplified in the following (77–79).

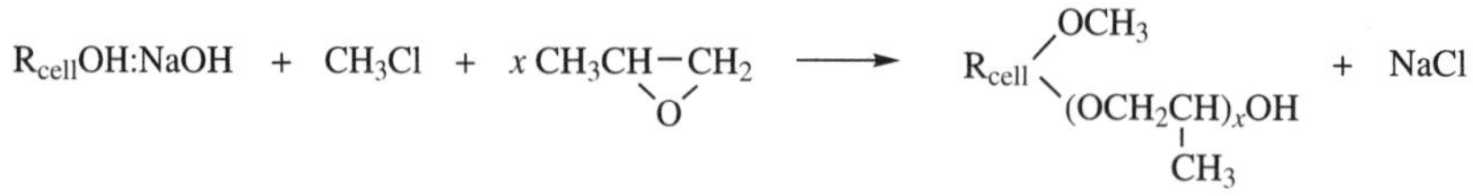

$$R_{cell}OH:NaOH \; + \; CH_3Cl \; + \; x\,CH_3CH{-}CH_2 \; \longrightarrow \; R_{cell} \begin{array}{l} {-}OCH_3 \\ {-}(OCH_2CH)_xOH \\ \qquad\qquad CH_3 \end{array} \; + \; NaCl$$

Similarly, ethylene oxide and 1,2-butylene oxide are used to make methyl-hydroxy-ethylcellulose and methylhydroxybutylcellulose, respectively.

Unlike HEC and CMC, which are purified by washing with aqueous organic solvents, methylcellulose and its hydroxyalkyl modifications are purified in hot water where they are insoluble. As with other cellulose ethers, drying and grinding complete the process.

Economic Aspects. A breakdown of salient figures in 1987 for the methylcelluloses is given in Table 9. The Dow Chemical Company is the only U.S. manufacturer. They produce and market methyl- (MC), hydroxypropyl-methyl-(HPMC), and hydroxybutylmethyl- (HBMC) celluloses. European producers include Aqualon Co. in Germany and Belgium, Hoechst AG in Germany,

Table 9. **Worldwide Supply and Demand for Methylcelluloses**[a]**, 10^3t**

Region	Capacity	Production	Consumption	Imports	Exports
United States	22	16	16	1	2
Western Europe	51	44	35.5		
Japan	7	7	5		2
Canada, Latin America, other					
Asia					
Total	*80*	*67*	*62*		

[a]In 1987 (1).

Dow Chemical GmbH in Germany, Wolff Walsrode AG in Germany, and Courtaulds Fibres Ltd. in the United Kingdom. Shin-Etsu Chemical Co., Ltd. and Matsumoto Yushi-Seiyaku Co., Ltd. are the two Japanese manufacturers.

Specifications and Standards; Test Methods. Premium grades of methylcellulose meet the requirements of U.S.P. XIX and *Food Chemicals Codex II*, and the *International Codex Alimentarius*. They are GRAS, meeting the requirements of Food Additives Regulation 182.1480 as multiple purpose food substances for nonstandardized foods. Premium grades of some hydroxypropylmethylcelluloses meet requirements of U.S.P. XIX, *Food Chemicals Codex II*, and Food Additives Regulation 172.874, which allows their use in nonstandardized foods. Methylcellulose and hydroxypropylmethylcellulose qualify as inert ingredients under CRF 180.1000 that may be used in formulations applied to growing crops or raw agricultural commodities after harvest.

Analytical procedures and test methods are described in manufacturers' technical bulletins (42,43).

Uses. There are numerous applications for methylcellulose and its derivatives. Some important ones are summarized in Table 10.

2.4. Ethylcellulose and Hydroxyethylethylcellulose. *Properties.* Ethyl cellulose [9004-57-3] (EC) is a nonionic, organo-soluble, thermoplastic cellulose ether, having an ethyl DS in the range of $\sim$2.2 – 2.7. Actually, EC is water-soluble at DS $\sim$1.2, but only those products that are thermoplastic and soluble in organic solvents are of commercial importance, because of their ability

Table 10. **Applications for Methylcellulose and its Derivatives**[a]

Industry	Application	Function
construction	cements, mortars	thickener, water-binder, workability
foods	mayonnaise, dressing	stabilizer, emulsifier
	desserts	thickener
pharmaceuticals	tablets	binder, granulation aid
	formulations	stabilizer, emulsifier
adhesives	wallpaper paste	adhesive
ceramics	slip casts	binder (promotes green strength)
coatings	latex paints	thickener
	paint removers	thickener
cosmetics	creams, lotions	stabilizer, thickener

[a]Refs. 42,43.

to form tough, stable films. Above a DS of about 2.5, EC is soluble in many non-polar solvents.

Film mechanical properties, such as tensile strength, elongation, and flexibility, depend more on the molecular weight (degree of polymerization) than on substitution. Elongation and tensile strength increase to a maximum with increasing molecular weight; flexibility increases linearly.

Ethylcellulose is subject to oxidative degradation in the presence of sun- or ultraviolet light, especially at elevated temperatures above the softening point. It must, therefore, be stabilized with antioxidants (44). EC is stable to concentrated alkali and brines but is sensitive to acids.

Organo-soluble hydroxyethylethylcellulose (HEEC) is highly ethoxylated with small amounts of hydroxyethyl substitution. It is used in coating applications that require solubility in fast-drying aliphatic hydrocarbons. These EHEC polymers are the only commercially available cellulosic polymers substantially soluble in low cost, low odor aliphatic hydrocarbon solvents. As a result, formulation costs are lowered and application conditions are simplified. Like ethylcellulose, HEEC is subject to oxidative degradation by the combination of heat and light (45). Degradation is more rapid above the melting point (175°C). An antioxidant, such as octylphenol, combined with an acid acceptor, such as an epoxy resin, provides protection against heat degradation. Benzophenone derivatives provide protection against sunlight. Primary uses for organic solvent-soluble EHECs are as additives in printing inks, clear lacquers, and other coatings such as alkyd, flat, and semigloss finishes. Table 11 gives typical properties for EC and HEEC.

Manufacture.   Ethyl chloride undergoes reaction with alkali cellulose in high pressure nickel-clad autoclaves. A large excess of sodium hydroxide and ethyl chloride and high reaction temperatures (up to 140°C) are needed to drive the reaction to the desired high DS values (≥2.0). In the absence of a diluent, reaction efficiencies in ethyl chloride range between 20 and 30%, the

Table 11. **Typical Properties of EC and HEEC**[a]

Property	EC	HEEC
Powder		
appearance	white	white
volatiles, %	2	
bulk density, g/cm^3	0.3–0.35	0.3–0.35
softening point, °C	152–162	
Film		
specific gravity	1.140	1.120
refractive index	1.470	1.47
tensile strength, MPa[b]	46–72	34–41
elongation, %	7–30	6–10
flexibility[c]	160–2000	500–900
dielectric constant, 60 Hz	2.5–4.0	

[a]Refs. 44,46.
[b]To convert mPa to psi, multiply by 145.
[c]MIT double folds.

majority of the rest being consumed to ethanol and diethyl ether by-products.

$$R_{cell}OH : NaOH + CH_3CH_2Cl \longrightarrow R_{cell}OCH_2CH_3 + NaCl$$
$$+ CH_3CH_2OH + (CH_3CH_2)_2O$$

Higher ethyl chloride efficiency is claimed for a process utilizing a hydrocarbon diluent coupled with stepwise addition of sodium hydroxide (80). Product work-up includes distillation to remove residual unreacted ethyl chloride, added diluent, methanol, and diethyl ether; neutralization of excess sodium hydroxide; washing in water to remove salts; drying; and grinding.

The manufacturing process for organo-soluble EHEC is similar to that for EC except that alkali cellulose reacts first with ethylene oxide to a low hydroxyethyl MS value of ~0.5 at a low temperature, ~50°C, followed by reaction of the ethyl chloride at a higher temperature. Additional by-products, which are removed during purification, include glycols and the reaction products of the glycols with ethyl chloride (glycol ethers).

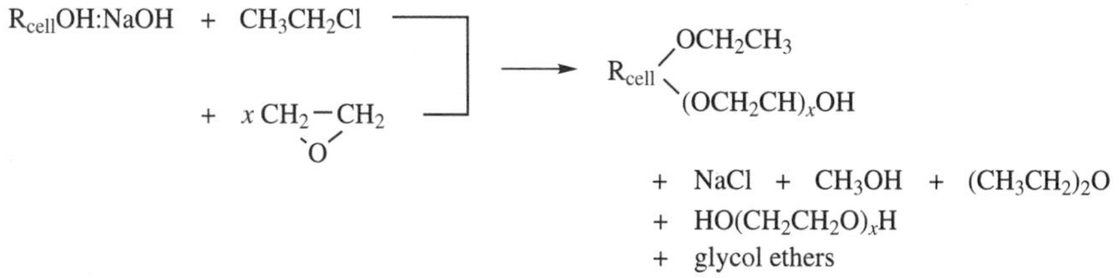

$$R_{cell}OH:NaOH + CH_3CH_2Cl$$
$$+ x\ CH_2-CH_2 \quad \underset{O}{\diagup}$$
$$\longrightarrow R_{cell} \underset{(OCH_2CH)_xOH}{\overset{OCH_2CH_3}{<}}$$

$$+ \quad NaCl \ + \ CH_3OH \ + \ (CH_3CH_2)_2O$$
$$+ \quad HO(CH_2CH_2O)_xH$$
$$+ \quad glycol\ ethers$$

Economic Aspects. The Dow Chemical Company and Aqualon Co. are the only listed principal producers of EC and HEEC products worldwide. Consumption has remained constant over the past several years, and the products are not expected to grow in the future. Production is estimated at 5000 t/yr, roughly equally divided between The Dow Chemical Company and Aqualon Co. As with other cellulose ethers, the price for EC and HEEC varies by grade.

Specifications and Standards; Test Methods. Ethylcellulose is cleared for many applications in food and food contact under the Federal Food, Drug, and Cosmetic Act, as amended. Examples include binder in dry vitamin preparations for animal feed, coatings and inks for paper and paperboard products used in food packaging, and closures with sealing gaskets for food containers (44). Methods of analyses are given in ASTM D914-72 (19), *National Formulary XIV*, and *Food Chemicals Codex II*.

Uses. A summary of the applications for ethylcellulose is given in Table 12.

Table 12. **Applications for EC and HEEC**[a]

Industry	Application	Function
coatings	lacquers, varnishes	protective film-former, additive to increase film toughness and durability, shorten drying time
printing	inks	film-former
adhesives	hot melts	additive to increase toughness

[a]Refs. 44,46.

2.5. Hydroxypropylcellulose.   *Properties.*   Hydroxypropylcellulose [9004-64-2] (HPC) is a thermoplastic, nonionic cellulose ether that is soluble in water and in many organic solvents. HPC combines organic solvent solubility, thermoplasticity, and surface activity with the aqueous thickening and stabilizing properties characteristic of other water-soluble cellulosic polymers described herein. Like the methylcelluloses, HPC exhibits a low critical solution temperature in water.

The substitution of HPC is defined by the MS. Molar substitutions higher than approximately 3.5 are needed for solubility in water and organic solvents.

HPC is available in a number of viscosity grades, ranging from about 3000 mPa · s(=cP) at 1% total solids in water to 150 mPa · s(=cP) at 10% total solids. HPC solutions are pseudoplastic and exceptionally smooth, exhibiting little or no structure or thixotropy. The viscosity of water solutions is not affected by changes in pH over the range of 2 to 11. Viscosities decrease as temperature is increased. HPC precipitates from water at temperatures between 40 and 45°C. Dissolved salts and other compounds can profoundly influence the precipitation temperature (50,81).

HPC is compatible with many natural and synthetic water-soluble polymers and gums (50). Generally, blends of HPC with another nonionic polymer such as HEC yield water solutions having viscosities in agreement with the calculated value. Blends of HPC and anionic CMC, however, produce solution viscosities greater than calculated. This synergistic effect may be reduced in the presence of dissolved salts or if the pH is below 3 or above 10.

Like the methylcelluloses, water solutions of HPC display greatly reduced surface tension. A 0.1% solution of HPC at 25°C has a surface tension of about 44 mN/m(=dyn/cm) (water is 74.1 mN/m) and interfacial tension of about 12.5 mN/m(=dyn/cm) against mineral oil. The molecular weight of the HPC has only a slight effect on the surface tension.

Examples of polar organic solvents that dissolve HPC are methanol, ethanol, propylene glycol, and chloroform. There is no tendency for HPC to precipitate as the temperature is raised. In fact, elevated temperatures improve the solvent power of organic liquids.

Some typical properties of commercial HPC are given in Table 13.

Manufacture.   HPC is manufactured by reaction of propylene oxide [75-56-9] with alkali cellulose.

$$R_{cell}OH{:}NaOH \; + \; x\,CH_3CH{-}CH_2 \; \longrightarrow \; R_{cell}(OCH_2CH)_xOH \; + \; HO(CH_2CHO)_xH$$

The reaction may be conducted in stirred autoclaves in the presence of hydrocarbon diluents (82,83). Like the methylcelluloses, advantage is taken of the low critical solution temperature of HPC and it is purified through multiple washings with hot water. Consequently, very low levels of residual salts and by-products are present in the final products.

Economic Aspects.   The Aqualon Co. is the only U.S. manufacturer. It is also produced in Japan by Nippon Soda Co., Ltd. Worldwide consumption in 1987 was estimated at 2300 metric tons.

Table 13. **Typical Properties of HPC**[a]

Property	Value
Powder	
appearance	off-white
volatiles, %	5 max
ash content (as Na_2SO_4), %	0.2–0.5
softening point, °C	100–150
molecular weight, M_w	$8.0 \times 10^4 - 1.15 \times 10^6$
Solution	
viscosity, Brookfield, 30 rpm, mPa · s(=cP)	
at 1% (high M_w)	2500
at 10% (low M_w)	100
surface tension, 0.1%, mN/m(=dyn/cm)	43.6
interfacial tension[b], 0.1%, mN/m	12.5
Film	
tensile strength, MPa[c]	14
elongation, %	50
flexibility[d] (50 µm film)	10,000
refractive index	1.559

[a]Ref. 50.
[b]Against mineral oil.
[c]To convert MPa to psi, multiply by 145.
[d]MIT double folds.

Specifications and Standards; Test Methods. Food-grade HPC products are manufactured for use in food and conform to the specifications for HPC set forth in CFR 21, Section 172.870. Food grades of HPC also conform to the specifications for HPC as listed in the current edition of the *Food Chemicals Codex*.

Pharmaceutical and cosmetic grades of HPC, as for example Klucel NF manufactured by Aqualon Co., conform to the requirements of the HPC monograph as listed in the current edition of the *National Formulary*.

Toxicity testing indicates that HPC is physiologically inert (50).

Procedures for determining the ash content and moisture level, solution preparation, and viscosity measurement techniques are given in the manufacturer's literature (50).

Uses. A summary of significant uses for HPC is given in Table 14.

Table 14. **Applications for HPC**[a]

Industry	Application	Function
polymerization	PVC suspension polymerization	protective colloid
pharmaceutical	tablets	binder, film-former
coatings	paint remover	thickener
foods	whipped toppings	stabilizer
	processed foods	extrusion aid
ceramics	slip casts	binder (promotes green strength)

[a]Ref. 50.

BIBLIOGRAPHY

"Cellulose Derivatives" in *ECT* 1st ed., Vol. 3, pp. 357–391; and in *ECT* 2nd ed., Vol. 4, pp. 616–652 both by E. D. Klug, Hercules Powder Company; "Cellulose Derivatives, Ethers" in *ECT* 3rd ed., Vol. 5, p. 143–163 by G. K. Greminger, Jr., Dow Chemical U.S.A.

CITED PUBLICATIONS

1. *Chemical Economics Handbook*, SRI International, Menlo Park, Calif., 1989, Section 581.5000A.
2. J. O. Warwicker, R. Jeffries, R. L. Colbran, and R. N. Robinson, *A Review of the Literature on the Effect of Caustic Soda and Other Swelling Agents on the Fine Structure of Cotton*, Shirley Institute Pamphlet No. 93, Shirley Institute, Didsbury, Manchester, UK, 1966.
3. T. Okano and A. Sarko, *J. Appl. Polym. Sci.* **29**, 4175–4182 (1984).
4. T. Okano and A. Sarko, *J. Appl. Polym. Sci.* **30**, 325–332 (1985).
5. H. Nishimura and A. Sarko, *J. Appl. Polym. Sci.* **33**, 855–866 (1987).
6. H. Nishimura and A. Sarko, *J. Appl. Polym. Sci.* **33**, 867–874 (1987).
7. D. Entwistle, E. H. Cole, and N. S. Wooding, *Text. Res. J.* **XIX**(9), 527–624 (1949).
8. R. I. C. Michie and S. M. Neale, *J. Polym. Sci. Part A* **2**, 2063–2083 (1964).
9. U.S. Pat. 2,512,338 (June 20, 1950), E. D. Klug and H. M. Spurlin (to Hercules Powder Co.).
10. U.S. Pat. 3,719,663 (Mar. 6, 1973), E. D. Klug (to Hercules Inc.).
11. U.S. Pat. 3,728,331 (Apr. 17, 1973), A. B. Savage (to The Dow Chemical Company).
12. U.S. Pat. 4,061,859 (Dec. 6, 1977), W-J. Cheng (to The Dow Chemical Company).
13. U.S. Pat. 3,391,135 (July 2, 1968), S. Ouno and co-workers (to Shin-Etsu Chemical Industry Co., Ltd.).
14. *Aqualon*™ *Cellulose Gum, Sodium Carboxymethylcellulose, Physical and Chemical Properties*, Aqualon Co., a Hercules Incorporated Company, Wilmington, Del., 1988.
15. D. J. Sikkema and H. Janssen, *Macromolecules* **22**, 364–366 (1989).
16. M. G. Wirick, *J. Polym. Sci. Part A-1* **6**, 1705–1718 (1968).
17. J. E. Glass, A. M. Buettner, R. G. Lowther, C. S. Young, and L. A. Cosby, *Carbohydr. Res.* **84**, 245–263 (1980).
18. U.S. Pat. 4,401,813 (Aug. 30, 1983), J. L. Lowell, M. J. Nevins, K. L. G. Reid, and K. L. Walter (to NL Industries, Inc.).
19. U.S. Pat. 4,525,585 (June 25, 1985), A. Taguchi and T. Ohmiya (to Diacel Chemical Industries, Ltd.).
20. S-I. Takahashi, T. Fujimoto, T. Miyamoto, and H. Inagaki, *J. Polym. Sci.: Part A* **25**, 987–994 (1987).
21. J. Reuben and H. T. Conner, *Carbohydr. Res.* **115**, 1–13 (1983).
22. J. Reuben, *Carbohydr. Res.* **157**, 201–213 (1986).
23. J. Reuben, *Carbohydr. Res.* **161**, 23–30 (1987).
24. J. Reuben and T. E. Casti, *Carbohydr. Res.* **163**, 91–98 (1987).
25. J. R. DeMember and co-workers, *J. Appl. Polym. Sci.* **21**, 621–627 (1977).
26. A. Parfondry and A. S. Perlin, *Carbohydr. Res.* **57**, 39–49 (1977).
27. Y. Tezuka, K. Imai, M. Oshima, and T. Chiba, *Polymer* **30**, 2288–2291 (1989).
28. B. Lindberg, U. Linquist, and O. Stenberg, *Carbohydr. Res.* **170**, 207–214 (1987).
29. S. P. Rowland, in N. M. Bikales, ed., *Encyclopedia of Polymer Science and Technology*, 1st ed., Suppl. Vol. 1, Wiley-Interscience, New York, 1976, 146–175.
30. J. Brandrup and E. H. Immergut, *Polymer Handbook*, 3rd ed., Section VII, John Wiley & Sons, Inc., 1989.

31. R. A. Gelman, *J. Appl. Polym. Sci.* **27**, 2957–2964 (1982).
32. R. A. Gelman and H. G. Barth, in J. E. Glass, ed., *Water-Soluble Polymers: Beauty with Performance, Advances In Chemistry Series No. 213*, American Chemical Society, Washington, D.C., 1986.
33. H. G. Barth, *J. Chromatogr. Sci.* **18**, 409–429 (1980).
34. E. Pfannkoch, K. C. Lu, F. E. Regnier, and H. G. Barth, *J. Chromatogr. Sci.* **18**, 430–441 (1980).
35. H. G. Barth and F. E. Regnier, *J. Chromatogr.* **192**, 275–293 (1980).
36. G. Holzwarth, L. Soni, and D. N. Schulz, *Macromolecules* **19**, 422–426 (1986).
37. L. M. Landoll, *J. Polym. Sci. Polym. Chem. Ed.* **20**, 443 (1982).
38. A. C. Sau and L. M. Landoll, in J. E. Glass, ed., *Polymers in Aqueous Media, Advances in Chemistry Series Vol. 223*, American Chemical Society, Washington, D.C., 1989, p. 343.
39. *Natrosol® Plus—Modified Hydroxyethylcellulose*, Aqualon Co., a Hercules Incorporated Company, Wilmington, Del., 1988.
40. *Natrosol® Hydroxyethylcellulose,* Physical and Chemical Properties, Aqualon Co., a Hercules Incorporated Company, Wilmington, Del., 1987.
41. *Cellosize® Hydroxyethylcellulose*, Union Carbide Corp., New York, 1981.
42. *Methocel®*, The Dow Chemical Company, Midland, Mich., 1978.
43. *Culminal® Methylcellulose, Physical and Chemical Properties*, Aqualon Co., a Hercules Incorporated Company, Wilmington, Del.
44. *Chemical and Physical Properties of Hercules® Ethylcellulose*, Hercules Inc., Wilmington, Del., 1982.
45. *Tough Ethocel® Ethyl Cellulose Resin*, The Dow Chemical Company, Midland, Mich., 1974.
46. *Chemical and Physical Properties of Hercules® EHEC Ethylhydroxyethylcellulose*, Hercules Inc., Wilmington, Del., 1981.
47. *Celquat®, Cationic Cellulosic Polymers For Cosmetics and Toiletries*, National Starch and Chemical Corp., Bridgewater, N.J.
48. *Polymer JR For Hair Care*, Union Carbide Corp., New York.
49. *Bermocoll® Cellulose Ethers*, Berol Kemi AB, Sweden, 1977.
50. *Klucel® Hydroxypropylcellulose, Physical and Chemical Properties*, Aqualon Co., a Hercules Incorporated Company, Wilmington, Del., 1987.
51. *National Fire Codes®: A Compilation of NFPA Codes, Standards, Recommended Practices, Manuals, and Guides*, Vol. II, National Fire Protection Association, Md., 1987.
52. U.S. Pat. 4,525,585 (June 25, 1985), A. Taguchi and T. Ohmiya (to Diacel Chemical Industries, Ltd.).
53. U.S. Pat. 2,517,577 (Aug. 8, 1950), E. D. Klug and J. S. Tinsley (to Hercules Powder Co.).
54. U.S. Pat. 2,976,278 (Mar. 21, 1961), O. H. Paddison and R. W. Somer (to E. I. du Pont de Nemours & Co., Inc.).
55. U.S. Pat. 4,460,766 (July 17, 1984), U-H Felcht and E. Perplies (to Hoechst Aktiengesellschaft).
56. U.S. Pat. 2,523,377 (Sept. 26, 1950), E. D. Klug (to Hercules Powder Co.).
57. U.S. Pat. 2,553,725 (May 22, 1951), L. N. Rogers, W. A. Mueller, and E. E. Hembree (to Buckeye Chemical Oil Co.).
58. R. N. Hadar, W. F. Waldeck, and F. W. Smith, *Ind. Eng. Chem.* **44**, 2803 (1952).
59. U.S. Pat. 2,879,268 (Mar. 24, 1959), E. I. Jullander (to Mo Och Domsjo Aktiebolag).
60. U.S. Pat. 3,072,635 (Jan. 8, 1963), J. H. Menkart and R. S. Allan (to Chemical Development of Canada).
61. U.S. Pat. 3,356,519 (Dec. 5, 1967), W. C. Chambers and M. Lee (to The Dow Chemical Company).

62. Can. Pat. 947,281 (May 14, 1974), W. Patten (Union Carbide Corp.).

63. U.S. Pat. 2,572,039 (Oct. 23, 1951), E. D. Klug and H. G. Tennent (to Hercules Powder Co.).

64. U.S. Pat. 2,682,535 (June 29, 1954), A. E. Broderick (to Union Carbide Corp.).

65. ASTM D2364-69, *Standard Methods for Testing Hydroxyethylcellulose*, American Society of Testing and Materials, Easton, Md.

66. U.S. Pat. 4,035,195 (July 12, 1977), T. J. Podlas (to Hercules Inc.).

67. U.S. Pat. 2,618,595 (Nov. 18, 1954), W. E. Gloor (to Hercules Powder Co.).

68. U.S. Pat. 3,284,353 (Nov. 18, 1966), J. B. Batdorf (to Hercules Powder Co.).

69. U.S. Pat. 4,433,731 (Feb. 28, 1984), J. Chatterji, B. G. Brake, and J. M. Tinsley (to Halliburton Co.).

70. U.S. Def. Publ. T103,401 (Sept. 6, 1983), T. G. Majewicz.

71. *Hercules CMHEC 37L, Technical Bulletin VC-402C*, Hercules Inc., Wilmington, Del.

72. U.S. Pat. 3,472,840 (Oct. 14, 1969), F. W. Stone and J. M. Rutherford, Jr. (to Union Carbide Corp.).

73. U.S. Pat. 4,464,523 (Aug. 7, 1984), D. Neigel and J. Kancylarz (to National Starch and Chemical Corp.).

74. U.S. Pat. 4,228,277 (Oct. 14, 1980), L. M. Landoll (to Hercules Inc.).

75. N. Sarkar, *J. Appl. Polym. Sci.* **24**, 1073–1087 (1979).

76. U.S. Pat. 4,117,223 (Sept. 26, 1978), W. Lodige, F. Lodige, J. Lucke, and E. Lipp.

77. U.S. Pat. 4,339,573 (July 13, 1982), W. Wust, H. Leischner, W. Rahse, F-J. Carduck, and N. Kune (to Henkel Kommanditgesellschaft Aktien).

78. U.S. Pat. 4,477,657 (Oct. 16, 1984), C. P. Strange, C. D. Messelt, and C. W. Gibson (to The Dow Chemical Company).

79. U.S. Pat. 4,456,751 (June 26, 1984), C. D. Messelt and G. P. Townsend (to The Dow Chemical Company).

80. U.S. Pat. 2,254,249 (Sept. 2, 1941), R. W. Swinehart and A. T. Maasberg (to The Dow Chemical Company).

81. E. D. Klug, *J. Polym. Sci.: Part C* **36**, 491–508 (1971).

82. U.S. Pat. 3,278,521 (Oct. 11, 1966), E. D. Klug (to Hercules Inc.).

83. U.S. Pat. 3,357,971 (Dec. 12, 1967), E. D. Klug (to Hercules Inc.).

GENERAL REFERENCES

E. K. Just and T. G. Majewicz, in J. I. Kroschwitz, ed., *Encyclopedia of Polymer Science and Engineering*, 2nd ed., Vol. 3, John Wiley & Sons, Inc., New York, 1985, 224–269.

E. Ott, M. Spurlin, and M. W. Graffin, ed., *Cellulose and Cellulose Derivatives, High Polymers*, Vol. V, Wiley-Interscience, New York, 1954–1955, Parts I–III.

N. M. Bikales and L. Segal, eds., *Cellulose and Cellulose Derivatives, High Polymers*, Vol. V, Wiley-Interscience, New York, 1971, Pts. IV–V.

R. L. Davidson, ed., *Handbook of Water-Soluble Gums and Resins*, McGraw-Hill, New York, 1980.

J. E. Glass, ed., *Water-Soluble Polymers—Beauty With Performance: Advances in Chemistry Series No. 213*, American Chemical Society, Washington, D.C., 1986.

J. E. Glass, ed., *Polymers In Aqueous Media: Performance Through Association: Advances In Chemistry Series No. 223*, American Chemical Society, Washington, D.C., 1989.

THOMAS G. MAJEWICZ
THOMAS J. PODLAS
Aqualon Company

CEMENT

1. Introduction

The term cement is used to designate many different kinds of substances that are used as binders or adhesives (qv). The cement produced in the greatest volume and most widely used in concrete for construction is portland cement. Masonry and oil well cements are produced for special purposes. Calcium aluminate cements are extensively used for refractory concretes (see ALUMINUM COMPOUNDS, ALUMINUM OXIDE; REFRACTORIES). Such cements are distinctly different from epoxies and other polymerizable organic materials. Portland cement is a hydraulic cement, ie, it sets, hardens, and does not disintegrate in water. Hence, it is suitable for construction of underground, marine, and hydraulic structures whereas gypsum plasters and lime mortars are not. Organic materials, such as latexes and water-soluble polymerizable monomers, are sometimes used as additives to impart special properties to concretes or mortars. The term cements as used herein is confined to inorganic hydraulic cements, principally portland and related cements. The essential feature of these cements is their ability to form on hydration with water relatively insoluble bonded aggregations of considerable strength and dimensional stability (see also BUILDING MATERIALS, SURVEY).

Hydraulic cements are manufactured by processing and proportioning suitable raw materials, burning (or clinkering at a suitable temperature), and grinding the resulting hard nodules called clinker to the fineness required for an adequate rate of hardening by reaction with water. Portland cement consists mainly of tricalcium silicate [12168-85-3], Ca_3SiO_5, and dicalcium silicate [10034-77-2], Ca_2SiO_4. Usually two types of raw materials are required: one rich in calcium, such as limestone, chalk, marl, or oyster or clam shells; the other rich in silica, such as clay or shale. The two other significant phases in portland cements are tricalcium aluminate [12042-78-3], $Ca_3Al_2O_6$, and ferrite phase (see FERRITES). A small amount of calcium sulfate [7778-18-9], $CaSO_4$, in the form of gypsum or anhydrite is also added during grinding to control the setting time and enhance strength development (see CALCIUM COMPOUNDS, CALCIUM SULFATE).

The demand for cement was stimulated by the growth of canal systems in United States during the nineteenth century. Process improvements were made in the calcination of certain limestones for the manufacture of natural cements, which were gradually displaced by portland cement. This latter was named in a 1824 patent because of its color and resemblance to a natural limestone quarried on the Isle of Portland in England. Research conducted since that time has provided a clear picture of the composition, properties, and fields of stability of the principal systems found in portland cement. These results led to the widely used Bogue calculation of composition based on oxide analysis (1). Details beyond the scope of this article may be found in the literature (2).

2. Clinker Chemistry

The conventional cement chemists' notation uses abbreviations for the most common constituents: calcium oxide [1305-78-8], CaO, = C; silicon dioxide [7631-86-9],

SiO_2 = S; aluminum oxide [1344-28-1], Al_2O_3 = A; ferric oxide [1309-37-1], Fe_2O_3, = F; magnesium oxide [1309-48-4], MgO, = M; sulfur trioxide [7446-11-9], SO_3, = $\bar{S}$; sodium oxide [1313-59-3], Na_2O, = N; potassium oxide [12136-45-7], K_2O, = K; carbon dioxide, CO_2, C; and water, H_2O, = H. Thus tricalcium silicate, Ca_3SiO_5, is denoted by C_3S.

Portland cement clinker is formed by the reactions of calcium oxide and acidic components to give C_3S, C_2S, C_3A, and a ferrite phase approximating C_4AF.

2.1. Phase Equilibria.   During burning in the kiln, ~20–30% of liquid forms in the mix at clinkering temperatures. Reactions occur at surfaces of solids and in the liquid. The crystalline silicate phases formed are separated by the interstitial liquid. The interstitial phases formed from the liquid in normal clinkers during cooling are also shown by X-ray diffraction (XRD) to be completely

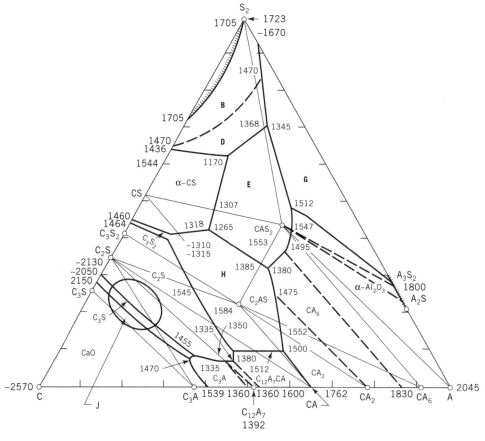

Fig. 1.   Phase equilibria in the C–A–S (CaO–Al_2O_3–SiO_2) system (3,4); temperatures are in °C. Shaded areas denote two liquids; compositional index marks on the triangle are indicated at 10% intervals; **B** denotes cristobalite [14464-46-1], and **D** denotes tridymite [15468-32-3], both of SiO_2 composition; E is anorthite [1302-54-1], $Al_2CaSi_2O_8$; G is mullite [55964-99-3]; **H**, gehlenite [1302-56-3], $Ca_2Al_2SiO_7$; and **J** is the area of portland cement compositions.

crystalline, although they may be so finely subdivided as to appear glassy (optically amorphous) under the microscope.

The high temperature phase equilibria governing the reactions in cement kilns have been studied, eg, in the $CaO-Al_2O_3-SiO_2$ system illustrated in Figure 1. In such a ternary diagram, the primary-phase fields are plotted, ie, the composition regions in which any one solid is the first to separate when a completely liquid mix is cooled to produce negligible supercooling. The primary-phase fields are separated by eutectic points on the sides of the triangle such as that at 1436°C between tridymite and α-CS.

In the relatively small portland cement zone, almost all modern cements fall in the high-lime portion ($\sim$65% CaO). Cements of lower lime content tend to be slow in hardening and may show trouble from dusting of the clinker by transformation of β- to γ-C_2S, especially if clinker cooling is very slow. The zone is limited on the high lime side by the need to keep the uncombined CaO to low enough values to prevent excessive expansion from hydration of the free lime. Commercial manufacture at compositions near the $CaO-SiO_2$ axis can present difficulties. If the lime content is high, the burning temperatures may be so high as to be impractical. If the lime content is low, the burning temperatures may even be low, but impurities must be present in the C_2S to prevent dusting. On the high alumina side the zone is limited by excessive liquid-phase formation that prevents proper clinker formation in rotary kilns.

The relations between the compositions of portland cements and some other common hydraulic cements are shown in the $CaO-SiO_2-Al_2O_3$ phase diagram of Figure 2 (5). In this diagram, Fe_2O_3 has been combined with Al_2O_3 to yield

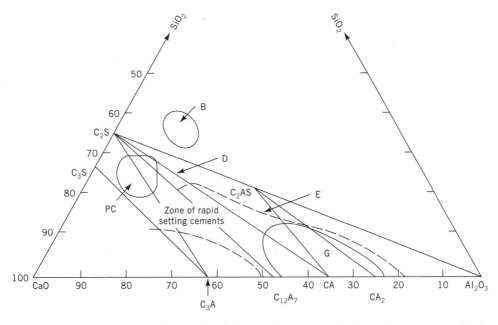

Fig. 2. Cement zones in the $CaO-Al_2O_3-SiO_2$ system (5) where B represents basic blast-furnace slag; D, cement compositions which dust on cooling; E, compositions showing no tendency to set; G, aluminous cement; and PC, portland cement.

the Al_3O_3 content used. This commonly applied approximation permits a two-dimensional representation of the real systems.

2.2. Clinker Formation.   Portland cements are ordinarily manufactured from raw mixes including components such as calcium carbonate, clay or shale, and sand. As the temperature of the materials increases during their passage through the kiln, the following reactions occur: evaporation of free water; release of combined water from the clay; decomposition of magnesium carbonate; decomposition of calcium carbonate (calcination); and combination of the lime and clay oxides. The course of these last reactions (6), which occur at the high temperature end of the kiln, just before and in the burning zone, is illustrated graphically in Figure 3 (7).

From the phase diagram of the $CaO-SiO_2-Al_2O_3$ system, the sequence of crystallization during cooling of the clinker can be derived if the cooling is slow enough to maintain equilibrium. For example, a mix at $1500°C$ of relatively low lime content, along the C_3S-C_2S eutectic line in Figure 1, is composed of solid C_3S and C_2S and a liquid along the C_3S-C_2S eutectic at the intersection with the $1500°C$ isotherm to the left of the 1470–1455 line. Upon cooling, this liquid deposits more C_3S and C_2S, moving the liquid composition down to the invariant point at $1455°C$, at which C_3A also separates until crystallization is complete. Although real cement clinkers contain more components, which alter the system and temperatures somewhat, the behavior is similar.

Cooling is ordinarily too rapid to maintain the phase equilibria. In the case in Figure 1, the lime-deficient liquid at $1455°C$ requires that some of the solid C_3S redissolve and that more C_2S crystallize during crystallization of the C_3A. During rapid cooling, there may be insufficient time for this reaction and the C_3S content is thus higher than when equilibrium conditions prevail. In this event, crystallization is not completed at $1455°C$, but continues along the C_3A-C_2S boundary until the invariant point at $1335°C$ is reached. Crystallization of

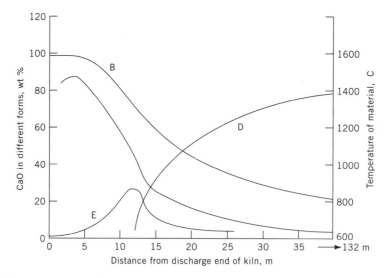

Fig. 3.   Temperatures and progress of reactions in a 132-meter wet-process kiln; area B represents proportion of total CaO in new compounds; D, CaO as $CaCO_3$; and E, free CaO.

C_2S, C_3A, and $C_{12}A_7$ then occurs to reach complete solidification. Such deviations from equilibrium conditions cause variations in the phase compositions that are estimated from the Bogue calculation, and cause variations in the amounts of dissolved substances such as MgO, alkalies, and the alumina content of the ferrite phase.

The theoretical energy requirement for the burning of portland cement clinker can be calculated from the heat requirements and energy recovery from the various stages of the process. Knowledge of the specific heats of the various phases, and the heats of decomposition, transformation, and reaction then permits calculation of the net theoretical energy requirement of 1760 kJ (420 kcal) for 1 kg of clinker from 1.55 kg of dry $CaCO_3$ and kaolin (see CLAYS) (8).

The kinetics of the reactions are strongly influenced by the temperature, mineralogical nature of the raw materials, fineness to which the raw material is ground, percentage of liquid phase formed, and viscosity of the liquid phase. The percentage of liquid formed depends on the alumina and iron oxides. When the sum of these oxides is low, the amount of liquid formed is insufficient to permit rapid combination of the remaining CaO. The viscosity of the liquid at clinkering temperature is reduced by increasing the amounts of oxides such as MnO, Fe_2O_3, MgO, CaO, and Na_2O (9).

The reaction of C_2S with CaO to form C_3S depends on dissolution of the lime in the clinker liquid. When sufficient liquid is present, the rate of solution is controlled by the size of the CaO particles, which depends in turn on the sizes of the particles of ground limestone. Coarse particles of calcite fail to react completely under commercial burning conditions. The reaction is governed by the rate of solution (10):

$$\log t = \log \frac{D}{A} + 0.43 \frac{E}{RT}$$

t is the time in minutes, D is the particle diameter in millimeters, A is a constant, T the absolute temperature, and E is the activation energy having a value of 607 kJ/mol (146 kcal/mol). For example, 0.05-mm particles require 59 min for solution at 1340°C but only 2.3 min at 1450°C. A similar relation applies for the rate of solution of quartz grains.

2.3. Phases Formed in Portland cements. Most clinker compounds take up small amounts of other components to form solid solutions (11). Best known of these phases is the C_3S solid solution called alite. Phases that may occur in portland cement clinker are given in Table 1. In addition, a variety of minor phases may occur in portland cement clinker when certain minor elements are present in quantities above that which can be dissolved in other phases. Under reducing conditions in the kiln, reduced phases, such as ferrous oxide [1345-25-1], FeO, and calcium sulfide [20548-54-3], CaS, may be formed.

The primary phases all contain impurities. In fact, these impurities stabilize the structures formed at high temperatures so that decomposition or transformations do not occur during cooling, as occurs with the pure compounds. For example, pure C_3S exists in at least six polymorphic forms each having a sharply defined temperature range of stability, whereas alite exists in three stabilized forms at room temperature depending on the impurities. Some properties of the more common phases in Portland clinkers are given in Table 2.

Table 1. **Phases in Portland Cement Clinker**[a]

Name of impure form	CAS Registry Number	Chemical name	Cement chemists' notation
free lime	[1305-78-8]	calcium oxide	C
periclase (magnesia)	[1309-48-4] and [1317-74-4]	magnesium oxide	M
alite	[12168-85-3]	tricalcium silicate	C_3S
belite	[10034-77-2]	dicalcium silicate	C_2S
C_3A	[12042-78-3]	tricalcium aluminate	C_3A
ferrite	[12612-16-7]	calcium aluminoferrite[b]	$C_2A_xF_{1-x}$
	[12068-35-8]	tetracalcium aluminoferrite	C_4AF
	[12013-62-6]	dicalcium ferrite[c]	C_2F
mayenite	[12005-57-1]	12-calcium-7-aluminate	$C_{12}A_7$
gehlenite	[1302-56-3]	dicalcium alumino monosilicate	C_2AS
aphthitalite	[12274-74-4] and [17926-93-1]	sodium, potassium sulfate[d]	$N_xK_y\bar{S}$
arcanite	[7778-80-5] and [14293-72-2]	potassium sulfate	$K\bar{S}$
metathenardite	[7757-82-6]	sodium sulfate form I	$N\bar{S}$
calcium langbeinite	[14977-32-8]	potassium calcium sulfate	$2X\bar{S} \cdot K\bar{S}$
anhydrite	[7778-18-9] and [14798-04-0]	calcium sulfate	$C\bar{S}$
calcium sulfoaluminate	[12005-25-3]	tetracalcium trialuminate-sulfate	$C_4A_3\bar{S}$
alkali belite	[15669-83-7]	α'- or β-dicalcium (potassium) silicate[e]	$K_xC_{23}S_{12}$
alkali aluminate	[12004-54-3]	8-calcium disodium trialuminate	NC_8A_3
	[65430-58-2]	5-calcium disilicate monosulfate	$2C2S \cdot C\bar{S}$
spurrite	[1319-44-42]	5-calcium disilicate monocarbonate	$2C2S \cdot C\bar{S}$
	[12043-73-1]	calcium aluminate chloride	$C_{11}A_7 \cdot CaCl_2$[f]
	[12305-57-6]	calcium aluminate fluoride	$C_{11}A_7 \cdot CaF_2$[f]

[a]Refs. (12–14).
[b]Solid solution series where $x = A/(A + F)$; $0 < x < 0.7$.
[c]End member of series.
[d]Solid solution series when $\frac{1}{3} \leq x/y$.
[e]Solid solution series when $x \leq 1$.
[f]Mixed notation.

2.4. Structure. Examination of thin sections of clinkers using transmitted light, and of polished sections by reflected light, reveals details of the structure. The polarizing microscope has been used to determine the size and birefringence of alite crystals, and the size and color of the belite to predict later age strength (17). The clinker phases are conveniently observed by examining polished sections selectively etched using special reagents as shown in Figure 4. The alite appears as clear euhedral crystalline grains, the belite as rounded striated grains, the C_3A as dark interstitial material, and the C_4AF as light interstitial material.

Table 2. **Properties of the More Common Phases in Portland Cement Clinker**[a]

Name	Crystal system	Density, g/L	Mohs' hardness
alite	triclinic		
	monoclinic	3.14–3.25	~ 4
	trigonal		
belite	hexagonal	3.04	
	orthorhombic	3.40	
	monoclinic	3.28	>4
	orthorhombic	2.97	
C_3A	cubic	3.04	<6
ferrite	orthorhombic	3.74–3.77	~ 5
free lime	cubic	3.08–3.32	3–4
magnesia	cubic	3.58	5.5–6

[a]Refs. (11–16).

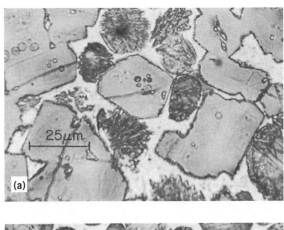

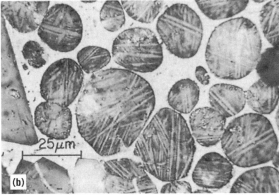

Fig. 4. Photomicrograph of polished and etched sections of portland cement clinkers. The C_3A appears as dark interstitial material, the C_4AF as light interstitial material. (**a**) Euhedral and subhedral alite crystals and rounded or ragged belite; (**b**) rounded and striated belite crystals.

Portland cement clinker structures (11,18,19) vary considerably with composition, particle size of raw materials, and burning conditions, resulting in variations of clinker porosity, crystallite sizes and forms, and aggregations of crystallites. Alite sizes range up to ∼80 μm or even larger, most being 15–40 μm.

2.5. Raw Material Proportions. The three main considerations in proportioning raw materials for cement clinker are the potential compound composition; the percentage of liquid phase at clinkering temperatures; and the burnability of the raw mix, ie, the relative ease, in terms of temperature, time, and fuel requirements, of combining the oxides into good quality clinker. The ratios of the oxides are related to clinker composition and burnability. For example, as the CaO content of the mix is increased, more C_3S can be formed, but certain limits cannot be exceeded under normal burning conditions. The lime saturation factor (LSF) is a measure of the amount of CaO that can be combined (20):

$$\mathrm{LSF} = \frac{\%\ \mathrm{CaO}}{2.8\ (\%\ \mathrm{SiO_2}) + 1.1(\%\ \mathrm{Al_2O_3}) + 0.7(\%\ \mathrm{Fe_2O_3})}$$

An LSF of 100 would indicate that the clinker can contain only C_3S and the ferrite solid solution. Lime saturation factors of 88–94 are frequently appropriate for reasonable burnability; low LSF indicates insufficient C_3S for acceptable early strengths, and higher values may render the mix very difficult to burn. Several other weight ratios such as the silica modulus and the iron modulus are also important (21).

The potential liquid-phase content at clinkering temperatures range from 18 to 25% and can be estimated from the oxide analysis of the raw mix. For example (22), for 1450°C:

$$\%\ \text{liquid phase} = 1.13(\%\ C_3A) + 1.35(\%\ C_4AF) + \%\ M + \%\ \text{alkalies}$$

The potential compound composition of a cement or cement clinker can be calculated from the oxide analyses of any given raw materials mixture, or from the oxide analyses of the cement clinker or finished cement. The simplest and most widely used method is the Bogue calculation (23). The ASTM C150 (24) calculation is somewhat modified.

The techniques of determining the proper proportions of raw materials to achieve a mix of good burnability and clinker composition may be determined by computer using an iterative program, starting with raw components of known composition. The concept of targets may be utilized, including fixed values of moduli, compound content, and amount of any raw material element in the final clinker. The number of targets that may be set is one less than the number of raw materials. The fuel ash must be considered as one of the raw materials. Representative chemical analyses of raw materials used in making portland and high alumina cements are given in Table 3, analyses of cements of various types appear in Table 4, along with their potential compound compositions.

Table 3. **Chemical Composition of Raw Materials**[a]**, wt** %

Type	SiO_2	Al_2O_3	Fe_2O_3	CaO	MgO	Loss on ignition
cement rock	13.4	3.5	1.7	42.9	1.0	37.2
limestone	1.2	0.2	0.4	53.4	1.3	43.4
dolomite	4.5	0.5	1.6	35.0	14.9	44.0
marl	6.0	0.6	2.3	49.1	0.4	40.4
oyster shells	1.5	0.4	1.2	52.3	0.7	41.8
shale	53.8	18.9	7.7	3.2	2.2	8.2
clay	67.8	14.3	4.5	0.9	1.2	8.0
mill scale			~100.0			
sandstone	76.6	5.3	3.1	4.7	1.7	6.6
bauxite	10.6	57.5	2.6			28.4

[a]Courtesy of the American Concrete Institute (25).

3. Hydration

3.1. Calcium Silicates. In hydrations at ordinary temperatures (27) pure C_3S and β-C_2S, corresponding to the alite and belite phases in portland cements, respectively, react with water to form calcium hydroxide and a single calcium silicate hydrate (C–S–H). Table 5 shows primary compound hydration reactions but not the many minor reactions.

These are the main reactions in portland cements. The two calcium silicates constitute ~75% of the cement. The average lime/silica ratio (C/S) in calcium silicate hydrate (C–S–H) may vary from ~1.5 ~2.0 or even higher, the average value is ~1.7. The water content varies with the ambient humidity, the 3 mol of water being estimated from measurements in the dry state and structural considerations. As the lime/silica ratio of the C–S–H increases, the amount of water increases on an equimolar basis, ie, the lime goes into the structure, resulting in less free calcium hydroxide.

Calcium silicate hydrate is not only variable in composition, but is very poorly crystallized, and archaically referred to as calcium silicate hydrate gel or tobermorite gel because of the colloidal sizes (<0.1 μm). The calcium silicate hydrates are layer minerals having many similarities to the limited swelling clay minerals found in nature. The layers are bonded together by excess lime and interlayer water to form individual gel particles only 2–3 layers thick. Surface forces, and excess lime on the particle surfaces, tend to bond these particles together into aggregations or stacks of the individual particles to form the porous gel structure.

Significant changes in the structure of the gel continue over very long periods. During the first month of hydration, appreciable quantities of the dimeric silicate anion $Si_2O_7^6$ are formed. These are reduced by later condensation to higher polysilicates, the amount of which together with the mean length of the metasilicate chains continues to increase for at least 15 years of moist curing. In one study, a mean length of 15.8 silica tetrahedra was found after such prolonged curing (28). These changes appear to have a positive effect on both strength development and reduction of drying shrinkage.

Table 4. **Chemical Composition, Compounds, and Fineness of Cements (51)**[a,b]

Type of Portland cement	Chemical composition, %							Potential compound composition, %				Blaine fineness m²/kg
	SiO_2	Al_2O_3	Fe_2O_3	CaO	MgO	SO_3	Na_2O eq	C_3S	C_2S	C_3A	C_4AF	
I (min–max)	18.7–22.0	4.7–6.3	1.6–4.4	60.6–66.3	0.7–4.2	1.8–4.6	0.11–1.20	40–63	9–31	6–14	5–13	300–421
I (mean)	20.5	5.4	2.6	63.9	2.1	3.0	0.61	54	18	10	8	369
II[c] (min–max)	20.0–23.2	3.4–5.5	2.4–4.8	60.2–65.9	0.6–4.8	2.1–4.0	0.05–1.12	37–68	6–32	2–8	7–15	318–480
II[c] (mean)	21.2	4.6	3.5	63.8	2.1	2.7	0.51	55	19	6	11	377
III (min–max)	18.6–22.2	2.8–6.3	1.3–4.9	60.6–65.9	0.6–4.6	2.5–4.6	0.14–1.20	46–71	4–27	0–13	4–14	390–644
III (mean)	20.6	4.9	2.8	63.4	2.2	3.5	0.56	55	17	9	8	548
IV (min–max)	21.5–22.8	3.5–5.3	3.7–5.9	62.0–63.4	1.0–3.8	1.7–2.5	0.29–0.42	37–49	27–36	3–4	11–18	319–362
IV (mean)	22.2	4.6	5.0	62.5	1.9	2.2	0.36	42	32	4	15	340
V (min–max)	20.3–23.4	2.4–5.5	3.2–6.1	61.8–66.3	0.6–4.6	1.8–3.6	0.24–0.76	43–70	11–31	0–5	10–19	275–430
V (mean)	21.9	3.9	4.2	63.8	2.2	2.3	0.48	54	22	4	13	373
White (min-max)	22.0–24.4	2.2–5.0	0.2–0.6	63.9–68.7	0.3–1.4	2.3–3.1	0.09–0.38	51–72	9–25	5–13	1–2	384–564
White (mean)	22.7	4.1	0.3	66.7	0.9	2.7	0.18	63	18	10	1	482

[a]Courtesy of the American Concrete Institute (51).
[b]Values represent a summary of combined statistics. Air-entraining cements are not included. For consistency in reporting elements are reported in a standard oxide form. This does not mean that the oxide form is present in the cement. For example, sulfur is reported an SO_3, sulfur trioxide, but portland cement does not have sulfur trioxide present. "Potential Compound Composition" refers to ASTM C 150 (AASHTO M 85) calculations using the chemical composition of the cement. The actual compound composition may be less due to incomplete or altered chemical reactions.
[c]Includes fine ground cements.

Table 5. **Portland Cement Compound Hydration Reactions (Oxide Notation)**

$2(3CaO \cdot SiO_2)$ Tricalcium silicate	$+ 11H_2O$ Water		$= 3CaO \cdot 2SiO_2 \cdot$ $8H_2O$ Calcium silicate hydrate (C-S-H)	$+ 3(CaO \cdot H_2O)$ Calcium hydroxide
$2(2CaO \cdot SiO_2)$ Dicalcium silicate	$+ 9H_2O$ Water		$= 3CaO \cdot 2SiO_2 \cdot$ $8H_2O$ Calcium silicate hydrate (C-S-H)	$+ CaO \cdot H_2O$ Calcium hydroxide
$3CaO \cdot Al_2O_3$ Tricalcium aluminate	$+ 3(CaO \cdot SO_3 \cdot$ $2H_2O)$ Gypsum	$+ 26H_2O$ Water	$= 6CaO \cdot Al_2O_3 \cdot$ $3SO_3 \cdot 32H_2O$ Ettringite	
$2(3CaO \cdot Al_2O_3)$ Tricalcium aluminate	$+ 6CaO \cdot Al_2O_3 \cdot$ $3SO_3 \cdot 32H_2O)$ Ettringite	$+ 4H_2O$ Water	$= 3(4CaO \cdot Al_2O_3 \cdot$ $SO_3 \cdot 12H_2O)$ Calcium mono- sulfoaluminate	
$3CaO \cdot Al_2O_3$ Tricalcium aluminate	$+ CaO \cdot H_2O$ Calcium hydroxide	$+ 12H_2O$ Water	$= 4CaO \cdot Al_2O_3 \cdot$ $13H_2O$ Tetra- calcium alumi- nate hydrate	
$4CaO \cdot Al_2O_3 \cdot Fe_2O_3$ Tricalcium aluminoferrite	$+ 10H_2O$ Water	$+ 2(CaO \cdot H_2O)$ Calcium hydroxide	$= 6CaO \cdot Al_2O_3 \cdot$ $Fe_2O_3 \cdot 12H_2O$ Calcium aluminoferrite hydrate	

Drying and other chemical processes can have significant effects on this structure, there being loss of hydrate water as well as physically adsorbed water, and collapse of the structure to form more stable aggregations of particles (29,30).

3.2. Tricalcium Aluminate and Ferrite. The hydration of the C_3A alone and in the presence of gypsum usually produces well-crystallized reaction products that can be identified by X-ray diffraction and other methods. C_3AH_6 is the cubic calcium aluminate hydrate; C_4AH_{19} and $C_4A\bar{S}H_{12}$ are hexagonal phases, the latter being commonly referred to as the monosulfate. The highly hydrated trisulfate, ettringite, $C_6A\bar{S}_3H_{32}$, occurs as needles, rods, or dense columnar aggregations. Its formation on the surfaces of anhydrous grains is responsible for the necessary retardation of hydration of the aluminates in Portland cements and the expansion process in expansive cements (31).

The early calcium aluminate hydration reactions in portland cements have been studied in simple mixtures of C_3A, gypsum, calcium hydroxide, and water (32). Figure 5 shows the progressive reaction of the gypsum, water, and C_3A as ettringite is formed, the reaction of the ettringite, calcium hydroxide, and water to form the monosulfate, and the solid of the monosulfate with C_4AH_{19}. These reactions are important in the portland cements to control the hydration of the C_3A, which otherwise might hydrate so rapidly as to cause flash set, or premature stiffening, in fresh concrete.

Other reactions taking place throughout the hardening period are substitution and addition reactions (29). Ferrite and sulfoferrite analogues of calcium monosulfoaluminate and ettringite form solid solutions in which iron oxide

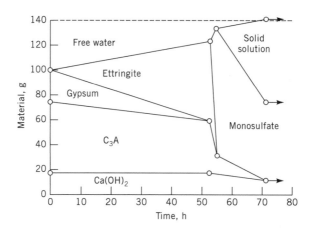

Fig. 5. The early hydration reactions of tricalcium aluminate in the presence of gypsum and calcium hydroxide. Initial molar proportions: $1\text{-}C_3A$; $1\text{-}Ca(OH)_2$; $3/4\text{-}CaSO_4 \cdot 2H_2O$; 0.4 water–solids ratio (32).

substitutes continuously for the alumina. Reactions with the calcium silicate hydrate result in the formation of additional substituted C–S–H hydrate at the expense of the crystalline aluminate, sulfate, and ferrite hydrate phases.

The hydration of the ferrite phase (C_4AF) is of greatest interest in mixtures containing lime and other cement compounds because of the strong tendency to form solid solutions. When the sulfate in solution is very low, solid solutions are formed between the cubic C_3AH_6 and analogous iron hydrate C_3FH_6. In the presence of water and silica, solid solutions such as $C_3ASH_4 \cdot C_3FSH_4$ may be formed (33). Table 6 lists some of the important phases formed in the hydration of mixtures of pure compounds.

3.3. Other Phases. In cements free lime, CaO, and periclase, MgO, hydrate to the hydroxides. The *in situ* reactions of larger particles of these phases can be rather slow and may not occur until the cement has hardened. These reactions then can cause deleterious expansions and even disruption of the concrete and the quantities of free CaO and MgO have to be limited. The soundness of the cement can be tested by the autoclave expansion test of portland cement ASTM C151 (24).

The expansive component $C_4A_3\bar{S}$ in Type K expansive cements hydrates in the presence of excess sulfate and lime to form ettringite is

$$C_4A_3\bar{S} + 8\ C\bar{S}H_2 + 6\ CH + 74\ H \longrightarrow 3\ C_6A\bar{S}_3H_{32}$$

The reactions in the regulated-set cements containing $C_{11}A_7 \cdot CF_2$ (note mixed notation) as a principal phase resemble those in ordinary portland cements. Initial reaction rates are controlled by ettringite formation. Setting occurs with formation of the monosulfate, along with some transitory lower-limed calcium aluminate hydrates that convert to the monosulfate within a few hours.

Pozzolans contain reactive silica that reacts with cement and water by combining with the calcium hydroxide released by the hydration of the calcium silicates to produce additional calcium silicate hydrate. If sufficient silica is added,

Table 6. Cement Phases Hydrated at Normal Temperatures[a]

Name	CAS Registry Number	Approximate composition[b]	Stability range rh, 25°C	Stability range Temp, °C	Crystal system	Density, kg/m³
calcium sulfate dihydrate (gypsum)	[10101-41-4] [13397-24-5]	$C\bar{S}H_2$	100–35	<100	monoclinic	2.32
calcium hydroxide (portlandite)	[1305-62-0]	CH	100–0	<512	trigonal–hexagonal	2.24
magnesium hydroxide (brucite)	[1309-42-8]	MH	100–0	<350	trigonal–hexagonal	2.37
calcium silicate hydrate gel (C–S–H gel)	[12323-54-5]	$C_xS_yH_z$[c]	indefinite	indefinite	indefinite	2.7[d]
tetracalcium aluminate,						
19-hydrate	[12042-86-3]	C_4AH_{19}	100–85	<15	trigonal–hexagonal	1.80
13-hydrate	[12042-85-2]	C_4AH_{13}	81–12		trigonal–hexagonal	2.02
7-hydrate	[12511-52-3]	C_4AH_7	2–0	to 120		
tetracalcium aluminate monosulfate,						
16-hydrate	[67523-83-5]	$C_4A\bar{S}H_{16}$	aq	<8	trigonal–hexagonal	
14-hydrate	[12421-30-6]	$C_4A\bar{S}H_{14}$	100–95	>9	trigonal–hexagonal	
12-hydrate	[12252-10-7]	$C_4A\bar{S}H_{12}$	95–12	>1	trigonal–hexagonal	1.95
10, 8, x-hydrate	[12252-09-4] [12445-38-4]	$C_4A\bar{S}H_x$	<12			
ettringite (6-calcium aluminate trisulfate, 32-hydrate)	[12252-15-2]	$C_6A\bar{S}_3H_{32}$	100–4	<60	trigonal–hexagonal	1.73–1.79
garnet-hydrogarnet solid solution series	[11070-82-9]	$C_6A\bar{S}_3H_8$ $C_3(F_{1-x}A_x)(S_{1-y}H_{2y})_3$[e]	4–2 stable	<110	cubic	
	[12042-80-7]	end member: C_3AH_6	100–0	>15	cubic	2.52

[a]Ref. 12,13, and 34.
[b]In cement chemists' notation.
[c]Where $1.3 < x/y < 2$ and probably $1 < z/y < 1.5$.
[d]Wet.
[e]$x = \dfrac{A}{A+F}$ and $y = \dfrac{2H}{2H+S}$.

479

~30% of the weight of cement, the calcium hydroxide can eventually be completely combined. Granulated blast-furnace slag is not ordinarily reactive in water, but in the presence of lime reactions occur with the silica framework. This breakdown of the slag releases other components so that a variety of crystalline hydrate phases can also form.

3.4. Hydration Process.   Portland cement is generally used at temperatures ordinarily encountered in construction, ie, from 5 to 40°C. Temperature extremes have to be avoided. The exothermic heat of the hydration reactions can play an important part in maintaining adequate temperatures in cold environments, and must be considered in massive concrete structures to prevent excessive temperature rise and cracking during subsequent cooling. Heat indued delayed expansion (delayed ettringite expansion) can also be controlled by keeping the concrete temperature <70°C.

The initial conditions for the hydration reactions are determined by the concentration of the cement particles (0.2–100 μm) in the mixing water (w/c = 0.3–0.7 on a wt% basis) and the fineness of the cement (300–600 m^2/kg). Upon mixing with water, the suspension of particles as shown in Figure 6 (35) is such that these particles are surrounded by films of water having an average thickness of ~1 μm. The anhydrous phases initially react by the formation of surface hydration products on each grain, and by dissolution into the liquid phase. The solution quickly becomes saturated with calcium and sulfate ions, and the

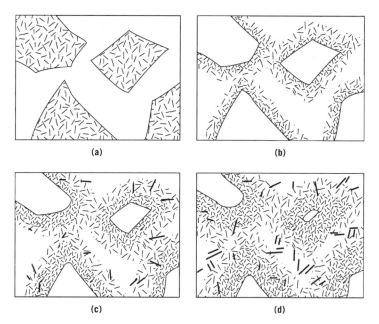

(a)

(b)

(c)

(d)

Fig. 6.   Four stages in the setting and hardening of portland cement: simplified representation of the sequence of changes. (**a**) Dispersion of unreacted clinker grains in water. (**b**) After a few minutes; hydration products eat into and grow out from the surface of each grain. (**c**) After a few hours; the coatings of different clinker grains have begun to join up, the hydration products thus becoming continuous (setting). (**d**) After a few days; further development of hydration products has occurred (hardening. Courtesy of Academic Press Inc. (London) Ltd. (35).

concentration of alkali cations increases rapidly. These reactions consume part of the anhydrous grains, but the reaction products tend to fill that space as well as some of the originally water-filled space. The porous hydration products occupy about twice the volume of the reacted anhydrous material (36). The hydration products at this stage are mostly colloidal (<0.1 μ) but some larger crystals of calcium aluminate hydrates, sulfoaluminate hydrate, and hydrogarnets form. As the reactions proceed the coatings increase in thickness and eventually form bridges between the original grains. This is the stage of setting. Despite the low solubility and mobility of the silicate anions, growths of the silicate hydrates also form on the crystalline phases formed from the solution and become incorporated into the calcium hydroxide and other phases. Upon further hydration the water-filled spaces become increasingly filled with reaction products to produce hardening and strength development.

The composition of the liquid phase during the early hydration of portland cements is controlled mainly by the solution of calcium, sulfate, sodium, and potassium ions. Very little alumina, silica, or iron are present in solution. Calcium hydroxide, as calcium oxide, and gypsum, as calcium sulfate, alone have solubilities of ~1.1 and 2.1 g/L at 25°C, respectively. In the presence of alkalies released in the first 7 min, the composition tends to be governed by the equilibrium:

$$CaSO_4 + 2\,MOH \rightleftharpoons M_2SO_4 + Ca(OH)_2$$

where M represents the alkalies. At advanced stages of hydration of low water/cement ratio pastes, the alkali solution concentration may exceed 0.4 N with a pH > 13. Saturated lime/water has a pH of 12.4 at 25°C.

The exact course of the early hydration reactions depends mainly on the C_3A, ferrite, and soluble alkali contents of the clinker, and the amount of gypsum in the cement. Following the rate of reaction by calorimetric measurements, at least two and sometimes three distinct peaks in the rate of heat liberation can be observed (33,37). A large initial peak lasts only a few minutes and may reach 6 J/(g · min)[1.4 cal/(g · min)] resulting mainly from the solution of soluble constituents and the surface reactions, especially the formation of the sulfoaluminate coating on the highly reactive C_3A phase. After the initial heat peak, the reactions are strongly retarded, producing a 1–2-h delay referred to as the induction period during which the cement/water paste remains plastic and the concrete is workable. The cause of the induction period is a subject of much debate. The most widely supported explanation of the induction period involves an initial reaction which forms a protective layer on the C_3S particles (38). Within 30 s, the C_3S is almost isolated from the solution. Eventually, C–S–H nucleates and grows. The protective layer is disrupted and increased access to the C_3S leads to an increasing rate of hydration, which produces the second heat peak. Around the time of the second heat peak the concrete sets and establishes its familiar monolithic structure. A third heat peak may be observed depending on the gypsum and C_3A contents. If present, the third peak corresponds to the exhaustion of the solid gypsum, a rapid decrease of sulfate in solution, conversion of the ettringite to monosulfate, and renewed rapid reaction of the remaining C_3A and ferrite phases. At optimum gypsum content the third peak ordinarily occurs between 18 and 24 h.

In these early reactions, the reactivities of the individual phases are important in determining the overall reaction rate. However, as the cement particles become more densely coated with reaction products, diffusion of water and ions in solution becomes increasingly impeded. The reactions then become diffusion controlled at some time depending on various factors such as temperature and water/cement ratio. After ~1 or 2 days, ie, at ~40% of complete reaction, the remaining unhydrated cement phases react more nearly uniformly.

Microscopic examination of sections of hardened cement paste show that the unhydrated cores of the larger cement particles can be distinguished from the hydrated portion or inner product, which is a pseudomorph of the original grain, and the outer product formed in the originally water-filled spaces. Measurements of these cores indicate the depth of penetration of the hydration reactions (39). The overall hydration rate increases with the temperature, the fineness of the cement, and to a lesser extent with the water/cement ratio; measurements of the activation energy indicate that the reaction becomes increasingly diffusion controlled (33). Although more finely ground cements hydrate more rapidly in the first month or more of hydration, these differences gradually disappear at later ages. After 1 year, most portland cements at usual water/cement ratios are >90% hydrated if continuously moist cured. At complete hydration the chemically combined water (the water retained after strong drying) is ~20–25% of the weight of the cement, depending on its composition. However, a minimum water/cement ratio of ~0.4 is required to provide enough space to permit complete hydration of the cement (36). If moist curing is stopped and the hardened cement is dried sufficiently, eg, to 80% relative humidity (rh), the hydration process stops.

4. Cement Paste Structure and Concrete Properties

The properties of both fresh and hardened mortars and concretes depend mainly on the cement/water paste properties. Practical engineering tests are usually made using concrete specimens because their properties also depend on the proportions, size gradation, and properties of the aggregates. Quality control testing and research on cement properties is usually done on cement pastes or cement mortars made with standard sands. The properties of hardened cement pastes, mortars, and concretes are similar functions of the water/cement ratio and degree of hydration of the cement. The properties of fresh concretes that determine the workability, or ease of mixing and placement into forms, also depend strongly on, but are not so simply related to, the cement paste rheological properties (40,41).

The fresh paste even in the dormant period is normally thixotropic, or shear thinning, indicating that the structure is being continuously broken down and re-formed during mixing. It is an approximately Bingham plastic body having a finite yield value and plastic viscosity from 5000 to 500 mPa $\cdot$ s($=$cP) as the water/cement ratio increases from 0.4 to 0.7 (42). The viscosity and yield values can be greatly reduced by the addition of certain organic water-reducing admixtures especially formulated for this purpose. Workability of concrete is measured by the slump of the concrete determined after removal of a standard slump cone (305 mm high) (43). Workable concretes have slumps of 75 mm or more.

After mixing and casting, sedimentation of the cement particles in the water results in bleeding of water to the top surface and reduction of water/ cement ratio in the paste. At high water/cement ratios, some of the very fine particles may be carried with the bleed water to the top resulting in laitance and perhaps the formation of flaws called bleeding channels. In concretes, sedimentation may cause flaws under the larger aggregate particles. If the fresh concrete is not protected from too rapid surface drying, capillary forces cause drying shrinkage that may cause plastic shrinkage cracks. Good construction practices are designed to minimize all of these flaws.

The engineering properties of the concrete, such as strength, elastic moduli, permeability to water and aggressive solutions, and frost resistance, depend strongly on the water/cement ratio and degree of hydration of the cement. A variety of empirical water/cement ratio laws express the strength as functions of water/cement ratio or porosity. The fraction of the original water-filled space, which is occupied by hydration products at any stage of hydration, is termed the gel-space ratio X. The compressive strength of hardened cement or mortar then approximately fits the power law:

$$\sigma = \sigma_o X^n$$

where n is ~3.0 and σ is the intrinsic strength of the densest ($X = 1$) gel produced by a given cement under normal hydrating conditions. Values of σ_o range upward from 100 MPa (15,000 psi), depending on the cement composition (44). Several direct relationships between porosity, p, and strength have been applied to cement pastes (45–48):

$$\sigma = \sigma_o (1 - p)^B$$

$$\sigma = D \ln (p/p_o)$$

$$\sigma = \sigma_o(1 - Ep)$$

where B, D, and E are constants. Whereas some of these equations break down at very low or very high porosities, they all provide reasonable estimates within an intermediate porosity range. One comparative study concluded that the last equation was the most satisfactory (49). Under extreme conditions, eg, hot pressing at very low water/cement ratios, strengths as high as 655 MPa (95,000 psi) have been reported (50). Tensile strengths and elastic moduli are similarly dependent on porosity or gel/space ratio, but the tensile strength is only about one-tenth of the compressive strength. The Young's modulus of the densest gels produced under normal hydrating conditions is about 34 GPa (5×10^6 psi) (30).

Under sustained loads, hardened cements and concrete creep or deform continuously with time, in addition to the initial elastic deformation. Under normal working loads this deformation may in time exceed the elastic deformation and must be considered in engineering design. This is especially true in prestressed concrete structural members in which steel tendons under high tensile stress maintain compressive stress in the concrete to prevent tensile cracking during bending. Both creep and drying shrinkage of the concrete may lead to loss of prestress. Some creep in ordinary concrete structures and in the cement paste between the aggregate particles can also be an advantage because it tends to reduce stress concentrations, cracking, and microcracking around aggregate particles.

Drying of hardened cements results in shrinkage of the paste structure and of concrete members. Linear shrinkage of hardened cements is ∼0.5% when dried to equilibrium at normal (∼50%) relative humidities. The cement gel structure is somewhat stabilized during drying so that upon subsequent wetting and drying smaller changes occur. Concretes shrink much less (∼0.05%), depending on the volume fractions of cement paste and aggregates, water/cement ratio, and other factors. Drying of concrete structural members proceeds very slowly and results in internal shrinkage stresses because of the moisture gradients during drying. Thick sections continue to dry and shrink for many years. Atmospheric carbon dioxide penetrates the partly dried concrete and reacts with the calcium silicate hydrate gel, as well as with calcium aluminate hydrates, releasing additional water and causing additional shrinkage. The density of the hydration products is increased, however, and the strength is actually increased. This reaction is sometimes used to advantage in the manufacture of precast concrete products to improve their ultimate strength and dimensional stability by precarbonation.

The slowness of drying and the penetration of the hardened cement by carbon dioxide or chemically aggressive solutions, eg, seawater or sulfate ground waters, is a result of the small sizes of the pores. Initially, the pores in the fresh paste are the water-filled spaces (capillaries) between cement particles. As these spaces become subdivided by the formation of the hydration products, the originally continuous pore system becomes one of more discrete pores or capillary cavities separated from each other by gel formations in which the remaining pores are very much smaller. These gel pores are so small (∼3-nm nominal diameter) that most of the water contained in them is strongly affected by the solid surface force four fields. These force fields are responsible for a large increase in the viscosity of the water and a decrease in mobility of ionic species in solution. Hence, the permeability of the paste to both water and dissolved substances is greatly reduced as hydration proceeds. This in part accounts for the great durability of concrete, especially when water/cement ratios are kept low and adequate moist curing ensures a high degree of hydration. High water/cement ratios result in large numbers of the capillary spaces (0.1 μm and larger) interconnected through capillaries that are 10 nm or larger. These capillaries not only lower the strength, but also permit the easy penetration of aggressive solutions. Furthermore, these capillary spaces may become filled with water which freezes < 0°C, resulting in destructive expansions and deterioration of the concrete.

5. Chemical Admixtures

Admixtures include materials other than cement, water, or aggregate added immediately before or during mixing. Admixtures serve a broad range of purposes including the control of setting, control of workability, and control of air content.

5.1. Retarders and Accelerators. Materials that control hardening of cement may be either organic or inorganic. Retarders are often incorporated in oil well cementing and hot-weather concrete applications, whereas accelerators may be useful for cold-weather concrete applications in which higher rates of reactivity are desirable. In most cases, these admixtures are used in low concentrations, suggesting that they act by adsorption.

5.2. Water-Reducers and Plasticizers. Common admixtures that improve fluidity of concrete mixes are often used in high strength concrete. These admixtures make possible the incorporation of silica fume while maintaining necessary workability. Three principal types of superplasticizers are in common use: salts of sulfonated melamine formaldehyde polymers; salts of sulfonated naphthalene formaldehyde polymers; and modified lignosulfonate materials (see AMINO RESINS; LIGNIN; PHENOLIC RESINS). One benefit of these admixtures is improved dispersion of cement grains in the mixing water.

5.3. Air-Entraining Admixtures. Materials that are used to improve the ability of concrete to resist damage from freezing are generally known as air-entraining admixtures. These surfactant admixtures (see SURFACTANTS) produce a foam that persists in the mixed concrete, and serves to entrain many small spherical air voids that measure from 10 to 250 μm in diameter. The air voids alleviate internal stresses in the concrete that may occur when the pore solution freezes. In practice, up to 10% air by volume may be entrained in concrete placed in severe environments.

6. Manufacture

6.1. Portland Cements. The process of portland cement manufacture consists of (1) quarrying and crushing the rock, (2) grinding the carefully proportioned materials to high fineness, (3) subjecting the raw mix to pyroprocessing in a rotary kiln, and (4) grinding the resulting clinker to a fine powder. A layout of a typical plant is shown in Figure 7 (51). Figure 8 gives the layout of a newer dry process plant (52). No new wet process plants are being built, and existing wet process plants are gradually being replaced by dry process plants (only 32.8% are wet process). The plants outlined are typical of installations producing approximately 1000 metric tons per day. Modern installations (53–55) are equipped with innovations such as suspension or grate preheaters, roller mills, or precalciner installations.

Because calcium oxide comprises ~65% of portland cement, these plants are frequently situated near the source of their calcareous material. The requisite silica and alumina may be derived from a clay, shale, or overburden from a limestone quarry. Such materials usually contain some of the required iron oxide, but many plants need to supplement the iron with mill scale, pyrite cinders, or iron ore. Silica may be supplemented by adding sand to the raw mix, whereas alumina can be furnished by bauxites and Al_2O_3-rich kaolinitic clays.

Industrial byproducts are becoming more widely used as raw materials for cement, eg, slags contain carbonate-free lime, as well as substantial levels of

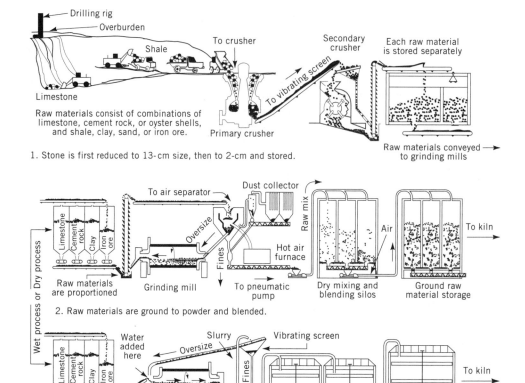

1. Stone is first reduced to 13-cm size, then to 2-cm and stored.

2. Raw materials are ground to powder and blended.

2. Raw materials are ground, mixed with water to form slurry, and blended.

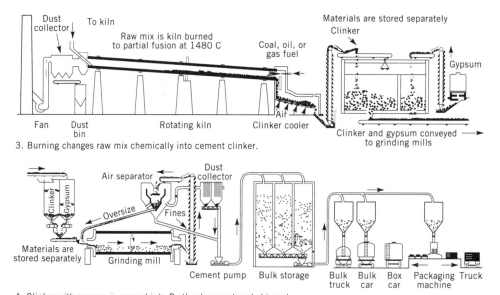

3. Burning changes raw mix chemically into cement clinker.

4. Clinker with gypsum is ground into Portland cement and shipped.

Fig. 7.   Steps in the manufacture of portland cement (51). Courtesy of Portland Cement Association.

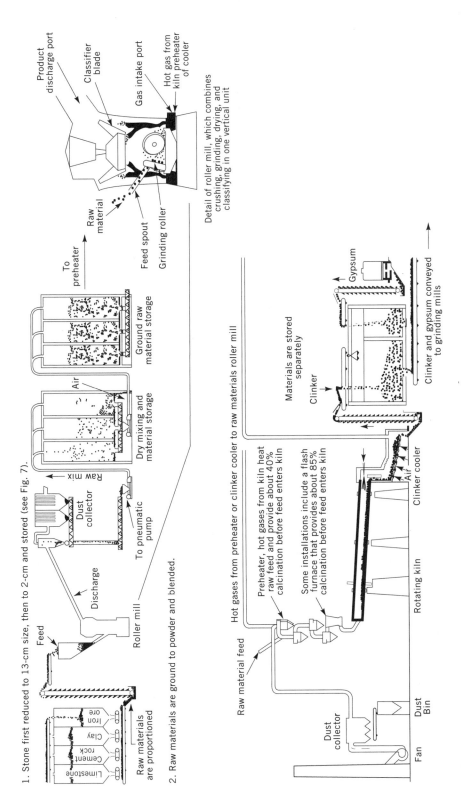

1. Stone first reduced to 13-cm size, then to 2-cm and stored (see Fig. 7).

2. Raw materials are ground to powder and blended.

Detail of roller mill, which combines crushing, grinding, drying, and classifying in one vertical unit

3. Burning changes raw mix chemically into cement clinker. Note four-stage preheater, flash furnaces, and shorter kiln.

4. Clinker with gypsum is ground into Portland cement and shipped (see Fig. 7).

Fig. 8. Technology in dry-process cement manufacture (52). Courtesy of Portland Cement Association.

silica and alumina. Fly ash from utility boilers can often be a suitable feed component, because it is already finely dispersed and provides silica and alumina. Even vegetable wastes, such as rice hull ash, provide a source of silica. Probably 50% of all industrial byproducts are potential raw materials for portland cement manufacture.

Clinker production requires large quantities of fuel. In the United States, coal (qv) and natural gas are the most widely used kiln fuels but fuels derived from waste materials, eg, tires, solvents, etc, are increasing in importance (53) (see FUELS FROM WASTE; GAS, NATURAL). In addition to the kiln fuel, electrical energy is required to power the equipment. This energy, however, amounts to only about one-ninth that of the kiln fuel. The cement industry carefully considers all measures that can reduce fuel demand.

Raw Materials Preparation.   The bulk of the raw material originates in the plant quarry when control of the clinker composition starts with systematic core drillings and selective quarrying. A primary jaw or roll crusher is frequently located within the quarry reducing the quarried limestone or shale to ∼100-mm top size. A secondary crusher, usually roll or hammer mills, gives a product of ∼10–25-mm top size. Clays may require treatment in a wash mill to separate sand and other high silica material. Combination crusher-dryers utilize exit gases from the kiln or clinker cooler to dry wet material during crushing.

Argillaceous, siliceous, and ferriferous raw mix components are added to the crusher product. At the grinding mills, the constituents are fed into the mill separately, using weigh feeders or volumetric measurements. Ball mills are used for wet and dry processes to grind the material to a fineness such that only 15–30 wt% is retained on a 74 μm (200 mesh) sieve. In the wet process the raw materials are ground with ∼30–40% water, producing a well-homogenized mixture called slurry. Low concentrations of slurry thinners may be added, such as sodium carbonates, silicates, and phosphates, as well as lignosulfonates and modified petrochemicals. Filter presses or other devices are sometimes used to remove water from slurries before feeding into the kiln.

Raw material for dry process plants is ground in closed-circuit ball mills with air separators, which may be set for any desired fineness. Autogenous mills, which operate without grinding media are not widely used. For suspension preheater-type kilns, a roller mill utilizes the exit gas from the preheater to dry the material in suspension in the mill.

A blending system provides the kiln with a homogeneous raw feed. In the wet process, the mill slurry is blended in a series of continuously agitated tanks in which the composition, usually the CaO content, is adjusted as required. These tanks may also serve as kiln feed tanks, or the slurry after agitation is pumped to large kiln feed basins. Dry-process blending is usually accomplished in a silo with compressed air.

Pyroprocessing.   Nearly all cement clinker is produced in large rotary kiln systems. The rotary kiln is a highly refractory-lined cylindrical steel shell (3–8 m dia, 50–230 m long) equipped with an electrical drive to rotate at 1–3 rpm. It is a countercurrent heating device slightly inclined to the horizontal so that material fed into the upper end travels slowly by gravity to be discharged onto the clinker cooler at the discharge end. The burners at the firing end produce a current of hot gases that heats the clinker and the calcined and raw materials in succession as it passes upward toward the feed end. Highly refractory bricks of magnesia or alumina (see REFRACTORIES) line the firing end, whereas in the less heat-intensive midsection of the kiln bricks of lower refractoriness and thermal conductivity can be used, changing to abrasion-resistant bricks or monolithic castable lining at the feed end. To prevent excessive thermal stresses and chemical reaction of the kiln refractory lining, it is necessary to form a protective coating of clinker minerals on the hot face of the burning zone brick. This coating also reduces kiln shell heat losses by lowering the effective thermal conductivity of the lining.

It is desirable to cool the clinker rapidly as it leaves the burning zone. This is best achieved by using a short, intense flame as close to the discharge as possible. Heat recovery, preheating of combustion air, and fast clinker cooling are achieved by clinker coolers of the traveling-grate, planetary, rotary, or shaft type. Most commonly used are grate coolers where the clinker is conveyed along the grate and subjected to cooling by ambient air, which passes through the clinker bed in crosscurrent heat exchange. The air is moved by a series of undergrate fans, and becomes preheated to 370–800°C at the hot end of the cooler. It then serves as secondary combustion air in the kiln; the primary air is that portion of the combustion air needed to carry the fuel into the kiln and disperse the fuel.

During the burning process, the high temperatures cause vaporization of alkalies, sulfur, and halides. These materials are carried by the combustion gases into the cooler portions of the kiln system where they condense, or they may be carried out to the kiln dust collector, usually a fabric filter or electrostatic precipitator, together with partially calcined feed and unprocessed raw feed. This kiln dust is reusable. However, the levels of total SO_3 and alkali content of cement are limited by considerations of product performance and some national specifications, so that it is not always feasible to recycle all the dust in the process. Other potential and actual uses of dust include fertilizer supplements (see FERTILIZERS), acid mine waste neutralization (see WASTES, INDUSTRIAL), boiler SO_2 control, and soil stabilization (qv).

Wet-Process Kilns. In a long wet-process kiln, the slurry introduced into the feed end first undergoes simultaneous heating and drying. The refractory lining is alternately heated by the gases when exposed and cooled by the slurry when immersed; thus the lining serves to transfer heat, as do the gases themselves. Large quantities of water must be evaporated, thus most wet kilns are equipped with chains to maximize heat transfer from the gases to the slurry. Large, dense chain systems permit energy savings of up to 1.7 MJ/kg (731 Btu lb) clinker in exceptionally favorable situations (53). After most of the moisture has been evaporated, nodules, which still contain combined water, move down the kiln and are gradually heated to ∼550°C. At this temperature, aluminosilicates decompose, evolving combined water, and leaving a mixture to highly reactive microcrystalline silica and alumina together with alkaki hydroxides.

At ∼650°C low temperature melts of alkali chlorides, sulfates and silicates start to form. These promote the decomposition of calcium carbonate. Initially, the decomposing calcium carbonate reacts directly with reactive silica and alumina to form the low-temperature forms of belite and a mixture of low lime aluminates. In these conditions, silica can be thought of as an "acid" chemically

decomposing the calcite. Little free calcium oxide forms. This reaction accelerates as the temperature reises, and when the equilibrium temperature of decomposition of calcium carbonate (898.6°C) is reached the combination of silica and
alumina is almost complete. This uses up ~70% of the available calcium carbonate and the initial endothermic silicate formation is as follows:

$$\text{CaCO}_3 + 0.5 \ \text{SiO}_2 \rightarrow 0.5 \ \text{Ca}_2\text{SiO}_4 + \text{CO}_2 \uparrow \quad (\Delta H = +115 \ \text{kJ} \cdot \text{mol}^{-1})$$

Above 900°C, thermal decomposition of the remaining calcite proceeds
rapidly with the formation of free calcium oxide in another endothermic reaction:

$$\text{CaCO}_3 \rightarrow \text{CaO} + \text{CO}_2 \uparrow \quad (\Delta H = +178 \ \text{kJ} \cdot \text{mol}^{-1})$$

Because these two reaction involve a large enthalpy transfer, a large proportion of the kiln system (the calcinations zone) is used for their completion.
Once the latter reaction is complete, at ~1050°C the temperature of the charge
rises rapidly. At 1300°C, the aluminates begin to melt. The resulting liquid then
acts as a solvent for the finishing reaction in which belite is converted to alite:

$$\text{Ca}_2\text{SiO}_4 + \text{CaO} \rightarrow \text{Ca}_3\text{SiO}_5 \quad (\Delta H = +13 \ \text{kJ} \cdot \text{mol}^{-1})$$

This occurs in the burning zone. The reaction proceeds at an economically
viable rate when the charge is raised to a peak temperature between 1400 and
1450°C in commercial kilns–the temperature needed depending on the burnability of the mix. Burnability depends on the fineness, homogeneity, and LSF of the
mix.

As the charge leaves the burning zone it begins to cool, and tricalcium aluminate and magnesia crystallize from the melt and the liquid phase finally solidifies to produce the ferrite phase. The material drops into the clinker cooler for
further cooling by air.

Dry-Process Kilns, Preheaters, and Precalciners. The dry process for
cement manufacture utilizes a dry kiln feed rather than a slurry. Early dry-process kilns were short, and the substantial quantities of waste heat in the exit
gases from such kilns were frequently used in boilers for electric power generation (qv); the power generated was frequently sufficient for all electrical needs of
the plant. In one modification, the kiln has been lengthened and chains have
been added; however, these serve almost exclusively a heat-exchange function
(see HEAT EXCHANGE TECHNOLOGY). Refractory heat-recuperative devices, such as
crosses, lifters, and trefoils, have also been installed so that the long dry kiln
is energy efficient. Other than the need for evaporation of water, its operation
is similar to that of a long wet kiln.

A second type of modern dry-process kiln is the suspension preheater system (56). The dry, pulverized feed passes through a series of cyclones where it is
separated and preheated several times. The partially calcined feed exits the preheater tower into the kiln at ~800–900°C. The kiln length required for completion of the process is considerably shorter than that of conventional kilns, and
heat exchange is very good. Suspension preheater kilns are very energy efficient:
as low as 3.1 MJ/kg (1334 Btu/lb) clinker in large installations. The intimate mixing of the hot gases and feed in the preheaters promotes condensation of alkalies
and sulfur on the feed, sometimes resulting in objectionably high alkali and

sulfur contents in the clinker. To alleviate this problem, some of the salt-laden kiln exit gases can be deverted to waste (bypassed) and fewer cyclone stages used in the preheater. The wasteage of the heat in these gases results in lower thermal efficiency.

The success of preheater kiln systems led to the development of precalciner kiln systems. These units utilize a second burner to carry out calcination in a separate vessel attached to the preheater. The flash furnace (57), eg, utilizes preheated combustion air drawn from the clinker cooler and kiln exit gases and is equipped with burner that burns ~60% of the total kiln fuel. The raw material is calcined almost 95%, and the gases continue their upward movement through successive preheater stages in the same manner as in an ordinary preheater.

The precalciner system permits the use of smaller kilns because only actual clinkering is carried out in the rotary kiln. Energy efficiency is comparable to that of a preheater kiln, except that the energy penalty for bypass of kiln exit gases is reduced because only ~40% of the fuel is being burned in the kiln. Precalciner kilns produce up to 10,000 metric tons of clinker per day; the largest long wet-process kiln, in the United States produces only 3270 t/day by comparison. The burning process and clinker cooling operations for the modern dry-process kiln systems are the same as for long wet kilns.

Finish Grinding. The cooled clinker is conveyed to clinker storage or mixed with 4–6% gypsum and introduced directly into the finish mills. The clinker and gypsum are ground to a fine, homogeneous powder having a surface area of ~300–600 m^2/kg. About 85–96% of the product is in particles having <45 μm dia. These objectives may be accomplished by two different mill systems. In open-circuit milling, the material passes directly through the mill without any separation. A wide particle size distribution range is usually obtained with substantial amounts of very fine and rather coarse particles. In closed-circuit grinding the mill product is carried to a cyclonic air separator in which the coarse particles are rejected from the product and returned to the mill for further grinding. Energy requirements for finish grinding vary in the range $33 - 77 \text{ kW} \cdot \text{h/t}$ cement, depending primarily on the required fineness, but also on the nature of the clinker.

Computer Control. Process computer control was introduced to the cement industry in the 1960s and a plant of a capacity of 1 million metric tons per year was built and placed in operation in 1973 having complete computer process and segmental control (58). Other plants have been built and some older plants computerized (59). Variables can be measured at intervals of 0.25 s and overall optimum response to operating problems is programmed, not always possible with manual operation. The rotary kiln is the largest and most difficult equipment to operate. Due to the hot, corrosive and abrasive condition. Temperature-sensing and gas-analyzing devices present special problems.

Quality Control. Beginning at the quarry operation, product quality is maintained by adjustments of composition, burning conditions, and finish grinding. Control checks are made for fineness of materials, chemical composition, and uniformity. Clinker burning is monitored by weighing a portion of sized clinker, known as the liter weight test, a free lime test, or checked by microscopic evaluation of the crystalline structure of the clinker compounds. Samples may be analyzed by X-ray fluorescence, atomic absorption, and flame photometry (see SPECTROSCOPY OPTICAL). Wet chemical analysis is described in ASTM C114 (24) and EN 680, but X-ray fluorescence analysis is usually used for quality control in most cement plants. Standard cement samples are available from the National Institute of Standards and Technology. Fineness of the cement is most commonly measured by the air permeability method. Finally, standardized performance tests are conducted on the finished cement (24).

Environmental Pollution Control.   The cement industry has had an intensive program of capital expenditure to install dust collection equipment on kilns and coolers since the 1970s (60). Modern equipment collects dust at 99.8% efficiency. Many smaller dust collectors are installed in new plants (61).

Government agencies have (62) established limits for cement plant effluents including water run-off from manufacturing facilities, quarrying, raw material storage piles, and wastewater. Compliance with these standards has required construction of diversion ditches for surface water, ponds for settling and clarification, dikes and containment structures for possible oil spills, and chemical water treatment in some cases. Since the cement industry obtains most of its raw material by quarrying, the standards for the mineral industry also apply.

One of the primary waste products of cement manufacturing is cement kiln dust (CKD). The CKD is collected from exhaust gases and either returned to the kiln with other raw materials or disposed of as landfill. CKD is also used in other applications, such as synthetic aggregate and tends to accumulate very low concentrations of heavy metals that originate in the fuels or in the raw materials. These metals volatilize in the high temperatures of the kiln and become associated with exhaust gases. Two significant studies of CKD (63,64) concluded that the environmental considerations are minor and that neither CKD nor cement have characteristics of hazardous waste as defined under the U.S. *Resource Conservation and Recovery Act*. The cement industry continues to search for new uses for CKD in construction applications.

6.2. Special Purpose and Blended Cements.   Special purpose and blended portland cements are manufactured essentially by the same processes as ordinary portland cements, but have specific compositional and process differences. White cements are made from raw materials of very low iron content. This type is often difficult to burn because almost the entire liquid phase must be furnished by calcium aluminates. As a consequence of the generally lower total liquid-phase content, high burning-zone temperatures may be necessary. Fast cooling (quenching) and occasionally supplementary reducing flame impinging on the charge at the kiln exit (bleaching) are needed to maintain both quality and color.

Regulated set cements are made using fluorite, CaF_2, additions which also act as fluxing agents, or mineralizers, to reduce burning temperatures. The clinker produced then contains $C_{11}A_7 \cdot CaF_2$, (mixed notation) as a principal phase. Another regulated set cement can be made in which the principal constituents are $C_4A_3\bar{S}$ and belite. Both cements rapidly harden and can reach modest compressive strengths within several hours. Final material properties are in most respects similar to comparable concretes made with portland cement.

Concern with regard to energy conservation has prompted the use of by-product materials in portland cement concrete. Blended hydraulic cements are produced by intimately and uniformly blending two or more types of fine materials. The primary blending materials are portland cement, ground granulated blast-furnace slag, fly ash, calcined clay, silica fume, and hydrated lime. Cement kiln dust and other materials are undergoing research for use in blended cements. Blended hydraulic cements can conform to the requirements of ASTM C 595, or ASTM C 1157. ASTM C 595 recognizes five classes of blended cements: portland blast-furnace slag cement, Type IS; portland–pozzolan cement, Type IP and Type P; slag cement, Type S; pozzolan-modified portland cement, Type I (PM); and slag-modified portland cement, Type I (SM). ASTM C 1157 cement types, which apply to both portland and blended cements, include general purpose cement, Type GU; high early strength cement, Type HE; moderate heat cement, Type MH; low heat cement, Type LH; moderate sulfate resistant cement, Type MS; and high sulfate resistant cement, Type HS.

Blended cements represent $\sim$2% of the cement shipped in the United States. In Europe, the use of blended cement is very common. Most of the blended cement used in the United States is Type IP and it is used in the same applications as that of regular Type I or II portland cement.

ASTM C845 Type E-I (K) expansive cement manufactured in the United States usually depends on aluminate and sulfate phases that result in more ettringite formation during hydration than in normal portland cements. Type K contains an anhydrous calcium sulfoaluminate. This cement can be made either by integrally burning to produce the desired phase composition, or by intergrinding a special component with ordinary portland cement clinkers and calcium sulfate.

Oil well cements are manufactured similarly to ordinary portland cements except that the goal is usually sluggish reactivity. For this reason, levels of C_3A, C_3S, and alkali sulfates are kept low. Hydration–retarding additives are also employed.

Pozzolans include natural materials such as diatomaceous earths (see DIATOMITE), opaline cherts, and shales, tuffs, and volcanic ashes or pumicites, and calcined materials such as some clays and shales. Byproducts such as fly ashes and silica fume are also employed. In the United States the proportion of pozzolan interground with clinker has varied from 15 to >30%, whereas in Italy, cements with a 30–40% pozzolan content are produced.

Portland cement clinker is also interground with 10–65% granulated blast-furnace slag to produce a Portland blast-furnace slag cement. The composition of the slag varies considerably but usually falls within the following wt% composition ranges: CaO, 40–50%, SiO_2, 30–40%, Al_2O_3, 8–18%; MgO, 0–8%; S (sulfide), 0–2%; and FeO, and MnO, 0–3%.

Masonry cements are used for making mortar for bricklaying. They are not suitable for use in concrete. Most masonry cements are finely interground mixtures where portland cement is a principal constituent. These cements also include finely ground limestones, hydrated lime, natural cement, pozzolans, clays, or air-entraining agents. Secondary materials are used to impart the required water retention and plasticity to mortars.

6.3. Non-Portland Cements. *Calcium Aluminate Cements.* These cements are manufactured by heating until molten or by sintering a mixture of limestone and a bauxite containing low amounts of SiO_2, FeO, and TiO_2 (see

ALUMINUM COMPOUNDS, ALUMINUM OXIDE, CALCINED, TABULAR, AND ALUMINATE CEMENTS).
The process is usually carried out in an open-hearth furnace having a long
vertical stack into which the mixture of raw materials is charged. The hot
gases produced by a blast of pulverized coal and air pass through the charge
and carry off the water and carbon dioxide. Fusion occurs when the charge
drops from the vertical stack onto the hearth at ~1425–1500°C. The molten
liquid runs out continuously into steel pans on an endless belt in which the
melt solidifies. Special rotary kilns, provided with a tap hole from which the molten liquid is drawn intermittently, and electric arc furnaces have also been used.

When calcium aluminate cements are made by the fusion process, the solidified melt must be crushed and then ground. The material is very hard to grind
and power consumption is high.

6.4. Hydraulic Limes. These materials are produced by heating below
sintering temperature a limestone containing considerable clay, during which

Table 7. **United States Portland Cement Consumption**[a]

U.S Cement Industry Consumption — Exports — Imports — Shipments[b]

| Year | Consumption[c] | | | Cement exports | Cement imports | Total shipments by domestic[d] products |
	Portland cement	Masonry cement	Total			
1979	75,523	3,343	78,866	244	4,101	75,032
1980	66,940	2,721	69,661	346	2,732	67,710
1981	63,456	2,440	65,896	614	2,287	64,338
1982	57.191	2,166	59,357	384	2,367	57,544
1983	62,918	2,615	65,533	292	2,522	63,584
1984	71,792	2,945	74,737	266	6,004	70,435
1985	74,434	2,960	77.394	258	8,845	70,235
1986	78,643	3,230	81,873	227	11,661	71,386
1987	80,291	3,375	83,666	271	13,184	71,243
1988	80,715	3,292	84,007	197	14,001	70,427
1989	79,155	3,071	82,226	300	12,547	70,321
1990	77,785	3,005	80,790	293	10,461	71,535
1991	69,098	2,495	71,593	272	7,215	64,842
1992	73,354	2,704	76,058	351	6,097	70,502
1993	76,566	3,014	79,580	398	6,151	73,934
1994	82,159	3,267	85,426	452	8,912	77,021
1995	82,825	3,160	85,985	485	11,625	74,936
1996	87,416	3,399	90,815	461	11,999	79,411
1997	92,708	3,458	96,166	519	13,814	82,978
1998	99,153	4,101	103,254	322	18,278	85,417
1999	104,074	4,352	108,426	315	22,534	86,328
2000	105,195	4,333	109,528	394	22,740	87,599
2001	108,090	4,475	112,565	443	22,397	91,097

[a]Ref. 69.
[b]Thousands of metric tons.
[c]Excludes Alaska and Puerto Rico.
[d]Excludes Puerto Rico.

Note: Domestic Shipments include cement shipments from domestic manufacturers and cement shipments ground from imported clinker, but exclude finished cement imports.

Source: U.S. Geological Survey/PCA Economic Research.

some combination takes place between the lime and the oxides of the clay to form hydraulic compounds.

7. Economic Aspects

From the beginning of the United States portland cement industry in 1872, annual cement consumption grew through 1970. From 1975 to 1990, cement consumption changed little. During the 1990s, cement consumption continued to grow. Table 7 gives United States' production figures and Table 8 gives the world production. China, having an annual output of >580 million metric tons in 2000, has emerged as the world's leading cement producer (68).

Since the 1940s, the cement industry reduced labor and energy costs by increased investment in capital equipment and larger plants to remain competitive with other building materials industries (see BUILDING MATERIALS, SURVEY; BUILDING MATERIALS, PLASTIC). The average plant size more than doubled between 1950 and 1990.

7.1. Energy Use. From 1972 to 2000, the cement industry has reduced unit energy usage by 32.9%. The wet process, used in 60% of the plants in the 1960s, was less labor-intensive than the dry process. However, as energy costs escalated in the early 1970s, the more energy efficient dry-process manufacturing was preferred. According to 1989 figures, wet-process plants consume 38% more energy per ton of cement than dry-process plants.

Coal is the primary kiln fuel as seen in Table 9. Energy from coal rose from 36 to 60% of the total energy required for cement production between 1972 and 2000.

7.2. Marketing Patterns. The cement industry reduced its dependence on bag (container) shipments (54.7% in 1950) and turned to the more labor-efficient bulk transport (96% in 2000). In addition, the amount of cement shipped by rail transportation declined from 75% of industry shipments in 1950 to <2% in 2000. Table 10 summarizes the shipment distribution by cement type.

In the past 30 years, the ready-mixed concrete industry became the primary customer for cement manufacturers. In 2000, >73.7% of the cement shipped was sold to the ready-mixed concrete industry, compared with 63% in 1975. The other primary uses are in building materials, concrete products, and highway construction.

8. Environmental Aspects

Cement plants in the United States are carefully monitored for compliance with Environmental Protection Agency (EPA) standards for emissions of particulates, SO_x, NO_x, and hydrocarbons. All plants incorporate particulate collection devices such as baghouses and electrostatic precipitators (see AIR POLLUTION CONTROL METHODS). The particulates removed from stack emissions are called cement kiln dust (CKD). It has been shown that CKD is characterized by low concentrations of metals which leach from the CKD at levels far below regulatory limits (63,64).

Table 8. **Top World Producers of Portland Cement**[a]

| | | | | Top World Producers of Hydraulic Cement[b] | | | | | |
	1991	1992	1993	1994	1995	1996	1997	1998	1999	2000[c]
China	252,610	308,220	367,880	421,180	475,910	491,190	511,730	536,000	573,000	583,190
India	51,000	50,000	53,812	57,000	62,000	75,000	80,000	85,000	90,000	95,000
United States[d]	68,465	70,883	75,117	79,353	78,320	80,818	84,255	85,522	87,777	89,510
Japan	89,564	88,253	88,046	91,624	90,474	94,492	91,938	81,328	80,120	81,300
Korea, Republic of	34,999	44,444	47,313	50,730	55,130	58,434	60,317	46,091	48,157	51,255
Brazil	27,490	23,903	24,843	25,230	28,256	34,597	38,096	39,942	40,270	39,208
Germany	34,396	37,529	36,649	36,130	33,302	31,533	35,945	36,610	38,099	38,000
Italy	40,806	41,347	33,771	32,713	33,715	33,327	33,721	35,512	34,000	36,000
Turkey	26,091	28,607	31,241	29,493	33,153	35,214	36,035	38,200	34,358	35,825
Russia						27,800	26,700	26,000	28,400	32,400
Thailand	18,054	21,832	26,870	29,900	34,900	38,749	37,086	30,000	34,000	32,000
Mexico	25,100	26,880	27,120	29,700	24,043	25,366	27,548	27,744	29,413	31,677
Spain	25,119	24,615	22,878	25,150	26,423	25,157	27,632	27,943	30,800	30,000
Total World[e]	1,181.8	1,231.1	1,290.9	1,370.0	1,445.0	1,493.0	1,547.0	1,547.0	1,603.0	1,643.0

[a]Ref. 69.
[b]Thousands of metric tons.
[c]estimate.
[d]Includes Puerto Rico.
[e]Million metric tons.
Source: U.S. Geological Survey.

496

Table 9. United States Portland Cement Industry Energy Consumption[a]

	1972	1995	1996	1997	1998	1999	2000	00/72	00/95
					Summary of Energy Consumption[b] % Change				
Gasoline	7.4	2.4	2.0	1.8	1.9	2.2	4.6	−37.8%	91.7%
Middle distillates	88.3	38.1	43.6	39.7	41.3	44.7	39.4	−55.4%	3.4%
Residual oil	808.9	17.1	6.0	8.1	6.9	3.6	3.9	−99.5%	−77.2%
LPG	1.4	0.6	0.6	0.4	0.4	0.3	0.3	−78.8%	−50.0%
Total Petroleum Products	*906.0*	*58.2*	*52.2*	*50.0*	*50.5*	*50.8*	*48.2*	*−94.7%*	*−17.2%*
Natural gas	3,347.1	564.0	374.6	305.2	372.4	327.7	261.6	−92.2%	−53.6%
Coal	2,639.0	2,887.2	3,052.6	3,110.2	2,998.6	2,984.0	2,984.1	13.1%	3.4%
Petroleum Coke	39.6	814.9	798.1	732.8	744.1	748.8	760.7	1821.0%	−6.7%
Total Coal and Coke	*2,678.6*	*3,702.1*	*3,850.7*	*3,843.0*	*3,742.7*	*3,732.8*	*3,744.8*	*39.8%*	*1.2%*
Waste fuel	—	416.2	389.5	444.4	413.5	407.8	402.5	—	−3.3%
Total Fossil Fuels	*6,931.7*	*4,740.5*	*4,667.0*	*4,642.6*	*4,579.1*	*4,519.1*	*4,457.1*	*−35.7%*	*−6.0%*
Electricity	488.6	520.9	533.7	528.7	528.1	520.8	524.6	7.4%	0.7%
Total Fuel and Power	*7,420.3*	*5,261.4*	*5,200.7*	*5,171.3*	*5,107.2*	*5,039.9*	*4,981.7*	*−32.9%*	*−5.3%*

[a]Ref. 69.

[b]Thousands of BTUs per equivalent metric ton. Weighted average of 92% clinker production plus 8% finished cement production.

Source: PCA U.S. and Canadian Labor-Energy Input Services.

Table 10. **United States Portland Cement Shipments by Type**[a]

Portland Cement Shipped from U.S. Plants by Type of Product[b]

	1998		1999		2000	
	Quantity	% of Total	Quantity	% of Total	Quantity	% of Total
General Use and Moderate Heat (Types I and II)	85,066	90.1	90,891	90.2	90,644	88.0
High Early Strength (Type III)	3,151	3.3	3,297	3.3	3,815	3.7
Sulfate-Resisting (Type V)	2,757	2.9	3,046	3.0	4,453	4.3
Block	594	0.6	632	0.6	636	0.6
Oil Well	797	0.8	578	0.6	1,039	1.0
White	790	0.8	848	0.8	894	0.9
Blended						
Portland Slag and Pozzolan	449	0.5	529	0.5	579	0.6
Portland Fly Ash and Other	672	0.7	664	0.7	718	0.7
Miscellaneous	132	0.1	260	0.3	171	0.2
Total Shipments	94,408		100,745		102,949	

[a]Refs. (66,69).
[b]Thousands of metric tons.
Source: U.S. Geological Survey.

9. Specifications and Types

Portland cements are manufactured to comply with specifications established in each country (70). In the United States, several different specifications are used, including those of the American Society for Testing and Materials and American Association of State Highway and Transportation Officials (AASHTO). The ASTM annually publishes test methods and standards (24), which are established on a consensus basis by its members which include consumers and producers.

In the United States, portland cement is classified in five general types designated by ASTM Specification C150 (24): Type I, when the special properties are not required; Type II, for general use, and especially when moderate sulfate resistance or moderate heat of hydration is desired; Type III, for high early strength; Type IV, for low heat of hydration; and Type V, for high sulfate resistance. Types I, II, and III may also be specified as air entraining. Chemical compositional, physical, and performance test requirements are specified for each type; optional requirements for particular uses may also be specified. Portland cement can also be specified under ASTM C 1157 as general purpose cement, Type GU; high early strength cement, Type HE; moderate heat cement, Type MH; low heat cement, Type LH; moderate sulfate resistant cement, Type MS; and high sulfate resistant cement, Type HS. Tables 11 and 12 list applications

Table 11. **Applications of Portland Cement Using U.S. Standards**

Cement specification	Applications[a]						
	General purpose	Moderate heat of hydration	High early strength	Low heat of hydration	Moderate sulfate resistance	High sulfate resistance	Resistance to alkali-silica reactivity (ASR)[b]
ASTM C 150 (AASHTO M B5) portland cements	I	II (moderate heat option)	III	IV	II	V	Low alkali option
ASTM C 595 (AASHTO M 240) blended hydraulic cements	IS, IP I (PM) I (SM), S, P	IS(MH) IP(MH) I(PM)(MH) I(SM)(MH)		P(LH)	IS(MS) IP(MS) P(MS) I(PM)(MS) I(SM)(MS)		Low reactivity option
ASTM C 1157 hydraulic cements[c]	GU	MH	HE	LH	MS	HS	Option R

[a]Check the local availability of specific cements as all cements are not available everywhere.
[b]The option for low reactivity with ASR susceptible aggregates can be applied to any cement type in the columns to the left.
[c]For ASTM C 1157 cements, the nomenclature of hydraulic cement, portland cement, air-entraining portland cement, modified portland cement, or blended hydraulic cement is used with the type desiganation.

Table 12. **Applications of Special Cements**

Special cements	Type	Application
White portland cements, ASTM C 150	I, II, III, V	White or colored concrete, masonry, mortar, grout, plaster, and stucco
White masonry cements, ASTM C 91	M, S, N	White mortar between masonry units
Masonry cements, ASTM C 91	M, S, N	Mortar between masonry units,[a] plaster, and stucco
Mortar cements, ASTM C 1329	M, S, N	Mortar between masonry units[a]
Plastic cements, ASTM C 1328	M, S	Plaster and stucco[b]
Expansive cements, ASTM C 845	E-1(K), E-1(M), E-1(S)	Shrinkage compensating concrete
Oil-well cements, API-10	A, B, C, D, E, F, G, H	Grouting wells
Water-repellent cements		Tile grout, paint, and stucco finish coats
Regulated-set cements		Early strength and repair[c]
Cements with functional additions ASTM C 595 (AASHTO M 240), ASTM C 1157		General concrete construction needing special characteristics such as; water-reducing retarding air entraining set control, and accelerating properties
Finely ground (ultrafine) cement		Geotechnical grouting[c]
Calcium aluminate cement		Repair, chemical resistance, high temperature exposures
Magnesium phosphate cement		Repair and chemical resistance
Geopolymer cement		General construction, repair, waste stabilization[c]
Ettringite cement		Waste stabilization[c]
Sulfur cements		Repair and chemical resistance
Rapid hardening hydraulic cement		General paying where very rapid ($\sim$4 h) strength development is required

[a]Portland cement Types I, II, and III and blended cement Types IS, IP, and I(PM) are also used in making mortar.
[b]Portland cement Types I, II, and III and blended cement Types IP, I(SM) and I(PM) are also used in making plaster.
[c]Portland and blended hydraulic cements are also used for three applications.

of portland and special cements. In Europe, cements are made to meet the requirements of EN 197. EN 197 cement Types CEM I, II, III, IV and V do not correspond to the cement types in ASTMC 150. CEM I is a portland cement and CEM II through V are blended (composite) cements. EN 197 also has strength classes and ranges (32.5, 42.5, and 52.5 MPa).

10. Uses

Hydraulic cements are intermediate products used for making concretes, mortars, grouts, and other composite materials (qv). High early strength cements may be required for precast concrete products or in high rise building frames to permit rapid removal of forms and early load carrying capacity. Cements of low heat of hydration may be required for use in massive structures, such as gravity dams, to prevent excessive temperature rise and thermal contraction and cracking during subsequent cooling. Concretes exposed to seawater or sulfate-containing ground waters require cements that are sulfate-resistant after hardening.

Air-entraining cements produce concretes that protect the concrete from frost damage. They are commonly used for concrete pavements subjected to wet and freezing conditions.

Low alkali cements may be used with certain concrete aggregates containing reactive silica to prevent deleterious expansions.

Expansive, or shrinkage-compensating cements cause slight expansion of the concrete during hardening. The expansion has to be elastically restrained so that compressive stress develops in the concrete (71). Subsequent drying and shrinkage reduces the compressive stresses but does not result in tensile stresses large enough to cause cracks. Special highly expansive cements have been used for demolition purposes.

Finely ground cements, often called ultrafine cements, having particles <10 μm and an average size of 4 μm are used to grout soils with fine pore spaces, such as fine sand with a permeability of 10^{-4} cm/s. These cements can be made with a wide combination of portland cement, slag, or silica fume (72).

Regulated-set cement, called jet cement in Japan, is formulated to yield a controlled short setting time, <1 h, and very early strength (73). It is a modified cement that can be manufactured in a conventional portland cement kiln. It incorporates set control and early strength development components.

Natural cements (74) that may be regarded as intermediate between portland cements and hydraulic limes in hydraulic activity, are no longer available in the United States.

Blended Cements. Portland cement clinker is also interground with suitable other materials such as granulated blast-furnace slags and natural or artificial pozzolans. These substances also show hydraulic activity when used with cements, and the blended cements (75) bear special designations such as portland blast-furnace slag cement or portland–pozzolan cement. Pozzolans are used in making concrete both as an interground or blended component of the cement or as a direct addition to the concrete mix. It is only when the two materials are supplied as an intimate blend that the mixture can be referred to as portland–pozzolan cement. Portland–pozzolan cements (76) were developed originally to provide concretes of improved durability in marine, hydraulic, and underground environments; they may also prevent deleterious alkali–aggregate reactions. Blast-furnace slag cements (77) may also reduce deleterious alkali–aggregate reactions and can be resistant to seawater if the slag and cement compositions are suitably restricted. Both cements hydrate and harden more slowly than portland cement. This can be an advantage in mass concrete structures where the lower rates of heat liberation may prevent excessive temperature rise, but when used at low temperatures the rate of hardening may be excessively slow. Portland blast-furnace slag cements may be used to advantage in steam-cured products which can have strengths as high as obtained with portland cement. Interest in the use of blended cements is stimulated by energy conservation and solid waste utilization considerations.

Oil well cements (78) are usually made from portland cement clinker and may also be blended cements. The American Petroleum Institute Specification for Materials and Testing for Well Cements (*API Specification 10*) (78) includes requirements for eight classes of oil well cements. They are specially produced for cementing the steel casing of gas and oil wells to the walls of the bore–hole and to seal porous formations (79). Under these high temperature and pressure conditions ordinary portland cements would not flow properly and would set prematurely. Oil well cements are more coarsely ground than normal, and contain special retarding admixtures.

Masonry cements (80) are cements for use in mortars for masonry construction. They are formulated to yield easily workable mortars and contain special additives that reduce the loss of water from the mortar to the porous masonry units.

Calcium aluminate cement (81) develops very high strengths at early ages. It attains nearly its maximum strength in one day, which is much higher than the strength developed by portland cement in that time. At higher temperatures, however, the strength drops off rapidly. Heat is also evolved rapidly on hydration and results in high temperatures; long exposures under moist warm conditions can lead to failure. Resistance to corrosion in sea or sulfate waters, as well as to weak solutions of mineral acids, is outstanding. This cement is attacked rapidly, however, by alkali carbonates. An important use of high alumina cement is in refractory concrete for withstanding temperatures up to 1500°C. White calcium aluminate cements, with a fused aggregate of pure alumina, withstand temperatures up to 1800°C.

Trief cements (83), manufactured in Belgium, are produced as a wet slurry of finely ground slag. When activators such as portland cement, lime, or sodium hydroxide are added in a concrete mixer, the slurry sets and hardens to produce concretes with good strength and durability.

Hydraulic limes (84) may be used for mortar, stucco, or the scratch coat for plaster.They harden slowly under water, whereas high calcium limes, after slaking with water, harden in air to form the carbonate but not under water at ordinary temperatures. However, at elevated temperatures achieved with steam curing, lime–silica sand mixtures do react to produce durable products such as sand–lime bricks.

Specialty Cements.   For special architectural applications, white portland cement with a very low iron oxide content can be produced. Colored cements are usually prepared by intergrinding 5–10% of pigment with white cement.

Numerous other specialty cements composed of various magnesium, barium, and strontium compounds as silicates, aluminates, and phosphates, as well as others, are also produced (85).

BIBLIOGRAPHY

"Cement, Structural" in *ECT* 1st ed., Vol. 3, pp. 411–438, by R. H. Bogue, Portland Cement Association Fellowship (Portland Cement); J. L. Miner and F. W. Ashton, Universal Atlas Cement Company (Calcium-Aluminate Cement); and G. J. Fink, Oxychloride Cement Association, Inc. (Magnesia Cement); "Cement" in *ECT* 2nd ed., Vol. 4, pp. 684–710, by Robert H. Bogue, Consultant to the Cement Industry; in *ECT* 3rd ed., Vol. 5, pp. 163–193, by R. H. Helmuth, F. M. Miller, T. R. O'Connor, and N. R. Greening, Portland Cement Association; in *ECT* 4th ed., Vol. 5, pp. 564–598, by R. H. Helmuth, F. M. Miller, N. R. Greening, E. Hognestad, S. H. Kosmatka, and D. Lang, Portland Cement Association; "Cement" in *ECT* (online), posting date: December 4, 2000, by Richard H. Helmuth, F. M. Miller, N. R. Greening, E. Hognestad, Steven H. Kosmatka, David Lang, Portland Cement Association.

CITED PUBLICATIONS

1. R. H. Bogue, *The Chemistry of Portland cement*, 2nd ed., Rheinhold Publishing Corp., New York, 1955.
2. H. F. W. Taylor, *Cement Chemistry*, Academic Press Ltd., London, 1997.
3. G. A. Rankin and F. E. Wright, *Am. J. Sci.* **39**, 1 (1915).
4. F. M. Lea, *The Chemistry of Cement and Concrete*, 3rd ed., Edward Arnold (Publishers) Ltd., London, 1971.
5. Ref. 4, p. 88.
6. Ref. 4, p. 122.
7. P. Weber, *Zem. Kalk Gips* Special Issue No. 9, (1963); Ref. 4, p. 130.
8. Ref. 4, p. 126.
9. K. Endell and G. Hendrickx, *Zement* **31**, 357, 416 (1942); Ref. 4, p. 128.
10. N. Toropov and P. Rumyantsev, *Zh. Prikl. Khim* **38**, 1614, 2115 (1965); Ref. 4, p. 135.
11. D. H. Campbell, *Microscopical Examination and Interpretation of Portland cement and Clinker*, SP030, Portland cement Association, Skokie, Ill., 1999.
12. *Guide to Compounds of Interest in Cement and Concrete Research, Special Report 127*, Highway Research Board, National Academy of Sciences, Washington, D.C., 1972.
13. H. F. W. Taylor, ed., *The Chemistry of Cements*, Vols. 1 and 2, Academic Press, Inc. (London) Ltd., London, 1964, Appendix 1.
14. Ref. 4, p. 121.
15. A. Guinier and M. Regourd, in *Proceedings of the 5th International Symposium on the Chemistry of Cement, Tokyo, 1968*, The Cement Association of Japan, Tokyo, Japan, 1969.
16. G. Yamaguchi and S. Takagi, in Ref. 15.
17. F. A. DeLisle, *Cement Technol.* **7**, 93 (1976); Y. Ono, S. Kamamura, and Y. Soda, in Ref. 15.
18. F. Gille and co-workers, *Microskopie des Zementklinkers, Bilderatlas*, Association of the German Cement Industry, Beton-Verlag, Dusseldorf, Germany, 1965.
19. L. S. Brown, *Proc. J. Am. Concr. Inst.* **44**, 877 (1948).
20. H. Kuhl, *Zement* **18**, 833 (1929); Ref. 4, p. 164.
21. Ref. 4, p. 166.
22. K. E. Peray and J. J. Waddell, *The Rotary Cement Kiln*, Chemical Publishing Co., New York, 1972, p. 65.
23. R. H. Bogue, *Ind. Eng. Chem. Anal. Ed.* **1**, 192 (1929); Ref. 1, p. 246.
24. *2002 Annual Book of ASTM Standards*, American Society for Testing and Materials, West Conshohocken, Pa., 2002.
25. F. R. McMillan and W. C. Hansen, *J. Am. Concr. Inst.* **44**, 553, 564, 565 (1948).
26. Ref. 4, p. 188.
27. S. Brunauer and D. K. Kantro, in Ref. 13.
28. C. W. Lentz, in *Special Report 90, Structure of Portland cement Paste and Concrete*, Highway Research Board, NRC-NAS, Washington, D.C., 1966.

29. L. E. Copeland and G. Verbeck, in *The 6th International Congress on the Chemistry of Cement, Moscow, 1974, English Preprints*, The Organizing Committee of the U.S.S.R., 1974.

30. G. Verbeck and R. A. Helmuth, in Ref. 15.

31. W. C. Hansen, in E. G. Swenson, ed., *Performance of Concrete*, University of Toronto Press, Toronto, Canada, 1968.

32. G. Verbeck, *Research Department Bulletin RX 189*, Portland cement Association, Skokie, Ill., 1965.

33. L. E. Copeland and D. L. Kantro, in Ref. 13.

34. F. M. Lea, *The Chemistry of Cement and Concrete*, 3rd ed., Edward Arnold (Publishers) Ltd., London, 1971.

35. H. F. W. Taylor, in Ref. 13, p. 21.

36. T. C. Powers, in Ref. 13, Chapt. 10.

37. W. Lerch, *ASTM Proc.* **46**, 1251 (1946).

38. E. M. Gardner and J. M. Gaidis, in J. P. Skalny, ed., *Materials Science of Concrete I*, American Ceramic Society, Westerville, Ohio, 1989, p. 95.

39. Ref. 4, p. 239.

40. G. H. Tattersall and P. F. G. Banfill, *The Rheology of Fresh Concrete*, Pitman Books, London, 1983.

41. T. C. Powers, *The Properties of Fresh Concrete*, John Wiley & Sons, Inc., New York, 1968.

42. E. M. Petrie, *Ind. Eng. Chem. Prod. Res. Dev.* **15**, 242 (1976).

43. *1992 Annual Book of ASTM Standards*, American Society for Testing and Materials, Philadelphia, Pa., 1992.

44. T. C. Powers, in *Proceedings of the 4th International Symposium on the Chemistry of Cement, Washington, D.C., 1960*, U.S. Government Printing Office, Washington, D.C., 1962.

45. M. Y. Bal'shin, *Dokl. Akad. Nauk SSSR* **69**, 831 (1949).

46. E. Ryshkewitch, *J. Am. Ceram. Soc.* **36**, 65 (1953).

47. K. K. Schiller, *Cem. Concr. Res.* **1**, 419 (1971).

48. D. P. H. Hasselman, *J. Am. Ceram. Soc.* **46**, 564.

49. M. Rössler and I. Odler, *Cem. Concr. Res.* **15**, 320 (1985).

50. D. M. Roy and G. R. Gouda, *Cement Concr. Res.* **5**, 153 (1975).

51. S. H. Kosmatka, B. Kerkhoff, and W. C. Panarese, *Design and Control of Concrete Mixtures*, EB001, Portland cement Association, 2002.

52. Ref. 51, p. 8.

53. *U.S. Cement Industry Fact Sheet*, Portland cement Association, Skokie, Ill., 2002.

54. W. H. Duda, *Cement Data Book*, Bauverlag GmbH, Wiesbaden, Germany, and Berlin, 1976.

55. K. E. Peray and J. J. Waddell, *The Rotary Cement Kiln*, Chemical Publishing Co., New York, 1972.

56. J. R. Tonry, *Report MP-96*, Portland cement Association, Skokie, Ill., 1961.

57. *Report: 1973 Technical Mission to Japan*, Portland cement Association, Skokie, Ill., 1973.

58. D. G. Courteney, *Rock Prod.* **78**(5), 75 (1975).

59. D. Grammes, in *Mill Session Papers M-195*, Portland cement Association, Skokie, Ill., 1969.

60. *Clean Air Act*, Public Law 88-206 (1963); Amendments: *Public Law* 89-675 (1966); 91-604 (1970); and 95-95 (1977).

61. W. E. Trauffer, *Pit Quarry* **67**(8), 52 (1975).

62. *Public Law 92-500*.

63. B. W. Haynes and G. W. Kramer, "Characterization of U.S. Cement Kiln Dust," *Bureau of Mines Information Circular (IC) 8885*, U.S. Dept. of the Interior, Bureau of Mines, Washington, D.C., 1991.

64. *An Analysis of Selected Trace Metals in Cement and Kiln Dust*, SP109, Portland cement Association, Skokie, Ill., 1992.

65. *Minerals Yearbook: Cement*, U.S. Bureau of Mines, Washington, D.C., 1990.

66. *Minerals Yearbook: Cement*, U.S. Bureau of Mines, Washington, D.C., 2001.

67. *Statistical Review No. 33, Production-Trade-Consumption 1974–1975*, Cembureau, Paris, France, Oct. 1976.

68. *Int. Cem. Rev.*, 29 (Apr. 1991).

69. *U.S. Cement Industry Fact Sheet*, Portland cement Association, Skokie, Ill., 2002.

70. *Cement Standards of the World*, Cembureau, Brussels, Belgium, 2002.

71. Ref. 24, ASTM C845.

72. S. H. Kosmatka, *Cementitious Grouts and Grouting*, EB111T, Portland cement Association, Skokie, Ill., 1990.

73. U.S. Pat. 3,628,973 (Dec. 21, 1971), N. R. Greening, L. E. Copeland, and G. J. Verbeck (to Portland cement Assoc.).

74. Ref. 24, ASTM C10.

75. Ref. 24, ASTM C595.

76. Ref. 4, Chapt. 14; R. Turriziani, in Ref. 13, Chapt. 14.

77. Ref. 4, Chapt. 15; R. W. Nurse, in Ref. 13, Chapt. 13.

78. *Specifications for Materials and Testing for Well Cements, API Standards 10*, API, Washington, D.C., 2002.

79. D. K. Smith, *Cementing*, Society of Petroleum Engineers of AIME, New York, 1987.

80. Ref. 24, ASTM C91.

81. Ref. 4, Chapt. 16; T. D. Robson, in Ref. 13, Chapt. 12.

82. Ref. 4, p. 481.

83. Ref. 4, p. 477.

84. Ref. 24, ASTM C141.

85. Ivan Odler, *Special Inorganic Cements*, E & FN Spon, New York, 2000.

STEVEN KOSMATKA
Portland Cement Association

CENTRIFUGAL SEPARATION

1. Introduction

Centrifugal separation is a mechanical means of separating the components of a mixture of liquids that are immiscible or of liquids and insoluble solid particles. The material is accelerated in a centrifugal field that acts upon the mixture in the same manner as a gravitational field. The centrifugal field can, however, be varied by changes in rotational speed and equipment dimensions, whereas gravity is essentially constant. Commercial centrifugal equipment can reach an acceleration of 20,000 times gravity (20,000 G); laboratory equipment can reach up to 360,000 G. The ultracentrifuge and gas centrifuge represent special cases that establish separation gradients on a molecular scale. The usual gravitational operations, such as sedimentation (qv) or flotation (qv) of solids in liquids, drainage or squeezing of liquids from solid particles, and stratification of liquids according to density, are accomplished more effectively in a centrifugal field (see SEPARATION, SIZE SEPARATION).

The development of theory for centrifugation equipment has been slow. Flow patterns in the centrifuge bowls are complex and difficult to model mathematically. The concept of a theoretical capacity factor for sedimentation depends only on equipment characteristics and is independent of the system (1,2). The theoretical effect of particle size distribution in single and multistage centrifugation has been demonstrated (3). Extensive application of centrifugation to dewatering (qv) and thickening of relatively soft solids and hydrogels associated with industrial and municipal waste treatment has resulted in changes in centrifuge design (see WASTES, INDUSTRIAL). A sound theoretical basis for centrifugal drainage or squeezing of liquids from solids has been only partially developed for compressible solids. Many aspects are in need of amplification.

Herein centrifugal separation in a liquid medium is discussed. For a brief discussion of centrifugal separation in a gaseous medium see previous editions of this article.

2. Theory

2.1. Separation by Density Difference.

A single solid particle or discrete liquid drop settling under the acceleration of gravity in a continuous liquid phase accelerates until a constant terminal velocity is reached. At this point the force resulting from gravitational acceleration and the opposing force resulting from frictional drag of the surrounding medium are equal in magnitude. The terminal velocity largely determines what is commonly known as the settling velocity of the particle, or drop under free-fall, or unhindered conditions. For a small spherical particle, it is given by Stokes' law:

$$v_g = \frac{\Delta\delta d^2 g}{18\ \mu} \tag{1}$$

where v_g = the settling velocity of a particle or drop in a gravitational field; $\Delta\delta = \delta_S - \delta_L$ = the difference between true mass density of the solid particle or liquid drop, and that of the surrounding liquid medium; d = the diameter of the solid particle or liquid drop; g = the acceleration of gravity; and μ = the absolute viscosity of the surrounding medium.

Stokes' law can be readily extended to a centrifugal field:

$$v_s = \frac{\Delta\delta d^2 \omega^2 r}{18\ \mu} = v_g\left(\frac{\omega^2 r}{g}\right) \tag{2}$$

where v_s = the settling velocity of a particle or drop in a centrifugal field; ω = the angular velocity of the particle in the settling zone; and r = the radius at which settling velocity is determined. Analogous equations describe the terminal velocity of a light particle or drop rising in a heavier continuous medium.

The settling velocity, v_s, is relative to the continuous liquid phase where the particle or drop is suspended. If the liquid medium exhibits a motion other than the rotational velocity, ω, the vector representing the liquid-phase velocity should be combined with the settling velocity (eq. 2) to obtain a complete description of the motion of the particle (or drop).

These concepts are used to analyze separations in the bottle centrifuge, the imperforate bowl centrifuge, and the disk centrifuge. Separation by density difference in other types of centrifuges can be analyzed by analogy.

The Bottle Centrifuge. Analysis of the performance of a bottle centrifuge is based on the model shown in Figure 1. A solid or liquid particle is considered in an initial position, X, at a radius, r, from the axis of rotation. If equation 2 is applied to this specific particle, assuming that $v_s = dr/dt$, then

$$\int_{r}^{r_C} \frac{dr}{r} = \int_{0}^{t} v_g \left(\frac{\omega^2}{g}\right) dt \tag{3}$$

where r_C = the radius of the sedimented cake, and t = the time during which the particle is subjected to centrifugal acceleration. Integration of equation 3 leads to the following:

$$\ln \frac{r_C}{r} = v_g \left(\frac{\omega^2}{g}\right) t \tag{4}$$

A radius, r', that divides the volume of supernatant into two equal parts can be defined as follows:

$$(r' - r_1) = (r_c - r') \quad \text{or} \quad r' = (r_c + r_1)/2 \tag{5}$$

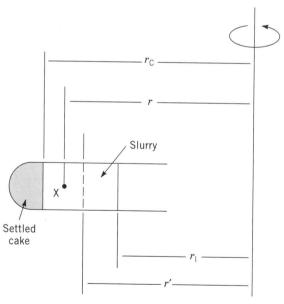

Fig. 1. Separation in a bottle centrifuge, where X is the initial position of a particle (drop). See text for definition of terms.

where r_1 is the radius of the free surface of the liquid. Assuming the presence of more than one particle in the liquid as well as a uniform initial particle distribution, then each of the two volumes defined by r'; initially contains the same number of particles. By making the further assumption that the particles in the suspension are identical, a settling time, t, is chosen so that those particles starting from radius r'; all reach the cake r_C after time t. Under these conditions, one-half of the particles that were in suspension at $t = 0$ are sedimented after the time, t, has elapsed. The other one-half, initially located above the level defined by the radius r', remain in suspension. If r is replaced in equation 4 by r', a sedimentation condition is established that is referred to as 50% cutoff. To determine 100% cutoff (complete solids capture) r should be replaced by r_1 in equation 4. An effective capacity, Q_0, for the bottle centrifuge is determined by the ratio between the volume, V, occupied by the slurry in the bottle, and the spinning time, t, calculated from equation 4:

$$Q_0 = V/t = 2v_g(\omega^2/g)V/(2 \ln(2r_c/(r_c + r_1))) \tag{6}$$

In equation 6, v_g characterizes the settling behavior of the solid particles or liquid drops in the suspension, whereas the second part of the right-hand side refers to speed and size of the bottle centrifuge and is expressed by the capacity factor Σ_B:

$$\Sigma_B = \omega^2V/2g \ln(2r_c/(r_c + r_1)) \tag{7}$$

This capacity factor has the dimension of an area and represents the area of a static gravity settling tank having a separation performance equal to that of the rotating bottle centrifuge handling the same particles. By combining equations 6 and 7, to eliminate volume, V, equation 8 is obtained:

$$\frac{Q_0}{\Sigma_B} = \frac{2g}{(\omega^2t)} \cdot \ln\left(\frac{2r_C}{r_C + r_1}\right) \tag{8}$$

Equation 8 provides the basis of comparison of the performance of various bottle centrifuges containing the same material, and also, under certain circumstances, of other types of sedimentation centrifuges, if geometric dissimilarities are also considered.

The capacity factor, Σ_B, defined by equation 7, is derived from a set of assumptions. An additional assumption is specific to the bottle centrifuge. Namely, a particle is considered sedimented when it reaches the surface of the cake without contacting the tube wall.

The Imperforate Bowl Centrifuge. In an imperforate bowl centrifuge the flow of the continuous liquid phase is nominally axial, except for areas immediately adjacent to the feed inlet and effluent outlet. Tubular bowl, imperforate basket and decanter centrifuges satisfy this definition.

The mathematical model chosen for this analysis is that of a cylinder rotating about its axis (Fig. 2). Suitable end caps are assumed. The liquid phase is introduced continuously at one end so that its angular velocity is identical everywhere with that of the cylinder. The flow is assumed to be uniform in the axial

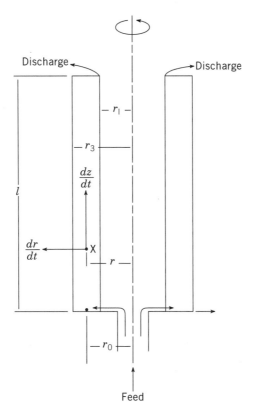

Fig. 2.   Separation in a basket or tubular centrifuge. Terms are defined in the text.

direction, forming a layer bound outwardly by the cylinder and inwardly by a free air–liquid surface. Initially the continuous liquid phase contains uniformly distributed spherical particles of a given size. The concentration of these particles is sufficiently low that their interaction during sedimentation is neglected.

Under these circumstances, the settling motion of the particles and the axial motion of the liquid phase are combined to determine the settling trajectory of these particles. The trajectory of particles just reaching the bowl wall near the point of liquid discharge defines a minimum particle size that starts from an initial radial location and is separated in the centrifuge. A radius r' is chosen to divide the liquid annulus in the bowl into two equal volumes initially containing the same number of particles. One-half of the particles of size d present in the suspension are separated; the other one-half escape, which is referred to as a 50% cutoff.

The feed rate corresponding to this condition is related to the bowl geometry r_1, r_3, and l; the bowl angular speed, ω and the Stokes' settling velocity, v_g (eq. 2).

$$Q_0 = V/t = 2\, v_g(\omega^2/g)\pi l(r_3^2 - r_1^2)/(\ln\,(2r_3^2/(r_3^2 + r_1^2))) \tag{9}$$

As an approximation with a maximum error of 4%:

$$Q_0 = 2v_g(\omega^2/g)2\pi l(3/4r_3^2 + 1/4r_1^2) \tag{10}$$

where v_g = the Stokes' settling velocity (see eq. 1); ω = the angular velocity of the centrifuge; g = the gravitational acceleration; 1 = the length of the settling zone; r_3 = the radius of the inside wall of the cylinder; and r_1 = the radius of the free surface of the liquid layer in the cylinder. Equation 9 can be rewritten as equations 11 and 12:

$$Q_0 = 2v_g\Sigma_T \tag{11}$$

where

$$\Sigma_T = 2\pi l\left(\frac{\omega^2}{g}\right)\left(\frac{3}{4}\,r_3^2 + \frac{1}{4}r_1^2\right) \tag{12}$$

Equation 11 estimates the flow or throughput rate, above which particles of size d are <50% sedimented, and below which >50% are collected. Equations 11 and 12 are also applicable to the light particles rising in a heavy phase liquid, provided that r_3 and r_1 are interchanged in equation 12.

The theoretical capacity factor, Σ_T, defined by equation 12 has the dimension of an area and can be interpreted as the area of a gravity settling tank where the separation performance is equal to the centrifuge provided that the factor v_g is the same for both. This restriction is required because the particles suspended in the continuous phase can be deaggregated and further dispersed by the vigorous shearing to which the feed is subjected during acceleration in the centrifuge. If this effect is not considered in comparison with that of the settling tank, centrifugal sedimentation might be less favorable in practice than is anticipated by theory.

Equation 12 can be further reduced to facilitate understanding of its use and application. If, instead of the radii of pond surface, r_1, and bowl wall, r_3, a mean radius, r_m, is introduced, equation 12 can be rewritten as follows:

$$\Sigma_T = 2\pi r_m l\left(\frac{\omega^2 r_m}{g}\right) \tag{13}$$

Equation 13 shows that Σ_T can be expressed as the product of a mean sedimentation area ($2\pi r_m l$) and the G level ($\omega^2 r_m/g$), and therefore reflects the increased sedimentation rate expected through a defined area having centrifugal acceleration instead of gravity.

Disk Centrifuge.   The separation of particles inside a disk stack is illustrated in Figure 3. The continuous liquid phase, containing solid or liquid particles to be separated, flows from the outside of the disk stack, having radius r_2, to the inside discharge opening, having radius r_1. Assuming that the liquid phase is evenly divided between the spaces formed by the disks, the flow in each disk space is Q_0/n, where Q_0 is the total flow through the entire disk stack and n is the number of spaces (disks). The flow of the continuous liquid phase is also assumed to be in a radial plane and parallel to the surfaces of the disks, and as having the same angular velocity as the stack.

Here again an equation is established (2) to describe the trajectory of a particle under the combined effect of the liquid-transport velocity acting in the x

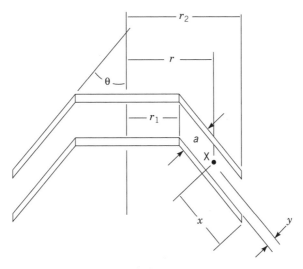

Fig. 3. Separation in a disk centrifuge, where X is the initial position of the particle. Other terms are defined in the text.

direction and the centrifugal settling velocity component in the y direction. Equation 14 determines the minimum particle size that originates from a position on the outer radius, r_2, and the midpoint of the space, a, between two adjacent disks, and just reaches the upper disk at the inner radius, r_1. Particles of this size initially located above the midpoint of space a are all collected on the underside of the upper disk; those particles initially located below the midpoint escape capture. This condition defines the throughput, Q_0, for which a 50% recovery of the entering particles is achieved. That is,

$$Q_0 = 2v_g \left(\frac{2\pi n \omega^2}{3 g} \cot \theta (r_2^3 - r_1^3) \right) \tag{14}$$

where v_g = the Stokes' settling velocity (see eq. 1); n = the number of spaces between disks in a stack; ω = the angular velocity of the centrifuge; θ = one-half the included angle of the disks; r_2 = the outer radius of the disks; and r_1 = the inner radius of the disks (see Fig. 3). If v_g, which describes the settling characteristics of the particles, is separated from parameters relating to the geometry and rotational speed of the disk stack, then,

$$Q_0 = 2v_g \Sigma_D \tag{15}$$

where

$$\Sigma_D = \frac{2\pi n \omega^2}{3 g} (\cot \theta (r_2^3 - r_1^3)) \tag{16}$$

Again, Σ_D corresponds to the area of a gravity settling tank capable of the same separation performance as the disk stack defined by the parameters included in equation 16.

2.2. Separation by Drainage. The theory covering drainage in a packed bed of particles is incomplete, and requires more development for a centrifugal field. Liquid is held within the bed by various forces. Removal involves several flow mechanisms. In addition, the centrifugal acceleration changes with radius in the bed, causing changes in packing tendencies of particles and accelerating forces on the residual liquid.

There are three types of liquid content in a packed bed: (1) in a submerged bed, there is liquid filling the larger channels, pores, and interstitial spaces; (2) in a drained bed, there is liquid held by capillary action and surface tension at points of particle contact, or near-contact, as well as a zone saturated with liquid corresponding to a capillary height in the bed at the liquid discharge face of the cake; and (3) essentially undrainable liquid exists within the body of each particle or in fine, deep pores without free access to the surface except perhaps by diffusion or compaction.

The last type of liquid can be removed by evaporation (qv) or displacement by another liquid but cannot be removed by simple flow in either a gravitational or centrifugal field. There is no sharp distinction between the first two types. The rate and extent of liquid removal from a submerged bed during drainage depends on the physical characteristics of the components, the force of the centrifugal field, and the time of exposure to the field. The residual liquid content of a drained cake consists largely of the capillary and irremovable types at the time of discharge from the separation equipment.

During cake formation and drainage the liquid moves into and through the bed in three different ways. During cake deposition, a continuous head of liquid ranging in composition from feed to clarified supernate may exist over the deposited cake. After feed is stopped, a layer of essentially clarified liquid may still exist over slow-draining cakes. Wash liquor, if it is used, may also create a liquid layer over the deposited cake. Drainage under these conditions requires continuous flow through the cake. The interstitial spaces are assumed to be full. When the free liquid layer no longer exists above the cake, the free liquid surface moves through the cake to an equilibrium position at the capillary height, leaving behind the larger voids filled with gas or vapor. Then, after bulk drainage of the larger voids, liquid still exists in the cake's upper zone in a film covering the surfaces of the solids and in partially filled voids having very restricted outlets. In time, some of this liquid flows as a film to the continuous liquid layer at the capillary height.

If a cake is sufficiently impermeable to permit the buildup of a feed or wash liquid head, flow through the cake approaches steady-state conditions except for changes in compaction or cake thickness as more feed is added, or in the liquid head if the drainage rate differs from the liquid rate addition. An equation for full-pore flow in a centrifugal field has been developed based on Darcy's equation. The hypotheses set forth and for the most part proved are as follows: flow radiates out from the rotation axis and the effect of the gravitational field is negligible; voids at all points are filled with liquid that moves in laminar flow through the cake; kinetic energy changes of the liquid in the cake may be neglected, ie, the filter medium is sufficiently permeable so that it does not run full of liquid; and ambient pressure exists at the outer face of the cake. Essentially incompressible solids produce very similar cake permeabilities in a vacuum, under pressure, and during centrifugal filtration, although significant local variations in permeability may occur because of irregularities in the feed and its distribution. When pores are filled, the flow rate of the supernatant liquid through the cake (4,5) is as follows:

$$Q = \left(\frac{\pi K \omega^2 h \delta_1}{\mu_l g}\right)\left(\frac{r_m^2 - r_l^2}{\ln\ (r_m/r_c)}\right) \tag{17}$$

where K is the permeability; h is the basket height; r_C and r_m are radii from the axis of rotation to the inner and outer faces of cake, respectively; ω is the angular velocity; and r_l is the radius from the axis to the inner face of the liquid layer. Comparison to the usual term α for cake resistance in pressure filtration shows that $K = \delta_l g/\alpha$. The functions related to cake thickness, (r_m/r_C); liquid layer thickness, $(r_m^2 - r_l^2)$; angular velocity, ω^2; and viscosity, μ_l, have been verified by experiment. The exponent on the angular velocity varies for materials that exhibit cake compression as speed is increased, but compression effects are minor on starch, chalk, and kieselguhr, ie, loose or porous diatomite (qv). In practice, the exponent of ω can probably be assumed to have a value of two and experimental variations in permeability can be absorbed by changes in the permeability coefficient. The real value of equation 17 lies in its use for estimating the effect of changes in operating variables for a material where the characteristics are already known from experimental data or plant operation.

Low permeability cakes draining under the conditions of equation 17 are usually handled in perforate basket centrifuges that have relatively long (20 min to several hours) cycles. Following elimination of the supernatant liquid, unsteady-state drainage of the cake may often be neglected. These slow-draining cakes often support such a large capillary height, eg, 90–99% of void volume for chalk (5), that little additional dewatering (qv) is obtained after completion of free-liquid drainage. Solids of larger particle size, or freer drainage characteristics, are handled in automatic perforate baskets or continuous screen centrifuges. Low final moistures are usually achieved. For cakes of high permeability, the period when a liquid layer exists above the cake is either short or nonexistent. The time cycle depends chiefly on the rate of film drainage at the completion of bulk liquid flow. Under these conditions film drainage and permanent residual moisture are most important.

The quantity of undrainable residual moisture cannot be predicted without the benefit of experimental data. Equation 18 (6) indicates the important parameters where the exponents were determined using limited experimentation. Introducing the approximation that $s\ \delta_s$ is proportional to $1/d$, where s is the specific surface area per weight of solid, the modified equation for undrainable liquid becomes

$$S^\infty = k\left(\frac{1-\epsilon}{\epsilon}\right)\left(\frac{1}{\bar{d}^2 G}\right)^{1/4}\left(\frac{\sigma \cos \Psi}{\delta_L}\right)^{1/4} \tag{18}$$

Where S^∞ is the fraction of void volume occupied by liquid after infinite drainage time, k is an experimental coefficient, ε is the void fraction, d is the mean particle diameter, G is $\omega^2 r/g$, σ is the surface tension, and ψ is the wetting angle. Appreciable internal porosity of the particles can badly distort an experimental value of S.

For cakes of high permeability, the capillary drain height may be an insignificant fraction of cake thickness, and film drainage becomes the controlling factor in a centrifugal field (7). Under unsteady-state conditions, equation 19 represents the drainable liquid left in the cake as a function of the centrifugal filtration parameters:

$$S - S^\infty = \frac{3}{\pi} \frac{s'}{\epsilon} \omega \left(\frac{\mu_2}{\delta_L}\right) \left(\frac{h}{2\,r_m - h}\right) \left(\frac{1}{t - t'}\right)^{1/2} \tag{19}$$

where S is the fraction of void volume occupied by liquid at time t, s'; is surface area/volume of cake, h is cake thickness, t'; is time at which free liquid surface enters the cake, and μ_2 is the viscosity of the surrounding medium.

It is difficult to obtain void volume data for a cake under drainage conditions in a centrifuge. Prediction of these values from filter cake data is uncertain because the compressive force in centrifugation increases with radius throughout the cake depth, and the effective mass of a particle, proportional to $(\delta_S - \delta_L)$ when the cake is submerged, becomes essentially δ_s after bulk drainage is completed. For engineering purposes, it is simpler to approximate the ratio of the volume of liquid to the volume of solid. Assuming that $G = \omega^2 (2r_m - h)/g$, and $s' \sim (1 - E)/d$ as for spheres, equation 19 becomes

$$q - q^\infty = \left(\frac{k'}{d}\right) \left(\frac{\mu_L}{\delta_2}\right)^{1/2} \left(\frac{h}{Gt}\right)^{1/2} 100 \tag{20}$$

where q is the ratio of the volume of liquid to the volume of solid. This measurement is readily obtained from the volume percentage of liquid in cake and is also easily converted by a density ratio to a weight ratio of liquid to solid. The value for q^∞ the ratio of the liquid volume to the volume of solid at infinite time, may also be applied when known.

Reasonable experimental agreement was obtained (7) using the exponents given in equation 20 for relatively slow-draining chalk and kieselguhr. Figure 4 (8) shows the effect of drain time after the disappearance of a free liquid head above the cake. The sharp break probably indicates completion of bulk drainage and start of drainage by film flow only. Drainage before the break is rapid and proportional to $t^{-1.9}$. After the break in the curve, the value $t^{-0.3}$ indicates that film drainage becomes controlling if low residual moisture is required. This exponent, -0.3, is appreciably lower than the -0.5 in theory but is close to -0.25, the value previously obtained (9). Considerable data indicate the validity of $G^{-1/2}$; limited data corroborate the theoretical exponent of $1/2$ for kinematic viscosity.

Experimental exponents for cake thickness vary from 0.5 to as much as 3.0. The theoretical value of $1/2$ may be approached only by incompressible cakes of a narrow range of sizes. The proper and characteristic value for the mean particle size, d, is difficult to ascertain. In practice, the most finely divided particles, eg, 10–15 wt% of solids, almost wholly determine the liquid content of a cake, regardless of the rest of the size distribution. It seems reasonable to use a d closely related to liquid content, eg, the 10% point on a cumulative weight-distribution curve.

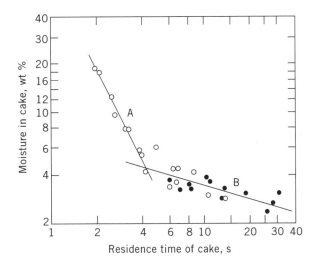

Fig. 4. Drainage of salt crystals in a cylindrical screen pusher-discharge centrifuge (8), where the cake thickness is 3.3 cm, the centrifugal field = 320 G, and the crystals 14 wt% <250 μm. (●) Represents moisture in the discharge cake, and (○) moisture in the cake by material balance with drainage flows; line A has slope = −1.9 and B, −0.3.

2.3. Separation by Compaction. In centrifuges used for dewatering compactable slurries, cake compaction follows the sedimentation of the suspended solids. Both solid bowl and perforated bowl centrifuges are used for dewatering compactable cakes. In a solid bowl centrifuge the expelled less dense liquid flows inwards countercurrent to the compacting cake, whereas it flows outwards through the compacting cake in a perforate bowl centrifuge, provided that the liquid pressure generated by rotation exceeds the capillary pressure. The bulk filtration rate of a filtering centrifuge can be estimated from equation 17 including accounting for the pressure drop over the filter medium and assuming an average cake permeability across the cake thickness.

In a decanter centrifuge the liquid released during compaction flows predominately radially inward and Darcy's equation can be used for estimating this liquid flow, with the following assumptions: (1) radial solids velocities are neglected, (2) cake motion due to scrolling is disregarded, and (3) the time required for full cake consolidation is much longer than the cake retention time in a centrifuge (10,11).

$$Q_0 = ((C_c - C_e)/(C_c - C_o))(K_m/\mu)(\delta_C - \delta_L)\pi l(r_3^2 - r_c^2)/\ln(r_3/r_c) \qquad (21)$$

$$Q_0 = \phi K_p \qquad (22)$$

Where K_p is a process constant and

$$\phi = \pi l(r_3^2 - r_c^2)/\ln(r_3/r_c) \qquad (23)$$

Comparing equation 23 with equation 9 and assuming that $R_c > R_1$, which is the normal operating mode for sludge compacting solid bowl centrifuges, one sees that the centrifuge related terms are identical, considering that equation 23 was developed for 100% solids capture, whereas, equation 9 assumes 50% cutoff. This in turn implies that scaling of solid bowl centrifuges based on sedimentation considerations (Σ concept) is identical to scaling based on cake compaction considerations.

2.4. Σ-Concept and Its Application. The assumptions and conditions for deriving equations 7, 12, and 16 impose limitations on the application of the Σ concept and fall into two groups. The first concerns the particulate material. Particles (or drops) are assumed to be spherical in shape and uniform in size. These should not deaggregate, deflocculate, coalesce, or flocculate during passage through the zone in which separation occurs. Initially, particles are evenly distributed in the continuous liquid phase, where their concentration is low enough for them to settle as individual particles without interaction. The settling velocity, v_s, of the particles is such that the Reynolds number does not exceed 1.0, thus ensuring that the deviation from Stokes' law does not exceed 10%. The settling velocity, v_g, in a gravity field, or v_s in a centrifugal field, is theoretically never reached because the accelerating time required for a particle to reach its terminal velocity is infinite. However, a particle of up to 100 µm approaches 90% of its terminal velocity in milliseconds. The time available for the particles to settle in a centrifuge is enough for each individual particle to be almost at its theoretical settling velocity, despite variation in G.

The second group of assumptions and conditions concerns flow conditions. Flow is assumed to be streamlined. Fresh feed is introduced uniformly into the full space available for its flow. In an imperforate bowl centrifuge, this condition requires that the continuous liquid phase immediately occupy the full liquid layer thickness between the free surface and the inside radius of the cylindrical wall. In a disk centrifuge, the continuous phase is assumed to divide evenly between all the disk spaces axially as well as circumferentially. In any imperforate bowl centrifuge, the continuous phase rotates everywhere at the same angular velocity as the bowl, ie, there is no forward or back swirl. The displacement of the flow pattern of the continuous phase by the layer of deposited material is neglected. Remixing at the interface of the separated material is negligible. Finally, the detrimental effect resulting from heavy separated material crossing the fresh feed stream outside of the disk stack is neglected.

Few of the assumed conditions are fully satisfied in practice. The last three items relate to potential interference between separated phases. Such interference can occur and leads to poor sedimentation performance if an excessive volume of the sedimented phase is retained in the centrifuge.

Excessive volume of solids may be retained in the bowl of conveyor centrifuges if (1) the conveyor volumetric displacement is not sufficient to handle the sedimentation rate of solids; (2) the sedimented solids cannot be successfully conveyed and discharged over the solids port until a sufficient layer has been built up inside the bowl; and (3) solids do not easily slide outwardly on the underside of the disk of a disk centrifuge.

In the case of the nozzle disk centrifuge, the flow of the solids phase through the discharge nozzles may be so restricted that an excessive layer can accumulate inside the bowl shell. When this layer reaches the zone utilized by the fresh feed stream entering the disk stack, reentrainment of the sedimented solids by the fresh feed may lead to poor sedimentation performance.

The sedimentation phenomenon that the Σ concept attempts to describe quantitatively is only part of the total task that the centrifuge has to accomplish. Thus, attempts to predict separation performance solely on the basis of Σ concepts have sometimes given disappointing results.

Nevertheless, the Σ concept is a valuable tool, allowing in theory a comparison between geometrically and hydrodynamically similar centrifuges operating on the same feed material. Equations 7, 12, and 16 show that the sedimentation performance of any two similar centrifuges having the same feed suspension is the same if the quantity Q_0/Σ; is the same for each. In practice, an efficiency factor, e, is often introduced to extend the use of Σ so as to compare dissimilar centrifuges. This factor takes into consideration differences in feeding, discharging, flow, turbulence, and remixing that exist in different types of centrifuges operating on the same feed material. The flow rate, $Q_{0\,2}$, of a No. 2 centrifuge can thus be compared to the rate, $Q_{0\,1}$, of a No. 1 centrifuge operating on the same feed. For equal sedimentation performance,

$$\frac{Q_{0_2}}{\Sigma_2 e_2} = \frac{Q_{0_1}}{\Sigma_1 e_1} \quad \text{or} \quad Q_{0_2} = Q_{0_1}\left(\frac{e_2}{e_1}\right)\left(\frac{\Sigma_2}{\Sigma_1}\right) \tag{24}$$

If the two centrifuges are geometrically and hydrodynamically similar, then $e_1 = e_2$ and equation 20 can be simplified to

$$Q_{0_2} = Q_{0_1}\frac{\Sigma_2}{\Sigma_1} \tag{25}$$

The Σ concept permits scale-up between similar centrifuges solely on the basis of sedimentation performance. Other criteria and limitations, however, should also be investigated. Scale-up analysis for a specified solids concentration, eg, requires knowledge of solids residence time, permissible accumulation of solids in the bowl, G level, solids conveyability, flowability, compressibility, limitations of torque, and solids loading. Extrapolation of data from one size centrifuge to another calls for the application of specific scale-up mechanisms for the particular type of centrifuge and performance requirement.

2.5. Other Sedimentation Scale-Up Equations. Some centrifuge suppliers use an area-equivalent, Ae, description instead of Σ; others use KQ or Lf_2 values. All of these are in units of area. For a disk centrifuge,

$$\Sigma_D = \frac{2\pi\eta\omega^2}{3g}\cot\theta(r_2^3 - r_1^3)$$

$$KQ = \frac{2\pi n\omega^{1.5}}{3g}\cot\theta(r_2^{2.75} - r_1^{2.75}) \tag{26}$$

$$Lf_2 = \frac{2\pi n\omega^2}{3g}\cot\theta\left(r_2^3 - \frac{r_1}{r_2} - \left(\frac{r_1}{r_2}\right)^2 - \left(\frac{r_1}{r_2}\right)^3\right) \tag{27}$$

For an imperforate tubular centrifuge,

$$Ae_{3/4} = 2\pi l\omega^2 (0.75r_3)^2 \tag{28}$$

$$\Sigma = \frac{2\pi l\omega^2}{g} \left(0.75 \ r_3^2 + 0.25 \ r_1^2\right) \tag{29}$$

All of these equations work in the scale-up of geometrically similar centrifuges. The KQ reduces the effect of rotational speed from ω^2 to $\omega^{1.5}$ and the disk radius from r^3 to $r^{2.75}$, based on empirical experience. The Lf_2 uses the projected cylinder area of the disks at their mean radius (see equation 13 of Σ_T for imperforate bowls. The parameter $Ae_{3/4}$ is the cylindrical area at three-quarter of the bowl wall radius).

When testing a new material for centrifugal separation, a bottle centrifuge is usually used to obtain the general G range needed, and to choose the centrifuge type and size. To estimate size, the Q_0/Σ_B must be determined for the bottle centrifuge (eq. 7), and then used in equation 25 to determine the Σ value of the centrifuge to be used. Efficiency factors for the various types of centrifuges (12) have been reported:

Disk bowl centrifuges	45–73 %
Scroll centrifuges	54–67 %
Tubular bowls	90–98 %

2.6. Factors Influencing Centrifugal Sedimentation.

The sedimentation velocity of a particle is defined by equations 1 and 2. Each of the terms therein effects separation.

Viscosity. Sedimentation rate increases with decreased viscosity, μ, and viscosity is dependent on temperature. Often mineral oils, which are highly viscous at room temperature, have a viscosity that is reduced by a factor of 10 at 70–80°C. Tar, solid at room temperature, is a low viscosity liquid at 150–200°C and can be clarified of inorganic solids at high flow rates. Even the viscosity of water changes significantly when the temperature changes between 10 and 35°C.

Density Difference Between Particle and Liquid. Separation cannot take place if $\Delta\delta = 0$. Some mineral oils have the same density as water at room temperature. When heated to 80°C, the reduction of the density of water is less than that of the mineral oil, resulting in the water becoming heavier. Therefore separation is possible. Dilution of a liquid by a solvent, eg, molasses by water or heavy oil by naphtha, results in lower density and lower viscosity of the liquid. Solvent stripping may be required at a later stage in the process.

Particle Size. Doubling particle size, d, increases sedimentation by a factor of 4. Thus, methods to increase size are important. Additives are commonly used to flocculate many fines into an agglomerate that acts as a single large particle. Many chemical companies offer a wide range of organic and inorganic flocculating products, in dry or liquid form. Bench tests are usually required to

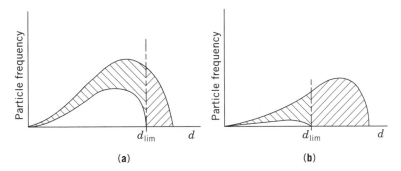

Fig. 5. Particle distribution (upper line) before and (lower line) after action of the separator where the cross-hatched areas represent the particles separated out. By definition, all particles of $d > d_{lim}$ are separated out. A number of particles having $d < d_{lim}$ are also separated. (**a**) Fine and (**b**) coarse particle dispersion (13).

determine best type and dose, ie, to optimize the flocculent choice. pH control, electrostatic devices, and mechanical coalascers are used to combine fine liquid drops in emulsions to produce larger particles. Special care must be exercised in pumping, and feeding mixtures of these easily breakable particle agglomerates, thus preventing the large particles from becoming fine before sedimentation.

    Particle Shape.   Whereas the Stokes' particle is assumed to be a sphere, very few real solids are actually spherical. Flat and elongated particles sediment slower than spheres. Normally an equivalent diameter would be used.

    Particle Size Distribution.   Almost every feed slurry is a mixture of fine and coarse particles. Performance depends on the frequency of distribution of particle size in the feed. Figure 5 shows that whereas all of the coarse particles having a diameter greater than some d_{lim} are separated, fewer of the very fine particles are, at any given feed rate. The size distribution frequency of particles in feed and centrate for a fine and coarse feed are quite different. More coarse particles separate out than fine ones. Classification of solids by size is often done by centrifugal sedimentation.

3. Liquid–Liquid-Phase Behavior

Liquid drops, suspended in a continuous liquid medium, separate according to the same laws as solid particles. After reaching a boundary, these drops coalesce to form a second continuous phase separated from the medium by an interface that may be well or ill defined. The discharge of these separated layers is controlled by the presence of dams in the flow paths of the phases. The relative radii of these dams can be shown by simple hydrostatic considerations to determine the radius of the interface between the two separated layers. The radius is defined by

$$r_i^2 - r_h^2 = \frac{\delta_l}{\delta_h} \left(r_i^2 - r_l^2 \right) \tag{30}$$

where r_i, r_h, and r_l are the radii of the interface, the liquid surface at the heavy discharge dam, and the liquid surface at the light discharge dam, respectively, and δ_h and δ_l are the densities of the heavy and light phases, respectively. Control of the interface radius, achieved by varying r_h or r_l for the desired ratio, is an important factor in liquid–liquid separation, as it determines whether the heavy or light phase is exposed to the greater separating effect (13).

Equation 30 is accurate only when the liquids rotate at the same angular velocity as the bowl. As the liquids move radially inward or outward these must be accelerated or decelerated as needed to maintain solid-body rotation. The radius of the interface, r_i, is also affected by the radial height of the liquid crest as it passes over the discharge dams, and these crests must be considered at higher flow rates.

4. Centrifuge Components

4.1. Power, Energy, and Drives.

Centrifuges accomplish their function by subjecting fluids and solids to centrifugal fields produced by rotation. Electric motors are the drive device most frequently used; however, hydraulic motors, internal combustion engines, and steam or air turbines are also used. One power equation applies to all types of centrifuges and drive devices.

The total power, P_T, needed to run a centrifuge, ie, delivered by the drive device, is equal to sum of all losses:

$$P_T = P_P + P_S + P_F + P_W + P_{BD} + P_{CP} \tag{31}$$

where P_P, the process power, $= Q\delta\omega^2 r^2$; Q is the flow rate of liquid or solids; $\delta =$ mass density; and $r =$ discharge radius of material being discharged. Power for each liquid and the solid phase must be added to get P_P. P_s, the solids process power $= k_p T_C \cdot \Delta N$ for scroll decanters, where $T_C =$ conveyor torque k_p is a conversion factor and $\Delta N =$ differential speed between bowl and conveyor. The parameter P_F is the friction power, ie, loss in bearings, seals, gears, belts, and fluid couplings. P_W, the windage power $= k_s \mu^{0.2} \rho^{0.8} N^3 D^{4.5}$ and $\mu =$ viscosity of surrounding gas; $\rho =$ density of gas; $D =$ rotor outside diameter; $N =$ bowl speed; and $k_s =$ shape constant. Increased density owing to gas pressure increases the windage power, and this may be very significant for high pressure applications. Also, many hydrocarbon gases are heaver than air, resulting in high windage power. Very high G centrifuges often operate in a vacuum to avoid excessive windage power. For constant G, the scale-up windage power $\sim D^3$; for constant rotor stress, the scale-up of windage power $\sim D^{1.5}$.

Windage power is a very important loss for large machines and must be determined. Whereas windage power can be calculated from drawing dimensions (14), it is preferable to measure the windage power for an actual rotor, and then extrapolate using the formula given for windage power for a geometrically similar (larger or smaller) size. Doubling the size of a rotor while maintaining the g level results in eight times the windage power loss.

P_{CP} is the friction power consumed by the centripetal pump. The centrate kinetic energy is partially recovered by the pump, which delivers the centrate flow at a positive pressure. The added power must be supplied by the centrifuge main drive, but use of a centripetal pump avoids the need for a separate centrate pump. The power required to bring the feed to the centrifuge is supplied by a feed pump at the feed tank, not by the centrifuge drive. This power may be significant where the feed pressure required due to flow rates is high.

For scroll centrifuges having back-drives, P_{BD} is the back-drive power:

$$P_{BD} = k_p(T_C/R)N_{BD} \tag{32}$$

where T_C = conveyor torque; R = gear box ratio; and N_{BD} = back-drive speed. Depending on the type of gearbox and wether the conveyor is leading or lagging the bowl, the backdrive may be a driver or a break. In the case of a driving backdrive, P_{BD} in equation 32 is provided by the backdrive driver, otherwise P_{BD} is provided by the centrifuge drive device. Braking backdrives can be regenerative or non regenerative. In the former case some of the braking power is regenerated using a motor as a generator, otherwise the energy is dissapated as heat. Regenerative back-drives reduce total power consumption. A high gear box ratio results in lower back-drive power. Direct hydraulic motor conveyor drive devices get their power from an external hydraulic power supply, not from the main drive motor, and must meet the direct conveyor torque demands. The equation for direct hydraulic conveyor power is $P_{BD} = k_p T_C \Delta N$; however, hydraulic losses in the power supply, rotary seals, and the hydraulic motor must be added. At least one manufacturer avoids these losses by using a direct drive planetary gear box powered by a variable frequency ac motor.

The choice of the main drive, usually an ac motor, must include starting specifications. Various methods are used to start centrifuges. These include mechanical or fluid couplings, as well as wye-delta electric motor starters or variable frequency drives. Centrifuges are high inertia rotating devices, sometimes taking up to 15 min to accelerate to operating speed. Details of the mass moment of inertia, friction, and windage losses must be considered to specify a drive device. The inertia seen by the drive device, when comparing centrifuges operating at constant G, is proportional to D^4: Small rotors are easy to start, but large rotors must be carefully reviewed.

Disk centrifuges having nozzles to discharge the solids slurry through small backward-pointing nozzles must have this power included in the calculations. Some thickening scroll centrifuges also use such nozzles, usually with an intermittent flow. The parameter P_N is the nozzle power:

$$P_N = Q_N \delta_N \omega r_N (\omega r_N - v_N \cos \phi) \tag{33}$$

where Q_N = the volume rate of the material discharged through the nozzles, δ_N = the mass density of the solids slurry discharged through the nozzles, ω = the angular velocity of the centrifuge, r_N = the radius at which the nozzles discharge, v_N = the linear discharge velocity out of the nozzles, and ϕ = the angle measured between the direction of the nozzle and the tangent to the circle of radius, r.

The discharge velocity out of the nozzles is given by equation 34:

$$v_N = C(2p_N/\delta_N)^{1/2} \tag{34}$$

where C = a discharge coefficient, and p_N = the hydraulic pressure at radius r_N resulting from the rotation of the bowl. The pressure p_N is given by equation 35:

$$p_N = \omega^2 \pi (r_N^2 - r_1^2)/2 \qquad (35)$$

where r_1 is the free-surface radius of the liquid phase, δ_p is the weighted average density of the process material in the bowl.

The coefficient C in equation 34 is a function of two phenomena. First, the presence of viscous friction accounts for a small loss of energy. Nozzle orifice contraction results in discharge coefficients which range from 0.5 to 0.85. The coefficient falls in the upper portion of this range when the length of the nozzle is two or three times its diameter, and in the lower end of the range where the nozzle diameter is more than five times its length.

Special vortex nozzle designs that deliver lower flows using a large opening have been used to reduce plugging problems. Also, viscosity-sensitive nozzles, where flow is increased as viscosity is increased, are used to control variation in solids concentrations.

Filtering centrifuges must consider the power needed to bring the solids to a final radius. The radius used to determine the centrate power consuption in equation 31 is the outside radius of the rotor supporting the filtering screen. Many filtering centrifuges discharge solids at a reduced bowl speed to avoid particle breakage. The main drive is often designed to recover the rotor kinetic energy during deceleration.

The energy absorbed by the liquid and solids stream, P_P, is transferred in the feed zone of the rotor or conveyor. One-half of the total energy is converted to the kinetic energy associated with the tangential velocity of the pond surface. This kinetic energy is dissipated as heat when the centrate is discharged to the stationary casing. The other one-half of the total energy is lost as turbulence in or near the feed zone. The intensity of this turbulence can break friable particles, or create tight emulsions of two immiscible liquids, both making the separation that follows more difficult. Reducing the pond radius reduces the total process power and particle degradation and thus improves total separation performance. Reducing the pond inside radius by 25% reduces power consumed by 44%.

4.2. Materials of Construction and Operational Stress. Before a centrifugal separation device is chosen, the corrosive characteristics of the liquid and solids as well as the cleaning and sanitizing solutions must be determined. A wide variety of materials may be used. Most centrifuges are austenitic or duplex stainless steels; however, many are made of ordinary steel, rubber or plastic coated steel, Monel, Hastelloy, titanium, and others. The solvents present and of course the temperature environment must be considered in elastomers and plastics, including composites.

Once the material choice based on corrosion is made, a careful analysis of the stresses produced by rotation for the particular type of centrifuge is required, so that for the given liquid and solids specific gravities a maximum operating speed can be determined. In general, the metals used are ductile and elastic in the operating speed range. The usual limits are the ultimate strength and yield strength (0.2% offset). The stresses of the centrifuge bowl are primarily tangential stress and axial stress. These are the result of the weight of the bowl material itself, and the need to contain the rotating process solids and liquids within the bowl.

The geometry of the bowl parts is important, and for intermittently discharging centrifuges, fatigue strength must be considered. In general the tangential stress in a bowl wall is σ_r.

$$\sigma_T = \sigma_{self} + \sigma_p \qquad (36)$$

where for a thin-shell bowl

$$\sigma_{self} = K'\delta_M\omega^2 r^2 \qquad (37)$$

$$\sigma_p = K'\delta_p\omega^2(r^2 - r_p^2) \qquad (38)$$

where δ_M = mass density of bowl material; r = bowl radius; δ_p = maximum density of process material; r_p = smallest radius of the process material (assuming that the bowl is full of process material); and K' is a function of bowl geometry and includes stress concentrations owing to changes in section, holes, slots, fillets, etc. The total stress increases with the square of bowl speed, ω^2, and the pond radius, and is directly proportional to the liquid and solid density.

Within the bowl, the pressure, p, developed also exerts axial separation forces. The internal pressure in a bowl is shown in equation 39:

$$p = \pi\omega^2(r^2 - r_p^2)/2 \qquad (39)$$

The axial projected area is $A = \pi(r^2 - r_p^2)$ and the average pressure is $1/2\,p$. The axial force $F_A = pA/2$, where

$$F_A = \delta_p\omega^2\pi(r^2 - r_p^2)/4 \qquad (40)$$

The stress resulting from the axial forces must be considered in analyzing the parts that resist axial separation, such as bolts, nuts, rings, etc. On scroll centrifuges the axial force owing to the axial component of the conveyor torque must also be considered.

Typically, the total stress on the bowl is 50–65% self-stress, and 35–50% process stress. The axial stress is usually 5–10% of the tangential stress. When the bowl material density is low (such as titanium), or where solids are heavy (as is coal), or when the pond surface is close to the bowl wall (as when using shallow pond scroll centrifuges), these proportions differ. All centrifuges must have a factor of safety against general yeilding and rupture. The ratio of the material strength to the actual stress is the factor of safety. Factor of safety is set by the manufacturer, and sometimes, especially in Europe, by government regulation (15). Every centrifuge supplier sets the limits of bowl speed, temperature, pressure and liquid, and solids density. On some very high G preparative or zonal centrifuges, the number of cycles of bowl use must be recorded and the rotor retired after a given number, because fatigue determines the bowl life.

Most disk and scroll centrifuges are made from forgings, centrifugal castings, or fabrications. Each manufacturer must carefully monitor the mechanical properties of the bowl material to ensure that the required factor of safety is maintained as well as the specified ductility. A rotor failure during operation is very serious, and can not only destroy the centrifuge, but also damage nearby equipment and injure operators.

Many of the centrifugal separation applications are abrasive and erosive to centrifuge parts because of the high relative velocity or high contact pressure between the particles and exposed parts. Areas of wear must be protected using materials that resist both mechanical and chemical attack. In general, the main areas of abrasive wear are the feed zone surfaces, where nonrotating process slurries are accelerated to speed; the tips and faces of the conveyor flights on decanter centrifuges; solids discharge openings; screens, where solids slide across the screen; and stationary collection surfaces that receive the impact of material being discharged from the bowl. Many abrasion resistant materials are used. Examples are sintered tungsten carbide bound with cobalt or nickel; sintered aluminum oxide ceramic; sprayed and fused or weld-deposited hard-surfacing alloys; and elastomeric coatings or inserts. There is great variation in the design and application of these materials that effect the actual working life in service and the cost to rebuild the worn areas or replace the worn insert.

Maintenance and operating costs are lower if replaceable wear protection is used, compared to the requirement to rebuild (and thus rebalance) the worn parts. In very severe applications, such as coal (qv) or coal refuse dewatering that use screens, wedge wire screens made from ceramic or carbide materials are required. The use of carbide or ceramic flight-tip protection on decanter centrifuges has permitted economic use in applications such as dewatering (qv) tar sands as well as coal and coal refuse, and dewatering and thickening mixed primary and secondary sewage sludges. Continuous nozzle discharge disk and decanter centrifuges would not be feasible without replaceable ceramic or carbide nozzle inserts. The use of relatively soft abrasion-resistant elastomers, such as urethanes, has been successfully applied in the solids receiver housing and the feed zone targets of scroll centrifuges. Hard chrome plating has been used on the first few disks of disk centrifuges. Stellite or carbide inserts are often used at the solids discharge opening.

4.3. Noise. Centrifuges, as do any rotating equipment, create noise. When the motion of air or gas entrained by a rotating bowl shell is deflected or otherwise disturbed, its energy is transferred to the environment through the casing or chutes. This mechanism suggests that the noise level created by the bowl is related to the surface linear speed of the rotor. High linear speed is important for maintaining separating capacity, so the noise level should be reduced without reducing the rotational speed, if possible.

In addition to surface speed, rotor imbalance, surface irregularities, clearance between covers and rotor, resonance in the supporting structure, conditions of installation, and particularly the drive motor contribute to centrifuge noise. An inadequate supporting platform can amplify centrifuge vibration. Open piping and venting and discharge connections allow noise generated inside the unit to escape. Discharge connections should be tight, yet flexible enough to prevent transfer of vibrational energy to plant piping. Size, spacing, materials of construction, and other properties of the centrifuge room also affect noise level. Sound-absorbing materials or enclosures should be provided when other means are inadequate.

The electric motors are often the noisiest component of the centrifuge assembly. Most standard motors in the 75–250 kW range develop noise levels of 85 dbA (weighted sound pressure level using filter A, per the ANSI standard). A quiet motor can reduce this level by 5 dbA and should be used whenever noise is of concern.

4.4. Equipment. Centrifugation equipment that separates by density difference is available in a variety of sizes and types (16,17) and can be categorized by capacity range and the theoretical settling velocities of the particles normally handled. Centrifuges that separate by filtration produce drained solids and can be categorized by final moisture, drainage time, G, and physical characteristics of the system, such as particle size and liquid viscosity.

For optimum results, a combination of several types of equipment may be used, eg, a gravity separator for oil recovery from sludge at a petroleum refinery. The sludge, an aqueous suspension of 1–5% oil and 5–30% solids <50 μm, is screened to remove trash, and degritted in a cyclone to eliminate the coarse solids that would cause excessive abrasion. A decanter centrifuge then removes 60–70% of the solids in oil-free condition. The resulting oil-in-water emulsion, stabilized by residual fine silt, is passed through a 0.25-mm (60-mesh) screen, and sent to a disk centrifuge that discharges an oil stream at 0.5–2% bottom sediment and water, and an oil-free peripheral nozzle discharge containing the remaining solids in water.

Equipment Materials and Abrasion Resistance. Stainless steel, especially Type 316, is the construction material of choice and can resist a variety of corrosive conditions and temperatures. Carbon steels are occasionally used. Rusting may, however, cause time-consuming maintenance and can damage mating locating surfaces, which increases the vibration and noise level. Titanium, Hastelloy, or high nickel alloys are used in special instances, at a considerable increase in capital cost.

Abrasion, a serious problem in some applications, requires the addition of hard-surfacing materials to points exposed to abrasive wear (18). The severity of wear depends on the nature, size, hardness, and shape of particles as well as the frequency of contact, the force exerted against the wearing parts, and solids loading as related to feed rate and solids concentration.

A wide range of abrasion-resistant materials is available. Nickel–chrome–boron and cobalt–chrome–tungsten hard-surfacing alloys have been used for many years. Composite coatings of nickel-base alloys containing crushed tungsten carbide particles, applied by flame spraying and fusing, are also used. Solid tungsten carbide, pressed to shape and sintered at high temperatures, provides the best protection. Tungsten carbide plates, previously induction brazed, or bonded are now in light of superior corrosion resistance more suitably vacuum brazed to stainless steel supports, which in turn can be easily welded to portions of a centrifuge such as conveyor flights. Ceramics have been used where minor impact and abrasive particle pressures are involved.

5. Centrifuges

Like other manufacturers, those building sedimentation and filtration centrifuges are subject to acquisition and merger. In order to find updated information it is perhaps best to check on line sources such as those listed in Reference 19.

5.1. Sedimentation Equipment. Centrifugal sedimentation equipment is usually characterized by limiting flow rates and theoretical settling capabilities. Feed rates in industrial applications may be dictated by liquid handling capacities, separating capacities, or physical characteristics of the solids. Sedimentation equipment performance is illustrated in Figure 6 on the basis of nominal clarified effluent flow rates and the applicable Q_0/Σ values. The latter are equivalent to twice the theoretical gravity settling velocities. In liquid–solid separations, the effluent rate represents the clarified stream of the liquid medium and does not include the volume of solids discharged or the volume of medium discharged with the solids. The effluent rate of liquid–liquid separation refers to the clarified, heavy, or light continuous phase that usually occupies the greater volume within the separating equipment. The flow range for a particular piece of equipment does not represent its absolute limitations, but the

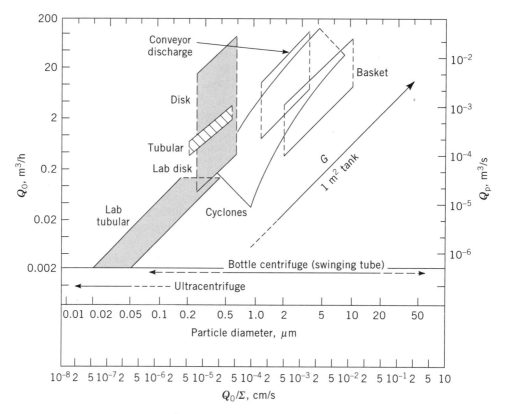

Fig. 6. Sedimentation equipment performance where the particles have a $\Delta\delta$ value of 1.0 g/cm^3 and a viscosity, μ, value of 1 mPa · s (=cP) . The value of Q_0/Σ is twice the settling velocity at $G = 1$, and $Q =$ overflow discharge rate in measurements given.

normal flows for good clarification in standard applications. Similarly, large particles can always be sedimented.

As an additional guide, the Q_0/Σ values are correlated with the equivalent spherical particle diameter by Stokes' law, as in equation 1. A density difference $\Delta\delta$ of 1.0 g/cm^3 and a viscosity of 1 mPa $\cdot$ s ($=$cP) are assumed, thus conversion to other physical characteristics of the system requires that the particle size scale be adjusted to equate a particle of 1.0-μm diameter to its Q_0/Σ in cm/s, according to the relationship $Q_0/\Sigma = 10^{-7} \times 1.09\ \Delta\delta/\mu$, for $\Delta\delta$ in g/cm^3, and viscosity μ in Pa $\cdot$ s. For interpretation of the particle sizes, the scale refers to the 50% cutoff particle size, and under actual centrifugation conditions the value of Σ, determined from Figure 6, must be increased by efficiency factors to give the theoretical value of Σ.

Figure 6 serves as a guide to the types of equipment that can handle a given separation. Other characteristics further narrow selection. For example, for the separation at $Q_0 = 3.5 \times 10^{-3}$ m^3/s (50 gpm) of kaolin clay solids from an aqueous suspension, where the particle density is 2.55 g/cm^3 and the size ranges from 0.25 to 30 μm with 55% $>$2 μm, the 1.0-μm point on the particle size scale would be equivalent to $Q_0/\Sigma = 1.69 \times 10^{-4}$ cm/s. Assuming that a high recovery is desired, a disk centrifuge is required, and recovery of most particles $>$0.4 μm is satisfactory, the Q_0/Σ equivalent to 0.4 μm on the adjusted scale is about 2.3×10^{-5} cm/s. Using a disk machine efficiency of 40%, the centrifuge needed would have a Σ value of

$$\Sigma = \frac{3.15 \times 10^{-3} \times 10^6}{0.40 \times 2.3 \times 10^{-5}} = 34.3 \times 10^7\ \text{cm}^2 \tag{41}$$

Because clay tends to pack hard, only the continuous nozzle discharge bowl would be satisfactory. Intermittently discharging disk centrifuges could not be used.

If classification of solids were desired, several other types of centrifuges could be used as well, assuming that only particles over $\sim$2 μm were to be removed from the suspension and that the oversize stream should be highly concentrated for disposal. Figure 6 shows that a decanter or basket centrifuge or standard cyclone could theoretically be used. Because cyclones cannot concentrate the oversize as much as the centrifuges, these are less satisfactory for this example. If the feed concentration were low, eg, less than a few percent solids, a basket centrifuge would be used with intermittent discharge of solids. A decanter centrifuge gives almost as good a concentration of oversize as the basket and is a more efficient classifier. The decanter centrifuge would thus be the better choice and is actually used in the kaolin industry.

In general, solids-retaining batch and batch automatic machines are limited to low feed concentrations to minimize the time required to unload the solids. Continuous disk centrifuges can have higher feed concentration. The limit is the underflow concentration. Conveyor discharge centrifuges can handle high feed concentration and are limited only by the volume of solids displacement, or torque capacity.

Using flocculent to create aggregated solids, varying degrees of deflocculation of the solid particles may occur during acceleration in the centrifugal field.

An additional problem is the removal of the resulting soft, slimy cakes under centrifugal force. Scroll centrifuges have been successfully used to discharge soft, slippery solids continuously. Specially modified, automated basket centrifuges can handle a broad range of soft sludges, often without polymer addition. Disk centrifuges are particularly well/suited for clarification of streams containing solids such as aluminum hydroxide or a secondary waste-activated sludge.

The disk centrifuge having high capacity and G level is normally used for separating a liquid–liquid mixture or for clarification of such a mixture containing fine solids (20). Specially modified conveyor centrifuges are also used for three-phase separations. Settling velocity is a criterion of selection, but the actual separating of emulsified liquid–liquid mixtures may not strictly follow a settling theory. To break an emulsion, a threshold level of centrifugal force may have to be exceeded. In addition, drainage of liquid from the continuous-phase film of the emulsion has a time factor. Centrifugal force and time cannot be calculated interchangeably as Σ theory would indicate. In centrifugation equipment, coalescence of the dispersed liquid occurs coincidentally with its separation because there is neither time nor space for appreciable interfacial retention of unbroken emulsion.

Batch equipment, such as the bottle centrifuge or ultracentrifuge, does not have a real throughput capacity. By increasing the time of operation, according to Figure 6, the smallest particle size of solids usually sedimented in the bottle centrifuge may be 0.1 µm, at which size Brownian movement controls. In the ultracentrifuge, separation can be achieved down to molecular size, perhaps 0.005 µm, where Brownian movement is controlling. There is clearly no limitation to the larger particles that may be settled in bottle centrifuges so that an arbitrary upper limit is indicated for practical minimum conditions of 1000 rpm and 10 s. Similarly, for the ultracentrifuge the upper limit for Q_0/Σ was estimated for minimum conditions of 1 h and 5000 rpm.

Commercial sedimentation centrifuges are characterized principally by how solids are discharged, and the general dryness of these solids. There are batch and automatic batch solid bowl machines which collect the solids at the bowl wall. Solids are removed very dry. Almost any solid is collectable, even those that are very soft and compressible.

Disk-type solid bowl machines are batch, batch automatic, and continuous. The solids are removed in many different ways, but are usually wet. Scroll centrifuges discharge solids continuously and usually drier than disk and imperforate batch types. Generally, disk centrifuges have the highest values of Σ or KQ for a given size and therefore the best ability to collect fine particles at a high rate.

Bottle Centrifuge. A bottle centrifuge is designed to handle small batches of material for laboratory separations, testing, and control. The basic structure is usually a motor-driven vertical spindle supporting various heads or rotors. A surrounding cover reduces windage, facilitates temperature control, and provides a safety shield. Accessories include timer, tachometer, and manual or automatic braking. Bench-top bottle centrifuges operate at 500–5000 rpm, producing centrifugal fields up to 3000 G in the lower speed range, and operate up to 20,000 rpm with 34,000 G in the high speed units. Larger models operate up to 6000 rpm and develop 8000 G, using special attachments that permit 40,000 G.

These models may also be equipped with automatic temperature control down to $-10°C$ and other programmable controls to manage the cycle.

There are three types of rotors: swinging bucket, fixed-angle head, or small perforate or imperforate baskets for larger quantities of material. In the swinging bucket type, the bottles are vertical at rest but swing to a horizontal radial position during acceleration so that solids are deposited in a pellet at the bottom of the tube. Although sedimenting particles must travel up to the full depth of the liquid layer, which requires appreciable time, the long path of travel and the perpendicularity of the sedimenting boundary to the axis of the tube are distinct advantages in effecting fractional sedimentation. Heads carrying fixed tubes at a 35–50° angle reduce centrifuging time because the maximum distance traveled by a particle is the secant of the tube angle times the diameter of the tube. Particles strike the wall and slide down the tube to collect near the bottom, but the angle makes it difficult to measure relative volumes of supernatant liquid and sedimented solids. Rotors carry 2–16 metal containers having tubes and bottles of various sizes and shapes. Containers range in capacity from capillaries for microanalysis to a 1 L maximum, limiting the batch capacity of this type of centrifuge to 4 L. Although glass bottles and tubes are generally used, plastic and metal containers are·available for high speed operation or corrosive liquids. Tubes are usually cylindrical, tapered, and graduated; special shapes for analytical work are available, including pear-shaped tubes having capillary tips for measuring small quantities of solids.

The bottle centrifuge is primarily used in the laboratory to separate small quantities of material. It is also used for standard analyses including many ASTM methods and in preliminary testing for scale-up to commercial centrifuges. The Σ_B value for free-swinging tubes is determined by equation 7 and the Q_0/Σ_B value by equation 8; Q_0/Σ_B data can be prepared by bottle centrifuge (21). The bottle centrifuge has been used to study waste treatment sludges for estimation of cake concentrations and feasibility of handling in a conveyor centrifuge (22). Because compaction of solids is largely a function of the centrifugal field force and the exposure time, the bottle centrifuge has been used to study these parameters. The results are not always in the range applicable to full scale industrial equipment which also utilizes the motion of the cake. Similarly, drainage of packed solids can be studied by using tubes with fretted glass or perforated metal bottoms. Closed containers should be used to prevent drying by windage.

Specialty rotors permit ordinary bottle centrifuges to achieve some of the results previously considered possible only in ultracentrifuges. A modified zonal rotor, shown in Figure 7, permits collection of sediment using continuous addition of feed and discharge of centrate.

Preparation Ultracentrifuge. Preparation ultracentrifuges are suitable for a range of applications, such as processing quantities of subcellular particles, viruses, and proteins (qv). Many design variations are available and only the common features are considered here. Preparation ultracentrifuges range in operating speed from 20,000 rpm, generating ~40,000 G, to 75,000 rpm and ~500,000 G. The rotor is surrounded by a high strength cylindrical casing and underdriven by an electric motor. To avoid overheating of the rotor by air friction at these speeds, the pressure in the casing is reduced to ~0.13 Pa (1 mm Hg).

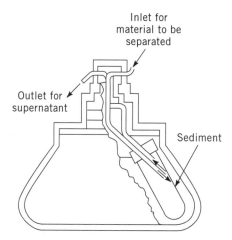

Fig. 7.   Tube-type continuous-flow rotor. Courtesy of Sorvall.

Sensors (qv) monitor the temperature and a cooling system controls the temperature in the range of -15 to $30°C$ within $\pm1°C$. Electronic controls maintain the rotor speed within a required narrow range and may be automatically programmed for sequential changes in speed, including control of the acceleration and deceleration (23).

Preparation ultracentrifuges are guaranteed for several billion revolutions and can be rebuilt using relatively few parts. Among the great number of rotors available are batch rotors and those accepting feed and discharging centrate continuously during rotation. Batch rotors include angle and swinging-bucket types as well as those having vertical tubes parallel to the axis of rotation, which present a very short sedimenting distance and time requirement. Swinging-bucket rotors are also used for density-gradient separations or volume evaluation of the settled cake.

Separation by selective sedimentation on the basis of size and density of the particles may be satisfactory for polydisperse particle systems. However, the cake contains a range of material depending on its starting position in the container. Selectivity of separation can be improved by introducing the sample near the surface after the container is up to speed. Reslurrying and recentrifuging may be necessary to achieve purer fractions. Isopycnic separation improves initial separation efficiency where particles differ in density. If the density of the medium is intermediate to the range of densities of particles, higher density particles settle, whereas others remain suspended or rise regardless of size.

Zonal Centrifuge.   The use of density gradients in centrifuge rotors greatly increases the sharpness of separations and the quantities of material that can be handled. In principle, the density gradient is established normal to the axis of rotation of the rotor and the highest density is located at the outer radius of the rotor. Low molecular weight solutes such as cesium chloride, sucrose, or potassium citrate, which are compatible with many systems in solution, are frequently used. A natural gradient may be formed by introducing a homogeneous solution and centrifuging for long periods of time. Continuous or step gradients may also be formed by introducing successive layers of solution,

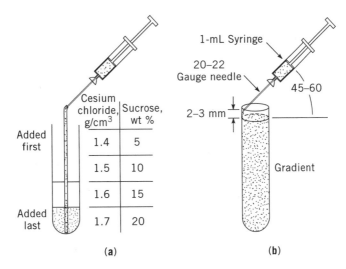

Cesium chloride, g/cm^3	Sucrose, wt %
1.4	5
1.5	10
1.6	15
1.7	20

(a)   (b)

Fig. 8. (a) Forming a gradient and (b) applying the sample to the gradient before inserting tubes in a centrifuge rotor.

the composition of which varies continuously or stepwise from low to high density, where the latter displaces the former toward the center of the rotor (17,18) (Fig. 8).

In the simpler rotors using batch containers having swinging, angle, or vertical tubes, the gradient is introduced while the rotor is at rest and then accelerated to speed. The gradient shows relatively little mixing. Slowing the rotor gradually at the end of the run allows retention of the gradient and permits collection of the material banded isopycnically (Fig. 9).

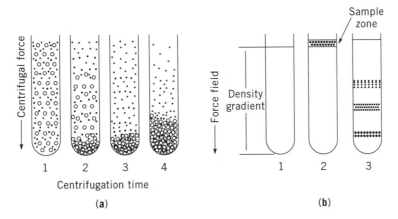

(a)   (b)

Fig. 9. (a) Differential centrifugation (pelleting), where time 1 < time 2 < time 3 < time. (b) Rate zonal separation in a swinging-bucket rotor, where tube 1 represents the density gradient solution, tube 2 the sample plus the gradient, and tube 3 the separation of sample particles under a centrifugal force, where the particles move at differing rates, depending on their mass.

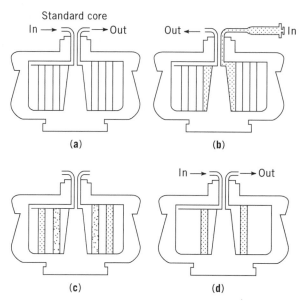

Fig. 10. Dynamic loading and unloading of a zonal rotor. (**a**) Gradient is loaded while rotor is spinning at 2000 rpm; (**b**) a sample is injected at 2000 rpm, followed by injection of overlay; (**c**) particles separated when the rotor is running at speed; and (**d**) contents are unloaded by introducing a dense solution at the rotor edge, displacing fractions at the center.

More sophisticated rotors can be loaded with gradient and sample while rotating. When the batch is finished or the bands are sufficiently loaded with material, the bowl may be stopped slowly and the reoriented layers displaced under static conditions. Rotors may also be designed to establish gradients and isopycnic bands of sample and then be unloaded dynamically by introducing a dense solution near the edge of the rotor as shown in Figure 10.

Particles in the gradient may be separated on the basis of sedimentation rate; a sample introduced at the top of the preformed gradient settles according to density and size of particles, but the run is terminated before the heaviest particles reach the bottom of the tube. If the density of all the particles lies within the range of the density limits of the gradient, and the run is not terminated until all particles have reached an equilibrium position in the density field, equilibrium separation takes place. The steepness of the gradient can be varied to match the breadth of particle densities in the sample.

Rotors are made of titanium or aluminum and may be cylindrical or bowl-shaped (see Fig. 10). Larger bowls reach 100,000 G; smaller units reach 250,000 G. The tubular rotors permit feed rates up to 60 l/h at 150,000 G or 120 l/h in a larger unit at 90,000 G. Such centrifuges may be used to separate relatively large quantities of viral material from larger quantities of cellular and subcellular matter, as, for example, in the production of vaccines (see VACCINE TECHNOLOGY).

Tubular Centrifuges. Tubular centrifuges (Fig. 11) separate liquid–liquid mixtures or clarify liquid–solid mixtures having less than 1% solids content and fine particles. Liquid is discharged continuously, whereas solids are removed

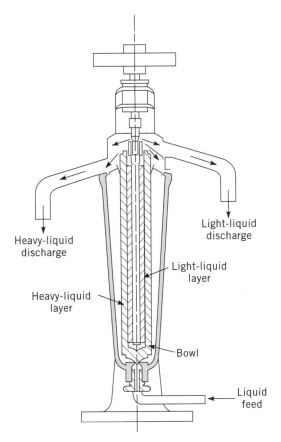

Fig. 11. A tubular centrifuge. Courtesy of Alfa Laval Inc.

manually when sufficient bowl cake has accumulated. For industrial use, the cylindrical bowls are 100–180 mm in diameter with length/diameter ratios ranging from 4 to 8. Bowl speeds up to 17,000 rpm generate centrifugal accelerations up to 20,000 G at the bowl wall (20). Because of the small bowl diameter, however, Σ_T values according to equation 11 result in flow rates in the range of 0.1–4 m³/h (0.5–16 gpm). The tubular centrifuge handles low to medium flows and theoretical particle settling velocities in the range of 5×10^{-6} to 5×10^{-5} cm/s. Clean in place (CIP) and sterilize in place (SIP) are available. SIP usually is done at 121–124°C, and slightly >1.5 bar. Drive motors of 1.5–7.5 kW are used.

The laboratory tubular centrifuge is similar to the industrial model. It operates with a motor or turbine drive at speeds to 50,000 rpm, generating 65,000 G at the latter speed in the 4.5-cm diameter bowl. The nominal capacity range is 30–2400 cm³/min. This centrifuge is uniquely capable of separating far finer particles than any other production centrifugation equipment except the bottle centrifuge. It is widely used in the production of flu virus.

A long, hollow, cylindrical bowl is suspended by a flexible spindle and driven from the top as shown in Figure 11. Axial ribs in the bowl ensure full acceleration of the liquid during its short time in the bowl. Feed is jetted into the bottom of the bowl and clarified liquid overflows at the top, leaving deposited solids as compacted cake on the bowl wall. The clarifying performance of the bowl is reduced as the deposited cake decreases the effective outer radius of the bowl in accordance with equation 12. Consequently, cake capacity of the industrial model is limited to 0.1–10 l. For liquid–liquid separation, the interface position (eq. 30) is determined by selection of ring-dam diameter or by the length of a hollow nozzle-type screw dam.

The tubular centrifuge was long used for the purification of contaminated lubricating oils because of the high centrifugal force developed and the simplicity of its operation. Colloidal carbon and moisture are removed from transformer oils to maintain dielectric strength; carbon and acid sludges are removed from diesel engine lubricating oils; and water and solid contaminants are removed from steam turbine lubricating oil. Polishing operations include the removal of small quantities of solids in the clarification of varnish, cider, fruit juices (qv), and even highly viscous chicle. In vegetable oil refining, oil losses in the semi-solid soap stock are kept low by compaction of the soap phase under high centri-fugal force. Automatic disk centrifuges which do not require manual solids unloading have largely replaced the batch-operating tubular. The laboratory tubular centrifuge is used to recover fine solids in batch preparations too large for bottle centrifuge separation, to estimate scale-up rates in larger centrifuges, and to analyze particle size distributions involving settling rates too low for feasible gravity sedimentation (24). Modern units having variable-frequency drives are available (20).

Disk Centrifuge.   Centrifuges that channel feed through a large number of conical disks to facilitate separation combine high flow rates with high theoreti-cal capacity factors (see Fig. 6). For industrial units flow rates up to 250 m^3/h (1100 gpm) can be obtained on easy separations, and theoretical settling velocities may range from 8×10^{-6} to $\sim 5 \times 10^{-5}$ cm/s. Both liquid–liquid and liquid–solid separations are performed using feed solids concentration <15% and small particle sizes. As seen from equation 15, the theoretical capacity factor depends on the number of disks, which is limited by the height of the disk stack. The performance is proportional to the cube of the disk diameter.

Several of the assumptions in the development of Σ_D do not apply in prac-tice (25), and mathematical representation of the actual flow pattern has been difficult to achieve. Computer studies of the flow to and between the disks, as well as experimental analysis of flow patterns within the disk stack, have improved the effectiveness of disks. Not all disk machine designs can be com-pared using Σ_D alone. Details such as type and size of disk spacer, number of spacers, and location of the feed holes with respect to the spacers and solids dis-charging ports are very important. Also the method of feed acceleration is espe-cially important in liquid–liquid separations and in the presence of fragile solids. If the disk centrifuge design is geometrically similar, scaling up by Σ_D from one speed and size of stack to another is reasonably accurate.

The outstanding feature of the disk bowl design is a stack of thin cones, commonly referred to as disks, which are separated by thin spacers. These are so arranged that the mixture to be clarified must pass through the disk stack before discharge. The resulting stratification of the liquid medium greatly reduces the sedimenting distance required before a particle reaches a solid surface and can be considered removed from the process stream. The angle of the cones to the axis of rotation is great enough to ensure that solid particles deposited on the surfaces slide, either individually, or as a concentrated phase according to the difference between their density and that of the medium.

The general flow patterns for a liquid–liquid separator and a recycle clarifier, respectively, are illustrated in Figure 12. Feed enters near the center of the bowl from either the top or the bottom, depending on the support, and is accelerated by vanes or disks (26) to the radius at which it enters the disk stack. When the disk stack is used for one phase, as in clarification or classification, the feed is distributed to the stock through the zone between the outer edges of the disks and the bowl wall. The clarified medium is discharged at a relatively small radius, generally at the top of the bowl. For a liquid–liquid separation, with or without solids, feed is distributed by a number of feed channels. The interface of

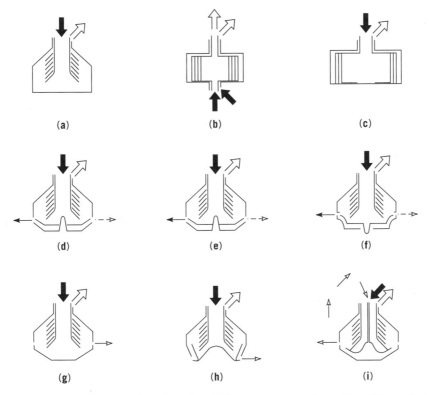

Fig. 12.   Disk centrifuge bowls, where bowl diameters range from 10 to 90 cm; feed flow rates from 0.06×10^{-3} to 38×10^{-3} m^3/s (1–600 gpm); and operating temperatures from 10 to 90°C, and (➡) represents the main inlet, (⇒) the main outlet, (◄–) continuous flow solids, (·–➤) intermittent flow solids, and (–➤) auxiliary liquid. (**a–c**) Solid wall bowl: base design, spiral cylinder inserts, and cylinder inserts, respectively; (**d–f**) bowl for intermittent solids discharges: base design, radial peripheral parts, and shoe; peripheral parts and shoe with nozzles for both continual and intermittent solids discharge; and axial peripheral parts and shoe, respectively; and (**g–i**) bowl with nozzles: base design and peripheral nozzles, nozzles at reduced diameter, and peripheral nozzles and solids circulation, respectively.

the two coalesced liquid phases is located at the disk feed holes, by appropriate selection (eq. 26) of the heavy- and light-phase discharge radii. During handling of the two liquids, the heavier moves toward the edge of the disks and the lighter moves inward. Separate channels in the bowl and separate cover compartments segregate the discharges.

Solids in either phase are sedimented to the underside of the disks and slide outward along the surfaces because of their higher density. The aggregated solids must move from the outer edges of the disks to the bowl wall; some may be reentrained into new feed material, and carried into the disk stack, which accounts in part for actual performance falling short of theoretical prediction.

Commonly used with disk centrifuges are centripetal pumps (Fig. 13) that discharge the clarified liquid phases under pressures up to 0.7 MPa (100 psi) at reduced aeration, and scoop the rotating liquid out by using a stationary impeller. Interface location can be altered by varying the centripetal pump-back pressure, allowing interface control without shutdown to change a discharge weir. Centripetal pumps are capable of discharging at rates to 250 m³/h (1100 gpm), and often eliminate the need for a tank and conventional pump (see PUMPS).

To maximize cake capacity, the simplest disk centrifuge bowl (Fig. 13a) is designed having a nonperforate bowl wall parallel to the axis. Feed solids should not exceed 0.5%. Bowl diameters of industrial units range from 180 to 600 mm with operating speeds from 8000 to 4500 rpm; the disks, between 30 and 200, are stacked at spacings of 0.3–6 mm; the half-angle is frequently 35–40° because solids handling is not critical. This type of centrifuge was originally employed for the separation of cream from milk and is still used widely in this field. Other uses include purification of fuel and lubricating oils having a low percen-

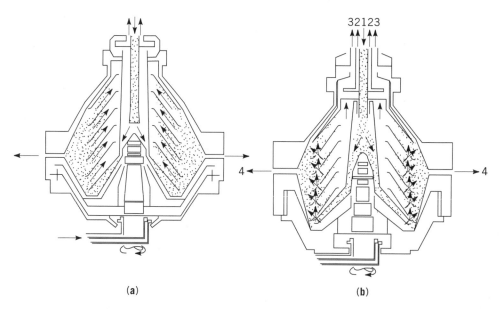

(a)   (b)

Fig. 13. (a) A clarifier and (b) a purifier of the paring disk-type design, where intermittent discharges of solids are designated by →, and 1 represents feed; 2, light phase; 3, heavy phase; and 4, solids.

tage of solids, separation of wash water from fats and vegetable or fish oils, and removal of moisture and solids from jet fuel. Solids that move readily in plastic flow can be continuously discharged as the heavy phase, eg, in the separation of soap stock from oil in vegetable oil refining.

Continuous discharge of solids, as a slurry, is achieved by sloping the inner walls of the bowl toward a peripheral zone containing between 8 and 24 discharge points containing nozzles, as shown in Figure 12**g**–**i**. The nozzles must be spaced closely enough so that the natural angle of repose of the solids deposited between the nozzles does not cause a buildup of cake to reach into the disk stack and interfere with clarification. The size of the nozzles is limited because the fluid pressure at the wall, which can be 6.9–13.8 MPa (1000–2000 psi), produces high nozzle velocities (see eq. 30). On the other hand, the nozzles must be large enough to prevent obstruction by individual particles; nozzle diameters at least four times the size of the largest particle are satisfactory. Replaceable, wear-resistant bushings in the nozzles provide orifices in the range of 0.8–2.5 mm. From 5 to 50% of the feed may be discharged with the solids through the nozzles. The upper limit of solids concentration depends on the particle packing characteristics but seldom exceeds 20 times that in the feed. The nozzle flow is directed backward with respect to rotation to recover much of the pressure energy and to reduce centrifuge power (see eq. 29). Bowl diameters range from 100 mm in laboratory units to 900 mm in industrial units and speeds from 12,000 to 3000 rpm, respectively; power requirements are between 10 and 2000 kW. Centrifugal accelerations to 12,000 G are obtained at the wall in the smaller bowls. Feed rates range from 1 to 136 m³/h (4–600 gpm).

Applications include kaolin clay dewatering, separation of fish oils from press liquor, starch and gluten concentration, clarification of wet-process phosphoric acid, tar sands, and concentrations of yeast, bacteria, and fungi from growth media in protein synthesis (20).

A variant of the continuous-discharge disk centrifuge provides for introduction of a recycle stream. Restrictions on number and size of nozzles sometimes prevent adequate concentration of the discharged solids to satisfy further process requirements. To increase this concentration, a portion of the discharged slurry is returned to the feed, thereby increasing the overall loading of solids in the bowl. A more efficient method, shown in Figure 12**i**, is to return some discharged slurry through a recycle system to the region of the nozzles where the higher density recycle stream preferentially joins the nozzle flow.

A further modification of the disk centrifuge provides peripheral ports that are opened only intermittently to discharge sludge (see Fig. 12**d**–**f**, and 13). This type is employed for medium quantities of solids (1–4%) for which neither continuous discharge nor batch operation is suitable, and for solids that break down under the shear forces of nozzle discharge and are therefore not suitable for nozzle recycle. Intermittent discharge provides longer holding time and better concentration of solids, often at the expense of decreased disk size and reduced bowl throughput. The frequency and duration of opening can be controlled to discharge very high concentrations of sludge by partial emptying of the bowl.

These centrifuges are available in 180-mm bowl laboratory and 460–600-mm bowl industrial units. Disks are mostly spaced at 0.6–1.0 mm and the half-angle is 40–45°. In the industrial units, bowls with 60–200 or more disks operate at maximum speeds of 4400–6500 rpm and require 10–50 kW. Feed flow rates range from 2.5–114 m³/h (10–500 gpm); temperatures from 30–90°C. Theoretical capacity factors are generally lower than in continuous discharge bowls of the same size. The disk outside diameter is smaller in order to provide solids-holding space of 4–20 L. Applications are limited to free-flowing solids that do not pack, and include recovery of wool grease from wool scouring liquor, orange juice clarification, recovery of soya protein, clarification of animal fats and food extracts, and purification of marine and jet engine fuels and lube oils as well as dairy applications.

Other modifications have special but more limited applications. A centrifugal bowl may contain, instead of disks, several annular baffles that take the liquid through a labyrinth path before discharge. The multiple cylinders increase cake capacity to as much as 70 L for easily sedimented solids. This centrifuge is used for clarification of food syrups and antibiotics (qv), and for recovery of heavy metallic salts and catalysts (see Fig. 12**c**).

Decanter Centrifuges.   A comprehesive discussion of the decanter centrifuge is availible (27). Decanter centrifuges collect solids by sedimentation and continuously discharge both liquid and solid material. These centrifuges have bowl diameters of 150–1400 mm and are essentially tubular shells with a length/diameter ratio of 1.5–5.2, as shown in Figure 14. Deposited solids are moved by a helical screw conveyor operating at a differential speed of 0.5–100 rpm with respect to the bowl. Centrifugal fields are lower than in disk or tubular centrifuges because of the conveyor and its associated mechanism. Maximum speeds range from 300–9000 rpm. Figure 8 shows that particles of intermediate settling velocities, such as 1.5 to ~15 × 10⁻⁴ cm/s, are handled at medium to large flow rates. For clarification, this type of centrifuge recovers medium and coarse particles from feeds at high or low solids concentration. Particle sizes less than ~2 μm are normally not collected without the addition of flocculating agents. For classification of solids, the flow rates are higher than for clarification and the overflow usually contains most of the finer solids. Feed flow rates range from 1 to 136 m³/h (4–600 gpm).

Incompressible solids discharged from the decanter, may not be as dry as those obtained by centrifugal filtration. Coarse crystals may discharge at 2–10% moisture, ground limestone at 15–20% moisture, and kaolin clay in the filler range (1–10 μm) at 30–35% moisture. It has been found that the decanter centrifuge is particularly well-suited for compacting low permeability compressible

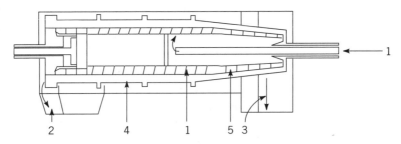

Fig. 14. A decanter centrifuge, where 1 corresponds to feed suspension; 2, to liquid phase; 3, to solid phase; 4, to liquid pool; and 5, to dry beach.

cakes, due to the beneficial motion and shearing of the compaction cake caused by the relative rotation of the conveyor.

Cake dryness's achieved with decanter centrifuges is much higher than what can be achieved with a bottle centrifuge operating at the same g-level, pond depth and with the same cake retention time (28). One manufacturer has recently launched a new series of municipal and industrial waste sludge decanters that optimises the ratio between sludge transportation and cake shearing thus further enhancing the cake dryness or feed rate of the centrifuges. Compressible, amorphous, and fibrous materials, such as sewage sludges, can be dewatered to 60–75% moisture, and meat rendering solids at 60–70% moisture plus 6–8% liquid fat. Operating pressures up to 1.03 MPa (150 psi) and temperatures up to 200°C are standard and 300°C is available.

Feed is introduced through an axial tube. Solids sedimenting to the bowl wall are conveyed along the cylindrical section and up a sloping beach. They are usually discharged at a radius smaller than that of the liquid discharge. Fine and flocculent solids compact under the liquid and show relatively little drainage on the beach. Coarse crystals and fibers do drain on the beach to a low residual moisture. The liquid level in the bowl is maintained by ports adjustable to the desired overflow radius. Considerable variation in design is available for the bowl shell, flight angle and pitch, beach angle and length, conveyor speed, feed position and type, and patterns of liquid and solids movement through the bowl. Bowl shapes having a high l/d exhibit high clarification capacity but may have wetter solids owing to the greater amount of fine particles recovered for discharge in the solids phase.

Bowl designs include countercurrent or cocurrent movement of the phases. In countercurrent flow, feed enters near the conical–cylindrical intersection; liquid flows toward the cylindrical end of the bowl to discharge over dams; while deposited solids are moved up the conical beach by the conveyor. In cocurrent flow, feed enters at the end away from the conical end. Liquid and settled solids move in the same axial direction toward the conical beach end. Axial conduits or a skimming device remove centrate from the pond surface, and solids move up the beach. The actual liquid flow between the helical flights is more tangential than axial so the cocurrent/countercurrent description is not particularly significant in regard to the actual movement of solids vs liquids within the bowl. In either type, solutions of flocculating agents may be introduced in the piping before entering the centrifuge, either in the feed tube, the feed acceleration zone, or in the pond after feed acceleration. Owing to the complexity of the chemistry of polyelectrolyte flocculation rate and efficiency, and floc damage owing to turbulence, the best solution for optimum polyelectrolyte use must be determined experimentally. A wash can be applied to the solids on the beach, but efficient rinsing depends on careful design to direct the rinse liquid to the solids. The wash is not collected separately but is discharged with the mother liquor.

A vertical decanter design using a bowl elastically suspended from a spindle with no bottom bearing is often used for high pressure and high temperature applications instead of horizontal axis decanters. Vertical units are easier to seal for pressure operation, and are well suited to accommodate bowl expansion at high temperatures. A vertical, elastically supported rotor as compared to a horizontal design does not offer separation advantages. There are mechanical advantages, however: the vertical design uses only one seal to three in the horizontal design. Moreover, the process connections are rigid (vertical) rather than flexible (horizontal), thermal expansion is not critical, and noise is lower.

Although Σ_T (eq. 12) indicates a reduced sedimentation performance level at increased liquid depth, this occurs only for coarse solids. The optimum pond depth varies with feed zone design, tendency of deposited solids to redisperse, conveyor differential speed, and particularly the depth of the cake layer required to produce a given solids concentration. Deep ponds are generally more effective for soft, slimy solids because conveying problems may reduce performance level and prevent complete clarification, even at low flow rates. The difficulty with which a decanter centrifuge is able to discharge soft, slippery solids has limited use in the past. To move the solids up the beach the solids must remain at bowl speed, with the helical flights moving the solids inward, against the high G. Often the soft solids slide back into the pond, building up in the bowl and ruining sedimentation performance. Setting the pond level inward of the solids discharge radius and separating the feed portion of the pond from the conical (beach) end using a deeply immersed baffle permits the centrifugal pressure head developed to assist the conveyor in discharging the solids. This method is used to thicken secondary sewage sludge from feed, ie, solids concentrations of 0.5–1% to solids concentrations of 5–10%. This thickening reduces total flow to the next stage of treatment by an order of magnitude. Flow rates up to 100 m³/h (440 gpm) have been achieved without the use of polyelectrolytes. A small amount of polyelectrolytes, however, can double this rate.

Dewatering sewage sludge to high solids levels has been achieved by the use of higher polyelectrolyte dosage (5–15 kg/t), increased G level (2000–4000 G), and longer solids residence time in the bowl. Longer solids residence has been achieved by mechanically restricting the solids near or at the beach and reducing the differential speed to 0.5–4 rpm. The torque between the bowl and conveyor is a good indication of solids dryness, and use of this factor has permitted automatic differential speed control to maintain a constant solid dryness at varying feed concentrations and feed rates. Differential speed control can be obtained by driving the conveyor by use of a planetary gearbox mounted on the bowl, the gear box receiving controlled torque, and speed from a variable speed motor or electrical brake or motor. Occasionally a hydraulic motor is mounted on the bowl and an external hydraulic power supply is used to deliver variable rates of high pressure oil to the rotor by means of a rotary union. All methods are increasing being controlled by programmable microprocessors. Higher torque differential speed controllers have permitted centrifuge capacity and solids dryness to achieve levels of 100 m³/h (440 gpm) at 35–40 wt% solids.

Final dewatering of sewage results from the solids almost filling the interior space between the conveyor hub and bowl shell interior. The use of polymers as flocculating agents results in easy separation in spite of a greatly reduced volume allowed to the centrate. The polymer also conditions the solids so that the pressure resulting from the deep layer of solids and the force needed to move the solids axially and inwardly toward discharge compresses and squeezes additional water from the space between and within the flocs. A drier discharged cake results. Compatibility studies (29) can be made to characterize particular solids. Improvements in dryness is proportional to $G^{1.5}t$, where G is the level at the mean solids radius and t is the solids residence time within the centrifuge. On a decanter having an $1/d = 4.2$, the solids have been found to occupy 75–80% of the bowl. The centrate clarification occurs just before discharging. Theoretical studies (30) comparing the ability to dewater compressible solids by sedimenting and filtering centrifuges to pressure filters, have shown that at high G levels, scroll decanters produce drier cakes than pressure filtration.

The capacity of decanters can be limited by any one of several factors (31): centrate clarity, usually a function of Σ; solids dryness, imposed by requirements of the next process step; conveyor torque, limited by rating of gear box, hydraulic motor, or back-drive capacity; chatter, by torsional instability resulting from stick-slip action of conveyor flights on fusible solids; swallowing capacity, by the ability of conveyor to accept feed without rejection; power, by the rating of drive motor; solids purity, where purity is obtained by rinsing; erosion rate, where abrasive particles are present; size purity, for particle classifications; solids volume, where solids feed concentration is high; fused solids, rate at which pasties are formed; and control, the maximum rate at which process control can be maintained.

Special designs have been offered which include one or more of the following features in vertical or horizontal construction: vapor-tight and pressure-tight enclosures; clean in place (CIP) and sterilize in place (SIP) rotors; three-phase separations having two immiscible liquids and solids; sanitary construction; a wide range of helices, pitches and leads, beach angles, compound beach angles, l/d ratios, pond depth/d ratios, and ribbed or grooved beaches or bowls; centrate centripetal pump devices; and abrasion protection systems, including complex flight tip geometries of ceramic, carbide, and conventional hard surfacing.

In clarifying operations, the conveyor discharge centrifuge recovers many types of crystals, meal from fish press liquor, and polymers, such as poly(vinyl chloride) and polyolefins. It is also used to dewater coal (32) and to concentrate solids from flue gas desulfurization sludges. Vertical designs, vapor-tight or under pressure, are applied to terephthalic acid, polypropylene, and catalyst recovery. Classification includes separation of particles ~2 μm from kaolin coating clay, and of particles over ~5 μm in the mill discharge of ground TiO_2; selective recovery of calcium carbonate from lime-treated waste sludges permits calcining and recycling of the lime without an overwhelming recycle load of inert material. The decanter centrifuge is frequently used to rough out medium and coarse solids before a second separation stage such as a disk centrifuge handling refinery sludges.

A varient of the decanter centrifuge, the screen bowl decanter uses both sedimentation and filtration primarily for the purposes of either drying or rinsing cake solids. The screen bowl and conventional decanter are shown in Figure 15-**a** and **b** respectively. Feed slurry is introduced conventionally into these imperforate scroll centrifuges. The dewatered solids are deposited onto the bowl wall and conveyed up a beach. The screen bowl decanter has a cylindrical screen with an inside diameter equal to the solids discharge diameter. The conveyor moves the solids axially over the bar screen area where further dewatering, fines loss and potentially rinsing takes place. Unlike the conventional decanter in which rinsing on the solid beach ultimately sends the rinse liquor to the centrate, the screen section allows rinse to be segragated from the

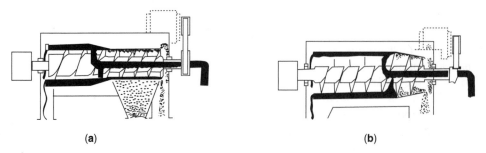

Fig. 15. Comparison of screen bowl (**a**) and conventional (**b**) decanters. Screen section can be used for purposes that include removal of additional fines (thru the screen) from cake solids, futher drying of cake solids or introducing rinse liquor that can be isolated from the centrate.

centrate. Screen bowls are used to dewater coarser solids to a very high degree. Typical screen bowl decanter would be used to dewater fine coal, while the conventional decanter whould more likely be used to capture the solids from a coal refuse stream. Screen bowls are also used in production of purified parazylene (18).

5.2. Centrifugal Filtration Equipment.   The important parameters of centrifugal filtration equipment (4,32) are screen area, level of centrifugal acceleration in the final drainage zone, and cake thickness. The latter affects both residence time and volumetric throughput rate. As indicated by equation 20, the particle size of the solids and the kinematic viscosity of the mother liquor also strongly affect the final moisture content. A limited correlation has been developed for the performance of perforate basket and conveyor discharge conical screen bowls, but the range of materials handled and the complexity of the drainage and washing operations do not lend themselves to broad correlations. The variables of correlation may be useful in a particular study, especially if more than one type of centrifuge is involved. An example of centrifugal filtration is the recovery of salt crystals from a mother liquor of $\sim 3 \times 10^{-6} \text{m}^2/\text{s}$ (3 cSt) viscosity; the crystal size is 170 μm at the 15% level and drained cake bulk density $\sim 1.5 \times 10^3$ kg/m^3 (95 lb/ft^3). The cone screen is 25.4 cm at the larger diameter and the automatic cycle basket centrifuge is 68.6 cm in diameter; other parameters for final values of $q = 7.6\%$ in both centrifuges are given in Table 1.

A conveniently expressed coordinate for plotting filtration performance is the drainage number, $d(G)^{1/2}/v$, where d is the mean particle diameter in micrometers, v is the kinematic viscosity of the mother liquor in m^2/s (Stokes $\times 10^{-4}$) at the drainage temperature, and G is $\omega^2 r/g$; r is the largest screen radius in a conical bowl. Because the final moisture content of a cake is closely related to the finest 10–15% fraction of the solids and is almost independent of the coarser material, it is suggested that d be used at the 15% cumulative weight level of the particle size distribution instead of the usual 50% point.

The other coordinate is $qt^{1/2}$, where q is the percentage of final moisture on the discharged cake, as the volume of mother liquor per unit volume of solids, and t is the drainage time in seconds. A weight ratio may be used for q, but the volume ratio makes the function more universal by eliminating densities.

Table 1. **Centrifuge Parameter Values for $q = 7.6\%$**

Properties	Centrifuge	
	Cone screen	Automatic basket
G	1440	865
bowl		
diameter, cm	25.4	68.6
speed, rpm	3180	1500
$\bar{d}(G)^{1/2}/\nu$	3.750×10^6	2.880×10^6
time, s		
feed and spin		12
spin after rinse		27
drain time	0.6^a	39
rinse^b		10
unload and screen rinse		4
cycle time		53
$qt^{1/2}, \% \cdot \mathrm{s}^{1/2}$	5.89	47.46
h, cm	0.51	3.81
screen area, cm^2	930	7675
approximate cake rate, kg/s	1.19	0.84

^{a}Differential speed = 60 rpm, 5/8 turn per helical flight.
^{b}Rinse is shown only in the basket centrifuge because rinse efficiency in the cone screen is fairly low compared to the basket. The latter may be selected for the application if good rinsing is needed, and the conveyor-discharge conical screen centrifuge selected if no rinsing is necessary.

For a conveyor discharge conical screen bowl, the helical conveyor is assumed to control the residence time of the solids, so that time becomes the number of turns of helical flight around the conveyor hub divided by the differential speed between the conveyor and the bowl. For a pusher centrifuge, t is the retention time on the screen as controlled by length and diameter of screen, thickness of cake, and frequency and length of stroke of the cake. In a basket centrifuge, dead and unload times should not be included in the calculation of drain time. Bulk drainage is completed so quickly that film drainage is usually controlling. Thus, drain time t is approximated by the sum of feeding time, spin time prior to rinsing, and spin time after rinsing but prior to unloading according to the theory of cyclical centrifuges (34). Filtration correlations generally show a spread as a function of cake thickness. Conical screen bowls characteristically have short residence times and achieve good drainage by maintaining thin layers of cake. Smaller perforate baskets operate having cakes 5–10 cm in thickness, whereas larger baskets may carry cakes up to 15-cm thick. Pusher centrifuges and high speed peeler baskets may handle cakes ranging in thickness from 5 to 20 cm.

    Perforate Basket Centrifuges. The simplest and most common form of centrifugal filter is a perforate-wall basket centrifuge, consisting of a cylindrical bowl having a diameter ranging from ~100–2400 mm and a diameter/height ratio ranging from 1 to 3. The wall is perforated with a large number of holes, more than adequate for the drainage of most liquid loads, and is lined with a filter medium. In the simplest case, the medium is a single layer of fabric or metal cloth or screen. In high speed basket centrifuges, one or more backup screens of relatively large mesh support a finer mesh filter surface. The method of discharging accumulated solids distinguishes three types of basket centrifuge: those that are stopped for discharge, those that are decelerated to a very low speed for discharge, and those that discharge at full speed (35).

Basket centrifuges that must be stopped for discharge are available in many sizes. The bowls are usually supported on a vertical spindle. Designs vary from a 300-mm diameter basket of 30 L of cake capacity (0.4-kW motor, 2100 rpm, and ~800 G) to a 1500-mm diameter basket having 500 L of cake capacity (14-kW motor, 600 rpm, and 300 G). Basket cake volumes are always nominal and must be modified according to cake density. Construction materials include carbon or stainless steel, Monel, Inconel, titanium, and a variety of rubber and plastic coatings. Normal operation includes pressures up to 35 kPa (5 psi) and temperatures up to 180°C. This type of centrifuge is used if a variety of materials must be filtered in small batches, if equipment must be sterilized between batches, or if the production rate is too low to warrant more automation. These centrifuges are also used in removing liquid from crystalline materials, in drying bulk materials such as raw leafy vegetables, and in clarifying process liquids and waste streams. Cycle times are >10 min. Large (dia = 1300–1500 mm) baskets operating at 700–800 rpm may use an inner-perforated container that mounts in the bowl but is removable for bulk loading outside the centrifuge. Such units are well suited to laundry and dyeing purposes and for dewatering of textiles, yarn, raw stock, feathers, and hair. Particle sizes range from very fine to 500 μm. Feed flow rates vary from 1 to 20 m³/h (4.5–90 gpm). A syphon to control the rate of filtrate removal improves dewatering (36).

Improved control systems and rising labor costs have led to manually or automatically controlled cycles with mechanical unloading at reduced speed. Baskets typically load at low to medium speed, accelerate to 900–1800 rpm for drainage and washing, decelerate to 35–75 rpm for mechanical unloading, and then start the cycle again. Cycle times range from 2 to 6 min. Cycles of 30 min for slow drainage or multiple rinses are common. Both a top-driven suspended bowl and underdriven bowl with three-point casing suspension designs are available. The cake is discharged in 20–120 s by means of a single or multiple plow that leaves a heel of cake on the filter medium. The heel can be completely removed with the help of a plastic-tipped plow and a perforated protecting plate. Filter media vary from perforated plates having 3-mm holes to 37-μm (400-mesh) Dutch twill. Vertical automatic dischare baskets always unload through a bottom discharge; a valve mechanism may be used to seal the bottom if the basket is fed with the whole charge at one time, so that an appreciable liquid layer develops.

The most fully automated basket is the horizontal basket or peeler centrifuge that feeds and discharges at spin speed, and normally operates at cycle times of <3 min at pressures up to 1.03 MPa (150 psi) and temperatures from −70 to 120°C, and in special cases to 350°C. It is primarily used for materials draining freely in a high centrifugal field, for medium tonnages, and for multiple rinses where nearly complete segregation of the rinses and motor liquor is desired. Ideally, the feed should have a constant composition and high concentration. For this purpose, a gravity slurry concentrating tank or a cyclone is often installed ahead of the centrifuge. The bowl rotates on a horizontal axis with a

Table 2. **Basket Centrifuge Operation on Free- and Slow-Draining Particles**

Operation	Type of crystal	
	Free draining[a]	Slow draining[b]
solids handling rate, t/h	20–24	1.5–2
conditioning rinse, s	1	0
feeding time, s	7	25
wash time, s	5	25
drain time, s	12	30
unloading time, s	2	6
total cycle, s	27	86

[a]For example, ammonium sulfate.
[b]Requiring several rinses, eg, polyolefins.

metal screen as the filter medium and discharges solids by cutting them out with a hydraulically operated knife. Bowls are 300–1200 mm in diameter and handle loads of 28–170 L (1–6 ft^3) of cake at bowl speeds of 1000–2500 rpm, producing maximum centrifugal fields of ~1250 G. The solids discharge through a front chute, which simplifies sealing against pressure. A distributor riding on the cake during feeding maintains uniform cake thickness, gives better bowl balance, and improves washing efficiency. Power requirements range from 15–150 kW. Materials of construction include carbon or stainless steel and Monel. The high speed during feeding and discharge may cause deformation of particles and breakage of crystals, whereas the heel of cake left on the screen may lose permeability through glazing or plugging with fractured fines. The heel can be conditioned by suitable washes after one or more cycles. Table 2 lists two process cycles.

This type of centrifuge is also used on borax and boric acid, p-xylene, sodium bicarbonate, and sodium chloride from glycerol or electrolytic caustic, in addition to dewatering of various products, eg, potato starch, or dewatering and washing slurries, eg, calcium hypochlorite.

Inverting Filter Centrifuge. Another batch automatic horizontal perforated bowl centrifuge inverts the flexible filter to discharge the solids. Feed slurry may be deposited on the inside surface of a cloth filter, with the bucket end completely closed. When the interior is full of dewatered solids, the bowl is decelerated to a slow speed and piston and closure plates move axially, inverting the filter cloth so that the solids reside on the outside diameter of the cloth. Very little residual material remains on the filter cloth surface for the next cycle. Centrifuges are available with diameters of 300–1300 mm, and operate at 720–1920 G. Filter cake rates of 75–250 kg/h are achieved at excellent rinsing efficiencies and low crystal breakage. Drives are electric or hydraulic, with the axial movement hydraulically actuated via a rotary seal. Power demands are low, from 3 kW for the smallest to 100 kW for the largest.

Continuous Cylindrical Screen Centrifuges. Continuous filtering centrifuges are used for very fast draining that do not require extremely dry final products. Rinsing efficiency varies considerably; power requirements are usually low; initial slurry concentrations can be somewhat more variable and not as high as for the high speed basket. Continuous centrifugal filters are equipped with either a cylindrical or a conical screen. Both types are made without a retaining lip on the solids discharge end of the bowl and employ various methods to move the solids through the bowl.

The cylindrical screen centrifuge deposits solids at one end of the bowl in a layer 6–80 mm thick and pushes the annular ring of cake axially through the bowl by means of a reciprocating piston (37,38). Washes are collected in separate sections of the casing but are not as distinctly separated as with basket centrifuge sequential operation. Drained solids at the end of the bowl are thrown into a casing that is separated by baffles from the liquid discharge zones. Bowls rotate on a horizontal axis, range in diameter from 200–1200 mm, and have capacities of 1–25 t of solids/h. Centrifugal fields of 300–600 G are common. To reduce the fines loss and to facilitate movement of the cake on the screen, maximum speeds are rarely used. Power requirements range from 4–60 kW. The cylindrical screens are generally of the bar screen type, with 0.1–0.5-mm spacing. The reciprocating piston operates at 20–100 strokes/min, with stroke lengths up to 80 mm. The thickness of the cake (max ∼80 mm) depends on the packing and draining characteristics, the bulking tendency of the layer in front of the pusher, and the frictional resistance of the cake on the screen. Friction also depends on the construction of the bar screen.

The feed slurry is introduced by gravity flow of 20–200 t/h or screw conveyor to an imperforate distributing cone that deposits the slurry at its original concentration immediately in front of the pusher, as shown in Figure 17. The cake must not buckle at this point, so slurries of 40% concentration are generally necessary. Because fast drainage is required, feed particle size should exceed 150 µm; medium and coarse crystals and granules or fibrous solids can be handled in this type of centrifuge. Crystal breakage is low within the basket but some breakage does occur during discharge.

To handle materials that form a soft or plastic cake or have a high frictional resistance, the cylindrical screen may consist of two to six steps with successively larger diameters, as shown in Figure 16. Alternate steps reciprocate with the piston. Thus, the cake is pushed across only a short length of screen before redistribution on the next step at a slightly larger diameter. Drainage and washing efficiency are increased by redistribution of the cake under these conditions. This type of centrifuge is used on sugar (qv), where the high viscosity of the mother liquor causes slow drainage, a high degree of plasticity in the partially drained cake, and poor penetration of wash liquor.

Another type of continuous, cylindrical screen centrifuge discharges the cake by moving it axially through the bowl with a helical conveyor. Crystal breakage through conveyor action is greater than with the pusher-type mechanism. Applications include the handling of copper as trisodium phosphate and purification of paraxylene.

Conical Screen Centrifuge.   In conical screen centrifuges the angle of the bowl causes or assists the cake to move axially and redistributes it in an increasingly thin layer which improves drainage characteristics. The feed slurry is deposited at the small end of the cone, where most drainage occurs. The drained solids are discharged from the large end, which has no retaining ring. Screens

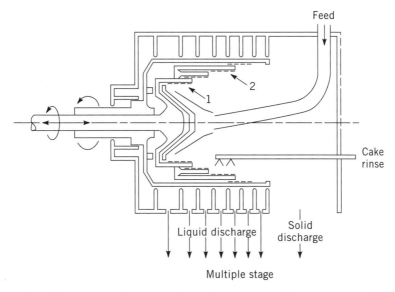

Fig. 16.   Multiple-stage pusher centrifuge. Screens 1 and 2 reciprocate with pusher.

generally have 0.08–1.5-mm slots or perforated plate holes. Screens less than ~0.25-mm thick require a backup screen to extend screen life. There are three types of conical screen centrifuges: those that are self-discharging, those that discharge by means of a helical conveyor (Fig. 17), and those that apply an axial vibration or oscillation to the bowl or the bowl and casing.

In its simplest form, the cone angle is slightly larger than the angles of repose of the solids at any stage in their drying cycle. Some bowls are made with two angles, such that the more shallow is at the small end, and the steeper, where solids concentrate, is higher and at the large end. Horizontal or vertical bowls of 500–1000-mm large-end diameter operate at speeds up to 2600 rpm, producing up to 2500 G; cone angles vary from 20–35° and selection of the proper cone angle and suitable screen surface is critical for each application. Temperature control of viscous feeds is necessary to maintain proper distribution and drainage. Feed slurry, usually under gravity flow, is introduced at the small end of the basket, where it is accelerated and spread evenly over the periphery of the screen. Viscous sugar massecuites are successfully handled at capacities of 2–7 t/h on a 750-mm diameter basket. Loss of fines is greater than on the automatic basket centrifuges, but improved rinse performance increases sugar purity. Rinsing efficiency is not generally high on this type of cone and segregation of rinse liquor is incomplete. This centrifuge is also used for the drying of crystalline materials such as ammonium sulfate and separation or dewatering of fine fiber in wet corn milling.

Conical screens with a helical conveyor turning slightly faster than the bowl can handle a variety of materials. The vertical baskets are underdriven and the larger diameter is downward, as shown in Figure 17. Horizontal axis designs are also available. Feed is introduced at the small end and the

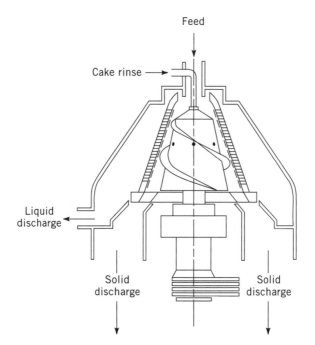

Fig. 17.   Conical screen conveyor discharge centrifuge.

rate at which solids move through the drainage zone is controlled to some degree by the differential speed of the conveyor. Bowls have large-end diameters of 250–500 mm, and lengths about two-thirds of the large diameter; bowl angles range from 10 to 20°, and operating speeds are 2500–3800 rpm, giving up to 3500 G. Solids capacities range from 1 to 30 t/h, with feed capacities of 1–15 m³/h (260–4000 gph). Centrifuge casings may be vapor-tight but are not intended for pressure operation; maximum temperature is about 150°C. Power requirements, low for the tonnages handled, range from 7 to 30 kW. Applications include dewatering of medium and coarse crystals, deoiling of proteinaceous solids, and removal of solids from fruit and vegetable pulps and other food slurries.

The third type of cone screen centrifuge operates with bowl angles of 13–18° and assists solids discharge by a vibratory motion of the bowl or bowl and casing. These units usually have underdriven bowls having 500–1100-mm larger diameter at the upper end; horizontal axis designs are also available where diameter/length ratios range from 1 to 2. Operating speeds are normally 300–500 rpm, and solids capacities range from 25 to 150 t/h. Pressurized units are not available, and operating temperatures range to 100°C. Power requirements are 15–35 kW. Bar screens are frequently used, and applications are largely on coal dewatering; particle sizes from 30 mm down to 0.25 mm (60 mesh) are easily handled. These centrifuges are also used in dewatering of potash and other crystalline solids. Where SW = separative work in kg/year.

6. Nomenclature

Symbol	Definition
$Ae_{3/4}$	the cylindrical area at 3/4 of the bowl radius
a	distance; space between adjacent disks
C	discharge coefficient
c	concentration
d	diameter of particle, particle size
$\bar{d}$	mean particle diameter
e	efficiency factor
E	energy
E_{rel}	relative performance
G	ratio of centrifugal acceleration to the acceleration of earth's gravity
g	acceleration of gravity
h	cake thickness, height
K	permeability
k	experimental coefficient
l	length, particularly from feed zone to centrate discharge
m	torque
N	bowl speed
n	number of disks
P	power
p	pressure
Q	Volumetric flow rate
q	volume of undrained liquid(mother liquir)/unit volume of solids, %
r	radius
S	Fraction of void volune occupied by liquid
s	external surface area/weight of solid
s'	surface area/volume of cake
T	torque
t	time
V	volume
v	velocity
α	cake resistance constance in pressure filtration, length/volume
Δ	difference, particularly density
δ	mass density, mass/unit volume
ε	voil fraction
μ	viscosity
ω	angular velosity
Ψ	wetting angle
Φ	angle between direction of nozzle and tangent to circle intersecting nozzle axis at discharge
Σ	theoretical capacity factor, length^2
σ	surface tension or material stress
θ	half angle of disks
ν	kinematic viscosity

Subscripts	
B	bottle centrifuge
BD	backdrive
C	cake or conveyor torque
F	friction factor
f	film flow
g	settling velocity of particle in gravity field
h	heavy phase

Symbol	Definition
i	interface
L	liquid medium
l	free surface of liquid or light phase
M	filter medium, or mass
m	mean value
N	nozzle
P	process power
p	process
S	solid medium
s	settling velocity of particle in centrifugal field or shape constant
W	windage pored

BIBLIOGRAPHY

"Centrifugal Separation" in *ECT* 1st ed., Vol. 3, pp. 501–521, by M. H. Hebb and F. M. Smith, The Sharples Corp.; in *ECT* 2nd ed., Vol. 4, pp. 710–758, by A. C. Lavanchy and F. W. Keith, Jr., The Sharples Co., and J. W. Beams (Gas Centrifugal Separation), University of Virginia; in *ECT* 3rd ed., Vol. 5, pp. 194–233, by A. C. Lavanchy and F. W. Keith, Jr., Pennwalt Corp. "Centrifugal Separation" in *ECT* 4th ed., Vol. 21, pp. 828–875, Alan Letki and R. T. Moll, Alfa Laval Separation Inc., and Leonard Shapiro, Consultant; "Separation, Centrifugal Separation" in *ECT* (online), posting date: December 4, 2000, by Alan Letki, R. T. Moll, Alfa Laval Separation Inc., Leonard Shapiro, Consultant.

CITED PUBLICATIONS

1. C. M. Ambler, *Chem. Eng. Prog.* **48**, 150 (1952).
2. F. W. Keith, Jr. and R. T. Moll, in R. A. Young and P. Cheremisinoff, eds., *Wastewater Physical Treatment Processes*, Ann Arbor Science Publishers, Ann Arbor, Mich., 1978.
3. J. Murkes, *Br. Chem. Eng.* **14**(12), 636 (1969).
4. H. P. Grace, *Chem. Eng. Prog.* **49**, 427 (1953).
5. J. A. Storrow, *AIChE J.* **3**, 528 (1957).
6. W. Batel, *Chem. Ing. Tech.* **33**, 541 (1961).
7. E. Nenninger, Jr. and J. A. Storrow, *AIChE J.* **4**, 305 (1958).
8. Technical data, Sharples Research Laboratory, 1961.
9. J. O. Maloney, *Ind. Eng. Chem.* **48**, 482 (1956).
10. F. M. Tiller and N. B. Hysung "Comparison of Compacted Cakes in Sedimenting and Filtering Centrifuges" *American Filtration Society*, Third Annual Meeting, Washington D.C., March 19–24, 1990.
11. N. Corner-Walker, "The Dry Solids Decanter Centrifuge: Capacity Scaling", Filtration-Separation May 2000, pp. 28–32.
12. H. Axelsson: "Centrifugation," in C. L. Cooney, A. E. Humphrey, eds., *Comprehensive Biotechnology*, Vol. 2, Chapt. 2, Pengamon Press, Oxford 1985, pp. 325–346.
13. Technical data, Alfa Laval Separation, Warminster, Pa., May 1993.
14. T. Theodorsen and A. Regier, *Experiments on Drag of Revolving Disk, Cylinders and Streamline Rods at High Speeds*, Report No. 793 National Advisory Committee for Areonautics 11944, pp. 367–384.
15. "Centrifuges—Common Safety Requirement" BSEN 12547:1999.
16. A. G. Letki, "Centrifuge Selection", *Chem. Proc. Eng.* **94**, 9 (Sept. 1998).

17. H. Axelsson "Cell Separation, Centrifugal" *Encyclopedia of Bioprocess Technology: Fermintation, Biocatalysis, and Bioseparation*, John Wiley & Sons, Inc., New York, 1999.

18. Technical data, Alfa Laval Separation, Warminster, Pa., 1984.

19. Thosmas Register http://www.thomasregister.com

20. H. Axelsson, *Centrifugal Separations—Principles and Techniques*, Alfa Laval Separation, Tumba, Sweden, presented at the Bioprocess Technology Program, University of Virginia, Charlottesville, Va., Oct. 17–25, 1991.

21. C. M. Ambler and F. W. Keith, Jr., in A. Weissberger, ed., *Techniques of Chemistry*, 3rd ed., Vol. XII, Wiley-Interscience, New York, 1978, Chapt. VI.

22. P. A. Vesilind, *Treatment and Disposal of Waste Water Sludges*, Ann Arbor Science Publishers, Ann Arbor, Mich., 1974.

23. O. M. Griffith, *Techniques of Preparative, Zonal and Continuous Flow Ultracentrifugation*, Spinco Division of Beckman Instruments, Inc., Palo Alto, Calif., 1975.

24. T. Lee and C. W. Weber, *Anal. Chem.* **39**, 620 (1967).

25. C. A. Willus and B. Fitch, *Chem. Eng. Prog.* **69**, 73 (Sept. 1973).

26. L. Borgstrom, C.-G. Carlsson, C. Inge, T. Lagerstedt, and H. Moberg, *Appl. Sci. Res.* **53**, 35–50 (1994).

27. F. A. Records and K. Sutherland, *Decanter Centrifuge Handbook*, 1st ed., Elsevier-Science Inc., New York, 2001.

28. B. Madsen, *J. Filt. Soc.* **2**(3), 30–32 (2000).

29. P. A. Vesilund and B. Zhang; *J. WPCF* **56**(12), 1231–1237 (Dec. 1984).

30. F. M. Tiller and N. B. Hsyung, *Chem. Eng. Prog.* 20–28 (Aug. 1993).

31. W. Gosile, *German Chem. Eng.* 3 (1980).

32. G. B. Cline, in E. S. Perry and C. F. van Oss, eds., *Progress in Separation and Purification*, Wiley-Interscience, New York, 1971, pp. 299–306.

33. R. Day, *Chem. Eng.*, 81 (May 13, 1974).

34. F. A. Records, *Chem. Proc. Eng.* **52**, 47 (Nov. 1971).

35. Technical data, Alfa Laval Separation, Warminster, Pa., Sept. 1991.

36. K. Lilley and G. Huhtsch, *Filtr. Sep.* **12**, 70 (Jan.–Feb. 1975).

37. D. K. Baumann and D. B. Todd, *Chem. Eng. Prog.* **69**, 62 (Sept. 1973).

38. P. M. T. Brown, *Chem. Proc. Eng.* **52**, 65 (Nov. 1971).

ALAN LETKI
Alfa Laval Inc.

NICK CORNER-WALKER
Alfa Laval AB

CERAMIC-MATRIX COMPOSITES

1. Introduction

With the development of emerging technologies such as aero-space, transportation and power generation, advanced materials are needed for components such as control surfaces, wing edges and nose cones; turbine blades and shrouds in more fuel efficient engines and heat exchanger elements. These structural components must operate in the temperature range between 1100 and 1650°C.

Ceramics, inorganic, nonmetallic, crystalline compounds with mixed ionic–covalent nature to their chemical bonds, have been the traditional candidate materials for such high temperature use. Their many desirable properties include high melting points, high chemical stability, high elastic modulus and hardness, high wear and creep resistance, and low mass density relative to metallic materials. Monolithic ceramics, however, are brittle and are thus very sensitive to intrinsic flaws and damage produced by use. Failure of these materials occurs in a catastrophic manner and at low strain-to-failure ratios. However, the problem can be alleviated by reinforcing monolithic ceramics with a second phase which is itself capable of operating at high temperatures. Such systems are designated as ceramic matrix composites (CMC).

The reinforcing phases in ceramic matrix composites are usually also ceramic and have many possible morphologies: particulate, platelet, whisker, short-fiber, or continuous-fiber. Reinforcing entities are typically added to ceramic matrices to produce tough composites. In comparison, high strength reinforcements are added to polymer-based composites to increase strength and stiffness. To enhance toughness high strength reinforcements with high elastic modulus and weak interfaces with the matrix are required; to produce high strength and stiffness, strong interfaces along with high stress transfer are needed to allow efficient load transfer or shedding from the matrix to the reinforcement.

2. Ceramic Composites Systems

With the appropriate choice of composite properties, such as reinforcement and matrix materials, reinforcing geometry and composite interface, an otherwise brittle mode of failure of a ceramic becomes more "ductile" and noncatastrophic in nature. Thus, the choice of the component materials is an important aspect of designing ceramic matrix composites. Two questions need to be addressed when making these choices. First, if a matrix crack encounters a potential bridging entity, will it deflect along the reinforcement/matrix interface or will fracture of the reinforcement occur? Second, if interface debonding occurs, will the interfacial sliding shear resistance, τ, be low enough to allow the bridge to slip in the matrix or will fracture of the bridging-reinforcement occur?

A partial answer to the first question has been provided by a theoretical treatment (1,2) that examines the conditions under which a matrix crack will deflect along the interface between the matrix and the reinforcement. This fracture–mechanics analysis links the condition for crack deflection to both the relative fracture resistance of the interface and the bridge and to the relative elastic mismatch between the reinforcement and the matrix. The calculations indicate that, for any elastic mismatch, interface failure will occur when the fracture resistance of the bridge is at least four times greater than that of the interface. For specific degrees of elastic mismatch, this condition can be a conservative lower estimate. This condition provides a guide for interfacial design of ceramic matrix composites.

About the second question, concerning the relative strengths of the bridge and the interfacial sliding resistance, little is known *a priori*. Some progress made for the system of continuous fiber-reinforced ceramic-matrix composites will be discussed later. The general recommendation is to have a high bridge strength and a low interfacial sliding shear resistance.

Various combinations of ceramic–matrix composites have been manufactured at the research level. Their properties are given in Table 1 for oxide-based matrices and in Table 2 for nonoxide matrices. Some commercial products are identified for information only. Such identification does not imply recommendation

Table 1. **Oxide-Based Ceramic-Matrix Composites**

	Reinforcement				Density, g/cm^3 or	
Type[a]	Amount, vol %	Strength,[b] MPa	Toughness,[c] MPa$\sqrt{m}$	Modulus, GPa[d]	% td[e]	Reference
Al_2O_3 matrix						
B_4C_p	50		4.5	380	3.28	(3,4)
SiC_w	20		2.5	400		5
SiC_w/SiO_{2i}[f]	20		6.0	420		6
SiC_w/Si_3N_{4w}	20	203	3.4		95% td	7
SiO_{2f}		6.3	28			8
TiC_p	30		4.0	400	4.26	4
BN_p		24.6		490	91.1% td	9
Al_p	20		8.4			10
$ZrO_2(t)_p$[g]		2000	5–8	333	4.54	11
Aluminosilicate glass matrix						
SiC_w			0.8	80		5
Al_2O_{3f}		311	3.3			12
Cordierite glass matrix						
SiC_f		128	1.6		2.44	9
Pyrex glass matrix						
Al_2O_{3f}		305	3.7			12
SiC_p	30	171	1.79			13
SiC_w	30	180	3.04			13
SiC_p/SiC_w[h]		159	2.73			13
Soda-lime silicate matrix						
SiC_w	20		0.7	72		5
LASIII glass ceramic matrix						
SiC_w	35	327	5.1			14
$3Al_2O_3 \cdot 2SiO_2$ matrix						
SiC_w	10	274	2.7	197	2.84	15
SiC_p		262	2.35	240		15
$ZrO_2(t)_p$[g]		250	4.0	150	98.9% td	16
ZrO_2 matrix						
$ZrO_2(t)_p$[g]		400–600	10	200	6.08	

[a] Subscripts denote reinforcement morphology; p = particulate, l = platelet, w = whisker, f = fiber, i = interlayer between reinforcement and matrix.
[b] Strength as measured in a four-point flexure test (modulus of rupture); to convert MPa to psi, multiply by 145.
[c] Fracture toughness; to convert MPa$\sqrt{m}$ to psi$\sqrt{in.}$, multiply by 910.048.
[d] To convert GPa to psi, multiply by 145,000.
[e] %td = percentage of theoretical density.
[f] 20% SiC_w.
[g] Tetragonal.
[h] 10% each.

Table 2. **Nonoxide-Based Ceramic-Matrix Composites**

Reinforcement						
Type[a]	Amount, vol %	Strength,[b] MPa	Toughness,[c] MPa$\sqrt{m}$	Modulus GPa[d]	Density, g/cm^3	Reference
AlN matrix						
BN$_p$		65.5		480		9
SiC matrix						
SiC$_w$			19.9	240		8
TiB$_{2p}$	16		4.5	430	3.30	4
TiC$_p$	25		6.0	450	3.36	4
Si$_3$N$_4$ matrix						
SiC$_w$	10	620	7.8			18
SiC$_w$	20		4.0	350		17
SiC$_w$	10	436	5.7			19
Si$_3$N$_{4p}$		680	7.6–8.6	160		19
TiC$_p$		578	7.2	328		17
TiC$_p$	30		4.5	350	3.7	20
TiN matrix						
Al$_2$O$_{3p}$/AlN[e]		229	10.2			21
WC matrix						
Co	20		16.9	442		10

[a] Subscripts denote reinforcement morphology; p = particulate, l = platelet, w = whisker, f = fiber, i = interlayer between reinforcement and matrix.
[b] Strength as measured in a four-point flexure test; to convert MPa to psi, multiply by 145.
[c] Fracture toughness; to convert MPa$\sqrt{m}$ to psi$\sqrt{in.}$, multiply by 910.048.
[d] To convert GPa to psi, multiply by 145,000.
[e] 30% each.

or endorsement by NIST, nor does it imply that the products are the best available for the purpose.

3. Composite Reinforcements

The structure of reinforcements can be either equiaxed or acicular. The nature of their placement within a composite, the composite architecture, is critical to the resultant composite properties. Possible architectures are summarized in Figure 1.

Equiaxed particles, which are well dispersed in the ceramic matrix, tend to produce isotropic composite behavior. The particles, either ceramic or metallic, may be single crystal or polycrystalline in nature.

Acicular reinforcements such as whiskers and platelets tend to produce rather more anisotropic composite properties. Whiskers and platelets are usually single crystals with aspect ratios up to 100 and with tensile strengths near their theoretical value. Composite processing can be tailored to produce either an aligned microstructure with the principal axis of all reinforcements lying in the same direction; a textured microstructure in which the principal axis is randomly arranged within a single plane; or an isotropic microstructure in which the reinforcements are randomly arranged in three dimensions. Aligned reinforcements produce a composite with highly unidirectional properties in the alignment direction, but with properties that are isotropic in the transverse

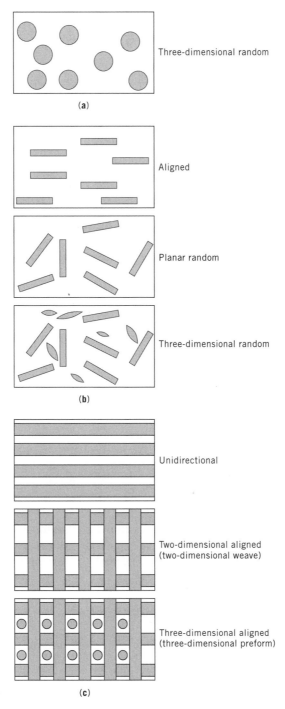

Fig. 1. Reinforcement architectures for ceramic–matrix composites and corresponding composite properties. (**a**) Spherical particles; (**b**) platelets, whiskers, short fibers; and (**c**) continuous fibers.

direction. Such microstructures are produced when whisker-reinforced composites are fabricated by extrusion or when platelet-reinforced composites are fabricated by tape-casting or hot-pressing techniques. Textured microstructures produce composites that have isotropic properties in the reinforcement plane. This tends to be the most common type of microstructure produced when a whisker-reinforced composite is fabricated by hot pressing or tape-casting techniques. Fabrication techniques required to produce a completely random microstructure with resulting isotropic properties are extremely difficult. Hence, most ceramic composites reinforced with acicular particles tend to have some form of texture and thus, some anisotropy of properties.

Fiber reinforcements can be amorphous, single crystal or polycrystalline in structure. They can be either short fibers producing similar composite architectures to those of whiskers or they can be continuous. Continuous fiber reinforced composites tend to have orthotropic properties. For unidirectional composites properties transverse to the fiber direction are significantly different from those parallel to the fiber direction. Some continuous fibers may be woven into two-dimensional weaves before being incorporated into a composite. Such composites have in-plane properties with fourfold symmetry and different properties out of plane. These two-dimensional fiber weaves are usually rotated through a fixed angle on stacking to produce more isotropic planar properties. A preform may also be woven in three dimensions to produce more isotropic properties in all directions, but this adds considerably to the overall cost of the resulting composite.

The role of reinforcements in a ceramic-matrix composite system is to transfer stress from the matrix to the reinforcement, thereby shielding the crack tip from the applied load and providing an additional dissipative energy sink to resist crack propagation. This function is usually achieved via strong reinforcements with a weak interface between the reinforcement and the matrix. This combination allows ligament debonding and energy dissipation via frictional sliding of the reinforcement in the matrix. The matrix and reinforcement are usually chosen to allow weak interface debond stress. However, in practice, it is difficult to achieve this state because most ceramic systems react chemically. In fiber and whisker reinforced ceramic composite systems an interlayer coating of pyrolitic carbon is usually incorporated at the interface to facilitate easy debonding. Alternative approaches to weaken the interface are given later.

Table 3. Platelet Reinforcements for Ceramic-Matrix Composites

Platelets	Tensile strength, GPa[a]	Modulus, GPa[a]	Density, g/cm³	Diameter, μm	Maximum use temperature, °C
Al_2O_3[b]		400	3.986	5–15/1	2040
SiC[c]	3	470	3.21	5–500/1–15	1600
SiC[d]	0.5	470	3.21	10–15	1600

[a] To convert GPa to psi, multiply by 145,000.
[b] Atochem, Centre de Recherche, France.
[c] C-Axis, Jonquiere, Quebec.
[d] Ref. 22.

Table 4. **Whisker Reinforcements for Ceramic-Matrix Composites**

Whiskers	Tensile strength, GPa[a]	Modulus, GPa[a]	Density, g/cm^3	Diameter, μm	Length, μm	Maximum use temperature, °C
Al$_2$O$_3$[b]	20	450	3.96	4–7	40–100	2040
B$_4$C[c]	14	490	2.52			2450
SiC						
Silar	7	340–690	3.2	0.6	900	1760
SC9[d]						
VLS[e]	8.3	580	3.2	4–7	5000	1400
Tokamax[f]		600	3.2	0.1–1.0	50–200	1400
SiC[g]		600	3.2	0.5–1	5–100	1400
Si$_3$N$_4$						
SNWB[h]	14	385	3.18	0.05–0.5	5–100	1900

[a] To convert GPa to psi, multiply by 145,000.
[b] Catapal XW, Vista Chemical Co., United States.
[c] Ref. 23.
[d] Advanced Composite Materials Corp. (ACMC), Greer, S.C.
[e] Los Alamos National Lab, Los Alamos, N. Mex.
[f] Tokai Carbon Co., Japan.
[g] J. M. Huber, Corp., Nacagdoches, Tex.
[h] UBE Industries, Japan.

A related and important issue in choosing a reinforcing material is the chemical compatibility of the reinforcement with the matrix. The reinforcement must also have high strength that is retained to elevated temperatures. If the environment has access to the reinforcement, either at the surface or through matrix cracking, then the reinforcement must be sufficiently chemically inert in the service-environment. Tables 3 through 5 present the properties of a few of the currently available platelets, whiskers, and fibers for use as reinforcements in ceramic composites.

4. Ceramic Matrices

Ceramic matrices are usually chosen on their merits as high temperature materials; reinforcements are added to improve their toughness, reliability, and damage tolerance. The matrix imparts protection to the reinforcements from chemical reaction with the high temperature environment. The principal concerns in choosing a matrix material are its high temperature properties, such as strength, oxidation resistance, and microstructural stability, and chemical compatibility with the reinforcement.

Another consideration is the difference in thermal expansion between the matrix and the reinforcement. Composites are usually manufactured at high temperatures. On cooling any mismatch in the thermal expansion between the reinforcement and the matrix results in residual mismatch stresses in the composite. These stresses can be either beneficial or detrimental: if they are tensile, they can aid debonding of the interface; if they are compressive, they can retard debonding, which can then lead to bridge failure (25).

Table 5. **Fiber Reinforcements for Ceramic-Matrix Composite**[a]

Fibers	Tensile strength, GPa[b]	Modulus, GPa[b]	Density, g/cm^3	Diameter, μm	Maximum use temperature, °C
Al$_2$O$_3$					
FP[c]	1.38	380	3.90	21	1316
PRD166[c]	2.07	380	4.20	21	1400
Sumitomo[d]	1.45	190	3.9	17	1249
Safimax[e]	2.0	300	3.30	3	1250
mullite					
Nextel312[f]	3.12	1.55	150	2.70	1204
Nextel440[f]	4.40	2.70	186	3.05	1426
Nextel480[f]	4.80	2.28	224	3.05	1200
SiC					
Nicalon[g]	2.62	193	2.55	10	1204
SCS[h]	2.80	280	3.05	6–10	1299
SCS6[h]	3.92	406	3.00	142	1299
Sigma[i]	3.45	410	3.40	100	1259
MPDZ[j]	1.75	175	2.30	10	
HPZ[j]	2.10	140	2.35	10	
MPS[j]	1.05	175	2.60	10	
SiTiCO					
Tyrrano[k]	2.76	193	2.5	10	1300
Si$_3$N$_4$					
TNSN[l]	3.3	296	2.5	10	1204
SiO$_2$					
Astroquartz	3.45	69	2.2	9	993
Graphite					
T300R[m]	2.76	2.76	1.8	10	1648
T40R[m]	3.45	276	1.8	10	1648

[a] Ref. 24.
[b] To convert GPa to psi, multiply by 145,000.
[c] E.I. du Pont de Nemours & Co. Inc., Wilmington, Del.
[d] Sumitomo Chemical America, New York.
[e] ICI Advanced Materials, Wilmington, Del.
[f] 3M Co., St. Paul, Minn.
[g] Nippon Carbon Co., Tokyo.
[h] AVCO Specialty Materials/Textron Inc., Lowell, Mass.
[i] Berghoff, Tubingen, Germany.
[j] Dow-Corning/Celanese, Midland, Mich.
[k] UBE Industries, Japan.
[l] Toa Nevyo Kogyo K. K. Tokyo.
[m] Amoco Performance Products, Ridgefield, Conn.

Compressive interfacial stresses increase the interfacial shear resistance. Although usually detrimental to toughening, these stresses can enhance toughening if bridge pullout is the operative toughening process.

5. Composite Interface

The prerequisite for tough, noncatastrophic failure of ceramic-matrix composite materials is that the interface between the reinforcement and the matrix is weak

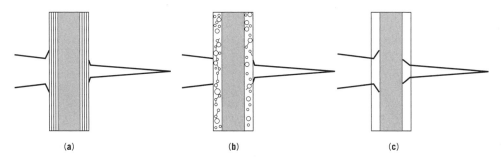

Fig. 2. Microstructural design approaches for composite interfaces: (**a**) mechanically weak coating; (**b**) porous interface and; (**c**) ductile interface.

enough to allow crack deflection along or around the reinforcement. For ceramic matrices and ceramic reinforcements this is not often the case as strong bonding can occur during composite fabrication. With ceramic materials primary chemical bonding occurs at the interface. One way to change the bonding is to change either the reinforcement or the matrix to materials that will not react chemically. More typically, the interface is weakened through the incorporation of a weak interlayer that is compatible with both the matrix and the reinforcements. There has been much interest in the development of weak reinforcement coatings. Alternative approaches have also been proposed; Figure 2 presents the state of the art in composite interfaces (26,27). Figure 2**a** represents a weak debond coating with a layer-type structure that does not react strongly with either the matrix or reinforcement, for example, pyrolitic carbon. Figure 2**b** presents a porous coating of the matrix material, which is much weaker owing to the presence of a large degree of porosity. Figure 2**c** presents a ductile interfacial layer, which has a low ductile yield stress and usually a lower elastic modulus than either the fiber or the matrix. These latter two coatings may be well-bonded to the matrix, the reinforcement, or both. These three approaches to a weak interface are discussed in more detail later.

A good example of the effectiveness of a weak debond coating is the Nicalon SiC–CVI SiC matrix composite fabricated at Oak Ridge National Laboratory (28) with varying amounts of pyrolitic carbon at the interface. The initial, as-manufactured, composite with no fiber coating produced a low strength, low toughness composite. Coating the fibers with a 0.17-μm-thick layer of carbon before deposition of the matrix dramatically changed the composite properties. The bend strength increased from 83 MPa (12,000 psi) to 383 MPa (55,000 psi), with an accompanying large strain-to-failure. The toughness, measured by a work of fracture, increased from 98 J/m^2 for the uncoated composite to 4110 J/m^2 ($\sim$2 ft lbf/in.^{2}) for the composite with the carbon-fiber coating. Figure 3 shows the corresponding load–deflection curves for the uncoated and coated composites.

Carbon is a commonly used and successful weak interfacial coating. For high temperature applications, however, carbon is not the best solution, because it oxidizes, leaving a physical gap between the reinforcement and the matrix or allowing interfacial reactions that result in a strong interface bond. Much

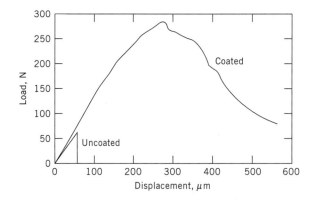

Fig. 3. Load–deflection curve for a SiC–C–SiC composite in four-point bending. Note the extreme change in behavior fora composite fabricated with a 0.17-μm carbon layer between the SiC fiber and the SiC matrix as compared with a composite with no interfacial layer (28).

research has been conducted to develop alternative high temperature debond coatings, with little success to date.

An alternative to the weak debond coatings is to create a mechanically weak debond interface (26). One that shows much promise is a porous coating of the matrix itself on the reinforcement (26). The coating is well-bonded to the reinforcement and the matrix but is mechanically much weaker than either because of its degree of porosity. A debond crack will thus run preferentially through the coating. Such a coating can be as temperature and oxidation resistant as the matrix itself, although the degree of oxidation protection rendered to the reinforcement phase may be reduced by any open porosity in the coating.

An alternative to the weak debond interface approach may lie in a ductile interface that is well-bonded to both the reinforcement and the matrix (27). Debonding of the interface then entails ductile yielding and shearing of the interface. Such a process potentially dissipates more energy than debonding and frictional interfacial sliding alone. The viability of such a coating has been explored for a model composite system of borosilicate glass reinforced with continuous SiC fiber (27). The coating used was electrolytically deposited copper up to 19 μm thick. The resulting composite showed an increase in toughness of 25% over the unreinforced matrix, compared with only 6% for the carbon-coated fiber composite. The disadvantage of such a system is that the degree of toughening changes with temperature, since ductility of the interlayer changes with temperature. Thus, although the high temperature toughness could be improved, the system might remain brittle at ambient temperature.

As discussed earlier, residual stresses can arise in composites when there is a mismatch between the linear thermal expansion coefficients of the reinforcement and the matrix. These stresses can be either deleterious or advantageous to the mechanical performance of the composite. Compressive stresses normal to the interface lead to enhanced interfacial sliding shear stresses. The corresponding hoop tensile stresses (and axial tensile stresses, if they arise) in the matrix reduce the strength of the matrix, and hence of the composite. Matching the thermal expansion coefficients of the reinforcement and the matrix is one solution to the problem, albeit a limited one. Another approach is the use of an interfacial layer to tailor the stresses, and the feasibility of such coatings has been proposed in a mathematical treatment of the problem (29). The results of this analysis predict that interlayer coatings with a higher expansion coefficient and a lower modulus than either the reinforcement or the matrix reduce the residual mismatch stresses. The analysis applies to the case when the thermal expansion of the matrix is greater than that of the reinforcement. The feasibility of this proposal has been demonstrated in a model composite of SiC continuous fiber reinforced borosilicate glass using a copper interlayer (27). Calculations using the theory (29) predict that a 10-μm thick copper layer will reduce the maximum hoop and radial stresses to one-third of the uncoated value. Stresses measured in the glass matrix immediately surrounding the fiber using a technique of stress-induced birefringence (30) indicate that the compressive stress of 80 MPa (11,600 psi) for the uncoated composite was reduced to a stress of less than 5 MPa (725 psi) for the copper interface composite.

6. Toughening Processes

The toughness induced in ceramic matrices reinforced with the various types of reinforcements, that is, particles, platelets, whiskers, or fibers, derives from two phenomena: crack deflection and crack-tip shielding. These phenomena usually operate in synergism in composite systems to give the resultant toughness and noncatastrophic mode of failure.

    6.1. Crack-Resistance Behavior. The goal of composite reinforcement is to produce tough, flaw-insensitive materials that fail in a "ductile" manner. Such materials are more damage tolerant than the monolithic ceramics because they can withstand larger cracks without fracture and the fracture strength may be independent of crack size within a certain flaw size range. This important property of flaw tolerance and stable crack growth results from a fracture resistance behavior known as $\Re$-curve or T-curve behavior, in which the fracture resistance rises with crack extension. Fracture resistance can be formulated in terms of either stress intensity factor T or strain energy release rate $\Re$ (or J). If stress intensity factor is used, then the ordinate is the square root of crack length and the plot is termed a T-curve. If, however, strain energy-release rate is used, $\Re$, (or J_c) is plotted directly as a function of crack length and the curve is termed an $\Re$-curve.

    Increasing fracture resistance with crack length is a phenomenon common to metals tested under plane-stress conditions. Ceramics also have been shown to have similar behavior, but the phenomenon arises from elements of their microstructure that resist crack propagation (31). Ceramic composites are a natural extension of this effect, where the reinforcements provide the elements that resist fracture. The shape and extent of the $\Re$-curve is dependent on the microstructural scale and its effectiveness at crack shielding. If a ceramic microstructure can be designed to have an $\Re$-curve with an extensive initial "knee" (or portion before the inflexion point), then there is stable crack growth before failure, and hence flaw tolerance. The material is thus tolerant to flaws of a size

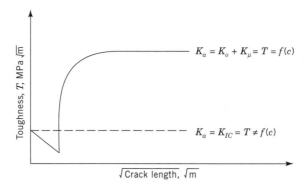

Fig. 4. Schematic representation of fracture resistance and its relation to crack length for a single-value toughness material and a material with a fracture resistance curve (T-curve). MPa = MN/m^2. To convert MPa$\sqrt{\text{m}}$ to psi$\sqrt{\text{in.}}$, multiply by 910.048.

larger than that predicted by a Griffith criterion to be unstable. The importance of toughness in engineering terms lies in the enhanced probability of detection of longer cracks via nondestructive evaluation techniques. Figure 4 is a schematic diagram of the toughness relation with crack length for a material with single-value toughness and for one with a T-curve. For a material with a single-value toughness, the toughness T can be represented by:

$$T = K_{IC} \tag{1}$$

which is not a function of crack length. For a material with a T-curve, the toughness can be represented by:

$$T = K_o + K_\mu \equiv T(c) \tag{2}$$

In this case K_o is the intrinsic toughness experienced by the crack tip (equal to K_{IC} for a single-value toughness material) and K_μ is the toughness associated with the microstructural shielding term (equal to zero for a single-value toughness material).

The propagation of a crack depends on the shape of the T-curve in relation to the scale of the initial crack length. Initially, when a flaw is small, it encounters very little of the microstructure along its length. Analyses (32,33) have suggested that the initial portion of the T-curve decreases as illustrated in Figure 4 owing to compressive microstructural elements. As the flaw extends, more of the microstructure is sampled and the crack becomes increasingly more difficult to propagate as the toughening contribution K_μ increases. Once the crack is long with respect to the shielding elements in the microstructure, the toughness saturates out and the crack grows as if it is sampling an average microstructure. The equilibrium condition for fracture is that the driving force is greater than or equal to the fracture resistance:

$$K_a \geq T \tag{3}$$

The stability condition,

$$\frac{dK_a}{dc} \geq \frac{dT}{dc} \tag{4}$$

determines the nature of the fracture, that is, either stable or unstable. Equation 4 is the condition for unstable fracture. At equality this is known as the tangency condition.

For a single-value toughness material, $dT/dc = 0$. Accordingly, if the applied stress intensity factor is always increasing with crack length, equation 4 is always satisfied. Thus, the condition for fracture is equation 5, where K_a is given by the applied loading conditions.

$$K_a = K_{IC} \tag{5}$$

For a nonsingle-value toughness material the equilibrium condition for fracture, equation 3, becomes

$$K_a \geq T \equiv K_o + K_\mu \tag{6}$$

The stability of crack extension in such materials depends on the rate of change of the applied driving force to that of the fracture resistance, equation 4.

$$\frac{dK_a}{dc} \geq \frac{dK_\mu}{dc} \tag{7}$$

Consider the toughness curve of Figure 5a. A preexisting flaw of length c_i will not extend until $K_a \geq K_o$. The dashed line represents a loading level that just satisfies the equilibrium condition for fracture, that is, equation 6. At this value of K_a, the crack propagates unstably until it arrests at a length $c'(c' > d$, where d is the length to the inflexion point on this simplistic T-curve). The arrest criterion is assumed to be the same as the propagation criterion, that is, equation 6, but alternative criteria have occasionally been used in the literature. Increasing K_a further, the crack propagates stably from c' to c^*, where the tangency condition, equation 7, is satisfied (see Fig. 5b) and spontaneous failure occurs. When the initial crack length $c_i > d$, there is no initial unstable crack growth, but rather the crack grows in a stable manner with increasing K_a until the crack length is equal to c^*. Preexisting flaws in the microstructure of size less than c^* but greater than a crack length c_o, as defined in Figure 5b behave in a flaw tolerant manner. The fracture strength in this initial flaw size regime is independent of crack length, as illustrated in Figure 5c. Theoretically, cracks smaller than c_o propagate unstably to failure in a high strength region. However, in practice this region is not observed experimentally because the smallest intrinsic flaws are typically of the order of d. Thus, a T-curve can produce a region of flaw sizes in which the material is flaw tolerant (3,34).

6.2. Crack Deflection Contribution to Toughening. Crack deflection is a phenomenon that leads both to toughening and to the formation of bridges that shield the crack tip from the applied stress. Little is known of the bridge

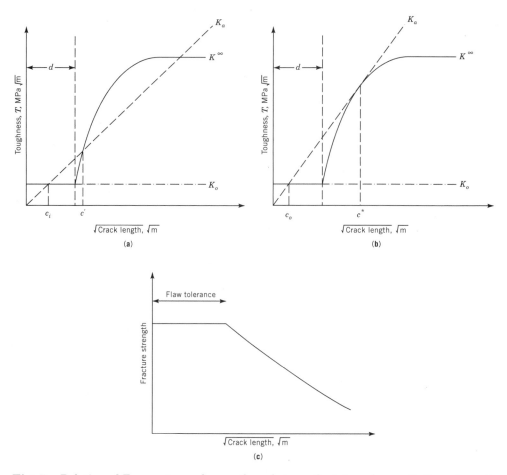

Fig. 5. Relation of T-curve to crack growth and strength: (**a**) crack growth for a flaw of size c_i; (**b**) tangency condition for crack growth for a flaw of initial size c_i.; (**c**) fracture strength as a function of crack size showing region of flaw size tolerance.

formation process, but its effect, that is, crack-tip shielding, is considered in the following section.

The condition for propagation of a mode I edge crack, that is, a crack that is subjected to pure opening (tensile) stresses applied perpendicular to the crack plane, is given by (35):

$$K_a = Y\sigma_a\sqrt{c} = K_{IC} \tag{8}$$

where Y is a dimensionless geometry term, σ_a is the applied stress, and c is the crack length. Once a crack is deflected from its original plane, further crack extension requires a higher driving force to accommodate the mode II (shear) or mode III (tearing shear) contribution to the stress intensity factor on the new crack plane. A schematic diagram of crack deflection with contributions from both modes is shown in Figure 6 (36). It has been shown (37) that the net

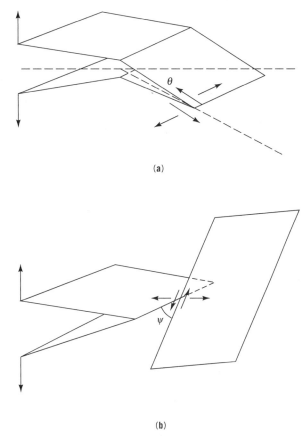

Fig. 6. (a) Crack deflection and propagation through a tilt angle $K(\theta) = K_{IC} \sec^2 (\theta/2)$. (b) Crack deflection and propagation through a twist angle, $K(\psi) = K_{IC} \sec^2 (\psi)$ (36).

applied stress intensity factor to drive a crack on a tilt plane of θ degrees from the original plane is given by equation 9.

$$K(\theta) = K_{IC}\sec^2(\theta/2) \tag{9}$$

Similarly, if a crack is deflected from its original plane through a twist angle of ψ degrees, a mode III tearing component must also be taken into account for further propagation of the crack. The resulting toughness is given by equation 10 (37).

$$K(\psi) = K_{IC}\sec^2(\psi) \tag{10}$$

Figure 7 shows these results schematically for both twist and tilt crack deflections. Thus, for the stress intensity factor required to drive a crack at a tilt or twist angle, the applied driving force must be increased over and above that required to propagate the crack under pure mode I loading conditions. Twist deflection out of plane is a more effective toughening mechanism than a simple tilt deflection out of plane.

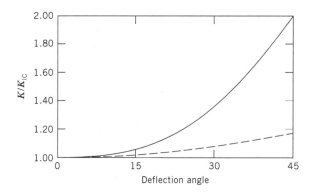

Fig. 7. Relative toughening owing to (– – –) tilt, θ, and (——) twist deflections, ψ.

A crack front may interact many times with inhomogeneities in the microstructure, causing multiple deflections. The resultant average driving force for a crack front that has undergone multiple deflection is an average of the many deflections of the crack tip. This problem has been treated mathematically (38–40). Having considered the effectiveness of various deflection morphologies for microstructural inhomogeneities, the authors came to the conclusion that the resultant toughening from deflection is dependent only on the shape and distribution of the inhomogeneities and not their size. The higher the density and aspect ratio of inhomogeneities, the higher the maximum toughening. Therefore, rodlike geometries produce higher toughening than either platelets or spheres. It was postulated that the toughening arises mainly from the twist component and the maximum possible toughening predicted by the model is approximately $2K_{IC}$.

Deflection rarely operates as the sole toughening mechanism in a system, although its contribution in some systems may be significant. Crack deflection, however, is a major aspect of bridge formation processes that leads to toughening via bridging ligaments.

6.3. Crack-Tip Shielding. Crack-tip shielding has two origins: process-zone shielding and crack-wake bridging. Process-zone shielding derives from mechanisms occurring in a zone around the crack tip which extend to the crack wake as the crack advances, indirectly applying closure forces to the crack flanks. Crack-wake bridging derives from intact bridging elements in the wake of the crack, directly applying closure forces to the crack flanks.

Process-Zone Shielding. An important mechanism that can lead to the phenomenon of crack-tip shielding is the development of a process zone around the crack. Stresses around the crack tip cause changes in the microstructure that can lead to process-zone shielding. In the presence of regions of nonuniform residual stresses in the microstructure, microcracks can open up, their dilation providing a back stress on the crack tip that retards further crack propagation (40). A similar dilatation can be produced by stress-induced phase transformations in which there is a volume change (41) or by yielding of ductile particles (10). Initially, a process zone is formed ahead of the crack tip, which may enhance crack growth, but as the crack advances, a fully developed process zone is formed in the wake of the crack, causing closure forces to be applied to the crack flanks. A

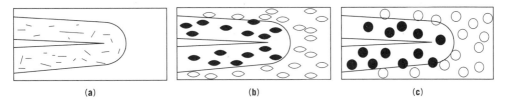

Fig. 8. Process-zone shielding mechanisms: (**a**) Microcrack cloud; (**b**) phase transformation; (**c**) yielding of ductile reinforcements.

schematic diagram presenting the identified process-zone shielding mechanisms in ceramic-matrix composites is given as Figure 8. The toughness produced by each type of zone has the same form in terms of the strain energy release rate approach (42):

$$\Re_\infty = \Re_o \left(1 + 2\eta\varepsilon V_f E'_c/\sigma_c\right) \tag{11}$$

where $\Re_\infty$ is the steady-state fracture resistance, (maximum in $\Re$-curve); $\Re_o$ is the crack-tip fracture resistance; η is a dimensionless zone-shape coefficient (≈ 0.03 to 0.06); ε is the dilatation strain per reinforcement (microcrack, transforming particle, or ductile yielding particle); V_f is the volume fraction of reinforcement; and E'_c is the plane strain composite modulus $[E_c/(1-\nu^2)]$ and σ_c the critical stress at which microcracks open up, particles transform, or particles yield.

The process zone toughening mechanism is seldom found to operate in isolation from other toughening mechanisms. The exception is the case of particles that undergo a phase transformation in which case the toughness is attributed to process zone shielding alone.

Crack-Wake Bridging. Crack-wake bridging occurs when matrix cracks, upon encountering a bridging entity, deflect around the entity, leaving it intact in the wake of the crack. Continued crack propagation requires further increases in the applied load to overcome the bridge closure forces. When the crack opening displacement at the bridging site is large enough to pull the bridge out of the matrix or to break the bridge, a steady state bridging zone develops, which then moves with the crack. New bridges are created at the crack front, and the bridges farthest away from the crack front become inactive. Such processes are energy dissipative and impart some degree of nonlinear stress–strain behavior to the composite; fracture becomes tougher with the possibility of developing large strains before final failure. The phenomenon of crack-wake bridging leads to T-curve behavior. In terms of a toughness approach, in which stress intensity factors can be linearly added, the far-field stress intensity factor K_a is equal to the stress intensity factor due to the bridging terms K_μ added to the stress intensity factor associated with the crack tip K_o:

$$K_a = K_\mu + K_o \tag{12}$$

The form of the solution for the bridging contribution is specific to the reinforcement and the crack geometry but is of the general form (35)

$$K_\mu = \sqrt{\frac{2}{\pi}}V_f \int_0^{x^*} \frac{\sigma_\mu[u(x)]}{\sqrt{x}} \, dx \tag{13}$$

where V_f is the volume fraction of bridging entities and $\sigma_\mu[u(x)]$ is the closure stress applied via a bridge to the flank of the crack as a function of crack opening $2[u(x)]$. The integral is evaluated over the bridging zone from the crack tip at $x = 0$ to the maximum extent of the bridging zone at $x = x^*$. Evaluation of the integral in equation 13 requires a knowledge of the crack profile with distance from the crack tip, $u = u(x)$. This function is obtained by evaluation of the integral equation (35,43):

$$2u(x) = \frac{8K_o}{E'}\sqrt{\frac{x}{2\pi}} + \left(\frac{4V_f}{\pi E'}\right)\int_0^{x^*}\sigma_\mu[u(x')]\left[2\sqrt{\frac{x}{x'}} - \ln\left|\frac{\sqrt{x} + \sqrt{x'}}{\sqrt{x} + \sqrt{x'}}\right|\right]dx' \qquad (14)$$

where E' is the plane strain elastic modulus. Solution of this integral equation is specific to the bridging relation and is not simple.

Toughness can also be calculated by considering a J-integral approach, where the equivalency with stress intensity factor is given by (44):

$$J_a = \frac{K_a^2}{E_c} \qquad (15)$$

$$J_o = \frac{K_o^2}{E_m} \qquad (16)$$

and

$$J_a = J_\mu + J_o \qquad (17)$$

where

$$J_\mu = 2V_f\int_0^{u^*}\sigma_b(u)\cdot du \qquad (18)$$

where $2u^*$ is the maximum crack opening displacement after which bridge failure occurs. Note that $2u^* = 2u(x = x^*)$. Using this approach the toughening increment, J_μ, can be calculated without having to solve the integral equation 14, however, the value obtained is only the steady-state toughening increment. No determination of the rising portion of the $\Re$-curve can be made using this approach. If the bridging relation includes dissipative processes, then the J-integral given in equation 18 is not strictly correct, because it assumes that the bridging process is nondissipative. This situation typically leads to an overestimate of the toughening increment (45).

7. Mechanical Performance

7.1. Particle Reinforcement. Particle reinforcement is an excellent method for toughening brittle ceramic matrices (3,41,46–48). The toughness imparted to such composites is due to multiple toughening mechanisms including crack deflection, crack pinning, microcracking, residual stress, frictional bridging, particle pullout, and transformation toughening. The mechanisms important to any specific system depend on the physical properties of the particles: size; morphology; thermal expansion mismatch with the matrix; and strength, toughness, and ductility.

Brittle Particles. Reinforcement via small brittle particles exploits the toughening mechanisms of crack deflection, microcracking, crack pinning, and crack bowing. The toughening contribution from the mechanisms of crack bridging and frictional pullout may be significant if the reinforcing particles are of the order of the matrix grain size or larger. All of these mechanisms arise from, or are strongly enhanced by, thermal expansion mismatch stresses in the composite. At the processing temperature the interface between the reinforcement and the matrix is stress free. On cooling to room temperature, thermal mismatch stresses can develop. If the matrix has a higher expansion coefficient than the reinforcement, the particle will be under compression and the matrix will have tensile hoop stresses and compressive radial stresses. The interface will be in compression, and the stresses in the matrix will decrease as the distance from the interface increases as a function of $1/r^3$, where r is the radial distance from the center of the particle. If, however, the particle has a higher expansion coefficient than the matrix, the nature of the residual stresses are reversed and the interface is in tension. For a particle the stresses are shown in Figure 9 (42).

These residual stresses play an important role in the mechanical response of ceramic matrix composites. Compressive stresses can lead to increased strength of the reinforcement, matrix or interface. Similarly tensile stresses act deleteriously. Even though tensile stresses reduce the apparent strength of the particles, the corresponding compressive hoop stresses in the matrix can act to deflect matrix cracks, thereby leading to toughening by crack deflection. Particles in compression have tensile hoop stresses in the matrix surrounding them. These stresses can cause microcracking at the particle. When the stress field of a matrix crack combines with tensile hoop stresses, microcrack

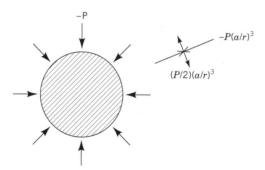

Fig. 9. Residual stresses owing to thermal expansion mismatch between a particle with radius a and thermal expansion coefficient α_p and a matrix with thermal expansion coefficient α_m. The stresses illustrated here are for $\alpha_m > \alpha_p$ and P is the interfacial pressure. $P = (\alpha_m - \alpha_p)(\Delta T)/[(1 + \nu_m)/2E_m + (1 + 2\nu_p)/E_p]$.

toughening is possible. In addition, tensile hoop stresses can attract matrix cracks to increase the probability of crack front pinning.

Crack pinning and the resultant crack bowing provide another toughening mechanism associated with brittle particle composites. As the driving force for the crack is increased, unpinned regions of the crack front tend to grow outward, retarded by the pinned regions. One description of the toughening that results from this phenomenon (46,49) proposes that a crack front possesses a line energy; accordingly, a crack that is pinned and bowed has a proportionally larger surface energy. The increment in fracture energy is proportional to the inverse average interparticle spacing (49) and is a function of the particle size (46). Fracture energy measurements for a composite system of alumina particles in a sodium–borosilicate glass are consistent with these functional dependencies, showing fracture energies as large as five times that of the unreinforced glass (46). (Of this increase, only a factor of less than two could be attributed to the enhanced surface roughness that results from the presence of the particles). This mechanism is not well understood and has not been widely verified; however, cracks can be pinned by inhomogeneities with a resulting tortured crack front. Figure 10 is a micrograph of crack pinning by a fiber.

A new form of particle-reinforced composite that derives its high toughness from particle bridging and pull-out is based on Al_2O_3–Al_2TiO_5 (3). In this system severe thermal expansion mismatch results in residual stresses in the microstructure that enhance grain-bridging toughening. The resulting composite, with steady state toughnesses of 8 MPa$\sqrt{m}$, a higher toughness and exhibits more flaw tolerance than the matrix alumina alone. The toughening mechanism has been identified as localized grain bridging (33) and bridges, labeled B, can be clearly seen in Figure 11.

Ductile Particles. Ductile particle reinforced ceramic composites show promise as composite material for high strength–high toughness applications.

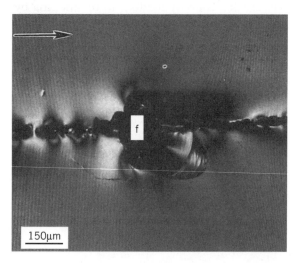

Fig. 10. Crack pinning by a SiC fiber in a glass matrix, photographed using an optical microscope and Nomarski contrast. Fiber lies perpendicular to plane of micrograph; lines represent crack position at fixed intervals of time, crack running left to right. Courtesy of T. Palamides.

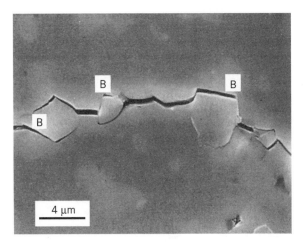

Fig. 11. Al_2O_3–Al_2TiO_5 typical microstructure showing grain bridging, B (SEM back-scattered electron image).Courtesy of L. Braun and S. J. Bennison.

Additions of up to 20 vol% aluminum particles to a matrix of alumina have shown toughness of up to 10 MPa$\sqrt{m}$ compared with typical alumina toughness of 2 MPa$\sqrt{m}$ (10). Figure 12 shows the toughening mechanisms schematically.

Ductile particles can act as bridging sites in the crack wake. Instead of fracturing in a brittle manner, they undergo plastic yielding as the crack opens up (10,47,50). The maximum strain energy release rate G^∞ in the steady-state region of the $\Re$-curve from ductile bridging alone for a bridged-edge crack is given by equation 19 (10):

$$G_\mu^\infty = aV_f\sigma_y R\varepsilon_y \tag{19}$$

where a is a geometrical constant, V_f is the volume fraction of bridging ductile particles, σ_y is the yield stress of particles in the matrix, ε_y is the rupture strain of the particles in the matrix and R is the particle radius. The solution for a bridged penny-shaped crack has been given (50). To enhance this mechanism large particles with a high flow stress are required.

A second toughening mechanism that operates simultaneously with crack bridging is the ductile yielding of particles in the crack-tip stress field within a process zone (10). To maximize this toughening mechanism requires a large volume fraction of particles of low yield strength.

Transforming Particles. A special type of particulate-composite are those based on the tetragonal form of zirconia. Tetragonal zirconia has the ability to undergo a stress-induced martensitic phase-transformation from its tetragonal crystal form to a monoclinic form with an accompanying dilatation of 4% unconstrained (41). Common examples of such ceramic systems include MgO or CaO partially stabilized zirconia (Mg-PSZ, Ca-PSZ, respectively), fine grain Y_2O_3-doped zirconia (known as tetragonal zirconia polycrystals, or Y-TZP), and composites of fine-grain zirconia with alumina or mullite, (zirconia toughened alumina, or ZTA and zirconia toughened mullite, ZTM, respectively). The toughening

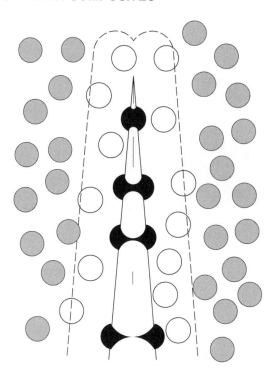

Fig. 12. Micromechanics of ductile reinforcement: particles yielding within process zone and particles bridging in crack wake.

owing to the phase-transformation-induced dilatation in terms of a stress intensity factor approach has been calculated to be that shown in equation 20,

$$K_a = K_o + K_p + T = K_o + 0.3\, Ee^T V_f w^{1/2} \tag{20}$$

where E is the modulus of the matrix, e^T is the transformation strain per particle, V_f is the volume fraction of transforming particles, and w is the width of the process zone (44). An equivalent solution in terms of a strain energy-release rate has been determined (51). This theoretical analysis indicates that a high modulus matrix and a large volume fraction of transforming particles are required to enhance the toughness. Such ceramic composites have been shown to have toughness as high as 15 MPa$\sqrt{\mathrm{m}}$ for Mg-PSZ with strengths of the order of 600 MPa (87,000 psi) (48). In contrast TZA composites have shown strengths as high as 2 GPa (290,000 psi) with toughness only in the range of 5 to 8 MPa$\sqrt{\mathrm{m}}$ (11). However, this particular toughening mechanism is restricted to zirconia particulate reinforcements and is only active below the phase-transition temperature, approximately 500°C.

7.2. Whisker Reinforcement. Toughening for whisker-reinforced composites has been shown to arise from two separate mechanisms: frictional

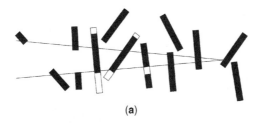

(a)

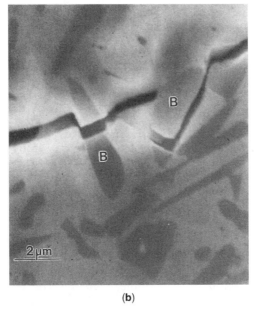

(b)

Fig. 13. Micromechanics of whisker toughening: (**a**) schematic diagram depicting frictional bridging, whisker fracture and pullout and (**b**) electron micrograph of whisker bridging in a SiC-reinforced alumina.

bridging of intact whiskers, and pullout of fractured whiskers, both of which are crack-wake phenomena. These bridging processes are shown schematically in Figure 13. The mechanics of whisker bridging have been addressed (52). The applied stress intensity factor is given by:

$$K_a = K_o + K_\mu \quad \text{and} \quad K_\mu = K_{fb} + K_{po} \tag{21}$$

K_{fb} is the toughness associated with the stretching of partially debonding whiskers in the wake of the crack, which includes frictional sliding of the debonded region of the whisker against the matrix. K_a for the single mechanism of frictional bridging is given by equation 22,

$$K_a = \left(E_c G_o + \frac{S_w^3 R_w V_w E_c}{6 E_w \tau} \right)^{1/2} \tag{22}$$

where E_c is the composite Young's modulus, G_o is the strain energy-release rate associated with the crack tip, S_w is the fracture strength of the whiskers, R_w is the radius of the whiskers, V_w is the volume fraction of bridging whiskers in the crack, E_w is the whisker modulus, and τ is the interfacial shear resistance of the sliding whisker–matrix interface (τ is proportional to the friction coefficient). This solution was determined assuming a linearly increasing constitutive relationship, $P_b(u)$, for the whisker bridging zone, where

$$G_{fb} \approx 2V_f \int_0^{u^*} P_b(u) \cdot du \tag{23}$$

The pullout regime assumes a linearly decreasing constitutive relationship as the crack opens up, and the whiskers pull out of the matrix with increasing ease. One solution (52) for frictional pullout alone is

$$K_{po} = \left(I_{po}/R_w\right)\left(E_c A_w \tau R_w\right)^{1/2} \tag{24}$$

where A_w is the area fraction of whiskers in the crack plane and I_{po} is the average pullout length given by

$$I_{po} \propto \left(\frac{R_w S_w}{2\tau}\right) \tag{25}$$

Experimentally it has been shown that both frictional bridging and whisker pullout play an important role in toughening industrially manufactured composites. Such investigations confirm that to maximize toughness via both mechanisms requires a high volume fraction of whiskers and a high composite modulus to whisker modulus ratio. For example, consider the effect of 20 vol% SiC whisker ($E_c = 500$ GPa) reinforcement of various matrices on the toughness as presented in Table 6 (53).

A high whisker strength combined with a low interfacial shear resistance τ enhances frictional bridging. Conversely, a high τ decreases the pullout contribution once whisker fracture has occurred. Increasing the whisker radius enhances both the frictional bridging and the whisker pullout contribution as evidenced by an increase in toughness of a 20 vol% SiC-reinforced alumina from 6.5 to 9 MPa $\cdot$ $\sqrt{m}$ when the radius of the bridging whiskers used were increased from 0.3–0.75 µm to 1.5 µm (53).

Whisker reinforcement is a viable method of toughening composites. However, health considerations associated with the aspiration of fine, high-aspect-ratio whiskers raise serious concern about their widespread use.

Table 6. **Whisker Reinforcement of Various Ceramic Matrices**[a]

Matrix	Matrix modulus,[a] GPa[b]	E_c/E_w[c]	Relative toughness, K_μ/K_m
glass	80	164	1.0
mullite	210	268	1.25
alumina	400	420	2.4

[a] Note the importance of matrix modulus on composite properties (54).
[b] To convert GPa to psi, multiply by 145,000.
[c] $E_c = E_m(1 - V_w) + E_w V_w$.

Table 7. **Effect of Platelet Size and Aspect Ratio on Toughness**[a]

Platelet size, μm	Aspect ratio	Bend Strength, MPa[b]	Fracture toughness, MPa$\sqrt{m}$
unreinforced		600	6.0
12	1.2	610	7.9
40	4	320	6.7
70	7	300	6.0

[a] For 20 vol% SiC platelet-reinforced RBSN (55).
[b] To convert MPa to psi, multiply by 145.

8. Platelet Reinforcement

Ceramic composites reinforced with crystalline platelets show similar values of toughness as whisker-reinforced ceramic matrices. Platelets have the additional advantages of being at least one tenth the cost of whiskers, easier to process, and have higher thermal stability and none of the health hazards associated with the aspiration of whiskers. Toughness comes from a combination of crack deflection, frictional bridging and platelet pullout. Whisker toughening models are applicable if the plate thickness is aligned perpendicular to the crack front. Platelets can be of the same material as the matrix. For example, an alumina with 5 μm equiaxed grains which is reinforced with 25 vol% of alumina platelets (100 – 200 μm × 10 μm thick) increased the toughness from 4 to 7 MPa$\sqrt{m}$ (54). The fracture toughness increases with platelet volume fraction; eg, for the 20 vol% SiC platelet reinforced reaction bonded silicon nitride (RBSN) (55). As the volume fraction increases from 5 to 30%, the bond strength changes from 400 MPa to 350 MPa and the fracture toughness increases from 4.9 MPa to 6.5 MPa$\sqrt{m}$. Platelet size also has an important influence on toughness and strength, which generally decrease with increasing size (Table 7).

A problem arises in using platelet reinforcements if their naturally mechanically weak crystallographic direction is aligned perpendicular to the crack front. The platelets easily fracture in this orientation. Further research is needed to grow platelets with favorable crystallographic orientations.

9. Fiber Reinforcement

9.1. Short Random Fibers. The whiskers bridging mechanics given in equations 21 through 25 apply also to short random fiber bridging mechanisms. The bridging terms come from (44):

$$G_a = G_o + G_\mu \qquad (26)$$

where

$$G_\mu = 2V_f \int_o^{u^*} P_b(u)du \qquad (27)$$

The integrand, $P_b(u)$, is the force-displacement relationship for the fibers pulling out of the matrix. This relationship is identical for fibers aligned parallel to one another; however, for randomly aligned fibers this constitutive relationship is a function of fiber alignment. Studies on continuous fibers misaligned to the crack front have been conducted (56,57) to determine the constitutive relationship as a function of misalignment angle. The greatest closure forces are applied to the crack flanks when the fibers are perpendicular to the crack front. As the fiber mismatch angle increases the closure forces are reduced. These results have not been incorporated into a new bridging model.

9.2. Continuous Fibers.   Composites reinforced by continuous fibers can fail in one of several possible modes depending on the interface properties and the fiber strength (58). A characterization of these modes by fiber strength and sliding shear resistance is depicted schematically in Figure 14. When the distribution of fiber strengths is broad (as characterized by a low Weibull modulus) in the regime of low fiber strength/high shear resistance, fiber fracture in the crack wake occurs away from the crack mid-plane and the fibers pullout. The majority of toughening is due to the frictional pullout mechanism, although, there may be some contribution from frictional bridging before fiber fracture occurs. In this regime, the composite fracture strength is a function of the crack length, and the matrix cracking stress is the ultimate tensile stress. The mechanics associated with failure in this regime are (9):

$$G_{fb} \propto \left[R(m-5)/\tau^{(m-2)} \right]^{1/(m+1)} \tag{28}$$

$$G_{po} \propto \left[R(m-3)/\tau^{(m-1)} \right] 1/(m+1) \tag{29}$$

where G_{fb} and G_{po} are the toughening contributions from frictional bridging and pullout, respectively. The dependencies of these contributions on the fiber radius R and the interfacial shear resistance τ are sensitive functions of the Weibull modulus m. The toughness increases with fiber radius when m is greater than 5 and decreases when m is less than 3. The toughness increases with a decreasing interfacial shear resistance when m is greater than 2 and decreases when m is less than or equal to 1.

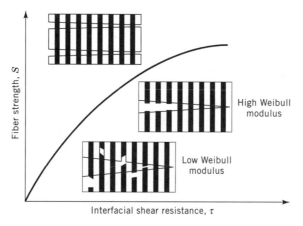

Fig. 14.   Failure mechanisms for continuous fiber reinforced ceramic matrices (58).

A transition to a different mode of composite failure, which is still within this low fiber strength/high shear resistance regime, occurs when the fiber strengths have a much tighter strength distribution as characterized by a high Weibull modulus. Initially the fiber strength is sufficient to allow the formation of a bridging zone in the crack wake before fiber fracture occurs at the point of highest stress in the fiber, that is, in the crack mid-plane. There is little to no fiber pullout contribution to toughening, and the contribution from fractional bridging predominates (45):

$$G_{fb} = \left(\frac{V_f V_m^2 E^{2m} R S^3}{3 E_f \tau E_c^2} \right) \qquad (30)$$

where S is the fiber strength. Accordingly, in this low fiber strength/high shear resistance regime, the fiber strength distribution, and in particular the Weibull modulus, are extremely important in tailoring composite properties to produce high toughness composites.

In the region of high fiber strength and low interfacial shear resistance, matrix cracks can propagate around the fibers leaving them intact in the wake of the crack. The matrix can be completely cracked through with the fibers supporting all the load before fiber failure begins. This regime of composite fracture leads to a very nonlinear stress–strain behavior, with a strong T-curve and crack length independent of strength at long crack lengths. Matrix cracks begin to propagate at the matrix cracking stress $\varsigma_{m\mu}$ (51,60).

$$\sigma_{m\mu} = \left(\frac{6 \tau G_{mc} E_f V_f^2 E_c^2}{R E_m^2 V_m} \right)^{1/3} \qquad (31)$$

In such a material the toughness is primarily due to bridging contribution, rather than fiber pullout.

10. Chemical and Thermal Stability

Ceramic-matrix composites are a class of materials designed for structural applications at elevated temperature. The response of the composites to the environment is an extremely important issue. The desired temperature range of use for many of these composites is 0.6 to 0.8 of their processing temperature. Exposure at these temperatures will be for many thousands of hours. Therefore, the composite microstructure must be stable to both temperature and environment. Relatively few studies have been conducted on the high temperature mechanical properties and thermal and chemical stability of ceramic composite materials.

10.1. Reinforcement Integrity. Strength degradation with increasing temperature occurs to a much greater extent with ceramic reinforcements, particularly those of continuous fibers, than it does with monolithic materials. Reinforcements have high surface areas to volume so that they are more susceptible to strength degradation resulting from surface reactions with the atmosphere.

These reactions can also decrease the toughness of the composite if crack-wake bridging and pullout are the predominant toughening mechanisms. This phenomenon results from the strong relationship between reinforcement strength and composite toughness predicted by the theoretical models.

Studies on the dependence of strength on temperature for Nicalon SiC fibers exposed to an oxidizing atmosphere have revealed that these fibers maintain their strength and stiffness up to 1000°C. By 1300°C, however, the tensile strength drops to 800 MPa (116,000 psi) (61). Mullite fibers (Nextel series) have a rapid fall off in elastic modulus at 900°C, but maintain a strength of at least 1 GPa (145,000 psi) up to 1400°C. Studies on strength degradation of alumina fibers (62) have revealed that these fibers maintain their strength and stiffness up to 800°C, with only 10% loss in both up to 1000°C. In general, nonoxide fibers tend to drop to half their 1000°C strength at 1400°C; oxide fibers drop to half their strength at 1100°C.

10.2. Composite Response. *Chemical Degradation.* A majority of ceramic-matrix composites show strong trends in the manner in which the mechanical properties are affected by temperature. If the interface degrades, allowing strong bonding to occur between the reinforcement and the composite matrix, the toughness is considerably reduced (58). If the interface remains weak enough to allow debonding and pullout, composite strength and elastic modulus are reduced. Usually, toughness increases with increasing temperature until a temperature is reached at which failure occurs by a different mechanism, such as creep, then toughness falls off rapidly (19,63). Composites that derive their toughness from stress-induced phase transformations, such as those based on zirconia, typically undergo a drastic reduction in toughness above 800°C, owing to the change in stable phase.

If reinforcements are nonoxides, then oxidation is a possibility at elevated temperatures. Oxidation of SiC and TiC to SiO_2 and TiO_2, respectively, may be rapid at 1200°C, leading to matrix cracking as a result of the volume expansion accompanying the oxidation. If the matrix is alumina, further reactions may take place between the alumina and the oxidation products, forming mullite $(3Al_2O_3 \cdot 2SiO_2)$ with SiO_2, and aluminum titanate, with TiO_2 (Al_2TiO_5).

A further problem is possible if the reinforcements are very small. Coarsening of the particles or whiskers may occur driven by Ostwald ripening, in which large particles grow through diffusional transport at the expense of smaller ones. This can be minimized by choosing matrices in which the reinforcement elements have very low solid solubilities and diffusion coefficients. Platelets, however, have been shown to be more resistant to coarsening than particles or whiskers.

Creep Resistance. Studies on creep resistanceof particulate reinforced composites seem to indicate that such composites are less creep resistant than are monolithic matrices. Silicon nitride reinforced with 40 vol% TiN has been found to have a higher creep rate and a reduced creep strength compared to that of unreinforced silicon nitride. Further reduction in properties have been observed with an increase in the volume fraction of particles and a decrease in the particle size (20). Similar results have been found for SiC particulate reinforced silicon nitride (64). Poor creep behavior has been attributed to the presence of glassy phases in the composite, and removal of these from the microstructure may improve the high temperature mechanical properties (64).

In contrast to the particulate-reinforced composites, all other reinforcement morphologies appear to provide enhanced creep resistance. The creep rate of Ce–TZP has been found to be reduced by a factor of 5 at 1250°C by the addition of 10 vol% Al_2O_3 platelets (55). The addition of 20 vol% SiC whiskers to mullite reduces the steady-state creep rate by a factor of 10 (65).

11. Conclusion

Ceramic matrix composites are candidate materials for high temperature structural applications. Ceramic matrices with properties of high strength, hardness, and thermal and chemical stability coupled with low density are reinforced with ceramic second phases that impart the high toughness and damage tolerance which is required of such structural materials. The varieties of reinforcements include particles, platelets, whiskers and continuous fibers. Placement of reinforcements within the matrix determines the isotropy of the composite properties.

The toughness of ceramic matrix composites derives from a combination of two phenomena: crack deflection and crack tip shielding mechanisms. Particulate reinforced matrices derive their toughness from processes such as crack deflection, crack pinning and bowing, microcracking and frictional bridging mechanisms. Whisker, platelet, and fiber-reinforced matrices derive their toughness from crack wake frictional bridging and pullout mechanisms. For the mechanism of crack-wake bridging the most important aspect of the composite is the interface between the reinforcement and the matrix. It must be weak enough to allow debonding around the reinforcement leaving it as an intact bridging element in the crack wake. Control of the interface between the matrix and reinforcements is extremely important in optimization of composite properties.

BIBLIOGRAPHY

"Composite Materials, Ceramic-Matrix", in *ECT* 4th ed., Vol. 7, pp. 77–108, by E.P. Butler and E.R. Fuller, Jr., National Institute of Standards and Technology.

CITED PUBLICATIONS

1. M.-Y. He and J. W. Hutchinson, *Int. J. Solids Structures* **25**, 1053–1067 (1989).
2. M.-Y. He and J. W. Hutchinson, *J. Appl. Mech.* **56**, 270–278 (1989).
3. J. L. Runyan and S. J. Bennison, *J. Eur. Ceram. Soc.* **1**, 93–99 (1991).
4. R. Warren, *Ceramic-Matrix Composites,* Blackie, Glasgow and London, 1992.
5. P. F. Becher, C.-H. Hseuh, P. Angelini, and T. N. Teigs, *J. Am. Ceram. Soc.* **71**, 1050–1061 (1988).
6. G. H. Campbell, M. Ruhle, B. Dalgleish, and A. G. Evans, *J. Am. Ceram. Soc.* **73**, (1990).
7. R. Chaim, L. Baum, and D. G. Brandon, *J. Am. Ceram. Soc.* **72**, 1636–1642 (1989).
8. F. K. Ko, *Am. Ceram. Bull.* **68**, 401–414 (1989).
9. L. M. Sheppard, *Am. Ceram. Bull.* **69**, 277 (1990).

10. L. S. Sigl, P. A. Mataga, B. J. Dagleish, R. M. McMeeking, and A. G. Evans, *Acta Metall.* **46**, 945–953 (1988).
11. K. Tsukuma, Y. Kubota, and T. Tsukidate, N. Claussen, M. Ruhle, and A. H. Heuer, eds., in *Advances in Ceramics: Science &* Technology of Zirconia II, ACerS, Columbus, Ohio, vol. 12, 382–390, 1984.
12. T. Michalaske and J. Hellmann, *J. Am. Ceram. Soc.* **71**, 725–731 (1988).
13. K. W. Lee and S. W. Sheargold, *Ceram. Eng. Sci. Proceedings* **8**, 702–711 (1987).
14. K. P. Gadkaree and K. Chyung, *Bull. Am. Ceram. Soc.* **65**, 370–376 (1986).
15. M. I. Osendi, B. A. Bender, and D. L. III, (1989).
16. R. Rundgren, P. Elfving, R. Pompe, K. P. D. Lagerlöf, and B. Larsson, in S. Somiya, N. Yamamoto, and H. Yanagida, eds., *Advances in Ceramics,* ACerS, Westerville, Ohio, vol. 24B, 1988, 1043–1052.
17. C. R. Blanchard and R. A. Page, *J. Am. Ceram. Soc.* **73**, 3442–3452 (1990).
18. V. J. Tennery, ACerS, Westerville, Ohio, 1989, 260–272.
19. J. Homeny and L. J. Neergaard, *J. Am. Ceram. Soc.* **73**, 3493–3496 (1990).
20. C. Blanchard-Arid and R. Page, *Ceram. Eng. Sci. Proc.* **9**, 1443–1452 (1988).
21. D. Lewis, *Am. Ceram. Soc. Bull.* **67**, 1349–1356 (1988).
22. W. Boecker and co-workers, in J. Cawley, ed., *Ceramic Transactions,* vol. 2, ACerS, Westerville, Ohio, 1988, 407–420.
23. A. R. Bunsell, in *Ceramic Matrix Composites,* Blackie, Glasgow and London, 1992, 12–33.
24. *Ceramic Source 1991–1992,* The American Ceramic Society.
25. M.-Y. He, A. Bartlett, A. G. Evans, and J. W. Hutchinson, *J. Am. Ceram. Soc.* **74**, 767–771 (1991).
26. H. W. Carpenter and J. W. Bohlen, *Ceram. Eng. Sci. Proc.* **13**, 238–256 (1992).
27. E. P. Butler, and E. R. Fuller, Jr., *"Ductile Coating for High Toughness Ceramic Matrix Composites,"* unpublished work, 1992.
28. R. A. Lowden, *4th Annual Conference on Fossil Energy Materials,* Fossil Energy AR&TD Materials Program, Oak Ridge National Laboratory, Oak Ridge, Tenn., 1990, 97–113.
29. C.-H. Hsueh, P. F. Becher, and P. Angelini, *J. Am. Ceram. Soc.* **71**, 929–33 (1988).
30. E. R. Fuller, Jr., E. P. Butler, and W. C. Carter, in *NATO Advanced Research Workshop on Toughening Mechanisms in Quasi-Brittle Materials,* Kluwer Academic Publishers, Dordrecht, The Netherlands, 1990.
31. R. M. Thomson, *Solid State Physics* **39**, 1 (1986).
32. D. B. Marshall and A. G. Evans, *Mater. Forum* **11**, 304 (1988).
33. S. J. Bennison and B. R. Lawn, *Acta Metall.* **37**, 26–59 (1989).
34. Y.-W. Mai and B. R. Lawn, *Ann. Rev. Mater. Sci.* **16**, 415 (1986).
35. H. Tada, P. C. Paris, and G. R. Irwin, *The Stress Analysis of Cracks Handbook,* Del Research Corp., St. Louis, Mo., 1985.
36. S. M. Wiederhorn, *Ann. Rev. Mater. Sci.* **14**, 373–403 (1984).
37. M. Gell and E. Smith, *Acta Metall.* **15**, 253–258 (1967).
38. K. T. Faber and A. G. Evans, *Acta Metall.* **31**, 565–576 (1983).
39. K. T. Faber and A. G. Evans, *Acta Metall.* **31**, 577–584 (1983).
40. D. R. Clarke and K. T. Faber, *J. Phys. Chem. Solids* **48**, 1115–1157 (1987).
41. R. C. Garvie, R. H. J. Hannink, and R. T. Pascoe, *Nature* **258**, 703–704 (1975).
42. B. R. Lawn, *Fracture of Brittle Solids,* The Cambridge Press, Cambridge, 1992.
43. I. N. Sneddon and M. Lowengrub, *Crack Problems in the Classical Theory of Elasticity,* John Wiley & Sons, Inc., New York, 1969.
44. R. M. McMeedking and A. G. Evans, *J. Am. Ceram. Soc.* **65**, 242–246 (1982).
45. M. D. Thouless, *Acta Metall. Mater.* **37**, 2297–2304 (1989).
46. F. F. Lange, *J. Am. Ceram. Soc.* **54**, 614–620 (1971).

47. K. S. Ravichandran, *Acta Metall.* **40**, 1009–1022 (1992).

48. D. B. Marshall and J. E. Ritter, *Am. Ceram. Soc. Bull.* **66**, 309–317 (1987).

49. F. F. Lange, *J. Am. Ceram. Soc.* **22**, 983–992 (1970).

50. F. Erdogan and P. F. Joseph, *J. Am. Ceram. Soc.* **72**, 262–270 (1989).

51. D. B. Marshall, B. N. Cox, and A. G. Evans, *Acta Metall.* **33**, 2013–2021 (1985).

52. P. F. Becher, *J. Am. Ceram. Soc.* **74**, 255–269 (1992).

53. P. F. Becher, C. H. Hsueh, P. Angelini, T. N. Tiegs, *J. Am. Ceram. Soc.* **71**, 1050–1061 (1988).

54. K. B. Alexander, P. F. Becher, S. B. Walters, in the *12th International Congress for Electron Microscopy,* San Francisco Press, San Francisco, 1990, 106–107.

55. N. Claussen in *11th Riso International Symposium on Metallurgy and Materials Science Structural Ceramics–Processing Microstructure and Properties,* Riso National Laboratories, Roskilde, Denmark, 1990, 1–12.

56. E. P. Butler, H. Cai, and E. R. Fuller, Jr., in *Engineering Ceramics Division, 16th Annual Conference on Composites & Advanced Ceramics,* ACerS, Cocoa Beach, Fla., 1992, 475–482.

57. H. Cai, K. Faber, and E. R. Fuller, Jr., *J. Am. Ceram. Soc.* **75**, 3111–3117 (1992).

58. E. Y. Luh and A. G. Evans, *J. Am. Ceram. Soc.* **70**, 466–469 (1987).

59. M. D. Thouless and A. G. Evans, *Acta Metallica* **36**, 517–522 (1988).

60. J. Aveston, G. A. Cooper, and A. Kelly, in *The Properties of Fiber Composites,* IPC Science & Technology Press, The National Physical Laboratory, 1971, 15–26.

61. D. J. Pysher, K. C. Goretta, R. S. Hodder, and R. E. Tressler, *J. Am. Ceram. Soc.* **71**, 284–288 (1989).

62. A. R. Bunsell, in *International Symposium on Composite Materials & Structures,* Beijing, 1986.

63. K. M. Prewo and J. J. Brennan, *J. Mater. Sci.* **17**, 1201–1206 (1982).

64. J. M. Birtch and B. Wilshire, *J. Mat. Sci.* **13**, 2627–2636 (1978).

65. R. D. Nixon, S. Chevacharoenkul, and R. F. Davis, in S. Somiya, R. F. Davis, and J. A. Pask, eds., *Mullite and Mullite Matrix Composites, Ceramic Transactions,* ACerS, Westerville, Ohio, vol. 6, 1990, 579–604.

E. P. Butler
E. R. Fuller, Jr.
National Institute of Standards
and Technology

CERAMICS AS ELECTRICAL MATERIALS

1. Introduction

Electronic ceramics materials contribute a wide range of functionalities as circuit components for microelectronics use (1,2). For most electroceramic applications the electrical conductivity, whether due to ionic, electronic, or mixed ionic-electronic conduction, is the dominant material property, that determines bulk insulation behavior. Even where the dominant material characteristic is, eg, magnetic, ferroelectric, piezoelectric, pyroelectric, electrooptic, or electrochemical, the underlying property of primary importance for device use of these materials is electrical conduction behavior (1–5). In oxide materials such as RuO_2 and $Bi_2Ru_2O_3$, metallic conduction does occur, making them ideal for use as components in thick-film pastes and in composite electrodes. Fast ion conduction in oxide materials such as $(Zr,Y)O_{2-}$ is made use of commercially in fuel cell applications (see FUEL CELLS). Superconductivity in ceramic oxides, based on the Y–Ba and Bi–Sr–Ca cuprate structures, is being exploited for use in microwave filters and for magnetic levitation. Other classes of ceramic materials that feature semiconducting properties are used in applications as varied as resistance heating elements, rectifiers, photocells, varistors, thermistors, and sensors. Ceramic materials (qv) serve equally important functions as electrical insulators (see INSULATION, ELECTRIC). For example, glasses (qv) and porcelains (see ENAMELS, PORCELAIN OR VITREOUS) are used in both low and high voltage insulation. Alumina [1314-28-1], Al_2O_3, beryllium oxide [1304-56-9], BeO, and aluminum nitride [24304-00-5], AlN ceramics are widely used as substrates in microelectronics packaging. These materials provide added benefits of excellent thermal conduction, high mechanical strength, high corrosion resistance, and compatible thermal expansion coefficients. As substrates for microelectronic packaging, good thermal expansion match with silicon [7440-21-3], Si in order to reduce thermal stresses, low dielectric constant and loss to enhance signal processing, and high thermal conductivity for heat dissipation, are all needed characteristics. Ceramics such as silicon nitride [12033-89-5], Si_3N_4, and silicon carbide [12504-67-5], SiC, are strong, less brittle insulators, which are useful in severe environments (see ADVANCED CERAMICS, ELECTRONIC CERAMICS). For these applications, the ceramic material must also exhibit high mechanical strength, and good thermal shock resistance. Ceramics are also widely used as thin-film insulators (see THIN FILMS). Such materials as silicon dioxide [7631-86-9], SiO_2, Al_2O_3, AlN, Si_3N_4, and diamond (see CARBON–DIAMOND, SYNTHETIC), as well as several type glasses, have been developed as thin-film interlayer dielectric insulation for passivation of integrated circuit devices (see INTEGRATED CIRCUITS).

An important category of ceramic materials is that which is developed through the aliovalent doping of the insulator material to give electron conducting behavior (3–5). With controlled processing, these materials typically develop large, n-type semiconducting grains separated by thin insulating grain boundaries. This type microstructure generally results in a space charge modulated system, which features high permittivity and barrier layer capacitive behavior (3–7). Control of charge flow across the grain boundaries, through variation in temperature, can suddenly transform the material from a semiconducting to insulating state, leading to the well-known (PTCR) effect. Likewise, accelerated charge transport across the grain boundaries on application of a critical electric field, can transform the material from insulating to conducting, giving the Varistor control effect. Grain boundaries typically do not play as significant a role in NTC type thermistors, which are mainly p-type conductors in the ternary $(Mn, Ni, Fe)_3O_4$ spinel system. In this system, p-type polaron conduction occurs due to the existence of acceptor cations on the octahedral B sites (8). The activation energy for polaron hopping, however, is high, leading to a correspondingly high dependence of the conductivity on temperature. These materials, therefore, exhibit very high temperature sensitivity and find wide usage as sensors, particularly near room temperature.

Barium titanate [12047-27-7], $BaTiO_3$ -based compounds exhibit high dielectric constants but typically must be specially formulated to satisfy specific capacitive and temperature compensating requirements (see BARIUM COMPOUNDS). They are the most widely used materials for disk, multilayer, and thick-film capacitors. More recently, relaxor-based formulations have been developed to serve both capacitive and actuator functions. These relaxors are based primarily on lead magnesium niobate (PMN), $Pb(Mg,Nb)O_3$, modified with Zn, Fe, or Ni as divalent cations and Ti, Nb, Ta, or W as higher valence cations. Important characteristics of the relaxors are their high, frequency-dependent dielectric permittivity, broad dielectric maxima, and low firing temperatures. This latter allows for cofiring in air ambient using less expensive Ag–Pd metallurgy. The trend in capacitor technology is increasingly toward integrated device packaging in which the capacitor, in thin- or thick-film form, may be embedded in the substrate (1,2). The development of ferroelectric thin films for nonvolatile memory (FeRAM) applications will likely accelerate this trend (see FERROELECTRICS). The memory films are based mainly on lead zirconate–titanate (PZT), $Pb(Ti,Zr)O_3$, formulations, either vapor deposited or sol-gel derived (see SOL–GEL TECHNOLOGY). However, other formulations, including lead titanate (PT), [12060-00-3], $PbTiO_3$, lead lanthanum zirconate titanate (PLZT), $(Pb,La)(Ti,Zr)O_3$, and relaxor compositions, have been developed for thin film use. In the bulk form, the materials are all strongly piezoelectric and rapid strides have been made in their development for use in actuator motors, ultrasonic transducers, and for electromechanical sensing (4,5). PLZT, in particular, has been widely used in electrooptic devices such as high-speed shutters, switches, light modulators and displays.

A wide range of other ceramic materials have been developed for sensor use, including gas, oxygen, temperature, voltage and humidity sensing. The sensor characteristic is typically based on the unique conduction response of oxides such as titanium dioxide [13463-67-7], TiO_2, zinc oxide [1314-13-2], ZnO, modified $BaTiO_3$, tin dioxide [18282-10-5], SnO_2, and chromium magnesium oxide [12053-26-8], $MgCr_2O_4$, to particular environmental stimulus. Because many of these sensors can provide a feedback signal for process control, ie, serve as smart sensors, these materials represent an expanding area for ceramic device application (1,2). For oxygen sensing, solid electrolyte conductors based on yttria [1314-36-9], Y_2O_3, stabilized ZrO_2 (YSZ) are the most widely used, particularly for automotive applications. Titanium dioxide is also used as an oxygen sensitive resistive sensor. Ceramic materials with mixed ionic–electronic conductivity can also be used to increase the electrocatalytic activity of the electrodes in solid oxide fuel cells (9–12). For these applications, the YSZ electrolytes are typically replaced by perovskite structure oxides, such as $LaGaO_3$, in which a wide range of oxygen stoichiometry can be developed through doping on either cation lattice site. The ionic or electronic conductivity can be readily controlled through careful processing. However, the presence of grain boundaries, which are often electronically active due to depletion effects (6), can significantly alter both the conduction and high permittivity characteristics.

Ferrites (qv) are ceramic oxides, that exhibit ferrimagnetic behavior by virtue of the opposed (A–B coupling), resulting from the presence of variable cations on the octahedral sites within the lattice. Ferrites are widely used as inductive circuit components, including use in recording heads, microwave filters, as permanent magnets and various type induction cores (see INFORMATION STORAGE MATERIALS; MICROWAVE TECHNOLOGY). Categories of ferrite include hard, soft, and microwave materials. All are based primarily on ferric oxide [1309-37-1], Fe_2O_3. The hard ferrites are of composition $MeO \cdot 6Fe_2O_3$, where $Me = Sr$, Ba, etc. This class of materials is used in permanent magnets and for motor applications. Soft ferrites, $MeO \cdot Fe_2O_3$, are based on the cubic spinel structure, where $Me = Ni$, Co, Cu, Zn, Li, and Mn. These cations can be readily interchanged so as to modify the magnetic permeability of the material. Ferrites are used in computer memory cores, deflection yokes, telecommunication switching and like applications. Microwave ferrites are based on the garnet structure and also contain iron oxide, eg, $3Y_2O_3 \cdot 5Fe_2O_3$. They feature high resistivity and high frequency capability. Electrical conduction can also be developed in spinel structures such as $CuCr_2O_4$, by doping with Mg^{2+} or Al^{3+} ions. These materials have been developed for use in thermoelectric power generation and the copper–nickel manganites $(Mn_{3-x-y} Ni_y Cu_x)O_4$ are used in low resistance thermistors (13).

In the oxide spinel structures cations of different valence may be distributed on octahedral or tetrahedral lattice sites, leading to electronic conduction by small polaron hopping mechanisms. In the nickel–manganites, eg, electrical conduction is due to electron hopping from Mn^{3+} to Mn^{4+} cation sites, and is controlled by dopant type and concentration (8). Similar polaron type conduction occur in the perovskite based titanates, which are used in thermistors and in high permittivity barrier layer capacitors (14–16). In these systems, cation stoichiometry is the critical factor that defines conductivity behavior, whether the mechanism be ionic, electronic or mixed (3–5).

High temperature ceramic superconductors are being developed as components in microelectronics circuitry, including thin-film devices for microwave applications, and bulk materials for various energy related storage devices (1). The structure of ceramic superconductors is based on an oxygen deficient, layered perovskite structure. Formulations in the Y–Ba–Cu–O system ($YBa_2 Cu_3O_{7-x}$) have been the most widely studied, but the processing aspects needed to achieve useful shapes, controlled microstructures and high current density, remains a challenge. Many theoretical views have been put forth to account for the superconduction in these materials (17–19). An important clue has been the recognition that these materials are antiferromagnetic at low dopant concentration. From the phase diagram, there is close proximity between the antiferroelectric and superconducting phases, with antiferroelectric spin fluctuations persisting into the superconducting phase (17). It is recognized that if the spin fluctuations are coupled to the charge carriers, a high temperature transition may occur. This theory differs from the original low temperature superconductivity (BCA) theories used to explain the phenomenon, which are based on phonon interactions.

In the broad range of ceramic materials that are used for electrical applications, each category exhibits unique property characteristics that directly reflects composition, processing, and microstructure. Detailed treatment is given here primarily to those characteristics relating to insulation behavior and electrical conduction processes. Further details concerning the more specialized electrical behavior in ceramic materials, eg, polarization, dielectric, ferroelectric, piezoelectric, electrooptic, and magnetic phenomena, are covered in (20–24).

2. Electrical Conduction

For electrical utilization, materials are usually classified according to their specific conduction mode (3–5,21,22). Insulators exhibit low carrier mobility and low conductivity, in contrast to metallic conductors, where both the carrier mobility and electrical conduction are high. Semiconductors are intermediate in conductivity, usually with exponential dependence of the conductivity on temperature. Figure 1 gives values of electrical conductivity at room temperature for a wide range of material systems. As seen from Figure 1, the electrical conductivity ranges over many orders of magnitude, from 10^5 $(\Omega\text{-cm})^{-1}$ for conducting oxides such as rhenium(VI) oxide [1314-28-9], ReO_3, and chromium(IV) [12018-01-8], CrO_2, to $10^{-14}(\Omega\text{-cm})^{-1}$ highly insulating materials such as steatite porcelains. Other compounds, such as TiO_2, may change conductivity by several orders of

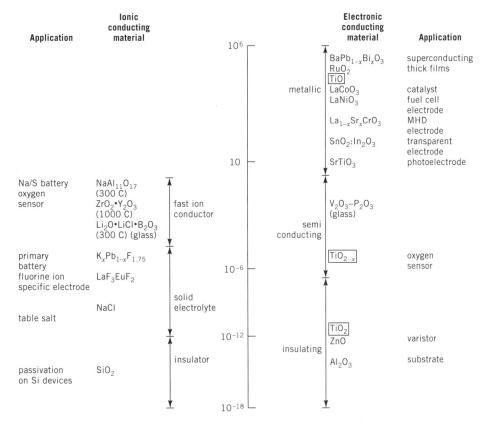

Fig. 1. Logarithmic scale of the electrical conductivities of materials categorized by magnitude and carrier type, ie, ionic and electronic, conductors. The various categories and applications are given. The wide conductivity range for the different valence–defect states of Ti oxide is highlighted. MHD is magnetic hydrodynamics.

magnitude as a result of aliovalent doping, or of high temperature heat treatment in controlled pO_2 ambient.

The charge transport mechanisms for the electrical conduction modes in ceramic materials vary greatly, since the transport of current may be due to the motion of electrons, electron holes, or ions. Crystal structure may also significantly affect the mobility of the charged species, as highlighted in Figure 1 for the various titanium oxides. The conductivity σ_i, which is associated with the mobility of the charged particles, is given as

$$\sigma_i = n_i z_i e(v_i/E) = n_i z_i^2 e^2 B_i \tag{1}$$

where n_i is the number of i particles per cubic meter, v_i is the average drift or net velocity in the direction of the applied field E, $e = 1.60 \times 10^{-19}$C, and ze is the charge on an ion of valence (z).

The conductivity term becomes complex if more than one type of charge carrier is present and involved in the conduction process. The total conductivity then becomes the sum of the fractional conductivities associated with each of the mobile carriers, be they electrons, holes or ions. This condition is represented by the expressions:

$$\sigma = \sigma_1 + \sigma_2 + \sigma_3 + \cdots + \sigma_n = \sum \sigma_i \ (i = 1 \cdots n) \tag{2}$$

The fraction of the total conductivity at a specific temperature and composition owing to the conduction of specie i is called the transference number (t):

$$t_i = \sigma_i/\sigma \tag{3}$$

where the value of t, defined as the transference number, lies between 0 and 1.0. Transference numbers for charged species in several compounds are given in Table 1, where t^+ and t^- represent the transference number for cation and

Table 1. **Transference Number of Cations t^+, Anions t^-, and Electrons or Holes $t_{e,h}$ in Several Compounds**[a]

Compound	Temperature, °C	t^+	t^-	$t_{e,h}$
AgCl	20–350	1.0		
CuCl	20	0.0		
	366	1.0		
KCl	435	0.96	0.04	
	600	0.88	0.12	
KCl + 0.02% CaCl$_2$	430	0.99	0.01	
	600	0.99	0.01	
NaCl	400	1.0		
	600	0.95	0.05	
Na$_2$O·11Al$_2$O$_3$	<800	1.0[b]		<10^{-6}
Na$_2$O·CaO·SiO$_2$		1.0[b]		
BaF$_2$	500		1.0	
ZrO$_2$ + 7% CaO	>700		1.0	10^{-4}
ZrO$_2$ + 18% CeO$_2$	1500		0.52	0.48
ZrO$_2$ + 50% CeO$_2$	1500		0.15	0.85
FeO	800	10^{-4}	0.0	1.0

[a] Ref. 1.
[b] Value is for sodium ion.

anion transport respectively, and $t_{e,h}$, that for electronic transport via electrons or holes. The compound CeO_2–ZrO_2, in the system cerium(IV) oxide [1306-38-3]– zirconium oxide [1314-23-4], is an example of variable transference number occurring in an ionic system. Here, the range of ionic transference numbers for this compound can vary from 0.03 to 1.0, depending on temperature, composition and the activities of the lattice components. The electronic transference numbers are also sensitive to these factors, including also oxygen partial pressure and defect concentration.

Electrical ceramic materials are commonly used in situations where refractoriness (see REFRACTORIES) or chemical resistance is needed, or where other environmental effects may be severe. Thus, it is important to understand the effects of temperature, chemical additives, gas-phase equilibration, and interfacial reactions on the conduction behavior.

3. Ionic Conduction

For an ion to move through a crystalline lattice, there must be an equivalent vacancy or interstitial lattice site available, and it must also acquire sufficient thermal energy to surmount the free energy barrier, ΔG, between the equivalent sites. Ionic conduction, which occurs through the transport of charge by mobile ions is, therefore, a diffusion activated process. From Fick's first law, the net flux, j in the direction x for a mobile specie in a concentration gradient (dn/dx) is given by $[j = -D \, (dn/dx)]$, where the diffusion coefficient D is given by

$$D = \alpha\lambda^2\nu = \alpha\lambda^2\nu_0 e^{-\Delta G^+/kT} \tag{4}$$

Here λ is the jump distance, approximately equal to the lattice parameter spacing $(a_o \sim 3 \times 10^{-8} \mathrm{cm})$ or to that of an adjacent plane; α is the number of possible jump sites, υ the jump frequency, υ_o the natural vibration frequency $\approx 10^{13}\mathrm{s}^{-1}$, and ΔG is the free energy of activation. The diffusion coefficient D_i for the ionic specie i, is related to the ionic mobility by the Einstein relationship:

$$D_i = \mu_i kT/ez_i = B_i kT \tag{5}$$

These equations generally are valid for all forms of conduction. For example, to determine the flux or conductivity of ions in a solid electrolyte, as compared to electrons in a semiconducting ceramic, the parameters of interest are the concentration of charge carriers and the carrier mobility. In crystalline ceramics the charge carriers are mainly mobile ions, with different mobility values for each specie. The mobility is related to the energetics of the site-to-site transport for each type ion, but this process can be enhanced significantly by forward biasing from an applied electric field. The effect of temperature, composition and structure on each of the terms in the general expression must be considered. Equation 6, which is the Nernst–Einstein relationship, gives the conductivity for an ionic specie i with transference number, t_i,

$$\sigma_i = t_i\sigma = \left(D_i n_i z_i^2 e^2\right)/kT \tag{6}$$

The expressions for extrinsic ionic conductivity of the material, as indicated by the shortened defect notation of equation 7, show that Na^+ ion vacancies are created mainly by Ca^{2+} ion doping and not primarily generated by thermal energy.

$$Ca^{2+} \xrightarrow{NaCl} Ca_{\dot{N}a} + V_{\dot{N}a} \tag{7}$$

In ionic compounds, intrinsic thermodynamic defects, known as Frenkel or Schottky defects, are present at concentrations determined by the entropy or energy state of the solid at the given temperature. The Frenkel defect pair consists of a lattice ion on an interstitial site with associated lattice ion vacancy. This type defect typically occurs in more open structures (CaF_2, AgI), in which interstitial ions can be accommodated. Conversely, Schottky-type defects occur in close-packed structures (MgO, Al_2O_3), where both cation and anion are considered to migrate to equivalent surface sites, leaving behind cation and anion vacancies. Creation of these thermal vacancies can be expressed in terms of an equilibrium constant and, therefore, as a free energy for the formation of the defects, ΔG_f. If the ion jump is to a lattice site, then a term for the creation of vacancies must be included. This results in an added exponential term in the conductivity expression of equation 6, and is valid when the concentration of defects caused by impurities is less than that caused by thermal energy (eq. 8).

$$\sigma_i = \frac{n_i z_i^2 e^2}{kT} e^{-\Delta G^+/kT} e^{-\Delta G_f/kT} \tag{8}$$

The term $e^{-\Delta G_f/kT}$ is the probability that a site is vacant.

In nonstoichiometric and doped materials the defect structure is generally more complex. In this extrinsic region the defect concentration is nearly temperature independent, and primarily controlled by the number and valence state of the solute atoms. For many materials, temperature ranges can be found in which only one type of charge carrier, intrinsic or extrinsic, is dominant. These regions can be distinguished by different activation energies as illustrated in Figure 2, where intrinsic conduction is dominant at elevated temperatures (25).

When aliovalent, ie, different valence, impurities are added to an ionic sites, the crystal lattice compensates by forming defects that maintain both electrical neutrality and the anion to cation ratio of the host lattice. For example, addition of x mols of CaO to ZrO_2 requires the formation of x mol of oxygen vacancies.

$$x\,(CaO) \xrightarrow{ZrO_2} x\,(Ca_{Zr}) + O_O + x\,(V_O) \tag{9}$$

If this concentration is larger than the oxygen vacancies created by thermal effects, then the conductivity from the motion of the doubly charged oxygen ions will be directly proportional to the concentration of the added Ca^{2+} ions (eq. 10).

$$\sigma_{O^{2-}} = \frac{\left([CaO](2)^2 e^2\right)}{kT} e^{-\Delta G^+/kT} \tag{10}$$

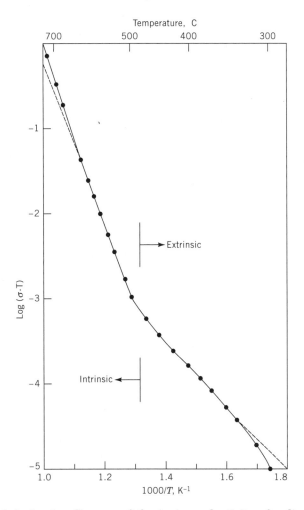

Fig. 2. Modified Arrhenius diagram of the ionic conductivity of sodium chloride. T is in kelvin, σ is in $(\Omega \cdot cm)^{-1}$. Over the temperature range shown the sodium ions are the primary charge carriers (25).

In polycrystalline materials, ion transport within the grain boundary must also be considered. For oxides with close-packed oxygen ions, the O^{2-} ion almost always diffuses much faster in the boundary region than in the bulk. This is due in part to second phases in the grain boundary region that are less close-packed, providing pathways for more rapid diffusion of ionic species. Thus the simplified picture of bulk ionic conduction becomes more complex when these additional effects are considered.

4. Fast-Ion Conductors

Inorganic compounds which exhibit exceptionally high ionic conductivity ($t = 1$), described as fast-ion conductors, are of technological interest for a variety of

Table 2. **Fast-Ion Conductors**

Compound	Temperature, °C	Conducting ion	σ_{ion}, $(\Omega \cdot m)^{-1}$
$NaAl_{11}O_{17}$	300	Na	35
$Na_3OZr_2PSi_2O_{12}$	300	Na	20
$CeO_2 + 12\,mol\%\,CaO$	700	O	4
$ZrO_2 + 12\,mol\%\,CaO$	1000	O	0.8
$K_{1.4}Fe_{11}O_{17}$	300	K	2
$Li_2B_4O_7{}^a$	150	Li	10^{-4}
	400	Li	0.1
$Li_4B_7O_{12}Cl^a$	300	Li	0.2
crystal	300	Li	0.8

^{a} Glass.

applications, including use as solid electrolytes. As shown in Table 2, compounds in this category include: (*1*) halides and chalcogenides of silver and copper, in which the metal atoms are disordered over a large number of interstitial sites; (*2*) oxides having the β-alumina [12005-48-0], $NaAl_{11}O_{17}$, structure in which migration of the monovalent cation is aided by the presence of conduction channels, leading to high mobilities; and, (*3*) oxides of the fluorite [14542-23-5], CaF_2, structure in which a large concentration of defects can be developed through incorporation of variable valence cations, or by solid state substitution of cations of lower valence, eg, $CaO \cdot ZrO_2$ or $Y_2O_3 \cdot ZrO_2$. Conductivity values in these type compounds are many orders of magnitude larger than for normal ionic compounds and are comparable to the conductivity of such liquid electrolytes as dilute solutions of sulfuric acid.

Owing to the precise relationship between the voltage and the chemical potential gradient across the electrolyte, these materials can be used in batteries (qv), fuel cells (qv), and as ion pumps or ion-activity probes. β-Alumina, eg, is used as an electrolyte in sodium–sulfur storage batteries. The β-aluminas exhibit an hexagonal structure with approximate composition $AM_{11}O_{17}$, where A, the mobile conducting ion, is monovalent Na and the M ion is trivalent Al. The crystal structure consists of four planes of oxygen ions in a cubic close-packed array, comprising a block wherein the aluminum ions occupy some of the octahedral and tetrahedral sites between each pair of the oxygen layers, similar to the site occupancy scheme in the spinel [1302-67-6], $MgAl_2O_4$ structure. These close-packed oxygen layers or spinel-like blocks are bound together by the Na^+ ions and AlO_4 tetrahedral units. They are separated by a more open and somewhat disordered basal plane layer, consisting of oxygen ions and loosely bound Na^+ ions. This layer provides a two-dimensional (2D) pathway for rapid migration of the Na^+ ions by greater than single jump distances which, coupled with their high concentration, lead to high ionic conduction in these structures.

Some other industrially important fast ion conductor materials display both mixed electronic and ionic conduction (6,9–12,26). This group includes the $LaGaO_3$ base system in which the A and B sites are doped with a variety of ions in order to impact the defect state, creating both extrinsic ionic and electronic states. The II–III–perovskites, such as $Sr(Co,Fe)O_3$, can be excellent mixed-ion conductors and have a two orders of magnitude higher oxygen ionic conductivity than the preferred III–IV–perovskites, such as $(LaSr)(MnCo)O_3$. These materials find use in the development of high temperature solid oxide fuel cells and sensors, where the combined effect of high electronic and ionic conduction can lead to improved electrochemical electrodes. Other uses include membranes for the production of high purity oxygen from air. The II–III based perovskite materials can develop a high concentration of ionic defects, but suffer from phase instability, which limits widespread usage (11). Doping on either the A or B site can easily change the phase transformation and temperature where electronic or ionic conductivity dominates, thereby controlling the property characteristics of these materials. Selective doping on the A or B site can retain the high electronic conductivity while increasing oxygen ion vacancy levels, and thus high ionic conductivity.

Typically, conductivity characteristics are dependent on temperature and oxygen partial pressure, hence will vary greatly depending on the degree of oxide nonstoichiometry and dopant level. The dopant level is critical since small fluctuations can lead to a decrease in electrical conductivity, oxygen nonstoichiometry, and oxygen permeation flux (10). For these mixed conducting systems, industrial applications require the avoidance of a decrease in oxygen permeation flux with time, typically associated with oxygen order–disorder phase transition in the oxygen sub–lattice. Likewise, chemical gradients can lead to strain and fracture, especially in fine separation membranes. For anode use in solid oxide fuel cells the material must exhibit high levels of mixed conduction under reducing conditions. Hence, the Ca doped $Gd_2Ti_2O_7$ pyrochlore system is attractive as an electrolyte material due to its high ionic conductivity and low electronic conductivity (12). Mo doping of this system leads to very high ionic and electronic conductivity under reducing conditions, making it suitable as an anode material. This behavior illustrates the importance of doping, which can drastically change property characteristics with only slight changes in processing parameters.

Other important fast ion conductors are the oxides having the fluorite structure, eg, ZrO_2. High (8–15 mol%) dopant levels of calcium oxide [1305-78-8] CaO or Y_2O_3 in solid solution with the ZrO_2 lead to large oxygen vacancy concentrations and vacancy ordering. The oxygen vacancies, which are produced during processing, are responsible for the high ionic conductivity in both the partially and fully stabilized zirconia. For example, for each Ca^{2+} that substitutes for a Zr^{4+} ion, an oxygen vacancy results, so that charge balance is maintained (eq. 9). Rapid oxygen migration occurs in these materials because of the high (15%) concentration of vacancies and correlated ion jumps over distances greater than the interionic separation. These structures support high oxygen ion mobility because of the low fourfold coordination of cations around the oxygens, coupled with the interconnected nature of the face-shared polyhedra that surround the oxygen sites (27). High ionic conduction results from the combination of the high oxygen vacancy concentration and high anion mobility (27). The doped ZrO_2 structures are widely used as potentiometric electrochemical sensors to detect oxygen in automotive exhausts (see EXHAUST CONTROL, AUTOMOTIVE). The sensor voltage is governed by the Nernst equation (eq. 11) where the activities are replaced by oxygen partial pressures and the air inside the chamber $(P_{O_2}^{II})$ is used as reference.

$$E = {}^-(RT/4F)\ln\left(P_{O_2}^{I}/P_{O_2}^{II}\right) \qquad (11)$$

Here, E is the voltage across the material, F is the Faraday constant (96,487 C), and R is the gas constant (8.314 VC/K). A porous platinum electrode is used so that the oxygen can pass freely through the electrode and react with the gas sensor material. Oxygen ions move rapidly through the $ZrO_2 \cdot Y_2O_3$ electrolyte and develop a potential difference as a result of the differential oxygen partial pressure (27,28).

Ceria (CeO_2) has the cubic fluorite structure as does ZrO_2, but it is a mixed conductor (9). In CeO_2, high electronic conductivity is observed at high temperatures and low partial pressure of oxygen, due to the formation of oxygen vacancies that controls the reduction of Ce^{4+} to Ce^{3+}, leading to n-type polaron conduction. In the ZrO_2–CeO_2 system, ionic and electronic conduction can vary over a wide concentration range, producing mixed to dominant ionic or electronic conductivity. Development of superstructures in these systems can lead to sharp changes in conductivity. Fast-ion conduction may, likewise, occur in certain glasses such as silver and alkali borates, phosphates, and molybdates (29). A condition for such conduction is the existence of channels in the highly ordered sublattice, within which the ions can easily move. Fast-ion conduction also occurs in glasses, where the mechanism can be attributed to other structural factors within the amorphous solid (27).

5. Conduction in Glasses

Electrical conduction in silicate glasses at ordinary temperatures can be attributed to the migration of univalent modifier ions such as Li^+, Na^+, K^+, H^+, and OH^-, under the influence of an applied field. At more elevated temperatures ($\sim$150°C), divalent ions, eg, Ca^{2+}, Mg^{2+}, and Pb^{2+} also contribute to conduction, although their mobility is generally low (30). Conduction in glass is an activated process and thus the number of ions contributing to conduction increases with both temperature and applied field. The temperature–resistivity dependence is given as

$$\rho = [6kT/\lambda\nu\alpha(ez)n]e^{E/kT} \tag{12}$$

where ν is the natural vibration frequency ($10^{13}s^{-1}$); λ is the average jump distance [$\sim$0.8–0.9 nm; α is the number of adjacent sites ($\sim$3); n is the concentration of charge carriers; and E is the activation energy].

As the concentration n of modifier ions increase, the effect is to loosen the structure, resulting in a lowering of the activation energy E and thus freer migration of the mobile ions. The jump distance and number of available jump sites can greatly impact the mobility of the ions. When two or more mobile ionic species are present, a condition commonly referred to as the mixed alkali effect causes a reduction in the conductivity, due to mutual blocking of the migration pathways, and thus to a mutual reduction in the ion mobility (30).

Glasses typically exhibit an amorphous or random network structure with characteristic short-range order of only a few atomic spacing. The creation of structural disorder occurs during fusion of the glass and, as in lithium borate glass, is presumed to result from the formation of a large excess of nearly equivalent sites, giving enhanced ionic mobility (29). Even though the glass structure can be described as open because of its random nature, there is no well-defined channel for conduction to occur as found in crystalline ceramics. With respect to a basic conduction mechanism, there is no general consensus that universally explains all aspects of ionic transport in glasses (31). However, application of percolation theory within the well-known random-energy model, leads to the most consistent explanation for both ac and dc conduction effects in glasses. Explanations for the observed strong dependence of ionic conductivity on composition have mainly been based on changes in the activation energy for electrical conduction (32). For alkali migration in oxide glasses this consists of two parts: the bonding energy between the mobile cation and its charge compensating center and the elastic strain energy associated with the distortion of the glass network as the ion moves from one site to another. For glasses with a high concentration of monovalent ions as charge carriers, ionic transport is determined not only by the interactions between the ions and the glass network but also between the ions themselves (33). These effects strongly influence the structure and the geometry of the transport pathways, leading to a lowering of the activation energy and increased conductivity.

6. Ceramic Insulators

Insulators are materials that offer effective resistance to current flow in an electric field, due to the very low concentration of mobile charge carriers. Important ceramic insulators such as SiO_2, Al_2O_3, mullite ($3Al_2O_3 \cdot 2SiO_2$), BeO, AlN, boron nitride (BN) [10043-11-5], and Si_3N_4 have resistivities $\sim 10^{14}$ (Ω-cm). These high values are the result of a large energy gap between a filled valence band and the next available energy level, where the promotion of an electron into a higher state is energetically unfavorable. The conductivity of these ceramics, therefore, is significantly influenced by both ionic and electronic defects. In insulating oxides, ionic defects arise from the presence of impurities of different valence from the host cation. An aluminum ion impurity substituting on a magnesium site in the MgO [1309-48-4] host lattice creates Mg vacancies.

$$x\,(Al^{3+}) \xrightarrow{\;\;MgO\;\;} x\,(Al_{Mg}^{\bullet}) \;+\; x/2\,(V_{Mg}'') \tag{13}$$

These vacancies facilitate ionic migration of Mg^{+2} ions under the influence of an electric field which is a high temperature process, however, and room temperature conductivities are very low (Table I).

Similarly, electronic conduction may arise in oxide materials from the natural loss of oxygen, which typically occur in oxides on heating to high temperatures.

$$O_O \longrightarrow \frac{1}{2}\,O_2(g) + [V_O + 2e'] \tag{14}$$

Electrons trapped at the vacancy can become partially or fully ionized, leading to weak n-type electronic conduction. Again, the conductivity is low. Conduction in materials such as AlN and SiC occur mainly through the presence of impurities having different valence states, leading to n- or p-type conduction.

$$Al^{3+} \xrightarrow{SiC} Al_{Si}^{\cdot} + h^{+} \qquad (\text{p-type}) \qquad (15)$$

$$Si^{4+} \xrightarrow{AlN} Si_{Al}^{\cdot} + e' \qquad (\text{n-type}) \qquad (16)$$

Similar conduction mechanisms can be expected for most nitride materials, including Si_3N_4 and BN, depending on the level of impurity doping SiC, which has only a moderate band gap ($E_g = 2.8$–3.2 eV), can become semiconducting and has been developed for device use. SiC is also widely employed as heating elements for furnace applications. Electrically insulating SiC can also be fabricated using BeO dopant additions. This is an important material for laser heat sink applications because of its high thermal conducting and electrical insulating properties (see LASERS) (34).

The primary function of insulation in electrical circuits is the physical separation of conductors and the regulation or prevention of current flow between them. Ceramic insulators are used in many demanding applications where high electrical resistance is a requirement, together with other important properties such as thermal conductivity, high operating temperatures, high dielectric strength, low dielectric loss, resistance to thermal shock, environmental resistance, thermal expansion, and long-life characteristics. Insulators of this type are known as linear dielectrics. The dielectric constant is a measure of the ability of the material to store charge relative to vacuum and is a characteristic material property. In an ac field, the electrical resistivity and dielectric constant are related by the dissipation factor, which measures the energy loss per cycle. This relationship is given by

$$\sigma = \omega \epsilon_0 k' \tan \delta = 1/\rho \qquad (17)$$

Where: $\sigma =$ conductivity; $\rho =$ resistivity; $\omega =$ frequency ($2\pi f$); ($\varepsilon_o =$ permittivity of vacuum (8.85×10^{-14} F/cm); ($\varepsilon_r =$ dielectric constant; $\tan \delta =$ dissipation factor; and $\varepsilon_r \tan \delta =$ dielectric loss factor $= \varepsilon_r''$. The dielectric strength (DS) is a measure of the maximum voltage gradient that can be impressed across the dielectric without physical degradation of its insulating properties. Ceramics that satisfy the following property criteria at 25°C are generally classified as good insulators: $\varepsilon_r \leq 30$; $\rho \geq 10^{12} \Omega$-cm; $\tan \delta \leq 0.001$; DS ≥ 5.0 kV/mm; $\varepsilon_r'' \leq 0.03$.

Silicon dioxide (SiO_2) has the lowest dielectric loss properties of any inorganic material. It is commonly used in insulating fibers and in the development of electrical porcelains ($R_2O \cdot Al_2O_3 \cdot SiO_2$). These materials have high dielectric strengths with low loss and are therefore suitable for high voltage applications such as transmission line insulators, high voltage circuit breakers, and cutouts. Mullite, $3Al_2O_3 \cdot 2SiO_2$, MgO, and steatite, $MgO \cdot SiO_2$, are extensively used for high temperature electrical insulation and for high frequency insulation because of their low loss characteristics. For electrical insulating applications and heat sinks, Al_2O_3, AlN, SiC, and Si_3N_4 are the most commonly used materials. Both SiC and $Si_3 N_4$ are also industrially valuable as high temperature heat exchangers because of high thermal conductivity and electrical insulating behavior, high hardness, durability, excellent high temperature, corrosion, and thermal shock resistance. Films of these materials, including diamond, have been developed, where the properties obtained are similar to that of the bulk materials. The conduction processes in the films mainly result from impurity and electrode injection effects, which degrades the high resistivity and dielectric properties of the materials.

7. Electronic Conduction

High electron and electron hole mobility in ceramic materials can contribute appreciably to electrical conductivity. In certain materials, metallic levels of conductivity can result while in others the electronic contribution can be very small. In all cases, the total electrical conductivity for all modes of conduction (electronic and ionic) is given by the general equation:

$$\sigma_i = \sigma(t_{\text{ionic}} + t_{\text{electronic}}) \tag{18}$$

where the total electronic conductivity is given as

$$\sigma - ne\mu_n + pe\mu_p \tag{19}$$

n and p denote the concentrations of electrons and holes, respectively, and μ_e and μ_h are the corresponding mobilities. The mobility of electrons and holes as charge carriers are generally much higher than for ionic carriers, because they are of lower mass and charge density and less confined to particular atomic sites (25). Scattering of electrons and holes occur by phonons and at point defects, dislocations, and grain boundaries. The conductivity is determined primarily by the concentration of electrons and holes, and can be described by an energy band structure (25). Figure 3 schematically illustrates the band energy configurations corresponding to metal, intrinsic semiconductor and insulator conduction. The highest energy band, which is completely filled at $T = 0$ K, is called the valence band, and the next higher band, being empty at this temperature, is the conduction band. These energy levels are separated by an energy or band gap (E_g), which normally is not occupied by electrons (25). The band gaps for several materials are given in Table 3. As shown in Figure 3, at $T = 0$ K, metallic conduction can occur in ceramic materials where there are partially filled valence bands and a corresponding overlap with the unoccupied conduction band states. When a small energy band gap is present with no overlap, semiconduction results. If a large energy gap exists the material is insulating, because the energy gap is too great to thermally promote electrons into the conduction band. Thus, with sufficient energy input, conduction occurs either by electrons being promoted into the conduction band or through the electron holes left behind in the valence band. If a semiconductor or insulator contains defects, ie, dopants, impurities, or vacancies, that have states within the band gap, as donor or acceptor levels, the result can be an increase in conduction.

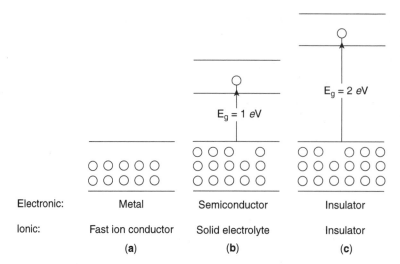

Fig. 3. Schematic of band gap energy, E_g, for the three types of electronic and ionic conductors. For electronic conductors the comparison is made of the relative occupancy of valence and conduction bands. For ionic conductors, the bands correspond to the relative occupancy of ionic sublattices. For (**a**), $n = 10^{22} \text{mL}^{-1}$ for (**b**), $n = 10^{10} \text{mL}^{-1}$; and for (**c**), $n = 1\text{mL}^{-1}$ (27).

In an ideal covalent semiconductor, electrons in the conduction band and holes in the valence band may be considered as quasifree particles. The carriers have high drift mobilities in the range of 10–$10^4 \text{ cm}^2/\text{Vs}$ at room temperature. As shown in Table 4, this is the case for both metallic oxides and covalent semiconductors at room temperature.

Two types of scattering may affect the motion of electrons and holes in a crystal lattice. One is lattice or phonon scattering, due to thermal vibrations within the lattice, where the vibration amplitude increases with temperature. With this enhanced vibration there is a corresponding decrease in mobility

Table 3. **Energy Gap (E_g) at Room Temperature for Intrinsic Semiconductors**

Crystal	E_g, eV	Crystal	E_g, eV
Cu_2O	2.1	Si	1.1
$BaTiO_3$	3.0–3.2	α-SiC	2.8–3.0
ZrO_2	3.2	BN	4.8
Fe_2O_3	3.4	C, diamond	5.2–5.6
ZnO	3.5		
TiO_2	3.05–3.8	AgI	2.8
CoO	4	KCl	7
BaO	5.5	CaF_2	12
CeO_2	5.5		
BeO	7.1	PbS	0.35
MgO	>7.8	GaAs	1.4
Al_2O_3	>8	GaP	2.25
SiO_2	~9	ZnS	3.6

Table 4. **Carrier Mobilities at Room Temperature**

Crystal	Mobility, cm^2/(V·s)		Crystal	Mobility, cm^2/(V·s)	
	Electrons	Holes		Electrons	Holes
Si	1500	450	CoFe$_2$O$_4$	10^{-4}	10^{-8}
diamond	1800	1200	CoO		~0.1
			NiO		~0.1
			Fe$_3$O$_4$		0.1
AgCl	50		Fe$_2$O$_3$	0.1	
KBr	100		TiO$_2$	0.2	
			(BaLa)TiO$_3$	0.5	0.1
AlN		10	BaO	5	
GaP	150	120	SrTiO$_3$	6	
PbS	600	200	ZnO	50	
GaAs	8500	450	SnO$_2$	160	

according to the relationship: $\mu \propto T^{-3/2}$. A second source of scattering arises is from impurities present at low (<100 K) temperature, which distort the periodicity of the lattice, giving a mobility relationship $\mu \propto T^{+3/2}$. The total mobility μ is proportional to the sum of these two terms. The temperature dependence of the mobility term for quasifree electrons and holes is much smaller than that for their concentration, which is exponentially determined. As a result, the conductivity has a temperature dependence determined mainly by this concentration.

Several theoretical models have been proposed to explain conduction in disordered materials, among which are the hopping and multiple trap models, as detailed by Blaise (14) and summarized briefly below. In the multiple trap model, charge transport occurs via extended states located above the mobility edge, which refers to the energy level separating localized states in the band gap from the extended or continuos states in the conduction or valence bands. In this model, the charge carriers are trapped within the band gap and released above the mobility edge, then recaptured and released many times as they progress through the material.

In the hopping model, charge transport occurs via impurity states located within the band gap. Hopping conduction is dominant when the trap energy levels are located well below the mobility edge and are populated at high concentration, such that there is an overlap of the wave functions of ions localized on the traps. The exact nature of the localized states depends on the type disorder resulting from the presence of defects such as vacancies, interstitial atoms and their ionized states. The distance between the states is in turn dependent upon the impurity concentration. A high concentration allows the localized states to overlap, with conduction occurring via thermally activated hopping between these states. The activation energy for conduction is typically < 1 eV, which allows conduction to occur at room temperature in these wide-band gap materials (14). This type conduction can be related also to charges moving from one impurity site to another via energy exchange with phonons (14–16).

In ionic host lattices where there is interaction between neighboring ions, polarization of the lattice due to the presence of an electron or hole as charge carrier, can occur. The resulting charged entity, consisting of the electronic carrier plus its polarization field, is referred to as a polaron (16). The presence of the electron or hole in the polar material also induces a dipole. If the extension of the resultant polarization field around the electron is large compared to the distance between the unit cells in the lattice, then the concept of a discrete dipole can be replaced by that of a polarization continuum, in which there is only weak-coupling between the charge carrier and the polarization field. This is referred to as a large polaron state, in which conductivity similar to a quasifree electron state results. Large polarons are typical of III–V compounds. An intermediate polaron state is one in which the polaron radius is comparable to the unit cell, and is typical of some perovskite materials (15). In contrast, when the lattice polarization, or polaron radius, is smaller than the lattice unit cell parameter (small polaron state), the mobility becomes strongly coupled to the lattice distortion, which must move along with the electronic carrier (15,16). This process is loosely referred to as a "hopping" mechanism. The localization of the polarons is a specific case in which the polaron has a very narrow bandwidth, where the spread of energy levels due to the lattice disorder is only a few tenths of electronics. This, however, is a simplification of the mechanism since the discrete energy levels can become localized states of lower energy within the band gap (16). Polar motion then becomes possible as a result of small overlap of the wave functions of neighboring ions. Electron mobility will increase according to the degree of overlap of the partially filled local states and not through conduction bands (16). Mobility value for the small polaron is typically less than 1 $cm^2/$Vs) but can be much lower.

8. Electronic Conducting Ceramics

8.1. Conducting Ceramics.   Metals are not the only materials that exhibit high conductivity. Metallic conduction occurs also in transition-metal oxides such as ReO_3, CrO_2, vanadium(II) oxide [12035-98-2], VO, titanium(II) oxide [12137-20-1], TiO, and rhenium(IV) oxide [12036-09-8], ReO_2 ; in doped perovskite structures such as lanthanum titanium oxide [12201-04-6], $LaTiO_3$, calcium vanadium(IV) oxide [12138-49-7], $CaVO_3$, $BaTiO_3$, strontium ferrate (1:1) [12022-69-4], $SrFeO_3$, lanthanum nickelite [12031-18-4], $LaNiO_3$, lanthanum colbaltite [12016-86-3], $LaCoO_3$, and lanthanum chromite [12017-94-6], $LaCrO_3$; in tungsten bronzes such as Na_xWO_3, La_xWO_3, and $Na_xNb_2O_5$, and in some spinels, $Li_{0.5} In_{0.5} Cr_2 S_4$, typically at specific dopant concentrations. These materials are all potentially important for fuel cell electrode use. In microelectronics, ReO_2 and RuO_2 ruthenium(IV) oxide [12036-10-1] compounds are used as the conducting phase in organic-based inks and pastes for the screen-printing of passive components in thick film circuits. The high conductivity in these transition-metal oxides typically result from overlap of unfilled d or f electron orbitals, forming energy bands with concentrations of quasifree electrons of the order of 10^{22}–$10^{23} m^{-1}$, equivalent to metallic conduction. The electronic conductivity in the perovskite and rutile oxide structures have been most studied. The electronic mobility in these simple transition-metal oxides is usually less than 1 $cm^2/$Vsec, leading to electrical conductivities in a broad range of [10^{-4}–10^4 $(\Omega\text{-cm})^{-1}$], with variable temperature dependencies.

Very high conduction also occurs in many ceramic materials represented by the group formula MeC_x or MeN_x, where Me is a metal, C and N are carbon and nitrogen respectively, and x is the carbon or nitrogen/metal ratio (35). These materials may exhibit covalent or ionic bonding, but with sufficient free electrons to display near-metallic behavior, including a positive and near-stable conduction with temperature. The high conductivity results from a redistribution of the orbitals and energy levels, resulting in band overlap at the Fermi energy level. TiC, for example, has no bandgap at the Fermi energy level and is, therefore, neither an insulator nor a semiconductor, but rather a metallic conductor, with resistivity values similar to the transition metals. Some of these materials, such as tantalum carbide and niobium carbide can become superconducting near 10 K (35).

8.2. Semiconducting Ceramics.   Defects can be created in many ceramic materials either by controlled doping, by heat-treated under conditions which create nonstoichiometric defects, or by uncontrolled impurity doping. Since mobility values for the defect transport mechanisms are often low and difficult to measure, the conductivity values that are reported for these materials are often at variance. This is in part due to the variable effects of impurities and thermal history which often overwhelm the expected dopant effects. Both the dopants and impurities strongly influence the properties of oxide semiconductors, since the substituted ions and created defects may introduce new localized energy levels that are intermediate between the valence and conduction bands (15,16). If the new energy levels are unoccupied and lie close to the top of the valence band, electrons can be excited out of the filled valence band into the new acceptor levels, leaving behind electron holes that contribute to p-type conductivity. These p-type oxides are commonly referred to as metal deficient conductors. Conversely, if the impurity additions have filled electron energy levels close to the conduction band, electrons may be excited from these donor levels into the conduction band as n-type charge carriers. The n-type conductivity in these oxides results from a so-called metal excess in these conductors. Similar impurity dopant effects occur also in non-oxide materials such as silicon carbide, which can be doped with boron [7440-42-8] to provide acceptor levels within the band gap (0.3 eV above the valence band), thus making it a p-type conductor. Conversely, nitrogen can be added to provide donor levels below the conduction band giving n-type conduction (0.07 eV).

Strontium titanate [12060-59-2], $SrTiO_3$, can become an n-type semiconductor by donor doping of the Ti lattice sites, or by heat treatment in a reducing atmosphere with resultant loss of oxygen. Electron mobility in the doped $SrTiO_3$ is $\sim$6 cm^2/Vs. ZnO also exhibits n-type conduction when sintered under reducing conditions, since the zinc interstitials formed become electron donors, each yielding one free electron. In contrast, Copper(I) oxide [1317-39-1] is a p-type semiconductor, in which the Cu^{2+} ions or vacancies act as acceptors for electron holes that conduct within a narrow band in the Cu d orbitals. Similar to Cu_2O, Nickel monoxide [1313-99-1], NiO, forms a metal deficient semiconductor in which vacancies occur on the cation lattice sites. For each cation vacancy two electron holes must be formed, which are assumed to be associated with the regular cations ($[Ni^{2+}]\rightarrow[Ni^{3+}]$). Transfer of the positive charges between the Ni cations leads to conduction within the lattice. This type conduction is similar to the polaron conduction described previously, where the localized states become polarized by electron hole motion. The charge-transfer process is of low mobility [0.1 cm^2 /V s], leading to relatively low conductivity of $\sim 10^{-13} (\Omega\text{-cm})^{-1}$ at 25°C, but the mobility value increases to $\sim 1 (\Omega\text{-cm})^{-1}$ on doping with Li^+ ions (25). The Li^+ ions stabilize the Ni^{3+} states at a higher concentration, resulting in higher and more stable conductivity. Similarly, the insulating characteristics of NiO can be improved by doping with trivalent ions such as Cr^{3+} or Al^{3+}, which decreases the fraction of Ni^{3+} ions formed and the overall conduction, since electron transfer between Ni^{2+} and Cr^{3+} or Al^{3+} does not occur.

Similar defect mechanisms can be expected in other oxide systems which exhibit both n- and p-type conduction. For example, Fe_2O_3 is an n-type semiconductor under conditions where oxygen ion vacancies are formed, with charge compensating electrons localized on the cation sites. This is the equivalent of forming Fe^{2+} ions on Fe^{3+} sites, resulting in electron polaron hopping and increased conduction. If Ti^{4+} is added to Fe_2O_3 in solid solution, an increased fraction of Fe^{3+} ions are forced into the Fe^{2+} state. As a result, the conductivity of the oxide substantially increases, ranging from 10^{-10} to $10^{-2} (\Omega\text{-cm})^{-1}$ for an added Ti^{4+} concentration of zero to 0.5 atom %. Again, conduction is determined primarily by the concentration of added titanium ions and is much less dependent on oxygen partial pressure and firing conditions. Tin oxide, SnO_2, and indium oxide [1312-43-2] In_2O_3, are other important oxides that are doped to increase conductivity. The SnO_2, In_2O_3, TiO_2, and in particular, $SrTiO_3$, are transparent to visible light and are often used as transparent electrodes, eg, on Vidicon tubes (Table 5).

Titanium dioxide as a semiconductor material is used as a gas sensor, and is sensitive to changes in ambient oxygen partial pressure. Whereas in the electrochemical sensor, $ZrO_2 \cdot Y_2O_3$, conduction is by oxygen ions, in TiO_2 sensors the conduction is by electrons (36). Pure TiO_2 has a band gap of 3.2 eV, and a fully occupied oxygen $2p$ valence band (27). Therefore, at room temperature it is a highly resistive material. At elevated (>500°C) temperatures, depending on oxygen partial pressure, it becomes oxygen deficient and Ti^{3+} ions are formed to compensate for the anion deficiency, along with mobile electrons. As more oxygen vacancies are created, the electron concentration increases and the resistivity decreases. Resistivity in these materials is, therefore, highly dependent on temperature, formation energy for the oxygen vacancies and on oxygen partial

Table 5. **Impurity Semiconductors**

		n-Type			
Cds	$BaTiO_3$	Nb_2O_5	Fe_2O_3	WO_3	GeO_2
CdSe	$SrTiO_3$	Ta_2O_5	Tl_2O_3	TiO_2	MnO_2
ZnF_2	$PbCrO_4$	Fe_3O_4	In_2O_3	SnO_2	ZnO
		p-Type			
Se	CuI	Ag_2O	Hg_2O	NiO	PdO
Te	SnS	Cu_2O	MnO	FeO	CoO
		Amphoteric			
Si	Sn	PbSe	Ti_2S	Al_2O_3	Mn_3O_4
Ge	PbS	SiC	PbTe	Co_3O_4	UO_3

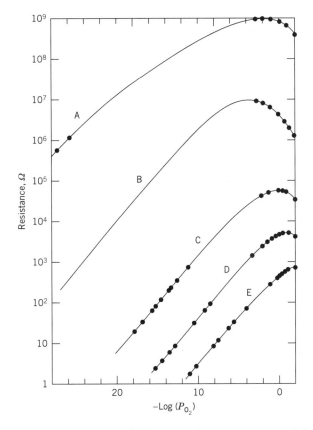

Fig. 4. Dependence of the resistance of TiO_2 ceramics on oxygen partial pressure in kPa. A, at 320°C; B, at 500°C; C, at 700°C; D at 850°C; and E, at 1000°C. To convert kPa to psi, multiply by 0.145 (36).

pressure, as shown in Figure 4 (36). This oxygen pressure dependence for many oxides is of the form: $\sigma = P_{O2}^{\pm\delta}$, where δ has values typically ranging from $\pm 1/8$ to 1/2. The sign of δ indicates p-type (+); or n-type (−) conduction.

Another method for obtaining semiconductors with controlled resistivities, which avoids the difficulties caused by stoichiometric deviations, is by solid solution of two or more compounds with widely different conductivities. Magnetite [1317-61-9], Fe_3O_4, is an excellent conductor having a specific resistivity of $\sim 10^{-4}$ Ω-cm at room temperature, compared to resistivities of $\sim 10^8$ Ω-cm for most stoichiometric transition-metal oxides. The high electrical conductivity of magnetite is a function of the random distribution of Fe^{2+} and Fe^{3+} ions on octahedral sites, allowing for easy electron transfer between cations. This is best illustrated by the order–disorder transformation occurring at ~ 120 K. Below this temperature the Fe^{2+} and Fe^{3+} ions are distributed in an ordered pattern on the octahedral sites. Above this temperature Fe^{2+} and Fe^{3+} positions become randomly distributed, resulting in a substantial increase in conductivity.

9. Spinels Ferrites

The spinel crystal structure (AB_2O_4) is based on the cubic close packing of oxygen ions in which the cations are situated on both the tetrahedral A and octahedral B sites (8,37). In spinel ferrites for magnetic applications, cations such Fe^{3+} or Mn^{2+} can occupy both A and B sites, giving an inverse [(AB)BO_4] structure and higher net magnetization. In the spinel lattice, the normal cations such as Ni, Co, Li, Mg, Zn, Cd, etc show a definite preference for either the tetrahedral or octahedral sites. The cation concentration and site preference essentially determines the properties of the ferrite. Ferrites need to have high electrical resistivities in order to eliminate eddy current and dielectric losses, as well as for allow full penetration of electromagnetic fields throughout the solid (38,39). High resistivity is obtained when a cation has only one valence state in the lattice. In the processing of these materials, therefore, high sintering temperatures and reducing atmospheres must be avoided since these conditions can produce variable valence states on some cations. The range of published resistivities for spinel and garnet [12178-41-5] ferrite materials is wide, from $\sim 10^{-4}$ to $10^9 \Omega$-cm at room temperature (38,39). The low conductivity is typically associated with the simultaneous presence of Fe^{2+} and Fe^{3+} ions on equivalent lattice sites. In general, a condition for appreciable conductivity in the ferrite structure is the presence of ions having multiple valence states on like crystallographic sites. Thus, the concentration of ions in Fe_3O_4 can be controlled by solid solution in which Fe^{2+} or Fe^{3+} are diluted by other ions that do not participate in the electron exchange.

There are three main commercial classes of ferrite spinels, namely: nickel–zinc ferrite, (NiZn)Fe_2O_4; manganese–zinc ferrite, (MnZn)Fe_2O_4; and, magnesium–manganese ferrites, (MnMg)Fe_2O_4. The electrical conduction is by small polaron mechanisms. Electrical resistivity primarily determines the utility of these materials in the high megahertz or microwave frequency ranges. Low frequency use requires a trade-off between high permeability and high resistivity. Nickel–zinc ferrites typically show an increase in permeability and a departure from stoichiometry in the iron-rich direction. Decreased resistivity results, therefore, when the formation of divalent iron becomes more probable, requiring close control of processing parameters is. Manganese–zinc and magnesium–manganese zinc ferrites are typically used in low frequency devices such as pulse transformers and memory-core devices. For higher frequency use, the high dc resistivity needed for full magnetic penetration and low eddy current losses can be obtained with an iron deficient oxide powder. However, a more complex processing situation arises because of the three possible valence states of manganese cation and the site preference for each. Sintering temperature and cooling rate can also impact the magnesium site locations. Increasing the amount of manganese increases the lattice constant, making it easier for the Mg ions to occupy both cation sites, such that divalent and trivalent ions are present on both sites, with adverse impact on the permeability, resistivity, and utility of the material (37,39).

Information regarding solid solutions of Fe_3O_4 and $MgCr_2O_4$, hercynite [1302-61-0], $FeAl_2O_4$, and ferrous chromite [1308-31-2], $FeCr_2O_4$, have been published (38). Semiconductor materials of this type, with controlled temperature coefficient of resistivity, have been prepared using materials such as $MgAl_2O_4$, $MgCr_2O_4$, and titanium zinc oxide [12036-69-0], Zn_2TiO_4, as the nonconducting component. Semiconductors made in this way are used as thermistors. Because the electrical conductivity the semiconductors made in this way have a negative temperature coefficient (NTC), these materials are typically utilized in temperature sensing and thermistors applications. Copper–nickel manganites, $Mn_{3-x-y}Ni_y Cu_xO_4$ spinels, are of technological importance for use in very low resistance thermistors (8). Through adjustment in stoichiometry, the electrical conductivity can increase to 0.1 $(\Omega = cm)^{-1}$ at room temperature. In nickel-manganites the conduction is primarily through electron hopping from Mn^{3+} to Mn^{4+} cations on the B sites. In the copper–nickel–manganites the situation is quite complex since there is the presence of Ni^{2+}, Mn^{2+}, Mn^{3+}, and Mn^{4+} ions as well, hence, there is uncertainty over distribution of copper cations and their ionic states.

10. Superconductivity

Oxide superconductors have been known since the 1960s. Compounds such as niobium oxide [12034-57-0], NbO, TiO, $SrTiO_{3-x}$ and AWO_3, where A is an alkali or alkaline earth cation, were found to be superconducting at 6 K or below. The highest T_c observed in oxides before 1986 was 13 K in the perovskite compound $BaPb_{1-x} Bi_xO_3$, for $x = 0.27$. Then in 1986 possible superconductivity at 35 K in the La–Ba–Cu–O compound was discovered (40). The compound was later determined to be $La_{1.85} Ba_{0.15} CuO_4$. Work on the Y–Ba–Cu–O system was published in 1987 with a reported a transition temperature above 90 K (41). This transition temperature was surpassed in early 1988 by the discovery of a series of bismuth-containing compounds, Bi–Sr–Ca–Cu–O, having the highest $T_c = 100$ K, and a $Tl_2 Ba_2 CaCu_2O_{8-y}$ compound having a $T_c = 120$ K (42,43) (see BISMUTH COMPOUNDS).

Superconductivity is partly typified by a perfect metallic conductor that has no resistance to current flow below a T_c. Besides the disappearance of electrical resistance, there is an expulsion of magnetic flux described as the Meissner effect. The widely acclaimed BCS theory explains traditional superconductivity based on the concept of Cooper pairs. The Cooper pairs are formed when electrons interact with phonons and attract each other, resulting in a combined energy state that is lower than the Fermi energy of the normal conduction electrons. These Cooper electron pairs move in such a way that at equilibrium the combined momentum is unchanged. Normal scattering effects, therefore, do not affect the forward momentum of the electrons accelerating in an electrical field. In spite of the success of the BCS theory for low temperature superconductors, the high temperature oxide superconductors do not conform well to the prediction of this model. Specifically, the BCS theory does not predict a $T_c >$ 30 K or the existence of a weak or no isotope effect. Any theory must consider the local charge inhomogeneities as well as the local electronic structure at distances of the order of the superconducting coherence length. Research studies aimed at understand the superconducting phenomenon in oxide systems and also increasing the transition temperature, have resulted in a large volume of literature on the subject.

A model by Emery and Kivelson (18) considers high temperature superconductors to be quasi-two dimensional doped insulators, obtained by chemically introducing charge carriers into a highly correlated antiferromagnetic insulating state. The model interprets the transition temperature as being inversely proportional to the spacing of one-dimensional arrays or stripes that are formed in the under-doped and optimally doped materials. In the under-doped insulating state, there exist charge inhomogeneities that are also antiphase domain walls for the background spins. The superconducting state is featured as a 2D array comprising rivers of charges flowing through the antiferromagnetic bulk, separating the background spins into loosely coupled regions (18). Another model proposed by Hirsch (19) states that only hole carriers in a solid can produce superconductivity, and that the negative charges will be expelled from the bulk and move to the surface, leaving the interior positively charged. In this theory the holes are expected to pair and have a smaller effective mass, becoming superfluid carriers. With this, there is an associated lowering of the kinetic energy that provides the superconducting condensation energy (19). Both the weak coupling and the strong coupling approaches described, can lead to the energy perturbation condition that triggers the superconducting state (17).

The $YBa_2Cu_3O_{7-x}$ compound is currently the most intensively investigated high temperature oxide superconductor. It has an oxygen deficient, distorted orthorhombic, 1:1:3 (ABO_3) perovskite-type structure, with the *Pmmm* space group (45). The crystal structure shows the Y and Ba ions located in the center of the unit cell, Cu ions on the corners, and O ions on the edges. Based on neutron diffraction studies, the structure shows two important features for the Cu ions: (*1*) nonplanar CuO_2 planes extend in the crystallographic *ab* planes at $z = 0.36$ and -0.36; and (*2*) fencelike, square planar CuO_3 linear chains extending along the *b* axis at $z = 0$. CuO_2 layers appear in all cuprate superconductors and seem to be a necessary but not sufficient condition for high temperature superconduction to occur (46). The $La_2SrCu_2O_{6.2}$ compound has CuO_2 layers but does not superconduct. Studies also indicate that T_c is proportional to the carrier density in the CuO_2 layer but not to the volume carrier density, which is further evidence that the $YBa_2Cu_3O_{7-x}$ is a 2D superconductor. In these materials, the electrical conductivity and critical fields are highly anisotropic. Figure 5 shows this for highly oriented films and the single crystal in two directions (47). Conductivity in the *ab* (CuO) planes is metallic, whereas along the *c* axis it is semimetallic. The conductivity can be greatly disturbed by defects such as vacancies and twin boundaries, in which the *a* and *b* axes are reversed, in particular by grain boundaries in the polycrystalline material. Because of the severe anisotropy, the critical current density is much higher in the plane than along the *c* axis.

The $YBa_2Cu_3O_{7-x}$ oxide superconductor is called an oxygen deficient perovskite because there are only seven oxygens per formula unit instead of the nine oxygens required for a perfect perovskite. The oxygen content in this compound is thus sensitive to processing and annealing conditions. When the *x* value is <0.5, an orthorhombic superconducting phase is present. As the compound loses oxygen and *x* becomes >0.5, an order–disorder phase transition occurs

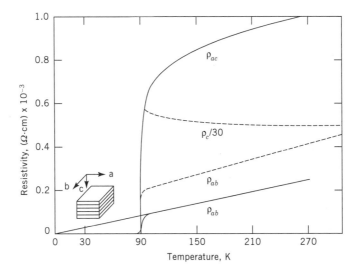

Fig. 5. Resistivity versus temperature showing onset of superconductivity for a single crystal (- - -) and for two highly oriented films (—), where ρ_{ab} denotes resistivity in the ab plane, ρ_{ac} in the ac plane, and $\rho_c/30$ represents 1/30 the resistivity in the c-direction to account for the much higher resistivity in this direction (47).

and the structure transforms to a tetragonal phase having semiconducting characteristics. The tetragonal phase has a $P4/mmm$ space group with oxygen disordered in the $z = 0$ planes, compared to the orthorhombic phase. There is a volume change of +0.2% associated with this phase transformation. Therefore, stresses are present resulting from the transformation and cause a twin structure formation, usually observed along the (110) reflection planes. The twin structure does not interrupt the continuity of the Cu–O superconducting planes, but does terminate the Cu–O chains between the Ba and Y planes.

The wide transition temperatures are found to be strongly dependent on oxygen stoichiometry. The high transition temperature at 92 K in this compound can only be obtained when the oxygen content is between 6.8 and 7.0. Although T_c generally decreases as the oxygen content is reduced, the exact relationship depends critically on the processing conditions. Applications for superconductors include magnetic resonance imaging for medical diagnostics (see IMAGING TECH-NOLOGY), particle accelerators for experimental physics, and superconducting quantum interference devices (SQUID).

11. Ferroelectrics

Barium titanate, $BaTiO_3$, has been a much studied ferroelectric, since it is the base material for large volume production of technologically important components such as disk and multilayer (MLC) capacitors; barrier layer (BL) and grain boundary barrier layer capacitors (GBBL); and, nonlinear PTCR. Barium titanate has a tetragonal structure over the range of ~5–125°C, with the dielectric constant rising to a peak near the critical transition temperature of ~125°C, defined as the Curie temperature, T_C. The ferroelectricity in $BaTiO_3$ and like (ABO_3) perovskite compounds arises from a dipolar shift in the B cations on cooling through the transition temperature. This results in a change in crystal symmetry from cubic to tetragonal and the appearance of spontaneous polarization charges (16), due to the development of a domain structure. Domains are unipolar regions within the crystal which, in the absence of an externally applied field, are randomly aligned so as to exhibit no net internal polarization. Above the Curie temperature, the material reverts to the nopolar state with disappearance of the spontaneous polarization (4,5,21,22).

$BaTiO_3$ has an intrinsic high resistivity of $\sim 10^{10}$ Ω-cm, when prepared in an oxidizing atmosphere. However, it can readily become semiconducting (48) through controlled doping of the A-site lattice using La^{3+}, Y^{3+}, or Nd^{3+}, or of the B-sites using Nb^{5+} or Ta^{5+} and like cations. A complex defect structure can result from the presence of different charge compensation mechanisms as the dopant level is increased. At low dopant concentration (0.1–0. 3 mol%), n-type conduction results, due to the formation of donor level Ti^{3+} ions with mobile electrons (48,49). At higher dopant levels, the mechanism changes from electronic to ionic compensation, due to cation vacancy formation, leading to a significant increase in resistivity (50,51). Low resistivities are obtained only within a very narrow dopant concentration range of $\sim$0.1–0.3 mol%. The formation of the Ti^{3+} ions in the doped $BaTiO_3$ lattice, leads to local distortion of the TiO_6 octahedra, causing a change in the polarization field and a lowering of the $Ti(3d)$ energy level associated with the Ti^{3+} state. Intermediate polaron conduction occurs due to the splitting of the $Ti(3d)$ orbitals (15,16). The mode of conduction results, therefore, from polarization field overlap of the localized Ti^{4+} and Ti^{3+} discrete energy levels, aided by the lattice distortion caused by the electron movement and change in Ti valence state.

Under controlled heat treatment, a thin, insulating grain boundary layer can be formed in the polycrystalline ceramic. In this grain boundary region oxygen is adsorbed during annealing, reducing the Ti^{3+} concentration (27). Oxygen vacancies may also diffuse from the interior to the grain boundary region where they act as electron traps. A space charge or barrier layer is thereby formed which repel electrons moving towards the grain boundary. Studies of this near-grain boundary region in doped and annealed $BaTiO_3$ samples show these to be regions of high strain, with defect segregation and a domain structure that is different from the interior (52,53). In contrast, the nonannealed materials feature only limited segregation, low strain and a near uniform domain structure. For the annealed material, a martensitic-type transformation or structure change within the narrow grain boundary region, coupled with the sudden release of stress, is considered to be the driving force for the abrupt change in grain boundary potential and resistivity that is observed (52,53). At temperatures below the critical ferroelectric phase transition, (+) spontaneous polarization charges can neutralize the (−) grain boundary charges in crystallographically coherent areas along the grain boundaries, creating thereby a low resistance pathways (52–55) and significantly lower ressistivity overall. In n-doped $BaTiO_3$, PTCR behavior is observed as a large increase in resistance (typically, several orders of magnitude) near the phase transition temperature over a narrow range of dopant concentration. Both the PTCR effect and room

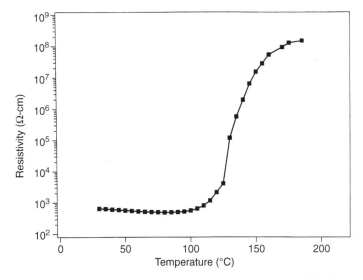

Fig. 6. PTCR resistivity–temperature characteristics for donor modified barium titanate.

temperature resistivities are highly dependent on dopant type and ionic radius. Figure 6 is a resistivity–temperature plot of a typical PTCR behavior, showing a low resistivity region and a sharp increase in resistivity near the ferroelectric phase transition. Materials with this type behavior are used for current limiting, temperature control, resistive heating, motor starters, and for sensor applications.

Barrier layer capacitors having high dielectric constants and low losses can be produced by increasing the grain boundary thickness through the use of additional dopants, such that the grain boundary barrier layer becomes impassable to electrons, by creating a large concentration of acceptor states that nullify the effects of the spontaneous polarization. This traps the conducting electrons by creating a space charge layer, which repels like charges. Several additional dopants, eg, Sr, Zr, Pb, Si, Cu, and Bi, to $BaTiO_3$-based systems are used in the manufacture of both capacitor and PTCR sensor devices, in order to adjust the switching temperature and other operating characteristics.

Perovskites with Pb on the A site are particularly important and show pronounced piezoelectric characteristics ($PbTiO_3$, PZT, PLZT). Different responses are found in $BaTiO_3$ and PZT to the addition of donor dopants such as La^{3+}. In PZT, lead monoxide [1317-36-8], PbO, which is partially lost by volatilization during sintering, can be replaced in the crystal by La_2O_3, where the excess positive charge of the La^{3+} is balanced by lead vacancies, giving ionic compensation and no generated electronic charge carriers (4). This results in a marked increase in resistivity of the material. Relaxors materials show a significant change in permittivity with frequency at temperatures near the Curie point, T_C. These materials also exhibit high permittivity values near T_C. Of particular interest for capacitors use are the lead-based relaxors having the general formula $Pb(B_1B_2)O_3$, where B_1 is typically a low valence cation (Mg, Zn, Fe, Ni, or Sc), and B_2 is a high valence cation (Ti, Nb, Ta, or W). The best known relaxor formulations are in the $Pb(Mg_{1/3}Nb_{2/3})O_3$ system (PMN), which shows excellent relaxor characteristics with a combined high dielectric permittivity and high resistivity. These lead-based materials (PZT, PLZT, PMN) form a class of ceramics with important dielectric, relaxor, piezoelectric, or electrooptic properties, and are thus used for actuator and sensor devices. Common problems associated with their use are low dielectric breakdown, increased aging and electrode injection, thereby decreasing the resistivity and degrading the dielectric properties (4).

12. Varistors

Varistors are primarily ZnO-based ceramic devices, which exhibit high nonlinear current–voltage behavior, which is ideal for use in protecting electronic equipment against voltage surges. To obtain the appropriate current–voltage characteristics, the polycrystalline ZnO is usually doped with antimony(III) oxide [1309-64-4], Sb_2O_3, bismuth(III) oxide [1332-64-5], Bi_2O_3, and like additives. The material is then processed so as to develop a microstructure consisting of conductive grains surrounded by a thin, resistive, Bi-rich second phase in the grain boundaries (56). This microstructure can be quite complex and is key to understanding the barrier layer conduction behavior in varistors. At low voltages the grain boundary resistance is sufficiently high that little current leaks from the circuit. At the higher breakdown voltages a tunneling process occurs, which allows the overvoltage pulse to be rapidly gated away from the device circuitry. This is because the varistor resistance to ground decreases several orders of magnitude (Fig. 7), similar to the PTCR devices previously described, although the actual operating mechanism is very different (56). Figure 7 shows the current density J versus applied field E for a typical varistor, in which the shape of the curve is largely defined by the slope **a**. A large variation in current density is seen for a comparatively small change in the field, which is a significant advantage for ZnO varistor use. Because the varistor action takes place across the ZnO

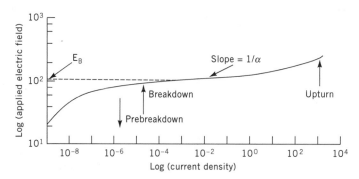

Fig. 7. Log–log plot of current density, J, versus applied electric field, E, for a ZnO varistor at room temperature, in which the breakdown field, E_B, is indicated. The exponent α equals the inverse slope of the curve, $\log(E/J) = 1/\alpha$, and is a measure of device nonlinearity. Units of current density and the electric field are A/cm^2 and V/cm, respectively (56).

grain boundaries, the tailoring of devices for specific breakdown voltages is done by fabricating the device with the appropriate number of grain boundaries in series between the electrodes. The current density /voltage behavior can be described by a power law:

$$I = kV^\alpha \ (1 \leq \alpha \leq \infty) \tag{20}$$

where I is the current through the device, V is the voltage, and the exponent α, which is a measure of device nonlinearity, is typically in the range 20–50. Ohmic-type behavior is usually found for low values of α (ie, for $\alpha \sim 1$), whereas varistor behavior is typically attained for values of $\alpha > 20$.

More recently, a different role has been ascribed to the grain boundary phase in current flow regulation. Here, the conductive grains are considered to be separated by a thin, adsorbed oxide layer that stores a negative charge equal in magnitude to the positive charge in the depletion layer, which lies wholly within each grain near the grain boundaries. These depletion regions are believed to create a barrier to electron flow at low voltages. Transport through the grain junction becomes then a compound step process. As the voltage increases, the depletion layer thins because there is a dramatic change in the conduction band shape and long-lived holes appear in the valence band. Electrons are able to tunnel through the interface because the tunneling barrier is now much thinner, and current flow becomes greatly enhanced (56).

In any event, it is clear that the state of the grain boundary region is the critical factor in determining device performance in both PTCR and varistor type devices. Depletion regions in the near grain boundary area are usually electonically active, since they act as barriers to charge flow. Thereby, it controls electronic and mixed conduction, as well as contributes to space charge polarization and high permittivity in these electroceramic systems. Understanding of the conduction phenomenon in these type materials is more complex than a simple description of depletion regions, however, since dopant effects, grain boundary and mechanical states also significantly affect conduction (6,57). The mechanical states can be described in terms of misorientation of adjacent grains and density of coinciding sites (sites that belong to the lattice of both grains). It affects, therefore, the coherency of the grain boundary and stress states. The combined mechanical and charge effects lead to depletion or accumulation effects for the mobile ionic and electronic charges. The grain boundary depletion or space charge layers are often described as back-to-back Schottky barriers, in which the width of the disruption region controls the properties (6,7,46,56,57). Since processing effects (including doping and heat treatment) have the potential to create and control grain boundary states, their impact on electrical properties can be substantial.

13. Processing Effects

Electrical ceramic properties are ultimately structure dependent, especially in those ceramic materials with variable valence cations. The processing of these materials, therefore, require close control of composition and heat treatment conditions. Microstructure and grain size control are also crucial since density changes and the presence of second phases can grossly affect properties. Thermal history during processing can affect properties in important ways as well. For example, in ferrite ceramics such as $MgFe_2O_4$, the distribution of cations between octahedral and tetrahedral sites determines magnetic behavior, but this site occupancy can be significantly altered by heat treatment and cooling rate. Rapid cooling from the sintering temperature, for instance, can cause development of a net magnetic moment in normally nonmagnetic Mg or even Zn ferrite.

Conducting ceramics based on nanocrystalline powders are of interest for use in electrochemical and catalytic systems and as photocatalysts. For many of these applications, the influence of density of grain boundaries on conductivity is not well understood, although it is known that grain size strongly influences conductivity in important material systems such as CeO_2, SnO_2, and ZrO_2 (59). The size dependence may be related to the development of space charge effects as the grain size approaches the Debye length (59). For example, the sensitivity of a porous SnO_2 gas sensor strongly increases when the grain size becomes comparable to the width of the space charge layer. In this state the entire grain can be considered depleted of electrons in the nonsensing condition. The dc conductivity in nanocrystalline TiO_2 is also enhanced relative to the coarse grain material. Nanocrystalline CeO_2 also shows a lower specific grain boundary resistivity than the coarse-grained material. The behavior, in this case, has been attributed to size dependent grain boundary impurity segregation. With increasing grain boundary area, due to smaller grains, the grain boundary impurity concentration decreases. This leads to a lower blocking effect on conduction electrons and ions and a correspondingly lower grain boundary resistivity (59). Grain boundary segregation and nonstoichiometry effects also impact electrical behavior in these nanostructured materials.

The effect of grain size on ionic and on mixed ionic-electronic conducting ceramics is also of interest, particularly for solid oxide fuel cells (SOFC) (57). Oxygen-ion conductors such as stabilized zirconia, ceria and lanthanum gallate, are promising systems for use as solid electrolytes. For long-term use and stability, the SOFCs need to operate at $\sim$750$°$C or below, but this requires a corresponding increase in electrolyte conductance and in gas–electrode reaction kinetics, parameters which typically decrease with operating temperature. The use of nanocrystalline materials can be advantageous for these type applications, if the ionic conduction along grain boundaries can be enhanced with the nanosized grains. This is because grain boundary diffusion is significantly greater than bulk diffusion, due to higher defect densities and mobilities in the disordered grain boundaries (57). Space charge effects may also develop, which repel like charges, leading to an accumulation of opposite charges along the grain boundary. If these opposite charges became mobile, ionic conduction along the grain boundary would be enhanced. Studies into enhanced ionic conductivity in nanosized crystalline materials have thus far been inconclusive, since enhanced electronic conductivity often masks changes in ionic conductivity, as, eg, in CeO_2. In designing electrical ceramics for particular applications, therefore, processing parameters must be carefully monitored and controlled.

BIBLIOGRAPHY

"Ceramics as Electrical Materials" in *ECT* 3rd ed., Vol. 5, pp. 290–314, by H. Kent Bowen, Massachusetts Institute of Technology; in *ECT* 4th ed., Vol. 5, pp. 698–728, by R. C. Buchanan and R. D. Roseman, University of Illinois, Urbana-Champaign; "Ceramics as Electrical Materials" in *ECT*, (online), posting date: December 4, 2000, by R. C. Buchanan and R. D. Roseman, University of Illinois, Urbana-Champaign.

CITED PUBLICATIONS

1. R. Waser, *J. Eur. Ceram. Soc.* **19**, 655 (1999).
2. R. E. Newnham, *Rep. Prog. Phys.* **52**, 123 (1989).
3. Y. Chiang, D. Birnie, and W. D. Kingery, *Physical Ceramics*, John Wiley & Sons, Inc., New York, 1997.
4. A. J. Moulson and J. M. Herbert, *Electroceramics*, Chapman and Hall, New York, 1990.
5. R. C. Buchanan, ed., *Ceramic Materials for Electronics*, 2nd ed., Marcel Dekker, Inc., New York, 1991.
6. R. Waser and R. Hagenbeck, *Acta Mater.* **48**, 797 (2000).
7. G. Garcia-Belmonte, J. Bisquert, and F. Fabregat-Santiago, *Solid State Electron.* **43**, 2123 (1999).
8. E. Elbadraoui, J. L. Baudour, F. Bouree, B. Gillot, S. Fritsch, and A. Rousset, *Solid State Ionics* **93**, 219 (1997).
9. G. Chiodelli, G. Flor, and M. Scagliotti, *Solid State Ionics* **91**, 109 (1996).
10. V. V. Kharton, A. P. Viskup, D. M. Bochkov, E. N. Naumovich, and O. P. Reut, *Solid State Ionics* **110**, 61 (1998).
11. H. Ullmann, N. Trofimenko, A. Naoumidis, and D. Stover, *J. Eur Ceram. Soc.* **19**, 791 (1999).
12. J. J. Sprague and H. L. Tuller, *J. Eur Ceram. Soc.* **19**, 803 (1999).
13. D. Basak and J. Ghose, *J. Solid State Chem.* **112**, 222 (1994).
14. G. Blaise, *J. Electrostat.* **50**, 69 (2001).
15. V. V. Paranjape and P. V. Panat, *J. Phys.: Condens. Matter.* **3**, 2319 (1991).
16. I. Bunget and M. Popescu, in C. Laird, ed., *Materials Science Monographs*, Vol. 19, Elsevier, Amsterdam, The Netherlands, 1978, Translation, V. Vasilescu, 1984.
17. J. R. Schrieffer, *Physica B*, 259–261 433 (1999).
18. V. J. Emery and S. A. Kivelson, *J. Phys./Chem. Solids* **61** 467 (2000).
19. J. E. Hirsch, *Phys. Lett. A* **281**, 44 (2001).
20. N. M. Tallen, ed., *Electrical Conductivity in Ceramics and Glasses*, Marcel Dekker, Inc., New York, 1974.
21. D. W. Richerson, *Modern Ceramic Engineering*, 2nd ed., Marcel Dekker, Inc., New York, 1992.
22. L. L. Hench and J. K. West, *Principles of Electronic Ceramics*, John Wiley & Sons, Inc., New York, 1990.
23. E. C. Subbarao, ed., *Solid Electrolytes*, Plenum Press, New York, 1980.
24. V. Z. Kresin and S. A. Wolf, eds., *Fundamentals of Superconductivity*, Plenum Press, New York, 1990.
25. R. Wernicke, in *Ceramic Monographs-Handbook of Ceramics*, Suppl. Interceram. Vol. 34(6), No. 3.1.5, 1985.
26. V. V. Kharton, A. P. Viskup, A. V. Kovalevsky, J. R. Jurado, E. N. Naumovich, A. A. Vecher, and J. R. Frade, *Solid State Ionics* **133**, 57 (2000).

27. D. C. Hill and H. L. Tuller, in R. C. Buchanan, ed., *Ceramic Materials for Electronics*, 2nd ed., Marcel Dekker, Inc., New York, 1991.
28. G. Fisher, *Am. Ceram. Soc. Bull.* **65** (4), 622 (1986).
29. H. L. Tuller, in Ref. 5.
30. R. C. Buchanan, in Ref. 5.
31. H. Cordes and S. D. Baranovskii, *Phys. Status. Solids B* **218**, 133 (2000).
32. C. H. Hsieh and H. Jain, *J. Non-Crystalline Solids* **183**, 1 (1995).
33. H. Kahnt, *J. Non-Crystalline Solids* **203**, 225 (1996).
34. M. F. Yan and A. H. Heuer, eds., in *Advances in Ceramics*, Vol. 7, American Ceramic Society, Columbus, Ohio, 1982.
35. W. Williams, *Inter. J. Refract. Metals Hard Mater.* **17**, 21 (1999).
36. E. M. Logothetis, *Ceram. Eng. Proc.* **1**, 281 (1980).
37. F. N. Bradley, in A. E. Javitz, ed., *Materials for Magnetic Functions*, Hayden Book Co., Inc., New York, 1971.
38. E. J. Verway, P. W. Haagman, and F. C. Romeijn, *J. Chem. Phys.* **15**, 18 (1947).
39. K. J. Standley, *Oxide Magnetic Materials*, 2nd ed., Clarendon Press, Oxford, UK, 1972.
40. J. G. Bednorz and K. A. Muller, *Z. Phys. B-Condensed Matter* **64**, 189 (1986).
41. M. K. Wu and co-workers, *Phys. Rev. Lett.* **58**(9), 908 (1987).
42. H. Maeda, Y. Tanaka, M. Fukutomi, and T. Asano, *Jpn. J. Appl. Phys.* **27**(2), L209 (1988).
43. Z. Z. Sheng and A. M. Hermann, *Nature. (London)* **332**, 138 (1988).
44. J. Bardeen, L. H. Cooper, and J. R. Schrieffer, *Phys. Rev.* **106**, 162 (1957).
45. M. A. Beno and co-workers, *Appl. Phys. Lett.* **51**(1), 57 (1987).
46. K. Mukae and A. Tanaka, *Ceramics Inter.* **26**, 645 (2000).
47. B. Kulwicki, in L. Levinson, ed., *Grain Boundary Phenomena in Electronic Ceramics, Advances in Ceramics.*, Vol. 1, 1981, pp. 138–154.
48. W. Heywang, *J. Mater. Sci.* **6**, 1214 (1971).
49. G. H. Jonker, *Solid State Electron..* **1**, 895 (1964).
50. K. Takada, E. Chang, and D. M. Smyth, in J. B. Blum and W. R. Cannon, eds., *Multilayer Ceramic Devices, Advances in Ceramics.*, Vol. 19, 1987 pp. 147–51.
51. J. Kim, R. D. Roseman, R. C. Buchanan, and *Ferroelectrics* **177**, 255 (1996).
52. R. D. Roseman, and J. Kim, R. C. Buchanan, *Ferroelectrics* **177**, 273 (1996).
53. R. D. Roseman, *Ferroelectrics* **215**, 31 (1998).
54. R. D. Roseman and G. Liu, *Ferroelectrics* **221**, 181 (1999).
55. L. M. Levinson and H. R. Philipp, in Ref. 12.
56. H. Tuller, *Solid State Ionics* **131**, 143 (2000).
57. T. H. Geballe and J. K. Hulm, *Science* **239**, 367 (Jan. 22, 1988).
58. M. Takemoto, T. Miyajima, K. Takayanagi, T. Ogawa, H. Ikawa, and T. Omata, *Solid State Ionics* **108**, 255 (1998).
59. C. Demetry and X. Shi, *Solid State Ionics* **118**, 271 (1999).

RELVA C. BUCHANAN
RODNEY D. ROSEMAN
University of Cincinnati

CERAMICS, MECHANICAL PROPERTIES

1. Introduction

Structural ceramics are used in applications such as gas turbines, advanced heat engines, semiconductor processing equipment, armor, thermal barrier coatings, medical implants, as thin films for wear and electronic applications, heat exchangers, aerospace and weapons components, in high temperature solid oxide fuel cells, and as bearings components. Advantages of ceramics over metals include dimensional stability; low densities, which translate into weight savings and increased fuel efficiencies; high temperature capabilities; and corrosion resistance. Although metals are ductile, and thus have greater damage tolerance than ceramics, the use of metals is limited to much lower temperatures. For example, the melting points of aluminum and 304 stainless steel are 660°C and from 1400–1450°C, respectively, whereas those of alumina, Al_2O_3, and stabilized zirconia are 2020°C and from 2500–2600°C, respectively.

The overriding concern with regard to the mechanical performance of ceramics is their brittleness and, hence, sensitivity to flaws. There is usually little or no warning that failure is imminent because deformation strain prior to failure is usually <0.1%. As a result, a primary thrust of structural ceramics research has been the development of tougher and stronger ceramics. Ceramics are now routinely available that have toughness values of 7–10 MPa·m$^{1/2}$ and strengths that exceed 1000 MPa (1.5×10^5psi) (1) (see also ADVANCED CERAMICS, STRUCTURAL CERAMICS). These values compare to toughness values of 120–153 MPa·m$^{1/2}$ and strengths of 1380–1790 MPa for structural metals such as AF1410 high strength steel (2).

The mechanical properties of ceramics are sensitive to the starting materials, forming processes, heat treatment conditions, and surface preparation. These properties are particularly sensitive to the microstructure, and vary with it in a complex manner. A unique aspect is the dependence of some of the mechanical properties on extremes rather than averages. For example, consider a tensile rod containing a significant volume of pores. In a metal, the remnant cross-sectional area would determine the load-bearing capacity. However, in a ceramic, the stress concentration from a single, large pore could determine the failure stress.

A much more detailed treatment of the mechanical properties of ceramics is available (3).

2. Properties and Behavior

2.1. Elastic Behavior. Elastic deformation is defined as the reversible deformation that occurs when a load is applied. Most ceramics deform in a linear-elastic fashion, ie, the amount of reversible deformation is a linear function of the applied stress up to a certain stress level. If the applied stress is increased any further, the ceramic fractures catastrophically. This is in contrast

Table 1. **Properties of Ceramics and Other Materials**

Material	Young's modulus, GPa[a]	Poisson's ratio
soft rubber	0.007–0.07	0.49
nylon	2.8	0.40
ice	9	0.2–0.88[b]
concrete	14	0.20
aluminum	70	0.34
plate glass	70	0.27
copper	110	0.34
ZrO_2, partially stabilized	205	0.23
Al_2O_3	380	0.26
SiC	207–483	0.14
diamond	1050	0.20

[a] To convert GPa to psi, multiply by 145×10^3.
[b] Values for the Poisson's ratio of ice are highly dependent on composition, structure, and strain rate (4).

to most metals, which initially respond elastically, and then begin to deform plastically. This plastic deformation allows stresses at stress concentrators to be dissipated rather than building to the point where bonds break irreversibly.

Elastic behavior is commonly quantified by the Young's modulus, E, the proportionality constant between the applied tensile stress σ and the tensile strain ε (Δ length/original length).

$$\sigma = E\varepsilon \tag{1}$$

Materials having high elastic moduli deform less for a given stress. Typical E values for several material categories, metals, ceramics, and polymers, are shown in Table 1. As ceramics have mostly covalent and ionic bonding, the bond strengths and consequently the elastic modulus values are high.

Crystals are often anisotropic because bond strengths and density are a function of direction. In single crystals this leads to anisotropy in the elastic moduli such that the strain depends on the stress application direction. Most polycrystalline materials are macroscopically isotropic because the anisotropies of randomly oriented, individual grains (single crystals) average to zero over all the grains. Glasses are elastically isotropic because of a random network structure.

Another commonly used elastic constant is the Poisson's ratio ν, which relates the lateral contraction to longitudinal extension in uniaxial tension. Typical Poisson's ratios are also given in Table 1. Other elastic moduli include the shear modulus G, which describes the amount of strain induced by a shear stress, and the bulk modulus K, which is a proportionality constant between hydrostatic pressure and the negative of the volume change ($-\Delta V/V$).

Porosity has a significant effect on elastic moduli. Empirical relations of the form

$$E = E_o e^{-bP} \tag{2}$$

where E_o = Young's modulus of dense material, P = relative volume of porosity, and b = constant, generally describe the elastic behavior of ceramics. Thermal expansion anisotropy between the various grains comprising a polycrystalline ceramic can result in tensile stresses high enough to produce intergranular cracks, with sizes on the order of the grain size. These cracks, called microcracks, lead to a decrease in the elastic modulus. The grain size of a ceramic has no effect on the modulus unless the material is very anisotropic and contains large grains that microcrack spontaneously.

Young's modulus can be determined by measuring the stress–strain response (static modulus), by measuring the resonant frequency of the body (resonant modulus), or by measuring the velocity of sound through the material (sonic modulus). The values of modulus obtained by static methods are less accurate (within 5–10%) than values using sonic techniques (<1% error).

2.2. Strength. *Measured Strength versus Theoretical Strength.* Strengths of ceramic materials depend both on the types of flaws present and the method of strength measurement. The elastic modulus describes how easily atoms in a solid can be moved together or apart for small deformations (higher values imply the lattice is stiffer). The shape of the interatomic potential is such that beyond a certain spacing, the atomic attraction is insufficient to hold the atoms together. If deformation of the material is continued beyond this point, entire planes of atoms separate and the ceramic fractures.

The theoretical tensile strength of the material, σ_{theor}, has been approximated by

$$\sigma_{\text{theor}} = \left(\frac{E\gamma}{a}\right)^{1/2} \approx \frac{E}{10} \tag{3}$$

where a = equilibrium spacing between planes of atoms, γ = fracture surface energy, and E = Young's modulus (5). If all bonds in a material were stressed equally up to the point of failure, the strength of a ceramic would be the theoretical strength. Large discrepancies between the theoretical and measured tensile strengths of ceramics result from the presence of imperfections. These imperfections or flaws can raise the local stress to the point that bonds in the immediate vicinity of the flaw can fail a few at a time, as opposed to every atom in the plane failing simultaneously. Regardless of the shape or orientation of a flaw, the energy required to break bonds in a given material is generally constant.

A flaw such as a simple spherical pore concentrates the stress on the bonds in the vicinity of the pore by a factor of 2 over the applied stress (6); however, most ceramics contain imperfections that enhance the stress to a much greater degree, leading to severe strength reductions. A typical ceramic such as alumina is as much as one hundred times weaker than the theoretical strength.

Stress concentration, such that the crack tip stress exceeds a material's theoretical strength, is a necessary condition for fracture; however, an energy requirement must also be satisfied. It has been recognized that crack growth could only occur if the total energy of the system was lowered by this extension. The energy requirements for crack growth have been postulated to be equal to the energy required to create two new surfaces. Thus

$$\sigma_{\text{fracture}} = \left(\frac{2E\gamma}{\pi c} \right)^{1/2} \tag{4}$$

where $c =$ crack dimension. The details of this formulation along with extensive information about brittle fracture are available (7).

A more practical approach for quantifying the conditions required for fracture uses a stress intensity criterion instead of an energy criterion. Using linear elastic theory, it has been shown that under an applied stress, σ_{applied}, when the stress intensity, K,

$$K = \sigma_{\text{applied}} Y(c)^{1/2} \tag{5}$$

where $Y =$ flaw and loading geometry factor (8), reached a critical value known as the fracture toughness, K_C, fracture would occur. The applied stress required to cause fracture, σ_{fracture}, can therefore be written as

$$\sigma_{\text{fracture}} = \frac{K_C}{Y(c)^{1/2}} \tag{6}$$

A very large proportion of failures in practical applications of ceramics results from applied tensile stresses. As a consequence, many published tables of properties of ceramic materials contain only tensile strength values. Ceramic strength is typically measured as a bending (flexural) strength because of the difficulties of gripping ceramic samples and achieving pure tension in tensile tests. In flexural tests, the reported strength is the value of the maximum stress at the tensile surface at failure. Guidelines for conducting strength tests are available (9). Measured strengths of ceramics in flexure and compression are shown in Table 2.

Flaws.   Pristine, undamaged ceramics (glass fibers and ceramic whiskers) exhibit strengths close to theoretical levels. Strengths of most ceramics, however, seldom approach theoretical levels because of processing flaws and damage introduced by handling, and during service. Examples of flaws commonly found in ceramics are shown in Figure 1.

Although each ceramic has a characteristic theoretical strength, the fracture toughness K_{IC} and the most severe flaw, ie, the flaw that produces the largest stress concentration, normally determine the actual strength. The larger the flaw, the greater the reduction in strength, and flaws in close proximity may act as a single, more severe flaw.

Microstructural effects on strength, such as those from porosity, impurities, grain size, and surface condition, are often difficult to identify because frequently several effects occur simultaneously. A common assumption for the relationship between strength and grain size is that flaw size scales directly with grain size. Hence the strength and grain size, G, are related in a similar manner to strength and flaw size (eq. 6). The strength is

$$\sigma \propto \frac{1}{G^{1/2}} \tag{7}$$

Table 2. **Measured Strengths of Common Ceramics**

| Material | Tensile strength | | | Compressive strength | |
	Measurement technique[a]	Value, MPa[b]	Reference	Value, MPa[b]	Reference
soda–lime silica (SLS) glass	ROR	79	10		
chemically strengthened SLS glass	ROR	293	10		
Al_2O_3	FP	370	11	4480	12
SiC	FP	345	13	3680	14
Si_3N_4	FP	1100	1	3450	12
SiAlON	FP	450	15		
SiC reinforced Al_2O_3 composite[c]	FP	640	16		
Y-TZP[d]	TT	745	17		
	FP	1630	17		
ZrO_2, CaO stabilized				2000	12
Al_2O_3/ZrO_2 composite	TP	2400	1		

[a] ROR = ring – on – ring bending; FP = four – point bending; TT = tensile test; and TP = three–point bending.
[b] To convert MPa to psi, multiply by 145.
[c] 30 vol% SiC whisker reinforced.
[d] 2 mol% yttria stabilized tetragonal zirconia polycrystal.

In some cases, it is not the average grain size that determines the strength. For example when exaggerated grain growth results in the formation of a few very large grains in an otherwise fine-grain material, one of those large grains can cause failure. That failure can occur at a stress much lower than expected for the fine-grain structure.

The relation between strength and impurities depends on the location and form of the impurities, and the failure mechanism. If impurities are present as discrete second phases, they can serve as obstacles to crack propagation; however, they can also act as stress concentrators, or as weak grain boundary fracture paths. An indirect effect of impurities on strength relates to their effect on the microstructure, eg, the grain size and shape, and porosity.

Statistical Variation in Strength. The wide variety of flaw types and sizes in ceramics produces the large (typically ±25%) variability in strength that has been one of the principal hurdles to the incorporation of ceramics in structural applications (18). This value compares unfavorably with the few percent for variability of the yield stress of a metal. The failure probability of a ceramic body at a given load depends on the probability of a flaw of a critical size being present in a location where it produces a stress concentration. Some of the strength variability reported in the literature can be attributed to inconsistencies in testing procedures; however, tests performed in accordance with recommended standards can give accurate and consistent results (9,11).

Many distribution functions can be applied to strength data of ceramics. The function that has been most widely applied is the Weibull function, which is based on the concept of failure at the weakest link in a body under simple tension. A normal distribution is inappropriate for ceramic strengths because

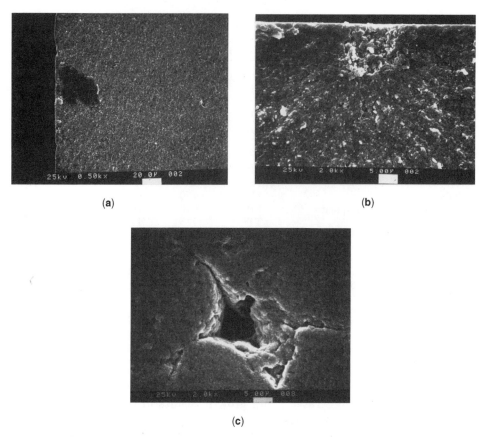

Fig. 1. Strength-reducing flaws commonly found in ceramics: (**a**) large grain surrounded by fine-grain matrix; (**b**) porous region resulting from incomplete densification; and (**c**) pore located where hard powder agglomerates did not sinter together.

extreme values of the flaw distribution, not the central tendency of the flaw distribution, determine the strength. One implication of Weibull statistics is that large bodies are weaker than small bodies because the number of flaws a body contains is proportional to its volume, and the likelihood of finding a large flaw increases with the volume.

The Weibull distribution function expresses the failure probability P_f as a function of the applied stress σ, and three Weibull parameters σ_μ, σ_o, and m

$$P_f = 1 - \exp\left[-\int_V \left(\frac{\sigma - \sigma_\mu}{\sigma_o}\right)^m dV\right] \tag{8}$$

where σ_μ = the threshold stress below which failure does not occur, ie, where $P_f = 0$; σ_o = normalizing strength for a unit volume of material loaded in tension; V = specimen volume, and m = measure of the spread in strength values and is called the Weibull modulus. The higher the m value, the tighter the strength distribution, ie, the lower the variability. For conservative design, σ_μ is set to zero, reducing the Weibull distribution to a two parameter description.

Under conditions other than tensile loading, the stress distribution in a body is nonuniform. To account for this, a loading factor k is used to calculate the effective volume under stress and kV replaces V. This analysis assumes that flaws are randomly distributed throughout the volume. In many ceramics, glass in particular, nearly all failures occur from surface flaws. If that is the case the volume term is replaced with an area term.

To obtain the Weibull parameters, the strengths of $N > 30$ samples are measured and ranked from lowest to highest. The failure probability is then calculated using an estimator:

$$P_f = \frac{n}{N+1} \tag{9}$$

where $N =$ number of specimens and $n =$ rank of sample. The data are plotted as $\ln \ln[1/(1-P_f)]$ versus $\ln \sigma$ according to equation 10, which is derived from equation 8.

$$\ln \ln\left(\frac{1}{1-P_f}\right) = \ln V - m \ln \sigma_o + m \ln \sigma \tag{10}$$

The terms V (or kV) and σ_o are usually grouped together as a single term σ_o' called the characteristic strength, which is specific to the test specimen geometry and volume. The parameters m and σ_o' can be obtained from the slope and intercept, respectively. Typical m values for ceramics range between 5 and 20 (18); 5 indicates a high variability in the strength and 20 a relatively low strength variability. Detailed information on the standard procedure for determining and reporting Weibull parameters is available (19).

Compressive Strength. Ceramics are much stronger in compression than in tension and are frequently used in applications where they bear compressive loads. Under excessive compressive loads ceramics fail in a brittle manner just as they do in tension; however, the measured compressive stresses are typically eight times greater than the measured tensile stresses for the same material (12). In compression, the failure process can begin with microplastic deformation, not the growth of preexisting flaws. Measured compressive strengths are shown in Table 2. Although there is limited data relating microstructure to compressive strength, compressive strength often shows an inverse relationship to the square root of the grain size. Porosity effects on compressive strength have been modeled using the same type of relation used for Young's modulus (eq. 2).

2.3. Fracture Toughness. The fracture criterion was defined by a critical value of the crack tip stress intensity, known as the fracture toughness, K_C. Ceramics often fail in pure tension, designated the mode I stress intensity, and K_{IC} replaces K_C in equation 6. Thus σ_{fracture}, the applied tensile stress at which fracture occurs, is a function of the flaw size and K_{IC}. A crack propagates catastrophically when the stress intensity factor at the crack tip reaches the critical value required for bond breakage, and when its rate of change with respect to the crack length is positive. Thus the critical stress intensity, K_{IC}, required for bond rupture is considered to be a material constant.

Under some circumstances the crack tip stress intensity factor is different than what the far-field stresses would indicate because of microstructural effects behind the crack tip. These include effects due to the presence of crack bridging fibers, whiskers, and grains. Often far-field values indicate the crack is propagating at a stress intensity value higher than K_{IC}, and this apparent value usually increases as crack length increases. Despite indications to the contrary, bonds continue to break at the same value of the stress intensity; however, the crack tip is being shielded from some of the applied stress intensity. To minimize confusion about K_{IC}, it has been suggested that the far-field value of the stress intensity be called $K_{applied}$. When there are no microstructural features that effectively reduce the crack tip stress intensity, such as in glasses, the crack tip stress intensity is accurately represented by the far-field values, and $K_{applied}$ equals K_{IC}. In this case, equation 6 can be used to predict what size flaw a material can tolerate at a given stress level.

K_{IC} measurements can be made by introducing a crack of a known size and a specific geometry in a body and then loading the body until catastrophic failure occurs. The shape factor Y, used to account for different flaw and loading geometries, can be found in reference handbooks (8). An alternative approach for determining K_{IC} consists of measuring the length of surface cracks introduced using a sharp indenter (20). The low fracture toughness, in units of $MPa \cdot m^{1/2}$, of ceramics is demonstrated by values of 0.75 for glass, 2.7–4.2 for alumina, 7–10 for Si_3N_4, and 8–9 for partially stabilized ZrO_2, as compared to 11–13 for common woods perpendicular to grains, 6–20 for cast iron, 46 for tool steel, 120–153 for high strength steel, and 100–350 for pure ductile metals, eg, Cu, Ni, and Ag (21).

R-Curve Behavior.   Ceramic toughening efforts have focused on the property of some ceramics to exhibit increased apparent fracture toughness as cracks grow. This increase is seen in terms of the far-field value of the stress intensity required to propagate the crack. An important consequence of this effect, which is commonly referred to as R- or T-curve behavior, is that the material has increased damage tolerance because there is a crack size regime in which the strength is independent of the crack size. The underlying basis of R-curve behavior is that the crack tip stress is redistributed, either to immediately adjoining material, as in the case of the process zone formed in transformation toughened materials, or to regions far removed from the crack tip, as in the case of fiber-reinforced ceramics. One reason for the interest in crack bridging mechanisms is that these toughening mechanisms should operate over a broad range of temperatures. Characteristics of R-curve mechanisms are that they are activated only as the crack advances, the number of activated shielding elements increases as the crack grows, and the measured fracture toughness saturates when the generation rate of shielding elements equals the rate at which elements become inactive.

One implication of R-curve behavior is that the measured fracture toughness is not a material constant equal to K_{IC}; rather it is a function of the crack length. The material can still be considered to have a constant toughness. However, there are microstructural influences that affect the effective stress intensity K at the crack tip. R-curves depend on crack size, loading conditions, and component geometry (22). Microstructural effects influencing R-curve behavior include grain size, shape, and orientation, grain boundary toughness, and thermal expansion anisotropy. In alumina, where R-curve behavior occurs because of ligamentary bridges formed by grains behind the crack tip, grain size determines the scale of grain pullout (23). At small grain sizes the bridging effect appears to be insignificant.

Transformation Toughening.   Transformation toughened materials exhibit enhanced toughness because of a process zone at the crack tip that consumes energy that would otherwise be used in the creation of fracture surface. Various processes at the crack tip have been postulated to consume energy such as the phase transformation from tetragonal to monoclinic zirconia, microcracking, and deviation of the primary crack around the transformed particles. The zone can also be thought of as a region that partially shields the crack tip from the far-field stresses. Significant toughening cannot occur until the process zone has grown sufficiently to extend behind an advancing crack tip as shown in Figure 2. A comprehensive review of transformation toughening of ceramics is provided (24).

The transformation toughening mechanism has been most successfully exploited in ZrO_2-based materials, where the phase transformation of interest is from tetragonal to monoclinic ZrO_2. Microcracking may also play a role in the transformation toughening of ZrO_2 materials because of the extra energy required to produce additional fracture surface. Toughness enhancements of almost 100% (from 5.2 to 9.5 MPa·m$^{1/2}$) have been achieved in ceramics such as Al_2O_3 by adding ZrO_2 as a second phase (24).

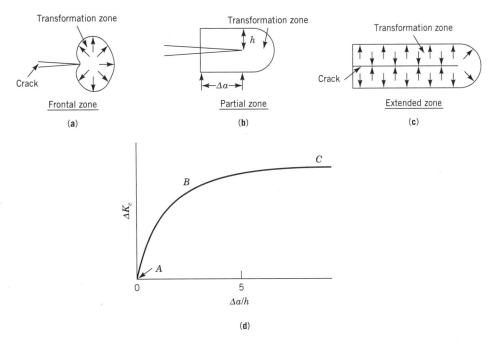

Fig. 2.   A schematic of a zirconia body containing a propagating crack: (**a**) the frontal zone of the transformed material; (**b**) the partial zone defining Δa and h; (**c**) the extended and the transformation and extended zones; and (**d**) change in fracture toughness as a function of growth of the transformation zone, where A represents the frontal zone; B, the partial zone; and C, the extended zone.

Oxides such as CaO, MgO, and Y_2O_3 are added to ZrO_2 to stabilize the tetragonal phase at temperatures below the tetragonal to monoclinic phase-transition temperature. Without stabilizer, the phase transition occurs spontaneously at temperatures below 850–1000°C, and no fracture toughness enhancement can occur (25).

Toughening Mechanisms in Composite Ceramics.   Significant toughening has been achieved by fabricating whisker- and fiber-reinforced ceramic composites, and metal–ceramic composites (see COMPOSITE MATERIALS, SURVEY). Toughening primarily results from crack bridging and/or crack deflection, both of which reduce the crack tip stress intensity. Crack deflection, which occurs when the reinforcement debonds from the matrix, changes the orientation of the crack relative to the stress application direction, changing the stress distribution around the crack such that the stress intensity at the crack tip is reduced. When a reinforcement phase bridges the crack behind the crack tip, it supports some of the applied load, and therefore shields the crack tip from some of the applied stress intensity. Crack tip shielding results in R-curve behavior and the ability of these materials to resist catastrophic failure as shown in Figure 3.

Crack deflection toughening resulting from debonding along interfaces between the matrix and reinforcement phase has been covered extensively in the literature (26–28). Crack bridging, which plays the largest role in enhancing the toughness, depends on intact whiskers or fibers being able to pull out of the matrix behind the crack tip. A very large amount of energy can be consumed in this process that would otherwise be used to propagate the crack. The stress–strain behavior of a fiber-reinforced composite in Figure 3 shows that during the initial stages of loading, the composite exhibits the same behavior as an unreinforced matrix, ie, it is linear elastic. At point A the matrix begins to crack and the slope of the stress–strain curve begins to decrease. At point B

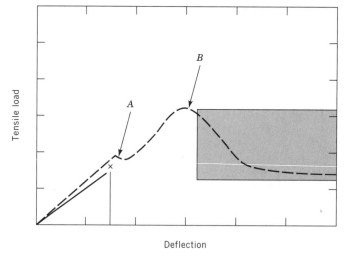

Fig. 3.   Tensile stress–strain curve for (——) unreinforced ceramic and (- - - -) fiber-reinforced ceramic composite. A represents the point where the matrix begins to crack; B, where the reinforcement begins to fail and the section where reinforcement pulls out of the matrix. X indicates the point of catastrophic failure.

the fibers begin to fail and the load the body can support diminishes. The area under the stress–strain curve, past the point where the matrix begins to crack, represents the energy absorbed during fiber debonding and pull-out. Many composites exhibit very large strains to failure and retain some load-bearing capability even after the integrity of the matrix has been lost.

Whiskers or fibers must be able to debond and pull out of the matrix in order for significant toughening to occur. Thus the parameters controlling the toughness of a composite are the interfacial strength and the interfacial frictional or pullout stress, τ, which has to be overcome before the reinforcement slides out of the matrix. If the interfacial strength is too high, the fibers break instead of debonding, and the value of the frictional stress is irrelevant. Methods for controlling the interfacial strength include choosing reinforcement and matrix materials that do not react, and coating the fiber with nonreactive or low strength porous coatings (29). Low τ values produce the greatest toughening with respect to fiber pullout.

Composites have also been made where strengthening is the goal. A high pullout stress, τ, is one of the important parameters for higher strength because high values allow load transfer from the matrix to the high strength fibers. The opposing effects of the pullout stress, τ, on the toughness and strength, highlight the importance of the design of the reinforcement-matrix interface in the overall design of composites. The role of the interface is covered (30).

Metal–ceramic composites, including those made by *in situ* oxidation of infiltrated molten metal (31), have high toughness values that result from residual ductile metal that bridges the crack behind the crack tip (32). Because of the large amount of plastic work required to cause these elements to fail, there is an increase in the amount of energy, or applied stress intensity, required to propagate a crack through the composite. Composites of this type exhibit fracture toughness values as great as three times the matrix toughness (33).

Ferroelasticity. Ferroelastic materials contain domains that can be switched by an applied stress (34) in a manner analogous to magnetic domain switching in ferromagnetic materials. A hysteresis loop between the applied field or stress and the induced state, and a permanent induced state when the applied field is removed, are characteristic of domain switching for both of these properties. The permanent state of a ferroelastic is the permanent strain induced by the applied stress. Many ceramics that exhibit ferroelasticity also exhibit ferromagnetism and/or ferroelectricity, such as $BaTiO_3$ and lead zirconate titanate (PZT). Toughening occurs because energy is absorbed in the switching of the domains.

Tetragonal zirconia is a structural ceramic that exhibits ferroelasticity and the toughness enhancement has been estimated to be as high as $5 \ MPa \cdot m^{1/2}$. An example of a partial hysteresis loop for this material is shown in Figure 4 (35). Domains do not have to be present prior to the stress application because stresses can also nucleate domains (36). Domain nucleation and additional fracture surface generated by fracture along domain boundaries appear to contribute to the toughening.

2.4. Plasticity. Although even at elevated temperatures [$>0.6 \ T_m$ (melting temperature)] mechanical failure of ceramics is dominated by brittle fracture, plastic deformation mechanisms often precede brittle fracture. Plastic

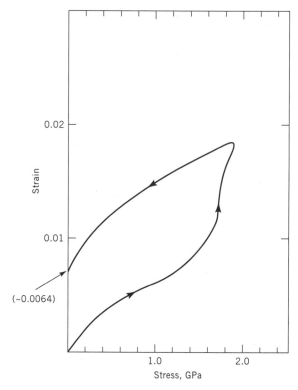

Fig. 4. Partial hysteresis loop for a ferroelastic material. After the applied stress is removed a permanent strain $\varepsilon \sim 0.0064$ remains (see eq. 1).

deformation is also important because of the role it plays in net shape forming operations such as extrusion, which require extremely high strain rates in a deformation regime known as superplasticity (37). In the lower range of the elevated temperature mentioned above, plastic deformation of crystalline ceramics can occur by slip. In a slightly higher temperature range ($>0.7\ T_m$), plastic deformation can also occur by grain boundary sliding and softening of secondary phases such as glass.

Plastic deformation by slip involves one atom plane sliding past another, or twinning by homogeneous shearing. Once slip occurs there is permanent deformation of the material. As the stresses at which slip is observed are significantly lower than those required to slide perfect planes of atoms past each other, defects that allow this motion to occur must be present. The primary mechanism of slip at relatively low stresses in polycrystalline materials is the motion or glide of line defects called dislocations. Dislocation motion allows bonds to break at relatively low applied stresses because dislocation motion occurs by sequential bond breakage. This is somewhat analogous to fracture. The primary difference between dislocation motion and fracture is that bonds immediately reform after a dislocation has passed, whereas they remain ruptured after a crack front passes.

Slip usually occurs on the planes that have the highest atom density and are the greatest distance from adjacent planes. Therefore the slip systems that can be activated in a crystal depend primarily on crystal structure, and those crystals with the highest symmetry, eg, cubic, have the most available slip systems. Slip systems and the temperatures at which they are activated are available (18).

If a ceramic single crystal is suitably oriented with respect to the applied stress direction, deformation can occur through dislocation motion. The rate of plastic deformation depends on the number of dislocations and their velocity. Although suitably oriented ceramic single crystals such as MgO deform plastically at low temperatures by slip, polycrystalline forms of the same material act in a brittle manner because of geometrical constraints. Plastic deformation in ceramics by slip is also suppressed by the large amount of energy required to move dislocations, especially for covalently bonded ceramics such as SiC and Si_3N_4. As temperature increases, plastic deformation by slip becomes easier because of the increased amount of energy available for dislocation glide. Other plasticity mechanisms also become activated at elevated temperature but despite the increased plasticity at these temperatures, failure ultimately occurs in a brittle manner when cracks nucleate at grain boundaries. The total strain prior to fracture in a tensile test conducted at elevated temperature is usually <1%.

The processes leading up to failure at elevated temperatures are known as creep. Figure 5 shows a typical creep curve divided into four regions. The first region, often ignored, represents the instantaneous deformation that occurs when a load is applied. The second region usually shows decreasing creep rate and is known as primary or transient creep. The third region, known as steady-state or secondary creep, is the most important for lifetime predictions. In tensile creep tests fracture often occurs in this region. The fourth region is known as tertiary creep and in this region the deformation rate accelerates just prior to complete failure. Fracture occurs as various types of creep damage accumulate. Damage includes loss of load bearing area from the formation of pores or cavities,

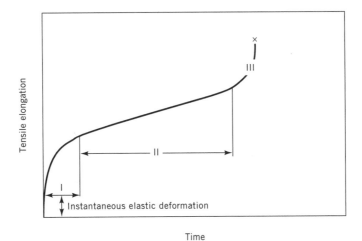

Fig. 5. Tensile elongation vs time demonstrating creep behavior of ceramics. Section I is primary creep; II, secondary or steady-state creep; III, tertiary creep; and X, fracture.

linking of the cavities to form cracks, and environmental degradation of the microstructure. Increased temperature or stress increases the creep rate and reduces the time to failure.

Polycrystalline ceramics exhibit diffusional creep, in which deformation occurs by diffusional flow of atoms to and from grain boundaries, either by diffusion through the lattice (Nabarro-Herring creep), or along the grain boundaries (Coble creep) (38,39). At temperatures close to the melting point, creep may also occur by dislocation motion, and this process is termed lattice creep . Creep can also occur by grain boundary sliding in materials with and without liquid grain boundary phases. Diffusional creep and creep by grain boundary sliding are categorized as boundary creep mechanisms. Detailed information on creep mechanisms can be found in the literature (38,40).

In the steady-state creep regime of ceramics, almost all creep mechanisms fit a strain rate dependence of the form (18):

$$\dot{\varepsilon} = \frac{A^*DGb}{kT}\left(\frac{b}{l_g}\right)^{m^*}\left(\frac{\sigma}{G}\right)^{n^*} \tag{11}$$

where $\dot{\varepsilon}$ = strain rate; A^* = dimensionless constant; D = diffusion coefficient (exponential temperature dependence); G = shear modulus; b = Burgers vector; k = Boltzmann's constant; T = temperature; l_g = grain size; σ = applied stress; n^* = stress exponent, which commonly has values of 1–5; and m^* = grain size exponent, which commonly has values of 0–3. Analysis of creep data using equation 11 usually allows a specific creep mechanism to be identified, based on the stress and grain size exponents. Once the creep mechanism has been identified the deformation kinetics can be determined and lifetime predictions can be made. Creep in some ceramics is sufficiently well understood over a broad range of conditions that the deformation behavior and theoretically predicted behavior can be displayed on maps known as Ashby plots (41).

Equation 11 shows that a larger grain size favors increased creep resistance. The potential detrimental effects of second phases are an important consideration for ceramics that require dopants as sintering aids. Impurities affect the diffusion coefficient and may end up as grain boundary phases, which are often glassy in nature.

Plastic deformation is commonly measured by measuring the strain as a function of time at a constant load and temperature. Deformation strain can be measured under many possible loading configurations. Because of problems associated with the preparation and gripping of tensile specimens, plastic deformation data are often collected using bend and compression tests.

2.5. Hardness. Although large-scale deformation in ceramics usually occurs only at elevated temperature, localized, constrained compressive loading can produce plastic deformation even at room temperature. Hardness (H) is a measure of the resistance of a material to deformation, in particular the resistance to plastic deformation during surface penetration. Hardness is related to the bond strength. Because covalent and multivalent ionic bonds are strong and highly directional in nature, slip is very difficult in these cases, and ceramics containing these bonds are generally the hardest materials (42). The ratio of the distance, a, between planes of atoms and the spacing, b, of the atoms in the plane also plays an important role in how easily planes of atoms slide past each other. Materials with small a/b ratios tend to possess a high resistance to slip, and are consequently harder (5). In contrast, glasses, which have strong covalent bonds and do not have any dislocation activity, are softer than crystalline ceramics. This is partially due to their low bond densities compared to polycrystalline ceramics. In some glasses, a significant portion of the deformation under compressive loading results from irreversible compaction of the glass, rather than from plastic flow.

Hardness is determined by measuring the penetration (depth or area) when a harder material, such as diamond, is pushed into the surface of the material of interest under a specified load. True hardness is defined as the force divided by the projected area. Vickers hardness tests, which employ a pyramid-shaped indenter, are frequently used to characterize ceramics; however, Vickers hardness calculations normally employ total surface area rather than projected area (43). Measurements are made on the diamond impression shown in Figure 6. Vickers hardness, H_v, is calculated using

$$H_{V, \text{totalarea}} = \frac{0.46\,P}{a^2} \tag{12}$$

where P = indentation load and a = half-length of Vickers impression diagonal. Hardness is normally expressed in units of GPa but Vickers hardness numbers are also expressed in units of kg/mm^2. Many other hardness scales are used; however, conversions between scales are not always possible (43). Hardness values of some ceramics are compared to common metals in the following table.

Material	Hardness, GPa
lead	0.07
copper	0.86
steel	2.23
cast iron	1–7
glass	6–7
zirconia (partially stabilized)	10–11
silicon nitride	8–19
alumina	18–23
silicon carbide	20–30
diamond	98

Hardness decreases with increasing porosity. Ceramics deform plastically more readily at higher temperatures and therefore hardness decreases with increasing temperature according to

$$H = H_o \left(1 - \frac{T}{T_o}\right) \tag{13}$$

where H_0 = hardness at 0 K and T_0 = temperature in Kelvin at which hardness goes to zero (44). Further information on hardness of ceramics is available (45,46).

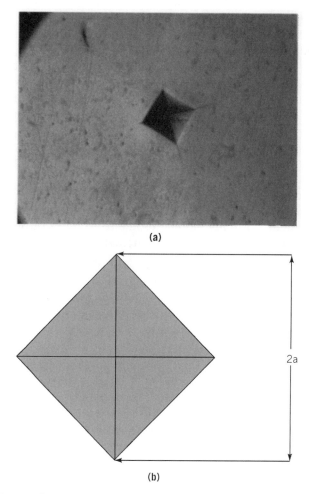

Fig. 6. (**a**) Vickers indentation in zirconia at 200 X. (**b**) Schematic of a Vickers indentation showing the dimension used in Vickers hardness determinations.

2.6. Subcritical Crack Growth. At low and modest temperatures, under certain environmental conditions, the strength of many ceramics, especially glasses, can degrade with time under mechanical loading. This phenomenon, referred to as static fatigue or delayed failure, is the result of subcritical crack growth (SCG). A preexisting crack grows slowly at an applied stress intensity lower than that necessary, ie, the critical value K_{IC}, to propagate a crack without environmental influences. SCG is pernicious because a flaw grows slowly at stresses far below the expected failure load until the flaw is large enough to satisfy the Griffith criterion, at which point failure occurs catastrophically.

The mechanism of subcritical crack growth is the reaction of the corrosive medium with highly stressed bonds at the crack tip. As such the mechanism is often referred to as stress corrosion cracking. In silica, in the absence of stressed bonds, the rate of the reaction between the bonds and corrosive media such as water is very low. The introduction of strain energy into crack tip bonds increases the activity of the bond. For silica glass in water, attack and bond breakage occurs by the following reaction (47):

$$H—O—H + \ —Si—O—Si— \ \longrightarrow \ —Si—OH + HO—Si—$$

(water)   (ionic/covalent bond   (2 silanol groups)
in glass)

in which weak bonds between silanol groups replace strong silicon–oxygen bonds. Similar reactions occur with ammonia, methanol, and other liquids (48).

The plot of crack velocity v, as a function of applied K_I, is usually divided into three regions as shown in Figure 7. In region I, v exhibits a power law dependence on K_I. The stress intensity determines the bond reactivity and therefore the kinetics of the chemical reaction at the crack tip and the rate of bond breakage. An increase in the reactive species concentration, eg, water, and increased temperature shift region I to lower K_I. Region II is determined by the rate at which the corrosive chemical species travels to the crack tip, and is insensitive to K_I. The Region II plateau velocity depends on the environment. For example, increasing relative humidity raises the plateau. In Region III, v is a strong function of K_I and believed to be independent of environment. Crack velocity in regions I and III can be approximated by an empirical relationship of the form

$$v = AK_I^n \tag{14}$$

where A = material constant and n = subcritical crack growth susceptibility constant. The relationship in equation 14 is most useful for region I because this

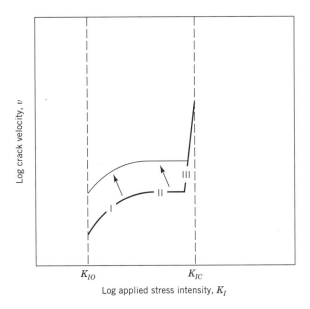

Fig. 7. Crack velocity, v, as a function of the applied stress intensity, K_I. Water and other corrosive species reduce the K_I required to propagate a crack at a given velocity. Increasing concentrations of reactant species shift the curve upward. Regions I, II, and III are discussed in text.

represents the majority of a ceramic component's lifetime. At K_{IC}, bonds break without environmental assistance and the crack becomes critical. There is also a threshold K_I, known as the stress corrosion limit or the fatigue limit, K_{IO}, below which crack growth ceases; however, K_{IO} is usually a practical limit dictated by how low a velocity can be measured, for example, 1×10^{-11}m/s. Information of the type presented in Figure 7 can be used to determine what applied K_I should be used to ensure that no crack propagation occurs, or to predict component lifetime. Materials with low n values are most susceptible to SCG. Values of n range from 15 for soda lime silica glass to 35 for silica glass, and from 90–100 for oxides such as MgO-stabilized ZrO_2.

2.7. Impact and Erosion.   Impact involves the rapid application of a load to a relatively small area. Two types of stress can arise during the impact; a localized stress at or near the impact point, and/or a macroscopic stress. Much of the kinetic energy from the impacting object may be transformed into strain energy for crack propagation. If the impact is from a blunt indenter, a crack called a Hertzian conoid usually forms from excessive tensile stresses around the point of contact. If, after the Hertzian conoid is formed, the impacting object still possesses a significant amount of kinetic energy, further damage may occur in the form of radial cracks and circumferential cracks. Sharp indenter impact can also produce Hertzian cone cracks as shown in Figure 8. If the impact load is relatively large and sustained, a macroscopic stress may be imposed on the target body causing it to bend. This may lead to excessive tensile stresses on the opposite face of the body, crack initiation, and catastrophic failure. Failure can also occur if erosion reduces the cross section and load-bearing capacity of the component, causes a loss of dimensional tolerance, or causes the loss of a protective coating. Detailed information on impact and erosion is available (49).

External factors affecting a ceramic's response to impact include the velocity, size, and shape of the impacting object, and its angle of incidence. Impacting particles do not have to be hard or tough materials to cause damage and erosion. For instance, water at high velocities, eg, rain on aircraft windshields, can have an extremely erosive effect because of the high localized stresses that it produces. Another mechanism of failure for liquid impact is the generation of shock waves that interact with preexisting flaws producing crack growth. Material and microstructural characteristics that influence a ceramic's impact resistance include density, elastic modulus, crystalline anisotropy, hardness, fracture toughness, strength, grain size, defects such as pores, and the presence of second phases. There are limited data quantifying the relationship between impact strength and microstructural parameters.

Impact resistance is determined using flyer plate impact tests, long rod impact tests, Hopkinson bar tests (50) , and the liquid jet technique (51). Impact damage resistance is often quantified by measuring the post-impact strength of the ceramic.

2.8. Tribiological Behavior.   Tribological performance of ceramics, which includes friction, adhesion, wear, and lubricated behavior of two solid materials in contact, has been reviewed (52). This topic is receiving increasing attention because of applications of ceramics such as bearings, gears, and seals, and because of ceramic coatings being evaluated for micro-electromechanical systems (MEMS).

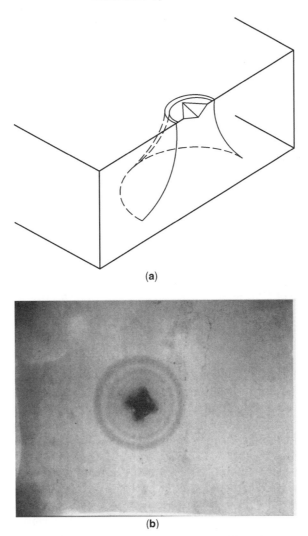

Fig. 8. (a) Schematic of a Vickers indentation-induced Hertzian cone crack. (b) View from the bottom of an aluminosilicate glass block of a Vickers indentation-induced Hertzian cone crack.

Friction and Adhesion. The coefficient of friction μ is the constant of proportionality between the normal force P between two materials in contact and the perpendicular force F required to move one of the materials relative to the other. Macroscopic friction occurs from the contact of asperities on opposing surfaces as they slide past each other. On the atomic level friction occurs from the formation of bonds between adjacent atoms as they slide past one another. Friction coefficients are usually measured using a sliding pin on a disk arrangement. Friction coefficients for ceramic fibers in a matrix have been measured using fiber pushout tests (53). For various material combinations (21):

Materials	μ
clean metals in air	0.8–2
steel on ceramics	0.1–0.5
ceramic on ceramic	0.05–0.5
high temperature lubricants (MoS_2, graphite)	0.05–0.2

Factors that affect μ include loading geometry, microstructure, crystal orientation, surface chemistry, environment, temperature, and the presence of lubricants.

Wear.   Ceramics generally exhibit excellent wear properties. Wear occurs by two mechanisms: adhesive wear and abrasive wear (21). Adhesive wear occurs when interfacial adhesion produces a localized K_I when the body on one side of the interface is moved relative to the other. If the strength of either of the materials is lower than the interfacial shear strength, fracture occurs. Lubricants (see LUBRICANTS AND LUBRICATION) minimize adhesion between adjacent surfaces by providing an interlayer that shears easily. Abrasive wear occurs when one material is softer than the other. Particles originating in the harder material are introduced into the interface between the two materials and plow into and remove material from the softer material (52). Hard particles from extrinsic sources can also cause abrasive wear, and wear may occur in both of the materials depending on the hardness of the particle.

Generally the harder the ceramic, the better its wear resistance; however, other properties such as fracture toughness may play the dominant role. If a ceramic is mated with a metal hardness is the determining factor, but when a ceramic is mated with another ceramic fracture toughness appears to determine the wear rate (54).

2.9. Thermal Stresses and Thermal Shock.   Thermal stresses arise when a body is heated or cooled and constrained from expanding or contracting. Thermal stresses can lead to fracture and catastrophic failure when the magnitude of the thermal stress exceeds the strength of the ceramic. Factors that contribute to the generation of large thermal stresses and the failure of ceramics under these stresses are low thermal conductivities, which produce large temperature gradients, and the lack of a stress relief mechanism such as plastic deformation. Approaches used to minimize thermal stresses include matching the expansion of ceramics with the expansions of the materials to which they are joined (55), minimizing temperature gradients, minimizing cross-sectional thickness changes to ensure that the body heats or cools uniformly, keeping the body at its operating temperature, and heating and cooling slowly.

Residual thermal stresses can occur in ceramics, especially in glasses, when they are cooled from elevated temperatures, and faster cooling of one region freezes in a structure that subsequently is unable to contract as much as another. This produces a situation wherein the regions that initially cooled more quickly are under compression and the slower cooling regions are under tension. Because tensile stresses are normally undesirable, annealing procedures are used to eliminate the residual stresses. Beneficial residual compressive surface stresses, which effectively strengthen the glass, are produced by a rapid cooling procedure known as thermal tempering.

When heating or cooling is extremely rapid, such as when a body is removed from a furnace and immersed in ice water, large thermal gradients produce very high, transient stresses. Rapid temperature changes, known as thermal shock, can lead to immediate failure of the body, or a degradation in the strength of the body resulting from the generation and propagation of cracks, or the growth of preexisting flaws. Thermal shock resistance (TSR) is the resistance to thermal shock-induced strength changes. TSR is usually quantified in terms of the temperature change, ΔT, below which no strength degradation occurs.

TSR depends on the conditions of thermal shock, the material, and the intended application of the material. TSR parameters are broadly divided into two groups, those based on conditions under which crack nucleation is favored, and those under which crack propagation is favored. For the former situation high TSRs are found for materials with low E and α, and high σ and thermal conductivity. For conditions favoring crack propagation, high TSRs are found for materials with high E and fracture surface energy, γ, and low σ. A compendium of TSRs is available (56). An example of a commonly used TSR parameter is

$$\Delta T = \frac{\sigma\,(1-\nu)}{\alpha E} \tag{15}$$

where σ = strength, ν = Poisson's ratio, α = coefficient of thermal expansion, and E = Young's modulus. TSR values, calculated using equation 15 and typical values of σ, ν, α, and E are shown in Table 3.

TSR is commonly measured by heating ceramics to various temperatures and then quenching them in a liquid medium. The critical temperature difference ΔT_{crit}, which causes severe damage, is used as a measure of the TSR. A common approach for quantifying the damage from thermal shock is to compare the strength of quenched and unquenched specimens. Another test utilizes thin circular disks heated rapidly with tungsten halogen lamps (57).

2.10. Cyclic Fatigue. Cyclic fatigue is the weakening and subsequent failure of a material during cyclic loading, often at stress levels significantly lower than those required to cause failure under static loading. Cyclic stresses

Table 3. **Thermal Shock Resistance Parameters**

Material	Strength, MPa[a]	Poisson's ratio	Coefficient of thermal expansion, $\times 10,^{-6}\,°C^{-1}$	E, GPa[a]	TSR parameter, °C
alumina	345	0.22	7.4	380	96
pyrex	70	0.20	4.6	70	170
silicon carbide	414	0.17	3.8	400	230
Y-TZP[b]	745	0.23	10.5	220	248
silicon nitride	310	0.24	2.4	172	650
LAS[c]	138	0.27	−0.3	70	4860

[a] To convert MPa to psi, multiply by 145.
[b] Y − TZP = yttria − stabilized tetragonal zirconia polycrystal.
[c] LAS = lithium aluminosilicate.

can be produced by repeated heating and cooling, by vibrations, and in applications in which the component is repeatedly loaded and unloaded. Ceramics were not recognized to exhibit cyclic fatigue (58) until the late 1980s when it was shown that cyclic fatigue occurs in ceramics under compressive loading (59). In ceramics that show crack bridging, cyclic fatigue is largely a result of the loss or destruction of ligamentary bridges when the crack closes. Some of the difficulties in identifying cyclic fatigue mechanisms in ceramics are related to the difficulties in obtaining reliable data using conventional cyclic fatigue testing, in which the failure stress is determined vs. the number of cycles to failure. An alternative approach for ceramic cyclic fatigue testing is based on the repeated indentation of a polished surface until chipping occurs (60).

3. Fracture Analysis

Fracture analysis, also known as fractography, plays an important role in understanding the relationships between the microstructure and mechanical properties, and the conditions that lead to failure (see FRACTURE MECHANICS). Systematic examination and interpretation of fracture markings and the crack path can often be used to reconstruct the sequence of events and stresses that led to failure. Fractography also plays an important role in the design and development of ceramic components because it helps differentiate whether failure occurred because the material was weakened by the introduction of processing and handling flaws, or because the applied stress exceeded the design stress.

Well-established fractographic techniques are available for determining crack propagation direction, failure origin location, estimating the failure stress, identifying what types of flaws are present, and for identifying local events that initiated failure (42,61,62). A significant amount of information about crack growth and failure can be determined using the fracture surface markings and the crack path. Some of the most useful fracture markings include crack branching patterns; twist hackle; Wallner lines; and arrest lines. At crack velocities approaching the terminal velocity, the fracture mirror, mist, and hackle, so called because these three terms aptly describe their appearance, are generated and can be seen on the fracture surface. These features are readily apparent for glasses as seen on the fracture surface in Figure 9. Fracture surface markings are influenced by, and therefore can provide information about the stress magnitude and orientation, crack velocity, interactions of the crack with microstructural inhomogeneities and stress pulses, flaw size and shape, and the test environment. The crack path provides information about the stress state at different positions in the body. The crack pattern can be a good indication of the conditions that gave rise to a certain stress state, such as thermal shock, impact, and twisting.

Fracture markings can be used to locate the failure origin, which is the discontinuity or flaw that caused the applied stress to be amplified locally. Once the failure origin has been located, the failure stress can be estimated using the flaw size and equation 6, or the distances to the boundaries of the mirror, mist, and hackle (whichever is most evident) and the following relation (63)

$$\sigma_{\text{fracture}} \sqrt{r_i} = A_i \qquad (16)$$

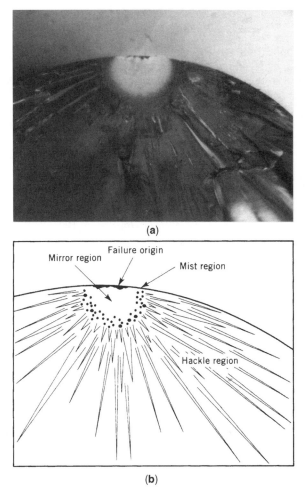

Fig. 9. (a) Optical micrograph of the fracture surface of a glass rod at 38 X. (b) Characteristic fracture markings such as mirror, mist, and hackle and the failure origin are indicated on the fracture surface schematic.

where r = radius; A = constant [values for various ceramics can be found in the literature (63)] and; $i = 1$, 2, or 3 refers to the mirror-mist, mist-hackle, and hackle-branching boundaries. If the ceramic body is small and the failure stress is extremely low, the mirror may extend over the entire fracture surface, and mist and hackle are absent.

A standard is available that describes an efficient and consistent methodology for conducting fractography (64). Simple instruments such as a pocket magnifying eyepiece or an optical microscope are often sufficient for obtaining critical information. As with any detective work, it is important to maintain careful records and to pay close attention to details in the reconstructions of the conditions under which fabrication and failure occurred. Seemingly unimportant details of fabrication, service, and/or the conditions under which failure occurred can frequently be the key to determining the cause of failure.

BIBLIOGRAPHY

"Properties and Applications of Ceramic Materials" under "Ceramics" in *ECT* 2nd ed., Vol. 4, pp. 793–832, by H. Palmour III, North Carolina State of The University of North Carolina, in *ECT* 3rd ed., Vol. 5, pp. 267–290, by T. M. Hare, North Carolina State University; "Mechanical Properties and Behavior" under "Ceramics", in *ECT* 4th ed., Vol. 5, pp. 634–658, by Jill Glass, Sandia National Laboratories; "Ceramics, Mechanical Properties and Behavior," in *ECT* (online), posting date: December 4, 2000, by Jill Glass, Sandia National Laboratories.

CITED PUBLICATIONS

1. F. L. Riley, *J. Am. Ceram. Soc.* **83**(2), 245 (2000).
2. J. Wolf, ed., *Aerospace Structural Metals Handbook*. Battelle Metals and Ceramics Information Center, Columbus, Ohio, 1988.
3. D. J. Green, *An Introduction to the Mechanical Properties of Ceramics*, Cambridge University Press, Cambridge, 1998.
4. N. K. Sinha, *J. Mater. Sci.* **21**(5), 1533 (1986).
5. A. H. Cottrell, *The Mechanical Properties of Matter*, Robert E. Krieger Publishing Co., Huntington, New York, 1981.
6. O. Vardar, I. Finnie, D. R. Biswas, and R. M. Fulrath, *Int. J. Fract.* **13**(2), 215 (1977).
7. B. R. Lawn, *Fracture of Brittle Solids*, 2nd ed., Cambridge University Press, Cambridge, U.K., 1993.
8. H. Tada, P. Paris, and G. Irwin, *The Stress Analysis of Cracks Handbook*, Del Research Corp., St. Louis, Mo., 1973.
9. ASTM C1161-02b, Standard Test Method for Flexural Strength of Advanced Ceramics at Ambient Temperature.
10. D. Connolly, A. C. Stockton, and T. C. O'Sullivan, *J. Am. Ceram. Soc.* **72**(5), 859 (1989).
11. G. Quinn, *J. Am. Ceram. Soc.* **73**(8), 2374 (1990).
12. R. W. Rice, in W. W. Kriegel and H. Palmour, III, eds., *Materials Science Research*, Vol. 5, Plenum Press, New York, 1971, pp. 195–229.
13. H. Kim and A. J. Moorhead, *J. Am. Ceram. Soc.* **73**(3), 694 (1990).
14. M. Srinivasan, in J. B. Wachtman, Jr., ed., *Treatise on Materials Science and Technology*, Vol. 29, Academic Press, Inc., New York, 1989, pp. 99–159.
15. K. Breder, T. Anderson, and K. Scholin, *J. Am. Ceram. Soc.* **73**(7), 2128 (1990).
16. J. Homeny and W. L. Vaughn, *J. Am. Ceram. Soc.* **73**(2), 394 (1990).
17. K. Noguchi, M. Fujita, T. Masaki, and M. Mizushina, *J. Am. Ceram. Soc.* **72**(7), 1305 (1989).
18. R. W. Davidge, *Mechanical Behaviour of Ceramics*, Cambridge University Press, Oxford, U.K., 1986.
19. ASTM C1239-00, Reporting Uniaxial Strength Data and Estimating Weibull Distribution Parameters for Advanced Ceramics.
20. G. R. Anstis, P. Chantikul, B. R. Lawn, and D. B. Marshall, *J. Am. Ceram. Soc.* **64**(9), 533 (1981).
21. M. F. Ashby and D. R. H. Jones, *Engineering Materials, An Introduction to their Properties and Applications*, Pergamon Press, Elmsford, N.Y., 1980.
22. Y.-M. Mai and B. R. Lawn, *Ann. Rev. Mater. Sci.* **16**, 415 (1986).
23. T. Fett and D. Munz, *J. Mater. Sci. Lett.* **9**, 1403 (1990).
24. D. J. Green, R. H. J. Hannink, and M. V. Swain, *Transformation Toughening of Ceramics*, CRC Press, Inc., Boca Raton, Fla., 1989.

25. E. C. Subbarao, in A. H. Heuer and L. W. Hobbs, eds., *Science and Technology of Zirconia, Advances in Ceramics*, Vol. 3, The American Ceramic Society, Inc., Columbus, Ohio, 1981.
26. K. T. Faber and A. G. Evans, *Acta Metall.* **31**(4), 565 (1983).
27. *Ibid.*, pp. 577–578.
28. H. Liu, K. L. Weisskopf, and G. Petzow, *J. Am. Ceram. Soc.* **72**(4), 559 (1989).
29. L. M. Sheppard, *Ceram. Bull.* **71**(4), 617 (1992).
30. R. J. Kerans, R. S. Hay, N. J. Pagano, and T. A. Parthasarathy, *Ceram. Bull.* **68**(2) (1989).
31. M. S. Newkirk, *Ceram. Eng. Sci. Proc.* **8**(7–8), 879 (1987).
32. F. F. Lange, B. V. Velamakanni, and A. G. Evans, *J. Am. Ceram. Soc.* **73**(2), 388 (1990).
33. B. D. Flinn, F. W. Zok, F. F. Lange, and A. G. Evans, *Mat. Sci. Eng.* **A144**, 143 (1991).
34. G. V. Srinivasan, J.-F. Jue, S.-Y. Kuo, and A. V. Virkar, *J. Am. Ceram. Soc.* **72**(11), 2098 (1989).
35. A. V. Virkar and R. L. K. Matsumoto, *J. Am. Ceram. Soc.* **69**(10), C-224 (1986).
36. J. F. Jue and A. V. Virkar, *J. Am. Ceram. Soc.* **73**(12), 3650 (1990).
37. I.-W. Chen and L. A. Xue, *J. Am. Ceram. Soc.* **73**, 2585 (1990).
38. J. Poirier, *Creep of Crystals*, Cambridge University Press, New York, 1985.
39. W. D. Kingery, H. K. Bowen, and D. R. Uhlmann, *Introduction to Ceramics*, John Wiley & Sons, Inc., New York, 1976.
40. W. R. Cannon and T. G. Langdon, *J. Mater. Sci.* **18**, 1 (1983).
41. H. J. Frost and M. F. Ashby, *Deformation Mechanism Maps*, Pergamon Press, New York, 1982.
42. D. W. Richerson, *Modern Ceramic Engineering: Properties, Processing and Use in Design*, 2nd ed., Marcel Dekker, Inc., New York, 1992.
43. ASTM E384-99 Standard Test Method for Microhardness of Materials; ASTM C1327-99 Standard Test Method for Vickers Indentation Hardness of Advanced Ceramics.
44. C. P. Alpert, H. M. Chan, S. J. Bennison, and B. R. Lawn, *J. Am. Ceram. Soc.* **71**(8), C-371 (1988).
45. D. Sherman and D. Brandon, in R. Riedel, ed., *Handbook of Ceramic Hard Materials*, Wiley-VCH, 2000, pp. 66–103.
46. B. R. Lawn and R. J. Blau, eds., *Microindentation Techniques in Materials Science and Engineering*, International Metallographic Society, ASTM STP 889, 1986.
47. K. M. Liang, R. Torrecillas, G. Orange, and G. Fantozzi, *J. Mater. Sci. Lett.* **25**, 5077 (1990).
48. T. A. Michalske and B. C. Bunker, *Sci. Am.* **255**(12), 122 (1987).
49. C. Preece, ed., *Treatise on Materials Science and Technology*, Vol. 16, Erosion, Academic Press, Inc., New York, 1979.
50. A. M. Rajendran and W. H. Cook, *Joint report between the University of Dayton Research and the Air Force Armament Laboratory*, AFATL-TR-88-143 SBI-AD-E801 843, Dec. 1988, 99 pp.
51. R. J. Hand and J. E. Field, *Eng. Fract. Mech.* **37**(2), 293 (1990).
52. D. H. Buckley and K. Miyoshi, in Ref. 14, pp. 293–365.
53. D. B. Marshall and W. C. Oliver, *J. Am. Ceram. Soc.* **70**(8), 542 (1987).
54. T. E. Fischer, M. P. Anderson, and S. Jahanmir, *J. Am. Ceram. Soc.* **72**(2), 252 (1989).
55. E. K. Beauchamp and S. N. Burchett, in S. J. Schneider, Jr., ed., *Engineered Materials Handbook*, Vol. 4, *Ceramics and Glasses*, ASM International, 1991, pp. 532–541.
56. D. P. H. Hasselman, *Bull. Am. Ceram. Soc.* **49**(12), 1033 (1970).
57. G. A. Schneider and G. Petzow, *J. Am. Ceram. Soc.* **74**(1), 98 (1991).
58. J. Tsai, C. Yu, and D. K. Shetty, *J. Am. Ceram. Soc.* **73**(10), 2992 (1990).

59. L. Ewart and S. Suresh, *J. Mater. Sci.* **22**(4), 1173 (1987).

60. M. J. Reece and F. Guiu, *J. Am. Ceram. Soc.* **73**(4), 1004 (1990).

61. V. D. Fréchette, *Failure Analysis of Brittle Materials, Advances in Ceramics*, Vol. 28, The American Ceramic Society, Inc., Westerville, Ohio, 1990.

62. V. D. Fréchette and J. R. Varner, eds., *Advances in Ceramics*, Vol. 22, The American Ceramic Society, Inc., Westerville, Ohio, 1988.

63. J. J. Mecholsky, Jr., S. W. Freiman, and R. W. Rice, *J. Mater. Sci.* **11**, 1310 (1976).

64. ASTM C1322-02, Standard Practice for Fractography and Characterization of Fracture Origins in Advanced Ceramics.

S. Jill Glass
Sandia National Laboratories

Rajan Tandon
Sandia National Laboratories

CERAMICS, PROCESSING

1. Introduction

Processing is key to the reproducible manufacture of ceramics. The tolerance of a finished ceramic to defects determines the raw materials selected, and the control that must be exercised during processing. More expensive advanced ceramics require higher quality, more expensive raw materials coupled with more carefully controlled manufacturing processes.

Ceramic processing is complicated both by the number of steps required in manufacture (Fig. 1), and by requirements to optimize the processing in the different steps. These factors are often opposed. For example, a fine particle size provides improved plasticity for forming and a higher thermodynamic driving force for sintering; however, electrostatic attraction and van der Waals forces promote agglomeration, ie, the formation of weakly bound particle clusters, and caking, making mixing and packing difficult. Submicrometer particles can be more easily mixed and packed in liquids, but the finer interparticle pore structure results in higher forming and drying stresses, as well as longer forming and drying times.

Ceramics are basically flaw intolerant materials. Consequently chemical and physical defects can severely degrade properties. Additionally, mistakes are cumulative in ceramic processing and these generally cannot be corrected during sintering and post-sintering processing (as they can be in metals processing, for example). The quality of the product is only as good as the quality of the raw materials used, and the control exercised in each of the process steps.

2. Raw Materials

Raw materials for ceramic processing range from relatively impure clay materials (see Clays) mined from natural mineral deposits, to ultrahigh purity powders

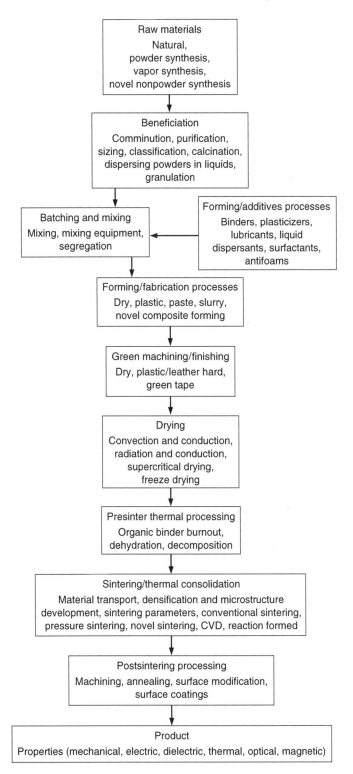

Fig. 1. Flow diagram of the steps and processes involved in manufacturing a ceramic.

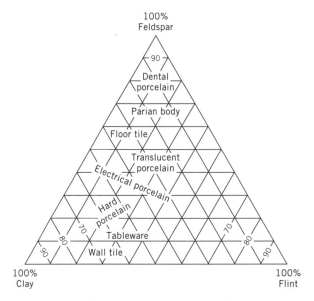

Fig. 2. Examples of traditional ceramics manufactured using traditional ceramic raw materials: clay, flint, SiO_2; and feldspar.

prepared by chemical synthesis. Inexpensive raw materials that cost a few cents per kilogram are typically used in manufacturing the traditional, high volume production ceramics shown in Figure 2, whereas synthesized ceramic powders, whiskers, and fibers costing hundreds of dollars per kilogram are used for advanced ceramics (qv). Chemically and physically beneficiated mined raw materials, eg, bauxite [1318-16-7] (see ALUMINUM COMPOUNDS), comprise the bulk of the oxide raw materials, eg, Al_2O_3, used in ceramic manufacturing, and make up the middle ground between the two extremes.

2.1. Natural. Naturally occurring ceramic raw materials (1–11) such as silica [7631-86-9] (qv), SiO_2, sand, quartz [14808-60-7], or flint [7631-86-9]; silicates eg, talc [14807-96-6], $Mg_3Si_4O_{10}\cdot OH_2$, wollastonite [14567-51-2], and asbestos [1332-21-4] (qv); and aluminosilicates, eg, clays, feldspars, pyrophyllite [12269-78-2], and sillimanite [12141-45-6], are used extensively as raw materials in the manufacture of traditional ceramics (see Fig. 2). Seawater and brine wells provide another source of ceramic raw materials, eg, magnesium oxide [1309-48-4], MgO (see CHEMICALS FROM BRINE). The principal advantage of naturally occurring raw materials is the low cost; the disadvantage is chemical impurity.

Naturally occurring raw materials are generally found in large, flat-lying mineral deposits near the earth's surface, and are mined using open-pit methods (8–12). Hydraulic mining techniques are also used to mine clays and glass (qv) sands (12) (see MINERAL RECOVERY AND PROCESSING).

2.2. Chemically Synthesized Powders. Chemical synthesis provides a means of producing powders for manufacturing advanced ceramics. Disadvantages of chemically synthesized raw materials are expense and difficulties in scale-up and availability. Additionally, ultrafine particle-size powders produced by chemical synthesis pose some unresolved processing problems in the areas of handling and mixing.

Solid State. Ceramic compounds can be formed by reacting constituent oxides and/or thermally decomposed salts at an elevated temperature in a solid-state process described as calcination (2,13–17). For example, spinel [1302-67-6], $MgAl_2O_4$, can be formed by reacting magnesia [1309-48-4], MgO, and alumina [1344-28-1], Al_2O_3. Because solid-state diffusion is inherently slow, fine, well-mixed powders are required to ensure that reactions go to completion and that chemical and structural homogeneity are achieved during calcination. Often multiple calcination, grinding, and mixing steps are performed to ensure homogeneity. Purity of the product is limited by the purity of the constituent raw materials, and by the impurities introduced during grinding. Ceramic powders formed by calcination are typically less expensive (<$1/kg) than those formed by liquid or vapor techniques ($\sim$$25 – 50/kg), but are also typically less pure and larger in size.

Solution Chemistry. Submicrometer particle size, high purity ceramic powders can be produced by precipitation, solvent evaporation, and solvent extraction from liquid solutions. The solid product is typically a salt that can be calcined at low temperatures, followed by light milling to produce reactive particles for sintering. Solution-derived powders have exceptionally high purity and homogeneity and can be extremely fine in size. Additionally, it is relatively easy to form multicomponent compounds and to add and disperse uniformly small concentrations of dopants, eg, sintering aids, by solution processing. However, chemical precursors are typically expensive, solution processes are not well suited to forming nonoxide ceramic powders, and surface contamination and microporosity common in solution-derived powders often make it difficult to produce dense, optical quality, ie, transparent, ceramics.

Precipitation (2, 13–17) techniques employ a combination of nucleation and growth induced by adding a chemical precipitant, or by changing the temperature and/or pressure of the solution. Chemical homogeneity is controlled by controlling the rate of precipitation. Heterogeneous precipitation involves the precipitation of a solid of different composition from the solution, and the composition of the precipitate may change as precipitation continues. Coprecipitation involves the simultaneous precipitation of similar size cations in a salt as a solid solution.

Precipitation is commonly performed using water-soluble salts; however, organometallics also provide a good source of high purity reactants for precipitation. Powders can also be formed by mixing reactant oxides in molten salts and precipitating. The calcination step for salt decomposition can be eliminated by precipitating powders in hot, high pressure water, using hydrothermal synthesis (see HYDROTHERMAL PROCESSING).

Solvent evaporation (13–15,17,18) techniques employ rapid drying to produce ceramic powders. Chemical homogeneity can be maintained by rapidly drying isolated, fine (10–100 μm) salt solution droplets produced by atomization, or by microwave or emulsion drying. Aerosol decomposition, also described as spray pyrolysis and evaporative decomposition of solutions, provides a means of enhancing powder production by combining rapid drying, precipitation, and decomposition in a single process similar to spray drying (see EVAPORATION).

Solvent extraction (2,13–15,17) by freeze drying provides another means of maintaining chemical homogeneity in solution-derived powders. Freezing solutions or emulsions prevents chemical segregation and the desired salt is obtained by removing the solvent by sublimation.

2.3. Sol-Gel Techniques.   Sol-gel powders (2,13,15,17) are produced as a suspension or sol of colloidal particles or polymer molecules mixed with a liquid that polymerizes to form a gel (see COLLOIDS; SOL-GEL TECHNOLOGY). Typically, formation of a sol is followed by hydrolysis, polymerization, nucleation, and growth. Drying, low temperature calcination, and light milling are subsequently required to produce a powder. Sol-gel synthesis yields fine, reactive, pseudo-crystalline powders that can be sintered at temperatures hundreds of degrees below conventionally prepared, crystalline powders.

2.4. Vapor-Phase Techniques.   Vapor-phase powder synthesis techniques, including vapor condensation, vapor decomposition, and vapor–vapor, vapor–liquid, and vapor–solid reactions, employ reactive vapors or gases to produce high purity, ultrafine, reactive ceramic powders. Many nonoxide powders, eg, nitrides and carbides, for advanced ceramics are prepared by vapor-phase synthesis.

Vapor-condensation (2,14,18) involves vaporizing a solid material at high temperature and precipitating a powder by condensation on cooling. Thermal plasmas that operate at several thousand degrees Kelvin at atmospheric pressure and higher can provide the vaporization energy (see PLASMA TECHNOLOGY). Only a very limited number of ceramic powders can be produced by this technique.

Vapor decomposition (14,15) involves drying, decomposing, and vaporizing a spray of salt precursor solution in a plasma, and subsequently nucleating and growing ceramic particles in the vapor. Silicon carbide [12504-67-5], SiC, powder is produced by this method.

Vapor–vapor reactions (14,16,17) are responsible for the majority of ceramic powders produced by vapor-phase synthesis. This process involves heating two or more vapor species which react to form the desired product powder. Reactant gases can be heated in a resistance furnace, in a glow discharge plasma at reduced pressure, or by a laser beam. Titania [13463-67-7], TiO_2, silica, silicon carbide, and silicon nitride, Si_3N_4, are among some of the technologically important ceramic powders produced by vapor–vapor reactions.

Vapor–solid reactions (13–17) are also commonly used in the synthesis of specialty ceramic powders. Carbothermic reduction of oxides, in which carbon (qv) black mixed with the appropriate reactant oxide is heated in nitrogen or an inert atmosphere, is a popular means of producing commercial SiC, Si_3N_4, aluminum nitride [24304-00-5], AlN, and sialon, ie, silicon aluminum oxynitride, powders.

2.5. Nonpowder Synthesis.   Many ceramic composites (qv) under investigation utilize reinforcing ceramic whiskers or fibers to achieve toughening (19). Whiskers (17,19,20) are produced by vapor-synthesis techniques. SiC whiskers can be produced by the rice hull or vapor–solid (VS) method whereby rice hulls are pyrolyzed to produce a mixture of carbon, C, and SiO_2, and whiskers are produced by directional growth by vapor deposition. The vapor–liquid solid (VLS) or Los Alamos process uses a liquid catalyzed vapor-deposition process to produce SiC whiskers from reagent-grade gases. Whiskers cost several hundreds of dollars per kilogram.

Fibers (17) can be formed by air drag or blow spinning molten precursors, and extruding or drawing gels through a spinneret. Air drag spinning is one of the oldest techniques for manufacturing fibers and uses air jets to draw a fiber from a melt reservoir and shape it. Blow spinning involves blowing a thin stream of compressed air or steam on a molten liquid as it passes through a small opening in the bottom of a melt reservoir. Al_2O_3–SiO_2 fibers are fabricated by blow spinning.

The extrusion process involves extruding, melt spinning, or drawing a controlled viscosity gel through a spinneret, followed by heating to effect gelation and crystallization. Al_2O_3 and Al_2O_3–ZrO_2 fibers are made this way. Nonoxide ceramic fibers have also been produced from amorphous preceramic polymer precursors by melt spinning. Fibers can also be formed indirectly by infiltrating an organic cloth fiber with a ceramic precursor solution, or by depositing a ceramic coating on a fiber preform using chemical vapor deposition (CVD). The desired ceramic fiber product is subsequently formed upon heating.

Although not as expensive as whiskers, fibers are considerably more expensive than powders. Extrusion processed Al_2O_3 and Al_2O_3–ZrO_2 fibers cost $100–200/kg.

3. Beneficiation

Beneficiation (2,11,12,21–27) involves a process or series of processes whereby the chemical and/or physical properties and characteristics of raw materials are modified to render the raw material more processable. The extent of beneficiation is determined by a combination of the starting raw materials, the processing scheme, the desired properties of the product, and economics. Powder cost increases with increased beneficiation; consequently, low value-added clay raw materials used to produce inexpensive structural clay products typically undergo a minimum of beneficiation, whereas higher value-added alumina powders undergo more extensive beneficiation.

Chemically synthesized materials also often undergo some form of beneficiation. Beneficiation processes for chemically synthesized ceramics are sometimes incorporated in the forming process such as in aerosol decomposition, where solvent evaporation and salt decomposition occur sequentially in a single operation.

3.1. Comminution. Particle size reduction by crushing and grinding/milling, otherwise known as comminution, is used extensively in ceramic processing to liberate impurities, break up aggregates, modify particle morphology, modify particle size distribution, improve processability during the mixing and forming operations, and produce more reactive powders for sintering (12,22,23). Crushing (11,12,21,22) operations are used for coarse grinding; milling (22–25) operations are used for fine grinding.

Primary crushers (11,12,21) are designed to reduce relatively large, up to 0.3 m diameter, mined material down to centimeter size particles. Secondary crushers (11,12,21) employ compression and impact compression to produce millimeter size particles. Natural raw materials processors typically supply material that has been processed through a primary and secondary crusher, and is often sized by screening.

Higher value-added, naturally occurring raw materials are generally ground to micrometer sizes by the raw materials supplier; however, additional grinding may be necessary prior to manufacturing. Micrometer size powders are produced by grinding millimeter and smaller size particles in a ball mill (2,22,23), vibratory mill (2,22), attrition mill (2,22), or fluid energy mill (2,23). Both particle size and size distribution decrease during grinding. In general, wet milling is more efficient than dry milling, producing a finer, typically submicrometer median particle size powder having a narrower size distribution. The practical lower particle size limit that can be achieved by milling is ~1.0 μm.

Ball milling (2,22,23), also referred to as jar or pebble milling, may be the most popular means of fine grinding ceramic powders. Ball milling can be performed wet or dry and as a batch or continuous process using spherical or cylindrical grinding media. Milling with media that is rod shaped is referred to as rod milling. Approximately 1 wt% grinding aid is generally added to minimize agglomeration during dry milling, and ≤1 wt% of a dispersant can be used to inhibit flocculation during wet milling. Water or alcohol are commonly used as the milling liquid.

3.2. Purification. Alumina, Al_2O_3, is produced by the Bayer process (1,9) (see ALUMINUM COMPOUNDS) which involves digestion followed by precipitation and calcination. High purity magnesia is extracted from natural brines and seawater by precipitation and calcination (1,9).

Soluble impurities can be extracted by washing with deionized or distilled water followed by filtration (1,12,26). Powders prepared by wet chemical synthesis are often washed and filtered for purification prior to use. The dewatering (qv) process can be enhanced by pressure filtration. Organic solvents can be used to remove water-insoluble impurities and wash-water sensitive materials.

Chemical leaching (1,12) with acids is used to extract metal contamination. High purity zirconia, ZrO_2, is produced by the caustic fusion of zircon [14475-73-1], $ZrSiO_4$, followed by the chemical removal of silica. Chemical leaching is generally followed by washing.

Magnetic separation (12,26) is used to extract magnetic impurities such as iron and iron minerals from raw minerals (see SEPARATION, MAGNETIC). Magnetic separation can be performed on dry powders and fluid slurries passing through an intense magnetic field. Froth flotation (qv) (12,26) can be used to separate intermixed raw materials and to separate the desirable material from the gangue.

3.3. Sizing. Sieve sizing (2,12,26,27) utilizes gravitational forces acting on particles translating across a screen (see SEPARATION, *SIZE*). The process can be enhanced by vibrating the sieves. Dry forced air and sonic sieving are used to size dry powders from 850 to 37 μm in size. Wet sieving eliminates electrostatic particle–particle and particle–sieve attractive forces to size particles down to ~5 μm.

3.4. Classification. Classification (2,12,26,28) or elutriation processes separate particles by the differences in how they settle in a liquid or moving gas stream. Classification can be used to eliminate fine or coarse particles, or to produce a narrow particle size distribution powder. Classification by sedimentation involves particle settling in a liquid for a predetermined time to achieve the desired particle size and size distribution or cut. Below ~10 μm, where interparticle forces can be significant, gravitational-induced separation becomes inefficient, and cyclone and centrifugation techniques must be used. Classification also separates particles by density and shape. Raw material separation by differential sedimentation is commonly used in mineral processing.

3.5. Calcination. Calcination (1,2,29) involves heat treating a powder or mixture of powders at a temperature well below its melting point to effect decomposition, ie, to liberate unwanted gases and/or chemically bound waste, solid-state reactions, and structural transformations to produce the desired composition and phase product. For example, water is liberated and the crystallographic structure changes when α-$Al_2O_3 \cdot 3H_2O$ is calcined to form α-Al_2O_3.

Calcination or dead burning is used extensively to dehydrate cements (qv) and hygroscopic materials such as MgO, and to produce a less water sensitive product. Calcination is also used to decompose metal salts to base oxides and to produce multicomponent or mixed oxide powders for manufacturing advanced ceramics. For example, barium carbonate, $BaCO_3$, is decomposed to form barium oxide, BaO, and fine particles of $BaCO_3$ and TiO_2 are intermixed and calcined to produce barium titanate [12047-27-7], $BaTiO_3$, powder for ceramic capacitors for solid-state electronics.

Calcination can be completed in an inclined rotary calciner, a heated fluidized bed, or by simply heating a static bed of powder in a refractory crucible.

3.6. Dispersing Powders in Liquids. The preparation of a slurry by dispersing a solid in a liquid entails (1) wetting the solid by the liquid; (2) breaking down flocs; and (3) stabilizing the system to prevent flocculation (30,31). Dispersing aids such as deflocculants and wetting agents are usually added to the liquid prior to mixing the powder with the liquid.

Slurry processing has a distinct advantage over dry, semi-dry, and plastic forming techniques in that the liquid provides an improved medium for mixing. Additionally, liquids eliminate electrostatic attraction forces between fine particles and particle surface charges and agglomeration can be controlled. The disadvantage is that the liquid has to be removed before firing. This increases the processing time and cost and can create problems with shape distortion and cracking during drying.

3.7. Granulation of Powders. To improve flow, handling, packing, and compaction, dry powders are typically granulated (2,32). Agglomerates or granules of controlled size, shape, and strength are formed by directly introducing a liquid/binder solution during powder stirring/mixing, or by spray drying.

Direct mixing (32) by compaction using a tableting die, briquetting rolls, a Muller mixer, extrusion, or spray granulation produces dense, hard, strong granules. An auger extruder or kneader is used to produce more plastic granules. Nearly spherical granules can be formed by spraying a liquid or binder solution into an agitating powder, such as a rotating pan, ribbon, double planetary, or V-cone mixer. The more intense the mixing action and higher the temperature, the smaller the granules.

Spray drying (2,32) provides a means of indirectly forming dense, homogeneous, nearly spherical granules from a slurry. The process involves atomizing a high solids content slurry through a pressure nozzle or rotary atomizer into a conical chamber with a cocurrent or countercurrent stream of heated air that dries the atomized droplets to form >20 μm, free-flowing granules. The product is collected in a container at the base of the spray dryer. Spray drying is widely used to prepare a variety of granulated, free-flowing ceramic powders with excellent compaction behavior for dry powder pressing operations.

4. Forming Additives/Processing Aids

Forming additives or processing aids (2,33–37) are commonly used to render ceramic powders more processable. Binders and plasticizers (qv) are typically added to improve or aid dry powder and plastic forming, whereas deflocculants, surfactants (qv), and antifoams are commonly used in slurry processing.

4.1. Binders. Binders (35,37,38) are used to impart strength to a green or unfired ceramic body for handling and green machining. Binders are polymer molecules or colloids (qv) that adsorb on particle surfaces and promote interparticle bridging or flocculation. Binders are used extensively in dry-pressing operations and are also added to plastic extrusion bodies and pastes. With the exception of tape casting, where a plasticized binder is required to produce flexible green tape, binders are generally avoided in slurry forming operations.

Binder selection depends on the ceramic powder, the size of the part, how it is formed, and the green density and strength required. Binder concentration is determined by these variables and the particle size, size distribution, and surface area of the ceramic powder. Three percent binder, based on dry weight, generally works for dry pressing and extrusion.

Binders such as poly(vinyl alcohol) (PVA) that form hard ceramic granules, produce free-flowing powders that are well-suited for automated pressing operations and yield strong bodies. Plasticizers and or lubricants (see LUBRICATION AND LUBRICANTS) can be added to hard binders to reduce the forming pressure and die wear. Wax binders such as poly(ethylene glycol) (PEG)form soft ceramic granules which yield lower strength, higher green density bodies on forming. Hard and soft binders can be combined to develop binder systems tailored to a given process and product.

Specialty binders may be used in novel or unique forming operations. Cross-linking polymers are used to induce gelation in gelcasting. Thermoplastic resins such as polyethylene and polystyrene often comprise the bonding matrix in injection-molded ceramic parts. Phosphate chemical bonds are used in forming specialty refractories (qv), and chemical hydration is employed in setting hydraulic cements. Preceramic polymers that decompose to a ceramic on heating serve as binders to provide green strength for handling and machining.

4.2. Plasticizers. Plasticizers (36–38) are often added to a binder to reduce cross-link density and increase flexibility. Plasticizers improve toughness, springback, and flexibility, but degrade overall green strength. Additionally, plasticizers can increase the sensitivity of a binder system to moisture.

Examples of plasticizers include adsorbed water and ethylene glycol for vinyl binders, stearic acid and oleic acid for wax binders, glycerine and ethylene glycol for clay bodies, and molten oils and waxes for thermoplastic polymers used in injection molding.

4.3. Lubricants. Lubricants (36,38) are added to lower frictional forces between particles, and between particles and die surfaces to improve compaction and minimize die wear. Typically ≤ 1 wt% of a lubricant is required for forming, and usually only with hard binders. Stearic and oleic acids are good lubricants for ceramics.

Individual particle surfaces can be lubricated by an adsorbed film that produces a smoother surface and/or decreases interparticle attraction. A plasticized binder may serve this purpose. Forming surfaces can be lubricated by coating with a film of low viscosity liquid such as water or oil. Die surfaces can also be coated with a solution of stearic acid dissolved in a volatile liquid that rapidly evaporates to leave a lubricating film.

Fine particles having a laminar structure and smooth surfaces are effective solid lubricants for rough surfaces and at high pressure. Graphite, boron nitride, BN, and talc are good solid lubricants; however, solid inorganic lubricants do not burn out during sintering and may affect the microstructure and properties of the finished product.

4.4. Liquids. Liquids (33) are common forming additives in plastic, paste, and slurry processing. In plastic forming operations, the liquid aids forming and serves as the binder/plasticizer for the system. In pastes and slurries, other additives are also dissolved or dispersed in the liquid/solvent. Water is a good, inexpensive solvent that can be recycled. Organic liquids such as alcohols are used to process water-sensitive materials and to dissolve water-insoluble forming additives, however, at considerably more expense.

4.5. Deflocculants. Deflocculants (34), dispersants (qv), or anticoagulants are added to slurries to improve dispersion and dispersion stability. Dispersants break up flocs in a slurry by lowering van der Waals interparticle forces. Deflocculants adsorb on particle surfaces and prevent the approach of particles either by electrostatic or steric stabilization. Deflocculation by electrostatic stabilization is common in clay slurries, as well as with ceramic particles dispersed in polar liquids such as water.

Monovalent cations are good deflocculants for clay–water slips and produce deflocculation by a cation exchange process, eg, Na^+ for Ca^{2+}. Low molecular weight polymer electrolytes and polyelectrolytes such as ammonium salts (see AMMONIUM COMPOUNDS) are also good deflocculants for polar liquids. Acids and bases can be used to control pH, surface charge, and the interparticle forces in most oxide ceramic–water suspensions.

Oleic acid is a good deflocculant for oxide ceramic powders in nonpolar liquids, where a stable dispersion is created primarily by steric stabilization. Tartaric acid, benzoic acid, stearic acid, and trichloroacetic acid are also deflocculants for oxide powders in nonpolar liquids.

4.6. Surfactants and Antifoams. Surfactants (33) or wetting agents can also be added to a slurry to improve dispersion. A wetting agent lowers the surface tension of the liquid, lowering the solid–liquid interfacial energy, which favors the liquid coating the solid. Antifoams (36) are added to slurries to remove trapped gas bubbles from the liquid (see DEFOAMERS). Commercial agents include fluorocarbons, dimethylsilicones, stearates, and high molecular weight alcohols and glycols. Typically, less than a few percent of an antifoam agent is required to prevent foaming.

5. Batching and Mixing

Mixing (24,39) is used to combine the constituents of a ceramic body to produce a more chemically and physically homogeneous system for forming (see MIXING AND BLENDING).

5.1. Convection, Shear, and Diffusion Mixing. Mixing is accomplished by material transport by convection, shear, and diffusion (39). Feeding devices, stirring, and baffles all produce material flow that contributes to convection. Material flow between surfaces moving at different velocities produces shear stresses that can break down agglomerates and mix viscous materials. Turbulent flow produced by eddy currents, cavitation, impact, and sonic vibration contribute to diffusive mixing, which also breaks down agglomerates.

5.2. Segregation. Segregation (24) or separation in mixtures of dissimilar particles can occur during storage and handling, resulting in macroscale heterogeneities in the product. Segregation can be minimized by minimizing size differences, storage and handling after mixing, and individual particle motion during mixing (see POWDERS, HANDLING).

6. Forming/Fabrication Processes

Ceramic forming involves consolidation and molding of ceramic powders to produce a cohesive body of the desired size and shape. Ceramic forming operations (38,40–66) are conducted with dry powders, plastic bodies, pastes, and slurries.

6.1. Dry Forming. Dry powders can be simultaneously compacted and shaped by pressing in a rigid die or flexible mold. Pressing (38,40–42,48–52) is the most widely used forming process in ceramics manufacturing. Microelectronic chip carriers, capacitors, spark plug bodies, cutting tools, and ceramic tiles are manufactured by powder pressing.

Powder pressing involves three general stages including: (1) filling the die or mold; (2) compacting and shaping the powder; and (3) extracting the pressed part from the die. When the pressing pressure is released, the dimensions of the powder compact and die increase or springback. Differential springback between the compact and mold aids extraction of the pressed part; however, excessive differential springback >0.75 linear% can generate catastrophic stresses that create defects or destroy the compact. Pressing defects can be minimized by using granulated powders and appropriate additives, and by controlling the pressing pressure, pressurization rate, and depressurization rate used in forming.

For reproducibility and rapid processing, pressing powders should be free-flowing, have a high bulk density, be comprised of deformable granules, cause minimal die wear, and not stick to die surfaces.

Dry Pressing. Dry or die pressing (38,40–42,45,48–50) which involves compacting a powder between two plungers in a die cavity, is a versatile technique for fabricating relatively uniform thickness and axial symmetric powder compacts. Manual dry pressing is used to form limited numbers of cylindrical pellets of single-phase and composite powders for sintering experiments. Automated dry pressing is capable of producing 5000 parts per minute.

Dies are made of hardened steel, specialty steels, and carbides (qv) or other structural ceramics, eg, SiC, Al_2O_3. Single-action, ie, moving top plunger, double-action, ie, moving top and bottom plungers, and floating die, (ie, moving top plunger and die body, pressing assemblies are used. In automated processes, the pressure is typically in the range of 20–100 MPa $(3 - 15 \times 10^6$ psi); however, manual die pressing operations may be carried out at pressures as high as 200 MPa $(29 \times 10^6$ psi). Die wear, ejection pressure, and differential springback all increase with pressing pressure.

Nonplastic ceramics are typically dry pressed using ≤4 vol% water. Plastic ceramics can be formed by semidry pressing using 10–15 vol %. Warm dry pressing with heated platens at ≤200°C is used to fabricate multilayer substrates from tape cast green tape for microelectronic packaging.

Isostatic Compaction. Isostatic pressing (38,40–42,45,51,52), which is also known as isopressing, hydrostatic pressing, and cold isostatic pressing (CIP), provides a means of manufacturing complex shapes such as tubes and spark plug bodies, and larger volume parts that are not easily dry pressed. Because the compacting pressure is applied uniformly over a larger free surface area, pressing gradients are typically less severe. Typically, a flexible mold or bag is filled with a granulated powder, sealed, and placed in a gas- or liquid-filled pressure chamber and pressurized to 5–200 MPa $(1 - 29 \times 10^6$ psi) to compact the powder. Because gas cannot escape during the isopressing operation, molds are typically deaired prior to pressing. Also, to improve the filling density, molds may be filled on a vibratory or tapping table. Both wet and dry bag techniques are employed in isopressing; although only the dry bag technique is suited to automation.

Isopressing molds are typically comprised of synthetic rubber, polyurethane, or silicone rubber with rigid steel fixtures and mandrels. Because of the relatively high springback of flexible mold materials, controlled, slow, depressurization is required below 2 MPa $(3 \times 10^5$ psi) to control mold expansion and to avoid damaging the compact on extraction. Isopressing is typically performed at ambient temperature; however, by heating the pressurizing liquid, warm isostatic pressing (WIP) can be accomplished. WIP is being investigated as an improved technique for laminating multilayers of tape cast green tape to produce microelectronic packages.

Vibratory Compaction. Vibratory compaction (38,53) provides a means of forming irregular shape compacts from ungranulated powders. Under agitation, a powder bed tends to rearrange to a configuration of closest packing. The amplitude and frequency of vibration required for compaction are dependent on the size and size distribution of the particles present, and the size of the compact, and are determined experimentally.

6.2. Plastic Forming. A plastic ceramic body deforms inelastically without rupture under a compressive load that produces a shear stress in excess of the shear strength of the body. Plastic forming processes (38,40–42,54–57) involve elastic–plastic behavior, whereby measurable elastic response occurs before and after plastic yielding. At pressures above the shear strength, the body deforms plastically by shear flow.

The cohesive strength required for plastic deformation is provided by capillary pore pressure and flocculation bonding. Plastic clay bodies are typically flocculated by the colloidal clay particles present in the system, whereas nonplastic ceramics can be flocculated with an organic binder. In clay bodies, submicrometer colloidal clay particles coat larger particles, and shear flow occurs between the larger particles in zones of oriented clay and finer nonplastic particles lubricated by interparticle water. In nonclay ceramic bodies plasticized with organic binders, the organic molecules play the role of the colloidal clay particles.

The dimensional tolerances of parts formed by plastic deformation are typically not as good as those in parts formed by dry powder pressing.

Extrusion.   Extrusion (38,40–42,54,55) is widely used in manufacturing structural clay products, including sewer pipe, chimney flues, bricks, and tiles. Additionally, refractory tubes and rods, and cordierite catalyst supports for catalytic converters in automobiles having 30–60 openings per square centimeter are formed by extrusion. Clay and talc bodies typically contain 12–20 vol% water, and can be extruded without a lubricant. Nonplastic ceramic powders require a plasticizer for extrusion. The higher the forming pressure, the lower the concentration of liquid required for extrusion, and the shorter the drying time and the lower the drying shrinkage. Extrusion pressures of up to 4 MPa (6×10^5 psi) are used to extrude porcelain bodies, whereas pressures up to 15 MPa (2×10^6 psi) are used to extrude ceramic powders having organic plasticizers.

Extrusion involves: (*1*) feeding plastic material into a chamber; (*2*) material consolidation and flow into the barrel, through the forming die and the finishing tube; and (*3*) ejection. A typical extruder may have a pug mill for additional mixing, and a vacuum chamber to deair the plastic body prior to extrusion. Dies are specially designed to generate sufficient back pressure to heal defects created during forming.

Both piston and auger type extruders are used in plastic forming. The former provide maximum control of the extrusion pressure and rate and are good for forming large parts. The latter provide a maximum extrusion rate without excess power requirements and are desirable in the continuous production of simple clay ware.

Defects in extruded bodies typically include: insufficient strength or rigidity from too much water or insufficient organic binder; cracks or delaminations because of excess differential springback, hard inclusions, or agglomerates from poor mixing; surface blisters from the expansion of dissolved air; surface delaminations created by slip-stick wall friction; nonuniform flow from improper die design; laminations produced by divided flow as a result of improper die design; gradients in extrudate stiffness; and curling on drying from differential drying.

Jiggering.   Jiggering (38,40,54) is one of the most widely used, cost effective, soft plastic forming techniques used in ceramic processing for manufacturing small, simple, axial–symmetrical whiteware ceramics including cookware and fine china, and electrical porcelain. Jiggering involves shaping a plastic clay body on a spinning porous plaster mold using a water-mist or steam-lubricated shaping tool. The shaping tool initially deforms the clay mass to the shape of the mold and subsequently creates a smooth surface finish. The mold and shaped part are then set aside and separation of the mold and part is accomplished by differential drying. Hand molding operations are conducted at 300–400 rpm; 500–1200 rpm are used in automated jiggering processes. Automated, highly mechanized jiggering operations can produce 1000 pieces per hour per line.

Powder Injection Molding.   Powder injection molding (PIM) (38,42,45,54, 56,57) provides an economical means of mass producing small, near-net, complex shapes. PIM involves injecting a hot ceramic and polymer binder mixture into a cooler die to form the desired shape, then extracting and deburring the part, and removing the binder. The binder concentrations for PIM range from 30 to 50 vol% and must be high enough for good flow during molding, but low enough to avoid debinding problems and excessive debinding times.

PIM is not cost competitive with die compaction for simple or axial–symmetric shapes. It is limited to small ($\leq$10 cm) shapes and is extremely process sensitive. Additionally, PIM tooling is expensive.

6.3. Paste Forming.   Ceramic pastes (58), often associated with thick-film printing operations, consist of a ceramic powder, a sintering aid such as borosilicate glass, and an organic vehicle comprised of a solvent, dispersant, plasticizer, and high molecular-weight binder. Ceramic thick-film pastes are used for decorating traditional ceramic tableware and to form capacitors and dielectric insulator layers on rigid substrates for microelectronic packaging.

6.4. Slurry Forming.   Slurries are used in ceramic casting operations (38,40–42,59–66) and slurry formation is perhaps the most important step. A good slurry is well dispersed and free of air bubbles and foaming; it has a high specific gravity; it has good rheological properties for casting; the solid particle settling rate in the slurry is low; and it is chemically stable. Additionally, when dry, the cast should possess sufficient strength for subsequent finishing and handling before firing.

The liquid vehicle in a slurry should: have a low vapor pressure for liquid extraction and drying; be compatible with the solids and casting mold; be inexpensive; and be capable of dissolving and dispersing deflocculants and other additives. Distilled or deionized water is generally used as the liquid vehicle, however, organic liquids must be used for such moisture sensitive oxide powders as CaO and MgO, and for oxidation sensitive nonoxide powders, eg, AlN.

A high solids concentration is desirable to minimize the amount of time and energy required for forming and to minimize drying shrinkage. Using deflocculants, fluid slurries can be made using as little as 15–20 vol% liquid.

Slip Casting.   Slip casting (38,40–42,45,59–62), the process in which a cast is formed from a slurry using a porous mold, is used to form sinks and other sanitary ware, figurines, refractory crucibles, porous thermal insulation, fine china, and complex shape structural ceramics such as multivane rotors.

Slip casting begins by pouring a stable slurry into a porous mold. The capillary suction of the porous mold draws the liquid from the slurry to form a higher solids content, close-packed cast on the inner surface of the mold. Wall thickness increases with the square root of forming time, and with sufficient time a completely solid cast can be formed. After a fixed time the excess slurry is drained from the cast (drain casting). Initial drying occurs in the mold, and the resultant cast shrinkage and strengthening aids in separating the cast from the mold. A mold-release agent, such as an alginate, graphite, a silicate, talc, silicone, or olive oil, can be used to enhance the removal of the cast from the mold. Inexpensive plaster molds comprised of 40–50 vol% porosity are typically used in slip casting.

Casting slips are typically pseudoplastic ($<$2 Pa·s at a shear rate of $1 - 10$ s^{-1}) for good mold filling and to allow the escape of gas bubbles. Additionally, the concentration of solids must be high enough to minimize settling during casting. Solid casting is generally conducted with slightly higher viscosity and higher solids content slips.

To control the concentration of submicrometer colloidal particles to $\leq$30%, clay slips are typically used in the partially deflocculated state. A coagulant can be added to a deflocculated clay slip immediately preceding casting to induce partial flocculation. Slurries for casting oxides, nitrides, and carbides are generally at least partially deflocculated. A fraction of a percent of organic binder increases the pseudoplasticity of the slurry and strength of the cast; too much binder can clog the plaster molds and decrease the rate of casting.

Casting rate can be enhanced through the use of vacuum casting (38,59), ie, by increasing the pressure gradient, or through pressure casting (38,59), ie, by decreasing the slurry viscosity, eg, by heating, or by centrifugal casting. Each enhance the rate of liquid flow from the slurry into the mold and decrease the casting time from hours to minutes, decrease the concentration of water in the cast and the shrinkage on drying from several percent to $\sim$1%, and yield higher green density casts that shrink less on firing. Additionally, centrifugal forces created by spinning the mold, and/or vibration can be used to enhance slurry flow and to ensure good filling of complex molds.

Disadvantages of vacuum, pressure, or centrifugal casting include higher costs because of specialty equipment and specialty molds, and density gradients in the casts that result from the higher pressure gradients during forming.

Gelcasting.   Gelcasting (63) is a novel method of fabricating particulate bodies from ceramic slurries. Gelcasting employs *in situ* polymerization of organic, eg, acrylamide (qv), monomers to produce a gel structure that binds individual particles together to form a cohesive body. Net and near-net complex shapes such as turbine rotors can be cast in a relatively short period of time, and the high (up to 62 vol%) solids concentrations improve the homogeneity of the cast and contribute to lower and more reproducible shrinkage during drying and sintering. Gelcast parts also have excellent strength for handling and green machining.

Tape Casting.   Tape casting (38,45,59,64–66), doctor-blading, or knife coating involves forming a thin film of slurry of controlled thickness by flowing a slurry beneath the knife edge of a doctor blade onto a support surface. Tape casting is an economical means of forming thin (0.02–2.0 mm), uniform thickness, smooth surface, flexible, green ceramic sheet or tape that can subsequently be cut and stacked to form multilayer ceramics for multilayer capacitors and dielectric insulator substrates (see also CERAMICS AS ELECTRICAL MATERIALS).

The thickness of the tape is controlled by the slip characteristics, the height of the doctor blade, the casting rate, and the pressure head of the slip reservoir behind the doctor blade. Slip viscosities in the range of $1-5$ Pa·s (10–50 P) are used to cast tapes at 5–100 cm/s. To achieve the desired strength and flexibility in the green tape, tape casting slurries contain more binder than those used in slip casting, as well as a plasticizer to ensure flexibility.

Tape can be cast on a stainless steel table or belt, glass plate, or a Mylar, Teflon, or cellulose acetate film carrier. The tape should adhere to the carrier sufficiently to prevent curling, but should be easily removable. In a continuous casting process, the tape is dried by air flowing 1–2 m/min counter to the casting direction. A typical dry green tape contains approximately 35 vol% organics, 50% ceramic powder, and 15% porosity.

Specialty Casting Techniques. Thixotropic casting (59) processes involve slurries comprised of a gelling, reactive, or hydraulic binder in which yield strength and viscosity increase with time. Thixotropic slurries are used for gunning or troweling castable refractory wall liners, preparing refractory shapes and molds for metal casting operations, obtaining dental impressions and filling cavities, and producing concrete structural materials.

Electrophoretic casting (38,59) is accomplished by inducing controlled migration of charged particles under an applied electric field to deposit on a mandrel. Thin tubular shapes and coatings of limited thickness are formed using this technique. Electrophoretic deposition (EPD) is also used to manufacture thin wall, solid β''-alumina [12005-16-2], $NaAl_5O_8$, electrolytes for sodium–sulfur batteries.

Freeze casting (59) is a hybrid of slip casting, gel casting, and freeze drying in which a slurry is poured into a rigid rubber mold, frozen, and the frozen liquid is removed by sublimation, ie, by freeze drying (see CRYOGENICS).

6.5. Novel Composite Forming Techniques. Particulates, whiskers, or fibers, can be used to improve the toughness of ceramics (19). Composites (qv) or multicomponent ceramic bodies containing second-phase reinforcing particulates, whiskers, or fibers can be processed by the dry powder, plastic, and slurry forming techniques described, but with considerably more difficulty. Challenges in composite processing include: uniformly dispersing the reinforcing phase; maintaining the desired orientation of the reinforcing phase within the matrix; and uniformly packing the matrix particles about the reinforcing phase.

Particulate Composites. In addition to the geometric obstacles to mixing dissimilar shape and size particles, there are also chemical barriers that must be overcome in processing composites, requiring novel techniques.

Coatings (qv) offer one means of reducing or eliminating the differences in dissimilar materials. For example, a silica coating on alumina allows alumina and silica particles to be dispersed together in water at the same pH. Likewise, oxide-resistant coatings on carbide and nitride powders render them compatible for coprocessing with otherwise incompatible oxide ceramics. Coatings can also be employed to aid the processing of whisker and fiber reinforced composites.

Infiltration (67) provides a unique means of fabricating ceramic composites. A ceramic compact is partially sintered to produce a porous body that is subsequently infiltrated with a low viscosity ceramic precursor solution. Advanced ceramic matrix composites such as alumina dispersed in zirconia [1314-23-4], ZrO_2, can be fabricated using this technique. Complete infiltration produces a homogeneous composite; partial infiltration produces a surface modified ceramic composite.

Preceramic polymer precursors (45,68) can be used to make ceramic composites from polymer ceramic mixtures that transform to the desired material when heated. Preceramic polymers have been used to produce oxide ceramics and are of considerable interest in nonoxide ceramic powder processing. Low ceramic yields and incomplete burnout currently limit the use of preceramic polymers in ceramics processing.

Whisker and Short Fiber Composites.   Whiskers and short fibers tend to align during forming, leading to anisotropic properties and large flaws in the product. Commercial whisker and fiber reinforced ceramics are formed by dry or isostatic pressing. Slip casting, tape casting, and injection molding processes are under development. Ceramic short fiber mat is manufactured in the dry mat process (42) by blowing a dispersion of 6–25 mm long staple fibers onto a moving belt by a stream of air and reacting it in a heated zone. The mats are resin reinforced and are limited to a thickness of 1–3 mm. The wet mat process (42) employs paper-making technology to fabricate denser, thicker web mat with staple fiber. Fibers and binder dispersed in water are cast, filtered, and dried in ovens (see PAPER).

Continuous Fiber Composites.   Textile manufacturing (42) techniques are used to fabricate continuous fiber ceramics and ceramic composites (see TEXTILES). Weaving (42) techniques can be used to manufacture ceramics and ceramic cloths having continuous monofilament fibers (qv) and fiber bundles (yarn). Complex parts can be woven with a high volume fraction of fiber; however, the process is slow and expensive. Complex, three-dimensional structures can be formed by multidirectional winding (42) continuous fiber on a mandrel. Rod, bar, tube, or sheet can be fabricated by pultrusion (42), in which a continuous fiber bundle is drawn through a bath of matrix solution that coats the fibers, through a card screen, and then through a heated die.

Preweave fabric to ceramic fabric (42) uses conventional cloth fiber as a preform of the desired shape that is formed by weaving, and then the weave is impregnated with a precursor solution of the desired ceramic. Zirconia ceramic cloth is formed this way. Impregnation (42) provides a means of fabricating polymer-filled, continuous ceramic fiber composites. Laminates of alumina fibers in a polymer resin matrix have been formed that have a thermal conductivity approaching that of alumina in the in-plane direction, and a dielectric constant approximating that of the polymer in the out-of-plane direction. Such composites have potential application in advanced microelectronic packaging.

7. Green Finishing/Machining

After forming, green finishing or machining is often required to eliminate rough surfaces, smooth forming seams, trim, and modify the size and shape of the green body in preparation for sintering (38,40,45,58).

7.1. Surface Grinding or Turning.   Surface grinding and turning (38,40,45,58) on a lathe is used to form contours in ceramic spark plug bodies and high tension porcelain insulators (qv), to machine threads, and to fabricate complex shape bioceramic implants. Typically, the strength of the green part must be >2 MPa ($>3 \times 10^5$ psi), and the tool force for grinding must be carefully controlled to avoid overstressing the relatively weak, green ceramic part during machining.

7.2. Blanking, Punching, and Laminating. Blanking, punching, and laminating (45,58,64–66) processes are used to form ceramic multilayers from tape cast green tape. Blanking, coining, or stamping is used to cut or form the desired substrate size and shape, large diameter holes or vias >0.5 mm, and cavities. Smaller vias are punched in a subsequent operation. To form a multilayer ceramic, several layers are stacked together and laminated by uniaxial or biaxial pressing at 3–30 MPa $(4 – 40 \times 10^5$ psi) between platens at 50–80°C. Warm isostatic pressing can also be used. To compensate for particle alignment and texture in the cast tape, and to minimize shrinkage anisotropy and warping during drying and sintering, alternating layers in the laminate are rotated 90° to the casting direction during stacking. An alignment die can be used to preserve via alignment during lamination. Final shaping and punching can be performed after lamination as necessary.

8. Drying

For economical reasons, drying (38,41,45,69–73) should be as fast as possible; however, to avoid differential shrinkage that can result in cracking, warping, and shape distortion, drying must be carefully controlled. Generally, larger parts containing more liquid require longer and costlier drying.

8.1. Convective And Conduction Drying. *Air Drying.* Air drying is the most common means of drying ceramics (69–73). Circulating hot air supplies heat to the ware to aid evaporation and compensate for evaporative cooling as it removes vapor from the ware surface. The drying rate is determined by two factors: the rate of liquid evaporation and the rate of liquid migration to the drying front.

In a saturated body, drying first occurs at a constant rate by the evaporation of the liquid film surrounding individual particles at the ware surface. If liquid migration cannot keep pace with evaporation, the drying front moves into the bulk of the ware, where drying continues by evaporation from the menisci of the liquid within the pores. During this later stage, falling rate period, the rate of drying continuously decreases with decreasing liquid content. During initial-stage drying, flowing air lowers the relative humidity and thins the boundary layer to enhance evaporation.

Most shrinkage occurs during initial-stage drying. Drying shrinkage can be reduced by lowering the concentration of liquid, decreasing the film thickness, decreasing the concentration of fines and plastic materials, and by adding nonplastics.

Controlled Humidity Drying. Controlled humidity drying (41,69–71) is used to minimize the moisture gradient and stresses during drying, and to optimize the drying rate. A high humidity atmosphere allows for rapid heating of moist ware without creating catastrophic drying stresses. After initial-stage drying, the humidity is rapidly decreased. Humidity controlled drying can reduce drying times from days to hours.

Air Drying Equipment.   Tunnel kiln dryers (70) are long furnaces comprised of several zones of different temperature, humidity, and air flow through which the ware travels on a moving car or belt. These kilns afford continuous processing. Periodic kiln cross-circulation dryers (70) are box furnaces in which ware is stacked on permanent racks or on a car that can be shuttled in and out of the furnace. Fans or jets are used to circulate heat uniformly through the ware. The process is not continuous, but production rates can be enhanced by shuttling multiple cars.

8.2. Radiation and Conduction Drying.   The rate of drying is often determined by the rate of forming and firing. The rate of drying ceramic ware processed with water can be enhanced by using microwave (69) or infrared (41,70) radiation. Radiation drying techniques can reduce drying times from hours to minutes.

Microwave Drying.   Microwave radiation drying (45,64) (see MICROWAVE TECHNOLOGY) is excellent for drying porous molds and porous refractories without shrinkage damage, for drying bulk powders without forming agglomerates, and for uniformly drying large volume bodies. Microwave drying is cost effective at or below 5–10 vol% water, and can be used in combination with conventional controlled humidity forced air drying to be economical at higher water concentrations.

Infrared Drying.   Infrared drying (41,70) using 4–8 μm electromagnetic radiation supplied by ir lamps is used to rapidly dry thin ceramic ware, such as china formed by jiggering, tape cast ceramic sheet, ceramic substrates, thick films, and coatings. The water content in a ceramic can be reduced from 25 to 5 vol% within 10–15 minutes, after which drying can be completed using conventional forced air techniques (see INFRARED TECHNOLOGY AND RAMAN SPECTROSCOPY).

8.3. Supercritical and Freeze Drying.   To eliminate surface tension related drying stresses in fine pore materials such as gels, ware can be heated in an autoclave until the liquid becomes a supercritical fluid, after which drying can be accomplished by isothermal depressurization to remove the fluid (45,69,72) (see SUPERCRITICAL FLUID). In materials that are heat sensitive, the ware can be frozen and the frozen liquid can be removed by sublimation (45,69).

9. Presinter Thermal Processing

A number of changes can occur on heating a ceramic prior to sintering, including additional drying, burnout of organic additives and impurities, removal of chemically bound water and water of crystallization, and decomposition of inorganic precursors or additives (29). These processes may be accomplished by a separate heat treatment well below the sintering temperature, or in a controlled series of ramps and isothermal holds in a single heat treatment process. For traditional clay ceramics, a separate heat treatment termed a bisque firing is often performed prior to glazing and sintering.

Thermal analysis using differential scanning calorimetry (dsc), thermogravimetric analysis (tga), and differential thermal analysis (dta) can provide useful information about organic burnout, dehydration, and decomposition.

9.1. Organic (Binder) Burnout. Organic decomposition (29,42,45,68), which occurs by dissociation or pyrolysis, produces gaseous products that may be hundreds of times the volume of the ceramic part. To preclude the development of catastrophic stresses from gases evolved within the ceramic body, organics must be removed gradually, prior to sintering to high relative density, when the permeability of the body is high. Organics are typically removed at $\leq$ the sintering temperature. Incomplete binder burnout can result in residual carbonaceous matter in the finished product that can degrade or alter its properties.

9.2. Dehydration. Residual liquid and physisorbed moisture on particle surfaces can be eliminated on heating to $\sim$200°C. Temperatures in excess of 1000°C may be required to eliminate chemisorbed water (29). Kaolin must be heated to 700°C to liberate the water of crystallization and produce the desired dehydrated aluminosilicate. As with binder burnout, rapid gas evolution from rapid dehydration can result in catastrophic stress development within a body.

9.3. Decomposition. Decomposition (29,68) processes are primarily accomplished in the beneficiation stage of processing by calcination; however, minor concentrations of sintering aids, which may be added as salt precursors, can be decomposed by heating prior to sintering.

10. Sintering/Thermal Consolidation

Most ceramics are thermally consolidated by a process described as sintering (29, 44,68,73–84), in which thermally activated material transport transforms loosely bound particles and whiskers or fibers into a dense, cohesive body.

10.1. Sintering. A ceramic densifies during sintering as the porosity or void space between particles is reduced. Additionally, the cohesiveness of the body increases as interparticle contact or grain boundary area increases. Both processes depend on and are controlled by material transport.

Thermodynamic Driving Force for Material Transport. Under the influence of elevated temperature and/or pressure, atoms move to lower energy, thermodynamically more stable positions within the system to decrease the volume-free energy of the system. In a powder compact, excess volume-free energy is present primarily in the form of excess pore surface or interfacial energy, ie, liquid–vapor and/or solid–vapor interfaces, and the driving force for sintering can be approximated by $dG = \Sigma - \gamma_i dA_i$, where γ_i is the interfacial energy and A_i is the interfacial area. The interfacial area of a powder compact decreases as material transport occurs during sintering.

The primary driving force for material transport comes from the chemical potential difference that exists between surfaces of dissimilar curvature within the system. The greater the curvature, ie, the finer the particle size, the greater the driving force for material transport and sintering.

Material Transport. During sintering, material transport can occur by solid-state, liquid-phase, and vapor-phase mechanisms individually or in combination. The consequences of material transport during sintering, and the different paths of transport are summarized in the two sphere model shown in Figure 3. Initially during sintering, material is transported from higher energy convex particle surfaces to the lower energy, concave grain boundary pore

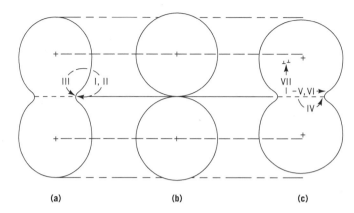

Fig. 3. The two-sphere model illustrating material transport paths I–VII during sintering where (**a**) represents coarsening; (**b**), the two spheres before sintering; and (**c**) densification. The + indicates the center of the sphere.

intersections to form necks between adjacent particles. As sintering progresses, interparticle necks grow, grain boundary area increases, interparticle contacts flatten, and the free energy of the system decreases. Densification occurs in a powder compact as the distance between particle centers and the volume of the compact decreases during sintering.

Material transport can also effect coarsening during sintering. Coarsening is the term used to describe material transport that changes the geometry of the system without resulting in densification, ie, the distance between particle centers remains constant. Coarsening decreases the driving force for material transport without affecting densification, and it is often considered to be an undesirable, competing process. Both densification and coarsening mechanism are generally operative at any given time during sintering.

10.2. Densification and Microstructure Development. The microstructure both of a sintered ceramic and development during sintering are also controlled by thermodynamics. The material transport that occurs during sintering manifests itself as grain boundary formation, interparticle pore shrinkage and annihilation, particle compact densification, and grain growth. Initially, interparticle necks and grain boundaries form, creating a three-dimensional array of approximately cylindrical, interconnected, ie, continuous pore channels. These pore channels shrink in diameter during intermediate-stage sintering, and ultimately pinch off to form closed, ie, isolated, approximately spherical pores for final-stage sintering. Additional pore shrinkage and grain growth occur during final-stage sintering.

The primary objective of sintering is to produce a cohesive body of the desired size and shape having a microstructure that optimizes the desired in-use properties. For ceramics, this is usually a theoretically or nearly theoretically dense body comprised of fine (micrometer size), uniform grains.

Physical Characteristics and Property Effects. Powder physical characteristics that can affect sintering include particle size, particle packing, ie, pore size and green density, and particle shape. Particle size has the most pronounced effect on sintering. In general, because material transport occurs faster over shorter distances and less material transport is required to eliminate smaller pores, the finer the particle size, the faster the sintering process, and the lower the sintering temperature. The larger grains grow at the expense of smaller ones, and the smaller pores are preferentially eliminated during sintering. Consequently, as sintering progresses, the average size of the grains and pores in the compact increase and the size distributions become narrower. Densification of large grain and pore size compacts, and compacts comprised of a distribution of grain and pore sizes during sintering, is often limited to <100% theoretical density. Uncontrolled grain growth during final-stage sintering can result in intragranular porosity that also limits densification.

Particle packing may be the single most important physical characteristic of a powder compact. Improved particle packing increases particle coordination and the relative density of the compact, increasing the number of material transport paths for densification, while decreasing the concentration, size, and size distribution of the pores present in the powder compact. Consequently, densification occurs faster, to higher end point densities, and with less overall volume shrinkage during sintering. Uniform particle packing in a green compact ensures uniform densification and shrinkage. Nonuniform particle packing resulting from the presence of agglomerates, ie, densely packed fine particle clusters, or density gradients is undesirable. Ceramic fabrication processes are usually designed and controlled with the intent of optimizing green density while minimizing density gradients during forming.

Particle shape also affects the sintering of a powder compact. Jagged or irregular shaped particles, which have a high surface area to volume ratio, have a higher driving force for densification and sinter faster than equiaxed particles. High aspect ratio platey particles, whiskers, and fibers, which pack poorly, sinter poorly.

Material properties, including surface energy, diffusion coefficients and fluid viscosity, and bond strength are also important factors in sintering. Surface energy anisotropy, which is common in crystalline ceramics, can lead to differential densification, nonuniform and exaggerated grain growth, and end point densities less than theoretical. Bond strength also affects sintering. Strongly bound covalent materials like SiC, Si_3N_4, and BN are typically difficult to sinter without a sintering aid. Weakly bound ionic materials like NaCl are also difficult to sinter because these tend to dissociate during sintering.

Phase transformations can also be an important factor in sintering. Volume changes in materials such as quartz and zirconia from polymorphic phase transformations on heating and cooling can generate catastrophic stresses that can destroy a ceramic body. Damage and loss can be minimized by carefully controlling the cooling rate, or by controlling the size and/or concentration of materials that undergo destructive phase transformations during processing. The polymorphic phase transformation in zirconia can be controlled and beneficially utilized to produce toughened ceramics.

Chemical impurities may form a solid solution that alters the rate and/or mechanism of material transport, second-phase precipitates that pin grain boundaries and impede grain growth, or a liquid phase that enhances densification relative to coarsening. Minor additions of appropriate impurities can provide improved control over microstructure development during sintering. MgO-doped

Al_2O_3 is the classic example in ceramics. Without MgO, the crystalline aniso-tropy in the Al_2O_3 system favors the formation of large lathlike grains; however, with as little as a few hundred parts per million MgO, fine, equiaxed, Al_2O_3 grains can be sintered to ~100% theoretical density in ≤1 h at ~1500°C. Chemical impurities that are intentionally added to a system to promote sinter-ing and/or control grain growth are referred to as dopants or sintering aids.

Sintering Parameters. For a given material system, densification and microstructure development are primarily controlled by the sintering tempera-ture, time, pressure, and atmosphere. Ceramic sintering is a thermally activated process that follows an Arrhenius-type behavior. Consequently, material trans-port and sintering occur faster at higher temperatures, but so does coarsening. Typically, ceramics are sintered at 0.50–0.75 of the absolute melting temperature.

Time is another important parameter. If the sintering time is short, a low density, porous ceramic body may be obtained. If the sintering time is long, coar-sening mechanisms may take over and produce a coarse grained microstructure. Short to intermediate sintering times are used to manufacture fine grain size, high strength ceramics such as alumina cutting tools, whereas longer times are used to manufacture larger grain size, more creep-resistant ceramics such as Si_3N_4 for advanced heat engines. Extended sintering times (>2 h) are used to manufacture large grain size, high permeability ferrites, and high permittivity ferroelectrics.

In contrast to conventional sintering, which is typically conducted at atmo-spheric pressure or in vacuum, pressure sintering is conducted using an exter-nally applied pressure. The applied pressure enhances the driving force for material transport during sintering, enhancing the rate of densification and making it possible to densify materials that are difficult to densify by conven-tional sintering. Oxide ceramics are typically sintered in an oxidizing or inert atmosphere to avoid reducing transition metals and degrading the optical prop-erties and/or aesthetics of the finished product. Depending on the oxygen partial pressure, lead zirconate titanate (PZT) can be densified by solid-state or liquid-phase sintering mechanisms; however, to optimize its piezoelectric properties, precautions must be taken to maintain a PbO overpressure to limit lead loss and maintain stoichiometry during sintering. Si_3N_4 and AlN are typically sin-tered in a nitrogen atmosphere to prevent nitrogen dissociation and/or oxidation during sintering. SiC must also be sintered in a nonoxidizing atmosphere to avoid oxidation. In processing ferrites (qv), the partial pressure of oxygen must be controlled to optimize magnetic properties. In special cases, such as in proces-sing $YBa_2Cu_3O_7$ ceramic superconductors, it is necessary to control and vary the sintering atmosphere during processing to optimize densification and properties.

10.3. Conventional Sintering. Ceramic sintering is usually accom-plished by heating a powder compact to ca two-thirds of its melting temperature at ambient pressure and holding for a given time. Densification can occur by solid-state, liquid-phase, or viscous sintering mechanisms.

In solid-state sintering (29,68,78–81) densification occurs by solid-state diffusion-controlled material support. Examples of two technologically important ceramics that density by solid-state sintering are MgO-doped Al_2O_3 and Y_2O_3-stabilized ZrO_2.

In general, material transport occurs faster through a liquid than through a solid. Thus densification generally occurs faster in the presence of a liquid phase as compared to solid-state sintering. The presence of a liquid during sintering can also lower the temperature required for densification. For example, polycrystalline alumina, which requires a sintering temperature of 1500–1600°C, can be densified faster and at temperatures hundreds of degrees lower using 0.5–6 wt% of an additive that forms a liquid phase during sintering (29). It is virtually impossible to densify materials such as SiC and Si_3N_4 without a liquid phase. When sintering with a liquid phase, trapped gases in closed pores often impede and limit densification. Additionally, a residual liquid or glass phase at the grain boundaries can degrade the properties of the finished product.

Liquid-phase sintering (LPS) (68,76,78–80,82) involves sintering with a liquid that is persistent throughout the sintering process. LPS is significantly more complex than solid-state sintering as there are more phases, interfaces, and material transport mechanisms to consider. The requirements for LPS are that the liquid wet the solid particles, that there is sufficient liquid present to wet the particles, and that the solid is soluble in the liquid. In general, the rate of densification increases with decreasing viscosity, increasing solubility, and increasing liquid concentration. Temperature is a critical processing parameter in LPS as the viscosity of the liquid decreases and the concentration and reactivity, ie, the solubility of the solid in the liquid, increase dramatically with increasing temperature above the eutectic temperature.

Reactive LPS (68,76,82) is similar to LPS, but differs by the fact that enhanced diffusion through a chemical reaction or phase transformation promotes faster densification. Examples of reactive liquid-phase sintered ceramics include MgO, Y_2O_3, and/or Al_2O_3 doped Si_3N_4. Transient LPS (76,82) takes advantage of LPS mechanisms and enhances densification in the presence of a liquid during sintering, after which the liquid vaporizes or transforms to a solid solution. Transient LPS produces ceramics devoid of a liquid grain boundary phase; consequently the finished product exhibits improved mechanical properties. LiF-doped MgO is an example of a transient liquid-phase sintered ceramic. Viscous sintering (83), which occurs by surface tension driven viscous flow, is widely used in the processing of glasses and glass-ceramics (qv) for optical and electronic applications. Nonreactive liquid-phase sintering (NLPS) (84) utilizes a combination of LPS and viscous sintering mechanisms to effect densification during sintering. Low temperature sintering ceramic-filled glass, ie, glass matrix, composites for electronic packaging, which typically contain ≥50 vol% glass, densify by NLPS. Densification and densification rate increase with liquid concentration and temperature during sintering.

Conventional Sintering Equipment. Like drying furnaces, sintering furnaces (29,76,85) can be periodic or continuous in nature. Periodic kilns offer greater flexibility; continuous tunnel kilns are more economical. Advanced ceramics are typically sintered in high purity, controlled atmosphere furnaces by electric resistance heating. Ceramic furnaces used to fire traditional ceramic ware are generally heated with inexpensive natural gas, oil, wood, or coal.

10.4. Pressure Sintering. Pressure sintering employs the simultaneous use of pressure and temperature to effect densification during sintering. The externally applied pressure supplements the existing surface tension driving force for material transport during conventional sintering, resulting in an increased driving force and faster densification at the same temperature as conventional sintering, or an equal driving force and densification rate at a lower temperature. Pressure sintering also enhances the rate of densification relative to coarsening, so dense, fine grain size ceramics can be fabricated with little or no sintering aid. The increased driving force for material transport and densification also makes it possible to eliminate larger pores.

Because of higher manufacturing costs relative to conventional sintering, pressure sintering is usually reserved for manufacturing ceramics that are difficult to sinter to high density by conventional sintering, such as nonoxide ceramics, and to produce specialty, pore-free ceramics with improved properties, such as CaF_2 radomes and hard ferrite recording heads.

Hot Pressing. Hot pressing (29,45,68,78–80,86), the oldest and most widely used pressure sintering technique, involves compressing a heated powder in a die between two forming rams. The process is similar to dry pressing, except that the pressing is conducted at the sintering temperature. Hydraulic presses are typically used to apply the pressure, and graphite dies, used to temperatures in excess of 2000°C, but limited to a maximum pressure of ~100 MPa, are commonly used.

Graphite powder impurities, metal oxide reduction, pressing gradients, and preferential grain alignment can be problems in hot pressed ceramics. Additionally, hot pressing is limited to forming simple shapes, and considerable machining may be required to produce the desired finished product.

Hot Isostatic Pressing. Hot isostatic pressing (HIP) (45,68,78,87) is similar to cold isostatic pressing used in green forming with the exceptions that pressing is conducted at the sintering temperature, and the pressing medium is typically a gas. Argon or nitrogen gas is generally used, however, helium, oxygen, and mixtures of argon and oxygen have also been used. HIP allows for net and near-net forming without pressing gradients. Complex shape rotor blades and axial rotors have been formed by HIP which can be conducted at temperatures up to 2000°C and ~300 MPa (3×10^6 atm). Sintered materials with impermeable surfaces can also be subjected to HIP to higher density, a process that has successfully been used to produce improved property $BaTiO_3$ multilayer capacitors and hard ferrites for magnetic recording heads.

Materials to be subjected to HIP must be impermeable to the pressurizing gas. Glasses and refractory metals have been used to "can" ceramics, or ceramics can be presintered to the closed pore stage, ie, $\geq 92\%$ theoretical density (TD), to produce an impermeable surface for HIP.

Hot Forging. Hot forging (78) takes advantage of the plastic or plasticlike properties of ultrafine grain crystalline ceramic compacts and crystalline ceramic compacts containing a fluid liquid or viscous glass phase at the sintering temperature. Plasticlike flow by grain boundary sliding under the influence of an applied stress at the sintering temperature results in simultaneous deformation and densification. By controlling deformation with forming dies, net and near-net shapes can be formed.

10.5. Novel Sintering Methods. Novel sintering methods employ combinations of radiation and conduction heating to directly heat ceramic parts rapidly in low mass furnaces. Rapid heating often promotes faster densification relative to coarsening, resulting in densification to high relative densities faster, with less grain growth. Additionally, direct sample heating minimizes the energy wasted heating the surrounding furnace and reduces the time required for heating and cooling.

Plasma Sintering. Plasma sintering (88) offers an economical means of rapidly sintering ceramics using rapid heating rates of up to $100°C/s$ to temperatures up to $10,000°C$. Sintering times, on the order of tens of seconds, produce up to 99.5% dense, fine grain size microstructures with improved mechanical properties. Rapidly heating, sintering, and cooling by plasma sintering also offers the possibility of producing unique microstructures comprised of metastable phases.

Microwave Sintering. Microwave sintering (89) offers the potential of uniformly processing large volume parts both rapidly and economically. Rapid heating at up to $600°C/min$ to temperatures up to $2000°C$ can be achieved by microwave heating. Microwave sintering is faster than conventional sintering and may favor densification at lower temperatures.

Rapid Thermal Processing (IR Sintering). Selective and preferential heating using ir radiation provides the potential for faster heating, shorter soak times, and faster cooling. Ir sintering can be 2–3 times faster than conventional sintering, and offers improved control over microstructure development and the benefit of simultaneously processing dissimilar materials.

10.6. Chemical Vapor Deposition. Chemical vapor deposition (CVD) (45,68,78,90) is a means of fabricating bulk ceramics, or depositing ceramic coatings (qv) on solid surfaces utilizing a homogeneous or heterogeneous surface vapor chemical reaction. CVD can be used to deposit high purity, fine grain size, controlled morphology ceramic films or monoliths at relatively low temperatures. CVD is used in advanced electronics to form dielectrics in metal oxide semiconductors (MOS), transistors and capacitors, insulators, and diffusion barriers. Additionally, in a process in which a ceramic is deposited on a mandrel, CVD is used to manufacture dense, high purity cubic boron nitride, BN, specialty refractories (qv).

10.7. Reaction Formed Ceramics. A variety of specialty ceramics are produced by a combination of a chemical reaction and growth, or by simultaneous chemical reaction and consolidation using relatively novel ceramic reaction forming and thermal consolidation processes. Reaction forming processes provide the potential of producing unique ceramics and ceramic composites and high purity ceramics for specialty applications.

Directed Oxidation of a Molten Metal. Directed oxidation of a molten metal or the Lanxide process (45,68,91) involves the reaction of a molten metal with a gaseous oxidant, eg, Al with O_2 in air, to form a porous three-dimensional oxide that grows outward from the metal/ceramic surface. The process proceeds via capillary action as the molten metal wicks into open pore channels in the oxide scale growth. Reinforced ceramic matrix composites can be formed by positioning inert filler materials, eg, fibers, whiskers, and/or particulates, in the path of the oxide scale growth. The resultant composite is comprised of both interconnected metal and ceramic. Typically 5–30 vol% metal remains after processing. The composite product maintains many of the desirable properties of a ceramic; however, the presence of the metal serves to increase the fracture toughness of the composite.

Self-Propagating High Temperature Synthesis. Self-propagating high temperature synthesis (SHS) (45,78,92) offers the potential for simultaneous synthesis and densification. Combustion front temperatures range from $\sim2200-2700°C$ and the combustion wave propagates at 0.1–100 cm/s. Sintered parts tend to be porous; however, higher densities can be achieved by applying pressure during the combustion process or by melt casting. SHS has been used successfully to produce corrosion-resistant coatings on pipes.

Vapor and Reactive Liquid Infiltration. Chemical vapor infiltration (CVI) (45,90,93,94) is a modified version of CVD whereby a matrix is chemically deposited within a porous preform by a solid–vapor chemical reaction. CVI provides a means of making fiber-reinforced ceramic matrix composites without chemically, thermally, or mechanically damaging the fragile fibers. Particulate composites are also formed by CVI which involves vapor-phase mass transport over distances corresponding to the size of the preform. The mass of the preform increases with time as the vapor deposits and reacts on free surfaces within the preform. CVI requires a porous preform and the reaction proceeds until the pore channels plug or close off at $\sim90\%$ TD and the reacting gas can no longer infiltrate the preform. CVI is a time-intensive process taking days to weeks to complete; however, an extremely high purity, fine grain matrix devoid of sintering aids or residual glassy phases is produced. Products manufactured by CVI include reaction bonded silicon nitride (RBSN) and reaction bonded silicon oxynitride that have good corrosion resistance to fluorides and chlorides. Additional products include carbon and silicon carbide fiber and matrix composites for aerospace applications and heat exchangers (see ABLATIVE MATERIALS).

Reactive liquid infiltration (45,68,90,93,94) is similar to the CVI process used to make RBSN. Driven by capillarity, a reactive liquid infiltrates a porous preform and reacts on free surfaces. Reactive liquid infiltration is used to make reaction bonded silicon carbide (RBSC), which is used in advanced heat engines and as diffusion furnace components for semiconductor wafer processing.

11. Postsintering Processes

Traditional ceramic processing is generally complete after forming and firing. In contrast, advanced ceramics used in engineering applications may require postsintering processing (95,96). Machining may be required to meet dimensional tolerances and surface finish requirements, thermal annealing may be required to optimize properties, or a surface coating or modification may be required to alter the properties of a ceramic surface for a given application.

11.1. Machining. Ceramics can be machined using mechanical, thermal, and chemical action. Techniques employed in machining ceramics include abrasive grinding, chemical polishing, electrical discharge machining, and laser machining (95,96).

11.2. Annealing. Oxygen sensitive ceramics often require a postsintering thermal anneal in a controlled partial pressure oxygen atmosphere to achieve the stoichiometry that optimizes properties (76). For example, the magnetic permeability of manganese zinc ferrite, $MnO_{0.5}ZnO_{0.5}Fe_2O_4$, and the electrical properties of $YBa_2Cu_3O_7$ ceramic superconductors are strongly dependent on oxygen concentration. Additionally, insulating metal oxides can be reduced in an oxygen deficient atmosphere to become electrical conductors, and electrically conducting metals in metal and ceramic co-fire microelectronic packaging systems can oxidize in an oxygen-rich atmosphere to become electrical insulators during sintering. Sometimes the oxygen partial pressure that favors a high sintered density is different from the one that yields optimum properties. In such cases sintering is initially conducted in one atmosphere and the properties are optimized by subsequently thermal annealing in another.

11.3. Surface Coatings. The surface properties and characteristics of a ceramic can be modified with coatings (58,68,78,97). Coatings are applied to traditional clay ware to create a stronger, impermeable surface, and for decoration. Coatings may be applied to advanced structural ceramics and ceramics used in corrosive environments to improve strength, abrasion resistance, ie, wear or erosion, and corrosion resistance. Coatings can be applied dry, as slurries, by spraying, or by vapor deposition.

Glaze coatings (58) are applied to dry or bisque-fired clay ceramics to form a strong, impermeable surface that is aesthetically pleasing. Protective ceramic coatings can also be deposited by CVD (68,90). Plasma activated CVD has been used extensively to produce diamond and diamondlike films. Diamond films can also be used to make optical coatings with a tailored refractive index.

Sol-gel coatings (97,98), which are used to produce reflective, colored, and anti-reflective coatings for glass, can also be used in optoelectronic and integrated optic applications as modulators and switches, eg, sol-gel polycrystalline ferroelectric ceramic films. Sol-gel coatings can also be used to form films of dielectric insulators (qv), high temperature superconductors, capacitors, and piezoelectrics for advanced microelectronic packaging. Common methods of applying sol-gel coatings include dipping and spin coating.

Plasma or flame spray coatings (76,95) are applied like spray glaze coatings except that on passing through the spray gun, the powder feed is superheated by a plasma or flame to produce molten droplets. Upon impinging the object surface, the molten droplets spread to form a continuous coating and are rapidly quenched and solidified. The rapid cooling of the droplets on the ware surface can lead to the formation of nonequilibrium phases with unique properties. Plasma spray coatings are used to form conductive, refractory, and erosion- and corrosion-resistant coatings.

Thin coatings can be formed by ion deposition or sputtering (95), whereby an electron beam strikes a target to produce ions that condense on the object surface. This process is used to alter the electrical properties of a surface by depositing either a conducting or insulating layer. Ion beam deposition is limited to forming thin coatings in a vacuum.

11.4. Surface Modification Techniques. Other means of modifying the surface of a ceramic include ion implantation (qv), ion exchange, and quenching. Ion implantation (95) is accomplished by physically bombarding and stuffing the surface of a ceramic with ions. Ion exchange involves the chemical exchange of a larger cation for a smaller one of equal valence in the surface. Thermal quenching by rapid cooling results in differential cooling between the surface and the bulk, freezing in a less ordered, higher volume structure at the surface relative to the bulk such as in tempered glass.

12. Processing–Microstructure–Properties

The processes employed in manufacturing a ceramic are defined and controlled to produce a product with properties suited to a specific application. Processing–microstructure–property relationships are determined by characterizing the ceramic raw materials, mixes, and the formed ceramic body intermittently during processing and after final thermal consolidation. It is possible to modify and optimize processes to optimize properties and to identify and correct processing deficiencies when less than optimal properties are obtained. Examples of some process–microstructure–property relations in advanced ceramics are outlined in Figure 4.

12.1. Characterization.
Ceramic bodies are characterized by density, mass, and physical dimensions. Other common techniques employed in characterizing include x-ray diffraction (XRD) and electron or petrographic microscopy to determine crystal species, structure, and size (100). Microscopy (qv) can be used to determine chemical constitution, crystal morphology, and pore size and morphology as well. Mercury porosimetry and gas adsorption are used to characterize pore size, pore size distribution, and surface area (100). A variety

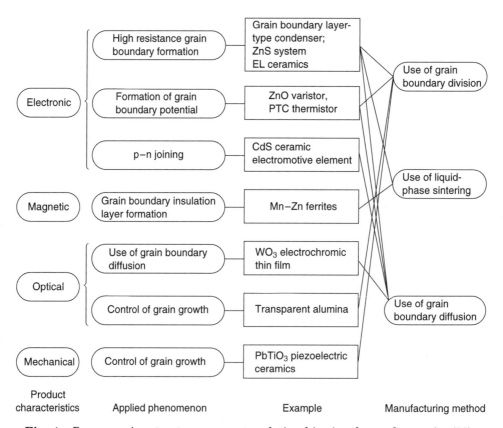

Fig. 4. Process–microstructure–property relationships in advanced ceramics (99).

of techniques can be employed to characterize bulk chemical composition and the physical characteristics of a powder (100,101).

BIBLIOGRAPHY

"Ceramic Forming Processes" under "Ceramics" in *ECT* 2nd ed., Vol. 4, pp. 776–783, by R. F. Stoops, North Carolina State of The University of North Carolina; in *ECT* 3rd ed., Vol. 5, pp. 253–259, by R. F. Stoops, North Carolina State University.

CITED PUBLICATIONS

1. J. S. Reed, *Introduction to the Principles of Ceramic Processing*, John Wiley & Sons, Inc., New York, 1988, pp. 31–46.
2. D. W. Richerson, *Modern Ceramic Engineering*, 2nd ed., Marcel Dekker, Inc., New York, 1992, pp. 374–417.
3. F. H. Norton, *Elements of Ceramics*, 2nd ed., Addison-Wesley Publishing Co., Reading, Mass., 1974, pp. 24–37.
4. F. H. Norton, in Ref. 3, pp. 38–41.
5. F. H. Norton, in Ref. 3, pp. 42–54.
6. W. D. Kingery, *Introduction to Ceramics*, John Wiley & Sons, Inc., New York, 1960, pp. 15–31.
7. J. T. Jones and M. F. Berard, *Ceramics: Industrial Processing and Testing*, The Iowa State University Press, Ames, Iowa, 1972, pp. 14–19.
8. W. E. Worrall, *Clays and Ceramic Raw Materials*, 2nd ed., Elsevier Applied Science Publishers, Ltd., New York, 1986, pp. 48–88.
9. *Ibid.*, pp. 204–219.
10. *Ibid.*, pp. 220–231.
11. W. E. Brownell, *Structural Clay Products, Applied Mineralogy*, Vol. 9, Springer-Verlag, New York, 1976, pp. 43–60.
12. F. H. Norton, in Ref. 3, pp. 55–71.
13. J. S. Reed, in Ref. 1, pp. 47–57.
14. D. W. Johnson, Jr., *Am. Ceram. Soc. Bull.* **60**(2), pp. 221–224, 243 (1981).
15. D. W. Johnson, Jr., in G. L. Messing, K. S. Mazdiyasni, J. W. McCauley, and R. A. Haber, eds., *Ceramic Powder Science, Advances in Ceramics*, Vol. 21, The American Ceramic Society, Westerville, Ohio, 1987, pp. 3–19.
16. W. H. Rhodes and S. Natansohn, *Am. Ceram. Soc. Bull.* **68**(10), pp. 1804–1812 (1989).
17. I. J. McColm and N. J. Clark, *Forming, Shaping and Working of High Performance Ceramics*, Chapman and Hall, New York, 1988, pp. 60–140.
18. H. Anderson, T. T. Kodas, and D. M. Smith, *Am. Ceram. Soc. Bull.* **68**(5) pp. 996–1000 (1989).
19. D. W. Richerson, in Ref. 2, pp. 731–807.
20. J. G. Lee and I. B. Cutler, *Am. Ceram. Soc. Bull.* **54**(2), pp. 195–198 (1975).
21. J. T. Jones and M. F. Berard, in Ref. 7, pp. 20–38.
22. J. S. Reed, in Ref. 1, pp. 255–276.
23. C. Greskovich, in F. F. Y. Wang, ed., *Ceramic Fabrication Processes, Treatise on Materials Science and Technology*, Vol. 9, Academic Press, New York, 1976.
24. R. Hogg, *Am. Ceram. Soc. Bull.* **60**(2), pp. 206–211, 220 (1981).
25. P. Somasundaran, in G. Y. Onada, Jr. and L. Hench, ed., *Ceramic Processing Before Firing*, John Wiley & Sons, Inc., New York, 1978, pp. 105–123.

26. J. S. Reed, in Ref. 1, pp. pp. 298–312.
27. T. Allen, *Particle Size Measurement*, 3rd ed., Chapman and Hall, New York, 1981, pp. 165–186.
28. T. Allen, in Ref. 27, pp. 325–349.
29. J. S. Reed, in Ref. 1, pp. 440–474.
30. T. Allen, in Ref. 27, pp. 246–266.
31. R. D. Nelson, in J. C. Williams and T. Allen, eds., *Dispersing Powders in Liquids, Handbook of Powder Technology*, Vol. 7, Elsevier Science Publishers, New York, 1988.
32. J. S. Reed, in Ref. 1, pp. 313–326.
33. *Ibid.*, pp. 123–131.
34. *Ibid.*, pp. 132–151.
35. *Ibid.*, pp. 152–173.
36. *Ibid.*, pp. 174–182.
37. T. Morse, *Handbook of Organic Additives for Use in Ceramic Body Formulation*, Montana Energy and MHD Research and Development Institute, Butte, Mont., 1979.
38. D. W. Richerson, in Ref. 2, pp. 418–518.
39. J. S. Reed, in Ref. 1, pp. 277–297.
40. F. H. Norton, in Ref. 3, pp. 92–113.
41. W. D. Kingery, in Ref. 6, pp. 33–77.
42. I. J. McColm and N. J. Clark, "Processing Stage 2—Greenbody Forming," in Ref. 17, pp. 141–207.
43. J. T. Jones and M. F. Berard, in Ref. 7, pp. 39–68.
44. J. M. Herbert, *Electrocomponent Science Monographs*, Vol. 6, Gordon and Breach Science Publishers, New York, 1985, pp. 63–94.
45. D. W. Richerson, ed., *Engineering Materials Handbook*, Vol. 4, *Ceramics and Glasses*, ASM International, Materials Park, Ohio, 1991.
46. A. L. Stuijts and G. J. Oudemans, *Proc. Brit. Ceram. Soc.* (3), pp. 81–99 (1965).
47. J. A. Mangels and G. L. Messing, *Forming of Ceramics, Advances in Ceramics*, Vol. 9, The American Ceramic Society, Westerville, Ohio, 1984.
48. J. S. Reed, in Ref. 1, pp. 329–354.
49. J. S. Reed and R. B. Runk, in Ref. 23, pp. 71–93.
50. H. Thurnauer, in W. D. Kingery, ed., *Ceramic Fabrication Processes*, John Wiley & Sons, Inc., New York, 1960, pp. 62–70.
51. G. F. Austin and G. D. McTaggart, in Ref. 23, pp. 135–151.
52. W. D. Kingery, in Ref. 50, pp. 70–73.
53. W. C. Bell, in Ref. 50, pp. 74–77.
54. J. S. Reed, in Ref. 1, pp. 355–379.
55. W. E. Brownell, in Ref. 11, pp. 61–100.
56. J. A. Mangels, *Ceram. Eng. Sci. Proc.* **3**(9–10), pp. 529–537 (1982).
57. R. M. German and K. F. Hens, *Bull. Am. Ceram. Soc.* **70**(8) pp. 1294–1302 (1991).
58. J. S. Reed, in Ref. 1, pp. 426–439.
59. *Ibid.*, pp. 380–402.
60. R. E. Cowan, in Ref. 23, pp. 153–171.
61. H. S. Magid, in Ref. 50, pp. 40–45.
62. P. D. S. St. Pierre, in Ref. 50, pp. 45–51.
63. A. C. Young, O. O. Omatete, M. A. Janney, and P. A. Menchhofer, *J. Am. Ceram. Soc.* **74**(3), pp. 612–618 (1991).
64. R. E. Mistler, D. J. Shanefield, and R. B. Runk, in Ref. 25, pp. 411–448.
65. J. C. Williams, in Ref. 23, pp. 173–197.
66. R. E. Mistler, *Am. Ceram. Soc. Bull.* **69**(6), pp. 1022–26 (1990).

67. S. J. Glass and D. J. Green, *Adv. Ceram. Mater.* **2**(2), pp. 129–131 (1987).

68. D. W. Richerson, in Ref. 2, pp. 519–595.

69. J. S. Reed, in Ref. 1, pp. 411–425.

70. F. H. Norton, in Ref. 3, pp. 114–125.

71. W. E. Brownell, in Ref. 11, pp. 101–125.

72. C. J. Brinker and G. W. Scherer, *Sol-Gel Science, The Physics and Chemistry of Sol-Gel Processing*, Academic Press, Inc., New York, 1990.

73. J. T. Jones and M. F. Berard, in Ref. 27, pp. 69–89.

74. F. H. Norton, in Ref. 3, pp. 126–153.

75. W. D. Kingery, H. K. Bowen, and D. R. Uhlmann, *Introduction to Ceramics*, 2nd ed., John Wiley & Sons, Inc., New York, 1967.

76. W. S. Coblenz, ed., in Ref. 45, pp. 242–312.

77. W. E. Brownell, in Ref. 11, pp. 126–164.

78. I. J. McColm and N. J. Clark, in Ref. 17, pp. 208–310.

79. R. L. Coble and J. E. Burke, in J. E. Burke, ed., *Progress in Ceramic Science*, Vol. 3, The MacMillan Co., New York, 1963.

80. F. Thümmler and W. Thomma, *J. Inst. Metals* **12**, pp. 69–108 (1967).

81. J. E. Burke and J. H. Rosolowski, in N. B. Hannay, ed., *Treatise on Solid State Chemistry*, Vol. 4, *Reactivity of Solids*, Plenum Press, New York, 1976.

82. R. M. German, *Liquid Phase Sintering*, Plenum Press, New York, 1985.

83. C. J. Brinker and G. W. Scherer, in Ref. 72, pp. 675–742.

84. K. G. Ewsuk, in K. M. Nair, P. Pohanka, and R. C. Buchanan, eds., *Ceramic Transactions*, Vol. 15, *Materials and Processes for Microelectronic Systems*, The American Ceramic Society, Westerville, Ohio, 1990.

85. F. H. Norton, in Ref. 3, pp. 154–162.

86. A. C. D. Chaklader, *J. Can. Ceram. Soc.* **40**, pp. 19–28 (1971).

87. H. T. Larker, in T. Garvare, ed., *Hot Isostatic Pressing; Theories and Applications*, Centek Publishers, Lulea, Sweden, 1988.

88. K. Upadhya, *Am. Ceram. Soc. Bull.* **67**(10), pp. 1691–1694 (1988).

89. W. H. Sutton, *Am. Ceram. Soc. Bull.* **68**(2), pp. 376–386 (1989).

90. D. P. Stinton, T. M. Besmann, and R. A. Lowden, *Am. Ceram. Soc. Bull.* **67**(2), 350–355 (1988).

91. M. S. Newkirk and co-workers, *Ceram. Eng. Sci. Proc.* **8**(7–8), pp. 879–885 (1987).

92. Z. A. Munir, *Am. Ceram. Soc. Bull.* **67**(2), pp. 342–349 (1988).

93. M. E. Washburn and W. S. Coblenze, *Am. Ceram. Soc. Bull.* **67**(2), pp. 356–363 (1988).

94. W. J. Lackey, *Ceram. Eng. Sci. Proc.* **10**(7–8), pp. 577–584 (1989).

95. D. W. Richerson, in Ref. 2, pp. 596–619.

96. K. Subramanian and R. F. Firestone, eds., in Ref. 45, pp. 313–376.

97. I. J. McColm and N. J. Clark, in Ref. 17, pp. 311–338.

98. C. J. Brinker and G. W. Scherer, in Ref. 72, pp. 839–880.

99. N. Ichinose, *Introduction to Fine Ceramics, Applications in Engineering*, Wiley, New York, 1987, p. 13.

100. H. D. Leigh, ed., in Ref. 45, pp. 547–627.

101. J. S. Reed, in Ref. 1, pp. 59–119.

KEVIN G. EWSUK
Sandia National Laboratories

CERIUM AND CERIUM COMPOUNDS

1. Introduction

Cerium [7440-45-1], Ce, at no. 58, is the most abundant member of the series of elements known as lanthanides. Lanthanide (Ln) is a collective name for the fifteen elements from at no. 57 (La) to 71 (Lu), also called the $4f$ elements. Rare-earth (RE) metal is the collective name for elements 21 (Sc), 39 (Y), plus 57 (La) to 71 (Lu). The label *light* is used herein for elements having atomic numbers from 57 to ~63 and the label *heavy* for numbers ~64 to 71.

Isolated first as an impure oxide in 1803, cerium was named after the earliest recognized asteroid Ceres. Cerium made its first contribution to chemical technology in the 1890s when, in Vienna, gas lights, using the Welsbach gas mantle based on a cerium-doped thorium oxide impregnated fabric, were introduced. Rapid widespread adoption of this form of illumination followed. Cerium still contributes to lighting in the 1990s but is now also to be found in automobiles, televisions, and other technologies.

Cerium, at wt 140.12; electron configuration [Xe] $4f^2 6s^2$; is characterized chemically by having two stable valence states, Ce^{3+}, cerous, and Ce^{4+}, ceric, for which the ionic radii are 114 pm and 97 pm, respectively. The easily accessible tetravalent ion is unique among the Ln series. Indeed, the ceric ion is a powerful oxidizing agent but when associated with the ligand oxygen, it is completely stabilized, and ceric oxide [1306-38-3], CeO_2, is the form of cerium most widely used. The most stable state for all lanthanides is a trivalent one, Ln^{3+}, having an electronic configuration of $[Xe]4f^n$; ie, for Ce^{3+}, $[Xe]4f^1$. The $4f$ electrons are in well-shielded inner orbitals not influenced by surrounding atoms, and hence the chemical behavior of all Ln^{3+} ions is very similar. The relative increased stability of empty $4f^0$, half-full $4f^7$, and completely full $4f^{14}$ shells, however, can, for certain elements, cause oxidation states other than three to be reasonably stable. Ce^{4+} has a $[Xe]4f^0$ configuration.

In bulk form cerium is a reactive metal that has a high affinity for oxygen and sulfur. It has a face centered cubic crystal structure, mp 798°C, bp 3443°C, density 6.77 g/mL, and a metallic radius of 182 pm. Detailed chemical and physical property information can be found in the literature (1,2).

Cerium ranks ca 25th in abundance in the earth's crust (3), and cerium, which occurs at 60 ppm crustal abundance, not lanthanum at 30 ppm, is the most abundant lanthanide.

2. Resources

There are few principal lanthanide deposits, and there are no minerals that are sources for cerium alone. All the lighter lanthanides occur together in any potential deposit, and processes separating the lanthanides are necessary to obtain pure cerium products.

Whereas certain rocks of igneous origin formed by melting and recrystallization can include minerals enriched in the lanthanides (4), cerium is usually present as a trace element rather than as an essential component. Only a few minerals in which cerium is an essential structure-defining component occur in economically significant deposits. Two minerals supply the world's cerium, bastnasite [68909-13-7], $LnFCO_3$, and monazite [1306-41-8], $(Ln,Th)PO_4$.

Bastnasite, a light Ln fluoride carbonate, occurs in an unusual type of magma-derived alkaline igneous rock in which the concentration of the Ln elements has been especially enhanced. These carbonatite magmas are produced when mantle rocks melt deep in the earth's crust in the presence of large amounts of carbonate. If fluoride ion is introduced during the ascent, the final stage of the emplacement can be specific lanthanide-containing fluorocarbonate minerals such as bastnasite.

Bastnasite has been identified in various locations on several continents. The largest recognized deposit occurs mixed with monazite and iron ores in a complex mineralization at Baiyunebo in Inner Mongolia, China. The mineral is obtained as a by-product of the iron ore mining. The other commercially viable bastnasite source is the Mountain Pass, California deposit where the average Ln oxide content of the ore is ca 9%. This U.S. deposit is the only resource in the world that is mined solely for its content of cerium and other lanthanides.

Monazite, a light lanthanide thorium phosphate, is found in many countries. It is an accessory mineral in granites, and because of a high specific gravity, on weathering of those primary rocks the mineral segregates out as placer deposits. Monazite-bearing beach and dune sand deposits occur in association with other heavy minerals such as ilmenite [12168-52-4], FeO_3Ti, zircon, and rutile [1317-80-2]. The other minerals are usually the economic driving force for exploiting the deposits and hence monazite is almost always derived as a by-product.

Several countries supply monazite concentrates for the world market. Extensive deposits along the coast of western Australia are worked for ilmenite and are the primary source of world monazite. Other regions of Australia, along with India and Brazil, also supply the mineral. Because monazite contains thorium [7440-29-1], India and Brazil have embargoed its export for many years. In the United States, commerce in the mineral is regulated by the Nuclear Regulatory Commission.

Phosphate rock, mined widely throughout the world for its fertilizer value (see FERTILIZERS), in certain regions contains a few percent of lanthanides. For example, the apatite deposits in the Kola peninsula on the Russian/Finnish border. The Ln content is recoverable from the various processing residues, and because other Ln-containing minerals, such as loparite [12173-83-0], are also found there, the location supplies a significant part of the demand in Eastern Europe.

World mine production of contained Ln oxide was estimated at 119,000 t with ca 68% of this in China, mostly at Baiyunebo, ca 4.2% in the United States, and ca 22.6% in India (5).

3. Production

The production of cerium derivatives begins with ore beneficiation and production of a mineral concentrate. Attack on that concentrate to create a suitable mixed lanthanide precursor for later separation processes follows. Then, depending on the relative market demand for different products, there is either direct production of a cerium-rich material, or separation of the mixed lanthanide precursor into individual pure lanthanide compounds including compounds of pure cerium, or both. The starting mineral determines how the suitable mixed lanthanide precursor is formed. In contrast the separation technology, which involves liquid–liquid countercurrent extraction or solvent extraction (SX), for preparing the individual lathanides is essentially independent of the starting mineral (see EXTRACTION, LIQUID–LIQUID). Thus different feedstocks can ultimately be processed by the same separation routines and equipment.

3.1. Processing of Bastnasite.

In the United States the ore is extracted by open-pit mining and the bastnasite fraction is separated by froth-flotation from other minerals including barite, calcite, strontium-containing materials, and quartz. The resulting bastnasite concentrate is a commercially available commodity. A typical analysis is given in Table 1. Bastnasite can be converted directly, without separating out the individual Ln, to other derivatives by dissolution in acid, such as sulfate or chloride, ie, $LnCl_3/RECl_3$. Bastnasite-derived rare-earth chlorides [68188-88-5] have a composition very similar to the analogous material produced from monazite.

The step used in California to crack or open the concentrate for further processing is to roast in air, whereby Ce^{3+} oxidizes to Ce^{4+}, then leach with HCl to produce an insoluble cerium-rich portion, cerium concentrate [68909-12-6], and a soluble cerium-poor lanthanum-rich fraction.

A typical analysis of the insoluble cerium concentrate portion is by wt%; CeO_2, ~62; other Ln oxides, ~10; CaO, 6; other oxides, ~4; and F^- ~10. The loss on ignition is about 8 wt%. Cerium oxide readily takes F^- ions into the lattice. The charge difference is then matched by some Ln^{3+} replacing Ce^{4+}.

The soluble fraction from the HCl leach, after a simple SX removal of heavy lanthanides, is either converted to a solid lanthanum concentrate [68188-83-0] or

Table 1. **Composition of Bastnasite Concentrate**[a]

Component	Composition, approx wt%
Lanthanides	
CeO_2	30
La_2O_3	20
Nd_2O_3	7
Pr_6O_{11}	2.4
other Ln	0.6
Non-Lanthanides	
SrO	5
CaO	4
BaO	1.5
F	5.5
SiO_2	1.5
PO_4	1
Fe_2O_3	0.5
SO_4	1

[a]Loss on ignition (carbonate) is 20%.

used as the feedstock for further separation processes to produce pure light lanthanide derivatives. The lanthanum concentrate is essentially a mixed lanthanide hydroxide chloride where the total Ln oxides is 80 wt%, Cl^- is 12 wt%, and the loss on ignition is 8 wt%. This material contains a small (12 wt%) percentage of cerium as CeO_2, and because lanthanum concentrate is consumed in quantity in fluid catalytic cracking (FCC) catalysts for gasoline production, it accounts for a portion of the net consumption of cerium. La_2O_3 is present at 46 wt%; the remaining 22 wt% consists of other lanthanides.

An alternative process for opening bastnasite is used in China: high temperature roasting with sulfuric acid followed by an aqueous leach produces a solution containing the Ln elements. Ln is then precipitated by addition of sodium chloride as a mixed sulfate. Controlled precipitation of hydroxide can remove impurities and the Ln content is eventually taken up in HCl. The initial cerium-containing product, once the heavy metals Sm and beyond have been removed, is a light lanthanide (La, Ce, Pr, and Nd) rare-earth chloride.

3.2. Processing of Monazite. Monazite, a by-product of the mineral-sands industry production of the economically more dominant minerals ilmenite, rutile, and zircon, is marketed worldwide as a concentrate in which the rare-earth oxide content is ~60%. The usual process to crack monazite is with alkali (6). After beneficiation, the monazite concentrate is finely ground and digested with an excess of caustic soda at ~150°C for several hours.

$$(Ln, Th)PO_4 + NaOH \longrightarrow Na_3PO_4 + Ln(OH)_3 + Th(OH)_4$$

A soluble sodium tripolyphosphate is produced as are insoluble lanthanide and thorium hydroxides (hydrated oxides).

The solids are treated with hydrochloric acid at 70°C at pH 3–4. The thorium hydroxide [13825-36-0] remains insoluble and can be filtered off. Small amounts of trace contaminants that carry through into solution, such as uranium and lead as well as some thorium, are removed by coprecipitation with barium sulfate in a deactivation step. The resulting product, after SX-removal of the heavy Ln fraction, is a rare-earth/lanthanide chloride, $LnCl_3 \cdot 6H_2O$. A typical analysis of this material gives from 22–25 wt% CeO_2, from 11–16 wt% La_2O_3, from 5–9 wt% Nd_2O_3, from 2–3 wt% Pr_6O_{11}, and from approximately 31 wt% chloride.

3.3. Production of Cerium Derivatives. Moderately pure (90–95%) cerium compounds can be made from rare-earth chloride through oxidation with, for example, hypochlorite to produce an insoluble cerium hydrate. The other lanthanides remain in solution. The hydrate, on calcination, converts to CeO_2.

$$2 Ce(OH)_4 \, nH_2O(s) \longrightarrow 2 CeO_2 + 4 \, nH_2O$$

The cerium concentrate derived from bastnasite can also be upgraded by dissolution and controlled reprecipitation.

Ce(IV) extracts more readily into organic solvents than do the trivalent Ln(III) ions providing a route to 99% and higher purity cerium compounds. Any Ce(III) content of mixed lanthanide aqueous systems can be oxidized to

Ce(IV) and the resulting solution, eg, of nitrates, contacted with an organic extractant such as tributyl phosphate dissolved in kerosene. The Ce(IV) preferentially transfers into the organic phase. In a separate step the cerium can be recovered by reduction to Ce(III) followed by extraction back into the aqueous phase. Cerium is then precipitated and calcined to produce the oxide.

Solvent extraction (SX) is also the process for the commercial separation of the individual light lanthanides (7). Each SX step, using nitrate or chloride solutions as the aqueous phase and kerosene as the organic phase and using chelating agents (qv) as extractants, cuts the Ln series into two parts. Consequently cerium can also be recovered from the second step in the general sequential SX processing of the Ln(III) series.

4. Cerium(IV) Chemistry

The fluorite structure, which has a large crystal lattice energy, is adopted by CeO_2 preferentially stabilizing this oxide of the tetravalent cation rather than Ce_2O_3. Compounds of cerium(IV) other than the oxide, ceric fluoride [10060-10-3], CeF_4, and related materials, although less stable can be prepared. For example ceric sulfate [13590-82-4], $Ce(SO_4)_2$, and certain double salts are known.

The tetravalent ceric ion [16065-90-0], Ce^{4+}, is the only nontrivalent lanthanide ion, apart from Eu^{2+}, stable in aqueous solution. As a result of the higher cation charge and smaller ionic size, ceric salts are much more hydrolyzed in aqueous solution than those of the trivalent lanthanides.

$$[CeL_m(H_2O)]^n + H_2O \longrightarrow [CeL_m(OH)]^{[n-1]+} + H_3O^+$$

Ceric salt solutions are strongly acidic, basic salts tend to form readily, and there are no stable simple salts of weak acids.

The double salts, ceric ammonium nitrate [16774-21-3], $(NH_4)_2[Ce(NO_3)_6]$, and ceric ammonium sulfate [19495-85-3], $(NH_4)_2[Ce(SO_4)_3]$, are stable orange compounds prepared by dissolving freshly prepared hydrated oxide in excess of the appropriate acid and adding the correct amount of ammonium salt. Cerium(IV) is present in the anion, not the cation, of these salts. The crystal structure shows that the formal name should, for example, be diammonium hexanitratocerate. The solid sulfate analogues are correctly sulfatocerates. These complex anions and other similar species are stable in aqueous solutions. Chlorocomplexes of cerium(IV) are not as stable as nitrato or sulfato derivatives; the simple cerium tetrachloride [14986-52-8], $CeCl_4$, is unstable, because Cl^- is oxidized by Ce(IV), but salts of the hexachlorocerate ion [35644-17-8], $[CeCl_6]^{2-}$, having large cations, do exist. Ceric salts range from yellow through orange to red in color, depending on the anionic form. Ceric ion in a glass matrix absorbs in the ultraviolet portion of the spectrum but not in the visible; the precise details of the spectra depend on glass composition.

Cerium in the tetravalent state is a strong oxidizing agent and can be reduced by, eg, oxalic acid, halogen acids, ferrous salts, and hydrogen peroxide. The exact oxidation potential for the one-electron reaction of Ce(IV) to Ce(III) ranges from ca 1.3 V in 1 N HCl to ca 1.8 V in 6 N HClO$_4$ and it depends on which small anionic groupings are complexed to the tetravalent cerium. Aqueous solutions containing Ce(IV) species are stable, despite the lower electrode potential for the H_2O/O_2 couple (1.23 V), probably for kinetic reasons. Cerium(IV) oxidation is a valuable tool in organic chemistry (8) and can be used for quantitative volumetric oxidation reactions called cerate oxidimetry in analytical chemistry (9).

4.1. Cerium Oxide. The most stable oxide of cerium is cerium dioxide [1306-38-3], CeO_2, also called ceria or ceric oxide. When cerium salts are calcined in air or if oxygen is present, this tetravalent Ce(IV) oxide is formed, cerium sesquioxide [1345-13-7] Ce_2O_3, can be prepared in strongly reducing conditions but is unstable in air and water, readily converting to the dioxide. Cerium has one of the highest free energies of formation for an oxide. The dioxide is soluble in mineral acids but can prove difficult to dissolve unless a trace of reducing agent such as hydrogen peroxide is added.

Ceria has the fluorite (CaF_2) structure having eight-coordinate cations and four-coordinate anions. Under some circumstances it can exhibit large deviations from stoichiometry giving CeO_{2-x}, where x can be up to 0.3. The color of the oxide is sensitive not only to stoichiometry but also to the presence of other lanthanides. Pure CeO_2 is a very pale yellow but a slight trace (ca 0.02 wt%) of Pr results in a buff color attributable to Ce(IV)–Pr(III) transitions. Grossly nonstoichiometric ceria samples are reported to be blue.

Solid solutions of ceria with trivalent ions, eg, Y and La, can readily be formed. The M^{3+} ions substitute for the tetravalent Ce and introduce one oxygen vacancy for every two Ln^{3+} ions. The dopant ions and the oxygen vacancies form charge associates. The resulting defect-fluorites have good oxide ion conductivity and are novel solid electrolytes for temperatures above ~600°C (10).

5. Cerium(III) Chemistry and Compounds

Cerium is strongly electropositive having a low ionization potential for the removal of the three most weakly bound electrons. The trivalent cerous ion [18923-26-7], Ce^{3+}, apart from its possible oxidation to Ce(IV), closely resembles, the other trivalent lanthanides in behavior.

The simple cerous salts can be prepared by dissolving the oxide, or preferably a more reactive precursor, in the appropriate acid or, when possible, produced by precipitation from solution. Upon crystallization a wide variety of hydrated species can result. These hydrates tend to be hygroscopic. Basic salts, eg, $Ce(OH)CO_3$, may be formed and these can be contaminants in the solid salts.

Ce(III) forms a water-insoluble hydroxide, carbonate, oxalate, phosphate, and fluoride; sparingly soluble sulfate and acetate; and soluble nitrate and chloride (and bromide). In solution the salts are only slightly hydrolyzed. The carbonate is readily prepared and is a convenient precursor for the preparation of other derivatives. The sparingly soluble sulfate and acetate decrease in solubility with an increase in temperature. Calcination of most Ce(III) salts results in CeO_2.

Cerous salts in general are colorless because Ce^{3+} has no absorption bands in the visible. Trivalent cerium, however, is one of the few lanthanide ions in which parity-allowed transitions between $4f$ and $5d$ configurations can take place and as a result Ce(III) compounds absorb in the ultraviolet region just outside the visible.

5.1. Hydroxide. Freshly precipitated cerous hydroxide [15785-09-8], $Ce(OH)_3$, is readily oxidized by air or oxygenated water, through poorly defined violet-tinged mixed valence intermediates, to the tetravalent buff colored ceric hydroxide [12014-56-1], $Ce(OH)_4$. The precipitate, which can prove difficult to filter, is amorphous and on drying converts to hydrated ceric oxide, $CeO_2 \cdot 2H_2O$. This commercial material, cerium hydrate [23322-64-7], behaves essentially as a reactive cerium oxide.

5.2. Carbonate. Hydrated cerous carbonate, $Ce_2(CO_3)_3 \cdot nH_2O$ when prepared under ideal conditions tends to crystallize as cerous carbonate octahydrate [16545-92-5], $n = 8$. The commercial product, prepared on a large scale by sodium carbonate addition, has a composition best represented as cerous carbonate trihydrate [64360-90-3], $n = 3$. Several other hydrates are known.

Temperature, pH, precipitation conditions, and the drying process determine the amount of water in the solid. In addition the species $Ce(OH)(CO_3) \cdot xH_2O$ and $Ce_2O_2(CO_3)$ may also be present. Thermal decomposition causes loss of crystallization water followed by the formation of ill-defined hydroxy- and oxy-species and eventually, at $\sim550°C$, CeO_2. Discrete intermediates are not seen because of the concomitant oxidation of Ce(III) to Ce(IV).

5.3. Nitrate. Cerium(III) nitrate hexahydrate [10294-41-4], $Ce(NO_3) \cdot 6H_2O$, is a commercially available soluble salt of cerium, and because of ready decomposition to the oxide, it is used, for example, when a porous solid is to be impregnated with cerium oxide. The nitrate is very soluble in water, up to about 65 wt%. It is also soluble in a wide range of polar organic solvents such as ketones, alcohols, and ethers.

5.4. Halides. Cerous chloride hydrate [19423-76-8], $CeCl_3 \cdot nH_2O$, usually with $n \sim 6$, on heating tends to form cerous oxychloride [15600-64-3], CeOCl. The anhydrous cerous chloride [7790-86-5] can be made from the hydrated salt by suppressing oxyhalide formation during thermal dehydration by the presence of hydrogen chloride or ammonium chloride. The anhydrous salt is soluble in a variety of organic solvents, eg, alcohols and ethers, has mp 817°C, and can be volatilized at high temperatures in vacuum.

Precipitation of cerous fluoride [7758-88-5], CF_3, from aqueous solution by HF addition produces $CeF_3 \cdot nH_2O$, $n \sim \frac{1}{2}$, from which the anhydrous salt can be prepared by controlled dehydration. CeF_3, mp 1432°C, can also be prepared by reaction of a suitable precursor, such as carbonate, and ammonium bifluoride, NH_4HF_2. The anhydrous chloride and fluoride are potential precursors for cerium metal production. Cerous oxyfluoride [20901-12-6], CeOF, materials have novel ionic conductivity. The oxyfluoride and oxychloride are both high melting species that often form during halide salt calcination.

5.5. Carboxylates. Cerium carboxylates, water-insoluble, can be made (11) by double decomposition and precipitation using water-soluble precursors, or by reaction of an insoluble precursor directly with the organic acid. Cerous oxalate [139-42-4], $Ce_2(C_2O_4)_3$, 2-ethylhexanoate (octanoate), naphthenate, and stearate are readily prepared.

5.6. Sulfides. Several cerium sulfides have been characterized (12) including the expected cerium(III) sulfide [12014-93-6], Ce_2S_3. Cerium monosulfide [12014-82-3], CeS, and tricerium tetrasulfide [12185-95-4], Ce_3S_4, are also known. CeS (13), bronze in color with a metallic lustre, adopts the NaCl structure with the ions Ce^{3+}, S^{2-}, and one electron in a conduction band. This sulfide has a high (in the metallic range) electrical conductivity, a high thermal conductivity, a high (ca 2715 K) melting point, and good thermal shock resistance.

Cerium(III) oxysulfide [12442-45-4], Ce_2O_2S, is a high melting stable compound that precipitates out when steel is treated with a cerium-based metal to control sulfide inclusions. Its thermodynamic properties are well known and the Ce–O–S phase diagram has been determined (14).

5.7. Miscellaneous Compounds. Among simple ionic salts cerium(III) acetate [17829-82-2] as commercially prepared, has $\sim 1\frac{1}{2}$ H_2O, has a moderate ($\sim$100 g/L) aqueous solubility that decreases with increased temperature, and is an attractive precursor to the oxide. Cerous sulfate [13454-94-9] can be made in a wide range of hydrated forms and has solubility behavior comparable to that of the acetate. Many double sulfates having alkali metal and/or ammonium cations, and varying degrees of aqueous solubility are known. Cerium(III) phosphate [13454-71-2], being equivalent to monazite, is very stable.

Derivatives such as borides, carbides, nitrides, and hydrides are best prepared by direct reaction between the elements. These metalloid-type compounds, which often show variable composition, are colored and sometimes semiconducting.

6. Metal

In bulk form cerium is a reactive metal. Pure metal is prepared by the calciothermic reduction of CeF_3.

$$2\,CeF_3 + 3\,Ca \longrightarrow 2\,Ce + 3\,CaF_2$$

A slight excess of calcium is used and the exothermic reaction, carried out in a tantalum crucible, is initiated at $\sim$900°C. After physical separation of the upper layer of immiscible fluoride slag, vacuum distillation removes unreacted volatile Ca. Cerium can also be made by the electrolytic reduction of fused chloride.

On a fresh surface the metal has a steely lustre but rapidly tarnishes in air as a result of surface formation of oxide and carbonate species. For protection against oxidation the metal is usually stored in a light mineral oil. When made finely divided, eg, on being cut, it can be strongly pyrophoric, and, for this reason is used, as the ferro-alloy mischmetal, in lighter flints and ordnance. Cerium reacts steadily with water, readily dissolves in mineral acids, and is also attacked by alkali; it reacts with most nonmetals on heating.

Cerium metal has unique solid-state properties and is the only material known to have a solid–solid critical point. Three allotropes, α, β, γ, are stable at or close to ambient conditions and have complex structural interrelationships.

6.1. Mischmetal. Mischmetal [62379-61-7] contains, in metallic form, the mixed light lanthanides in the same or slightly modified ratio as occurs in the resource minerals. It is produced by the electrolysis of fused mixed lanthanide chloride prepared from either bastnasite or monazite. Although the precise composition of the resulting metal depends on the composition of chloride used, the cerium content of most grades is always close to 50 wt%.

An alternative commercial form of a metallic mixed lanthanide-containing material is rare-earth silicide [68476-89-1], produced in a submerged electric-arc furnace by the direct reduction of ore concentrate, bastnasite, iron ore, and quartz. The resulting alloy is approximately 1/3 mischmetal, 1/3 silicon, and 1/3 iron. In addition there are some ferro-alloys, such as magnesium–ferrosilicons, derived from cerium concentrate, that contain a few percent of cerium. The consumption of metallic cerium is overwhelmingly in the mixed lanthanide form in ferrous metallurgy.

7. Economic Aspects

The worldwide production of lanthanides in 2002 was 85,500 t (5), measured as contained Ln oxide. The primary sources are given in Table 2. The rise of Chinese lanthanide production during the 1980s has become a significant factor in the global market picture.

Forecast demand for rare earths is expected to grow at the rate of 4–9%/yr through 2007 (15).

Cerium is used in several forms other than as the pure oxide. Only a small fraction of the 80,000 ton Ln total is produced as separated, relatively pure individual Ln derivatives, cerium included. The bulk of the material is consumed as concentrates, cerium included.

Table 2. **2002 Lanthanide Production, t[a]**

Brazil	200
China	75,000
India	2,700
Malaysia	450
Sri Lanka	120
United States	5,000
former USSR	2,000

[a]As contained Ln oxide, wt approximate.

Table 3. **Commercially Available Cerium-Containing Materials and Uses**

Type	Material	Use	Cerium content
A	rare-earth chloride, mischmetal	FCC[a] catalysts iron metallurgy.	principal component[b]
B	lanthanum concentrate, La–Ln chloride	FCC[a] catalysts	minor component[b]
C	cerium concentrate	glass polishing, glass decolorizing	dominant element[c]
D	oxide, nitrate, metal	autoemission catalysts, etc	> ca 90 wt %
E	oxide, salts	luminescence, catalysts	> ca 99 wt %

[a]FCC = fluid catalytic cracking.
[b]Of mixed-lanthanide composition.
[c]In oxide-type compound.

Table 4. **U.S. Exports of Cerium Compounds by Country**[a,b]

Country[c]	2000		2001	
	Gross weight, kg	Value, $	Gross weight, kg	Value, $
Australia	5,630	39,000	2,740	15,400
Belgium	86,000	1,580,000	104,000	211,000
Brazil	205,000	337,000	241,000	486,000
Canada	201,000	1,520,000	300,000	2,640,000
France	124,000	515,000	121,000	401,000
Germany	832,000	2,460,000	518,000	1,900,000
Hong Kong	60,100	359,000	35,700	357,000
India	4,070	37,800	89,400	557,000
Japan	213,000	1,550,000	462,000	2,580,000
Korea, Republic of	1,150,000	4,950,000	1,080,000	4,900,000
Malaysia	178,000	889,000	122,000	594,000
Mexico	216,000	1,490,000	232,000	1,640,000
Netherlands	206,000	697,000	11,100	96,200
Singapore	15,100	83,100	13,600	69,900
South Africa	6,000	108,000	988	10,400
Taiwan	237,000	976,000	286,000	1,260,000
United Kingdom	133,000	400,000	386,000	703,000
other	172,000	997,000	477,000	1,700,000
Total	*4,050,000*	*19,000,000*	*4,490,000*	*20,100,000*
Total estimated equivalent rare-earth oxide (REO) content	*4,050,000*	*19,000,000*	*4,490,000*	*20,100,000*

[a]Ref. 16.
[b]Data are rounded to no more than three significant digits; may not add to totals shown.
[c]Harmonized Tariff Schedule of the United States category numbers.

The various cerium-containing derivatives available commercially are summarized in Table 3.

The average prices for imported cerium compounds, excluding cerium chloride increased to $4.92/kg in 2001 from $4.57/kg in 2000 (16).

U.S. exports of cerium compounds by country are listed in Table 4. U.S. imports for consumption of cerium products by country are listed in Table 5.

8. Analytical Methods

Preliminary separation of the lanthanides as a complete group is often possible by oxalate precipitation at low pH. In most cases direct calcination, at ~1000°C, of oxalate to the oxide provides a gravimetric determination of total Ln oxide content. If necessary controlled hydroxide precipitation can reject alkaline earths. The Ln oxide can be redissolved in strong acid and titrated against EDTA or other complexing agents. Depending on analytical procedures and on the sample it may be necessary to ensure that all the Ce is in one oxidation state, eg, 3+, by

Table 5. U.S. Imports for Consumption of Cerium Compounds by Country[a,b]

Country[c]	2000 Gross weight, kg	Value, $	2001 Gross weight, kg	Value, $
cerium compounds, including oxides, hydroxides, nitrates, sulfate chlorides, oxalates:				
Austria	49,600	468,000	59,000	439,000
China	3,470,000	13,500,000	4,060,000	14,000,000
France	2,390,000	7,130,000	1,240,000	6,650,000
Japan	410,000	7,820,000	288,000	6,500,000
other	134,000	561,000	115,000	737,000
Total	*6,450,000*	*29,400,000*	*5,760,000*	*28,300,000*
Total estimated equivalent rare-earth oxide (REO) content	*4,310,000*	*29,400,000*	*3,870,000*	*28,300,000*

[a]Ref. 16.
[b]Data are rounded to no more than three significant digits; may not add to totals shown.
[c]Harmonized Tariff Schedule of the United States category numbers.

using a reducing agent such as ascorbic acid. The ceric content can be determined by oxidation of Ce^{3+} to the tetravalent state by persulfate or bismuthate then titrating the resulting Ce^{4+} with standard ferrous ammonium sulfate. Qualitative colorimetric tests are also possible.

Several instrumental methods are available for quantitative estimation of from moderate to trace amounts of cerium in other materials. X-ray fluorescence is widely available, versatile, and suitable for determinations of Ce, and any other Ln, at percent levels and lower in minerals and purer materials. The uv-excited visible luminescence of cerium is characteristic and can be used to estimate Ce content, at ppm levels, in a nonluminescing host. X-ray excited optical luminescence (17), a technique especially appropriate for Ln elements including cerium, relies on emissions in the visible, and also measures ppm values. Atomic emission spectrometry is applicable to most lanthanides, including Ce (18). The precise lines used for quantitative measurement must be chosen with care, but once set-up the technique is suitable for routine analyses.

9. Health and Safety Factors

In general the lanthanides, including cerium, have a low toxicity rating (19), especially when they are present in material having low aqueous solubility. Cesium resembles aluminum in its pharmacological action as well as in its chemical properties. The insoluble salts such as the oxalates are stated to be non-toxic even in large doses. The greatest exposures are likely to be during manufacture of cerium. Exposed workers have experienced sensitivity to heat, itching, and skin lesions (20). When orally administered poor absorption from the gastrointestinal tract tends to result in the lanthanides generally having little effect. The anion is often an important determinant in toxicity.

Cerium is a strong reducing agent. There is a moderate explosion hazard when cerium in the form of dust is exposed to flame. There is also a moderate fire hazard since cesium ignites spontaneously in air at 150–180°C (20).

Historically the use of monazite, a thorium-containing mineral, as the principal lanthanide resource led to confusion regarding the relation between radioactivity and the lanthanides. Inadequate separations produced Th-contaminated Ln products. Modern processing technology results in products that meet all regulatory requirements.

10. Uses

The technological applications of cerium rely predominantly on its high thermodynamic affinity for oxygen and sulfur, potential redox chemistry involving cerium(III) and cerium(IV), and the absorption/excitation energy bands associated with its electronic structure. Uses may be categorized as occurring in metallurgy (qv), glass (qv) and ceramics (qv), catalysis (qv), and chemicals, plus phosphors/luminescence.

The purity of the cerium-containing materials depends on the application as indicated in Table 3, and purity can mean not only percentage of cerium content but also absence of unwanted components. For some uses, eg, gasoline production catalysts, the lanthanides are often used in the natural-ratio without separation and source literature for these applications often does not explicitly mention cerium. Conversely, particularly in ferrous metallurgy, cerium is often assumed to be synonymous with rare-earth or lanthanide and these terms are used somewhat interchangeably.

10.1. Metallurgical Applications. *Steel.* Mischmetal (MM) and other cerium-containing ferro-alloys are used to improve the physical properties of high strength/low alloy (HSLA) steels (21). Cerium is added primarily to provide sulfide shape control but also as a deoxidizer or desulfurizer because steel quality is improved when the oxygen and sulfur content is minimal. Nonmetallic inclusions, particularly of certain sulfides, can deform at high working temperatures. These platelike inclusions produce layers of weakness and hence undesirable mechanical properties in the final steel. When added to the molten steel, cerium and other lanthanides combine avidly with oxygen and sulfur, even reducing other oxides and sulfides originally present, to form high melting, hard Ln oxysulfides and oxides, which do not deform later during rolling of the steel. In addition these same inclusions can provide nuclei during solidification that promote a fine-grained final product.

Modern steelmaking technology, with an emphasis during production on low levels of oxygen and sulfur, results in cleaner steels. This, together with the general downturn in steel production in the western economies, has resulted in a significantly reduced demand for mischmetal in the steel industry. In contrast, in China, which has abundant Ln reserves and less advanced steelmaking technology, mischmetal consumption in steel treatment amounts to about 75% of their Ln production.

Cast Iron. Cast irons contain carbon (qv) as the main alloying element, are heterogeneous in microstructure, and form an extensive family of materials that includes gray iron, compacted graphite, and ductile iron. The key to obtaining distinctive differences in properties between individual cast-iron types is the control of the carbon content and especially control of the morphology that graphitic carbon precipitates assume in the final product. Crystal morphology depends on tramp elements bound to the growing crystallite surfaces.

The lanthanides, and in particular cerium, are used (22) to provide this graphite morphology control, for example to produce spherulitic or vermicular crystallites. The elements are usually added as silicides or in various ferro-alloys with cerium as the principal lanthanide. The function of cerium and all the lanthanides is probably to remove free oxygen and sulfur from the melt through formation of stable lanthanide oxysulfides, akin to the role in steel technology; to initiate the special carbon crystal growth by nucleation on those oxysulfide compounds; and to tie up undesirable tramp elements, eg, Pb and Sb, as intermetallics.

Lighter Flints and Getters. Traditionally the item most widely associated with cerium has been the pyrophoric iron-mischmetal (~60%) alloy for lighter flints, in limited use. Similar low vapor pressure reactive alloys based on cerium, such as Th_2Al-MM, can also be used as getters for electronic equipment and vacuum tubes (see ELECTRONIC MATERIALS; VACUUM TECHNOLOGY).

Super Alloys. Super alloys are nickel- or cobalt-based alloys that are exceptionally heat-resistant and are used, for example, in gas turbine engines in aircraft (see COBALT AND COBALT ALLOYS; NICKEL AND NICKEL ALLOYS). One operating problem, caused by the repetitive cycling from ambient up to high temperature, is the tendency for the essential protective oxide skin on the metal surface to spall off. Several commercial alloys have ~0.05 wt% pure cerium that significantly improves this oxidation resistance, provides creep resistance, and confers a longer operating life (23). This alloy property improvement probably comes from the trapping of trace unwanted sulfur impurities from the metal crystallite boundaries plus a modification to the diffusion mechanism for oxide skin growth. The oxide skin formed at high temperatures shows less tendency to spall off when cerium is present in the alloy.

Aluminum Alloys. Aluminum alloy systems, under development for use at higher temperatures than is normally possible with aluminum, can be made by rapid solidification powder metallurgy processes (see ALUMINUM AND ALUMINUM ALLOYS; HIGH TEMPERATURE ALLOYS). The novel compositions produced have additive element concentrations, eg, of cerium, beyond those possible using conventional ingot metallurgy. One of the most promising of the lightweight alloys is an Al-8.3Fe-4.0Ce (wt%) material, having excellent properties in the 230°C to 340°C range (24).

The technique of rapid solidification enables relatively large amounts of insoluble metallic elements to be finely dispersed within atomized powders. Upon freezing very small intermetallic particles are formed, resulting, after further processing, in a high volume fraction of finely dispersed particles within the aluminum matrix and hence a dispersion strengthened alloy. The intermetallic phases, or possibly oxidic species, responsible for the dispersion strengthening are probably binary Al–Fe and ternary Al–Fe–Ce compounds.

Chromium Plating.   Chromium is deposited onto many consumer articles and industrial items to provide decoration as well as corrosion resistance. The chromium coating is produced by an electroplating (qv) process from an aqueous solution containing a chromium salt. Chromium cannot easily be plated directly from Cr(III), however, because of the high stability of the aquo-ion $[Cr(H_2O)_6]^{3+}$. It needs to be present as Cr(VI), which can be reduced to Cr(O), ie, chromium metal, through the intermediary of a protected Cr(III) species avoiding the formation of the stable hexaaquochromium(III) ion. The plating solutions must contain an anion, such as fluoride, to stabilize the active Cr(III) intermediate and prevent the aquo-ion from forming. It has proved difficult to maintain a stable F^- concentration. The addition of cerous ions, Ce^{3+}, as cerium fluoride added directly to the plating bath, gives a self-regulating electrolyte (25). The solubility of cerous fluoride is nearly independent of the temperature of the plating solution within the practical range. By ensuring that excess, undissolved CeF_3 is always present, the fluoride ion concentration can be closely controlled and quality plating achieved.

Welding Electrodes.   The electrodes used in certain welding (qv) technologies, such as inert gas tungsten-arc welding, contain a finely dispersed oxide distributed throughout the tungsten matrix. These oxide particles give an arc-strike reliability to the electrodes at lower voltages than tungsten alone can provide. Cerium oxide, at 2 wt% loading, provides an alternative to thorium oxide, a common additive that is being phased out for environmental reasons (26).

10.2. Glass and Ceramic Applications.   *Glass Polishing.*   The most efficient polishing agent for most commercial glass compositions is cerium oxide (27). This application consumes, either as a moderately pure oxide or as a cerium oxide-dominated concentrate, a significant portion of the cerium products produced annually. Commercial glass polishes (qv) are based on cerium oxide powders having defined particle sizes and controlled dispersibility in aqueous systems. The polishing process requires water, and it is a softer hydrated surface layer that is removed or reformed. In general, the polishing agent should have a Mohs' hardness of ~6.5, close to the hardness of most glasses. Cerium oxide slurried in water contains the potential polyvalent cerium atom, and redox reactions from the Ce(IV)/Ce(III) couple may well provide chemical assistance in breaking up the silicate lattice. Transient formation of complexed groupings consisting of $\cdots$Ce–O–Si$\cdots$ has been suggested.

The cerium concentrate derived from bastnasite is an excellent polish base, and the oxide derived directly from the natural ratio rare-earth chloride, as long as the cerium oxide content is near or above 50 wt%, provides an adequate glass polish. The polishing activity of the latter is better than the CeO_2:LnO ratio suggests. Materials prepared prior to any Ln purification steps are sources for the lowest cost polishes available used to treat TV face plates, mirrors, and the like. For precision optical polishing the higher purity materials are preferred.

A cerium-based abrasive material for producing a highly accurate polished surface and improved grindability has been reported (28).

Glass Decolorization.   An important use for cerium compounds is the decolorization of glass. The dominant glass produced is the soda lime-type made from inexpensive raw materials, the purity of which determines the color of the finished product. One common impurity from silica sand, iron, is a moderately strong colorant in glass and as little as 0.01 wt% can be visually detected. The coloration caused by iron results from absorption of both the ferric and ferrous ions. Glass can be decolorized without changing the total iron content by keeping all the iron in the Fe^{3+} state by addition of Ce^{4+} to the glass bath.

Cerium(IV) oxidizes ferrous ion to ferric and the cerium ions are stable under the conditions of a molten silicate–glass bath. Furthermore, cerium itself has no absorption in the visible region. Economical additions of cerium, as cerium concentrate, enable the efficient use of raw materials containing trace quantities of iron (29).

Ultraviolet Absorption. The damage sunlight causes materials from near uv (300–400 nm range) radiation, can be limited by screening out the damaging radiation through the incorporation of components that absorb at those wavelengths. Cerium(IV) in particular makes glass opaque to near uv radiation but shows no absorption in the visible; cerium(III) has similar behavior (30). Cerium-doped glass has applications in several areas, eg, medical glassware, display case glass windows, etc.

The photostability of pigments can be enhanced by surface additives that increase the provision of recombination centers for the photoproduced charge carriers that would otherwise cause chemical damage. The surface additive must be a one-electron redox couple and be present in both valence states. Cerium, through the Ce(III)/Ce(IV) interconversion, can provide this protective process, giving pigments lightfastness. The rate at which certain pigments, such as titanium dioxide, darken on exposure to light can be reduced by producing a precipitated cerium salt coating on the pigment particles (31).

Radiation-Resistant Glass. Television glass faceplates are subjected to electron bombardment by high energy electrons, particularly with the high tube voltages needed for color displays. Over time this bombardment causes discoloration or browning of the glass because of the creation of color centers, an effect which is suppressed by the addition of up to 1 wt% or so of cerium oxide to the glass. A similar suppression of gamma-ray induced discoloration is also possible and cerium-containing glasses are used in the construction of viewing windows in hot-cells in the nuclear industry. The suppression mechanism in both instances is believed to depend on the presence of both Ce^{3+} and Ce^{4+} ions within the glass lattice (32).

Photo-Sensitive Glass. On exposure to strong light, certain glass formulations develop a latent image that can, in a later step, be converted into a permanent structural or color change. This type of glass contains cerium ions that absorb ultraviolet radiation and release electrons into the glass matrix. Heat treatment causes these electrons to migrate to silver ions, also present, that initially form silver specks that in turn nucleate the crystal growth of other compounds within the glass (33). Highly detailed patterns can be produced that are not only decorative but also can help create masks, spacers, and the like for electronic uses.

Opacifier in Enamels. Cerium oxide has a high (~2.2) refractive index and is a potential opacifying agent (34) in enamel compositions used as protective coatings on metals. In addition to a high thermal stability, ceria has only one crystallographic form throughout the range of temperatures met during enameling. Opacity in porcelain enamels refers to the desired white cover-coat surface and is created when the opacifier, the oxide, precipitates out as micrometer-sized crystals during the firing on of the enamel (see ENAMELS, PORCELAIN, OR VITREOUS).

Zirconia Ceramics. Zirconia, ZrO_2, the high temperature engineering ceramic, needs an additive to produce components having high strength and toughness (see ADVANCED CERAMICS). A so-called phase stabilizer is added in order to maintain a portion of the ZrO_2 as dispersed particulates of the tetragonal phase within the matrix cubic phase at ambient temperatures. Addition of ~12 mol% of CeO_2, for example, to zirconia produces a material having exceptional toughness and good strength (35,36). Cerium oxide-doped zirconia is also used in thermal barrier spray coatings on metal surfaces (37) such as aircraft engine parts in order to reduce the high temperatures to which the metal substrate would be exposed (see METAL SURFACE TREATMENTS).

Optical Coatings. Thin surface coatings are applied to optical components to improve performance. Wideband antireflection coatings for the visible and ir regions need materials with a refractive index of ~1.5 for the best efficiency. Cerium fluoride, a stable material resistant to humidity damage, has a suitable index, 1.63 in the visible, 1.59 in the infrared, and is transparent over the range 0.5 μm to 5 μm. It is one of the compounds used to build up the multilayers deposited on lenses, sensors, and the like.

10.3. Catalytic and Chemical Applications. *Cracking Catalysts* Several catalysts used to convert crude oil to lower molecular-weight fractions, such as gasoline, contain lanthanides including cerium (38). Within the United States alone ~500 of FCC are consumed per day. An FCC unit has a lower reactor temperature and a higher regenerator temperature; the catalyst circulates between the two. FCC catalysts contain crystalline zeolites and additives embedded in an inert matrix. The zeolite, a special alumino–silicate with organic molecule-sized pores, requires cations within those pores for charge neutrality, to give catalytic reactivity in the reactor and to provide thermal stability in the regenerator.

Large highly charged ions, such as La^{3+} or Ce^{3+}, bound within the zeolite pores create a high electric field gradient strong enough to dissociate adsorbed water and provide a high surface acidity. The Ln content can reach up to 10 wt% or higher by weight of the zeolite, but not all FCC catalysts contain lanthanides. Catalysts with a range of Ln content are made, each designed to meet specific refinery needs. Cerium, because of the potential availability of the Ce^{4+} state that tends to hydrolyze at the ion-exchange pH used, is often partially removed from the precursor solutions.

The lanthanides, be they La, Ce, or others, are used to give high cracking activity to the catalysts, especially to produce low octane fuel from heavy crude oil feedstocks. Consumption of lanthanides, and hence of cerium, in FCC catalysts has altered during the decade of the 1980s because of increased demand for high octane fuels, greater feedstock availability of lighter crudes, and changes in catalyst technology. FCC Ln demand reached a peak in ~1984 when one-third of all Ln consumption was in FCC.

This trend has influenced the supply and availability of cerium, particularly in comparison to the availability of lanthanum-rich cerium-poor materials. The increase in Ln demand for FCC catalysts up to the mid-1980s, together with the need to separate out cerium in order to make the La-rich Ce-poor compositions increasingly preferred, led to a glut of Ce-based raw materials at that time.

Emission Control Catalysts.   There is strong demand for cerium as one of the catalytically active components used to remove pollutants from vehicle (auto-exhaust) emissions (16,39). The active form of cerium is the oxide that can be formed *in situ* by calcination of a soluble salt such as nitrate or by deposition of slurried oxide (see EXHAUST CONTROL, AUTOMOTIVE).

The most widely used exhaust control device consists of a ceramic monolith with a thin-walled open honeycomb structure. The accessible surface of this monolith system is increased by applying a separate coating, a wash coat, of a high surface area material such as gamma-alumina with the catalytically active species impregnated into this washcoat. The catalyst needs to oxidize hydrocarbons, convert CO to CO_2, and reduce NO_x. The whole system forms a catalytic converter that, suitably encased, is placed between the engine and the muffler/silencer unit.

In addition to platinum and related metals, the principal active component in the multifunctional systems is cerium oxide. Each catalytic converter contains 50–100 g of finely divided ceria dispersed within the washcoat. Elucidation of the detailed behavior of cerium is difficult and complicated by the presence of other additives, eg, lanthanum oxide, that perform related functions. Ceria acts as a stabilizer for the high surface area alumina, as a promoter of the water gas shift reaction, as an oxygen storage component, and as an enhancer of the NO_x reduction capability of rhodium.

The tendency for high surface area gamma-alumina to sinter and lose that crucial area during high temperature operation is retarded by the intimate addition of several percent of cerium oxide. The mechanism is still under debate but may involve a surface LN–aluminate species on the alumina.

An oxygen storage component stores oxygen under lean operating conditions, ie, fuel-poor/air-rich, and releases it under fuel-rich, air-poor conditions to continue the oxidation of unburnt hydrocarbons and the removal of carbon monoxide even when there is insufficient gaseous oxygen. CeO_2 readily provides elemental oxygen by going nonstoichiometric to CeO_{2-x} in those air-poor portions of the exhaust cycle and then reoxidizing to CeO_2 during the air-rich period. The role of the cerium oxide however is more complex than just this oxygen storage capability. Overall ceria in autoexhaust catalysts provides better low temperature performance.

A process for the treatment of gases with a high oxygen content, with a view to controlling nitrogen oxide emissions, using a catalytic composition comprising cerium oxide and/or zirconium oxide has been reported (40).

Combusion Additives.   The ability of cerium oxide to act as an oxidizing agent underlies the potential use of various cerium derivatives as additives to aid combustion (see COMBUSTION TECHNOLOGY). Diesel exhaust often contains unburnt carbonaceous material as particulate matter, and in order to reduce particle emissions the exhaust can be passed through a ceramic trap, a closed honeycomb. In order to extend the lifetime of these traps and to reduce the temperature needed for regeneration, a cerium carboxylate, in particular cerium naphthenate, can be used as an additive to the fuel (41). The cerium compound, dissolved in the fuel at a concentration equivalent to around 25–50 pmm by weight of oxide, is transformed within the engine into CeO_2, which in turn is trapped in the so-called trap oxidizer. This finely divided oxide, thoroughly dispersed throughout the trap, produces conditions under which continuous regeneration effectively occurs; burn-off happens at a lower temperature because CeO_2 catalyzes the carbon combustion.

Sulfur Oxide Removal.   In the refinery catalytic cracking process (FCC) sulfur-containing crude oil fractions can give rise to sulfur oxide in the gases emitted under the oxidizing conditions in the high temperature (750°C) regenerator unit. An additive to the actual FCC catalyst can capture this regenerator-SO_x as sulfate and later release, in the cracking, (reducing) region, a more easily trapped form of sulfur, H_2S. Several catalyst additives containing cerium and/or lanthanides can act as the SO_x control agent by forming a stable sulfate that is reducible at the operating temperatures of the riser reactor.

Cerium oxide acts as a catalytic oxidizer in a spinel-based additive (42) that aids SO_2 to SO_3 conversion and promotes the required sulfate formation. Bastnasite itself is the most economical source of cerium and can be used directly at ~1% as the capture additive (43).

Polymerization Initiator.   Some unsaturated monomers can be polymerized through the aid of free radicals generated, as transient intermediates, in the course of a redox reaction. The electron-transfer step during the redox process causes the scission of an intermediate to produce an active free radical. The ceric ion, Ce^{4+}, is a strong one-electron oxidizing agent that can readily initiate the redox polymerization of, for example, vinyl monomers in aqueous media at near ambient temperatures (44). The reaction scheme is

$$\textit{Initiation} \quad RH + Ce^{4+} \longrightarrow complex \longrightarrow R + H^+ + Ce^{3+}$$

$$\textit{Propagation} \quad R \longrightarrow R-M\cdot \longrightarrow R-M-M\cdot \text{ etc}$$

where the monomer M can be methylmethacrylate, acrylamide, etc (see ACRYLAMIDE).

Cellulose and similar materials are polyhydric alcohols having hydroxyl groups that can react with ceric ions. The resulting macroradicals provide active sites for the polymerization of monomer with the special advantage that the radicals remain attached to the backbone polymer and hence copolymerization can occur without homopolymer being formed. Ceric ions can initiate graft polymerization of vinyl monomers onto wool, starch, cotton, and the like thereby modifying the properties of the natural polymer in order to, for example, improve mechanical strength.

Dehydrogenation, Ammoxidation, and Other Heterogeneous Catalysts. Cerium has minor uses in other commercial catalysts (45) where the element's role is probably related to Ce(III)/Ce(IV) chemistry. Styrene is made from ethylbenzene by an alkali-promoted iron oxide-based catalyst. The addition of a few percent of cerium oxide improves this catalyst's activity for styrene formation presumably because of a beneficial interaction between the Fe(II)/Fe(III) and Ce(III)/Ce(IV) redox couples. The ammoxidation of propylene to produce acrylonitrile is carried out over catalytically active complex molybdates. Cerium, a component of several patented compositions (46), functions as an oxygen and electron transfer through its redox couple.

Lubrication Additive.   Cerium fluoride, CeF_3, can be used as an additive to lubricant formulations to improve extreme pressure and antiwear behavior (47). The white solid has a crystal structure that can be pictured as [CeF] layers separated by [F] atom sheets, a layer structure analogous to that of MoS_2, a material that CeF_3 resembles in properties.

Carbon Arcs.   An electric arc struck between two rods of carbon can be a source of very intense light approximating sunlight in quality. The efficiency with which the input electrical energy is converted to the radiant visible energy can be enhanced by making the electrode rod with a core of lanthanide fluoride, $(Ce,Ln)F_3$. The function of the cerium and other lanthanides is to increase the light's intensity by the absorption of energy resulting in excitation of the Ln atoms to higher energy states. These excited atoms then emit light, falling back to their normal ground energy states. The atomic emission spectra of cerium and other lanthanides have many lines in the visible.

Paint Driers and Polymer Additives.   Paints based on alkyd resins (qv) dry by the oxidation and cross-linking of unsaturated side chains. Metal catalysts are included in paint formulations to promote this drying. Cerium carboxylates, eg, the naphthenate, are used as through driers, ie, to promote drying in the body of the paint film rather than at the film's surface (48).

Silicones are oligomers and polymers, with good high temperature stability, based on the siloxane $-O-Si-O-Si-$ backbone that itself is insensitive to oxidative scission, a degradation mechanism common to $-C-C-C-$ backbone polymers. The oxidative stability, however, of the side chain groupings can remain a weak point for the molecule. Additives, in particular cerium derivatives (49), can improve properties. Metal soaps, such as the cerium octanoate, blended into the polymer can enable, for example, silicone fluids to be used at 250°C. Thermal degradation proceeds through free-radical reactions and the function of the cerium is, through the potential one-electron redox Ce(IV)/Ce(III) system, to mop up these radicals. Low levels of additive are effective because the behavior of cerium appears to be catalytic. Any Ce(III) formed is reoxidized back to the active Ce(IV) by the slow diffusion of oxygen/air through the silicone.

10.4. Phosphor/Luminescence Applications.   *Fluorescent Lighting Phosphors.*   Fluorescent lighting relies on phosphors to convert the efficient low pressure mercury-arc emission at 254 nm in the ultraviolet into energy emitted within the visible spectrum, 400 nm to 700 nm. Until the mid-1970s the phosphors of choice were halophosphates and others that produced an essentially continuous broad white-light spectrum. It was shown then that, for the visual perception of white, three line-emitting phosphors would suffice. Such a blend of three phosphors, with narrow line emissions centered at ~450, ~550, and ~610 nm, is not only more efficient in converting input electrical energy to output luminance but can also provide excellent color rendition. Cerium is an essential component in these new generation phosphors that have made possible the so-called tricolor lamps, essential for energy efficient and compact fluorescent lighting (50).

The role of cerium in these lighting phosphors is not as the emitting atom but rather as the sensitizer. The initial step in the lighting process is the efficient absorption of the 254 nm emission; Ce^{3+}, with broad absorption bands in the uv, is very suitable. This absorbed energy is then transferred to the sublattice within the crystalline phosphor; eventually the activator ion is fed and emission results. Cerium, as a sensitizer ion, is compatible in crystal lattices with other lanthanide ions, such as Eu and Tb, the usual activator atoms.

The precise choice of compound for the phosphor depends on a complex interplay of symmetry, interatomic spacing, stability, etc in the crystal structure. The initial green emitting phosphor was the aluminate $Ce_{0.67}Tb_{0.33}MgAl_{11}O_{19}$. More recently the choice has been a $(La_{0.4}Ce_{0.45}Tb_{0.15})PO_4$ compound and in addition a borate, $(Ce,Gd,Tb)MgB_5O_{10}$, is also now used.

Another Ce-containing fluorescent lighting phosphor is $Y_3Al_5O_{12}$:Ce used in some blends to convert energy from the otherwise unwanted 435 nm mercury line from the blue into the yellow, lowering the color temperature of the light. This same garnet phosphor is also used in high pressure mercury discharge lamps for the same effect.

Cathode Ray Tube Phosphors. The cerium atom, upon excitation by energetic cathode ray electrons, produces a characteristic emission (luminescence) that is usually in the blue to ultraviolet region, the precise wavelength depending on the symmetry and nature of the ions immediately surrounding the Ce atom in the host lattice. The Ce^{3+} emission is highly efficient, is broadband in character, and corresponds to a $5d-4f$ transition. Because the transition is allowed the emitting energy level has a very short ($\sim$50 ns) lifetime and the luminescence decays very rapidly. This property underlies the use of some cerium-containing phosphors in specialized cathode-ray tube applications (51).

Beam-indexing display tubes require phosphors that emit ultraviolet when struck by an electron beam. The phosphor needs a high efficiency and an extremely fast decay time; the preferred material is cerium doped ($\sim$2 atomic%) yttrium diorthosilicate, $Ce:Y_2Si_2O_7$, having a peak emission at 380 nm. A flying-spot scanner images a transparent film and converts the image to an electronic signal with a very fast phosphor covering the whole visible spectrum. Two cerium-containing components, a garnet, $Ce:Y_3Al_5O_{12}$, and a silicate, $Ce:Y_2SiO_5$, the first emitting in the range 500–650 nm and the second over the range 370–500 nm, are combined to give the correct phosphor properties.

Cando-Luminescence, Gas Mantles. The role of a gas mantle is to produce visible radiation, in excess of the expected black-body thermal radiation, from an impinging gas flame. The closely woven fabric is impregnated with the nitrate solution that decomposes on heating to leave the oxide mantle adopting the fine structure of the textile yet having reasonable mechanical stability. The resulting oxide $Ce_{0.01}Th_{0.99}O_2$ has a broad emission band centered around 500 nm covering most of the visible; the trace of cerium moves the emission from the violet end into the visible and hence the emission becomes more pleasing to the eye. In addition it is probable that ceria is acting catalytically to ensure complete combustion.

10.5. Electrorheological Fluids. Researchers have used cerium and titanium dioxide to make an improved electrorheological (ER) fluid. Using an electric field, ER fluids have variable viscosity, stiffness, and heat transference. Under an electric field, the particles in the ER fluid are polarized and organize into chain structures. This increases its viscosity. Cerium-doped titanium dioxide in dimethylsilicone oil had a shear stress five to six times higher than pure titanium dioxide. Potential applications of these fluids are in viscous clutches, variable-cushion shock absorbers, and other variable coupling devices (52).

BIBLIOGRAPHY

"Cerium" in *ECT* 1st ed., Vol. 3, pp. 634–647, by H. E. Kremers, Lindsay Light and Chemical Co.; "Cerium and Cerium Compounds" in *ECT* 2nd ed., Vol. 4, pp. 840–854, by W. L. Silvernail and R. M. Healy, American Potash & Chemical Corp.; in *ECT* 3rd ed., Vol. 5, pp. 315–327, by W. L. Silvernail, consultant; in *ECT* 4th ed., Vol. 5, pp. 728–749, by Barry T. Kilbourn, Molycorp Inc.; "Cerium and Cerium Compounds" in *ECT* (online), posting date: December 4, 2000, by Barry T. Kilbourn, Molycorp Inc.

CITED PUBLICATIONS

1. J. W. Mellor, *A Comprehensive Treatise on Inorganic and Theoretical Chemistry*, Vol. 5, Longmans, UK, 1924, Chapt. 38; R. C. Vickery, *Chemistry of the Lanthanons*, Academic Press, New York, 1953.
2. K. A. Gschneidner and L. Eyring, *Handbook on the Physics and Chemistry of Rare Earths*, Vols. 1–13, North-Holland, 1992; *Handbook of Inorganic Chemistry*, Gmelin, System 39, various dates.
3. S. R. Taylor, *Geochim. Cosmochim. Acta* **28**, 1973 (1964).
4. A. M. Clark, in P. Henderson, ed., *Rare-Earth Element Geochemistry*, Elsevier Science Publishing Co., New York, 1984, p. 33.
5. J. B. Hedrick, "Rare Earths," *Mineral Commodity Summaries*, U.S. Geological Survey, Reston, Va., Jan. 2003.
6. N. S. Narayanan and co-workers, *Mater. Sci. Forum* **30**, 45 (1988).
7. J. Kaczmarek, in K. A. Gschneidner, ed., *Industrial Applications of the Rare Earths*, ACS Symposium Series 164, 1981, p. 135.
8. W. H. Richardson, in K. B. Wiberg, ed., *Ceric Ion Oxidation of Organic Compounds*, Academic Press, New York, 1965, p. 243; G. A. Molander, *Chem. Rev.* **92**, 29 (1992).
9. G. F. Smith, *Cerate Oxidimetry*, G. Frederick Smith Chemical Co., Columbus, Ohio, 1964.
10. H. Yahiro and co-workers, *J. Appl. Electrochem.* **18**, 527 (1988).
11. K. W. Bagnall, *MTP Int. Rev. Sci., Inorg. Chem. Ser. 2* **7**, 41 (1975).
12. J. Flahaut, in Ref. 2, Vol. 4, Chapt. 31, p. 1.
13. O. H. Krikorian and P. G. Curtis, *High Temp. High Press.* **20**, 9 (1988).
14. A. Vahed and D. A. R. Kay, *Met. Trans. B* **7B**, 375 (1976).
15. "Rare Earths and Yttrium," *Roskill Metals and Minerals Reports*, 2003.
16. J. B. Hedrick, "Rare Earths," *Minerals Yearbook*, U.S. Geological Survey, Reston, Va., 2001.
17. A. P. D'Silva and V. A. Fassel, in Ref. 2, Vol. 4, Chapt. 37E.
18. G. W. Johnson and T. E. Sisneros, in G. McCarthy and co-eds., *The Rare Earths in Modern Science and Technology*, Vol. 3, Plenum Press, 1982, p. 525; K. Jyrkas and M. Leskela, *J. Less Common Metals* **126**, 291 (1986).
19. T. J. Haley, *J. Pharm. Sci.* **54**, 663 (1965); T. J. Haley, in Ref. 2, Chapt. 40; P. Arvela, *Prog. Pharmacology* **2**(3), 69 (1979).
20. R. J. Lewis, Sr., *Sax's Dangerous Properties of Industrial Materials*, 10th ed., Vol. 2, John Wiley & Sons, Inc., New York, 2000.

21. P. E. Waudby, *Int. Metals Rev.* **2**, 74 (1978); L. A. Luyckx, in Ref. 7, p. 43.

22. M. J. Lalich, *Foundry Met. Treat. (3)*, 118 (1978); H. F. Linebarger and co-workers, in Ref. 7, p. 20.

23. F. Cosandey, *Met. Trans.* **14A**, 611 (1983).

24. U.S. Pat. 4,464,199 (Aug. 7, 1984), G. J. Hildeman and R. E. Sanders; R. A. Rainen and J. C. Ekvall, *J. Metals (5)*, 16 (1988).

25. U.S. Pat. (Jan. 11, 1972), E. J. Seyb, Jr. (to M & T Chemicals Inc.); M. A. Schluger and co-workers, *Zh. Prikl. Khim. (Leningrad)* **51**(9), 2105 (1978).

26. M. Uchio, *Pure Appl. Chem.* **60**(5), 809 (1988); A. A. Sadek and co-workers, *Met. Trans. A* **21A**, 3221 (1990).

27. T. Izumitani and Sh. Harada, *Wiss. Z. Friedrich-Schiller-Univ. Jena Math.-Naturwiss. Reihe* **2–3**, 389 (1979); W. L. Silvernail and co-workers, *Optical World* (Aug. 7, 1980); R. V. Horrigan, in Ref. 7, p. 95.

28. U. S. Pat. 6,585,787 (July 1, 2003), H. Yamasaki, Y. Uchino, A. K. Takahashi (to Mitsu Mining & Smelting Co.).

29. A. P. Herring and co-workers, *Glass Ind.* **51**(7), 316 (1970); **51**(8), 350 (1970); **51**(9), 394 (1970); T. C. Shutt and co-workers, *Ceram. Bull.* **51**(2), 155 (1972).

30. A. Paul and co-workers, *J. Mater. Sci. (England)* **11**, 2082 (1976).

31. U.S. Pat. 2,365,171 (Dec. 19, 1944), E. C. Botti (to E. I. du Pont de Nemours & Co., Inc.); U.S. Pat. 4,022,632 (May 10, 1977), G. C. Newland and co-workers, (to Eastman Kodak); U.S. Pat. 4,461,810 (July 24, 1984), H. W. Jacobson (to E. I. du Pont de Nemours & Co., Inc.).

32. A. M. Bishay, *J. Am. Ceram. Soc.* **45**(8), 389 (1962); H.-G. Byhan, *Silikattechnik* **33**(12), 359 (1982).

33. S. D. Stookey and co-workers, *J. Appl. Phys.* **49**(10), 5114 (1978).

34. A. I. Nedeljikovic and R. L. Cook, *Vitreous Enameller* **26**(1–2), 2 (1975).

35. S. Meriani, *Mat. Sci. Eng.* **A109**, 121 (1989).

36. *Ceram. Bull.* **65**(10), 1386 (1986).

37. J. W. Holmes and B. H. Pilsner, *Thermal Spray, Proceedings of the National Thermal Spray Conference 1987*, ASM, 1988, p. 259.

38. J. Scherzer, in R. G. Bautista and M. M. Wong, eds., *Rare Earths, Extraction, Preparation, and Applications*, TMS, 1988, p. 317.

39. G. Kim, *Ind. Eng. Chem. Prod. Res. Dev.* **21**, 267 (1982); H. C. Yao and co-workers, *J. Cat.* **86**, 254 (1984); B. Harrison and co-workers, *Platinum Met. Rev.* **32**(2), 73 (1988); M. Funabiki and co-workers, *Catal. Today* **10**, 33 (1991).

40. U.S. Pat. 6,548,032 (April 15, 2003), P. Bazthe, C. Hedovin, and T. Seguelong (to Rhodia Chemie).

41. K. Pattas and co-workers, *Cordierite Filter Durability with Cerium Fuel Additive: 100,000 km of Revenue Service in Athens*, SAE Technical Paper 920363, Society of Automotive Engineers, Warrendale, Pa., 1992; U.S. Pat. 4,522,631 (June 11, 1985), A. M. Mourao and C. H. Falst.

42. A. A. Bhattacharyya and co-workers, *Ind. Eng. Chem. Res.* **27**(8), 1356 (1988).

43. U.S. Pat. 4,341,661 (July 27, 1982), K. Baron and D. P. McArthur.

44. D. J. McDowall and co-workers, *Prog. Polym. Sci.* **10**, 1 (1984).

45. B. T. Kilbourn, *J. Less. Comm. Metals.* **126**, 101 (1986).

46. J. F. Brazdil and R. K. Graselli, *J. Catal.* **79**, 104 (1983).

47. J. M. Dumdum and co-workers, "Lubricant Grade Cerium Fluoride: A New Solid Lubricant Additive for Greases, Pastes and Suspensions, paper presented at *Annual Meeting National Lubrication Grease Institute, Oct. 23–26, 1983*, Kansas City, Mo.

48. P. Ducros and J. Less, *Comm. Metals* **11**, 37 (1985).

49. U.S. Pat. 3,142,655 (1964), W. J. Bobear (to General Electric); M. Heidingsfeldova and co-workers, *Kautsch. Gummi Kunstst.* **37**(8), 694 (1984).

50. B. M. J. Smets, *Mat. Chem. Phys.* **16**, 283 (1987).
51. A. Bril and co-workers, *Philips Tech. Rev.* **332**, 125 (1971).
52. J. Yin and X. Zhao, *J. Physics, D, Appl. Physics* **34** 2063–2067 (2001).

BARRY T. KILBOURN
Molycorp Inc.

CESIUM AND CESIUM COMPOUNDS

1. Introduction

Cesium [7440-46-2], Cs, is a member of the Group 1 (IA) alkali metals. It resembles potassium and rubidium in the metallic state, and the chemistry of cesium is more like that of these two elements than like that of the lighter alkali metals.

Cesium, first discovered by Bunsen and Kirchoff in 1860 while examining spring water, was the first element discovered spectroscopically (1). The name, comes from the Latin *caesius,* sky blue, and refers to the characteristic blue spectral lines of the element. Cesium salts were not successfully reduced to metal until 1881. Electrolysis of the molten chloride did not yield cesium metal under the same conditions that led to the reduction of the other alkali metal chlorides.

Cesium was first produced in the metallic state by electrolysis of a molten mixture of cesium and barium cyanides (2). Subsequently the more common thermochemical–reduction techniques were developed (3,4). There were essentially no industrial uses for cesium until 1926, when it was used for a few years as a getter and as an effective agent in reducing the electron work function on coated tungsten filaments in radio tubes. Development of photoelectric cells a few years later resulted in a small but steady consumption of cesium and other applications for cesium in photosensing elements followed.

Until the late 1970s, cesium continued to be little more than a research element. Much of its limited production was for research into thermionic power conversion, magnetohydrodynamics (qv), and ion propulsion. Although the potential for these applications has not materialized, cesium chemical usage has increased significantly as catalysts in the chemical and petrochemical industries and in biotechnical engineering (see GENETIC ENGINEERING; NUCLEIC ACIDS).

2. Physical Properties

Pure cesium is a silvery white, soft, ductile metal. Surface alteration by minute traces of oxygen result in the metal taking on a golden hue. Of the stable alkali metals, ie, excluding francium, cesium has the lowest boiling and melting point,

Table 1. **Physical Properties of Cesium**

Property	Value
atomic number	55
atomic weight	132.905
melting point, °C	29
boiling point, °C	685
specific gravity, kg/m^3	
solid at 17°C	1892
liquid at 40°C	1827
atomic radius,[a] nm	0.274
ionic radius,[b] nm	0.165
viscosity at mp, mPa · s (=cP)	0.686
surface tension at mp, mN/m	39.4
heat of fusion, ΔH_{fus}, kJ/mol[c]	2.13
heat of vaporization, ΔH_{vap}, at 0.1 MPa, J/mol[c,d]	65.9
specific heat, J/g · °C[c]	
solid at 20°C	0.217
liquid at bp	0.239
vapor at bp	0.156
thermal conductivity, W/(m · °C)	
liquid at mp	18.4
vapor at bp	4.6×10^{-3}
ionization potential, eV	3.893
work function, eV	1.91
standard electrode potential, V	−2.923
electrical conductivity, $(\Omega \cdot m)^{-1}$	
solid at mp	4.9×10^6
vapor at 1250°C	2.0×10^4
Moh's hardness	0.2
Brinell hardness, kg/mm^2	0.015

[a]The metal is 12 coordinate.
[b]The ion is usually six coordinate.
[c]To convert J to cal, divide by 4.184.
[d]To convert MPa to psi, multiply by 145.

the highest vapor pressure, the highest density, and the lowest ionization potential. These properties and the large radius of the monovalent cesium ion have important consequences directly related to applications (5). Selected physical properties are given in Table 1.

3. Chemical Properties

The ionization potential of the alkali metals decreases with increasing atomic number; consequently cesium is generally far more reactive than the lower members of the alkali metal group. When cesium is exposed to air, the metal ignites spontaneously and burns vigorously producing a reddish violet flame to form a mixture of cesium oxides. Similarly cesium reacts vigorously with water to form cesium hydroxide, the strongest base known, as well as hydrogen; together with air and water a hydrogen explosion usually occurs as the burning cesium readily ignites the liberated hydrogen gas. Cesium, the most active of the alkali metals toward oxygen and the halogens, is the least reactive toward nitrogen, carbon, and hydrogen.

Cesium salts are, in general, chemically similar to other alkali metal salts. The solubility of alkali metal salts of simple anions generally increases with the atomic weight of the alkali ion; in contrast the solubility of the alkali metal salts of complex anions generally decreases with increasing atomic weight. The salts of cesium and simple anions are usually hygroscopic as well as very soluble, but the sparingly soluble salts of cesium and complex anions are seldom hydrated and are usually not hygroscopic.

Cesium forms simple alkyl and aryl compounds that are similar to those of the other alkali metals (6). They are colorless, solid, amorphous, nonvolatile, and insoluble, except by decomposition, in most solvents except diethylzinc. As a result of exceptional reactivity, cesium aryls should be effective in alkylations wherever other alkaline alkyls or Grignard reagents have failed (see GRIGNARD REACTIONS). Cesium reacts with hydrocarbons in which the activity of a C–H link is increased by attachment to the carbon atom of doubly linked or aromatic radicals. A brown, solid addition product is formed when cesium reacts with ethylene, and a very reactive dark red powder, triphenylmethylcesium [76-83-5], $(C_6H_5)_3CCs$, is formed by the reaction of cesium amalgam and a solution of triphenylmethyl chloride in anhydrous ether.

4. Occurrence

Cesium is the rarest of the naturally occurring alkali metals, ranking fortieth in elemental prevalence. Nevertheless, it is widely distributed in the earth's crust at very low concentrations. Granites contain an average of ~1 ppm (7), sedimentary rocks ~4 ppm (8), and seawater ~0.2 ppm (9). Higher concentrations are found in lepidolite [1317-64-2], a lithium mica containing ~0.5% Cs_2O but reaching 1.9% on rare occasions (10), in carnallite [1318-27-0], $KMgCl_3 \cdot 6H_2O$, a double salt of potassium and magnesium chlorides containing 10–40 ppm Cs_2O; in the rare mixed-cesium–antimony–tantalum oxide cestibtantite (11); as well as in muscovite, beryl, spodumene, potassium feldspars, leucite, petalite, and related minerals. Both lepidolite and carnallite have yielded commercial quantities of cesium (12) and a process for its recovery from muscovite micas has been developed (13).

By far, the most important commercial cesium source is pollucite [1308-53-8], ideally $Cs_2O \cdot Al_2O_3 \cdot 4SiO_2$. The theoretical cesium content of pure pollucite is 45 wt% Cs_2O; however, natural pollucite usually contains 5–32% Cs_2O because of other minerals intimately associated with the pollucite. Additionally, the Cs is often replaced in the crystal lattice by varying amounts of Rb, K, or Na plus H_2O. Natural pollucite can be regarded as an intermediate mineral in the isomorphous series analcime [1318-10-1], $Na_2O \cdot Al_2O_3 \cdot 4SiO_2 \cdot 2H_2O$, and theoretical pollucite (14). Pollucite is a clear to milky greyish mineral with uneven fracture, similar to quartz but from which it can generally be distinguished by the ubiquitous veinlets of alteration products that occur in the pollucite (14). Alternatively, a qualitative field test that produces a bright red reaction stain (15) can be used to ascertain its presence.

Table 2. **World Mine Production, Reserves, and Reserve Base of Cesium**[a]

Country	Reserves[b], kg	Reserve base[b], kg
Canada	70,000,000	73,000,000
Namibia		9,000,000
Zimbabwe	23,000,000	23,000,000
other countries	na	na
World total (rounded)	*100,000,000*	*110,000,000*

[a]From Ref. 18, estimated, based on the occurrences of pollucite.
[b]na = not available.

Economic concentrations of pollucite usually occur in highly zoned complex pegmatites, associated with lepidolite, petalite, and spodumene. The Bernic Lake orebody of Tantalum Mining Corp. of Canada Ltd. (Tanco) in southeastern Manitoba, Canada, is the world's largest cesium source containing approximately two-thirds of the known ore. The pollucite occurs as essentially monomineralic zones within a flat lying pegmatite having a spatial extent of some 3 km^2. The main zones contain over 400,000 tons of pollucite grading ~24% Cs_2O with another zone of 100,000 tons of 5% ore (16). Tanco's production accounts for about two-thirds of the world's requirements. Other significant ore deposits are the Bikita pegmatite in Zimbabwe, and in the Karibib desert of Namibia (17). Other, smaller concentrations occur in China and Brazil, where some production takes place, as well as Scandinavia, Czechoslovakia, Afghanistan, and the United States. Some ore was mined from a deposit in Maine between World Wars I and II, but there has been no U.S. production for several decades. Russia has some low-grade resources, but the evidence points to them not being exploited, presumably because of low grade. The known high grade reserves of pollucite are sufficient to supply a steady demand for well over a century, and since 1960, cesium has essentially been produced from pollucite.

The estimates of reserves and reverse base are based upon occurrences of pollucite and are shown in Table 2 (18).

5. Processing of Pollucite

Pollucite preparation consists simply of mining the ore, crushing it to required size, followed in some instances by hand picking. No other concentration is required. Although flotation processes for the concentration of low grade ores have been developed in both the United States and Russia (19), these have commercially lagged because of the availability of high grade ores (20). Chemtall 6 mbh is the predominant processor worldwide. The only significant U.S. producer is Fluids (Woodland, Texas). Their production plant is in Bernic Lake, Manitoba, Canada. Canada is the major source of cesium ores. Other sources are found in United Kingdom (18)

There are three basic methods of converting pollucite to cesium metal or compounds: direct reduction with metals; decomposition with bases; and acid digestion. In each case, grinding of the ore to 75 µm precedes conversion.

5.1. Direct Reduction with Metals. Pollucite can be directly reduced by heating the ore in the presence of calcium to 950°C in a vacuum (21), or in the presence of either sodium or potassium to 750°C in an inert atmosphere (22). Extraction is not complete. Excessive amounts of the reducing metal is required and the resultant cesium metal is impure except when extensive distillation purification is carried out. Engineering difficulties in this process are significant, hence, this method is not commercially used.

5.2. Decomposition with Bases. Alkaline decomposition of pollucite can be carried out by roasting pollucite with either a calcium carbonate–calcium chloride mix at 800–900°C or a sodium carbonate–sodium chloride mix at 600–800°C followed by a water leach of the roasted mass, to give an impure cesium chloride solution that is separated from the gangue by filtration (23). The solution can then be converted to cesium alum [7784-17-0], $Cs_2SO_4 \cdot Al_2(SO_4)_3 \cdot 24H_2O$. Extraction of cesium from the pollucite is almost complete. Solvent extraction of cesium carbonate from the cesium chloride solution using a phenol in kerosene has also been developed (24).

5.3. Acid Digestion. Acid digestion of pollucite is the primary commercial process for cesium production. Hydrofluoric, hydrobromic, hydrochloric, and sulfuric acids can be used. Hydrofluoric acid has been used in Germany (25); it gives the most complete cesium recovery, but the inherent difficulties with its use eliminate any current advantage. Likewise, the proposed hydrobromic acid process (26), which converts pollucite to the bromide that is precipitated using isopropyl alcohol from which the cesium is removed by liquid bromine, is not in use in part because of the engineering problems associated with the hot acids.

Hydrochloric acid digestion takes place at elevated temperatures and produces a solution of the mixed chlorides of cesium, aluminum, and other alkali metals separated from the siliceous residue by filtration. The impure cesium chloride can be purified as cesium chloride double salts such as cesium antimony chloride [14590-08-0], $4CsCl \cdot SbCl_3$, cesium iodine chloride [15605-42-2], Cs_2Cl_2I, or cesium hexachlorocerate [19153-44-7], $Cs_2[CeCl_6]$ (27). Such salts are recrystallized and the purified double salts decomposed to cesium chloride by hydrolysis, or precipitated with hydrogen sulfide. Alternatively, solvent extraction of cesium chloride direct from the hydrochloric acid leach liquor can be used.

Sulfuric acid digestion has been investigated by several laboratories, including the Canadian Mines Branch (CANMET), Ottawa, Canada (28). Pollucite is digested at 110°C, close to the boiling point of 35–40% sulfuric acid, followed by a hot water wash and filtration. Cesium alum is crystallized from the leach filtrate by stage cooling to 50°C and then to 20°C, and roasted in the presence of 4% carbon

$$Cs_2SO_4 \cdot Al_2(SO_4) \cdot 4H_2O + 1.5\,O_2 + 3\,C \longrightarrow 24\,H_2O + Cs_2SO_4$$
$$+ 3\,SO_2 + 3\,CO_2 + Al_2O_3$$

The residue is leached to give cesium sulfate solution, which can be converted to cesium chloride by ion exchange on Dowex 50 resin and elution with 10% HCl, treatment using ammonia or lime, to precipitate the aluminum, or by solvent extraction, followed by purification at neutral pH using hydrogen peroxide or ammonia.

In one process developed by Carus Corp., Illinois, pollucite is digested with sulfuric acid to cesium alum that is dissolved in an aqueous hydroxide solution to form cesium alum hydroxide and potassium sulfate, from which cesium permanganate is directly precipitated by addition of potassium permanganate (29).

$$CsAl(SO_4)_2 \cdot 2\,H_2O + 4\,KOH \longrightarrow CsAl(OH)_4 + 2\,K_2SO_4 + 12\,H_2O$$

$$CsAl(OH)_4 + KMnO_4 \longrightarrow CsMnO_4(s) + KAl(OH)_4$$

Alternatively, permanganate can be added to the cesium chloride resulting from hydrochloric acid digestion, after removal of excess iron and alumina by precipitation as hydroxides, followed by centrifugation and filtration (30) to pure cesium permanganate. The resultant cesium permanganate can be converted to the carbonate or chloride by reduction using an agent such as methanol.

6. Production of Cesium Metal

6.1. Thermochemical Methods.   Cesium halides can readily be reduced using calcium or barium, but not magnesium. Purified cesium chloride and calcium in roughly equal proportions are heated together to 700–800°C under a vacuum or in an atmosphere of an inert gas such as argon or helium, and 90–95% of the cesium is distilled as metal (31), although lower temperatures (300–400°C) and higher vacuum achieve the same result but with higher purity (32).

$$2\,CsCl + Ca \longrightarrow CaCl_2 + 2\,Cs$$

Magnesium is used to obtain cesium metal from cesium hydroxide [21351-79-1], Cs(OH), cesium carbonate [534-17-8], Cs_2CO_3, or cesium aluminate [20281-00-9], $Cs_2O \cdot Al_2O_3$, according to the following equations (32):

$$2\,CsOH + 2\,Mg \longrightarrow 2\,MgO + H_2 + 2\,Cs$$

$$Cs_2CO_3 + 3\,Mg \longrightarrow 3\,MgO + C + 2\,Cs$$

$$CsO \cdot Al_2O_3 + Mg \longrightarrow MgO \cdot Al_2O_3 + 2\,Cs$$

Vacuum redistillation at low temperature is used for final purification of the cesium metal, if required.

6.2. Thermal Decomposition.   Cesium azide [22750-57-8], CsN_3, which is prepared by reacting aqueous solutions of cesium sulfate and barium azide, melts at 326°C and decomposes at 390°C to cesium metal (33):

$$2\,CsN_3 \longrightarrow 3\,N_2 + 2\,Cs$$

6.3. Electrolytic Reduction.   The extreme reactivity and relatively high volatility of cesium metal combine to make conventional fused-salt electrolysis of cesium salts impractical for the direct production of cesium metal, although it can be used as one stage in metal production. For example, electrolysis of fused cesium chloride using a molten lead cathode at 700°C results in a cesium–lead alloy containing ~8.5% Cs, from which cesium metal can be distilled at 600–700°C under vacuum (28). Alternatively, electrolysis of concentrated aqueous solutions using a mercury cathode can be followed by distillation of the amalgam resulting in cesium metal (34).

7. Cesium Alloys

Eutectics melting at about −30, −47, and −40°C are formed in the binary systems, cesium–sodium at about 9% sodium, cesium–potassium at ~25% potassium, and cesium–rubidium at about 14% rubidium (35). A ternary eutectic with a melting point of about −72°C has the composition 73% cesium, 24% potassium, and 3% sodium. Cesium and lithium are essentially completely immiscible in all proportions.

Cesium does not alloy with or attack cobalt, iron, molybdenum, nickel, platinum, tantalum, or tungsten at temperatures up to 650°C (36).

8. Cesium Compounds

Cesium compounds are manufactured and distributed by a comparatively large number of companies, considering the size of the total cesium market. Those companies that process pollucite produce their own range of products, some of which are then reprocessed and refined by other, smaller, specialty companies, many of which are located in the United States.

8.1. Carbonates.   Cesium carbonate [534-17-8], Cs_2CO_3, mol wt 325.82, specific gravity 4072 kg/m^3, has theoretical cesium content of 81.58%. It is a colorless, very hygroscopic, crystalline solid, which is stable up to its melting point of 610°C at which temperature it decomposes. The carbonate is prepared from the hydroxide by the addition of carbon dioxide, but it can also be prepared by decomposing the nitrate with excess oxalic acid to form the oxalate and igniting and decomposing the cesium oxalate to the carbonate.

Cesium hydrogen carbonate [15519-28-5], $CsHCO_3$, mol wt 193.92, theoretical cesium content 68.54%, is a colorless, slightly hygroscopic, crystalline solid having a specific gravity of ~1400 kg/m^3, which decomposes at 175°C. It has a solubility of 2.1 kg/L.

8.2. Cesium Chromate.   Cesium chromate [13454-78-9], Cs_2CrO_4, has a mol wt 381.80, and a theoretical cesium content of 69.62 wt%.

8.3. Cesium Halides.   Cesium bromide, [7787-69-1], CsBr, mol wt 212.82, theoretical cesium content 62.45%, is a colorless crystalline solid, having a melting point of 636°C, a specific gravity of 4433 kg/m^3, and a solubility of 1.23 kg/L of water at 25°C. It is usually prepared by neutralizing the carbonate or hydroxide with HBr, but it is also the primary product of the Dow process (26) for pollucite processing.

Cesium chloride, [7647-17-8], CsCl, mol wt 168.36, theoretical cesium content 78.9%, has a melting point of 646°C, a boiling point 1290°C, and a specific gravity of 3983 kg/m^3. Cesium chloride is a primary product of pollucite processing using hydrochloric acid digestion, and it is usually purified by precipitation as a complex double salt, which is then decomposed by hydrolysis or sulfide precipitation, leaving purified cesium chloride in solution. It crystallizes readily from water in well-defined, colorless, cubic crystals; its solubility is 2.7 kg/L of water at 100°C, 1.86 kg/L at 20°C, and 1.62 kg/L at 0°C. It can be formed by neutralization of the carbonate or hydroxide with hydrochloric acid.

Cesium perchlorate [13454-84-7], $CsClO_4$, mol wt 232.35 and theoretical cesium content 57.2%, is a crystalline powder that decomposes at 250°C.

Cesium fluoride [13400-13-0], CsF, mol wt 151.90, theoretical cesium content 87.49%, has a melting point of 682–703°C and a boiling point of 1253°C. Cesium fluoride is an extremely hygroscopic, colorless, crystalline solid; it has a solubility of 3.665 kg/L of water at 18°C. Cesium fluoride is made by exactly neutralizing cesium hydroxide with hydrofluoric acid and evaporating the resultant solution to dryness at 400°C. Excess HF results in a bifluoride salt that does not decompose at 400°C, and carbonate in the starting material gives an alkaline product.

Cesium iodide [7789-17-5], CsI, mol wt 259.81, theoretical cesium content 51.2%, has a melting point of 621°C and a specific gravity of 4510 kg/m^3. It is colorless, crystalline, hygroscopic, and has a solubility of 0.74 kg/L of water at 20°C, and 1.6 kg/L at 61°C. It is formed by the neutralization of the hydroxide or carbonate using hydriodic acid.

8.4. Cesium Hydroxide.   Cesium hydroxide [21351-79-1], CsOH, mol wt 149.91, theoretical cesium content 88.66%, is a colorless, crystalline, hygroscopic, anhydrous, lumpy solid, having a melting point of 272°C and a specific gravity of 3680 kg/m^3. It has a solubility of about 4 kg/L of water at 15°C and is sold both in the solid form and as a 50% solution. It is the strongest base known, and hot, concentrated cesium hydroxide rapidly attacks nickel and silver, which are often used as container materials for less reactive hydroxides. Cesium hydroxide solutions can be dehydrated in platinum at 180°C to give the monohydrate, and further dehydration at 400°C, in a vacuum, results in the anhydrous solid being formed. Carbon dioxide (qv) is absorbed rapidly from the air by both the solid and aqueous solutions of cesium hydroxide. Reaction of the solid hydroxide and carbon monoxide at atmospheric pressure and elevated temperature results in the formation of cesium formate [3495-36-1], $CsHCO_2$, cesium oxalate [18365-41-8], $Cs_2C_2O_4$, and the carbonate.

Cesium hydroxide monohydrate [35103-79-8], $CsOH \cdot H_2O$, mol wt 167.93, theoretical cesium content 79.14 wt%, is a colorless, hygroscopic, crystalline powder, having a melting point of 205–208°C and a specific gravity of 3500 kg/m^3. It is highly soluble, 8.6 kg/L of water at 15°C; similar to the anhydrous hydroxide, it is an extremely strong base.

8.5. Cesium Nitrate.   Cesium nitrate [7789-18-6], $CsNO_3$, mol wt 194.91, theoretical cesium content 68.19 wt%, crystallizes from solution in well-defined, glittering, colorless, hexagonal prisms. It has a melting point of 414°C, a specific gravity of 3685 kg/m^3, and a solubility of 0.09 kg/L of water at 0°C, and 1.97 kg/L at 100°C. Cesium nitrate is prepared from the chloride by heating with excess nitric acid until a negative test for chlorides is obtained; the reverse conversion to chloride using excess hydrochloric acid is also possible. Cesium nitrate is an ideal salt for obtaining cesium free from other alkalies, as fractional crystallization from water effectively eliminates traces of lithium, sodium, potassium, and rubidium.

8.6. Cesium Oxides. Cesium forms a series of oxides, including cesium monoxide [20281-00-9], Cs_2O, mol wt 281.81, theoretical cesium content 94.32 wt%; the suboxides: cesium heptaoxide [12433-62-4], CsO_7, tetracesium oxide [12433-60-2], Cs_4O, heptacesium dioxide [12433-63-5], Cs_7O_2, and tricesium oxide [12433-59-9], Cs_3O; cesium peroxide [12053-70-2], Cs_2O_2; and cesium superoxide [12018-61-0], CsO_2. The suboxides are formed by incomplete oxidation of cesium metal or by treating cesium monoxide with cesium metal. Cesium monoxide can be formed by direct combination of the elements or by thermal decomposition of the suboxides in the form of polycrystalline laminated plates, which are lemon yellow at $-80°C$, orange-yellow at room temperature, and cherry red $>180°C$. The crystal structures of both Cs_2O and Cs_3O have been determined by single-crystal X-ray studies (37).

Partial oxidation of cesium metal changes its color from silver-white to golden yellow; further oxidation using dry oxygen gives a black reaction product which, in the presence of oxygen at 330°C, changes to the bright yellow superoxide, CsO_2. Thermal decomposition of this product at 280–360°C yields the peroxide, without the formation of intermediate sesquioxide; further decomposition results in the monoxide (38).

8.7. Cesium Permanganate. Cesium permanganate [13456-28-5], $CsMnO_4$, mol wt 251.84, theoretical cesium content 52.77 wt%, has a specific gravity of 3597 kg/m^3, decomposes at 320°C, and is relatively insoluble at ~ 1 g/L of water. It is prepared in the Carus process by precipitation through the reaction of hydrated cesium alum with potassium permanganate (29). It can be reduced to cesium chloride or cesium carbonate using a reducing agent such as methanol.

8.8. Cesium Sulfates. Cesium sulfate [10294-54-9], Cs_2SO_4, mol wt 361.87, theoretical cesium content 73.46 wt%, forms colorless, rhombic, or hexagonal crystals and has a melting point of 1010°C and a specific gravity of 4243 kg/m^3. It can be obtained by adding a hot solution of barium hydroxide to a boiling solution of cesium alum until all the aluminum is precipitated. This equivalence point is well indicated by spot-testing for alkalinity using bromthymol blue, which has pH 7.6. Filtration yields a filtrate barren of aluminum hydroxide and barium sulfate, and the cesium sulfate is then obtained by concentration and recrystallization from water.

Cesium aluminum sulfate [7784-17-0] (cesium alum) $Cs_2SO_4 \cdot Al_2(SO_4)_3 \cdot 24H_2O$, mol wt 1136.39, theoretical cesium content 23.39 wt%, has a melting point of 117°C and a specific gravity of 1970 kg/m^3. It is the least soluble of the alkali alums, having a solubility of ~ 0.32 kg/L of water at 100°C but only 0.015 kg/L at 50°C. It therefore crystallizes first from aqueous mixed solutions of the alkali alums, as colorless, octahedral crystals and can readily be separated completely from lithium, sodium, and potassium alums by fractional distillation; the relative solubility of rubidium alum is such, however, that multiple-stage fractional crystallization is required to effect a separation. Cesium alum is the digestion product of pollucite in sulfuric acid and is one of the starting points for the production of cesium compounds.

8.9. Other Cesium Compounds. Cesium acetate [3396-11-0], $CsOOCCH_3$, mol wt 191.95, theoretical cesium content 69.24 wt%; cesium trifluoroacetate, CF_3COOCs, mol wt 245.93, theoretical cesium content 54.04 wt%; cesium–precious metal compounds such as cesium dicarbonyltetrachlororuthenium, [22594-81-6] $Cs_2RuCl_4(CO)_2$, mol wt 564.71, ruthenium content of 17.9 wt%, a yellow crystalline powder; and cesium tetrachlorogold [13682-60-5], $CsAuCl_4$, mol wt 471.7, gold content of 41.8 wt% a yellow powder; are all known.

9. Handling, Storage, and Shipment

Because of the high reactivity of cesium metal, special precautions are required for its storage, transportation, and use. Small quantities are usually contained in evacuated glass ampuls, larger quantities in stainless steel containers that are themselves contained in an outer packing, ensuring that the metal is kept from moisture or air. Most cesium compounds are hygroscopic, especially the halides, and must therefore be stored dry. Other precautions that must be taken depend on the anion. Most products are sold in polyethylene bottles inside of clamping ring steel drums.

The toxicology, occupational health hazards, and transportation regulations of cesium compounds result from the anion rather than the cesium cation. Producers and distributors provide an MSDS as well as detailed shipping requirements for each product.

10. Economic Aspects

The cesium market is very small. Most current applications require relatively small quantities of cesium and hence annual world production of cesium and compounds is estimated to be on the order of 375 t of cesium chloride equivalents. As a result, there is no official market price. However, several companies publish prices for cesium and cesium compounds. The prices have remained stable for several years (18). The per unit prices for the metal or compounds purchased from these companies varies with the quantity and quality of material purchased. As an example, one company offered 1-g ampuls of 99.98%-grade cesium metal at $50.00. The price for 100 g of the same material was $1370 (18). Technical grade compounds were priced in the range of $20–30/kg.

The United States depends on imports for 100% of its needs (18).

11. Standards and Analytical Procedures

The determination of cesium in minerals can be accomplished by X-ray fluorescence spectrometry or for low ranges associated with geochemical exploration, by atomic absorption, using comparative standards. For low levels of cesium in medical research, the proton induced X-ray emission technique has been developed (39).

Cesium metal and cesium compounds are produced and marketed in a variety of grades, eg, from 99% for technical grades to as high as 99.999% for ultrapure compounds.

Analysis and purities of the metal or compounds are determined by difference, subtracting the sum of the analyzed levels of all impurities from 100%. Analysis of impurity levels is carried out by the most appropriate technique, which may include spectroscopy, atomic absorption, and photometry.

12. Health and Safety Aspects

The cesium ion is more toxic than the sodium ion but less toxic than the potassium, lithium, or rubidium ion. No threshold limit value is stated for cesium or cesium chloride; the TLV for cesium hydroxide is 2 mg/m^3. The oral LD$_{50}$ of cesium chloride for mice is 2300 mg/kg, and for cesium flouride is 400–700 mg/kg (40).

The hydroxide, which is one of the strongest know bases, can be formed from moisture resulting in alkaline toxic effects. Cesium has been studied as indicated medically in depressive disorders (41). Because of the small-scale production of cesium products, no significant environmental problems have been encountered (18).

13. Uses

The number of commercial uses of both cesium metal and its compounds has grown significantly since the early 1980s. Cesium compounds are used in research and development, and commercially in electronic, photoelectric, and medical applications (18).

13.1. Electronic Applications.   Electronic applications make up a significant sector of the cesium market. The main applications are in vacuum tubes, photoemissive devices, and scintillation counters (see ELECTRONIC MATERIALS).

Vacuum Tubes.   In the manufacture of vacuum tubes for use in polarized ion sources, vaporized cesium is used as a getter for residual gaseous impurities in the tube and as a coating to reduce the work function of the tungsten filaments or cathodes of the tube. The cesium vapor is generated by firing, at ~850°C within the sealed and evacuated tube, a cesium chromate pellet and zirconium (12) (see VACUUM TECHNOLOGY).

$$4\,Cs_2CrO_4 + 5\,Zr \longrightarrow 5\,ZrO_2 + 8\,Cs + 2\,Cr_2O_3$$

Photoemissive Devices.   The development of the silver–oxygen–cesium photoemitter, which converts photons into free electrons, resulted in commercial exploitation of photoemissive devices; the first important use was the reproduction of sound from film, followed by such devices as the photomultiplier, iconoscope, image orthicon, and optical character recognition devices (41) (see IMAGING TECHNOLOGY; PHOTODETECTORS). The Ag–O–Cs photoemitter has relatively low sensitivity, and most photoemitters are constructed using intermetallic compounds such as antimony tricesium [12018-68-7], $SbCs_3$, and the bialkali K–Cs–Sb, rather than alloys. In the photomultiplier tube, cesium is used both in the photocathode and also as a secondary emission material in the dynode. Photomultipliers are used in a wide range of equipment, including pollution and radiation monitoring equipment; scientific, military, and medical equipment; and gamma-ray cameras.

Scintillation Counters. Cesium iodide and cesium fluoride are used in scintillation counters, which convert energy from ionizing radiation into pulses of visible light. Such units have special application in the fields of medical diagnostics, oil and mineral exploration, analysis, and space, military, and nuclear physics research (42). Thallium activated cesium iodide monocrystals have been incorporated in a synchrotron for detecting high energy gamma rays (43). Cesium iodide and cesium bromide are used for the preparation of lenses, prisms, and cuvettes for use in infrared spectrometers, especially in the 500–550 nm range (see INFRARED AND RAMAN SPECTROSCOPY).

Other. Alkali chlorochromate compounds, including cesium chlorochromate, $CsCrCl_4$, are ferromagnetic substances being studied for potential application in optically read computer memory devices. Cesium has also been used in vapor glow lamps (44), vapor rectifiers, and high energy lasers (qv) (45).

13.2. Biotechnology and Medical Applications. Cesium chloride, and to a lesser extent the other halides, cesium trifluoracetate and cesium sulfate, are used in the purification of nucleic acids (qv), ie, RNA and DNA, viruses, and other macromolecules (46). Molecules are separated according to density after being subjected to a centrifugal density gradient. In medicine, cesium salts have been considered both as an antishock reagent following the administration of arsenical drugs, though a contraindication is the disturbance to heart rhythm (47), and for the treatment of epilepsy.

The isotope ^{134}Cs, $t_{1/2} = 2.05$ year, emits a β particle and is useful in radio-autography (48). Research into the use of the stable ^{133}Cs isotope and the positron-emitter ^{132}Cs, $t_{1/2} = 6.47$ day, as well as ^{131}Cs, $t_{1/2} = 9.7$ day, for cancer treatment has been carried out (49), although these processes have not yet been commercialized.

13.3. Chemical Applications. Cesium metal is used in carbon dioxide purification as an adsorbent of impurities; in ferrous and nonferrous metallurgy (qv) it can be used as a scavenger of gases and other impurities.

The performance of many metal-ion catalysts can be enhanced by doping with cesium compounds. This is a result both of the low ionization potential of cesium and its ability to stabilize high oxidation states of transition-metal oxo anions (50). Catalyst doping is one of the principal commercial uses of cesium. Cesium is a more powerful oxidant than potassium, which it can replace. The amount of replacement is often a matter of economic benefit. Cesium-doped catalysts are used for the production of styrene monomer from ethyl benzene at metal oxide contacts or from toluene and methanol as cesium-exchanged zeolites; ethylene oxide; ammonoxidation, acrolein (methacrolein); acrylic acid (methacrylic acid); methyl methacrylate monomer; methanol; phthalic anhydride; anthraquinone; various olefins; chlorinations; in low pressure ammonia synthesis; and in the conversion of SO_2 to SO_3 in sulfuric acid production.

A growing use of cesium compounds is in the field of organic synthesis, replacing sodium or potassium salts. Various cesium compounds are highly soluble in polar solvents. These compounds do not decompose as do many organic compounds, thus avoiding undesirable by-product formation. Additionally, the cesium component can be recovered and recycled. The use of cesium fluoride in esterifications is typical, eg, the synthesis of phenacyl esters (51); in the production of trialkyl phosphates, ie, a plasticiser (52); in polymerizations; and in the preparation of organoflourine compounds, such as ring fluorinated aromatics, by halogen exchange. In this latter case, it can be used either alone, as a catalyst to accelerate KF reactions, or in combination with calcium fluoride as a support reagent (53). Cesium carbonate can be used in intramolecular cyclizations; eg, cyclic oligoethers can be prepared utilizing the template effect of the cesium cation (54). Cesium hydroxide can be used for the synthesis of insoluble fatty acid esters and polyesters (52).

Molten (177–343°C) cesium hydroxide can be used in desulfurizing heavy oils. The hydroxide may be recycled by steam hydrolysis (55).

13.4. Energy Related Applications. Much research, with regard to the use of cesium in energy related processes, has resulted in little commercial application. The heightened awareness of the environmental degradation caused by fossil fuel power stations has resulted in increased research both into efficiency improvements for existing plants and into alternative power generation (qv) methods.

Cesium is ideally suited for use in magnetohydrodynamic (MHD) power generation. The metal can be used as the plasma seeding agent in closed-cycle MHD generators using high temperature nuclear reactors (qv) as the primary heat source. However, open-cycle MHD offers considerable potential for increasing the efficiency of fossil fuel fired power plants from 30–35% to 45–50%. Hot combustion gases are seeded using cesium oxide or cesium carbonate, potassium carbonate, or a mixture to form a highly conductive plasma that is accelerated through a magnetic field channel, ideally a superconducting magnet, and an electric current is generated at right angles to both the flow of plasma and the magnetic field. The off-gases thereafter pass to a conventional power generator. One of the significant potential side benefits of this process is the scrubbing of sulfur from the off-gases by the seeding material. Potassium carbonate is considerably cheaper but also much less effective than the cesium compounds; the use of a mixture of the salts has been proposed to be the best choice (56) (see PLASMA TECHNOLOGY).

MHD generated considerable interest in the 1960s and 1970s and by 1971, a 20 MW unit had been constructed in Russia. By 1986 a commercial-scale 500 MW plant was built at Ryazan. Interest elsewhere waned, but since the mid-1980s interest in the United States has again increased, resulting in plans to build pilot-plant versions (57).

One alternative method of generating electricity directly from a heat source is by the use of a cesium vapor thermionic convertor, which uses cesium to neutralize the space charge above a hot cathode that is emitting electrons toward a cooler anode. A nuclear reactor is required as the heat source because temperatures of ∼1900°C are necessary (58). Cesium has also been considered as a working fluid for high temperature Rankine-cycle, turboelectric generators (59).

Cesium oxide has been mentioned as a coating in solar photovoltaic cells (qv) (see SOLAR ENERGY), and cesium hydroxide has been considered as a partial replacement of sodium or potassium hydroxide in alkaline storage batteries (qv) especially for use at low temperatures (60), although lithium is now preferred.

Ion engines are used in satellites for orientation control. Cesium is vaporized in a vacuum and ionized as it passes through a heated porous tungsten disk, the ions are accelerated by an electric field to ~135 km/s and are neutralized by the injection of electrons and exhausted from the thruster. However, mercury-, xenon-, and argon-based ion engines are preferred.

13.5. Other Applications. The refractive index of silicate or borosilicate glass can be modified by the addition of cesium oxide, introduced as cesium nitrate or carbonate. Glass surfaces can be made resistant to corrosion or breakage by surface ion exchange with cesium compound melts or solutions. This process can also be used for the production of optical wave guides (61). A cesium–lithium–borate crystal can be used for frequency conversion of laser light (62).

Cesium metal is used for time standards based on the natural vibration of the ^{133}Cs atom, which oscillates 9,192,631,770 times/s, and in high precision oscillators to synchronize fiber optic telecommunication.

It has been suggested that cesium may be useful in the fixation of radioactive waste in a cesium-based glass and in detoxification procedures for fugitive ^{137}Cs emissions, such as at Chernobyl, Ukraine. Methods for the removal of cesium from radioactive waste liquids have been patented (63,64).

14. Cesium Isotopes

Naturally occurring cesium and cesium minerals consist only of the stable isotope ^{133}Cs. The radioactive cesium isotopes such as ^{137}Cs are generated in fuel rods in nuclear power plants (65).

Cesium isotopes can be recovered from fission products by digestion in nitric acid, and after filtration of waste the radioactive cesium phosphotungstate is precipitated using phosphotungstic acid. This technique can be used to prepare radioactive cesium metal or compounds. Various processes for removal of ^{137}Cs isotopes from radioactive waste have been developed including solvent extraction using macrocyclic polyethers (66) or crown ethers (67) and coprecipitation with sodium tetraphenylboron (68).

The radioactive isotope ^{137}Cs is important commercially in process controlinstruments and for sewage sludge sterilization. The isotope has a long half-life ($t_{1/2} = 30$ year); however, it must be well-shielded because of the high biological hazard. A method of separation of ^{131}Cs from barium has been patented (69).

BIBLIOGRAPHY

"Cesium" under "Alkali Metals and Alkali Metal Alloys" in *ECT* 1st ed., Vol. 1, pp. 453–458, by E. H. Burk5b- J. A. Morrow, and M. S. Andrew, E. I. du Pont de Nemours & Co., Inc.; "Cesium Compounds" in *ECT* 1st ed., Vol. 3, pp. 648–651, by J. J. Kennedy, Maywood

Chemical Works; "Cesium and Cesium Compounds" in *ECT* 1st ed., Suppl. 2, pp. 190–192, by J. N. Hinvard, American Potash & Chemical Corp.; in *ECT* 2nd ed., Vol. 4, pp. 855–868, by R. E. Davis, American Potash & Chemical Corp.; in *ECT* 3rd ed., Vol. 5, pp. 327–339, by C. T. Williams, Tantalum Mining Corp; in *ECT* 4th ed., Vol. 5, pp. 728–764, by Richard O. Burt, Tantalum Mining Corp.; "Cesium and Cesium Compounds" in *ECT* (online), posting date: December 4, 2000, by Richard O. Burt, Tantalum Mining Corp.

CITED PUBLICATIONS

1. W. P. Barton, *U.S. Bur. Mines Bull.* **585** (1960).
2. C. Setterberg, *Ann. Chem.* **211**, 100 (1882).
3. E. Graefe and M. Eckhardt, *Z. Anorg. Allgem. Chem.* **22**, 158 (1900).
4. L. Hackspill, *Compt. Rend.* **141**, 106 (1905).
5. M. Bick, *Cesium Chemicals from the World's Leading Producer*, Chemetall, Frankfurt, Germany.
6. N. V. Sidgwick, in *The Chemical Elements and Their Compounds*. Vol. 1, Oxford University Press, London, 1950, pp. 59–102.
7. E. L. Horstman, *Geochim. Cosmochim. Acta* **12**, 1 (1957).
8. A. A. Smales and L. Salmon, *Analyst* **80**, 37 (1955).
9. R. Greenwood, *Min. Eng.* **12**, 482 (1960).
10. P. Quensel, *Ark. Mineral. Geol.* **2**, 9 (1956).
11. A. V. Voloshin, Yu. P. Menshikov, Ya. A. Pakhomovskyi, and L. I. Poleezhaeva, *Zapiski Vses Mineral. Obshtch.* **110**, 345 (1981).
12. V. E. Plyushchev and I. V. Shakhno, *Khim. Nauka Promst.* **1**, 534 (1956).
13. V. V. Starotsin, N. V. Petrova, T. G. Gladkova, I. Yu. Chueva, and M. A. Kolenkova, *Izv. Vyssh. Uchebn. Zaved., Tsvetn. Metall.* **6**, 54 (1960).
14. P. Černy, *Short Course in Granitic Pegmatites in Science and Industry*, Mineralogical Association, Winnipeg, Canada, 1982, p. 150.
15. K. C. Dean, I. L. Nichols, and B. H. Clemmons, *J. Metals* **19**, 1198 (1966).
16. R. A. Crouse, P. Černy, D. L. Trueman, and R. O. Burt, *Can. Min. Metall. Bull.*, 142 (Feb. 1979).
17. J. J. Hogan, *Canadian Minerals Yearbook* Department of Energy, Mines and Resources, Ottawa, Canada, 1981.
18. R. G. Reese Jr., "Cesium", *Mineral Commodity Summaries*, U.S. Geological Survey, Jan. 2002.
19. Yu. F. Petrova and co-workers, *I.P. Bardin Otechestvennaya Metall*, 77 (1983).
20. K. C. Dean and I. L. Nichols, *Bur. Mines Rep. Invest.* **5940** (1962); U.S. Pat. 3,107,215 (Oct. 15, 1963), K. C. Dean (to the United States of America).
21. L. Hackspill and G. Thomas, *Compt. Rend.* **230**, 1119 (1950).
22. U.S. Pat. 3,207,598 (1965), C. E. Berthold.
23. R. E. Davis and R. E. Jones, *J. Metals* **18**, 1203 (Nov. 1966).
24. W. D. Arnold, D. J. Crouse, and B. Brown, *I EC Proc. Des. Dev.* **4**, 249 (1965).
25. F. M. Perel'man, *Cesium and Rubidium*, MacMillan, New York, 1953, pp. 5–11.
26. U.S. Pat. 2,481,455 (1949), V. A. Stenger.
27. A. P. Bayanov, Z. A. Temerdashev, and B. P. Burylev, *Zh. Prikl. Khim.* **56**, 11 (1983).
28. H. W. Parsons, A. Vezina, R. Simard, and H. W. Smith, *Mines Branch Technical Bulletin TB 50*, Department of Mines, Ottawa, Canada, 1963.
29. Can. Pat. 1,222,377 (1984), P. G. Mein (to Carus Corp.).

30. U.S. Pat. 4,447,406 (1984), P. G. Mein (to Carus Corp.).

31. L. Huntspill, *Bull. Soc. Chim.* **9**, 466 (1911).

32. P. H. Schmidt, *J. Electrochem. Soc.* **116**, 1279 (1969).

33. Eastman Kodak Co., *Chem. Week.* **93**, 42 (1963).

34. R. E. Davis, in C. A. Hempel, ed., *Encyclopedia of Electrochemistry*, Reinhold Publishing Corp., 1964.

35. C. Goria, *Gazz. Chim. Ital.* **65**, 1226 (1935).

36. J. M. Lamberti and N. D. Saunders, *NASA Technical Note D-1739*, Office of Technical Services, Washington, D.C., 1963, p. 43.

37. K.-R. Tsai, P. M. Harris, and E. M. Lassettre, *J. Phys. Chem.* **60**, 338, 345 (1956).

38. G. V. Morris, *The Thermal Decomposition of Cesium Superoxide*, Ph.D. dissertation, University of Rhode Island, Kingston, R.I., 1962.

39. J. S. C. McKee and co-workers, *J. Environ. Sci. Health, Part A* **16**, 5 (1981).

40. R. Bose and C. Pinsky, *Pharmacol. Biochem. Behav.* **18** (1983).

41. S. W. Pierce, in E. Bingham, B. Cohrssens, and C. H. Powell, eds., *Patty*'s Toxicology, 5th ed., Vol. 3, John Wiley & Sons, Inc., New York, 2001, Chapt. 46, p. 601.

42. S. Kubota and co-workers, *Nucl. Instrum. Methods Phys. Res.* **268**, 275 (1988).

43. *Am. Met. Mark.* (Dec. 3, 1985).

44. N. C. Beese, *J. Opt. Soc. Am.* **36**, 555 (1946).

45. P. Sorokin and J. R. Lankard, *J. Chem. Phys.* **54**, 2184 (1971).

46. J. Vinograd and J. E. Hearst, *Equilibrium Separation of Macromolecules and Viruses in a Density Gradient*, Springer-Verlag, Wien, Austria, 1962.

47. *The Economics of Cesium and Rubidium*, 2nd ed., Roskill Information Services, Ltd., London, U.K., 1984.

48. L. Szentkuti and W. Giese, *Hisochemie* **34**, 211 (1972).

49. C. Pinsky and co-workers, *Can. Nuclear Soc. Conf.* **2** (1983).

50. H. Prinz, *Chemspec. Eur. 90 Symp.*, 1990.

51. J. H. Clark, *Chem. Rev.* **80**, 429 (1980).

52. K. K. Ogilvie and S. L. Beaucage, *J. Chem. Soc. Chem. Commun.*, 443 (1976).

53. J. H. Clark, A. J. Hyde, and D. K. Smith, *J. Chem. Soc. Chem. Commun.*, 791 (1986).

54. B. Klieser, B. Rossa, and F. Vögtle, *Kontakte* **1**, 3 (1984).

55. Brit. Pat. 913,730 (1962) (to Esso Research & Engineering Co.).

56. P. D. Bergman and D. Beinstock, *Report of Investigations 7717*, U.S. Department of the Interior, Bureau of Mines, Washington, D.C., 1972.

57. D. Fishlock, *Financial Times*, 14 (Nov. 8, 1988).

58. J. Raloff, *Sci. News* **113**, 13 (1978).

59. T. P. Moffitt and W. Klag, *Analytical Investigation of Cycle Characteristics for Advanced Turboelectric Space Power Systems*, NASA TN D-472, Office of Technical Services, Washington, D.C., 1960.

60. U.S. Pat. 2,683,102 (1954), R. S. Coolidge.

61. Ger. Pat. 35 01 898, L. Ross and J. Rasper (to Schott Glaswerke).

62. U.S. Pat. 6,296,784 (Oct. 2, 2001), T. Sasaki and co-workers (to Research Development Corporation of Japan).

63. U.S. Pat. 6,214,234 (April. 10, 2001), R. Harjula and J. Lehto (to Ivo Power Engineering Oy).

64. U.S. Pat. 6,270,737 (Aug. 7, 2001), B. N. Zaitsev and co-workers (to the United States of America).

65. M. Bick, *Ullmans Encyclopedia of Industrial Chemistry*, 5th ed., Vol. 6, VCH Verlagsgellschaft mbH, 1986.

66. Eur. Pat. App. 73,262 (Mar. 1983), E. Blasius and K. H. Nilles (to Kernforshungszentrum Karsruhe GmbH).

67. I. H. Gerow, J. E. Smith, and W. Davis, *Sep. Sci. Technol.* **16**, 5 (1981).

68. W. H. Bond, M. K. Williams, M. C. Colvin, and G. L. Silver, *Treat. Handl. Radioact. Wastes*, 385 (1982).

69. U.S. Pat. 6,066,302 (May 23, 2000), L. A. Bray.

WILLIAM FERGUSON
DENA GORRIE
Tanco Lac du Bonnet

CHELATING AGENTS

1. Introduction

A chelating agent, or chelant, contains two or more electron donor atoms that can form coordinate bonds to a single metal atom. After the first such coordinate bond, each successive donor atom that binds creates a ring containing the metal atom. This cyclic structure is called a chelation complex or chelate, the name deriving from the Greek word chela for the great claw of the lobster (1).

Chelation is an equilibrium system involving the chelant, the metal, and the chelate. Equilibrium constants of chelation are usually orders of magnitude greater than are those involving the complexation of metal atoms by molecules having only one donor atom.

Chelating agents may be used to control metal ion concentrations. Chelation complexes usually have properties that are markedly different from both the free metal ion and the chelating agent. Consequently, chelating agents provide a means of manipulating metal ions through the reduction of undesirable effects by sequestration or through creating desirable effects such as in metal buffering, corrosion inhibition, solubilization, and cancer therapy.

Chelates and chelation reactions are abundant in nature, ranging from delicately balanced life processes depending on traces of metal ions to extremely stable metal chelates in crude petroleums. Examples of biochemical processes involving chelates include photosynthesis, oxygen transport by blood, certain enzyme reactions, ion transport through membranes, and muscle contraction. Technological applications include scale removal from steam boilers, water softening, ore leaching, textile processing, food preservation, treatment of lead poisoning, chemical analysis, tissue specific medical procedures, and micronutrient fertilization of agricultural crops.

2. Structure and Terminology

The structural essentials of a chelate are coordinate bonds between a metal atom or a stable oxo cation, M, which serves as an electron acceptor, and two or more

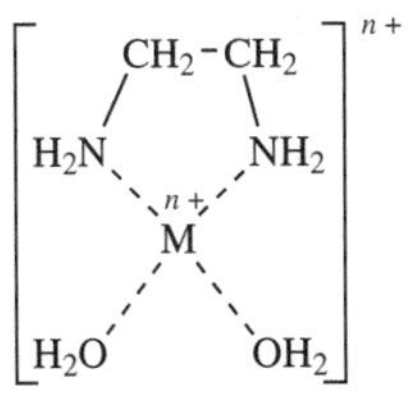

Fig. 1. Types of chelates where (**1**) represents a tetracoordinate metal having the bidentate chelant ethylenediamine and monodentate water; (**2**), a hexacoordinate metal bound to two diethylenetriamines, tridentate chelants; (**3**), a hexacoordinate metal having triethylenetetramine, a tetradentate chelant, and monodentate water; and (**4**), a porphine chelate. The dashed lines indicate coordinate bonds.

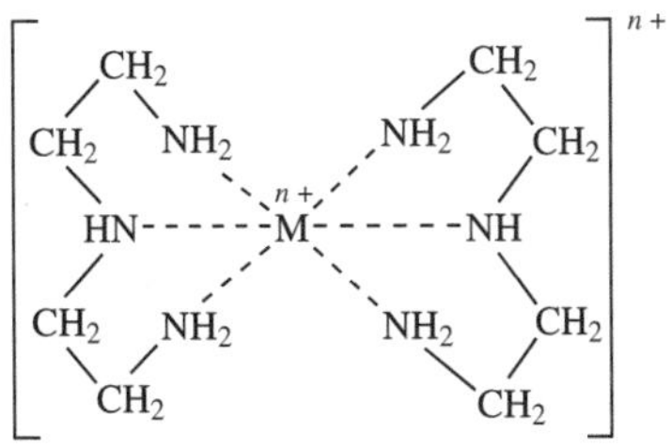

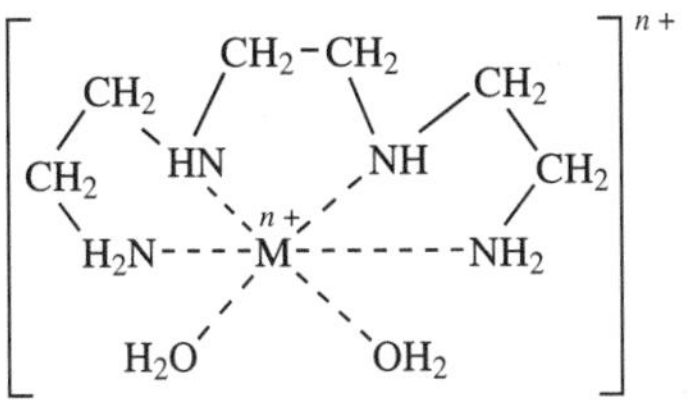

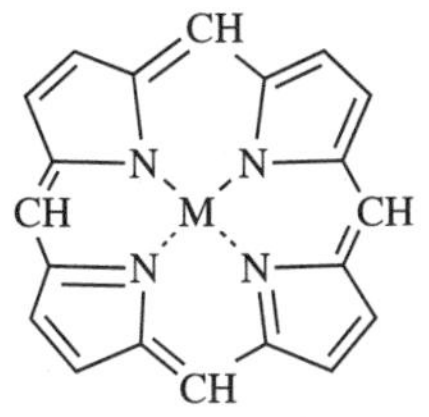

atoms in the molecule of the chelating agent, or ligand, L, which serve as the electron donors. A chelating agent may be bidentate, tridentate, tetradentate, and so on, according to whether it contains two, three, four, or more donor atoms capable of simultaneously complexing with the metal atom. Examples are shown in Figure 1. Molecules having only one donor atom, such as water and dimethylamine, are monodentates and form coordination complexes but not chelates. The principal donor atoms in practical use are N, O, and S, but P, As, and Se also form chelates. Metals are characterized by coordination numbers which correspond to the number of donor atoms bound to the central metal atom in a particular compound. The most common coordination numbers are four and six. Some metals have more than one coordination number, depending on the valence state. The term coordination number also refers, albeit less commonly, to the maximum number of donor atoms to which the metal atom can coordinate.

If the coordination number of M is greater than the number of donor atoms in the ligand L, more than one ligand molecule may combine with the metal to form the complex ML_n as in structure (**2**) where $n = 2$. Moreover, different chelating molecules can combine with the same metal atom to form species such as $L_mML'_n$. Remaining vacant coordination sites of the metal may also bind monodentate molecules as illustrated in (**1**) and (**3**). Solvated metal ions in water or other monodentate solvents are generally solvent coordinated and therefore, chelate formation should be regarded as a displacement of solvent molecules by the donor atoms of the chelating ligand.

Just as a metal can coordinate with more than one chelating molecule, a ligand having enough donor atoms in the proper configuration can bind more than one metal. These metal atoms may be the same or different.

A chelate compound may be either a neutral molecule or a complex ion associated with the appropriate counterions to produce electroneutrality. The formal charge on the complex is the algebraic sum of the charges resulting from any charge on the ligand or ligands and the charge of the metal. The ligand may be neutral or completely or partially ionized before chelation occurs. Changes in the charge of either or both metal atom and chelating agent may occur during the chelating reaction as, eg, by the displacement of hydrogen atoms or ions from the ligand donor atoms, or by an oxidation–reduction reaction between the metal and the chelant. After the chelate is formed, the charge of the complex can also change. Charge change often involves ionization of groups on the ligand that are not involved in the chelate structure, usually as a result of changes in the pH, or by oxidation or reduction of the chelated metal atom.

Most chelating agents are linear or branched chains where the donor atoms are separated by suitable numbers of other atoms to allow the formation of the chelate rings. However, there are also classes of chelating agents where the donor atoms are contained within macrocyclic structures. The spacer atoms complete the chelate rings between pairs of coordinated donor atoms, forming a pattern of fused rings centered about the metal. The porphyrins are examples of this type of chelate, and structure (4) represents chelates of porphine [101-60-0], the parent compound of the porphyrins.

Another group of macrocyclic ligands that have been extensively studied are the cyclic polyethers, such as dibenzo-[18]-crown-6 (5), in which the donor atoms are ether oxygen functions separated by two or three carbon atoms. The name crown ethers has been proposed (2) for this class of compounds because of the resemblance of their molecular models to a crown. Sandwich structures are also known in which the metal atom is coordinated with the oxygen atoms of two crown molecules.

(5)   (6)   (7)

Related to the crown ethers are compounds, such as hexamethyl-[14]-4,11-diene N_4 (6), which differ by the replacement of one or more of the oxygen atoms by other kinds of donor atoms, particularly N or S. Macrocyclic amine and thioether compounds have been synthesized. Compounds having more than one kind of heteroatom in the ring are called mixed-donor macrocycles. The naturally occurring metabolites nonactin [6833-84-7] and monactin [7182-54-9] have both ether and ester groups incorporated in the macrocyclic structure.

Three-dimensional polymacrocyclic chelating agents are formed by joining bridgehead structures with chains that contain properly spaced donor atoms. For example, bicyclic molecules result from joining nitrogen bridgeheads with chains of ($-OCH_2CH_2-$) groups as in 2.2.2-cryptate (**7**) (3). Such bicyclic structures form a cavity that holds a metal coordinated to the donor atoms in the surrounding chains. Other groups that are at least trifunctional can serve as bridgeheads, eg, pentaerythritol [115-77-5] (see ALCOHOLS, POLYHYDRIC). The donor atoms of the bridges may all be O, N, or S, or the compounds may be mixed donor macrocycles in which the bridge strands contain combinations of these donor atoms. Synthesis, metal binding, and thermodynamic properties of synthetic multidentate macrocyclic complexing agents have been reviewed (4).

Incorporating ligand groups into a cross-linked polymer structure gives the chelate-forming resins that perform ion-exchange functions by chelation. The ligand groups may either be present in the monomer before polymerization, or they may be attached to a preformed polymer by appropriate reactions. Several types are commercially available. Cross-linked styrenedivinylbenzene bonded at the nitrogen atoms to iminodiacetic acid groups is such a polymer.

3. Compounds Having Chelating Properties

Compounds with chelating properties can be found in almost any class of structures containing two or more donor atoms spatially situated so that they can coordinate with the same metal atom. The chelate rings formed contain four or more members, but for the same donor atoms, the five- or six-membered chelate rings are usually the most stable and most useful. Complexes having chelate rings of more than six members are rare. In the macrocyclic molecules, the stability of the metal complex depends strongly on the relationship between the size of the metal ion and the size of the opening within the crown or the crypt.

Chelating agents may be either organic or inorganic compounds, but the number of inorganic agents is very small. The best known inorganic chelants are polyphosphates. The annual consumption of these compounds exceeds that of all the organic chelating agents combined. Polyphosphates are less expensive than the organics but are hydrolytically unstable at high temperature and pH. Although many hundreds of organic chelating agents are known, only a few members of a few classes of compounds find extensive industrial use. One important class of organic chelating agents is the group of phosphonic acids analogous to the amino- and hydroxycarboxylic acids. These phosphonate chelants possess many of the complexing properties of the inorganic polyphosphates, particularly threshold-scale inhibition, effective at much less than stoichiometric ratios of chelant to metal ion, but unlike the polyphosphates, the phosphonates are stable in water at high temperature and pH.

Table 1 lists a number of chelating agents, grouped according to recognized structural classes. Because systematic nomenclature of chelating agents is frequently cumbersome, chelants are commonly referred to by common names and abbreviations. For the macrocyclic complexing agents, special systems of abbreviated nomenclature have been devised and are widely used. Some of the donor atoms involved in chelation and the many forms in which they can occur have been reviewed (5).

Table 1. **Classes of Chelating Agents**

Chelating agent	CAS Registry Number	Molecular formula	Abbreviation
Polyphosphates			
sodium tripolyphosphate	[7758-29-4]	$Na_5P_3O_{10}$	STPP
hexametaphosphoric acid	[18694-07-0]	$H_6O_{18}P_6$	
Aminocarboxylic acids			
ethylenediaminetetraacetic acid	[60-00-4]	$C_{10}H_{16}N_2O_8$	EDTA
hydroxyethylethylenediaminetriacetic acid	[150-39-0]	$C_{10}H_{18}N_2O_7$	HEDTA
nitrilotriacetic acid	[139-13-9]	$C_6H_9NO_6$	NTA
N-dihydroxyethylglycine	[150-25-4]	$C_6H_{13}NO_4$	2-HxG
ethylenebis(hydroxyphenylglycine)	[1170-02-1]	$C_{18}H_{20}N_2O_6$	EHPG
1,3-Diketones			
acetylacetone	[123-54-6]	$C_5H_8O_2$	acac
trifluoroacetylacetone	[367-57-7]	$C_5H_5F_3O_2$	tfa
thenoyltrifluoroacetone	[326-91-0]	$C_8H_5F_3O_2S$	TTA
Hydroxycarboxylic acids			
tartaric acid	[526-83-0]	$C_4H_6O_6$	
citric acid	[77-92-9]	$C_6H_8O_7$	cit
gluconic acid	[133-42-6]	$C_6H_{12}O_7$	
5-sulfosalicylic acid	[97-05-2]	$C_7H_6O_6S$	5-SSA
Polyamines			
ethylenediamine	[107-15-3]	$C_2H_8N_2$	en
diethylenetriamine	[111-40-0]	$C_4H_{13}N_3$	dien
triethylenetetramine	[112-24-3]	$C_6H_{18}N_4$	trien
triaminotriethylamine	[4097-89-6]	$C_6H_{18}N_4$	tren
Aminoalcohols			
triethanolamine	[102-71-6]	$C_6H_{15}NO_3$	TEA
N-hydroxyethylethylenediamine	[111-41-1]	$C_4H_{12}N_2O$	hen
Aromatic heterocyclic bases			
dipyridyl	[366-18-7]	$C_{10}H_8N_2$	dipy, bipy
o-phenanthroline	[66-71-7]	$C_{12}H_8N_2$	phen
Phenols			
salicylaldehyde	[90-02-8]	$C_7H_6O_2$	
disulfopyrocatechol	[149-46-2]	$C_6H_6O_8S_2$	Tiron, PDS
chromotropic acid	[148-25-4]	$C_{10}H_8O_8S_2$	DNS
Aminophenols			
oxine, 8-hydroxyquinoline	[148-24-3]	C_9H_7NO	Q, ox
oxinesulfonic acid	[84-88-8]	$C_9H_7NO_4S$	
Oximes			
dimethylglyoxime	[95-45-4]	$C_4H_8N_2O_2$	
salicylaldoxime	[94-67-7]	$C_7H_7NO_2$	
Schiff bases			
disalicylaldehyde 1,2-propylenediimine	[94-91-7]	$C_{17}H_{18}N_2O_2$	
Tetrapyrroles			
tetraphenylporphin	[917-23-7]	$C_{44}H_{30}N_4$	
phthalocyanine	[574-93-6]	$C_{32}H_{18}N_8$	
Sulfur compounds			
toluenedithiol (Dithiol)	[496-74-2]	$C_7H_8S_2$	tdth
dimercaptopropanol	[59-52-9]	$C_3H_8OS_2$	BAL

Table 1 (*Continued*)

Chelating agent	CAS Registry Number	Molecular formula	Abbreviation
thioglycolic acid	[68-11-1]	$C_2H_4O_2S$	
potassium ethyl xanthate	[140-89-6]	$C_3H_6OS_2 \cdot K$	
sodium diethyldithiocarbamate	[148-18-5]	$C_5H_{11}NS_2 \cdot NA$	
dithizone	[60-10-6]	$C_{13}H_{12}N_4S$	dz
diethyl dithiophosphoric acid	[298-06-6]	$C_4H_{11}O_2PS_2$	
thiourea	[62-56-6]	CH_4N_2S	
Synthetic macrocyclic compounds			
dibenzo-[18]-crown-6	[14187-32-7]	$C_{20}H_{24}O_6$	
hexamethyl-[14]-4,11-dieneN_4	[29419-92-9]	$C_{16}H_{32}N_4$	
(2.2.2-cryptate)	[23978-09-8]	$C_{18}H_{36}N_2O_6$	
Polymers			
polyethyleneimines			PEI
	[9002-98-6]	$(C_2H_5N)_x$	
	[25988-99-2]	$(C_2H_5N)_x$	
	[32167-41-2]	$(C_2H_5N)_n +$ $C_8HF_{17}O_2S$	
polymethacryloylacetone	[25120-51-8]	$(C_7H_{10}O_2)_x$	
poly(*p*-vinylbenzyliminodiacetic acid)	[30395-28-9]	$(C_{13}H_{15}NO_4)_x$	
Phosphonic acids			
nitrilotrimethylenephosphonic acid	[6419-19-8]	$C_3H_{12}NO_9P_3$	NTP, ATMP
ethylenediaminetetra (methylenephosphonic acid)	[1429-50-1]	$C_6H_{20}N_2O_{12}P_4$	EDTMP
hydroxyethylidenediphosphonic acid	[2809-21-4]	$C_2H_8O_7P_2$	HEDP

4. Nomenclature and Structural Representation

4.1. Chelating Agents. Besides the conventional empirical and structural formulas, chelating compounds and chelates are often represented by type formulas, ie, formulas that show only generalized types of structural features. Chelants having proton acid groups may be shown as H_nA or, if partially dissociated, as H_mA^{n-}. Alcohol or phenol groups that lose protons on chelation are shown as $A(OH)_n$. The letter A may be used to represent an entire multidentate ligand molecule or to show only a donor atom as in A–A–A–A for a tetradentate ligand.

For many macrocyclic ligands, simplified names are in common use. For example, crown ether nomenclature consists of four parts: (*1*) the number and type of fused rings on the polyether ring; (*2*) in square brackets the number of atoms in the polyether ring; (*3*) the word crown; and (*4*) the number of oxygen atoms in the macro ring (2). Ligand structures may be represented by any of the conventional means for depicting structure, eg, see structures (5–7).

4.2. Chelates. Because of length and complexity, systematic names of chelates are little used except for special purposes, such as where unequivocal referencing is essential. Chelates are named in the literature in a variety of ways. The name of the ligand in a chelate is usually given a suffix -o or -ato if it is a negative group but remains unchanged if the ligand is electrically neutral. Prefixes indicate the number of bound ligand molecules. The central atom is given the name of the metal, or a derivative name having the suffix -ate, eg, cuprate and ferrate, if the complex is negatively charged. Oxidation states of the metals are indicated by Roman numerals, eg, iron(III), and ionic charges are shown as part of the name by Ewens-Bassett numbers, eg, (2+) or (1−).

Chelates are often named merely as a complex, eg, cadmium complex with acetylacetone. A common practice in the literature is to give the symbol of the central atom and an abbreviation for the ligand with or without an indication of ionic charges, oxidation states, structure, or counterions, as in the following: Pb-EDTA, Cacit⁻, Cu(en)₂, Co(II)-(phen), [Cu(dipy)₃]SO₄, [Ru(dipy)₂(en)]²⁺, and Na[Co(acac)₃]. Ligand abbreviations are given in Table 1.

Several ways of representing chelates are shown in Figure 2. Structures (**8**) and (**9**) represent bidentates; structures (**10**) and (**11**), tetradentates; and structures (**12**) and (**13**) the same hexadentate. Square brackets, evident in structures

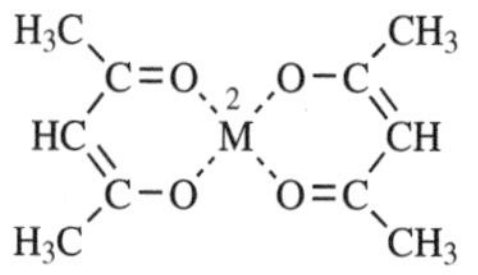

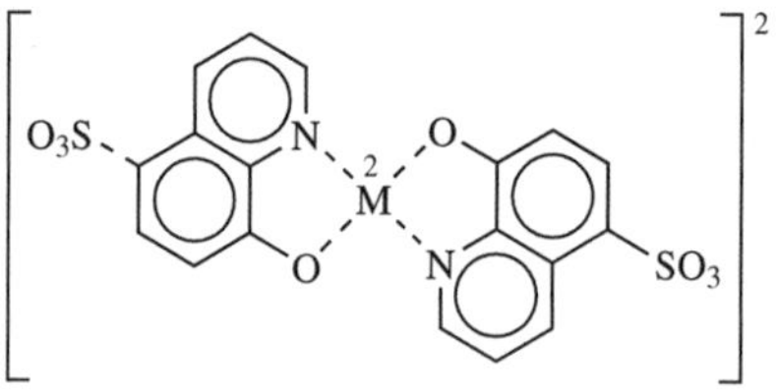

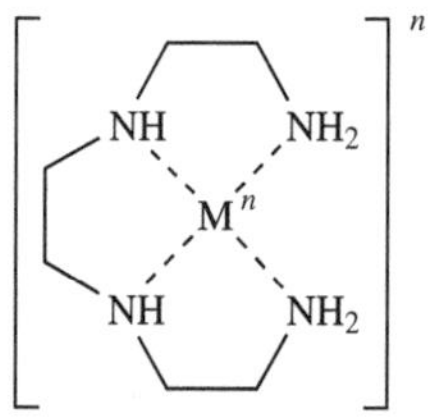

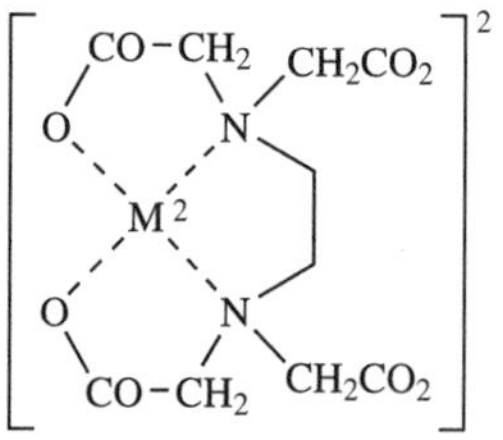

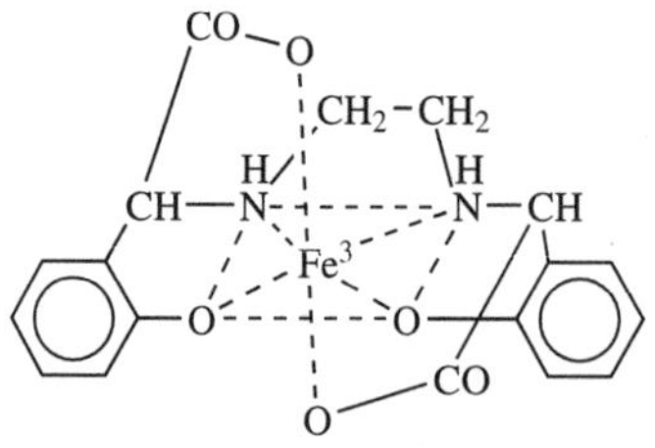

Fig. 2. Structural representations of chelates where (**8**) corresponds to M(acac)₂; (**9**) to ML_2^{2-}, L = 5-sulfo-8-hydroxyquinoline; (**10**) to M-Trien; (**11**) to a tetracoordinate metal bound to the hexadentate ligand EDTA; and (**12**) to the Fe(III) chelate of EHPG. Structure (**13**), which emphasizes the spatial arrangement of the donor atoms, also represents the chelate in (**12**).

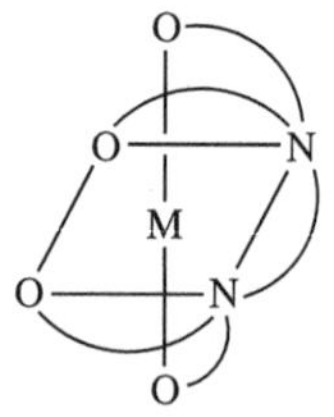

(**9–11**), may be used to emphasize that the structure is a complex ion, but brackets are not a requirement. The sum of charges shows that structure (**12**) is also a charged (1−) complex; structure (**8**) is neutral. In structure (**13**), which is used to show spatial arrangements, the curved lines represent chains of unspecified length connecting the donor atoms. Note that the aromatic rings in (**9**) sterically prevent the sulfo groups from chelating with the central metal.

For the many details of constructing or interpreting structures and systematic names, the literature on nomenclature and indexing (6) can be consulted. Systematic nomenclature is illustrated by the *Chemical Abstracts* name of the sodium iron(III) EHPG chelate: sodium [[*N*,*N'*-1,2-ethanediylbis[2-(2-hydroxyphenyl)glycinatol]](4-)-*N*,*N'*,*O*,*O'*,*O²*,*O²'*]ferrate(1-) [16455-61-1]. The ferrate anion (**12**) [20250-28-6] and the potassium salt [22569-56-8] are also listed in *Chemical Abstracts* (7).

5. The Chelation Reaction

5.1. Chelate Formation Equilibria. In homogeneous solution, the equilibrium constant for the formation of the chelate complex from the solvated metal ion and the ligand in its fully dissociated form is called the formation or stability constant. Whereas the ligand displaces solvent molecules coordinated to the metal, these solvent molecules do not generally enter into the equations. When more than one ligand molecule complexes with a metal atom, the reaction usually proceeds stepwise. For a metal having a coordination number of six and a bidentate chelating agent, the equations representing the equilibria are

$$\text{M} + \text{L} \rightleftharpoons \text{ML} \qquad K_1 = \frac{[\text{ML}]}{[\text{M}][\text{L}]} \tag{1}$$

$$\text{ML} + \text{L} \rightleftharpoons \text{ML}_2 \qquad K_2 = \frac{[\text{ML}_2]}{[\text{ML}][\text{L}]} \tag{2}$$

$$\text{ML}_2 + \text{L} \rightleftharpoons \text{ML}_3 \qquad K_3 = \frac{[\text{ML}_3]}{[\text{ML}_2][\text{L}]} \tag{3}$$

$$overall: \text{M} + 3\,\text{L} \rightleftharpoons \text{ML}_3 \qquad K = \frac{[\text{ML}_3]}{[\text{M}][\text{L}]^3} \tag{4}$$

where the square brackets represent concentrations in units of molarity. The overall stability constant is the product of the step stability constants, ie, $K = K_1 K_2 K_3$, and is often designated by β. Protons displaced from a ligand in the chelation reaction may be shown, as in equation 5, where H_2A^{2-} represents the divalent anion of a multidentate ligand such as EDTA.

$$\text{M}^{2+} + \text{H}_2\text{A}^{2-} \rightleftharpoons \text{MA}^{2-} + 2\,\text{H}^+ \qquad k = \frac{[\text{MA}^{2-}][\text{H}^+]^2}{[\text{M}^{2+}][\text{H}_2\text{A}^{2-}]} \tag{5}$$

Such a reaction is the sum of an acid dissociation reaction

$$H_2A^{2-} \rightleftharpoons A^{4-} + 2\,H^+ \qquad K_a = \frac{[A^{4-}][H^+]^2}{[H_2A^{2-}]} \qquad (6)$$

and the reaction of chelate formation from the fully dissociated form of the ligand

$$M^{2+} + A^{4-} \rightleftharpoons MA^{2-} \qquad K = \frac{[Ma^{2-}]}{[M^{2+}][A^{4-}]} \qquad (7)$$

where K is the chelate formation constant, K_a is the overall acid dissociation constant for the two stages, and $k = KK_a$.

Experimentally determined equilibrium constants are usually calculated from concentrations rather than from the activities of the species involved. Thermodynamic constants, based on ion activities, require activity coefficients. Because of the inadequacy of present theory for either calculating or determining activity coefficients for the complicated ionic structures involved, the relatively few known thermodynamic constants have usually been obtained by extrapolation of results to infinite dilution. The constants based on concentration have usually been determined in dilute solution in the presence of excess inert ions to maintain constant ionic strength. Thus concentration constants are accurate only under conditions reasonably close to those used for their determination. Beyond these conditions, concentration constants may be useful in estimating probable effects and relative behaviors, and chelation process designers need to make allowances for these differences in conditions.

Stability constants for a number of industrially important metals and some representative chelating agents are given in Table 2. Extensive listings of stability constants are available (8). The practical significance of formation constants is that a high log K value means a large ratio of chelated to unchelated or free metal when equivalent amounts of metal and ligand are present. From the values shown in Table 2, it is apparent that the concentration of free metal ion can range from relatively large to very low depending on the chelant.

Many experimental approaches have been applied to the determination of stability constants. Techniques include pH titrations, ion exchange, spectrophotometry, measurement of redox potentials, polarimetry, conductometric titrations, solubility determinations, and biological assay. Details of these methods can be found in the literature (9,10).

5.2. Displacement Equilibria.   Species in solution are generally in formation—dissociation equilibrium, and displacement reactions of any given metal or ligand by another are possible. Thus,

$$ML + M' \rightleftharpoons M'L + M \qquad (8)$$

or

$$ML + L' \rightleftharpoons ML' + L \qquad (9)$$

Table 2. **Concentration Formation Constants of Metal Chelates**

Metal ion	STPP	Citric acid	EDTA	EDTMP	NTA $\text{Log } K_1$	$\text{Log } K_2$	$\text{Log } K_1 K_2$
V(III)[a]			25.9				
Fe(III)		10.9	25.1		15.9	9.9	25.8
In(III)			25.0		15.0	9.6	24.6
Th(IV)			23.2		12.4		
Hg(II)			21.8		12.7		
Cu(II)	8.7	6.1	18.8	23.21	12.7	3.6	16.3
Ni(II)	6.7	4.8	18.6	16.38	11.3	4.5	15.8
Y(III)			18.1		11.4	9.1	20.5
Pb(II)		5.7	18.0		11.8		
Zn(II)	7.6	4.5	16.5	18.76	10.5		
Cd(II)		4.2	16.5		10.1	4.4	14.5
Co(II)	6.9	4.4	16.3	17.11	10.6		
Fe(II)	2.5	3.2	14.3		8.8		
Mn(II)	7.2	3.4	14.0		7.4		
V(II)			12.7				
Ca(II)	5.2	3.5	10.7	9.36	6.4		
Mg(II)	5.7	2.8	8.7	8.43	5.4		
Sr(II)	4.4		8.6		5.0		
Ba(II)	3.0		7.8		4.8		
rare earths			15.1–20.0		10.4–12.5		

[a] The oxovanadium(IV) ion, VO^{2+}, forms a complex with EDTA having $\log K = 18.8$.

If the stability constants for ML and M′L are K and K', respectively, then for the exchange shown in equation 8, the equilibrium constant is K_x.

$$K_x = \frac{[\text{M}][\text{M}'\text{L}]}{[\text{ML}][\text{M}']} = \frac{K'}{K} \qquad (10)$$

The extent of displacement depends on the relative stabilities of the complexes and the mass action effect of an excess of M′. For equivalent total amounts of M and M′, K_x must be on the order of 10^4 for 99% complete displacement to occur. Similar considerations apply for the displacement of L from ML by L′. The situation is quite analogous to the familiar competition of two bases for the hydrogen ion.

If the metals or ligands involved in a displacement reaction form chelates where type formulas are different, the exchange equilibrium constant is the simple ratio of the formation constants of the chelates. Rather, for the reaction

$$2\,\text{ML} + \text{M}' \Longleftrightarrow 2\,\text{M} + \text{M}'\text{L}_2 \qquad (11)$$

the equilibrium constant $K_x = K'/K^2$ assuming K and K' are the respective formation constants of ML and $\text{M}'\text{L}_2$. The proper evaluation of K_x can be derived for each particular case.

Metal exchange is the mechanism by which many foods, such as shortenings, shellfish, and dairy products, are stabilized against deleterious effects of trace metals by the addition of $Na_2CaEDTA$ ($\log K = 10.7$). Copper ($\log K = 18.8$) and iron ($\log K = 25.1$) displace calcium and become sequestered so that the remaining concentration of free iron or copper ions is too low for catalytic effects to occur at significant rates (11). Ligand exchange occurs when ascorbic acid [50-81-7] bound to copper ($\log K = 1.57$) is displaced by EDTA, stabilizing this vitamin by disrupting an oxidation mechanism (12). Dyes and bleaches are similarly protected in the textile industry (13) (see FOOD ADDITIVES; DYES AND DYE INTERMEDIATES; BLEACHING AGENTS).

The addition of a chelating agent to a solution of two or more metal ions leads to an order of metal ion complexation that is regulated by the displacement equilibrium constants. If the objective is to bind only a particular ion, then enough chelant to combine with the target ion and all the other ions that are capable of displacing the target ion should be added. For any particular chelating agent under similar solution conditions, a displacement series of metal ions can be assembled by calculating the K_x values from series of stability constants such as those in Table 2. For selective complexation of one metal in the presence of another, a chelating agent with sufficiently different stability constants for the two metals is necessary so that K_x becomes large. For example, beryllium occurs at the bottom of a displacement series with NTA allowing this metal to be recovered as the hydroxide by pH adjustment of an ore processing solution; all of the interfering metals remain sequestered by chelation (14). Additionally, because other metals present cannot displace iron in an iron–EHPG chelate, the chelate can be used in highly calcareous soils to supply iron as a trace nutrient in agriculture (15).

Selectivity for a single metal of a group is the basis of a solvent extraction process for the recovery of copper (qv) from low concentration ore leach solutions containing high levels of iron (qv) and other interfering metals (16).

5.3. Rates of Reaction. The rates of formation and dissociation of displacement reactions are important in the practical applications of chelation. Complexation of many metal ions, particularly the divalent ones, is almost instantaneous, but reaction rates of many higher valence ions are slow enough to measure by ordinary kinetic techniques. Rates with some ions, notably Cr(III) and Co(III), may be very slow. Systems that equilibrate rapidly are termed kinetically labile, and those that are slow are called kinetically inert. Inertness may give the appearance of stability, but a complex that is apparently stable because of kinetic inertness may be unstable in the thermodynamic equilibrium sense.

5.4. Factors Affecting Stability. A characteristic of chelation distinguishing it from monodentate metal coordination is the increased stability from ring formation of the resultant chelate complex. For equal numbers of similar coordinated donor atoms, as in amine complexes compared to chelates of ethylenediamine, eg, $M(RNH_2)_2$ vs $M(en)$ or $M(RNH_2)_4$ vs $M(en)_2$, the stability constants of the chelates are from one to several orders of magnitude greater than those of the monodentate complexes. The greater stability of the chelates is largely the result of an increase in entropy resulting from an increase in the number of free molecules, usually solvent or other monodentate ligand, liberated as the chelate is formed. This extra stabilization produced by the ring formation is called the chelate effect.

Many parameters influence the stability of chelates. Several of the stability factors common to all chelate systems are the size and number of rings, substituents on the rings, and the nature of the metal and donor atoms. In the macrocyclic complexes, the degree to which the size of the metal ion fits the space enclosed by the macro rings is a significant factor. In chelation, five- and six-membered rings are most stable: coordination angles on the metal atoms prohibit the formation of three-membered rings, and ring closure is improbable for rings having more than seven members. In these latter systems, coordination in linear chains is a competing reaction. Formation of each additional ring by the same ligand contributes extra stability from the entropy effect of displacing coordinated solvent molecules. Substituents on a ring may also produce steric hindrance, or otherwise alter the availability of the donor atom electrons for coordination.

The alkaline and rare-earth metals, and positive actinide ions, generally have greater affinity for $-O^-$ groups as electron donors. Many transition metals complex preferentially with enolic $-O^-$ and some nitrogen functions. Polarizability of the donor atoms correlates with stability of complexes of the heavier transition metals and the more noble metal ions.

In any series of chelates, the stability constants are usually influenced by more than one of the parameters that are known to affect chelate stability. The data in Table 3 illustrate some of these relationships: the calcium complexes containing EDTA and homologues decrease in stability as the size of the chelate ring formed by the metal and the two coordinating nitrogen atoms increases. The aminoacetic acid series shows the stability gained from the formation of additional rings on a single-ligand molecule. And the copper–polyamine series shows the combined effects of ring size, number of rings for similar donor atoms, and whether the rings are formed by one or more ligand molecules.

5.5. pH Effects. Being Lewis bases, the donor atoms of chelating agents react with Lewis acids such as metal and hydrogen ions. In the pH range of aqueous solutions, most of the well-known chelating agents exist as an equilibrium mixture of both protonated and unprotonated forms. Metal ions compete with hydrogen ions for the available donor atoms, and therefore, simultaneous equilibria exist that are treated mathematically by the simultaneous equations for the formation constants of the chelates and the acid dissociation constants of the chelating agents.

In aqueous systems, water is a competing ligand, and its dissociation into hydrogen and hydroxyl ions must often be considered in the system of simultaneous equilibria (see HYDROGEN-ION ACTIVITY). In nonaqueous solvents, similar treatment is possible with appropriate modifications for the acidity in those systems. The pH leveling effect of water affects and limits the acid dissociation behavior of chelating agents in aqueous systems. However, coordination with certain metals in aqueous solution can result in loss of a proton from aliphatic $-OH$ and $-NH_2$ groups.

Consider the equilibria in an aqueous system composed of a bidentate ligand HA, eg, the enol form of acetylacetone, and a tetracoordinate metal

Table 3. **Ring Effects on Complex Stability**

Ligand, L	Complex formula	Ring size	Number of rings	Number of coordinated donor atoms	Log K^a		
EDTA, $n = 2^b$	CaL	5			10.5		
homologue, $n = 3^{b,c}$	CaL	6			7.1		
homologue, $n = 4^{b,d}$	CaL	7			5.2		
homologue, $n = 5^{b,e}$	CaL	8			4.6		
					Cu(II)	Ni(II)	Co(II)
$H_2NCH_2COOH^f$	ML	5	1	2	8.6	6.2	5.2
$HN(CH_2COOH)_2{}^g$	ML	5	2	3	10.6	8.2	7.0
$N(CH_2COOH)_3$	ML	5	3	4	12.7	11.3	10.6
					Log K_1	Log K_2	Log K^h
$H_2NC_2H_4NH_2$	CuL	5	1	2	10.72		
	CuL$_2$	5	2	4		9.31	20.0
$H_2N(CH_2)_3NH_2{}^i$	CuL	6	1	2	9.77^j		
	CuL$_2$	6	2	4		7.1	16.9^j
$HN(C_2H_4NH_2)_2{}^k$	CuL	5	2	3	15.9		
	CuL$_2$	5	2	4		5.4	21.3
$(H_2NC_2H_4NHCH_2)_2$	CuL	5	3	4	20.5		

^{a} Ref. 9.
^{b} Formula is $(HOOCCH_2)_2 N(CH_2)_n N(CH_2COOH)_2$.
^{c} Propylenedinitrilotetraacetic acid [1939-36-2], $C_{11}H_{18}N_2O_8$.
^{d} Tetramethylenedinitrilotetraacetic acid [1798-13-6], $C_{12}H_{20}N_2O_8$.
^{e} Pentamethylenedinitrilotetraacetic acid [1798-14-7], $C_{13}H_{22}N_2O_8$.
^{f} Glycine [56-40-6], $C_2H_5NO_2$.
^{g} Iminodiacetic acid [142-73-4], $C_4H_7NO_4$.
^{h} Log $K = K_1K_2$.
^{i} 1,3-Propanediamine [109-76-2], $C_3H_{10}N_2$.
^{j} Ref. 10.
^{k} N-2-Aminoethyl-1,2-ethanediamine [111-40-0], $C_4H_{13}N_3$.

M^{2+}, structure (**8**). The equations are

$$HA \rightleftharpoons A^- + H^+ \qquad K_a = \frac{[H^+][A^-]}{[HA]} \tag{12}$$

$$M^{2+} + 2\,A^- \rightleftharpoons MA_2 \qquad K = \frac{[MA_2]}{[M^{2+}][A^-]^2} \tag{13}$$

giving the relation

$$[M^{2+}] = [H^+]^2 \times \frac{[MA_2]}{[HA]^2} \times \frac{1}{KK_a^2} \tag{14}$$

Equation 14 shows that an increase in acidity of the solution increases the concentration of uncomplexed metal, which must result from the displacement of

M^{2+} from MA_2, causing a simultaneous decrease in the ratio of complexed to free ligand $[MA_2]/[HA]$. The opposite effects result upon decreasing the acidity. This behavior occurs in the pH range where appreciable amounts of both HA and A^- coexist. Outside this range the ligand is present almost entirely as either HA or MA_2 and A^-, and the system is essentially independent of pH.

Chelating agents that are polybasic acids give two or more hydrogen ions per molecule. The four stages of dissociation of EDTA, eg, are represented by the equations

$$H_4A \xrightleftharpoons{-H^+} H_3A^- \xrightleftharpoons{-H^+} H_2A^{2-} \xrightleftharpoons{-H^+} HA^{3-} \xrightleftharpoons{-H^+} A^{4-} \qquad (15)$$

These equilibrium constants K_1, K_2, K_3, and K_4 are known (10). The pK values for the four dissociation steps as well as the proportions of the species present in aqueous solution as a function of pH are shown in Figure 3. The reaction of Na_2EDTA and M^{2+}, represented by equation 5 and noting that $KK_3K_4 = k$, give

$$[M^{2+}] = [H^+]^2 \times \frac{[MA^{2-}]}{[H_2A^{2-}]} \times \frac{1}{KK_3K_4} \qquad (16)$$

which is of the same form as equation 14. In general, for the reaction

$$[M^{x+}] + H_nA^{y-} \rightleftharpoons [MA^{x-y-n}] + n\, H^+ \qquad (17)$$

the equation for the concentration of free metal is

$$[M^{x+}] = [H^+]^n \times \frac{[MA^{x-y-n}]}{[H_nA^{y-}]} \times \frac{1}{KC} \qquad (18)$$

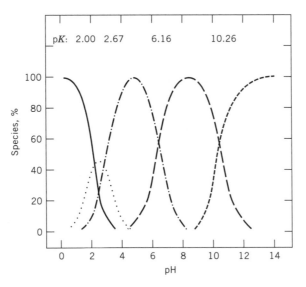

Fig. 3. Distribution of ionic species of EDTA as a function of pH where (—) represents H_4A, (···) H_3A^-, (–·–·–) H_2A^{2-}, (— —) HA^{3-}, and (---) A^{4-}.

where K is the formation constant (eq. 7) of MA^{x-y-n} and C is the product of the n stepwise acid dissociation constants involved. Taking the negative logarithm of both sides of the equation and letting $-\log [M^{x+}] = pM$, the equation becomes

$$pM = n\,pH + \log \frac{[H_nA^{y-}]}{[MA^{x-y-n}]} + \log KC \qquad (19)$$

The corresponding generalized form for equation 14 is

$$pM = n\,pH + \log \frac{[HA]^n}{[MA_n^{x-n}]} + \log KK_{a^n} \qquad (20)$$

In both cases n is the number of hydrogen ions displaced in the formation of the complex. In solutions where the ratio of free chelating agent to complex, $[H_nA^{y}-]$ $[MA^{x-y-n}]$, is held constant, the slopes of the curves pM vs pH are equal to n in the region where H_nA^{y-} is the principal form of the chelating agent.

The displacement of hydrogen ions by a metal ion from a protonated form of the chelating agent (eq. 17) generates an autogenous pH that depends on the base strength of the counterions of the metal salt. The pH of the solution can become quite low if these counterions are those of a strong acid, eg, Cl^- or SO_4^{2-}. However the pH of chelate solutions can be controlled by the use of compatible buffers, and the chelating agent itself can sometimes serve as the pH buffer if one of its acid dissociation stages occurs in the desired pH range.

The variation of pM with pH according to equation 17 is shown in Figure 4 for EDTA and Cu(II) and EDTA and Mn(II) at three different ratios of free chelant to metal chelate, $[H_nA^{y-}]/[MA^{x-y-n}]$. The pM value at any pH indicates the concentration of the free aqua metal ion in equilibrium with the chelate. From pH 2–6, the free form of the EDTA is H_2A^{2-}, and because two protons are displaced by the metal on chelation, the slope of the curves is two. The slope changes to one from pH 6 to 10 where the EDTA is present as HA^{3-} and only one proton is displaced. Above pH 10, the chelates are formed from the fully dissociated chelant, A^{4-}, no hydrogen ions are displaced, the slope of the curves is zero, and pM is independent of pH up to the pH of the intersection of the solid lines with the dashed lines for the same metal. Up to pH 10, the rise of pM with pH shows the increasing degree of metal binding as competition by hydrogen ions for the chelant is decreased. Cu(II) and Mn(II) form insoluble metal hydroxides to which the following applies at equilibrium in aqueous solution:

$$M^{2+} + 2\,OH^- \underset{}{\overset{K_{sp}}{\rightleftharpoons}} M(OH)_2\,(s)$$
$$K_{sp} = [M^{2+}][OH^-]^2, \quad K_w = [H^+][OH^-] \qquad (21)$$

and from these

$$pM = 2\,pH + \log(K_w^2/K_{sp}) \qquad (22)$$

The dashed lines in Figure 4 are plots of equation 22 for Cu^{2+} and Mn^{2+} and indicate the concentration of the aqua metal ions in equilibrium with the solid

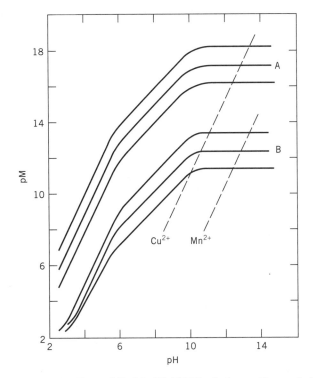

Fig. 4.   pM vs pH for A: Cu(II), and B: Mn(II) EDTA chelates. For each family of curves, the lowest curve represents 1%; the second, 10%; and the top curve, 100% of free ligand species in excess of the amount needed to form the metal chelate. Broken lines represent solid–solution equilibria for corresponding metal hydroxides. See equation 22.

hydroxides as function of pH. At any pH where the solid curve is above the dashed line for the same metal, the EDTA is holding the unchelated metal ion concentration at a value too low for the precipitation of the solid hydroxide. Relatively large quantities of the metal can thus be maintained in solution as the chelate at pH values where otherwise all but trace quantities of the metal would be precipitated. In Figure 4, this corresponds to pH values where pM of the dashed curves is 4 or greater. At the pH of intersection of the solid and dashed lines for the same metal, the free metal ion is in equilibrium with both the solid hydroxide and the chelate. At higher pH the hydroxyl ion competes more effectively than the chelant for the metal, and only a trace of either the chelate or the aqua metal ion can exist in solution. Any excess metal is present as solid hydroxide.

The more stable the chelate, the higher the pM that it can maintain, and the higher the pH required to precipitate the metal hydroxide. From equation 22 it can be seen that the smaller the solubility product K_{sp}, ie, the more insoluble the metal hydroxide, the higher the pM that a chelant must maintain to prevent precipitation. The stability constant of the Fe(III)–EHPG complex (**12**), is so large (10^{35}) that iron is not precipitated even in strongly alkaline solutions.

If instead of hydroxyl ion the precipitating agent is the anion of an acid stronger than or comparable in strength to the chelating agent, the metal salt

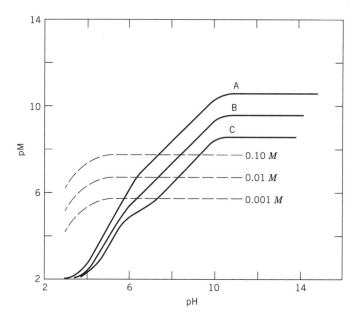

Fig. 5. pM vs pH for M = Ca(II), L = EDTA, in the presence of excess oxalate. Solid lines A, B, C represent 100, 10, and 1% excess EDTA, respectively. Broken lines indicate solid–solution equilibria of calcium oxalate in the presence of dissolved oxalate.

may be insoluble at low pH where the chelant is protonated but not the precipitant anions, and soluble at higher pH where the complexing form of the ligand is relatively more available. Solutions of oxalate, EDTA, and Ca(II) show this behavior; Figure 5 shows the relationships of pM to pH. The solid lines give pM for the Ca–EDTA system, and the dashed lines represent the calcium oxalate [563-72-4] CaC_2O_4, solubility. As in Figure 4, at pH values where the dashed lines are above the solid lines, the metal is present almost entirely as the insoluble salt. To the right of the intersections of dashed and solid lines the metal is almost entirely chelated, and the solid salt phase cannot exist.

5.6. Titration Behavior.   Protonated chelating agents exhibit titration behavior typical of their respective acidic groups, eg, carboxyl phenolic hydroxyl, ammonium, or sulfhydryl moieties, if they are titrated with bases where the cations have a very weak or no tendency to form chelates. In the presence of a metal ion that coordinates with the donor atom of one of these acidic groups, hydrogen is displaced by the metal, and the acid strength of the group appears to be enhanced. The hydrogen ion concentration is then higher than in the absence of the metal. Strongly chelated metal ions can increase the acidity of an acidic group by several orders of magnitude. With A representing donor groups, which may be the same or different, the release of hydrogen ions is shown schematically by

$$M^{n+} + AH \overset{\frown}{} \overline{AH}\,\overline{AH}\,\overline{AH} \rightleftharpoons \left[\begin{smallmatrix} A & & A \\ & M & \\ A & & A \end{smallmatrix} \right]^{n-4} + 4\,H^+ \qquad (23)$$

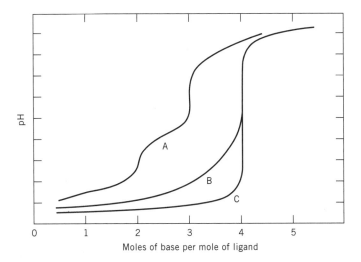

Fig. 6. Base titration of H_4A chelant: A, free acid without coordinating metal; B, in the presence of a metal of intermediate coordinate strength; and C, in the presence of a strongly chelated metal.

In the absence of metal coordination the compound H_4A_4 would titrate as a typical tetrabasic acid having a stepwise titration curve as shown in Figure 6. Upon strong chelation, all four protons are displaced and base titration resembles that of a typical strong acid at four times the equivalent concentration. This statement is in agreement with equation 19, which shows that pM can be large (low concentration of free metal) at low pH if K is large (strong chelation).

If metal chelation is intermediate in strength, the chelation of the groups that are last to coordinate to the metal may not do so to completion at the low pH generated by the hydrogen ions released from the first groups to coordinate. Then, in accordance with equation 19, because K is smaller, the completion of the chelation reaction, as shown by the reduction of the metal ion concentration to a low value (high pM), may not occur until a higher pH is attained where more A-groups are available to the metal. The curve of pH vs base added thus rises sooner than the curve for strong chelation, resembling the titration curve of a weak acid and reflecting the lower apparent acid strength resulting from weaker coordination. The intermediate curve of Figure 6 shows this effect of weaker chelation.

Titration of the hydrogen ion liberated from a strong chelating agent is used to determine the concentration of metal ions in solution. The strength of chelation can also be determined from these data.

Deprotonation of enols of β-diketones, not considered unusual at moderate pH because of their acidity, is facilitated at lower pH by chelate formation. Chelation can lead to the dissociation of a proton from as weak an acid as an aliphatic amino alcohol in aqueous alkali. Coordination of the O atom of triethanolamine to Fe(III) is an example of this effect and results in the sequestration of iron in 1 to 18% sodium hydroxide solution (Fig. 7). Even more striking is the loss of a proton from the amino group of a gold chelate of ethylenediamine in aqueous solution (17).

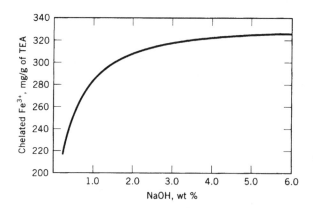

Fig. 7. Chelation of iron by TEA in aqueous sodium hydroxide. Courtesy of The Dow Chemical Company.

Another group of chelants that form stable chelates at high pH because of metal–alkoxide coordination are the sugar acids, such as gluconic acid [526-95-4], $C_6H_{12}O_7$ (18). Utility for this group is found in high alkalinity bottle washes and other cleansers (19).

5.7. Metal Buffering. The equation for the formation constant of the reaction

$$M^{n+} + L^{m-} \overset{K}{\rightleftharpoons} ML^{n-m} \tag{24}$$

can be rearranged to give

$$\frac{1}{[M^{n+}]} = \frac{K[L^{m-}]}{[ML^{n-m}]} \tag{25}$$

from which, on taking logarithms and defining $-\log[M^{n+}]$,

$$pM = \log\frac{[L^{m-}]}{[ML^{n-m}]} + \log K \tag{26}$$

The concentration of the metal ion can be controlled by adjusting the ratio of the concentrations of free ligand and metal chelate. If both species are present in appreciable amounts, moderate changes in either concentration have little effect on the ratio. The concentration of the metal ion can thus be buffered in a manner analogous to the buffering of pH by the presence of a weak acid and its anion

$$pH = \log\frac{[A^-]}{[HA]} + pK_a \tag{27}$$

In the equation for pM, $\log K$ appears instead of pK because K is a formation constant, the reciprocal of the chelate dissociation constant, which is analogous to the acid dissociation constant K_a.

By choice of chelating agents, and thus log K, pM can be regulated over a wide range. Two or more metals may be selectively buffered at different concentrations by a single chelating agent having different stability constants for the metals. Selective buffering of one metal to a low concentration in the presence of other metals is termed masking. It is the ability to maintain a nearly constant concentration of metal ions at almost any level of concentration that is the basis of many of the commercial uses of chelating agents. The buffering capacity may be used to supply metal ions at a definite concentration as in electroplating (qv) and in nutrient media (see MINERAL NUTRIENTS), or to remove or sequester metal ions in cleaning baths where the fresh stock entering the bath continually introduces additional amounts of metals.

The effect of pH on metal buffering is shown by equations 19 and 20. If a constant pH is imposed on a system by a hydrogen ion buffer, variations in pM are controlled only by variations in the ratio of the free and metal-bound forms of the ligand, and of course by the characteristics of the ligand. The free form of the ligand is the acid form its acidic dissociation stage at the imposed pH, ie, HA or H_nA^{y-}. If the acid groups of the chelating agent are fully dissociated at the pH of the buffer, no hydrogen ions are displaced when chelation with the metal occurs, no dissociation constants of the ligand are involved, n in the equations is essentially zero, and pM is independent of pH. Equations 19 and 20 then reduce to the form of equation 26.

5.8. Solubilization. The solubility product of a slightly soluble salt determines the concentration of metal ion that can be present in solution with the anion of that salt. For the salt MX the solubility product is

$$K_{sp} = [M^{n+}][X^{n-}] \tag{28}$$

from which is obtained

$$pM = \log[X^{n-}] - \log K_{sp} \tag{29}$$

The presence of a sufficiently strong chelating agent, ie, one where K in equation 26 is large, keeps the concentration of free metal ion suppressed so that pM is larger than the saturation pM given by the solubility product relation (eq. 29) and no solid phase of MX can form even in the presence of relatively high anion concentrations. The metal is thus sequestered with respect to precipitation by the anion, such as in the prevention of the formation of insoluble soaps in hard water.

Deposits of an insoluble salt can be dissolved as a salt of the metal chelate.

$$MX\ (s) + L^{m-} \xrightleftharpoons{\quad K_s \quad} [ML^{n-m}] + X^{n-} \tag{30}$$

In the presence of the chelating agent and the insoluble salt, MX, pM of the solution is subject to both the metal buffering and the solubility equilibria. Equating the right-hand sides of the equations 26 and 29 and rearranging gives

$$\log[X^{n-}] = \log \frac{[L^{m-}]}{[ML^{n-m}]} + \log KK_{sp} \tag{31}$$

As the dissolving of the salt progresses $[X^{n-}]$ is approximately equal to $[ML^{n-m}]$, and both represent the amount of MX dissolved. Substituting $[ML^{n-m}]$ for $[X^{n-}]$ in equation 31 gives

$$2 \log[ML^{n-m}] = \log[L^{m-}] + \log KK_{sp} \tag{32}$$

for the equilibrium concentrations for the process. If $KK_{sp} = 1$, $\log KK_{sp} = 0$, then $[ML^{n-m}] = [L^{m-}]^{1/2}$, and the amount of MX solubilized is equal to the square root of the amount of excess chelating agent required, which is an amount that is in the practical range. A dilution–efficiency effect can be calculated from this relationship. If the amount of MX dissolved gives only a 0.1 M solution of $[ML^{n-m}]$, the excess ligand concentration is 0.01 M and almost 90% of the total amount of ligand is effective in solubilizing salt deposit. However, for $[ML^{n-m}] = 1.0\,M$ an equal concentration of excess ligand is required and solubilization to 2.0 M requires 4.0 M excess ligand, giving efficiencies of 50 and 33%, respectively. If for economic reasons the chelating agent must be recovered, dilution is a disadvantage, and dilution and efficiency must be compromised. If the stability constant K is large enough, equation 32 shows that only small amounts of the chelating agent in excess of that bound to the metal are required to dissolve a given amount of the deposit.

The product KK_{sp} is equal to the equilibrium constant K_s for the reaction shown in equation 30. It is generally considered that a salt is soluble if $K_s > 1$. Thus sequestration or solubilization of moderate amounts of metal ion usually becomes practical as K_s approaches or exceeds one. For smaller values of K_s, the cost of the required amount of chelating agent may be prohibitive. However, the dilution effect may allow economical sequestration, or solubilization of small amounts of deposits, at K_s values considerably less than one. In practical applications, calculations based on concentration equilibrium constants can be used as a guide for experimental studies that are usually necessary to determine the actual behavior of particular systems.

The K_s values for some common scale deposits and NTA, which is an effective agent for solubilizing $CaSO_4$ and $CaSiO_3$ $(K_s > 1)$, are shown in Table 4. For removal of $CaCO_3$ deposits, a stronger Ca(II) chelating agent would be required. A large cleaning business that services industry is based on the aminocarboxylic acid chelants.

5.9. Electrochemical Potentials.   The oxidation potential of a solution containing a metal in two of its valence states, M^{x+} and M^{x+n}, is given by

$$E = E^0 - \frac{RT}{nF} \ln \frac{[M^{x+n}]}{[M^{x+}]} \tag{33}$$

In the presence of a chelating agent, the concentrations of the two forms of the metal are buffered according to the simultaneous equations

$$[M^{x+n}] = \frac{[M_{ox}L]}{K_{ox}[L]} \quad \text{and} \quad [M^{x+}] = \frac{[M_{red}L]}{K_{red}[L]} \tag{34}$$

where $M_{ox}L$ and $M_{red}L$ are the chelates of the oxidized and reduced forms of the ions and K_{ox} and K_{red} are the respective formation constants. Substituting these

Table 4. K_s **Values for NTA and Calcium Salts**[a]

Salt	K_{sp}	K_s
$CaSO_4$	6.1×10^{-5}	152
$CaSiO_3$	6.6×10^{-7}	1.65
$CaCO_3$	8.7×10^{-9}	0.022

[a] Formation constant $K = 2.5 \times 10^6$.

values in the potential equation gives

$$E = E^0 + \frac{RT}{nF} \ln \frac{K_{ox}}{K_{red}} - \frac{RT}{nF} \ln \frac{[M_{ox}L]}{[M_{red}L]} \tag{35}$$

The first two terms of the right-hand side of the equation are sometimes combined and expressed as $E^{0\prime}$, which is called the standard oxidation potential for the chelate system. If the chelation is strong and the ligand is in excess, the metal would be almost entirely in the chelated forms, and $[M_{ox}L]$ and $[M_{red}L]$ would essentially be equal to the total concentrations of the oxidized and reduced forms of the metal. If, as is usual, the oxidized form is the more strongly chelated ($K_{ox} > K_{red}$), the oxidation potential of a system is increased by the addition of the chelant.

In electrodeposition, the reduced form of the metal is the elemental form M, $x = 0$, and there is no chelated M in solution. Neglecting activity coefficients, the reversible potential is

$$E = E^0 - \frac{RT}{nF} \ln[M^{x+}] \tag{36}$$

By buffering the metal ion concentration using a chelant, E can be adjusted to and stabilized at values that give desirable properties to the deposit. Selective buffering can sequester the properties of interfering ions or can be used to regulate the potentials of two or more ions to approximately the same value in order to effect codeposition.

6. Economic Aspects

Production and price estimates for the principal industrial chelating agents are given in Table 5. The list is dominated by STPP, which can be used in certain chelation applications (see PHOSPHORIC ACID AND THE PHOSPHATES). Primary chelating uses are in household cleaners and detergents, industrial and institutional cleaners, and in water treatment formulations (21). Most of the citric acid (qv) is consumed in nonchelant applications in the food and beverage industry. Citric acid is used for its chelating properties mainly in household detergents and cleaning applications where STPP is not suitable (22).

The aminopolycarboxylic acids are used principally as chelating agents, and a large proportion is used in pulp and paper, water treatment, photographic

Table 5. **United States Production of Industrial Chelating Agents**

Agent	Production	Price	Producers[a]
EDTA	132[b]		Akzo, BASF, Dow
DTPA	46[b]		Akzo, BASF, Dow
HEDTA	13[b]		Akzo, BASF, Dow
NTA	49[b]		Akzo, BASF, Dow, Sol
EDTA(Na)$_4$ solution (38–39%)		0.39–0.42[c]	Akzo, BASF, Dow
DTPA(Na)$_5$ solution (40%)		0.50–0.51[c]	Akzo, BASF, Dow
HEDTA(Na)$_3$ solution (41%)		0.58[c]	Akzo, BASF, Dow
NTA(Na)$_3$ solution (40%)		0.31[c]	Akzo, BASF, Dow, Sol
Gluconates	30[b]		Glu, PMP
Gluconic acid solution (50%)		0.50[c]	Glu, PMP
Glucoheptonates	21[b]		Cal, En
Sodium glucoheptonate solution (50%)		0.20–0.21[c]	Cal, En
Organophosphonates	55[b]		Cal, Rh, Sol
ATMP(H)$_6$ solution (50%)		0.80–0.88[c]	Cal, Rh
HEDP(H)$_4$ solution (60%)		1.21–1.23[c]	Cal, Rh
STPP	200[d]	0.46[e]	As, Pr, Rh
Citric Acid	460[f]	0.75[g]	C, ADM, TL

[a] AKZO = Akzo Nobel Chemicals, Inc.; BASF = BASF Corporation; DOW = The Dow Chemical Co.; Sol = Solutia Inc.; Glu = Glucona America, Inc.; PMP = PMP Fermentation Products, Inc.; Cal = Callaway Chemical Co.; En = EnCee Chemicals, Inc.; Rh = Rhodia, Inc.; As = Astaris, LLC; Pr = Prayon Inc.; C = Cargill; ADM = Archer Daniels Midland Co.; TL = Tate & Lyle (20–22).
[b] Millions of pounds, 100% dry basis; estimated for 1998 (20).
[c] Dollars per pound; estimated for 1990 (20).
[d] Thousands of short tons; estimated for 2000 (21).
[e] Dollars per pound; estimated for 2000 (21).
[f] Millions of pounds, 100% dry basis; estimated for 1998 (22).
[g] Dollars per pound; estimated for 1998 (22).

processing, and cleaning formulations. The data of Table 5 include NTA, most of which was exported.

Gluconates and glucoheptonates are largely interchangeable except in foods where the heptonates are prohibited. These compounds chelate polyvalent metals in strongly alkaline solution and are used in many ways, including metal cleaning, bottle washing, food service cleaning applications, electroplating, derusting, aluminum etching, and in concrete. Alkaline gluconate solutions dissolve ferric oxide.

Organophosphonates are similar to polyphosphates in chelation properties, but they are stable to hydrolysis and replace the phosphates where persistence in aqueous solution is necessary. They are used as scale and corrosion inhibitors (23) where they function via the threshold effect, a mechanism requiring far less than the stoichiometric amounts for chelation of the detrimental ions present. Threshold inhibition in cooling water treatment is the largest market for organophosphonates, but there is a wide variety of other uses (20).

7. Environmental, Health, and Safety Factors

The primary industrial chelating agents are essentially environmentally benign and nontoxic under the conditions incidental to normal handling and use. With these, eye irritation is mainly a function of the acidity or alkalinity of the form of the product and its solubility. However, use of NTA is regulated in some jurisdictions. In medical uses the effects of the chelating agents on metal ion balances in the body tissues must be accommodated. Some of the commercial compounds used in smaller amounts as chelants, such as oxalic acid [144-62-7], are toxic, however. The hazards of using any chelant should be determined prior to use.

Solutions of iron chelates can be used to remove hydrogen sulfide and oxides of sulfur and nitrogen in industrial gas scrubbing processes (20,24,25) before flue gases are released to the atmosphere.

There is an increased interest in the development of chelating agents that have improved biodegradable characteristics.

8. Applications

Three features of chelation chemistry are fundamental to most of the applications of the chelating agents. The first and probably the most extensively used feature is the control of free metal ion concentration by means of the binding–dissociation equilibria. The second is where the special properties of the chelate itself provide the basis of the application. The third feature comprises displacement reactions: metal by other metal ions, chelant by chelant, and chelant by other ligands or ions.

8.1. Concentration Control. Sequestration, solubilization, and buffering depend on the concentration control feature of chelation. Traces of metal ions are almost universally present in liquid systems, often arising from the materials of the handling equipment if not introduced by the process materials. Despite very low concentrations, some trace metals produce undesirable effects such as coloration or instability.

Sequestration. The suppression of certain properties of a metal without removing it from the system or phase is called sequestration. Sequestration is invaluable in controlling trace ion effects. Chelation produces sequestration mainly by reducing the concentration of free metal ion to a very low value by converting most of the metal to a soluble chelate that does not possess the properties to be suppressed. A sufficiently large stability constant in the medium of the application is required, and the sequestering agent must not cause any undesirable change that would render the system unsuitable for its intended purpose.

The largest single use for sequestration is probably the control of water (qv) hardness. Chelating agents are used to prevent the formation of precipitates in a wide variety of aqueous systems such as washing solutions of soaps (qv) and detergents, boiler feed water (26), textiles (27) and paper (qv) processing solutions (28), preparations of cosmetics (qv) and pharmaceuticals (qv), photographic developing solutions, chemical process water, beverages, and foodstuffs (18).

Oil-soluble sequestrants suppress metal-catalyzed development of rancidity, gum formation in fuels, and other oxidative degradations (29). In textile bleaching, catalytic decomposition of bleaching agents (qv) is reduced, and degradation of fabrics is minimized by the sequestration of metals that catalyze the reaction of the bleach with the material (30). In dyeing, metal contaminants that cause spotting, off-color shades, and decomposition of the dyes are sequestered (31). Brightness reversion in wood pulp used to make paper is diminished (32), and iron is removed in the alkaline washing step of the preparation of high brightness grades of paper. Discoloration of leather (qv) by metal-tannin complexes is prevented. Chelants are used in many metal treating operations such as phosphatizing, alkaline derusting, and etching (see METAL SURFACE TREATMENTS). Poisoning by metal ions in mammals is also treated by sequestration. Chelating agents have been used extensively in oil/gas field treatments. Primary uses include stabilization of iron, scale removal, and well stimulation. Other examples can be found in the literature on sequestration and chelation.

Solubilization.   Causing the constituents of a phase that is normally insoluble to dissolve in the medium is termed solubilization. Chelation solubilization depends on the formation of a chelate having groups that confer solubility in the medium and a stability sufficient to sequester the metal ion to a concentration (pM) that can exist in the presence of the associated counterions. Usually solubilization into an aqueous phase is thought of in connection with chelation. The donor atoms involved in the chelation may be sufficiently hydrophilic to produce a soluble species, as in structures (**10**) and (**12**). If the organic group is large, more hydrophilic groups may be required for water solubility. Oxine is a well-known precipitant, but its sulfonated derivative, shown in structure (**9**) (Fig. 1), is a solubilizer in water. A neutral chelation complex, such as structure (**8**), may be solubilized in organic media. The macrocyclic chelates have solubility in both aqueous and organic media as a result of their ionic nature and the largely organic character of the complex.

Dissolving hardness scale from the surfaces of boilers, heat exchangers, and piping (see PIPING SYSTEMS) is probably the largest industrial example of solubilization (33). The cleaning of films from dairy equipment and reusable beverage bottles is also a significant use (19). Deposits on processing tanks that are unique to a particular industry, such as paper, textiles (qv), metal treating, or photography (qv), are often removed by solubilization. Prevention of metal deposition by sequestration is usually preferred where possible because solubilization is sometimes slow and can lead to costly down time. Chelants are used in recovery of metals from ores (34) and in cleaning oxide films from metals in preparation for surface treatments (35). Chelation solubilization is especially useful for cleaning up radioactive contamination.

Macrotetrolides of the valinomycin group of electrically neutral antibiotics form stable 1:1 complexes with alkali metal ions that increase the cation permeability of some biological and artificial lipophilic membranes. This solubilization process appears to have implications in membrane transport research (36) (see ANTIBIOTICS, PEPTIDES).

Buffering.   If addition or removal of an appreciable amount of a metal ion produces only a relatively small change in the concentration of that ion is a solution, the solution is buffered with respect to the ion. Metal ions are buffered by chelants of various strengths, ie, stability constants, in a manner exactly analogous to the buffering of hydrogen ions by bases of various strengths.

Chelation buffering is particularly useful in supplying micronutrient metal ions to biological growth systems at controlled, very low concentrations (37). At the very small subtoxic concentrations required for some metals, the amount present would ordinarily soon be depleted, but using the buffer, a reserve supply of the metal as its chelate is available over long periods with automatic control of the concentration. Examples of this kind of use are found in microorganism cultures in closed, controlled systems and in field use in agriculture in open environments (38). The metal concentration can be held at optimum values in electroplating, and chelates have supplied the metal in electroless depositions.

Control of concentrations enables simultaneous deposition of metals in alloy plating. Increasing attention is being given to the use of chelants to replace cyanide in electroplating baths. Chelants are used to control the activity of redox polymerization catalysts by buffering the metal ions participating in the mechanism. Buffering of the metal is produced by having appreciable amounts of both the chelant and the chelate present simultaneously. The pM is determined mainly by the stability constant of the chelate, and within the region of this primary regulation over a small practical range, by the ratio of chelant and chelate concentration.

8.2. Special Properties of Chelates. Some of the principal applications of the preparative feature of chelation depend on the solubility properties, color, or catalytic effects of the chelates. Selection of a chelant to form a chelate that is soluble in the medium enables the solubilization of mineral deposits, pipe and boiler scale, films on surfaces, constituents of ores, and similar insoluble materials. Chelates having suitable solubilities can be designed to concentrate a metal into a particular phase by extraction from water into an organic solvent, by binding to a liquid but water-insoluble chelant, by precipitation of the chelate as a solid phase, or by ion exchange onto an insoluble, solid, chelating resin. Color and color fastness in some dyeing processes depend on the properties of chelates. The phthalocyanine [574-93-6] pigments (qv) are intensely colored, insoluble chelates. The color of chelates is the basis for many analytical procedures. Catalytic effects may result from a chelate, or chelation of the reactant may itself be part of the mechanism of catalysis by a metal. In biological systems the properties of some enzymes and vitamins (qv) involve chelation, and the activities of chlorophyll and hemoglobin are associated with their chelate structures.

In photography, specially designed chelates suppress or release metal ions to start or stop reactions at appropriate stages in processing sequences, sensitize or desensitize substances to radiation, function in optical and multiplex recording systems, and replace the less environmentally suitable ferricyanide for photographic bleaching (39). Chelates have been used in the preparation of superconducting compounds (40) and as cross-linking agents in fracturing fluids and plugging gels for subterranean formations (41).

Catalysis. In catalysis (qv), the importance of coordination between ligands and metals has long been recognized. The special properties of chelating ligands are especially evident in asymmetric syntheses catalyzed by chelates of an asymmetric ligand, such as in the homogeneous hydrogenation of double-bond functions by a chelate of cobalt and the chiral ligand quinine [130-95-0], $C_{20}H_{24}N_2O_2$ (42). In another application, a cobalt chelate is used as an oxygen carrier in the sweetening of gasoline by oxidation of mercaptans.

Chelation is a feature of much research on the development and mechanism of action of catalysts. For example, enzyme chemistry is aided by the study of reactions of simpler chelates that are models of enzyme reactions. Certain enzymes, coenzymes, and vitamins possess chelate structures that must be involved in the mechanism of their action. The activation of many enzymes by metal ions most likely involves chelation, probably bridging the enzyme and substrate through the metal atom. Enzyme inhibition may often result from the formation by the inhibitor of a chelate with a greater stability constant than that of the substrate or the enzyme for a necessary metal ion.

Many reactions catalyzed by the addition of simple metal ions involve chelation of the metal. The familiar autocatalysis of the oxidation of oxalate by permanganate results from the chelation of the oxalate and Mn(III) from the permanganate. Oxidation of ascorbic acid [50-81-7], $C_6H_8O_6$, is catalyzed by copper (12). The stabilization of preparations containing ascorbic acid by the addition of a chelant appears to be due to the inhibition of catalytic oxidation which results from the sequestration of the copper. Many such inhibitions are the result of sequestration. Catalysis by chelation of metal ions with a reactant is usually accomplished by polarization of the molecule, facilitation of electron transfer by the metal, or orientation of reactants.

Chelation itself is sometimes useful in directing the course of synthesis and is called the template effect (43). The presence of a suitable metal ion facilitates the preparation of the crown ethers, porphyrins, and similar heteroatom macrocyclic compounds. Coordination of the heteroatoms about the metal orients the end groups of the reactants for ring closure. The product is the chelate from which the metal may be removed by a suitable method. In other catalytic effects, reactive centers may be brought into close proximity, charge or bond strain effects may be created, or electron transfers may be made possible.

The crown ethers and cryptates are able to complex the alkali metals very strongly (44). Applications of these agents depend on the appreciable solubility of the chelates in a wide range of solvents and the increase in activity of the coanion in nonaqueous systems. For example, potassium hydroxide or permanganate can be solubilized in benzene [71-43-2] by dicyclohexano-[18]-crown-6 [16069-36-6]. In nonpolar solvents the anions are neither extensively solvated nor strongly paired with the complexed cation, and they behave as naked or bare anions with enhanced activity. Small amounts of the macrocyclic compounds can serve as phase-transfer agents, and they may be more effective than tetrabutyl-ammonium ion for the purpose. The cost of these macrocyclic agents limits industrial use however.

Precipitation and Extraction.   The processes of extraction and precipitation comprise transferring the metal into another phase. If the ligand charges neutralize those of the metal ion, the complex becomes a neutral molecule. As the size of the hydrophobic part of the ligands increases, the neutral chelates become less soluble in water and precipitate when enough chelate is present to exceed the solubility. Some ligands precipitate certain metals essentially quantitatively. These materials are used for analytical methods and for recovery of metals from ores or from waste streams. Oxine (8-hydroxyquinoline) is a well-known precipitating agent. Selective and successive precipitations are used in the separation and recovery of the rare-earth elements. Passivation of metals by many organic corrosion inhibitors may involve the formation of an insoluble chelate film with the oxide on the surface or with the metal itself (45) (see CORROSION AND CORROSION INHIBITORS).

A special kind of transfer of metal ions to a solid phase is found in chelating resins, which are similar to ordinary cation-exchange resins except that they have chelating groups in place of the salt-forming moieties. The behavior of the two kinds of resins is similar except that the special effects of the chelation equilibria must be considered. An important use of chelating resins is for the preconcentration of metal ions from such media as seawater, body fluids, and geological materials in which concentration is exceedingly small, so that these ions may be detected or determined analytically (46). Chelating resins may be used functionally in process streams (24).

If a neutral chelate formed from a ligand such as acetylacetone is sufficiently soluble in water not to precipitate, it may still be extracted into an immiscible solvent and thus separated from the other constituents of the water phase. Metal recovery processes (see MINERAL RECOVERY AND PROCESSING), such as from dilute leach dump liquors, and analytical procedures are based on this phase-transfer process, as with precipitation. Solvent extraction theory and many separation systems have been reviewed (47).

Displacement. In many of the applications of chelating agents, the overall effect appears to be a displacement reaction, although the mechanism probably comprises dissociations and recombinations. The basis for many analytical titrations is the displacement of hydrogen ions by a metal, and the displacement of metal by hydrogen ions or other metal ions is a step in metal recovery processes. Some analytical pM indicators function by changing color as one chelant is displaced from its metal by another.

The pH effect in chelation is utilized to liberate metals from their chelates that have participated in another stage of a process, so that the metal or chelant or both can be separately recovered. Hydrogen ion at low pH displaces copper, eg, which is recovered from the acid bath by electrolysis while the hydrogen form of the chelant is recycled (48). Precipitation of the displaced metal by anions such as oxalate as the pH is lowered (Fig. 4) is utilized in separations of rare earths. Metals can also be displaced as insoluble salts or hydroxides in high pH domains where the pM that can be maintained by the chelate is less than that allowed by the insoluble species (Fig. 3).

Rare earths have been separated by elution from ion-exchange resins with chelates of iron, manganese, and cadmium. In another separation, a band of rare earths on an ion-exchange resin was eluted with a chelant and the eluate was passed over an ion-exchange bed loaded with copper. The copper displaced the rare-earth metals that deposited on the second bed. In a solution mining process, a leach solution of $Na_2CaEDTA$ and sodium bicarbonate exchanges Ca for Cu, and the copper is then displaced by lime and the leach solution is regenerated (28). In using chelated chromium in leather tanning, the chromium is captured from the chelant by the collagen in the hide.

The calcium form of EDTA instead of free EDTA is used in many food preparations to stabilize against such deleterious effects as rancidity, loss of ascorbic acid, loss of flavor, development of cloudiness, and discoloration. The causative metal ions are sequestered by displacing calcium from the chelate, and possible problems, such as depletion of body calcium from ingestion of any excess of the free chelant, had it been used, are avoided.

8.3. Medical Uses. A significant usage of chelation is in the reduction of metal ion concentrations to such a level that the properties may be considered to be negligible, as in the treatment of lead poisoning. However, the nuclear properties of metals may retain their full effect under these conditions, eg, in nuclear magnetic resonance or radiation imaging and in localizing radioactivity.

In the treatment of poisoning by lead or other metal ions, higher concentrations of chelant can be obtained in humans by administering $Na_2CaEDTA$ rather than Na_4EDTA. The metal ion is bound by displacing small amounts of Ca^{2+} that the body can tolerate. Use of Na_4EDTA would result in calcium chelation and thus serious depletion of calcium in the body fluids (49). Removal of iron in Cooley's anemia is accomplished by using chelants that are relatively specific for iron (50).

Bifunctional chelating agents are capable of being covalently attached to an antibody having specificity for cancer or tumor cell epitopes or antigens (see Chemotherapeutics, Anticancer). Radioactive metal complexes of such antibody–chelant conjugates are useful in diagnostic, eg, imaging, and therapeutic (irradiation) applications as a means of delivering the radioactive metal to a cancer or tumor cell or to a specific tissue or location (51). Technetium-99m chelates are the most widely used agents in nuclear medicine for diagnostic imaging of the brain, liver, kidneys, and skeleton. Chelants commonly used include diethylenetriaminepentaacetic acid [67-43-6], 1-hydroxyethylidene-1,1-diphosphonic acid [2809-21-4], and glucoheptonate (52). Organophosphonic acid chelates of the radioactive isotopes samarium-153 and rhenium-186 have shown promise in bone cancer therapy (53). Gadolinium(III) complexes have been used to enhance the nuclear magnetic resonance images of cerebral tumors (54).

BIBLIOGRAPHY

"Sequestering Agents" in *ECT* 1st ed., Vol. 12, pp. 164–181, by H. Kroll and M. Knell, Alrose Chemical Co.; "Complexing Agents" in *ECT* 2nd ed., Vol. 6, pp. 1–24, by A. E. Martell, Illinois Institute of Technology; "Chelating Agents" in *ECT* 3rd ed., Vol. 5, pp. 339–368, by A. L. McCrary and W. L. Howard, Dow Chemical USA; in *ECT* 4th ed., Vol. 5, pp. 764–765, by William L. Howard, Consultant and David A. Wilson, The Dow Chemical Company; "Chelating Agents" in *ECT* (online), posting date: December 4, 2000, William L. Howard, Consultant and David A. Wilson, The Dow Chemical Company.

CITED PUBLICATIONS

1. G. T. Morgan and H. D. Drew, *J. Chem. Soc.* **117**, 1456 (1920).
2. C. J. Pedersen, *J. Am. Chem. Soc.* **89**, 7017 (1967).

3. B. Dietrich, J. M. Lehn, and J. P. Sauvage, *Tetrahedron Lett.*, 2889 (1969).

4. S. Quici and P. L. Anelli, *Chim. Oggi* **7**(10), 49 (1989) in English; A. V. Bajaj and N. S. Poonia, *Coord. Chem. Rev.* **87**, 55 (1988), in English; C. J. Pedersen, *Synth. Multident. Macrocyclic Compd.*, 1 (1978); G. W. Liesegang and E. M. Eyring, *Synth. Multident. Macrocyclic Compd.*, 245 (1978); J. J. Christensen, D. J. Eatough, and R. M. Izatt, *Chem. Rev.* **74**, 351 (1974).

5. H. Diehl, *Chem. Rev.* **21**, 39 (1937).

6. *Chemical Abstracts Ninth Collective Index, Index Guide*, Appendix IV F. Specialized Substances, par. 215, pp. 202I–205I (1972–1976). In the *Tenth and Eleventh Collective Indexes*, the corresponding page numbers are 192I–195I and 193I–197I, respectively.

7. *Chemical Abstracts Eighth Collective Index*, pp. 7004F and 12481S (1967–1971); *Ninth Collective Index*, p. 9668F (1972–1976).

8. J. Bjerrum, G. Schwarzenbach, and L. G. Sillen, *Stability Constants*, Pt. I, Special Publication No. 6, 1957, and Pt. II, Special Publication No. 7, 1958, The Chemical Society, London, U.K.;L. G. Sillen andA. E. Martell, eds., *Stability Constants*, 2nd ed., Special Publication No. 17, The Chemical Society, London, UK, 1964; R. M. Smith and A. E. Martell, *Critical Stability Constants*, Vols. 1–6, Plenum Press, New York, 1974–1989; J. R. Van Wazer and C. F. Callis, *Chem. Rev.* **58**, 1011 (1958); D. T. Sawyer, *Chem. Rev.* **64**, 633 (1964); J. Kragten, *Atlas of Metal Ligand Equilibria in Aqueous Solution*, John Wiley & Sons, Inc., New York, 1978.

9. S. Chaberek and A. E. Martell, *Organic Sequestering Agents*, John Wiley & Sons, Inc., New York, 1959, pp. 126–130; F. J. C. Rossotti and H. Rossotti, *The Determination of Stability Constants*, McGraw-Hill Book Co., Inc., New York, 1960.

10. A. E. Martell and M. Calvin, *Chemistry of Metal Chelate Compounds*, Prentice-Hall, Inc., Englewood Cliffs, N.J., 1952, pp. 522, 537.

11. T. E. Furia, *Food Technol.* **18**, 1874 (1964).

12. E. Niadas and L. Robert, *Experientia* **14**, 399 (1958).

13. H. W. Zussman, *Am. Dyestuff Reptr.* **38**, *Proc. Am. Assoc. Text. Chem. Color.*, P500-4 (1949).

14. K. Vetejska and J. Mazacek, *Czech.* **101**, 864 (Dec. 15, 1961).

15. U.S. Pat. 2,921,847(Jan. 19, 1960), M. Knell and H. Kroll (to Geigy Chemical Corp.).

16. P. J. Bailes, C. Hanson, and M. A. Hughes, *Chem. Eng. (N.Y.)* **83**(2), 86 (1976).

17. B. P. Block and J. C. Bailar Jr., *J. Am. Chem. Soc.* **73**, 4722 (1951).

18. D. T. Sawyer, *Chem. Rev.* **64**, 633 (1964).

19. U.S. Pat. 2,584,017 (Jan. 29, 1952), V. Dvorkovitz and T. G. Hawley (to The Diversey Corp.).

20. R. Davenport with F. Dubois, A. DeBoo, and A. Kishi, in *Chemical Economics Handbook*, SRI International, Menlo Park, Calif., 2002.

21. B. Suresh with M. Yoneyama, in *Chemical Economics Handbook*, SRI International, Menlo Park, Calif., 2002.

22. T. Esker with H. Janshekar, and Y. Sakurma in *Chemical Economics Handbook*, SRI International, Menlo Park, Calif., 1999.

23. U.S. Pat. 4,872,996 (Oct. 10, 1989), J. G. Grierson, D. A. Wilson, and D. K. Crump (to The Dow Chemical Company); U.S. Pat. 4,640,818 (Feb. 2, 1987), J. G. Grierson, C. A. Jones, and W. D. Spears (to The Dow Chemical Company).

24. H. Asanuma, A. Takemura, N. Toshima, and H. Hirai, *Ind. Eng. Chem. Res.* **29**(11), 2267 (1990).

25. R. W. Kuhr, C. P. Wedig, and L. N. Davidson, "The Status of New Developments in Flue Gas NO_x and Simultaneous NO_x/SO_x Cleanup," *1988 Joint Power Generation Conference*, Philadelphia, Pa., Sept. 26–28, 1988; S.-M. Yih and C.-W. Lii, *Chem. Eng. J.* **42**, 145 (1989).

26. J. C. Edwards and E. A. Rozas, *Proc. Am. Power Conf.* **23**, 575 (1961).

27. J. H. Wood, *Am. Dyestuff Reptr.* **65**(11), 32 (1976).

28. V. N. Gupta and D. B. Mutton, *Pulp Pap. Mag. Can.* **70**(1), T174 (1969).

29. U.S. Pat. 2,181,121 (Nov. 28, 1939), F. B. Downing and C. J. Pedersen (to E. I. du Pont de Nemours & Co., Inc.).

30. E. P. Bayha, L. R. Hubbard, and W. H. Martin, *Intern. Dyer* **131**, 529 (1964).

31. H. E. Millson, Jr., *Am. Dyestuff Reptr.* **45**, *Proc. Am. Assoc. Text. Chem. Color.*, P66 (1956).

32. R. H. Dick and D. H. Andrews, *Pulp Pap. Mag. Can.* **66**(3), T201 (1965).

33. J. R. Metcalf, *Ind. Water Eng.* **8**(1), 16 (1971).

34. D. J. Baure and R. E. Lindstrom, *J. Metals* **23**(5), 31 (1971).

35. J. K. Aiken and C. Garnett, *Electroplat. and Met. Finish.* **10**(2), 31 (1957); *Eng. Index*, 667 (1957).

36. W. Simon, W. E. Morf, and P. Ch. Meier, *Struct. Bonding (Berlin)* **16**, 113 (1973).

37. J. J. Mortvedt and co-eds., *Micronutrients in Agriculture*, Soil Science Society of America, Madison, Wis., 1972.

38. A. Wallace, *J. Plant Nutr.* **6**(6), 429 (1983).

39. D. D. Chapman and E. R. Schmittou, in G. Wilkinson, R. D. Gillard, and J. A. McCleverty, eds., *Comprehensive Coordination Chemistry*, Vol. 6, Pergamon Press, New York, 1987, pp. 95–132; U.S. Pat. 4,717,647 (Jan. 5, 1988), A. Abe and J. Nakajima (to Fuji Photo Film Co., Ltd.); Jpn. Kokai Tokkyo Koho JP 63,231,742 [88 231,742] (Sept. 27, 1988) A. Taomoto, K. Waratani, K. Nichogi, I. Machida, and S. Asakawa (to Matsushita Electric Industrial Co., Ltd.); Jpn. Kokai Tokkyo Koho JP 63,165,181 [88 165,181] (July 8, 1988) T. Fukui, K. Miura, Y. Oguchi, and Y. Takasu, (to Canon K. K.); Jpn. Kokai Tokkyo Koho JP 02 120,084 [90 120,084] (May 8, 1990) N. Yokoyama, T. Noda, and T. Kitao to Daihachi Chemical Industry Co., Ltd.); Jpn. Kokai Tokkyo Koho JP 02 69,739 [90 69,739] (Mar. 8, 1990) T. Tamaoki and K. Ichimura (to Agency of Industrial Sciences and Technology).

40. T. Fujisawa, A. Takagi, K. Okuyama, and S. Ohshima, *Jpn. J. Appl. Phys., Pt. 1* **29**(10), 1914 (1990) (Eng.); T. Fujisawa and co-workers, *Jpn. J. Appl. Phys., Pt. 1* **28**(8), 1358 (1989) (Eng.).

41. Eur. Pat. Appl. 278684 (Aug. 17, 1988), D. E. Putzig (to E. I. du Pont de Nemours & Co., Inc.).

42. L. Marko and B. Heil, *Catal. Rev.* **8**, 269 (1973).

43. C. A. Vitali and B. Masci, *Tetrahedron* **45**(7), 2213 (1989).

44. C. J. Pedersen and H. K. Frensdorff, *Angew. Chem. Int. Engl. Ed.* **11**, (1972); A. C. Knipe, *J. Chem. Ed.* **53**, 619 (1976).

45. D. C. Zecher, *Mater. Perform.* **15**(4), 33 (1976).

46. G. Schmuckler, in N. Bikales, ed., *Encyclopedia of Polymer Science and Technology*, Suppl. Vol. 2, John Wiley & Sons, Inc., New York, 1979, p. 197.

47. J. Stary, *The Solvent Extraction of Metal Chelates*, The Macmillan Co., New York, 1964.

48. M. A. Hughes, *Chem. Ind. (London)* (24), 1042 (1975).

49. M. Rubin, S. Gignac, S. P. Bessman, and E. L. Belknap, *Science* **117**, 659 (1953).

50. A. Cerami, R. W. Grady, C. M. Peterson, and K. K. Bhargava, in *4th Cooley's Anemia Symposium*, 1979; *Ann. N.Y. Acad. Sci.* **344**, 425 (1980).

51. C. F. Meares and co-workers, *Anal. Biochem.* **142**, 687 (1984); G. E. Krejcarek and K. L. Tucker, *Biochem. Biophys. Res. Commun.* **77**, 581 (1977); D. Parker, *Chem. Br.* **26**, 642 (1990); D. A. Goodwin and C. F. Meares, "Bifunctional Chelates for Radiopharmaceutical Labeling," in R. P. Spencer, ed., *Radiopharmaceuticals: Structure—Activity Relationships*, Grune & Stratton, Inc., New York, 1981, pp. 281–306.

52. W. C. Eckelman and S. M. Levenson, *Int. J. Appl. Radiat. Isot.* **28**, 67 (1977); T. C. Pinkerton, C. P. Desletts, D. J. Hoch, M. V. Mikelsons, and G. M. Wilson, *J. Chem. Educ.* **62**, 965 (1985).
53. W. A. Volkert and co-workers, *Drugs Future* **14**(8), 799 (1989); W. F. Goeckeler and co-workers, *J. Nucl. Med.* **28**, 495 (1987); U.S. Pat. 4,898,724 (Feb. 6, 1990), J. Simon, D. A. Wilson, W. A. Volkert, D. E. Troutner, and W. F. Goeckeler (to The Dow Chemical Company).
54. R. B. Lauffer, *Chem. Rev.* **87**, 901 (1987).

GENERAL REFERENCES

G. Wilkinson, R. D. Gillard, and J. A. McCleverty, *Comprehensive Coordination Chemistry, The Synthesis, Reactions, Properties & Applications of Coordination Compounds*, Vols. 1–7, Pergamon Press, Oxford, New York, Beijing, Frankfurt, São Paulo, Sydney, Tokyo, Toronto, 1987.
Chemical Economics Handbook Marketing Research Reports, SRI International, Menlo Park, Calif.
A. Catsch, A. E. Harmuth-Hoene, and D. P. Mellor, "The Chelation of Heavy Metals," in *International Encyclopedia of Pharmacology and Therapeutics*, Section 70, Pergamon Press, Oxford, UK, 1979.

WILLIAM L. HOWARD
Consultant

DAVID WILSON
The Dow Chemical Company

CHEMICAL METHODS IN ARCHAEOLOGY

1. Introduction

Archaeology is the study of past human life, culture, and activities, which is shown by material evidence in the form of surviving artifacts, biological and organic remains, and a variety of other evidence recovered by archaeological excavation (1). It differs from the other historical disciplines since these reconstruct the past on the basis of documentary sources. Documentary sources mostly testify to important events and can be affected by the human tendency of a writer to represent his/her own subjective reality. Archaeology, by contrast, tells about the past by studying materials that entered into the life of common people.

In its endeavour to study material evidence, archaeology has strongly interacted with almost all scientific disciplines. Among these, chemistry has certainly played an important role. Chemistry has developed methods to date archaeological material and has also allowed us to infer trade routes by studying ancient artifacts and to shed light on the technology used to make them. The diet and customs of ancient peoples have also been discovered by applying chemical methods. Chemistry intervenes in the understanding of the mechanisms that cause archaeological material to degrade in order to set up procedures aimed at stabilizing decay and preventing further deterioration. It also tries to find the best way to restore ancient artifacts.

Today, a number of resources are available to researchers involved in studies dealing with chemistry and archaeology. *Archaeometry* and the *Journal of Archaeological Science* are two specific journals devoted to the interactions between the sciences and archaeology. A variety of different conference proceedings contain results from chemical studies in archaeology. Between them those published by the American Chemical Society (2–6) and the Materials Research Society (7–12) are of particular relevance. Several books cover topics of relevance to archaeological chemistry. The comprehensive book edited by Brothwell and Pollard (13) represents a landmark in the field of archaeological sciences and replaces the outdated book edited by Brothwell and Higgs (14,15). A broad introduction to archaeological chemistry was provided by Goffer (16) in 1980 while the most recent book from Pollard and Heron (17) concentrates on a series of case studies. The books from Lambert (18) and Henderson (19) are based on the study of a series of archaeological materials. Techniques of analytical chemistry applicable to archaeology have been discussed in the book edited by Ciliberto and Spoto (20). Methods used to date archaeological materials are discussed in the books by Aitken (21) and Taylor and Aitken (22).

2. Chemical Methods in Archaeology Over the Last Three Centuries

The chemical methods used in the study of antiquities go back to the eighteenth and nineteenth century. Eminent scientists such as Humphry Davy (17 Dec. 1778–29 May 1829), Jöns Jakob Berzelius (20 Aug. 1779–7 Aug. 1848), Michael Faraday (22 Sept. 1791–25 Aug. 1827), Marcelin Berthelot (25 Oct. 1827–18 Mar. 1907), Friedrich August Kekulé von Stradonitz (7 Sept. 1829–13 Jul. 1896), and Wilhelm Conrad Röntgen (27 Mar. 1845–10 Feb. 1923) turned their attention to ancient coins, glasses, pigments, pottery, and other remnants of the past during the course of their studies. Similar investigations continued throughout the nineteenth century thanks to a number of other investigators, with most of them operating in isolation (23–25). Important basic concepts in the field started to be introduced at that time when European chemists suggested that chemical composition could be used to identify the source of archaeological materials (25). The concept of *provenance* is still active in the field (26) and the development of instrumental methods and of new ways to mathematically treat data sets (27–29) have allowed us to better define the potential and the limitation of scientific provenance of archaeological material.

The use of scientific examination to investigate the past was greatly fostered when major museums began to establish laboratories for that purpose on their premises. After a scientific laboratory was established in 1888 at the Königlichen Museen in Berlin, the director Friedrich Rathgen (30) provided the first manual dealing with practical procedures for the conservation and restoration of antiquities (31). This handbook remained for many years the only collected scientific documentation in the field. It was in 1920, as a direct result of the First World War, that one of the leading laboratories in the field was established at the British Museum. The discovery of the alterations suffered by many of the objects stored in 1918 in the Holborn Post Office tunnel as protection against possible war damage (32) moved the Trustees to invite Alexander Scott to carry out an investigation. On the basis of his report, an emergency laboratory was set up in 1920 within the precincts of the British Museum. At this time, Alexander Scott was aged 70 and was a senior fellow of the Royal Society, Superintendent of the Davy-Faraday Laboratory of the Royal Institution and President of the Chemical Society. His interests, ranging from sciences to arts, led to the systematic application of chemical methods to the study of ancient objects held in museums (33). Just a few years later the first research facility devoted to the study and the conservation of archaeological and artistic objects was established in the United States. In fact, in 1928 the Center for Conservation and Technical Studies was opened at the Fogg Art Museum of Harvard University by its Director Edward W. Forbes.

The pioneering approach developed in museum science laboratories was followed by the involvement of university laboratories that ensured the continuous development of new chemical methodologies for the investigation of remnants of the past. In spite of this continuous evolution, only in the past four decades has the use of advanced analytical instrumentation, together with the increased knowledge of statistical methods for the elaboration of coherent data-sets, established a fundamental link between instrumental analytical chemistry, art, and archaeology. The opening in 1955 of the Research Laboratory for Archaeology and the History of Art at Oxford University was certainly a starting point for this process. Today, science departments are normally active within the major museums while departments and institutes devoted to studies of relevance for chemistry applied to archaeology are present in most of the world's university institution.

3. Materials Study

Analytical chemistry plays an important role in the study of archaeological materials and has a variety of goals. When artifacts are investigated, some of the most important aims concern the study of the technology used to produce them, to reconstruct their distribution from the production areas, and to understand the use to which they were put in the past (34). By interpreting such information, it is possible to better understand the behavior of ancient people. Long-term storage often tends to obscure chemical information that contribute to the above mentioned aims. It is thus important to understand, at the deepest possible level, all the altering processes that intervene in the life cycle of the archaeological material.

3.1. Instrumental Methods for Chemical Analysis. Archaeological materials are studied from a chemical point of view by using the variety of instrumental methods available today to chemists (20). However, all chemists involved in studies concerning samples of archaeological interest will recognize how important it is to make an appropriate selection of the analytical method to use in their studies. The main questions raised usually concern how the proposed analytical procedure will affect the integrity of the object under examination. From this point of view, only those techniques that do not alter the integrity and appearance of archaeological objects can be considered as "ideal" techniques. Techniques that operate *in situ*, making sample-taking unnecessary, come close to this ideal.

In the attempt to find a balance between the requirements of scientific methods and the need to maintain the integrity of the object under study, the only alternatives to *in situ* analysis require the object itself to be placed in the analyzing chambers of the equipment or tiny fragments of samples to be scraped from its surface. The former approach cannot be applied in all cases, since only small objects such as coins, certain jewelry, and statuettes are of a size and shape that will fit into common analytical tools. The destructive approach thus remains as a last resort for the extraction of chemical information from archaeological samples.

Tremendous improvements have been made with regard to the scope and efficiency of today's chemical instrumental methods, which has led to the development of new methodologies that satisfy specific requirements to a greater degree such as micro-destructiveness or nondestructiveness of the sample to be analyzed. A wider range of information is now available and the greater sensitivity and reproducibility of analyses is thus ensured. In this context, the use of spatially resolved analytical techniques have provided new opportunities for micro-destructive and, at times, completely nondestructive analyses, opening up new diagnostic approaches for the study of archaeological samples (35).

Instrumental chemical methods used to study archaeological materials include elemental analytical techniques (36) and molecular analytical techniques (37).

X-ray fluorescence (XRF) has been popular for elemental analysis since the 1960s. Major and minor element composition of a great variety of solid archaeological materials including stones, metals, glasses, ceramics, bones, paintings and other materials is daily obtained in most archaeometry laboratories making use of energy dispersive XRF (EDXRF) or wavelength dispersive XRF (WDXRF) equipment (38–40). Synchrotron radiation XRF (SRXRF) has been recently proposed as a technique that significantly improves the performance of the standard XRF (41). However, its use is limited by the small number of synchrotron radiation facilities available. Particle-induced X-ray emission (PIXE) also has been widely used to study archaeological materials by using multiproject facilities (42,43). A facility dedicated to the study of museum artifacts has been developed at the Research Laboratory of the Museums of France in Paris (44).

Trace element analyses contribute to chemically fingerprint archaeological material. Provenance is often inferred by evaluating results from trace element analysis. Optical emission spectroscopy (OES) has been used for analysis of elements present at concentration up to ~0.001% since the 1930s. It was largely replaced by atomic absorption spectroscopy (AAS) during the 1970s and by the inductively coupled plasma (ICP) techniques during the 1980s (45). Instrumental neutron activation analysis (INAA) (46) has been the technique of choice in provenance investigation for a long time, largely due to the high sensitivity to many trace elements along multiple dimensions of element concentration (47). It also ensures good precision and accuracy of data compared to other techniques (46). However, ICP–mass spectrometry (ICP–MS), less costly equipment, today rivals INAA in provenance investigations (45,48).

The above mentioned methods for trace element analysis require samples, often in the range of tenths of milligrams, to be properly treated. Inappropriate procedures have been shown to affect the precision of the results (49). Efforts are placed today in trying to minimize the destructive impact of trace element analysis. In this perspective, the use of the laser ablation–ICP–MS has been investigated (50–53). Laser-induced breakdown spectroscopy (LIBS), also known as laser-induced plasma spectroscopy (LIPS) or laser ablation spectroscopy (LAS), has been recently proposed as a new micro-destructive method for major, minor, and trace element analysis in the study of archaeological material (54).

Isotope analysis plays an important role in the study of archaeological material. The quantitative determination of the relative amount of isotopes of interest allows us to hypothesize on the provenance of metal objects and also to support paleodietary research. Thermal ionization mass spectrometry (TIMS) has been used largely for the analysis of heavy metal isotopes for provenance purposes (55). More recently, the multi-collector–ICP–MS (MC–ICP–MS) has been shown to attain the required reproducibility for lead isotope analysis (56,57). Archaeological investigations based on the study of stable isotopes such as ^{13}C, ^{15}N, or ^{18}O are carried out by making use of isotope ratio mass spectrometry (IRMS) based techniques (58,59).

Molecular analytical techniques are used in archaeological science with a variety of purposes. The identification of organic materials is one of them. These organic materials are for the most part natural products and, hence, composed of complex mixtures of biochemical components. Detailed compositional information from such materials are derived making use of a wealth of instrumental methodology, most of which are based on organic mass spectrometry (MS) methods (60,61).

Gas chromatography mass spectrometry (GC–MS) has become the workhorse of organic analysis in archaeology thanks to its ability to separate and analyze mixtures of thermally stable volatile compounds or compounds that can be volatilized by the application of heat. Liquid chromatography mass spectrometry (LC–MS), more often in the form of high performance liquid chromatography mass spectrometry (HPLC–MS), is used when thermally unstable or not volatile organic mixtures are going to be analyzed. The use of different mass analyzers allows for improved performance of the above mentioned techniques. Tandem mass spectrometry (MS^n) has also been shown to contribute to the study of organic remains. In fact, GC–MS–MS has been used to confirm the presence of DNA residue in ancient seeds (62).

However, the possibility of studying solid samples offered by cross-polarization magic angle spinning NMR (CP–MAS–NMR), coupled with the ability to analyze less sensitive elements such as silicon and aluminium, has recently allowed the demonstration of the potential of this technique in studying ancient organic residues, archaeological bone and wood, fossilized resins, and a range of other archaeological materials (63).

Both Raman spectroscopy (64) and Fourier transform infrared spectroscopy (FT–IR) (65) have been used increasingly in archaeology to study a wide range of both inorganic and organic archaeological materials. Modern applications tend to favor the microdestructive approach guaranteed by the use of micro-Raman spectroscopy and FT–IR microspectroscopy (35).

A variety of other analytical techniques have been used to solve specific archaeometric questions. Among them surface sensitive analytical techniques (see SURFACE AND INTERFACE ANALYSIS) such as X-ray photoelectron spectroscopy (XPS), Auger electron spectroscopy (AES) and secondary ion mass spectrometry (SIMS) have been sparingly used, mostly to investigate degradation processes and to microdestructively characterize both inorganic and organic material (66–69).

The fundamental contribution of statistical methods in the understanding of data generated by the variety of analytical methods used to investigate archaeological material should not be overlooked. The exploratory multivariate methodologies have been extensively used in interpreting analytical data from archaeological materials (27,29). Recently, the Bayesian approach has been used as a statistical modeling of data (28,29).

3.2. Inorganic Archaeological Materials.
It has been clarified that *"The primary aim of materials studies in archaeology is to contribute to the investigation of the overall life cycle or chaîne opératoire of surviving artifacts.... This life cycle starts with production that includes the procurement and processing of the raw materials through to the fabrication and decoration of the artifacts. It then continues through distribution of the artifacts to their use, re-use and ultimate discard"* (34). Inorganic materials better survive the degradation processes that increase with time and thus have more easily been subjected to such investigations.

Stone. Stone is certainly one of the earliest inorganic materials used by humans (70). In particular, flint was used from the Paleolithic period onward for a variety of purposes such as cutting and pounding thanks to its properties to produce sharp blades with characteristic conchoidal fractures when worked. The advent of farming during the Neolithic period expanded the need for flint, which was extracted from mines or quarries and transported to different regions. Chemical studies aimed at establishing what distance the flint traveled from its source are based on the premise that it is possible to source the flint chemically to a particular location (19). Earliest works were able to distinguish between broad geological areas in western Europe in which flint occurs by using the amount of a single element (thorium) as a discriminator (71–73). Successive studies demonstrated that a more precise sourcing was possible by detecting aluminium, iron, magnesium, potassium, sodium, calcium, and lithium at levels >50 µg/g (74). In such a way, the chemical composition of flint was linked to specific areas in England.

Obsidian is certainly the lithic material providing archaeologists with the clearest evidence of contact between different cultures. In fact, obsidian is almost the ideal material for source characterization by elemental analysis and was the material of choice for the manufacture of a variety of cutting tools. Moreover, it is shiny and attractive, and for this reason was used in the past for ornamental purposes. Obsidian is a glass formed when highly viscous volcanic lava of high silicon and aluminium content cools rapidly, usually at the margins of a lava flow, such that the process of mineral crystallization is precluded. The presence of obsidian far from any source of volcanic activity represented an intriguing puzzle to the archaeologists (18,19). Now we know that the acquisition of obsidian developed in different ways, including local collection over land or sea. For this reason, sourcing archaeological obsidian is of great assistance in the investigation of the cultural, social, and economic development of ancient societies.

Early works based on chemical elemental analysis carried out with optical emission spectroscopy showed that it was not until 7500 BC that obsidian artifacts were moved beyond their immediate environs of source areas in the Near East (75–77). The advent of LA–ICP–MS has allowed us to improve our ability to precisely source obsidian. Nine separate Cappadocian sources have been clearly separated with a minimal impact of the destructiveness of the analysis (78). The Mediterranean area also represented an important source for obsidian (17,79). Elemental chemical analyses carried out by OES in early works and successively by INAA, XRF, and ICP–MS have identified the island of Melos as the source for obsidian for Greece, Crete, and the Aegean islands. Northern Italy and Macedonia were supplied by Carpathian sources (80,81). Central Mediterranean regions were mainly supplied by the Italian islands of Lipari, Sardinia, Palmarola, and Pantelleria (79). The obsidian source located in the area between east-central Mexico and Guatemala, called Mesoamerica, have been studied in an attempt to shed light on ancient trade routes set up when important cultures such as Aztecs and Maya were present in the area (82,83).

Ceramics. [See FINE ART EXAMINATION AND CONSERVATION (19).] The discovery of fire allowed humans to process natural materials to improve or simply change their characteristics (~1,600,000 years ago) (84). One of the earliest uses of fire was for cooking. Food became safer and tasted better after cooking. Later, stones were heated to improve their hardness (~80,000 BC). The complex technology required for making pottery was not developed until thousands of years after fire had been discovered. Paleolithic objects from Dolni Vestonice in the Czech Republic are probably pottery's earliest ancestors and can be dated to 24,000 BC (85). The development of pottery is still, however, a subject for debate and its origins have been placed between 12,000 and 10,000 BC (18,86). Pottery technology evolved with time and more and more complex procedures allowed the production of finely finished ceramic artifacts that combined aesthetics with usefulness. The introduction of glaze certainly improved both the appearance and utility of pottery, especially when the use of proper fluxes provided suitable glazes for curved surfaces such as those of pots (87,88). Flat surfaces were in fact already being glazed in Egypt in ~5000 BC but the glazed pots found in Syria have been dated to a later period, ~1700 BC.

Different methods for finishing pottery required control of the oxidizing/reducing atmosphere during firing. The most famous and striking examples of this procedure were provided by the Greeks after the invention of the so-called "red-figures" technique (89). On the basis of documented evidence, most attribute this technique to the painter Andokides, dating it to ~530 BC. It involved producing red and black paints on pottery as a result of switching from an oxidizing firing atmosphere to one that was reducing to form Fe_2O_3 (red) or Fe_3O_4 (black) on the surface of the pottery previously covered with suitable clay materials. The complex decorations developed during the Middle Ages and the Renaissance bear witness to the tremendous evolution of pottery decoration techniques through successive ages.

Ceramics are synthetic materials whose production is affected by choices and actions taken by humans during each stage of production that reflects their cultural symbolism, tradition, and individual preferences. Their study can thus improve our knowledge of past societies. The complex range of parameters that led to the various modes of pottery production, distribution, and consumption recently has been discussed (90). Pottery has been the biggest class of material to be studied for provenance purposes. In the simplest approach, the chemical composition of the fired ceramic is considered indicative of the composition of the raw clay material. However, a number of factors could influence the final composition of the final products and thus it is a normal procedure to compare the finished pottery composition with that of fired pottery of certain provenance (17). Since the 1970s INAA has been the preferred analytical technique for pottery trace analysis (46,91). More recently INAA has been rivaled by ICP–MS in this field. The limited length of this article does not allow us to account for the enormous literature developed in an attempt to provenance archaeological ceramics. The reader is addressed to further readings for a detailed list of case studies (17–19,46,90–92).

Glass. (See GLASS.) The development of the technology necessary to obtain glass could be linked to the smelting of metal ores or to the manufacturing of glazed pottery. The earliest known glass material are supposed to be linked to smelting technology and have been dated to ~2000 BC (93,94). Glazing technology may, however, have anticipated the origin of glass (18,88). In fact, the first vitreous materials were glazed stones and ground quartz bodies coated with a glaze called faience. Mesopotamia was probably the region in which glass production was first established but it was in Egypt after 1500 BC under the XVIII dynasty that glass production found its first prominence.

The early chemical analyses carried out with the aim to create ancient glass composition data sets began in the 1950s (95). However, it was only in 1961 that the first report as to where ancient glasses were grouped in term of chemical composition and correlated to both geographical and chronological criteria were published (96). Five elements were determined by using INAA and expressed in term of oxides: magnesium, potassium, manganese, antimony, and lead. Magnesium was an impurity in the alkali used as flux, potassium was both an impurity in the alkali and an alkali itself. Both manganese and antimony were used to eliminate the pale green color of glass with iron impurity and obtained under oxidizing conditions (17,19). Ancient soda-lime glasses dated between 1500 and 800 BC and 800 and 1000 AD were categorized as being high magnesium (HMG) and low magnesium (LMG) containing glasses. The amount of magnesium put in relation to the potassium contents reflected the use of mineral (natron) or plant-ash sources of alkali. High antimony soda-lime glasses produced between 600 and 200 BC were identified as a separate group. Islamic glasses were grouped as high magnesium containing glass produced between 840 and 1400 AD and high lead glasses produced between 1000 and 1400 AD.

Today, a number of other ancient glass composition groups have been identified. Each of them reflecting changes in the raw material or in the technology used to produce the glass. Low magnesia, high potassium oxide glasses produced in Europe between 1150 and 700 BC were obtained by innovating the raw materials used (97). Later in Medieval northern Europe high potash glasses were used to produce church windows and vessels easily subject to degradation (17). High potassium and barium oxide glasses were produced under the Chinese Han Dynasty (206 BC–221 AD) (98). New materials were also used in India from the first millennium AD to produce high alumina glasses (99).

Metals. [See FINE ART EXAMINATION AND CONSERVATION (18,19).] The advantages offered by metals compared to other materials used by ancient people, such as stone or wood, were discovered 10,000–12,000 years ago in Southwest Asia. The advent of metallurgy with the development of farming and domestication of animals allowed the rise of urban civilizations. The exploitation of metals enhanced previously existing trade routes and the specialization required by metal working encouraged social stratification.

Copper in its native state is believed to be the earliest metal used by humans even though native gold could have preceded its use due to its beauty and resistance to corrosion. The production of copper and copper alloy objects developed alongside the evolution of the technology linked to the extraction of the metal from ores—smelting, the improved ability to work the metal and, finally, the ability to form alloys (100). Much of this evolution was a consequence of the ability of metalworkers to discover and control processes that allowed them to heat the metal, in its native state or in ores containing copper compounds, at increasingly high temperatures so as to reach the temperature necessary to melt copper (1083°C).

The shaping of native copper was a well-established custom in Southwest Asia from ~10,000 BC onward. This area was by far the most advanced in copper-work technology. In fact, evidence for the smelting of copper-based ores, which leads us to suppose that casting skills were already established, has been dated to 7000–6000 BC (Çatal Hüyük, Anatolia).

The addition of elements other than copper to form alloys with better properties in terms of castability, hardness, and appearance may originally have been accidental. Arsenic, the first element used to form copper alloys, was alloyed with copper during Chalcolithic times, possibly using arsenic-containing copper ores. Tin was the most important alloying element in the Old World after 4000 BC until the resulting alloy being *bronze*. Lead and zinc were two other important elements in forming or modifying the characteristics of copper alloys. The former was used especially by the Greeks who intentionally added lead to the tin-bronzes used to make statues both to improve the fluidity of the molten alloy and to enhance appearance. The intentional alloying of zinc—thus forming *brass*—was certainly lead to the use of the mentioned elements in copper alloy preparation. The earliest brass objects were probably manufactured in eastern Turkey during the first millennium BC. With the *calamine process* been established and perfected, brass production increased rapidly thanks to its use in minting coins during the Roman Empire (17).

The smelting of lead was assumed to precede that of copper. The lower temperature required for lead smelting (800°C) is the most obvious reason for this assumption. Lead use was at its height when the Greeks used it with copper-tin alloys to make statues. However, during the Roman Empire lead production increased tremendously and reached levels of production that were to be reproduced only after the Industrial Revolution (101). Under the Romans, lead was used to produce a great variety of objects ranging from coins to pipes for the distribution of water.

The beginning of the Early Iron Age (102)—fixed as being 1200 BC—coincides with the ability of people from western Asia to smelt iron and alloy carbon so as to obtain steel. Up to 1500 BC, the Hittites had the best developed technology for working iron. The advantages of iron over copper and its alloys had, however, been known since the Bronze Age. Evidence of uneven uses of iron, sometimes in its native state, have been dated to 2500 BC. Iron smelting technology presumably dates back to the Bronze Age as the temperature required for smelting iron (1100–1150°C) is similar to that required for copper. However, iron smelting necessitated a more accurate control of the carbon and oxygen present in the furnace so as to maximize the percentage of iron present in the spongy mass, or *bloom*, obtained after the furnace was cooled (103). Glass-like materials, or *slag*, were formed during smelting due to a reaction between the silica impurities present in the ore and fluxes. The formation of slag, inevitably containing certain amounts of iron, was sometimes due to the intentional addition of fluxes to the furnace to help rid the ore of silica impurities.

Lead isotope analysis has been used for provenancing metal objects (104). The applicability and the use of the method has caused discussions mainly centred on the evaluation of the correct interpretation of the use and limitation of lead isotope analysis in the investigation of ancient metal production (105–107). Nevertheless, lead isotope analysis has allowed better identification of sources of copper used in the Mediterranean Bronze Age. Mines at Laurion were the largest sources of lead in the late Bronze Age. Lead found at Knossos on Crete from Middle Minoan to Mycenaean times was excavated at Laurion. Also, Egyptian artifacts from Amarna and Abydos have proven to be derive from metals excavated at Laurion.

3.3. Organic and Biomolecular Archaeological Materials. Most of the research studies carried out in archaeological science over the last 50 years have been devoted to the investigation of inorganic material. This situation was due to the idea that biological and organic material can only survive in the archaeological record under exceptional circumstances. After a few pioneering investigations in the 1980s, the use of increasingly sophisticated organic techniques have since demonstrated how a variety of organic and biomolecular archaeological residues can be studied. This approach consists in identifying molecular markers capable of identifying unknown organic samples on the basis of their presence in contemporary natural substances (108). Lipids, in particular, have been shown to be particularly important as biomolecular markers (109).

Archaeological Lipids. Lipids occur ubiquitously in plants and animals and preserve under favorable conditions in association with a range of different classes of archaeological materials ranging from unglazed pottery, soil, human and animal remains, resins, and a range of other amorphous materials. The use of modern chromatographic techniques coupled with mass spectrometric analyzers (60) has contributed to studies of artifact use patterns (110) and food consumption (111) through the identification of lipid residues. Lipids are extracted from the powdered original matrices by using organic solvents. They are properly derivatized and then analyzed by GC or GC/MS techniques or by GC–C–IRMS for isotope ratio studies (112).

Degradation processes cause lipids to be hydrolyzed or oxidated. Secondary ketones are commonly found in the lipid extracts of ancient cooking vessels. The contribution of isotope ratio studies carried out by GC–C–IRMS in differentiating ancient lipid residues has been clearly demonstrated, eg, by distinguishing between cow milk and adipose fats using the $\delta^{13}C$ values of their $C_{16:0}$ and $C_{18:0}$ fatty acids (113). The identification of lipid biomarkers also provides insights into ancient anthropogenic activities. In fact, soil lipid profile is affected by different agricultural practices, while detection of ancient faecal inputs to the soil may allow the location of ancient cesspits (114,115). In such studies 5β-stanols, bile acids, and n-alkyl lipids act as useful biomarkers. Lipids help also in studying decay processes associated with human remains. Lipid analysis of skin tissue from the Tyrolean Ice Man showed that some acyl lipids were preserved. However, it was noted that all triacylglycerols with more than one double bond were completely degraded. The combined histological evidence of loss of epidermis with chemical evidence of the transformation of fats into adipocere indicates submersion of the body in water for several months before its freeze–drying (116). Cholesterol is another lipid that persists in long-buried bones of humans and animals (117) and its evidence can be used as a source of paelodietary information (118). Recently, lipid biomarkers have helped to shed light on chemical treatments used in ancient Egyptian mummification (119). Organic material from Egyptian mummies dating from 1900 BC to 395 AD have been studied by GC–MS, thermal desorption (TD) and pyrolysis (Py)–GC–MS. In Egyptian mummification, natron was used as the desiccant while a variety of organic materials were used to prevent the decomposition of the body. Plant oils and, to a lesser extent, animal fats were used to protect tissues from degradation thanks to the cross-linked network they generate after polymerization. Both coniferous resin, identified by the presence of diterpenoid components, and beeswax, identified by the presence of alkanes, wax esters, and hydroxy wax esters, were increasingly used with time for their antimicrobial and antibacterial properties. A number of other plant derived components of the embalming mixture also have been identified while no component coming from petroleum bitumens were detected. The latter evidence contradicted the previously supposed use of natural bitumens in mummification (120).

Proteins.   Proteins have rarely survived to the archaeological record (121). Only under unusual burial environments have they survived microbial degradation and proteins in hard tissues such as tooth, bone, and shell, are prevalently protected (122). Temperature plays the main role in protein preservation, however, deposition within small pores whose dimensions physically excludes enzymes and close interaction with minerals have been proposed as situations that enhance protein preservation (121,123). As a consequence, we should expect that it is possible to obtain protein residues from ancient ceramics that may have been in contact with protein-rich foodstuff for prolonged periods of time. However, protein extraction from mineral and ceramic surfaces is difficult. Most of the proposed methods disrupt the macromolecular structure of the protein residue (124–126). Immunological methods have also been used as extraction methods of protein from mineral surfaces, however, a yield of ~0.0025% was evaluated for the proposed methods (127). Recently, a new immunological method that allows protein extraction yield up to 0.1% was proposed (128,129).

Other Organic Residues.   Organic residues have been identified by making use of FT–IR. In particular, this technique, in combination with other spectroscopic and chromatographic analytical techniques, has provided chemical evidence of ancient food and beverages (130,131).

Food is fundamental for all human society, not only for nutrition and health but also in economic, social, and ritual life. Information about prepared food is therefore critical for an understanding of ancient cultures but unfortunately examining the food of the past is extremely difficult. Food is normally consumed and leftovers usually decay. These problems are of particular relevance when residues of ancient beverages are to be studied. Fermented beverages are made from a variety of sugar-containing materials. Barley (*Hordeum vulgare* L.) was probably the first to be domesticated and used to make beer. This beverage first appeared in Mesopotamia ~6000 BC (18). Wine followed this early beer production.

A small fragment of the yellowish residue present in a Neolithic jar found in the Hajji Firuz Tepe village (North of Iran) and dated 5400–5000 BC was analyzed by means of diffuse-reflectance FT–IR spectroscopy (130). The spectra showed features attributed to calcium tartrate. After comparison with ancient and modern reference samples, this result, confirmed by means of chromatographic and ultraviolet (UV) spectroscopic methods, was the earliest chemical evidence for wine since tartaric acid occurs naturally in significant amounts (~1%) only in grapes (*Vitis vinifera* L.). The calcareous environment of the site had converted the acid into calcium tartrate. Moreover, the shape of the jar confirmed its use as a liquid holder. A detailed study of the spectra (with peaks centred at 2926 and 2858 cm^{-1}) and their comparison with reference samples, together with the analytical evidence obtained from chromatographic and UV spectroscopic investigations, also confirmed the presence of an oleoresin extracted from the *Pistacia atlantica* Desf. terebinth tree. Alcohol soluble resin was used to inhibit bacterial growth and improve the wine in taste and odor.

A similar analytical approach also led to the identification of the ancient organic material present in vessels found at Gordion (Turkey) in a tomb dated ~700 BC (131). The tomb is assumed to be that of the Phrygian King, Midas. The study of the region at 1420, 1390, 1170, and 1120 cm^{-1} in the diffuse-reflectance FT–IR spectra of the methanol extracts of 14 food samples indicated that a mutton or goatmeat-based dish had been left in the vessels. The contemporary presence of bands due to long-chain esters of beeswax, calcium oxalate, and calcium tartrate in the diffuse-reflectance FT–IR spectra of the methanol extracts of 16 beverage samples instead provided evidence of a mixed fermented beverage of grape wine (calcium tartrate), barley beer (calcium oxalate, also called "beerstone"), and honey mead (beeswax).

This study, which revealed one of the most comprehensive Iron Age drinking sets ever found, suggested that such food and drink was eaten at a feast before the interment of the king, thus providing fundamental chemical evidence for ancient cuisine in the Mediterranean area.

Ancient DNA.   DNA entered the archaeological record from the second half of the 1980s. In fact, before then it was not imagined that long-term preservation of DNA was possible. A breakthrough in the field was a study published in 1985 where the successful detection of intact genetic information in a 4000-year old Egyptian mummy was presented (132). Ancient DNA studies were boosted by the invention of the polymerase chain reaction (PCR) (133), which allows a targeted stretch of DNA to be amplified millions of times so as to be properly sequenced. Unfortunately, the high sensitivity of the method renders contamination from modern DNA highly probable if appropriate procedures are not set up (134). Moreover, a deep understanding of the degradation processes that concern postmortem DNA and of the conditions under which DNA preserves is required (134,135).

DNA is a record of ancestry. For this reason, ancient DNA can be used to determine kin relationship within a group of specimens (136). Moreover, ancient DNA can express some of the biological characteristics of an archaeological specimen. Biological sex (137) or genetic diseases (138) can be inferred by studying archaeological DNA. Studies carried out on DNA sequences older than 1 million years ago (antediluvian DNA) have concluded that such ancient sequences cannot be reproduced or derive from contaminations (139). A variety of studies on DNA sequences dated up to 100,000 years ago from extinct animals have revealed the phylogenetic relationships of extinct animals (134). For example, the extinct *moas* of New Zealand have been shown to be related to flightless birds in Australia rather than extant kiwis in New Zealand (140). The study of ancient human DNA sequences opened up a new and exciting view of our ancestry (134,141). It is today known that Neanderthal hominids that lived in Europe and western Asia until ~30,000 years ago, were not directly related to modern Europeans (142). The common ancestor of modern Europeans lived ~170,000 years ago, possibly in Africa (143). However, a mixture of modern humans and Neanderthals coming to Europe from Africa ~40,000 years ago cannot be excluded.

Amber Provenance. *"It will, of course, for ever remain a secret to us whether this amber is derived from the coast of the Baltic or from Italy, where it is found in several places, but particularly on the east coast of Sicily."* With this sentence of his book "Mycenae: a narrative of researches and discoveries at Mycenae and Tiryns" the German archaeologist Heinrich Schliemann (1822–1890), discoverer of the ruins of Troy and Mycenae, seemed to be challenging scientists to solve the puzzling question of the provenance of amber. Amber is a fossil resin, derived from coniferous trees. It comprises a complex mixture of molecules based primarily on monoterpenoid and diterpenoid structures. It has been used for ornamental purposes since prehistoric times when it was believed that amber was sunlight solidified by sea waves. Understanding the provenance of amber made it possible to establish the earliest known trade routes that involved its transportation from northern to southern Europe ~5000 BC. In the 1960s, IR spectroscopy contributed greatly to this discovery by providing evidence of differences in composition between Baltic amber and Sicilian amber. Transmittance IR spectra acquired from hundreds of amber samples made it clear that the vast majority of amber from prehistoric Europe derives from material originating in the Baltic coastal region (144). Differences in the absorption patterns generated by the vibrational stretching of C–O bonds (1110–1250 cm^{-1}) provided the analytical evidence of Baltic or non-Baltic provenance.

CP–MAS–NMR has also been shown to be able to characterize both modern and fossil amber on a worldwide basis by distinguishing them in both their botanic as well as geographical differences (145,146).

Amino Acid Racemization Dating.   Amino acids are the "building blocks" of proteins and up to several hundred of such building blocks can be contained in a protein. All of the amino acids that occur in proteins, except glycine, have at least an asymmetric carbon atom and thus can occur in two optical isomers called D and L. In life, the amino acids making up the proteins of higher eukaryotes consist solely of the L form. Metabolically active tissues contain specialized enzymes known as racemases that maintain a disequilibrium in our cells of only the L isomers. After death, the enzymes in living organism cease their activity and amino acids undergo racemization thus interconverting the L isomers in to the D isomers at a time dependent rate. In the late 1960s, the dating method based on the racemization of the amino acids in fossil organisms was first announced (147,148). The method was expected to extend beyond the radiocarbon range (40,000–150,000 years BP) (for an overview on dating see FINE ART EXAMINATION AND CONSERVATION) but acquired a controversial reputation after errors made in some of the dating carried out in the mid-1970s. The dating of Californian paleoindian skeletons carried out by quantifying the aspartic acid racemization led to an estimated age for the fossil bones of ~50,000 and 60,000 years (149). Such result suggested an early colonization of North America. Later, after more precise calibration by AMS radiocarbon dating, the bones were redated to 5000–6000 years (150). [A review on amino acid racemization of Californian paleoindian remains may be found in (17).] Today, it is clear that fluctuation in the burial environments and the degree of degradation of the protein can affect the dating and that studies carried out on protein extracted from well-preserved bones can led to more reliable dating (151,152) even though a variety of factors may render the method prone to contamination by exogenous or degraded proteins (153).

4. Degradation of Archaeological Materials

Most of the materials studied by the archaeology have survived for long time in the ground and to a variety of degradation processes. Degradation processes affect different materials to a different extent and follow different paths. For this reason, certain materials entered the archaeological records more often than others. Stone survives almost unaltered while materials such as metal, glass, and certain organic material such as amber, undergo some degradation but often survive in a recognizable form. Biological materials such as skin and hair survive only under exceptional condition such those that preserved the Tyrolean Ice Man in the Alps on the Austrian–Italian border (154). Biological hard tissues such as bone, tooth, and shell undergo complex degradation processes.

The overall degradation processes that act on organic remains after death are studied by *taphonomy*. The term was first introduced in 1940s (155) and comes from the Greek word ταφοϛ (taphos, burial). Taphonomy studies all the natural and anthropogenic processes that affect the organism in its transferral from the living word (biosphere) to the sedimentary record (lithosphere). Taphonomy includes two different stages. The first one, *biostratinomy*, includes all the interactions involved in the transferral of the living organism from the living world to the inorganic world, including burial. *Diagenesis* includes all the transformation occurring after burial (156,157).

More recently, these concepts, referring only to living organisms, have been broadened and diagenesis is now "*... the cumulative physical, chemical and biological processes that alter all archaeological materials in the burial environment, and is consequently a fundamental characteristic of the archaeological record*" (156). Diagenetic studies thus also involves postdepositional changes that affect the structure of metal, glass or ceramic during burial (158,159). In this perspective, geochemical modeling was used to provide a deeper understanding of the complex variety of postdepositional processes affecting inorganic materials such as ceramics (156). Also, great progress has been made in understanding taphonomic processes affecting bone (160–162). Bone is an important component of the archaeological record due to the wide range of information its organic and inorganic components carry. Paleodietary information is obtained by the elemental and isotopic analysis of bone components while a variety of other information, partially described above, can be obtained from lipids, proteins, and DNA, which are often preserved in bone. For these reasons, attention is increasingly being placed on understanding all taphonomic processes involving bone.

BIBLIOGRAPHY

1. C. Renfrew and P. Bahn, *Archaeology: Theories, Methods, Practice*, 3rd ed., Thames and Hudson, London, 2000.
2. C. W. Beck, ed., *Archaeological Chemistry*, Advances in Chemistry Series No. 138, American Chemical Society, Washington, D.C., 1975.
3. G. F. Carter, ed., *Archaeological Chemistry II*, Advances in Chemistry Series No. 171, American Chemical Society, Washington, D.C., 1978.
4. J. B. Lambert, ed., *Archaeological Chemistry III*, Advances in Chemistry Series No. 205, American Chemical Society, Washington, D.C., 1984.
5. R. O. Allen, ed., *Archaeological Chemistry IV*, Advances in Chemistry Series No. 220, American Chemical Society, Washington, D.C., 1989.
6. M. V. Orna, ed., *Archaeological Chemistry V*, Advances in Chemistry Series No. 625, American Chemical Society, Washington, D.C., 1996.
7. E. V. Sayre, P. B. Vandiver, J. Druzik, and C. Stevenson, eds., *Materials Issues in Art and Archaeology*, Materials Research Society Symposium Proceedings, Vol. 123, Materials Research Society, Pittsburgh, Pa., 1988.
8. P. B. Vandiver, J. Druzik, and G. S. Wheeler, eds., *Materials Issues in Art and Archaeology II*, Materials Research Society Symposium Proceedings, Vol. 185, Materials Research Society, Pittsburgh, Pa., 1991.
9. P. B. Vandiver, J. R. Druzik, G. S. Wheeler, and I. C. Freestone, eds., *Materials Issues in Art and Archaeology III*, Materials Research Society Symposium Proceedings Vol. 267, Materials Research Society, Pittsburgh, Pa., 1993.
10. P. B. Vandiver, J. R. Druzik, J. L. G. Madrid, I. C. Freestone, and G. S. Wheeler, eds., *Materials Issues in Art and Archaeology IV*, Materials Research Society Symposium Proceedings Vol. 352, Materials Research Society, Pittsburgh, Pa., 1995.

11. P. B. Vandiver, J. R. Druzik, J. Merkel, and J. Stewart, eds., *Materials Issues in Art and Archaeology V*, Materials Research Society Symposium Proceedings, Vol. 462, Materials Research Society, Pittsburgh, Pa., 1997.

12. P. B. Vandiver, M. Goodway, J. R. Druzik, and J. L. Mass, eds., *Materials Issues in Art and Archaeology VI*, Materials Research Society Symposium Proceedings, Vol. 712, Materials Research Society, Pittsburgh, Pa., 2001.

13. D. R. Brothwell and A. M. Pollard, eds., *Handbook of Archaeological Sciences*, John Wiley & Sons, Inc., New York, 2001.

14. D. R. Brothwell and E. Higgs, eds., *Science in Archaeology: A Survey of Progress and Research*, Thames & Hudson, London, 1963.

15. D. R. Brothwell and E. Higgs, eds., *Science in Archaeology: A Survey of Progress and Research*, 2nd edition, Thames & Hudson, London, 1969.

16. Z. Goffer, *Archaeological Chemistry: A Sourcebook on Applications of Chemistry to Archaeology*, John Wiley & Sons, Inc., New York, 1980.

17. A. M. Pollard and C. Heron, *Archaeological Chemistry*, The Royal Society of Chemistry, Cambridge, UK, 1996.

18. J. B. Lambert, *Traces of the Past: Unraveling the Secrets of Archaeology Through Chemistry*, Addison-Wesley, Reading, Massachusetts, 1997.

19. J. Henderson, *The Science and Archaeology of Materials*, Routledge, London, 2000.

20. E. Ciliberto and G. Spoto, eds., *Modern Analytical Methods in Art and Archaeology*, John Wiley & Sons, Inc., New York, 2000.

21. M. J. Aitken, *Science-based Dating in Archaeology*, Longman, London, 1990.

22. R. E. Taylor and M. J. Aitken, eds., *Chronometric Dating in Archaeology*, Plenum Press, New York, 1997.

23. E. R. Caley, *J. Chem. Educ.* **28**, 64 (1951).

24. E. R. Caley, *J. Chem. Educ.* **44**, 120 (1967).

25. Ref. 17, pp. 1–19.

26. L. Wilson and A. M. Pollard, in Ref. 13, pp. 507–517.

27. M. J. Baxter, *Exploratory Multivariate Analysis in Archaeology*, Edinburgh University Press, Edinburgh, 1994.

28. C. E. Buck, W. G. Cavanagh, and C. D. Litton, *The Bayesian Approach to Archaeological Data Interpretation*, John Wiley & Sons, Inc., Chichester, 1996.

29. M. J. Baxter and C. E. Buck, in Ref. 20, pp. 681–746.

30. M. Gilberg, *J. Am. Inst. Cons.* **26**, 105 (1987).

31. F. Rathgen, *The Preservation of Antiquities: a Handbook for Curators*, Cambridge University Press, Cambridge, UK, 1905.

32. M. L. Caygill, in P. B. Vandiver, J. R. Druzik, G. S. Wheeler, and I. C. Freestone, eds., *Materials Issues in Art and Archaeology III*, Material Research Society, Pittsburgh, 1992, pp. 29–99.

33. A unique overview of the history of scientific conservation was provided by Harold J. Plenderleith with a lecture held at the British Museum in 1978. The text of the Plenderleith's lecture has been recently published in H. J. Plenderleith, *Stud. Conserv.* **43**, 129 (1998).

34. M. S. Tite, in Ref. 13, pp. 443–448.

35. G. Spoto, A. Torrisi, and A. Contino, *Chem. Soc. Rev.* **29**, 429 (2000).

36. R. G. V. Hancock, in Ref. 20, pp. 11–20.

37. G. Spoto, in Ref. 20, pp. 167–175.

38. L. Moens, A. von Bohlen, and P. Vandenabeele, in Ref. 20, pp. 55–79.

39. E. S. Lindgren, ed., Special Millennium Issue on Cultural Heritage, *X-Ray Spectr.* **29** (2000).

40. M. Mantler, M. Schreiner, *J. Radioanal. Nucl. Chem.* **247**, 635 (2001).

41. A list of paper on SR in archaeology may be found in http://srs.dl.ac.uk/arch/publications.html
42. G. Demortier and A. Adriaens, eds., *Ion Beam Study of Art and Archaeology Objects*, European Commission, Directorate-General of Research, Brussels, 2000.
43. J. C. Dran, T. Calligaro, and J. Salomon, in Ref. 20, pp. 135–166.
44. T. Calligaro, J. C. Dran, H. Hamon, B. Moignard, and J. Salomon, *Nucl. Instr. Meth. B* **136–138**, 339 (1998).
45. S. M. M. Young and A. M. Pollard, in Ref. 20, pp. 21–53.
46. H. Neff, in Ref. 20, pp. 81–134.
47. R. L. Bishop, V. Canouts, P. L. Crown, and S. P. De Atley, *Am. Antiq.* **55**, 537 (1990).
48. R. H. Tykot and S. M. M. Young, in Ref. 6, pp. 116–130.
49. H. Neff, M. D. Glascock, R. L. Bishop, and M. J. Blackman, *Am. Antiq.* **61**, 389 (1996).
50. C. E. B. Pereira, N. Miekeley, G. Poupeau, and I. L. Kuchler, *Spectrochim. Acta B* **56**, 1927 (2001).
51. W. Devos, M. Senn-Luder, C. Moor, and C. Salter, *Fresenius J. Anal. Chem.* **366**, 873 (2000).
52. J. D. Robertson, H. Neff, and B. Higgins, *Nucl. Instr. Meth. Phys. Res. B* **189**, 378 (2002).
53. S. A. Junk, *Nucl. Instr. Meth. Phys. Res. B* **181**, 723 (2001).
54. A. Demetrios, *Appl. Spectr.* **55**, 186A (2001).
55. N. H. Gale and Z. Stos-Gale, in Ref. 20, pp. 503–584.
56. A. N. Halliday, D. C. Lee, J. N. Christensen, M. Rehkamper, W. Yi, X. Luo, C. M. Hall, C. J. Ballentine, T. Pettke, and C. Stirling, *Geochim. Cosmochim. Acta* **62**, 919 (1998).
57. T. Hirata, *Analyst* **121**, 1407 (1996).
58. M. J. De Niro, *Am. Sci.* **75**, 182 (1987).
59. I. T. Platzner, *Modern Isotope Ratio Mass Spectrometry*, John Wiley & Sons, Inc., New York, 1997.
60. R. P. Evershed, in Ref. 20, pp. 177–239.
61. J. R. Chapman, *Practical Organic Mass Spectrometry: A Guide for Chemical and Biochemical Analysis*, 2nd ed., John Wiley & Sons, Inc., New York, 1995.
62. K. O'Donoghue, T. A. Brown, J. Carter, and R. P. Evershed, *Rapid Commun. Mass Spectrosc.* **8**, 503 (1994).
63. J. B. Lambert, C. E. Shawl, and J. A. Stearns, *Chem. Soc. Rev.* **29**, 175 (2000).
64. F. Cariati and S. Bruni, in Ref. 20, pp. 255–278.
65. M. Bacci, in Ref. 20, pp. 321–361.
66. G. Spoto and E. Ciliberto, in Ref. 20, pp. 363–404.
67. G. Spoto, *Acc. Chem. Res.* **35**, 652 (2002).
68. G. Spoto, *Thermochim. Acta* **365**, 157 (2000).
69. A. Adriaens, in D. C. Creagh and D. A. Bradley, eds., *Radiation in Art and Archaeology*, Elsevier Science, Amsterdam, The Netherlands, 2000, pp. 180–201.
70. M. Edmonds, in Ref. 13, pp. 461–470.
71. G. de G. Sieveking, P. T. Craddock, M. J. Hughes, P. Bush, and J. Ferguson, *Nature (London)* **228**, 251 (1970).
72. A. Aspinal and S. W. Feather, *Archaeometry* **14**, 41 (1972).
73. M. De Bruin, P. J. M. Korthoven, C. C. Bakels, and F. C. A. Groen, *Archaeometry* **14**, 55 (1972).
74. P. T. Craddock, M. R. Cowell, M. N. Leese, and M. J. Hughes, *Archaeometry* **25**, 135 (1983).
75. C. Renfrew, J. R. Cann, and J. E. Dixon, *Ann. Br. School Athens* **60**, 225 (1965).
76. C. Renfrew, J. E. Dixon, and J. R. Cann, *Proc. Prehistoric Soc.* **32**, 30 (1966).

77. C. Renfrew, J. E. Dixon, and J. R. Cann, *Proc. Prehistoric Soc.* **30**, 319 (1968).
78. B. Gratuze, *J. Archeol. Sci.* **26**, 869 (1999).
79. R. H. Tykot, *Acc. Chem. Res.* **35**, 618 (2002).
80. O. Williams-Thorpe, S. E. Warren, and J. Nandris, *J. Archaeol. Sci.* **11**, 183 (1984).
81. T. K. Biro, I. Pozsgai, and A. Vlader, *Acta Arcaheol. Acad. Sci. Hung.* **38**, 257 (1986).
82. M. D. Glascock, *Acc. Chem. Res.* **35**, 611 (2002).
83. P. F. Healy, H. I. McKillop, and G. B. Walsh, *Science* **225**, 414 (1984).
84. R. M. Rowlett, *Science* **284**, 741 (1999).
85. P. V. Vandiver, O. Soffer, B. Klima, and J. Svoboda, *Science* **246**, 1002 (1989).
86. E. Copper, *A History of Pottery*, Longman, London, 1972.
87. M. S. Tite, I. Freestone, R. Mason, J. Molera, M. Vendrell-Saz, and N. Wood, *Archaeometry* **40**(2), 241 (1998).
88. M. Tite, A. Shortland, S. Paynter, *Acc. Chem. Res.* **35**, 585 (2002).
89. J. Boardman, *Athenian Red-Figure Wares: the Archaic Period, a Handbook*, Thames and Hudson, London, 1976.
90. M. S. Tite, *J. Archeol. Meth. Theory* **6**, 181 (1999).
91. R. L. Bishop and M. J. Blackman, *Acc. Chem. Res.* **35**, 603 (2002).
92. I. K. Whitbread, in Ref. 13, pp. 449–459.
93. T. A. Wertime, *Am. Sci.* **61**, 670 (1973).
94. R. H. Brill, *Sci. Am.* **209**, 120 (1963).
95. W. E. S. Turner, *J. Soc. Glass Technol.* **40**, 162 (1956).
96. E. V. Sayre and R. W. Smith, *Science* **133**, 1824 (1961).
97. J. Henderson, *Archaeometry* **30**, 77 (1988).
98. L. Jiazhi and C. Xianqiu, in H. C. Bhardwaj, ed., *Archaeometry of Glass*, Indian Ceramic Society, Calcutta, 1987, pp. 21–26.
99. R. H. Brill, in H. C. Bhardwaj, ed., Archaeometry of Glass, Indian Ceramic Society, Calcutta, 1987, pp. 1–25.
100. R. F. Tylecote, *A History of Metallurgy*, Institute of Metals, 2nd ed., London, 1992.
101. S. Hong, J. P. Candelone, C. C. Patterson, and C. F. Boutron, *Science* **265**, 1841 (1994).
102. R. Maddin, J. D. Muhly, and T. S. Wheeler, *Sci. Am.* **237**(4), 122 (1977).
103. H. Hodges, *Artifacts: an Introduction To Early Materials and Technology*, Gerald Duckworth & Co., Ltd., London, 1989.
104. N. H. Gale and Z. Stos-Gale, in Ref. 20, pp. 503–584.
105. E. V. Sayre, K. A. Yenner, E. C. Joel, and I. L. Barnes, *Archaeometry* **34**, 73 (1992).
106. N. H. Gale and Z. Stos-Gale, in A. M. Pollard, ed., *New Developments in Archaeological Science*, Oxford University Press, Oxford, 1992, pp. 63–108.
107. E. Pernicka, *Archaeometry* **35**, 259 (1993).
108. R. P. Evershed, *World Archaeol.* **25**, 74 (1993).
109. R. P. Evershed, S. N. Dudd, M. J. Lockheart, and S. Jim, in Ref. 13, pp. 331–349.
110. C. Heron and R. P. Evershed, in M. Schiffer, ed., *Archaeological Method and Theory V*, University of Arizona Press, Arizona, 1993, pp. 247–284.
111. R. P. Evershed, C. Heron, S. Charters, and L. J. Goad, in A. M. Pollard, ed., *New developments in Archaeological Science*, Oxford University Press, Oxford, 1992, pp. 187–202.
112. R. P. Evershed, S. N. Dudd, M. S. Copley, R. Berstan, A. W. Stott, H. Mottram, S. A. Buckley, and Z. Crossman, *Acc. Chem. Res.* **35**, 660 (2002).
113. S. N. Dudd and R. P. Evershed, *Science* **282**, 1478 (1998).
114. P. F. van Bergen, I. D. Bull, P. R. Poulton, and R. P. Evershed, *Org. Geochem.* **26**, 117 (1997).
115. I. D. Bull, I. A. Simpson, P. F. van Bergen, and R. P. Evershed, *Antiquity* **73**, 86 (1999).

116. B. X. Mayer, C. Reiter, and T. L. Bereuter, *J. Chromatog. B* **692**, 1 (1997).
117. R. P. Evershed, G. Turner-Walker, R. E. M. Hedges, N. Tuross, and A. Leyden, *J. Archaeol. Sci.* **22**, 277 (1995).
118. R. P. Evershed and A. W. Stott, *Anal. Chem.* **68**, 4402 (1996).
119. S. A. Buckley and R. P. Evershed, *Nature (London)* **413**, 837 (2001).
120. P. G. Bahn, *Nature (London)* **356**, 109 (1992).
121. A. M. Gernaey, E. R. Waite, M. J. Collins, O. E. Craig, and R. J. Sokol, in Ref. 13, pp. 323–329.
122. A. M. Child, *Stud. Cons.* **40**, 19 (1995).
123. L. M. Mayer, *Geochim. Cosmochim. Acta* **58**, 1271 (1994).
124. R. P. Evershed and N. Tuross, *J. Archaeol. Sci.* **23**, 429 (1996).
125. T. M. Oudemans and J. J. Boon, *J. Anal. Appl. Pyrolysis* **20**, 197 (1991).
126. P. F. VanBergen, C. J. Nott, I. D. Bull, and P. R. Poulton, *Org. Geochem.* **29**, 1779 (1998).
127. O. E. Craig and M. J. Collins, *J. Archaeol. Sci.* **29**, 1077 (2002).
128. O. E. Craig and M. J. Collins, *J. Immunol. Methods* **236**, 89 (2000).
129. O. Craig, J. Mulville, M. P. Pearson, R. Sokol, K. Gelsthorpe, R. Stacey, and M. Collins, *Nature (London)* **408**, 312 (2000).
130. P. E. McGovern, D. L. Glusker, L. J. Exner, and M. M. Voigt, *Nature (London)* **381**, 480 (1996).
131. P. E. McGovern, D. L. Glusker, R. A. Moreau, A. Nuñez, C. W. Beck, E. Simpson, E. D. Butrym, L. J. Exner, and E. C. Stout, *Nature (London)* **402**, 863 (1999).
132. S. Pääbo, *Nature (London)* **314**, 644 (1985).
133. R. K. Saiki, D. H. Gelfand, S. Stoffel, S. J. Scharf, R. Higuchi, G. T. Horn, K. B. Mullis, and H. A. Erlich, *Science* **239**, 487 (1988).
134. M. Hofreiter, D. Serre, H. N. Poinar, M. Kuch, and S. Pääbo, *Nature Rev. Gen.* **2**, 353 (2001).
135. H. N. Poinar, *Acc. Chem. Res.* **35**, 676 (2002).
136. S. Hummel and B. Herrmann, *Homo* **47**, 215 (1996).
137. K. A. Brown, *Ancient Biomol.* **2**, 3 (1998).
138. D. Filon, M. Faerman, P. Smith, and A. Oppenheim, *Nature Genet.* **9**, 365 (1995).
139. T. Lindahl, *Nature (London)* **365**, 700 (1993).
140. A. Cooper, C. Mourer-Chauvire, G. K. Chambers, A. von Haeseler, A. C. Wilson, and S. Pääbo, *Proc. Natl. Acad. Sci. USA* **89**, 8741 (1992).
141. I. V. Ovchinnikov, A. Gotherstrom, G. P. Romanova, V. M. Kharitonov, K. Liden, and W. Goodwin, *Nature (London)* **404**, 490 (2000).
142. M. Krings, A. Stone, R. W. Schmitz, H. Krainitzki, M. Stoneking, and S. Pääbo, *Cell* **90**, 19 (1997).
143. M. Ingman, H. Kaessmann, S. Pääbo, and U. Gyllensten, *Nature (London)* **408**, 708 (2000).
144. C. W. Beck, *Appl. Spectr. Rev.* **22**, 55 (1986).
145. J. B. Lambert and J. S. Frye, *Science* **217**, 55 (1982).
146. J. B. Lambert and G. O. Poinar, *Acc. Chem. Res.* **35**, 628 (2002).
147. P. E. Hare and P. H. Abelson, *Yearbook Carnegie Inst. Washington* **66**, 526 (1967).
148. J. L. Bada, B. P. Luyendyk, and J. B. Maynard, *Science* **170**, 730 (1970).
149. J. L. Bada, R. A. Schroeder, and G. F. Carter, *Science* **184**, 791 (1974).
150. J. L. Bada, *Am. Antiq.* **50**, 645 (1985).
151. B. J. Johnson and G. H. Miller, *Archaeometry* **39**, 265 (1997).
152. J. Csapó, Z. Csapó-Kiss, and J. Csapó Jr., *Trends Anal. Chem.* **17**, 140 (1998).
153. M. J. Collins, E. R. Waite, and A. C. T. van Duin, *Proc. R. Soc. London* **B354**, 51 (1999).

154. K. Spindler, *The Man in the Ice. The Preserved body of a Neolithic Man Reveals the Secrets of the Stone Age*, Phoenix, London, 1995.
155. I. A. Efremov, *Pan-Am. Geol.* **74**, 81 (1940).
156. L. Wilson and A. M. Pollard, *Acc. Chem. Res.* **35**, 644 (2002).
157. C. Caple, in Ref. 13, pp. 587–593.
158. M. McNeil and L. S. Selwyn, in Ref. 13, pp. 605–614.
159. I. C. Freestone, in Ref. 13, pp. 615–625.
160. R. E. M. Hedges, *Archaeometry* **44**, 319 (2002).
161. M. J. Collins, C. M. Nielsen-Marsh, J. Hiller, C. I. Smith, J. P. Roberts, R. V. Prigodich, T. J. Wess, J. Csapó, A. R. Millard, G. Turner-Walker, *Archaeometry* **44**, 383 (2002).
162. A. Millard, in Ref. 13, pp. 637–647.

Giuseppe Spoto
Università di Catania

CHEMICAL PRODUCT DESIGN

1. Introduction

Chemical product design is the process by which we choose what product we will make. It precedes chemical process design, which deals with how we plan to make the product chosen. In the past, most of those involved in the chemical industry emphasized process design, because this enterprise was focused on perhaps 50 commodity products. For these commodities, price was the key, and efficient production was the route to low prices. This strategy correctly dominated the efforts of the chemical industry in the twentieth century.

Now the goals of this chemical enterprise have become much broader. These goals include not only the 50 or so commodity chemicals, but several thousand high value added chemicals. Many of these are pharmaceuticals. The broader goals of the chemical industry also include chemical mixtures, often with specific microstructures, like detergents, and coatings. The goals include polymers processed to have characteristics desired in filters or films. These broader goals mean that chemical professionals now participate in a wider variety of business decisions. These include deciding what to make.

In this article, we suggest a four step template by which we can decide which chemical products we want to make. Because of the enormous variety of products which are possible, we should not expect this template to work perfectly in every case. Instead, we offer the template as a mental checklist for organizing our thinking. We ourselves use this checklist again and again as we refine our efforts at product design. The four steps are as follows:

1. *Needs*: We must decide what need our product will fill.
2. *Ideas*: We must generate chemical ideas that could satisfy this need.

3. *Selection*: We must efficiently select the best ideas.

4. *Manufacture*: We must design the process for making our product.

The first three steps of this template are unique to Product Design, but the fourth step includes the more familiar topics of Process Design.

We will give details of each of these four steps in the body of this chapter. Before we give these details, we can gain an overview from quickly applying this template to three possible products: an improved amine for gas scrubbing, a pollution preventing ink, and a battery separator.

Current gas scrubbing often uses aqueous solutions of amine like monoethanol amine to absorb acid gases like carbon dioxide and hydrogen sulphide. Once saturated, or loaded, the resulting solutions are warmed to strip out the absorbed acid gases. Because the solutions are dilute, the warming takes a lot of energy. Thus the need is for an amine that can be more easily regenerated. Ideas to do this could include amines that are used as concentrated solutions, or those that undergo phase transitions and, hence, require a smaller temperature increase than currently dictated. The idea actually selected was to search for a pressure sensitive, hindered amine, one that complexed acid gases at moderate pressure, but released them under partial vacuum. Those involved synthesized and tested several hundred hindered amines to find attractive candidates. They eventually synthesized one of the best in industrially useful quantities. Note that the important steps were the definition of the need and the creation of ideas, even though most of the work was in the chemical synthesis of candidate amines.

The second example, the pollution preventing ink, originates in the use of methylene chloride in printing. This carcinogenic solvent is used to adjust the ink's viscosity and to clean the presses between runs. The need to reduce methylene chloride use led to ideas like using less solvent, capturing and condensing the solvent's vapor, using toluene as a solvent, and modifying the ink's chemistry. Many of these ideas were tried in turn; the one selected eventually was to change the ink's chemistry. In particular, the ink contains a low molecular weight polymeric resin. By changing the molecular weight distribution of the resin, we can make an ink that does not require adding solvent to adjust its viscosity. By adding pendant carboxylic acid groups to the resin, the ink can be washed off the presses with aqueous base: The ink essentially becomes its own soap. Manufacturing the new ink turned out to be very similar to manufacturing the old ink. The key was inventing the new ink chemistry.

The third example, a shutdown battery separator, is used in the high energy batteries in laptop computers and other portable electronic devices. These batteries store chemical energy in an anode and a cathode separated by a microporous polymer sheet. This microporous separator is often a polyolefin made by stretching to form small pores, followed by a careful warming with the film under tension to relax the polymer and freeze the pores in place. Unfortunately, if these batteries short out, they can explode. Thus the need is for a separator that will allow unrestricted ionic transport under normal operating conditions, but which will stop this transport if the current is excessive.

Ideas to meet this need ranged widely, from pedestrian fuses to silly fancies. The idea selected depended on the fact that a shorted battery gets hot. This warmth softens the microporous polymer separator. If the heat treatment in making the separator was changed, then this softening would close the pores in the separator. In developing this product, the engineers involved used their experience in making the original film with pores that always stayed open. In these original efforts, the engineers frequently had failures, when the pores would not stay open. Now, with a new product need, these failures became the key to a success.

In the rest of this article, we will examine in detail this template of needs, ideas, selection, and manufacture. We will illustrate each step with additional examples, but we want to stress the steps independently, rather than to list case studies. Using the template this way, we believe, best illustrates how products are designed.

2. Needs

The first step in designing a chemical product is defining what it is that a successful product should achieve, which is the Needs stage of product design.

Before considering how to define product needs, it is worth thinking about what it is that stimulates product development. We broadly identify two driving forces to product design. The first is the pull of the market, where a market opening is identified and then a product designed to fill this opening. This route is typical for new consumer products. A company will formulate a new soft drink in response to the marketing department identifying a market opportunity. Examples of market pull products are a deodorizing fabric for sports, a needle-free injector for medical applications, and an improved dusting cloth.

The second driving force to product development is technology push. Here the initial stimulant is not a perceived consumer opportunity, but an advance in technology, a new invention looking for an application. For example, during the Cold War there was a major effort to produce strong, lightweight materials for military applications. Having developed these high performance materials, companies began to look for other uses and found squash rackets and golf clubs. These are examples of technology push products.

As a second example, in 1775, shortly after discovering oxygen, Priestly said, "Who can tell but that, in time, this pure air may become a fashionable article in luxury. Hitherto only two mice and myself have the privilege of breathing it". By 2001, small bottles of oxygen were being marketed as OPUR and used by celebrities. Perhaps this is the slowest technology push product design in history.

The two driving forces of market pull and technology push result in rather different statements of product needs. In the case of a market pull product, we wish to define exactly what the market opportunity is. Product design then identifies an appropriate technology to exploit this opportunity. In the case of a technology push product, it is the other way around: the needs stage involves specifying the new technology and identifying areas of superiority to existing technology. The product design then consists of identifying markets where this new technology can be advantageously exploited.

While the form of the needs specification will vary depending on the spur to product development, the process by which this specification can be established is similar. This topic is what we now discuss.

2.1. Needs Identification. In defining product needs, we first remember that the product is not for ourselves. We must make sure that the needs we define reflect the requirements of those who will ultimately use the product and not simply our own prejudices. This means that needs identification will always begin by asking others what it is that they require: we must find our "customers" and identify their needs. We use the term customers in a loose sense here. We do not necessarily mean those who will buy our product, rather those who will benefit from its chemistry, which may be organizations such as companies or government agencies rather than individuals.

The consensus among marketing organizations is that the best way to get this sort of information is by face-to-face interviews. Fewer than 10 such interviews risk missing significant information, while more than 50 simply leads to duplication. If organizations are involved, we must of course interview individuals within those organizations. It is worth talking separately to several; it is always surprising the degree of disagreement present within a supposedly homogeneous organization.

One group merits particular attention, the "lead users". These are the people most expert in the product and those who will benefit most by its improvement. In the case of a market pull product, the lead users are those who very much depend on existing and competing products. In the case of a technology push product, the lead users are those most responsible for the technological advance that has stimulated the product development.

In addition to individual interviews, test panels and focus groups are sometimes used to help identify needs, particularly for consumer products.

Example 2A: Nonionic Surfactants. A typical washing powder for clothes cleaning made by a major manufacturer contains ~5% of nonionic surfactants, the remainder being anionic surfactant, mainly linear alkyl sulphonates. Nonionic surfactants are particularly effective for the removal of greasy stains. The company would like to increase their loading to ~10% to improve washing performance. The problem is that nonionic surfactants are sticky liquids, while washing powders must be free flowing granules. Simply increasing the nonionic surfactant loading using current technology results in agglomeration of the powder into a sticky mess. How can the nonionic surfactant loading be increased whilst maintaining the physical properties of the washing powder?

Solution. We must first identify the customers, so that we can ask what is required. It is tempting to think of the customers as those who buy and use the washing powder, but in this case that is not appropriate. They will simply tell you they want an inexpensive product that works well. As a result of numerous interviews with users, the marketing department has already decided that an increase in nonionic surfactant loading is desirable. The relevant people to talk to in this case are within the company, those who are going to have to implement changes in formulation: the engineers who run the powder manufacturing plants. We should aim to talk to 5–10 such people, to give us a good coverage of the issues involved and highlight differences of opinion. A typical interview with a plant engineer might go as follows:

What Do You Do Now?   We blend the anionic detergent and solid additives in an extrusion process. This makes the detergent go white and springy, which we want. We granulate to get the particle size right, and finally spray on nonionics and perfumes.

What Happens If You Spray on More Nonionic?   It is hopeless. You end up with a big sticky lump of detergent—its not a powder any more.

What Is Good about What You Do Now?   It gives a high density product—better than spray drying the detergent. It is efficient—we get the solids mixing and physical structuring of the anionic detergent at the same time. Its easy to change the additives at the end for different products just by spraying different amounts.

What Is Wrong with What You Do Now?   Nothing. We make five different brands this way. Have done for years. They clean people's clothes OK. I do not see the problem. Why do marketing want something new anyhow?

Where Do You Get Your Materials From?   We buy them from a subcontractor—a specialist manufacturer. They supply the anionic detergent—it's a yellow sticky paste when we get it—and the nonionic—it's a thick liquid. Solids come from another company and perfume from a third. The formulation chemists tell us what to put in and then we have to find it. We always like to have at least two suppliers of each material.

This interview starts to show us the needs—how any solution must fit into the existing process. It also tells us other people to talk to—the formulation chemists and the subcontractors who supply the detergents. We will then get a fuller picture of what is needed. Our next task will be to organize these interviews into a list of more coherent and specific needs.

2.2. Interpretation of Needs.   The needs recorded from the information gathering just described will be a hotchpotch of conflicting and incomplete statements, of varying relevance and practicality. Our task is to organize these needs as groups and to edit them into a cogent list. We will drop some stated needs, either because they appear impractical or are beyond our company's expertise.

It is also useful to rank the needs, for example, as essential, desirable and useful. The essential needs are those without which the product cannot succeed. We will aim to achieve as many desirable needs as possible, particularly if competitor products fail to do so. We are unlikely to design explicitly for the useful needs, although we will keep in mind that it will be a bonus if our product can fulfill these, too.

The way in which needs are grouped and organized will depend on the product being considered. It will usually be an easy task if we aim to improve an existing product. The more innovative the proposal product, the harder it is to satisfactorily define the needs. We may need to modify the needs statement later as our ideas take shape. It may also be useful to return to the customers, perhaps a different group, to further explore our marked list of needs.

Example 2B: A Fishy Business.   Fish farming is now a major international business. One company supplies food for salmon and other farmed species to markets in Scotland, Norway, and Chile. They wish to improve the performance of their product in order to secure these growing markets. A group of engineering consultants has been hired to recommend improvements. The current product consists of cylindrical pellets made by extrusion of a mixture of fish meal, fish oil and wheat. An interview with a fish farmer might go as follows:

What Do You Do Now? *We use a water cannon to spray the pellets over the fish pens. We feed once a day.*

What Is Good About the Current Product? *It's really easy to deliver—the water cannon is no work. The fish love it. They grow really fast.*

What Is Bad About It? *It is oily. You always get an oil film. This seems a waste and is not good for the fish. I also worry about the oil breeding fungi, which gets into the gills of the fish. The big fish always get to the top first and eat most of the food. We would like to give more to the smaller fish.*

Where Do You Buy Your Food From? *We get 70% from your company and 30% is local waste. The waste is not so good; it is messy and the fish will not eat it all—they seem to prefer food with oil in. But it is cheap.*

Interpret these needs into a form useful for product specification.

Solution. We can easily classify the needs inferred from this interview.

Essential

Good nutritional value.

Mechanical strength.

Fast dispersion to get food to smaller fish.

Desirable

Inexpensive.

Good oil retention.

Useful

Ease of manufacture using current process.

Hygienic appearance.

We are now ready to continue development.

2.3. Quantification of Needs. Our aim now is to convert our qualitative list of needs into specifications, including as much quantitative and chemical detail as possible. In doing this, it is useful to consider three steps:

1. Write complete chemical reactions for any chemistry involved.
2. Make mass and energy balances important to the product's use.
3. Estimate the rates of any important changes that occur during the product's use.

Having produced a set of ideal product specifications in as much detail as possible, we should examine these carefully. Being a result of individuals' wish lists, they will often be entirely impractical involving huge flows, enormous concentrations or massive costs. If this is the case, we must revise our specifications to be more realistic. This may lead us to abandon the project altogether. If the only way of meeting customer requirements is to break a law of thermodynamics, we should stop product development now. This type of critical examination is sometimes known as a "chicken test" after a Canadian method for testing aero engines for their capacity to fly through flocks of geese. We are asking ourselves if the project is obviously unrealistic, before committing large resources to it.

Table 1. **Heats of Combustion of Alternative Fuels (kJ)**[a]

Fuel	per mol	per g	per cm^3
H_2	286	143	0.013 (at 1 atm)
CH_3OH	726	23	18
$Li \rightarrow Li^+ + e^-$	293	42	23
octane	5470	48	34
toluene	3910	43	37

[a] The comparison shown is sometimes obscured by the way data are presented.

Setting specifications often involves estimating physical properties of the final product. These estimates depend on two types of knowledge. First, they depend on the so-called "structure–property relations", which relate macroscopic properties like scratch resistance or creaminess to chemical and physical structure at much smaller size scales. Most obviously, this may be molecular structures, which dominate properties like drug efficacy or product acidity. Surprisingly often, "structure–property relations" involve structures of nanometer or micrometer size. Meat tenderness, paint "hiding power", and polymer transparency are examples where such microstructure are key.

The second type of knowledge useful for setting specifications involves qualitative generalizations of physical properties. These generalizations are often obscured by differences between results per mole, per volume, and per mass. As an illustration, consider the alternative fuels shown in Table 1. On a molar basis, octane is the most powerful fuel listed; on a mass basis, hydrogen is; on a volume basis, toluene is the choice. Because we sensibly describe chemical reactions with stoichiometry, generalizations like this are often obscured. This type of judgment is a key to setting product specifications.

The final step in quantification of needs is to specify a benchmark, which is an existing or idealized product against which we can measure the performance of our development. If we cannot beat the benchmark, the product is not worth developing. In some cases, when an innovative new product is being contemplated, no obvious benchmark will be available. We may be able to envisage an idealized benchmark by analogy with similar products. Sometimes, however, we must do without the security of a benchmark for comparison.

Example 2C: Water Purification for a Family. The WHO estimates that 1.7 billion people do not have access to a clean water supply (12,000 children per year die from diarrhoeal diseases—the world's biggest cause of child death). In the absence of major civil engineering projects, the point of demand is the family unit. A large non governmental organization has decided to design a water purification unit suitable for use by individual families in third world countries. Interviews with governmental and nongovernmental agencies working in such countries, and most importantly the lead users, usually the mother, reveal the following list of needs:

Essential
Supply a family with safe drinking water.
Inexpensive.

Desirable
  Rural focus.
  Simple to operate.
Useful
  Environmentally sustainable.

Assess these needs, quantifying where possible.

  Solution. The first question is how much water we need to purify. A person requires about 5 L of drinking water per day and typically uses ~50 L for cooking, cleaning, and washing. A household might be 10 people. So a minimum reasonable requirement would be 100 L/day, up to 500 L/day being desirable. A 5-year lifetime will be desirable, so a total lifetime throughout of 200,000 L will be necessary.

  What do we need to remove? Waterbourne diseases are the greatest threat—we must remove protozoa, bacteria, and viruses. Toxic materials are also sometimes a problem, but it is probably not useful to focus on these—the variety is very great, the problem usually local and a better solution is often preventing discharge. So we will focus on disease organisms. There are WHO guidelines for microbial removal and it would be useful to have these in mind, but to see these high standards as an absolute requirement might be a mistake. After all, an affordable but slightly less effective device will be more useful than one which no one can buy.

  Getting the cost right will be critical. What is affordable is clearly variable, but we must be aware that we are considering some of the poorest people in the world. We might hope for some support from aid agencies, but ideally a device affordable by the users without aid is best. Interviews in Nepal have revealed that a cost of >$10–15, with a annual running cost of $3–5, will be beyond the reach of most of the rural population. This requirement is clearly stringent and one we may not be able to meet and should not be seen as an absolute limitation: What we are really saying is the cheaper the better, this limit being the target.

  The rural focus means that the device must operate in the absence of a power supply. Any consumable should be locally available. The product must be easy to explain and to use by those with little experience of technology.

  Environmental sustainability is a rather vague need. It is best rephrased as the need to avoid the consumption of source local resources or the production of damaging waste.

  So our final needs list will be

Remove viruses, bacteria, and protozoa from 100 L/day.

Total lifetime at least 200,000 L.

Equipment as cheap as possible, ideally $10–15; annual operating cost under $5.

Operate in the absence of power, using locally available materials.

Simple to operate.

No consumption of source resources or discharge of toxic waste.

Chlorination is a sound benchmark, whis is simple, cheap, and well established, but has problems in terms of supply, use, and discharge of chemicals. Our device must be more attractive than local chlorination if it is to succeed.

We have now completed the Needs stage of product design. We have produced a ranked list of what our product needs to achieve and put this into quantitative and scientific terms as far as possible. We have also made a check that our aims are not wholly unreasonable and ideally we have a benchmark by which to judge the success of our product development. Up to this point, we have consciously avoided trying to think of solutions for our product needs. We want to define what we wish to achieve without prejudice caused by a preconception of what the product will look like. If we do already have an idea of the product's nature, we should try to keep it out of our considerations until the end of the Needs stage. Only now that we have well-established criteria for the success of any product should we begin to develop ideas for the product itself.

3. Ideas

Once we have chosen specifications for our target product, we need some good product ideas. In principle, we only need one idea, the one that we will manufacture. In practice, product development requires up to 100 ideas in order to find one truly worth pursuing. In these paragraphs, we describe how we get these ideas, how we organize them, and how we choose our best candidates for further effort.

3.1. Idea Generation.   To get our 100 or so product ideas, we will depend on people more than publications. The most important people are those on the team responsible for developing the specific product. We will normally assemble this team for free-ranging discussions that aim at generating possible answers. How to run such "brainstorming" discussions is carefully described in the literature, and so is not detailed here. We mention only that these discussions should initially be noncritical, and that all participants should be treated as equals. Discussions of new chemical products are sometimes curtailed for reasons as trivial as that the boss' spouse disagrees with some of the ideas suggested.

In addition to depending on the product team, we should pay special attention to customers who already are using existing, similar products. Some of these customers, called "lead users" in the business literature, may have already adapted our existing products for their own uses. These lead users often have excellent suggestions. Other human sources—consultants, private inventors, and the like—are often less useful. Literature has widely ranging value. Patents and trade information from competitors is often more useful than archival literature. Other methods for ideas use forms of chemical synthesis, as detailed elsewhere. Still, in most cases, the key is most often the product development team.

3.2. Ideas Sorting.   We now have our 100 or so ideas of widely varying quality. We must somehow sort through these ideas to locate the best five or so for further developments. Evaluating all to the same degree will normally take more resources than we will ever have, and will take much longer than we will even want. Thus we must find fast ways to find the best ideas.

We suggest proceeding with two stages. First, without quantification, we should try to sort the ideas on completely qualitative grounds, reducing the number to perhaps twenty. Later, with a bare minimum of quantification, we should try to screen the surviving ideas, aiming to get just the five or so we think are best. For the present, we will talk about how to get from 100 to 20; later, we will describe ways to go from 20 to 5.

To reduce the number of ideas to 20, we just make a list of all the ideas. We can then easily remove redundancy. Often this redundancy will occur because some ideas are special cases of others. For example, in a discussion of better barriers for landfills, one idea could be "The barrier should capture mercury."

A second idea might be "The barrier should adsorb all heavy metals except calcium."

The first idea is just a special case of the second.

In addition to removing redundancy, we want to drop ideas that are obvious folly. In doing so, we should be cautious, because some silly ideas may contain dreams. Sometimes, we can benefit from keeping a separate list of these flawed dreams, just to serve as a stimulus to later development. Normally, the efforts to remove redundancy and folly will still leave us up to 70 ideas.

To reduce the number of ideas further, we should try to organize them into categories, in a type of outline. How this should be done depends on the ideas generated: There seems to be no general strategy. Once this outline is made, it may expose gaps, which may imply repeating the brainstorming. More often, we will find that large groups of ideas will be inconsistent with our organisation's objectives or its strengths. These groups of ideas can be dropped, a major simplification. These last steps commonly cut the number of ideas to meet our target of twenty. One note of caution, however, many of our best ideas will often cluster under a single heading. Because we do not want to overspecialize too soon, we should consider choosing at least one idea beyond this cluster for the next stage of product development.

Example 3A: Multilayered Polymer Films for Secure Documents. Counterfeiting documents is big business. In the United States alone, >$250 million of counterfeit currency is recovered each year. To reduce this problem, some nations have gone to composite polymer films produced by multilayer extrusion as an alternative to paper for printing money. Australia has led the way. Their experience suggests that small denomination bills circulate so rapidly that paper bills wear out in <1 year, but large denomination bills circulate much more slowly. Because composite polymer bills last four times longer, their higher cost—twice that of paper—can be recovered in a couple of years.

Our company hopes to build on our experience in multilayer extrusion to make new films for all types of secure documents. We are part of a project team charged with identifying these documents. What should we recommend?

Solution. Our team quickly identified >200 possible products in one of the fastest, most straightforward efforts that our company has seen. After the redundancy and folly is removed, the ideas can be organized under the four headings shown in Table 2. Because this list is so broad and often so vague, it should be further edited to represent our company's strengths.

3.3. Idea Screening. We must now reduce our 20 surviving idea down to a still smaller number, normally 5 or fewer. We will still not have the

Table 2. **Possible Products Based on Multilayered Films**

I. Minor Improvements in Existing Products
    These improve or modify the existing paper document.
    A. Currency
       1. Paper currency with irremovable polymer strip.
       2. Machine-readable currency to tell denomination or lifetime.
    B. Passports and Corporate Identification
       1. Identification page of a passport made of multilayered polymer.
       2. Fingerprints added with optical ink.
    C. Credit Cards
       Polymer patch that takes signature.
    D. Surface Treatments
       1. Antidestructive.
          a. Multilayer polymer window that diffracts light.
          b. Polymer lamination.
       2. Antigraffiti coating.
       3. Miscellaneous.
          a. Fluorescent patches.
          b. Antibacterial agents.
          c. Waterproof coatings.

II. Change in Substrate
    These replace the paper with a polymer composite.
    A. Synthetic Polymers
       1. Currency, stock certificates, and Traveler's checks.
          a. Synthetic plastic currency stamped with optical ink.
          b. Polymer currency with different colours for the different denominations
             using dyed injected polymers.
          c. Metal coins that have a polymer core.
       2. Identification cards.
          a. Polymer passports.
          b. Polymer social security card.
          c. Oriented polymer paper.
          d. Polymer currency with layers of different orientation.
       3. Memorabilia like baseball cards made of polymer.
    B. Optically Unique Synthetic Polymers
       1. Currency, stock certificates and traveller's cheques.
          a. Polymer films with a multi-layer window.
          b. Polymer films with different thickness to induce colour.
          c. Polymer films with partial burnouts to generate patches of colour.
          d. Machine readable strips for use in vending machines.
       2. Credit Cards made with a homogenous polymer backing.
       3. Identification Cards, including driver's licences.
          a. Cards.
          b. Cards that change colours with light.
          c. Cards with burn out patches that change colour with light.
    C. Combinations of Synthetic and Optically Unique Polymers

III. Adhesive Additions
     These products use adhesive to attach them to existing documents.
     A. Currency
     B. Documents
        1. Multilayer optical films to authenticate legal documents.
        2. Multilayer films to replace notary stamps.
     C. Identification
        1. Multilayer optical polymer films adhered to passports.
        2. Multilayer optical polymer films adhered to student IDs with school logo.

Table 2 (*Continued*)

    D. Credit Cards
       1. Multilayer optical polymer films placed on credit cards.
    E. Checks
       1. Multilayer optical polymer films with adhesive backing used on checks.

  IV. Environmentally Benign Products
     This smaller set of ideas involves recycling, and is partially redundant.
    A. Secure Products from Recycled Plastic
    B. Secure Products with Later Uses
     Polymer currency that can be recycled into other products.

resources to make more detailed calculations for the 20 survivors, so we need approximate but quantitative tools that let us continue the screening, but on a still more rational basis.

One commonly effective method for this screening is to choose five or fewer key attributes shared by most of the surviving ideas. These attributes will include factors like scientific maturity, ease of engineering, risk of failure, and cost. We should choose factors that are different for different products. For example, even if safety is the most important product attribute, we gain nothing by choosing safety if all our potential products are equally safe.

Once these key attributes are chosen, we need to assign weighting factors to each. Normally, we will normalize these weighting factors, assuming values that sum to 1. Note that this implies that all products that we are still considering are capable of satisfying all attributes at least to a limited extent. If there is one attribute which is truly essential, we should drop all ideas that cannot satisfy this constraint, and continue our evaluation for the survivors.

On the basis of these attributes and their weighting factors, we now score all of our ideas relative to a convenient benchmark. We find it convenient to assign the benchmark scores of 5, and then to choose scores from each new product between one (poor) and 10 (excellent). We then calculate an average weighted score for each product. The potential products with the highest scores are those that we choose for further development. This screening method is partially illustrated by the following example.

Example 3B: Lab-on-a-Chip. This example is of a technology push project. Developments in the microelectronics industry have resulted in the fabrication of very small devices becoming standard practice. Channels down to 100 μm can be cut, with flow control devices and separation stages on a comparable scale, which means that it is possible to produce miniature chemical plants on a scale of a few centimetres. While the potential of such devices as detectors or reactors is large, few commercial products have yet been produced. Prototypes for deoxyribonucleic acid (DNA) sequencing, blood testing and a handful of other applications have been made. A company holding patents in nanofabrication is looking at the best areas in which to launch its technology in the marketplace. The following needs have been established:

Table 3. **Grouping of Selected Ideas for Lab-on-a-Chip**

Detectors/Analysers
   A. Medical/Laboratory
      1. Compact analytical instrument.
      2. Drug development.
      3. Screening of blood/vaccines/food/water
         supplies for infectious agents.
      4. Diagnosis of disease, eg, acquired immune deficiency
         syndrome (AIDS), hepatitis, cancers, and Alzheimer's disease
   B. Industrial
      1. Chemical plant on-line testing.
      2. Effluent testing (chemical plant/sewage plant).
   C. Commercial End User
      1. Air pollution detector (smoke/CO/H_2S).
      2. Food testing (nut/GM/rot/animal products).
      3. Accurate breathalyzer.
      4. Blood sugar testing for athletes/diabetics.
      5. Home pregnancy tests.
      6. Check blood–iron levels for anaemics.
      7. Allergy testing at restaurants.
      8. Alerting asthma.
      9. Check pollution levels for surfers.
   D. Regulatory/Police Use
      1. DNA analysis and fingerprinting (spit, urine, hair).
      2. Drug detection—sniffer chip.

Producers
   E. Medical
      1. Hormone production.
      2. Making insulin.
      3. Wasp sting remedy.
      4. Timed production and dose of drugs, especially for the elderly.
      5. On-the-spot antidotes for biocides or rabies etc.
      6. Production of drugs with very short shelf life in, eg, first aid kit.
      7. Internal surgery—nutrients for tissue growth.
   F. Industrial
      1. Production of speciality chemicals as/when needed.
      2. Chips to spin fibres—tiny spinnerettes for polymers.
      3. Manufacture of dangerous chemicals in small
         quantities (eg, phosgene/hydrogen cyanide).
   G. Commercial
      1. Make-up manufacture.
      2. Night-clubs—smoke/foam production.

Combined Detectors and Producers
   H. Commercial End User
      1. Suncream detector and applicator.
      2. Variable strength deodorant—alter strength.
      3. Air freshener—variable strength.
      4. Worktop coating that tests and releases antibacterial agent.

Cleaners/Removers
   I. Medical
      1. Minidialysis—remove alcohol from system to sober up quickly.
      2. Artery cleaner.
   J. Industrial
      1. Critters-on-a-chip—detect, map and
         digest environmental pollutants.
      2. Cleaning water supplies via ion exchange.
      3. Fouling/corrosion detection/remediation.
   K. Commercial
      1. Stain removal for washing—seek out and remove dirt.

Essential

Use of current technology for a more cost effective product or a completely new product.

Quick to market—relies on using existing technology.

Desirable

Use company's technology to its full potential.

Useful

Environmentally benign.

Easily marketable.

Explore ideas for devices that could be produced using existing technology.

Solution. After brainstorming, the design team came up with 73 ideas; this list was reduced to 39 by the removal of redundancy, folly, and excessive requirements. The remaining ideas were then organised as shown in Table 3.

We must reduce this list to a handful for detailed consideration in the Selection stage. This reduction can be done in two stages. First, one major identified need is using technology to ensure a swift entry into the market, which is desirable both for the usual reason that the first product into a market usually takes the lion's share of sales, even in the absence of other competitive advantages. Also, in this particular case, the product being considered is intended not just as a device in its own right, but also as an exemplar of the many other potential applications of the company's nanotechnology—a flagship for the company's future. It is doubly important to minimize the risk of technological failure or delay. The company's technology is currently not developed in the areas of detection and separations. The whole categories of E, F, G, H, and I are therefore rejected not because they are inherently bad ideas, but because they will not provide the most risk-free and speedy route to a successful first product. We may well return to these ideas in future years.

To achieve an initial screening of the remaining 21 ideas we use the criteria developed in the Needs section. We very crudely assess each idea on three criteria: market size; maturity, and reliability of technology; likely time to market. Only those ideas that look promising and all three of these vital criteria will be considered further. An example screening of the first four ideas is shown in Table 4. Of these only the first is taken forward to the Selection stage, as are B1, C1, D1 and J3.

Table 4. **Screening of Four Ideas for Lab-on-a-Chip**

	Market size	Technology	Time
compact analytical instrument	++	+	+
drug development	++	+	−
screening of liquids for infectious agents	+	+	−
diagnosis of disease	++	−	−

4. Selection

With a handful of good ideas remaining, we must next choose from these the best one to take forward for product development. Because all the remaining ideas are promising, this decision involves considerably more effort than we put into cutting the number of ideas down in the screening just described under Ideas. As far as possible, we must quantify how each idea will measure up to the criteria we set for a successful product at the Needs stage. This quantification will involve making estimates based on chemistry and engineering, and perhaps doing some simple experiments. At the same time, we still wish to develop our product as quickly as possible and do not want to put resources into exploring products we will end up rejecting. The key to success in the Selection stage is to make reliable choices with minimal effort.

4.1. Assessment.   The Needs stage provided a list of criteria by which to judge the success of our product development. We emphasized the importance of quantifying the needs as far as possible. This approach will now show its value. The first step in selecting between the remaining ideas is to estimate how each will perform relative to our criteria. In order to do this, we must gather more information about each idea. This process will involve firming up exactly how each idea will work; we may need to do some simple experiments to achieve this, we will certainly need to explore the literature some more. As we generate more detailed information on each idea, the idea itself will change and become clearer. Thus there is an iteration between the Ideas and Selection stages—as we explore an idea in more detail, the idea evolves and new ideas may emerge. Thus, although we present product design as a linear four step procedure, we understand that this is a simplification.

4.2. Comparison.   Having made an assessment of each idea against each criterion, we must now make an overall comparison among the ideas. In some cases, particularly where the identified needs are primarily technical, this will be easy once good estimates of performance are available. In other cases, subjective judgements will be necessary, either in comparing unlike criteria or inherent in the criteria themselves. In such cases, we often proceed by drawing up a decision matrix in the same way as was described for screening at the Ideas stage. Previously, we were interested in eliminating the weak ideas and the emphasis was on making quick decisions. Now, we are considering strong ideas and the emphasis must be on making a good decision. While the methodology of the decision matrix remains the same, we must now put a lot more effort in to evaluating each idea against each criterion. We should recognize that the decision matrix is a highly imperfect tool for making a complex decision where numerous subjective and objective criteria are involved. We believe it has value because it ensures that all important factors are explicitly examined and because it allows more detailed research to be conducted where the decision remains inconclusive. Also, for the technically trained, the decision matrix is a useful way of ensuring that subjective factors, such as the feel, look or taste of a product, are not overlooked in favor of the more objective factors that, we can calculate and feel more comfortable with.

Example 4A: Selecting a Polymer Film that Stops Ultraviolet (UV) Light.   On summer days, cars get hot because UV sunlight passes through the windows and is absorbed by the car's interior. Because our company has considerable expertise in thin film technology, we are interested in making a film which could be attached to these windows, and which would block light. Such a defect-free film should have four key properties

1. It should be transparent to visible light.
2. It should block 99% of UV light.
3. It should be 70–150-μm thick, but cost less than current competitors.
4. It should be easy to apply.

These properties derive partially from a competitive benchmark, a clear plastic film with a vapor-deposited metal coating. This benchmark, available in a variety of colours, costs $8–12 m².

After idea generation, sorting and screening, we came up with two alternative products. One is a multilayered polymer composite with an internal UV-absorbing layer and an adhesive backing. The second, more imaginative product is also a multilayered film, but the internal layer has a slight electrical conductivity. When no current is flowing through this conducting layer, it is opaque; but when a current is flowing, it becomes transparent.

Which product should be developed further?

Solution. After considerable discussions, our project team decides on four attributes for evaluating these products: cost, engineering, ease of application, and aesthetics. "Cost" should include manufacturing and development expenses. The attribute "engineering" reflects the ease of manufacturing the product. Ease of application includes the effort of installation and maintenance required by the customer. Aesthetics includes both quality and market appeal.

Using these criteria, our project team comes up with the selection matrix shown in Table 5. The UV absorbing polymer is cheap and easily made with our existing technology, but the product is not otherwise much different to the benchmark. The conducting polymer is costly, hard to make, and difficult to install. While it is superior aesthetically, it ranks below the benchmark that it is designed to replace. We should make the product which builds on our current skills, if we decide to make any product at all.

4.3. Getting Close to a Decision. At this point we should have a good indication of which idea looks the most promising in terms of fulfilling the needs defined in the first stage of product development. However, before proceeding with the development of this idea, we pause to consider two important factors, which so far have been largely overlooked: intellectual property and risk.

Table 5. **Selecting a UV Barrier Film**

Attribute	Attribute importance	Benchmark product	UV absorber with adhesive backing	Electrically activated absorber
cost	0.25	5	7	2
engineering	0.25	5	9	2
ease of applications	0.25	5	5	3
aesthetics	0.25	5	5	9
score		5	6.5	4.0

Intellectual property is a complex area and should be referred to an expert. The important point here is to make sure that the ownership of the intellectual property is clear before large resources are invested. Often the profitability of a product (notably a pharmaceutical) depends on the exclusive license granted by patent protection. In such cases we must ensure there is at least a good prospect of obtaining such protection before proceeding. In all cases, we must at least ensure that our activities will not be restricted by any intellectual property held by others.

In assessing how well our product ideas measure up to the criteria set by the defined needs, we have largely ignored the issue of risk. However, the ideas we are choosing among may range from minor developments to an existing product to risky and untested new technology. We need to factor this into our decision. Risk may take three forms: the product may not work; the product may take a long time to develop; and external problems may occur because of local politics, fashion or a changing economic situation. The first of these, product function, should be unlikely; by this stage we should have eliminated product ideas that are likely to fail. The second, development time, is largely a technical issue that we should be able to make a good stab at predicting. It can often be translated into a financial risk—the longer a development programme is likely to be, the greater the uncertainty in cost and the larger the return must be. The third, external problems, is the hardest to estimate, and ultimately comes down to a matter of judgment.

Nevertheless, at this point, we should at least think hard about what factors could compromise the success of the product. In thinking about risk, one needs to consider both the probability of an event and the seriousness of its consequences. It is often useful to do this by drawing up a table giving total risk as a product of probability and consequence. Just as when using decision matrices, the numbers should not be treated with excessive reverence—the procedure is really a means of ensuring all factors are carefully considered.

Having identified the risks involved in our favored product, we have three possible responses. We might decide an idea, while attractive, is too risky to merit time and effort. This realization might lead us to select a different idea of smaller potential, or to abandon the project altogether. It is essential that we carefully consider this latter option before proceeding to the manufacturing stage, when large amounts of money must be invested. Most practitioners of product design will say that the most common mistake is to abandon projects too late. Alternatively, we may decide simply to accept the risks and to proceed. This minimizes the time to product launch and is often the appropriate strategy if financial risks can be offset against the advantage of getting to market earlier. Proceeding in spite of risk is particularly apt where we hold a patent. We should be more circumspect in dealing with risks of a safety or environmental nature. Third, we may decide to do a little more research, perhaps including an experimental programme, before committing resources to product manufacture, which is closer to the traditional approach of prototyping: It will often result in a better product, but also in a longer (and more expensive) development period.

Example 4B: Moderate Scale Oxygen Production.   Oxygen, one of the largest commodity chemicals, is often made at the site where it is needed. The traditional manufacturing method, dating from Von Linde's 1905 process, is cryogenic distillation. This technique is extremely effective, but requires a large capital investment, and so is most suitable for large scale. In the last few decades, pressure swing absorption has been developed to produce oxygen at smaller scales, capturing this part of the market from distillation. For example, pressure swing adsorption units not much larger than a beer can are commercially available to produce oxygen enriched air for patients with emphysema.

As a manufacturer of distillation equipment, we are considering new technologies for making 85% oxygen at a rate of 6000 scfh. Our current technology, based on high capacity trays, is not economic at this small scale. As alternatives, we are interested either in structured packing or in hollow fibre membranes. The structured packing, which recently came off patent, consists of metal guides, mounted crossways, looking like stacks of stainless steel venetian blinds. It is an established technology now supported by both the original manufacturers and new entrants.

Hollow fiber membranes are much more speculative. Selective membranes are commercially used to produce nitrogen enriched air. Selective membranes, which retain a nitrogen enriched waste and permeate an oxygen enriched product are not commercially attractive at this scale. Porous, nonselective membranes are not used to effect selectivity but to control a condensate flow moving countercurrently to the vapor, in a configuration like that of a shell-and-tube heat exchanger. Such membranes are completely untried, though academic reports promise productivity increases of several orders of magnitude. It is the porous, nonselective membranes used as a form of structured packing which are then focused here.

Which if any technology merits development?

Solution. The cryogenic distillation of air involves three major types of equipment: compressors, heat exchangers, and distillation columns. The compressors are the biggest capital expense, perhaps one-half of the total; and the heat exchangers are another 30%. Thus we could conclude that distillation design is not important anyway. However, the compressors and heat exchangers will be standard to any distillation process, and margins for a commodity like oxygen will always be small. Thus cutting the size of the distillation columns could be a significant gain.

To examine the effect of column size in more detail, we compare equipment capacity and cost for trays, structured packing, and membranes in Table 6. The first column of figures in this table gives the relative cost of the three internals. In the past, structured packing has been expensive; but with patents expiring, costs have dropped significantly. Membrane costs are a complete guess: while

Table 6. Cost Estimates for Different Distillation Internals

Column internals	Cost per column volume	Column capacity	Column size	Column cost
Sieve trays	1	1	1	1
Structured packing	2	3	1/3	2/3
Nonselective membranes	10	50	1/50	1/5

Table 7. **Risks for Alternative Distillation Internals**

Structured packing	Probability	Consequence	Risk
competition from other manufacturers	0.9	0.3	0.27
performance failures	0.1	0.5	0.05
nonselective membranes competition from other manufacturers	0.1	0.5	0.05
performance failures	0.7	0.7	0.49

in principle there is no reason that membranes should be more expensive than trays, we expect that they will need frequent replacement.

The other columns in the table give the capacity, the size, and the column cost. The capacity of the membranes is extremely high, a consequence of the fact that membrane units are largely unaffected by flooding. The result is that the cost of the membrane-based column is potentially much less than those of the other internals.

The difficulty with the membrane alternative is that it is so risky. To understand the origin of this risk, we can compare structured packing with membranes using the risk estimates in Table 7. Structured packing is going to work. If we have any trouble, we know that we have not assembled it correctly, and we can imitate existing products to overcome these shortcomings. However, the expiration of patents in this area means that many competitors are now scrambling for new business. Thus the risk for structured packing comes from the marketplace, not from the technology.

As Table 7 shows, the risk for membranes has a different origin. There is no current competition. Even if competitors appear, they are unlikely to have more experience than we, so we will not have any implicit disadvantages. At the same time, we may have major difficulty in achieving the performance that we expect from the literature. If we do have trouble with membrane performance, we have little experience of how to fix this, so the consequences will be major.

In this situation, our selection is unclear. We can do nothing, accepting that the market for our trays may get smaller in the future. We can decide that we are efficient manufacturers, and enter the competitive structured packing market, recognising both the risk and the fact that we cannot do much to reduce it. Finally, we can gamble on the membrane market, trying through further technical development to reduce the probability of performance failures from the current value of 0.7 to a reasonable goal of 0.2. Then—and only then—membranes may make sense as an alternative to structured packing.

5. Product Manufacture

By this stage, we have decided which chemical product we want to manufacture. We have identified a customer and that customer's product needs. We have generated ideas to fill that need, and we have selected the best idea. We are ready to decide how we will make the product.

5.1. Preparation. The very wide range of chemical products possible means that we will need to consider a wide range of manufacturing methods.

Table 8. **Different Chemical Products are Manufactured Differently**

	Commodity chemicals	High value chemicals	Microstructured products
key factor	cost of product	speed of manufacture	function, from microstructure
examples	ethylene, ammonia	penicillin, viagra	paint, detergent
amount made	$>10^7$ kg/year^{-1}	$<10^4$ kg/year^{-1}	$<10^6$ kg/year^{-1}
typical molecular weight	100	600	>100; often $>10^4$
phase during synthesis	gas	dilute liquid solution	concentrated solution or melt
chemical reactor	dedicated, continuous, plug flow	generic, batch, stirred tank	ranges widely
common separation	distillation, absorption	extraction, adsorption	often not required

To provide some initial guidance for manufacture, we find it useful to think of four types of products. The first type is devices, especially for medical applications. Examples include the artificial kidney and the osmotic pump for drug delivery. The manufacture of many devices depends on mechanical engineering more than on chemistry and chemical engineering. We do not discuss devices, because this topic is thoroughly described elsewhere.

The other types of chemical products can be roughly classified as shown in Table 8. The first type, commodity chemicals, made in amounts $>10^7$ kg/year^{-1}, normally have molecular weights <100. As a result, their manufacture uses gaseous reagents, supported catalysts, and purification via distillation. This manufacture always has product cost as its primary focus. Commodity chemicals were the source of chemical industry growth in the last one-half century.

The two other types of chemical products shown in Table 8 are much harder to describe. The more obvious type, 10 thousand high value added chemicals, are exemplified by drugs. These compounds have molecular weights typically in the range of 500–700 Da. Their chemical structure is normally well defined, one of many available isomers. The finished products are crystalline, of exceptionally high purity.

Once a high value added product is identified, it is normally manufactured as quickly as possible, because the first manufacturer will usually command two-thirds of the market. This need for purity and speed, plus the small amounts to be manufactured, dictate the use of dilute solutions and generic equipment. Manufacture is more like gourmet cooking than like petroleum refining.

The third type of chemical product, which is the most heterogeneous, often has a useful microstructure. This large group of products includes most "speciality chemicals". Manufacturing ranges widely, and product properties are often a function of product history. For example, polymer properties are affected by molecular weight distribution, and paint quality changes if the paint is frozen. The process engineering required is normally a compromise between the first two types.

At the end of the Selection stage, we had sufficient information to convince ourselves that we had selected the best idea and that it had a high probability of success. Inevitably, there will have been gaps in our knowledge of the details of exactly how the product would work. In order to make the product, we not only need to be confident that the product can be made to work, but we must also establish in detail exactly how it is going to fulfil our aims. Before going further, we must fill in this missing information. This missing information may be obtained by literature searches, by consulting experts and by conducting experiments. It may be tedious, time consuming and expensive. That is why we put it off until after we had decided to proceed with product development. We do not wish to put this level of effort into more than one product idea or into an idea we later abandon if we can possibly avoid it.

Example 5A: Corrosion Resistant Paints. For many years, marine paints have contained high concentrations of hexavalent chromium to reduce corrosion. The advantage of such a paint is that it is "self-healing": When the paint is scratched, an electrochemical cell is established between the newly exposed metal and the chromium ions which, after an intricate series of chemical steps, results in the exposed metal being coated with a layer of pentavalent chromium gel. This new "self-healed" layer provides considerable protection against further corrosion. Unfortunately, even without being scratched, these marine paints also leach chromium into the water, causing considerable damage to marine life.

Efforts to replace chromium have been unsuccessful, but they have spurred development of a wide variety of paints with the potential to reduce corrosion. Most obviously, these have centred on the development of new polymers. Epoxy-based resins have been particularly successful. However, these still provide less protection than desired.

One method to further improve epoxies is to incorporate flakes of clay or mica into the resin. If these inorganic materials are exfoliated and pretreated so that they are wet by the resin, they can offer dramatically enhanced corrosion protection. However, the improvements reported vary widely, probably because the effect of the flakes depends dramatically on their alignment. If they are aligned randomly, the effect of flakes on corrosion is minor, a reduction of perhaps 10%. If the same flakes are aligned parallel to the surface, then the effect is predicted and observed to be much larger, a reduction of at least a factor of 10—over 1000%. Such a corrosion resistant paint is an example of a microstructured product, a product whose performance is dictated not so much by its chemical composition as by its microstructure, in this case the aspect ratio and alignment of the flakes. A patent search shows no significant conflicts, though patents exist for flake filled plastic bottles. Curiously, there are a large number of patents and publications that mention human skin, which has a similar geometry of impermeable protein flakes.

How should we study corrosion resistance in paints with aligned flakes?

Solution.   Answering this question requires a method of evaluating corrosion and a technique for aligning flakes. The standard method of evaluating corrosion is microscopic evaluation of metal coupons after exposure to a salt fog. While this method is well established and reliable, it is tedious, effective for a final evaluation but not for a preliminary screening. As a result, we decide to defer using the salt fog and just to put cooled coupons into boiling 0.1 N sodium chloride solution. We will examine them microscopically.

Aligning flakes can be done in at least three ways. First and most directly, we can use a suspension of flakes in a mixture of polymer and solvent. Such a solution—a lacquer—is spread onto the metal coupons; and evaporation of the solvent aligns the flakes. However, the solvents used are often chlorinated, so their evaporation is environmentally abusive. Second, we can spread a hot melt of polymer containing flakes over a metal coupon, using the shear to align the flakes. However, applying such a shear dependent coating to an entire ship will be extremely tedious.

The third method of aligning flakes is the most attractive alternative, but the one about which the least is known. It uses a water borne latex containing suspended flakes. If the flakes are mica, they are larger than the latex droplets; if the flakes are clay, they are smaller. We should plan experiments with both sizes of flakes to measure the alignment induced by the fusion of the latex droplets into a film.

5.2. Final Product Specification. Before designing the processing route, we need to produce a final specification for the product we have decided to make. We should by now have obtained all the information required to do this. For a typical chemical product, we need to define the physical structure, the chemical composition, the chemical reactions that occur during the product's operation and the thermodynamics of the product (including the microstructure). Clearly which of these dominates depends on the type of product being made, and this was the purpose of considering this in the preparation step just outlined.

The key to a product is often how it responds to a change, which may involve dissolution (or precipitation) in a solvent, response to a temperature or other physical change, or a chemical reaction (pH change being the most common initiator). It is important to specify these chemical changes carefully as well as the nature of the product itself.

Example 5B: A Self-Warming Baby Milk Bottle. A baby milk manufacturer wishes to market a self-warming bottle for feeding babies in the absence of a easily accessible power supply. It has been decided to achieve this by using a double skinned bottle, the inner space being filled by a material which will undergo an exothermic reaction.

Set final specifications for such a product.

Solution. We must first think what to use for the exothermic reaction. Obviously, it is much better if it can be regenerable. A quick literature search reveals the crystallization of sodium acetate trihydrate from water to be ideal. This crystallization can be triggered in a variety of ways, the liquid phase is stable in the absence of triggering, and can be regenerated by putting the empty bottle in boiling water. Recent advances in triggering make the reaction much more reliable and also allow regeneration by microwave heating, ideal for a baby's bottle. The maximum temperature reached is $\sim$50°C, hot enough to get the milk to the required 37°C, but without posing any safety hazard. We need to determine two things for our final specification: the mass of sodium acetate solution required and the characteristic time for warming. First, we look at the overall heat balance

$$\left[\begin{pmatrix} \text{mass of} \\ \text{sodium acetate} \end{pmatrix} + \begin{pmatrix} \text{mass of} \\ \text{milk} \end{pmatrix}\right] C_p(T_{\text{final}} - T_{\text{initial}}) = -\begin{pmatrix} \text{mass of} \\ \text{sodium acetate} \end{pmatrix} \Delta H_{\text{rxn}}$$

The enthalpy of reaction ΔH_{rxn} is -125 kJ kg^{-1}, and the heat capacities C_P for both milk and sodium acetate are assumed to be 4.2 kJ kg^{-1} K^{-1}. To heat 0.4 kg of milk from a $T_{initial}$ of 15°C to a T_{final} of 37°C, we need 1.13 kg of sodium acetate. This result is quite high, but not impractical.

Next we must look at the rate of heat transfer:

$$-\frac{d}{dt}(\Delta T) = \frac{UA}{C_p}\left(\frac{1}{M_{milk}} + \frac{1}{M_{acetate}}\right)\Delta T$$

where U is the overall heat transfer coefficient from acetate to milk, around 50 W m^{-2} K^{-1}; ΔT is the difference between the temperature of the acetate and that of the milk; M_{milk} and $M_{acetate}$ are the masses of milk and sodium acetate, respectively; and A is the surface area between milk and acetate.

We want to find this interfacial area A. To do so we integrate the above equation, giving

$$\frac{\Delta T}{\Delta T_0} = \exp\left\{-\frac{UAt}{C_p}\left(\frac{1}{M_{milk}} + \frac{1}{M_{acetate}}\right)\right\}$$

where ΔT_0 is the initial temperature difference between the room temperature milk, at perhaps 15°C, and the acetate, at $\sim$50°C after its rapid crystallisation. We want the milk to reach 37°C after perhaps five minutes, so we should use $\sim$1.5 kg of sodium acetate, giving a final temperature of 43°C. The milk will be at 37°C when $\Delta T = 7$°C, and so

$$\frac{7}{35} = \exp\left\{-\frac{50 \times A \times 300 \times \left(1 + \frac{0.4}{1.5}\right)}{4.2 \times 10^3 \times 0.4}\right\}$$

$$A = 140 \text{ cm}^2$$

Thus the final specification is a double skinned bottle, containing 1.5 kg of sodium acetate in the core surrounded by 0.4 kg of milk, with an interfacial area between the sodium acetate and the milk of at least 140 cm^2.

5.3. The Manufacturing Process. Finally, we must specify the process by which we will achieve the specifications we have just set. This is where chemical product design and traditional process design finally merge. It is worth noting that there are significant differences of emphasis in the process design for a chemical product as compared to commodity chemicals. In commodity chemical manufacture, minimizing cost and maximizing efficiency are usually the goal. Economies of scale, continuous processing and good heat integration are normally required. In the case of products, we are usually producing much smaller quantities of a higher value material. The emphasis tends to be on speed, rather than optimization, and batch processing in standard or shared units is the norm. Also, chemical products usually involve larger, more delicate molecules or solid materials. Separations therefore tend to focus on extraction, adsorption and crystallisation, rather than the distillations which typify commodity manufacture.

The details of the process design will of course vary as widely as the nature of the products we have been discussing. This was highlighted in Table 8, where different categories of chemical products were summarized. Here, we simply give one example of how a chemical product might be manufactured. This example highlights both the similarity to traditional process design and the different emphasis required in producing in small volumes large delicate molecules of high value.

Example 5C: Key Manufacturing Steps in Hormone Replacement Therapy. To relieve the symptoms of the menopause, women are often given estrogens prepared from a variety of natural sources. One such preparation is an extract of the urine of pregnant mares. While other domesticated animals will excrete similar hormone mixtures in their urine, horses are preferred because they evolved in the desert, where water is in short supply. As a result, the physiology of horses aims to limit water loss: their faeces are much drier than, for example, those of cows; and their urine is more concentrated.

Our company is interested in a generic equivalent to this extract because it sells well and because the patents for its production have long since expired. These patents suggest one possible recovery route of the estrogens is by extraction with hexanol. The active materials in the hexanol extract are concentrated by evaporating some of the hexanol. Hexane is then added to the hexanol concentrate, so that the active estrogens can be easily back extracted into water. The result is an aqueous solution not that different to the starting urine, but with a much higher concentration of active estrogens. According to the patents, this concentrate is then purified by washing with various solvents and dried to a powder. The powder is the crude product.

What steps of this preparation are likely to be critical to its manufacture?

Solution. The manufacture of drugs typically involves a reactor and a separation sequence, just like other chemical syntheses. When the drugs are obtained from natural sources, the reactor is a plant, an animal, or a microorganism. Here, the horse is the reactor, and the urine processing is the separation sequence. Such separation sequences tend to involve four steps, sometimes called a RIPP sequence, after the first letter of each of the four steps. The first, "R" step, Removal of Insolubles, is normally a filtration or centrifugation. In this case, a filtration will remove straw and other suspended solids from the urine. The second, "I" step, Isolation, concentrates the active material, most often by extraction. For Premarin, this step is the hexanol extraction and the hexane-driven back extraction. The third, "P" step, is Purification, most often based on chromatography but in this case accomplished by solvent washing. The fourth step, Polishing, normally includes crystallizations and drying. For this case, the product is a mixture of estrogens and, hence, cannot be crystallized. It is dried.

We are interested in which step in this RIPP sequence will be most important for manufacturing. In most cases, this surprisingly turns out to be the Isolation step, not the Purification step. Isolation tends to dominate the cost of the product. As evidence of this, we note that, a plot of the logithrim of the product cost versus the logarithm of the concentration entering the separation sequence is linear, with a slope of (-1), over a range of 10 orders of magnitude in feed concentration. Thus if one product has a feed concentration ten times smaller than a second product, then we should expect its price to be 10 times higher.

For the case considered here, this implies that the hexane–hexanol extraction steps will be the key to the cost-effective production of a generic form of this drug.

6. Conclusion

During the twentieth century, the chemical industry focused on commodity chemicals, especially those derived from petroleum. Profits depended on the efficient, large-scale production of a small number of these chemicals. In recent decades the emphasis shifted toward the manufacture of higher added value products. These chemical products are designed and manufactured to achieve a specific effect, in which the crucial element is a chemical or a chemical transformation. The effectiveness of the product allows it to be sold at a premium price, often under patent protection. In contrast, the traditional commodity products are specified chemically, and sold into a highly competitive global market for a wide range of uses. For chemical products, function is key. For commodities, price is key.

These differences between chemical products and commodities imply important differences in the way they are designed. Because chemical products are defined by function, the design procedure must start earlier: We cannot decide how to make something until we have decided what to make. A chemist or engineer involved in chemical product design must expect to participate in the identification of market needs, the generation of possible solutions and the selection of a product, in addition to the manufacturing decisions, which have been the traditional role of the engineer. This holistic approach to design is in sharp contrast to traditional process design, in which a specification is dictated to the chemist or engineer who then optimizes a process. Product design involves participation at a much earlier stage of product development, usually in a multidisciplinary team.

Since high value added chemical products are normally produced in low volume, they are typically manufactured by batch processing, often in campaigns in generic equipment used for many different products. As a result, much traditional process design becomes irrelevant. Heat integration is harder if several different products are to be made in batch in the same vessel. Control will often be rudimentary. Because margins are high, process optimisation is less of an issue. Instead, time to market is critical. Because products usually have a short lifetime, being first in the marketplace represents a major advantage. Product design therefore tends to focus on speed, not optimization.

Process design is a well-established subject and very effective heuristics exist to aid practitioners. As chemical product design increases in importance, it needs similar heuristics. As we have outlined, there are significant differences in design of chemical products and design of processes for commodity manufacture. For this reason, a simple mapping of conventional process design onto product design is unlikely to be effective. In this chapter, we have outlined heuristics for tackling the design of chemical products. Chemical products are of course immensely varied, ranging from dialysis machines, through herbicides, to ice cream. Inevitably no design scheme offers a panacea. Nonetheless we now have a heuristic available to provide an intellectual framework within which we can start thinking.

BIBLIOGRAPHY

GENERAL REFERENCES

M. F. Ashby and K. Johnson, *Materials Selection in Mechanical Design*, Oxford, 1992, ISBN 0080419062.

M. F. Ashby and K. Johnson, *Materials and Design. The Art and Science of Material Selection in Product Design*, Butterworth-Heinemann, Oxford, 2002, ISBN 0750655542.

P. A. Belter, E. L. Cussler, and W. S. Hu, *Bioseparations*, John Wiley & Sons, Inc., New York, 1988, ISBN 0471847372.

L. T. M. Blessing, *A Process-Based Approach to Computer-Supported Engineering Design*, Ph.D. Thesis, University of Twente, Netherlands, 1994, ISBN 0952350408.

R. G. Cooper, *Winning at New Products, Accelerating the Process from Idea to Launch*, 2nd ed., Addison-Wesley Publishing Company, New York, ISBN 0-201-56381-9.

C. B. Cobb, "Prepare for a Different Future", *Chem. Eng. Prog.*, 69 (Feb. 2001).

E. L. Cussler and G. D. Moggridge, *Chemical Product Design*, CUP, New York, 2001, ISBN 0521791839.

E. L. Cussler and J. Wei, "Chemical Product Engineering", *AIChEJ* **49**(5), 1072 (2003).

P. V. Danckwerts, "Science in Chemical Engineering", *The Chem. Eng.*, 155 (July/August 1966).

J. M. Douglas, *Conceptual Design of Chemical Processes*, McGraw-Hill, New York, 1988, ISBN 0070177627.

M. F. Edwards, *The Importance of Chemical Engineering in Delivering Products with Controlled Microstructure to Customers*, Institute of Chemical Engineers, North Western Branch Papers No. 9, 1998.

H. S. Fogler and S. E. LeBlanc, *Strategies for Creative Problem Solving*, Prentice Hall, New Jersey, 1994, ISBN 0131793187.

T. E. Graedel and B. R. Allenby, *Design for Environment*, Prentice Hall, Inc, ISBN 0-13-531682-0.

H. Grabowski, S. Rude, and G. Grein, eds., *Universal Design Theory*, Shaker Verlag, Aachen, ISBN 3-8265-4265-7.

S. A. Gregory, ed., *The Design Method*, Butterworths, London, 1966.

S. A. Gregory, ed., *Creativity and Innovation in Engineering*, Butterworths, London, 1972.

R. M. Kanter, J. Kao, and F. Wiersema, *Breakthrough Thinking at 3M*, DuPont, GE, Pfizer, and Rubbermaid Innovation, Haper Collins, 1997, ISBN 0-88730-771-X.

J. F. Louvar and B. D. Louvar, *Health and Environmental Risk Analysis*, Prentice-Hall, New York, 1997, ISBN 0131277391.

M. E. McGrath, *Setting the Pace in Product Development, A Guide to Product and Cycle-Time Excellence*, Revised edition, Butterworth-Heinemann, London, ISBN 0-7506-9789-Xproduct.

J. McMillan, *Games, Strategies, and Managers, How Managers Can Use Game Theory to Make Better Business Decisions*, Oxford University Press, Oxford, 1992, ISBN 0-19-507403-3.

G. Pahl and W. Beitz, *Engineering Design, A Systematic Approach*, 2nd ed., Springer, London, 1996, ISBN 3-540-19917-9.

W. Rahse and S. Hoffmann, Product Design, *Chem. Ingenieur Technik*, 1220 (2002).

R. M. D. Rosenau, Jr., A. Griffin, G. A. Catellion, and N. F. Anschuetz, *The PDMA Handbook of New Product Development*, Published by John Wiley & Sons, Inc., New York, ISBN 0-471-14189-5.

R. H. Schlosberg, "Identification, Evaluation and Development of New Products for Environmental Benefit", *AIChEJ* **46**(4), 672 (2000).

P. Spitz, *Petrochemicals: The Rise of an Industry*, John Wiley & Sons, Inc., New York, 1988.

S. S. Stevens, *Psychophysics: Introduction to its Perceptual, Neural and Social Prospects, Transaction Books*, New Brunswick, USA, 1985, ISBN 0887386431.

K. T. Ulrich and S. D. Eppinger, *Product Design and Development*, McGraw-Hill, New York, 2000, ISBN 0071169938.

J. Wei, "A Century of Changing Paradigms in Chemical Engineering", *Chemtech* **26**(5), 16 (1996).

J. Wei, "Molecular Structure and Property: Product Engineering", *Ind. Eng. Chem. Res.* **41**(8) 1917 (2002).

A. W. Westerberg and E. Subrahmanian, "Product Design," *Comp. Chem. Eng.* **24**(2–7) 959 (2000).

P. R. Whitfield, *Creativity in Industry*, Penguin Books Ltd., Harmondsworth, 1975, ISBN 0140219013.

J. A. Wesselingh, and L. P. B. M. Janssen, "Teaching Product Engineering", *Proceedings of the 2nd European Congress of Chemical Engineering, Montpellier*, 1999.

G. D. MOGGRIDGE
University of Cambridge

E. L. CUSSLER
University of Minnesota

CHEMICALS FROM BRINE

1. Introduction

Nearly every country in the world has a source of brine containing useful minerals. Many have underground ore bodies that may be turned into brine by solution mining. Oceans and seas of the world are the largest sources of brine. Hundreds of companies extract sodium chloride, commonly called salt or halite, from the sea. Exportadora Del Sal is the largest producer of solar salt. They process over 8,000,000 tons each year at their facility in Baja California, Mexico. Many other small and large salt producing companies can be found along the coasts of Africa, India, China, Australia, South America, and coasts of the Mediterranean and the Aegean Sea. Salt is produced in the largest quantities, but there are other compounds of sodium, magnesium, potassium, and calcium made from sea brine.

A second, and significant source of brine, is found in terminal lakes. The Dead Sea in Israel and Jordan is an example of a large terminal lake with almost unlimited supplies of magnesium chloride, potassium chloride, and sodium chloride. Over two and one-half million tons of potassium chloride are extracted from the Dead Sea each year. Magnesium compounds and bromine are also extracted.

Great Salt Lake, Utah, is the largest terminal lake in the United States. From its brine, salt, elemental magnesium, magnesium chloride, sodium sulfate, and potassium sulfate are produced. Other well-known terminal lakes are Qinghai Lake in China, Tuz Golu in Turkey, the Caspian Sea and Aral'skoje in the states of the former Soviet Union, and Urmia in Iran. There are thousands of small terminal lakes spread across most countries of the world. Most of these

Table 1. Some Common Evaporite Salts

Mineral name	CAS Registry Number	Other names	Formula
anhydrite	[7778-18-9]	muriazite	$CaSO_4$
antarcticite			$CaCl_2 \cdot 6H_2O$
aragonite	[14791-73-2]	calcite	$CaCO_3$
arcanite	[14293-72-2]	sulfate of potash (SOP)	K_2SO_4
bischofite	[13778-96-6]		$MgCl_2 \cdot 6H_2O$
bloedite	[15083-77-9]	astrakanite	$Na_2SO_4 \cdot MgSO_4 \cdot 4H_2O$
borax	[1303-96-4]		$Na_2B_4O_7 \cdot 10H_2O$
burkeite	[12179-88-3]		$Na_2CO_3 \cdot 2Na_2SO_4$
caliche		mixture of nitrate, chloride, and sulfate salts	
carnallite	[1318-27-0]	crackel salt	$MgCl_2 \cdot KCl \cdot 6H_2O$
colemanite	[12291-65-5]		$Ca_2B_6O_{11} \cdot 5H_2O$
darapskite	[12196-75-7]		$NaNO_3 \cdot Na_2SO_4 \cdot H_2O$
epsomite	[14457-55-7]	epsom salts, pickrite	$MgSO_4 \cdot 7H_2O$
gaylusite			$Na_2CO_3 \cdot CaCO_3 \cdot 5H_2O$
glaserite	[16349-83-0]	aphthitalite	$3K_2SO_4 \cdot Na_2SO_4$
gypsum	[13397-24-5]	karstenite	$CaSO_4 \cdot 2H_2O$
halite	[14762-51-7]	salt	$NaCl$
hanksite	[12180-10-8]		$9Na_2SO_4 \cdot 2Na_2CO_3 \cdot KCl$
hexahydrate	[17830-18-1]		$MgSO_4 \cdot 6H_2O$
hydrophilite			$CaCl_2$
kainite	[1318-72-5]		$4KCl \cdot 4MgSO_4 \cdot 11H_2O$
kieserite	[14567-64-7]	wathlingenite	$MgSO_4 \cdot H_2O$
kernite	[12045-87-3]		$Na_2B_4O_7 \cdot 4H_2O$
langbeinite	[14977-37-8]		$K_2SO_4 \cdot 2MgSO_4$
leonite	[15650-69-8]		$K_2SO_4 \cdot MgSO_4 \cdot 4H_2O$
loweite	[16633-52-6]		$Na_2SO_4 \cdot MgSO_4 \cdot 2.5H_2O$
mirabilite	[14567-58-9]	Glauber's salt	$Na_2SO_4 \cdot 10H_2O$
nahcolite	[15752-47-3]		$NaHCO_3$
niter	[7757-79-1]		KNO_3
polyhalite	[15278-29-2]	mamanite	$K_2SO_4 \cdot MgSO_4 \cdot 2CaSO_4 \cdot 2H_2O$
pinsonite			$NaCO_3 \cdot CaCO_3 \cdot 2H_2O$
schoenite	[15491-86-8]	picromerite	$K_2SO_4 \cdot MgSO_4 \cdot 6H_2O$
silvinite	[12174-64-0]		$KCl + NaCl$
sylvite	[14336-88-0]	muriate of potash (MOP)	KCl
syngenite		kaluszite	$K_2SO_4 \cdot CaSO_4 \cdot H_2O$
tachyhdrite	[12194-70-6]	tachydrite	$CaCl_2 \cdot 2MgCl_2 \cdot 12H_2O$
thenardite	[7757-82-6]	salt cake	Na_2SO_4
trona	[15243-87-5]		$Na_2CO_3 \cdot NaHCO_3 \cdot 2H_2O$
ulexite	[1319-33-1]		$NaCaB_5O_9 \cdot 8H_2O$
vanthoffite	[15557-33-2]		$3Na_2SO_4 \cdot MgSO_4$

lakes contain sodium chloride, but many contain ions of magnesium, calcium, potassium, boron, lithium, sulfates, carbonates, and nitrates. Some terminal lakes have dried up leaving large expanses of salt called salars and salt flats. One such flat is in Wendover, Utah. This large expanse of salt is used as a race-way where many land speed records have been recorded. The largest salar in the world is in Uyuni, Bolivia. It is 100-km wide.

A third source of brine is found underground. Underground brines are primarily the result of ancient terminal lakes that have dried up and left brine entrained in their salt beds. These deposits may be completely underground or start at the surface. Some beds are thousands of meters thick. The salt bed at the Salar de Atacama in Chile is >300-m thick. Its bed is impregnated with brine that is being pumped to solar ponds and serves as feed stock to produce lithium chloride, potassium chloride, potassium sulfate, borax, and magnesium chloride. Searles Lake in California is a similar ancient terminal lake. Brine from its deposit has been used to recover soda ash, borax, sodium sulfate, sodium bicarbonate, potassium chloride, potassium sulfate, and lithium compounds.

A fourth source of brine is obtained through solution mining. Potash (KCl) is mined at Kane Creek, in Moab, Utah by solution mining. Much of the food grade sodium chloride in the United States, Europe, and other parts of the world is solution mined. Large beds of potassium salts in Canada and trona beds in Wyoming and California are being solution mined.

The main metals in brines throughout the world are sodium, magnesium, calcium, and potassium. Other metals are found in lesser amounts such as lithium, and boron. The nonmetals are chloride, sulfate and carbonate, with nitrate occurring in a few isolated areas. A major fraction of sodium nitrate and potassium nitrate comes from these isolated deposits. Other nonmetals produced from brine are bromine and iodine.

All of these metallic and nonmetallic ions join together in a complicated array of salts and minerals called evaporites. Several evaporites usually crystallize simultaneously in a mixture, which often makes separation into pure chemicals difficult. A list of some of the more common evaporites is shown in Table 1. This table also shows the chemical formula, and other mineral names.

2. Recovery Process

2.1. Solar Evaporation.
Recovery of salts by solar evaporation is favored in hot dry climates. Solar evaporation is also used in temperate zones where evaporation exceeds rainfall and in areas where seasons of hot and dry weather occur. Other factors affecting solar pond selection are wind, humidity, cloud cover, and land terrain.

Large solar salt operations in the United States can be found along the shores of Great Salt Lake (1) and in the San Francisco Bay area (2). Salt production from solar ponds represents 14% of the total salt produced in the United States.

Total salt (NaCl) produced and consumed from all sources in the world is near 225 million tons. Large quantities of unusable salt are produced in preconcentration ponds as an intermediate step in the production of other chemicals such as potassium chloride. For example, the Dead Sea facilities produce 50 million tons of salt each year but sell <0.5 of 1% because of the high cost of transportation to markets. Great Salt Lake Minerals Corporation also deposits >15 million tons a year of by-product salt, but can market only 1 million tons.

Brine depth in solar ponds is typically 15–50-cm deep. Ponds are usually built over flat areas where silts and clays have settled to make a tight soil base to prevent leakage through the bottom of the pond. In areas where soils are not tight, artificial liners of rubber, poly(vinyl chloride) (PVC) or high density poly ethylene (HDPE) are used.

Until the 1970s, solar ponds were constructed and operated as more of an art than a science. Since then, rising land value, environmental conscientiousness, limited space, and rising costs have forced a scientific approach to solar pond optimization, design, and operation, to make ponds more productive.

Where possible, solar salt is replacing vacuum salt because of rising energy costs. For example, at noon in July, the 162 km^2 (40,000 acres) of solar ponds at Great Salt Lake Minerals Corporation evaporate >907 million kg (2 billion lb) of water each day. This would require an equivalent of 100,000 tons of coal to supply the same energy each day.

Salts formed in saturated terminal lakes or in manmade solar ponds and lakes are called evaporites. A list of major evaporites is shown in Table 1.

2.2. Seawater. Salt extraction from seawater is done in most countries having coast lines and weather conducive to evaporation. Seawater is evaporated in a series of concentration ponds until it is saturated with sodium chloride. At this point >90% of the water has been removed along with some impurities, $CaSO_4$ and $CaCO_3$, which crystallize at the bottom of the ponds. This brine, now saturated in NaCl, is transferred to ponds, called crystallizers, where salt precipitates on the floor of the pond as more water evaporates. Brine left over from the crystallizers is called bitterns because of its bitter taste. Bitterns is high in $MgCl_2$, $MgSO_4$, and KCl. In some isolated cases (India and China), magnesium and potassium compounds have been commercially extracted, but these represent only a small fraction of total world production.

Bays in San Francisco and San Diego are used to make salt from sea water. Worldwide, ~50 million tons of salt is produced from ocean water. Salt made from the seas is 97 to 98% pure (unwashed). The deposit is harvested in the fall after the weather turns cool and evaporation ceases (3). After harvesting, some of the salt may be washed to obtain salt >99% pure. Much of this salt is sold in bulk but some is rewashed and dried in rotary kilns or fluidized bed dryers to obtain salt of 99.8 and higher purity.

2.3. The Great Salt Lake. The Great Salt Lake, located in Northern Utah, is the largest lake in the western hemisphere that does not drain into an ocean. The level of the lake fluctuates depending on the weather and so does the mineral concentration. In 1983, annual rainfall was double the normal and runoff caused the lake to rise 5 ft. Again, in 1984, rain was above normal and the lake rose another 5 ft. causing hundreds of millions of dollars in flood damage and diluting the mineral concentration to half its preflood value.

In 1981, seven facilities extracted minerals from Great Salt Lake brine but flooding in 1983 and 1984 reduced the number to five. By 1992, four companies were operating and are still in operation in 2003. All Great Salt Lake mineral extracting facilities have solar ponds as the first stage in processing minerals from brine.

The first salt to saturate and crystallize is halite. This salt is successively followed by epsomite, schoenite, kainite, carnallite, and finally bischofite. See Table 1 for the chemical composition of these minerals. In the winter, freezing temperatures cools the brine causing mirabilite (Glauber's salt) to crystallize. Solar pond end brine, called bitterns, contains 30–35% magnesium chloride. These bitterns are used as a feed stock to make magnesium metal, bischofite flake, dust suppressants, freeze prevention, fertilizer sprays, and in ion exchange resins. See Figure 1 for a multiproduct facility that uses Great Salt Lake Brine.

2.4. Solution Mining. Solution mining, also known as brining, is the recovery of sodium chloride (or any soluble salt) in an underground deposit by dissolving it *in situ* and forcing the resultant solution to the surface.

Solution mining produced nearly 20.5 million metric tons of salt in 2001 representing about one-half of the total U.S. salt production (4). Salt brine is made from bedded salt at >18 different locations and from 17 salt domes. Bedded salt of the Salina formation is the most widely and intensively exploited by solution mining. Enormous reserves of Salina salt are available. Cost of solution mining salt is usually less than the cost of salt produced by dry mining. The method is particularly good where salt deposits are deep and dry mining would not be feasible.

The essentials of solution mining a salt dome have not changed over the years. The method is shown in Figure 2. Production rate of any given mine is limited to the dissolving rate of the salt to bring it close to saturation. The flow rates of the injected water controls the rate of out flowing brine. Adjustment of this injected brine therefore controls the degree of saturation of recovered brine. Since energy must be used to remove all the water to recover sodium chloride, it is important to keep brine near saturation. Frequently, in thin bed salt strata, two wells (5) are used. The wells communicate through hydraulically formed fracture channels.

An environmental risk in solution mining is surface subsidence. This risk is greatest with embedded salt. No cases of salt subsidence have been reported in mining domes that have been mined according to standard industry approved practice in the United States, but some has been seen in other countries. One side benefit of dome solution mining is use of the cavities later for storage of industrial fluids, chiefly petroleum and natural gas.

Solution mining for sodium chloride is extensively used in Europe and the states of the former Soviet Union.

2.5. General Economics and Uses. Demands and prices have a wide swing for some evaporites. New technology and development of brine reserves are increasing each year in the United States and abroad. This affects the uses and price of brine chemicals. Development of the Salar de Atacama in Chile in the 1980s as the largest producer of brine lithium in the world has affected lithium production and prices world wide. Figure 3 illustrates some major brine evaporites and their derivative products. Some of these chemicals find application in thousands of household objects. Sodium chloride alone is reported

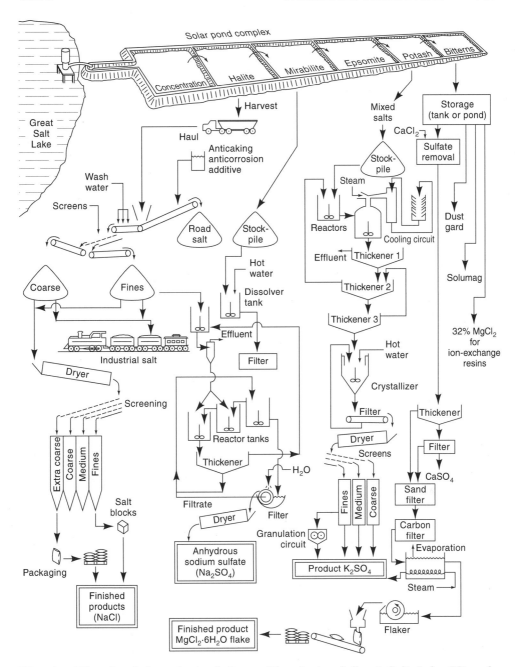

Fig. 1. Abbreviated flow sheet of the multiproducts of Great Salt Lake Minerals Corporation.

to have over 14,000 different uses (6). The United States Geological Survey, USGS, has a web site (7) and a list of most evaporites showing pricing and production rates in the United States and abroad. Much information may be obtained about these chemicals from this web site.

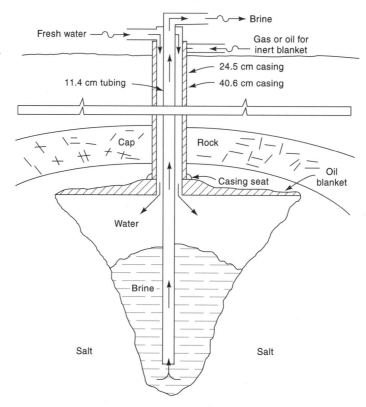

Fig. 2. Typical solution mining operation in a salt dome (9).

There has been much interest in making chemicals from brine because of the low expense as compared to alternative methods. Lithium, eg, had been mostly produced from spodumene ore in the 1970s but is now produced from brine. Similarly, solar salt has cost advantages over mined rock salt. Potassium chloride produced from brine has more than doubled from 1980 to 2000.

3. Minerals from Brine

A list of common minerals and their use is shown in Figure 3. This is only a small fraction of the tens of thousands of uses of these minerals extracted from brine.

3.1. Boron Compounds. *Occurrence.* Brine found in Searles Lake, California is the only major brine source where boron from brine is produced commercially in the United States. Brines at the Salar De Atacama in Chile also contain boron and extraction began in the 1990s.

Boron is found in two underground ores, ulexite, and colemanite. Research and pilot plant studies were completed in the last decade to solution mine these ores with water. Yield with water was low and replaced with a dilute solution of sulfuric acid that gave much higher yields. Boron is found in over 200 minerals but only four are of major importance. Borax, kernite, colemanite, and ulexite. These mineral are extracted mostly in California, and Turkey. To a lesser extent

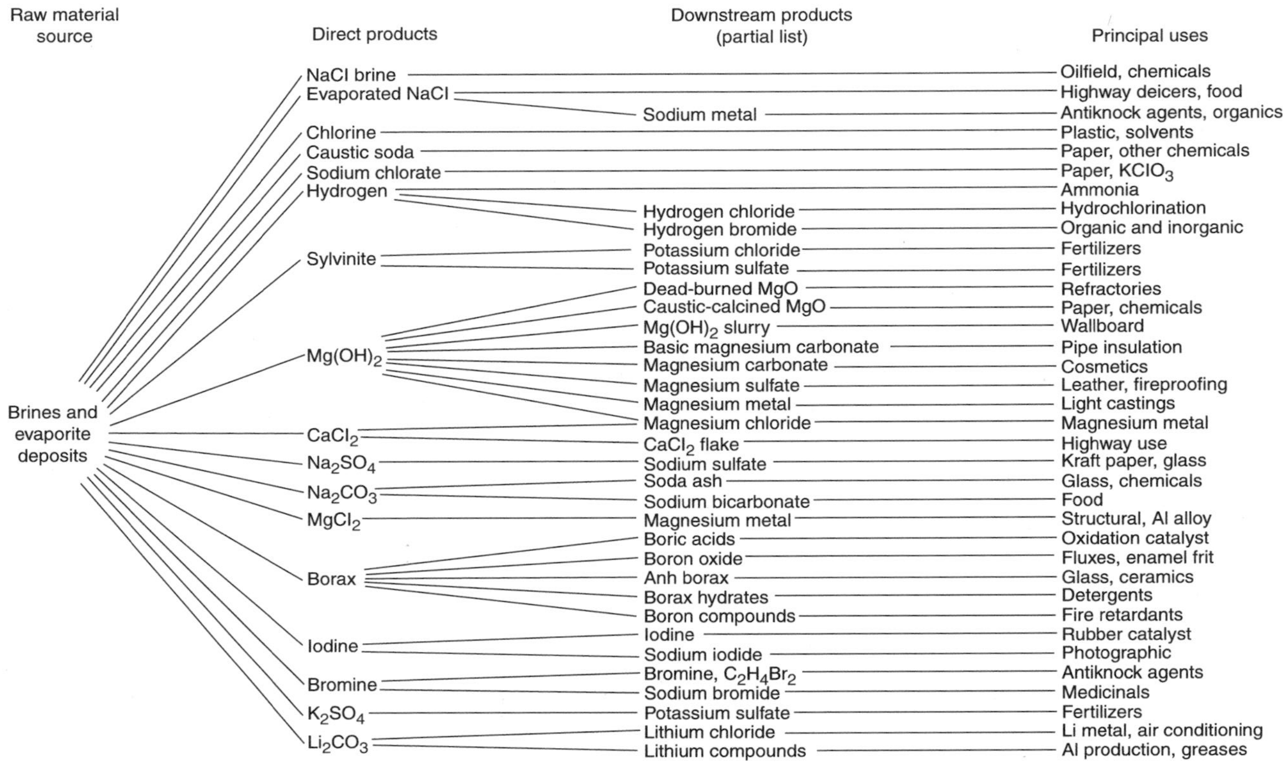

Fig. 3. The brine chemical industry and some of its products.

Argentina, Bolivia, Chile, China, and Peru also produce boron compounds from these minerals.

Recovery Process.   Boron values are recovered from brine of Searles Lake by North American Chemicals Corp. In one process, the brine is heated to remove some water and burkeite. The remaining brine is cooled to remove potassium chloride. This cooled brine is then transferred to another crystallizer where borax pentahydrate, $NaB_4O_7 \cdot 5H_2O$, precipitates. In a separate process, boron is removed by liquid–liquid extraction, followed by stripping with dilute sulfuric acid. Evaporator-crystallizers are used to recover boric acid. In a third process, borax is recovered by refrigerating carbonated brine. The most recent production is by Fort Cady Minerals Corp near Hectar, California. A solution of dilute sulfuric acid is used to solution mine a bed of borate ore 427 m underground. The boron enriched solution is then reacted with lime to precipitate a pure calcium borate product (8).

Economics and Uses.   The principal producers in the United States are U.S. Borax and Chemical Corporation, North American Chemical, American Borate Corporation, and recently, Fort Cady.

Minerals.   Their combined annual capacity in 2001 was reported to be 536,000 metric ton of equivalent B_2O_3. Of this tonnage, about one-half is exported. About 41% of boron compounds are used in glass fiber insulation, and heat resistant glass. Another 13% is used in enamel frits and glazes. Soaps and personal care products take 12%; Agriculture uses another 6% and the remaining 28% is consumed in all other uses. The United States is the largest producer of boron compounds in 2002.

3.2. Bromine.   *Occurrence.*   Bromine is found in sea water and in underground brine deposits of marine origin (9). Bromine is also found in Dead Sea brine and is currently being produced there by the Dead Sea Works and the Jordan Bromine Co.

The earliest commercial production of Bromine in the United States, in the mid-1800s, made use of a brine well in Freeport, Pennsylvania. Later, recovery of bromine from brine wells in Midland County, Michigan was developed. Production from brines in Michigan, Ohio, and West Virginia supplied the major portion of production in the United States until 1935. Michigan brines are still a source of bromine today. A major source of bromine comes from wells in Arkansas. Albermarle Corp opened up new wells in Union county, Arkansas July 2000. They now have a total of 31 wells.

Bromine is found in Searles Lake brine and was produced there at one time, but commercial extraction has been discontinued. The United States produces 40% of the world bromine production.

Recovery Process.   Commercial processes depend on the oxidation of bromide to bromine. Most of the liberated bromine remains dissolved in the brine. The brine is then stripped of bromine, followed by recovery of bromine from the stripping agent. Subsequent purification by distillation is often a final step.

Direct electrolysis was used at one time for the oxidation step. Manganese dioxide has also been used as an oxidant. Most present day processes use chlorine:

$$Cl_2 \,(gas) + 2\,NaBr \,(liq) \longrightarrow 2\,NaCl \,(liq) + Br_2 \,(gas)$$

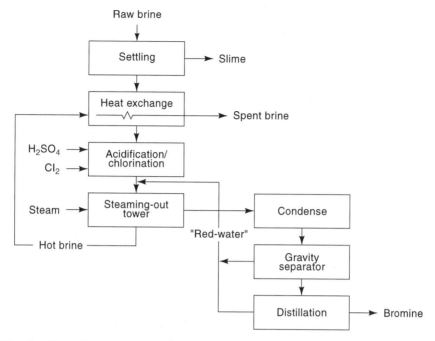

Fig. 4. Steaming-out process for recovery of bromine from high bromide brines.

Steam stripping or steaming-out (Fig. 4), is a method used in the United States to remove elemental bromine from the brine. This method is economical for brines containing bromine >1000 ppm (10).

Older methods of stripping used with concentrations below 1000 ppm utilize a stream of air flowing counter current to the brine stream. The bromine is then recovered from the air with wet scrap iron, ammonia, sodium carbonate, or sulfur dioxide (11,12).

Economics and Uses. The United States produces ~225,000 tons of bromine per year at an estimated value of 200 million dollars. All bromine is extracted from brine wells. Abermarle Corp. reported in 2000 that 31% of bromine production was used for flame retardants; pharmaceuticals and agriculture, 18%; catalysts and additives, 21%; and performance chemicals, 30%.

Ethylene dibromide is used in gasoline. It is also used as a soil fumigant (methyl bromide). Other commercial forms are alkali metal bromides, ammonium bromide, and hydrobromatic acid. About 100 kg of plastic is used per car. These plastics use bromine flame retardant (BFR). The BFRs are used in oil filters, battery cases, bumper fascias, carpeting, connectors, corrugated wire conduits, dash board covers, electric insulation, and many more automotive parts.

4. Calcium Chloride

4.1. Occurrence. Brines are the main commercial source of calcium chloride. Some brines of Michigan, Ohio, West Virginia, Utah, and California contain >4% calcium. Michigan is the leading state in natural calcium chloride production with California a distant second.

A former commercially important source of calcium chloride was a by-product of the Solvay Process used to produce soda ash. Because of environmental concerns and high energy costs, the Solvay Process has been discontinued in the United States, but it is still used extensively elsewhere in the world.

4.2. Recovery Process.   Because of its high solubility compared to that of other brine constituents, calcium chloride is the final constituent recovered in a multiproduct brine processing operation. Magnesium chloride is another highly soluble component that may occur with the calcium chloride. Separation of magnesium ions may be accomplished by precipitating magnesium hydroxide, or by crystallizing tachyhydrite, $CaCl_2 \cdot 2MgCl_2 \cdot 12H_2O$ or $2CaCl_2 \cdot MgCl_2 \cdot 2H_2O$. Both yield additional calcium chloride liquor. Calcium chloride flake is made similar to magnesium chloride flake (see Fig. 5).

4.3. Economics and Uses.   Most production of calcium chloride is from Michigan brines. The principle use of calcium chloride is to melt snow and ice from roads. It is also used in dust control, concrete setting control, waste water treatment, and various industrial uses. As a dust suppressant, it is simply

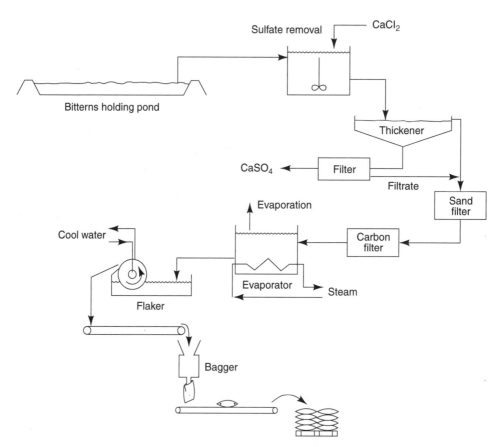

Fig. 5.   Magnesium chloride flaking circuit $MgCl_2 \cdot 6H_2O$.

sprayed over dusty areas or on unpaved roads where its dust wetting and binding properties can last for months in dry arid regions.

5. Iodine

5.1. Occurrence. Iodine is widely distributed in the lithosphere at low concentrations ($\sim$0.3 ppm). It is present in sea water at a concentration of 0.05 ppm. Three companies in Oklahoma accounted for all of the Iodine made in the United States in 2002. Certain marine plants concentrate iodine to higher levels than occur in the sea brine and these plants have been used for their iodine content. A major source of iodine is caliche deposits of the Atacama Desert, Chile. About 32% of the world's iodine is produced in Japan from natural gas wells. This is down from 40% over the last 10 years because of lower production costs from Atacama Desert caliche. By 1992, Chile was the major world producer and by 2002, Chile was producing 55% of the world's Iodine. In the United States, underground brine is the sole commercial source of Iodine. Such brine can be found in the northern Oklahoma oil fields originating in the Mississippian Geological System.

5.2. Recovery Process. Prior to 1966, iodine was recovered at Long Beach, California from oil field brine and from natural brines near Shreveport, Louisiana. The silver process was used. Silver nitrate was reacted with sodium iodide to precipitate silver iodide. Iron was then added to form ferrous iodide and free silver. The ferrous iodide was then reacted with chlorine gas to release free iodine. After 1966, the silver process was replaced with the "blowing out" process similar to the bromine process.

Later, more concentrated brines of the Midland, Michigan producers displaced the California producers. In 1976, Houston Chemicals began recovery using the blowing out process from underground brines of the Anadarko Basin in northwestern Oklahoma. Annual capacity was 900 metric tons. By 2003, all of the iodine produced was made from Oklahoma brines by a blowing out process.

Japan was the leading producer of iodine in the 1980s, producing nearly 7000 metric tons/year. Production in 2002 was 6,100 tons. Elemental iodine is released into brine by treatment with sodium nitrate or chlorine. The free iodine is then absorbed on activated carbon, and stripped from the carbon with sodium hydroxide followed by acidification to form a slurry of elemental iodine:

$$I_2 \text{ (soln)} \longrightarrow I_2 \text{ (absorbed)}$$

$$3\,I_2 + 6\,NaOH \longrightarrow 5\,NaCl + NaIO_3 + 3\,H_2O$$

$$5\,NaI + NaIO_3 + 3\,H_2SO_4 \longrightarrow 3\,Na_2SO_4 + 3\,H_2O + 3\,I_2$$

A large reserve of caliche ore-bearing iodine is being processed in the Atacama Desert. About 55% of the world supply of iodine is made from these Chilean deposits. The process consists of leaching the caliche with water. Brine is stripped of iodine using an organic solvent. The iodine is then removed from the solvent to form a slurry. Solid-phase iodine is separated from the slurry in conventional flotation cells, dried and packaged. Details of the process are proprietary.

5.3. Economics and Uses.   Most of the iodine used in the United States comes from Japan and Chile. The United States produces 7% of the world supply but consumes 28%. Iodine is produced in Woodward, and Vici Oklahoma. These two locations produced 1290 tons in 2001, and an estimated 1700 tons in 2002. Total world consumption is approximately 19,000 tons. Price per kilogram in 2002 was ~$13.

In 2001, estimated uses for Iodine were as follows: sanitation 45%; animal feed, 27%; pharmaceuticals, 10%; catalysts, 8%; heat stabilizers, 5%; and other, 5%. These other uses include inks and colorants, photographic chemicals, laboratory reagents, batteries, motor fuels, and lubricants (13).

6. Lithium

6.1. Occurrence.   Numerous brines contain lithium in minor concentrations. Commercially valuable natural brines are located at Silver Peak, Nevada (300 ppm), and at Searles Lake, California (50 ppm). Great Salt Lake brine contains 40 ppm and is a source not yet exploited. Sea Water contains <0.2 ppm. Lithium production started at Silver Peak in the 1970s, but concentrations of lithium in the brine have diminished by 25% and now major production occurs from brine in the Salar de Atacama, Chile. Lithium brines with commercial potential are found in the Altiplano of Bolivia and Argentina, in salt beds of Chile, and in several salt beds in central and western China. Lithium is presently being extracted from the Salar de Hombre Muerto in Argentina.

6.2. Recovery Process.   Lithium is extracted from brine at Silver Peak Marsh, Nevada, and at the Salar De Atacama, Chile. Both processes were developed by Foote Mineral Corporation. The process at Silver Peak consists of pumping shallow underground wells to solar ponds where brines are concentrated to 6000 ppm. Lithium is then removed by precipitation with soda ash to form lithium carbonate [554-13-2]. At the Atacama, virgin brine with nearly 3000 ppm lithium is concentrated to near saturation in lithium chloride. This brine is then shipped to Antofagasta, Chile, where it is combined with soda ash to form lithium carbonate.

6.3. Economics and Uses.   In 1976, one-third of the lithium produced in the United States was extracted from brines of Searles Lake and Silver Peak. Since then, Lithium production at Searles Lake has been discontinued and the lithium concentration at Silver Peak is decreasing. During the 1980s lithium extraction was started at the Salar De Atacama, Chile, and expanded in the 1990s. The Chilean operation is the largest lithium production in the world. Prior to 1980 most of the lithium was produced from Spodumene ore. It still is in some countries, but now nearly all lithium is produced from brine.

The United States produces and consumes about one-half of all the world production. Imports of brine based lithium carbonate are increasing. FMC, Lithium Division, now has long term contracts with Producers of lithium carbonate in Chile (14).

Over one-half of the lithium production is used as a cell-bath additive in aluminum production and in ceramics and glass. Almost all major battery manufactures produce some types of lithium batteries. Other uses are in lubricants, synthetic rubber, plastics, and pharmaceuticals. Since lithium is a light, strong metal, it finds applications in aerospace metals and alloys where a light metal is needed.

Lithium carbonate prices are steady at $4.47/kg from 1990 to 2000, but pricing for 2001 and 2002 is not available (15).

7. Magnesium Compounds

7.1. Occurrence. Magnesium hydroxide and magnesium chloride are two commercially important magnesium compounds recovered directly from natural brines. From these compounds many other compounds of magnesium are made such as elemental magnesium and magnesia. Other important compounds containing magnesium are epsomite, schoenite, kainite, and carnallite (see Table 1). Major magnesium sources are sea water (1300 ppm Mg), Great Salt Lake (1.1% Mg), underground brines near the surface east of Wendover, Utah (1%), subterranean brines in Michigan (0.7–2.5%), and brine from the Yates formation in the Midland Basin of West Texas (3%). Besides the oceans, there are vast reserves of magnesium chloride in the Dead Sea, Qaidam Basin, China, and many salars of South America.

7.2. Recovery Process. Magnesium hydroxide [1309 42-8] can be recovered in relatively pure form either from the brine or from an intermediate plant liquor by increasing alkalinity. For example:

$$MgCl_2 \text{ (liq)} + Ca(OH)_2 \text{ (solid)} \longrightarrow CaCl_2 \text{ (liq)} + Mg(OH)_2 \text{ (solid)}$$

Better recoveries can be made by replacing lime with dolomite [39445-23-3] that has been calcined. Sometimes NaOH is used when low calcium in the product is required.

Recovery of magnesium chloride [7786-30-3] as a direct product is usually economically feasible only as a by-product. Like calcium chloride, magnesium chloride is highly soluble and is extracted at the end of a series of processes that first removes the less soluble products as the magnesium ion concentrates. In Michigan brines, NaCl, Br_2, I_2, $CaCl_2$, $Mg(OH)_2$, and KCl are coproducts. At Great Salt Lake, NaCl, Na_2SO_4, and K_2SO_4 are coproducts and $MgCl_2$ is the last to be recovered. In a current process for Michigan Basin brines, pure $MgCl_2$ liquor is formed by reacting previously precipitated $Mg(OH)_2$ with stack gas and concentrated brine:

$$Mg(OH)_2 + CaCl_2 + CO_2 \longrightarrow MgCl_2 + H_2O + CaCO_3$$

End brines at Wendover Utah and the Great Salt Lake have concentrations up to 36% $MgCl_2$. At Great Salt Lake Minerals Corporation, this highly concentrated brine is artificially evaporated to remove 15% of the water. The hot solution is then cooled to form bischofite ($MgCl_2 \cdot 6H_2O$). Figure 5 shows the process.

Magnesium metal is produced by the electrolysis of molten $MgCl_2$ by two different processes. In the old Dow sea water process, $MgCl_2$ liquor is formed by reacting lime or dolomite with sea water to form $Mg(OH)_2$ and then neutralized with HCl to form $MgCl_2$. The slurry is dried to a hydrous cell feed containing ~70% $MgCl_2$. This process was ended in 2001 when Alcoa Inc. stopped production of its 43,000 tons/year facility. The other process used by MagCorp with Great Salt Lake brines, recovers anhydrous $MgCl_2$ directly from concentrated brine by spray drying. In both cases, elemental magnesium and chlorine gas are formed in the cells. MagCorp's facility is the only major producer of magnesium, metal in 2003. They produce ~45,000 t/yr magnesium from brine.

The Dead Sea has an unlimited reserve of $MgCl_2$. The Dead Sea Works, produces about 30,000 tons/year of MgO equivalent from some of their concentrated brines. The Arab Potash Company is also beginning to produce MgO at its new bromine plant.

MgO has also been produced from concentrated ocean brines that have first been used to make sodium chloride in solar ponds. The process is basically to form $Mg(OH)_2$ by adding dolomite or lime to the brine and then to burn $Mg(OH)_2$ to MgO.

7.3. Economics and Uses. Magnesium is removed from brines of the Great Salt Lake in the form of magnesium chloride. This is then used to make elemental magnesium, dust suppressants, and bischofite flake. Magnesium chloride is also used in drilling mud, ion exchange resins, oxy-chloral cements, fertilizers, and miscellaneous industrial uses. Magnesium hydroxide and magnesium chloride are used as a basic feed stock to make elemental magnesium, refractories (MgO), and reactive chemicals. Over 1,000,000 tons of magnesium and magnesium compounds were made in the United States in 2001 (16). Twelve million tons were made world wide. Prices vary widely depending on the product. MgO (dead burnt) sold for $363/ton in march of 2003. Hydrous magnesium chloride sold for $0.145/lb (17). The metal price is ~1.2/lb (18).

8. Potassium Compounds

8.1. Occurrence. There are two forms potassium compounds in brines of the world. Muriate of potash (KCl) and sulfate of potash (K_2SO_4). These brine potash operations are located in Utah (Moab, Ogden, Wendover) and one in California (Searles Lake). Operations in Searles Lake have produced both muriate and sulfate of potash. The Ogden operation produces sulfate of potash. The others produce muriate.

8.2. Recovery Process. Moab Salt, LLD, Cane Creek potash operation of Moab, Utah, produces KCl by solution mining. Brine is pumped from underground to 400 acres of solar ponds where a mixture of KCl and NaCl is crystallized in a salt mass called sylvinite. Sylvinite is removed from the ponds with scraper-loaders and hauled to a central pit. The salts are then transported to the refinery in a slurry line. KCl is separated from the NaCl by flotation. The flotation process is standard throughout the industry and is the same process used to separate KCl from impurities in a carnallite decomposition process explained later. An amine collector is used as one of the flotation reagents.

Production of KCl at the Wendover, Utah operation employs a large 7000-acre complex of solar ponds. Both shallow brine wells and deeper wells are used to pump brine into the pond complex. Surface canals 7-m deep have also been dug out into the ore body. These canals fill with brine which is pumped to nearby ponds. In the preconcentration ponds, water is evaporated and sodium chloride is crystallized. Later the brine is transferred to production ponds where sylvinite is deposited. Brine is then transferred to other ponds where carnallite is crystallized. Sylvinite is removed from drained ponds with self-loading scrapers and taken to the plant where KCl is separated by flotation with an amine oil collector. The carnallite ($KCl \cdot MgCl_2 \cdot 6H_2O$) is decomposed by leaching with water. The water dissolves $MgCl_2$ and leaves KCl and impurities, mainly NaCl. KCl is then separated from impurities by conventional flotation.

Great Salt Lake Minerals Corporation (GSL) near Ogden, Utah, produces potassium sulfate and several other products from Great Salt Lake brines. Presently 35,000 acres is divided into 85 solar ponds. Two years are required to process brine through the ponds. During the 2 years, salts crystallize in the following sequence: NaCl, $MgSO_4 \cdot 7H_2O$, $MgSO_4 \cdot K_2SO_4 \cdot 6H_2O$, $MgSO_4 \cdot KCl \cdot 2.75H_2O$, $MgCl_2 \cdot KCl \cdot 6H_2O$, and $MgCl_2 \cdot 6H_2O$. In the winter $Na_2SO_4 \cdot 10H_2O$ crystallizes. All the salts containing potassium are harvested from the ponds and hauled into the refinery where impurities are leached out and remaining potassium is converted to K_2SO_4. In addition to potassium, three other major products are made as shown in the flow of Figure 1.

8.3. Economics and Uses. Total world production of potassium products is 27,000,000 ton/year (19). Potassium chloride is removed from brine at Moab; Wendover, Utah; and at Searles Lake, California. Potassium sulfate is made from Great Salt Lake brine by Great Salt Lake Minerals Corporation who is the largest producer of solar potassium sulfate in the world. Combined, these U.S. facilities still produce a relatively small percentage of potash fertilizers in the world. Production from the Dead Sea, eg, is 10 times greater than production of potassium from brines in the United States. More than 95% of all the potassium produced is used in fertilizer blends. The remainder is converted to other potassium chemicals for industrial use.

Domestic potash production supplies one-third of the U.S. consumption. Price of both KCl and K_2SO_4 fluctuates. KCl ranges between 105 and 125 dollars a ton. Sulfate of potash sells for \$190–200/ ton (20).

9. Sodium Carbonate

9.1. Occurrence. The brines of Searles Lake, California are the sole brine source of sodium carbonate (soda ash) production in the United States. There is a large underground deposit of sodium carbonate brine in the Sua Pan area of Botswana, Africa where 270,000 tons were produced in 2001. Another potential source is Owens Lake, California. Owens Lake brines were used to produce soda ash but discontinued in 1967. There are many other deposits in the world where there are deposits of trona and nahcolite (21).

9.2. Recovery Process. Presently North American Chemicals Co. at Searles Lake is the only significant producer in the United States of sodium carbonate from brine (abt. 1,000,000 tpy). The process is based on converting all sodium carbonate in the brine to sodium bicarbonate by adding CO_2:

$$2\,Na^+ + CO_3^{2-} + CO_2 \longrightarrow 2\,NaHCO_3$$

Impurities remain in the brine solution and the sodium bicarbonate crystallizes and is filtered from the liquor. The bicarbonate is then converted to carbonate by heating:

$$2\,NaHCO_3 + Heat \longrightarrow Na_2CO_3 + CO_2 + H_2O$$

carbondioxide is recycled back again to the carbonation vessels.

In 1976, >3,000,000 ton of soda ash was produced in the United States using a procedure called the Solvay Process. This process has been discontinued in the United States because of pollution problems and high processing costs. It is still an important process in other countries.

9.3. Economics and Uses. North American Chemical Company at Searles Lake is now the only producer of soda ash from naturally occurring brine in the United States. Production from brine represents ~10% of U.S. production. Total 2001 production in the United States was 10 million tons. Total world production was 35 million. Price in 2003 was $120/ ton bulk.

One-half of all sodium carbonate is used in glass production. Production of other chemicals use another 25%. Soap and detergents use 11% and Flue gas desulferization, pulp and paper, water treatment consume what is left.

10. Sodium Chloride

10.1. Occurrence. About one-half of all the sodium chloride produced in the world is from brine. Approximately 100 million tons a year are produced from brines of the ocean, terminal lakes, subterranean aquifers and solution mining. Sodium is found in large quantities in most areas of the world. Its quantity is so large that prices in some locations are only $25 a ton. Many areas have millions of tons but prices are so low that it is not economical to mine or process the salt. The largest exposed sodium chloride bed is at the Salar de Uyuni, Bolivia, but Bolivia is land locked and very little of the salt can be processed and sold at a profit.

10.2. Recovery Process. Figure 1 shows a typical process for producing sodium chloride (halite). There are two main processes. One is to flood solar ponds with brine and evaporate the water leaving sodium chloride crystallized on the pond floor. The other is to artificially evaporate the brine in evaporator-crystallizers. Industrial salt is made from solar ponds where food grade salt made for human consumption is mostly made in the evaporator-crystallizers.

10.3. Economics and Uses. The United States is the largest producer of salt in the world. In 2001. Some 45 million tons was produced compared to 180 million tons made in the rest of the world. Salt continues to be one of the most heavily traded chemical ores in the world representing >50% of all seaborn mineral trade. World consumption is >200 million tons/year. Most of this salt is made from brines. Salt is used directly or indirectly in 14,000 different products (22).

Salt has had a major impact on the economies of some countries and in some places it still does. Salt was the first chemical recognized and used in ancient times (23). It was first used in food for taste enhancement and as a preservative.

Today most salt is used to make caustic soda and chlorine. These chemicals are used in thousands of household products. The next biggest use is for highway deicing. Other uses are for water softening, livestock feeds, meat packing and foodstuffs. Chemical use in 2001 was 41%; ice control, 34%; retail distribution, 8%; general industrial, 6%; agricultural, 4%; food processing, 4%; all other, 5%. Price of salt ranges from \$20/ton raw bulk, to \$122 processed and packaged.

11. Sodium Sulfate

11.1. Occurrence. In the United States, natural sodium sulfate brines are found in California at Searles Lake, at the shallow Castile formation underlying Terry and Gains counties, Texas, and at the Great Salt Lake, Utah.

Other natural sodium sulfate brines of commercial importance are found in dry lake beds of southwestern Saskatchewan, Canada; Laguna Del Ray in Coahuila, Mexico; the Gulf of Kara-bogaz, Russia, and in western China. Some lagoons and salars around the world have brines rich in sulfates such as the Huyamampa Lagoon in Santiago, Argentina and Aguas Blancas, Chile.

11.2. Recovery Process. The process for making sodium sulfate is different at each facility where it is produced. One step common to all facilities is a cooling step to form Glauber's salt followed by a purification and recrystallization step to form anhydrous sodium sulfate.

In Texas, brine is pumped from underground deposits. Sodium chloride is added to bring the brine near saturation. This solution is then chilled to $-8°C$ to crystallize Glauber's Salt (24). Anhydrous Na_2SO_4 [7757-82-6] is recovered by artificially evaporating the liquor formed after re-melting the Glauber's salt.

At Searles Lake, sodium sulfate is recovered as one of three coproducts in a series of complex operations where soda ash and borax are also recovered from the brine. Anhydrous sodium sulfate is recovered by artificially evaporating the water from melted Glauber's Salt.

In processing Great Salt Lake brine, Glauber's salt is crystallized in solar ponds by cooling during the winter. This salt is harvested and stockpiled when abient temperatures are still cold. Once the stockpile has drained, the Glauber's salt is stable in the solid phase throughout the year and can be reclaimed at any time. Reclaimed Glauber's salt is dissolved in hot water at the refining plant. The resulting liquor is filtered to remove insolubles and then reacted with sodium chloride to form anhydrous sodium sulfate according to the reaction below. The process is shown in Figure 1.

$$Na_2SO_4 \text{ (liq)} + NaCl \text{ (solid)} \longrightarrow Na_2SO_4 \text{ (solid)} + NaCl \text{ (liq)}$$

11.3. Economics and Uses. About one-half of all sodium sulfate produced in the United States is from brine. In 2002, 500,000 ton was produced.

All of the production is from North American Chemicals (Searles Lake) and Ozark-Mahoning at Brownfield and Seagraves, Texas. Great Salt Lake Mineral Corp has discontinued production.

Of the sodium sulfate produced in the United States, the paper industry consumes 13%.; detergents and soap, 46%; glass, 11%; textiles, 12%; carpet fresheners, 7%; and miscellaneous, 11% (25). Powdered detergents are on the decline in favor of liquids that do not use sodium sulfate. Since, the pulp and paper industry is also using less, the price of sodium sulfate has been declining.

BIBLIOGRAPHY

"Great Salt Lake Chemicals" in *ECT* 2nd ed., Suppl. Vol., pp. 438–467, by G. Flint, Great Salt Lake Minerals and Chemical Corp.; "Chemicals from Brine" in *ECT* 3rd ed., Vol. 5, pp. 375–393, by R. B. Tippin, Great Salt Lake Minerals and Chemicals Corp., and P. E. Muehlberg, The Dow Chemical Company; in *ECT* 4th ed., Vol. 5, pp. 817–837, by David Butts, Great Salt Lake Minerals Corp.; "Chemicals from Brine" in *ECT* (online), posting date: December 4, 2000, by David Butts, Great Salt Lake Minerals Corp.

CITED PUBLICATIONS

1. J. Wallace Gwynn, ed., *Great Salt Lake*, Utah Geological Survey Special Publication, Department of Natural Resources, 2002, pp. 201–233.
2. W. E. VerPlank and R. F. Heiser, *Salt in California*, Bull. 175, California Department of Natural Resources, San Francisco, Calif., Mar. 1958.
3. E. H. Rivera, *Forth Symposium on Salt*. Vol. II, Northern Ohio Geological Society, Cleveland, Ohio, 1974, pp. 411–413.
4. *U.S. Geological Survey Minerals Yearbook*, 2001. p. 64.6. www.minerals.usgs.gov.
5. K. Henderson, in Ref. 3, Vol. II, pp. 211–218.
6. H. L. Bradley, *Guidelines for the Establishment of Solar Salt Facilities from Seawater*, Underground Brine and Salted Lakes, United Nations Publication, UNIDO Dept., Vienna, Austria, 1983, p. 1.
7. See Ref. 4.
8. See Ref. 4, p. 13.3.
9. V. M. Goldschmidt and A. Muir, eds., *Geochemistry*, Oxford University Press, London, 1954, p. 607.
10. F. Yaron, in Z. E. Jolles, ed., *Bromine and its Compounds*, Ernest Benn Ltd., London 1966, pp. 3–41.
11. U.S. Pat. 2,143,223, S. B. Heath (to Dow Chemical Co.).
12. U.S. Pat. 2,143,224, G. W. Hooker (to Dow Chemical Co.).
13. Phyllis A. Lyday, see Ref. 4, pp. 38.1–38.7.
14. Joyce A. Ober, see Ref. 4, pp. 46.1–46.8.
15. *U.S. Geological Survey*, Mineral Commodity Summaries, January 2003.
16. D. A. Kramer, Magnesium and Magnesium compounds Compounds, *U.S. Gerological Survey mineral yearbook*, 2001.
17. Chemical Market Reporter, March 17, 2003. www.chemicalmarketreporter.com.
18. D. A. Kramer, *U.S. Geological Survey*, Mineral Commodity Summaries, January 2003.
19. J. P. Searls, *Potash,U.S. Geological Survey*, Minerals Commodity Summaries, Jan. 2003.

20. See Ref. 17.
21. D. E. Garrett, *Natural Soda Ash*, Van Nostrand Reinhold Co., New York, 1992.
22. H. L. Bradley, Guidelines for the Establishment of Solar Salt Facilities from Sea Water, Underground Brines and Salted Lakes, United Nations Publication, UNIDO Dept., Vienna, Austria, 1983, p. 1.
23. *Holy Bible*, Leviticus, King James Version, published by the Church of Jesus Christ of Latter-Day Saints, 1979, p. 149, chap. 2, ver. 13.
24. W. I. Weisman and R. C. Anderson, *Mining Eng.* **5**(7) 1953.
25. Dennis Kostick, *U.S. Geological Survey*, Mineral Commodity Summaries, January 2003.

GENERAL REFERENCES

A very good source of information for most minerals from brine is The United States Geological Survey, where information, pricing, production, and other satistics are compiled by assigned comodity experts. This information can be acessed on the internet at www.minerals.usgs.gov/minerals.

W. J. Schlitt, ed., *Salts & Brines '85*, Port City Press, Baltimore, Md. 1985, 209 pp. This reference lists 501 other references concerning minerals from brines.

Economics and up to date pricing are available at www.chemicalmarketreporter.com, however, there is a charge for this service.

DAVID BUTTS
DSB International Inc.

CHEMICAL VAPOR DEPOSITION

1. Introduction

Vapor deposition is a process that transfers gaseous molecules into solid-state materials (Fig. 1). From this process, thin films can be grown on substrates and fine powders can be formed. There are two major types of vapor depositions, physical vapor deposition (PVD, see THIN FILMS, FILM FORMATION TECHNIQUES) (1) and chemical vapor deposition (CVD) (2–4). The difference is that in a PVD process, a source is vaporized to deposit the film. Usually, the deposited film is the same as the source material. This vaporization–deposition process is a physical process that involves phase change but does not typically involve a chemical reaction. Major PVD techniques are vacuum evaporation, sputtering, ion plating, and molecular beam epitaxy (MBE) (see THIN FILMS, FILM FORMATION TECHNIQUES). In the CVD process (Fig. 2), a molecular source, frequently called a precursor, is vaporized into a flow reactor. The precursor is activated by a form of energy and decomposed through steps of chemical reactions to deposit thin films. The

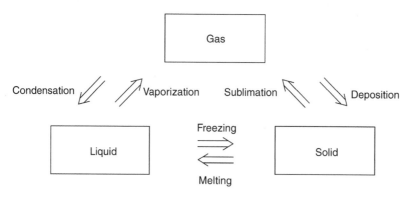

Fig. 1. Definition of deposition.

CVD processes are complex because there are many variables, such as temperature, pressure, chemical nature of the precursor, physical shape of the reactor, etc, to be controlled. Consequencetly, the physical and chemical properties of the films grown may vary widely from one reactor system to another. Using CVD technology, it is possible to grow thin films of metals (5), semiconductors (6), and insulators (7). This process is used extensively to produce semiconductor devices (see INTEGRATED CIRCUITS; LIGHT-EMITTING DIODES; LIGHT GENERATION SEMICON-DUCTOR LASERS) (8), and protective coatings on precision machine parts (see METALLIC COATINGS, SURVEY) such as diamond films (9). Recently, CVD has been used to grow nanotubes, nanorods, and nanowires (see NANOTECHNOLOGY) of many materials (10–12).

In a typical CVD system, there are three major components to be considered: the precursors, the substrate, and the reactor. The precursors are the volatile molecules that will deposit thin film materials inside the reactor. The substrate provides a foundation for the film to grow upon. The reactor, usually a flow reactor (see REACTOR TECHNOLOGY), allows the deposition reaction to proceed. The precursors are supplied into the reaction zone of the reactor through a delivery system. Upon arrival, the precursors are activated and decomposed to deposit thin films on the substrates placed inside the reactor. Vapor-phase by-products are removed and treated if necessary (see EXHAUST CONTROL, INDUSTRIAL). More descriptions are provided below.

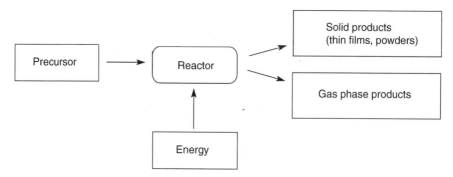

Fig. 2. Schematic diagram of CVD Process.

Table 1. Some Common Precursors and Corresponding Thin Films Grown by CVD

Precursor[a]	Thin films by CVD
hydrides, MH$_x$	
(see HYDRIDES)	
SiH$_4$	Si and Si containing materials
	(see SILICON COMPOUNDS, SILANES; SILICON, PURE)
AlH$_3$(NMe$_3$)$_2$	Al and Al containing materials
	(see ALUMINUM AND ALUMINUM ALLOYS)
NH$_3$	nitrides (see AMMONIA; NITRIDES)
PH$_3$	phosphides (see PHOSPHINE AND ITS DERIVATIVES)
halides, MX$_y$	
SiCl$_4$	Si and Si containing materials
	(see SILICON COMPOUNDS, SILICON HALIDES; SILICON, PURE)
GeCl$_4$	Ge and Ge containing materials
	(see GERMANIUM AND GERMANIUM COMPOUNDS)
TiCl$_4$	Ti and Ti containing materials
	(see TITANIUM AND TITANIUM ALLOYS)
TaCl$_5$	Ta and Ta containing materials
	(see TANTALUM AND TANTALUM COMPOUNDS)
MoF$_6$	Mo and Mo containing materials
	(see MOLYBDENUM AND MOLYBDENUM ALLOYS; MOLYBDENUM COMPOUNDS)
WF$_6$	W and W containing materials
	(see TUNGSTEN AND TUNGSTEN ALLOYS; TUNGSTEN COMPOUNDS)
metal-organics metal alkyls	
AlMe$_3$	Al and Al containing materials
	(see ALUMINUM AND ALUMINUM ALLOYS)
Ali-Bu$_3$	Al and Al containing materials
	(see ALUMINUM AND ALUMINUM ALLOYS)
Ti(CH$_2$$t$-Bu)$_4$	Titanium carbide and other Ti containing materials
	(see TITANIUM AND TITANIUM ALLOYS)
metal alkoxides	
(see ALKOXIDES, METAL)	
Ti(Oi-Pr)$_4$	TiO$_2$ (see TITANIUM COMPOUNDS, INORGANIC)
[Cu(Ot-Bu)]$_4$	Cu and Copper Oxides (see COPPER; COPPER COMPOUNDS)
metal dialkylamides	
Ti(NMe$_2$)$_4$	titanium nitride and carbonitride
	(see TITANIUM COMPOUNDS, INORGANIC; NITRIDES; CARBIDES, INDUSTRIAL HARD)
Cr(NEt$_2$)$_4$	chromium nitride and carbonitride
	(see CHROMIUM AND CHROMIUM ALLOYS; NITRIDES; CARBIDES, INDUSTRIAL HARD)
metal diketonates	
Cu(acac)$_2$	Cu and Copper Oxides (see COPPER; COPPER COMPOUNDS)
Pt(hfac)$_2$	Pt (see PLATINUM-GROUP METALS; PLATINUM-GROUP METALS, compounds)
metal carbonyls	
(see METAL CARBONYLS)	
Fe(CO)$_5$	Fe (see METAL CARBONYLS; IRON COMPOUNDS; IRON)
Ni(CO)$_4$	Ni (see METAL CARBONYLS; NICKEL AND NICKEL ALLOYS; NICKEL COMPOUNDS)

[a]acac = Acetylacetonato and hfac = hexafluoroacetylacetonato.

2. Precursors and Delivery System

Traditionally, volatility is the most important factor for selecting a CVD precursor. However, suitable simple molecules with high vapor pressure are relatively rare. Thus, only limited types of materials can be deposited by CVD.

Recently, the search for new CVD precursors has been an active research area. Many new precursors have been synthesized. Generally, the volatility of a molecule is determined by its molecular weight and molecularity (degree of polymerization). The volatility, the result of basic molecular structure and bonding, is a property difficult to predict and adjust. Other characteristics to be considered for choosing a suitable CVD precursor are the stability and reactivity, usually defined by the bond strength. An ideal precursor needs to be stable in storage and in delivery into the reactor. Also, it needs to be reactive enough to deposit thin films at low temperatures so that more substrates can be chosen. However, these two characteristics are in conflict. For example, high stability frequently means low reactivity; consequently, a high temperature of deposition is required. This limits the selection of the substrate materials to be deposited on. Thus, a compromise between the stability and the reactivity is frequently needed. Safety is another important issue. Many CVD precursors are toxic, poisonous, or flammable in the air. Their volatility suggests that fast diffusion of the precursor molecules in the space is likely. Thus, great caution should be paid to handling of CVD precursors. Some common precursors and the deposited materials are given in Table 1. Frequently, more than one precursor is needed to grow materials containing mixed elements. In addition to the precursors listed in Table 1, there are many other precursors with mixed ligands.

The precursor delivery system usually has flow controllers (see FLOW MEASUREMENT) to monitor and to adjust the quantities of the gaseous reactants and the carrier gases supplied. For liquid precursors, temperature-controlled bubblers are used. Solid precursors can be delivered using temperature-controlled evaporators. Solid precursors with extremely low vapor pressures can be dissolved in solutions and supplied to the reactor through a solution delivery system.

3. CVD Reactors

CVD reactors are constructed from temperature and chemical resistant materials such as quartz and stainless steel. A source of energy is provided and controlled to supply thermal (see TEMPERATURE MEASUREMENT), plasma (see PLASMA TECHNOLOGY) or photo (see PHOTOCHEMICAL TECHNOLOGY SURVEY) energy to the reactor. The thermal energy can be supplied from resistively heated elements (see FURNACES, ELECTRIC, RESISTANCE FURNACES) inside or outside the reactor, called cold-wall and hot-wall reactors, respectively. The reactors can be heated inductively (see FURNACES, ELECTRIC, INDUCTION FURNACES) also. In addition, a focused laser beam can be used to heat a small area. This method is called laser CVD (LCVD) (13). Depending on the thermal stability of the precursor and the substrate, the temperature of deposition of thermal CVD can be anywhere between

350 and 2000 K. Room temperature deposition is possible with some processes, particularly with plasmas.

The pressure of deposition can vary widely. When the CVD process is carried out under atmospheric pressure, it is frequently called "atmospheric pressure CVD" (APCVD). In many cases, a vacuum system (see VACUUM TECHNOLOGY; PUMPS) is attached to the reactor to maintain a suitable deposition pressure (see PRESSURE MEASUREMENT) usually <100 Torr (1.33×10^4 Pa). This technique is classified as "low pressure CVD" (LPCVD). Chemical beam epitaxy (CBE) is a technique similar to MBE. It uses molecular chemicals precursors instead of elemental solids to generate molecular beams (or chemical beams) to deposit thin films under extremely low pressures ($< 10^{-9}$ Torr or 1.33×10^{-7} Pa).

Plasma (glow discharge) is another form of energy frequently used in CVD. The technique is called plasma-enhanced CVD (PECVD) or plasma-assisted CVD (PACVD) (14). Inside the discharge zone, partially ionized gas composed of ions, electrons, and neutral species are generated. Chemical interactions among these active species and the substrate frequently deposit non-stoichiometric thin films with unique properties at low temperatures. PECVD is a more complex technology than the thermal CVD mentioned above. Usually, PECVD is carried out at low pressures to generate low temperature plasma. In addition to the factors, such as pressure and temperature, mentioned for the thermal CVD, many other parameters have to be considered. These include, eg, the frequency and the power supplied to induce the glow discharge process, and the shape and position of the electrodes and coils. Photon energy from light sources, such as discharge tubes and lasers, is another form of energy that can be used for low temperature CVD processes. The technique is called photo CVD (PCVD) (15). Several types of reactors are given as examples.

3.1. Horizontal Hot-Wall Reactor. The simplest type of CVD reactor is shown in Figure 3. It can be constructed easily from a tube furnace and a quartz tube. It is called "hot-wall" because the furnace outside heats the tube wall. The flow pattern is horizontal and parallels the way the substrates are arranged.

3.2. Vertical Cold-Wall Reactor. As shown in Fig. 4, the substrates are heated by the heating stage inside the reactor. The reactor is called "cold-wall" because the reactor wall has a temperature lower than the central heating stage. The flow pattern is vertical and perpendicular to the substrates.

3.3. Parallel Plate Plasma Reactor. In addition to the basic features of the cold-wall reactor discussed above, this reactor (Fig. 5) has a radio frequency

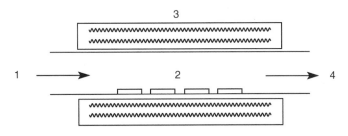

Fig. 3. Horizontal hot-wall reactor. 1: precursor in; 2: substrates; 3: heater or furnace; 4: exhaust.

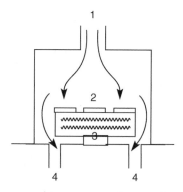

Fig. 4.   Vertical cold-wall reactor. 1: precursor in; 2: substrates; 3: heater; 4: exhaust.

(rf) power supply. (Other frequency ranges are also possible.) It provides an electric field through a pair of parallel electrodes to generate glow discharge under low pressure. The substrates, placed between the electrodes, are in the glow discharge zone. Because the substrates are placed in the plasma zone, possible radiation damage from the active species in the plasma to the deposited films is a drawback.

3.4. Inductive Tube Remote Plasma Reactor.   This reactor (see Fig. 6) has similar features the parallel plate plasma reactor; however, the plasma is generated by an induction coil outside the reactor. The substrates are placed away from the discharge zone, which is why it is called "remote plasma reactor". In this way, radiation damage can be significantly reduced.

In many CVD processes, the precursors and the vapor-phase by-products are frequently hazardous chemicals. They can be poisonous, flammable, and corrosive. Thus, proper treatment is necessary before they can be discharged to the environment. The most common methods are to decompose, adsorb, or remove them using catalytic crackers, traps and scrubbers (see EXHAUST CONTROL, INDUSTRIAL).

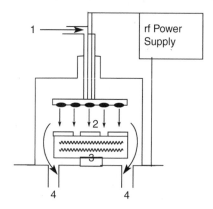

Fig. 5.   Parallel plate plasma reactor. 1: precursor in; 2: substrates; 3: heater; 4: exhaust.

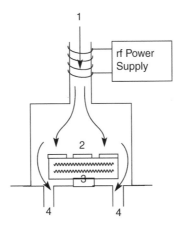

Fig. 6. Inductive tube remote plasma reactor. 1: precursor in; 2: substrates; 3: heater; 4: exhaust.

4. Substrates

One advantage of CVD is that thin films can grow on complicated shapes, which is due to the fact that if the molecular diffusion rate is faster than the kinetic decomposition rate the precursor molecules can diffuse from the flow stream to the surface of the substrates uniformly. Since the thin film grows directly on the surface of the substrates, the surface preparation is very important. The substrate can be cleaned either by physical methods or chemical methods. Physical methods include high pressure scrubbing and liquid spraying, ultrasonic cleaning, centrifugal cleaning, solvent treatment, extraction and vapor degreasing, thermal desorption, sublimation, and sputtering. Examples of chemical methods are oxidation (thermal and wet) and stripping, chelating and complexion, solution (wet) and vapor etching, plasma etching, and ultraviolet (uv) irradiation. It is better to clean the substrates right before deposition. If they are not used immediately, they should be stored in a clean environment to avoid recontamination. The selection of substrate materials in a thermal CVD process is limited by their high temperature stability and compatibility with the deposited film. On the other hand, PECVD can deposit thin films at much lower temperatures, which offers more substrate selections. However, at much lower temperatures often the quality of the films is poorer; eg, adherence and crystallinity.

5. Fundamental Principles

5.1. Thermodynamics. Like most chemical reactions, it is possible to establish theoretical feasibility for a CVD reaction by calculating the Gibbs energy of reaction and the equilibrium constant (see THERMODYNAMICS). In general, an ideal gas mixture in contact with pure condensed phases is assumed for the calculations. However, the prediction is of somewhat limited use, as the vast majority of CVD reactions are inherently nonequilibrium because they are

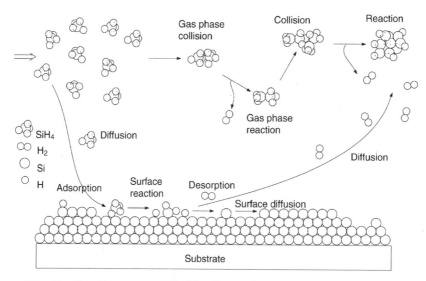

Fig. 7.   Reaction steps of Si thin-film and particle growth from SiH_4.

carried out in flow reactors under conditions where there are significant growth rates and volatile by-products are removed constantly. However, CVD kinetics (see KINETIC MEASUREMENTS) is far more important than the thermodynamics.

5.2. Kinetics.   Growth of the thin film in a CVD process, after the precursor molecules are delivered into the reactor, is the result of several basic steps. These steps are presented in Figure 7, using CVD of Si from SiH_4 as an example.

- Diffusion of precursor to the surface.
- Gas-phase reactions of the precursor.
- Adsorption–desorption of precursor at the surface.
- Reaction of adsorbed precursor on the surface.
- Incorporation of surface atoms into the bulk via surface diffusion.
- Desorption–adsorption and diffusion of the surface reaction by-products.

If the precursor concentration is high so that the diffusion rate is low, or if the reaction temperature is high, powders could form through a gas phase nucleation process. Collisions of the precursor molecules and gas-phase reactions might then take place resulting in the enlargement of particles' size.

In a CVD system, the overall reaction rate is influenced by the rates of precursor's feed, diffusion, surface as well as gas-phase reactions. The rate limiting cases are discussed below.

Feed-Rate-Limited CVD.   In this case, the film growth rate is limited by the amount of the precursor molecules delivered into the reactor. This is frequently encountered for precursors with low vapor pressure and fast reaction on the surfaces. In the extreme case, the quantity of the reacted precursor equals the quantity of the precursor introduced. The film growth rate appears to be almost independent of deposition temperature, which is true for diffusion controlled

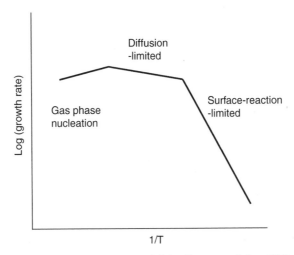

Fig. 8. Arrhenius plot of thin-film growth by CVD.

processes too, because the surface reaction rate is much higher than the feed rate.

Diffusion and Surface-Reaction-Limited CVD. These are dominant when the precursor feed rate is sufficiently high. Diffusion limited CVD, or as it frequently called diffusion-controlled or mass-transfer-controlled, is rate limiting when the surface reaction rate is high enough so that the growth rate of the film is limited by the diffusion rate of the reactants to the surface. This usually proceeds at high pressure, typically atmospheric pressure, and higher temperatures. On the other hand, the growth rate of the film is limited by the surface reaction rate when the precursor diffusion (transport) to the surface is relatively high. It is then referred to as surface reaction limited CVD, surface-reaction-controlled or kinetic-controlled. This usually occurs at low pressure, typically <10 Torr (1.33×10^3 Pa), and lower temperatures. An Arrhenius plot (see KINETIC MEASUREMENTS) is sometimes useful to differentiate the two cases (Fig. 8). The apparent energy of activation (E_a) can be derived from the slop of the plot to distinguish them. The surface-reaction-limited cases show steeper slope and larger E_a than the diffusion-limited cases. At high temperature (small $1/T$), it is common to observe decreased growth rate due to the increased rate of nucleation in the gas phase, which depletes the reactant concentration and lowers the thin-film growth rate.

In general, mass-transfer rate of the reactant gases to the substrate surface is more influential for APCVD than other CVD techniques. Mass transfer of gases involves the diffusion of molecules across the slow-moving boundary layer above substrate surface. The mass-transfer rate depends on the reactant concentration, the boundary layer thickness, and the diffusivity of the gases. In APCVD, mass-transfer rate and surface reaction rate are of the same order of magnitude. The substrates must lay side by side in the reactor to attain uniform film thickness.

In LPCVD, the mass-transfer rate of the reactant gases to the substrate surface is greatly enhanced due to the increased mean-free path. (Diffusivity is proportional to $1/P$) Here, the surface reaction rate, which is dependent on the reactant concentration and the temperature, controls the rate of deposition. In contrast to APCVD, the substrates can be arranged close-spaced and vertically in LPCVD and still be coated uniformly.

Surface Reactions.   The basic CVD reaction steps on a surface can be described by surface adsorption (see ADSORPTION, FUNDAMENTALS), reaction and desorption, as shown in Figure 7. The model is closely related to the Langmuir-Hinshelwood Mechanism frequently used to describe heterogeneous catalysis on the surfaces (see CATALYSIS) (16–18). The composition of the thin films and the fundamental surface reaction steps are studied using surface science analysis techniques (see SURFACE AND INTERFACE ANALYSIS) (18).

6. Conclusions

Just like PVD, CVD is an important and versatile technique to grow high quality thin solid films for many applications. In general, PVD is the preferred method of deposition in many industrial settings. The reason is that CVD is a more complicated process than PVD because the equipment setup is more complex and the chemical reaction steps are more difficult to control. However, the capability of conformal deposition of thin films on complex structures by CVD is the most important advantage of CVD over PVD. Thus, CVD and PVD methods complement each other.

CVD is a bottom-up method of fabrication. It converts small molecules, or building blocks, into extended structures such as thin films. Recently, it has been found that in addition, many nano-sized materials in other shapes, such as particles, tubes, wires, and rods, can be grown effectively by CVD. Thus, CVD is expected to play an important role in the development of the ever-important area of nanotechnology (see NANOTECHNOLOGY).

BIBLIOGRAPHY

1. J. E. Mahan, *Physical Vapor Deposition of Thin Films*, Wiley-Interscience, New York, 2000.
2. M. L. Hitchman and K. F. Jensen, eds., *Chemical Vapor Deposition: Principles and Applications*, Academic Press, London, 1993.
3. H. O. Pierson, *Handbook of Chemical Vapor Deposition: Principles, Technology and Applications*, 2nd ed., Noyes Publications, Norwich, New York, 1999.
4. C. E. Morosanu, *Thin Films by Chemical Vapour Deposition*, Elsevier, Amsterdam, The Netherlands, 1990.
5. T. T. Kodas and M. J. Hampden-Smith, *The Chemistry of Metal CVD*, VCH, Weinheim, 1994.
6. A. C. Jones and P. O'Brien, *CVD of Compound Semiconductors: Precursor Synthesis, Development and Applications*, VCH, Weinheim, 1997.
7. W. S. Rees, ed., *CVD of Nonmetals*, VCH, Weinheim, 1996.
8. A. Sherman, *Chemical Vapor Deposition for Microelectronics: Principles, Technology, and Applications*, Noyes Publications, Park Ridge, New Jersey, 1987.
9. K. E. Spear and J. P. Dismukes, eds., *Synthetic Diamond: Emerging CVD Science and Technology*, Wiley-Interscience, New York, 1994.

10. W. Z. Li, S. S. Xie, L. X. Qian, B. H. Chang, B. S. Zou, W. Y. Zhou, R. A. Zhao, and G. Wang, *Science* **274**, 1701 (1996).
11. X. T. Zhou, N. Wang, H. L. Lai, H. Y. Peng, I. Bello, N. B. Wong, C. S. Lee, and S. T. Lee, *Appl. Phys. Lett.* **74**, 3942 (1999).
12. T. I. Kamins, R. Stanley Williams, D. P. Basile, T. Hesjedal, and J. S. Harris, *J. Appl. Phys.* **89**, 1008 (2001).
13. J. Mazumder and A. Kar, *Theory and Application of Laser Chemical Vapor Deposition*, Plenum Press, New York, 1995.
14. S. Sivaram, *Chemical Vapor Deposition: Thermal and Plasma Deposition of Electronic Materials*, Van Nostrand Reinhold, New York, 1995.
15. J. G. Eden, *Photochemical Vapor Deposition*, Wiley-Interscience, New York, 1992.
16. R. P. H. Gasser, *An Introduction to Chemisorption and Catalysis by Metals*, Oxford University Press, New York, 1985.
17. G. A. Somorjai, *Chemistry in Two Dimensions: Surfaces*, Cornell University Press, Ithaca, 1981.
18. G. A. Somorjai, *Introduction to Surface Chemistry and Catalysis*, Wiley-Interscience, New York, 1994.

HSIN-TIEN CHIU
National Chiao Tung University

CHEMICAL WARFARE

1. Chemical Warfare

Weapons of mass destruction have received increased attention recently. Terrorist groups, rogue states, and foreign governments possessing these weapons have raised major concerns to stability in different regions across the globe, and about homeland security in the United States. Weapons of mass destruction are defined as nuclear, chemical, or biological weapons. While categorized together due to their potential destructive power, the nature of the threat of each of these types of weapons is very different, and defensive measures taken in response to each varies greatly. This article focuses on the details of chemical warfare, and deals primarily with U.S. capabilities. Concerning the role of chemical warfare in U.S. strategic planning, all modern chemical warfare capabilities are defensive in nature. Due to the fact that the first step of defending against any threat is understanding the threat, a discussion of chemical agents follows. However, from there discussion turns to classes of chemicals that, among other uses, can be utilized to thwart attempts to detect the presence of chemical agents. Rapid and accurate determination of the presence of exceedingly small amounts of a chemical substance on a battlefield, an environment that is far from pristine, is a difficult task. However, certain chemicals can be used to further complicate detection by orders of magnitude, and a sample of those are be included in this work. A short discussion describing contamination avoidance, protection, and decontamination materials is also included. None of the material covered in this article should be considered to be comprehensive, even within the framework of what constitutes the U.S. militaries equipment or capabilities.

Chemicals used in the context of chemical warfare fit into five categories: toxic agents, riot control agents, flame agents, incendiaries, and smokes and obscurants. Toxic chemical agents, used to achieve military objectives by producing casualties and discouraging enemy troops from certain areas of the battlefield (terrain denial), may be incapacitating or lethal. Army Field Manual 8-285 defines a chemical agent as "a chemical substance... intended for use in military operations to kill, seriously injure, or incapacitate humans (or animals) through its toxicological effects." Army and Department of Defense doctrine specifically excludes from this definition riot-control agents, which are considered to be legitimate law-enforcement tools. Also excluded are chemical herbicides and smoke and flame materials, all of which are considered legitimate military assets (1). Riot control agents are nonlethal tear agents most effective against unprotected personnel. Flame and incendiary agents can be used to harass and inflict casualties, and to destroy structures and material. Smokes and obscurants are employed for screening, signaling, and target marking, in both offensive and defensive applications.

The use of chemical agents and weapons in war is commonly thought to be a modern military technique. However, several historic uses are reported (2) including the contamination of drinking water and an attack against ships with earthen pots filled with live serpents, both before the birth of Christ.

Chemical warfare, a term used since 1917, is of vital interest not only to the world powers, but also to many developing countries. The relative simplicity of production and ease with which existing pesticide or other chemical plants can be converted to a weapons facility, make chemical weapons a continuing threat. Iraq and Libya, each of which possesses chemical weapon production facilities, have been known to use chemicals in war during the latter part of the twentieth century. The Geneva Convention of 1925 prohibited the use of chemical agents in war. The United States did not sign the Geneva Convention. The Paris Conference on Chemical Weapons in 1989 reaffirmed the prohibition on nerve agents (3–5). Thus, chemical weapons are not, as of this writing, banned. However, whether employed or not, these materials exist as a potential weapon in any conflict. Toxic chemical agents may exist as chemical substances in gaseous, liquid, or solid state intended to produce casualty effects ranging from harassment through varying degrees of incapacitation to death. A few such agents are true gases but most are solids or liquids that are converted in use into a gaseous state or disseminated as aerosols (qv). For contamination of terrain the agent can also be disseminated in bulk form with or without additives to modify physical properties.

Since 1944, the DoD Nuclear Biological and Chemical (NBC) Defense Program has been overseen by the Office of the Secretary of Defense, Chemical and Biological Defense Steering Committee, and the Joint NBC Defense Board. The Joint Service Integration Group generates joint NBC requirements, including inputs of the geographic commanders in chief. The Joint Service Material Group takes those priorities and coordinates planning for all NBC nonmedical research. Development and Acquisition (RDS). Medical RDA is coordinated through the Armed Services Biomedical Research and Evaluation Management Committee (6).

According to the U.S. General Accounting Office, in its report 'Chemical and Biological Defense: Coordination to Nonmedical Chemical and Biological R&D Programs', August 1999, the United States budgeted ~$1 billion/year on military defense against chemical and biological weapons in 2000, with an additional $1.4 billion for civilian preparedness (6).

2. Lethal Agents and Incapacitants

The modern history of the military use of toxic chemical agents (2,6–8) dates from the first full-scale (chlorine) gas attack on April 22, 1915, near Ypres, Belgium in World War I. Some estimates put the number of casualties of World War I to chemical warfare agents at 5 million (1). There have been a few reports of the limited use of toxic chemicals since that time. Historically, the most important of the tactical agents, and the one still used most often in battle, is sulfur mustard. Although it is chemically similar to the cytotoxic nitrogen mustards used in contemporary medicine, it has little commercial value other than its role in warfare. The first heavy military use of mustard occurred in the summer of 1917 (9). It is estimated that during 10 days of a German attack on the French town of Armentieres in September of that year, more than 1 million mustard shells were fired (10). While sulfur mustard was stockpiled by the warring nations of World War II, there is little evidence of its deliberate use in combat (9). In December 1943, an Allied ship carrying large quantities of mustard and other munitions was attacked by German planes and exploded in the harbor or Bari, Italy, disseminating mustard over a wide area. The incident resulted in more than 600 casualties due to mustard exposure (11). The Italians employed mustard, a blister agent, during the Ethiopian war in 1935 and 1936: the Japanese used toxic chemicals in a number of small-scale engagements in the early years of their war with China: and Iraq purpotedly employed both mustard and nerve gases in the 1980s. The Iran-Iraq war produced at least 45,000 casualties to chemical agents (1).

Research on chemical agents after World War I led to the elimination of all but a handful of chemicals as being of practical battlefield significance. Some of the criteria used in the selection of a suitable agent are effectiveness in extremely small concentrations; time to onset of action; effectiveness through various routes of entry into the body, such as the respiratory tract, eyes, and skin; stability in long-term storage; and ease of dissemination in feasible munitions. At the time of World War II, the only chemicals considered to be of practical significance included the mustard gases and phosgene.

Discovery of nerve agents in Germany led to the availability of a class of compounds at least one order of magnitude more lethal than previously known where death might occur in a matter of minutes instead of hours.

An excellent survey of the many factors involved in the use of toxic chemicals in war, including historical background, legal aspects, attempts at international agreement, and reasons why toxic chemicals were not employed in World War II or the Korean Conflict, is available (12).

2.1. Mustard and Related Vesicants.   Mustard, bis(2-chloroethyl) sulfide [505-60-2] (Chemical Agent Symbol HD), $Cl(CH_2)_2S(CH_2)_2Cl$, is a colorless, oily liquid when pure. Most samples have a characteristic garliclike odor. Known as the "king of war gases," it is a potent alkylating agent that causes severe vesication of epidermal surface. It is favored for it's ability to incapacitate opponents, lower their mobility, restrict the use of terrain, and *reduce* their ability to flight. At high-dose levels, it exerts systemic cytotoxic effects, in particular involving the rapidly proliferative cells of the hematological system and intestinal mucosa. It is also genotoxic, mutagenic, and a dose-related carcinogen (8). Sulfur mustard in an oily liquid that becomes aerosolized when dispersed by spraying or by explosive blast from a shell or bomb. In temperate climates, it vaporizes slowly, posing particular risk in prolonged, closed-space, or below-grade exposures. At higher temperatures, vaporization increases markedly and contributes to more severe clinical effects. Because of its low volatility, mustard is persistant. At temperate climates, in open areas with little wind, it is expected to persist for >1 week. By contrast, persistence for only a day is likely at temperatures >37.7°C in the Saudi desert (9).

Battlefield air concentrations during World War I mustard gas attacks were estimated in the range of 19–33 mg/m^3 (13). At such concentrations, exposure for several minutes causes skin and eye injury, and exposure for 30–60 min can result in severe respiratory injury, systemic poisoning, and death (42).

In the period following World War I and during World War II, a wide variety of sulfur analogues of mustard were investigated and many potent vesicants were discovered. Each had two 2-chloroethyl groups attached to a sulfur atom. Examples of such compounds are 1,2-bis(2-chloroethylthio)ethane [3563-36-8] (Chemical Agent Symbol Q), $Cl(CH_2)_2S(CH_2)_2S(CH_2)_2Cl$; and bis(2-chloroethylthioethyl) ether [63918-89-8] (T), $Cl(CH_2)_2S(CH)_2O(CH_2)_2S(CH_2)_2Cl$. It is noteworthy that each 2-chloroethyl group is associated with a separate sulfur atom, not with the same one as for mustard. Both Q and T are more vesicant than mustard, but neither is as volatile. For that reason, neither is as effective by the vapor route. One disadvantage of sulfur mustard, however, is that it freezes at ~10°C, so that solidification in airplane spray tanks presents a serious problem.

The procedure by which mustard is manufactured can be modified to yield either a mixture of mustard and Q (HQ) or a mixture of mustard and T (HT). These mixtures have several advantages over mustard alone, unless the agent is used only for vapor effects. HQ and HT are more toxic, more vesicant, more persistent, and have lower melting points than mustard alone.

In 1935, nitrogen analogues of sulfur mustards were first synthesized and found to have marked vesicant action (14). These are tertiary amines containing at least two 2-chloroethyl groups, $RN(CH_2CH_2Cl)_2$. The most important of these compounds were tris(2-chloroethyl)amine [555-77-1] (Chemical Agent Symbol HN3), $N(CH_2CH_2Cl)_3$; *N*-methyl-2,2'-dichlorodiethylamine [51-75-2] (HN2), $CH_3N(CH_2CH_2Cl)_2$; and 2,2'-dichlorotriethylamine [538-07-8] (HN1), $CH_3CH_2N(CH_2CH_2Cl)_2$. The nitrogen mustards are colorless when pure but turn yellow to amber in storage. These materials have faint odors varying from fishy or soft-soaplike to practically odorless, and they act on the body in a manner similar to sulfur mustard.

Table 1. **Properties of Mustard Gases**

Property	HD	Q	T	HN1	HN2	HN3
mol wt	159.08	219.08	263.25	170.08	156.07	204.54
bp, °C^a	$80^{0.67}$		$120^{0.003}$	$85^{1.3}$	$87^{2.4}$	$144^{2.0}$
mp, °C	14.5	56	10	−34	−60	−4
density at 25°C, g/mL	1.2682		1.24	1.086	1.118	1.2347
volatility at 25°C, mg/m^3	925	0.4	2.8	2s.29	3.581	0.120

^{a} Pressure in kPa at which boiling point was determined is given as a superscript. To convert kPa to mm Hg, multiply by 7.5.

2.2. Properties. The physical properties of the mustards are summarized in Table 1. The sulfur mustards are only slightly soluble in water, whereas the nitrogen mustards are slightly soluble at neutral pH, but form water-soluble salts under acid conditions. Both sulfur and nitrogen mustards are extremely soluble in most organic solvents.

Although sulfur and nitrogen mustards have limited solubility in water at neutral pH, the small quantity that dissolves is extremely reactive. The reaction proceeds via a cyclic sulfonium or imonium intermediate

$$Cl(CH_2)_2S(CH_2)_2Cl \;\rightleftharpoons\; Cl(CH_2)_2-\overset{+}{S}\underset{CH_2}{\overset{CH_2}{\diagup\!\!\!\diagdown}}\; Cl^-$$

This intermediate attacks compounds containing a variety of functional groups, such as primary, secondary, and tertiary amino nitrogen atoms, carboxyl groups, and sulfhydryl groups (15).

With nitrogen mustards, the imonium ion, which apparently forms even in the absence of any solvent, readily attacks another molecule to form a dimer. For this reason, the nitrogen mustards are less stable than sulfur mustards in long-term storage.

An excellent reagent for detection and quantitative estimation of the mustards is p-nitrobenzylpyridine (16). On treatment of the reaction product with alkali, a blue color appears, which detects as little as 0.1 µg of mustard.

The mustards readily alkylate inorganic thiosulfates to form Bunte salt anions (17,18). For sulfur mustard, the product is $Cl(CH_2)_2S(CH_2)_2SSO_3^-$. Phosphates react in a similar manner and the product isolated from sulfur mustard is $Cl(CH_2)_2S(CH_2)_2OPO_3^{2-}$.

Sulfur mustard reacts rapidly with chlorine or with bleach, and this reaction is a suitable means of decontamination. Nitrogen mustards, however, chlorinate extremely slowly; thus chlorination is not suitable for their decontamination. The formation of water-soluble salts, such as by neutralization with sodium bisulfate, is the usual method for nitrogen mustard removal from contaminated surfaces. The mustard salts are much less vesicant than the corresponding free bases.

The mustards can be oxidized by such oxidizing agents as hydrogen peroxide or potassium bichromate in sulfuric acid. Oxidation occurs at the sulfur atom of sulfur mustard and at the nitrogen atom of nitrogen mustard. The product formed on strong oxidation of sulfur mustard, bis(2-chloroethyl)sulfone, [471-03-4], $(CLCH_2CH_2)_2SO_2$, exhibits vesicant properties.

2.3. Physiological Effects. The sulfur and nitrogen mustards act first as cell irritants·and finally as a cell poison on all tissue surfaces contacted. The first symptoms usually appear in 4–6 h (7). The higher the concentration, the shorter the interval of time between the exposure to the agent and the first symptoms. Local action of the mustards results in conjunctivitis (inflammation of the eyes); erythema (redness of the skin), which may be followed by blistering or ulceration; and an inflammatory reaction of the nose, throat, trachea, bronchi, and lung tissue. Injuries produced by mustard heal much more slowly and are much more liable to infection than burns of similar intensity produced by physical means or by other chemicals.

Mustard's most important metabolic effect, at least for acute toxicity, is inhibition of cellular glycolysis. Disruption of that metabolic pathway is found after exposure to nearly all vesicant chemicals, usually due to interference with sulfhydryl-rich hexokinase enzyme systems (19). In the case of mustard, however, inhibition of glycolosis is an indirect result to nucleic acid damage and repair. Initially, sulfur mustard rapidly alkylates the purine bases (guanine and adenine) of DNA. Activation of endonucleases then leads to removal (depurination) of alkylated bases, leaving apurinic sites where DNA breaks occur readily. In turn, need for DNA repair activates poly(ADP-ribose) polymerase, an enzyme that rapidly depletes cellular NAD^+ (20–22).

Depletion of cellular NAD^+, which begins within 1 h of exposure and is maximal after ~4 h, inhibits glycolysis, leads to releas of tissue proteases, and ultimately results in cell necrosis. These biochemical changes parallel the development of tissue injury (19,23). In animals, treatment with polymerase inhibitors (3-methoxybenzamide or nicotinamide) or NAD^+ precursors (niacin or nicotinamide) increased cellular NAD^+ levels, decreased the severity of skin damage, and delayed the onset of toxic effects after cutaneous exposure (20).

The rate of detoxification is slow, and the effects of even very small repeated exposures are cumulative or more than cumulative owing to sensitization.

2.4. Uses. The nitrogen mustards are used clinically in the treatment of certain neoplasms (21). They have been used in treatment of Hodgkin's disease, lymphosarcoma, and leukemia (see CHEMOTHERAPEUTICS, ANTICANCER).

2.5. Nerve Agents. Discovery of nerve agents in Germany led to the availability of a class of compounds at least one order of magnitude more lethal than previously known where death might occur in a matter of minutes instead of hours. Nerve agents are part of a class known as organophosphorus compounds, and were first synthesized in 1854, but their development as warfare agents did not occur until 80 years later (23). The first military nerve agent, ethyl phosphorodimethylamidocyanidate, known as tabun, or chemical agent symobol GA (Fig. 1), was synthesized in 1936 by Schrader while searching for

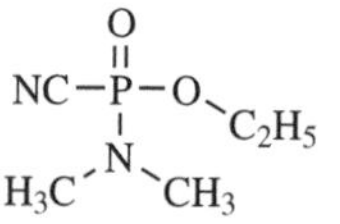

Fig. 1. Tabum.

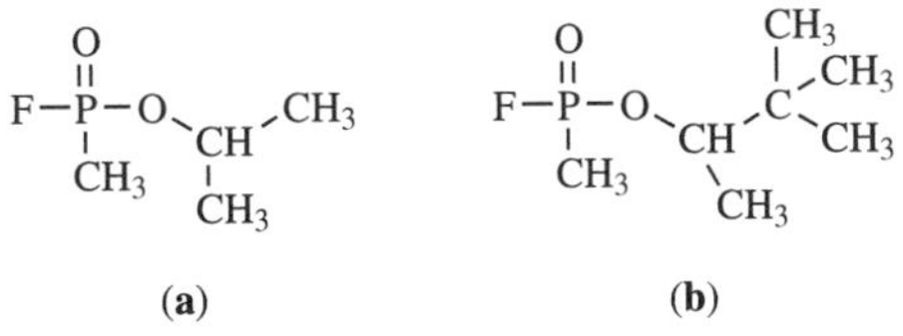

Fig. 2. (a) Sarin (b) Soman.

more effective agricultural organophosphorate insecticides at the German chemical firm of IG Farbenindustrie.

Two years later, he synthesized a second nerve agent, isopropyl methylphosphonofluoridate, now known as Sarin or GB [Fig. 2(**a**)]. In 1944, a third nerve agent, pinacolyl methyl-phosphonofuoridate, now known as Soman or GD [Fig. 2(**b**)], was synthesized (14).

By the end of World War II. Germany had large stockpiles of nerve agents. Although estimates vary widely, as much as 30,000 tons of Tabun and 500 tons of Sarin may have been prepared in munitions ready for battlefield deployment (24). It is uncertain why these agents never were used by the German military; the Allies were not aware that Germany had synthesized these agents for use in battle and did not have antidotes (25). During the waning stages of war. Russian forces captured the German nerve agent-manufacturing facilities at Duhernfurt and moved the buildings and staff to Russia, where production and stockpiling of these agents continued (26). At approximately the same time. Allied forces discovered munitions containing nerve agents, and scientists elucidated the pharmacology and the antidotal activity of atropine. An excellent survey of the many factors involved in the use of toxic chemicals in war, including historical background, legal aspects, attempts at international agreement, and reasons why toxic chemicals were not employed in World War II or the Korean Conflict, is available (12).

In the early 1950s, a fourth agent of current military importance. *O*-ethyl-*S*-[2-(diisopropylamino)ethyl]-methylphosphonothioate, now known as VX (Fig. 3), was synthesized in England, again in a search for better insecticides (25). This and related compounds are a great deal more toxic than either mustard or chlorine and are capable of being disseminated by munition systems including bombs, artillery rounds, rockets, grenades, missiles, and aerial spray (27,28).

Nerve agent refers to two groups of highly toxic chemical compounds that generally are organic esters of substituted phosphoric acid (8,11,27). The nerve

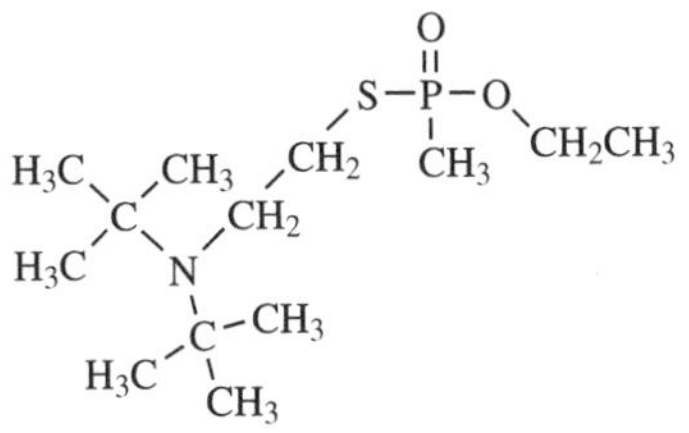

Fig. 3. VX.

agents inhibit cholinesterase enzymes and thus come within the category of anticholinesterase agents (see ENZYME INHIBITORS). The three most active G-agents are tabun, ethyl phosphorodimethylamidocyanidate [77-81-6] (Chemical Agent Symbol GA), $(CH_3)_2NPOCNOC_2H_5$; sarin, isopropyl methylphosphono-fluoridate [107-44-8] (GB), $CH_3POFOCH(CH_3)_2$; and soman, pinacolyl methyl-phosphonofluoridate [96-64-0] (GD), $CH_3POFOCHOCH_3C(CH_3)_3$.

The G-agent liquids under ordinary atmospheric conditions have sufficiently high volatility to permit dissemination in vapor form. They are generally colorless, odorless or nearly so, and are readily absorbable through not only the lungs and eyes but also the skin and intestinal tract without producing any irritation or other sensation on the part of the exposed individual. These agents are sufficiently potent so that even a brief exposure may be fatal. Death may occur in 1–10 min, or be delayed for 1–2 h depending on the concentration of the agent.

Another class of nerve agents, discovered after World War II, is the V-agents. Nerve agent vapors are four to six times denser than air. As a result, they tend to remain close to the ground and pose a particular risk to people in low areas and below-ground shelters (25). These materials are generally colorless and odorless liquids that do not evaporate rapidly at normal temperatures. In liquid or aerosol form, V-agents affect the body in a manner similar to the G-agents. Although all four nerve agents are significantly hazardous by all routes and are percutaneous hazards, VX is least volatile and most efficiently absorbed through the skin (29,30). Their relative lethality as determined in animal studies is VX > Soman > Sarin > Tabun (30). Estimates of lethal doses vary. For inhalation exposure in a 70-kg adult breathing 15 L/min, approximate mean lethal doses are 150–400 mg min/m^3 for Tabun, 75–100 mg min/m^3 for Sarin, 35–50 mg min/m^3 for Soman and 10 mg min/m^3 for VX. Mean lethal doses due to skin exposure in a 70-kg adult are 1700 mg for Sarin, 1000 mg for Tabun, 100 mg for Soman, and 6 mg for VX (24,31).

2.6. Properties.   Some physical properties of nerve agents are given in Table 2. The G-agents, miscible in both polar and nonpolar solvents, hydrolyze slowly in water at neutral or slightly acid pH and more rapidly under strong acid or alkaline conditions. The hydrolysis products are considerably less toxic than the original agent.

GB is unstable in the presence of water. Maximum stability in aqueous solutions occurs from pH 4.0–6.5 with the hydrolysis rate increasing as the pH increases. The half-life in distilled water at 25°C is ~36 h, but hydrolysis is accelerated in the presence of acids or bases. Because bases are far more effective in this respect than acids, caustic solutions are useful for decontamination.

Table 2. **Properties of Nerve Agents**

Property	GA	GB	GD	VX^a
formula wt	162.13	140.10	182.18	267.38
bp, °C	246	147	167	298
mp, °C	−50	−56	unknown	below −51
density at 25°C, g/mL	1.073	1.0887	1.0222	1.0083
volatility at 25°C, mg/m^3	610	21,900	3,060	10.5

^{a} VX [50782-69-9], $C_{11}H_{26}NO_2PS$, is a phosphonothioic acid ester.

GB decomposes thermally to form a variety of phosphorus-containing products as well as propylene. The rate of decomposition increases with increase in temperature and in the presence of acids. At the boiling point of GB, under atmospheric conditions, decomposition is fairly rapid.

GB and other G-agents react with perhydryl ions at pH 9–10 to form a perphosphonate ion, $CH_3P(O)(OC_3H_7)OO^-$, which has a sufficiently high redox potential to oxidize indole or o-dianisidine to produce colored products. This reaction is thus useful as a method of detection, and <1 μg of GB can be detected in this manner (32).

Another useful reagent for detection and estimation of G-agents is diisonitrosoacetone (33). A magenta color is produced with 1 μg of GB at pH 8.5. Coupling agents, such as p-phenylenediamine, increase the reaction rate.

2.7. Physiological Effects. Inhalation of G-agent vapor at realizable field concentrations is immediately incapacitating. The symptoms in normal order of appearance are runny nose; tightness of chest; dimming of vision and pinpointing of the eye pupils (miosis); drooling and excessive sweating; nausea and vomiting; cramps and involuntary defecation or urination; twitching, jerking, and staggering; and headache, confusion, drowsiness, coma, and convulsion. These symptoms are followed by cessation of breathing and death.

Although GB is also effective by penetration through the skin, the dose required to produce toxic effects by this route is high. Thus a person wearing skin covering and a gas mask is reasonably well protected.

The toxicity of nerve agents is due primarily to irreversible inactivation of acetylcholinesterase leading to accumulation of toxic levels of acetylcholine. Like other organophosphorous compounds (pesticides), these agents act by binding to a serine residue at the active site of the cholinesterase molecule. The resulting accumulation of toxic levels of acetylcholine at the synapse initially stimulates and then paralyzes cholinergic synaptic transmission. Cholinergic synapses are found in the central nervous system, at the termination of somantic nerves, in ganglionic synapses of autonomic nerves, at parasympathetic nerve endings, and at some sympathetic nerve endings, such as those in the sweat glands (25,34). The reaction between organophosphourous compounds and acetylcholinesterase occurs in a three-step process (Fig. 4).

In step 1, a reversible enzyme-inhibitor complex is formed. The rate of complex formation depends on the structure of the organophosphorous compound, particularly the size and shape of R_1 and R_2. In step 2, which results in phosphorylation and inactivation of the acetylcholinesterase molecule, the X moiety is displaced from the phosphorous atom by the hydroxyl group of the enzyme's target serine residue. The kinetics of this reaction is determined by the X moiety, which is referred to as the leaving group. Step 3 is a time-dependent reaction that modifies the enzyme-inhibitor complex in such a way that the organophosphorous compound and enzyme cannot be separated and the enzyme cannot be reactivated by oximes. This reaction, called "aging," involves the formation of a monophosphoric acid residue that is bound to the enzyme (25).

2.8. Binary Munitions. Binary munitions contain two nonlethal components that are mixed during flight to form a nerve agent. Each component is manufactured separately, remaining in its own container until the munition components are assembled just prior to use. Mixing and subsequent agent

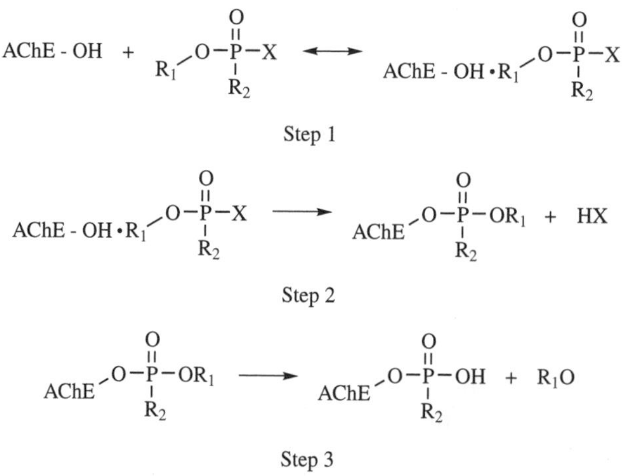

Fig. 4.   Activity of oganophosphorus nerve agents.

formation occurs after firing or launch of the munition. In addition to greatly reduced storage and handling hazards, the binary components can be manufactured in ordinary chemical facilities, which need not be equipped with the stringent safety and environmental controls required for the older nerve agent munitions. The binary technology also overcomes the long-term storage problems associated with the unitary nerve agents.

2.9. Other Lethal Agents.   There are a number of substances, many found in nature, which are known to be more toxic than nerve agents (27). None has been weaponized. Examples of these toxic natural products include shellfish poison, isolated from toxic clams; puffer fish poison, isolated from the viscera of the puffer fish; the active principle of curare; "heart poisons" of the digitalis type; the active principle of the sea cucumber; active principles of snake venom; and the protein ricin, obtained from castor beans (see CASTOR OIL).

2.10. Incapacitants.   Incapacitating agents, or incapacitants, are just what the name implies. In wartime, soldiers and civilians must be physiologically, physically, and mentally able to perform their jobs. Thus, an agent rendering an individual incapable of job performance may be classified as an incapacitating agent (27,35).

Incapacitants are most suitable for consideration in limited warfare situations, eg, when enemy troops are intermingled with a friendly population, or in a city that is a key military objective. The purpose is to capture the enemy without killing the civilians. Incapacitating agents should produce no permanent aftereffects and allow for complete recovery.

Agent BZ, 3-quinuclidinyl benzilate [6581-06-2], $C_{21}H_{23}NO_3$, is a typical incapacitant. BZ is one of a group of substances, many of them glycolate esters, sometimes known as atropinemimetics. Their action on the central and peripheral nervous systems resembles that of atropine [51-55-8], $C_{17}H_{23}NO_3$. The effects of BZ are those of an anticholinergic psychotomimetic drug (27). These effects follow ~1/2 h after exposure to BZ aerosol, reach a peak in 4–8 hs, and may then take up to 4 days to pass. Effects include disorientation with visual and auditory hallucinations. The agent disturbs the higher integrative functions of memory, problem solving, attention, and comprehension. There is a gradual return to normalcy.

By U.S. Army criteria, incapacitation agents do not include (35): (1) lethal agents that are incapacitating at sublethal doses, such as nerve agents; (2) substances that cause permanent or long-lasting injury, such as blister agents, choking agents, and those causing eye injury; (3) drugs that exert marked effects on the central nervous system, such as barbiturates, belladonna alkaloids (qv), tranquilizers, and many of the hallucinogens. These drugs are logistically infeasible for large-scale use because of the high doses required (see HYPNOTICS, SEDATIVES, AND ANTICONVULSANTS AND ANTIOLYTIC; NEUROREGULATORS; PSYCHOPHARMACOLOGICAL AGENTS); (4) agents of temporary effectiveness that produce reflex responses interfering with performance of duty. These include skin and eye irritants causing pain or itching (vesicants or urticants), vomiting- or cough-producing compounds (sternutators), and tear compounds (lacrimators); and (−5) agents that disrupt basic life-sustaining systems of the body and thus prevent physical activity. These include agents that lower blood pressure, paralyzing agents such as curare, fever-producing agents, and blood poisons. Such agents almost invariably have a narrow margin of safety between the effective and possibly lethal doses, and the basic purpose of an incapacitating agent is to reduce military effectiveness without endangering life.

2.11. Use.   Owing to inherent physical characteristics, chemical agents can be adapted to a variety of munitions, including grenades, mines, artillery shells, bombs, bomblets, spray tanks, rockets, and missiles. Tactically, chemical agents have defensive and offensive capabilities in limited or general wars. Toxic chemical agents may be used alone or in conjunction with other types of weapons. Chemical weapons do not destroy matériel but allow physical preservation of industrial complexes and other facilities. Incapacitating agents may also be used to preserve life and avoid permanent injury.

The use of chemical agents in battle imposes a significant burden on troops because of the cumbersome nature of the protective clothing and the attendant heat load in hot climate situations. This factor alone imposes a burden on potential target personnel, lowering their effectiveness. U.S. troops in the 1991 Mideast war Desert Storm were provided with protective gear that did not deter them with regard to the outcome of the action.

3. Irritants

Irritant compounds such as the lacrimators and sternutators used in World War I are traditional examples of harassing agents, the effects of which are reversible and briefly incapacitating. Riot control agent CS, a modern irritant compound (27,35), causes physiological effects that include extreme burning of the eyes, accompanied by a copious flow of tears; coughing; difficulty in breathing; chest tightness; involuntary closing of the eyes; stinging sensation of moist skin; runny nose; and dizziness or "swimming" of the head. Heavy concentrations also cause nausea and vomiting.

The effects of agent CS are immediate, even in extremely low concentrations. The median concentration for respiratory effects is 12–20 mg/m^3; for eye effects it is 1–5 mg/m^3. The onset of maximum effects is 20–60 s, and the duration is 5–10 min after the individual has been removed to fresh air.

A water-soluble white crystalline solid, CS is disseminated as a spray, as a cloud of dust or powder, or as an aerosol generated thermally from pyrotechnic compositions. The formulation designated CS1 is CS mixed with an antiagglomerant; when dusted on the ground, it may remain active for as long as 5 days. CS2 formulated from CS1 and a silicone water repellent, may persist for as long as 45 days (12).

The principal uses of CS are in riot control and training; it has limited tactical use in defensive military modes (36).

4. Flame

In the modern weapons arsenal flame agents are defined as various hydrocarbons, blends of hydrocarbons, and other readily flammable liquids, usually thickened with additives, that are easily ignited and can be projected to military targets (35). Although flame agents may be employed against buildings and other flammable targets, their primary role is against personnel in hardened structures or emplacements. In the United States, the principal application of flame agents is now in flame throwers and flame projectors, including flame rockets. The firebomb is becoming obsolete. The replacement for the portable flame-thrower is the multishot flame rocket, which delivers an encapsulated flame warhead directly to the target from a significantly great and safe range. The large-caliber flame rounds could become all-weather, highly aimable replacements for the obsolete air-deliverable fire-bombs and the mechanized flame-throwers.

4.1. Flame Throwers and Projectors.   One advance in flame-throwers since World War II was a mechanized flame-thrower kit for a variety of armored vehicles other than the main battle tank. The multishot, lightweight, shoulder-fired, four-tube flame system capable of firing one to four flame rounds semiautomatically is replacing the portable flame-thrower. Indeed the mechanized flame-thrower is expected to become obsolete because of the family of large-caliber flame rounds.

The U.S. Army's M202/M74 flame system is a shoulder-fired, four-tube launcher equipped with front and rear hinged protective covers. A folding sight and trigger handle assembly provide compact carrying and storage capabilities. An adjustable sling is used to carry the launcher over the shoulder. The rocket system is aimed and fired from the right shoulder from the standing, kneeling, sitting, or prone position. Ammunition for the launcher is provided in rocket clips preloaded with four rockets that slip-fit into the four launcher tubes. The user can fire from one to four rockets semiautomatically at a rate of one per second; the launcher can be reloaded with a new clip repeatedly. The flame agent payload is thickened pyrophoric agent TPA, a polyisobutylene-thickened metal alkyl formulation (37).

4.2. Fire Bombs. After World War II, the fire-bomb became a standard item of military equipment. A cigar-shaped, thin-cased tank similar in appearance to the aircraft fuel tank from which it evolved, the typical fire-bomb consisted of the basic tank, two igniters, two fuses, an arming wire, and the flame agent payload. Most fighter and fighter-bomber aircraft carried two, one under each wing. Ideal delivery was at low altitude, almost level flight, with the fire-bomb impacting along the aircraft flight path and producing a rolling wall of flame covering an area ~35 m wide and 100 m in length. The initial fireball, a direct function of flame agent quality, burned for ~10 s with intense heat. The dispersed particles could burn for up to 10 min but at greatly reduced intensity.

Upon the advent of high performance jet-powered aircraft, the fire-bomb became obsolete. When delivered at speeds approaching Mach one (345 m/s), the design characteristics were often grossly exceeded and many units broke up or functioned while still on the aircraft. Delivery from high altitude created craters and deposited most of the flame agent payload in those craters.

Studies of fire-bomb modifications and the corresponding flame agent payloads were terminated once the controlled fireball damage mechanism was developed. In field firings it was shown that large-caliber flame rounds could produce equivalent effects. The rolling wall of reacting flame could be generated, accompanied by a highly aimable, all-weather flame system capable of sustained fire.

4.3. Flame Agents. For some time after World War II, effort was expended in improvement of naplam-thickened hydrocarbons as the standard flame agents. The problem was the breakdown of the flame agent in the presence of traces of water and the need for peptizers in cold-weather applications. More recently, thickened pyrophoric flame agents (35) have been deployed in the field and as the payload for the U.S. Army's flame rocket system. Advantages of this newer flame agent include: the ability to prepackage warheads and other containers and store them for indefinite periods with no deterioration; and the fact that these pyrophoric flame agents do not require an ignition system to function on the target.

Computer-aided research by army personnel has resulted in development of the controlled fireball damage mechanism for efficient and effective coupling of heat energy to effect maximum thermal damage. Moreover, flame agents may be tailored to each candidate flame system. Optimization of several low viscosity metal−alkyl formulations has also removed the problem of temperature by making these flame agents relatively independent of temperature variations.

5. Incendiaries

Incendiary agents are designed for use in the planned destruction of buildings, property, and matériel by fire (35). Incendiaries burn with an intense, localized heat. They are very difficult to extinguish and are capable of setting fire to materials that normally do not ignite and burn readily. Although there are tactical applications for incendiary agents and munitions, they have played primarily a strategic role in modern warfare.

5.1. Incendiary Requirements. The mechanics of starting fires using incendiary agents involve a source of heat to act as a match to initiate combustion in a larger mass; combustible material to serve as kindling; and fuel. The match and the kindling are provided by the incendiary munition; the target is the fuel. All incendiary munitions, except for those containing materials that are spontaneously combustible, must have some sort of initiator such as a fuse or an ignition cup. The second element of the incendiary munition, the kindling, is the important factor, and both the amount and the nature of the combustible material in the munition have been the subject of much research and development.

The maximum total heat output of an incendiary agent can be readily calculated and it is obviously desirable to use a filling that has a high heat evolution. The rate of the heat release varies with the agent and depends on: flash, fire, or decomposition temperature; particle size of the agent after ejection from the munition, which controls the surface/volume ratio of the agent; and oxidizing agents blended with the combustible material to increase the rate of heat evolution. The incendiary agent must be capable of heating the target or fuel until the ignition temperature, which can vary from 200–400°C, is reached. To be really effective, the incendiary agent must generate at least four times as much heat as is necessary to raise the temperature to this point.

5.2. Metal Incendiaries. Metal incendiaries include those of magnesium in various forms, and powdered or granular aluminum mixed with powdered iron(III) oxide. Magnesium is a soft metal that, when raised to its ignition temperature, burns vigorously in air. It is used in either solid or powdered form as an incendiary filling, and in alloyed form as the casing for small incendiary bombs.

Magnesium has an ignition temperature of 623°C and a burning temperature of ~1982°C. The burning temperature is variable; it depends on the rate of heat dissipation, rate of burning, and other factors. Magnesium, burning with a blinding white flame, melts as it burns and the burning liquid metal drops to lower levels, igniting all combustible materials in its path. Burning stops if oxygen is prevented from reaching the metal or if the metal is cooled to a point below its ignition temperature. Whereas magnesium does not have the highest heat of combustion of the metals, none of the other metals have been successfully used as air-combustible incendiaries. Some metals may be alloyed with magnesium without affecting its ignitability. The alloyed metal has the strength to withstand distortion, whereas pure magnesium does not. In massive form, magnesium is difficult to ignite. This problem is overcome by packing a hollow core in a bomb with thermite [8049-32-9]. This results in an easily ignited mixture that supplies its own oxygen and burns at a very high temperature.

Thermite is essentially a mixture of ~73 wt% powdered iron(III) oxide, Fe_2O_3, and 27 wt% powdered or granular aluminum. The aluminum has a higher affinity for oxygen than iron, and if a mixture of iron oxide and aluminum powder is raised to the combustion temperature of aluminum, an intense reaction occurs

$$Fe_2O_3 + 2\,Al \longrightarrow Al_2O_3 + 2\,Fe \qquad \Delta H = -3.35 \text{ kJ/g } (-800 \text{ cal/g})$$

Under favorable conditions the thermite produces temperatures of ~2200°C, high enough to turn the newly formed metallic iron into a white-hot liquid that acts as a heat reservoir to prolong and spread the heat or igniting action.

The thermate mixture, composed of thermite and various additives, is used in igniter compositions for magnesium bombs. A number of such compositions have been developed. Three of these were Therm-8, Thermate-TH2 (formerly Therm-8-2), and Thermate-TH3 (formerly Therm-64-C). Therm-8 was the precursor of later, improved igniting formulations TH2 differs from Therm-8 in that TH2 contains no sulfur and slightly less thermite. TH3 was found to be superior to the others and thus adopted for use in the incendiary magnesium bomb. The wt% composition of TH3 is thermite, 68.7; barium nitrate, 29.0; sulfur, 2.0; and as a binder, oil, 0.3.

The TH3 core is ignited by the primer, and the burning core then melts and ignites the magnesium alloy body of the bomb. The incendiary action on the target is localized. There is little scattering of the incendiary material.

5.3. Oil and Metal Incendiary Mixtures. PT1 is a complex mixture composed of magnesium dust, magnesium oxide, and carbon (qv), along with an adequate amount of petroleum (qv) and asphalt (qv) to form the paste (37). The U.S. developers have adopted the formula: type c paste (goop), -49.0 ± 1.0; IM polymer AE, 3.0 ± 1.0; coarse magnesium, 10.0 ± 1.0; petroleum oil extract, 3.0 ± 0.2; gasoline, 30.0 ± 3.0; and sodium nitrate, 5.0 ± 0.5.

PTV is an improved oil and metal incendiary mixture having the composition: polybutadiene, 5.0 ± 0.1; gasoline, 60.0 ± 1.0; magnesium, 28.0 ± 1.0; sodium nitrate, 6.0 ± 0.1; and p-aminophenol, 0.1.

Incendiary bombs containing PT1 or PTV mixtures are easily ignited by nose or tail fuses because they contain many combustible ingredients. These formulations contain both metal and an oxidizer, eg, sodium nitrate; thus condensed phase reaction products are obtained. The resulting heat flux conducted to the target is increased, resulting in greater potential target damage.

6. Smokes

Military smokes are aerosols (qv) of gaseous, liquid, or particulate matter that are tactically employed to defeat enemy surveillance, target acquisition, and weapons guidance devices. The traditional battlefield applications of smoke are screening and marking. Screening smokes are normally employed to obscure a military objective from enemy observation by creating an aerial blanket, vertical curtain, or ground haze. Signaling smokes are pyrotechnic mixtures that incorporate an organic compound dyed for target marking or for transmitting messages by prearranged color code (see PYROTECHNICS).

Battlefield smoke has a great significance in the countermeasure role. Threat weapons no longer rely simply on optical devices that operate in the visible spectrum. Modern lethal weapons are augmented with sophisticated surveillance, target acquisition, and guidance devices that operate throughout the visible, infrared (ir), and microwave (radar) regions of the electromagnetic spectrum. Accordingly, research and development is concentrated on nonvisual multispectral screening systems designed to defeat modern weaponry such as the antitank guided missile (ATGM), laser-guided munitions, and heat-seeking missiles.

6.1. Screening Smokes. Military smoke screens are produced by dispersing either finely divided solids or minute liquid droplets in air. To be useful, a smoke screen must be sufficiently opaque to provide the desired screening power and long-lasting enough to achieve effective military results. In designing a screen for the modern battlefield, it is necessary to defeat sophisticated surveillance technologies that operate in the near-, mid-, and far-ir as well as in the millimeter wavelength frequencies. Other factors considered in the evaluation of potential screening agents include cost, ease of dispersion, and efficiency of dispersion. Agents used for screening friendly areas must be as nonirritating as possible.

Both the opacity and persistency of smoke screens are largely dependent on the nature of the individual smoke particles. All present U.S. smoke agents produce aerosols. The principal mechanism of obscuration is the scattering effect. To optimize its effectiveness, the aerosol's particle size should be approximately in the same size range as the wavelength of the light to be screened. Many U.S. smoke agents produce aerosols having particle-size distribution that is log-normal having a mean diameter of ~ 0.6 µm, allowing maximum effectiveness in the visible and near-ir wavelength regions. More recently, fine fibers have been used as screens for longer wavelengths and in mixtures or combined usage to screen the entire region of concern.

The obscuring action of visual screening smokes is largely caused by reflection and refraction of light by the individual suspended particles of which the smoke is composed (35). Because this obscuring action occurs to the greatest extent in the absence of light-absorbing particles such as carbon, white smokes have the greatest screening action.

Chiefly wind and convection currents in the air determine the persistency of a smoke cloud. Ambient temperature also plays a part in the continuance or disappearance of fog oil smokes. Water vapor in the air has an important role in the formation of most chemically generated smokes, and high relative humidity improves the performance of these smokes. The water vapor not only exerts effects through hydrolysis, but it also assists the growth of hygroscopic (deliquescent) smoke particles to an effective size by a process of hydration. Smoke may be generated by mechanical, thermal, or chemical means, or by a combination of these processes (35).

6.2. Types of Screening Smokes. The generation of oil smoke is based on the production of minute oil droplets by purely physical means. The most desirable droplet size is 0.5–1.0 µm. The tiny droplets of oil scatter light rays and produce a smoke that appears to be white, and any individual droplet would be transparent under magnification. These droplets are produced as the vaporized oil passes through the nozzle of a generator and is subsequently cooled by the surrounding air. The air cools the oil vapor so quickly that only very small droplets are able to form. The process thus depends on a high oil temperature followed by quick cooling. Proper selection of the smoke oil ensures that the final smoke cloud is stable, and the life of the cloud is determined almost entirely by meteorological conditions. The smoke generator uses a low viscosity petroleum oil (U.S. Army designation: SGF No. 2 smoke generator fog). SGF corresponds somewhat to an SAE 10 (light) motor oil in viscosity. Below 0°C, a mixture of SGF No. 2 and a paraffin-free kerosene is used.

Another type of smoke mixture, a volatile hygroscopic chloride for thermal generation, has the U.S. Army designation HC, type C. It is composed of ~6.7 wt% grained aluminum, 46.7 wt% zinc oxide ZnO, and 46.7 wt% hexachloroethane [67-72-1], C_2Cl_6. The ratio of zinc oxide to hexachloroethane is held between 1.04 and 1.00, but the aluminum may be varied slightly to regulate the burning time. Because this mixture is composed of solids, it can be compressed to provide high payloads in small volumes, and it is used as a filling for smoke grenades, smoke pots, and artillery shells. The initial heat needed to start the burning of HC is provided by a starter mixture, typically silicon, potassium nitrate, charcoal, iron oxide, grained aluminum, cellulose nitrate, and acetone. A burning HC mixture produces zinc chloride which, in turn, is volatilized by reaction heat and condenses to form a zinc chloride–water smoke.

A third screening smoke-type is white phosphorus [7723-14-0] (WP), -P_4 (see PHOSPHORUS AND THE PHOSPHIDES), which reacts spontaneously with air and water vapor to produce a dense cloud of phosphorus pentoxide [1314-56-3]. An effective screen is obtained as the P_2O_5 hydrolyzes to form droplets of dilute phosphoric acid aerosol. WP produces smoke in great quantity, but it has certain disadvantages. Because WP has such a high heat of combustion, the smoke it produces from bulk-filled munitions has a tendency to rise in pillarlike mass. This behavior too often nullifies the screening effect, particularly in still air. Also, WP is very brittle, and the exploding munitions in which it is used break it into very small particles that burn rapidly.

The disadvantages of WP have been overcome to some degree by absorbing WP into felt wedge submunitions that burn and release smoke at a lower rate and by the development of plasticized white phosphorus (PWP). PWP is produced by melting WP and stirring it into cold water, which results in a slurry of WP granules of ~0.5-mm diameter. The slurry is mixed with a viscous solution of synthetic rubber, so that the granules are coated with a film of rubber and thus separated from each other. When PWP is dispersed by an exploding munition, it does not break into such small particles; the burning rate is slowed, and the tendency of the smoke to pillar is reduced. WP and PWP are used as fill for grenades, artillery shells, mortar shells, bombs, and rockets.

Developments in several countries have resulted in red phosphorus (RP) as a screening smoke agent having less performance problems. RP is an allotropic form of elemental phosphorus, which is made by heating white phosphorus at high temperatures in the absence of air. RP is less reactive than WP and thus lends itself to the manufacture of presized subunits that can be packaged in artillery and mortar shells. These subunits, which are dispersed as multiple smoke-producing sources, enhance target effectiveness by the rapid formation of a large homogeneous and persistent screen. The pillaring phenomenon is thus minimized. Another advantage of RP is that it does not undergo a change of state at operational temperatures thereby precluding the munition instability problems that sometimes occur with WP, which liquefies $> 43°C$.

A solution of sulfur trioxide [7446-11-9], SO_3, dissolved in chlorosulfonic acid [7990-94-5], $ClSO_3H$, has been used as a smoke (U.S. designation FS) but it is not a U.S. standard agent (see CHLOROSULFURIC ACID; SULFURIC ACID AND SULFUR TRIOXIDE). When FS is atomized in air, the sulfur trioxide evaporates from the small droplets and reacts with atmospheric moisture to form sulfuric acid vapor. This vapor condenses into minute droplets that form a dense white cloud. FS produces its effect almost instantaneously upon mechanical atomization into the atmosphere, except at very low temperatures. At such temperatures, the small amount of moisture normally present in the atmosphere, requires that FS be thermally generated with the addition of steam to be effective. FS can be used as a fill for artillery and mortar shells and bombs and can be effectively dispersed from low performance aircraft spray tanks. FS is both corrosive and toxic in the presence of moisture, which imposes limitations on its storage, handling, and use.

6.3. Signaling Smokes.   Screening smokes also can be adapted for signaling purposes. For example, phosphorus-filled artillery and mortar rounds can be used to mark targets and determine range corrections. However, a good signaling smoke must be clearly distinguished from the smoke incident to battle. The standard signaling smokes, described in Table 3, afford good visibility and unmistakable identity. Unfortunately, all colors become gray and indistinguishable at great distances, and even excellent signaling smokes have a maximum effective visual range that varies from ~1–3 km, depending on the color intensity of the smoke and the type and size of the munition.

Volatilizing and condensing a mixture containing an organic dye produce colored signaling smokes. Of the dyes tested by the U.S. Army, the most satisfactory ones are the azo dyes (qv), the azine dyes (qv), the diphenylmethane dyes, and those of anthraquinone (qv) (35) (see DYES AND DYE INTERMEDIATES). The filling for a colored smoke munition is essentially a pyrotechnic mixture of fuel and dye, with a cooling agent sometimes added to prevent excessive decomposition of the dye. The heat produced by the fuel volatilizes the dye, and the dye condenses outside the munition to form the colored smoke. In U.S. munitions, the fuel is made up of a mixture of an oxidizing agent such as potassium chlorate, $KClO_3$, and a combustible material such as sulfur or sugar. The burning time can be regulated by adjusting the proportions of oxidant and combustible material and by use of coolants such as baking soda. A typical starter mixture is given in Table 3. Colored smoke mixtures are used in hand and rifle grenades and in canisters for use with larger projectiles.

7. Defense Against Toxic Agents

Defensive measures against toxic agents may be divided into four categories: agent detection and identification, individual and collective protection, decontamination, and medical defense. To these may be added a high degree of training in defensive measures and discipline in using them.

7.1. Detection.   The utilization of various technolgies to sample the environment in search of chemical weapons agents is known as detection, and two major classifications of detection exist. If the environment being probed for the existence of chemical agent concentrations is in the immediate vicinity, it is known as point detection. If the search for chemical agents is occurring at some distance from the detector, it is referred to as remote or stand-off detection.

The United States has deployed several chemical sensors. The Automatic Chemical Agent Detector and Alarm is a man-portable vapor alarm designed

Table 3. **Colored Smoke Fillings**

Ingredient	CAS Registry number	Molecular formula	Composition, wt%
Red smoke			
dye:			
85% 1-*N*-methylaminoanthraquinone;	[82-38-2]	$C_{15}H_{11}NO_2$	
15% dextrin	[9004-53-9]		42
sodium bicarbonate	[144-55-8]	$NaCHO_3$	19
potassium chlorate	[3811-04-9]	$KClO_3$	28
sulfur	[7704-34-9]	S	11
Green smoke			
dye:			
70% 1,4,di-*p*-toluidinoanthraquinone;	[128-80-3]	$C_{28}H_{22}N_2O_2$	
10.5% indanthrene golden yellow;	[128-66-5]	$C_{24}H_{12}O_2$	
19.5% benzanthrone	[82-05-3]	$C_{17}H_{10}O$	40.0
sodium bicarbonate			22.6
potassium chlorate			27.0
sulfur			10.4
Yellow smoke			
dye:			
65% benzanthrone;	[82-05-3]	$C_{17}H_{10}O$	
35% indanthrene golden yellow	[128-66-5]	$C_{24}H_{12}O_2$	38.5
sodium bicarbonate			33.0
potassium chlorate			20.0
sulfur			8.5
Violet smoke			
dye:			
80% 1,4-diamino-2,3-dihydroanthraquinone;	[83-61-0]	$C_{14}H_{12}N_2O_2$	
20% 1-*N*-methylaminoanthraquinone	[82-38-2]	$C_{15}H_{11}NO_2$	42.0
sodium bicarbonate			18.0
potassium chlorate			28.8
sulfur			11.2
Starter			
potassium nitrate	[7757-79-1]	KNO_3	37.8
silicon	[7440-21-3]	Si	28.0
charcoal	[7440-44-0]	C	4.2
cellulose nitrate	[9004-70-0]		1.2
acetone	[67-64-1]	C_3H_6O	28.8

to have greater capability than the current M8A1 detector against standard blister and nerve agents. The first U.S. Army unit was equipped in 1998, and it is starting to be procured in numbers. The Improved Chemical Agent Monitor (ICAM) is a hand-held, real-time detector of nerve and mustard agents in vapor phase on personnel or equipment. The program is also based on a product developed by Graseby in the U.K., and is intended to replace the CAM, significantly improving that system's reliability and maintainability. The United States has bought ~5000 ICAMs (38).

The Joint Chemical Agent Detector (JCAD) is a pocket-sized detector for use especially on aircraft and ships. It is entering into the operational test and evaluation phase, with production of more than 250,000 expected to begin in

Fiscal Year 2003 (FY03). Manufactured by BAE Systems of Austin, Texas, it is expected eventually to replace all current U.S. point detectors (38).

In the area of stand-off chemical detection, The Joint Service Lightweight Nuclear, Biological, Chemical Reconnaissance System (JSLNBCRS) is a vehicle-mounted system designed to sample, detect and identify threats within a unit's area of responsibility. It is intended to replace the Fox Nuclear, Biological and Chemical Reconnaissance System vehicle made by General Dynamics Land Systems and Henschel Wehrtechnik (Germany). The Fox uses the M21 Remote Sensing Chemical Agent Alarm, e.g., while the JSLNBCRS will incorporate the Joint Service Standoff Chemical Agent Detector (JSLSCAD) that improves on the M21. Some advanced detectors are also planned to be integrated into the Block II Fox NCBRS. JSLSCAD prototypes for High-Mobility Multipurpose Wheeled Vehicles (HMMWVs) and Light Armored Vehicles are under going development and initial operational testing, with full-scale production of 700 vehicles scheduled to begin in FY03 (38).

Individual chemical sensors would be much more effective if integrated into a network. Multiple sensors would provide mutual support, and rapid attack characterization, as well as minimizing false alarms. The Joint Warning and Reporting Network (JWARN) is a command and control information system that will be used by NBC specialists in various headquarters units to analyze data and support commanders' decisions, while extending warning down to individual battlefield units. It will seek to integrate legacy systems and future detectors. Block I software has been developed. A new Block II Operational Requirements Document has just been issued, and a solicitation is expected from the U.S. Marine Corps Systems Command (38).

7.2. Individual and Collective Protection. The United States has in the past insisted its troops be able to "fight dirty," in a contaminated environment. The United States sets a high limit for chemical agent protection (10 g/m^2 for liquids; 5000–10,000 mg min/m^3 for vapors). This was designated in light of the threat when the former Soviet Union stockpiled chemical munitions by the ton and a possible struggle for societal survival was envisaged. Current potential enemies have less well-developed toxic capabilities, and there are suggestions that the high standards for protection need to be cased. This would also facilitate contaminated environments, in that heat stress and cumbersome equipment could be diminished. While current and planned protective equipment is designed to meet the high standard, there has been an increased emphasis on trying to reduce the significant troop performance decrement associated with it. This includes developing decontaminant creams to wear on the skin under seams and gloves (38).

The primary item of individual protection is the protective or gas mask. The U.S. Army standard mask is the M17A2, which has as its basic components a molded rubber facepiece with large cheek pouches that hold filter elements. These elements consist of six sheets of core laminated between two sheets of backing layer. The backing layers are composed of vinyl chloride–vinyl acetate copolymer, cellulose, and glass fibers. The core layers are composed of these same materials plus 75% Whetlerite absorbent, a finely ground activated carbon impregnated by immersion in an ammoniacal solution of silver, copper, chromium, and carbon dioxide, and then dried at temperatures sufficient to expel substantially all ammonia from the granules. Inhalation valves are attached to the filter element caps through the cheek-pouch portions of the mask faceblank. A voicemitter-outlet assembly, incorporating a speech diaphragm and an outlet valve, is attached to the faceblank at the mouth position. Large eyepieces in the faceblank provide a wide field of vision. The M17A2 mask provides complete respiratory protection against all known military toxic chemical agents, but it does not provide protection against some industrial toxics such as ammonia and carbon monoxide.

A newer improved mask developed by the U.S. Army is in production and is expected to be fielded in the 1990s. The M40 (Figs. 5 and 6) and M42 chemical-biological masks are to replace the M17A2 and several other specialty masks, making for greater logistical simplicity as well as better respiratory, eye, and face protection against field concentrations of chemical and biological agents. Improvements include an externally mounted NATO standard canister, front and side voicemitters, microphone compatibility, drinking capability for all personnel, compatibility with combat spectacles, and better fit and comfort.

The mask alone, however, does not provide protection from substances such as nerve and blister agents that penetrate through the skin. Thus protection for the entire body should be provided. Airtight, impermeable clothing is available for personnel who must enter heavily contaminated areas. Such clothing is cumbersome and enervating because it retards release of body heat and moisture, and personnel efficiency is drastically lowered when it is worn for a long time. Although resistant to liquid chemical agents, impermeable clothing may be penetrated after a few hours exposure to heavy concentrations of agent. Consequently, liquid contamination must be neutralized or removed as soon as possible.

Fig. 5. The M40 chemical-biological mask shown with hood that increases protection against liquid contamination.

Fig. 6. Overgarment issued to troops operating in forward areas (19) that do not have access to or time to use any type of decontamination procedures.

For field use, combat troops use the battledress overgarment (BDO) shown in Figure 6. This overgarment protects against skin contact with chemical agent vapors, aerosols, and droplets of liquid. It consists of a two-piece suit in a camouflage pattern: the jacket has a zippered front, and trousers have a fly front and zippered legs. The BDO, supplied in a vapor-barrier bag that protects it from rain, moisture, and sunlight, consists of a water-repellent outer layer of nylon–cotton and an inner layer of charcoal impregnated polyurethane foam. The overgarment is discarded once contaminated.

The BDO is being replaced with an effective lighterweight suit, the Joint Service Lightweight Integrated Suit Technology (JSLIST). The BDO generated controversy previously as quality-control issues came to light. The BDO is being supplanted by JSLIST, as the military is procuring 300,000 of these suits annually. One of the great improvements is that soldiers are supposed to be able to work ∼45 min/h while wearing this suit, as compared to only 15 min/h in the old one. There is also a preplanned improvement for JSLIST under development (38).

The U.S. approach is to enable donning increasing levels of protection over the uniform, going from Military Oriented Protective Posture (MOPP)-ready (having gear available) to fully protected (MOPP 4) in ∼8 min. However, adding garments when one is already masked often causes the mask to break its seal temporarily but significantly. The new joint service general purpose mask is more comfortable, but may not remedy this problem. Other countries, such as

France, have a different approach, and have lighterweight suits worn directly on the skin rather than over the uniform (39).

The protective clothing issued to military personnel provides protection against high concentrations for prolonged periods, but is constraining and, in a hot climate, increases the risk of heat stroke, especially in individuals treated with atropine (39).

The U.S. military now has a mission to support homeland defense. However, its Personal Protective Equipment is not compatible with that of civilian responders. National Institute for Occupational Safety and Health does not accept military MOPP gear for civilian first responders, mandating a self-contained breathing apparatus rather than air purifying respirators. The DoD may want to reconsider whether it should purchase some civilian protection equipment to meet some of its needs, in addition to its military-specific developments (38).

Collective protection enclosures are required for groups of personnel. Such enclosures must be airtight to prevent inward seepage of contamination. They can be independent units or can be formed by adequately treating the interior walls of structures, tents, airplanes, or vehicles. A supply of uncontaminated air, provided by passing ambient air through high efficiency aerosol and carbon filters, must be provided.

Simplified collection protection equipment (SCPE) (U.S. designation M20) provides such protection using lightweight elements consisting of an inflatable enclosure, a hermetically sealed filter canister, motor blower, protective entrance, and a support kit. The SCPE is designed to be used inside another structure or tent. The modular collective protection equipment (MCPE) is designed to support enclosures such as rigid wall shelters or vans. The MCPE comprises a family of items, consisting of four filter units of different sizes, external and internal protective entrances, and a motor controller. The entire system is flexible by choice of items to fit a large variety of adaptations.

7.3. Decontamination. If contaminated equipment or material does not have to be used immediately, natural aeration is an effective decontaminant procedure, as most chemical agents, including the blister and V-agents, are volatile to a certain degree. Wind accelerates their evaporation and hastens their dissipation. Rain and dew may also cause sufficient hydrolysis of some agents. Sunlight increases the surface temperatures of military equipment and thus accelerates agent evaporation.

If decontamination cannot be left to natural processes, chemical neutralizers or means of physical removal must be employed. In general, the neutralizers are of two types: chlorine-based oxidants or strong bases. Some neutralizers have been especially developed for the decontamination of chemical agents.

One such decontaminant is supertropical bleach (STB). STB is a mixture of chlorinated lime and calcium oxide containing $\sim 30\%$ available chlorine. It can be used either as a dry mix or as a slurry to decontaminate some equipment surfaces and terrain. The dry mix is prepared with two parts bleach to three parts earth by volume. A slurry typically consists of 40 parts STB to 60 parts by weight of water. This material is then sprayed or swabbed on the contaminated surface (see BLEACHING AGENTS). STB is an effective decontaminant for mustard, lewisite, and VX. It is less effective against nerve agents other than VX.

Decontaminating agent DS2 is a general purpose equipment decontaminant, consisting of 70 wt% diethylenetriamine, 28 wt% -ethylene glycol monomethyl ether, and 2 wt% sodium hydroxide. DS2 reacts with nerve agents and blister agents to effectively reduce their hazards within 5 min. Important limitations in the use of DS2 include: irritation to the eyes and skin, and inhalation of the vapors may be harmful; it is a combustible liquid, therefore it must not be used on hot metal surfaces such as running engines or exhaust pipes; it is not suitable for use on personal equipment such as mask or clothing or on electrical or electronic equipment; in the pure state, DS2 is noncorrosive to most metals but it may cause corrosion of aluminum, cadmium, tin, or zinc after prolonged contact; and whereas it can be used on surfaces coated with polyurethane paint, it may remove other paints, especially fairly new ones.

The M258A1 decontaminant kit, personal, is used by the individual soldier for emergency decontamination of skin and partial decontamination of personal equipment. The kit consists of three Decon 1 towelette wipes and three Decon 2 wipes, which are individually sealed in impermeable foil packets. The Decon 1 packet contains a towellette prewetted with a solution of 72 wt% ethanol, 10 wt% phenol, 5 wt% sodium hydroxide, 0.5 wt% ammonia, and the rest water. The Decon 2 packet contains a towellette impregnated with dry chloramine B and glass ampuls filled with a solution 44 wt% ethanol, 5 wt% zinc chloride, and the remainder water. The glass ampuls are contained in plastic mesh bags to prevent injury when the ampuls are broken. The M258A1 is effective against all threat agents including biological agents and toxins. It is capable of three complete decontaminations of exposed skin (hand, neck, face) and partial decontamination of personal equipment. Storage and handling of the kits require special consideration. Each of the solutions are poisonous and present a caustic burn hazard. The wipes are not to be used on eyes, wounds, or mouth. Also the kits must be stored in a dry place away from direct sunlight and flammable materials.

The M25A1 is being replaced through attrition by the M291 skin decontamination kit. This item contains six individual skin decontamination packets for the emergency decontamination of skin. Each packet contains a nonwoven fiber polybacked applicator pad impregnated with 2.8 g of Ambergard XE-555 resin. Applicator pads have strap handles for ease of use. Individual pads are hermetically packaged in olive drab-colored, heat-sealed, polyester-foil laminate material, which is notched on each corner for ease of opening. The six packets are carried in a flexible, walletlike carrying pouch constructed of nonwoven fabric. Total weight is 45 g, and it fits easily into the trousers pocket of the battledress uniform (BDU) and the protective BDO.

Besides these specially designed decontaminants and kits there are a number of commercially available materials that will partially decontaminate chemical agents. Such materials are listed in the *U.S. Army Field Manual FM 3–5, NBC Decontamination.*

7.4. Medical Defense. The most important items of U.S. medical defense against organophosphorus nerve agents are atropine and pralidoxime chloride [51-15-0], (2-PAM), $C_7H_9ClN_2O_2$. These agents neutralize the effects of the anticholinesterase compounds and are capable of reactivating the inhibited enzymes. If adequate emergency treatment is immediately available, it is theoretically possible to save a high percentage of those affected by the agent. Oximes such as pralidoxime displace nerve agents bound to AchE, allowing

AchE to resume hydrolyzing ACh (40). Before aging has occurred, administration of nucleophilic oximes, such as pralidoxime chloride, can reactivate the phosphorylated acetylcholinesterase, but such treatment is ineffective after aging (34). Oximes are ineffective after aging has occurred (41). American soldiers carry three 600-mg autoinjectors for concurrent intramuscular use with atropine (39). Atropine competitively inhibits muscarinic effects and has a central role in the treatment of nerve agent poisoning. Soldiers carry three 2-mg autoinjectors for use when first symptoms appear (41). More recently the use of a pretreatment has been developed using pyridostigmine bromide [101-26-8], $C_9H_{13}BrN_2O_2$, when the threat or possibility of exposure occurs. During the 1991 Desert Storm action, over 40,000 U.S. troops were so pretreated. Presently, American soldiers are issued 30-mg tablets of pyridostigmine bromide to be taken, on order, every 8 h for up to 14 days when at risk of nerve agent exposure (42). Similar pretreatment compounds are in use in other countries (13,23).

Immediate treatment of an exposed individual is essential. The U.S. regimen includes the pretreatment, and after exposure atropine and 2-PAM are self-administered. Further treatment includes up to two additional doses, followed by the tranquilizer Valium. As required, artificial respiration is instituted, clearing the airway if necessary. The current standard U.S. Army atropine item is the automatic injector, Atropen, designed for self-administration by the individual in the field.

Research on other protective measures includes the use of excess circulating acetyl- and butyryl- cholinesterases as scavengers. Monoclonal antibodies and carboxylesterase are being investigated as means to deactivate nerve agents in blood before they reach synapses (29).

Vesicant agents, such as mustard, require no special treatment once the burns have occurred. Copious washing is quite effective when used early for liquid contamination of the eyes, and soap and water removes the liquid agent from the skin. Burns resulting from mustard agent are treated like any other severe burn. The pulmonary injuries are treated symptomatically; antibiotics are used only if indicated for the control of infection.

BIBLIOGRAPHY

"Chemicals in War" under "Gas Warfare Agents" in *ECT* 1st ed., Vol. 7, pp. 117–145, by R. Macy and co-workers, Chemical Corps, Chemical and Radiological Laboratories, Army Chemical Center, and H. A. Charipper, New York University; "Chemical Warfare" in *ECT* 2nd ed. Vol. 4, pp. 869–907, by T. E. Puro, United States Army, Edgewood Arsenal; "Chemicals in War" in *ECT* 3rd ed., Vol. 5, pp. 393–416, by B. L. Harris, F. Shanty, and W. J. Wiseman, Chemical Systems Laboratory, U.S. Department of Defense; in *ECT* 4th ed., Vol. 5, pp. 795–816, by B. L. Harris, Consultant, U.S. Chemical Systems Laboratory; "Chemicals in War" in *ECT* (online), posting date: December 4, 2000, by B. L. Harris, Consultant, U.S. Army Chemical Systems Laboratory.

CITED PUBLICATIONS

1. J. Newmark, *Military Med.* **166**, 9 (2001).
2. J. H. Rothschild, *Tomorrow*'s Weapons, McGraw-Hill Book Co., New York, 1964.

3. S. A. Carnes and A. P. Watson, *JAMA* **262**, 653 (1989).
4. H. R. Hu, Cook-Deegan, and A. Shukri, *JAMA* **262**, 640 (1989).
5. J. M. Orient, *JAMA* **262**, 644 (1989).
6. B. H. Liddell-Hart, *A History of the World War, 1914–1918*, Little. Brown & Co., Boston, Mass., 1948.
7. *Chemistry*, in W. A. Noyes, Jr., ed., *Science in World War II*, Little, Brown & Co., Boston, Mass., 1948.
8. *The Problem of Chemical and Biological Warfare*, Vol. LL. Stockholm International Peace Research Institute (SIPRI), Humanities Press, New York, 1973.
9. J. Borak and F. R. Sidell, *Ann. Emerg. Med.* **21**, 303 (1992).
10. W. K. Blewets, *NBC Defense Technol. Int.* **1**, 64 (1985).
11. S. F. Alexander, *Milit. Surg.* **101**, 1 (1947).
12. J. B. Kelly, *Gas Warfare in International Law*, Master's Thesis, Georgetown University, Washington, D.C., June 1960.
13. J. F. Thorpe, *Thorpe's Dictionary of Applied Chemistry*. Vol. 3, New York: Longmans, Green, 1939, p. 8.
14. K. Ward, Jr., *J. Am. Chem. Soc.* **57**, 914 (1935).
15. S. B. Davis and W. F. Ross, *J. Am. Chem. Soc.* **69**, 1177 (1947).
16. B. Gehauf, *CWS Field Lab Memo 1-2-8*, National Technical Information Service, U.S. Department of Commerce, Washington, D.C., April 1943.
17. W. H. Stein and co-workers, *J. Org. Chem.* **11**, 664 (1946).
18. C. Golumbic and co-workers, *J. Org. Chem.* **11**, 518 (1946).
19. M. Dixon and D. M., *Nature (London)* **158**, 432 (1946).
20. B. Paplmeister and co-workers, *Fund. Appl. Toxicol* **5**, S134 (1985).
21. P. W. Rankin and co-workers, *Cancer Res.* **4C**, 1803 (1988).
22. T. Lindahl, *Prog. Nucleic Acid Res. Mol. Biol.* **22**, 135 (1979).
23. B. Holmstedt, ed. "Structure-Activity Relationships of the Organophosphorous Anticholinesterase Agents", in G. Koelle, ed., *Cholinesterase and Anticholinesterase Agents*, Springer-Verlag, Berlin, 1963, pp. 428–485.
24. J. P. Robinson, *The Problem of Chemical and Biological Warfare; Vol I. The Rise of CB Weapons*, Humanities Press, New York, 1971, pp. 72–87.
25. F. R. Sidell and J. Borak, *Ann. Emerg. Med.* **21**(7), 128 (1992).
26. G. B. Koelle, *Fund. Appl. Toxicol.* **1**, 129 (1981).
27. *Military Chemistry and Chemical Compounds*. U.S. Army Field Manual 3-9/U.S. Air Forces Field Manual 355-7, U.S. Government Printing Office, Washington, D.C., October, 1975.
28. *Commander's* Call, Chemical Warfare, Department of the Army Pamphlet, 360–831, January–February 1977.
29. M. A. Dunn and F. R. Siddell, *JAMA* **262**, 649 (1989).
30. D. J. Hickett, J. F. Glenn, and W. E. Houston, *Milit. Med.* **152**, 35 (1987).
31. Demilitarization, P.E.O.-P.M.f.C., *Chemical Stockpile Disposal Program: Final Programmatic Environmental Impact Statment*, Aberdeen Proving Ground, Maryland, 1988.
32. B. Gehauf and co-workers, *Anal. Chem.* **29**, 278 (1957).
33. S. Sass and co-workers, *Anal. Chem.* **29**, 1346 (1957).
34. P. Taylor, ed. *Anticholinesterase Agents*, 8 ed, in A. G. Goodman, R. T. W, and A. S. Nies and co-workers, eds., *The Pharmacological Basis of Therapeutics*, Pergamon Press: New York, 1990, pp. 131–140.
35. R. N. Sterlin, V. I. Yemel'yanov, and V. I. Zimin, *Khim. Oruzhive Zashchita Nego* (1975).
36. *Renunciation of Certain Uses in War of Chemical Herbicides and Riot Control Agents*, Executive Order 11850, April 8, 1975.

37. *Launcher, Rocket: 66MM.* 4-Tube, M202A1 (NSN 1055-00-021-3909) U.S. Army Technical Manual 3-1055-456-12. U.S. Government Printing Office, Washington, D.C., October 1969.
38. D. Siegrist, *Janes Defense Weekly* **37**(16), 22 (2002).
39. U.S. Departments of the Army, N., and Air Force., *Field Manual 8-285: Treatment of Chemical Agent Casualties and Conventional Military Chemical Injuries. ed.* H. (DASG-HCD). 5109 Leesburg Pike, Falls Church, Va-22041-3258.
40. I. B. Wilson, *J. Biol. Chem.* **190**, 111 (1951).
41. C. H. Gunderson and co-workers, *Neurology* **42**, 946 (1992).
42. J. R. Keeler, C. G. Hurst, and M. A. Dunn, *JAMA* **266**, 693 (1991).
43. B. Ranshaw, *Mechanisms in Production of Cutaneous Injuries by Sulfur and Nitrogen Mustards*, Chemical Warfare Agents and Related Chemical Problems. Vol. 1. Washington, D.C. : US Office of Scientific Research and Development, 1946, pp. 479–518.

GENERAL REFERENCES

A. Goodman and H. Martens, *Studies on the Use of Electric Eel Acetylcholinesterase for Anticholinesterase Agent Detection*, Edgewood Arsenal Report No. Ed-TR-74096, Feb. 1975*.

A. Silvestri and co-workers, *Development of a Kit for Detection of Hazardous Material Spills into Waterways*, Edgewood Arsenal Special Publication No. ED-SP-76023, Aug. 1976*.

J. S. Parsons and S. Mitzner, *Environ. Sci. Technol.* **9**, 1053 (Nov. 1975).

F. W. Karasek, *Detection Limits in Instrumental Analysis, Research/Development*, July 1975.

J. E. Estes, *Remote Sensing Techniques for Environmental Analysis*, Hamilton Publishing Co., 1974.

G. P. Wright, *Designing Water Pollution Detection Systems*, Ballinger Publishing Co., Cambridge, Mass., 1974.

W. S. C. Chang, *Lasers and Applications*, Ohio University, Athens, Ohio, 1963.

R. M. Gamson, D. W. Robinson, and A. Goodman, *Environ. Sci. Technol.* **7**, 1137 (1973).

L. H. Goodson and W. B. Jacobs, *Real Time Monitor, Immobilized Enzyme Alarm and Spare Parts*, Edgewood Arsenal Report No. ED-CR-77015, February, 1977*.

J. P. Mieure and W. M. Dietrich, *J. Chrom. Sci.* **13**, 559 (Nov. 1973).

B. J. Ehrlich and S. F. Spencer, *Development of an Automated Mustard Stock Monitor*, Edgewood Arsenal Report No. ED-CR-76084, Tracor, Inc., June 1976*.

L. Schwartz and co-workers, *Evaluation of M15/M18 Enzyme Detector Ticket System with Low Concentration of GB*, Edgewood Arsenal Report No. ED-TR-74018, June 1974*.

L. H. Goodson, *Feasibility Studies on Enzyme System for Detector Kits*, Edgewood Arsenal Report No. ED-CR-77019, December, 1976*.

H. W. Levin and E. S. Erenrich, *Enzyme Immobilization Alternatives for the Enzyme Alarm*, Edgewood Arsenal Report No. ED-CR-76005, August, 1975*.

R. M. Gamson and co-workers, *Detection of GB, VX and Parathion in Water*, Edgewood Arsenal Report No. ED-TR-74015, June 1974*.

D. P. Soule, *Agent Concentractor Feasibility Studies*, Edgewood Arsenal Report No. ED-CR-76075*.

J. W. Scales, *Air Quality Instrumentation*, Vols. I and II, Instrument Society of America, 1974.

H. Tannenbaum, *Proc. Soc. Photo-Opt. Instrum. Eng.* **49** (1975).

T. Hirschfeld and co-workers, "Remote Spectrosoopic Analysis of Parts-Per-Million-Level Air Pollutants by Raman Spectroscopy," *Appl. Phys. Lett.* **22**(1) (Jan. 1973).

H. A. Walter, Jr. and D. F. Flanigan, "Detection of Atmosphere Poliutants: A Correlation Technique," *Appl. Opt.* **14** (June 1975). ED: the DTD does not allow Footnote within citation Available from National Technical Information Service, U.S. Department of Commerce, Springfield, Va.

Tom J. Evans
Cubic Defense Applications Group

CHEMILUMINESCENCE, ANALYTICAL APPLICATIONS

1. Introduction

Chemiluminescence (CL) has many advantageous features as a tool of detection in instrumental analysis, including sensitivity, selectivity and simplicity. A CL reaction needs no excitation light source, thus, it is not accompanied with any scattering light. This results in a large signal-to-noise (S/N) ratio and consequently, a substantial increase in detector's sensitivity. To date, several kinds of CL reactions have been clarified on their reaction mechanisms. Generally, concentrations of the substrates or catalysts in each CL reaction can be determined by measuring the CL generated. Among the CL reactions, luminol (5-amino-2,3-dihydro-1,4-phthalazinedione) derivatives, acridine derivatives, aryloxalate derivatives, and ruthenium derivatives have frequently been utilized in analytical applications (1–4). In this article, representative analytical applications of CL in liquid phase for the determination of a variety of compounds are described.

2. Luminol Derivatives

2.1. Luminol as a CL Reagent. Luminol is generally used for the determination of compounds that catalyze the luminol CL reaction. Aliphatic alcohols, aldehydes, ethers, and sugars containing an oxygen atom in their molecules can be converted into H_2O_2 by oxygenation in photochemical reaction using anthraquinone disulfonate as a sensitizer, which can then be detected by a CL reaction with Co(II) and luminol (5). The lower limit of detection is in the picogram level. After separation on a cation-exchange column, Co(II) in rice powder was determined by high performance liquid chromatography (HPLC) in the same manner. The sensitivity is very high with the detection limit of 0.5 ng/L (6). Although Cu(II) can catalyze the luminol–hydrogen peroxide CL, the coexisting proteins can quench the yield of emission. Based on this phenomenon, as low as 50 ng of ovalbumin were determined by HPLC (7). Hydroperoxides, the primary lipid peroxidation products, were separated on a normal-phase HPLC and selectively was determined by reaction with a mixture of luminol and cytochrome c as hydroperoxide-specific postcolumn CL reagents (8,9). Phosphatidylcholine derivatives are important biological phospholipids, which were separated on a normal-phase column and determined by the same CL reaction system; the amounts of peroxides were 28–431 pmol hydroperoxide–O_2/1 mL plasma (10). Phosphatidylcholine hydroperoxide in low density lipoprotein (LDL) in human plasma was determined by the same method, and revealed that the level is higher in patients with atherosclerosis and hyperlipemia compared to normal patients (11). Quantification of phospholipid hydroperoxides in biological tissues is important in order to know the degree of peroxidation damage of membrane lipids. Hydroperoxides of phosphatidiylcholine and phosphatidylethanolamine in rat liver and brain tissue were determined also (12). Determination of lipid peroxides in native LDL is clinically important because these peroxides are considered to cause the pathogenesis of atherosclerosis. A rapid flow injection method was developed by utilizing the luminol–microperoxidase CL reaction for the determination of peroxides in LDL (13). By using the same CL system, it was confirmed that the lipid peroxides in triacylglycerol contained in butter or spreads increased during the storage (14). The column switching HPLC method for simultaneous determination of hydroperoxides in phosphatidylcholine and cholesteryl ester has been developed; lipid hydroperoxide levels of phosphatidylcholine and cholesteryl ester in human plasma were 36.0 ± 4.0 and 12.3 ± 3.1 nM, respectively (15). Cholesterol hydroperoxide in human red cell membrane was also determined by a similar luminol–microperoxidase CL method (16). The unique CL method utilizing isoluminol and microperoxidase was developed for the measurement of lipid peroxidation in the cultivated medium, which can be used to directly measure *in vitro* lipid peroxidation in cells (17). The use of m-chloroperoxybenzoic acid instead of H_2O_2 causes an increase in luminol CL catalyzed by microperoxidase 8 by an order of magnitude. The enhancement of CL intensity is pH dependent (18). The manganese–tetrasulfonatophthalocyanine catalyzed by a luminol–H_2O_2 CL system, can be quenched in the presence of proteins. By using this quenching reaction, a highly sensitive method has been developed for the determination of proteins, such as human serum albumin, human γ-IgG, and bovine serum albumin (BSA) with the detection limits of 1.9, 2.7, and 3.4 ng/ml, respectively (19). Phenanthroline–H_2O_2 CL or luminol–H_2O_2 CL were used for the photographic determination of 1,10-phenanthroline and Fe(II) complex with the detection limit of 6×10^{-5} and 2×10^{-15} M, respectively (20).

2.2. Luminol Derivatives for the Measurement of Hydrogen Peroxide and Substrates that Produce Hydrogen Peroxide in Enzyme Reactions.

Hydrogen peroxide is an excellent oxidizing agent for luminol; the CL increases with the increase of H_2O_2 concentration. Therefore H_2O_2 or substrates that produce it in an enzyme reaction can be determined by measuring the CL yielded; needless to say, enzyme activity can also be estimated. Glucose, cholesterol, and xanthines were determined by measuring H_2O_2 produced by enzyme reactions with glucose oxidase, cholesterol oxidase, and xanthine oxidase, respectively. Adenosine is converted to inosine by adenosine deaminase, which is further converted to hypoxanthine by nucleoside phosphorylase. The H_2O_2 formed by the reaction of hypoxanthine and xanthine oxidase can be determined by the luminol–peroxidase CL reaction. This reaction system was applied to the automatic determination of purine derivatives; it is very rapid and 200 samples per day can be assayed (21). A flow injection (FIA)–CL method for the convenient determination of H_2O_2 was devised. In this system, an immobilized enzyme reactor (IMER) was placed in a flow line, and substrate was indirectly assayed by measuring the H_2O_2 yield. Moreover, FIA and HPLC methods for the determination of β, D-glucosides were developed; phenyl-β-D-glucoside, p-nitrophenyl-β-D-glucoside, and salicyl-β-D-glucoside are all converted to β, D-glucose with β, glucosidase, which was further converted into H_2O_2 with β-glucose oxidase. The H_2O_2 produced was determined by the luminol–CL reaction. Under the HPLC conditions employed, although the mobile phase contained 30% acetonitrile, no remarkable influence on the activity of IMER was observed. The sensitivity of the method was 2 pmol on column as the detection limit (22).

The luminol CL reaction was also applied to immunoassays and DNA hybridization assays (23,24); by using microperoxidase and luminol as CL reagents, α-fetoprotein or ferritin as a tumor marker, thyroid-stimulating hormone (TSH), luteinizing hormone (LH), and follicle-stimulating hormone (FSH) were assayed. The HRP–luminol CL was utilized in hybridization assays of DNA on membrane, magnetic bead, polymer, etc. (25,26).

2.3. Isoluminol Derivatives as CL Reagents. Isoluminol has a higher luminescence efficiency than luminol. The 1-methoxy-5-methylphenadinium methylsulfate-isoluminol-microperoxidase CL reaction system has been utilized for the sensitive determination of 1-β- and 6-α-hydroxybile acids and bile acids in human urine samples. The detection limits of the method for bile acids were 8–250 pmol per assay. Among newborn babies, women, and women at a late pregnancy stage, the concentrations and conjugated forms of bile acids in their urine are markedly different, and the metabolic reaction of fetus and newborn babies are clearly different (27). A series of isoluminol derivatives has been synthesized for the chemiluminescent immunoassay (CLIA) by Schroeder and co-worker (28). These CL reagents have been applied to the assays of biotin and thyroxine and steroid hormones (estradiol, estriol, progesterone, etc). The CL reaction scheme is shown in Fig. 1. The labeling reaction of the analytes with these reagents needs condensation agents such as N,N'-disuccinimidyl carbonate (DSC), N-ethyl-N'-(3-dimethylaminopropyl)carbodiimide, and N,N'-dicyclohexylcarbodiimide. N-(4-Aminobutyl)-N-ethylisoluminol (ABEI) has been used as a precolumn labeling reagent in HPLC. For example, an unsaturated fatty acid, eicosapentaenoic acid, was labeled with ABEI using DSC as a condensing agent, and the derivative obtained was separated with a reversed-phase HPLC followed by CL detection with microperoxidase and H_2O_2 as postcolumn reagents. The determination range was 2 pmol–2 nmol on column with the detection limit being 200 fmol (29). Polyunsaturated fatty acids were determined by HPLC–CL using 6-[N-(4-aminobutyl)methylamino]-2,3-dihydro-1,4-phthaladinedione as a CL labeling reagent, EDC as a condensing agent, and a mixture of peroxidase–H_2O_2 as a postcolumn CL reagent. The DSC also can be used to condensate amino groups of reagent and analyte, which was used for the determination of amine compounds (30) and methamphetamine (31). N-(4-Aminohexyl)-N-ethylisoluminol (AHEI) was used for quantification of the N-terminal

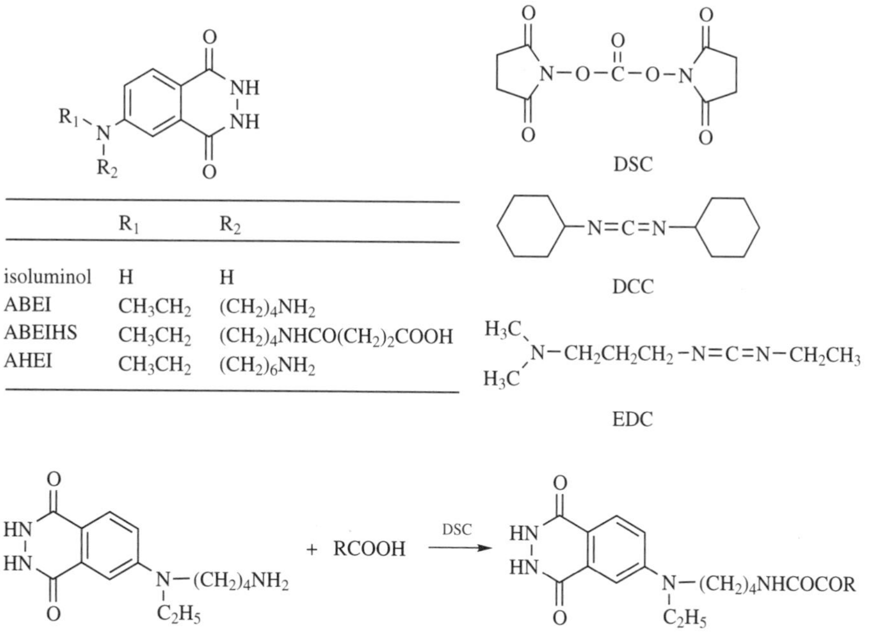

	R_1	R_2
isoluminol	H	H
ABEI	CH_3CH_2	$(CH_2)_4NH_2$
ABEIHS	CH_3CH_2	$(CH_2)_4NHCO(CH_2)_2COOH$
AHEI	CH_3CH_2	$(CH_2)_6NH_2$

Fig. 1. Structures of isoluminol derivatives, condensation agents, and derivatization reactions of ABEI with carboxylic acid.

amino acid of peptides; 500 amol of the derivative can be detected (32). The distinct characteristics of ABEI analogues are that their reactions can be performed under the mild conditions, ie, at room temperature in polar solvents.

2.4. Other Luminol Derivatives.   4-Isothiocyanatophthalhydrazide and 6-isothiocyanobenzo[g]phthalazine-1,4-(2H,3H)-dione (IPO) were used as highly sensitive CL labels of amino compounds. Twelve amino acids were sensitively determined by HPLC with 4-isothiocyanatophthalhydrazide and a detection limit of 10 fmol on column was acheived (33). The antidepressant maprotiline in plasma could be detected at a very low level of 0.1 ng/ml by HPLC with IPO (34). Primary amines such as n-hexylamine, n-butylamine, and n-octylamine were sensitively detected using IPO as a label over the range from 30 to 120 fmol at a S/N of 3. Secondary amines such as N-methyloctylamine, di-n-amylamine, di-n-hexylamine of at levels of 0.8–3 fmol (S/N = 3) were determined using the same approach (35). A new luminol-type reagent, 6-aminomethylphthalhydrazide (6-AMP), was developed for the selective CL labeling of 5-hydroxyindole derivatives. As shown in Fig. 2, 5-hydroxyindoles were selectively reacted with 6-AMP in the presence of potassium hexacyanoferrate(III) to yield highly CL compounds. These compounds were detected after separation with a reversed-phase column by a postcolumn reaction with H_2O_2 and potassium hexacyanoferrate(III) in sodium hydroxide solution. The detection limits were as low as 0.7–4 fmol on column (S/N = 3) (36).

The principle that analytes of interest can be determined after being converted to luminol derivatives, was successfully introduced to the measurement of some biological components. For instance, 4,5-diaminophthalhydrazide

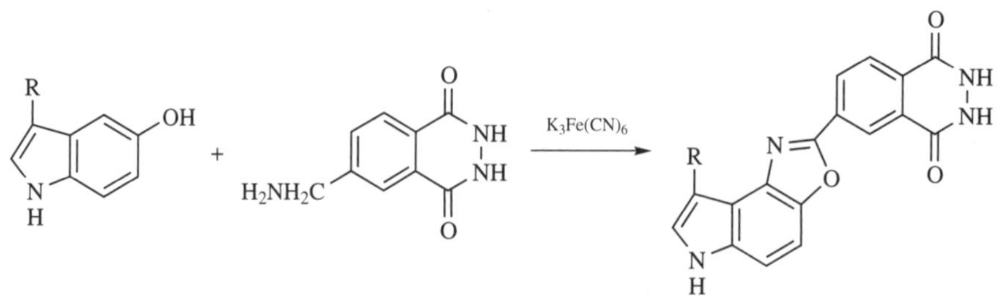

Fig. 2. Derivatization reaction of 5-hydroxyindole with 6-AMP.

(DPH) gives CL compounds by reaction with α-keto acids, which are important intermediates in the biosynthesis of amino acids, carboxylic acids, and sugars. Nine DPH derivatives of α-keto acids were separated by a reversed-phase column within 50 min followed by a postcolumn CL reaction with a mixture of H_2O_2 and potassium hexacyanoferrate(III) in sodium hydroxide solution; detection limits were in the range of 4–50 fmol on column. This method was applied to the assay of α-keto acids including phenylpyruvic acid; the detection limits were 9–92 pmol/mL plasma (S/N = 3) (37). The DPH was also applied to the quantification of N-acetylneuraminic acid derived from hydrolysis of sugar protein or sugar lipids in serum or urine; the detection limit was 9 fmol (S/N = 3) using as little as 10 μL of serum or 50 μL of urine (38). By modifying the reaction conditions, DPH can react with α-dicarboxylic acids to give CL quinoxaline derivatives. Phenylglyoxal, diacetyl, 2,3-pentanedione, 2,3-hexanedione and 3,4-hexanedione were examined and determined sensitively; the detection limits were 1.1–8.7 fmol except for 3,4-hexanedione whose limits were 300 fmol (39). 3α,5β-Tetrahydroaldosterone and dexamethasone have an α-keto moiety in their structure, which can be converted into a α-dicarbonyl moiety with copper(II) acetate and determined with HPLC–CL using DPH as a derivatizing reagent (40). 5-Amino-4-sulfanilphthalhydrazide was synthesized as a CL labeling reagent for aromatic aldehydes in HPLC. Benzaldehyde, 4-tolualdehyde, 4-chlorobenzaldehyde, 4-formylbenzoic acid, 4-hydroxybenzaldehyde, and vanillin were selected as model aromatic aldehydes. The detection limits of these compounds were in the range of 0.2–4.0 fmol on column (41). 4-(6,7-Dihydro-5,8-dioxothiazolo[4,5-g]phtalazin-2-yl)benzoic acid N-hydroxysuccinimide ester as a sensitive CL reagent for amines was developed. Model amines such as methyl-n-octylamine, n-nonylamine, and n-decylamine were derivatized with this reagent followed by separation with a reversed-phase column and CL detection. The detection limits at S/N = 3 were subfmol levels on column (42).

2.5. Utilization of Enhancers for Luminol CL Reaction. In spite of the continuous efforts by analytical chemists to develop useful CL reagents, only a few new ones have been successfully developed. Thus, trials to enhance the efficiency of known CL reagents, were introduced in CL analyses. In the luminol–peroxidase–H_2O_2 CL system, phenol compounds such as 4-iodophenol (43), aromatic compounds such as hydroxybenzothiazoles and dehydroluciferins (44,45), and phenylboronic acids (46) have been found to serve as strong enhancers. Competitive CL immunoassay of s-Triazine pesticides with horseradish

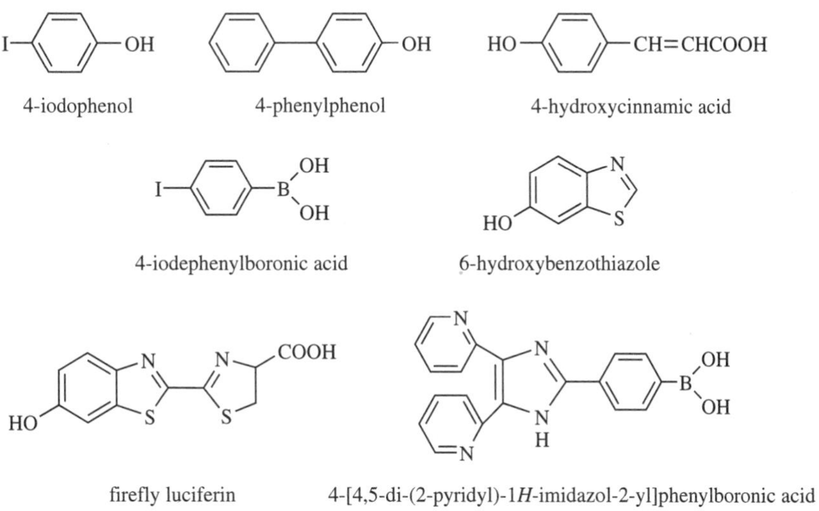

4-iodophenol   4-phenylphenol   4-hydroxycinnamic acid

4-iodephenylboronic acid   6-hydroxybenzothiazole

firefly luciferin   4-[4,5-di-(2-pyridyl)-1*H*-imidazol-2-yl]phenylboronic acid

Fig. 3. Representative enhancers for luminol CL.

peroxidase as a label was developed by using 4-iodophenol as an enhancer. *s*-Triazine classification of environmental samples containing various analyte mixtures was correct in 79–100% of all cases depending on the type of analyte (47). Representative enhancers are summarized in Fig. 3. In the CL system using glucose oxidase as a labeled enzyme to produce H_2O_2, the CL efficiency can be increased by this enhancing system. Although the mechanism has not been clarified in details yet, it is proposed as follows where HRP is used as the enzyme. First, HRP accepts oxygen from hydrogen peroxide to form HRP-I and HRP-II, which withdraw hydrogen from luminol to give luminol radical; in this system, the enhancer accelerates the radical formation. The resultant radical may convert into the diazaquinone intermediate that produces an aminophthalic acid dianion and an emission of light by the reaction with H_2O_2. Substrates having 5-bromo-4-chloro-3-indolyl group (BCI) were synthesized for alkaline phosphatase (ALP), which were combined with this enhanced luminol CL. For instance, hybridization assays using ALP as a marker enzyme have been developed (48).

3. Acridine Derivatives

Lucigenin (*N,N'*-dimethyl-9,9'-biacridinium dinitrate), a representative chemiluminogenic acridine derivative, is one of the oldest CL reagents and has been used for analytical purposes. Lucigenin reacts with H_2O_2 to yield an emission of light in a similar manner to that of luminol, and it also can produce emission with organic reductants such as reducing sugars. This finding is based on the reaction of lucigenin with the α-hydroxycarbonyl group of the organic reductants. Therefore, compounds bearing an α-hydroxycarbonyl group such as glyceraldehydes, cortisols, phenacylalcohols, or pehancylesters can be determined by lucigenin

CL. A FIA method was developed for the determination of glucuronides; herein, IMER of β-glucuronidase hydrolyzed glucuronides to glucuronic acid, which was detected by the lucigenin CL reaction. Phenyl, nitrophenyl, methylumbelliferil, bromonaphthyl, estradiolglucuronide, and androsterone glucuronides were determined with the lower detection limits of 5–10 μM. The lucigenin CL detection system can be applied to HPLC with an ion-exchange column, where the disturbing effects of biological substances on quantification could be omitted (49). Corticosteroids and p-nitrophenacyl esters of 0.5-pmol levels on column were quantified by HPLC with lucigenin CL (50). Furthermore, lucigenin was found to yield an emission of light by reaction with catecholamines. By utilizing this phenomenon, catecholamines were quantified over the range of 1×10^{-7}–1×10^{-4} M (51). The lucigenin CL was applied for the investigation of human seminal plasma, which has powerful antioxidant and immunosuppressive activities (52). A sensitive and simple CL assay for alkaline phosphatase (ALP) was developed by using dihydroxyacetone phosphate or its ketal as a substrate. Both substrates were transformed to dihydroxyacetone by hydrolysis with ALP, followed by a CL reaction with lucigenin with the detection limits of 3.8×10^{-19} and 1.5×10^{-18} mol of ALP, respectively (53). Quantification of epinephrine with lucigenin CL was achieved by an FIA system fabricated on a microchip. Two types of flow cells were used; (1) two streams entered through separate inlet ports and merged to flow adjacently, and (2) a lucigenin solution-containing epinephrine was split up to 36 partial flows by passage through the nozzles, and was injected into the alkaline solution. The CL intensity in cell 2 was six-times larger than that in cell 1, with the detection limit of 8.0×10^{-7} M (54).

Acridinium esters or acid chlorides were developed as CL labeling reagents to be applied to immunoassays and HPLC of trace amounts of biological components. Acridinium labeled compounds have 100 times stronger CL intensity compared to luminol labeled ones, and acridinium esters have the dominant feature that they do not loss the CL efficiency even after binding to antigen or antibody. Acridinium esters, for instance, were applied to compensative DNA probe (cDNA) in hybridization assay (55), human gonadotropin and TSH (56), virus, and immunoassay of a cancer marker. By using a commercially available kit (TSA), as low as 0.01 mU/mL of TSH can be determined. This sensitivity makes it possible to clearly distinguish the TSH levels in normal human blood (0.53–3.05 μU) and Basedow's disease patients' blood (<0.10 μU) (57). 10-Methyl-9-acridinium carboxylate was used for HPLC–CL determination of the important environmental pollutants chlorophenols; the detection limits were very low of 300 amol–1.25 fmol (S/N = 3) (58). Several activated derivatives of 9-acridinecarboxylic acids were synthesized and their utility for H_2O_2 detection was investigated. Among these, 9-acridinecarbonylimidazole was found as a useful reagent for measuring the activity of a number of enzymes that directly produce peroxide, including glucose oxidase, with the detection limits in the 1–10-amol range (59). Brain natriuretic peptide (BNP) is a potent vasodilator and natriuretic factor regulating salt and water homeostasis. A nonradioactive sensitive and specific assay for N-terminus of the precursor of BNP was developed using 4-(2-succinimidyl-oxycarbonylethyl)phenyl-10-methylacridium 9-carboxylate fluorosulphonate as a CL label (60).

4. Aryloxalate Derivatives

4.1. Aryloxalates and Fluorescent Compounds as Chemiluminescent Reagents.

Chemiluminescent produced by the reaction of aryloxalate (or oxamides), H_2O_2, and fluorescent compounds is known as peroxyoxalate chemiluminescence (PO–CL). Many kinds of aryloxalates have been synthesized and evaluated as PO–CL reagents. Among these, bis(2,4,6-trichlorophenyl)oxalate (TCPO) is the most popular. The TCPO can be easily synthesized and is inexpensive. The solubility of this compound is low in organic solvents such as acetonitrile and methanol commonly used in HPLC, which may limit its use in some cases. Meanwhile, bis[2-(3,6,9-trioxadecyloxycarbonyl)-4-phenyl]oxalate (TDPO) was found to be useful because of its higher CL efficiency and greater solubility in organic solvents compared to TCPO. In PO–CL, fluorescent compounds having large fluorescent quantum yields and lower oxidative electric potentials to yield efficient CL intensities are preferable. Representative aryloxalates are presented in Fig. 4.

4.2. Determination of Fluorescent Compounds.

Generally, highly fluorescent compounds, which should be suitable to PO–CL detection, are rare among biologically active components, environmental pollutants, etc. Therefore, derivatization of non-fluorescent or weakly fluorescent compounds to yield highly fluorescent ones is commonly adopted practice in PO-CL. Many known fluorescence derivatization methods can be used to this aim. Dipyridamole, a blood platelet improving agent, and benzydamine, an antiinflammatory agent, in rat plasma samples were sensitively determined by PO–CL without derivatization, because they have strong native fluorescence in themselves (61). Phenothiazine antipsychotic agents are also determined without derivatization (62). Many aromatic compounds such as air pollutants have strong native fluorescence, and

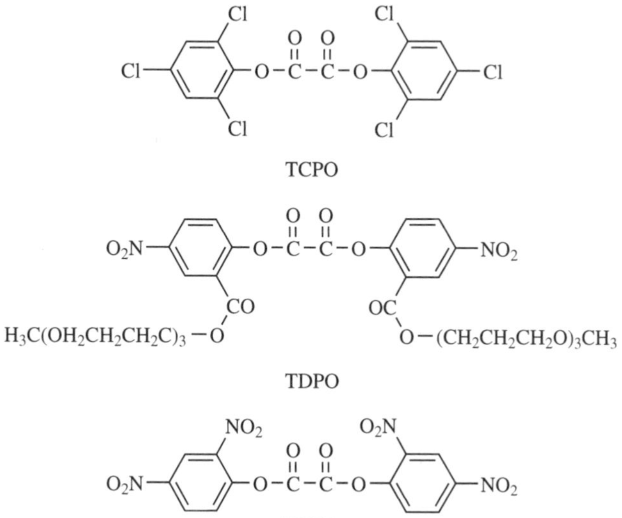

TCPO

TDPO

DNPO

Fig. 4. Representative aryloxalates are TCPO, TDPO, and bis(2,4-dinitrophenyl)oxalate (DNPO).

thus, can also be determined with PO–CL without a derivatization step. A FIA method for determining perylene, anthracene, and pyrene was developed using TCPO and H_2O_2 in acetonitrile as CL reagents with the detection limits of 0.05, 65, and 75 pg (S/N = 5), respectively (63). Nitroarenes derived from diesel engine exhaust are known as carcinogens. In spite of being non-fluorescence, nitroarenes can be determined after reduction to highly fluorescent aminoarens. Air pollutants such as polycyclic aromatic hydrocarbons (PAHs) and their nitrated derivatives were determined over a period of 12 months by HPLC–PO–CL. The toxic equivalency factors adjusted concentration of total PAHs determined was 2.33 ng/m^3 in Nagasaki city area (64). Fluorescent compounds having their absorption wavelengths in the infrared (ir) region are preferable as fluorophores, because they can be excited with low excitation energy that generally the PO–CL reaction can produce. Methylene blue, pyridine 1, oxazine 1, and 3,3'-diethylthiadicarbocyanine iodide (DTDCI) were evaluated as fluorophores in HPLC–PO–CL, and were sensitively detected with the detection limits of 120, 27, 31, and 0.19 fmol on column, respectively (65). Among these, DTDCI is a potential candidate as a lead compound for preparing fluorophores in PO–CL.

A photographic method was devised for the detection of FL compounds and H_2O_2. Pyrimido[5,4-d]pyrimidine derivatives and rhodamine B can be detected as colored spots on an instant photographic film. Glucose in serum was semi-quantified by the method (66). The CL immunoassay of antihuman T-cell virus antibody was developed by a photographic method (67).

4.3. Determination of Amino Compounds. Dansyl chloride (Dns-Cl) was first used for FL labeling of amino acids in thin-layer chromatography (TLC) (68). Excitation energy for Dns-derivatives is usually covered with that of PO–CL reaction. Sensitivities for determining several kinds of primary alkyl amines in PO–CL were compared using some labeling reagents, ie, Dns-Cl, 4-chloro-7-nitrobenzo-1,2,5-oxazole (NBD-Cl) and o-phtalaldehyde (OPA). The detection limits of DNS-, NBD- and OPA-amines were 0.8–14, 19–270, and 94–580 fmol, respectively (69). Dns-amino acids were also quantified by HPCL–PO–CL using TCPO as an oxalate reagent with the detection limit of 10 fmol (70). Sixteen kinds of Dns-amino acids were separated within 30 min by a gradient elution, and detected over the range from 2 to 5 fmol (S/N = 2). By applying a microbore column in this method, as low as 0.2 fmol of dansylated Ala, Val, Ileu, and Phe could be detected (71).

Furthermore, Dns derivatization was applied to quantifiy abuse drugs (amphetamines) in urine samples with the detection limit of 2×10^{-10} M (72). Methamphetamines in a single hair sample were detected at very low level of 20 pg (73).

Mexiletine, an antiarrhythmic drug, was assayed in rat plasma using TDPO as an oxalate over the calibration range of 20–100 ng/mL with the detection limit of 1.0 fmol on column (74). A highly sensitive HPLC–PO–CL method for catecholamines, which act as neurotransmitters, was developed using fluorescemine as a label with the detection limit of 25 fmol (75). The sensitivity is 20-times higher than that of FL detection and only 10 μL of urine sample were needed for quantification. Meanwhile, it was reported that FL detection was superior than CL detection in HPLC for histamine. Fluorescamine derivatives of histamine can be detected with FL 100 times more sensitively than

CL (76). This large difference may be due to the limitation of solvents to separate fluorescamine derivatives in HPLC–PO–CL, where solvents seriously affect the CL efficiency. Fluorescamine has also been utilized to quantify sulfamethazine (an antibacterial drug) in chicken eggs and serum; 1 ng/mL of standard sulfamethazine can be detected (77). 4-(N,N-Dimethylaminosulfonyl)-7-fluoro-2,1,3-benzoxadiazole (DBD-F) is non-fluorescent, but gives strong fluorescent derivatives by the reaction with primary and secondary amines. Methamphetamine and its related compounds were simultaneously assayed by HPLC with TDPO and H_2O_2 as the postcolumn CL reagents. Six DBD-derivatives were separated by a reversed-phase column with a gradient elution, and their CL intensities were monitored over the range from 25 to 133 fmol (S/N = 3) by using DBD-N-ethylbenzylamine as an internal standard (78). The DBD-F was also utilized for determining metoprolol in serum (79). Naphthalene-2,3-dicarboxaldehyde (NDA) reacts with primary amines to yield fluorescent cyanobenz[f]isoindole (CBI) derivatives (80). By utilizing this labeling reaction, dopamine and norepinephrine in urine were determined at sub–low femtomole levels with a sample size of only 20 μL (81). Fluvoxamine, an antidepressant, was also determined with the same manner (82). Ethylenediamine was applied for the on-line postcolumn derivatization approach of catecholamines following PO–CL detection with TDPO and H_2O_2 as the postcolumn reagents. The method is very sensitive and allows the detection of catecholamines as low as 1 fmol on column (S/N = 2). Furthermore, the method was applied to develop an automatic analysis of catecholamines in rat plasma (83). The automatic analytical method was applied to determining (–)-isoproterenol and (–)-(R)-1-(3,4-dihydroxyphenyl)-2-[(3,4-dimethoxyphenethyl)amino]ethanol in rat plasma with the detection limits of 1.3 and 0.9 fmol on column, respectively (84). 1,2-Diarylethylenediamine derivatives can also be used for precolumn labeling of catecholamines. The labeled compounds were separated on a reversed-phase column and detected by PO–CL with TDPO and H_2O_2 as CL reagents; amol levels of catecholamines could be detected at S/N = 3 (85). Luminarin 1 bearing quinolizino coumarin skeleton as a fluorophore and N-hydroxysuccinimide ester as a reactive group, reacts with amines to give highly fluorescent compounds. Luminarin derivatives of pentylamine, pyrrolidine, thiamine, and proline were separated on a reversed-phase column and detected by PO–CL with the detection limits of 15–100 fmol (S/N = 3), which were 3–10 times more sensitive than those obtained by fluorescence detection (86). Representative derivatization reagents for amines are summarized in Fig. 5.

    Sensitizing effects of polyamines on CL reaction was utilized to their sensitive determination in tomatoes. A FIA method was developed for determining 55 kinds of amines by using TDPO and sulforhodamine as CL reagents; aliphatic amines of $\sim 1.2 \times 10^{-8}$ M and polyamines of $\sim 7 \times 10^{-10}$ M were detected. Histamine in fish meat samples was determined by the same method (87).

    4.4. Carbonyl   Compounds.   5-N,N'-Dimethylaminonaphthalene-1-sulfonohydrazide (Dns-H) was used as a labeling reagent for carbonyl compounds. Oxo-steroids and oxo-bile acid ethyl esters in serum samples were derivatized to Dns-hydrazones, which were purified by gel-permeation chromatography, separated on a reversed-phase columns, and detected with TDPO–PO–CL. Corticosterone, testosterone, and progesterone were quantified with the detection limits of 3, 2, and 4 fmol at S/N = 2, respectively (88). 7α-Hydroxy-3-oxo-5β-cholanic

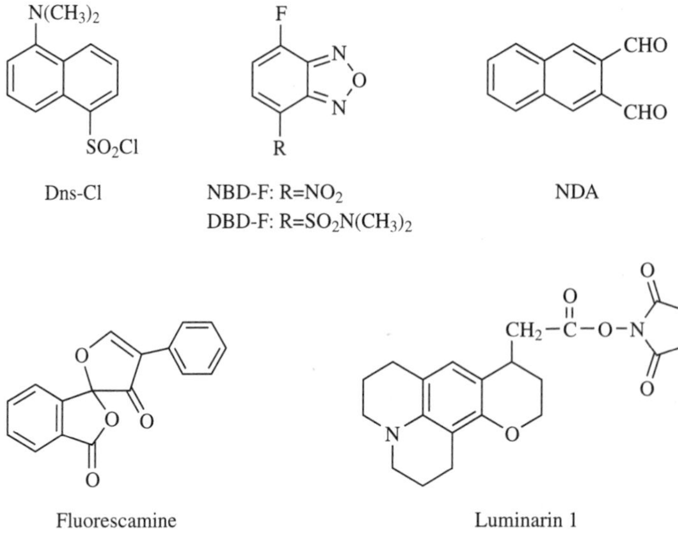

Dns-Cl   NBD-F: R=NO$_2$ NDA
           DBD-F: R=SO$_2$N(CH$_3$)$_2$

Fluorescamine   Luminarin 1

Fig. 5. Labeling reagents for amines in CL detection.

acid, an unusual oxo-bile acid, in cholestatic hepatic disease patient's urine sample was determined at nmol/L level. 3α- or 3β-Hydroxysteroide, 3β-hydroxy-5-cholenic acid, pregnanediol, 5-pregnene-3β,20β-diol, 5-pregnene-3β,20α-diol were converted into 3-oxo-steroide with the IMER of hydroxysteroidedehydrogenase, which was derivatized with Dns-H followed by reversed-phase separation and CL detection at a few femtomole level (89). Dns-H was also applied to the labeling of sugars; hyaluronic acid, chondroitin sulfate, dermatan sulfate were determined with the detection limit of 100 fmol (S/N = 3). Hyaluronic acid was converted with hyaluronidase SD to an unsaturated disaccharide, ΔdiHA, and labeled with Dns-H to give a FL compound. Disaccharides in rat's mast cells can be determined (90). Carbonyl compounds in air such as formaldehyde, acetaldehyde, and acetone were determined with HPLC–PO–CL after labeling with Dns-H (91). These compounds were derivatized by drawing an air sample through a small glass cartridge packed with porous glass particles impregnated with Dns-H; this method is very simple and can detect amounts at sub-ppb concentrations of carbonyl compounds. 4-(N,N-Dimethylaminosulfonyl)-7-hydrazino-2,1,3-benzoxadiazole (DBD-H) showing very weak FL, reacts with carbonyl compounds to yield strongly FL derivatives. Medroxyprogesterone acetate, a synthetic progesterone, reacts with DBD-H to form a fluorescence compound, whose determination in serum was performed in the calibration range of 15.6–96.6 ng/mL using 100 µL of sample (92). Propentofylline, which has a nervous cell protecting effect, was quantified by TDPO–PO–CL with 0.031 ng on column (93). Microdialysate obtained from rat hippocampus was successfully applied to this study. 3-Aminofluoranthene was applicable to labeling of aldehydes and ketones (94). Malondialdehyde is well known as a degradation product of polyunsaturated lipids, and has been used as a marker of biological lipid-peroxidation. 1,3-Diphenyl-2-thiobarbituric acid (DPTBA) and 1,3-diethyl-2-thiobarbituric acid (DETBA) were proposed as useful labeling reagent candidates

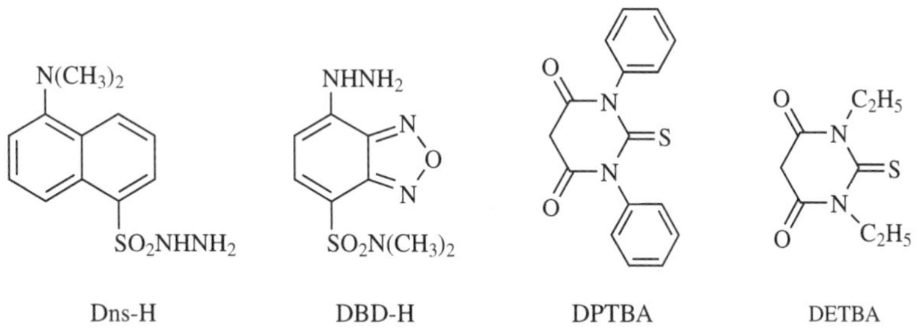

Fig. 6. Labeling reagents for carbonyl compounds in CL detection.

for malondialdehyde (MDA). By using DETBA, MDA in fractions of rat brain tissues was first determined with $TCPO-H_2O_2$ CL (95). Labeling reagents for carbonyl compounds used in PO–CL are summarized in Fig. 6.

4.5. Determination of Carboxylic Acids. Coumarin derivatives, ie, 4-(bromomethyl)-7-methoxycoumarin (Br-Mmc), 7-(diethylamino)cumarin-3-carbohydrazide (DCCH), 7-(diethylamino)-3-[(4-iodoacetylamino)phenyl]-4-methylcumarine (DCIA), have been used for labeling of carboxylic acids. These reagents were evaluated as PO–CL using a $TCPO-H_2O_2$ CL reagent. However, Br–Mmc was found to give non-CL derivatives, because the energy of the PO–CL system used was not enough to excite them. Meanwhile, DCIA derivatives of linear-chain fatty acids showed good CL with low femtomole levels of detection limits (96). N-(Bromoacetyl)-N'-[5-(dimethylamino)naphthalene-1-sulfonyl]piperadine (Dns-BAP) was developed as a labeling reagent for carboxylic acids, and applied to their HPLC–PO–CL assays. For instance, carboxylic acids were derivatized in aprotic solvents at 55°C for 30 min under coexistence of 18-crown-6 and potassium bicarbonate, followed by separation on a reversed-phase column and detection with bis(2-nitrophenyl)oxalate and H_2O_2 as CL reagents. Retinein acid, a precursor of vitamin A, was also detected at 25 fmol (97). 6,7-Dimethoxy-1-methyl-2(1H)-qinoxalinone-3-propionyl carboxylic acid hydrazide was applied to determine metabolites of arachidic acids with EDC and DCC as condensing reagents. The derivatives were determined by HPLC with $TCPO-H_2O_2$ as CL reagents at 500 amol (S/N = 3) (98). 2-(4-Hydrazinocarbonylphenyl)-4,5-diphenyl-imidazole (HCPI) can be applied to fluorescent labeling of fatty acids. The HCPI derivatives of fatty acids in human serum were detected by HPLC–PO–CL over the range of 12–18 fmol (S/N = 2) (99). 7-(Diethylamino)coumarin-3-carbohydrazide and luminarin 4 can also be applied to determine sensitively fatty acids at femtomol levels (100).

Derivatizing reagents for carboxylic acids used in PO–CL are shown in Fig. 7.

4.6. Determination of Thiols, Alcohols, Phenols and Others. Usable fluorescent derivatizing reagents for thiols and phenols are limited. N-[4-(6-Dimethylamino-2-benzofuranyl)phenyl]maleimide (DBPM), a maleimide-type fluorescent reagent, was used for thiol compounds. The DBPM as well as other maleimide-type fluorescent labeling reagents is natively non-fluorescent, but gives strong fluorescent products upon reaction with thiol compounds. The DBPM derivatives of biologically important thiols such as glutathione, cysteine, N-acetylcysteine, cysteamine, and antirhumatide D-penicillamine were separated

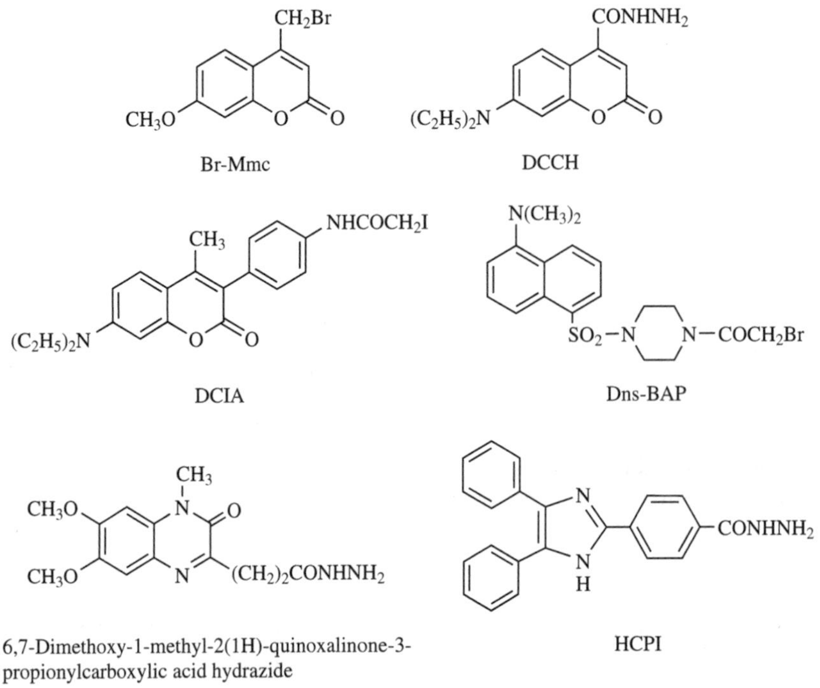

Fig. 7. Labeling reagents for carboxylic acids in CL.

on a reversed-phase column and detected by PO–CL using TCPO and H_2O_2 as the postcolumn reagents. The derivatives were separated within 12 min, and detected in the range from 7 to 113 fmol (S/N = 2) on column (101). The method was applied to the analysis of glutathione and cysteine in rat liver tissue.

Normal phase separation is preferable to a PO–CL detection system, because PO–CL can yield a stronger emission of light in nonaqueous solvents than in aqueous solutions. However, due to the limited usability of normal-phase separation for biological samples, only a few reports have appeared. Estradiol (one of the major estrogens secreted by the human ovary) in serum was determined after extraction by a solid-phase column followed by Dns-derivatization and separation on a normal-phase column. The method can detect as low as 50 pg of estradiol (102). Estradiol in plasma sample was also determined by HPLC–PO–CL combined with liquid–liquid extraction using ethyl acetate. The detection limit of Dns–estradiol was 15 fmol (4 pg) in the standard solution and 44 fmol (12 pg) in the rat plasma (103).

Phenols labeled with lissamine rhodamine B sulfonyl chloride show high quantum yield and emission at long wavelength (>550 nm), which permits a considerable reduction of the background emission. The derivatization can be performed under mild conditions, ie, at room temperature for 1 min. Lissamine Rhodamine B derivatives of chlorophenol was detected in the low picogram range after normal or reversed-phase column separation following PO–CL detection. Pentachlorophenol in river water could be detected as low as 0.8 ppb without pretreatment (104). DBPM and lissamin rhodamine B sulfonyl chloride are shown in Fig. 8.

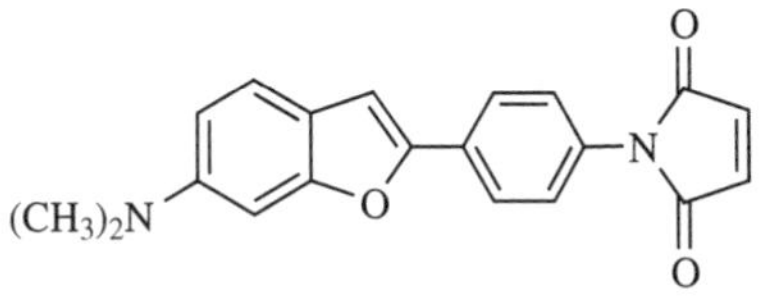

Fig. 8. N-[4-(6-dimethylamino-2-benzofuranyl)phenyl]maleimide (DBMP) and lissamin rhodamine B sulfonyl chloride.

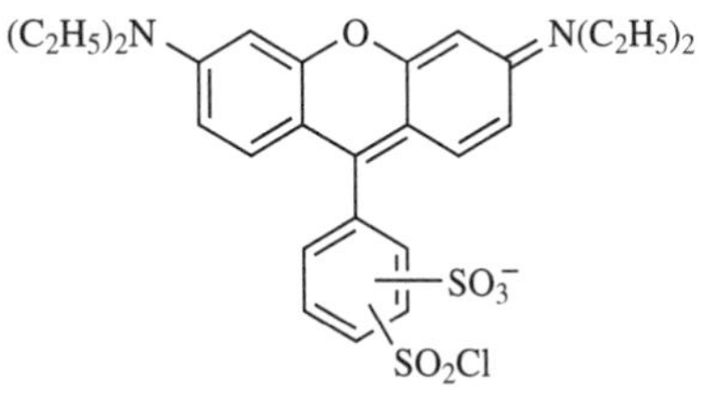

The DCIA, a labeling reagent for carboxylic acid, can be also applied to fluorescent labeling of fluoropyrimidines anticancer drugs using a crown ether–potassium complex as a catalyst. The DCIA derivatives of 5-fluorouracil, 5-fluorouridine, 5-fluoro-5′-deoxyuridine, and 1-(tetrahydro-2-furanyl)-5-fluoro-uracil were detected at several tens of a femtomole level (105,106). Metal ions can be determined after formation of chelates with 8-hydroxyquinoline followed by FIA or HPLC with PO–CL detection. The ions Al(III), Zn(II), Cd(II) and In(III) were determined with the detection limit from 20 to 70 ppb (107).

PO–CL can also be applied to the detection of oxalic acid and porphyrins (108,109). Oxalic acid emits light in ethanol solution by the reaction with H_2O_2, N,N'-dicyclohexylcarbodiimide, and diphenylanthracene at pH 1. Oxalic acid in serum was also quantified. Enhancement of this PO–CL by porphyrins was applied to their determination in urine.

A capillary electrophoresis (CE) with PO–CL detection was developed for Dns–amino acids, which is 35 times more sensitive than ultraviolet (uv) detection (110). Bovine serum albumin forms complexes with eosin and rose bengal, which can be separated by CE. These complexes were analyzed after separation by CE followed by PO–CL detection; albumin, from 5×10^{-7} to 1×10^{-4} M, can be determined with the detection limit of 2×10^{-7} M (4 fmol) (111). The sensitivity of the method was further improved to be 1.7 fmol (112). The CE, thus, will become a useful tool for analysis of macromolecular biological compounds (113–115).

4.7. Determination of Hydrogen Peroxide and Substrates that Produce Hydrogen Peroxide in Enzyme Reactions. Hydrogen peroxide and substrates from which H_2O_2 is produced in enzymatic oxidation can be sensitively determined by using a PO–CL system. Several enzymatic oxidation reactions are summarized in Fig. 9. For these aims, batch, FIA, HPLC, photographic methods, etc, have been developed. For the determination of H_2O_2 by PO–CL, pyrimido[5,4-d]pyrimidines were synthesized and evaluated. Among them, 2,4,6,8-tetrathiomorphorinopyrimido[5,4-d]pyrimidine (TMP) was found to give the largest CL (116). Structure of TMP is given in Fig. 10. As low as 1×10^{-8} M of H_2O_2 was quantified by a batch method with TCPO and TMP as reagents (117). The H_2O_2 was first determined by HPLC after separation on reversed-phase column followed by CL detection with TCPO and TMP as reagents with the detection limit of 188 fmol (9.9×10^{-9} M) at S/N = 3. Hydrogen peroxide in

$$\text{Uric acid} + O_2 \xrightarrow{\text{Uricase}} \text{Allantoin} + CO_2 + H_2O_2$$

$$\text{Choline} + O_2 \xrightarrow{\text{Choline oxidase}} \text{Betain aldehyde} + H_2O_2$$

$$\text{Cholesterol} + O_2 \xrightarrow{\text{Cholesterol oxidase}} \text{4-Cholesten-3-one} + H_2O_2$$

$$\text{Xanthine} + O_2 \xrightarrow{\text{Xanthine oxidase}} \text{Uric acid} + H_2O_2$$

$$\text{Glucose} + O_2 \xrightarrow{\text{Glucose oxidase}} \text{Gluconolactone} + H_2O_2$$

Fig. 9. Representative enzyme reactions that produce hydrogen peroxide.

coke drinks and commercially available hydroperoxide reagents were success-fully determined by this method (118). Rhodamine derivatives can increase the PO–CL efficiency. A FIA method using sulforhodamine 101 as a sensitizer can detect as low as $3 \times 10^{-9}\ M$ H_2O_2 with TDPO as an oxalate (119). A fluorescent reactor of 3-aminofluoranthene immobilized on pore glass beads was used in FIA, and H_2O_2 could be detected at $1 \times 10^{-8}\ M$ (120).

An IMER can be used as an online reagent for the determination of sub-strates in FIA or HPLC system. However, several conditions should be optimized to obtain a satisfactory CL yield, ie, nature, pH, and flow rate of carrier, reaction temperature, catalysts, CL reagents, organic solvents, etc. Eight kinds of L-amino acids were enantio-selectively determined by using an IMER of L-amino acid oxidase in urine and beer (121). The IMERs of glucose oxidase and uricase were applied for quantification of glucose and uric acid, respectively. These methods are simple and can be applied to human serum without pretreat-ment except a simple dilution. An IMER immobilized with acetylcholine esterase and choline oxidase was applied to determine simultaneously acetylcholine, an important neurotransmitter, and choline. The determination range is from

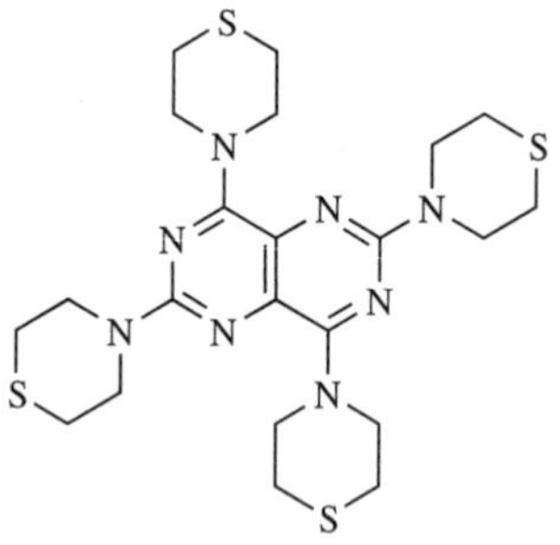

TMP

Fig. 10. Structure of TMP.

10 pmol to 10 nmol on column (122). Phospholipids containing choline were determined by a hyphenated HPLC–FIA method. Phospholipids are highly lipophilic compounds, and thus, a special column such as an aminopropyl-modified silica gel column is required for their separation. Several kinds of choline containing phospholipids were separated and fractionated. Each fraction was dried *in vacuo* and redissolved in Triton-X aqueous solution, and the resulted solution was injected into FIA–PO–CL system equipped with the IMER of phospholipase D and choline oxidase. Choline-containing phospholipids in human serum can be determined by this method (123). Polyamines are regarded as cancer markers, which can be determined by HPLC–PO–CL. After separation on a reversed-phase column, polyamines were converted into H_2O_2 by the IMER of polyamine oxidase and putrescine oxidase. Putrescine was detected with DNPO and 8-anilinonaphthalene sulfonic acid as CL reagents with the detection limit of 5 pmol (124). In the same manner, polyamines and their monoacetylated derivatives can be determined by using the IMER of acetylpolyamine amidohydrolase and putrescine oxidase with TCPO and TMP as CL reagents. The method was applied to determine polyamines in potatoes (125). Pyruvic acid can be converted to H_2O_2 by the IMER of pyruvic acid oxidase, which was determined by PO–CL with TCPO and perylene as the postcolumn reagents. Pyruvic acid in serum can be quantified over the range of 1×10^{-6}–1×10^{-2} M by this method.

5. Dioxetane Derivatives

Adamantyl dioxetanes have been used as analytical reagents of CL. 3-(2′-Spiroadamantane)-4-methoxy-4-(3″-phosphoryloxy)-1,2-Dioxetane (AMPPD) emits of light by the decomposition with alkaline phosphatase (ALP). By utilizing this reaction, as low as sub-attomol levels of ALP were assayed (126). The AMPPD can be successfully applied for enzyme immunoassay with LAP as a label by an automatic analyzer. In the same manner, 3-(2-spiroadamantane)-4-methoxy-4-(3″-β-D-galactopyranosyl-oxyphenyl)-1,2-dioxetane (AMPGD) is decomposed with β-D-galactosidase (β-Gal) to emit light, by which low levels of β-Gal can be determined. Several kinds of enzymatic substrates having adamantly dioxetane as an emitter have been developed for β-glucuronidase, β-glucosidase, and phospholipase. Owing to the very long lifetime of CL produced by adamantly dioxetanes, they are recommended to photographic detection. Reverse transcriptase (RT) assays are very important for the detection of retroviruses including human immunodeficiency virus (HIV) and for the development of new antiretroviral substances. A highly sensitive RT assay was developed by using 0.25 mM disodium 3-(4-methoxyspiro[1,2-dioxetane-3,2′-(5′-chloro)-tricyclo[3.3.1.1.3,7]decan]-4-yl)phenyl phosphate (CSPD) and 0.1 mg/mL poly(benzyltributyl)ammonium chloride in 1 M diethanolamine buffer (pH 9.5) containing 1 mM MgCl$_2$ as a CL reagent and enhancer solution. The assay can detect the RT activity in serum, plasma, and cell culture medium directly without preconcentration or extraction of enzyme (127). A sensitive CL enzyme-linked immunosorbent assay (ELISA) for the bioanalysis of carboxy-terminal B-chain analogues of human insulin was developed using only 10 μL of serum sample (128).

6. Lophine and Indole Derivatives

Lophine (2,4,5-triphenylimidazole) is the first synthetic CL compound. However, its analytical applications are relatively rare. Lophine produces an emission of light by the reaction with H_2O_2, which is sensitized by some ions such as Co(II), ClO^-, Cr(III), and Cu(II). These ions, thus, can be determined by measuring the CL produced at the concentration of $\sim10^{-6}\,M$. Hydroxylamine was found to enhance the CL of lophines-H_2O_2-Co(II) reaction, and by a FIA method with this enhancing reaction, CL intensities of 18 kinds of lophine derivatives were evaluated (129). Furthermore, this enhancement was utilized for quantification of Co(II) with the determination limit of $4.5 \times 10^{-8}\,M$ (130). Lophines show very strong FL, and thus they have been applied to FL detection rather than CL detection.

Indole derivatives have been much studied on their CL mechanisms because their structures are well correlated with luciferin of *Cypridina hilgendorfi*, but studies on their analytical characteristics are rather few. Indoxyl β-D-galactose was synthesized as a substrate of β-Gal, which allowed the assay of as low as 3 amol of β-Gal (131). Superoxide-triggered CL of *Cypridina* luciferin analogue, 2-methyl-6-phenyl-3,7-dehydroimidazo[1,2-α]pyrazin-3-one, is enhanced by nonionic detergents, such as Tween 20, Tween 80, and Triton X-100. By utilizing this detergent-amplified CL reaction, xanthine and xanthine oxidase were determined as low as 5 nmol/L and 3.85×10^{-7} U/mL, respectively (132).

7. Ruthenium(II) Complex

Tris(2,2′-bipyridyl)ruthenium(II) complex (bpy = 2,2′-bipyridyl) is a strongly fluorescent compound, which emits at 620 nm with excitation of 450 nm. As shown in Fig. 11, CL is produced by the oxidation of ruthenium(II) complex and reducing compound on the surface of the electrode followed by the excitation of oxidized ruthenium(III) complex with a radical derived from reducing compound. Therefore, reducing compounds can be determined after HPLC separation and following oxidation. Primary amines yield little CL on this system, but secondary and tertiary amines, especially cyclic aliphatic amines can

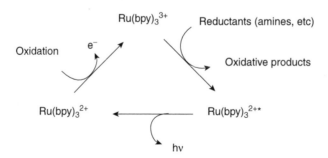

Fig. 11. Electrogenerated chemiluminescence (ECL) reaction of ruthenium(II) complex with reductants.

produce strong CL. The CL intensities of aliphatic amines are ordered as follows: tertiary > secondary > primary, and thus, many kinds of medicines containing a tertiary aliphatic amine moiety were determined by this system. Primary amines can be sensitively determined after conversion into tertiary amines; propylamine and 3-aminopentane were determined after conversion to the corresponding cyclic tertiary amines with divinyl sulfone followed by reversed-phase separation and ECL detection with the detection limits of 30 pmol and 1 pmol on column, respectively. In the same manner, a simultaneous determination of primary, secondary, and tertiary amines was performed; after an HPLC separation, primary and secondary amines were converted to tertiary amines with acrylonitrile and detected by ECL (133). A new electrogenerated chemiluminescent immunoassay (ECLIA) was developed for the determination of cytokeratin 19 (CYFRA 21-1), a marker of epithelial differentiation, in serum and urine samples. The clinical value of urinary CYFRA 21-1 for the detection of bladder cancer was evaluated (134). D- and L-Tryptophan were determined after a ligand-exchange HPLC followed by ECL detection with 0.2 pmol on column (S/N = 2) (135). Dns-amino acids were also determined by ruthenium(II) complex CL; Dns–Glu was quantified with the detection limit of 0.1 μM (2 pmol, S/N = 2), which is 1000 times more sensitive than that of nonlabeled Glu (136). Oxalic acid in urine and blood samples was assayed after a separation with a reversed-phase ion-paired column followed by ECL detection with the detection limit of 1 nmol/mL (25 pmol on column) (137). By directly adding ruthenium(II) complex to the mobile phase, the detection limit can be improved to 0.1 μM, which was applied to the assay of erythromycin in urine and plasma samples with the detection limit of 50 fmol on column (S/N = 3) (138). Similarly, a new type of erythromycins, de(N-methyl)-N-ethyl-8,9-anhydroerythromycin A 6,9-hemiacetal and its three metabolites in plasma and urine samples were quantified with the determination limits of 1 and 10 ng/mL, respectively (139). Glyphosate, a herbicide, and its structural analogues were determined after HPLC separation and ECL detection, and their CL intensities were ordered as follows: glycine < diethanol amine < hydroxyethyl glycine < iminodiacetic acid < glyphosate. The detection limit of glyphosate was 0.01 μM (140). Yohimbine (an indole alkaloid with α_2-adrenergic blocker) in human serum was separated with an ion-paired column and determined by ECL with the detection limit of 30 ng/mL (S/N = 3) (141). Tris(1,10-phenanthroline)ruthenium(II) complex (phen = 1,10-phenanthroline) was used as a CL reagent and applied to determine oxalic acid in tea (142).

Labeling reagent was prepared by introducing N-hydroxysuccinimide into ruthenium(II) complex, which can be used for ECL immunoassay of antigens, antibodies, and nucleic acids. An automatic analyzer using ECL immunoassay is commercially available, and utilized for clinical analysis of markers of hepatic cancer or infectious diseases.

8. Others

In spite of its long history, gallic acid has a few analytical applications. Recently, catechin and (−)-epigallocatechin 3-gallate, which are focused as a radical quencher, were found to yield CL by the reaction with acetaldehyde, horseradish peroxidase and H_2O_2. Catechin in rat and human plasma samples was assayed by this method (143). Pyrogallol is also known as a potential CL compound. Carix[4]allenes having pyrogallol as a structural unit were synthesized, and their CL properties were examined (144). The analytical applications of these compounds are anticipated.

Adenine and guanidine compounds can produce an emission of light by reaction with 9,10-phenanthrenequinone, which could be utilized for HPLC–CL of guanidino compounds (145). As well as adenine and its nucleosides and nucleotides (146,147), guanine and its nucleosides and nucleotides (148) react with glyoxal derivatives to give CL compounds, which can intensely emit under alkaline conditions in aprotic solvents such as dimethylformamide (DMF). For instance, adenyl compounds react with dimethylglyoxal dimethylacetal to give strong CL in the coexisting of tungustic acid and propanol; as low as 3×10^{-8} M of adenine, which corresponds to a few nanogram of DNA can be detected (149). 3′,4′,5′-Trimethoxyphenylglyoxal can yield intense CL by the reaction with DNA or guanidine nucleotide. The DNA adsorbed on a nylon membrane was sensitively detected at zmol by a CCD camera at $-25°C$ (150).

The CL of nitrogen monoxide (NO) occurs with the reaction of ozone in the range of 600–3200 nm. Nitrogen monoxide produced from NO_2^- by the reduction with potassium iodide was analyzed by FIA or HPLC with the detection limit of 0.1 ng (151). After nitrosation of glyphosate in grain, the resultant N-nitrosoglyphosate was separated by an ion-exchange column, and NO derived by a denitrosation reaction was detected. The method was applied to the determination of glyphosate in water, beer, ale, lentils and beans, and a few cereals (152). Purity of the pharmaceutical-grade synthetic peptides was evaluated by using a nitrogen–CL detector (153). N-Nitroso compounds (NOC) comprise nitrosoamines and nitrosamides, and were produced by the nitrosation of secondary amines and N-alkyl amides, respectively. Total NOC and their precursors were found in extracts of food and tobacco products. After these compounds were decomposed to NO with refluxing HBr/HCl/HOAc/EtOAc, NO was measured by CL (154).

Morphine, a narcotic analgesic, can yield CL by reaction with potassium permanganate in polyphosphoric acid solution (pH 1–2). After separation with a stylene–divinylbenzene column, CL produced from on-column reaction of morphine with potassium permanganate was assayed with the detection limit of 0.5 ng on column (155). Similarly, morphine and monoacetylmorphine were simultaneously determined with the detection limits of 20 and 300 pg on column, respectively (156). The 1,10-phenanthroline–H_2O_2–Cu(II) CL system was applied for assay of BSA. Gradient elution as a function of pH with a metal-chelate affinity column was examined for the separation of BSA, lysozyme, and bovine serum γ-globulin in the range from 1×10^{-4} to 1×10^{-1} g/L; the sensitivity is 200 times higher than that with uv detection (157).

In a similar manner, ovalbumin-trypsin inhibitor and BSA-trypsin inhibitor were detected (158).

The CL of sulfur-containing compounds is also reported (159). A FIA for thiopronine and its metabolite, 2-mercaptopropionic acid, was developed by utilizing the phenomenon that Ce(IV) CL can be enhanced by quinine (160). In the same manner, rhodamine B was used as an enhancer for the determination of preparations of captopril and hydrochlorothiazide (161,162).

As mentioned above, CL is a versatile tool for sensitive and selective determination of trace amounts of organic and inorganic compounds including biologically important components such as bioactive amines, nucleic acids, sugars, etc. Therefore, it is expected that CL will be utilized to postgenome sciences such as proteome, metabolome, etc.

BIBLIOGRAPHY

"Chemiluminescence" in *ECT* 3rd ed., Vol. 5, pp. 416–450, by M. M. Rauhut, American Cynamid Co.; "Luminescent Materials, Chemiluminescence" in *ECT* 4th ed., Irena Bronstein, Tropix, Inc., Larry J. Kricka, University of Pennsylvania, and Richard S. Givens, University of Kansas; "Chemiluminescence" in *ECT* (online), posting date: December 4, 2000, by Irena Bronstein, Tropix, Inc., Larry J. Kricka, University of Pennsylvania, and Richard S. Givens, University of Kansas.

CITED PUBLICATIONS

1. F. Li, C. Zhang, X. Guo and W. Feng, *Biomed. Chromatogr.* **17**, 96 (2003).
2. A. M. Garcia Campana and W. R. G. Baeyens, eds., *Chemiluminescence in Analytical Chemistry*, Marcel Dekker, New York, 2001.
3. K. Nakashima, *BUNSEKI KAGAKU* **49**, 135 (2000).
4. N. Kuroda and K. Nakashima, in T. Toyo'oka, ed., *Modern Derivatization Methods for Separation Sciences*, John Wiley & Sons, Inc., Chichester, 1999.
5. M. S. Gandelman and J. W. Birks, *J. Chromatogr.* **242**, 21 (1982).
6. P. Jones, T. Williams, and L. Ebdon, *Anal. Chim. Acta* **217**, 157 (1989).
7. T. Hara, M. Toriyama, and T. Ebuchi, *Bull. Chem. Soc. Jpn.* **58**, 109 (1985).
8. T. Miyazawa, *Free Radical Biol. Med.* **7**, 209 (1989).
9. T. Miyazawa, K. Fujimoto, T. Suzuki, and K. Yasuda, *Methods Enzymol.* **233**, 324 (1994).
10. T. Miyazawa, K. Yasuda, K. Fujimoto, and T. Kaneda, *Anal. Lett.* **21**, 1033 (1988).
11. T. Miyazawa, K. Fujimoto, and S. Okikawa, *Biomed. Chromatogr.* **4**, 131 (1990).
12. T. Miyazawa, T. Suzuki, and K. Fujimoto, K. Yasuda, *J. Lipid Res.* **3**, 1051 (1992).
13. L. Cominicacini, A. M. Pastrino, A. McCarthy, M. Campagno, U. Garbin, A. Davoli, A. DeSantis, and V. LoCacio, *Biochim. Biophys. Acta* **1165**, 279 (1993).
14. T. C. Christensen and G. Hoelmer, *J. Food Sci.* **61**, 486 (1996).
15. M. Yasuda and S. Narita, *J. Chromatogr. B* **693**, 211 (1997).
16. J. Adachi, M. Asano, T. Naito, Y. Ueno, and Y. Tatsuno, *Lipids* **33**, 1235 (1998).
17. G. T. Shwaery, J. M. Samii, B. Frei, and J. F. Keaney, Jr., *Methods Enzymol.* **300**, 51 (1999).
18. H.-C. Yeh and W.-Y. Lin, *Anal. Bioanal. Chem.* **372**, 525 (2002).
19. Y. Li, D. Zhao, C. Zhu, and L. Wang, J. Xu, *Anal. Bioanal. Chem.* **374**, 395 (2002).
20. K. Ueno, F. Sagara, I. Yoshida, H. Enami, S. Etoh, H. Miyazaki, A. Sonoda, and M. Saito, *Anal. Sci.* **4**, 477 (1988).
21. H. Kather, E. Wieland, and W. Wass, *Anal. Biochem.* **163**, 45 (1987).
22. P. J. Koerner, Jr. and T. A. Nieman, *J. Chromatogr.* **449**, 217 (1988).
23. J. A. Mattews, A. Batiki, C. Hynds, and L. J. Kricka, *Anal. Biochem.* **151**, 205 (1985).
24. T. Segawa, T. Kamidate, and H. Watanabe, *Anal. Sci.* **4**, 659 (1988).
25. T. Stone and I. Durrant, *Mol. Biotechnol.* **6**, 69 (1996).

26. J. Stevens, F. S. Yu, P. M. Hassoun, and J. J. Lanzillo, *Mol. Cell. Prob.* **10**, 31 (1996).
27. S. Ikegawa, N. Hirabayashi, T. Yoshimura, M. Tohma, M. Maeda, and A. Tsuji, *J. Chromatogr.* **577**, 229 (1992).
28. H. R. Schroeder and F. M. Yeager, *Anal. Chem.* **50**, 1114 (1978).
29. H. Yuki, Y. Azuma, N. Maeda, and H. Kawasaki, *Chem. Pharm. Bull.* **36**, 1905 (1988).
30. T. Kawasaki, M. Maeda, and A. Tsuji, *J. Chromatogr.* **328**, 121 (1985).
31. K. Nakashima, K. Suetsugu, S. Akiyama, and M. Yoshida, *J. Chromatogr.* **530**, 154 (1990).
32. Y. Hasegawa, D. C. Jette, A. Miyamoto, H. Kawasaki, and H. Yuki, *Anal. Sci.* **7**, 945 (1991).
33. S. R. Spurlin and M. M. Cooper, *Anal. Lett.* **19**, 2277 (1986).
34. J. Ishida, N. Horike, and M. Yamaguchi, *J. Chromatogr. B* **669**, 390 (1995).
35. J. Ishida, N. Horike, and M. Yamaguchi, *Anal. Chim. Acta* **302**, 61 (1995).
36. J. Ishida, T. Yakabe, H. Nohta, and M. Yamaguchi, *Anal. Chim. Acta* **346**, 175 (1997).
37. T. Nakahara, J. Ishida, and M. Yamaguchi, M. Nakamura, *Anal. Biochem.* **190**, 309 (1990).
38. J. Ishida, T. Nakahara, and M. Yamaguchi, *Biomed. Chromatogr.* **6**, 135 (1992).
39. J. Ishida, S. Sonezaki, and M. Yamaguchi, *J. Chromatogr.* **598**, 203 (1992).
40. J. Ishida, S. Sonezaki, M. Yamaguchi, and T. Yoshitake, *Anal. Sci.* **9**, 319 (1993).
41. H. Yoshida, R. Nakao, H. Nohta, and M. Yamaguchi, *J. Chromatogr. A*, **898**, 1 (2000).
42. H. Yoshida, R. Nakao, T. Matsuo, H. Nohta, and M. Yamaguchi, *J. Chromatogr. A* **907**, 39 (2001).
43. G. H. G. Thorpe and L. J. Kricka, *Methods Enzymol.* **133**, 331 (1986).
44. T. P. Whitehead, G. H. G. Thorpe, T. J. N. Carter, C. Croucutt, and L. J. Kricka, *Nature (London)*, **305**, 158 (1983).
45. G. H. Thorpe, L. J. Kricka, E. Gillespie, S. Moseley, R. Amess, N. Baggett, and T. P. Whitehead, *Anal. Biochem.* **145**, 96 (1985).
46. L. J. Kricka, M. Cooper, and X. Jin, *Anal. Biochem.* **240**, 119 (1996).
47. J. V. Samsonova, M. Yu. Rubtsova, A. V. Kiseleva, A. A. Ezhov, and A. M. Egorov, *Biosen. Bioelect.* **14**, 273 (1999).
48. H. Arakawa, M. Maeda, and A. Tsuji, *Anal. Biochem.* **199**, 238 (1991).
49. L. L. Klopf and T. A. Nieman, *Anal. Chem.* **57**, 46 (1985).
50. M. Maeda and A. Tsuji, *J. Chromatogr.* **352**, 213 (1986).
51. T. Kamidate, K. Yoshida, T. Segawa, and H. Watanabe, *Anal. Sci.* **5**, 359 (1989).
52. F. R. Ochsendorf, C. Goy, J. Fuchs, W. Morke, H. A. Beschmann, and H. Bromme, *Free Rad. Res.* **34**, 153 (2001).
53. A. Kokado, H. Arakawa, and M. Maeda, *Luminescence* **17**, 5 (2002).
54. T. Kamidate, T. Kaide, H. Tani, E. Makino, and T. Shibata, *Anal. Sci.* **17**, 951 (2001).
55. N. C. Nelson, A. B. Cheikh, E. Matsuda, and M. M. Becker, *Biochem.* **35**, 8429 (1996).
56. N. Sato, K. Shirakawa, Y. Kakihara, H. Mochizuki, and T. Kanamori, *Anal. Sci.* **12**, 853 (1996).
57. T. Nakai and K. Isobe, *Rinsyo Kensa* **42**, 283 (1998).
58. T. J. Novak and M. L. Grayeski, *Microchem. J.* **50**, 151 (1994).
59. A. A. Waldrop III, J. Fellers, and C. P. H. Vary, *Luminescence* **15**, 169 (2000).
60. D. Hughes, S. Talwar, I. B. Squire, J. E. Davies, and L. L. NG, *Clin. Sci.* **96**, 373 (1999).
61. A. Nishitani, Y. Tsukamoto, S. Kanda, and K. Imai, *Anal. Chim. Acta* **251**, 247 (1991).
62. B. Mann and M. L. Grayeski, *Biomed. Chromatogr.* **5**, 47 (1991).
63. M. Maeda, K. Tsukagoshi, M. Murata, and M. Takagi, *Anal. Sci.* **10**, 583 (1994).

64. M. Wada, H. Kido, N. Kishikawa, T. Tou, M. Tanaka, J. Tsubokura, M. Shironita, M. Matsui, N. Kuroda, and K. Nakashima, *Environ. Poll.* **115**, 139 (2001).
65. K. Hayakawa, N. Terai, P. G. Dinning, K. Akutsu, Y. Iwamoto, R. Etoh, and T. Murahashi, *Biomed. Chromatogr.* **7**, 262 (1993).
66. K. Nakashima, S. Kawaguchi, R. S. Givens, and S. Akiyama, *Anal. Sci.* **6**, 833 (1990).
67. N. Kuroda, S. Hosoki, K. Nakashima, S. Akiyama, and R. S. Givens, *J. Biolumin. Chemilumin.* **13**, 101 (1998).
68. T. G. Curtis and W. Seitz, *J. Chromatogr.* **134**, 343 (1977).
69. G. Mellbin and B. E. F. Smith, *J. Chromatogr.* **312**, 203 (1984).
70. S. Kobayashi and K. Imai, *Anal. Chem.* **52**, 424 (1980).
71. K. Miyaguchi, K. Honda, and K. Imai, *J. Chromatogr.* **316**, 501 (1984).
72. K. Hayakawa, Y. Miyoshi, H. Kurimoto, Y. Matsushima, N. Takayama, S. Tanaka, and M. Miyazaki, *Biol. Pharm. Bull.* **16**, 817 (1993).
73. N. Takayama, S. Tanaka, and K. Hayakawa, *Biomed. Chromatogr.* **11**, 25 (1997).
74. A. Nishitani, S. Kanda, and K. Imai, *Biomed. Chromatogr.* **6**, 124 (1992).
75. S. Kobayashi, J. Sekino, K. Honda, and K. Imai, *Anal. Biochem.* **112**, 99 (1981).
76. D. L. Walters, J. E. James, F. B. Vest, and H. T. Karnes, *Biomed. Chromatogr.* **8**, 207 (1994).
77. C-E. Tsai, F. Kondo, Y. Ueyama, and J. Azama, *J. Chromatogr. Sci.* **33**, 365 (1995).
78. K. Nakashima, K. Suetsugu, K. Yoshida, S. Akiyama, S. Uzu, and K. Imai, *Biomed. Chromatogr.* **6**, 149 (1992).
79. S. Uzu, K. Imai, K. Nakashima, and S. Akiyama, *Analyst (London)* **116**, 1353 (1991).
80. P. de Montigny, F. Stobaugh, R. S. Givens, R. G. Carlson, K. Srinivasacher, L. A. Sterson, and T. Higuchi, *Anal. Chem.* **59**, 1096 (1987).
81. T. Kawasaki, K. Imai, T. Higuchi, and O. S. Wong, *Biomed. Chromatogr.* **4**, 113 (1990).
82. P. J. K. Kwakman, H. Koelewijin, I. Kool, U. A. Th. Brinkman, and G. J. de Jong, *J. Chromatogr.* **511**, 155 (1990).
83. P. Prados, S. Higashidate, and K. Imai, *Biomed. Chromatogr.* **8**, 1 (1994).
84. P. Prados, S. Higashidate, K. Imai, Y. Sato, and T. Nagao, *Biomed. Chromatogr.* **8**, 49 (1994).
85. G. H. Ragab, H. Nohta, M. Kai, Y. Ohkura, and K. Zaitsu, *J. Pharm. Biomed. Anal.* **13**, 645 (1995).
86. H. Kouwatli, J. Chalom, M. Tod, R. Farinotti, and G. Mahuzier, *Anal. Chim. Acta* **266**, 243 (1992).
87. M. Katayama, H. Takeuchi, and H. Taniguchi, *Anal. Chim. Acta* **281**, 111 (1993).
88. K. Imai, S. Higashidate, A. Nishitani, Y. Tsukamoto, M. Ishibashi, J. Shoda, and T. Osuga, *Anal. Chim. Acta* **227**, 21 (1989).
89. S. Higashidate, K. Hibi, M. Senda, S. Kanda, and K. Imai, *J. Chromatogr.* **515**, 577 (1990).
90. H. Akiyama, S. Shidawara, A. Maeda, H. Toyoda, T. Toida, and T. Imanari, *J. Chromatogr.* **579**, 203 (1992).
91. L. Nondek, R. E. Milofsky, and J. W. Birks, *Chromatographia* **32**, 33 (1991).
92. S. Uzu, K. Imai, K. Nakashima, and S. Akiyama, *J. Pharm. Biomed. Anal.* **10**, 979 (1992).
93. Y. Hamachi, M. N. Nakashima, and K. Nakashima, *J. Chromatogr. B.* **724**, 189 (1999).
94. B. Mann and M. L. Grayesky, *J. Chromatogr.* **386**, 149 (1987).
95. K. Nakashima, M. Nagata, M. Takahashi, and S. Akiyama, *Biomed. Chromatogr.* **6**, 55 (1992).
96. M. L. Grayeski and J. K. De Vasto, *Anal. Chem.* **59**, 1203 (1987).
97. P. J. M. Kwakman, H. P. Van Schaik, and U. A. T. Brinkman, G. J. De Jong, *Analyst (London)* **116**, 1385 (1991).

98. B. W. Sandmann and M. L. Grayeski, *J. Chormatogr. B* **653**, 123 (1994).
99. G-L. Duan, K. Nakashima, N. Kuoda, and S. Akiyama, *J. Chin. Pharm. Sci.* **4**, 22 (1995).
100. M. Tod, M. Prevot, J. Chalom, R. Farinotti, and G. Mahuzier, *J. Chromatogr.* **542**, 295 (1991).
101. K. Nakashima, C. Umekawa, S. Nakatsuji, S. Akiyama, and R. S. Given, *Biomed. Chromatogr.* **3**, 39 (1989).
102. O. Nozaki, Y. Ohba, and K. Imai, *Anal. Chim. Acta* **205**, 255 (1988).
103. H. Yamada, Y. Kuwahara, Y. Takamatsu, and T. Hayase, *Biomed. Chromatogr.* **14**, 333 (2000).
104. P. J. M. Kwakman, J. G. J. Mol, D. A. Kamminga, R. W. Frei, U. A. Th. Brinkman, and G. J. de Jong, *J. Chromatogr.* **459**, 139 (1988).
105. S. Yoshida, K. Urakami, M. Kito, S. Takeshima, and S. Hirose, *J. Chromatogr.* **530**, 57 (1990).
106. S. Yoshida, K. Urakami, M. Kito, S. Takeshima, and S. Hirose, *Anal. Chim. Acta* **239**, 181 (1990).
107. K. Sato and S. Tanaka, *Microchem. J.* **53**, 93 (1996).
108. S. Albrecht, H. Brandl, W. D. Bohm, R. Beckert, H. Kroschwitz, and V. Neumeister, *Anal. Chim. Acta* **255**, 413 (1991).
109. S. Albrecht, H. Hornak, T. Freidt, W. D. Bohm, K. Weis, and A. Reinschke, *J. Biol. Chem.* **8**, 21 (1993).
110. N. Wu and C. W. Huie, *J. Chromatogr.* **634**, 309 (1993).
111. T. Hara, J. Yokogi, S. Okamura, S. Kato, and R. Nakajima, *J. Chromatogr.* **652**, 361 (1993).
112. T. Hara, S. Kayama, H. Nishida, and R. Nakajima, *Anal. Sci.* **10**, 223 (1994).
113. W. R. G. Baeyens, B. L. Ling, K. Imai, A. C. Calokerinos, and S. G. Schulman, *J. Microcol. Sep.* **6**, 195 (1994).
114. X.-J. Huang, Z.-L. Fang, *Anal. Chim. Acts* **414**, 1 (2000).
115. A. M. Garcia Campana, W. R. G. Baeyens, and N. A. Guzman, *Biomed. Chromatogr.* **12**, 172 (1998).
116. K. Nakashima, K. Maki, S. Akiyama, and K. Imai, *Biomed. Chromatogr.* **4**, 105 (1990).
117. K. Nakashima, K. Maki, S. Kawaguchi, S. Akiyama, Y. Tsukamoto, and K. Imai, *Anal. Sci.* **7**, 709 (1991).
118. K. Nakashima, M. Wada, N. Kuroda, S. Akiyama, and K. Imai, *J. Liq. Chromatogr.* **17**, 2111 (1994).
119. M. Katayama, H. Takeuchi, and H. Taniguchi, *Anal. Lett.* **24**, 1005 (1991).
120. G. Gubitz, P. Zoonen, C. Goojer, N. H. Velthorst, and R. W. Frei, *Anal. Chem.* **57**, 2071 (1985).
121. H. Jansen, U. A. T. Brinkman, and R. W. Frei, *J. Chromatogr.* **440**, 217 (1988).
122. K. Honda, K. Miyaguchi, H. Nishino, H. Tanaka, Y. Yao, and K. Imai, *Anal. Biochem.* **153**, 50 (1986).
123. M. Wada, K. Nakashima, N. Kuroda, S. Akiyama, and K. Imai, *J. Chromatogr. B* **678**, 129 (1996).
124. S. Kamei, A. Ohtsubo, S. Saito, and S. Takagi, *Anal. Chem.* **61**, 1921 (1989).
125. M. Wada, N. Kuroda, T. Ikenaga, S. Akiyama, and K. Nakashima, *Anal. Sci.* **12**, 807 (1996).
126. L. J. Kricka, *Clin. Chem.* **37**, 1472 (1991).
127. F. Odawara, H. Abe, T. Kohno, Y.-N. Fujii, K. Arai, S. Imamura, H. Misaki, H. Azuma, K. Ikebuchi, H. Ikeda, S. Mohan, and K. Sano, *J. Virol. Meth.* **106**, 115 (2002).
128. Y. Cao, W. C. Smith, and R. R. Bowsher, *J. Pharm. Biomed. Anal.* **26**, 53 (2001).

129. K. Nakashima, H. Yamasaki, N. Kuroda, and S. Akiyama, *Anal. Chim. Acta* **303**, 103 (1995).

130. K. Nakashima, H. Yamasaki, R. Shimoda, N. Kuroda, and S. Akiyama, *Biomed. Chromatogr.* **11**, 63 (1997).

131. M. Maeda, *Rinsyo Kensa* **42**, 263 (1998).

132. A. M. Osman, C. Laane, and R. Hilhorst, *Luminescence* **16**, 45 (2001).

133. S. Yamazaki, R. Chiba, K. Uchikura, and T. Tanimura, *Chromatography* **16**, 322 (1995).

134. M. Sanchez-Carbayo, A. Espasa, V. Chinchilla, E. Herrero, J. Megias, A. Mira, and F. Soria, *Clin. Chem.* **45**, 1944 (1999).

135. K. Uchikura and M. Kurisawa, *Anal. Sci.* **7**, 971 (1991).

136. W-Y. Lee and T. A. Nieman, *J. Chromatogr. A* **659**, 111 (1994).

137. D. R. Skotty and T. A. Nieman, *J. Chromatogr. B* **665**, 27 (1995).

138. J. S. Ridlen, D. R. Skotty, P. T. Kissinger, and T. A. Nieman, *J. Chromatogr. B* **694**, 394 (1997).

139. H. Monji, M. Yamaguchi, I. Aoki, and H. Ueno, *J. Chromatogr. B* **690**, 305 (1997).

140. J. S. Rinden, G. J. Klopf, and T. A. Nieman, *Anal. Chim. Acta* **341**, 195 (1997).

141. R. Chiba, M. Shinriki, Y. Ishii, and A. Tanaka, *Anal. Sci.* **14**, 975 (1998).

142. F. Wu, Z. He, O. Luo, and Y. Zeng, *Anal. Sci.* **14**, 971 (1998).

143. K. Nakagawa and T. Miyazawa, *Anal. Biochem.* **248**, 41 (1997).

144. M. Nakazono, Y. Ohba, and K. Zaitsu, *Chem. Pharm. Bull.* **47**, 569 (1999).

145. I. Furukawa, H. Hosotsubo, C. Hayashi, and Y. Ishida, *Jpn. J. Clin. Chem.* **9**, 279 (1980).

146. N. Kuroda, K. Nakashima, and S. Akiyama, *Anal. Chim. Acta* **278**, 275 (1993).

147. N. Sato, K. Shirakawa, K. Sugihara, and T. Kanamori, *Anal. Sci.* **13**, 59 (1997).

148. M. Kai, Y. Ohkura, S. Yonekura, and M. Iwasaki, *Anal. Chim. Acta* **287**, 75 (1994).

149. N. Kuroda, K. Nakashima, S. Akiyama, N. Sato, N. Imi, K. Shirakawa, and A. Uemura, *J. Biolumin. Chemilumin.* **13**, 25 (1998).

150. M. Kai, S. Kishida, and K. Sakai, *Anal. Chim. Acta* **381**, 155 (1999).

151. N. P. Sen, P. A. Baddoo, and S. W. Seaman, *J. Chromatogr. A* **673**, 77 (1994).

152. N. P. Sen and P. A. Baddoo, *Int. J. Environ. Anal. Chem.* **63**, 107 (1996).

153. E. M. Fujinari, J. D. Manes, and R. Bizanek, *J. Chromatogr. A* **743**, 85 (1996).

154. J. Haorah, L. Zhou, X. Wang, G. Xu, and S. S. Mirvish, *J. Agric. Food Chem.* **49**, 6068 (2001).

155. R. W. Abbott, A. Townshend, and R. Gill, *Analyst (London)* **112**, 397 (1987).

156. E. Amiott and A. R. J. Andrews, *J. Liq. Chrom. Rel. Technol.* **20**, 311 (1997).

157. T. Hara, K. Tsukagoshi, and T. Yoshida, *Bull. Chem. Soc. Jpn.* **61**, 2779 (1988).

158. T. Hara, K. Tsukagoshi, and H. Tsuji, *Bull. Chem. Soc. Jpn.* **63**, 770 (1990).

159. T. B. Ryerson and R. E. Sievers, *Chem. Anal. (N.Y.)* **131**, 1 (1995).

160. Y. N. Zhao, W. R. G. Baeyens, X. R. Zhang, A. G. Calokerinos, K. Nakashima, G. Van der Weken, and A. Van Overbeke, *Chromatographia* **44**, 31 (1997).

161. X. R. Zhang, W. R. G. Baeyens, G. Van der Weken, A. C. Calokerinos, and K. Nakashima, *Anal. Chim. Acta* **303**, 121 (1995).

162. J. Ouyang, W. R. G. Baeyens, J. Delanghe, G. Van der Weken, W. Van Dael, D. De Keukeleire, and A. M. Garcia Campana, *Anal. Chim. Acta* **386**, 257 (1999).

KENICHIRO NAKASHIMA
Nagasaki University